Selected Papers on

Optical Storage

Books in SPIE's Milestone Series
Brian J. Thompson, *General Editor*

Published

Title	Editor
Acousto-Optics (1990)	Adrian Korpel
Automatic Object Recognition (1991)	Hatem Nasr
CO_2 Lasers (1990)	James D. Evans
Coherence and Fluctuations of Light (1990)	Emil Wolf Leonard Mandel
Computer-Controlled Optical Surfacing (1991)	Robert A. Jones
Computer-Generated Holograms and Diffractive Optics (1992)	Sing H. Lee
Deposition of Optical Coatings (1989)	Michael Ray Jacobson
Design of Optical Coatings (1990)	Michael Ray Jacobson
Digital Image Processing (1990)	Mohan M. Trivedi
Digital Image Restoration (1992)	M. Ibrahim Sezan
Dye Lasers (1992)	F. J. Duarte
Ellipsometry (1991)	R. M. A. Azzam
Fiber Optic Gyroscopes (1989)	Robert B. Smith
Free-Space Laser Communications (1991)	David L. Begley
Holographic and Diffractive Lenses and Mirrors (1991)	Thomas W. Stone Brian J. Thompson
High-Power Lasers (1991)	John M. Soures
Holographic Particle Diagnostics (1990)	Chandra S. Vikram
Image Tubes (1990)	Illes P. Csorba
Infrared Design (1985)	R. Barry Johnson William L. Wolfe
Infrared Fiber Optics (1990)	James A. Harrington
Interferometry (1991)	P. Hariharan
Laser Damage in Optical Materials (1990)	Roger M. Wood
Laser Design (1991)	Hugo Weichel
Laser Scanning and Recording (1985)	Leo Beiser
Light Scattering (1988)	Milton Kerker
Liquid Crystals for Optics (1992)	Stephen D. Jacobs
Multiple Scattering in Plane Parallel Atmospheres and Oceans: Methods (1991)	George W. Kattawar
Natural Optical Activity (1990)	Akhlesh Lakhtakia
Nonlinear Optics (1991)	Howard E. Brandt
Optical Computing (1989)	H. John Caulfield Gregory Gheen
Optical Fibers in Medicine (1990)	Abraham Katzir
Optical Shop Metrology (1990)	Daniel Malacara
Optical Storage (1992)	Glenn T. Sincerbox James M. Zavislan
Optical Tolerancing (1991)	Gary E. Wiese
Optomechanical Design (1988)	Donald C. O'Shea
Photon Statistics and Coherence in Nonlinear Optics (1991)	Jan Peřina
Polarization (1990)	Bruce H. Billings
Quasielastic Light Scattering (1990)	Benjamin Chu
Radiometry (1990)	Irving J. Spiro
Rare-Earth-Doped Fiber Laser Sources and Amplifiers (1992)	Michel J. F. Digonnet
Scalar Wave Diffraction (1992)	Kurt Edmund Oughstun
Scattering in the Atmosphere (1989)	Craig F. Bohren
Semiconductor Diode Lasers (1992)	James J. Coleman
Solid State Lasers (1991)	Richard C. Powell
Speckle Metrology (1991)	Rajpal S. Sirohi
Surface-Enhanced Raman Scattering (1990)	Milton Kerker
Turbulence in a Refractive Medium (1990)	Edgar L Andreas
Ultrafast Laser Technology (1991)	Timothy R. Gosnell Antoinette J. Taylor
Visual Communication (1990)	T. Russell Hsing Andrew G. Tescher

Forthcoming

Title	Editor
Architectural Lighting	Mark S. Rae
Applications of Polarized Light	Bruce H. Billings
Astronomical Optics	Daniel J. Schroeder
Atmospheric Optics	Vladimir E. Zuev
Characterization of Optical Coatings	Michael Ray Jacobson
Coherence and Radiometry	Ari T. Friberg
Coherent Optical Processing	Francis T. S. Yu Shizhuo Yin
Color—Fundamentals	David L. MacAdam
Effects of Aberrations in Optical Imaging	Virendra N. Mahajan
Electron Optics	Peter W. Hawkes
Fourier Optics	Mustafa A. G. Abushagur H. John Caulfield
Fundamentals of Lasers	William T. Silfvast
Gaussian Beam Mode Optics for MM-Wave and Terahertz Systems	James C. G. Lesurf
Gradient Index Optics	Duncan Moore
Image Analysis and Machine Vision	Mohan M. Trivedi
Image Compression	Majid Rabbani
Laser Isotope Separation	Jeff W. Eerkens
Laser Safety	D. H. Sliney
Model-Based Vision	Hatem Nasr
Optical Fiber Technology	Donald B. Keck
Optical Microlithography	Harry Stover
Optical Moiré and Applications	Guy Indebetouw Robert Czarnek
Optical Transfer Function: Foundation and Theory	Lionel Baker
Optical Transfer Function: Measurement	Lionel Baker
Optometry	William M. Rosenblum William J. Benjamin
Photochemistry	Douglas Neckers
Photoconductivity	Narahari V. Joshi
Scanning Acoustic Microscopy	Butrus Khuri-Yakub Calvin Quate
Schlieren Optics	Jurgen R. Meyer-Arendt
Semiconductor Infrared Detectors	Antoni Rogalski
Solar Radiation and Solar Thermal Systems	Donald Osborn
X-Ray Optics	Richard B. Hoover

SPIE Milestone Series
Volume MS 49

Selected Papers on
Optical Storage

Glenn T. Sincerbox
James M. Zavislan, *Editors*
IBM Research Division
Almaden Research Center

Brian J. Thompson
General Editor, SPIE Milestone Series

SPIE OPTICAL ENGINEERING PRESS

A Publication of SPIE—The International Society for Optical Engineering
Bellingham, Washington USA

Library of Congress Cataloging-in-Publication Data

Selected papers on optical storage / editors, Glenn T. Sincerbox,
James M. Zavislan.
p. cm. — (SPIE milestone series ; v. MS 49)
Includes index.
ISBN 0-8194-0890-5 (hardbound). — ISBN 0-8194-0891-3 (softbound)
1. Optical storage devices. I. Sincerbox, Glenn T., 1937- .
II. Zavislan, James M., 1960- . III. Series.
TK7895.M4S39 1992
621.382`7—dc20 92-7650
CIP

ISBN 0-8194-0890-5 (hardbound)
ISBN 0-8194-0891-3 (softbound)

Published by
SPIE—The International Society for Optical Engineering
P.O. Box 10, Bellingham, Washington 98227-0010 USA
Telephone 206/676-3290 (Pacific Time) • Telex 46-7053

This book is a compilation of outstanding papers selected from the world literature on optical and optoelectronic science, engineering, and technology. SPIE Optical Engineering Press acknowledges with appreciation the authors and publishers who have given their permission to reprint material included in this volume. An earnest effort has been made to contact the copyright holders of all papers reproduced herein.

Printed in the United States of America

Introduction to the Series

There is no substitute for reading the original papers on any subject even if that subject is mature enough to be critically written up in a textbook or a monograph. Reading a well-written book only serves as a further stimulus to drive the reader to seek the original publications. The problems are, which papers and in what order?

As a serious student of a field, do you really have to search through all the material for yourself, and read the good with the not-so-good, the important with the not-so-important, and the milestone papers with the merely pedestrian offerings? The answer to all these questions is usually yes, unless the authors of the textbooks or monographs that you study have been very selective in their choices of references and bibliographic listings. Even in that all-too-rare circumstance, the reader is then faced with finding the original publications, many of which may be in obscure or not widely held journals.

From time to time and in many disparate fields, volumes appear that are collections of reprints that represent the milestone papers in the particular field. Some of these volumes have been produced for specific topics in optical science, but as yet no systematic set of volumes has been produced that covers connected areas of optical science and engineering.

The editors of each individual volume in the series have been chosen for their deep knowledge of the world literature in their fields; hence, the selection of reprints chosen for each book has been made with authority and care.

On behalf of SPIE, I thank the individual editors for their diligence, and we all hope that you, the reader, will find these volumes invaluable additions to your own working library.

Brian J. Thompson
University of Rochester

Selected Papers on

Optical Storage

Contents

Section Six
Head Integration/ Miniaturization

Section Seven
Servos—Focus and Tracking

Introduction

This volume on optical storage is an attempt to assemble, under one cover, the key papers that have been published in the field of optical storage technology over the past 25 years. Almost since the discovery of the laser itself, this field has progressed through numerous research and development cycles, some more successful than others, to reach the level it is at today. Indeed, a wide range of optical products are now available that span levels of capability from read-only to write-once to erasable, as well as enabling a new class of applications based on media removability and mass replication.

In spite of occasional setbacks, each period has provided important technical contributions that have been passed forward to enhance the next generation of development; in many cases, this obviated the need to completely reinvent the wheel in each new cycle. It is in this spirit, therefore, that we have attempted to identify those papers that represent the critical underpinnings and important building blocks in the evolution of this technology. They are presented in this context to provide historical perspective and, hopefully, some degree of future guidance in the continued advancement of this field.

Whereas a true representation of the history would argue for arranging the papers in chronological order, we note the great diversity of fields that contribute to today's optical storage systems and have chosen to organize this volume according to subfields. Within a subfield, we have preserved chronology.

The enormous size and age of this field prompts us to comment on the task that was set before us and the results of our labor--in other words, our disclaimer. Great difficulty is encountered when one attempts to identify and label papers as milestones. Are they the first observation or understanding of a process, often published as a letter to the editor, or are they in-depth analysis and experimental results requiring months, even years, of preparation and work? We phrase this question rhetorically in order to identify an important issue, and no attempt will be made to provide an answer. Perhaps it is a candidate for some future editorial debate. Needless to say, it complicates our task. Selection is further compounded by the fact that since 1969, more than 3200 papers have been published in this field. We have selected 92 and make no claim that we have found them all nor, for that matter, that all of these assembled papers satisfy everybody's definition of a milestone. We wish to thank those authors who have given permission to reprint their papers and apologize in advance to those authors whom we have overlooked.

The future for optical storage is bright. Optical libraries and individual drives are available in quantities. Advances in the many constituent technologies continue to push the performance envelope. Hard limits are not yet in sight. Also, new applications such as multimedia information processing and interactive television rely on optical storage technology. Librarians now use optical disks to store book catalogs and journal references. We hope that periodicals and journals soon will be distributed on optical disks and that in the future, you will be reading the introductions to volumes similar to this from an optical disk.

Glenn T. Sincerbox
James M. Zavislan
IBM Research Division
Almaden Research Center

March 1992

Section One
System Modeling and Analysis

Reprinted with permission from *Journal of Applied Physics,* Vol. 38(3), pp. 1192-1196 (March 1967). ©1967 American Institute of Physics.

Magneto-Optic Detection of High-Density Recordings

D. TREVES*

Ampex Corporation, Redwood City, California

The various noise sources in a high-density magneto-optic readout system are evaluated. It is shown that shot noise is not severe even at interrogation rates of 10^7 bit/sec and bit areas of 10^{-6} cm^2. Various techniques used to reduce the noise originating from surface imperfections are described and their effectiveness demonstrated by experimental results.

INTRODUCTION

THE utilization of the magneto-optic effects for reading out information stored in magnetic media was considered in the past by several authors.[1–5]

Because of the smallness of the magneto-optic effects and because of the low power density obtainable with conventional light sources this method of information retrieval was often thought impractical.[2] The situation drastically changed with the advent of high-power c.w. lasers, and a second look was taken at the feasibility of readout of high-density recording (bit area of 10^{-5} cm^2 or less) using magneto-optic effects. In this paper the discussion is centered on the longitudinal Kerr effect; however, most of the problems and methods described are equally well suited for other magneto-optic effects.

The shot noise is evaluated taking into account the limitations due to the temperature rise in the illuminated area. With available lasers, large signal-to-noise ratios are predicted for wideband and high density readout systems, probably making this noise source less important than the other noise sources.

After a discussion of the effects of surface imperfections on the light reflected from the magnetic medium, three techniques developed in order to reduce these effects are described. Experimental results demonstrating the effectiveness of these techniques are presented.

THE MAGNETO-OPTIC SIGNAL

It will be assumed that the information is in digital form, that is, the information is stored in bits which are magnetized to saturation in either direction. In a readout system that utilized the longitudinal Kerr effect, linearily polarized light is reflected by the magnetic medium. The magnetic information is transferred to the reflected light beam essentially as a change in the direction of polarization. Let the difference of this direction for the two states of magnetization be denoted by the Kerr angle ϕ. If a polarizing crystal (the analyzer) is introduced in the path of the reflected beam, then the change in transmitted power for light reflected from two oppositely magnetized areas will be

$$\Delta W = W_0 \left[\sin^2(\theta+\phi) - \sin^2\theta\right] \quad (1)$$

where θ is the angle of rotation of the analyzer from its position of minimum transmittance and W_0 is the light power reflected by the magnetic medium. The angle ϕ is usually very small, of the order of a few milliradians. Also, when signal and not contrast is important, one usually will set $\theta \gg \phi$. Under these conditions

$$\Delta W = W_0 \phi \sin 2\theta, \quad (2)$$

and the average light power reaching the photocathode will be

$$W = W_0 \sin^2\theta. \quad (3)$$

NOISE

Having found the magnitude of the signal that one expects, it is now appropriate to deal with the various noise sources inherent in such a system. The three major ones are: shot noise in the photodetector due to the quantized nature of light; surface noise due to imperfections in the surface of the magnetic media, such as pinholes, dust and scratches; light level fluctuations due to arc instability, plasma noise in gas lasers, variations in reflectivity from region to region of the magnetic media, and slight changes in the alignment of the optical system during the process of readout.

Shot Noise

The shot noise current, I_n, in photodetectors is given by[2]

$$I_n = (2efPW_0)^{1/2} \sin\theta, \quad (4)$$

and the signal current, I_s, is given by

$$I_s = PW_0 \phi \sin 2\theta, \quad (5)$$

where $e = 1.6\times10^{-19}$ C is the electron charge, f is the bandwidth, and P is the conversion coefficient of the photocathode, a typical value of which is 3×10^{-2} A/W.

From these two equations one finds the signal-to-noise ratio

$$I_s/I_n = (2PW_0/ef)^{1/2} \phi \cos\theta. \quad (6)$$

It is interesting that this ratio is essentially independent of θ, the analyzer angle, in the whole region of interest,

---

* On leave from the Weizmann Institute of Science, Rehovoth, Israel.

[1] T. Lentz and J. Miyata, Electronics **34**, 36 (1961).

[2] G. Fan, E. Donath, E. S. Barrekette, and A. Wirgin, IEEE EC12, 3 (1963).

[3] B. Edwards and O. Kornei, Electronics, **36**, 48 (1963).

[4] D. B. Dove, J. Appl. Phys. **35**, 1991 (1964).

[5] J. T. Chang, J. F. Dillon, Jr., and U. F. Gianola, J. Appl. Phys. **36**, 1110 (1965).

which rarely will exceed 10°, for reasons which will be apparent later. This ratio increases with the light power, W_0; it is therefore worthwhile at this point to compare conventional light sources to coherent sources.

For an incoherent source the maximum brightness of an illuminated spot is the brightness b of the source. The highest values obtainable for continuous illumination are in the order of $b=250$ W/cm² (PEK Lab. 107/109 mercury-arc lamp). If a light spot of area A (small compared to the source dimensions) is produced by an incoherent source, the power impinging on it will be of the order of Ab, and is thus proportional to the area of the spot. On the other hand, the total power of a coherent source can be focussed on any spot larger than the diffraction-limited spot of the optical system. Therefore, comparing a 1-W coherent source with an incoherent source with $b=250$ W/cm², one finds that for spots of an area of 10^{-5} cm² the coherent source will supply optical power W_0 several orders of magnitude larger than the incoherent source, and thus will yield a considerably larger signal-to-noise ratio.

From Eq. (6) it may appear that shot noise can be always eliminated as a problem simply by using a high enough power W_0. However, the temperature rise of the magnetic medium puts an upper limit to the amount of power one can use. If the amount of power is determined only by the allowed temperature rise Δt, the $f^{-1/2}$ bandwidth dependence given in Eq. (6) may change considerably. In general, at higher bandwidths each bit will remain exposed to the light beam a shorter time, allowing a higher light power for the same Δt, resulting in a slower dependence of the signal to shot noise ratio on frequency. For example, in a situation in which no heat diffusion effects occur (approximated by a thin metallic film on a good thermal insulator substrate), Δt is given by

$$\Delta t = W_a/JGC, \tag{7}$$

where G is the weight of the illuminated portion of the magnetic film, C is the specific heat and W_a is the power absorbed in the magnetic medium. Assuming that all the nonreflected power is represented by W_a, one gets

$$W_a = W_0(1-R)/R, \tag{8}$$

where R is the reflectance. Substituting (7) and (8) in (6), one finds

$$I_s/I_n = [2P\Delta tGCR/e(1-R)]^{1/2}\phi \cos\theta \tag{9}$$

which is independent of f. If heat diffusion into the substrate occurs, the heat is spread on a larger amount of material and more power, W_a, can be absorbed for the same temperature rise, Δt. Therefore, (9) describes the worst-case situation.

A better approximation to the heat-limited case is to assume a thin magnetic film deposited on a thick substrate of diffusivity, α, and thermal conductivity, k. If the linear dimensions of the illuminated spot are large compared to the heat penetration depth, heat diffusion occurs only in a direction perpendicular to the surface and the problem is one dimensional. The surface temperature rise, Δt, and the penetration depth, s, are then[6]

$$\Delta t = (\alpha T)^{1/2} 2q/k\pi^{1/2}, \qquad s = (\alpha T)^{1/2}, \tag{10}$$

where q is the power density absorbed (assumed here to be constant over the spot), and T is the time the power is on, which is essentially the inverse of f.

Using (10) and substituting $T=f^{-1}$, $W_a=qA$ and (8) in (6), one gets

$$I_s/I_n = \{(P/e)[R/(1-R)]A\Delta t\pi^{1/2}\}^{1/2}[(k^2/\alpha)T]^{1/4}\phi \cos\theta. \tag{11}$$

It is instructive to compare the signal-to-noise ratio derived from (9) and (11) for $A=10^{-6}$ cm², $\Delta t=100$°C, $P=3\times10^{-2}$ A/W, $R=0.5$ and $\phi=10^{-3}$ rad. For a film thickness of 10^{-5} cm, a specific heat $C=0.5$ J/g·deg, and a specific gravity of 8 g/cm³, Eq. (9) yields $I_s/I_n \approx 30$ or 30 dB, independent of frequency. For a copper substrate with $\alpha=1$ cm²/sec and $k=4$ J/sec·cm·deg, Eq. (11) yields $I_s/I_n \approx 10^4\, T^{1/4}$. For $T=10^{-8}$ sec (100-MHz bandwidth) the signal-to-noise ratio is 40 dB and the penetration depth calculated from (10) is 10^{-4} cm, showing that at this frequency the one-dimensional approximation is valid. From (10) the power W_a is found to be 4 W. At 1 MHz the signal-to-noise ratio is 50 dB, $s=10^{-3}$ cm, and $W_a=0.4$ W. Here s is already the same as the linear dimensions of the spot, and heat will then flow also parallel to the surface. Therefore, more power can be absorbed for the same temperature rise and the value of 50 dB is an underestimation.

It is evident from (9) and (11) that enhancement of the Kerr rotation by a dielectric layer, which simultaneously increases ϕ and decreases R, may have only a limited effect on the signal-to-noise ratio. On the other hand, with the availability of high-power coherent sources, shot noise in readout of high-density information is not a severe problem and will usually be considerably smaller than other sources of noise.

Surface Noise and Light Fluctuations

While shot noise can be evaluated analytically, surface noise by its very nature is unpredictable. Observation of a large number of thin films deposited on microscope cover glass under laboratory conditions revealed always a considerable amount of surface imperfections. Figure 1(a) is a photograph of a typical film, taken with the analyzer close to extinction. The bright band in the center is a 50-μ wide bit. The bits in this and all subsequent cases were recorded on the

[6] M. Jakob, *Heat Transfer* (John Wiley & Sons, Inc., New York, 1953), Vol. I.

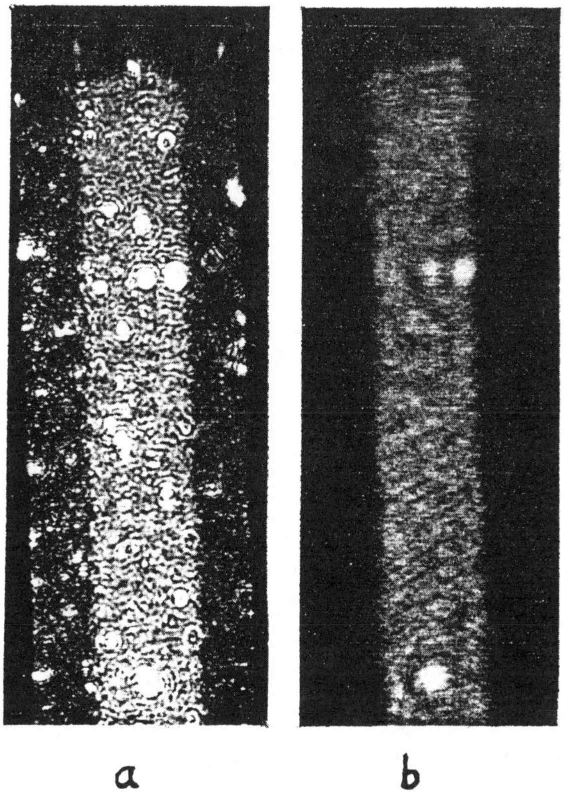

FIG. 1. Photograph of a 50-μ wide bit recorded on a typical film. (a) Without spatial filter. (b) With spatial filter.

films by contact printing from an oxide tape. On the assumption that some surface noise is inevitable even with the most careful preparation methods, several optical techniques aimed at reducing the deleterious effect of this noise were developed. It was found that the majority of the surface imperfections have dimensions, d, of less than a few microns; they scatter light

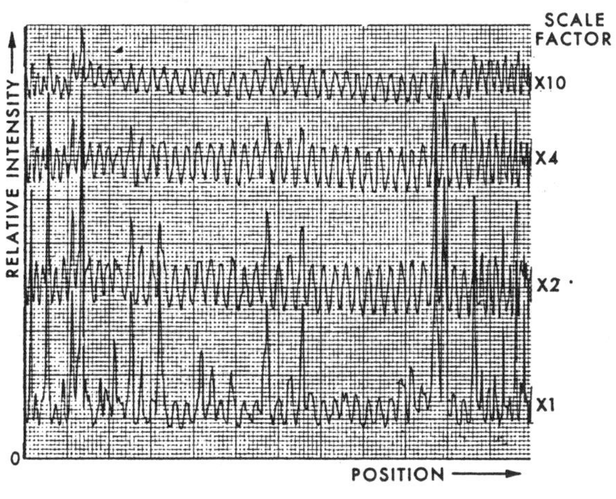

FIG. 2. Scans of 12.5-μ alternate bits for different analyzer angles θ. Bottom trace—$\theta=5'$. Top trace—$\theta=1.9°$. Double Kerr angle $\varphi=3.5'$. Illuminating spot: $7\ \mu\times200\ \mu$.

roughly as if they were pinholes in the path of a plane wave, and thus the diffracted light originating from them spreads with a cone angle, β, of the order of $\beta\approx\lambda/d$ where λ is the wavelength of light. Although the incident light is linearily polarized, the light diffracted by the imperfections is essentially completely depolarized.

The first noise reduction method is based on this last property. As the diffracted light is depolarized, its fraction transmitted by the analyzer is independent of the analyzer angle θ. On the other hand, Eq. (2) shows that the signal is proportional to $\sin 2\theta$. Therefore, by increasing θ, one will increase the signal-to-noise ratio due to surface imperfections. In order to observe this effect, alternate bits of information were recorded on a film, mounted on a micrometer-driven stage so that it could be scanned. The photomultiplier output and a voltage proportional to the position of the stage were fed to an X–Y recorder. S-polarization and 30° angle of incidence were used throughout. Figure 2, which shows scans of 12.5-μ bits of information for different angles θ, clearly demonstrates the

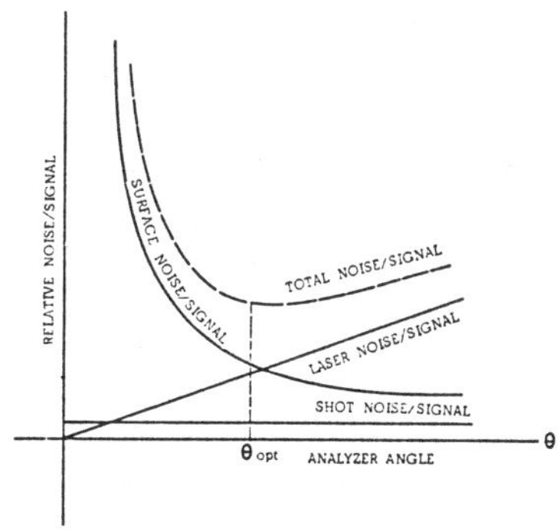

FIG. 3. Schematic representation of the dependence of noise on the analyzer angle θ.

effective reduction in surface noise as θ increases. On the other hand, as θ increases, the contrast decreases, and the effect of light level fluctuations becomes more and more pronounced as seen by the fluctuation of the average light level of the top trace. Following Eq. (3), this fluctuation is proportional to $\sin^2\theta$; thus, the ratio of the signal to this kind of noise is proportional to $\cot\theta$ and decreases with increasing θ. The relative size of these two noise sources will determine the optimum angle θ for minimum total noise. This balance between noise sources is schematically shown in Fig. 3. Curves depicting noise-to-signal ratios, N/S, are used so that they can be added up in order to find the total noise to signal of the system.

The next noise reducing technique takes advantage of the diffracted nature of the light scattered by the surface imperfections and the small dimensions of the latter. If the illuminating spot is a diffraction-limited spot equal in dimensions to the bit size, D, the ratio of the cone angles in the far field region for the light diffracted by the imperfection and the main beam that carries the information is of the order of D/d. By inserting an appropriate iris in the far field, one can then stop most of the light diffracted by the imperfection, while letting through the main beam. This simple form of spatial filtering is effective only if the ratio of D/d is large. Figure 4 shows two scans of the same region of a film on which 6.25 μ bits were recorded. The top and bottom traces were taken respectively with and without the spatial filter. Note that in the filtered scan there is no loss in signal while the noise is reduced by almost two orders of magnitude. The effect of the spatial filter can be well observed photographically, as shown in Fig. 1(b), which is identical to Fig. 1(a) except that the filter was inserted. Most of the short-wavelength surface noise is eliminated. Figure 5 shows scans of 6.25-μ bits, taken with the spatial filter and optimum analyzer angle θ, and of 3.75 μ bits taken near extinction.

As a considerable fraction of the residual light that reaches the photodetector at extinction originates from surface imperfections, the introduction of the spatial filter will greatly improve extinction and thus increase the contrast. Figure 6 shows short scans of 50 μ bits taken with spatial filter and at various small angles θ. The base line is obtained by completely obstructing the light beam. The lowest scans have intensity ratios for the two states of magnetization of 10. As the double Kerr angle ϕ was in this case 3.3′, the extinction ratio in the system is better than 10^{-7}, which is an exceptionally low value.

The last noise reducing method is a differential scheme based on the fact that the surface noise and the signal are carried in a different way by the beam of light; the noise is essentially an amplitude modulation, while the signal is a modulation of the direction of polarization of the light. Because of this difference, the beam can be split into two information channels in which the noise appears in common mode while the signal appears in opposite polarity so that the useful information can be added, and the noise cancelled out by feeding the two channels to a differential amplifier. The beam reflected by the magnetic medium is amplitude split by a nonpolarizing beam splitter into two beams identical in every respect except di-

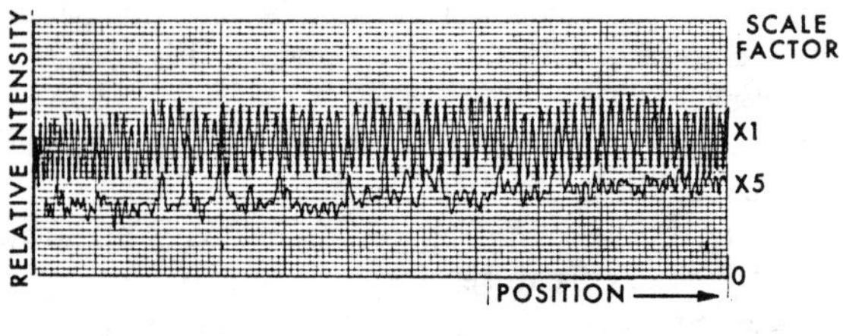

FIG. 4. Scans of 6.25-μ alternate bits, close to extinction. Illuminating spot: 7 $\mu\times$200 μ. Top trace—with spatial filter. Bottom trace without spatial filter and attenuated by a factor of 5.

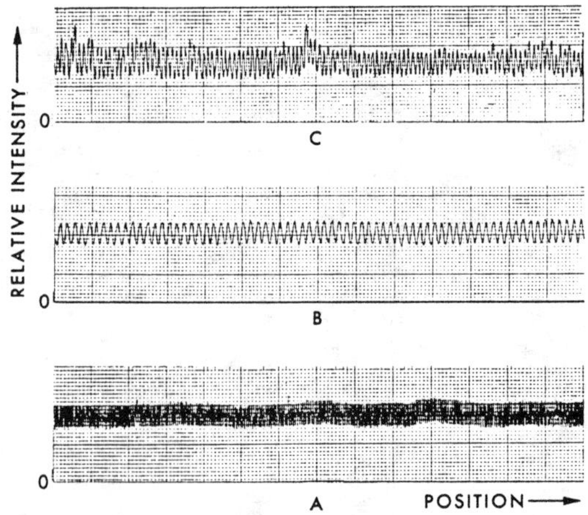

FIG. 5. (A) Scans of 6.25-μ alternate bits, with spatial filter and optimum analyzer angle θ. (B) Same as (A) but on expanded scale. (C) Scans of 3.75-μ alternate bits with spatial filter and close to extinction.

rection of propagation. Two analyzers are used for the two beams, set at angles $+\theta$ and $-\theta$ to the minimum transmittance position, respectively. This setting is made for the medium magnetized in "zero" state. The light is fed to two photomultipliers and to a difference amplifier. As the two beams reflected exactly from the same area of the medium have the same direction of polarization and angle of incidence and come from the same light source, all the fluctuations that come from surface imperfections, changes in reflectivity, and source intensity will cancel out. If the medium is now magnetized in the "one" state, one analyzer will make an angle $\theta+\phi$ and the other $-\theta+\phi$ with the minimum transmittance position, and the difference in light power that the two photomultipliers will receive is given by Eq. (2). The advantage of this scheme with respect to that of Lentz and Miyata[1] is that here the two beams are completely symmetrical, and that there is complete freedom in the choice of θ. Figures 7 and 8 show the effectiveness of this technique. In each figure there are four traces: a baseline which is the zero light level; A and −B which are the output of the individual channels (the B channel is inverted); and A−B which is their difference. In Fig. 7, $\theta=8'$ and in Fig. 8, $\theta=130'$. The bits are 25 μ and no spatial filter was used. Figure 7 shows the remarkable reduction in surface noise, and Fig. 8 shows that light level fluctuations, which in the bottom trace are larger than the signal, are practically eliminated. The only remaining effect due to these fluctuations is a small amplitude modulation of the signal that can be observed in the A−B trace of Fig. 8.

CONCLUSION

The analysis of shot noise made, taking into account the limitations due to the temperature rise in the

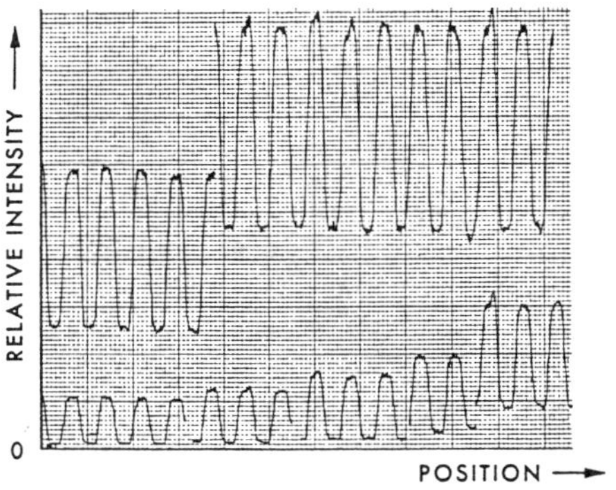

FIG. 6. Short scans of 50-μ alternate bits at various small analyzer angles θ, with spatial filter. Baseline obtained by obstructing the light beam. Double Kerr angle $\varphi = 3.3'$.

illuminated area, indicates that bits with area of 10^{-6} cm^2 can be interrogated at a repetition rate of 10^7 bits/sec with a signal-to-noise ratio larger than 40 dB. The power of the source required is a few watts, a reasonable value for continuous gas lasers. This

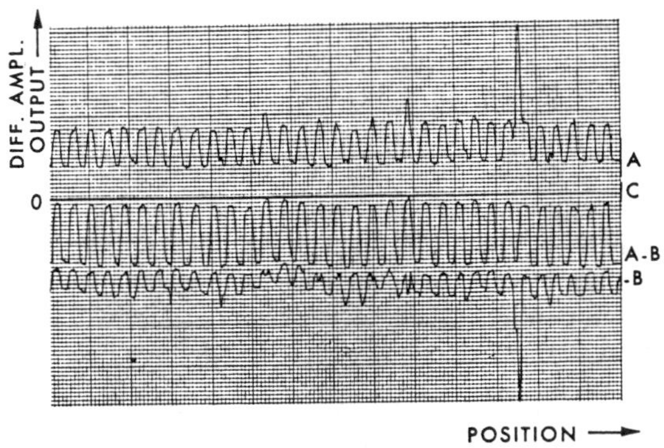

FIG. 7. Scans of 25-μ alternate bits, in differential scheme. No spatial filter, and analyzer angle $\theta = 8'$. Traces A, B, and A−B are of the individual channels and their difference, respectively. Trace C is with light off.

calculation is based on a double Kerr angle of 10^{-3} rad, which is rather conservative, and one should remember that the signal-to-noise ratio is proportional to this angle. On the basis of observations carried out on a large number of thin films prepared under laboratory conditions, it is found that, in general, the noise introduced by surface imperfection in the magnetic medium is large compared to the shot noise. However,

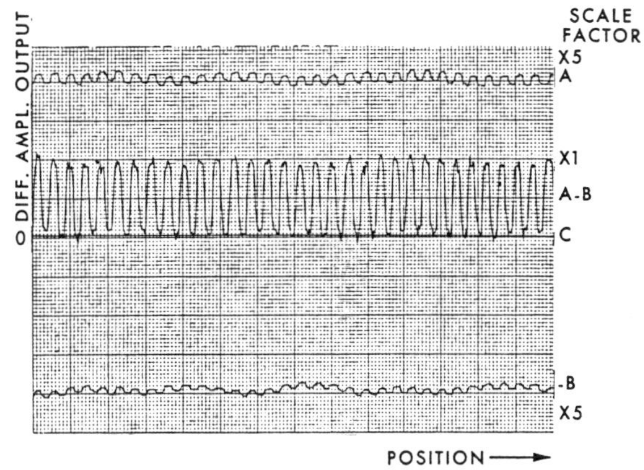

FIG. 8. As Fig. 7, except $\theta = 130'$.

the three methods described above, namely, proper adjustment of the analyzer angle, spatial filtering, and the differential technique are very effective in reducing the surface noise. It is felt that actually these methods, if they can be applied simultaneously, can reduce surface noise to the extent of making a magneto-optic readout system of practical value.

ACKNOWLEDGMENTS

The author is greatly indebted to I. W. Wolf, H. M. A. El-Sum, P. Smaller, and R. P. Hunt for many stimulating discussions.

Reprinted with permission from *Journal of the Optical Society of America,* Vol. 69(1), pp. 4-24 (January 1979).

Diffraction theory of laser read-out systems for optical video discs

H. H. Hopkins

Department of Physics, University of Reading, Reading, England

(Received 19 August 1978)

The form of discs considered has $\lambda/4$-deep pits along the track, the spacings and lengths of which are modulated. A laser beam is focused on the track by means of a high-aperture objective, and the light which is scattered back through the objective is detected. The diffraction theory of this system is treated, both for simple periodic and for more complicated forms of disc. Formulas are obtained for the fundamental and harmonics of the read-out signal. The influences of focus error and aberrations are studied, for certain cases of which numerical results are given. Also treated are cross-talk between adjacent tracks and intermodulation between the pit-spacing frequencies and those of the length modulation of the pits.

I. INTRODUCTION

The Video Long Play (VLP) system of Philips[1,2] allows the recording and reproduction of color television programs. The recording is made on a conventional size 30-cm diameter gramophone disc, and has a playing time of about 30 min. In such a system it is necessary to record a luminance signal and also chrominance and sound signals. The time-bandwidth product is thus large, and accordingly the signals must be encoded on a very small physical scale. This is achieved by using small "pits" along a spiral track in the disc and an optical read-out system, as indicated schematically in Fig. 1. The appearance of three adjacent tracks is shown schematically in Fig. 2. The repetition frequency of the pits along the track determines the luminance signal, whilst the chrominance and sound signals can be encoded as a modulation of the length of the pits. The optical read-out system is illustrated in Fig. 1.

Light from a small He-Ne laser B passes through a grating G and produces three spots of light at O. The center spot at O is imaged on the track at O', after passing through a polarizing beamsplitter P, reflection at the mirror M, and traversal of the $\lambda/4$-plate W. The role of the polarizer and $\lambda/4$-plate is two-fold: it gives an efficient use of the polarized beam of light from the laser, and also avoids feedback to the laser of light reflected from the disc at O'. With no pits on the disc all of the light is reflected back to the detector at D. If the light spot at O' falls on a pit, some light is diffracted outside the aperture of the objective and so a reduced signal is given by the detector. This signal is thereby modulated as the disc rotates, and the signal from the detector then contains the necessary frequencies for the reproduction of the picture and sound.

The two first-order diffracted beams from the grating are imaged as spots on either side of the tracks being read, as shown diagrammatically at A_1 and A_2 in Fig. 2. The light reflected from these spots falls on two additional detectors at D, and these signals are employed for tracking. If the spots move off-center, that is upwards or downwards in Fig. 2, the signal from one will increase and that from the other will decrease in strength. In the same manner the amount of undiffracted light from the read-out spot at O' will also change if it moves off-center. The study of these factors, namely the required precision of tracking and the sensitivity of the error signals from A_1 and A_2, are all made possible by the detailed diffraction analysis to be described here.

Each pit has a width of about 0.5 μm and adjacent tracks are spaced center-to-center typically by 1.7 μm. Using an objective of numerical aperture NA = 0.4 and a He-Ne laser, $\lambda = 632.8$ nm, the diameter of the Airy disc is 1.93 μm, and this size is indicated by the three circles centered on A_1, O', and A_2 in Fig. 2. The optical depth of each pit is $\lambda/4$, and thus imposes a phase change of π on light which falls inside it. In practice the sides of the pits are sloping, but this is also taken into account in the analysis to be described.

In addition to tracking errors, the signals received at the detectors will also be influenced by focus error of the objective and by its aberrations. It was the purpose of the diffraction theory of the read-out system to make possible the study of the influence of all of these factors on the detector signals. The analysis will first be developed for the special case shown in Fig. 2, in which all pits have the same length, and where the pits in adjacent tracks have no "stagger" relative to each other. It will then be indicated how the relative "staggering" of adjacent tracks and the duty cycle of length modulation of the pits may be included. The case of adjacent tracks having different repetition frequencies for the pits is also treated.

Scalar diffraction theory is employed in the analysis, and this calls for two comments. In the first place, the objective used has a numerical aperture which can be as high as 0.5. It is known, however, that the vector diffraction theory of the Airy disc gives results which do not differ very significantly from those given by the scalar theory provided the numerical aperture does not exceed about 0.5. The second consideration is that the size of detail on the video disc is of the same order of magnitude as the wavelength of the light employed. In consequence the appropriate boundary-value problem should, strictly speaking, be solved in order to find the variation of the reflection coefficient over the disc. The structure on the disc has, however, a periodic or quasi-periodic nature so that the theory given here would still apply, given only that the coefficients in the Fourier series employed are suitably changed to take account of any differences between their values given by the simpler treatment employed here and a full boundary-value calculation. Account would also need to be taken of differences between directions of polarization.

II. COORDINATE SYSTEMS EMPLOYED

It is convenient here, as in other cases in both the geometrical and physical treatments of image formation, to use canonical coordinates[3] for the object and image planes, and also for the entrance and exit pupils. In Fig. 3, (ξ,η) are real-space

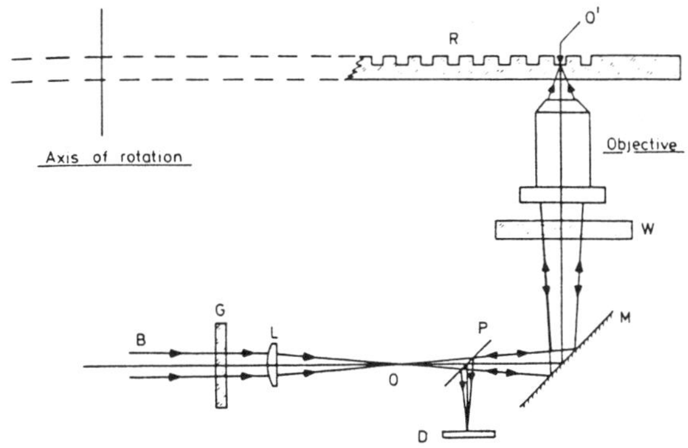

FIG. 1. Optical scanning system.

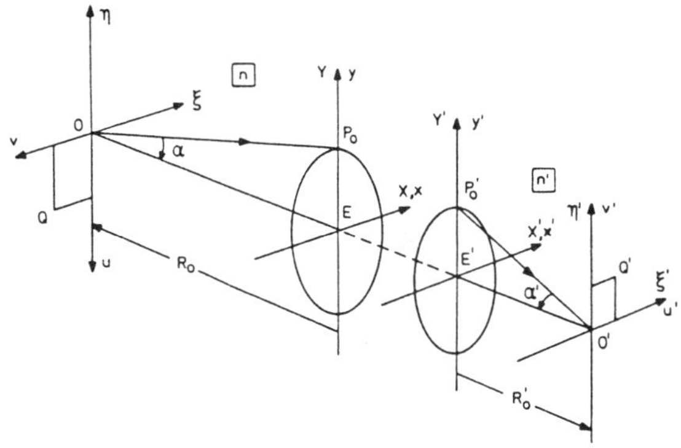

FIG. 3. Object and image-plane coordinates.

coordinates in the object plane at O, similarly (ξ',η') are real-space coordinates for the image plane at O'. The incident and emergent portions of the marginal ray are shown as OP_0 and $P_0'O'$ respectively. The numerical apertures in the object and image spaces are respectively $(n\ \sin\alpha)$ and $(n'\ \sin\alpha')$, where n and n' are the refractive indices of the object and image spaces. The reduced coordinates of an object point Q, having real-space coordinates (ξ,η), are defined by

$$u = (n\ \sin\alpha/\lambda)\xi \quad v = (n\ \sin\alpha/\lambda)\eta \tag{2.1}$$

and the coordinates for an image point Q', real-space coordinates (ξ',η'), are similarly defined to be

$$u' = (n'\ \sin\alpha')/\lambda)\xi' \quad v' = (n'\ \sin\alpha'/\lambda)\eta', \tag{2.2}$$

where λ is the vacuum wavelength of the light employed. The angles α and α' are respectively negative and positive for the case shown in Fig. 3, so that (u,v) and (u',v') are oriented as indicated in the diagram. If Q' is the geometrical image of Q, its coordinates are given by $\xi' = M\xi, \eta' = M\eta$, where

$$M = \eta\ \sin\alpha/\eta'\ \sin\alpha' \tag{2.3}$$

is the geometrical magnification. Thus, the geometrical image of the point (u,v) has coordinates $(u' = u,\ v' = v)$.

The entrance and exit pupils are shown in Figs. 3 and 4 at E and E'. The marginal ray intersects the pupil planes at P_0 and P_0'. It is, however, essential in the theory of image formation to identify the pupil surfaces with the pupil spheres, rather than the pupil planes, if Fourier transform theory is to be applied. Thus, in Fig. 4, the pupil spheres are EBP and $E'B'P'$ with centers at O and O' respectively. Any general aperture ray, $OB\ldots B'O'$, intersecting the pupil spheres in B and B', having real-space coordinates (X,Y) and (X',Y'), has reduced coordinates defined by

$$x = X/h, \quad y = Y/h, \qquad x' = X'h', \quad y' = Y'/h', \tag{2.4}$$

so that $x^2 + y^2 = 1$ and $x'^2 + y'^2 = 1$ are the peripheries of the entrance and exit pupils.

For a system satisfying the sine condition, the angles θ and θ' in Fig. 4, which is shown in two dimensions merely for convenience, are such that

$$\sin\theta'/\sin\alpha' = \sin\theta/\sin\alpha$$

and, multiplying the two sides by R_0'/R_0' and R_0/R_0 respectively, gives

$$X'/h' = X/h, \quad Y'/h' = Y/h \tag{2.5}$$

for any ray. Thus a ray entering with entrance pupil coordinates (x,y) emerges from the point $(x' = x, y' = y)$ of the exit pupil sphere.

In the present context the use of the above canonical coordinates is important from a number of points of view. Of particular importance is their relevance to diffraction theory. Thus, if $f(x',y')$ is the distribution of complex amplitude over the pupil sphere $E'B'P'$, of Fig. 4, the complex amplitude at any point D', with reduced coordinates (u',v'), in the diffraction image formed at O', is given, apart from an irrelevant constant, very exactly by

$$U_{D'} = \exp\left[-ik\left(\frac{\xi'^2 + \eta'^2}{2R_0'}\right)\right] F(u',v'), \tag{2.6}$$

where

$$F(u',v') = \int_{-\infty}^{+\infty}\int_{-\infty}^{+\infty} f(x',y') \times \exp\left[i2\pi(u'x' + v'y')\right] dx'\,dy' \tag{2.7}$$

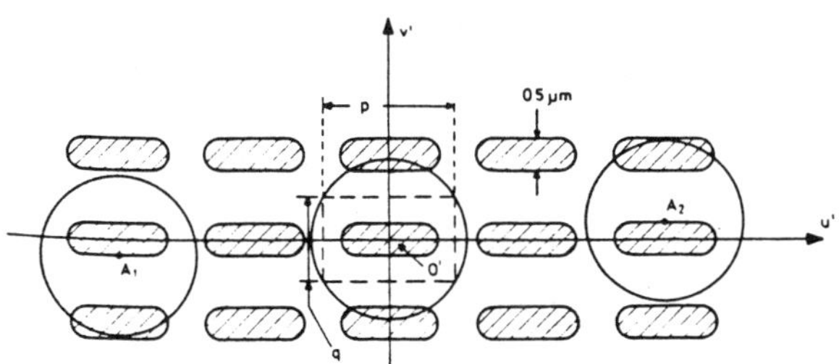

FIG. 2. Read-out and tracking spots.

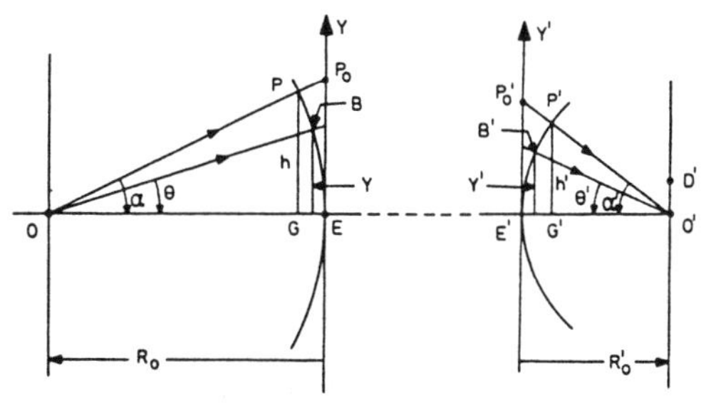

FIG. 4. Pupil sphere coordinates.

is the Fourier transform of $f(x',y')$, and (ξ',η') are the real-space coordinates of D'. In the case to be considered the exit pupil E' is at infinity, so that $R_0' = \infty$ and

$$U_{D'} = F(u',v') \tag{2.8}$$

which is the basic Fourier transform relation between the aperture and the image distributions of complex amplitude. The relations (2.6) and (2.8) only hold accurately if the function $f(x',y')$ expresses the complex amplitude over the pupil sphere as a function of the rectangular coordinates (x',y') of points on this sphere.

When the exit pupil is at infinity, both h' and (X',Y') become infinite, but the exit pupil is still the unit circle $x'^2 + y'^2 \leq 1$ when the relative coordinates (2.4) are employed. It is also useful to note that the dimensionless coordinates (u,v) and (u',v') both have the values 0.61 corresponding to the Airy disc radii in the object and image spaces respectively, independently of the numerical aperture of the optical system and of the wavelengths of the light used.

III. GENERAL FORMULAS FOR THE LIGHT RECEIVED BY THE DETECTOR

Light from the focal point O of the laser beam, Fig. 1, will give a diverging spherical wave at the entrance pupil of the objective, but it is a wave which is inhomogeneous in real amplitude. This we shall denote by $\tau(x,y)$, where for the particular laser mode

$$\tau(x,y) = \exp[-(1/2)\sigma(x^2+y^2)] \tag{3.1}$$

corresponding to an intensity distribution

$$I(x,y) = \exp[-\sigma(x^2+y^2)] \tag{3.2}$$

so that the intensity at the center of the wave is normalized to unity, and

$$I_m/I_0 = \exp(-\sigma) \tag{3.3}$$

is the ratio of the intensity at the edge of the pupil to that at the midpoint. The subsequent analysis is not restricted to the above case; and accordingly the nonuniform amplitude of the wave at the entrance pupil will be denoted simply by $\tau(x,y)$. Since, for corresponding rays in the image and object spaces we have $(x' = x, y' = y)$ for the exit pupil coordinates, the real amplitude over the exit pupil sphere will be given by $\tau(x',y')$. It is assumed here that the pupil is of uniform transmittance.

If the objective has focus error and/or aberration the wavefront at the exit pupil of the objective will be specified by its wavefront aberration $W(x',y')$. The distribution of complex amplitude over the exit pupil sphere will then be given by

$$\begin{aligned} f(x,y) &= \tau(x,y)\exp[i2\pi W(x,y)] \quad (x^2+y^2) \leq 1, \\ &= 0 \quad\quad (x^2+y^2) > 1, \end{aligned} \tag{3.4}$$

where the primes may be omitted since $x' = x, y' = y$ for any given ray. In (3.4) the wavefront aberration and/or focus error will be expressed using the wavelength of the light as the unit of length. With the entrance pupil at the front focus of the objective, the exit pupil is at infinity; in accordance with (2.8), the complex amplitude in the point-spread function, formed at O' in Fig. 1, is given by

$$F(u',v') = \int_{-\infty}^{+\infty}\int_{-\infty}^{+\infty} f(x,y)\exp[i2\pi(u'x+v'y)]dx\,dy; \tag{3.5}$$

i.e., the Fourier transform of $f(x,y)$. This light is diffracted by the pits of the disc recording, and it is required to calculate how much of this diffracted light falls inside the exit pupil of the objective. This is the light which eventually falls on the detector at D in Fig. 1.

The upper surface of the disc, R in Fig. 1, is metallized. We shall assume its real amplitude reflection coefficient to be uniform and equal to unity. This last merely omits an irrelevant scaling factor. The complex amplitude reflection coefficient we shall denote by $R(u',v')$. Thus, outside a pit we shall have $R(u',v') = 1$; and inside a pit $R(u',v') = \exp[i\phi(u',v')]$, where $\phi(u',v')$ describes the phase profile of the pit. The light leaving the disc has thus a complex amplitude distribution $F(u',v')R(u',v')$; and, by the converse of (2.8), the complex amplitude diffracted back to the point (x,y) of the exit pupil will be given by

$$a'(x,y) = \int_{-\infty}^{+\infty}\int_{-\infty}^{+\infty} F(u',v')R(u',v') \times \exp[-i2\pi(u'x+v'y)]dx\,dy$$

which is simply the (inverse) Fourier transform of the product $F(u',v')R(u',v')$. Hence

$$a'(x,y) = \int_{-\infty}^{+\infty}\int_{-\infty}^{+\infty} f(x-x',y-y')r(x',y')dx'\,dy' \tag{3.6}$$

by a standard theorem in Fourier analysis.

We shall initially assume the simple form of disc depicted in Fig. 2. The position of the disc is there shown with no tracking error and with a pit of the center track at the origin at time $t = 0$. We denote by $R_0(u',v')$ the complex reflection coefficient of the disc in this position. If ω is the angular velocity of the disc, and a the radius of the given track, the disc will have moved a reduced distance

$$u = (n'\sin\alpha'/\lambda)\omega a t \tag{3.7}$$

at time t. If there is then also a tracking error of reduced amount v, the complex reflection coefficient of the disc at time t will be

$$R(u',v') = R_0(u'-u, v'-v), \tag{3.8}$$

where, for the case in question, the function $R_0(u',v')$ is doubly periodic with respect to the rectangular axes (u',v').

Thus, if p and q are the reduced center-to-center separations of the pits along each track and the center-to-center spacings of the tracks respectively, the "unit cell" for $R_0(u',v')$ is the rectangle bounded by $u' = \pm p/2$, $v' = \pm q/2$ as indicated in Fig. 2. The function $R_0(u',v')$ may thus be written as a complex Fourier series,

$$R_0(u',v') = \sum_{m=-\infty}^{+\infty}\sum_{n=-\infty}^{+\infty} R(m,n) \times \exp\left\{i2\pi\left[\left(\frac{m}{p}\right)u' + \left(\frac{n}{q}\right)v'\right]\right\}, \tag{3.9}$$

where

$$R(m,n) = \frac{1}{pq} \int_{-p/2}^{+p/2} \int_{-q/2}^{+q/2} R_0(u',v') \times \exp\left\{-i2\pi\left[\left(\frac{m}{p}\right)u' + \left(\frac{n}{q}\right)v'\right]\right\} du'\, dv' \quad (3.10)$$

are the coefficients. By (3.8) and (3.9) the reflection coefficient of the disc at time t will be of the form

$$R(u',v') = \sum_{m=-\infty}^{+\infty} \sum_{n=-\infty}^{+\infty} \exp\left\{-i2\pi\left[\left(\frac{m}{p}\right)u + \left(\frac{n}{q}\right)v\right]\right\} \times R(m,n) \exp\left\{i2\pi\left\{\left(\frac{m}{p}\right)u' + \left(\frac{n}{q}\right)v'\right]\right\}$$

whose inverse Fourier transform is

$$r(x',y') = \sum_{m=-\infty}^{+\infty} \sum_{n=-\infty}^{+\infty} \exp\left\{-i2\pi\left[\left(\frac{m}{p}\right)u + \left(\frac{n}{q}\right)v\right]\right\} \times R(m,n)\delta\left(x' - \frac{m}{p}, y' - \frac{n}{q}\right), \quad (3.11)$$

where $\delta(\ )$ is a delta function of two variables. We now substitute (3.11) in (3.6), and evaluate the integrals using the sifting property of the δ function. This gives

$$a'(x,y) = \sum_{m=-\infty}^{+\infty} \sum_{n=-\infty}^{+\infty} \exp\left\{-i2\pi\left[\left(\frac{m}{p}\right)u + \left(\frac{n}{q}\right)v\right]\right\} \times R(m,n)\, f\left(x - \frac{m}{p}, y - \frac{n}{q}\right), \quad (3.12)$$

for the complex amplitude diffracted backwards to the point (x,y) of the exit pupil. The intensity falling at this point is given by the squared modulus of (3.12); that is, by

$$I(x,y) = \sum_m \sum_n \sum_{m'} \sum_{n'} \exp\left\{-i2\pi\left[\left(\frac{m-m'}{p}\right)u + \left(\frac{n-n'}{q}\right)v\right]\right\} \times R(m,n)R^*(m',n') f\left(x - \frac{m}{p}, y - \frac{n}{q}\right) f^*\left(x - \frac{m'}{p}, y - \frac{n'}{q}\right), \quad (3.13)$$

where m, n, m', and n' each take the values $0, \pm 1, \pm 2, \ldots$, and $*$ denotes the complex conjugate.

The total light diffracted back to the detector is found by integrating $I(x,y)$ over the pupil area; that is, by

$$I = \sum_m \sum_n \sum_{m'} \sum_{n'} \exp\left\{-i2\pi\left[\left(\frac{m-m'}{p}\right)u + \left(\frac{n-n'}{q}\right)v\right]\right\} \times R(m,n)R^*(m',n')D(m,n;m',n'), \quad (3.14)$$

where

$$D(m,n;m',n') = \iint_{(x^2+y^2)\le 1} f\left(x - \frac{m}{p}, y - \frac{n}{q}\right) \times f^*\left(x - \frac{m'}{p}, y - \frac{n'}{q}\right) dx\, dy, \quad (3.15)$$

the pupil being bounded by the circle $x^2 + y^2 = 1$. It is to be noted that in (3.14) the light flux I is expressed as a function of u and v, the former of which is related to time through (3.7) whilst v is the reduced magnitude of the tracking error. The ranges of values of m, n, m', and n' in the summations of (3.14) are restricted to those values for which the integral $D(m,n,m',n)$ of (3.15) has nonzero values. This will be considered further in the next section, in which (3.14) is used to find the dc and ac components of the detector light flux I considered as a function of time.

IV. THE FOURIER COMPONENTS OF THE TIME VARIATION OF THE LIGHT FLUX REACHING THE DETECTOR

We put in (3.14),

$$(m' - m) = \mu, \quad (n' - n) = \nu \quad (4.1)$$

and, writing $I(t)$ for I, we then obtain

$$I(t) = \sum_m \sum_n \sum_\mu \sum_\nu \exp\left\{i2\pi\left\{\left(\frac{\mu}{p}\right)u + \left(\frac{\nu}{q}\right)v\right]\right\} R(m,n) \times R^*(m+\mu, n+\nu)D(m,n;m+\mu,n+\nu), \quad (4.2)$$

in which μ and ν now also take the values $0, \pm 1, \pm 2, \ldots$, subject to $D(m,n,m+\mu,n+\nu)$ being nonzero. Using (3.7) for u, we write

$$\omega_\mu t = 2\pi(\mu/p)u = \mu\{2\pi(n' \sin\alpha'/\lambda)\, \omega a/p\}t \quad (4.3)$$

or, since the reduced period p is given by

$$p = (n' \sin\alpha'/\lambda)\xi' \quad (4.4)$$

where ξ' is the real-space center-to-center distance between the pits, (4.3) gives

$$\omega_\mu = \mu\{2\pi(\omega a/\xi')\} = \mu\omega_1,$$
$$\omega_1 = 2\pi(\omega a/\xi'). \quad (4.5)$$

Inverting the order of the summations, (4.2) is now written

$$I(t) = \sum_\mu \left\{\sum_\nu \sum_m \sum_n \exp\left[i2\pi\left(\frac{\nu}{q}\right)v\right] R(m,n)R^*(m+\mu,n+\nu) \times D(m,n;m+\mu,n+\nu)\right\}\exp(i\omega_\mu t) \quad (4.6)$$

which, being the integral of the squared modulus of $a'(x,y)$ in (3.12), is essentially real.

The expression (4.6), considered as a function of t, is the complex Fourier series representing the time-varying signal $I(t)$. If we write (4.6) as

$$I(t) = \sum_{\mu=-\infty}^{+\infty} I_\mu \exp\{i\mu\omega_1 t\} \quad (4.8)$$

the coefficients

$$I_\mu = \sum_\nu \sum_m \sum_n \exp\left[i2\pi\left(\frac{\nu}{q}\right)v\right] R(m,n)R^*(m+\mu,n+\nu) \times D(m,n;m+\mu,n+\nu) \quad (4.9)$$

must be such that I_0 = real, and $I_{-\mu} = I_\mu^*$ for $\mu = +1, +2, +3, \ldots$, since $I(t)$ is a wholly real function. We may, therefore, write from (4.8)

$$I(t) = I_0 + \sum_{\mu=1}^{\infty} [I_\mu \exp(i\,\mu\omega,t) + I_\mu^* \exp[-i\mu\omega,t)]$$

and only non-negative values of μ now occur. Replacing $\mu\omega_1$ by ω_μ, this formula gives

$$I(t) = A_0 + \sum_{\mu=1}^{\infty} [A_\mu \cos(\omega_\mu t) + B_\mu \sin(\omega_\mu t)] \quad (4.10)$$

where

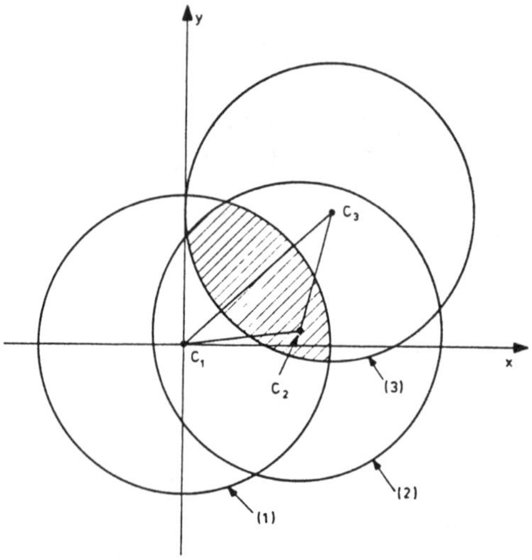

FIG. 5. Region of integration for $D(m,n;m+\mu,n+\nu)$.

$$A_0 = I_0,\quad A_\mu = 2\mathrm{Re}(I_\mu),\quad B_\mu = -2\,\mathrm{Im}(I_\mu) \tag{4.11}$$

in which Re ( ) and Im ( ) denote respectively the real and imaginary parts. In (4.10) A_0 gives, apart from a constant factor, the dc component of the time-varying light flux falling on the detector, whilst A_μ, B_μ (with the same constant factor omitted) give the amplitudes of the two quadrature components of the fundamental and harmonics of the ac components.

If the disc has no pits, $R_0(u',v') = 1$ and the coefficients in the Fourier series (3.9) are all zero except for $m = n = 0$, when $R(0,0) = 1$. (4.9) is then also zero except for $m = n = 0$, when it has the value

$$I_\mu = \sum_\nu \exp\left[i2\pi\left(\frac{\nu}{q}\right)v\right] R(0,0)R^*(\mu,\nu)D(0,0;\mu,\nu)$$

and, since the factor $R^*(\mu,\nu) = 0$ except when $\mu = \nu = 0$, this reduces to

$$I_0 = |R(0,0)|^2 D(0,0;0,0). \tag{4.12}$$

The disc is assumed to have unit real amplitude reflection coefficient, so that, using (3.15), this expression becomes

$$I_0 = \iint_{(x^2+y^2)\le 1} |f(x,y)|^2\,dx\,dy, \tag{4.13}$$

which is simply the total flux falling on the entrance pupil, as was to be expected. We shall divide (4.10) by this quantity in order to normalize $I(t)$ such that the detector signal is unity when there are no pits on the disc.

We shall denote the quantity (4.13) by A. For the present case, $|f(x,y)| = \tau(x,y)$ has the Gaussian form (3.1). The normalizing factor (4.13) is then given by

$$A = \iint_{(x^2+y^2)\le 1} \exp[-\sigma(x^2+y^2)]\,dx\,dy,$$

which is easily integrated to give

$$A = (\pi/\sigma)[1-\exp(-\sigma)] \tag{4.14}$$

and, after computing the coefficients A_0, A_μ, and B_μ in (4.10), they have all finally to be divided by this quantity A in order to give $I(t)$ in the suitably normalized form. It will be seen that, in this form, the assumption of 100% reflection coefficient for the disc is merely one of convenience, which has no influence on the normalized forms of the coefficients A_0, A_μ, and B_μ.

V. FORMULAS FOR THE FOURIER COEFFICIENTS OF THE DETECTOR SIGNAL

To evaluate the Fourier coefficients it is necessary to find the real and imaginary parts of the coefficients I_μ of (4.9), in which ν, m, and n take the values 0, ±1, ±2, . . . The ranges of values of μ, ν, m, and n in the summations of (4.10) and (4.9) are determined by those values for which the integral (3.15), with $m' = m + \mu$, and $n' = n + \nu$, namely

$$D(m,n;m+\mu,n+\nu) = \iint_{x^2+y^2\le 1} f\left(x-\frac{m}{p},\, y-\frac{n}{q}\right) \times f^*\left(x-\frac{m+\mu}{p},\, y-\frac{n+\nu}{q}\right)dx\,dy, \tag{5.1}$$

has nonzero values. The region of integration for (5.1) is shown in Fig. 5. The circle (1) in the diagram is the region $x^2 + y^2 \le 1$, with center at the origin: this is the formal region of integration. However, $f(x,y) = 0$ for $x^2 + y^2 > 1$, so that the integrand in (5.1) is zero except in the region of overlap of the circles

$$(x - m/p)^2 + (y - n/q)^2 = 1 \tag{5.2}$$

and

$$[x - (m+\mu)/p]^2 + [y - (n+\nu)/q]^2 = 1 \tag{5.3}$$

with centers C_2 and C_3 at $(x = m/p, y = n/q)$ and $[x = (m + \mu)/p, y = (n + \nu)/q]$ respectively. These circles are denoted by (2) and (3) respectively in Fig. 5. The effective region of integration is thus the region common to the three circles (1), (2), and (3), which is shown shaded in the figure. Each of these circles is of unit radius.

The integrand in (5.1) is thus zero unless at least the circles (2) and (3) overlap each other; that is, we require the distance (C_2C_3) to be numerically less than 2. The coordinates of C_2 are $x = m/p, y = n/q$, and those of C_3 are $x = (m + \mu)/p, y = (n + \nu)/q$: one necessary condition for $D(m,n;m + \mu,n + \nu) \neq 0$ is thus

$$(\mu/p)^2 + (\nu/q)^2 < 4 \tag{5.4}$$

and μ will have its numerically largest permissible value when $\nu = 0$. The range of values of the integer μ in (4.10) is, therefore, restricted to

$$0 \le \mu \le \mu_0, \tag{5.5}$$

where μ_0 is the largest integer for which

$$\mu_0 < 2p \tag{5.6}$$

in accordance with (5.4) when $\nu = 0$.

We may note that the highest temporal frequency that can be present in the detector signal will be given by (4.3) with $\mu = 2p$. These temporal frequencies will thus never exceed the value

$$f = 2(n' \sin\alpha'/\lambda)\omega a \tag{5.7}$$

since $f = \omega_\mu/2\pi$, and μ in (4.3) may never exceed the value $2p$.

We now consider the range of values assumed by the integer ν in the summations (4.9) for the coefficients I_μ, when μ has any given value. It follows at once from (5.4) that ν must lie in the range

$$-q[4-(\mu/p)^2]^{1/2} < \nu < q[4-(\mu/p)^2]^{1/2} \tag{5.8}$$

so that, if ν_0 is the largest positive integer for which

$$\nu_0 < q[4-(\mu/p)^2]^{1/2} \tag{5.9}$$

the values taken by ν will be

$$\nu = 0, \pm 1, \pm 2, \ldots, \pm \nu_0, \tag{5.10}$$

and, for each positive value of ν there will be a corresponding negative value.

We may now write (4.10) as

$$I(t) = A_0 + \sum_{\mu=1}^{\mu_0} [A_\mu \cos(\omega_\mu t) + B_\mu \sin(\omega_\mu t)] \tag{5.11}$$

with, as in (4.4),

$$A_0 = I_0, \quad A_\mu = 2\,\mathrm{Re}(I_\mu), \quad B_\mu = -2\,\mathrm{Im}(I_\mu), \tag{5.12}$$

and (4.9) has the form

$$\begin{aligned} I_\mu = {} & \sum_m \sum_n R(m,n)R^*(m+\mu,n)D(m,n;m+\mu,n) \\ & + \sum_{\nu=1}^{\nu_0} \exp\left[i2\pi\left(\frac{\nu}{q}\right)v\right] \sum_m \sum_n R(m,n)R^*(m+\mu,n+\nu) \\ & \qquad \times D(m,n;m+\mu,n+\nu) \\ & + \sum_{\nu=1}^{\nu_0} \exp\left[-i2\pi\left(\frac{\nu}{q}\right)v\right] \sum_m \sum_n R(m,n)R^*(m+\mu,n-\nu) \\ & \qquad \times D(m,n;m+\mu,n-\nu), \end{aligned} \tag{5.13}$$

where μ_0 and ν_0 are given by the conditions (5.6) and (5.9) respectively. It remains to find the ranges of the inner summations in (5.13), that is, over m and n.

We have so far used only the condition that the circles (2) and (3) overlap. Two further necessary conditions are that the circles (1) and (3), and also the circles (1) and (2), overlap; that is, that $(C_1C_3) < 2$ and $(C_1C_2) < 2$. These two conditions are

$$[(m+\mu)/p]^2 + [(n \pm \nu)/q] < 4 \tag{5.14}$$

and

$$(m/p)^2 + (n/q)^2 < 4. \tag{5.15}$$

In (5.14), $\nu = 0$ for the first series in (5.13) whilst the positive and negative signs apply to the second and third series respectively. Since μ takes only values greater than or equal to 0, the maximum numerical values of m permitted by (5.14) and (5.15) will be when $(n \pm \nu) = 0$ and $n = 0$ respectively. (5.14) then gives

$$(-2p) < (m+\mu) < (+2p)$$

or

$$(-2p-\mu) < m < (2p-\mu) \tag{5.16}$$

and (5.15) similarly requires

$$-2p < m < +2p. \tag{5.17}$$

Since μ is never negative, we shall always have $(-2p) > (-2p - \mu)$ and $(2p-\mu) < 2p$. It is thus (5.17) which determines the lower limit for m, and (5.16) which determines the upper limit. Hence, if m_1 and m_2 are respectively the algebraically smallest and largest integers for which

$$\begin{aligned} m_1 &> (-2p) \\ m_2 &< (2p-\mu) \end{aligned} \tag{5.18}$$

the range of values of m will be

$$m_1 \le m \le m_2 \tag{5.19}$$

for all three series in (5.13).

The above considerations serve to impose necessary conditions on the values of μ, ν, and m. For given values of these integers, (5.14) imposes necessary conditions on the values of n. With the positive sign, and $\nu > 0$, (5.14) applies to the first two series in (5.13): the values of n must now satisfy

$$-\left[4-\left(\frac{m+\mu}{p}\right)^2\right]^{1/2} < \left(\frac{n+\nu}{q}\right) < +\left[4-\left(\frac{m+\mu}{p}\right)^2\right]^{1/2};$$

that is,

$$\left\{-q\left[4-\left(\frac{m+\mu}{p}\right)^2\right]^{1/2} - \nu\right\} < n < \left\{+q\left[4-\left(\frac{m+\mu}{p}\right)^2\right]^{1/2} - \nu\right\}$$

and similarly (5.15) requires that

$$\left\{-q\left[4-\left(\frac{m}{p}\right)^2\right]^{1/2}\right\} < n < \left\{+q\left[4-\left(\frac{m}{p}\right)^2\right]^{1/2}\right\}$$

in which both μ and ν are greater than 0. Now $\mu > 0$ will make $[4-(m+\mu)^2/p^2]^{1/2} < [4-m^2/p^2]^{1/2}$, so that the upper limit for n will always be the algebraically largest integer n_2 for which

$$n_2 < q\{4-[(m+\mu)/p]^2\}^{1/2} - \nu, \tag{5.20}$$

but the lower limit will depend on the particular values of μ and ν. We thus choose the algebraically lowest value n_1 such that both the conditions

$$\begin{aligned} n_1 &> -q\{4-[(m+\mu)/p]^2\}^{1/2} - \nu, \\ n_1 &> -q[4-(m/p)^2]^{1/2}, \end{aligned} \tag{5.21}$$

are satisfied. The ranges of values of n in the first and second series of (5.13) are then

$$n_1 \le n \le n_2 \tag{5.22}$$

with n_2 given by (5.20) and n_1 by one or other of the expressions in (5.21). For the special case $\mu = 0$, for example, the limits will be given by

$$\begin{aligned} n_1 &> -q[4-(m/p)^2]^{1/2}, \\ & \qquad\qquad (\mu = 0) \\ n_2 &< +q[4-(m/p)^2]^{1/2} - \nu, \end{aligned} \tag{5.23}$$

for the first and second series of (5.13).

The third series in (5.13) has the negative sign in (5.14). The conditions to be imposed on n are now

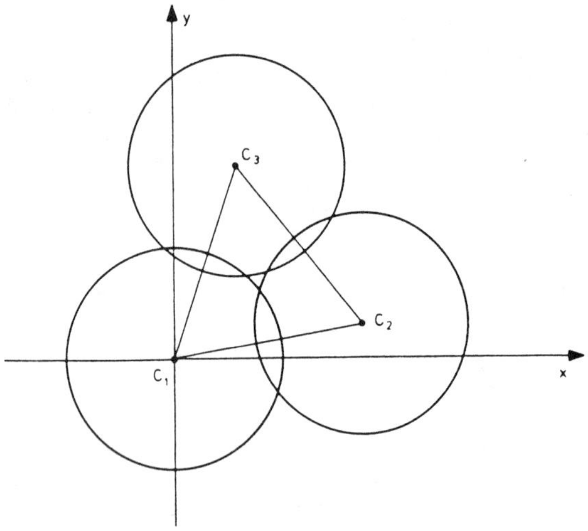

FIG. 6. Special case of the circles C_1, C_2, and C_3.

$$\left\{-q\left[4-\left(\frac{m+\mu}{p}\right)^2\right]^{1/2}+\nu\right\} < n < \left\{+q\left[4-\left(\frac{m+\mu}{p}\right)^2\right]^{1/2}+\nu\right\}$$

and

$$\{-q\,[4-(m/p)^2]^{1/2}\} < n < \{+q[4-(m/p)^2]^{1/2}\}$$

giving for the lower limit

$$n_3 > -q\{4-[(m+\mu)]^2/p\}^{1/2}+\nu \qquad (5.24)$$

since again μ and ν are both greater than 0. The upper limit is given by the algebraically largest integer n_4 such that both the conditions

$$n_4 < +q\{4-[(m+\mu)/p]^2\}^{1/2}+\nu,$$
$$n_4 < +q[4-(m/p)^2]^{1/2}, \qquad (5.25)$$

are satisifed and the range of values of n is then

$$n_3 \le n \le n_4 \qquad (5.26)$$

for the third series of (5.13). For the special case $\mu = 0$, the limits will be given by

$$\left.\begin{aligned} n_3 &> -q[4-(m/p)^2]^{1/2}+\nu \\ n_4 &< +q[4-(m/p)^2]^{1/2} \end{aligned}\right\} \quad (\mu = 0) \qquad (5.27)$$

for the third series.

The complex coefficient I_μ, given by (5.13), is now written

$$\begin{aligned} I_\mu = &\sum_{m=m_1}^{m_2}\sum_{n=n_1}^{n_2} R(m,n)R^*(m+\mu,n)D(m,n;m+\mu,n) \\ &+\sum_{\nu=1}^{\nu_0}\exp\left\{i2\pi\left(\frac{\nu}{q}\right)v\right\}\sum_{m=m_1}^{m_2}\sum_{n=n_1}^{n_2} R(m,n) \\ &\times R^*(m+\mu,n+\nu)D(m,n;m+\mu,n+\nu) \\ &+\sum_{\nu=1}^{\nu_0}\exp\left\{-i2\pi\left(\frac{\nu}{q}\right)v\right\}\sum_{m=m_1}^{m_2}\sum_{n=n_1}^{n_4} R(m,n) \\ &\times R^*(m+\mu,n-\nu)D(m,n;m+\mu,n-\nu) \end{aligned} \qquad (5.28)$$

with ν_0, m_1, m_2, n_1, n_2, n_3, and n_4 given by (5.9), (5.18), (5.21), (5.20), (5.24), and (5.25) respectively. Consideration of these conditions and of the different values of μ and ν shows that the maximum numbers of terms in the various summations in (5.28) are always limited by

$$\begin{aligned} (n_2-n_1+2) &\le 4q+2 \\ (n_4-n_3+2) &\le 4q+2 \\ (m_2-m_1+2) &\le 4p+2 \end{aligned} \qquad (5.29)$$

so that the total number of terms is never excessively large.

It should be emphasized that the above are only necessary conditions, namely that each of the circles (1), (2), and (3) overlap the other two: they do not ensure that all three circles have a region in common. In a few instances the special case illustrated in Fig. 6 may arise, where the circles all overlap in pairs, but where there is no common overlap of the three circles. Such cases are included formally in the summations, but they are automatically excluded at the stage of numerical evaluation of the integrals $D(m,n;m+\mu,n+\infty)$.

We are now in a position to obtain explicit formulas for the coefficients A_0, A_μ, and B_μ, in the expression (5.11) for the time-varying light flux $I(t)$ falling on the detector. From (5.12), we have $A_0 = I_0$; or, using (5.28)

$$\begin{aligned} A_0 = &\sum_{m_1}^{m_2}\sum_{n_1}^{n_2} R(m,n)R^*(m,n)D(m,n;m,n) \\ &+\sum_{\nu=1}^{\nu_0}\exp\left\{i2\pi\left(\frac{\nu}{q}\right)v\right\}\sum_{m_1}^{m_2}\sum_{n_1}^{v_2} R(m,n) \\ &\times R^*(m,n+\nu)D(m,n;m,n+\nu) \\ &+\sum_{\nu=1}^{\nu_0}\exp\left\{-i2\pi\left(\frac{\nu}{q}\right)v\right\}\sum_{m_1}^{m_2}\sum_{n_3}^{n_4} R(m,n) \\ &\times R^*(m,n-\nu)D(m,n;m,n-\nu) \end{aligned} \qquad (5.30)$$

which, we shall now show, is always wholly real.

From the definition (3.15), we have

$$D(m,n;m',n') = \iint_{(x^2+y^2)\le 1} f\left(x-\frac{m}{p},y-\frac{n}{q}\right) \times f^*\left(x-\frac{m'}{p},y-\frac{n'}{q}\right)dx\,dy, \qquad (5.31)$$

so that

$$D(m,n;m,n) = \iint_{(x^2+y^2)\le 1}\left|f\left(x-\frac{m}{p},y-\frac{n}{q}\right)\right|^2 dx\,dy$$

is wholly real. Thus, each term in the first series of (5.30) is necessarily real. In the second and third series, the ranges of the parameters ν and m are identical. For the case $\mu = 0$, (5.23) and (5.27) give, for the ranges of n,

$$\begin{aligned} n_1 &> -q[4-(m/p)^2]^{1/2} & n_2 &< +q[4-(m/p)^2]^{1/2}-\nu \\ n_3 &> -q[4-(m/p)^2]^{1/2}+\nu & n_4 &< +q[4-(m/p)^2]^{1/2} \end{aligned} \qquad (5.32)$$

so that, for each term $n = \tilde{n}$ in the second series there will be a term $n = (\tilde{n}+\nu)$ in the third series. These corresponding terms will be

$$R(m,\tilde{n})R^*(m,\tilde{n}+\nu)D(m,\tilde{n};m,\tilde{n}+\nu)$$

and

$$R(m,\tilde{n}+\nu)R^*(m,\tilde{n})D(m,\tilde{n}+\nu;m,\tilde{n})$$

respectively. From (5.31) it is seen that

$$D(m,\bar{n}+\nu;m,\bar{n}) = D^*(m,\bar{n};m,\bar{n}+\nu) \quad (5.33)$$

and it follows that to each term in the second series there will exist a term in the third series which is its complex conjugate. We have, therefore, shown that A_0 is always wholly real. (5.30) may thus be written

$$A_0 = \sum_{m_1}^{m_2}\sum_{n_1}^{n_2} R(m,n)R^*(m,n)D(m,n;m,n) + 2\,\mathrm{Re}\left\{\sum_{\nu=1}^{\nu_0} \exp\left[i2\pi\left(\frac{\nu}{q}\right)v\right]\sum_{m_1}^{m_2}\sum_{n_1}^{n_2} R(m,n) \times R^*(m,n+\nu)D(m,n;m,n+\nu)\right\} \quad (5.34)$$

where Re{ } denotes the real part.

To separate the real and imaginary parts for use in computation, the following general notation will be used. Let

$$R(m,n) = G(m,n) + iK(m,n), \quad (5.35)$$

where $G(m,n)$ and $K(m,n)$ are the real and imaginary parts, respectively, of $R(m,n)$. We now write

$$R(m,n)R^*(m',n') = M(m,n;m',n') + iN(m,n;m',n'), \quad (5.36)$$

where

$$M(m,n;m',n') = G(m,n)G(m',n') + K(m,n)K(m',n') \quad (5.37)$$

and

$$N(m,n;m',n') = K(m,n)G(m',n') - G(m,n)K(m',n') \quad (5.38)$$

are the real and imaginary parts of the product $R(m,n)$ $R^*(m',n')$. We shall also write

$$D(m,n;m',n') = C(m,n;m',n') + iS(m,n;m',n'), \quad (5.39)$$

where $C(\ )$ and $S(\ )$ are the real and imaginary parts respectively of the integral (5.31). For $m' = m$, $n' = n$, the imaginary parts $N(m,n;m,n)$ and $S(m,n;m,n)$ are identically zero. Substitution of (5.36) and (5.39) in (5.34) now easily gives

$$\begin{aligned} A_0 = &\sum_{m_1}^{m_2}\sum_{n_1}^{n_2} M(m,n;m,n)\,C(m,n;m,n) \\ &+ 2\sum_{\nu=1}^{\nu_0} \cos\left[2\pi\left(\frac{\nu}{q}\right)v\right]\sum_{m_1}^{m_2}\sum_{n_1}^{n_2}\{M(m,n;m,n+\nu) \\ &\times C(m,n;m,n+\nu) - N(m,n;m,n+\nu)S(m,n;m,n+\nu)\} \\ &+ 2\sum_{\nu=1}^{\nu_0} \sin\left[2\pi\left(\frac{\nu}{q}\right)v\right]\sum_{m_1}^{m_2}\sum_{n_1}^{n_2}\{-M(m,n;m,n+\nu) \\ &\times S(m,n;m,n+\nu) - N(m,n;m,n+\nu)C(m,n;m,n+\nu)\} \end{aligned} \quad (5.40)$$

for the dc component of $I(t)$.

The coefficients A_μ and B_μ of the quadrature components of the ac components are given by $2\,\mathrm{Re}(I_\mu)$ and $-2\,\mathrm{Im}(I_\mu)$ respectively, as in (4.11). Substitution of (5.36) and (5.39) in the formula (5.28) for I_μ, gives

$$\begin{aligned} A_\mu = &2\sum_{m_1}^{m_2}\sum_{n_1}^{n_2}[M(m,n;m+\mu,n)C(m,n;m+\mu,n) \\ &- N(m,n;m+\mu,n)\,S(m,n;m+\mu,n)] \\ &+ 2\sum_{\nu=1}^{\nu_0} \cos\left[2\pi\left(\frac{\nu}{q}\right)v\right]\sum_{m_1}^{m_2}\sum_{n_1}^{n_2}[M(m,n;m+\mu,n+\nu) \\ &\times C(m,n;m+\mu,n+\nu) - N(m,n;m+\mu,n+\nu) \\ &\times S(m,n;m+\mu,n+\nu)] \\ &+ 2\sum_{\nu=1}^{\nu_0} \sin\left[2\pi\left(\frac{\nu}{q}\right)v\right]\sum_{m_1}^{m_2}\sum_{n_1}^{n_2}[-M(m,n;m+\mu,n+\nu) \\ &\times S(m,n;m+\mu,n+\nu) - N(m,n;m+\mu,n+\nu) \\ &\times C(m,n;m+\mu,n+\nu)] \\ &+ 2\sum_{\nu=1}^{\nu_0} \cos\left[2\pi\left(\frac{\nu}{q}\right)v\right]\sum_{m_1}^{m_2}\sum_{n_3}^{n_4}[M(m,n;m+\mu,n-\nu) \\ &\times C(m,n;m+\mu,n-\nu)N(m,n;m+\mu,n-\nu) \\ &\times S(m,n;m+\mu,n-\nu)] \\ &+ 2\sum_{\nu=1}^{\nu_0} \sin\left[2\pi\left(\frac{\nu}{q}\right)v\right]\sum_{m_1}^{m_2}\sum_{n_3}^{n_4}[M(m,n;m+\mu,n-\nu) \\ &\times S(m,n;m+\mu,n-\nu) + N(m,n;m+\mu,n-\nu) \\ &\times C(m,n;m+\nu)] \end{aligned} \quad 5.41$$

for twice its real part, and

$$\begin{aligned} B_\mu = &2\sum_{m_1}^{m_2}\sum_{n_1}^{n_2}[-M(m,n;m+\mu,n)S(m,n;m+\mu,n) \\ &- N(m,n;m+\mu,n)C(m,n;m+\mu,n)] \\ &+ 2\sum_{\nu=1}^{\nu_0} \cos\left[2\pi\left(\frac{\nu}{q}\right)v\right]\sum_{m_1}^{m_2}\sum_{n_1}^{v_2}[-M(m,n;m+\mu,n+\nu) \\ &\times S(m,n;m+\mu,n+\nu) - N(m,n;m+\mu,n+\nu) \\ &\times C(m,n;m+\mu,n+\nu)] \\ &+ 2\sum_{\nu=1}^{\nu_0} \sin\left[2\pi\left(\frac{\nu}{q}\right)v\right]\sum_{m_1}^{m_2}\sum_{n_1}^{n_2}[-M(m,n;m+\mu,n+\nu) \\ &\times C(m,n;m+\mu,n+\nu) + N(m,n;m+\mu,n+\nu) \\ &\times S(m,n;m+\mu,n+\nu)] \\ &+ 2\sum_{\nu=1}^{\nu_0} \cos\left[2\pi\left(\frac{\nu}{q}\right)v\right]\sum_{m_1}^{m_2}\sum_{n_3}^{n_4}[-M(m,n;m+\mu,n-\nu) \\ &\times S(m,n;m+\mu,n-\nu) - N(m,n;m+\mu,n-\nu) \\ &\times C(m,n;m+\mu,n-\nu)] \\ &+ 2\sum_{\nu=1}^{\nu_0} \sin\left[2\pi\left(\frac{\nu}{q}\right)v\right]\sum_{m_1}^{m_2}\sum_{n_3}^{n_4}[M(m,n;m+\mu,n-\nu) \\ &\times C(m,n;m+\mu,n-\nu) - N(m,n;m+\mu,n-\nu) \\ &\times S(m,n;m+\mu,n-\nu)] \end{aligned} \quad (5.42)$$

for minus twice the imaginary part of I_μ. In the above formulas, the quantities $M(\ )$ and $N(\ )$ depend only on the coefficients $R(m,n)$ of the complex Fourier series representing the complex amplitude reflection coefficient $R_0(u',v')$ of the disc. Also the quantities $C(\ )$ and $S(\ )$, being the real and imaginary parts of $D(\ )$ defined by (5.31), depend only on the real amplitude $\tau(x,y)$ of the laser beam and the aberration and/or focus error of the objective. It is accordingly convenient to condense the above notation in the following manner.

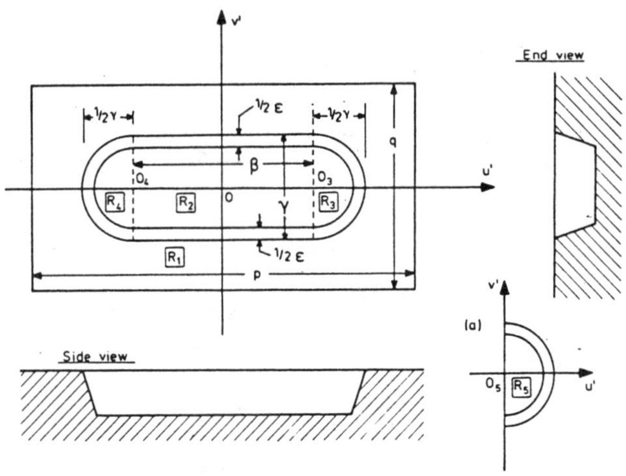

FIG. 7. Form of each pit.

We represent the various series in (5.40), (5.41) and (5.42) by

$$P_1(\mu,\nu) = \sum_{m_1}^{m_2}\sum_{n_1}^{n_2} [M(m,n;m+\mu,n+\nu)C(m,n;m+\mu,n+\nu) - N(m,n;m+\mu,n+\nu)S(m,n;m+\mu,n+\nu)], \quad (5.43)$$

$$P_2(\mu,\nu) = \sum_{m_1}^{m_2}\sum_{n_1}^{n_2} [M(m,n;m+\mu,n+\nu)S(m,n;m+\mu,n+\nu) + N(m,n;m+\mu,n+\nu)C(m,n;m+\mu,n+\nu)], \quad (5.44)$$

and

$$Q_1(\mu,\nu) = \sum_{m_1}^{m_2}\sum_{n_0}^{n_4} [M(m,n;m+\mu,n-\nu)C(m,n;m+\mu,n-\nu) - N(m,n;m+\mu,n-\nu)S(m,n;m+\mu,n-\nu)], \quad (5.45)$$

$$Q_2(\mu,\nu) = \sum_{m_1}^{m_2}\sum_{n_3}^{n_4} [M(m,n;m+\mu,n-\nu)S(m,n;m+\mu,n-\nu) + N(m,n;m+\mu,n-\nu)C(m,n;m+\mu,n-\nu)], \quad (5.46)$$

respectively. Using this notation we now write for (5.40)

$$A_0 = P_1(0,0) + 2\sum_{\nu=1}^{\nu_0}\left\{P_1(0,\nu)\cos\left[2\pi\left(\frac{\nu}{q}\right)v\right] - P_2(0,\nu)\sin\left[2\pi\left(\frac{\nu}{q}\right)v\right]\right\} \quad (5.47)$$

and, similarly for (5.41),

$$A_\nu = 2P_1(\mu,0) + 2\sum_{\nu=1}^{\nu_0}\left\{[P_1(\mu,\nu)+Q_1(\mu,\nu)]\cos\left[2\pi\left(\frac{\nu}{q}\right)v\right] + [-P_2(\mu,\nu)+Q_2(\mu,\nu)]\sin\left[2\pi\left(\frac{\nu}{q}\right)v\right]\right\}, \quad (5.48)$$

whilst (5.42) gives

$$B_\mu = -2P_2(\mu,0) + 2\sum_{\nu=1}^{\nu_0}\left\{-P_2(\mu,\nu) - Q_2(\mu,\nu)]\cos\left[2\pi\left(\frac{\nu}{q}\right)v\right] + [-P_1(\mu,\nu)+Q_1(\mu,\nu)]\sin\left[2\pi\left(\frac{\nu}{q}\right)v\right]\right\} \quad (5.49)$$

and the influence of tracking error may now very easily be studied, once the values of $P_1(\mu,\nu)$, $P_2(\mu,\nu)$, $Q_1(\mu,\nu)$, and $Q_2(\mu,\nu)$, are known, merely by inserting suitable values of v.

VI. COMPUTING PROCEDURES

For any given disc the Fourier coefficients $R(m,n)$ are calculated, giving the real and imaginary parts $G(m,n)$ and $K(m,n)$, from which the quantities $M(m,n;m',n')$ and $N(m,n;m',n')$ defined in (5.37) and (5.38) are found. The real and imaginary parts, $C(m,n;m',n')$ and $S(m,n;m',n')$, of $D(m,n;m',n')$ defined by the integral (5.31) depend only on the aberrations and/or focus error of the objective. We shall first describe the calculation of the Fourier coefficients $R(m,n)$ of the complex reflection function of the disc, and then outline the procedure used for the evaluation of the integrals $D(m,n;m'n')$.

A. Calculation of the Fourier coefficients $R(m,n)$

The formula (3.10) gives

$$R(m,n) = \frac{1}{pq}\int_{-1/2p}^{+1/2p}\int_{-1/2q}^{+1/2q} R_0(u',v') \times \exp\left\{-i2\pi\left[\left(\frac{m}{p}\right)u' + \left(\frac{n}{q}\right)v'\right]\right\} du'\,dv' \quad (6.1)$$

for the coefficients in the Fourier series representing the complex reflection function $R_0(u',v')$ of the disc. The "unit cell" of this function is of reduced dimensions (p,q), as shown in Figs. 2 and 7. To find the values of the coefficients $R(m,n)$ it is necessary to specify the form of each pit.

Arising from the production technique employed, each pit is of the general form suggested in Fig. 7: the two ends are rounded, and the side walls slope inwards. In the diagram the side walls are depicted as inclined straight sides, which meet the flat floor of each pit in a sharp angle. This is merely for the purposes of illustration, and the analysis to be described is in no way restricted to this simple form.

For the purposes of illustration we also assume here that the real amplitude reflection coefficient is the same at all points, both in the interior of the pits and on the surface of the disc, and further that the effect of a pit on the phase of the reflected light is merely to impose a phase difference corresponding to twice the optical depth. It will be appreciated that, as mentioned in the introduction, a full boundary-value solution of the problem would merely alter the values of the Fourier coefficients, but the general analysis would no way be invalidated.

We regard each pit as composed of a rectangle, of reduced length β and reduced width γ, together with two semicircles of radius (1/2) γ as indicated in Fig. 7. For pits of different lengths it is merely the length β of the rectangular portion which changes; and it is, accordingly, useful to compute the coefficients $R(m,n)$ in such a way that the dependance on β is isolated from those factors which remain unchanged.

It is of importance to note that in all cases the pits have symmetry about the u' and v' axes: that is $R_0(-u',v') = R_0(u',v')$ and $R_0(u',-v') = R_0(u',v')$. It then follows from (6.1) that

$$R(-m,n) = R(m,n),$$
$$R(m,-n) = R(m,n), \quad (6.2)$$
$$R(-m,-n) = R(m,n),$$

so that it is only necessary to tabulate $R(m,n)$ for the values

$m = 0, +1, +2, \ldots, m_2$ and $n = 0, +1, +2, \ldots, n_2$ where m_2 and n_2 are respectively the values of m_2 and n_2 defined by (5.18) with $\mu = 0$ and by (5.20) with $m = \mu = \nu = 0$. By (5.25) with $m = 0$ the maximum value of n_4 is never greater than that of n_2. There will thus be no more than $2p$ values of m and so no more than $2q$ values for n.

For the moment, and for later reference, we shall regard $R_0(u',v')$ as zero outside the unit cell at the origin, and then denote it by $R_{00}(u',v')$. The inverse Fourier transform of this function is

$$r_{00}(x',y') = \int_{-1/2p}^{+1/2p} \int_{-1/2q}^{+1/2q} R_{00}(u',v') \times \exp\{-i2\pi(x'u' + y'v')\}\, du'\, dv' \quad (6.3)$$

and comparison with (6.1) then shows that

$$R(m,n) = (1/pq) r_{00}(m/p, n/q) \quad (6.4)$$

gives the Fourier coefficients of the reflection function for the whole disc.

We now let the unit cell be divided into the four regions R_1, R_2, R_3, and R_4 shown in Fig. 7. If $R_1(u',v')$, $R_2(u',v')$, $R_3(u',v')$, and $R_4(u',v')$, are the complex reflection functions for each of these regions, we have

$$R_{00}(u',v') = R_1(u',v') + R_2(u',v') + R_3(u',v') + R_4(u',v') \quad (6.5)$$

for the whole region of the unit cell. To deal with $R_3(u',v')$ and $R_4(u',v')$, relating to the semicircular regions R_3 and R_4, we consider first the region R_5 shown at (a) in Fig. 7. If $R_5(u',v')$ is the reflection function for R_5, we shall have in place of (6.5)

$$R_{00}(u',v') = R_1(u',v') + R_2(u',v') + R_5(u' - \beta/2, v') + R_5(-u' - \beta/2, v'), \quad (6.6)$$

and it remains to specify $R_1(u',v')$, $R_2(u',v')$, and $R_5(u',v')$.

Let the following "window" functions be defined:

$$G_1(u';p) = 1, \quad |u'| \le (1/2)p; \qquad 0, \quad |u'| > (1/2)p \quad (6.8)$$

$$G_2(u',v';\gamma) = 1, \quad u' \ge 0; (u'^2 + v'^2) \le (\gamma/2)^2; \qquad = 0, \quad u' < 0; (u'^2 + v'^2) > (\gamma/2)^2 \quad (6.9)$$

so that $G_2(u',v';\gamma) = 1$ inside the region R_5 and is zero outside this region. In terms of (6.8) and (6.9) we may write

$$R_1(u',v') = G_1(u';p)G_1(v';q) - G_1(u';\beta)G_1(v';\gamma) - G_2(u' - \beta/2, v';\gamma) - G_2(-u' - \beta/2, v';\gamma), \quad (6.10)$$

which is equal to 1 for all points of the unit cell excluding the region occupied by the pit.

If $H(v')$ is the "phase-profile" across the rectangular region of the pit, the reflection function for this region is

$$R_2(u',v') = G_1(u';\beta)G_1(v';\gamma)\exp[iH(v')], \quad (6.11)$$

where, it should be noted, $H(v')$ is twice the phase depth of the pit. In a similar manner

$$R_5(u',v') = G_2(u',v';\gamma)\exp\{iH[(u'^2 + v'^2)^{1/2}]\} \quad (6.12)$$

gives the reflection function for the region R_5 of Fig. 7(a).

Using (6.10), (6.11), and (6.12) in (6.6), the reflection function for the whole unit cell becomes

$$\begin{aligned} R_{00}(u',v') = {} & G_1(u';p)G_1(v';q) \\ & + G_1(u';\beta)G_1(v';\gamma)\{\exp[iH(v')] - 1\} \\ & + G_2(u' - \beta/2, v';\gamma)[\exp\{iH[(u' - \beta/2)^2 + v'^2)\} - 1) \\ & + G_2(-u' - \beta/2, v';\gamma)[\exp\{iH(V(-u' - \beta/2)^2 \\ & + v'^2)\} - 1) \end{aligned} \quad (6.13)$$

whose inverse Fourier transform is given by

$$\begin{aligned} r_{00}(x',y') = {} & [\sin(\pi p x')/\pi x'][\sin(\pi q y')/\pi y'] \\ & + [\sin(\pi\beta x')/\pi x']\, I_1(y';\gamma) + \exp(-i\pi\beta x') I_2(x',y';\gamma) \\ & + \exp(+i\pi\beta x') I_2(-x',y';\gamma), \end{aligned} \quad (6.14)$$

where

$$I_1(y';\gamma) = \int_{-\gamma/2}^{+\gamma/2} \exp[iH(v')]\exp\{-i2\pi y'v'\}\, dv' \quad (6.15)$$

and

$$I_2(x',y';\gamma) = \iint_{R_5} [\exp\{iH[u'^2 + v'^2)^{1/2}\} - 1] \times \exp\{-i2\pi(u'x' + v'y')\}\, du'\, dv' \quad (6.16)$$

in which R_5 is the region shown in Fig. 7(a). Using polar coordinates (ρ', θ'), with origin at O_5, (6.16) becomes

$$I_2(x',y';\gamma) = \int_0^{\gamma/2} \int_{\pi/2}^{+\pi/2} \{\exp[iH(\rho')] - 1\} \times \exp[-i2\pi(x'\rho'\cos\theta' + y'\rho'\sin\theta')]\rho\, d\rho'\, d\theta', \quad (6.17)$$

which, like the integral $I_1(y';\gamma)$, depends only on the width of the pit and the phase profile across it.

Both $I_1(y';\gamma)$ and $I_2(x',y';\gamma)$ are thus quite independent of the length of the pit, and also do not depend on the spacings p and q of the pits and tracks respectively. (6.14) is usefully written

$$\begin{aligned} r_{00}(x',y') = {} & [\sin(\pi p x')/\pi x']\,[\sin(\pi q y')/\pi y'] \\ & + \left[\frac{i}{(2\pi x')} I_1(y';\gamma) + I_2(x',y';\gamma)\right]\exp(-i\pi\beta x') \\ & + \left[\frac{-i}{(2\pi x')} I_1(y';\gamma) + I_2(x',y';\gamma)\right]\exp(+i\pi\beta x'), \end{aligned} \quad (6.18)$$

where the dependence on β is now determined solely by the two exponential functions. For the present case of a simply periodic disc, the Fourier coefficients of the reflection function for the whole disc are given by (6.4): that is, putting $x' = m/p$ and $y' = n/q$ in (6.18), .

$$\begin{aligned} R(m,n) = {} & \frac{\sin(\pi m)}{\pi m}\frac{\sin(\pi n)}{\pi n} \\ & + \left[\left(\frac{i}{2\pi m q}\right) I_1\left(\frac{n}{q};\gamma\right) + \left(\frac{1}{pq}\right) I_2\left(\frac{m}{p},\frac{n}{q};\gamma\right)\right] \\ & \times \left[\exp[-i\pi\left(\frac{m}{p}\right)\beta\right] + \left[\left(\frac{-i}{2\pi m q}\right) I_1\left(\frac{n}{q};\gamma\right)\right. \\ & \left. + \left(\frac{1}{pq}\right) I_2\left(-\frac{m}{p},\frac{n}{q};\gamma\right)\right] \exp\left[+i\pi\left(\frac{m}{p}\right)\beta\right], \end{aligned} \quad (6.19)$$

the first term of which is zero except for $m = n = 0$ when it has the value +1. The integrals $I_1(n/q;\gamma)$ and $I_2(m/p,n/q;\gamma)$ may be evaluated numerically using (6.15) and (6.17) for any given form of the phase profile $H(v')$ and pit width γ; $R(m,n)$ is then found for any value of β using (6.19). In this manner the values of the real and imaginary parts are found and tabulated for the requisite values of m and n.

B. Evaluation of the integrals $D(m,n;m',n')$

The integrals to be evaluated are of the form (5.31), namely

$$D(m,n;m',n') = \iint_{(x^2+y^2)\leq 1} \times f\left(x-\frac{m}{p}, y-\frac{n}{q}\right) f^*\left(x-\frac{m'}{p}, y-\frac{n'}{q}\right) dx\, dy \quad (6.20)$$

with, from (3.4)

$$\begin{aligned} f(x,y) &= \tau(x,y)\exp[i2\pi W(x,y)], \quad (x^2+y^2)\leq 1 \\ &= 0 \qquad\qquad (x^2+y^2)>1 \end{aligned} \quad (6.21)$$

and for $m' = m+\mu$, $n' = n \pm \nu$, where μ and ν take the non-negative values 0, 1, 2, ... The function $\tau(x,y)$ is of the form (3.1), namely

$$\tau(x,y) = \exp[-(\sigma/2)(x^2+y^2)] \quad (6.22)$$

and describes the variation of real amplitude over the incident laser beam.

It simplifies the tabulation of the integrand, and also reduces the numerical work, if the origin in (6.20) is moved to the point $(x = m/p,\ y = n/y)$. The integral then becomes, with $m' = m+\mu$, $n' = n\pm\nu$,

$$D(m,n;m+\mu,n\pm\nu) = \iint_{(x-m/p)^2+(y-n/q)^2\leq 1} \times f(x,y) f^*\left(x-\frac{\mu}{p}, y\pm\frac{\nu}{q}\right) dx\, dy \quad (6.23)$$

and the regions of integration are now those shown shaded in Fig. 8. For the numerical integration the values of the integrand are tabulated at a suitable set of mesh-points in the circle $x^2+y^2 = 1$: this is the base tabulation.

The regions of integration shown in Figs. 8(a) and 8(b) are for the cases $D(m,n;m+\mu,n+\nu)$ and $D(m,n;m+\mu,n-\nu)$ respectively. Only positive or zero values of μ and ν occur. The three circles, with centers C_1, C_2, and C_3, correspond to those of Fig. 5. C_1 now has coordinates $(x = m/p, y = n/q)$; and C_3 has coordinates $(x = \mu/p, y = \nu/q)$. Thus, for any given values of (μ,ν) the region of summation includes only points

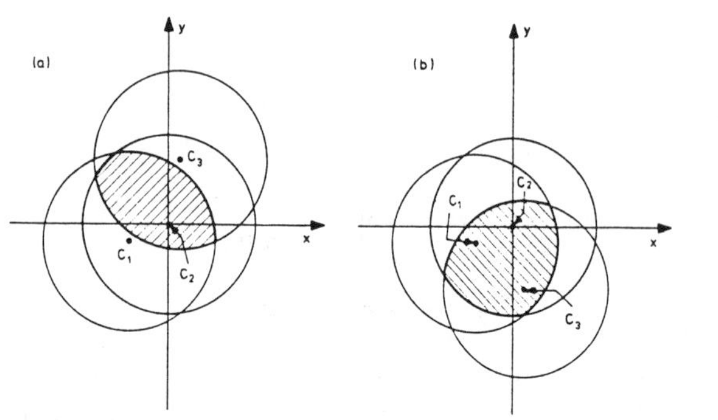

FIG. 8. Regions of integration for (a) $D(m,n;m+\mu,n+\nu)$ and (b) $D(m,n;m+\mu,n-\nu)$.

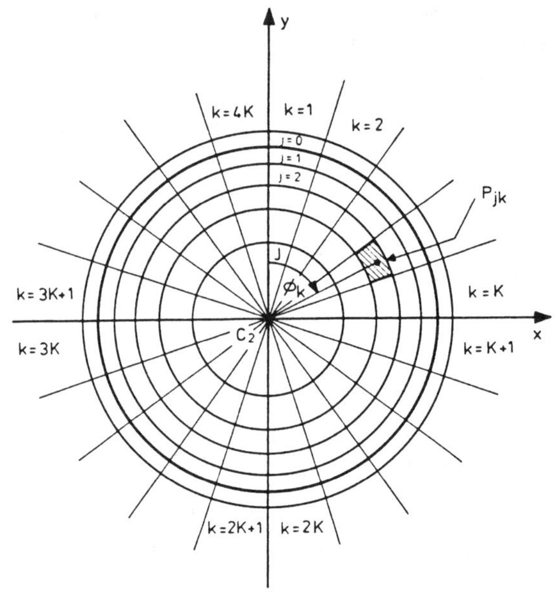

FIG. 9. Mesh points for the base tabulation.

in the region of overlap of the circles (2) and (3), and the integrand in (6.23), that is

$$f(x,y)f^*(x-\mu/p, y-\nu/q)$$

has therefore to be subtabulated only at points in this region, no matter what the values of m and n.

It is convenient to employ polar coordinates for (6.23). We use (r,ϕ), where $\phi = 0$ is along the positive y axis and ϕ increases in a clockwise direction. Thus (r_j,ϕ_k) are the polar coordinates of the point P_{jk} in Fig. 9. Using these coordinates we have, for (6.23),

$$D(m,n;m+\mu,n\pm\nu) = \iint_D f(r,\varphi)f^*(r',\varphi')r\, dr\, d\phi, \quad (6.24)$$

where D denotes the region of integration, and (r',ϕ') are the polar coordinates corresponding to the rectangular coordinates $(x-\mu/p, y\mp\nu/q)$; that is,

$$r' = \{[(r\sin\phi-\mu/p)^2 + (r\cos\alpha \mp \nu/q)^2]\}^{1/2}, \quad (6.25)$$

$$\cos\phi' = \frac{r\cos\phi \mp (\nu/q)}{r'}$$

give the values of (r',ϕ') for $f^*(r',\phi')$ corresponding to any values of (r,ϕ) in $f(r,\phi)$.

Writing (6.21) and (6.22) in polar coordinates gives

$$f(r,\phi) = \exp[-\sigma/2)r^2]\exp[i2\pi W(r,\varphi)], \quad (6.26)$$

where the wave front aberration, $W(r,\phi)$, is a function of the variables in the combinations r^2 and $r\cos\phi$. For an off-axis image formed by a microscope objective, a typical form of $W(r,\phi)$ will be

$$\begin{aligned} W(r,\phi) = {} & (W_{20}r^2 + W_{22}r^2\cos^2\varphi) + (W_{40}r^6 \\ & + W_{60}r^6 + W_{80}r^8 + W_{100}r^{10}) + (W_{31}r^3\cos\phi + W_{51}r^5\cos\varphi \\ & + W_{71}r^7\cos\varphi + W_{91}n^9\cos\varphi), \end{aligned} \quad (6.27)$$

where W_{20} represents the focal shift from the chosen plane of focus to the sagittal focal line, W_{22} is the focal shift between the sagittal and tangential focal lines, and the next two brackets represent different orders of spherical aberration and coma respectively.

For the purpose of numerical integration the variables ($t = r^2, \phi$) will be used. We then have, for (6.24)

$$D(m,n;m+\mu,n\pm\nu) = \frac{1}{2}\iint_D f(t,\varphi) f^*(t',\varphi')\, dt\, d\phi, \tag{6.28}$$

where, from (6.25),

$$r' = [(\sqrt{t}\sin\varphi - \mu/p)^2 + (\sqrt{t}\cos\phi - \nu/q)^2]^{1/2},$$

$$\cos\phi' = \frac{\sqrt{t}\cos\phi \mp (\nu/q)}{r'}, \tag{6.29}$$

for any given values of (μ,ν). We also have, for (6.26),

$$f(t,\phi) = \exp[-(1/2)\sigma t]\exp[-i2\pi W(\sqrt{t},\phi)], \tag{6.30}$$

whilst the wave-front aberration, (6.27), is written

$$W(t,\varphi) = U(t,\varphi) + V(t,\varphi),$$

where

$$U(t,\varphi) = W_{20}t + W_{22}t\cos^2\phi + W_{40}t^2 + W_{60}t^3 + W_{80}t^4 + W_{100}t^5 \tag{6.31}$$

and

$$V(t,\varphi) = (W_{31}t + W_{51}t^2 + W_{71}t^3 + W_{91}t^4)\sqrt{t}\cos\phi \tag{6.32}$$

are, respectively, the even and odd parts of the wave front aberration.

Because of the form of $f(t,\phi)$ the quantities

$$T(t) = \exp[-(\sigma/2)t]$$

$$\Phi(t,\varphi) = 2\pi W(t,\varphi) = 2\pi U(t,\varphi) + 2\pi V(t,\varphi) \tag{6.33}$$

need to be tabulated only in the first quadrant. This is divided into K equal sectors and by J circles equally spaced in $t = r^2$. The coordinates of the center of the mesh point P_{jk}, Fig. 9, are then

$$t_j = 1 - (2j-1)/2J, \qquad \varphi_k = [(2k-1)/4K]\pi, \tag{6.34}$$

$j = 1$ being the outermost annulus bounded by the pupil periphery $t = 1$. The mesh points in other quadrants are labeled as shown in Fig. 9, and the values of $\Phi(t,\phi)$ in the other quadrants are given by the rules

$$\begin{aligned} \Phi(t_j,\phi_{2K-k}) &= 2\pi U(t_j,\phi_k) - 2\pi V(t_j,\phi_k),\\ \Phi(t_j,\phi_{2K+k}) &= 2\pi U(t_j,\varphi_k) - 2\pi V(t_j,\varphi_k),\\ \Phi(t_j,\phi_{4K-k}) &= 2\pi U(t_j,\phi_k) + 2\pi V(t_j,\varphi_k), \end{aligned} \tag{6.35}$$

for any value of j and for $1 \le k \le K$. These rules follow from the forms of the dependence of $U(t,\phi)$ and $V(t,\phi)$ on ϕ.

With the particular form of mesh employed, the mesh elements all have the same area

$$\delta D = \pi/(2JK) \tag{6.36}$$

and the value of the integral (6.28) is then approximated by the sum

$$D(m,n;m+\mu,n\pm\nu) = D\left[\frac{1}{D}\sum_j\sum_k T(t_j)T(t'_j).\times\exp\{i[\Phi(t_j,\phi_k) - \Phi(t_j',\phi'_k)]\}\frac{\pi}{2JK}\right], \tag{6.37}$$

where each factor in the integrand has been written using (6.33), and the values of (t_j',ϕ'_k) are related to those of (t_j,ϕ_k) by means of (6.29). The summations extend over all mesh elements whose midpoints fall in the region of overlap, if any, of the three circles (1), (2), and (3). The quantity D is the area of this region of overlap, so that the quantity in square brackets in (6.37) represents the mean value of the integrand over this region. Thus if N is the total number of such mesh elements, $D = N\delta D = N\pi/2JK$ is the area of the region covered by the mesh elements, which will not coincide exactly with the boundary of the region of overlap. We, therefore, write (6.37) as

$$D(m,n;m+\mu,n\pm\nu) = D\left[\frac{1}{N}\sum_j\sum_k T(t_j)T(t'_j).\times\exp\{i[\Phi(t_j,\phi_k) - \Phi(t'_j,\phi_k')]\}\right] \tag{6.38}$$

and the area D may be found exactly, so that any error lies only in the difference between the mean value of the integrand over the approximation to the region D and its value for the exact region. The values of $\Phi(t_j,\phi_k)$ and $T(t_j)$ are given directly by the base tabulation, whilst those of $T(t_j')$ and $\Phi(t'_j,\phi'_k)$ are found by linear interpolation between adjacent points in the base tabulation.

C. Computation of the Fourier coefficients A_0, A_μ, and B_μ of the detector signal

These Fourier coefficients are given by (5.47), (5.48), and (5.49), and each requires the evaluation of the series, $P_1(\mu,\nu)$, $P_2(\mu,\nu)$, $Q_1(\mu,\nu)$, and $Q_2(\mu,\nu)$ for the given value of μ and the appropriate range of values $\nu = 0, 1, 2, \ldots, \nu_0$. For each value of ν there will be ranges of m and n as specified earlier. The subtabulation described in the previous section is carried out for each value of ν in turn, and for each ν the set of different values of m and n give different centers for the circle C_1 in Fig. 8, and thus different ranges of (j,k) in the sum (6.38). In this manner the real and imaginary parts, $C(m,n;m+\mu,n\pm\nu)$ and $S(m,n;m+\mu,n\pm\nu)$, of $D(m,n;m+\mu,n\pm\nu)$ are calculated. The corresponding values of the real and imaginary parts, $M(m,n;m+\mu,n\pm\nu)$ and $N(m,n;m+\mu,n\pm\nu)$, of the product $R(m,n)\,R^*(m+\mu,n\pm\nu)$ are then found using the tabulated values of $G(m,n)$ and $K(m,n)$ in (5.37) and (5.38).

In the above manner, the values of the coefficients $P_1(\mu,\nu)$, $P_2(\mu,\nu)$, $Q_1(\mu,\nu)$, and $Q_2(\mu,\nu)$ are easily obtained, whence A_0, A_μ, and B_μ are found as simple functions of the tracking error υ by (5.47), (5.48), and (5.49), respectively.

VII. THE CASE OF OFF-AXIS BEAMS

The above treatment has assumed the entrance pupil of the objective in Fig. 1 to be at the front focal plane of the objective, and also that the incident laser beam enters along the optical axis. If these conditions are not satisfied it requires only simple modifications to the above formulas.

In Fig. 10 the entrance pupil has its axial point at E, and the central ray from the extra-axial object point Q is QE. This

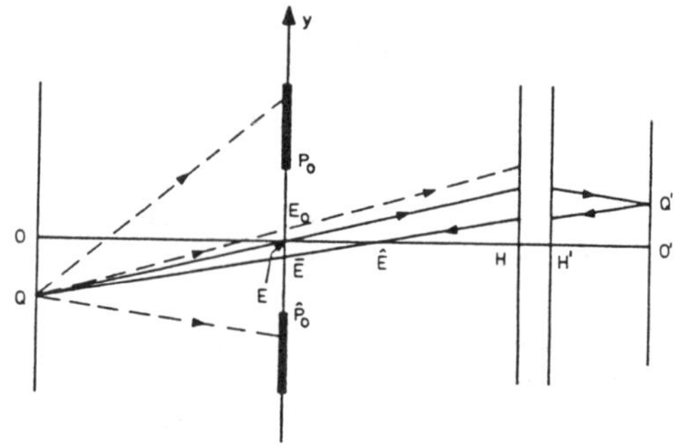

FIG. 10. Ray paths for the general case.

ray meets the image plane OQ' at Q and the ray has non-normal incidence when E is not at the front focal plane of the objective, which is represented in Fig. 10 by its principal planes H and H'. The ray reflected at Q' returns to give an image of the pupil at $\hat{E}$, and then cuts the entrance pupil at $\overline{E}$. The zero-order undiffracted beam thus focuses at $\hat{E}$, and proceeds to $\overline{E}$ where the limiting aperture is a circle of center E.

Let $\overline{E}$ have real-space coordinates $(\overline{X},\overline{Y})$, and let h be the radius of the entrance pupil as in Fig. 4. If we note that the ray from O through P_0 returns after reflection at O' to cut the entrance pupil plane at $\hat{P}_0$, we see that the normalizing radius is $-h$, and

$$\bar{x} = -\,\overline{X}/h, \quad \bar{y} = -\,\overline{Y}/h \tag{7.1}$$

are the reduced coordinates of the point $\overline{E}$ of the ray returned from Q'. The returned diffracted distribution of light at the pupil plane E is centered in the point $\overline{E}$, so we have the same situation as moving the center of the circular aperture E to the point $(-\bar{x},-\bar{y})$ and the integrals $D(m,n;m+\mu,n\pm\nu)$ of (6.20) then become

$$D(m,n;m+\mu,n\pm\nu) = \iint_{(x-\bar{x})^2+(y-\bar{y})^2\le 1,} f\left(x-\frac{m}{p},y-\frac{n}{q}\right) \times f^*\left(x-\frac{m+\mu}{p},y-\frac{n\pm\nu}{q}\right)dx\,dy,$$

or, shifting the origin to $(x = m/p,\ y = n/q)$,

$$D(m,n;m+\mu,n\pm\nu) = \iint f(x,y)f^*\left(x-\frac{\mu}{p},y-\frac{\pm\nu}{q}\right)dx\,dy,$$

$$\left[x-\left\{\frac{m}{p}-\bar{x}\right\}\right]^2+\left[y-\left\{\frac{n}{q}-\bar{y}\right\}\right]^2 \le 1, \tag{7.2}$$

so that the only effect is that the circle (1) now has its center at the point $(x = m/p - \bar{x}, y = n/q - \bar{y})$. Apart from this the evaluation of the integrals is exactly as described above.

If the laser beam focuses at Q, but is not aimed symmetrically at the pupil, the axis ray of the beam will cut the pupil at a point such as E_0 in Fig. 10. If this point has reduced coordinates

$$x_0 = X_0/h \quad y_0 = Y_0/h \tag{7.3}$$

the factor $\tau(x,y)$ in the function $f(x,y)$ has to be written

$$\tau(x,y) = \exp\{-(1/2)\sigma[(x-x_0)^2+(y-y_0)^2]\} \tag{7.4}$$

in place of (3.1), and the previous analysis then applies without change.

VIII. MORE COMPLICATED FORMS OF DISC: INTERMODULATION AND CROSS-TALK

The first arrangement of pits to be considered here is shown in Fig. 11, in which the tracks are numbered $n = 0, \pm1, \pm2, \ldots$, and the pits along each track have the numbers $m = 0, \pm1, \pm2, \ldots$. In each track the center-to-center separation between adjacent pits is constant, with reduced value p. The tracks are "staggered" in the sense that the center points of corresponding pits in adjacent tracks are relatively displaced along the direction of u'. The track n is thus the same as the track $n = 0$ but displaced by ns where s is the reduced value of the stagger.

In each track the pit length is modulated, and such that the mth pit of each track has a reduced length β_m for the rectangular portion of the pit. For the central track $n = 0$, the mth pit has coordinate $u' = mp$ and its length parameter is taken to have the form

$$\beta_m = \beta[1 + M\cos(2\pi mp/P)], \tag{8.1}$$

where β is the mean value, M is the amplitude of modulation of β, and P is the reduced period of this duty cycle.

It will be shown that the preceding analysis for the case of the simply periodic disc of Fig. 2 is readily adapted to the form of disc shown in Fig. 11. We let $\hat{R}_{00}(u',v')$ be the reflection function for the mth pit of the central track, $n = 0$, with the origin taken at the midpoint $(u' = mp, v' = 0)$ of this pit. We define $\hat{R}_{00}(u',v')$ to be zero outside the region of the pit. The reflection function for the pits of the central track when the origin is taken to be at the midpoint of the pit $m = 0$, is then given by

$$\sum_{m=-\infty}^{+\infty} \hat{R}_{00}(u'-mp,v') \tag{8.2}$$

and the nth track will be of the form

$$\sum_{m=-\infty}^{+\infty} \hat{R}_{00}(u'-mp-ns,v'-nq) \tag{8.3}$$

since the nth track is merely the track $n = 0$ displaced a distance ns along u' and nq along v', where q is the reduced spacing of the tracks. The reflection function for the whole disc is obtained by summing the expression (8.3) over n; that is,

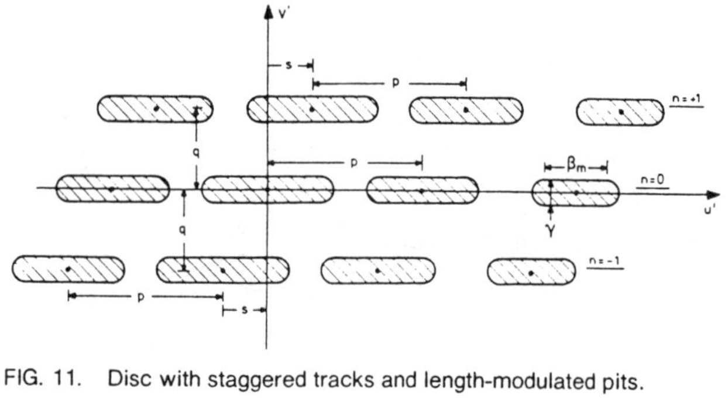

FIG. 11. Disc with staggered tracks and length-modulated pits.

$$\hat{R}_0(u',v') = \sum_{m=-\infty}^{+\infty} \sum_{n=-\infty}^{+\infty} \hat{R}_{00}(u' - mp - ns, v' - nq) \tag{8.4}$$

is the reflection function when the center of the pit $m = 0$ of the track $n = 0$ is at the origin. That is, $R_0(u',v')$ corresponds to the reflection function $R_0(u',v')$ for the simply periodic track of the earlier analysis.

As before, we need to find the inverse Fourier transform of $\hat{R}_0(u',v')$. Using (8.4), this is given by

$$\hat{r}_0(x',y') = \iint_{-\infty}^{+\infty} \sum_{m=-\infty}^{+\infty} \sum_{n=-\infty}^{+\infty} \hat{R}_{00}(u' - mp - ns, v' - nq) \times \exp\{-i2\pi(u'x' + v'y')\}\, du'\, dv'$$

or, with $u' - mp - ns = \alpha$, $v' - nq = \beta$,

$$\hat{r}_0(x',y') = \sum_{m=-\infty}^{+\infty} \sum_{n=-\infty}^{+\infty} \iint_{-\infty}^{+\infty} \hat{R}_{00}(\alpha,\beta) \times \exp[-i2\pi(\alpha x' + \beta y')]\, d\alpha\, d\beta \times \exp\{-i2\pi[(mp + ns)x' + nqy']\};$$

that is,

$$\hat{r}_0(x',y') = \sum_{m=-\infty}^{+\infty} \sum_{n=-\infty}^{+\infty} \hat{r}_{00}(x',y') \times \exp\{-i2\pi[m(px') + n(qy' + sx')]\}, \tag{8.5}$$

where $\hat{r}_{00}(x',y')$ is the inverse Fourier transform of the function $\hat{R}_{00}(u',v')$.

The function $\hat{R}_{00}(u',v')$ is identical with the function $R_{00}(u',v')$ of the earlier analysis in Sec. VI, except that β is now replaced by the form (8.1). Thus, using this in (6.18), we obtain

$$\begin{aligned}\hat{r}_{00}(x',y') = {} & E_1(x',y') + E_2(x',y') \\ & \times \exp\{-i\pi\beta[1 + M\cos(2\pi mp/p)]x'\} \\ & + E_3(x',y')\exp\{+i\pi\beta[1 + M\cos(2\pi mp/p)]x'\},\end{aligned} \tag{8.6}$$

where

$$\begin{aligned} E_1(x',y') &= [\sin(\pi px')/(\pi x')][\sin(\pi qy')/(\pi y')], \\ E_2(x',y') &= (i/2\pi x')I_1(y';\gamma) + I_2(x',y';\gamma), \\ E_3(x',y') &= -(i/2\pi x')I_1(y';\gamma) + I_2(-x',y';\gamma), \end{aligned} \tag{8.7}$$

denote the expressions occurring in (6.18).

Using the expansion

$$\exp(\pm ir\cos\phi) = \sum_{r=-\infty}^{+\infty} (\pm i)^r J_r(r)\exp(\pm ir\phi), \tag{8.8}$$

where $J_r(z)$ denotes the Bessel function of the first kind of order r, (8.6) may be written

$$\begin{aligned}\hat{r}_{00}(x',y') = {} & E_1(x',y') + E_2(x',y')\exp(-i\pi\beta x') \\ & \times \sum_{r=-\infty}^{+\infty} (-i)^r J_r(\pi\beta Mx')\exp\left[-ir\left(\frac{2\pi mp}{p}\right)\right] \\ & + E_3(x',y')\exp(+i\pi\beta x') \\ & \times \sum_{r=-\infty}^{+\infty} (+i)^r J_r(\pi\beta Mx')\exp\left[+ir\left(\frac{2\pi mp}{p}\right)\right],\end{aligned} \tag{8.9}$$

which reduces to $r_{00}(x',y')$ of (6.18) if $M = 0$.

Substitution of (8.9) in (8.5) gives

$$\begin{aligned}\hat{r}_0(x',y') = {} & E_1(x',y') \sum_{m=-\infty}^{+\infty} \exp[-i2\pi m(px')] \\ & \times \sum_{n=-\infty}^{+\infty} \exp[-i2\pi n(qy' + sx')] \\ & + E_2(x',y')\exp(-i\pi\beta x') \sum_{r=-\infty}^{+\infty} (-i)^r J_r(\pi\beta Mx') \sum_{m=-\infty}^{+\infty} \\ & \times \exp\left[-i2\pi m\left(px' + \frac{rp}{p}\right)\right] \sum_{n=-\infty}^{+\infty} \exp[-i2\pi(qy' + sx')] \\ & + E_3(x',y')\exp(+i\pi\beta x') \sum_{r=-\infty}^{+\infty} (+i)^r J_r(\pi\beta Mx') \sum_{m=-\infty}^{+\infty} \\ & \times \exp\left[-i2\pi m\left(px' - \frac{rp}{P}\right)\right] \sum_{n=-\infty}^{+\infty} \exp[-i2\pi n(qy' + sx')],\end{aligned} \tag{8.10}$$

or, replacing r by $-r$ in the third term, and noting that $J_{-r}(z) = (-1)^r J_r(z)$,

$$\begin{aligned}\hat{r}_0(x',y') = {} & E_1(x',y') \sum_{m=-\infty}^{+\infty} \exp[-i2\pi m(px')] \\ & \times \sum_{n=-\infty}^{+\infty} \exp[-i2\pi n(qy' + sx')] \\ & + \sum_{r=-\infty}^{+\infty} [E_2(x',y')\exp(-i\pi\beta x')(-i)^r \\ & + E_3(x',y')\exp(+i\pi\beta x')(i)^r] \\ & \times J_r(\pi\beta Mx') \sum_{m=-\infty}^{+\infty} \exp\left[-i2\pi m\left(px' + \frac{rp}{P}\right)\right] \\ & \times \sum_{n=-\infty}^{+\infty} \exp[-i2\pi n(qy' + sx')],\end{aligned} \tag{8.11}$$

having combined the final two series of (8.10).

The summations over m and n in (8.11) may be converted to sums of δ functions in the following manner: Consider the series

$$\begin{aligned}\sum_{n=-\infty}^{+\infty} \exp[\pm in(2\pi\Delta)] &= \lim_{N\to\infty} \sum_{n=-N}^{+N} \exp[\pm in(2\pi\Delta)] \\ &= \lim_{N\to\infty} \left\{1 + 2\sum_{n=1}^{N} \cos[n(2\pi\Delta)]\right\}\end{aligned}$$

or, using the known sum of the series of cosines,

$$\sum_{n=-\infty}^{+\infty} \exp\{\pm in(2\pi\Delta)\} = \lim_{N\to\infty} \frac{\sin(N+1)\pi\Delta}{\sin(\pi\Delta)}, \tag{8.12}$$

in which the denominator is zero for $\pi\Delta = n\pi$, where $n = 0, \pm 1, \pm 2, \ldots$. For each of these values of Δ the right-hand side of (8.12) defines a δ function. We thus have

$$\sum_{n=-\infty}^{+\infty} \exp\{\pm in(2\pi\Delta)\} = \sum_{n=-\infty}^{+\infty} \delta(\Delta - n), \tag{8.13}$$

which may now be used for the summations over m and n in (8.11).

Using (8.13) in (8.11) gives

$$\hat{r}_{00}(x',y') = E_1(x',y') \sum_{m=-\infty}^{+\infty} \delta(px' - m)$$

$$\times \sum_{n=-\infty}^{+\infty} \delta(qy' + sx' - n)$$

$$+ \sum_{r=-\infty}^{+\infty} \{E_2(x',y') \exp\{-i\pi\beta x'\}(-i)^r$$

$$+ E_3(x',y') \exp\{+i\pi\beta x'\}(i)^r\}.$$

$$J_r(\pi\beta M x') \sum_{m=-\infty}^{+\infty} \delta\left(px' + \frac{rp}{P} - m\right) \sum_{n=-\infty}^{+\infty} \delta(qy' + sx' - n)$$

or, with the identity $\delta(az) = \delta(r)/|a|$,

$$\hat{r}_0(x',y') = E_1(x',y') \left(\frac{1}{pq}\right) \sum_{m=-\infty}^{+\infty} \delta\left(x' - \frac{m}{p}\right) \sum_{n=-\infty}^{+\infty}$$

$$\times\, \delta\left(y' + \frac{sx'}{q} - \frac{n}{q}\right) + \sum_{r=-\infty}^{+\infty} [E_2(x',y')$$

$$\times \exp(-i\pi\beta x')(-i)^r + E_3(x',y') \exp(+i\pi\beta x')(i)^r]$$

$$\times J_r(\pi\beta M x') \left(\frac{1}{pq}\right) \sum_{m=-\infty}^{+\infty} \delta\left(x' + \frac{r}{P} - \frac{m}{p}\right) \sum_{n=-\infty}^{+\infty}$$

$$\times\, \delta\left(y' + \frac{sx'}{q} - \frac{n}{q}\right), \tag{8.14}$$

which is the inverse Fourier transform of the reflection function for a disc with staggered tracks and pits length-modulated in accordance with (8.1) above.

The complex amplitude diffracted back to the point (x,y) of the pupil is given by (3.6) with $r(x',y')$ now replaced by $\hat{r}(x',y')$, which is the inverse transform of

$$\hat{R}(u',v') = \hat{R}_0(u' - u, v' - v); \tag{8.15}$$

that is, the reflection function for the disc of Fig. 11 with the disc displaced by amounts u and v from the position in defining $\hat{R}_0(u',v')$. We thus have

$$\hat{r}(x',y') = \exp[-i2\pi(ux' + vy')]\hat{r}_0(x',y')$$

and (3.6) takes the form

$$a'(x,y) = \iint_{-\infty}^{+\infty} f(x - x', y - y') \exp[-i2\pi(ux' + vy')]$$

$$\times\, \hat{r}_0(x',y')\, dx'\, dy' \tag{8.16}$$

with $\hat{r}_0(x',y')$ given by (8.14).

Substitution from (8.14) in (8.16) gives integrals which may be evaluated using the sifting property of the δ function. This requires the replacement of x' and y' in the integrand by

$$x' = m/p - r/P, \quad y' = n/q - sx'/q,$$

or

$$x' = \hat{m}/p, \quad y' = \hat{n}/q, \tag{8.17}$$

where

$$\hat{m} = m - p\,(r/P), \quad \hat{n} = n - s(m/p - r/P) \tag{8.18}$$

with $r = 0$ for the first series in (8.14). If there is no stagger between tracks and no length modulation, that is, $s = M = 0$, (8.18) reduces to $\hat{m} = m$ and $\hat{n} = n$ with only the term $r = 0$ occurring in the second series of (8.14).

With the above notation we now find, using (8.14) in (8.16),

$$a'(x,y) = \sum_m \sum_n \exp\left\{-i2\pi\left[\left(\frac{\hat{m}}{p}\right)u + \left(\frac{\hat{n}}{q}\right)v\right]\right\}$$

$$\times \left(\frac{1}{pq}\right) E_1\left(\frac{\hat{m}}{p}, \frac{\hat{n}}{q}\right) f\left(x - \frac{\hat{m}}{p}, y - \frac{\hat{n}}{q}\right)$$

$$+ \sum_r \sum_m \sum_n \exp\left\{-i2\pi\left[\left(\frac{\hat{m}}{p}\right)u + \left(\frac{\hat{n}}{q}\right)v\right]\right\}\left(\frac{1}{pq}\right)$$

$$\times \left[E_2\left(\frac{\hat{m}}{p}, \frac{\hat{n}}{q}\right) \exp\left(-i\pi\beta\frac{\hat{m}}{p}\right)(-i)^r\right.$$

$$\left. + E_3\left(\frac{\hat{m}}{p}, \frac{\hat{n}}{q}\right) \exp\left(+i\pi\beta\frac{\hat{m}}{p}\right)(i)^r\right]$$

$$\times J_r\left(\pi\beta M\frac{\hat{m}}{p}\right) f\left(x - \frac{\hat{m}}{p}, y - \frac{\hat{n}}{q}\right), \tag{8.19}$$

again with the understanding that $r = 0$ in the values of $\hat{m},\hat{n}$ when used in the first of these series. Using the symbol $\delta_{mn} = 1$ for $m = n = 0$ and $\delta_{mn} = 0$ for other values of m and n, we may combine the series in (8.19) to give

$$a'(x,y) = \sum_r \sum_m \sum_n \exp\left\{-i2\pi\left[\left(\frac{\hat{m}}{p}\right)u + \left(\frac{\hat{n}}{q}\right)v\right]\right\}\left(\frac{1}{pq}\right)$$

$$\times \left\{\delta_{mn}E_1\left(\frac{\hat{m}}{p}, \frac{\hat{n}}{q}\right) + \left[E_2\left(\frac{\hat{m}}{p}, \frac{\hat{n}}{q}\right) \exp\left(-i\pi\beta\frac{\hat{m}}{p}\right)(-i)^r\right.\right.$$

$$\left. + E_3\left(\frac{\hat{m}}{p}, \frac{\hat{n}}{q}\right) \exp\left(+i\pi\beta\frac{\hat{m}}{p}\right)(i)^r\right] J_r\left((\pi\beta M\frac{\hat{m}}{p}\right)$$

$$\left. \times f\left(x - \frac{\hat{m}}{p}, y - \frac{\hat{n}}{q}\right)\right\}, \tag{8.20}$$

with r,m,n each taking the values $0, \pm1, \pm2, \ldots$. This reduces to (3.12) for the case $s = M = 0$.

We shall not here carry the analysis through completely, but merely note that the total light flux reaching the detector is given by

$$I(t) = \iint_{(x^2+y^2)\leq 1} |a'(x,y)|^2\, dx\, dy \tag{8.21}$$

and that, as in (4.2) above, the time enters by way of a term

$$\exp\left[i2\pi\left(\frac{\hat{m}' - \hat{m}}{p}\right)u\right]$$

in the squared modulus of $a'(x,y)$; and this, as seen from (8.20), will have a sixfold summation, with indices (ν,m,n) and (ν',m',n'). With $\hat{\nu} = (\hat{m} - \hat{m})$ corresponding to $\mu = m' - m$ in (4.1), the angular frequencies found in the detector signal will be given, as in (4.3), by

$$\omega_{\hat{\mu}}t = 2\pi(\hat{\mu}/p)u, \tag{8.22}$$

where

$$\hat{\mu} = \hat{m}' - \hat{m} = (m' - m) - p/P(r' - r)$$

from the defining formula (8.18). We now write

$$\mu = m' - m,$$

$$\rho = r' - r, \tag{8.23}$$

$$\hat{\mu} = (\mu - (p/P)\rho),$$

and (8.22) then becomes

$$\omega_{\hat{\mu}}t = 2\pi(\mu/p)u - 2\pi(\rho/P)u, \tag{8.24}$$

where μ has exactly the same significance as in (4.5) for the case of a simple periodic disc. In the same manner as in (4.3) above, (8.24) may be written as

$$\omega_{\hat{\mu}} t = \mu \left[2\pi \left(\frac{n' \sin\alpha'}{\lambda} \right) \frac{\omega a}{p} \right] t + \rho \left[2\pi \left(\frac{n' \sin\alpha'}{\lambda} \right) \frac{\omega a}{P} \right] t$$

using (3.7) for u. The reduced period p is given by (4.4), and P is given by

$$P = \left(\frac{n' \sin\alpha'}{\lambda} \right) \tilde{\xi}' \tag{8.25}$$

where $\tilde{\xi}'$ is the real-space period of the duty cycle of length modulation. We thus find

$$\omega_{\hat{\mu}} = \mu\omega_1 + \rho\tilde{\omega}_1 \tag{8.26}$$

where, as in (4.5)

$$\omega_1 = 2\pi(\omega a/\xi') \tag{8.27}$$

is the fundamental angular frequency corresponding to the center-to-center spacing of the pits, and

$$\tilde{\omega}_1 = 2\pi(\omega a/\tilde{\xi}') \tag{8.28}$$

is the fundamental angular frequency in $I(t)$ corresponding to the periodic modulation of the length of the pits.

As in the case of a simple periodic disc, it is found that μ takes on the values $\mu = 0, +1, +2, \ldots$ in the series for $I(t)$, whereas $\rho = (r' - r)$ may take the values $\rho = 0, \pm 1, \pm 2, \ldots$. The term $\mu = \rho = 0$ gives the dc component, whilst the terms $\rho = 0$, $\mu = +1, +2, \ldots$ give the fundamental and harmonic components corresponding to the center-to-center spacing of the pits. The terms having $\mu = 0$, $\rho \neq 0$ similarly give the fundamental and harmonic components corresponding to the periodic length modulation of the pits. When $\mu = 0$, it is found that the terms $\pm\rho$ combine such that there are effectively only the terms $\rho = +1, +2, \ldots$. When $\mu \neq 0$, we find that ρ takes the values $\pm 1, \pm 2, \ldots$, so that there are sums and differences of the two basic sets of frequencies. These terms thus represent intermodulation between the two desired signals.

To find the amplitudes of any of the above terms fortunately only requires the evaluation of integrals of the form (3.15) obtained for the simple case. As may be seen from (8.20) and (8.21), the required integrals are

$$D(\hat{m},\hat{n};\hat{m}',\hat{n}') = \iint_{(x^2+y^2)\leq 1} f\left(x - \frac{\hat{m}}{p}, y - \frac{\hat{n}}{q}\right) \times f^*\left(x - \frac{\hat{m}}{p}, y - \frac{\hat{n}'}{q}\right) dx\, dy \tag{8.29}$$

which is precisely the form (3.15), or (6.20), but with $\hat{m},\hat{n}$ and $\hat{m}',\hat{n}'$ in place of m, n and m', n'. With

$$\hat{\mu} = (\hat{m}' - \hat{m}) = \mu - (p/P)\rho \tag{8.30}$$

as in (8.23), and from (8.18)

$$\hat{\nu} = (\hat{n}' - \hat{n}) = (n' - n) - (s/p)[(m' - m) - (p/P)(r' - r)]$$

that is

$$\hat{\nu} = \nu - (s/p)[\mu - (p/P)\rho] \tag{8.31}$$

and using (8.18) for m and n, the integral (8.29) takes the form

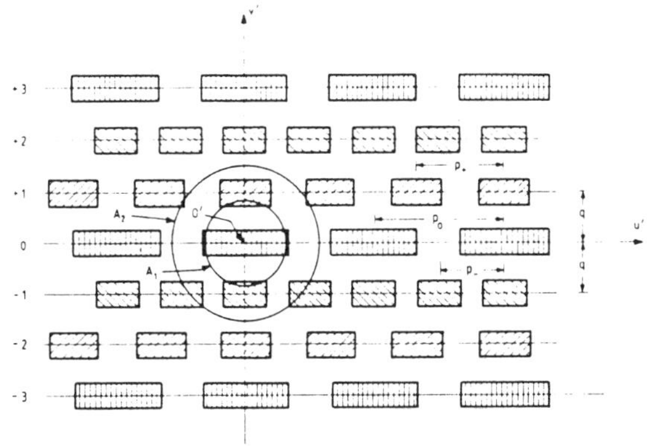

FIG. 12. Disc with adjacent tracks of different period.

$$D(\hat{m},\hat{n};\hat{m} + \hat{\mu},\hat{n} + \hat{\nu}) = \iint_{(x-\hat{m}/p)^2+(y-\hat{n}/q)^2\leq 1} f(x,y) f^*\left(x - \frac{\hat{\mu}}{p}, y - \frac{\hat{\nu}}{q}\right) dx\, dy \tag{8.32}$$

as in (6.23); and, for any values of μ and ρ in (8.30) and (8.31), $D(\hat{m},\hat{n};\hat{m} + \hat{\mu},\hat{n} + \hat{\nu})$ may be evaluated using the same computer program as for the case of a simple periodic disc.

In addition to the above intermodulation terms there may be cross-talk between adjacent tracks. This arises from light diffracted by pits in the neighboring tracks on which the outer parts of the diffraction image falls. A high density of information on a disc demands that the tracks be as closely spaced as possible; and, for example, in the case illustrated in Fig. 12, not only do the outer edges of the Airy disc just fall on adjacent tracks, but parts of the surrounding diffraction ring fall on them. In Fig. 12, A_1 and A_2 are respectively the first and second dark rings of the Airy disc formed at O'. A simple way to study this effect is to calculate the amplitudes of the dc and ac components in the time-varying light flux $I(t)$ for the correct track spacing q, and then to calculate the same amplitudes for, say, track-spacings $2q$, $3q$, $4q$. The differences between the amplitudes for the spacing q and a sufficiently greater spacing will then indicate the amount of cross-talk present.

The above calculations would assume the center-to-center spacings of the pits to be the same for all tracks, and a case of obvious interest is when the track being read and the two adjacent tracks have three different center-to-center spacings for the pits. In this case, illustrated schematically in Fig. 12, there might be expected to be spurious frequencies in the signal arising from cross-talk, rather than merely changes in the amplitudes and phases of those frequencies present in the absence of cross-talk.

The case of different periods on adjacent tracks may clearly be studied using the general formalism of Sec. III, provided the reflection coefficient of the disc may be represented by a Fourier series. It proves possible, as will be seen, to represent the reflection coefficient of the disc as the sum of three Fourier series, and in this manner to study the question of cross-talk between tracks having different periods.

The form of disc is now assumed to be as shown schematically in Fig. 12. The pits along tracks $n = 0, \pm 3, \pm 6, \ldots$ are assumed to be of identical form. The adjacent track $n = +1$ differs from that of $n = 0$, but is assumed to be identical with the tracks $n = \ldots -5, -2, +4; +7. \ldots$ Similarly, the track $n = -1$ is assumed to be different from both the tracks $n = 0$ and $n = +1$, but it is taken to be identical with the tracks $n = \ldots -7, -4, +2, +5, \ldots$ The center-to-center spacings of the pits along the tracks will be denoted as follows:

$$\begin{aligned} &\text{tracks } (+1\ +3k)\ p_+, \\ &\text{tracks } (0\ \ +3k)\ p_0, \\ &\text{tracks } (-1\ +3k)\ p_-, \end{aligned} \tag{8.33}$$

with k taking the values $0, \pm 1, \pm 2, \ldots$ All tracks are taken to have the same separation q, where q (as also p_+, p_0, p_-) is the reduced value.

If the tracks $(-1+k)$ and $(+1+k)$ of (8.33) are omitted, the remaining array of pits along the tracks $(0 + k)$ is simply periodic, with reduced periods p_0 and $3q$ along the directions of u' and v' respectively. Let the corresponding reflection coefficient be denoted by $R_0^0(u',v')$ which may be written as the Fourier series

$$R_0^0(u',v') = \sum_{m_0} \sum_{n_0} R_{m_0 n_0}^0 \exp\left\{i2\pi\left[\left(\frac{m_0}{p_0}\right)u' + \left(\frac{n_0}{3q}\right)v'\right]\right\} \tag{8.34}$$

with coefficients given by

$$R_{m_0 n_0}^0 = \left(\frac{1}{3qp_0}\right)\int_{-(p_0/2)}^{+(p_0/2)}\int_{-(q/2}^{+(q/2)} R_0^0(u',v') \times \exp\left\{-i2\pi\left[\left(\frac{m_0}{p_0}\right)u' + \left(\frac{n_0}{3q}\right)v'\right]\right\} du'\, dv', \tag{8.35}$$

the limits for v' being $\pm q/2$, not $\pm 3q/2$, since $R_0^0(u',v') = 0$ for $q/2 < |v'| < 3q/2$. The coefficients (8.35) are equal to the corresponding coefficients for a simple periodic disc with track spacings equal to $3q$. They may therefore be calculated for any practical case as described in Sec. VI (A), above.

If we now omit tracks $(-1+k)$ and $(0+k)$, there will remain only the track $(+1+k)$, whose pits again form a simple periodic array, with periods p_+ and $3q$. Let the corresponding reflection function, with the origin at the center of the pit $m = 0$ of track $n = +1$, be denoted by $R_0^+(u',v')$, which may be written as a Fourier series similar to (8.34), namely

$$R_0(u',v') = \sum_{m_+} \sum_{n_+} R_{m_+,n_+}^+ \exp\left\{i2\pi\left[\left(\frac{m_+}{p_+}\right)u' + \left(\frac{n_+}{3q}\right)v'\right]\right\} \tag{8.36}$$

with coefficients given by

$$R_{m_+ n_+}^+ = \left(\frac{1}{3qp_+}\right)\int_{-(p_+/2)}^{+(p_+/2)}\int_{-(q/2)}^{+(q/2)} R_0^+(u',v') \times \exp\left\{-i2\pi\left[\left(\frac{m_+}{p_+}\right)u' + \left(\frac{n_+}{3q}\right)v'\right]\right\} du'\, dv' \tag{8.37}$$

and, with the origin at the center of the pit $m = 0$ of track $n = 0$, the reflection function for the tracks $(+1+k)$ will be $R_0^+(u',v'-q)$.

In a similar manner the reflection function for the tracks $(-1+k)$ will be $R_0^-(u',v'+q)$, where

$$R_0^-(u',v') = \sum_{m_-} \sum_{n_-} R_{m_- n_-}^- \exp\left\{i2\pi\left[\left(\frac{m_-}{p_-}\right)u' + \left(\frac{n_-}{3q}\right)v'\right]\right\} \tag{8.38}$$

is the reflection function for these tracks with the origin at the center of the pit $m = 0$ of track $n = -1$. The coefficients in (8.38) are given by

$$R_{m_- n_-}^- = \left(\frac{1}{3qp}\right)\int_{-(p_-/2)}^{+(p_-/2)}\int_{-(q/2)}^{+(q/2)} R_0^-(u',v') \times \exp\left\{-i2\pi\left[\left(\frac{m_-}{p_-}\right)u' + \left(\frac{n_-}{3q}\right)v'\right]\right\} du'\, dv' \tag{8.39}$$

and these may again be calculated by the method of Sec. VI A, above.

The reflection function for the whole disc is now given by

$$R_0(u',v') = R_0^-(u',v'+q) + R_0^0(u',v') + R_0^+(u',v'-q) \tag{8.40}$$

or, using the series (8.39), (8.34), and (8.36), and grouping the terms having $m_- = m_0 = m_+ = m$ and $n_- = n_0 = n_+ = n$ together,

$$\begin{aligned} R_0(u',v') = \sum_m \sum_n \Bigg(& R_{mn}^- \exp\left(\frac{i2\pi n}{3}\right) \times \exp\left\{i2\pi\left[\left(\frac{m}{p_-}\right)u' + \left(\frac{n}{3q}\right)v'\right]\right\} \\ & + R_{mn}^0 \exp\left\{2\pi\left[\left(\frac{m}{p_0}\right)u' + \left(\frac{n}{3q}\right)v'\right]\right\} \\ & + R_{mn}^+ \exp\left\{-\frac{i2\pi n}{3}\right\} \exp\left\{i2\pi\left[\left(\frac{m}{p_+}\right)u' + \left(\frac{n}{3q}\right)v'\right]\right\}\Bigg) \end{aligned} \tag{8.41}$$

which is, of course, simply the sum of three Fourier series, and has the three different fundamental spatial frequencies $(1/p_-)$, $(1/p_0)$, and $(1/p_+)$ in the u'-direction, and a single fundamental spatial frequency $(1/3q)$ in the v' direction.

The reflection function (8.41) corresponds to that of (3.9) for the case of a simple periodic disc. When the disc considered here is moved distances u and v along the u' and v' axes, the reflection coefficient is given by $R(u',v') = R_0(u' - u, v' - v)$, exactly as in (3.8) for the simple disc. The inverse Fourier transform of $R(u,v)$ is thus given by

$$\begin{aligned} r(x',y') = & \sum_m \sum_n \exp\left\{-i2\pi\left[\left(\frac{m}{p_-}\right)u + \left(\frac{n}{3q}\right)v\right]\right\} \times R_{mn}^- \exp\left(i\frac{2\pi n}{3}\right)\delta\left(x' - \frac{m}{p_-}, y' - \frac{n}{3q}\right) \\ & + \sum_m \sum_n \exp\left\{-i2\pi\left[\left(\frac{m}{p_0}\right)u + \left(\frac{n}{3q}\right)v\right]\right\} \times R_{mn}^0\, \delta\left(x' - \frac{m}{p_0}, y' - \frac{n}{3q}\right) \\ & + \sum_m \sum_n \exp\left\{-i2\pi\left[\left(\frac{m}{p_+}\right)u + \left(\frac{n}{3q}\right)v\right]\right\} \times R_{mn}^+ \exp\left(-i\frac{2\pi n}{3}\right)\delta\left(x' - \frac{m}{p_+}, y' - \frac{n}{3q}\right) \end{aligned} \tag{8.42}$$

corresponding to (3.11).

If $r(x',y')$ given by (8.42) is now substituted in the general formula (3.6), the complex amplitude diffracted back to the point (x,y) of the pupil will be given by

$$
\begin{aligned}
a'(x,y) = \sum_m \sum_n & \exp\left\{-i2\pi\left[\left(\frac{m}{p_-}\right)u + \left(\frac{n}{3q}\right)v\right]\right\} \\
& \times R^-_{mn} \exp\left(i\,\frac{2\pi n}{3}\right) f\left(x - \frac{m}{p_-}, y - \frac{n}{3q}\right) \\
& + \sum_m \sum_n \exp\left\{-i2\pi\left[\left(\frac{m}{p_0}\right)u + \left(\frac{n}{3q}\right)v\right]\right\} \\
& \times R^0_{mn} f\left(x - \frac{m}{p_0}, y - \frac{n}{3q}\right) \\
& + \sum_m \sum_n \exp\left\{-i2\pi\left[\left(\frac{m}{p_-}\right)u + \left(\frac{n}{3q}\right)v\right]\right\} \\
& \times R^+_{mn} \exp\left\{-\,i\,\frac{2\pi n}{3}\right\} f\left(x - \frac{m}{p_+}, y - \frac{n}{3q}\right), \qquad (8.43)
\end{aligned}
$$

corresponding to (3.12) for the case of a simple disc.

We shall again not carry the analysis through completely, but merely note that the total light flux reaching the detector is again given by (4.2) above, namely

$$
I(t) = \iint_{(x^2+y^2)\leq 1} |a'(x,y)|^2\, dx\, dy, \qquad (8.44)
$$

with $a'(x,y)$ now given by (8.43). Clearly, there will be cross-talk terms with frequencies given by sums and differences of the frequencies associated with each of the three sets of tracks. The amplitudes of these cross-talk terms again only require the coefficients R^-_{mn}, R^0_{mn} and R^+_{mn}, to be known and the evaluation of integrals of precisely the same form, namely $D(m,n;m',n')$ of (6.20), as those for the case of a simple periodic disc.

The method described here could clearly be extended to cases where more than the three tracks n = 1, 0, and +1, have different center-to-center spacings for the pits. This was not considered necessary in the present investigations because, for a diffraction image centered on the track n = 0, the central disc and first ring of the Airy diffraction pattern fall entirely within the circle A_2 of Fig. 12. Only a very small outer part of the second diffraction ring overlaps the inner halves of the pits of the tracks $n = \pm 2$, so that the particular forms of the pits in these tracks would not be expected to have any significant effect.

Although there would be no difficulty in principle to introducing "stagger" between adjacent tracks, this was also considered unnecessary; the pits m = 0 shown in Fig. 12 have no relative "stagger", but as the disc rotates and the pits $m \neq 0$ are read there are time-varying "staggers" between the track n = 0 and the two adjacent tracks. For this reason, it was felt that having relative "stagger" between the initial pits on the tracks would not make any material difference to the final numerical results.

It would again be possible, using the methods described in the first part of the present section, to treat the present case but with duty cycles of length modulation of the pits along each of the tracks n = −1, 0, and +1. The theory presents no difficulty, but there are then inevitably a very large number of different frequency components in the detected light flux $I(t)$, with a consequent considerable increase in the numerical work.

IX. NUMERICAL RESULTS AND DISCUSSION

The above formulas have been programmed and used extensively to study the dependence of the performance of the read-out system on the different parameters. Results predicted by the analysis have shown good agreement with practical measurements. Only a small number of results will be quoted here, with emphasis being given to the influence of focus error and aberrations on the read-out signal.

The total light flux reaching the photodiode is given by the formula (4.10), namely

$$
I(t) = A_0 + \sum_{\mu=1}^{\infty} \{A_\mu \cos(\omega_\mu t) + B_\mu \sin(\omega_\mu t)\} \qquad (9.1)
$$

with the coefficients A_0,A_μ, and B_μ, given as functions of the reduced tracking error v by (5.47), (5.48), and (5.49). The angular temporal frequencies are given by

$$
\omega_\mu = \mu[2\pi(\omega a/\xi')] \qquad (9.2)
$$

where ω is the angular velocity of the disc, a is the radius of the track being read, and ξ' is the real-space center-to-center separation of the pits. The dc component is given by A_0, and $(A^2_\mu + B^2_\mu)^{1/2}$ is the amplitude of the component of frequency ω_μ; we shall denote this latter by

$$
I_\mu = (A^2_\mu + B^2_\mu)^{1/2} \qquad (9.3)
$$

in the figures. The real-space tracking error will be denoted by the same symbol v which is used for the reduced-space value of this quantity in the analysis. In each of the figures the values used for the wavelength of the light and the numerical aperture of the objective were: λ = 0.63 μm and NA = 0.45.

Figure 13 shows the variation of the amplitude I_1, of the

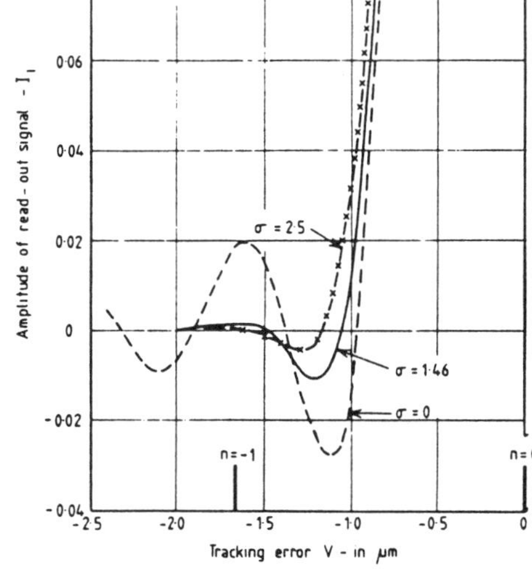

FIG. 13. Influence of nonuniformity of amplitude on the read-out signal.

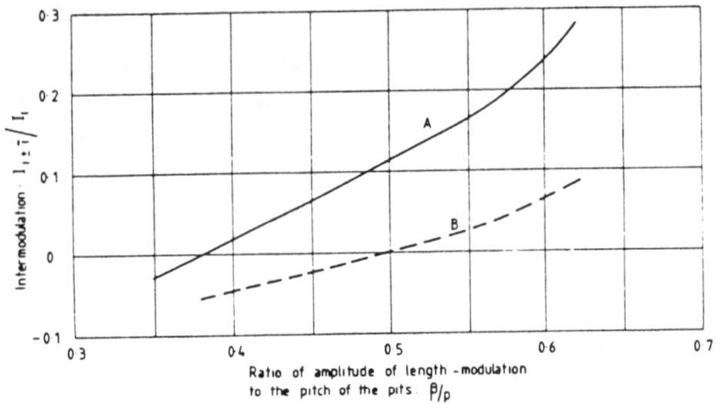

FIG. 14. Intermodulation between length-modulation and fundamental read-out frequencies. Curve A: full aperture, $\phi = \pi$; curve B: push-pull operation, $\phi = \pi/2$.

fundamental in the read-out signal as a function of the tracking error for different amounts of nonuniformity of amplitude over that part of the incident laser beam which is accepted by the pupil of the objective. The quantity σ is the parameter in (3.2) specifying the Gaussian variation of intensity across the incident laser beam. The more this is expanded before entering the objective, the smaller the decrease in intensity between the center and the edge of the pupil. The latter is given by $I_m = \exp(-\sigma)I_0$, as in (3.3), where I_0 is the intensity at the center. Thus, $\sigma = 2.50$ and $\sigma = 1.46$ correspond to $I_m = 0.08\, I_0$ and $I_m = 0.23\, I_0$ respectively. The case $\sigma = 0$, shown for comparison, is that of uniform intensity, $I_m = I_0$. In the curves of subsequent figures the value $\sigma = 1.46$ has been employed.

The curves of Fig. 13 are normalized to make $I_1 = 1$ when there is no tracking error, $v = 0$. The read-out signal decreases more rapidly the more uniform the intensity over the incident beam, and the spurious signals for larger tracking errors are also greater. This is consistent with the known fact that, with $I_m < I_0$, the central disc of the Airy pattern is slightly broadened but less and less energy appears in the surrounding diffraction rings.

The curves shown were calculated for a case in which the tracks adjacent to that being read have different pit spacings, p_+ and p_-, from that of the center track p_0. This is the notation of (8.33), and the formulas used were those of the analysis of Sec. VIII, leading to (8.43) and (8.44). Each pit was assumed to be of the form shown in Fig. 7; that is, having sloping walls and rounded ends, with $(1/2)\,\epsilon = 0.05\ \mu$m. The width of each pit is $\gamma = 0.5\ \mu$m and the spacing between tracks is $q = 1.67\ \mu$m. The read-out signal shown, I_1, is the fundamental of the temporal frequency corresponding to the pit spacing p_0 of the center track. The marks $n = 0$ and $n = 1$ show respectively the positions of the center lines of the track being read and an adjacent track.

An important aspect of the problem studied here is that of intermodulation between the temporal frequencies associated with the center-to-center spacing ξ' of the pits and those associated with the period $\bar{\xi}'$ of the length modulation of the pits along the track. The study of this problem has been given in Sec. VIII above, where the length of the mth pit is given by

$$\beta_m = \beta[1 + M\cos(2\pi mp/P)], \qquad (9.4)$$

where β is the mean length of the pits, M_β is the amplitude of length modulation, and p and P are the reduced values of ξ' and $\bar{\xi}'$ respectively. The form of disc is now as shown in Fig. 11, and the basic formulas for this case are those of (8.20) and (8.21) above.

The read-out signal now contains terms whose temporal frequencies are given by

$$\omega\hat{\mu} = \mu[2\pi(\omega a/\xi')] + \rho[2\pi(\omega a/\bar{\xi}')] \qquad (9.5)$$

as shown in (8.26), (8.27), and (8.28) above. In (9.5) the first term gives the temporal frequencies associated with the center-to-center spacing of the pits, and the second term gives those arising from the length modulation. The dominant intermodulation terms are the sum and difference frequencies $\mu = 1, \rho = \pm 1$, and the ratio $I_{1\pm\bar{1}}/I_1$ is a measure of the degree of intermodulation between the signals from the periods ξ' and $\bar{\xi}'$.

The ratio $I_{1\pm\bar{1}}/I_1$ is shown in Fig. 14 as a function of β/p, for the case of simple rectangular pits. The amplitude of length modulation is $M = 0.2$, and the other parameters have the values $p = 1.5\ \mu$m, $q = 1.67\ \mu$m, $P = 1000\, p$, and $\gamma = 0.5\ \mu$m. Curve A is for pits having a depth $\lambda/4$, giving a phase-difference $\phi = \pi$, and detecting all of the light diffracted back through the pupil of the objective. Curve B shows the case $\phi = \pi/2$, but employing a "push-pull" mode of operation, in which the light diffracted back to the objective is detected separately for the two halves of the pupil, and the two resulting signals are differenced.

The remaining diagrams, Figs. 15–18 show the effects of errors of focusing of the objective on the disc and of aberrations in the objective. Both focus error and aberrations will

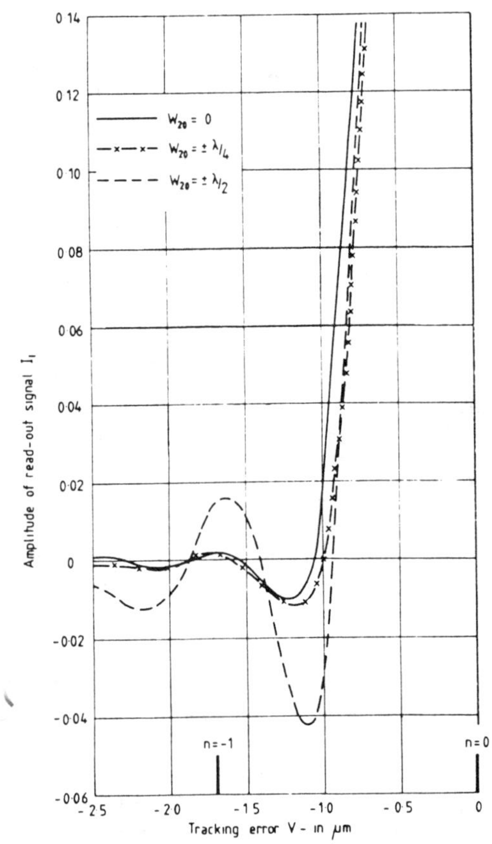

FIG. 15. Influence of focus error on the read-out signal.

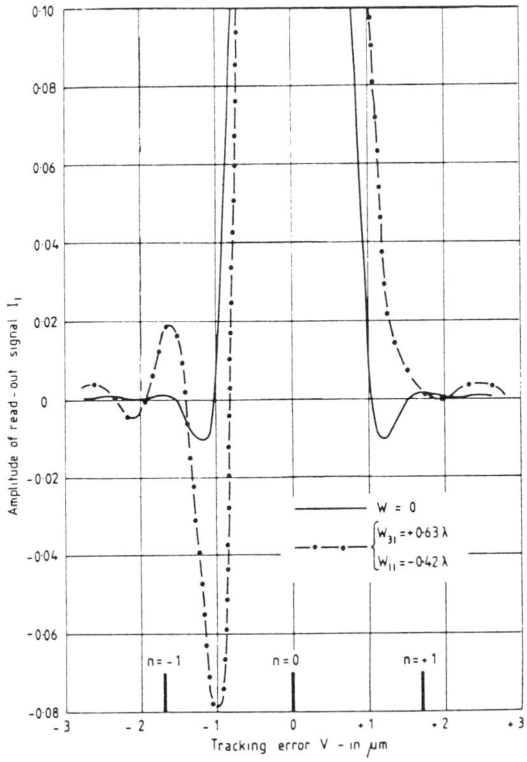

FIG. 16. Influence of coma on the read-out signal.

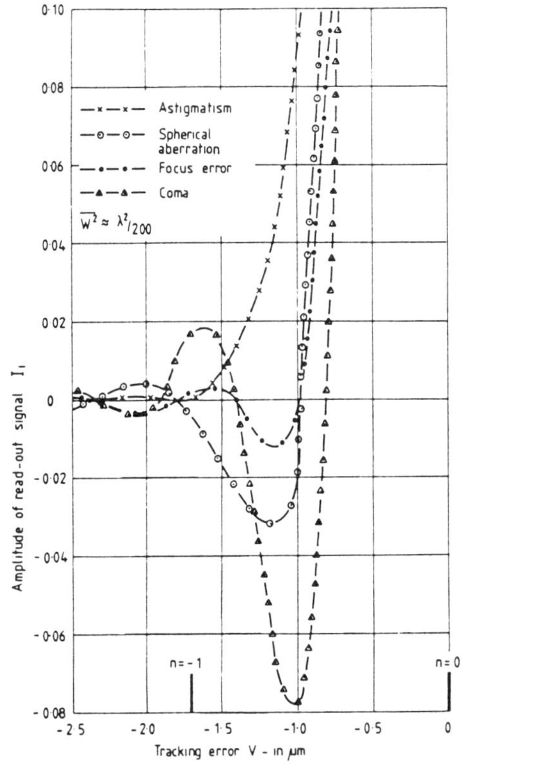

FIG. 18. Influence of focus error, coma, spherical aberration, and astigmatism on the read-out signal.

cause the diffraction image formed on the disc to be of greater size and will also modify the distribution of complex amplitude in it. The form of disc assumed in each of the figures is where adjacent tracks have slightly different center-to-center spacings for the pits. The pits are all assumed to have sloping walls and rounded ends. The lengths of the rectangular parts of the pits are assumed to be equal to half the center-to-center spacings.

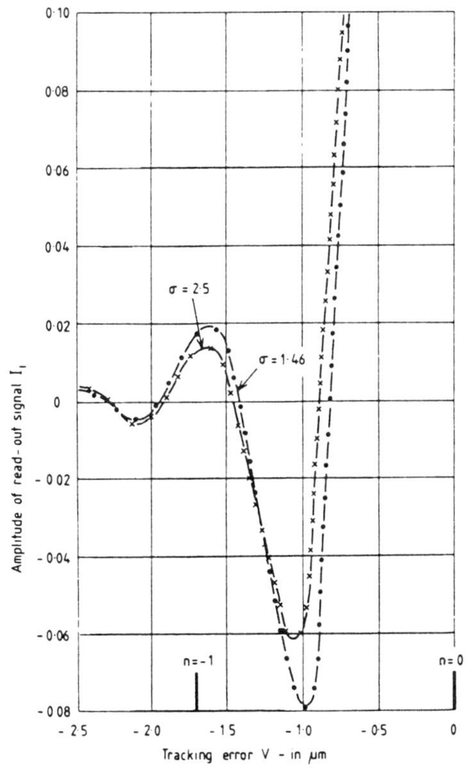

FIG. 17. Influence of nonuniformity of amplitude on the read-out signal in the presence of coma.

Since the tracks n = 1, 0, and +1, all have different center-to-center spacings they contribute different temporal frequencies to the total read-out signal. The signal, in each of these figures, is the amplitude of the fundamental frequency corresponding to the track $n = 0$ shown as a function of tracking error. Because of the different frequencies associated with the adjacent tracks, the change in the read-out signal for the track $n = 0$ with increasing tracking error is unaffected by the signals from the adjacent tracks, even when the tracking error is so large that the read-out spot falls on one of the adjacent tracks.

Figure 15 shows the effect of focus on the amplitude of the read-out signal I_1 as a function of tracking error. The form of the aberration function for this case is, in polar coordinates,

$$W(r,\phi) = W_{20}r^2, \tag{9.6}$$

where $r = 1$ is the periphery of the pupil. The Rayleigh criterion, and equally the Strehl criterion, for the ordinarily permissible focus error is $|W_{20}| \leq \lambda/4$. The curves for $W_{20} = 0$ and $W_{20} = \pm\lambda/4$, for the present case, show very little difference. By contrast, when W_{20} reaches the value $\lambda/2$ there is an appreciable effect on the read-out signal.

Figure 16 shows the influence of primary (3rd-order) coma on the read-out signal. The aberration function now has the form

$$W(r,\phi) = W_{31}r^3 \cos\phi + W_{11}r \cos\phi \tag{9.7}$$

where W_{11} corresponds to a transverse shift of the reference focus. The case shown in Fig. 16 is when $W_{21} = 0.63\ \lambda$, which is when the Strehl intensity ratio has the value 0.8. The shift of focus $W_{11} = -0.42\ \lambda$ brings one from the geometrical image point to the principal maximum in the diffraction pattern. Thus, in the figure, no tracking error ($\upsilon = 0$) is when this maximum is on the center of the track $n = 0$. It is seen that the effects of tracking error are asymmetrical, and they are very marked when the tracking error is to the same side as the "coma flare."

The nonuniformity of amplitude assumed in Fig. 16 is that corresponding to $\sigma = 1.46$, for which $I_m = 0.23\ I_0$. It seemed useful to investigate whether the value $\sigma = 2.50$, for which $I_m = 0.08\ I_0$, would reduce the effects of coma because of the smaller amplitude at the edge of the aperture, where the aberration has its maximum value. The curves of Fig. 17 show this comparison, in which a small but not very significant improvement is seen.

The curves of Fig. 18 show, for purposes of intercomparison, the separate effects of focus error, coma, spherical aberration, and astigmatism respectively. The magnitudes of the aberrations are in each case such as to give $\overline{W^2} \approx \lambda^2/200$ for the mean-square values. This, as Maréchal has shown, is the tolerance in aberration needed to ensure a value greater than or equal to 0.80 for the Strehl ratio. The aberration functions for the cases of focus error and of coma are precisely the same as used for the curves of Figs. 15 and 16 respectively. For spherical aberration the aberration function has the form

$$W(r,\phi) = W_{40}r^4 + W_{20}r^2 \tag{9.8}$$

with $W_{40} = -1\lambda$ and $W_{20} = +1\lambda$, the latter being the longitudinal shift from the paraxial image plane to the plane of best focus. For astigmatism, the aberration function is

$$W(r,\phi) = W_{22}r^2 \cos^2\phi, \tag{9.9}$$

which corresponds to the focus being chosen with the focal line perpendicular to the tracks. The coefficient has the value $W_{22} = 0.35\lambda$.

The above examples have considered only cases of primary (3rd-order) aberrations present separately, although the analysis is in no way restricted as to the form of complete aberration of the system. They, nevertheless, serve to show the high degree of correction of aberration and precision of focusing which are necessary if very high packing densities of information are to be achieved.

ACKNOWLEDGMENTS

The analysis given here was checked by Dr. J. Braat, who also wrote all the computer programs and supplied the numerical results. I am also grateful to Philips Research Laboratories, Eindhoven, for agreeing to publication of this paper.

[1]K. Compaan and P. Kramer, "The Philips VLP System," Philips Tech. Rev., **33**, 178–180 (1973).

[2]G. Bouwhuis and P. Burgstede, "The Optical Scanning System of the Philips VLP Record Player," Philips Tech. Rev., **33**, 186–189 (1973).

[3]H. H. Hopkins and Maria J. Izuel, Opt. Acta, **17**, 157–182 (1970).

Reprinted with permission from *Journal of Applied Physics*, Vol. 53(6), pp. 4485-4494 (June 1982). ©1982 American Institute of Physics.

Signal and noise in magneto-optical readout

Masud Mansuripur and G. A. N. Connell
Xerox Corporation, Palo Alto Research Center, 3333 Coyote Hill Road, Palo Alto, California 94304
Joseph W. Goodman
Department of Electrical Engineering, Stanford University, Stanford, California 94305

For a magneto-optical readout system which uses the Kerr effect, the power spectral density of the read signal has been derived. The noise components originating in the photodetection and amplification stages appear as additive terms in the expression for the overall spectral density, enabling us to define a signal to noise ratio for the readout system. Adjustable parameters of the system are then optimized and a multilayer disk structure is introduced which enhances the Kerr effect and, consequently, results in an increased signal to noise ratio.

I. INTRODUCTION

Presently up to a million bits of information can be stored per square inch on conventional magnetic disk media. The upper limit on the storage density is set by the sensitivity of the read-write head rather than by the magnetic properties of the media, particularly when vertical as opposed to in-plane recording is considered. Thus, new methods providing access to smaller regions of magnetic film, could provide marked improvements in storage density. In principle, the ability to focus laser light to sufficiently small dimensions offers an attractive solution to this problem, as well as a way of exploiting the optical and thermal properties of perpendicular media to store, erase, and retrieve information at densities approaching the media limits. The possibilities and realization of this approach are therefore of considerable interest.[1,2]

In this paper, we will discuss the read signal from a magneto-optical disk when the polar Kerr effect is used, reserving the problem of the writing and erasure of data for a later publication. The polar Kerr effect itself is observed by the changes in the state of polarization of light that occur after normal reflection from a perpendicularly magnetized medium. Figure 1, therefore, represents the simplest system that can use this effect to extract data from a magneto-optical disk. The magnetic film has perpendicular anisotropy, and the direction of magnetization (either up or down) in any region determines whether a zero or one is encoded there. When linearly polarized light is normally incident on the film, the rotation of the major axis from the original polarization direction and the ellipticity of the polarized reflected light both have signs that depend on the direction of magnetization. Thus, the polarization analyzer shown converts major axis rotations to changes in light intensity which are then converted to electric current variations by a photodetector. The addition of a quarter wave plate at Q would permit changes of ellipticity to be detected in the same way.

The polar Kerr effect is usually very small. Typically the rotation angle and the ellipticity of the reflected light are of the order of 0.1 degrees. Thus, with the simple detector of Fig. 1, the information-carrying signal would be but a small fraction of the detected signal. To improve on this situation, two approaches have been taken. The first is to use the differential detection scheme of Fig. 2.[3] In this scheme, the reflected light is split in two halves and each half is detected by a separate detector. The detectors are identical except for the setting of the analyzers. The null axis of one analyzer is set at angle $+\Theta$ with respect to the unrotated polarization vector, while the other is set at $-\Theta$. When the two outputs are fed to a differential amplifier, the dc components cancel each other and the amplified signal is received at the output. This signal is processed by a filter before reaching the decision point. The second approach is to enhance the rotation angle or the ellipticity through interference effects that occur when, for example, the sample is overcoated with a thin dielectric film. Our aim is to find the conditions that make best use of these elements when used together. Thus, in this paper we first focus on the performance of the differential detection scheme itself. Then, we consider how interference effects enhance the magneto-optic response and so find that

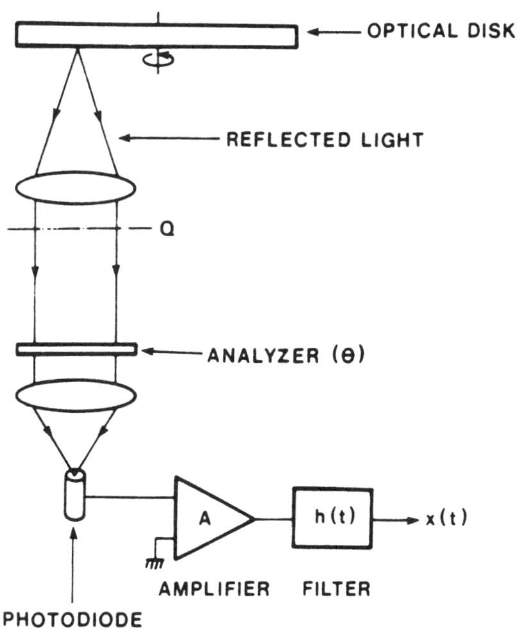

FIG. 1. Schematic diagram of a simple readout system.

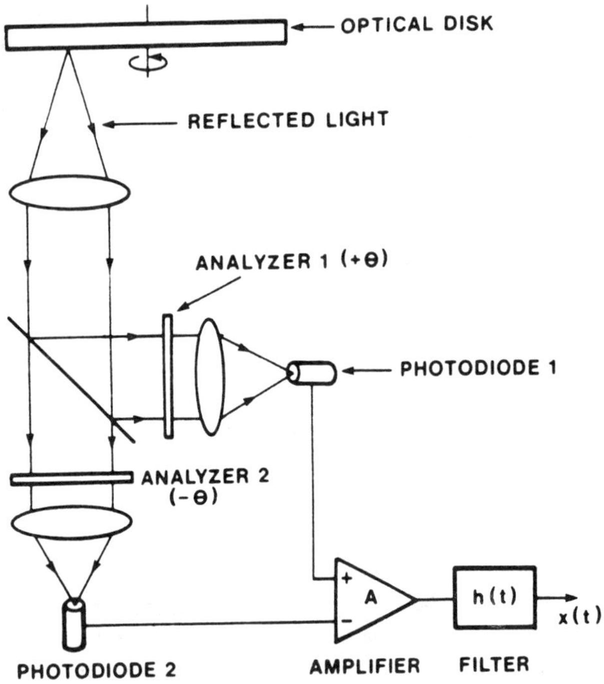

FIG. 2. Schematic diagram of a differential readout system.

multilayer device that results in maximum signal to noise ratio when used with the differential detection scheme.

The paper is organized as follows. Section II is a brief account of the polar Kerr effect that provides both an intuitive view of how interference schemes can improve the response of the magneto-optic medium and the mathematical framework for magneto-optic calculations on multilayer devices. This is a basis for the derivation of the power spectral density of the read signal from the differential detection system in Sec. III and of the signal to noise ratio in Sec. IV. The signal to noise ratio in a digital detection system is directly related to error rates only when the noise is additive and Gaussian, which is not exactly the case in this problem. An exact evaluation of the system should therefore be based on a more complicated analysis. We have shown elsewhere,[4] however, that in a conventional photodetection system the assumption of Gaussian statistics yields reasonably accurate predictions. Therefore the signal to noise ratio is a good measure of achievable error rates. Results of Sec. IV are then used in Sec. V to obtain the optimum disc structure and maximum signal to noise ratio for the differential detection system. Our conclusions are summarized in Sec. VI.

II. THE POLAR KERR EFFECT

The polar Kerr effect occurs when light is reflected normally from a perpendicularly magnetized medium. Under these conditions, it is readily shown[5] that the normal modes of propagation are right (a^-) and left (a^+) circularly polarized with eigenvalues:

$$n^{\pm} = (\epsilon \pm i\epsilon')^{1/2}. \tag{1}$$

In this equation, ϵ and ϵ' are the complex diagonal and off-diagonal elements of the dielectric tensor, which are even and odd functions of magnetization, respectively. For isotropic, non-magnetic materials, it therefore follows that $\epsilon' = 0$ and $n^+ = n^- = \epsilon^{1/2}$, i.e., the usual relationship between refractive index and dielectric constant.

It is often convenient experimentally to use linearly polarized light for the observation of the Kerr effect. In this case, with the incident light linearly polarized along the X direction, the reflected light has the regular component r_x and a magneto-optically induced component r_y, where r_x and r_y are complex. The magnitude of the polar Kerr effect, $|r_y|$, is determined both by the internal conversivity[6] of the medium set by ϵ', and by the ability to introduce light into and remove magneto-optic radiation from the medium such that maximum conversion, as measured outside the medium, occurs. This latter aspect is very much affected by interference effects within the device structure and considerable enhancement of $|r_y|$ over its value for a bulk sample can be obtained. For example, the simplest case of a dielectric antireflection coating on a bulk material produces a small gain because more light enters the magnetic medium. It is not effective, however, either in increasing the efficiency of conversion within the magnetic medium itself or in increasing the collection of the magneto-optic radiation. Consider, however, the device shown in Fig. 3 in which the magnetic thin film is deposited on a transparent dielectric film that itself rests on an opaque reflector, to form what may be called a trilayer device. The thicknesses of the sample and transparent dielectric films are adjusted so that $r_x = 0$. This occurs because the ray reflected directly by the sample is exactly cancelled by light which has suffered at least one reflection from the back reflector. To achieve this, the optical thickness of the dielectric layer must be approximately $(2m + 1)\lambda/4$, where λ is the wavelength of the light used in the magneto-optic experiment and $m = 0, 1$, etc. (In actual practice, $m = 0$ is adequate for most purposes.) Therefore, when a perfect opaque reflector is used, all of the incident intensity is absorbed in the thin absorbing sample. Furthermore, magneto-optically induced light $(E \perp X)$, which is emitted both forwards and backwards from the sample, suffers relatively little reabsorption in the thin absorbing film, and the interference conditions that have been set up to create total absorption of the incident beam are exactly those required to maximize $|r_y|$ by in-phase addition of the exiting

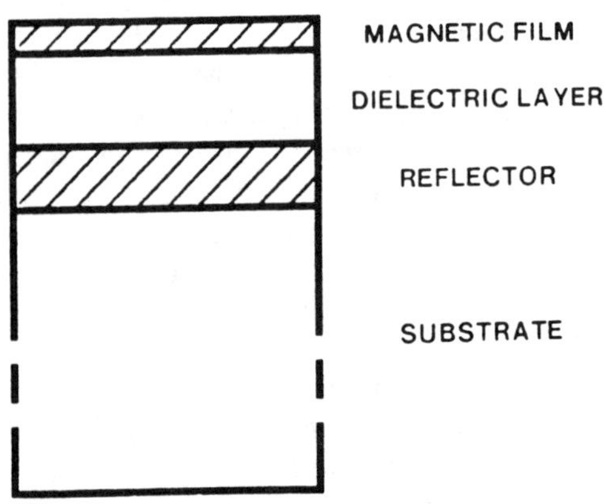

FIG. 3. Trilayer structure for magneto-optic device.

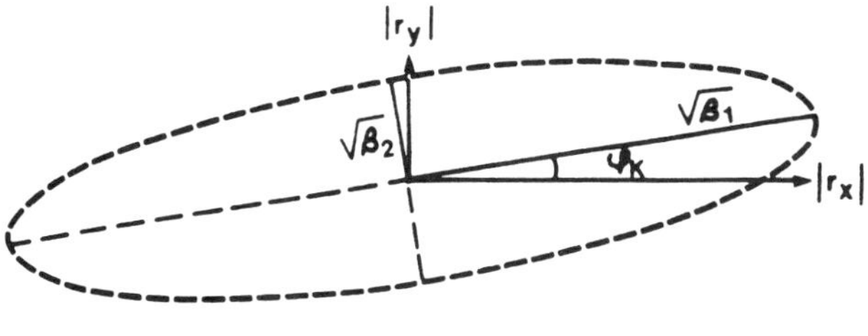

FIG. 4. Polarization of reflected light from the disk surface.

rays. Thus, a large interference enhancement of $|r_y|$ occurs. It has been shown[7] that the approximate enhancement of the polar Kerr intensity $|r_y|^2$ over its value in a semi-infinite sample is $E \sim 8\alpha t/(1-R)^2$ where R and α are the reflectance and absorption coefficient of semi-infinite sample, respectively, and t, the thickness of the thin absorbing layer, is typically less than 100 Å. Thus, for reasonable values of these parameters, $E \sim 10$–10^2, and the importance of device structure for obtaining maximum signal to noise during magneto-optic readout is intuitively obvious.

The mathematical framework needed to describe these effects is set up by decomposing the incident linearly polarized light into the two circularly polarized normal modes.

$$a_x = \tfrac{1}{2}(a_x + ia_y) + \tfrac{1}{2}(a_x - ia_y) = \tfrac{1}{2}(a^+ + a^-). \tag{2}$$

Here a_x and a_y are unit vectors along the X and Y directions. On reflection, each of the above components will be attenuated by the corresponding reflectivity. Let us use complex numbers r^+ and r^- to identify the amplitude reflectivities for left and right circularly polarized components, respectively. These of course are readily obtained, as discussed in the appendix, from $n^\pm$ of the materials that constitute the layers of the device. The reflected field can then be written as

$$r_x a_x + r_y a_y = \tfrac{1}{2}r^+(a_x + ia_y) + \tfrac{1}{2}r^-(a_x - ia_y). \tag{3}$$

In general, the reflected light is elliptically polarized (see Fig. 4). We define the Kerr angle φ_k as the angle between the major axis of the ellipse and the X axis. If the field component polarized along the major axis has intensity β_1, and that along the minor axis has intensity β_2, one can write the reflected amplitude in the following equivalent form:

$$\begin{aligned} r_x a_x + r_y a_y = {} & [\beta_1^{1/2}(\cos\varphi_k a_x + \sin\varphi_k a_y) \\ & + \beta_2^{1/2}\exp(\pm i\pi/2)(\sin\varphi_k a_x \\ & - \cos\varphi_k a_y)]\exp(i\delta). \end{aligned} \tag{4}$$

Here δ is a constant phase angle and can be determined, along with β_1, β_2, and φ_k, by equating the right hand sides of Eqs. (3) and (4). The sign of $\pi/2$ in Eq. (4) determines the direction of rotation around the ellipse and must be deduced likewise.

All we need in our later analysis is β_1, β_2, and $\varphi = |\varphi_k|$. These parameters can be expressed in terms of r^+ and r^-, independent of δ and the sign of $\pi/2$. The following results can be easily verified:

$$r_x = \tfrac{1}{2}(r^+ + r^-), \tag{5a}$$

$$r_y = \tfrac{1}{2}i(r^+ - r^-), \tag{5b}$$

$$\beta_1 = \tfrac{1}{2}(|r_x|^2 + |r_y|^2 + |r_x^2 + r_y^2|), \tag{5c}$$

$$\beta_2 = \tfrac{1}{2}(|r_x|^2 + |r_y|^2 - |r_x^2 + r_y^2|), \tag{5d}$$

$$\cos(2\varphi) = (|r_x|^2 - |r_y|^2)/|r_x^2 + r_y^2|. \tag{5e}$$

In the following sections we will refer to β_1, β_2, and φ as disk parameters, keeping in mind the fact that they can always be determined from the more basic optical parameters of the constituent layers.

III. SPECTRAL ANALYSIS OF THE READ SIGNAL

Figure 5 shows a portion of a magneto-optical disk which is located in the XY plane; the presence of reverse magnetized domains is also indicated. A narrow beam of linearly polarized light, propagating in the Z direction and moving with velocity v along the X direction, is used for the retrieval of data. The beam is Gaussian and its intensity distribution in the XY plane can be expressed as

$$I(x,y,t) = (P_0/\pi r_0^2)\exp\{-[(x-vt)^2 + (y-y_0)^2]/r_0^2\}. \tag{6}$$

Here P_0 is the beam's total power, r_0 is the beam's radius at the e^{-1} point, and y_0 is the y coordinate of the track being read.

If the disk is uniform, its parameters β_1, β_2, and φ will be constant over the entire XY plane. Define a function $Z(x,y)$ with only two values: "+ 1" when the magnetization at (x,y) points up and "− 1" when the magnetization points down. One can then write

$$\varphi_k(x,y) = Z(x,y)\varphi. \tag{7}$$

With this notation the optical power collected at one of the photodetectors in Fig. 2 is

$$\begin{aligned} P_1(t) = \tfrac{1}{2}\int\int_{-\infty}^{\infty} & I(x,y,t)\{\beta_1\sin^2[\Theta + Z(x,y)\varphi] \\ & + \beta_2\cos^2[\Theta + Z(x,y)\varphi]\}\,dx\,dy. \end{aligned} \tag{8a}$$

The power collected at the other detector will then be

$$\begin{aligned} P_2(t) = \tfrac{1}{2}\int\int_{-\infty}^{\infty} & I(x,y,t)\{\beta_1\sin^2[\Theta - Z(x,y)\varphi] \\ & + \beta_2\cos^2[\Theta - Z(x,y)\varphi]\}\,dx\,dy. \end{aligned} \tag{8b}$$

Knowing the optical energy delivered to the detectors, one must in principle be able to determine the power spectral

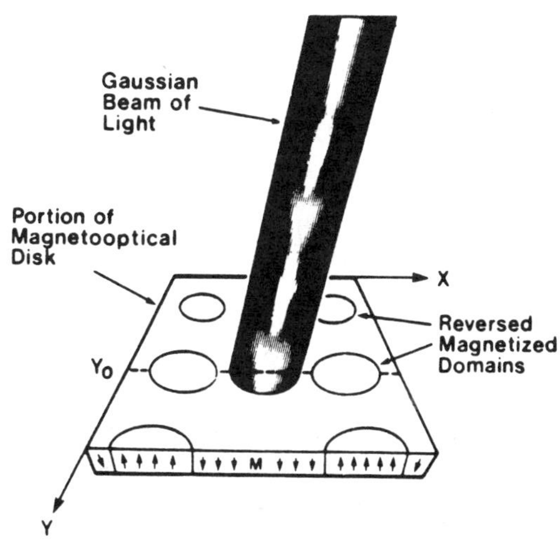

FIG. 5. Coordinate system for the spectral analysis of the read signal.

density of the output signal (in volt^2/Hz), which is defined as

$$S_x(\omega) = \lim_{T\to\infty} \mathbf{E}\left[(1/T)\left|\int_{-T/2}^{T/2} x(t)\exp(-i\omega t)\,dt\right|^2\right]. \quad (9)$$

The expectation in Eq. (9) is taken over the space of random functions $x(t)$. The signal $x(t)$ is the difference between the outputs of the two photodetectors and can be written as

$$\begin{aligned} x(t) &= x_1(t) - x_2(t) \\ &= \sum_{k_1} (qRA) G_{k_1} h(t - t_{k_1}) - \sum_{k_2} (qRA) G_{k_2} h(t - t_{k_2}). \quad (10) \end{aligned}$$

The subscripts 1 and 2 identify the corresponding photodetectors. In the above equation q is the electronic charge; R is the resistance in each photodetector branch; A is the amplifier's gain; G_k is the internal gain of the photodiode associated with the k'th released electron; $h(t)$ is the impulse response of the filter, and t_k is the instant at which the k'th electron has been released. The summations in Eq. (10) are over all electrons that contribute to the signal at time t. By substituting for $x(t)$ in Eq. (9) one arrives at

$$\begin{aligned} S_x(\omega) = \lim_{T\to\infty} \mathbf{E}\Bigg[(1/T)\Bigg| &\sum_{k_1=1}^{K_1} (qRA) G_{k_1} H(\omega)\exp(-i\omega t_{k_1}) \\ &- \sum_{k_2=1}^{K_2} (qRA) G_{k_2} H(\omega)\exp(-i\omega t_{k_2})\Bigg|^2\Bigg]. \end{aligned} \quad (11)$$

Here $H(\omega) = \int_{-\infty}^{\infty} h(t)\exp(-i\omega t)\,dt$, and K_1 and K_2 are the total number of electrons released in the interval $[-T/2, +T/2]$ in photodetectors 1 and 2, respectively. T is assumed to be very large compared to the width of $h(t)$. Equation (11) can be expanded as

$$\begin{aligned} S_x(\omega) = (qRA)^2 |H(\omega)|^2 \lim_{T\to\infty} \mathbf{E}\Bigg[(1/T)\Bigg(&\sum_{k_1=1}^{K_1} G_{k_1}^2 + \sum_{k_2=1}^{K_2} G_{k_2}^2 \\ &+ \sum_{k_1\neq k_1'}\sum G_{k_1} G_{k_1'} \exp[-i\omega(t_{k_1} - t_{k_1'})] \\ &+ \sum_{k_2\neq k_2'}\sum G_{k_2} G_{k_2'} \exp[-i\omega(t_{k_2} - t_{k_2'})] \\ &- \sum_{k_1=1}^{K_1}\sum_{k_2=1}^{K_2} G_{k_1} G_{k_2} \{\exp[-i\omega(t_{k_1} - t_{k_2})] \\ &+ \exp[-i\omega(t_{k_2} - t_{k_1})]\}\Bigg)\Bigg] \end{aligned} \quad (12)$$

The expectation in Eq. (12) is over K_1, K_2, and all t_{k_1}, G_{k_1}, t_{k_2}, and G_{k_2}'s. We determine this expectation in several steps, fixing some of the random variables in each step and taking the conditional expectation over the remaining ones. First, let us fix K_1, K_2, and all t_{k_1}'s and t_{k_2}'s. Since the photodetector gain for one electron is independent of that for other electrons, the result of taking conditional expectations over G_{k_1}'s and G_{k_2}'s will be

$$\begin{aligned} S_x(\omega) = (qRA)^2 |H(\omega)|^2 \lim_{T\to\infty} \mathbf{E}\Bigg[(1/T)\Bigg(&\langle G^2\rangle (K_1 + K_2) \\ &+ \langle G\rangle^2 \sum_{k_1\neq k_1'}\sum \exp[-i\omega(t_{k_1} - t_{k_1'})] \\ &+ \langle G\rangle^2 \sum_{k_2\neq k_2'}\sum \exp[-i\omega(t_{k_2} - t_{k_2'})] \\ &- \langle G\rangle^2 \sum_{k_1=1}^{K_1}\sum_{k_2=1}^{K_2} \{\exp[-i\omega(t_{k_1} - t_{k_2})] \\ &+ \exp[-i\omega(t_{k_2} - t_{k_1})]\}\Bigg)\Bigg]. \end{aligned} \quad (13)$$

Here $\langle G\rangle$ and $\langle G^2\rangle$ are the first two moments of photodetector gain. In the absence of photomultiplication both moments are equal to 1.

In Eq. (13) we now fix K_1 and K_2 and determine the conditional expectation over the t_{k_1}'s and the t_{k_2}'s. Due to the Poisson nature of photodetection, all t_{k_1}'s and t_{k_2}'s are conditionally independent[8] and their probability density functions are proportional to $P_1(t)$ and $P_2(t)$ [See Eq. (8)]. Calculation of these conditional expectations leads to

$$\begin{aligned} S_x(\omega) = (qRA)^2 |H(\omega)|^2 \lim_{T\to\infty} \mathbf{E}\Bigg((1/T)\Bigg\{ &\langle G^2\rangle (K_1 + K_2) \\ &+ \langle G\rangle^2 K_1(K_1 - 1) \int\int P_1(t) P_1(t') \\ &\times \exp[-i\omega(t - t')]\,dt\,dt' \Big/ \Big[\int P_1(t)\,dt\Big]^2 \\ &+ \langle G\rangle^2 K_2(K_2 - 1) \int\int P_2(t) P_2(t') \\ &\times \exp[-i\omega(t - t')]\,dt\,dt' \Big/ \Big[\int P_2(t)\,dt\Big]^2 \\ &- \langle G\rangle^2 K_1 K_2 \int\int [P_1(t)P_2(t') + P_1(t')P_2(t)] \\ &\times \exp[-i\omega(t - t')]\,dt\,dt' \Big/ \\ &\Big[\int P_1(t)\,dt \int P_2(t)\,dt\Big]\Bigg\}\Bigg), \end{aligned} \quad (14)$$

where the limits of integrations are from $-T/2$ to $+T/2$. Finally, the expectation will be taken over K_1 and K_2 which are Poisson random variables with parameters $\lambda_1 = (\eta/h\nu)\int P_1(t)\,dt$ and $\lambda_2 = (\eta/h\nu)\int P_2(t)\,dt$, respectively. [$\eta$ is the quantum efficiency of photodetectors and $h\nu$ is the photon energy.] For a Poisson random variable K with parameter λ it can be shown that $\mathbf{E}[K] = \lambda$ and $\mathbf{E}[K(K-1)] = \lambda^2$; therefore

$$\begin{aligned} S_x(\omega) = (qRA)^2 |H(\omega)|^2 &\times \lim_{T\to\infty} \mathbf{E}\Bigg((\eta/h\nu)\langle G^2\rangle \Big\{(1/T)\int [P_1(t) + P_2(t)]\,dt\Big\} \\ &+ (\eta/h\nu)^2 \langle G\rangle^2 \\ &\times \Big\{(1/T)\int\int [P_1(t) - P_2(t)][P_1(t') - P_2(t')] \\ &\times \exp[-i\omega(t - t')]\,dt\,dt'\Big\}\Bigg). \end{aligned} \quad (15)$$

$P_1(t)$ and $P_2(t)$ can now be replaced in the above equation to yield

$$
\begin{aligned}
S_x(\omega) = (qRA)^2|H(\omega)|^2\Big(&\tfrac{1}{2}(\eta/h\nu)(\beta_1-\beta_2)\langle G^2\rangle \\
&\times\{[(\beta_1+\beta_2)/(\beta_1-\beta_2)]-\cos(2\Theta)\cos(2\varphi)\} \\
&\times\lim_{T\to\infty}\left[(1/T)\int_{-T/2}^{T/2}dt\int_{-\infty}^{\infty}\int I(x,y,t)\,dx\,dy\right] \\
&+\tfrac{1}{4}(\eta/h\nu)^2(\beta_1-\beta_2)^2\langle G\rangle^2\sin^2(2\Theta)\sin^2(2\varphi) \\
&\times\lim_{T\to\infty}\mathbf{E}\left[(1/T)\left|\int_{-T/2}^{T/2}\exp(-i\omega t)\right.\right. \\
&\left.\left.\times\int\int_{-\infty}^{\infty}Z(x,y)I(x,y,t)\,dx\,dy\,dt\right|^2\right]\Big). \qquad (16)
\end{aligned}
$$

Substituting for $I(x,y,t)$ from Eq. (6) and defining the function $Z^{\circ}(x,y)$ as

$$
Z^{\circ}(x,y_0) = (\pi r_o^2)^{-1/2}\int_{-\infty}^{\infty}Z(x,y)\exp[-(y-y_0)^2/r_o^2]\,dy, \qquad (17)
$$

one arrives at the following equation for the power spectral density:

$$
\begin{aligned}
S_x(\omega) = (qRA)^2|H(\omega)|^2\{&(\eta P_0/2h\nu)(\beta_1-\beta_2)\langle G^2\rangle \\
&\times\{[(\beta_1+\beta_2)/(\beta_1-\beta_2)]-\cos(2\Theta)\cos(2\varphi)\} \\
&+(\eta P_0/2h\nu)^2(\beta_1-\beta_2)^2\langle G\rangle^2\sin^2(2\Theta)\sin^2(2\varphi) \\
&\times\exp[-\tfrac{1}{2}(\omega r_0/v)^2]S_{Z^{\circ}}(\omega)\} \qquad (18)
\end{aligned}
$$

in which $S_{Z^{\circ}}(\omega)$ is the power spectral density of the desired signal

$$
S_{Z^{\circ}}(\omega) = \lim_{T\to\infty}\mathbf{E}\left[(1/T)\left|\int_{-T/2}^{T/2}Z^{\circ}(vt,y_0)\exp(-i\omega t)\,dt\right|^2\right]. \qquad (19)
$$

Note that the spread of the Gaussian beam in the Y direction affects the ideal signal, $Z(x,y)$, through Eq. (17). In the limit when r_0 approaches zero we will have $Z^{\circ}(x,y_0) = Z(x,y_0)$ and neither smoothing nor cross talk between tracks occurs. The spread of the beam in the X direction, however, affects the signal through the factor exp $[-\tfrac{1}{2}(\omega r_0/v)^2]$ in Eq. (18). This is a bell-shaped function that attenuates the high frequency components of the signal more than the low frequency ones. Roughly speaking frequencies above $\omega = v/r_0$ will be strongly affected. Thus a smaller r_0 will reduce this effect while a larger v does not change it. This is due to the fact that the bandwidth of the signal itself increases linearly with v [see Eq. (19)].

In addition to the spectrum of the useful signal, the power spectral density in Eq. (18) contains a constant shot noise level. This is similar to the white noise spectral density usually generated by thermal fluctuations in the electronic circuitry. In fact, because of the independence of thermal fluctuations from the signal and because of their additive nature, one can add their spectral density (a constant level N_0) to the shot noise density, and lump them together in the expression for the power spectral density of total noise

$$
\begin{aligned}
S_N(\omega) = |H(\omega)|^2[&N_0+(qRA)^2(\eta P_0/2h\nu) \\
&\times(\beta_1-\beta_2)\langle G^2\rangle \\
&\times[\{[(\beta_1+\beta_2)/(\beta_1-\beta_2)]-\cos(2\Theta)\cos(2\varphi)\}]. \qquad (20)
\end{aligned}
$$

The filter $H(\omega)$ can be designed to minimize the effects of smoothing and/or optimize the signal to noise ratio. Its primary task, however, must be to eliminate the components of noise which fall outside the signal bandwidth.

IV. SIGNAL TO NOISE RATIO

In the previous section we derived expressions for the spectral densities of signal and noise. By integrating over all frequencies, one obtains the following expressions for the total signal power S and the total noise power N:

$$
\begin{aligned}
S = [&(\eta P_0/2h\nu)(qRA)(\beta_1-\beta_2)\langle G\rangle\sin(2\Theta)\sin(2\varphi)]^2 \\
&\times\int_{-\infty}^{\infty}\exp[-\tfrac{1}{2}(\omega r_0/v)^2]\,|H(\omega)|^2S_{Z^{\circ}}(\omega)\,d\omega, \qquad (21)
\end{aligned}
$$

$$
\begin{aligned}
N = \Big\{&N_0+(qRA)^2(\eta P_0/2h\nu)(\beta_1-\beta_2)\langle G^2\rangle \\
&\times\Big[[(\beta_1+\beta_2)/(\beta_1-\beta_2)]-\cos(2\Theta)\cos(2\varphi)\Big]\Big\} \\
&\times\int_{-\infty}^{\infty}|H(\omega)|^2d\omega. \qquad (22)
\end{aligned}
$$

We now define the following parameters:

$$
2B = \int_{-\infty}^{\infty}|H(\omega)|^2d\omega, \qquad (23)
$$

$$
n_{eq} = N_0/(qRA)^2, \qquad (24)
$$

$$
F_G = \langle G^2\rangle/\langle G\rangle^2, \qquad (25)
$$

where B can be interpreted as the system's bandwidth, n_{eq} as the equivalent rate of generation of thermal electrons prior to amplification, and F_G as the excess noise factor of the diodes. For avalanche detectors F_G obeys the following relation[9]:

$$
F_G = k\langle G\rangle + (1-k)(2-\langle G\rangle^{-1}), \qquad (26)
$$

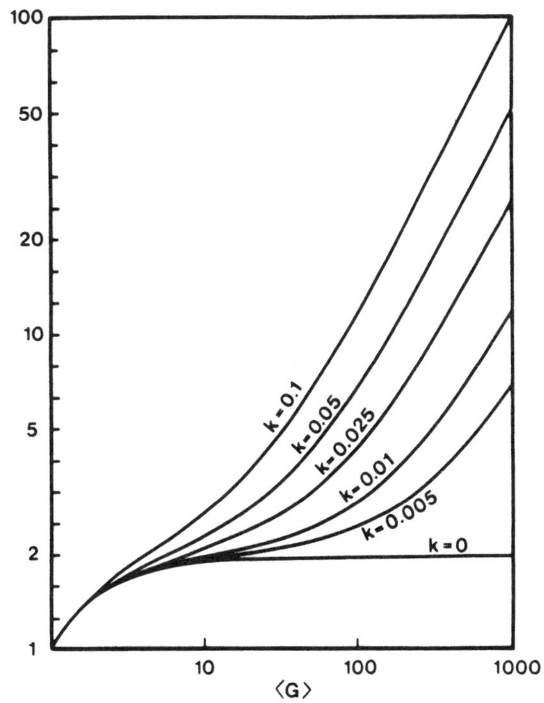

FIG. 6. Noise factor vs average gain for avalanche photodiodes.

where κ is a constant parameter known as the ionization ratio of the diode. Figure 6 shows the dependence of F_G on $\langle G \rangle$ for different values of k. Note in particular the increasing nature of F_G with the average gain.

The signal to noise ratio can now be expressed as

$$\mathrm{SNR} = \frac{\frac{1}{2}(\eta P_0/h\nu)(\beta_1 - \beta_2)\sin^2(2\Theta)\sin^2(2\varphi)}{2B\,(F_G\{[(\beta_1+\beta_2)/(\beta_1-\beta_2)] - \cos(2\Theta)\cos(2\varphi)\} + 2\langle G\rangle^{-2}\{n_{eq}/[(\eta P_0/h\nu)(\beta_1-\beta_2)]\})} \tag{27}$$

The factor $\int\exp[-\frac{1}{2}(\omega r_0/v)^2]|H(\omega)|^2\ S_{Z'}(\omega)\,d\omega$ has been set equal to unity in the above equation for the following reason. In the ideal case, where $r_0 = 0$ and $H(\omega)$ does not distort the signal, the value of the integral is equal to 1. In a practical case, however, although this value might slightly differ from unity, it would be constant and would not vary with the parameters we seek to otpimize. Therefore, as long as we are interested in the relative values of SNR, it makes sense to set this constant equal to 1.

The signal to noise ratio can now be maximized with respect to Θ. Equation (27) can be written for this purpose as

$$\mathrm{SNR}(\Theta) = \tfrac{1}{2}\gamma\sin^2(2\Theta)/[1 - \alpha\cos(2\Theta)], \tag{28}$$

where

$$\gamma = \frac{(\eta P_0/h\nu)(\beta_1-\beta_2)\sin^2(2\varphi)}{2B\,(F_G[(\beta_1+\beta_2)/(\beta_1-\beta_2)] + 2\langle G\rangle^{-2}\{n_{eq}/[(\eta P_0/h\nu)(\beta_1-\beta_2)]\})}, \tag{29}$$

$$\alpha = \frac{F_G\cos(2\varphi)}{F_G[(\beta_1+\beta_2)/(\beta_1-\beta_2)] + 2\langle G\rangle^{-2}\{n_{eq}/[\eta \mathrm{P}_0/\mathrm{h}\nu)(\beta_1-\beta_2)]\}}, \tag{30}$$

and clearly $\gamma > 0$ and $|\alpha| < 1$. It is easy to show that the optimum value of Θ is given by

$$\cos(2\Theta) = \alpha/[1 + (1-\alpha^2)^{1/2}], \tag{31}$$

and that the maximum value of SNR, obtained by replacing Eq. (31) in Eq. (28), can be expressed as

$$\mathrm{SNR} = \gamma/[1 + (1-\alpha^2)^{1/2}]. \tag{32}$$

A modified version of the basic differential detection scheme which is used in practice replaces the beam splitter and the two analyzers of the system of Fig. 2 with a single polarizing beam splitter (See Fig. 7). The advantage of this new version is that the reflected light need not be split before reaching the analyzers and this results in a more efficient use of the read power P_0. The disadvantage is that Θ is now fixed at 45° and cannot be optimally adjusted. The signal to noise ratio for the modified system is obtained from Eq. (27) by setting $\Theta = 45°$ and replacing P_0 with 2 P_0, i.e.,

$$\mathrm{SNR}^\circ = \frac{(\eta P_0/h\nu)(\beta_1-\beta_2)\sin^2(2\varphi)}{2B\,(F_G[(\beta_1+\beta_2)/(\beta_1-\beta_2)] + \langle G\rangle^{-2}\{n_{eq}/[(\eta \mathrm{P}_0/\mathrm{h}\nu)(\beta_1-\beta_2)]\})}. \tag{33}$$

SNR° in the above equation is clearly greater than γ in Eq. (29) which is in turn greater than SNR in Eq. (32). The modified version is therefore superior to the basic system in terms of signal to noise ratio and our attention will thus be restricted to that in the remainder of this paper.

Finally, by replacing for β_1, β_2, and φ from Eqs. (5) in Eq. (33) we obtain

$$\mathrm{SNR}^\circ = \frac{4(\eta P_0/h\nu)|r_x|^2|r_y|^2\cos^2(\Phi_{rx} - \Phi_{ry})}{2B\{F_G(|r_x|^2 + |r_y|^2) + \langle G\rangle^{-2}[n_{eq}/(\eta P_0/h\nu)]\}}, \tag{34}$$

where Φ_{rx} and Φ_{ry} are the phase angles of the complex numbers r_x and r_y. For a maximum signal to noise ratio Eq. (34) immediately suggests several system requirements. First, the read power P_0 should be as large as possible. Obviously, however, P_0 cannot be made very large because the heat induced in the disk will then reduce the signal sharply and may even erase the recorded spots. For a given disk structure, therefore, P_0 has an upper limit. Second, the phase difference between r_x and r_y should be eliminated. This can be done by the introduction of a phase plate with appropriate relative phase delay between the x and y directions of polarization in the path of the reflected light. An arrangement such as that shown in Fig. 7 leaves the magnitude of r_x and r_y unperturbed but eliminates the phase difference between them. For the combination of disk and the phase plate, therefore, $\beta_2 = 0$ or equivalently $\cos^2[\Phi_{rx} - \Phi_{ry}] = 1$. The presence of this phase plate will be assumed in the remainder of this article. Third, the avalanche gain $\langle G \rangle$ should be optimized. The denominator of Eq.(34) is a function of $\langle G \rangle$ and can be minimized with respect to it. The optimum gain can be shown to be

$$\langle G\rangle_{opt} = \begin{cases} 1 & \text{if } n_{eq} < \frac{1}{2}(\eta P_0/h\nu)(|r_x|^2 + |r_y|^2)\,, \\ [(a^2+b^3)^{1/2} + a]^{1/3} - [(a^2+b^3)^{1/2} - a]^{1/3} & \text{if } n_{eq} > \frac{1}{2}(\eta P_0/h\nu)(|r_x|^2 + |r_y|^2)\,, \end{cases} \tag{35}$$

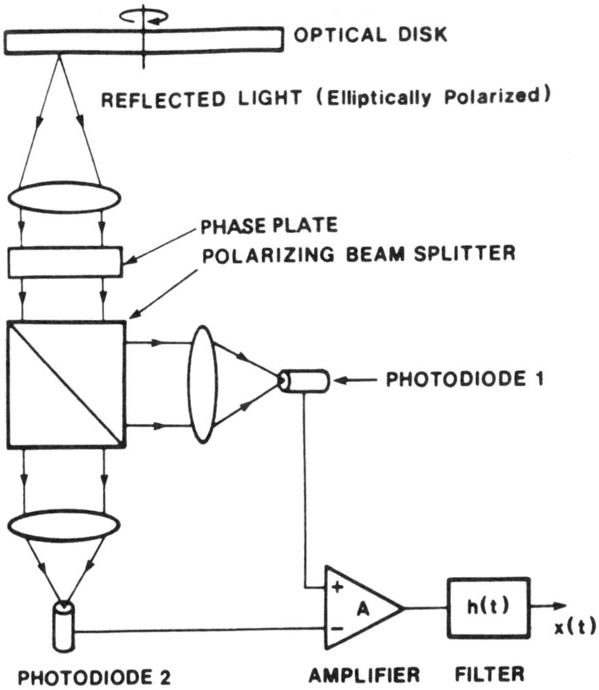

FIG. 7. Schematic diagram of the modified differential readout system.

where $a = n_{eq}/[\kappa(\eta P_0/h\nu)(|r_x|^2 + |r_y|^2)]$ and $b = (1-\kappa)/3\kappa$. Finally, the structure of the disk should be optimized. With the aid of an appropriate interference device, it is possible to increase $|r_y|$ at the expense of $|r_x|$. The optimum structure then corresponds to a compromise between $|r_x|$ and $|r_y|$ that results in the maximum signal to noise ratio.

V. INTERFERENCE ENHANCEMENT OF THE KERR EFFECT

We now apply the foregoing results to examine the role of disk structure in the overall performance of the differential readout system. For concreteness, we take the system of Fig. 7 characterized by the parameters given in Table I. We also assume that each interference device is deposited on a glass substrate and has a typical magneto-optic material such as MnBi as one layer, SiO_2 or SiO for thin dielectric layers, and Al for the reflecting layer. The optical constants used are shown in Table II. The quantitative results are therefore quite specific. Nevertheless, many of the trends observed are independent of the parameters chosen, and this allows us to make numerous general conclusions.

As a base line for discussion of our later data, we show in Fig. 8 our results for a single magnetic layer on a glass substrate. We have found the values of the avalanche gain $\langle G \rangle$ and the phase retardation required for the phase plate, represented by $(\Phi_{r_x} - \Phi_{r_y})$, that are required to maximize SNR for each magnetic film thickness. The resulting values of $|r_x|^2$ and $|r_y|^2$ are also shown.

There are numerous points to note. At thicknesses above 1600 Å, where the MnBi layer is optically thick, all the parameters are constant as expected, and SNR is about 14 db. Moreover, the phase difference between r_x and r_y is quite small, and the removal of the phase plate would not reduce SNR appreciably. At smaller thicknesses, interference effects within the metal layer become important and lead to a maximum SNR of 16.5 db, although in contrast to the thick film result, an appropriately chosen phase plate is essential. The origin of this increase, relative to the value for a thick film, may be traced directly to the enhancement of $|r_y|^2$ since from Eq. (34) SNR $\sim |r_y|^2$ for $|r_y|^2 \ll |r_x|^2$. We may therefore expect to obtain significant improvements in SNR with multilayer interference devices specifically designed to increase $|r_y|^2$, and our earlier qualitative arguments suggest that a trilayer-like device would be most effective from this point of view.

In Fig. 9 we show the quadrilayer device that has been studied. The overcoating has been added both to make it possible to fabricate the device (since oxidation of the very thin magnetic layer must be prevented) and to lend a further degree of flexibility (to be discussed later) that is absent for the trilayer. However, the intuitive ideas given earlier that suggest exceptional performance for the trilayer carry over naturally to this quadrilayer case. Figure 10 then shows the results for a device in which the intermediate and overcoat layers are SiO_2.A computer search was used to establish the values of the intermediate layer and overlayer thicknesses, t_i and t_c, respectively, that maximize SNR for each thickness of the magnetic film. The corresponding optimum values of $\langle G \rangle$ and $(\Phi_{r_x} - \Phi_{r_y})$ were calculated as before. Again at large thicknesses beyond 1600 Å, SNR, $|r_x|^2$ and $|r_y|^2$ are constant, and the increase of 4db in SNR over its corresponding value in the single film represents the effect of the (antireflecting) overlayer on the size of $|r_y|^2$. The intermediate and reflecting layers in this thick-film region are inconsequential since they are optically inaccessible. At smaller thicknesses, however, they make a significant contribution

TABLE I. Numerical values for the parameters of the differential detection system.

Parameter	Definition	Numerical value
η	Quantum efficiency of photodiodes	0.85
P_0	Laser read power	2 μW
λ	Wavelength of light	8400 Å
B	Bandwidth	10 MHz
k	Ionization ratio of photodiodes	0.02
n_{eq}	Equivalent rate of generation of thermal electrons	10^{15}sec^{-1}

TABLE II. Optical constants of the materials.

Material	n^+	n^-
SiO_2	1.50	1.50
SiO	2.00	2.00
Al	$2.00 + 7.10i$	$2.00 + 7.10i$
MnBi	$3.77 + 3.92i$	$3.56 + 3.79i$

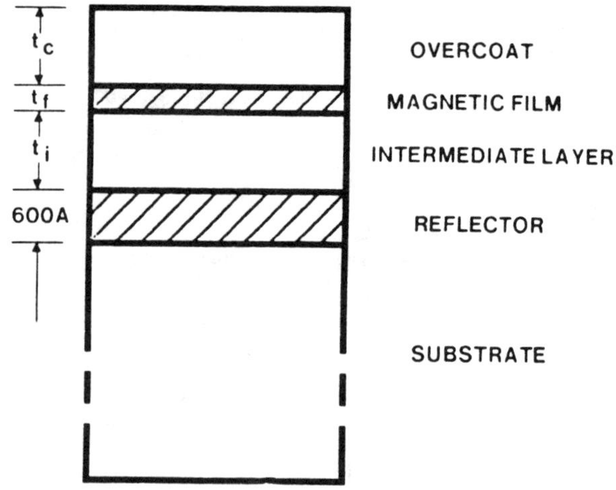

FIG. 9. Structure of the quadrilayer disk.

to the performance of the structure as can be seen from the increase in $|r_y|^2$ and SNR near 800 Å where t_c is constant. Thus the improvement in the performance of the device is the result of the intermediate layer appropriately coupling the light reflected from the Al layer back into the magnetic film. At even smaller thicknesses, where the qualitative arguments for the enchancement of r_y become appropriate, SNR climbs to a maximum of 22.5 db, an increase of 8.5 db over the uncoated, thick film case, and of 4.5 db over the overcoated, thick film case. The maximum gain in $|r_y|^2$ relative to its uncoated bulk value is 9, which compares reasonably with the estimate $E \approx 8\alpha t/(1-R)^2 \approx 16$ obtained earlier. This agreement in fact would be closer if $|r_y|^2$ were maximized rather than SNR, because in this regime $|r_x|^2$ is of order $|r_y|^2$ when $|r_y|^2$ is maximized, and the compromise between them is significant in maximizing SNR in Eq. (34).

A parenthetical remark should be made about the discontinuities in the slopes of the curves of Fig. 10 at around $t_f = 170$ Å. This seemingly curious behavior is merely a consequence of the nature of the function being optimized and bears no physical significance. In fact, a multitude of analytic functions can be found which exhibit a similar feature. As an illustrative example, consider the simple case of the polynominal function

$$P(X,Y) = 12X^2Y(Y-1) + 2XY^2(2Y-15) - Y^3(3Y+16)$$

for $0<X<4$. Then as shown in Fig. 11, the value of Y that maximizes the polynominal at each X clearly shows a discontinuity in slope at $X = 1$. Below $X = 1$, there are three extrema with one an absolute maximum, while for $1<X<4$ there is but one maximum. The discontinuity in slope is caused by the transition from the maximum in one region to the maximum in the other. Numerically we find just this behavior in the more complicated function that was examined in determining the results in Fig. 10.

The results of optimization for a quadrilayer which has

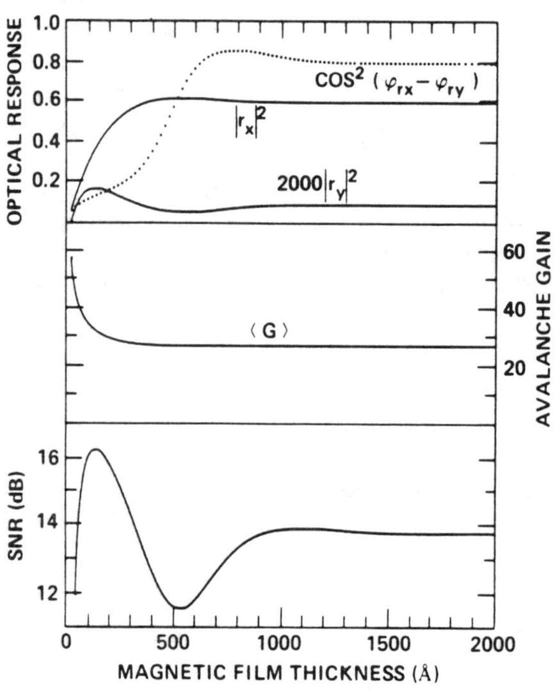

FIG. 8. Characteristics of readout: Film on glass substrate.

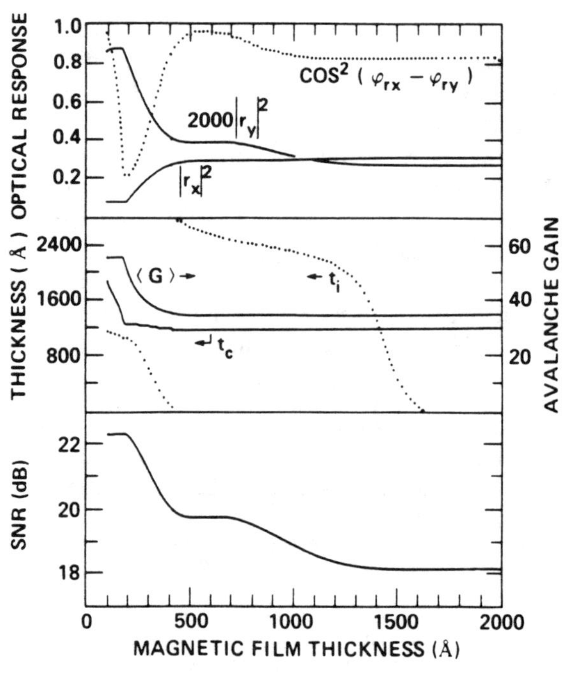

FIG. 10. Characteristics of readout: quadrilayer with SiO_2 for dielectric layers.

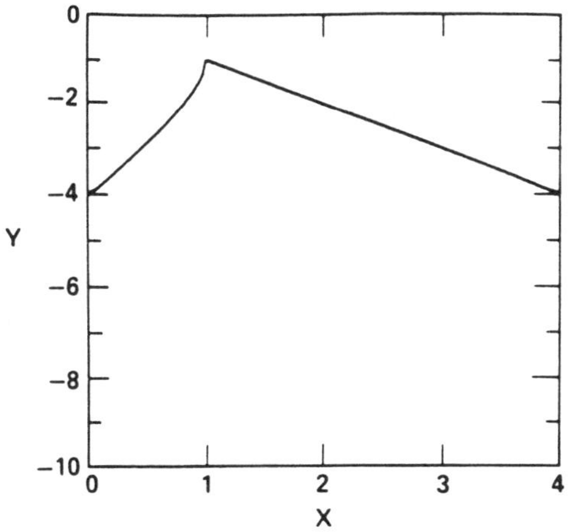

FIG. 11. Plot of the values of Y that maximize the polynominal $P(X,Y)$ at each X.

$$P(X,Y) = 12X^2Y(Y-1) + 2XY^2(2Y-15) - Y^3(3Y+16).$$

SiO for intermediate and overlayers are presented in Fig. 12. Qualitatively, the results are identical to those of the previous case, but there are a few important quantitative differences. For instance, the plateau in SNR between 400 and 700 Å magnetic film thickness is now raised by about 2 db so that the difference between the absolute maximum, and this level is now only 1 db. This effect is mainly due to the change in the refractive index of the overcoat; using SiO_2 or any other dielectric as the intermediate layer has no significant effect on SNR, presumably because this layer merely creates an appropriate phase retardation. There is, therefore, much more flexibility in the design of this structure which could be of value in optimizing its performance as far as other system parameters are concerned.

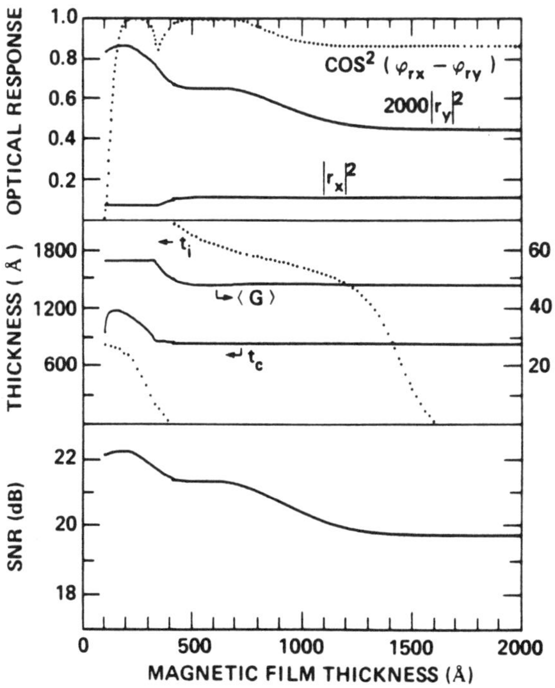

FIG. 12. Characteristics of readout: quadrilayer with SiO for dielectric layers.

VI. CONCLUDING REMARKS

In this paper we have begun a theoretical study of the readout system from a magneto-optical disk, but much is still to be done. For example, our analysis has been restricted to the most fundamental sources of noise, namely, shot noise in photodetection and thermal noise in electronic circuitry. Nevertheless, our method for the power spectral density derived in Sec. III should offer a tractable approach to considering the effects of other sources of noise that may be important in any given system. Some examples might be laser instabilities, nonuniformities of the disk structure, and nonuniformities in the pattern of recorded data.

Our optimization procedures for the disk structure might also be carried out to favor aspects of the system other than signal to noise in conditions of limited laser power. This approach was taken because at the present stage of GaAs laser development, we were concerned about the availability of power for writing, and assumed that this therefore set the limit on read power. If this situation changes to one of essentially unlimited power, either through laser or media improvements, then the figure of merit for best performance must be changed to one which considers the variation of the maximum read power with structure. Nevertheless, we expect that this can be handled within the present framework.

APPPENDIX: SURFACE REFLECTIVITY OF MULTILAYERS

Consider the N layer of Fig. 13 with its surface located at $Z = 0$. Starting at the substrate with $\kappa = 1$, the layers are numbered in increasing order. For the κ'th layer we will denote the thickness by Z_k and the complex refractive index by $n^{(\kappa)} = n_r^{(\kappa)} + in_{im}^{(\kappa)}$. When an optical plane wave propa-

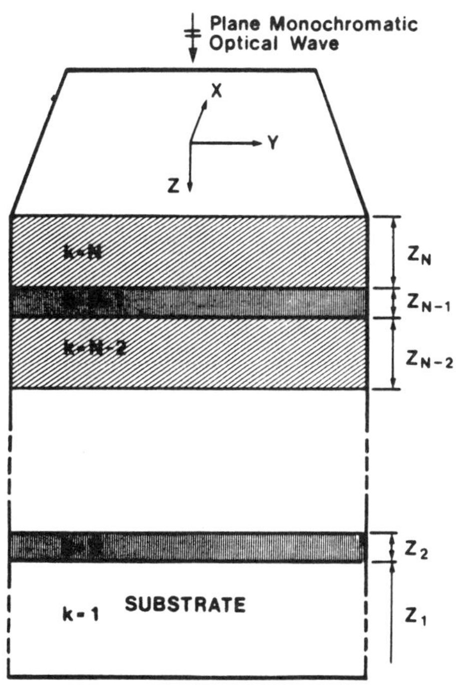

FIG. 13. Coordinate system for the calculation of reflectivity of a multilayer.

gating in the Z direction with wavelength λ_0 illuminates the surface of the multilayer, one can write the following Maxwell equations for the electric and magnetic field components in the κ'th layer[10]:

$$[(d^2/dZ^2) + (2\pi n^{(\kappa)}/\lambda_0)^2]\mathbf{E}_k(Z) = 0, \quad \text{(A1a)}$$

$$(d/dZ)\mathbf{E}_k(Z) - i(2\pi/\lambda_0)\mathbf{H}_k(Z) = 0. \quad \text{(A1b)}$$

The general solution to these equations can be written as

$$\mathbf{E}_k(Z) = A_1^{(k)}\left\{\exp\left[+i(2\pi n^{(k)}/\lambda_0)\left(Z - \sum_{j=k+1}^{N+1} z_j\right)\right] + A_2^{(k)}\exp\left[-i(2\pi n^{(k)}/\lambda_0)\left(Z - \sum_{j=k+1}^{N+1} z_j\right)\right]\right\}, \quad \text{(A2a)}$$

$$\mathbf{H}_k(Z) = n^{(k)}A_1^{(k)}\left\{\exp\left[+i(2\pi n^{(k)}/\lambda_0)\left(Z - \sum_{j=k+1}^{N+1} z_j\right)\right] - A_2^{(k)}\exp\left[-i(2\pi n^{(k)}/\lambda_0)\left(Z - \sum_{j=k+1}^{N+1} z_j\right)\right]\right\}. \quad \text{(A2b)}$$

In the above equations $Z_{N+1} = 0$. The coefficients A_1 and A_2 must be determined in view of the continuity requirements at the boundaries. The continuity of H/E is assured if the A_2's satisfy the following recursive relation:

$$A_2^{(k)} = \left(\frac{[n^{(k)} - n^{(k-1)}]/[n^{(k)} + n^{(k-1)}] + A_2^{(k-1)}}{1 + \{[n^{(k)} - n^{(k-1)}]/[n^{(k)} + n^{(k-1)}]\}A_2^{(k-1)}}\right) \times \exp[+i(4\pi n^{(k)}/\lambda_0)z_k]; \; 2 \leqslant k \leqslant N+1. \quad \text{(A3)}$$

Since the reflected component of light in the substrate is zero one can start with $A_2^{(1)} = 0$ and determine all other A_2's from Eq. (A3). Note that $A_2^{(N+1)}$ is the amplitude reflectivity of the surface; n_{N+1} is the refractive index of air, through which light travels to reach the multilayer, and $Z_{N+1} = 0$.

[1] R. P. Hunt, IEEE Trans. Magn. **MAG-5**, 700 (1969).
[2] D. Chen, G. N. Otto, and F. M. Schmit, IEEE Trans. Magn. **MAG-9**, 66 (1973).
[3] D. Treves, J. Appl. Phys. **38**, 1192 (1967).
[4] M. Mansuripur, J. W. Goodman, E. G. Rawson, and R. E. Norton IEEE Trans. Commun. **COM-28**, 402 (1980).
[5] P. S. Pershan, J. Appl. Phys. **38**, 1482 (1967).
[6] D. O. Smith, Opt. Acta **12**, 13 (1965).
[7] G. A. N. Connell, Appl. Phys. Lett. **40**, 212 (1982).
[8] E. Cinlar, *Introduction to Stochastic Processes* (Prentice Hall, Englewood Cliffs, New Jersey, 1975).
[9] S. D. Personick, Bell Syst. Tech. J. **50**, 3075 (1971).
[10] J. D. Jackson, *Classical Electrodynamics* (Wiley, New York, 1967).

Reprinted with permission from *Applied Optics*, Vol. 22(14), pp. 2202-2209 (July 15, 1983). ©1983 Optical Society of America.

Two-dimensional modeling of an optical disk readout

V. B. Jipson and C. C. Williams

The design of high density digital optical storage systems requires a thorough understanding of the various interactions affecting the integrity of both data and servo signals under various readout conditions. Here, a 2-D Fourier optic model of the readout system is presented along with an efficient calculational procedure based on 2-D fast Fourier transforms (FFT). The model is utilized to examine the effects of data density, focus errors, tracking errors, lens fill conditions, phase pit depth, etc. on the quality of the readout signal. The derivation of servo signals is also studied through the modeling of a simple diffraction based scheme. It is shown that a reasonably constant servo signal (small variations with focus errors, track depth, or added amplitude data) can be generated by this technique provided that a $\lambda/8$ phase groove is present. The same technique can be shown to be unacceptable if an amplitude or $\lambda/4$ phase groove is utilized.

I. Introduction

There has been extensive activity over a 10-year period[1] aimed at utilizing the laser as a primary element in a write-once, read-only digital optical storage system. Recent demonstrations by several authors have shown the potential of these systems.[2,3] To optimize these systems, it is necessary to understand the details of both the recording and readout processes. Here, we present a detailed 2-D model of the readout process.

Hopkins[4] previously modeled the optics of the Philips video disk system, and our analysis follows similar lines. However, since we are interested in ultimately studying random patterns, we adopt a different calculational scheme from the one employed by Hopkins. Conceptually, the problem of calculating the output voltage from an optical disk as a function of disk position is straightforward; however, the calculation itself is time-consuming due to the necessity of calculating 2-D transforms of relatively large arrays. Here, we employ a 2-D FFT algorithm to reduce the calculational time.

The optical system employed in an optical disk reader is directly analogous to a Type 1 scanning optical microscope,[5,6] and thus the model described here could be used to calculate the resultant image from such a microscope. A schematic diagram of the optical system studied is shown in Fig. 1. A collimated laser source with a field distribution given by $u_i^+(x,y)$ is incident onto an objective lens through a beam splitter. This lens acts to focus the radiation to a fine spot at the object plane. For generality we assume that the object is a distance ε removed from the lens focal plane. The incident light interacts with the object being imaged, and the reflected light that falls within the numerical aperture of the objective lens is collimated and directed to the output detector through the input beam splitter. In general, the beam splitter would be a polarizing one, and a quarterwave plate would be inserted between the beam splitter and the objective lens. However, for this analysis we neglect the effects of polarization (i.e., a scalar calculation), and thus, the omission of these components is not important. This is a reasonable approximation for numerical apertures (N.A.) <0.6, but it becomes questionable for numerical apertures significantly >0.6. A second approximation employed is that the interaction between the object and the incident field distribution can be described through a simple spatial multiplication of the field by the possibly complex reflectivity function of the object. While this is a good approximation for most situations of interest here, it is less valid for structures that exhibit a large change in reflectivity with angle of incidence, since these effects are ignored. Thus, one would expect the model to show better agreement when modeling the readout from holes written in a thin metal film than for a similar pattern on a highly tuned structure such as a trilayer.[7] Experimental results, however, indicate that, at least for first minimum trilayers (i.e., spacer layers $\approx \lambda/4$ thick), this problem is not significant.

II. Theory

In this section the theoretical model for the system shown in Fig. 1 is outlined. An angular spectrum ap-

V. B. Jipson is with IBM Watson Research Center, P.O. Box 218, Yorktown Heights, New York 10598, and C. C. Williams is with Stanford University, Stanford, California 94305.

Received 17 March 1983.
0003-6935/83/142202-08$01.00/0.

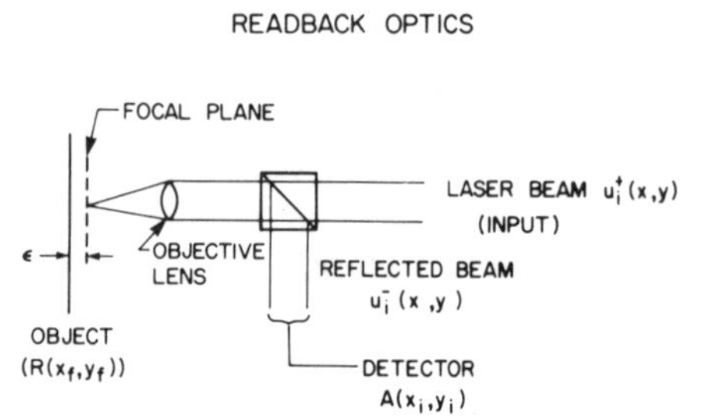

Fig. 1. Simplified schematic of the readout optics for an optical disk system.

proach is utilized, since it allows one to easily incorporate the effects of focus errors and substrate thickness variations within the model. We chose to use a plane-by-plane calculational procedure rather than a direct imaging calculation,[8] since we are interested in the light intensity at planes other than the output plane. A basic assumption of the model is that the objective lens performs a good Fourier transform; however, lenses that exhibit aberrations can be modeled by including the appropriate phase errors within the lens pupil function.

The details of the theoretical model are included in the Appendix. Conceptually, the approach is straightforward. An input field distribution to the objective lens is first either calculated or measured. In the case of no focus error ($\varepsilon = 0$), this field is used to calculate the field at the object by performing a 2-D Fourier transform of the input field distribution after multiplication by the lens pupil function. At the object plane the reflected field is found through multiplication with the object reflectivity. The reflected field is then utilized to calcualte the field at the output plane through another 2-D Fourier transform. Again, the finite aperture of the objective lens is taken into account through the lens pupil function. The detected output voltage is then found by integrating the field intensity over the detector area. For cases with focus errors ($\varepsilon \neq 0$), the problem becomes somewhat more complicated. However, this problem can be easily handled by utilizing an angular spectrum approach[9,10] which is described in detail in the Appendix.

III. Calculational Procedure

While the theoretical model outlined in the previous section is straightforward, it does involve a large amount of computation. To study the interaction of neighboring tracks with the readout from a given track it is desirable that the object reflectivity function contain at least a 3×3 matrix of written spots with a 5×5 matrix more desirable. Furthermore, to insure adequate sampling of the recorded spot represented in the array, it is desirable that of the order of 100 or more resolution elements be used to represent each spot. Finally, to reduce computation time we use the fast Fourier transform (FFT) algorithm to calculate the necessary 2-D Fourier transforms. With the above constraints in mind, we represent the object reflectivity function by a 128×128 matrix. This matrix can be made up of real (pure amplitude) or complex (amplitude and phase) values.

The calculational procedure follows the Appendix directly with 128×128 point FFTs replacing the 2-D Fourier transforms. A lens distribution and fill factor are selected, and the field distribution at the object plane is calculated using Eqs. (A3)–(A7). Note that this distribution will not change until the lens illumination or the object plane changes, and thus, the result is stored for reuse. The output voltage for object position x_s, y_s is now calculated through the use of Eqs. (A8)–(A12). Additional points of the image are calculated by shifting the position of the reflectivity function and repeating the steps starting at Eq. (A8). Each additional image point requires the calculation of one 2-D FFT.

We note that the calculation time is dominated by calculation of the 2-D Fourier transform. Using the FFT algorithm requires $n^2 \log_2 n$ complex multiplies and adds per transform ($n = 128$ for our case), compared with the n^4 computations needed for straightforward numerical integration. For $n = 128$ the reduction in computation time is >1000 to 1.

IV. Results

A. Readout Analysis

In Eq. (A2) a generalized pupil function was introduced for the objective lens with little comment. This function can, however, serve many functions. First it serves as the limiting aperture for the objective lens with respect to the input light distribution, and, thus, it determines the optical fill conditions of the objective lens. For example, if the objective lens is uniformly illuminated (i.e., overfilled), a relatively narrow focal spot with high sidelobes will be obtained. If the aperture size is such that the lens is underfilled, most of the available laser power will pass through the objective, but the effective numerical aperture will be reduced, and the focal spot will be broadened. Proper selection of the lens fill condition will result in a maximum peak intensity of the focal spot (for an assumed Gaussian input distribution). This is a usually desirable if reading and writing are to be performed with the same beam. It should be noted that, if two lasers are being used (one for writing and one for reading), it may be desirable to use different fill conditions for each. In Fig. 2 the calculated focal distribution is shown for three different fill conditions. Figure 3 gives the relative peak intensity at the center of the focal spot as a function of the lens fill condition, assuming a Gaussian input beam. It is seen that the peak intensity is obtained when the incident field is reduced to $e^{-1.25}$ at the lens edge. This, however, results in an ~10% increase in the full width to half-maximum (FWHM) focal spot that is achievable with uniform illumination (Fig. 2). For the results given below, the lens fill condition was selected such that the relative field amplitude was reduced to 0.37 at the lens edge.

A second function of the pupil function is to account for less than diffraction-limited performance of the objective lens. This is readily accomplished by intro-

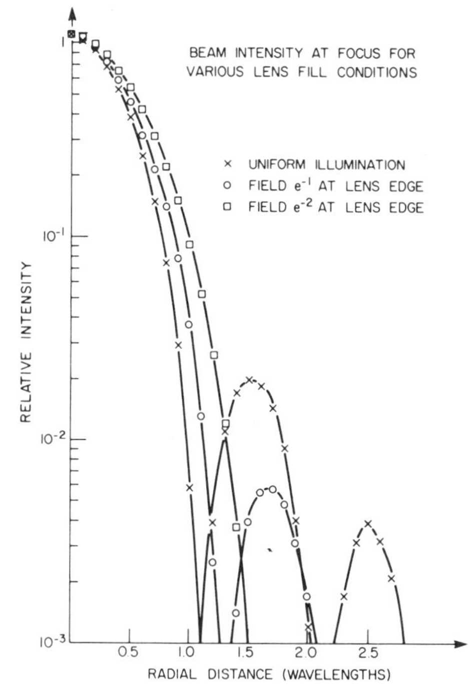

Fig. 2. Normalized focal plane light intensity distribution for three lens fill conditions.

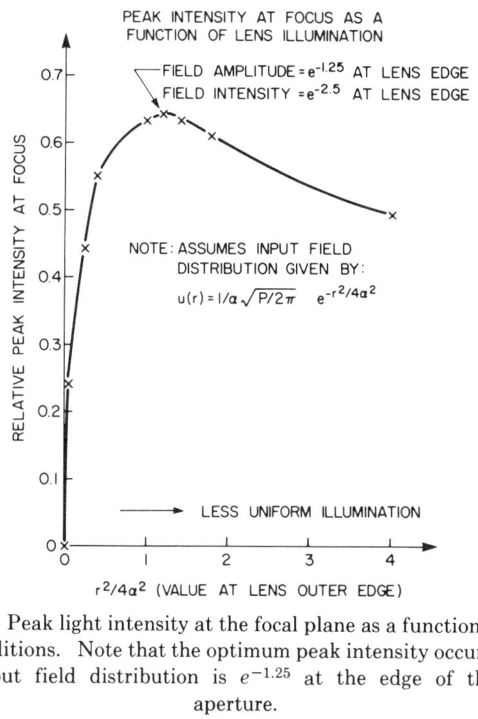

Fig. 3. Peak light intensity at the focal plane as a function of lens fill conditions. Note that the optimum peak intensity occurs when the input field distribution is $e^{-1.25}$ at the edge of the lens aperture.

ducing within the pupil function phase or amplitude errors that model the aberration of interest.

In Fig. 4 a section of a typical disk reflectivity function is shown. In general the object consists of a 2-D array of elliptical objects with x and y dimensions given by d_x and d_y. The total disk function consists of a 128 × 128 complex array, while only a 55 × 55 array is displayed in Fig. 4. The distance from mark edge to mark edge along the track is taken to be s_x and the track-

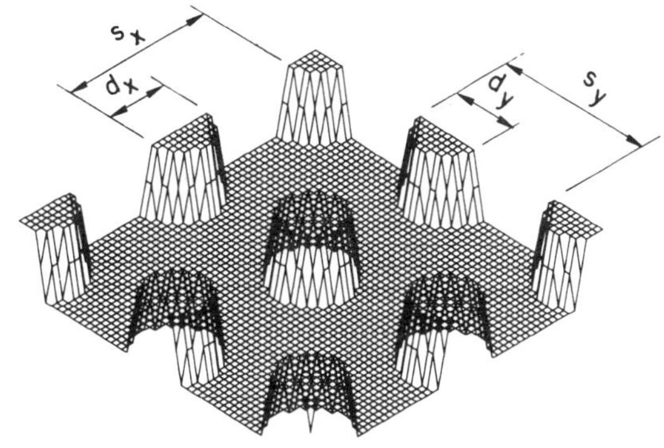

Fig. 4. Section of a typical disk reflectivity function. d_x and d_y represent the axes of spots while s_x and s_y correspond to the mark spacing along the track and perpendicular to the track, respectively. Note that in general the spots can be elliptical, but the ones shown here are circular.

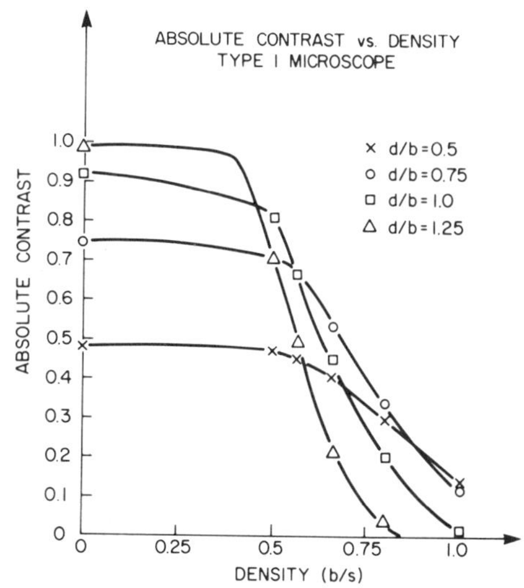

Fig. 5. Optical contrast vs linear spot density with normalized mark size as a parameter.

to-track spacing is given as s_y. To normalize dimensions, we represent the FWHM of the reading beam by b and then divide all quantities by it. In this way the numerical aperture and wavelength are easily scaled for a given calculation. In fact, these quantities do not have to be known until the final step for most cases. One exception to this is calculations involving the effects of defocus. In these cases the numerical aperture and wavelength must be known in order to account for these errors correctly. It should be noted that, while the disk reflectivity function shown in Fig. 4 is periodic and real, the model is capable of analyzing any complex reflectivity function.

In Fig. 5 the calculated peak optical contrast for a linear array of circular amplitude spots is shown as a function of the spot spacing. The ratio of the recorded spot diameter ($d = d_x = d_y$) to the beam diameter (b) is treated as a parameter, and four different spot diameters are examined. As expected, large isolated spots yield greater contrast than smaller isolated spots but, if uniformity of contrast as a function of packing density is desired, smaller written spots are desirable. With smaller recorded spots it is possible to get to higher densities while maintaining adequate readout contrast. In a write-once, read-only file this parameter is generally not under the control of the designer but is instead a function of the medium. Typically, d/b will be of the order of 1.0 for these applications. However, in a read-only device where the information is being mastered onto the disk, it may be desirable to write smaller spots in order to maximize the storage density while minimizing the readout contrast variations with density. It will be seen later that, in the case where information is recorded as phase pits (video disk or digital audio disk), it is desirable to work with d/b ratios significantly below 1.0.

In Figure 6 the measured contrast vs linear density is shown for a thin film metal disk as measured on a static test stand. The conditions of the test were: (1) N.A. = 0.6; (2) wavelength = 647.1 nm; (3) FWHM = 600 nm (measured—note that this is slightly less than diffraction-limited and is most likely due to less than perfect optics); and (4) $d_x = d_y = 600$ nm. A theoretical response is also shown and is seen to be in reasonable agreement. This is particularly true when one considers the relative accuracy of measuring the laser profile and the written spot size.

The effects of tracking errors on the readout contrast are shown in Fig. 7 for several different relative mark sizes. Here, the normalized readout contrast from a single track is shown as a function of the normalized tracking error. As one would expect, smaller mark sizes exhibit greater sensitivity to tracking errors than do larger marks. If one arbitrarily sets a 5% threshold for contrast changes due to tracking errors, it can be seen from Fig. 7 that, depending on the relative pit size, one must servo to the track with an error of <0.1–0.2 beam diameters. In Fig. 8 the normalized readout signal in decibels is shown as a function of tracking error for the case of $d/b = 1.0$. The most interesting point of this result is that the signal from an adjacent track is down by more than 50 dB when it is 1.3 beam diameters away. For digital storage applications this is more than adequate, and it would seem reasonable to pack adjacent tracks at least this closely. In the case of the video disk,[11] this distance is somewhat greater due to the necessity of reducing the cross talk between adjacent frames by more than 40 dB for all conditions. For digital applications, cross talk suppression of 23–26 dB for worst-case conditions should be adequate.

In Fig. 9 the effects of lens focus error are shown for errors of one and two wavelengths for a 0.5-N.A. focusing objective. It can be seen that focus errors of one wavelength result in only a minor loss in optical contrast for the particular case shown ($d/b = 1.0$ and $s/b = 1.5$). However, a defocus error of two wavelengths results in a significant loss in optical contrast. It must be remembered that this problem will get worse if the data are more densely packed or the numerical aperture of the lens is increased. In fact, this analysis greatly underestimates the problem of focus error in a write-once, read-only file. The constraints on focus during the

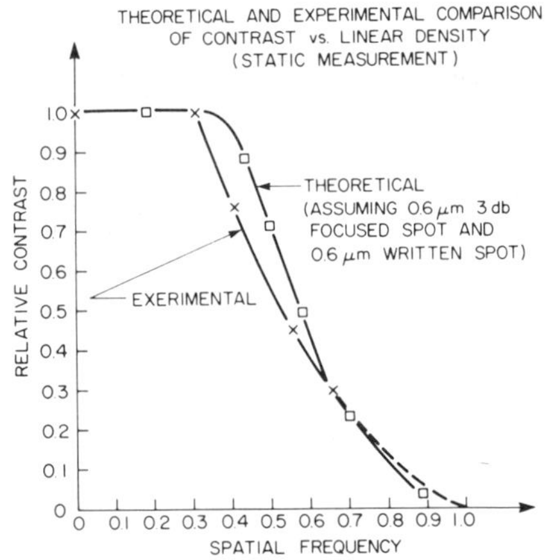

Fig. 6. Measured and theoretical contrast vs linear density for a thin film metal disk: N.A. = 0.6; λ = 647.1 nm; b = 600 nm (measured), $d_x = d_y$ = 600 nm (measured).

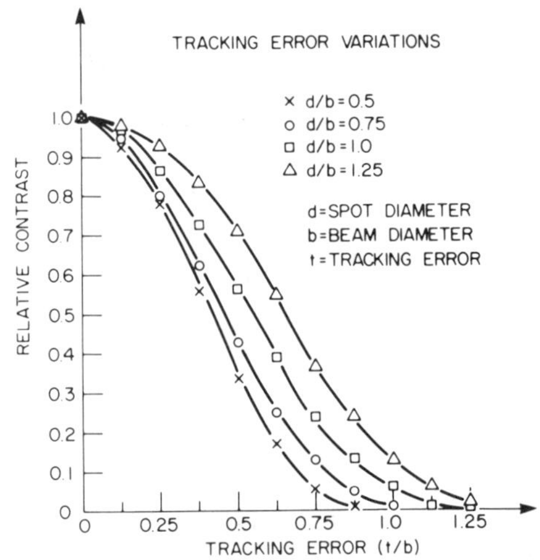

Fig. 7. Optical contrast vs tracking error with normalized mark size as a parameter.

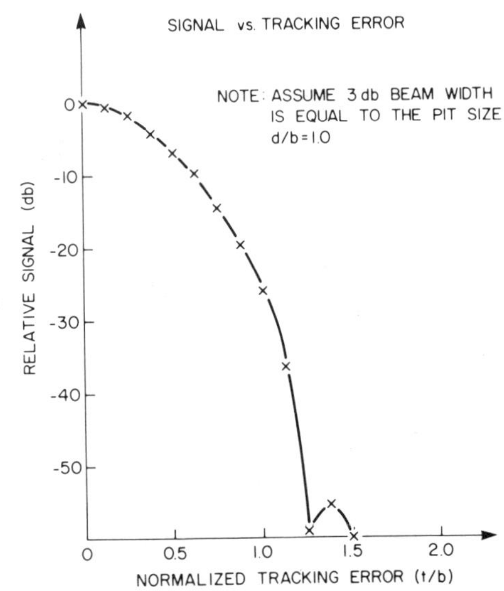

Fig. 8. Optical contrast vs tracking error for $d/b = 1.0$. Of note is the fact that the track-to-track cross talk is reduced by >50 dB when $s_y = 1.3\ b$. This is excessive for digital storage applications.

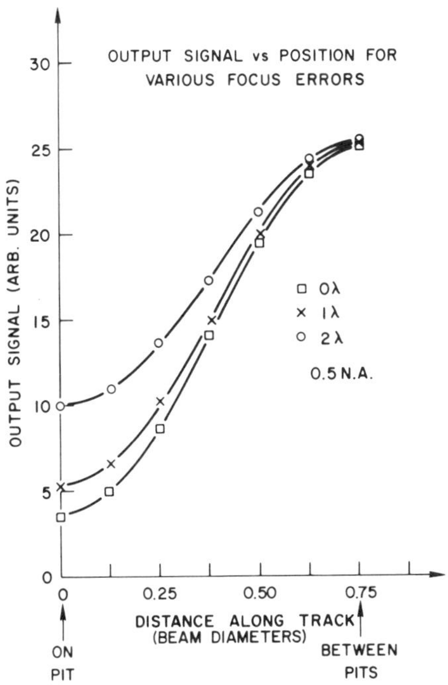

Fig. 9. Readout signal vs position for three focal positions: d/b = 1.0; s/b = 1.5, N.A. = 0.5.

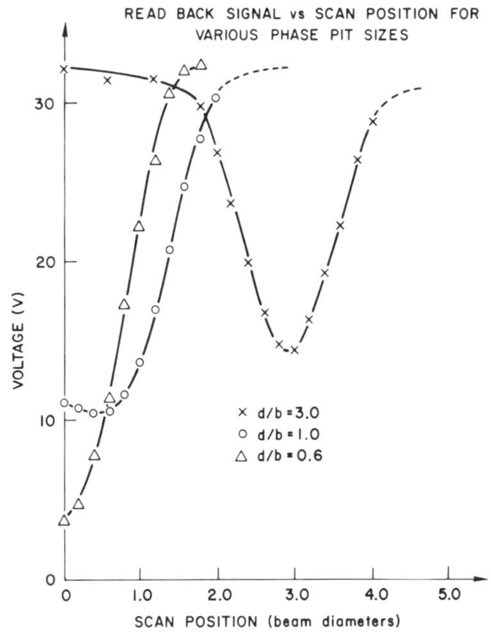

Fig. 10. Readout signal vs position for pure phase pits (λ/4 deep) of various sizes.

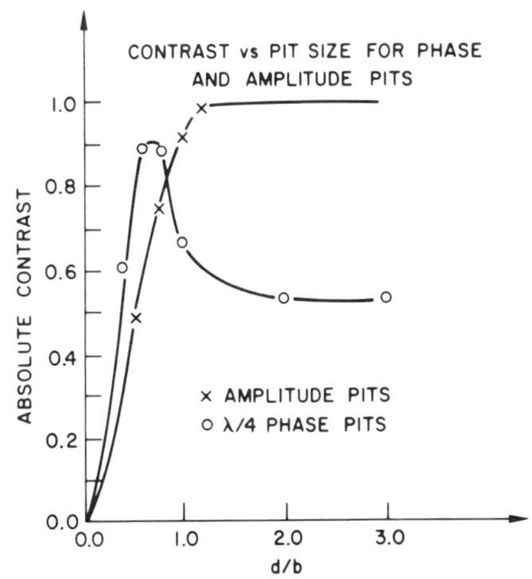

Fig. 11. Peak readout contrast vs phase pit diameter and peak readout contrast vs amplitude pit diameter. Note that the phase pit has an optimum size from a contrast point of view.

writing process are more stringent than during reading, since it is the distribution of the light during writing that will determine the size and the shape of the written spot. Depending on the exact writing characteristics of the material, a defocus error above some level may indeed result in no mark being written. For example, if a medium with a sharp writing threshold is utilized and it is written at 20% above its writing threshold, defocus errors that result in a FWHM increase of 10% would result in no mark being written.

Up to now all readout signals have been calculated assuming that the written spot is an amplitude only spot. This is a reasonable assumption for most write-once materials but not all. For example, trilayer structures[11] inherently have a phase component to their optical response. This is particularly true if the trilayer is not well tuned. Furthermore, certain organic materials that are written by melting a pit in the material[12,13] contain phase as well as amplitude information. Also, all prerecorded and pregrooved information is typically in the form of pure phase information. For these reasons it is desirable to understand the difference between reading phase and amplitude information. In Fig. 10 we show the relative readout signal from isolated phase pits (one-quarter wavelength deep). For a large pit, contrast occurs only as the pit edge is reached. It is interesting that the peak contrast is not greater. This has to do with the details of how the scattered light is detected. When the phase pit is reduced in size to d/b = 1.0, it is seen that the peak contrast is greater, but it is still evident that the peak contrast does not occur at the center of the pit. Finally, when the pit size is reduced to d/b = 0.6, we see still greater peak contrast, and now it occurs at the pit center. This is an interesting result in that it suggests that there is an optimum pit size if one wishes to maximize the readout contrast from phase pits. Such a calculation is shown in Fig. 11 where we plot the peak readout contrast as a function of the normalized pit diameter. The peak contrast is seen to occur for d/b about 0.6. It should be noted that this value is very close to the value of d_y/b used in the digital audio disk where information is recorded as λ/4 phase pits of various lengths. A similar result is also shown for amplitude pits, and, as expected, the contrast monotonically increases to unity as the pit size increases.

The model has also been used to study the effects of phase pit depth on signal readout strength. For example, Fig. 12 gives the relative readout as a function of depth for circular pits with d/b = 0.6 and s/b = 1.5. It is interesting to note that there is considerable con-

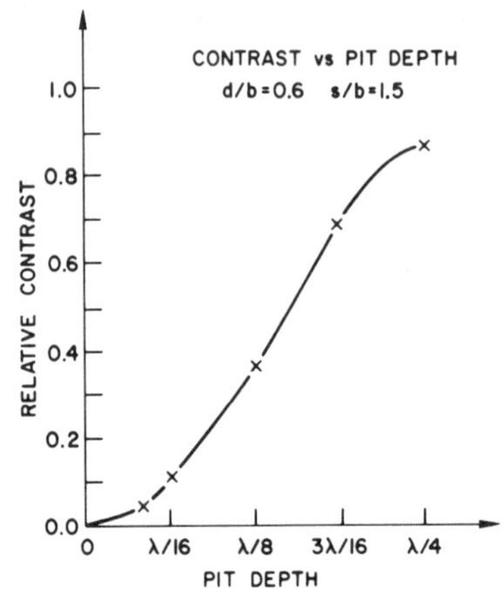

Fig. 12. Peak readout contrast vs phase pit depth: $d/b = 0.6$, $s/b = 1.5$.

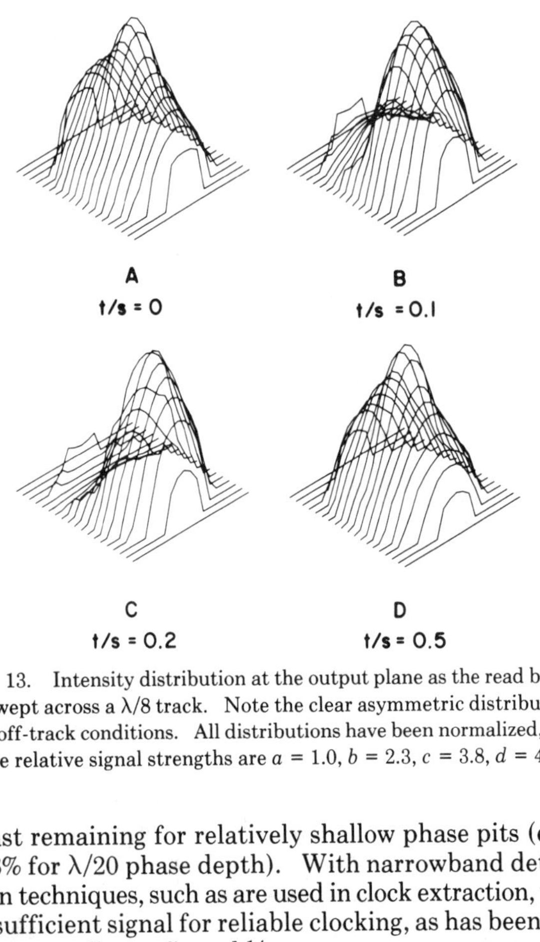

Fig. 13. Intensity distribution at the output plane as the read beam is swept across a λ/8 track. Note the clear asymmetric distribution for off-track conditions. All distributions have been normalized, but the relative signal strengths are $a = 1.0$, $b = 2.3$, $c = 3.8$, $d = 4.0$.

trast remaining for relatively shallow phase pits (e.g., ~8% for λ/20 phase depth). With narrowband detection techniques, such as are used in clock extraction, this is sufficient signal for reliable clocking, as has been experimentally confirmed.[14]

B. Servo Analysis

Up to now the model has been utilized to examine the changes in readout signal as a function of various parameters. However, it can also be used to study various servoing schemes and their interaction with each other and the recorded data. As an example of this, we consider the use of a single beam tracking scheme based on diffraction from a λ/8 phase groove that is embossed into the disk surface. This scheme utilizes a split de-

The authors gratefully acknowledge helpful discussions with T. DiStefano and N. J. Mazzeo.

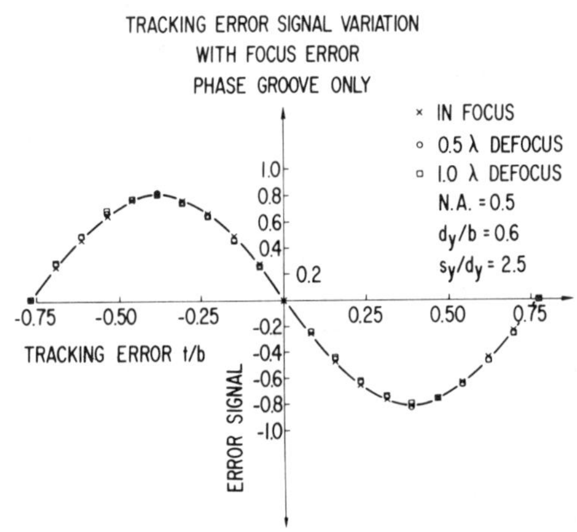

Fig. 14. Calculated tracking error signal as a function of track position for a λ/8 groove at three focal positions. Note that the error signal is not dramatically affected by small focus errors.

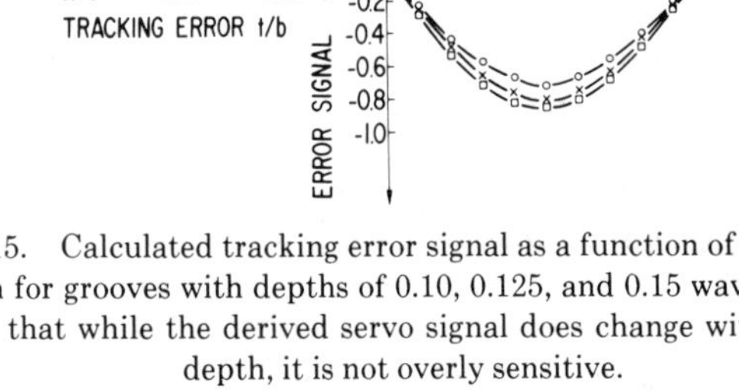

Fig. 15. Calculated tracking error signal as a function of track position for grooves with depths of 0.10, 0.125, and 0.15 wavelengths. Note that while the derived servo signal does change with groove depth, it is not overly sensitive.

tector at the output plane and derives the necessary tracking signal by measuring the difference signal from the split detector. In Fig. 13 the field intensity at the output plane is shown as a function of tracking error. It can be seen that as the beam moves off the track the amount of light striking one-half of the detector increases while the other half decreases, thus providing the necessary servo signal. In Fig. 14 the calculated error signal is given as a function of tracking error for three different focal positions. It can be seen that focus errors up to 0.5 wavelength (0.5 N.A.) result in only minor changes in the shape of the servo response. Finally, in Fig. 15 the effect of groove depth is examined. Grooves 0.10, 0.125, and 0.150 wavelengths are modeled. Again, there is obviously a change in the servo response as a function of groove depth, but it is not great and thus should not present a problem if reasonable groove depth control is achieved.

It is not possible to use the above tracking scheme on a $\lambda/4$ groove or a pure amplitude groove. For these conditions the model predicts (and experiments confirm) that while in focus there is no tracking error signal. Furthermore, when slightly out of focus an error signal is generated, but its magnitude and polarity are dependent on the magnitude and the polarity of the focus error. This is obviously not a very stable servo signal.

This result suggests that, while the tracking scheme described above for a simple $\lambda/8$ groove may be adequate when no amplitude data are present, it might not be when amplitude data are included. To investigate this, the tracking error signal was calculated for the disk reflectivity function shown in Fig. 16. The disk in this case consists of amplitude data pits written onto $\lambda/8$ tracking grooves. The background reflectivity is taken to be 1.0 while the written spots have a reflectivity of 0.1. The resultant tracking error signal is given in Fig. 17. (This error signal is taken as a time average as the beam moves along the track at various off-track positions.) It can be seen that, while the variation of error signal with focus is greater than it was without the amplitude spots (Fig. 14), it is still not excessive. A similar analysis for written spots of higher reflectivity (trilayer case) yields comparable results, and thus it appears that the addition of data to the tracks does not affect the servo signal to a significant extent.

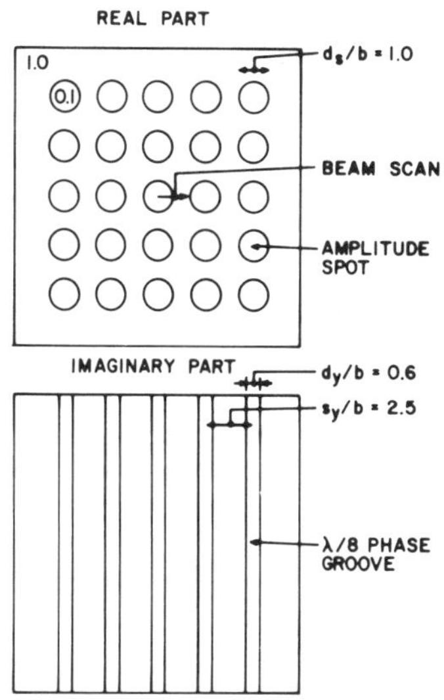

Fig. 16. Complex disk reflectivity function for an object consisting of amplitude pits written onto $\lambda/8$ grooves.

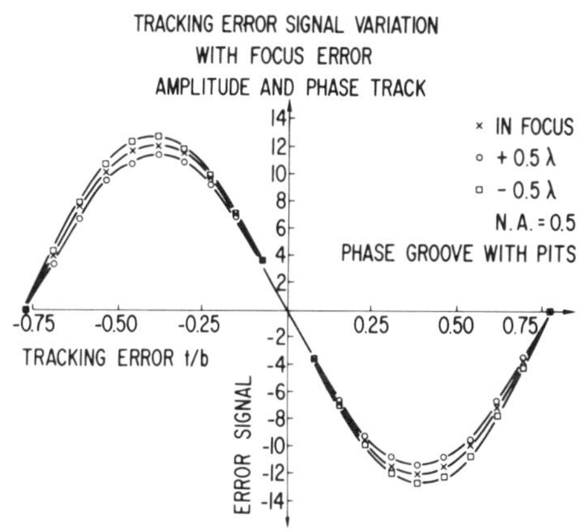

Fig. 17. Calculated tracking error signal as a function of track position for the object of Fig. 16. Results are shown for three focal positions.

V. Conclusion

A detailed Fourier optic model of an optical disk reader has been presented, and a calculational procedure outlined. The model has been utilized to examine the effects of data density, focus errors, tracking errors, etc. on the final readout data signal. Furthermore, the model was used to evaluate a simple track following scheme based on diffraction from a phase groove. It was demonstrated that this scheme is fairly stable with variations in groove depth and focus position provided that a $\lambda/8$ phase groove is utilized. However, it was theoretically determined that a $\lambda/4$ phase groove or an amplitude groove could not be used with this scheme. It was further demonstrated that the addition of amplitude data to a $\lambda/8$ phase groove results in additional tracking servo signal variation with focus, but this additional variation is not excessive.

While the choice of data density and servo schemes associated with an optical storage device will be determined by several variables (some of which are not addressed here), it is essential that detailed modeling of the various interactions be performed. This allows one to more readily evaluate the alternatives while independently varying different parameters. This is not easily accomplished with an experimental setup, and thus the model described here provides an important design aid.

Appendix

With reference to Fig. 1, the output of the laser after appropriate expansion and collimation is described by the function $u_i^+(x_i,y_i)$ where the + sign indicates a field traveling in the $+z$ direction. This field will typically be of the form

$$u_i^+(x_i,y_i) = c \exp - \left(\frac{x_i^2}{a_x^2} + \frac{y_i^2}{a_y^2}\right), \tag{A1}$$

i.e., a 2-D Gaussian where c is an irrelevant constant, a_x is the standard deviation in the x direction, and a_y is the standard deviation in the y direction. For gas lasers a_x and a_y are usually equal, but this is not necessarily true for laser diodes that exhibit greatly different angles of divergence in the plane of the junction vs the plane orthogonal to the junction.

The field at the focal plane of the objective lens can be easily calculated once the lens illumination is specified and the lens pupil function is known. For a diffraction-limited lens this field is given by

$$u_f^+(x_f,y_f) = \mathcal{F}^2\{P(x_i,y_i)u_i^+(x_i,y_i)\}|_{S_x=x_f/\lambda F, S_y=y_f/\lambda F}, \tag{A2}$$

where $P(x_i,y_i)$ represents the generalized pupil function of the objective lens (note that in the absence of aberration this function will simply be unity within the lens aperture and zero elsewhere), F is the lens focal length, and $\mathcal{F}^2\{\ \}$ represents a 2-D Fourier transform defined in the usual way:

$$\mathcal{F}^2\{u(x,y)\} = \int_{-\infty}^{\infty}\int_{-\infty}^{\infty} u(x,y)\exp\{i2\pi(S_x x + S_y y)\}dxdy. \tag{A3}$$

Typically, it will be the field at the focal plane of the lens that is of interest (i.e., in the absence of focus errors), but for generality we assume that the object plane is ε away from the lens focal plane. The field at this plane can be easily found by utilizing an angular spectrum approach.[9,10] The angular spectrum of the field $u_f^+(x_f,y_f)$ is given by

$$A^+\left(\frac{x_i}{\lambda F}, \frac{y_i}{\lambda F}, 0\right) = \mathcal{F}^2\{u_f^+(x_f,y_f)\}|_{S_x=x_i/\lambda F, S_y=y_i/\lambda F}, \tag{A4}$$

or using Eqs. (A2) and (A3):

$$A^+\left(\frac{x_i}{\lambda F}, \frac{y_i}{\lambda F}, 0\right) = P(-x_i,-y_i)u_i^+(-x_i,-y_i), \tag{A5}$$

where $x_i = \alpha F, y_i = \beta F$ and α and β refer to the appropriate direction cosines.[9] Note that the angular spectrum at the focal plane can be found by appropriate scaling of the input distribution without performing the time-consuming 2-D transforms. The angular spectrum is spatially inverted due to the repetitive forward transforms.

Once in the angular spectrum domain, the spectrum is transferred to the object plane through a complex multiplication[9]:

$$A^+\left(\frac{x_i}{\lambda F}, \frac{y_i}{\lambda F}, \varepsilon\right) = A^+\left(\frac{x_i}{\lambda F}, \frac{y_i}{\lambda F}, 0\right) \times \exp\left\{\frac{i2\pi\varepsilon}{\lambda}\sqrt{1 - \left(\frac{x_i}{F}\right)^2 - \left(\frac{y_i}{F}\right)^2}\right\}. \tag{A6}$$

The field at this plane is found by returning to the spatial domain through an inverse transform, i.e.,

$$u_f^+\{x_f,y_f,\varepsilon\} = \mathcal{F}^{-2}\left\{A^+\left(\frac{x_i}{\lambda F}, \frac{y_i}{\lambda F}, \varepsilon\right)\right\}. \tag{A7}$$

At this plane the field is reflected by the object to yield a reflected field distribution given by

$$u_f^-(x_f,y_f,\varepsilon) = u_f^+(x_f,y_f,\varepsilon) \times R(x_f - x_s, y_f - y_s), \tag{A8}$$

where $R(x_f - x_s, y_f - y_s)$ represents the object reflectivity function for scan position x_s,y_s. Note that we have assumed that the reflectivity operation is given by a simple point-by-point multiplication of the object reflectivity function with the input field distribution. This distribution is now transformed to the angular spectrum domain:

$$A^-\left(\frac{x_i}{\lambda F}, \frac{y_i}{\lambda F}, \varepsilon\right) = \mathcal{F}^2\{u^-(x_f,y_f,\varepsilon)\} \tag{A9}$$

and then propagated to the lens focal plane

$$A^-\left(\frac{x_i}{\lambda F}, \frac{y_i}{\lambda F}, 0\right) = A^-\left(\frac{x_i}{\lambda F}, \frac{y_i}{\lambda F}, \varepsilon\right) \times \exp\left\{-\frac{i2\pi\varepsilon}{\lambda}\sqrt{1 - \left(\frac{x_i}{F}\right)^2 - \left(\frac{y_i}{F}\right)^2}\right\}. \tag{A10}$$

The field at the output plane can now be found through an inverse transform (to return to the field domain) followed by a forward transform (to return through the lens), or we can use the symmetry of multiple transforms to write

$$u_i^-(x_i,y_i) = A^-\left(\frac{x_i}{\lambda F}, \frac{y_i}{\lambda F}, 0\right) P(-x_i,-y_i), \tag{A11}$$

which is the field at the output of the objective lens. Here the pupil function is included primarily in recognition of the fact that any spatial frequencies that fall outside the collection aperture of the objective lens must be discarded. The voltage of the readout detector is now calculated through a 2-D integration of the intensity at the detector plane, i.e.,

$$v(x_s,y_s) = \int_{-\infty}^{\infty}\int_{-\infty}^{\infty} u_i^-(x_i,y_i)u_i^{-*}(x_i,y_i)D(x_i,y_i)dx_i dy_i, \tag{A12}$$

where $D(x_i,y_i)$ is the detector intensity response and $v(x_s,y_s)$ is the detected signal for the object position x_s,y_s. Additional image points are now readily calculated by returning to Eq. (A8), changing x_s,y_s, and repeating steps (A8)–(A12).

References

1. D. Maydan, Bell Syst. Tech. J. **50**, 1761 (1971).
2. K. Bulthuis *et al.*, IEEE Spectrum **26** (Aug. 1979).
3. G. Kenney *et al.*, IEEE Spectrum **33** (Feb. 1979).
4. H. H. Hopkins, J. Opt. Soc. Am. **69**, 4 (1979).
5. C. J. R. Sheppard and A. Choudhury, Opt. Acta **24**, 1051 (1977).
6. T. Wilson, Appl. Phys. **22**, 119 (1980).
7. A. E. Bell and F. W. Spong, IEEE J. Quantum Electron. **14**, 487 (1978).
8. D. Goodman, IBM Research; private communicatins.
9. J. W. Goodman, *Introduction to Fourier Optics* (McGraw-Hill, New York, 1968), p. 48.
10. A. Atalar, J. Appl. Phys. **49**, 5130 (1978).
11. D. G. Howe and J. J. Wrobel, J. Vac. Sci. Technol. **18**(1), 92 (1981).
12. J. Drexler, J. Vac. Sci. Technol. **18**(1), 87 (1981).
13. V. Jipson and C. Jones, J. Vac. Sci. Technol. **18**(1), 105 (1981).
14. M. G. Carasso and A. Huijser, in *Technical Digest, Conference on Lasers and Electrooptic Systems* (Optical Society of America, Washington, D.C., 1982), paper WH1.

Reprinted with permission from *Journal of Applied Physics*, Vol. 56(4), pp. 1165-1170 (August 15, 1984). ©1984 American Institute of Physics.

High-density magneto-optical disk storage and the effect of finite beam size in readout

M. Mansuripur
Boston University, College of Engineering, Boston, Massachussetts

A method of data reconstruction from a digital magneto-optical disk system with diffraction-limited spot size is described. Using a linear filter with high-frequency enhancement we show that significant improvements in the storage capacity can be achieved at a reasonable cost in terms of the signal-to-noise ratio. Performance is evaluated against an optimum but impractical method as well as a technique for the reconstruction of the analog waveforms.

I. INTRODUCTION

In comparison with the conventional magnetic disk systems, erasable optical storage has the advantages of high capacity, remote read/write/erase heads, and removability.[1,2] Among those media suggested for erasable optical storage to date, amorphous rare-earth-transition metal alloys with perpendicular magnetization have shown great promises.[3] With thermomagnetic recording and magneto-optical readout at red or near infrared wavelengths, densities in excess of 10^8 bits/cm^2 have been achieved.

A significant difference between magnetic and optical recording techniques stems from the fact that in the former the readout energy is derived from the magnetic domains themselves, whereas in the latter the domains act as gates, directing the energy of the reading light beam into separate channels for detection. This means that the magnetic recording density is inherently limited while the optical recording density is only limited by the wavelength of the available lasers.

Optical storage is presently limited by the wavelength of the read beam. The diffraction limited spot formed by focusing a gaussian beam of wavelength λ through an objective lens of numerical aperature NA is gaussian, and has a diameter proportional to λ / NA. Increasing the storage density for a given spot size results in reduced signals from individual domains and overlapped signals from neighboring domains, a phenomenon referred to as the intersymbol interference (ISI). Both reduced signal and overlap degrade the signal-to-noise ratio (SNR) and impose limitations on achievable storage densities. It is important to emphasize here that ISI results from the read beam overlapping more than one domain and not magnetic coupling between domains. In fact, since most media of interest for erasable optical storage are ferrimagnetic with compensation point in the vicinity of room temperature, magnetic coupling between neighboring domains should be insignificant.

The aim of this paper is to discuss different signal processing schemes that can recover data from a high density storage medium in the presence of ISI. In Sec. II the various components and parameters of the proposed readout system are described. Section III is devoted to the analysis of the three suggested techniques for data reconstruction. Section IV is a summary of the results and contains a few general comments and conclusions.

II. MAGNETO-OPTICAL READOUT AND DIFFERENTIAL DETECTION SCHEME

The schematic diagram of the readout system is shown in Fig. 1; this section describes its components and defines various parameters and functions that will be used throughout the paper.

A. The storage format

The disk contains a magnetic layer with perpendicular magnetization on which the information is recorded in the form of small magnetic domains. Assuming that the magnetization of the film at a point (ρ,θ) is saturated in either of the two possible directions along the Z axis, the function $z(\rho,\theta)$ is defined as follows: $z(\rho,\theta) = \pm 1$ if the magnetization at (ρ,θ) is in the $\pm Z$ direction. The read beam reads only one particular track at any given time, so only the neighborhood near a fixed radius, say $\rho = \rho_0$, is of interest. The sequence of data on this track can then be represented as $z(x)$ where $x = \rho_0\theta$ is

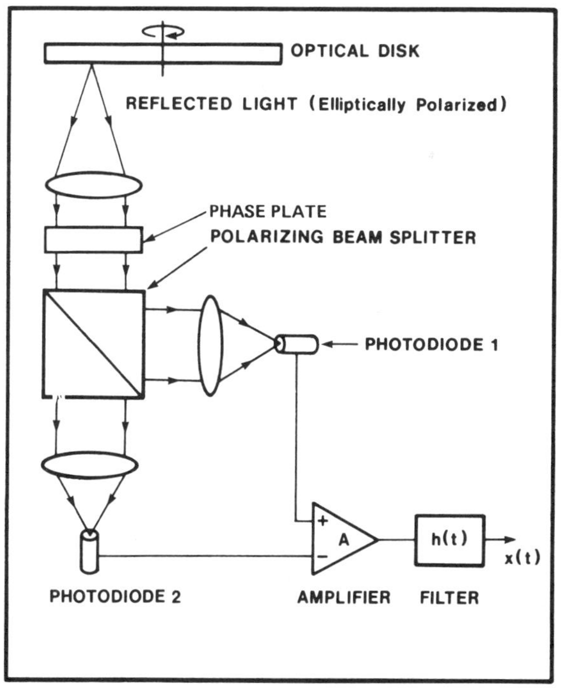

FIG. 1. Schematic diagram of the magneto-optical readout system.

the distance along the track from an arbitrary reference angle. Assume that the length of the domains in the x direction is a constant γ. If the linear velocity of the track is v, then the time each domain spends under the beam will be $T = \gamma/v$. It is assumed that T is a fixed system parameter, so that comparisons of the next section between different techniques are made at fixed data rates.

The ideal signal, expected from the system in the absence of noise and ISI, is $z_0(t) = z(vt)$. Using the rectangular pulse function $s(t)$ of unit height and duration T, this signal can be written as

$$z_0(t) = \sum_n a_n s(t - nT), \tag{1}$$

where $a_n = \pm 1$ and $\{a_n\}$ represents the binary sequence of data recorded on the track.

B. The read beam

The beam is focused on the disk surface through the objective lens. The intensity distribution on the film is assumed to be gaussian with

$$I(x,y) = \frac{P_0}{\pi r_0 r_1} \exp\{-[(x/r_0)^2 + (y/r_1)^2]\}. \tag{2}$$

r_0 is the beam radius in the X direction (along the track), r_1 is the radius in the Y direction (perpendicular to the track), and P_0 is the total incident power. Throughout the paper it is assumed that in the Y direction the domain width is much larger than r_1 so that the intensity distribution in this direction becomes irrelevant. One can then use the function

$$g(x) = \frac{P_0}{\sqrt{\pi} r_0} \exp[-(x/r_0)^2] \tag{3}$$

to represent the incident power density on the track.

The read beam is linearly polarized. Upon reflection from the magnetic material its state of polarization will change. This is the well-known polar Kerr effect. In general, two parameters $r_\parallel$ and $r_\perp$ describe the new polarization state. $r_\parallel$ is the amplitude reflectivity for the component polarized in the direction of incident polarization and $r_\perp$ is the amplitude reflectivity for the component perpendicular to it. Both $r_\parallel$ and $r_\perp$ are complex numbers and the phase difference between them determines the amount of ellipticity. Nonetheless, with proper adjustment of the phase plate (see Fig. 1) this phase difference can be eliminated, converting ellipticity to pure rotation of the direction of linear polarization. Since the system does not detect ellipticity the above conversion improves performance.[4,5] $r_\parallel$ and $r_\perp$ are thus assumed to be real.

C. The optical system

The axis of the polarizing beam splitter is at 45 deg with respect to the unrotated polarization so that in the absence of the polar Kerr effect the photodiodes receive identical amounts of energy $(0.5P_0r_\parallel^2)$. When the Kerr effect is present the energy is split as $0.5P_0(r_\parallel \pm r_\perp)^2$ between the two detectors, with the state of magnetization determining the detector which receives the larger amount. The difference between these energies, $2P_0r_\parallel r_\perp$, gives the magneto-optic signal amplitude while their sum, $P_0(r_\parallel^2 + r_\perp^2)$, is proportional to the total shot-noise density.[4]

D. The electronic circuitry

The photodiodes are identical light detectors with conversion factor η. Each photocurrent goes through a parallel RC circuit before being amplified. The RC circuit represents the impedances of the detector, the amplifier, and the bias circuitry. Its frequency response is $RH_0(f)$ where

$$H_0(f) = 1/(1 + j2\pi RCf). \tag{4}$$

The effect of thermal noise from the electronic circuitry is minimized if avalanche detectors with internal gain mechanism are employed. SNR analysis is thus confined to shot noise although thermal noise can be easily included if so desired. The shot-noise variance is independent of the state of magnetization and is given by[4]

$$\sigma^2 = \eta e P_0(r_\parallel^2 + r_\perp^2)\int_{-\infty}^{\infty} |H_{eq}(f)|^2 df, \tag{5}$$

where e is the electronic charge and $H_{eq}(f)$ is the frequency response of the equivalent filter. If $H(f)$ represents the frequency response of the filter $h(t)$ in Fig. 1 then

$$H_{eq}(f) = H_0(f)\,H(f). \tag{6}$$

Equation (5) does not include gain factors such as R, amplifier gain, and avalanche gain that cancel out in the SNR. When avalanche photodiodes are used, however, a noise factor due to the random nature of the multiplication process must be included.

III. EFFECT OF INTERSYMBOL INTERFERENCE ON READOUT

As mentioned in the introduction, the nonzero read beam size is responsible for ISI. When the beam travels along the track, it illuminates an area whose length is approximately $4r_0$ (about 99.5% of the beam's energy is contained in this region). If the domain size γ is smaller than this length there will be some degree of overlap between the signals from neighboring domains. Moreover, when the beam crosses the boundary between adjacent domains it sends back a signal which represents both domains, producing ISI. The degrading effect of this phenomenon as well as three different methods of eliminating the problem are discussed in this section.

A. Elimination of ISI by spatial filtering: An optimum technique

This approach is based on a method which could eliminate ISI in principle, but requires an objective lens with an impractically large numerical aperture. Besides, if a large numerical aperture were available in the first place, the spot size would have been reduced and the problem eliminated. The technique nonetheless provides a yardstick for the evaluation of other methods.

The time allocated to each domain for readout is T and the read beam in an optimum system must spend *all* of this time on the domain. In other words, the beam must jump from the center of one domain to the center of the next domain on the track every T seconds and remain stationary in the meantime. Now, consider the pattern of polarization at the disk during each T second interval. With a high quality imaging system this pattern can be exactly reconstructed in another plane. To eliminate ISI, a spatial filter is placed in this plane, transmitting the image of the domain under consideration and blocking the image of neighboring domains. The method is optimal since all the information about a domain is contained in the light reflected from the domain itself and by blocking the neighboring domains no information is being lost.

Beyond the spatial filter, the optimum detection scheme will be the same as the differential detector of Fig. 1. The output signal amplitude is thus given by

$$\pm 2\eta r_{\parallel} r_{\perp} \int_{-\gamma/2}^{\gamma/2} g(x)dx = \pm 4\eta P_0 r_{\parallel} r_{\perp} \operatorname{erf}\left(\frac{\gamma}{\sqrt{2}r_0}\right). \quad (7)$$

The spectral density of the accompanying shot noise is given by

$$2\eta e P_0 (r_{\parallel}^2 + r_{\perp}^2) \operatorname{erf}\left(\frac{\gamma}{\sqrt{2}r_0}\right). \quad (8)$$

For the sake of optimum detection, the system impulse response $h_{eq}(t)$ must be a rectangle of duration T. Since the bandwidth of this filter is $2B = 1/T$ the signal-to-noise ratio is given by

$$\mathrm{SNR} = \frac{4\eta P_0 T r_{\parallel}^2 r_{\perp}^2}{e(r_{\parallel}^2 + r_{\perp}^2)} \left[2\operatorname{erf}\left(\frac{\gamma}{\sqrt{2}r_0}\right)\right]. \quad (9)$$

A logarithmic plot of the function $2\operatorname{erf}(\gamma/\sqrt{2}r_0)$ vs γ/r_0 is shown in Fig. 2. For large γ/r_0 the SNR is at its maximum possible value. This is because the beam completely covers the domain under consideration while no time is spent crossing the boundary between domains. At $\gamma = 3r_0$ where 96.5% of energy covers the domain the SNR deterioration is only 0.15 dB. At $\gamma = r_0$ the coverage is about 50% and the SNR drop is about 3 dB.

B. Reconstruction of the pattern of magnetization reversals: The analog signal approach

In a previous paper[4] the power spectral density of the signal $x(t)$ in the system of Fig. 1 was derived. It is given by

$$S(f) = (2\eta P_0 r_{\parallel} r_{\perp})^2 \exp[-2(\pi f r_0/v)^2] \times S_{z0}(f)\,|H_{eq}(f)|^2, \quad (10)$$

where $S_{z0}(f)$ is the spectrum of the ideal signal $z_0(t)$ in Eq. (1) and is given by[6]

$$S_{z0}(f) = T\left(\frac{\sin(\pi f T)}{\pi f T}\right)^2. \quad (11)$$

Let us choose $h(t)$ to make the equivalent filter $h_{eq}(t)$ an ideal low pass filter with bandwidth $2B = \alpha/T$ where α is an arbitrary constant. Then

$$\mathrm{SNR} = \frac{4\eta P_0 T r_{\parallel}^2 r_{\perp}^2}{e(r_{\parallel}^2 + r_{\perp}^2)} \left\{(2/\pi\alpha) \int_0^{\pi\alpha/2} \exp[-2(r_0 x/\gamma)^2] \times (\sin x/x)^2 dx\right\}. \quad (12)$$

Figure 3 is a logarithmic plot of the bracketed term in Eq. (12) versus γ/r_0 for several values of α. For large γ/r_0 the ISI is negligible and the signal reconstruction is possible through the use of a sufficiently wide band filter (i.e., large α). However, as α increases the noise level rises and consequently the SNR drops. For small γ/r_0 the signal at any instant of time contains information about several domains and identification of individual data bits is impossible without further equalization. The issue of equalization, however, will not be addressed here since the subject of analog signal reconstruction and the enhancement of storage capacity using various encoding schemes is beyond the scope of this paper and will be taken up in a separate publication. The point of the discussion here is that for large γ/r_0 the loss of SNR in the analog approach is signficant and it is reasonable therefore to expect that such inefficiency applies to small γ/r_0 as well. For example, the SNR loss of the analog method for $\alpha = 2$ and large γ/r_0 is about 3.5 dB compared to only 0.5 dB for the digital method described below.

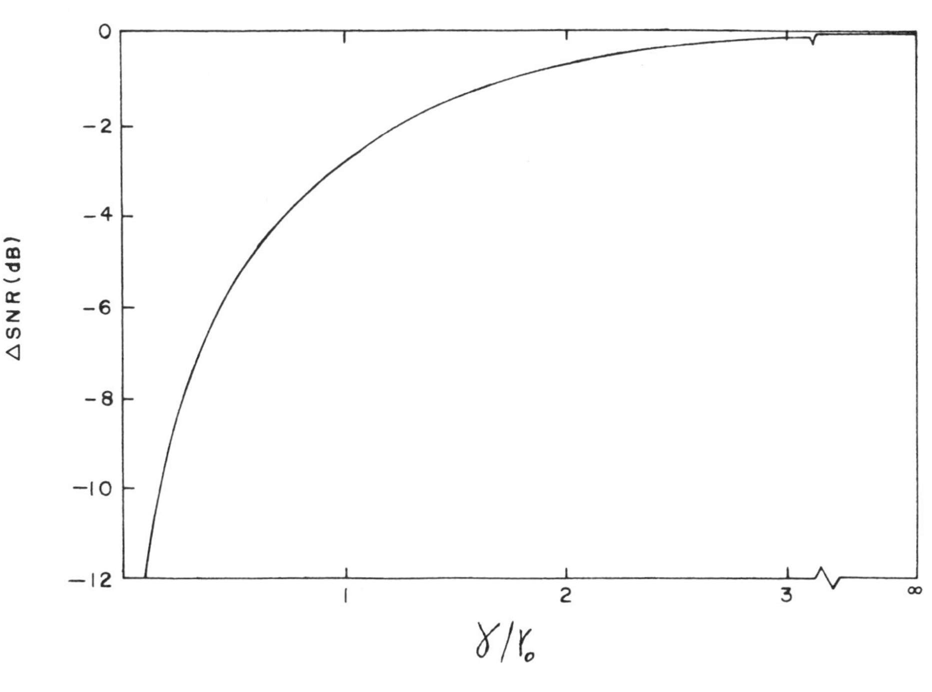

FIG. 2. Logarithmic plot of the function $2\operatorname{erf}(\gamma/\sqrt{2}r_0)$ vs γ/r_0.

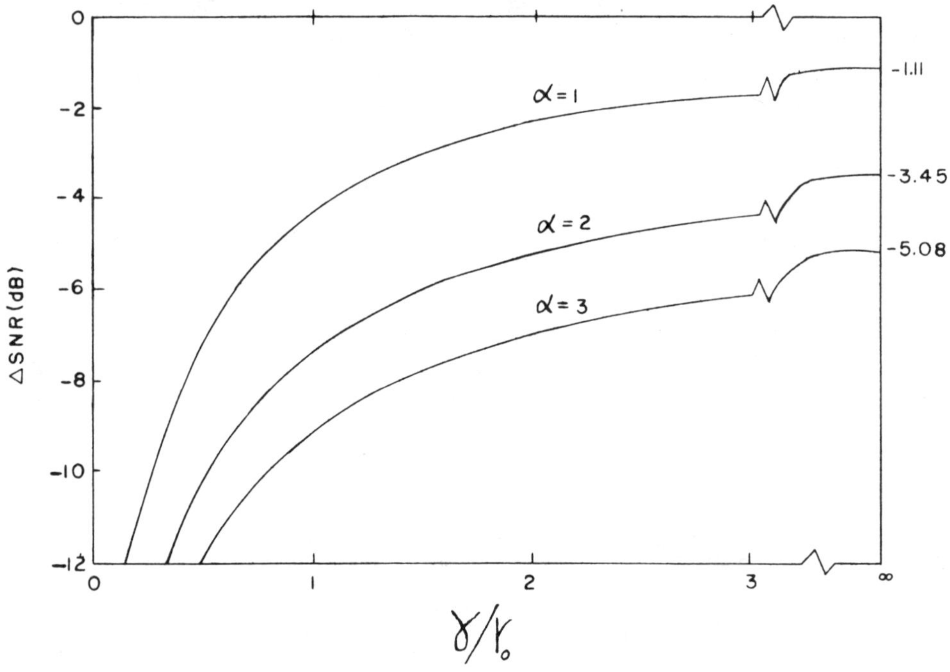

FIG. 3. Logarithmic plot of the function $(2/\pi\alpha)\int_0^{\pi\alpha/2}\exp[-2(r_0x/\gamma)^2] \times(\sin x/x)^2dx$ vs γ/r_0 for several values of α.

C. The digital signal approach to the elimination of ISI

In this approach the filter $h(t)$ is designed to eliminate interference only at instants of time which are multiples of T, namely $t=nT$. This allows a simple threshold device to sample $x(t)$ at $t=nT$ and reconstruct the sequence of data $\{a_n\}$. SNR at $t=nT$ is then a direct measure of the error rates involved.

Consider the instantaneous power $P(t)$ incident on the detectors

$$P(t)=\frac{1}{2}P_0(r_\parallel^2+r_\perp^2)\pm r_\parallel r_\perp\int_{-\infty}^{\infty}g(vt-x)z(x)dx, \quad (13)$$

where $g(x)$ is defined in Eq. (3). Defining

$$g_0(t)=\frac{\gamma}{\sqrt{\pi}r_0T}\exp\left[-\left(\frac{\gamma t}{r_0T}\right)^2\right] \quad (14)$$

the integral in Eq. (13) can be written as $P_0g_0(t)*z_0(t)$, where * stands for convolution. The output signal will then be

$$\begin{aligned}x(t)&=2\eta P_0r_\parallel r_\perp g_0(t)*z_0(t)*h_{eq}(t)\\&=2\eta P_0r_\parallel r_\perp\sum_n a_nq(t-nT),\end{aligned} \quad (15)$$

in which $q(t)$ is given by

$$q(t)=s(t)*g_0(t)*h_0(t)*h(t). \quad (16)$$

If $q(t)$ is now chosen to be equal to 1 at $t=0$ and equal to zero at $t=mT$ for all $m\neq 0$, then for $t=nT$ we have

$$x(nT)=2\eta P_0r_\parallel r_\perp a_n. \quad (17)$$

This signal clearly contains information about a_n without interference from the rest of the sequence.

A proper choice of $q(t)$ would be the following family of the so called raised cosine functions[7]

$$q(t)=\frac{\sin(\pi t/T)}{\pi t/T}\,\frac{\cos(\pi\beta t/T)}{1-(2\beta t/T)^2}, \quad (18)$$

where β is a parameter in the interval [0,1]. The Fourier transforms of the functions $s(t)$, $g_0(t)$, and $q(t)$ are given in Table I. Using these transforms in Eq. (16) yields

$$H(f)=\frac{Q(f)}{S(f)\,G_0(f)\,H_0(f)}. \quad (19)$$

This is the frequency response of the appropriate filter. A plot of $H(f)$ for typical parameter values is shown in Fig. 4.

Using the noise variance given in Eq. (5), the signal-to-noise ratio at $t=nT$ is calculated as

TABLE I. Fourier transforms of $s(t)$, $g_0(t)$, and $q(t)$.

Function	Fourier transform	
$s(t)$	$S(f)=T\dfrac{\sin(\pi Tf)}{\pi Tf}$	
$g_0(t)$	$G_0(f)=\exp\left[-\left(\dfrac{\pi Tf}{\gamma/r_0}\right)^2\right]$	
$q(t)$	$Q(f)=T$	$0\leqslant\lvert f\rvert\leqslant\dfrac{1-\beta}{2T}$
	$\dfrac{T}{2}\{1-\sin[\pi(Tf-0.5)/\beta]\}$,	$\dfrac{1-\beta}{2T}\leqslant\lvert f\rvert\leqslant\dfrac{1+\beta}{2T}$
	0	Otherwise

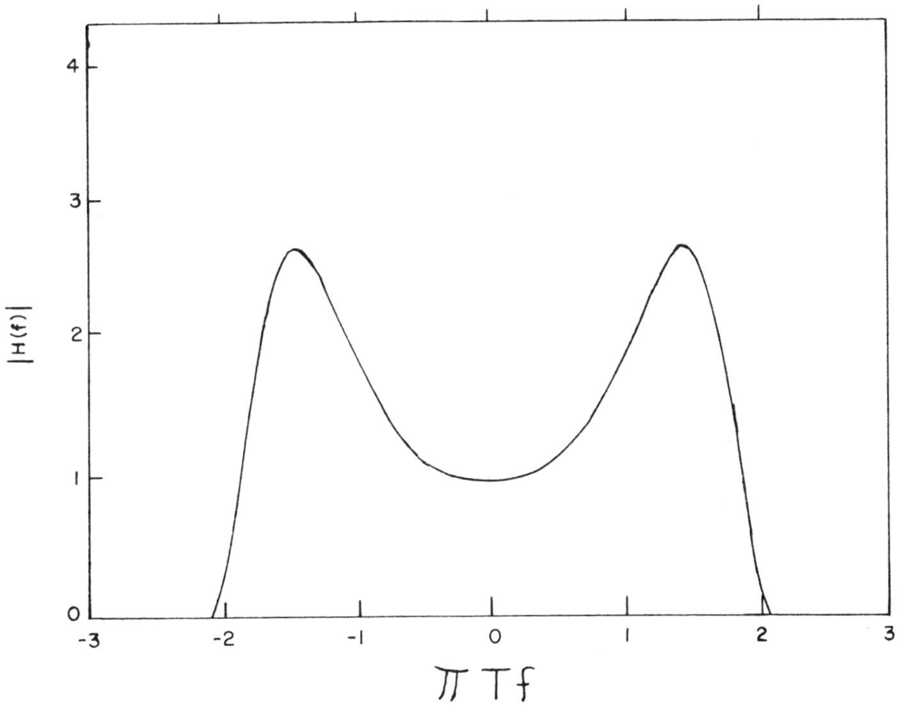

FIG. 4. Plot of the function $|H(f)|$ vs f for $\gamma = 1.5r_0$, $\beta = 0.3$ and $RC = T/10$.

$$\mathrm{SNR} = \frac{4\eta P_0 T r_{\parallel}^2 r_{\perp}^2}{e(r_{\parallel}^2 + r_{\perp}^2)} (1/\xi), \tag{20}$$

where

$$\xi = \frac{1}{\pi} \int_0^{\pi(1+\beta)/2} \frac{\exp[2(r_0 x/\gamma)^2]}{(\sin x/x)^2} \times \begin{Bmatrix} 2 & x \leqslant \frac{\pi}{2}(1-\beta) \\ \frac{1}{2}\left\{1 - \sin\left[\left(x - \frac{\pi}{2}\right)/\beta\right]\right\}^2, & x > \frac{\pi}{2}(1-\beta) \end{Bmatrix} dx.$$

For a given set of system parameters β can be chosen to minimize ξ. The minimum value of ξ for large γ/r_0 is 1.12 (at $\beta = 0.87$). In the limit of large γ/r_0 therefore the SNR in Eq. (20) is only 0.5 dB below the optimum SNR in Eq. (9). A logarithmic plot of $1/\xi$ vs γ/r_0 for the optimum β is shown in Fig. 5. (The optimum β is also shown in the figure.) According to this curve one loses about 1.2 dB in going from large γ to $\gamma = 3r_0$. At the cost of another 4 dB one can further reduce γ to $1.5r_0$. Compare these figures with the corresponding numbers obtained from the optimum curve of Fig. 2, namely 0.15 and 1.3 dB, respectively.

IV. CONCLUDING REMARKS

Three methods of data reconstruction from a magneto-optical storage medium with intersymbol interference were discussed. The performance of each method at different storage densities was evaluated on the basis of the signal-to-noise ratio degradation below the maximum level. It was found, for example, that a linear filter with high frequency enhance-

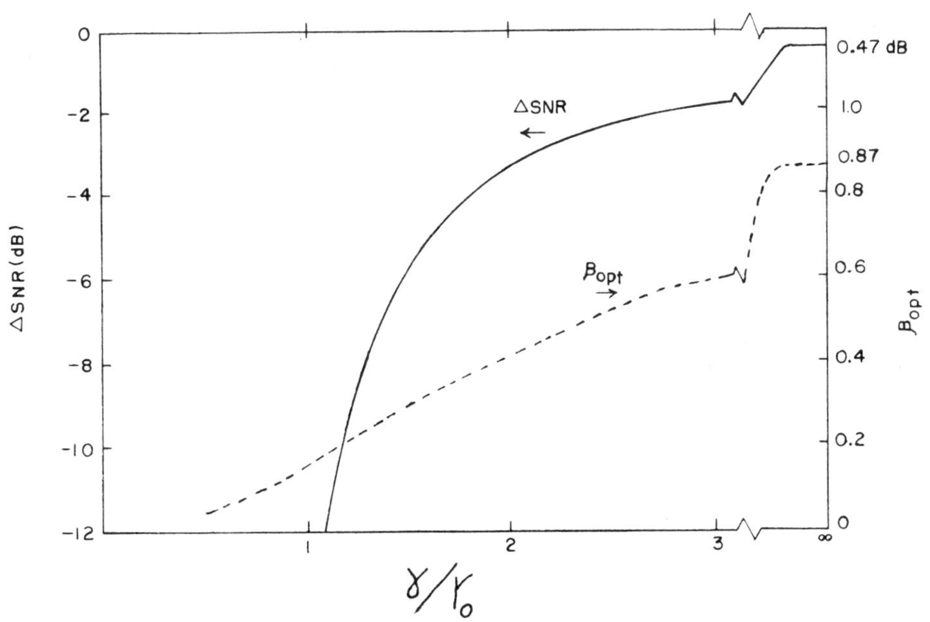

FIG. 5. Logarithmic plot of the function $1/\xi$ vs γ/r_0 for optimum β. The broken curve is a plot of β_{opt}.

ment can be used to reconstruct data stored at $\gamma = 1.5r_0$ at a cost of 5.5 dB in SNR. This is only 4 dB below the SNR of the optimum system. At smaller values of γ reconstruction is still possible although SNR degradation become quite severe. At $\gamma = r_0$, for example, the cost in SNR is 14 dB, which is 11 dB below optimum.

ACKNOWLEDGMENTS

The author is grateful to Dr. Neville Connell of Xerox Corporation and Dr. David Treves of the Weizmann Institute of Technology for many helpful discussions. The project is supported by the National Science Foundation grant number ECS-8307928, and in part by the IBM corporation.

[1]R. M. White, IEEE Spectrum **12**, 32 (1983).
[2]A. E. Bell, Comput. Des. **8**, 133 (1983).
[3]P. Chaudhari, J. J. Cuomo, and R. J. Gambino, Appl. Phys. Lett. **22**, 337 (1973).
[4]M. Mansuripur, G. A. N. Connell, and J. W. Goodman, J. Appl. Phys. **53**, 4485 (1982).
[5]M. Mansuripur, G. A. N. Connell, and J. W. Goodman, Proc. SPIE Conf. **329**, 215 (1982).
[6]A. Papoulis, *Probability, Random Variables, and Stochastic Processes* (McGraw-Hill, New York, 1965).
[7]S. D. Personick, Bell Syst. Tech. J. **52**, 843 (1973).

Reprinted with permission from *Applied Optics*, Vol. 25(4), pp. 483-489 (February 15, 1986). ©1986 Optical Society of America.

Compact magnetooptical disk for coded data storage

Masahiro Ojima, Atsushi Saito, Toshimitsu Kaku, Masaru Ito, Yoshito Tsunoda, Shinji Takayama, and Yutaka Sugita

A compact magnetooptical disk having 500 times the capacity of a 13.34-cm (5.25-in.) floppy disk has been developed for use in coded data storage applications. Low error rates are attained by the use of advanced highly sensitive polarization detection optics and high-quality recording media. Progress in these fields has been based on noise analysis of readout signals. The disk drive can write and erase data by sectors and can also play current write-once-type optical disks. This disk/disk drive unit promises to lead to enhanced storage capacity and to support the current drive to enhance office automation and other information systems.

I. Introduction

Both floppy and rigid magnetic disks have come into common use for data storage and retrieval in office automation or other smaller information systems. Until now, these devices provided users with adequate data access and storage functions. However, demands for higher storage capacities have greatly increased. To meet these demands, magnetooptical disks have been developed.

The idea of thermomagnetic writing and magnetooptical readout using a laser source was first proposed about 20 years ago.[1] Since then, progress in developing amorphous magnetic materials,[2] optical disk recording technology, and semiconductor diode lasers has been such that a magnetooptical disk is now feasible using these technologies.[3] Such a disk offers benefits like erasability, large capacity, and wide spacing between the optical head and disk. Use of this disk in coded data storage is particularly promising, since disk capacity is much greater than for floppy or rigid magnetic disks.

This paper reports on the development by the authors of a compact magnetooptical disk. Furthermore, a compact magnetooptical disk drive has also been developed, which can play conventional write-once optical disks in addition to the magnetooptical disks.

The authors are with Hitachi, Ltd., Central Research Laboratory, 1-280 Higashi-koigakubo, Kokubunji, Tokyo 185, Japan.

Received 3 September 1985.
0003-6935/86/040483-07$02.00/0.

II. Improvements in Signal-to-Noise Ratio

The most critical problem to be faced in developing a practical magnetooptical disk is the signal-to-noise ratio. Low values occur since the magnetic recording material has a Kerr rotation angle of $<1°$. To maintain a low error rate, both SNR increase and disk-defect reduction are necessary. This section describes noise analysis and other techniques for increasing the SNR in both the optical head and the recording medium.

A. Noise Analysis

The optics configuration for a magnetooptical disk is shown in Fig. 1. The diode laser is Hitachi's HL8312E or HL8314E with 830-nm wavelength. Beam splitters 1 and 2 transmit partially to achieve p polarization and reflect totally to achieve s polarization of light. Beam splitter 2 divides the laser beam reflected from the disk into two parts. One is channeled into an analyzer for magnetooptical readout of user data, and detected by a single PIN or APD (avalanche photodiode) detector. Differential detection is not adopted. The other beam is used to achieve focusing or tracking servo signals. Focusing/tracking error detection optics are the same as optics for a write-once optical disk[4] to permit compatibility between write-once and magnetooptical disks within the same system. The tracking detector also picks up sector header signals prerecorded in the form of pits or user data from the write-once optical disk.

The causes of noise in the magnetooptical disk were investigated to determine approaches for increasing SNR. Experimental results and calculated curves for carrier C, noise N, and carrier-to-noise ratio C/N are shown in Fig. 2 as a function of the analyzer angle, θ_A. Magnetooptical signals are read out by a PIN detector (left half of the figure) and an APD detector with a

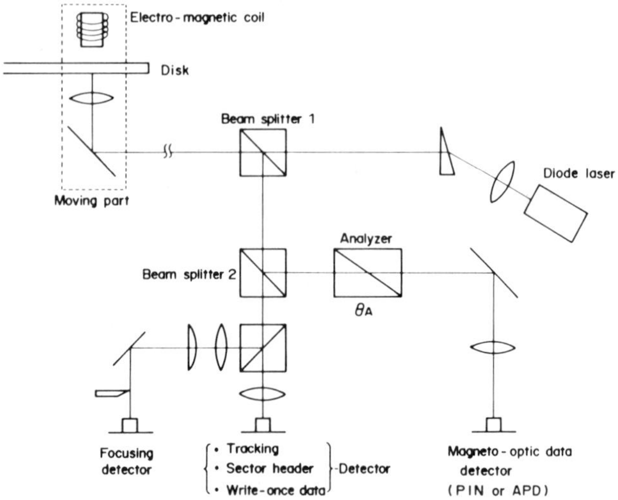

Fig. 1. Optical configuration of a magnetooptical disk.

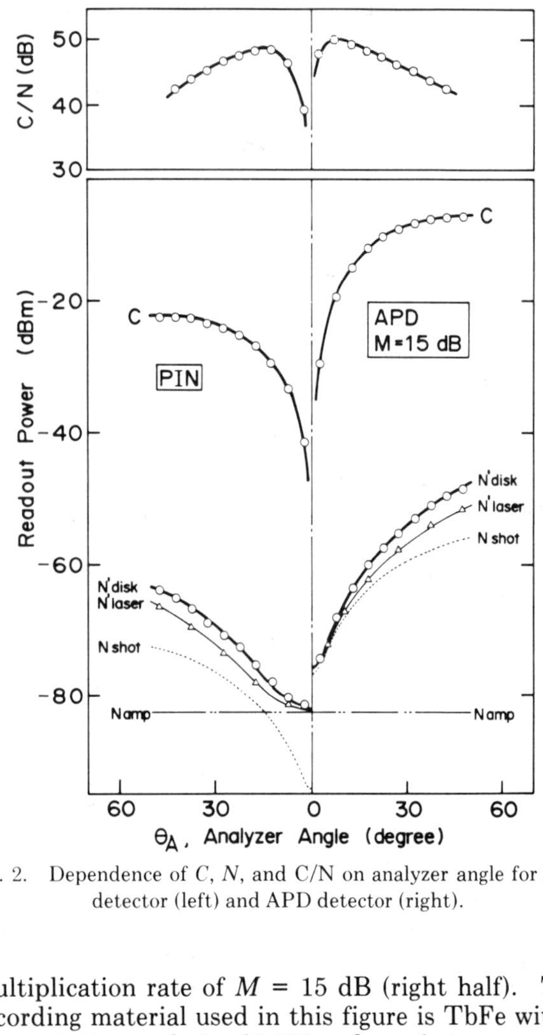

Fig. 2. Dependence of C, N, and C/N on analyzer angle for PIN detector (left) and APD detector (right).

multiplication rate of M = 15 dB (right half). The recording material used in this figure is TbFe with a Kerr rotation angle θ_K of 0.25° and a curie temperature T_C of 140°C. Carrier frequency f is 0.9 MHz and noise power bandwidth Δf is 30 kHz.

A 12-cm diam disk was rotated at 1200 rpm. Disk readout power is 1 mW (Fig. 2). Disk noise, N_{disk}, was measured during rotation, laser noise, N_{laser}, at a time of no rotation, and amplifier circuit noise, N_{amp}, with the laser turned off. Shot noise, N_{shot}, generated in the detector was calculated as

$$N_{\text{shot}} = [2eP_0\alpha M^{2.25}\Delta f(\sin^2\theta_A + \epsilon)]^{1/2} \cdot R_G. \tag{1}$$

Here, e is the electronic charge, P_0 is the laser power before entering the analyzer, α = 0.45 A/W is the detector efficiency, ϵ = 1/225 is the extinction ratio, and R_G = 430 kΩ is the amplifier gain resistance. The C, N_{disk}, and N_{laser} are calculated as

$$C = P_0\alpha MR_G \cdot \sin(2\theta_A)\sin(2\theta_K{}'')m, \tag{2}$$

$$N_{\text{disk}} = \eta_{\text{disk}} \cdot \text{dc} = \eta_{\text{disk}}P_0\alpha MR_G(\sin^2\theta_A + \epsilon), \tag{3}$$

$$N_{\text{laser}} = \eta_{\text{laser}} \cdot \text{dc} = \eta_{\text{laser}}P_0\alpha MR_G(\sin^2\theta_A + \epsilon). \tag{4}$$

Variable m is the modulation factor determined by the written domain size and the laser spot diameter. The effective Kerr rotation angle, $\theta_K{}''$, is[5]

$$\theta_K{}'' = \theta_K/(R_{p1}R_{p2})^{1/2}. \tag{5}$$

Here, R_{p1} and R_{p2} are the reflection coefficients for p polarization of beam splitters 1 and 2, respectively, η_{disk} and η_{laser} are relative noise factors compared with dc power level for disk noise and laser noise, respectively. Curves representing $N_{\text{disk}}{}'$ and $N_{\text{laser}}{}'$ in the figure are calculated as

$$N_{\text{disk}}{}' = (N_{\text{disk}}{}^2 + N_{\text{laser}}{}^2 + N_{\text{shot}}{}^2 + N_{\text{amp}}{}^2)^{1/2}, \tag{6}$$

$$N_{\text{laser}}{}' = (N_{\text{laser}}{}^2 + N_{\text{shot}}{}^2 + N_{\text{amp}}{}^2)^{1/2}. \tag{7}$$

It can be seen that N_{disk} and N_{laser} are dominant for larger θ_A values, because the laser power before entering the detector is relatively high. For $\theta_A = 0$, the noise is limited by N_{shot} for APD and N_{amp} for PIN. Since C/N reaches its maximum near $\theta_A = 0$, all these noises contribute to the total noise.

B. Highly Sensitive Polarization Detection Optics

Amplifier noise, N_{amp}, can be defined like this[6]:

$$N_{\text{amp}} = R_G\left\{4kT\Delta f\left[\frac{1}{R_G} + \frac{(2\pi f)^2}{3}(C_o + C_i)^2R_{\text{eq}}\right]\right\}^{1/2}. \tag{8}$$

Here, R_G is the gain resistance, k is Boltzmann's constant, T is the absolute temperature, Δf is the frequency bandwidth, C_o is the output capacitance of the photodetector, and C_i and R_{eq} are input capacitance and equivalent noise resistance of the first-stage FET. In the amplifier circuit used, the observed value of N_{amp} = −85 dBm is almost equal to that expected from Eq. (8).

To eliminate the influence of N_{amp}, all signals and noises except N_{amp} need to be augmented through increasing readout power P_0 or by multiplying photocurrent using an APD with $M > 1$. The first of these two approaches is limited by the maximum power so as not to disturb the written domain (more discussion of this point later). As regards the second approach, it can generally be stated that APD use is preferable to PIN use (also discussed later).

The increase in C/N compared to PIN, ΔC/N, is calculated as a function of M in Fig. 3. Here, it is assumed that P_0, η_{disk}, η_{laser}, ϵ, and N_{amp} have the same

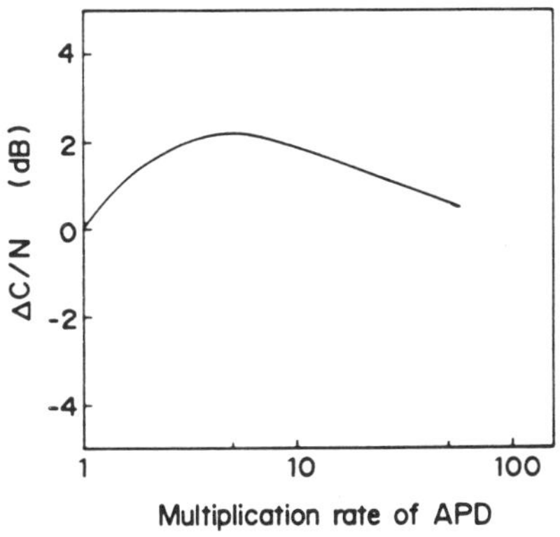

Fig. 3. Relationship between the increase of C/N and multiplication rate of APD (calculated).

values as in Fig. 2. The ΔC/N increases with M for smaller M values and reaches its maximum of ~2 dB at $M \sim 5$ (20 log$M$ = 15 dB). For $M > 6$, C/N decreases with M, since N_{shot} increases with $M^{1.13}$ more rapidly than do C, N_{disk}, and N_{laser}.

Increasing the laser beam readout power is effective in lessening the influence of N_{shot}, because N_{shot} is proportional to the square root of P_0, while N_{disk} and N_{laser} are proportional to P_0. A high readout power, however, requires high T_C values so that recorded data are not disturbed.

Diode laser noise, N_{laser}, varies from diode laser sample to sample and with output power. It also depends on such factors as optical feedback from the disk to laser, the distance between the disk and the laser, and temperature. It is common knowledge that an index-guided single-mode laser tends to oscillate in multiple longitudinal modes under substantial optical feedback and that laser noise increases with feedback.[7] In the case illustrated in Fig. 2, η_{laser} was found to be −78 dB (20 logη). In terms of RIN (relative intensity noise),

$$\text{RIN} = 10^{(-1/10 \log dc/N_{\text{laser}})}/\Delta f = 10^{(2\log \eta_{\text{laser}})}/\Delta f. \tag{9}$$

This value corresponds to RIN = 5×10^{-13} Hz^{-1}, which is expected as a typical value from past data on diode laser noise.[7] As for the magnetooptical readout, RIN should be less than this value, which is ~10^{-2} of the noise specification for a write-once disk to be used in digital recording and about 10 times that for an optical video disk.[7]

The extinction ratio in polarization detection optics plays an essential role in noise reduction for small θ_A values. The relationships of C, N, and C/N vs θ_A calculated for various extinction ratios are shown in Fig. 4. Here, P_0, η_{disk}, η_{laser}, N_{amp}, and M = 15 dB are assumed to be the same as in Fig. 2. When ϵ decreases, N decreases and C/N increases. As ϵ approaches 0, N_{disk} and N_{laser} decrease in parallel with $\theta_A{}^2$, while C and N_{shot} are proportional to θ_A for small θ_A values. Therefore, when θ_A decreases, N_{disk} and N_{laser} decrease more rapidly than do C and N_{shot}. At $\theta_A = 0$, C/N is determined by C/N_{shot} if N_{amp} can be neglected.

The decrease in C/N is related to the extinction ratio in Fig. 5. Here, M is changed to give the maximum

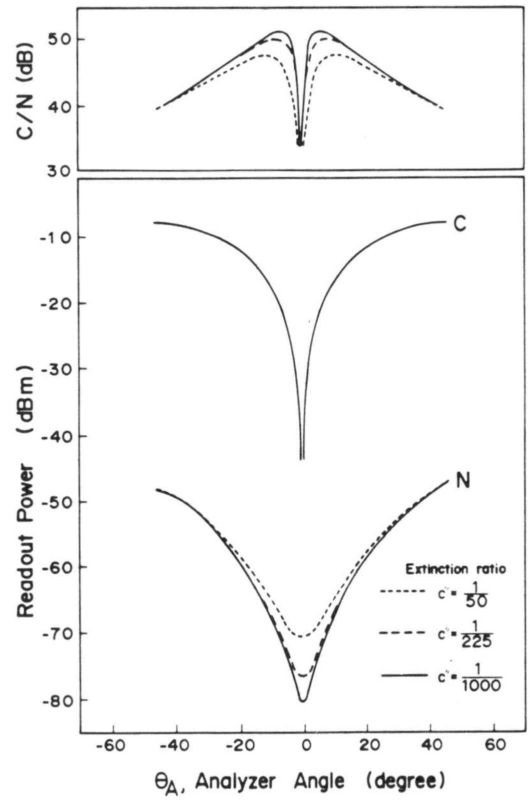

Fig. 4. Dependence of C, N, and C/N on the extinction ratio of polarization detection optics.

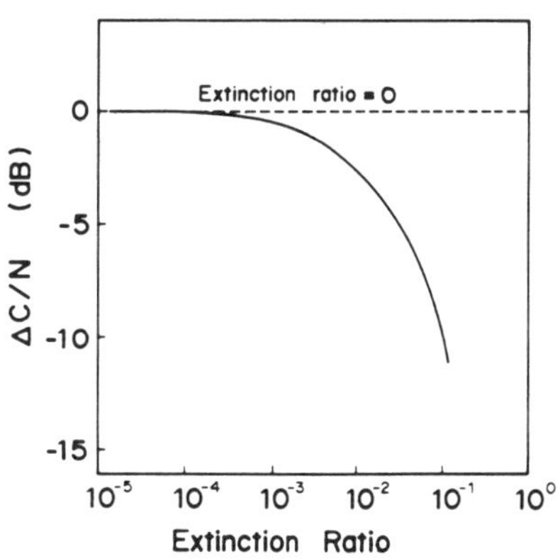

Fig. 5. Relationship between ΔC/N and extinction ratio (calculated).

C/N at a certain extinction ratio. Therefore, as ϵ approaches 0, C/N tends to be determined by C/N_{shot} in Fig. 5. For our optics of $\epsilon = 1/225$, Δ C/N is calculated as −2 dB. This suggests that a further improvement in extinction ratio assures a C/N increase of 2 dB at the most. To improve the extinction ratio, polarization properties for optical components must be carefully designed.

C. High SNR Recording Media

Disk noise has a major influence on magnetooptical disks. This noise is mainly due to fine roughnesses in the guide groove which reflect slight fluctuations in

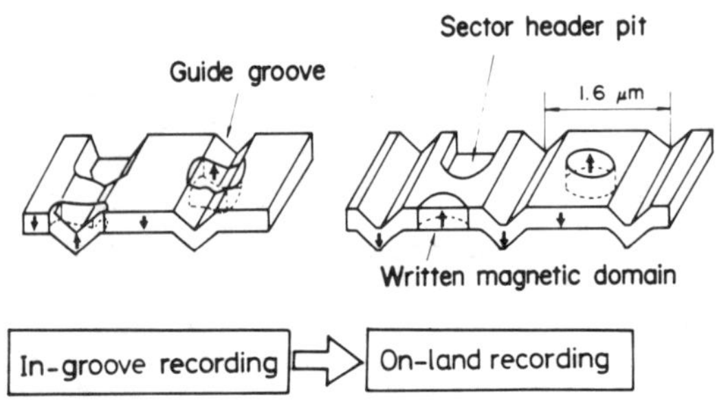

Fig. 6. Surface section of enlarged recording medium.

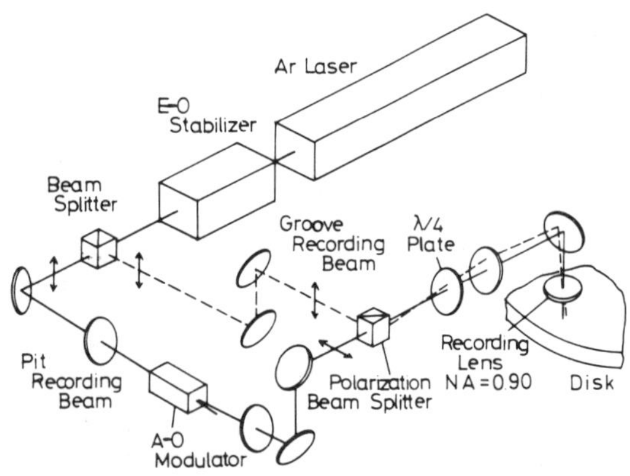

Fig. 7. Laser mastering optics for cutting a between-the-grooves recording disk.

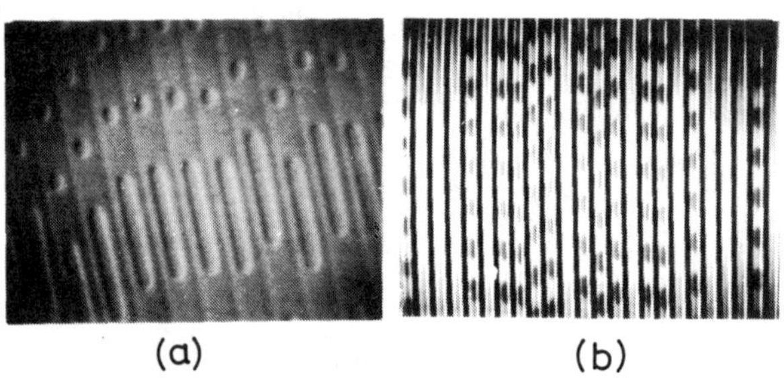

Fig. 8. (a) SEM photograph of sector headers and (b) polarization microscope photograph of written magnetic domains.

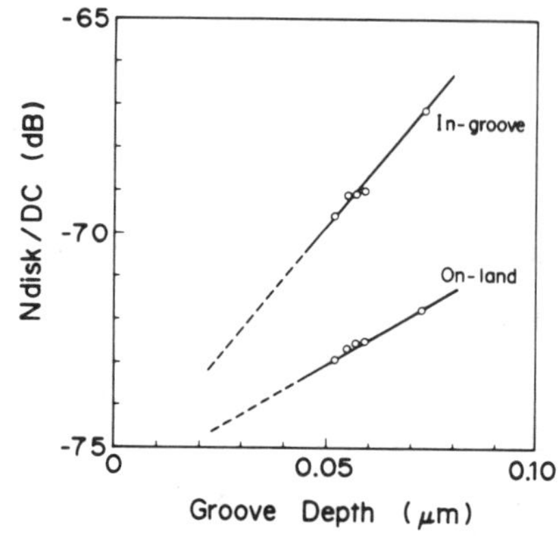

Fig. 9. Comparison of disk noise between in-groove recording and on-land (between-the-grooves) recording.

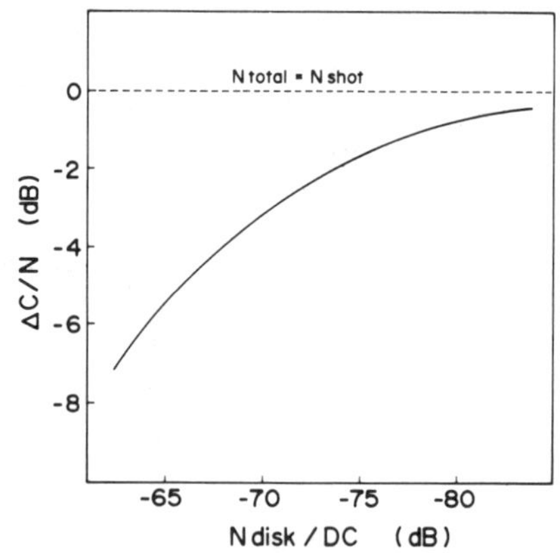

Fig. 10. Relationship between ΔC/N and disk noise (calculated).

laser light power during groove cutting. To avoid this noise, data are written on areas (referred to as plateaus) between this guide's grooves instead of in them.

Enlarged disk surface sections for in-groove and on-land (plateau) recording are shown schematically in Fig. 6. Sector header pits are also prerecorded between the grooves for the on-land recording disk using the newly developed mastering optics shown in Fig. 7. In these optics, the Ar laser beam is split into two beams. One is used for dc groove cutting and the other for pit recording. The two beams are combined by a polarization beam splitter and focused by a lens of N.A. = 0.9. The position of the dc groove recording spot is adjusted by a pair of mirrors placed in front of the polarization beam splitter. A scanning electron microscope (SEM) photograph of sector headers and a polarization microscope photograph of the written domains are shown in Figs. 8(a) and (b). Groove pitch is 1.6 μm.

A comparison between in-groove and on-land disk noise as a function of groove depth is shown in Fig. 9. In this figure, N_{disk} is measured at $\theta_A = 45°$ to eliminate the influence of N_{shot} and is normalized by dc power. The vertical axis represents 20 log η_{disk}. On-land N_{disk} is 4 or 5 dB lower than in-groove N_{disk} for a depth of ~λ/8. This leads to a 2- or 3-dB C/N increase at small values of θ_A.

The decrease in C/N (a comparison is made with the case without N_{disk}) is shown in Fig. 10. Here, P_0, ϵ, and N_{amp} are assumed to be the same as in Fig. 2. The N_{laser} is neglected and M is changed to give the maximum C/N at a certain η_{disk} value. As N_{disk} approaches 0, C/N tends to be determined by C/N_{shot}, as mentioned earlier. From this figure, a C/N increase of 2 dB may be expected from η_{disk} = −69-dB in-groove to −73-dB on-land. This agrees with the experimental results. On-land (plateau) disk noise for our replicated disk is still larger than that for a pure glass substrate, where the disk noise level is about −80 dB. Figure 10 suggests that further improvement in disk surface noise leads to a C/N increase of ~2 dB at most.

Recording material was developed using TbFeCo amorphous magnetic alloy originally studied by KDD.[8] By optimizing the composition of the three elements, high curie temperatures could be attained keeping θ_k = 0.35°. High T_C values allow a high power readout.

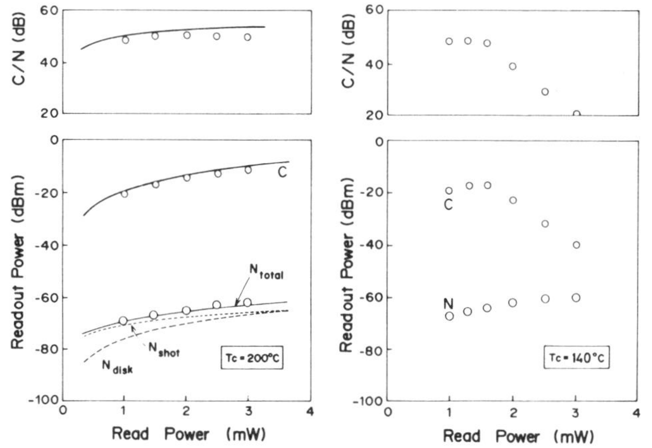

Fig. 11. Dependence of C, N, and C/N on reading laser power for T_C = 200°C (left) and 140°C (right).

This improves C/N by effectively reducing the shot noise. Dependences of C, N, and C/N on the reading laser power for the disk are shown in Fig. 11 for T_C = 200°C (left) and 140°C (right). The C/N increases ~2 dB with an increase of the reading laser power from 1 to 2 mW for recording material with a T_C of 200°C. However, the carrier level drops by more than a 2-mW reading power for material where T_C = 140°C, since written domains are disturbed by the readout laser power.

The disk substrate was also developed by choosing an epoxy resin whose thermal and optical properties are superior to those of PMMA or polycarbonate. Pregrooves and sector header pits were fabricated using the 2P (photopolymerization) method.

Typical examples of a readout signal and its associated frequency spectrum are shown in Fig. 12. A C/N value of 57 dB is obtained for f = 0.9 MHz using the technology described in this paper. This C/N value is sufficient for maintenance of a low error rate in coded data applications. Laser writing onto a disk requires ~8 mW, which is the same for write-once-type optical disks. This also facilitates compatibility.

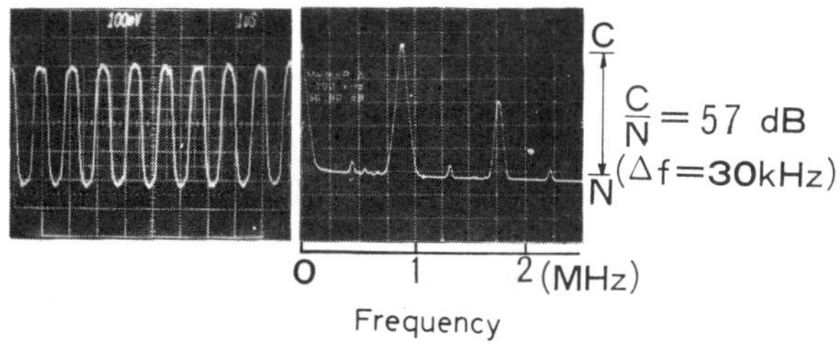

Fig. 12. Readout carrier waveform and corresponding spectrum.

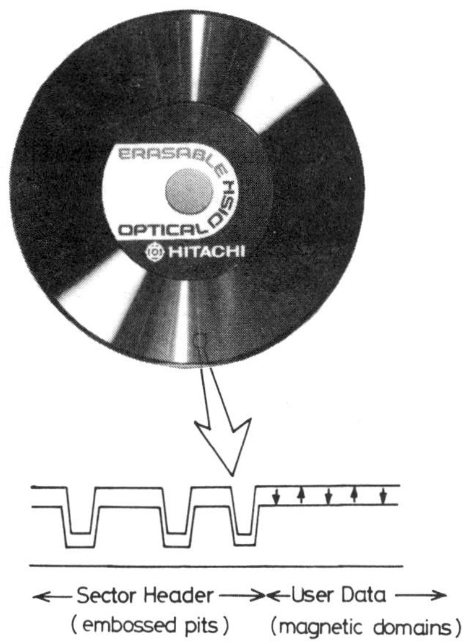

Fig. 13. Magnetooptical disk with sector header pits.

III. Writing/Erasing Data by Sectors

It is necessary to write or erase data by sector in coded data applications. The disk being described here has thirty-two sectors per track, as shown in Fig. 13. Each sector has header pits embossed on the area between the grooves. The optical head must pick up both sector header signals as light intensity modulation and user data as polarization rotation modulation.

A. Sector Header/User Data Detection Optics

Sector header signals and user data are separately picked up by two different detectors (Fig. 1). The laser beam passing through the analyzer contains both sector header signals and magnetooptical data, permitting one detector to pick up both components. However, the analyzer angle for maximum C/N is different from the angle needed to balance the amplitude and polarity between header and data. To solve this problem, the optical head combines the two signals

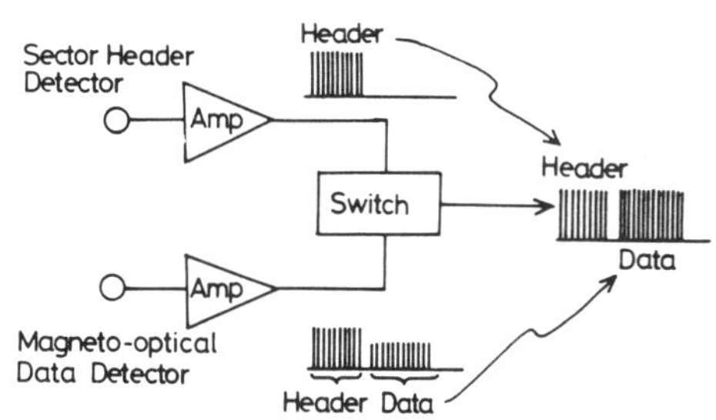

Fig. 14. Sector header/magnetooptical data detection and processing configuration.

using a switching circuit after separate detection. This is illustrated schematically in Fig. 14.

B. Magnetic Field Switching

A magnetic field is applied to the recording medium during laser pulse irradiation to write or erase data. This field generated by the electromagnetic coil must switch polarity between writing and erasing.

A time history for the magnetic field and laser pulses is shown in Fig. 15. Write/erase status signals are also presented. Rise time τ for the magnetic field is ~3 ms, which is determined from a coil inductance L of 300 mH and resistance R of 150 Ω. Since τ is larger than the time interval for a sector (1.6 ms), the magnetic field must exist prior to the laser pulses. Therefore,

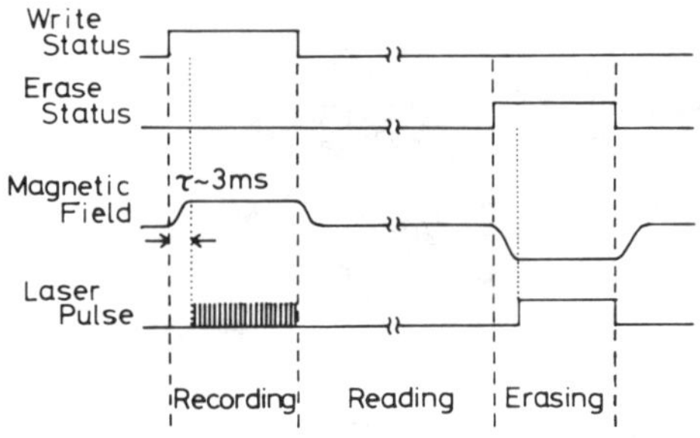

Fig. 15. Time chart pattern for magnetic field and laser pulses.

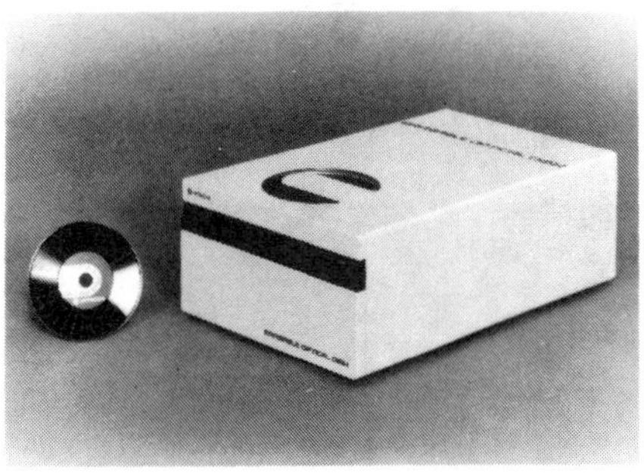
Fig. 16. Prototype compact magnetooptic disk and drive unit.

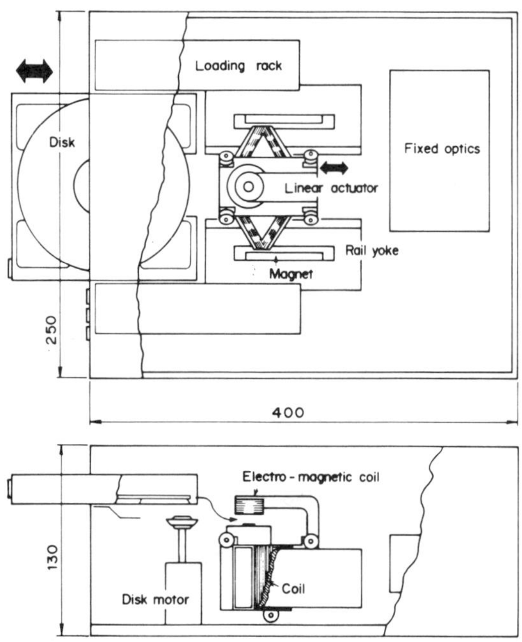

Fig. 17. Top view (top) and side view (bottom) block diagrams of the drive unit.

write/erase status signals are switched on two sectors (3.2 ms) before data can be written or erased. The magnetic field acting on the recording media is effectively the sum of the field generated by the external electromagnetic coil and a leakage field from the lens actuator. The initial magnetization direction of the recording layer must be set the same as the direction of the leakage field from the actuator. When using both sides of a disk, the initial magnetization directions for two recording layers should be opposite each other.

Table I. Prototype Specifications

Disk/drive comb.	Disk diameter	12 cm (~5 in.)
	Storage capacity	550 MB/both sides
	Data transfer rate	440 KB/s
	Average access time	100 ms
	C/N ratio (bandwidth 30 kHz)	57 dB
	Error rate (without ECC)	$\sim 10^{-6}$
	Error rate (with ECC)	$\sim 10^{-12}$
Disk only	Track number/side	17,000
	Track pitch	1.6 μm
	Sector number/track	32/track
	Total data capacity/sector	690 bytes/sector
	User data capacity/sector	512 bytes/sector
	Recording material	TbFeCo
	Curie temperature, T_C	>200°C
	Kerr rotation angle, θ_K	0.35°
	Disk substrate material	Epoxy resin
Drive only	Disk rotation speed	1200 rpm
	Error correcting code	BCH code
	Laser output power	20 mW
	Recording laser power (on disk)	7–8 mW
	Reading laser power (on disk)	1–2 mW

Two single-side recording disks were thus combined by adhesive after separately being started in terms of magnetization direction.

IV. Prototype Development

Prototypes of a compact magnetooptical disk and drive unit were developed from the technology described previously. Photographs of these devices are shown in Fig. 16.

Top and side views of the drive unit are shown in Fig. 17. The disk is loaded from the front; the electromagnetic coil plus a focusing lens is moved by the newly developed linear actuator. Other major optics are fixed to the base. By separating the optics into two parts, high access speed is achieved; average access time is 100 ms.

Prototype specifications are listed in Table I. The disk diameter is the same as for currently available compact digital audio disks. The storage capacity of 550 M-bytes was obtained by using both sides of the disk. This value is 500 times the capacity of 13.34-cm floppy disks now in use. Initial error rate without correction is $\sim 10^{-6}$, which is almost the same rate for a current write-once optical disk. With error correction, however, a value of 10^{-12} is attained, which is adequate for coded data applications.

V. Conclusions

Noise in a magnetooptical disk has been experimentally analyzed into disk noise, diode laser noise, shot noise, and amplifier noise to determine approaches effective in increasing the SNR. Highly sensitive polarization detection optics could be constructed using an APD detector, high extinction-ratio optical components, and a high-reading-power laser. To obtain recording disk media with high-level SNR characteristics, data have been recorded between the guide grooves, and a perpendicular magnetic recording film with a high curie temperature (>200°C) has been developed by optimizing the composition of the TbFeCo alloy. Furthermore, to assure sector-by-sector writing and erasure of data, it has been made possible for sector header signals to be detected separately from magnetooptical user data and for the magnetic field to switch on prior to laser pulse irradiation.

Successful prototypes of a compact magnetooptical disk and its drive unit have been developed for coded data storage. The drive unit, as developed, can also play write-once optical disks in current use. The 12-cm diam disk developed can hold 550 Mbytes of data or 500 times the storage capacity of an ordinary 13.34-cm floppy disk. The bit error rate was improved to 10^{-12} with error correction coding, making the system as reliable as magnetic disks and tapes in current use. The systems compact size and large storage capacity make it an ideal storage device for personal computer and office automation applications when the need arises to upgrade from a floppy disk or small rigid disk drive.

The authors wish to express their sincere thanks to Tsuyoshi Kato and Masafumi Yoshihiro for help with experiments and calculations, Yoshinori Miyamura for disk preparation, and to members of the optical disk laboratory group for their assistance in construction of the prototypes.

References

1. Honeywell, U.S. Patent 3368209 (22 Oct. 1964).
2. P. Chaudhari, J. J. Cuomo, and R. J. Gambino, "Amorphous Metallic Films for Bubble Domain Applications," IBM J. Res. Dev. **17,** 66 (1973).
3. N. Imamura and C. Ota, "Experimental Study on Magneto-Optical Disk Exerciser with the Laser Diode and Amorphous Magnetic Thin Films," Jpn. J. Appl. Phys. **19,** L731 (1980).
4. Y. Tsunoda, S. Horigome, and Z. Tsutsumi, "Optical Digital Data Storage Technologies with Semiconductor Laser Head," Proc. Soc. Photo-Opt. Instrum. Eng. **382,** 24 (1983).
5. K. Ohta, A. Takahashi, T. Deguchi, T. Hyuga, S. Kobayashi, and H. Yamaoka, "Magneto Optical Disk with Reflecting Layers," Proc. Soc. Photo-Opt. Instrum. Eng. **382,** 252 (1983).
6. K. Horita, S. Kubo, and K. Takahashi, "Design of Junction FET for Video Camera," Telev. Eng. Jpn. **TEBS74-1,** 1 (1980).
7. M. Ojima, A. Arimoto, N. Chinone, T. Goto, and K. Aiki, "Diode Laser Noise at Video Frequencies in Video Disk Players," Appl. Opt. **25,** (1986), to be published.
8. F. Tanaka, Y. Nagao, and N. Imamura, "Dynamic Read/Write Characteristics of Magneto-Optical TbFeCo and DyFeCo Disk," IEEE Trans. Magn. **MAG-20,** 1033 (1984).

Reprinted with permission from *Applied Optics*, Vol. 25(9), pp. 1398-1403 (May 1, 1986). ©1986 Optical Society of America.

Optimum conditions for the high frequency noise reduction method in optical videodisc players

Akira Arimoto, Masahiro Ojima, Naoki Chinone, Akio Oishi, Toshihoko Gotoh, and Nobutaka Ohnuki

In a previous paper, the authors demonstrated the usefulness of the high frequency (HF) current superposition noise reduction method. This paper discusses optimum conditions for the HF method. The main results obtained are that modulation current frequency must be higher than 600 MHz and modulation depth must be >120%. The optical path length between laser and disk and the removal for interference in the signal detectors must also be taken into consideration.

I. Introduction

The diode laser used in optical videodisc players causes serious noise in the video frequency range[1] as well as in the servo control frequency range.[2]

In the authors' previous paper,[3] the problem of diode laser noise in optical videodisc players and the behavior of the diode laser were discussed. It was shown that the high frequency (HF) noise reduction method is the best way to solve the noise problem, and that it can be successfully used in optical videodisc players.

In this paper, the optimum conditions for reducing the noise and the mode behavior of the diode laser in optical videodisc players will be discussed.

II. Optimum Conditions for the High Frequency Current Superposition Noise Reduction Method

We have reported[3] that the HF method can be used in optical videodisc players. In this section, we examine the optimum conditions for suppressing the noise to less than RIN $< 3 \times 10^{-14}$ (Hz^{-1}) by the HF method.

The allowable noise upper limit of the diode laser itself can be assumed to be 3×10^{-14} (Hz^{-1}) as a result of the playback experiments conducted on videodisc players.[3]

All the authors are with Hitachi, Ltd., Japan; T. Gotoh is at Consumer Research Center, Totsukaku, Yokohama, Kanagawa 244; N. Ohnuki is at Yokohama Works, Totsukaku, Yokohama, Kanagawa 244; the other authors are with Central Research Laboratory, Kokubunji, Tokyo 185.

Received 2 August 1985

0003-6935/86/091398-06$02.00/0.

A. Drive Current Frequency Properties

First, let us examine the drive current frequency dependence properties of the laser noise in the HF method. Figure 1 shows the noise level vs the temperature when the drive current frequency is changed. When the drive current frequency is low, the noise level has sharp peaks. By increasing the frequency, the noise level becomes stable and flat. Figure 2 represents the drive current frequency properties of the noise when the optical length between laser and optical disk is 70 and 85 mm. The noise level values in the figure indicate the worst level obtained in the temperature range between 10 and 50°C.

In the drive frequency range of <500 MHz, the noise defined as RIN cannot be reduced sufficiently, and it is sensitive to optical feedback. By increasing the drive current frequency to more than 600 MHz, the RIN can be kept to $<3 \times 10^{-14}$ (Hz^{-1}), and it stabilizes against the optical feedback.

Figure 2 shows that the noise increases again near 1 GHz. The reasons for this will be discussed in Sec. II.D.

Next, let us observe the longitudinal modes of the diode laser (Fig. 3) for HF modulation. At 300 MHz, the diode laser is already oscillating in a multimode state with about ten modes oscillating. By increasing the drive frequency, the number of oscillating modes increases and the mode profile envelope acquires a quasi-Gaussian shape. None of the modes oscillates more strongly at 600 and 800 MHz.

B. Modulation Depth Properties

The relationship between the HF current modulation depth and the noise is presented in this section.

The modulation depth (MD) in this paper is defined as

$$MD = \frac{\text{average laser power with HF current}}{\text{laser power with dc drive}} \times 100 \qquad (>100\%), \tag{1a}$$

$$MD = \frac{\text{lowered average power when the HF current reaches the threshold}}{\text{original average laser power}} \times 100 \qquad (<100\%). \tag{1b}$$

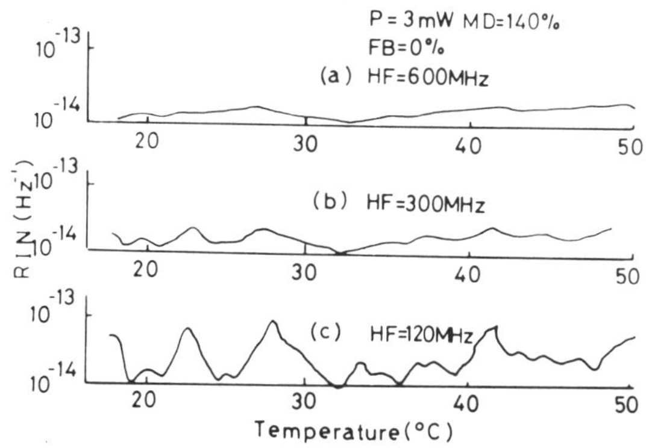

Fig. 1. Temperature dependence properties of the high frequency current superimposition noise reduction method: (a) HF = 600 MHz, (b) HF = 300 MHz, and (c) HF = 120 MHz.

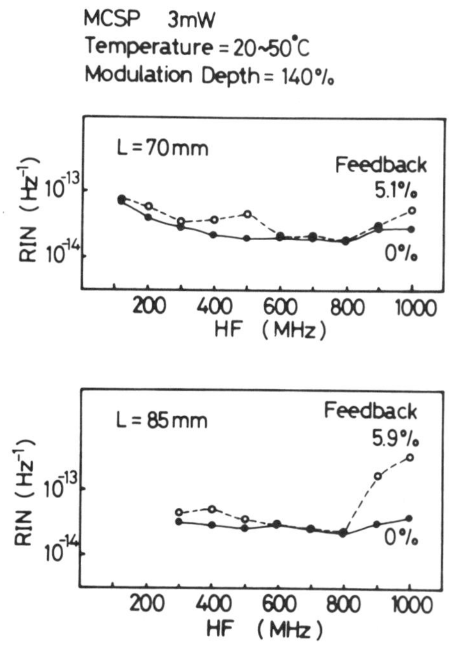

Fig. 2. Drive current frequency dependence of diode laser noise.

The diode laser is driven with HF current as shown in Fig. 4. If the HF current drives the laser below its threshold, a pulsed laser putput is obtained. In this case, the average output power is larger than the laser power without HF modulation. This means that the MD is over 100%, and Eq. (1a) applies.

If the HF current is adjusted so that the threshold current is never crossed, a sine waveform output is obtained and the average output power does not change as the HF current varies. We then lower the bias point of the current until the lowest value of the HF drive current reaches the threshold value. The average power is now lower than the original level. We define the MD according to Eq. (1b) when it is <100%.

Figure 5 shows the relationship between the MD and the corresponding laser noise. When the MD is <100%, the noise cannot be reduced sufficiently; the RIN exceeds 3×10^{-14} (Hz^{-14}). However, in the case of more than 120% modulation, the RIN becomes lower than 3×10^{-14} (Hz^{-1}) and is insensitive to optical feedback within the 10–15°C temperature range.

Next, let us examine the mode behavior (Fig. 6).

At 50% MD, one particular mode oscillates strongly, and it oscillates even stronger when optical feedback is applied. At 100% MD, the diode laser oscillates in a

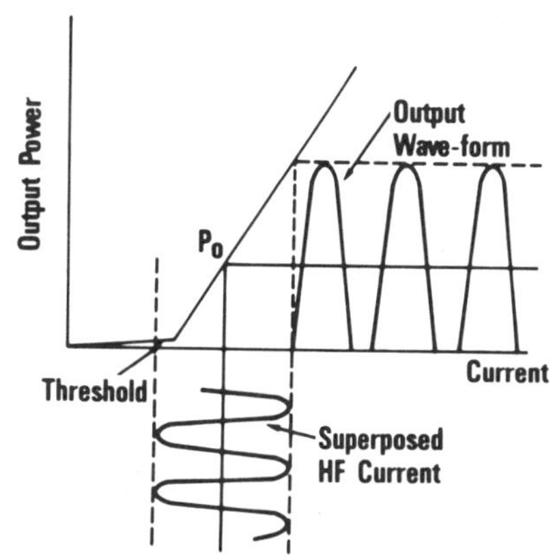

Fig. 4. Driving method of a diode laser with a high frequency current superimposition.

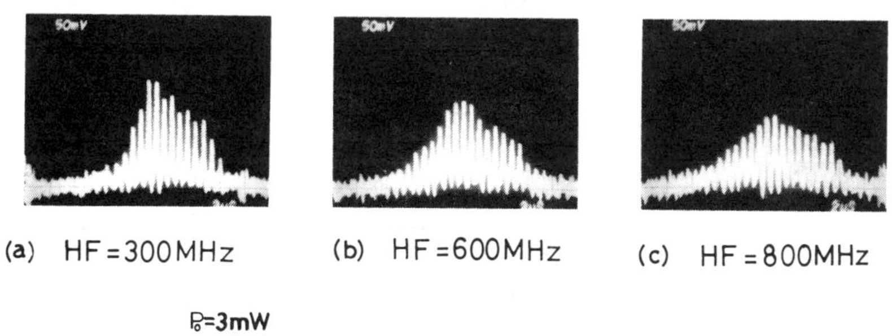

Fig. 3. Mode profile of diode laser vs superimposed drive current frequency: (a) HF = 300 MHz, (b) HF = 600 MHz, and (c) HF = 800 MHz.

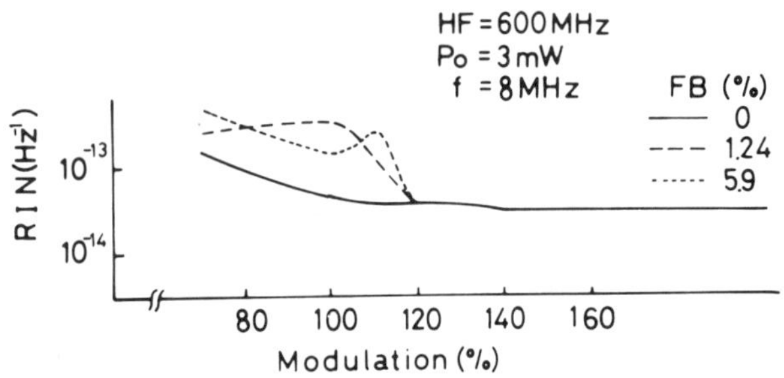

Fig. 5. Modulation dependence of diode laser noise: (a) M = 50%, (b) M = 80%, and (c) M = 120%.

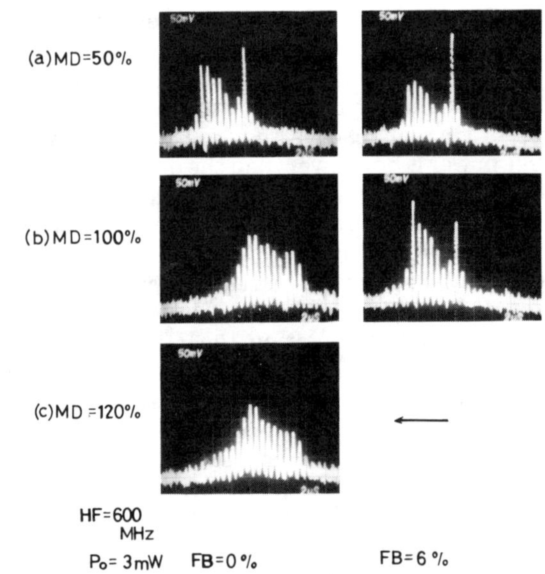

Fig. 6. Mode profile of diode laser vs modulation depth: (a) M = 50%, (b) M = 100%, and (c) M = 120%.

multimode state without optical feedback. With optical feedback, some modes oscillate more strongly and they are sensitive to optical feedback. This suggests that mode competition noise is induced by the optical feedback. At 120% MD, the diode-laser mode profile does not change with the addition of optical feedback. More than fifteen modes oscillate stably. This indicates that more than 120% MD is necessary to obtain reduction in the noise level. It is well known that, at the oscillation threshold level, diode lasers oscillate in a multimode state. It is therefore important to make sure that the drive current crosses the threshold level.

Mode observation will be discussed in more detail in Sec. III.

C. Optical Path Length Dependence Properties

Here we discuss the optical path length dependence of this noise reduction method.

In Sec. II, it was shown that the noise increases near 1-GHz frequency due to the relationship of the optical path length between laser and disk vs the applied drive current frequency. Figure 7 shows the optical path dependence in the cases of 600- and 915-MHz HF current modulation.

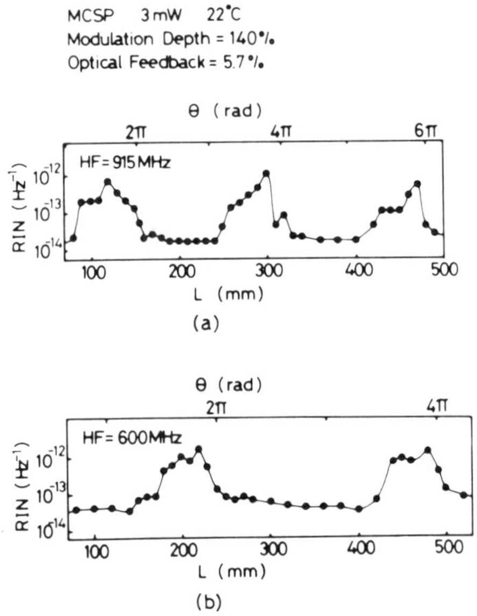

Fig. 7. Optical path dependence of diode laser noise: (a) HF = 915 MHz and (b) HF = 600 MHz.

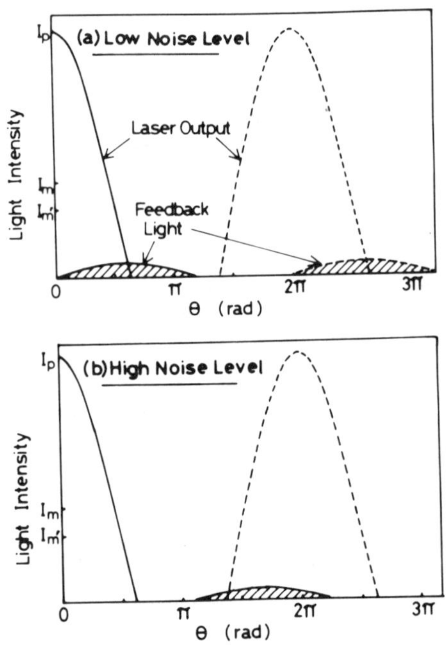

Fig. 8. Diode laser noise vs normalized optical path length: (a) low noise level and (b) high noise level.

The peak of the noise level appears cyclically in both cases. θ, defined as $2\pi[(\mathrm{HF} \cdot 2L)/c]$, is the normalized optical path length between the laser and the optical disk. 2π in the figure corresponds to the pulse repetition period of the diode laser output. These figures indicate that the high noise level appears at the $2m\pi$ point, where m is an integer. This means that the noise increases when the laser beam reflected from the optical disk coincides with the leading edge of succeeding output pulses at the resonator. Figure 8 clarifies this relationship.

To reduce the noise, the optical path must be carefully chosen in accordance with the drive current fre-

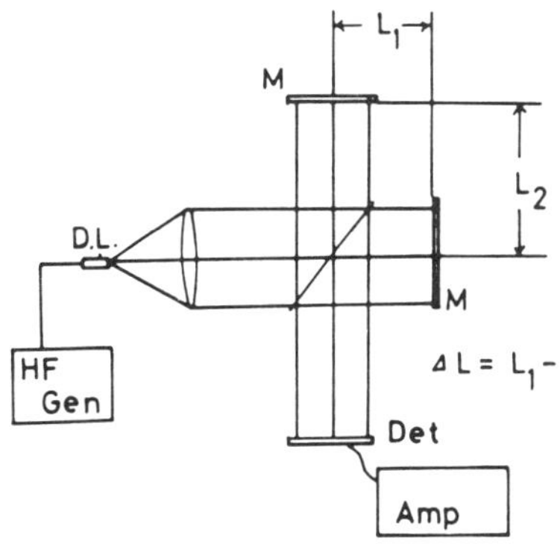

Fig. 9. Schematic of measuring the interference effect of diode laser noise.

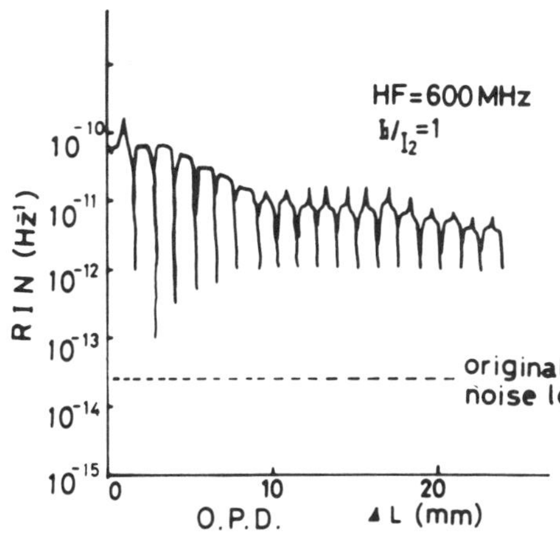

Fig. 10. Optical path difference properties of diode laser noise under interference. The intensity ratio of two beams is unity. The dashed line shows the laser noise without interference.

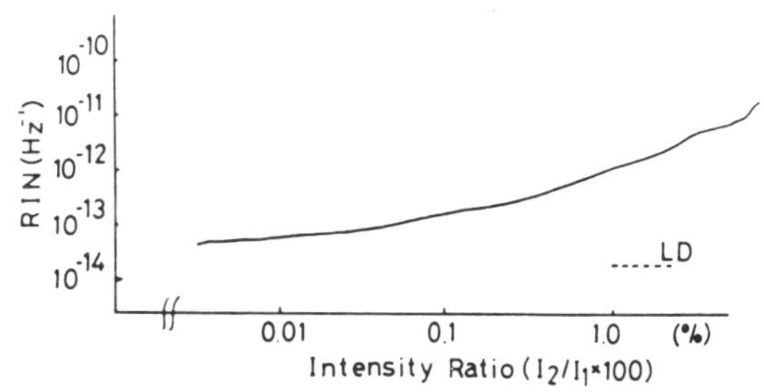

Fig. 11. Diode laser noise properties under interference vs the intensity ratio of two beams. The dashed line shows the diode laser noise without interference.

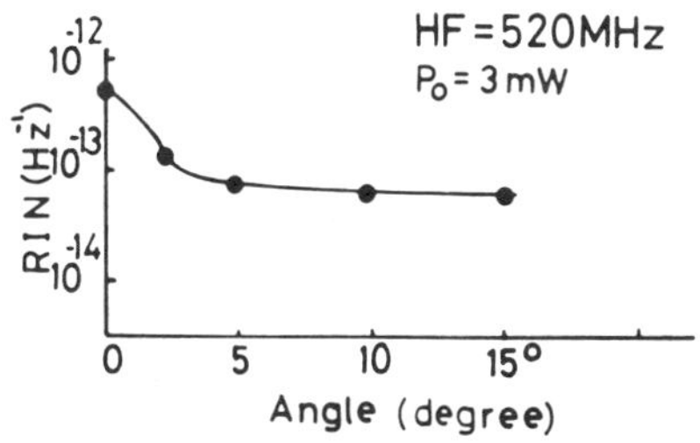

Fig. 12. Interference effect of diode laser noise due to the glass cover on the detector.

quency to prevent the reflected beam from coinciding with the succeeding optical pulses.

D. Interference Properties

One of the important reasons this HF method can be used in videodisc players is that the coherence of the emitted laser beam is lower than that of a single-mode laser, which means that this method is insensitive to optical feedback and does not cause control frequency noise.[2] However, we found that the interference effect in the video frequency range must be taken into consideration when using this method. The interference effect of gain-guided lasers has been reported by Petermann and Weidel.[4] Their paper stated that noise level increases due to mode partition and phase laser intrinsic noise.

We examined the interference properties of this HF method by using a Michelson interferometer as shown in Fig. 9. The diode laser is driven by HF current. Figure 10 shows the relationship between the optical path difference of two arms ($\Delta L = L_1 - L_2$) and the noise.

In the presence of the interference, the noise level changes cyclically and drastically. The cycle of the sharp peak of the noise level coincides with the resonator length of the diode laser which is 1.2 mm in our case. Our result is very similar to Petermann's for gain-guided lasers, i.e., the HF method is also influenced by interference.

Figure 11 shows the noise level under interference when the intensity ratio of the two beams in the interferometer varies. In this experiment, the path difference between two beams is set so as not to coincide with multiples of the laser resonator length. As the ratio increases, the noise level increases. These two results indicate that interference must be removed even for videodisc players using the HF noise reduction method.

Next let us examine the negative effect which interference has on videodisc players.

We tilted the glass covered signal detector away from the perpendicular position. The relationship between the tilt angle and the noise is shown in Fig. 12. At a 0 angle (i.e., the detector is perpendicular to the optical axis), the noise level is highest, because the beam reflected from the glass cover reaches the signal detector and causes interference between this beam and the signal light. This interference raises the noise level. To avoid this noise increase, the detector must be tilted toward the optical axis more than 7°.

However, no noise is caused in the control frequency range at any angle with this method, because visibility is very low, so no interference pattern can be observed on the signal detectors due to multimode oscillation.[2]

III. Precise Observation of the Mode Spectra

In Sec. II, we showed the optimum conditions for the HF noise reduction method in videodisc players and said that the HF method modulation current forces the diode laser to oscillate in a multimode state. In this section, we investigate and observe how these multimode oscillations are generated.

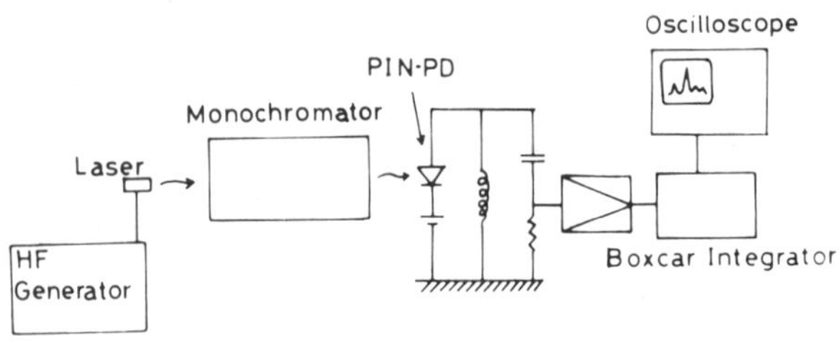

Fig. 13. Schematic of measuring longitudinal modes.

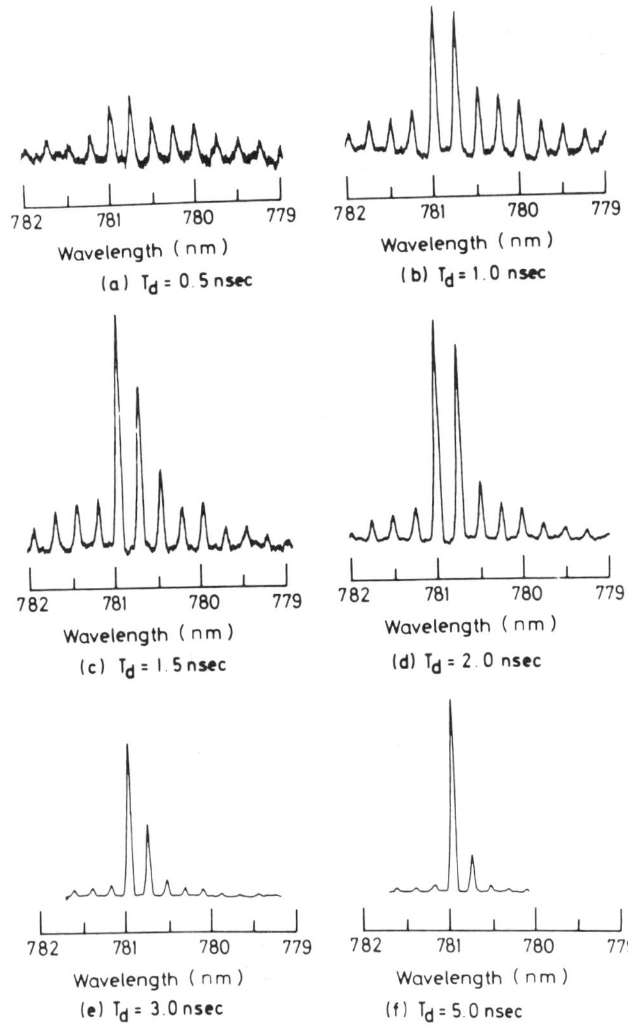

Fig. 14. Longitudinal mode behavior when short pulses are applied. T_d denotes the period between oscillation start time and observation time.

First, let us examine the mode behavior when driving current is applied. Figure 13 is a block diagram of the apparatus. It is well known that the diode laser oscillates in a multimode state at the threshold current level.[5] Rectangular current pulses are superimposed on the dc current which is set equal to 0.9 times the threshold value. The pulse width is 20 ns and the pulse height is 20 mA. T_d denotes the time of which the oscillation of the diode laser starts. Several modes oscillate with nearly equal intensity when T_d is 0.5 ns as in Fig. 14(a). As the time (T_d) passes (i.e., as the current increases), one or two particular modes oscillate strongly [Figs. 14(b)–(e)]. Finally, the laser oscillates in nearly a single mode, and other modes are suppressed [Fig. 14(f)].

When $T_d < 1.0$ ns, we assume that the laser oscillates in the multimode state. It was shown that more than 600-MHz modulation frequency is necessary for sufficient noise reduction. One must assume that these two facts are strongly related to each other.

Next, let us examine how each mode oscillates by the superimposed current pulse. Each mode is resolved by the grating monochromator and observed with an oscilloscope. The boxcar integrator in Fig. 13 is removed in this experiment, and the PIN detector output is fed directly to the oscilloscope. Figure 15(a) shows the real-time total output laser pulse when the

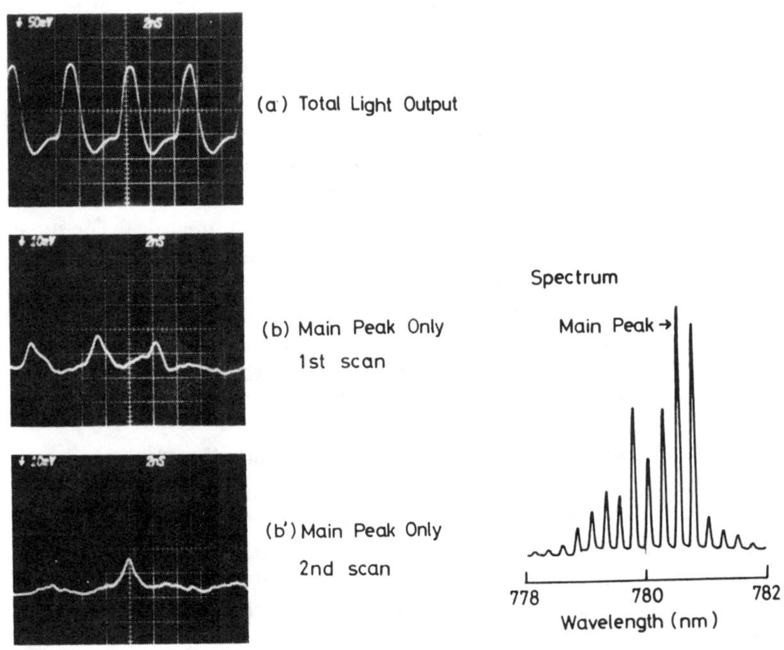

Fig. 15. Output response of main mode: (a) total output, (b) main mode (first scan), and (b′) main mode (second scan).

driving pulse current is applied. In each driving pulse these waveforms are constant. However, we observe one particular mode that oscillates most strongly in Fig. 15. Figure 15(b) shows the output waveform of this mode as well as its behavior. The observation time is different. In this figure, three optical pulses are observed when four current pulses are applied. In Fig. 15(b′) only one optical pulse is observed. This behavior indicates that even the strongest mode does not always oscillate when current pulses are applied. These characteristics are quite different from those of a gain-guided multimode laser.

The above observation shows that every pulse oscillates independently and each mode in a pulse has no correlation with that mode in other pulses. Therefore, the oscillating mode profile is stable at video frequency and mode competition cannot occur at video frequency in a HF-driven laser unless the feedback pulse coincides with the leading edge of succeeding output pulses. The reflected beam pulse at the start of the laser oscillation tends to lock the oscillation spectrum so that every pulse tends to oscillate in the same spectrum for a certain amount of time. However, the oscillation mode spectrum can be changed at video frequency in a way similar to the mode competition in a multimode laser. Therefore, optical feedback can induce mode competition noise in a HF-driven laser if the feedback pulse coincides with the start of the pulsed oscillation.

IV. Conclusion

This paper discusses how a high frequency current superimposition noise reduction method can be used in videodisc players. The results obtained are:

(1) Drive current frequency must be larger than 600 MHz.

(2) Modulation depth (MD) must be higher than 120% to obtain pulse laser output.

(3) Due to variations in the drive current frequency, the optical path between diode laser and disk must be carefully chosen to avoid overlap between the returning laser beam and the succeeding output pulse.

(4) Interference must be removed from the detectors.

We would like to thank M. Nakamura, Z. Tsutsumi, K. Akatsu, and T. Chiba of Hitachi, Ltd. for their useful comments and constant encouragement.

References

1. T. Gotoh, A. Arimoto, M. Ojima, and H. Chinone, "Characteristics of Laser Diodes and Picture Quality," Proc. Soc. Photo-Opt. Instrum. Eng. **329,** 11 (1982).
2. A. Arimoto and M. Ojima, "Diode Laser Noise at Control Frequencies in Optical Videodisc Players," Appl. Opt. **23,** 2913 (1984).
3. M. Ojima, A. Arimoto, N. Chinone, T. Gotoh, and K. Aiki, "Diode Laser Noise at Video Frequencies in Optical Videodisc Players," Appl. Opt. **25,** 1404 (1986).
4. K. Petermann and E. Weidel, "Semiconductor Laser Noise in an Interferometer System," IEEE J. Quantum Electron. **QE-17,** 1252 (1981).
5. T. Ikegami, "Spectral Broadening and Trailing Effect in Directly Modulated Injection Lasers," in *Technical Digest, First European Conference on Optical Communication*, London (1975), p. 111.

Reprinted with permission from *Optical Engineering*, Vol. 25(7), pp. 881-891 (July 1986). ©1986 Society of Photo-Optical Instrumentation Engineers.

Signal, noise, and codes in optical memories

David Treves, MEMBER SPIE
Weizmann Institute of Science
Department of Electronics
Rehovot, Israel 76100

Dan S. Bloomberg, MEMBER SPIE
Xerox Corporation
Palo Alto Research Center
Palo Alto, California 94304

Abstract. An analysis is given of the effect of additive and multiplicative noise in the data and servo channels of both write-once and magneto-optic recording systems. Electronic (additive) noise is most severe in the magneto-optic data channel, and laser noise is of particular concern in both the sum tracking channel and the write-once data channel. In a typical optical memory system the maximum laser light fluctuation should never exceed 10% of the average light level. In the differential channel of such a system, the laser noise should be further reduced by another order of magnitude. In the data channel, nonreturn-to-zero (NRZ) encoding has the greatest immunity to noise. Analysis is made of the extent to which the S/N margin is lost when various run-length-limited codes are used, for both additive and multiplicative noise.

Subject terms: magneto-optic recording; magneto-optics; optical recording; optical memory.

Optical Engineering 25(7), 881-891 (July 1986).

CONTENTS

Paper 2203 received Sept. 19, 1985; revised manuscript received March 29, 1986; accepted for publication March 29, 1986; received by Managing Editor April 16, 1986.

1. INTRODUCTION

The purpose of this paper is to clarify allowable noise levels in the data, focus, and tracking channels of digital optical disk memories. Write-once (WO) and erasable magneto-optic (MO) systems have much in common and are therefore treated together.

The various signals may appear as a sum or as a difference of optical power on photodetectors. For example, the data signal is given as the sum in WO,[1] whereas in MO it is usually the difference.[2] The focus channel is given as the difference for both WO and MO,[1,3] while for the tracking channel either the sum[4,5] or the difference[1,6] may be used. The noise components are either additive or multiplicative, and their effect on the channel depends on whether a sum or difference is used. Examples are given in Secs. 2 and 3.

In optical recording, there are three major types of noise: electronic, medium, and laser noise.

Electronic noise, which includes Johnson and excess electronic noise in the amplifiers and shot noise in the detectors, is Gaussian and additive. With the exception of the shot component, which is proportional to the square root of the optical power,[2] electronic noise is independent of the signal.

Medium noise is primarily additive, stationary (unchanging for any place on the disk), non-Gaussian, and again exists independently of the signal. It depends on the quality of the film(s), on the smoothness of the surface, and, for recording through the substrate, on the substrate material as well (polycarbonate, methyl methacrylate, or glass). Experience with WO media shows that although there is little rms noise, the amplitude distribution is anomalous, resulting in raw error rates of 10^{-5} to 10^{-4}. Although there is presently insufficient information about MO media to give meaningful error rates, it appears that medium noise is no greater for MO than for WO.

Laser noise is due essentially to fluctuations of the light intensity, and it is usually multiplicative. However, there may be a nonmultiplicative type of laser noise associated with interference of light beams that reach the detectors via different paths. Fast fluctuations of the phase difference of the light beams may result in intensity noise in the frequency band of the data.[7] Laser noise, like medium noise, does not have a Gaussian distribution at large amplitudes. Rather, it probably has a sharp cutoff at some maximum amplitude.

Laser noise is caused by light reflected from optical components or the film medium back into the laser. The physics of this effect, although not completely understood, has been amply discussed in the literature. We shall briefly describe here only those observations that are directly relevant to optical memory systems.

In the absence of reflected light, the laser usually oscillates in a single longitudinal mode.[8] The reflected light causes the laser to oscillate in several modes, and its intensity fluctuates in a noisy manner, presumably due to a fast energy transfer between modes. In addition to the light level fluctuations, one can also expect a frequency variation of the light because the modes differ in wavelength from each other.

In general, with increasing laser current and for a given fraction of reflected light, the noise increases until it reaches a maximum somewhere between threshold and about 2 mW. The noise then goes through a minimum and rises again at high power. For a given laser power, say 3 mW, the following is observed as the fraction of reflected light is increased: At well under 1% reflectivity, the noise is non-Gaussian, composed mainly of large amplitude spikes occurring at low frequency and probably due to mode hopping between a small number of modes. Between 1% and 10% reflectivity, the noise is usually Gaussian-like, with a steady average value. At higher reflectivities, the average light level often has two metastable levels, presumably associated with different mode structures. These states are possibly induced by fluctuations in the phase of the reflected light. The laser is sometimes noisy at one metastable level and quiet at the other. In the former state, the noise does not appear at all Gaussian, being composed primarily of large spikes. In WO systems, one can use an optical isolator to optimize the amount of light feedback. In MO systems, however, optical isolation is impracticable, and the amount of feedback is usually large, determined by medium reflectivity. For such systems, it may be necessary to superimpose an rf current with a frequency of several hundred MHz on the dc current driving the laser.[9] Turning the laser off once during each cycle forces it into many modes, reducing the noise to an acceptable level.

The WO data, focus, and tracking servo signals are derived from a relatively large modulation of the incident optical power and are usually not affected by electronic noise. On the other hand, the MO signal is due to the magneto-optic Kerr effect, and because it is small, electronic noise must be considered.

The requirements on the signal-to-noise ratio (S/N) depend on the type of encoding that is implemented. For a given minimum feature size, determined by the optics, run-length-limited (RLL) codes[10,11] can increase the encoded data density at the cost of decreasing the size of the timing window. When timing errors due to electronic noise and transition and clocking jitter are small compared to the window, performance is indeed limited by the minimum feature size, and RLL codes can be very useful.

Electronic noise is relevant mainly to MO and is therefore treated in Sec. 2, together with the MO signal. Laser noise, on the other hand, affects all data and servo channels. It is discussed in Sec. 3 for nonreturn-to-zero (NRZ) encoding. The interdependence of timing window, bit density, and required S/N for various codes, in the presence of both additive and multiplicative noise, is covered in Sec. 4.

2. ELECTRONIC NOISE AND SIGNAL IN A MAGNETO-OPTIC DATA CHANNEL

In this section, electronic noise and signal in an MO system are analyzed, showing the interrelation of the reflectivity, signal, and contrast. The basic figure of merit for MO media is explained, and a convenient method is given for its measurement. The light level that reaches the detector in an MO system is more than an order of magnitude smaller than in a WO system. Therefore, the front-end detector amplifier must have very low noise, and it should approach as closely as possible a shot-noise-limited situation.

2.1. Basic noise relations

In order to relate all noise expressions and the signal in common units, it is convenient to convert all noises to an equivalent optical power at the input to the photodetector (PD). This is the commonly used noise equivalent power (NEP) that often specifies photodetectors. In the following equations,

e = 1.6×10^{-19} C, the electron charge,
η $\simeq$ 0.5, the quantum efficiency of the PD (A/W),
P = the optical average power on each PD (W),
f = the bandwidth of the electronics (Hz),
kT = 4.2×10^{-21} (J),
r = the load resistor of the PD (Ω),
G = the PD photocurrent-to-voltage gain of each amplifier (V/A),
U_n = the noise voltage at the output of each amplifier (V),
P_i = the optical power incident on the disk (W),
T = the optical transmission from disk to photodiodes,
R = the reflectivity of the disk,
ρ = the fraction of p (parallel) component of electric field diverted by the polarizing beam splitter,
P_s = peak-to-peak signal on each diode for bulk switching (W),
I_s = signal current in the PD module (A),
I_{sh} = shot noise current in the PD module (A),
E_p = the p (parallel) component of electric field at the input of the PD module ($W^{0.5}$), and
E_s = the s (perpendicular) component of electric field at the input of the PD module ($W^{0.5}$).

The shot noise is given by Ref. 2 as

$$\text{NEP (shot)} = (2e\eta Pf)^{1/2}\,\eta^{-1} \quad \text{(W)} , \tag{1}$$

where the term in parentheses is the shot noise current I_{sh}.

The Johnson noise in the photodetector load resistor is given by Ref. 2 as

$$\text{NEP (Johnson)} = \left(\frac{4kTf}{r}\right)^{1/2} \eta^{-1} \quad \text{(W)} . \tag{2}$$

The term in parentheses is the Johnson noise current.

The *measured* electronic noise includes the Johnson noise. The latter has to be subtracted in order to obtain the *excess* electronic noise of the amplifier:

$$\text{NEP (electronic)} = U_n(G\eta)^{-1} \quad (W) \ . \tag{3}$$

Because the various noise contributions are uncorrelated, their powers are additive. It is useful to calculate in relative dB terms how two noise sources add up. Assume a noise source P_1 with rms value $<P_1>_{rms}$ at 0 dB, and a second noise source P_2 with $<P_2>_{rms}$ at some number of dB relative to $<P_1>_{rms}$. The rms value of the sum noise is given by

$$<P_1 + P_2>_{rms} = \sqrt{<P_1^2>_{rms} + <P_2^2>_{rms}} \ . \tag{4}$$

The table below shows how this works for $<P_1>_{rms} = 0$ dB:

$<P_2>_{rms}$	0	−1	−2	−3	−5	−10	(dB)
$<P_1 + P_2>_{rms}$	3	2.5	2.1	1.75	1.2	0.4	(dB)

A practical test to determine if the predominant noise source is electronic, shot, or laser noise is to measure the noise, insert a filter (say, 50%) in front of the detector, and measure the noise again. If the noise is unchanged, it is primarily electronic noise. If the noise is reduced by the square root of the power, it is primarily shot noise. And if the noise is reduced proportionally to the power, it is mostly laser noise. To discriminate between laser and medium noise, one can compare the noise from a static disk with that from a spinning disk.

2.2. The magneto-optic signal

2.2.1. Physical basis

The MO effect in an isotropic medium is due to a dielectric tensor of the form[12]

$$\varepsilon = \begin{bmatrix} \varepsilon_0 & \varepsilon_1 \\ -\varepsilon_1 & \varepsilon_0 \end{bmatrix} \ , \tag{5}$$

where ε_1 is proportional to the magnetization M. A linearly polarized field $E_{\parallel}$ incident on such a material will, after the interaction, have a component $E_{\perp}$ in the orthogonal direction. This component is proportional to M. Because ε_1 is in general complex, $E_{\perp}$ will not be in phase with $E_{\parallel}$. This holds for any complex structure containing a magnetic layer, such as sandwiches between various dielectric or metallic films.

The elliptically polarized reflected light can be pictorially described as in Fig. 1. A variable phase compensator with its principal axes parallel to the $E_{\perp}$ and $E_{\parallel}$ axes can simultaneously convert the two elliptically polarized fields to linear ones (dashed lines). This will maximize the MO signal. Polarizers are ther used to convert the linearized E(±M) into distinct intensities P(+M) and P(−M) impinging on a photodetector. The MO signal is then defined as P(+M) − P(−M).

2.2.2. Detection methods

After interaction with the magnetic material and the phase compensator (and appropriate losses in the optical path), the field reaches the MO detector module with two in-phase components: E_p parallel to the incident field and E_s perpendicular to it. E_s changes sign with the magnetization direction.

Three different detection schemes[2] are shown in Fig. 2. In

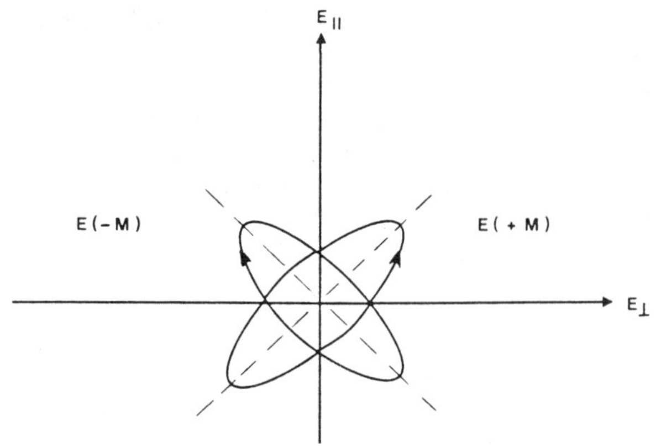

Fig. 1. The magneto-optic effect. The field $E_{\parallel}$ is incident on the medium, and the fields E(±M) are produced after interaction with the two states of magnetization.

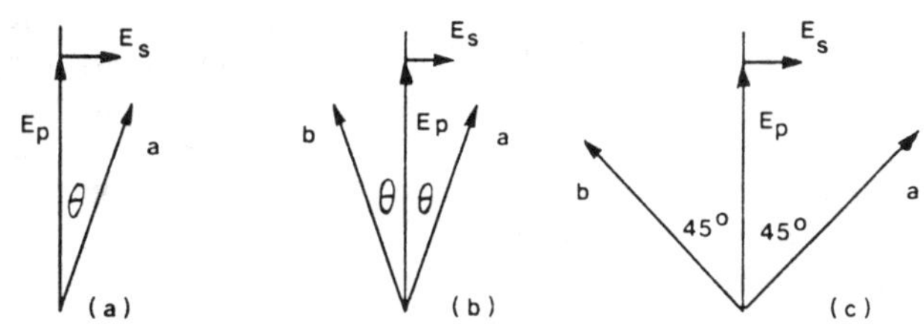

Fig. 2. Three different detector schemes. In (a), a single analyzer is placed at an angle θ to E_p. In (b), the beam is split, and two analyzers are placed at angles ±θ in front of two photodiodes. In (c), a polarizing beam splitter is used with axes at ±45° to E_p. Two photodiodes are again used.

(a), the single analyzer at an angle θ to E_p transmits only the projection of the field along its axis:

$$E_a = E_p\cos\theta \pm E_s\sin\theta \ , \tag{6}$$

$$\frac{I_s}{\eta} = P_s = P(+M) - P(-M) = 2E_pE_s\sin 2\theta \ . \tag{7}$$

E_p should be larger than E_s to maintain the signal at a reasonable level. The value of E_p can be controlled by an antireflection (AR) coating on the disk or by a suitable optical system. Control by AR coating has the advantages that it improves the signal-to-shot-noise ratio and reduces the optical feedback into the laser. Too large a value of E_p will result in low contrast, thus requiring a large dynamic range of the electronics and making the system more sensitive to laser noise. Using Eq. (1), the peak-to-peak signal current to rms shot noise for the case $E_p \gg E_s$ ($P = E_p^2\cos^2\theta$) is given by

$$\frac{I_s}{I_{sh}} = 4CE_s\sin\theta \ , \tag{8}$$

with

$$C = \left(\frac{\eta}{2ef}\right)^{1/2} \ . \tag{9}$$

Note that the maximum S/N is obtained in the limiting case of θ = 90°, where there is no signal! When θ = 45°, the signal is maximum and the S/N is a factor $\sqrt{2}$ below its maximum value. This scheme is simple, but it is very sensitive to laser noise and reflectivity fluctuations.

Identical results for signal and signal/shot noise ratio are obtained when the beam is split and two analyzers are used differentially, as in Fig. 2(b). However, this method has the advantage that is much less sensitive to laser noise and reflectivity fluctuations.

When a polarizing beam splitter is used instead of a normal beam splitter followed by two analyzers, as in Fig. 2(c), the results for signal and shot noise are

$$\frac{I_s}{\eta} = 2P_s = 4E_pE_s , \quad (10)$$

$$\frac{I_s}{I_{sh}} = 4CE_s . \quad (11)$$

The S/N of this scheme is equal to $(S/N)_{max}$ of the other methods, and it has the advantage that the signal is *twice* as large as the maximum signal in the other methods. The power on each PD is approximately 0.5 E_p^2.

2.2.3. Magneto-optic signal and contrast

In the following we assume that a polarizing beam splitter (PBS) is used for differential detection. A typical MO optical module is shown schematically in Fig. 3. A second PBS in the main beam diverts a fraction ρ of the parallel component of the optical electric field and, essentially, all of the orthogonal component to the differential detector. In this case, the optical power on each detector is

$$P = 0.5P_iTR\rho^2 . \quad (12)$$

As the magnetization state of the storage medium is reversed, P on each diode changes by an amount P_s:

$$P_s = 2KTP_i\rho\sqrt{R} , \quad (13)$$

where

$$K^2 = \frac{E_s^2}{TP_i} . \quad (14)$$

K is the intrinsic MO parameter of the disk structure. It describes the fraction of the optical field transformed from the incident polarization to the orthogonal polarization. (The transmission T is included because E_s is defined at the input to the PD module.) K is related to the Kerr rotation angle after reflection, Θ, by

$$K = \Theta\sqrt{R} . \quad (15)$$

From Eq. (11) it follows that in a shot-noise-limited case, S/N is directly proportional to K. Remember that P_s is the peak-to-peak signal on each diode for the bulk switching of the magnetization. For high resolution recording, due to the convolution of the written bit with the resolving beam spot, P_s will drop by roughly a factor of 2, resulting in a signal loss of about 6 dB.*

The contrast P_s/P, given by

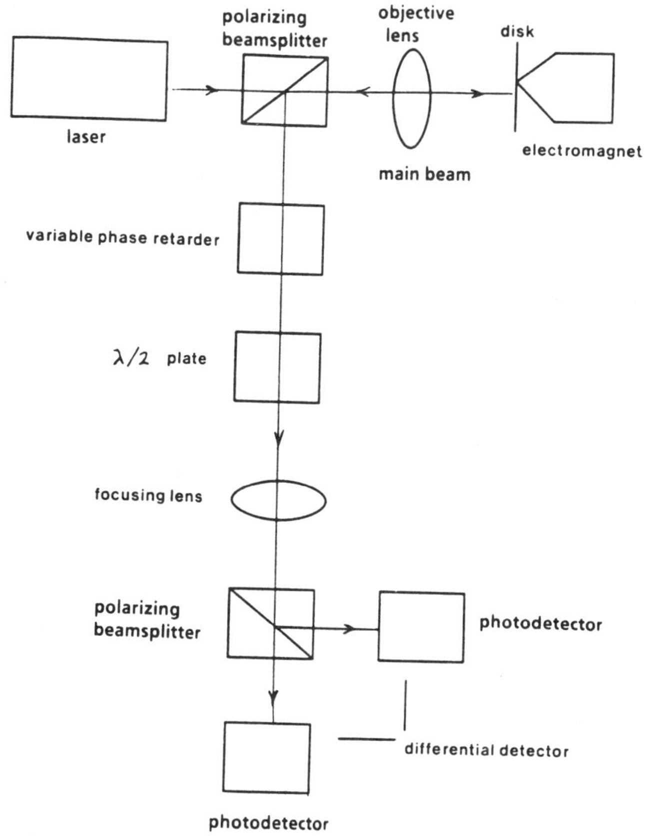

Fig. 3. A typical optical module for a magneto-optic memory system. The λ/2 plate is used to equalize the intensities on the two photodetectors.

$$\frac{P_s}{P} = \frac{4K}{\rho\sqrt{R}} , \quad (16)$$

is readily measured in a conventional MO detection module by dividing the peak-to-peak signal by the average power that reaches the detectors. This measurement does not require the knowledge of optical power or losses within the system. The contrast, together with the measurement of R and ρ, allows a convenient determination of K. A good MO medium has been achieved with $K = 8\times10^{-3}$.[13] A value of $K = 17\times10^{-3}$ has also been claimed.[14] Note that K depends to some extent on the multilayer structure of the disk. The improvement in K between the simplest structure and an optimized quadrilayer is about a factor of 2.[13]

2.2.4. Effect of the contrast ratio

In a shot-noise-limited case, such as with an avalanche PD, the S/N depends only on K for contrasts up to about 0.6. However, too small a contrast will require a high dynamic range of the amplifier, and the system will be more sensitive to alignment and laser noise. If there is an appreciable amount of electronic and Johnson noise, it may be worthwhile to increase P_s by increasing ρ (decrease the contrast). This will increase the signal, thus improving S/N. However, one must remember that the amount of power absorbed by the disk is proportional to $(1 - R)(1 - \rho^2)$, and it will therefore decrease as ρ increases. This is important for the writing process. For a constant signal P_s, the absorbed power is a maximum for $\rho^2 = R$.

*Photodetectors convert optical power into electrical current. Thus, a reduction by a factor of 2 in optical signal power results in a 6 dB drop in S/N.

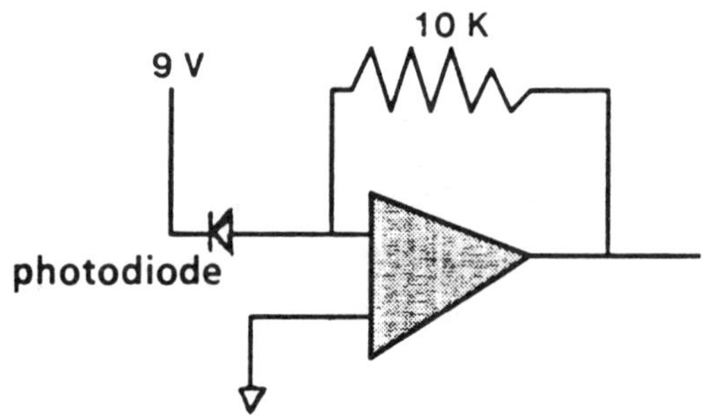

Fig. 4. Signal detection using a transimpedance amplifier (NE 5539 operational amplifier).

2.3. Avalanche versus PIN photodetectors

Avalanche diodes have built-in gain with typically a shot-noise-limited S/N. However, they are expensive, require high voltages, and are not easy to integrate. It is therefore of interest to consider the S/N obtainable with a PIN diode and a transimpedance amplifier, as shown in Fig. 4. The feedback resistor of 10 kΩ allows operation up to 20 MHz, the assumed bandwidth in the following calculation. A good amplifier will have a total input spectral noise current of 3 $pA/\sqrt{Hz}$.* Using a detector quantum efficiency $\eta = 0.5$, bandwidth of 20 MHz, and Eqs. (1) to (3),

$$\begin{aligned} &\text{NEP (shot, P} = 10\ \mu\text{W)} &&= 0.011\ \mu\text{W} \\ &\text{NEP (Johnson, 10 k}\Omega\text{)} &&= 0.012\ \mu\text{W} \\ &\text{NEP (electronic, excess)} &&= 0.028\ \mu\text{W} \\ &\text{NEP (electronic, total)} &&= 0.030\ \mu\text{W}\ . \end{aligned} \tag{17}$$

The S/N for different values of ρ and incident optical power P_i can now be calculated using the equation

$$S/N\ (dB) = 10\log\left(\frac{P_s^2}{NEP_{shot}^2 + NEP_{electronic}^2}\right) . \tag{18}$$

From this value we (a) subtract 9 dB to convert the p-p signal to rms signal, (b) subtract another 6 dB to convert the saturation signal to signal at moderate density, and (c) add 3 dB for combining the two differential channels. P_s is calculated from Eq. (13). Results obtained assuming R = 0.2 and $K = 7.1\times10^{-3}$ are given in Table I. The shot-noise-limited case was added for comparison. An avalanche PD will be a few dB worse than shot-noise-limited. For convenience, both P, the average light power on each detector, and P_s, the peak-to-peak saturation signal, are also given in Table I.

3. NOISE REQUIREMENTS FOR DATA, TRACKING, AND FOCUS CHANNELS

In this section we consider the requirements on laser noise, both for data channels that use threshold detection and for servo channels. We also briefly discuss the effect of electronic noise. To specify a maximum allowed noise level in a given channel, it is necessary to know both the signal level and the required S/N ratio. Although only specific examples are analyzed here for illustration, the method is general and can be applied to other situations.

*See for example the data sheets of the Intersil IH5101 high frequency amplifier.

TABLE I. $(S/N)_{rms}$ for high resolution recording for electronic NEP = 0.03 μW in a bandwidth of 20 MHz. Note that the power delivered to the disk in the writing process is proportional to $1-\rho^2$. P and P_s have been calculated assuming R = 0.2 and K = 7.1×10^{-3}.

P_iT (mW)	ρ^2	P (μW)	P_s(p-p) (μW)	S/N (shot) (dB)	S/N (total) (dB)
1	0.1	10	2	33	24
1	0.2	20	$2\sqrt{2}$	33	26.4
1	0.3	30	$2\sqrt{3}$	33	27.7
2	0.1	20	4	36	29.4
2	0.2	40	$4\sqrt{2}$	36	31.7
2	0.3	60	$4\sqrt{3}$	36	32.7
4	0.1	40	8	39	34.7
4	0.2	80	$8\sqrt{2}$	39	36.3
4	0.3	120	$8\sqrt{3}$	39	37.1

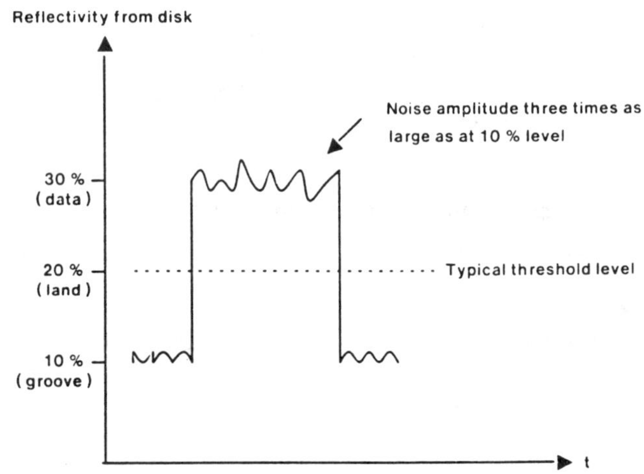

Fig. 5. Multiplicative laser noise in sum data channel (write-once).

3.1. Data channel

3.1.1. Sum channel (write-once)

The data signal in a WO memory is due to a change in reflectivity. A typical example of the signal with its associated multiplicative laser noise is shown in Fig. 5. Data are written within preformatted grooves, which are separated from adjacent tracks by land. If the threshold is set at the midpoint between a "0" (groove) and a "1" (data), it is obvious that the probability of an error is largest in the data region. An error will occur when a fluctuation from the 30% reflected level (data) crosses the threshold at about the 20% level. Thus, a peak fluctuation of about 30% of the average light power will cause an error. It is therefore required that the relative fluctuation γ not exceed 0.3. Note that because γ is a function of the reflectivity, this requirement applies to the situation with 30% reflectivity.

3.1.2. Difference channel (magneto-optic)

Assuming use of the differential detection method† described in Sec. 2.2.2, it is necessary to distingush among cases that differ in the alignment of the PBS in front of the photodetectors. In the symmetrically balanced mode (Fig. 6), the output

†If in an MO system a nondifferential detection method is used, such as that shown in Fig. 2(a), the data channel would be described analogously to the sum channel, as in Sec. 3.1.1.

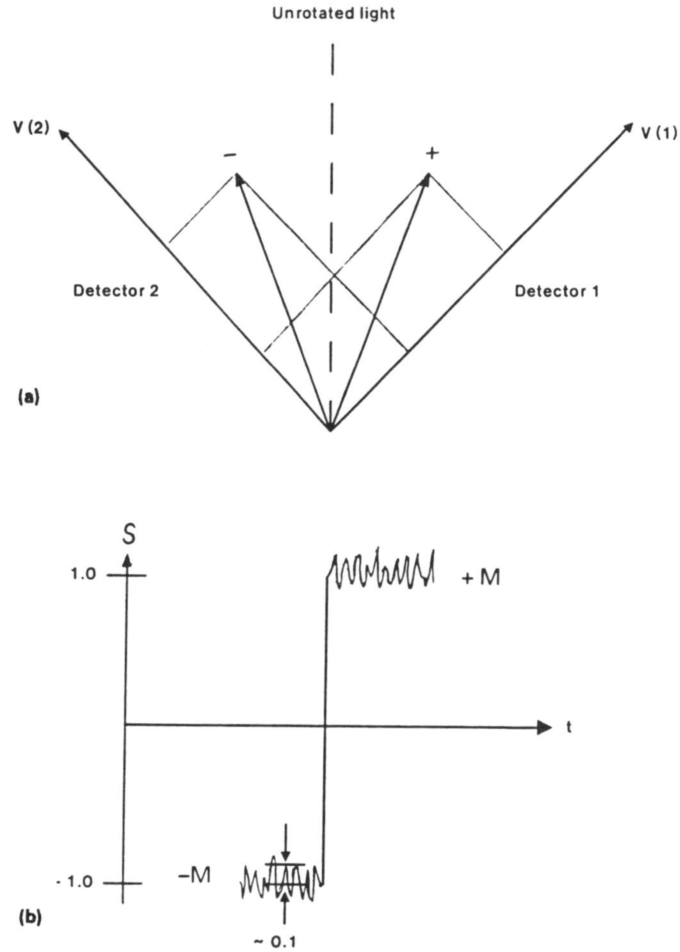

Fig. 6. ***Balanced*** **detectors. (a) The magnitude of the differential signal is the same in the + and − states. (b) Multiplicative laser noise in a differential (magneto-optic) data channel. The noise amplitudes are equal in the two states, and the threshold is set halfway between the − and + signal levels. The effect of laser noise with balanced detectors is to reduce momentarily the effective signal.**

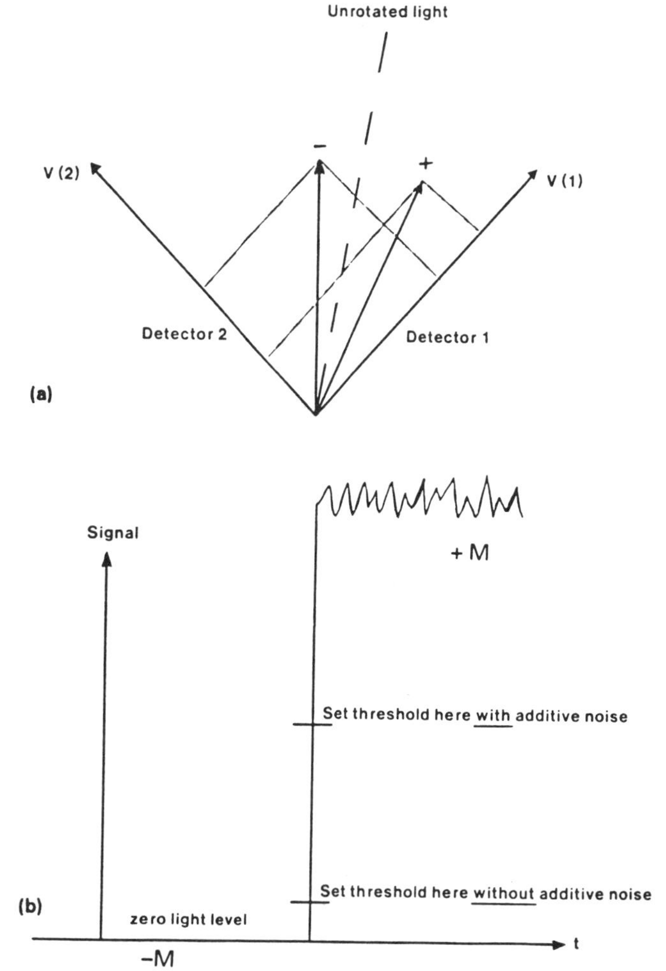

Fig. 7. ***Unbalanced*** **detectors. (a) The differential signal V(1) − V(2) is zero in the − state. (b) Multiplicative laser noise in a differential (magneto-optic) data channel. There is no laser noise in the − state.**

difference signal is of equal magnitude and opposite sign for the two magnetization directions. Therefore, the multiplicative laser noise is equal in the two states, and the threshold level is set at zero differential output. In the unbalanced mode (Fig. 7), both the output difference signal and noise are zero when the magnetization is bulk-switched in the (−) direction.

In the absence of other additive noise, laser noise will not cause errors because the threshold can be placed, even in the unbalanced case, near the zero light level. The effect of laser noise is to reduce the ratio S/N_{other} due to additive noise. For a relative laser noise γ, the total S/N will be

$$S/N = (1 - \gamma)S/N_{other} \tag{19}$$

because the presence of the multiplicative laser noise is equivalent to reducing the signal by a factor of $1 - \gamma$. If, in the unbalanced case, the threshold is set at the midpoint [Fig. 7(b)], the reduced S/N will be

$$S/N = (1 - 2\gamma)S/N_{other} \ . \tag{20}$$

3.2. Tracking servo channel (sum)

The tracking method chosen here for discussion is based on sampling the sum signal on two consecutive offset flags.[4,5] The servo error signal is given by the difference of the two sum signals. A typical error sensitivity is such that 0.1 μm tracking error will cause the difference of the two sums to be about 10% of the average sum. Because usual specifications require a tracking precision of ±0.1 μm under all conditions, the allowed laser noise γ should be appreciably below 10% within the bandwidth of this channel, which is about 2 MHz for typical flag dimensions.

The electronic noise is negligible even in a magneto-optic system. It was shown in Sec. 2 that the light power into each detector is at least 10 μW and that in a 20-MHz bandwidth the NEP of the electronics, using a PIN diode with a transimpedance amplifier, is about 0.03 μW. This is at least 30 dB smaller than a fluctuation that would cause the maximum allowable tracking error. To this we must add 10 dB because only noise within the tracking servo bandwidth need be considered, and a futher 3 dB can be added because the noise is incoherent in the two detectors. Thus, the NEP is about 43 dB below a fluctuation that would cause a maximum allowable tracking error.

3.3. Focus servo channel (difference)

A popular focus error detection method uses astigmatic optics.[3] This is a differential technique with a typical sensitivity β of 5% change in light level on each detector for each 1 μm out of focus at the disk (Fig. 8). Denote the optical power on the detectors by P_1 and P_2, the distance out of focus by d, the fluctuation in the difference channel and in the estimated distance out of focus due to laser noise by $<P_1 - P_2>$ and $<d>$, respectively, and the fractional light level fluctuation in the difference channel due to laser noise by ε. Then,

$$<P_1 + P_2> = \gamma(P_1 + P_2) \ , \tag{21}$$

$$<P_1 - P_2> = \varepsilon(P_1 + P_2) \ , \tag{22}$$

$$<d> = \frac{<P_1 - P_2>}{\beta(P_1 + P_2)} \ . \tag{23}$$

Because the maximum allowed value of $<d>$ is about ± 0.2 μm, one requires

$$\varepsilon < 0.01 \ . \tag{24}$$

The usual assumption is that laser noise fluctuations on the two detectors are completely correlated, and hence

$$\varepsilon = \frac{\gamma(P_1 - P_2)}{P_1 + P_2} \ . \tag{25}$$

Thus, at the balance point there should be no laser noise. However, as mentioned in the introduction, there may be noise associated with interference of stray light on the detectors,[7] and one must ensure that in fact $\varepsilon < 0.01$.

As with the tracking channel, the effect of electronic noise on the focus channel is negligible. Repeating the calculations of Sec. 3.2, we find a margin of 23 dB.

4. ENCODING METHOD: TIMING ERRORS, NOISE, AND ERROR RATES

Among the various ways in which data can be encoded in the recording process, NRZ has the largest tolerance to noise. If noise and timing errors are sufficiently low that the minimum feature size is limited by the optics, then the bit density can be increased by using more sophisticated codes[10,11] that yield a higher data bit density per minimum feature length. However, this is always done at the expense of a narrower timing window and an associated increase in the required S/N.

In this section we discuss the sources of timing error in an optical recording system. We then briefly describe a few run-length-limited codes, and derive the dependence of the error rate for these codes on both *additive* and *multiplicative* noise. The former is appropriate for electronic noise in an MO differential channel. The latter is of most concern for laser noise in a WO sum channel.

4.1. Timing errors

With NRZ encoding, one could use either *threshold* detection or *zero-crossing* detection in the center of the data cell. In the former method, an error would occur if either the signal incorrectly crossed the threshold or the clocking error became greater than half the data cell. In the latter method, a bit would be encoded by a transition that must be detected within half the data cell width from its center. With zero-crossing detection, the effect of noise is converted into a *timing error* due to a shift in position (or time) of the zero-crossing.

Included in the various contributions to timing error in the data channel are (1) electronic (including Johnson and shot) noise, (2) laser noise, (3) medium noise, (4) jitter in the pos tion of the written bit, (5) linear readback bit shift, (6) interfe ence from data in adjacent tracks, (7) interference fro overwritten data on the same track, and (8) clock jitter.

For convenience, we consider separately the additive ar multiplicative noise contributions to (1) and (2) above. Th remaining six contributions to timing shift are assumed occupy a fraction α of the NRZ window. The combined effe of RLL encoding and additive noise is calculated in Sec. 4. and the effect of multiplicative laser noise in the sum chann with NRZ and 2,7 codes is given in Sec. 4.4.

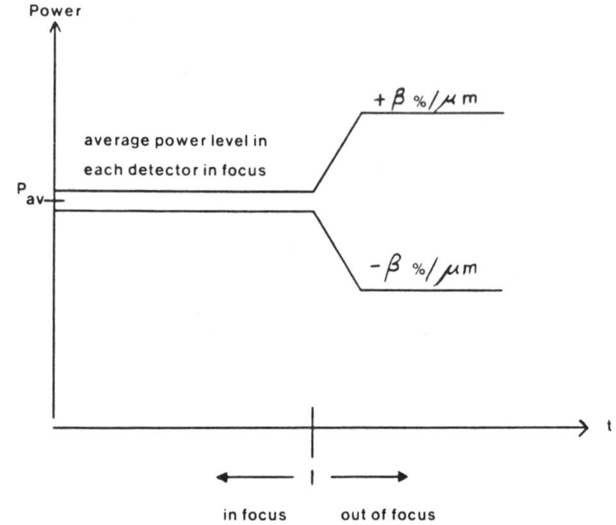

Fig. 8. Schematic of power in the focus channel, both in and o of focus.

TABLE II. The window width, data bit size, minimum feature lengt and (number of data bits)/(minimum feature distance) for three ru length-limited codes, as compared with NRZ. Numbers are given f both compressed (com) and uncompressed (uncom) 1,7 ar 2,7 codes.

	Fraction of NRZ					
		(MFM)	1,7,2,3		2,7,1,2	
Code	NRZ	1,3,1,2	com	uncom	com	uncor
Window width	1	0.5	0.5	0.67	0.33	0.5
Data bit size	1	1	0.75	1	0.67	1
Minimum feature	1	1	1	1.33	1	1.5
No. data bits/ minimum feature	1	1	1.33	1.33	1.5	1.5

4.2. Codes

Compared with NRZ, which is not self-clocking unless au mented with a k-constraint, all the d = 1 RLL codes (lik MFM and the 1,7,2,3 code) have windows that are only ha as large when the minimum distances between transitions a made equal.[10] In the case of MFM, the sacrifice of half th window yields only self-clocking, since the data rate unchanged from NRZ. For the 1,7,2,3 code, the data densi is increased to 4/3 of the MFM or NRZ data density, again the expense of 1/2 the NRZ window. For the 2,7,1,2 cod the window width is further decreased, to 1/3 the NRZ wi dow, and the data density rises to 3/2 the NRZ density.

Table II gives the window width, data bit size, minimu

feature length, and (number of data bits)/(minimum feature length) for these four codes. The 1,7 and 2,7 codes are given both uncompressed and compressed. By *uncompressed*, we mean that the *data bit* size is the same as NRZ, and the minimum feature size (i.e., minimum distance between transitions) can be greater than for NRZ. After *compression*, the *minimum feature* size is the same as for NRZ, and the size of the data bit can be smaller than for NRZ.

4.3. Error rate due to additive noise

What is the requirement on S/N such that a given error rate R results when a fixed fraction α *of the NRZ window* is allocated to timing error other than electronic or laser noise? It is important to note that this allocated period of time is subtracted from the window *of the particular encoding method under consideration.*

Two calculations are given below. The first is a simple model that assumes the high density readback waveform is a sawtooth pattern with linear transitions extending over the width of the data cell. The second model is more accurate and assumes that the high density readback waveform is sinusoidal. Similar results are obtained from the two models, which demonstrates *little sensitivity to the details of each model.*

4.3.1. Linear transitions

Figure 9 shows a linear transition extending the width of the data cell. At high density, the waveform would be a sawtooth. The NRZ window is the full width of the linear transition. The width of the compressed 2,7 window is 1/3 of the NRZ window. For a given amount of noise, the rectangular construction (dashed line) gives the error in position of the zero-crossing.

Assume first that there are no other sources of noise, so that the entire window widths are available for Gaussian noise. Then the amount of noise that would cause an error in the 2,7 code is 1/3 the noise necessary to exceed the NRZ window. For comparable error rates with NRZ and 2,7 codes, we have

$$(S/N)_{2,7} = 3(S/N)_{NRZ} \ . \tag{26}$$

To use a 2,7 code, about 10 dB more S/N is required than for NRZ with the same minimum feature size.

There always do exist other sources of noise to be accounted for. In system design, a "budget" is created that allocates each of the noise sources some amount of the window. Suppose a fraction α of the NRZ window is allocated to *all sources of timing error other than electronic and laser noise:*

$$w^{other} = \alpha w_{NRZ} \ . \tag{27}$$

The NRZ window allocated to noise is

$$w^{n}_{NRZ} = (1 - \alpha) w_{NRZ} \ , \tag{28}$$

and the 2,7 window allocated to noise is

$$w^{n}_{2,7} = \frac{1}{3} w_{NRZ} - w^{other} = \left(\frac{1}{3} - \alpha\right) w_{NRZ} \ . \tag{29}$$

If S_o is the magnitude of the signal, then its slope at the zero-crossing is approximately

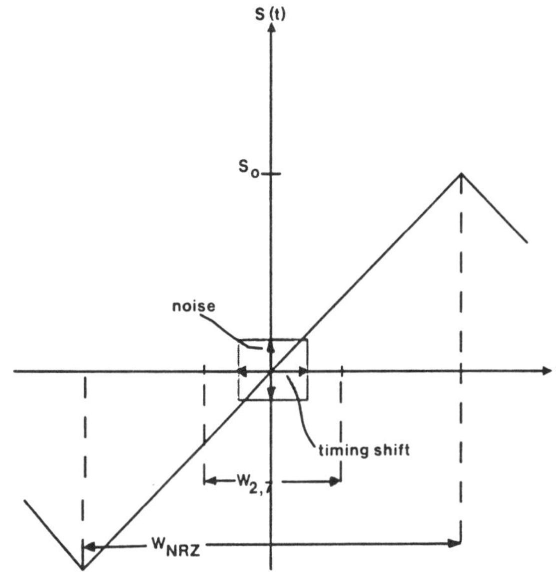

Fig. 9. Simple model of a *linear* transition, and the effect of noise on the zero-crossing shift.

$$\text{slope} = \frac{2S_o}{w_{NRZ}} \ , \tag{30}$$

and the signal at the edge of the NRZ noise window, given by *half* of Eq. (28), is then

$$(S_{edge})_{NRZ} = \frac{S_o w^{n}_{NRZ}}{w_{NRZ}} = (1 - \alpha) S_o \ . \tag{31}$$

Let the rms noise level be N. Then an error will occur when a noise fluctuation of magnitude βN ($\beta > 1$) exceeds S_{edge}:

$$\beta N > (1 - \alpha) S_o \quad \text{(for NRZ)} \ . \tag{32}$$

If the fluctuations are normally distributed, the probability of occurrence of an error can be found from a table of the complementary error function.[15] In terms of the rms signal $S = S_o/\sqrt{2}$, the S/N for NRZ, corresponding to a fluctuation given by β, is simply

$$(S/N)_{NRZ} = \frac{\beta/\sqrt{2}}{1 - \alpha} \ . \tag{33}$$

The signal at the edge of the 2,7 noise window, given by *half* of Eq. (29), is

$$(S_{edge})_{2,7} = \frac{S_o w^{n}_{2,7}}{w_{NRZ}} = S_o\left(\frac{1}{3} - \alpha\right) . \tag{34}$$

The noise βN will equal this value when

$$(S/N)_{2,7} = \frac{\beta/\sqrt{2}}{(1/3) - \alpha} \ . \tag{35}$$

This ratio becomes infinite when $\alpha = 1/3$ because there is no window left for the noise.

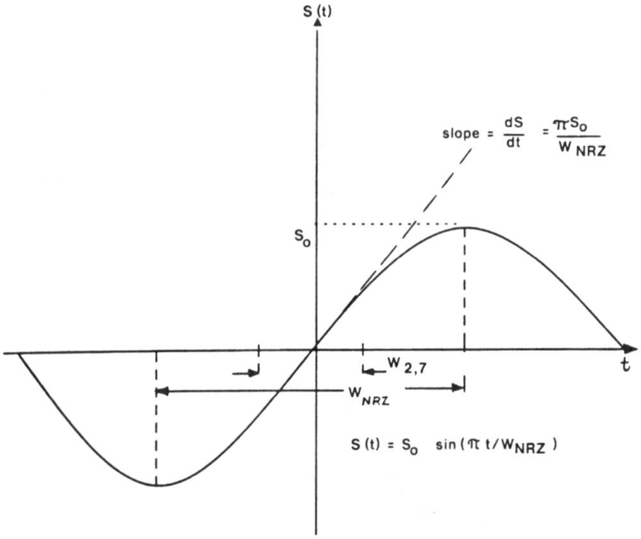

Fig. 10. Signal at high density (approximately sinusoidal), showing the approximation for threshold detection of NRZ (dashed line $S = S_o$) and the approximation for a narrow window 2,7 code near the zero-crossing.

4.3.2. Sinusoidal waveform

A portion of the signal at high transition density is shown in Fig. 10. Assuming a sine wave, the NRZ window is one-half the period, and the slope of the signal at the zero-crossing is

$$\frac{dS}{dt} = \frac{\pi S_o}{w_{NRZ}} . \tag{36}$$

Equations (28) and (29) can again be used for the windows allocated to noise for NRZ and 2,7 codes, respectively. For NRZ encoding and for $\alpha \leq 1/3$, an error will occur when the noise βN exceeds approximately the maximum signal S_o. (Because the sine wave in Fig. 10 approximates the horizontal dotted line near $t = w_{NRZ}/2$, NRZ essentially uses threshold detection.) In terms of the rms signal $S = S_o/\sqrt{2}$, the signal/noise ratio for NRZ, corresponding to β, is simply

$$(S/N)_{NRZ} = \frac{\beta}{\sqrt{2}} . \tag{37}$$

Note that Eq. (33) reduces to this when $\alpha = 0$.

For 2,7 encoding, the waveform is well approximated by the dashed line through the zero-crossing. With peak signal S_o and noise βN, the timing error is

$$\text{shift}_{2,7} = \frac{\beta N w_{NRZ}}{\pi S_o} . \tag{38}$$

Setting this equal to *half* of the allocated 2,7 window,

$$(S_o/N)_{2,7} = \frac{2\beta/\pi}{(1/3) - \alpha} . \tag{39}$$

In terms of the rms signal S, we have finally

$$(S/N)_{2,7} = \frac{\sqrt{2}\beta/\pi}{(1/3) - \alpha} . \tag{40}$$

TABLE III. For *additive* noise, $(S/N)_{rms}$ required for immunity against fluctuation βN with $\beta = 7$ for NRZ and 2,7,1,2 codes. Results for both linear transitions and sinusoidal waveforms are shown, with a fraction α of NRZ window allocated to other sources of noise.

	$(S/N)_{rms}$ in dB			
	Linear transition		Sinewave	
α	NRZ[a]	2,7,1,2[b]	NRZ	2,7,1,2[d]
0	14	24	14[c]	20
0.20	16	32	14	28
0.33	17.5	∞	15	∞

[a]Use Eq. (33).
[b]Use Eq. (35).
[c]Use Eq. (37).
[d]Use Eq. (40).

Analogously, for the 1,7,2,3 code, the required rms S/N is approximately

$$(S/N)_{1,7} = \frac{\sqrt{2}\beta/\pi}{(1/2) - \alpha} . \tag{41}$$

If $\alpha = 0$, the signal/noise ratio for the 2,7 code must be twice as large (6 dB) as for NRZ to get the same error rate (as determined by β). But if α is greater than zero, the S/N required for the 2,7 code rapidly becomes much larger than that required for NRZ and goes asymptotically to infinity when the fraction α budgeted to *all* other sources of timing error reaches *one-third* of the NRZ window.

To illustrate these results, suppose we accept an error rate of 10^{-12} ($\beta = 7$ for normally distributed fluctuations) due to electronic and laser noise, with an additional noise contribution α defined as before. Table III gives the required rms signal/noise for NRZ and 2,7,1,2 codes for different values of α.

4.4. Error rate due to multiplicative noise: laser noise requirement in sum data channel (WO)

We give a derivation of the effect of laser noise on the WO data channel for both NRZ and 2,7 RLL codes. Figure 11 shows both a linear and a sinusoidal transition from the data level of about 30% reflectivity to the groove level of about 10% reflectivity. The zero-crossing is detected halfway between, at the 20% level. Multiplicative noise fluctuations, which are larger on the upper side of the zero-crossing, will shift the zero-crossing to the left. The narrowing of the window in going from NRZ to 2,7 encoding makes the laser noise more critical. We shall calculate the effect on allowable light fluctuations and include for comparison the results for NRZ.

In the presence of other noise, the window is narrowed by a fraction α of the NRZ window, as discussed above. This is shown schematically in Fig. 11, where both the NRZ and 2,7 window edges (solid vertical lines) are narrowed (dashed lines).

The worst multiplicative light fluctuations, from the point of view of causing errors, will occur at the window edge. An error occurs when the fluctuation is downward and equal to the difference between the value at the window edge and the 20% level. The four points at window edges (linear and sinusoidal; NRZ and 2,7) are given by

$$P_{NRZ;sine} = 20 + 10\sin\left[\frac{\pi}{2}(1 - \alpha)\right] , \tag{42}$$

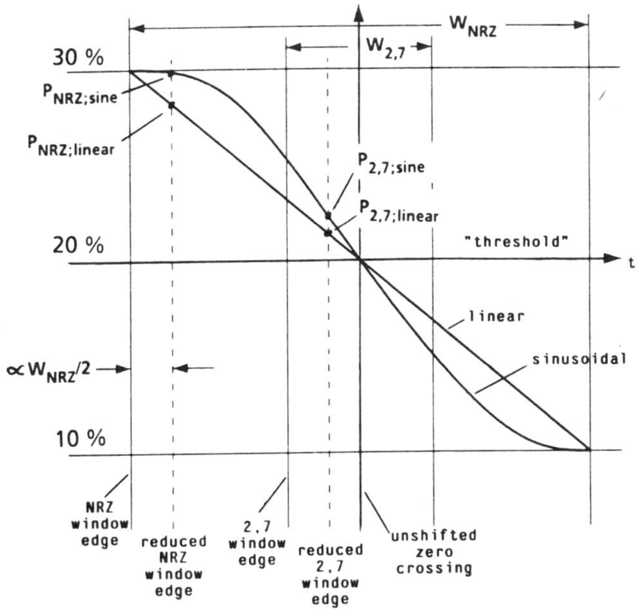

Fig. 11. Reflectivity of sinusoidal and linear transitions in a write-once sum data channel. Percentages are representative. Window edges for NRZ and 2,7 codes are shown, both full and reduced by a margin due to additional noise. The magnitude of the waveform at window edge, P, determines the fluctuation, P − 20, required to cause an error. The fluctuations most likely to cause an error are those at the window *edge*, not those far outside the window.

$$P_{NRZ:linear} = 20 + 10(1-\alpha) , \tag{43}$$

$$P_{2,7:sine} = 20 + 10\sin\left[\frac{\pi}{2}\left(\frac{1}{3}-\alpha\right)\right], \tag{44}$$

$$P_{2,7:linear} = 20 + 10\left(\frac{1}{3}-\alpha\right) . \tag{45}$$

Using these values of P, the *maximum* allowable fractional fluctuation is

$$\gamma_{max} = \frac{P-20}{P} . \tag{46}$$

This maximum fluctuation is given for the four cases in Table IV. As in Sec. 4.3, no noise can be tolerated when α approaches 1/3 because the timing margin allocated to other noise sources is $w_{NRZ}/6$, which takes up the entire 2,7 half-window.

We note that these timing shifts (which are due to multiplicative noise) depend only on γ and hence are *independent of signal amplitude.*

5. CONCLUSION

Laser noise can cause both a loss of tracking or focus and errors in the data channel. In a typical optical memory system, the maximum laser light fluctuation should never exceed 10% of the average light level. In the differential channel of such a system, the laser noise should be further reduced by another order of magnitude.

Run-length-limited codes are useful because they allow self-clocking and can permit a greater packing of data bits into the minimum feature size distance. However, the timing window must be reduced to accommodate the restrictions on the encoded transitions. Consequently, there is always a price to pay in required S/N when going from NRZ to an RLL code. If the only source of zero-crossing shift is the noise being considered, the price for typical RLL codes is an increase in S/N of between 6 and 10 dB. In an actual system, there are other sources of zero-crossing shift, for which some fraction of the window must always be allocated. The increase in required S/N can consequently be much larger than 10 dB. When the fraction of the window that is allocated to timing jitter *other than Gaussian noise* is comparable to the reduced window for the RLL code, the system will fail (it will not tolerate *any* noise). *Write-once* systems are sensitive to multiplicative laser noise in the data channel, because the fluctuations in the sum channel are proportional to the signal. The *timing shift* that results from such fluctuations is *independent of signal amplitude*. Additionally, some fraction of the reduced RLL window must be allocated to other sources of timing jitter, including the nonrepeatable jitter (edge acuity) that is produced during writing. Consequently, when an RLL code is used on write-once media, timing shifts due to laser noise can bring the zero-crossings outside the window. The data channel in a *magneto-optic* system is not expected to be as severely impacted by laser noise because of some (generally unknown) degree of cancellation of laser noise in the difference channel. To predict the actual performance of an RLL code, an estimate must be obtained of zero-crossing shifts due to factors other than laser and electronic noise, such as edge acuity, medium noise, clock jitter, and overwrite. Additionally, the amplitude distribution of different noise sources must be known for the system.

For the MO data channel, a comparison of Tables I and III shows that the electronic noise leaves little margin, in particular for a 2,7 code, and the system must be designed carefully to avoid either loss of signal or increase in noise much beyond the estimates derived here.

TABLE IV. For *multiplicative* noise, the ratio γ_{max} (maximum fluctuation/power in sum channel) that will cause an error for NRZ and 2,7 codes with a fraction α of NRZ window allocated to other sources of noise. Results for both a linear transition and sinusoidal waveform are shown. The required γ (rms fluctuation/power in sum channel) for $\beta = 7$ (which corresponds to an error rate of 10^{-12} if the fluctuations are distributed normally) is found by subtracting 17 dB from each of these numbers. Values in parentheses are in dB.

	Maximum fluctuation/dc power			
	Linear transition		Sinusoidal waveform	
α	NRZ	2,7	NRZ	2,7
0	0.33 (−10)	0.14 (−17)	0.33 (−10)	0.20 (−14)
0.1	0.31 (−10)	0.103 (−20)	0.33 (−10)	0.15 (−16)
0.2	0.27 (−11)	0.061 (−24)	0.32 (−10)	0.092 (−21)
0.3	0.26 (−12)	0.015 (−36)	0.31 (−10)	0.023 (−33)
0.33	0.25 (−12)	0 (−∞)	0.30 (−10)	0 (−∞)

6. ACKNOWLEDGMENTS

It is a pleasure to acknowledge many helpful discussions with Neville Connell and Dan Kowalski.

7. REFERENCES

1. G. Bouwhuis and J. J. M. Braat, "Recording and reading of information on optical disks," in *Applied Optics and Optical Engineering*, R. R.

Shannon and J. C. Wyant, eds., p. 73ff., Academic Press, New York (1983).
2. J. W. Beck, "Noise considerations of optical beam recording," Appl. Opt. 9, 2559 (1970).
3. D. K. Cohen, W. H. Gee, M. Ludeke, and J. Lewkowicz, "Automatic focus control: the astigmatic lens approach," Appl. Opt. 23, 565 (1984).
4. S. Yonezawa and T. Tsuyoshi, "Sector servo data file optical disk memory," in *Third International Conference on Optical Mass Data Storage*, R. A. Sprague, ed., Proc. SPIE 529, 84 (1985).
5. R. Millar and C. Abate, "Servo and clock sampling in the Optimem 1000," in *Third International Conference on Optical Mass Data Storage*, R. A. Sprague, ed., Proc. SPIE 529, 140 (1985).
6. J. J. M. Braat, "Optically read disks with increased information density," Appl. Opt. 22, 2196 (1983).
7. A. Arimoto and M. Ojima, "Diode laser noise at control frequencies in optical videodisc players," Appl. Opt. 23, 2913 (1984).
8. K. E. Stubkjaer, "Noise properties of semiconductor lasers due to optical feedback," IEEE J. Quantum Electron. QE-20, 472 (1984).
9. A. Ohishi, N. Chinone, M. Ojima, and A. Arimoto, "Noise characteristics of high-frequency superimposed laser diodes for optical disc systems," Electron. Lett. 20, 821 (1984).
10. P. A. Franaszek, "Sequence-state methods for run-length-limited coding," IBM J. Res. Dev. 14, 376 (1970).
11. K. Norris and D. S. Bloomberg, "Channel capacity of charge-constrained run-length limited codes," IEEE Trans. Magn. MAG-17, 3452 (1981).
12. R. P. Hunt, "Magneto-optic scattering from thin solid films," J. Appl. Phys. 38, 1652 (1967).
13. M. Hartmann, J. Braat, and B. Jacobs, "Erasable magneto-optical recording media," IEEE Trans. Magn. MAG-20, 1013 (1984).
14. R. N. Gardner, T. A. Rinehart, L. H. Johnson, R. P. Freese, and R. A. Lund, "Characteristics of a new high c/n magneto-optic media," in *Optical Storage Media*, A. E. Bell and A. A. Jamberdino, eds., Proc. SPIE 420, 242 (1983).
15. A. Bruce Carlson, *Communication Systems*, 2nd edition, pp. 474–475, McGraw-Hill, New York (1975).

David Treves, born on June 28, 1930, in Ital received the B.Sc. degree in 1955, the M.S degree in 1956, and the D.Sc. degree in ele trical engineering in 1958, all from the Techn on—The Israel Institute of Technology in Haif Since 1953, he has been associated with Th Weizmann Institute of Science and has bee involved in the study of magnetization pr cesses, magnetic materials, magneto-opt memories, lasers, coherent optics, and micr wave antennas. He spent several years in th U.S.A. as a research associate at Pomona College, Claremont, Calif.; a member of the technical staff at Bell Laboratories, Murray Hill, N.J.; the Ampex Research Laboratory at Redwood City, Calif.; as World Trad Exchange Fellow at IBM, San Jose, Calif.; and at the Xerox Palo Al Research Center. Presently he is the Chairman of the Electroni Department at The Weizmann Institute of Science. Dr. Treves is a fello of the IEEE and a member of the Israel Physical Society, the Optic Society of America, the American Physical Society, and SPIE.

Dan S. Bloomberg was born in Clevelan Ohio, on Jan. 24, 1946. He received the A. degree in physics from the University of Cal fornia at Berkeley in 1967 and the Ph. degree in solid state physics from Simon Fras University, British Columbia, in 1973. He wa a teaching postdoctoral fellow at Simon Fras University and at the University of Britis Columbia from 1973 to 1975. Since joining th Xerox Corporation Palo Alto Research Cent in 1976, he has been involved with theoretic and experimental aspects of digital magnetic recording systems, ma neto-optic recording technology, and network-based software for eng neering and scientific workstations. Dr. Bloomberg is a member of th Federation of American Scientists, Computer Professionals for Soci Responsibility, the American Association for the Advancement Science, and the IEEE Magnetics and Computer societies.

Reprinted from *Optical Mass Data Storage II*, Proc. SPIE Vol. 695, pp. 72-78 (1986). ©1986 Society of Photo-Optical Instrumentation Engineers.

Measurement and modeling of optical disk noise

A. G. Dewey

IBM Research
Almaden Research Center, Dept. K67/802
650 Harry Road, San Jose, California 95120-6099

Abstract

This paper discusses the various components of the noise in the readback signal from an optical disk and suggests methods for measuring them. An analysis is presented of the relationship between optical efficiency of an M-O head, light intensity at the detector(s), choice between PIN and avalanche photodetectors and the resulting noise levels.

A number of measured noise spectra from grooved and ungrooved substrates and disks are presented and mathematical models are used to explain the results. In particular a simple model of an M-O system with differential detectors is included.

Introduction

It is essential to understand and control the sources of noise in an optical disk system in order to achieve high S/N ratio and eventually a low BER[1]. Most of the results reported in this paper have been generated in the course of studying the performance of experimental magneto-optic (M-O) disks. While some of the sections are specific to M-O systems, there are others, such as the study of substrate noise, that would be applicable to all types of optical disk systems. One approach to disk evaluation used in this work has been measurement of carrier-to-noise ratio (CNR) using the standard 30 kHz resolution bandwidth. This same bandwidth is therefore used in all discussions of noise levels throughout the paper. The primary instrument used for CNR evaluation of disks has been a modern RF Spectrum Analyzer, the printout from which appears as most of the figures in the paper.

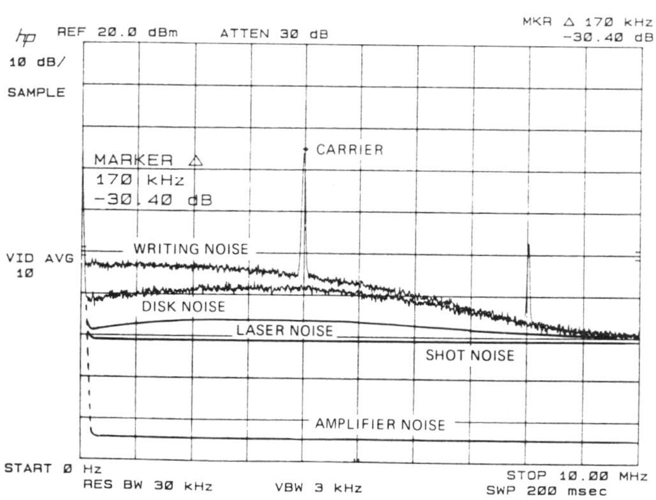

Figure 1. CNR measurement.

As an aid to the discussion of noise components, the results of a CNR test on a grooved M-O disk are presented in Figure 1. The track was first erased and the resulting spectrum was recorded. This trace is labeled "disk noise". Then a 4 MHz carrier was written on the track and a second spectrum which has been labeled "writing noise" was recorded. For this experiment both the disk and the recording conditions were deliberately chosen to emphasize the noise levels. The other three traces are schematic and were added to the chart to represent (a) the amplifier noise, defined as the noise level when no light is incident on the detector(s), (b) the shot noise associated with the photocurrent in the detector(s), and (c) the laser noise.

The labeling of the traces in Figure 1 is not strictly correct in that each noise level includes the combined effect of all the levels below it. The optical system provides adequate light levels at the photodetectors to ensure that the *amplifier noise* is much lower than the shot noise. *Shot noise* is white and the level may be calculated knowing the light intensity on the detectors. Alternatively one could illuminate the detectors with a low-noise light source at the same intensity and wave length as the laser, and measure the response. The *laser noise* spectrum may be measured in many cases by stopping the disk, provided that the stationary focussed beam does not create any damage. Such damage is likely however with a plastic substrate. Laser noise may be white (frequency independent) or, as has been sketched in Figure 1, it may have some frequency dependence. In either case one has to determine the shot noise level using one of the methods described above in order to evaluate the laser noise.

The *disk noise* may contain contributions from the substrate material, from tracking grooves and from the thin-film coatings. There are two basic types of noise which may be called amplitude noise and polarization noise. The former results from local variations in the reflectivity of the disk or in light scattering or diffraction. Polarization noise is a local variation in the polarization state of the light reflected from the disk. It is particularly important in an M-O disk system and will be discussed further in later sections. The shape of the disk noise spectrum is determined by the disk itself, by the size of the focussed laser beam, and by the media velocity. There is a cut-off frequency, related to beam size and velocity which in this case is about 10 MHz. These issues will be discussed in detail in a later section.

The final component of the noise is *writing noise* created when a sequence of spots on the disk have their magnetization reversed. There are at least two components to this writing noise: one is caused by irregular shape, size and magnetization of the written spots. The other occurs in differential detection systems which can be balanced only for one state of magnetization of the disk. In this state amplitude noise and laser noise are cancelled, but when a track is written the detectors are unbalanced and amplitude noise is no longer cancelled.

When there are insignificant levels of laser, disk and writing noise, and sufficient light P on the detectors, the shot noise which is proportional to $\sqrt{P}$ is dominant. Since the signal is proportional to P, the shot-noise-limited CNR is proportional to $\sqrt{P}$ and is limited only by the amount of read power that the disk can handle. The other noise sources are all proportional to P and hence the appearance of more than 3 dB of noise above the shot noise floor means that no significant increase in CNR can be achieved by increasing the read level. Thus in the experiment depicted in Figure 1 where the noise at 4 MHz is 15 dB above the shot noise, there is no point in increasing the read power.

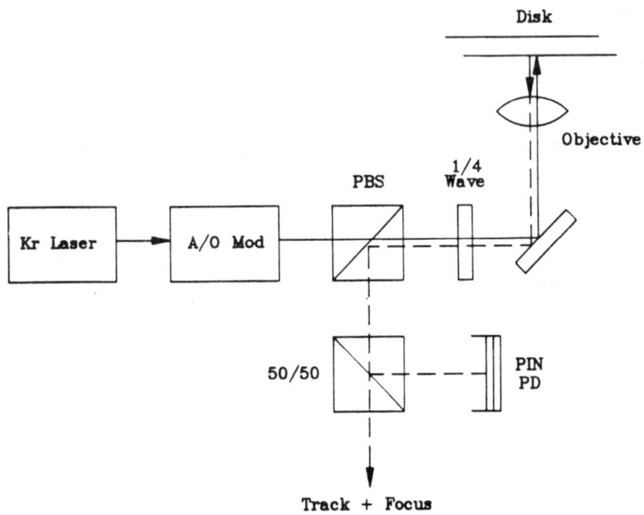

Figure 2. Kr laser test system.

Measurement of Disk Noise

Much of our substrate and disk noise evaluation has been performed on the system shown in Figure 2 which measures primarily amplitude noise. The A/O modulator is not an essential part of the system for this application but is included because the system is also used for testing phase change disks. The advantage of the Kr laser is that the power level P at the disk is not limited to the 10 to 15 mW typical of a GaAs laser system. When testing an uncoated plastic or glass substrate the reflected light collected by the objective is only a few percent of P, and in order to have sufficient light on the detectors, the 15 mW read level may be marginal. Another advantage is that the laser noise level is very low. However, the Kr laser wavelength and spot size are smaller than that for an equivalent GaAs system and if the primary interest is in data for a GaAs system it would be preferable to make all measurements using a diode laser. A system similar to that of Figure 2 could be built with a 40 mW laser diode that would be perfectly adequate for measuring uncoated substrates.

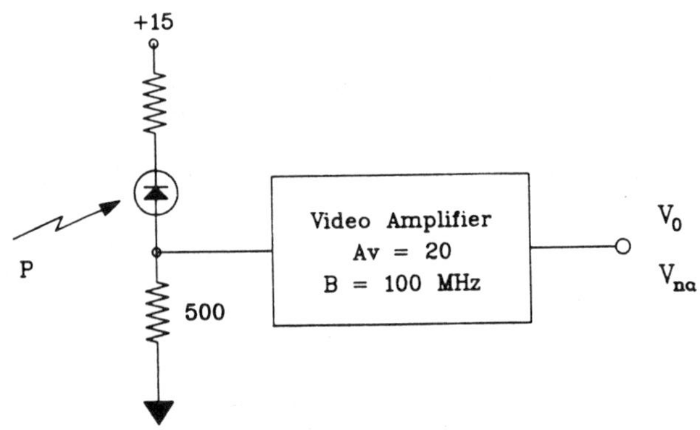

Figure 3. PIN diode detector.

The detection system used in this tester is shown in Figure 3. It was designed as a general purpose high speed detector and uses a fast PIN photodiode. For noise testing the frequency response should be reasonably flat up to the cut off frequency of the optics. We chose to do most of our testing at a media velocity of 3.14 m/s (100mm ϕ, 10 r/s) for which a bandwidth of 5 MHz is adequate. The amplifier noise V_{na} (in a bandwidth Δf = 30 kHz) is 14 μV. Using this value, the calculated CNR performance of the detector as a function of light intensity is shown in Figure 4. If the photocurrent is I amps then the diode shot noise current is given by

$$I_{ns} = \sqrt{2eI\Delta f} = 10^{-7}\sqrt{I} \tag{1}$$

The shot_noise voltage $V_{ns} = R\ I_{ns} \times A_v = 500\ I_{ns} \times 20 = 10^{-3}\sqrt{I}$. If the power incident on the detector is P μW and the responsivity is 0.5 A/W, then I = 0.5P μA and

$$V_{ns} = \sqrt{0.5P}\ \mu V \tag{2}$$

The noise voltage V_n is given by

$$V_n^2 = V_{na}^2 + V_{ns}^2 = (14^2 + 0.5P)\ \mu V^2 \qquad (3)$$

The d.c. output of the detector V_0 is given by

$$V_0 = 500 \times 0.5P \times 20 = 5 \times 10^3 P\ \mu V \qquad (4)$$

and an idealized sinusoidal readback signal that varied from V_0 down to $V_0/2$ would have an r.m.s. value of $V_s = V_0/4\sqrt{2}$. V_s, V_{ns} and V_n are plotted in Figure 4 which shows that with only 1 μW on the detector the CNR would be limited to 36 dB. As the power is increased the shot noise limited CNR increases by 3 dB/octave, but in addition, the effect of the amplifier noise is reduced. At 100 μW the CNR of our ideal signal would be 75 dB which, in a real system, would almost certainly be reduced by laser and disk noise. In an efficient M-O head where the light level on the detector is at most a few μW, the performance of this detection system would be poor.

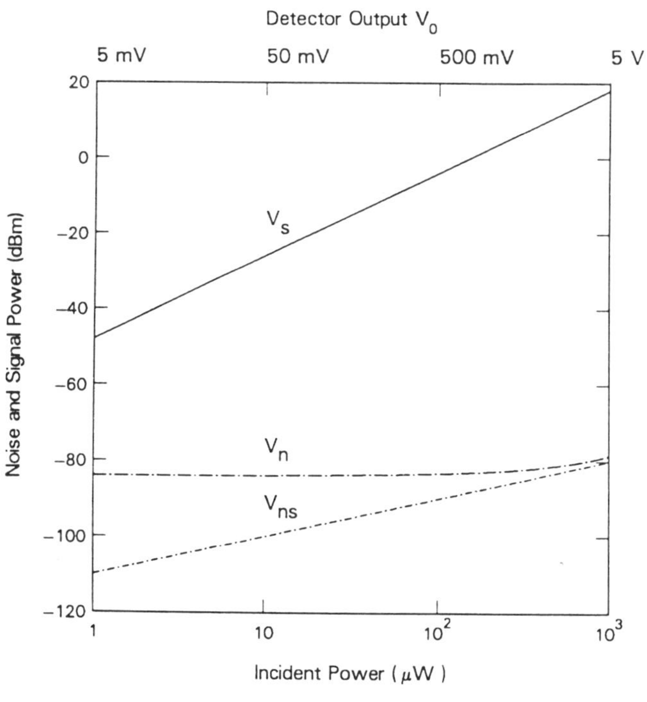

Figure 4. Dependence of signal and noise on light intensity.

If an avalanche detector with a gain G were used, it would boost both the signal current and I_{ns} by this factor. With G = 100 the shot noise would equal the amplifier noise at a light level of only 4 μW. There is , however, an additional noise associated with an avalanche photodiode[2] which must be taken into account. Future research will be directed toward understanding and quantifying this avalanche noise.

The Kr laser noise was measured by focusing the laser on an Al mirror. The power incident on the detector was adjusted to be 0.8 mW which gives V_0 = 4V (+25 dBm). The shot noise level is -80 dBm and the measured laser noise at 1 MHz as shown in Figure 5, is -68 dBm. This is 93 dB down from the d.c. level and is thus negligible when measuring disk noise.

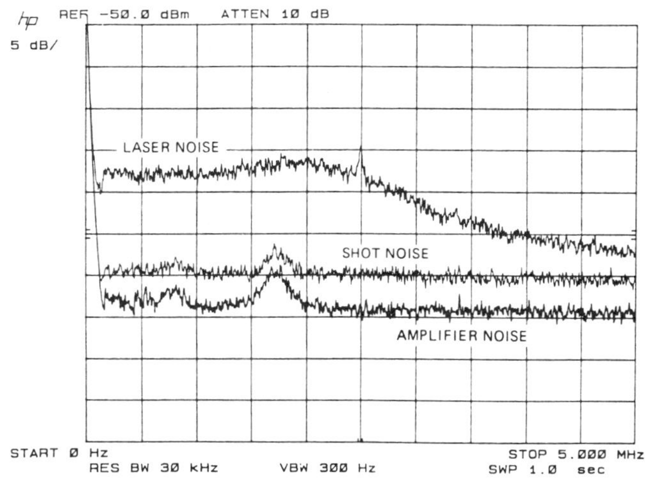

Figure 5. Kr laser noise.

Disk Noise

Disk noise may be caused by the surface roughness of the substrate and, in the case of a grooved substrate, by the roughness of the grooves. Defects on the surface such as dust particles will, if they are sufficiently dense, appear on the spectrum analyzer as noise, especially if averaging is employed. At some point a distinction has to be made between noise and defects: in this paper it is assumed that the surface features are sufficiently dense and random to create a smooth noise spectrum. The other sources of disk noise are the thin-film coatings that are deposited on the substrate. These are all intended to be uniform in composition and amorphous. In this case the reflectivity is constant and the only source of noise would be polarization noise from an M-O medium that was not uniformly magnetized. Some materials are difficult to deposit in an amorphous state and become polycrystalline. Random orientation of the crystallites with an associated change in refractive index, plus scattering and diffraction at the grain boundaries can cause high noise levels. Included in this class are certain metals such as Cr and Ni that are used in the fabrication of master disks and stampers for the injection molding or 2P process. Testing of these disks for groove noise may be impossible because of the noise created by the coating. Some metal oxides that are used as dielectric layers are difficult to deposit as amorphous films and others are subject to stress cracking. All of these sources of noise need to be controlled in order to produce a finished low-noise disk.

Figures 6 and 7 show results of disk noise evaluation on the Kr test system. In most cases the data was taken at 3.14 m/s (100mm ϕ, 10 r/s) with a light level at the detector of 40 μW (V_0 = 200 mV = -1 dBm). Thus the system is capable of detecting noise levels up to 83 dB below the d.c. level. Figure 6 is for an injection molded plastic substrate with a groove width of 0.8 μm and a pitch of 2.5 μm. The groove noise is much higher than the land noise which implies that noise is dominated by the roughness of the groove edges as described by Nakajima et al.[3] The diameter of the beam is about 0.75 μm (FWHM) and when it is tracking in the center of a land, only the tails of the beam probe the groove edges. For a similar disk with a pitch of 1.6 μm the land noise is much higher and if land and groove width are equal, the noise spectra may be identical.

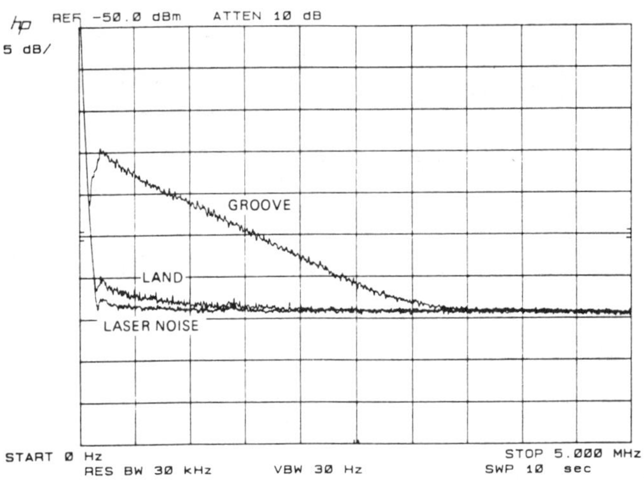

Figure 6. Plastic substrate with 2.5 μm pitch.

One of the problems of measuring groove noise when it is dominated by the groove edges is focussing the beam. If the groove width is comparable to that of the beam then the noise will be maximized when the beam is slightly out of focus. If the grooves have sector marks or track headers, these may be used for focussing. If not, the track crossing signal, when the track servo is unlocked, may be maximized but this will only be valid if the astigmatizm in the beam is small compared with the depth of focus. The method will definitely not work with heads that use an astigmatic lens for the servo system.

Figure 7 shows tests of a grooved disk with a pitch of about 1.6 μm on the Kr test system and on a similar system that employs a GaAs laser. The detectors are identical but the spot size of the GaAs laser is about 1 μm. The differences between the two results are (a) about 6 dB of noise from the GaAs laser, (b) a lower cutoff frequency for the GaAs laser due to the larger beam diameter and (c) less discrimination between land and groove noise, again due to the larger beam. The magnitude of the noise at low frequency is almost identical; this is somewhat surprising considering the difference in beam size and wavelength.

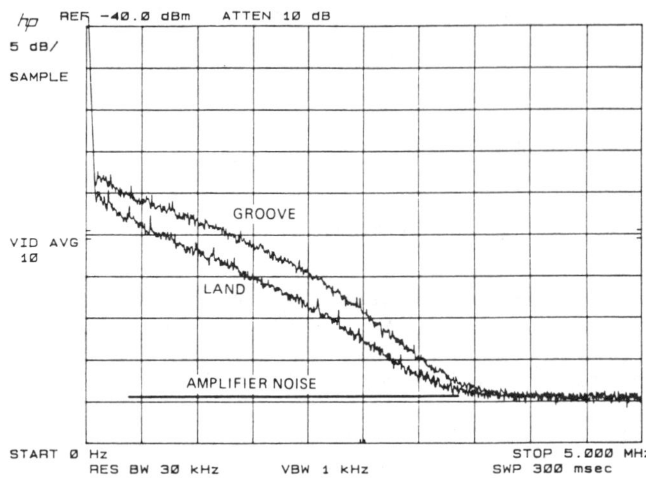

Figure 7a. Grooved disk with 1.6 μm pitch (Kr laser).

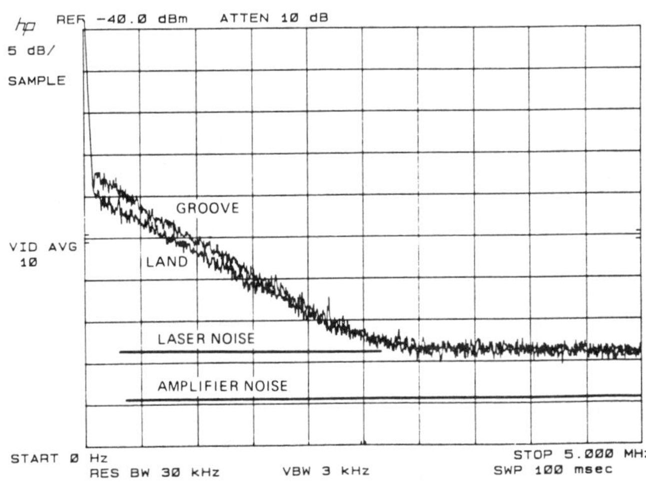

Figure 7b. Grooved disk with 1.6 μm pitch (GaAs laser).

All the previous measurements have been made with a fixed media velocity of 3.14 m/s and the question arises as to the behavior of the disk noise as a function of velocity. If the profile of the focussed beam in the scan direction x is given by

$$h(x) = I_0 \exp(-x^2/2\sigma^2) \qquad (5)$$

and the reflectivity of the disk is r(x), then the read signal g(x) is the convolution of h and r. In the spatial frequency domain, G(k) = R(k) H(k) where

$$H(k) = \mathfrak{F}\{h(x)\} = I_0 \sqrt{2\pi}\, \sigma \exp(-k^2\sigma^2/2) \qquad (6)$$

If the statistical variation of r(x) may be characterized as "white noise", then R(k) is a constant and G(k) has the same form as H(k). The spectrum analyzer displays the power spectral density function $P(\omega)$ which is proportional to $|G(\omega)|^2$.

If the disk velocity is v, then we can find $P(\omega)$ by substituting $\omega = vk$ and obtain the expression

$$P(\omega) = K \frac{\sigma^2}{v} \exp(-\omega^2\sigma^2/v^2) \qquad (7)$$

The term v in the denominator is included to account for the fact that the total noise power is constant, and hence for a given resolution bandwidth B, $P(\omega)$ must be inversely proportional to v.

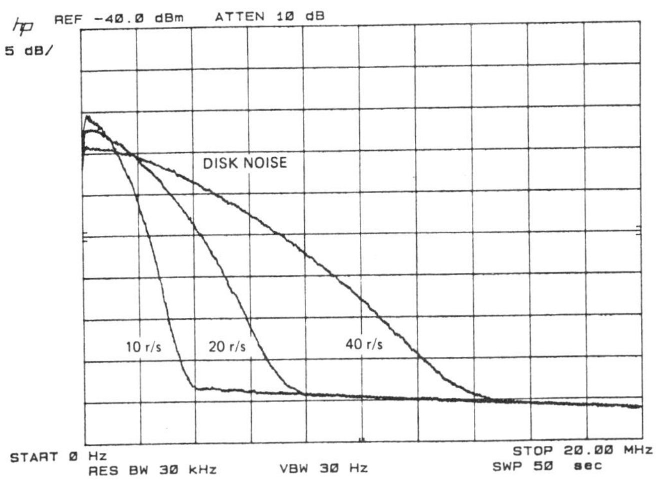

Figure 8. Disk noise at different media velocities.

Figure 8 shows noise measurements taken at velocities of π, 2π and 4π m/s. The d.c. values differ by approximately 3 dB as predicted by the model. We can use the data to estimate σ by measuring the frequency at the e^{-3} points.

$$P(\omega)/P(0) = e^{-3} \quad \text{when} \quad \omega^2\sigma^2 = 3v^2 \qquad (8)$$

That is, $\sigma = \sqrt{3}\ v/\omega$ and W, the width of the laser beam at half its maximum intensity, is given by

$$W = 2.35\sigma \simeq 4v/\omega = 2v/\pi f \qquad (9)$$

Taking the -13 dB (e^{-3}) points from Figure 8 we get W = 0.87, 0.93 and 1.0 μm. These values are larger than that measured by a knife edge scan of the Kr beam (0.75μm) which is most probably due to the disk noise reflectivity r(x) not being "white". Note that the groove noise spectrum shown in Figure 6 has a different shape from the curves of Figure 8. This indicates a different spectral content of the disk noise.

Magneto-Optic Test System

The optical system used for testing M-O media is shown schematically in Figure 9. Two of the polarizing beam splitters PBS1 and PBS2 are rotated slightly so that they reflect not only all of the s-polarized light (rotated light from the M-O disk), but also a controlled fraction of the un-rotated p-polarized light. Note that whatever fraction of the p-light is sent to the detectors (track, focus and data) the same fraction is diverted on the inward path from laser to disk. Thus the desire to have plenty of light at the detectors, for reliable servoing and high S/N with PIN detectors, is in direct conflict with the goal of having efficient delivery of light from the laser facet to the disk. Note also that a large fraction of the light reflected by the disk is returned to the laser facet, causing undesirable laser noise. One solution to this problem employing a Faraday rotator was recently discussed by Yamanaka et al.[4]

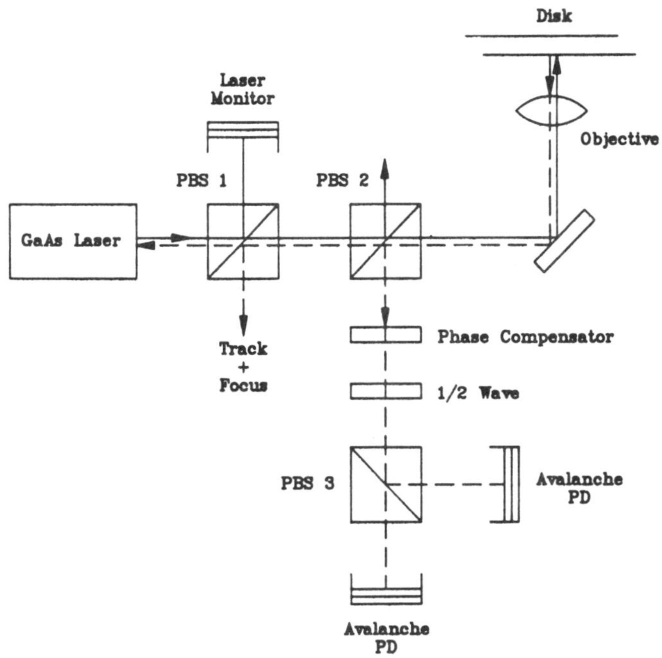

Figure 9. M-O test system.

The outputs of the avalanche detector/preamplifiers are a.c. coupled to a differential amplifier whose output is connected to a spectrum analyzer. In a typical CNR measurement a track of the disk is first beam erased, then the bias field is reversed and a carrier is written at a frequency f with a certain laser power and pulse width. If the read power incident on the disk is P watts, the reflectivity is R and the Kerr rotation is Θ_k, then the power levels at the two detectors are given approximately by:

$$P_{1,2} = 0.5\eta PR(1+\delta m)\{\gamma^2 \pm 2\gamma M(\delta\Theta + \Theta_k \cos 2\pi ft)\} \qquad (10)$$

where η is the efficiency with which light is delivered from the disk to the detectors, δm represents the amplitude noise components of laser noise and disk noise, $\delta\Theta$ is the polarization noise of the disk, γ^2 is the fraction of p-light that is deflected by PBS2, and M is the MTF of the optical system at the recording frequency f.

The efficiency η depends not only on the transmission and reflection of the optical components, but also on the disk groove structure. The apparent reflectivity when tracking in a groove or on a land may be significantly less than R, measured on a flat area of the disk. The differences, caused by light diffraction and scattering are accounted for by the parameter η. The term γ^2 in Equation (10) represents the d.c. light level at the detectors and determines the shot noise. The amplitude noise from this d.c. term is cancelled by the differential amplifier but the noise from the modulation term is additive and will show up as a component of the writing noise if the detection system is balanced for an erased track.

In the analysis that follows the voltage at the output of the differential amplifier is calculated assuming a responsivity of the detectors equal to
S A/W, and the combined effect of the avalanche gain and that of the amplifier is represented by the effective impedance Z. Avalanche noise[2] is not included in this model. Thus, the shot noise V_{ns}, r.m.s. signal voltage V_s, amplitude noise V_{nm} and polarization noise V_{np} are given by

$$< V_{ns} >^2 = 2eS\eta PR\gamma^2 \Delta f Z^2 \tag{11}$$

$$< V_s >^2 = 2(SZ\eta PR\gamma M\Theta_k)^2 \tag{12}$$

$$< V_{nm} >^2 = 2(SZ\eta PR\delta m\gamma M\Theta_k)^2 \tag{13}$$

$$< V_{np} >^2 = (2SZ\eta PR\gamma M\delta\Theta)^2 \tag{14}$$

The shot noise limited CNR is given by:

$$\mathrm{CNR} = 10\log_{10} < V_s >^2 / < V_{ns} >^2$$

$$= 10\log_{10}(S\eta PRM^2\Theta_k^2/e\Delta f) \tag{15}$$

Choosing the following values for these parameters: S=0.5, η=0.6, R=0.2, M=0.6, Θ_k=0.01 rad and P = .002 W, we calculate a CNR of 61.3 dB. This value is close to that actually measured on an un-grooved, low-noise disk having similar parameters.

The presence of laser or disk amplitude noise will show up as writing noise and $V_{nm} = V_{ns}$ when

$$S\eta PR < \delta m >^2 M^2 \Theta_k^2 = e\Delta f \tag{16}$$

Using the above parameters this implies that there will be 3 dB of writing noise at the recording frequency when $< \delta m >^2 = 1.1 \times 10^{-6} \simeq$ -60 dB.

Polarization noise is not cancelled by the differential detectors, so if it is the dominant noise source, there is no justification for the added expense of the differential system. The polarization noise $V_{np} = V_{ns}$ when

$$2S\eta PRM^2 < \delta\Theta >^2 = e\Delta f \tag{17}$$

Using the above parameters again, there will be 3 dB of disk noise at the recording frequency when $< \delta\Theta > = 7\mu$rad. This noise may be caused by poor erasure of a track or it may be associated with the roughness of the groove edges. If a track is beam erased the width of the erased region will depend on a number of parameters including the beam power and the bias field. There will always be regions outside of the erased track that are seen by the tails of the read beam. If the magnetization state of these regions is random, this will show up as polarization noise.

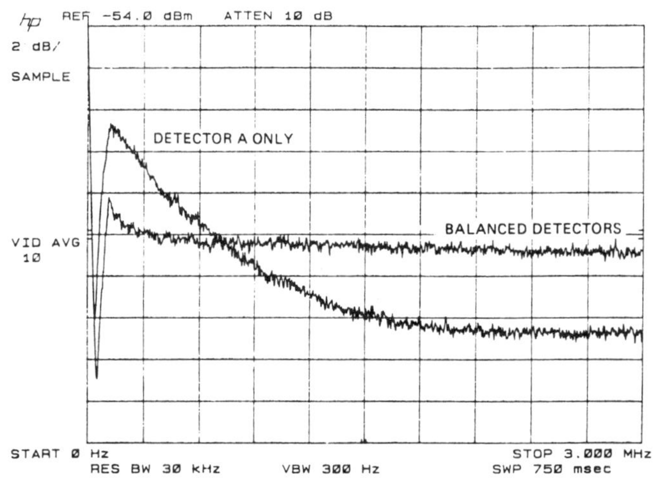

Figure 10. Disk noise cancelled by differential detectors.

As an example of these different disk noise types, Figures 10 and 11 show results from the M-O test system on two different disks. In both cases the detectors were balanced for minimum noise and then the light incident on one of the detectors was blocked. The noise level

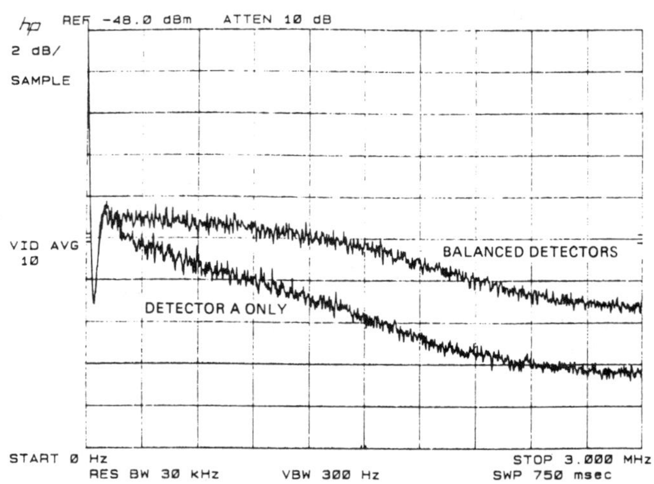

Figure 11. Disk noise not cancelled by differential detectors.

from both detectors includes an additional 3 dB of shot noise which explains the noise increase at 3 MHz. The disk in Figure 10 has about 10 dB of amplitude noise which is almost completely cancelled by the differential amplifier. The disk in Figure 11 has both amplitude and polarization noise. The amplitude noise is subtracted but the polarization noise is added; the net result is a residual 4 dB of disk noise.

Conclusions

Means of identifying and measuring the noise components in the readback signal from an optical disk have been examined. Some of these may be minimized through the design of the readback system but there are tradeoffs with complexity and optical efficiency that must be made. In an M-O head it has been shown that differential detection can reduce the level of amplitude noise, such as that from the laser, but not eliminate it completely. Polarization noise, on the other hand, is not cancelled by differential optics and may be the limiting factor in obtaining high CNR from a grooved disk. Finally some simple models have been presented that help to explain the disk noise spectrum and the performance of an M-O head.

Acknowledgement

The author wishes to acknowledge the technical assistance of U. G. Gerber and several useful discussions with M. Chen, R. T. Lynch and D. Rugar.

References

1. Chen, M., Bell, A. E., and Marrello, V., "Characterization of Optical Recording Disk Noise", SPIE Vol 420 - Optical Storage Media (1983), p. 306-12.

2. Mansuripur, M, Connell, G. A. N. and Goodman, J. W., "Signal and Noise in Magneto-optic Readout", J. Appl. Phys. 53(6), June 1982, p. 4485 -94.

3. Nakajima, M., Hamada, M., Moribe, M., Hirano, H., Itoh, K. and Ogawa, S., "Reduction of Media Noise in Optical Disks", Optical Data Storage, paper ThCC5, Washington, DC, Oct. 15, 1985.

4. Yamanaka, Y., Katayama, R., Itoh, M., Ohta, Y., Hibiya, T. and Okada, M., "Magneto-optical Disk Head With Optical Circulator Structure", CLEO '86, p. 342-3.

Reprinted from *Optical Mass Data Storage II*, Proc. SPIE Vol. 695, pp. 270-276 (1986). ©1986 Society of Photo-Optical Instrumentation Engineers.

Retardation effects in magneto-optic readout

Alan B. Marchant

Verbatim Corporation, 323 Soquel Way, Sunnyvale, California 94086

Abstract

The performance of a magneto-optic readout system is easily degraded by optical retardations, or polarization-dependent phase shifts. Optical interference structures meant to enhance the magneto-optic effect can produce retardation which reduces the readout signal. Polarization-dependent optical elements (prisms, mirrors, etc.) can either exacerbate or compensate for this problem. Retardation due to substrate birefringence is an even more serious problem because it can cause relatively large intensity fluctuations at the photodetectors. Injection-molding techniques which minimize normal-incidence birefringence in polycarbonate substrates have been demonstrated; however vertical birefringence, which causes large retardation for marginal light rays, remains an unsolved problem.

Introduction

Magneto-optic (M-O) readout is accomplished by detecting the slight polarization rotation (Kerr effect) of a linearly polarized beam reflected off the recording film. An analyzer mounted at some angle to the incident polarization direction (say, $90^0-\theta$) passes more/less light to a photodetector, depending on whether the rotated beam polarization is more/less aligned with the analyzer axis. The direction of rotation (clockwise or counterclockwise) depends on the magnetization of the recording film in the illuminated region. As the beam scans along a data track on an M-O disk the modulated part of the photocurrent presents an analog representation of the magnetization pattern previously recorded on the track.

This idealized M-O readout system can be understood using Jones Matrix formalism to show how the light beam is affected by the M-O medium and the analyzer. The x-axis is defined as the direction of incident polarization, so the field-representation of the incident beam, normalized with respect to the incident power, is (1 , 0). At the photodetector, the field is transformed to the following:

$$(1, 0) \begin{pmatrix} 1 & \pm k \\ \mp k & 1 \end{pmatrix} \begin{pmatrix} \cos\theta & \sin\theta \\ -\sin\theta & \cos\theta \end{pmatrix} \begin{pmatrix} 0 & 0 \\ 0 & 1 \end{pmatrix} = (0 , \sin\theta \pm k \cos\theta). \tag{1}$$

The first transfer matrix represents Kerr rotation, divided by the media reflectance. In the ideal case of pure rotation, $\pm k$ is the tangent of the Kerr rotation angle. Since this is always quite small ($<1^0$ in current systems) k is approximately equal to the Kerr angle. The second transfer matrix rotates the coordinate system to align it perpendicular to the analyzer. The third matrix represents the ideal polarizing function of an analyzer. The light intensity at the photodetector and the resultant photocurrent are proportional to the square of the field:

$$I \propto |(0, \sin\theta \pm k \cos\theta)|^2 = \sin^2\theta + k^2\cos^2\theta \pm k \sin 2\theta \tag{2}$$

The photocurrent has a small signal ($S = \pm k\sin 2\theta$) superimposed on a relatively large DC level ($DC = \sin^2\theta + k^2\cos^2\theta$) which is essentially independent of the magnetization pattern. Note that the signal is maximized when $\theta = 45^0$, but at this point the DC is quite large. The ratio of signal divided by media noise or laser noise (S/DC) is maximized when the analyzer is nearly perpendicular to the incident polarization ($\theta = k$). The ratio of signal to shot noise ($S/\sqrt{DC}$) is maximized at an alignment angle $\theta \simeq \sqrt{k}$ (about 10^0).

Two detection schemes are in common use. "Single detection" uses one detector and one analyzer, oriented at an intermediate small angle to maximize the overall signal-to-noise-ratio (SNR). In "differential detection", two detectors are used to increase the signal and remove the DC offset. The analyzer is a polarizing beam splitter which divides the light into orthogonally polarized beams directed at the two detectors. The effective analyzer orientations for the two detectors are $\pm 45^0$. Their photocurrents are differenced, leading to a net signal, $S=\pm 2k$, cancellation of the DC, and a substantial increase in SNR.[1]

So far, this discussion has assumed perfect rotation in the Kerr effect and no phase shifts or polarization rotation in the readout optics. But phase shifts leading to significant variations in the signal and/or DC levels are not unusual. The second section of this paper describes the effects of media-induced polarization ellipticity in the Kerr effect, which tends to reduce the signal. The third section shows how ordinary optical components used in the readout system can either exacerbate or compensate for the Kerr ellipticity problem. The final section describes the effects of substrate birefringence, which can lead to serious fluctuations in the DC level.

Kerr Ellipticity

From a simplistic optical point of view, the polar Kerr effect arises because vertical magnetization (parallel to the direction of beam propagation) alters the refractive index for left-hand and right-hand circularly polarized light. For example, refractive indices $n_r = 3.24 + 3.58$ and $n_l = 3.16 + 3.52$ have been measured on a particular magnetized sample of TbFeCo.[2] Upon reflection, circularly polarized light remains circular, with a change in field amplitude and phase given by the complex reflection coefficient, r_r or r_l, which is a function of n_r or n_l, respectively. E.g. for normal air-incidence on a thick magneto-optic film we calculate $r_{r,l}=(n_{r,l}-1)/(n_{r,l}+1)$. In a practical M-O disk, other thin films and optical interfaces must be considered; standard multilayer calculations must be used to compute r_r and r_l. Note that (by symmetry) the values of r_r and r_l are reversed when the magnetization switches.

Given that circularly polarized beams are the eigen-states for reflection from an M-O medium, the Kerr effect Jones matrix can be derived as

$$\bar{r}\begin{pmatrix} 1 & \pm i\Delta r/\bar{r} \\ \mp i\Delta r/\bar{r} & 1 \end{pmatrix} \equiv \bar{r}\begin{pmatrix} 1 & \pm k\cdot e^{i\kappa} \\ \mp k\cdot e^{i\kappa} & 1 \end{pmatrix} \tag{3}$$

where $\bar{r} \equiv (r_r+r_l)/2$, $\Delta r \equiv (r_r - r_l)/2$, and k and κ are the amplitude and phase of $i\Delta r/\bar{r}$. The average media reflectance, $\bar{r}$, serves as an overall scale factor and will be ignored in the remaining discussions. For $\kappa=0$, the Kerr matrix achieves the ideal condition of pure rotation. This occurs when r_r and r_l have equal magnitudes. But circular dichroism ($|r_r| = |r_l|$) is usually present, especially when multilayer interference structures are used to enhance the weak M-O signals, and κ may take on any value whatever.

A more realistic model of the M-O readout process should include the effect of κ. It should also account for the beam splitter which deflects the beam to the detectors (instead of back to the laser). This is usually a polarizing beam splitter cube (PBSC) modified to transmit most of the incident polarized light to the medium, and deflect a small fraction (f^2) of the reflected light including nearly all of the orthogonal polarization created by the Kerr effect. The transfer formula for the optical field at the detector (eq. 1) now becomes

$$(1, 0)\begin{pmatrix} 1 & \pm ke^{i\kappa} \\ \mp ke^{i\kappa} & 1 \end{pmatrix}\begin{pmatrix} f & 0 \\ 0 & 1 \end{pmatrix}\begin{pmatrix} \cos\theta & \sin\theta \\ -\sin\theta & \cos\theta \end{pmatrix}\begin{pmatrix} 0 & 0 \\ 0 & 1 \end{pmatrix} \tag{4}$$

$$= (0, f\sin\theta \pm k\, e^{i\kappa}\cos\theta)$$

and the photocurrent (eq. 2) becomes

$$I \propto f^2\sin^2\theta \pm f\, k\cos\kappa\, \sin 2\theta . \tag{5}$$

It has been noted that a small value of f effectively increases the Kerr rotation, increasing the SNR without requiring careful alignment of the analyzer.[3] However, the effective reflectivity of the disk and the maximum achievable signal (at $\theta = 45^0$) are significantly reduced. Very small values of f may be counterproductive in systems employing differential detection.

Equation 5 also shows that the M-O signal is not dependent only on the reflectivity of the medium orthogonal to the incident polarization (which equals $\bar{r}^2k^2$). Rather, the signal is proportional to $k \cos\kappa$. In general, the reflected light is elliptically polarized with the major axis rotated an angle θ_k, where $\tan 2\theta_k = 2\cos\kappa \;/\; (1-k^2)$. So in the small-angle limit, the signal is proportional to θ_k.

The importance of κ in even the most mundane situations is illustrated in Figure 1. θ_k is plotted as a function of overcoat thickness (AlN, n=2.1) for first-surface readout of a TbFeCo film. θ_k reaches a maximum when the overcoat is about 800 Å thick, at which point the reflectivity is nearly minimized. At this point, κ is close to 0^0. But if a thinner overcoat is employed (e.g. to achieve higher reflectivity) then κ will be large enough to significantly reduce the signal. If an optical element, such as a Babinet compensator oriented parallel to the x-axis, is used to modify κ, the Kerr angle can be increased markedly (c.f. the $+40^0$ curve). Mansuripur has suggested that this sort of technique should be used to remove all ellipticity from the reflected beam.[1] In the case of conventional differential detection, variable compensation can be achieved more cheaply using a quarter-wave-plate (QWP). If the QWP is oriented at 45^0 to the x-axis, rotation of the analyzing beam splitter or of an intermediate half-wave-plate will permit perfect optimization of the signal.

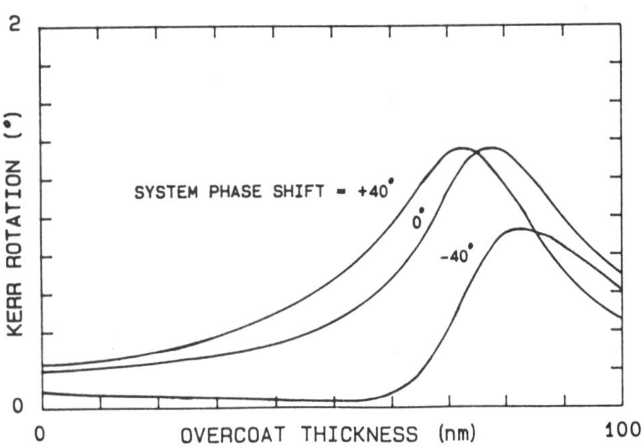

Figure 1

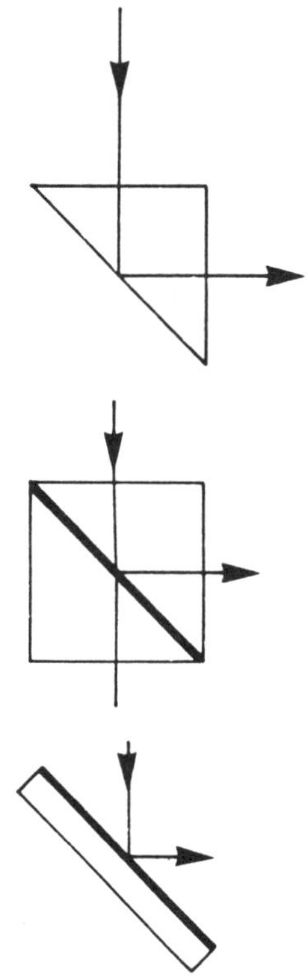

Figure 2

Optical elements

Figure 1 also demonstrates the devastating impact which an inappropriate system phase shift can have (c.f. the -40^0 curve). Any beam-turning element, or even a beam splitter used in transmission, can introduce phase differences between the p- and s-polarizations. Such phase shifts may not affect the incident beam, which should be completely polarized, but they must be taken into account in the design of any practical M-O readout system.

Figure 2 illustrates the phase-shifts expected from several types of beam-turning optical elements. A 45^0 right-angle prism can be used to bend the beam 90^0 without any loss in power. However, the resultant phase shift is very large - 37^0 for n=1.5. A half-silvered beam splitting cube (n_{glass}=1.5, Ag thickness >50Å) introduces a lesser phase-shift, but still $>20^0$. Perhaps the best simple optical element is a front-surface Aluminum mirror, with a phase-shift of only 9^0. But note that dielectric coatings used to protect and enhance such mirrors can increase the phase shift.

In most M-O heads, a PBSC is used to divide the illumination and readout beams. The PBSC includes a complex multilayer film with a detailed structure dependent on the desired reflectivities, wavelength range, and other factors. Clearly, the phase characteristics must also be taken into account in designing and testing this beam splitter. In a mature head design, where the media characteristics (i.e. κ) are well established, this component should be designed to compensate for phase-shifts from the media and other components.

Birefringence

The most important - and problematical - source of retardation in M-O systems is substrate birefringence. From most points of view - (Tg, stability, clarity, moldability, etc.) bisphenol-a polycarbonate is a good candidate for M-O disk substrates. However, it is prone to birefringence (a difference in refractive index for different polarizations) and resultant retardation (Δn times the distance traversed). Several years of process development have resulted in substrates with very low retardation for normally-incident light. But, as Figure 3 shows, the real problem has not been solved. The single-pass retardation for a particular injection-molded polycarbonate disk was less than 10nm at normal incidence, but increased to as much as 80nm ($\lambda/10$) at an incidence angle of 30°. Note that for an objective lens with a numerical aperture of 0.5, nearly half the light strikes the disk at incidence angles greater than 20°. This pattern of high retardation at moderate incidence angle is typical of all injection-molded polycarbonate disks we have examined. Intrinsically non-birefringent substrate materials such as glass or PMMA do not show this effect.

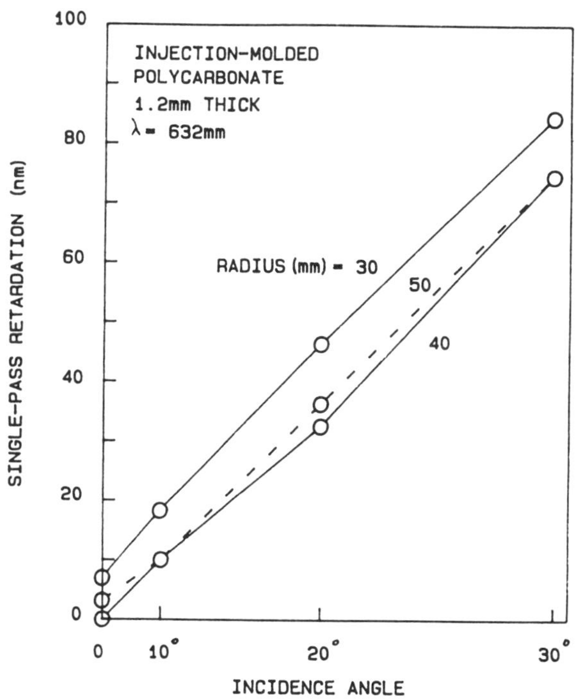

Figure 3

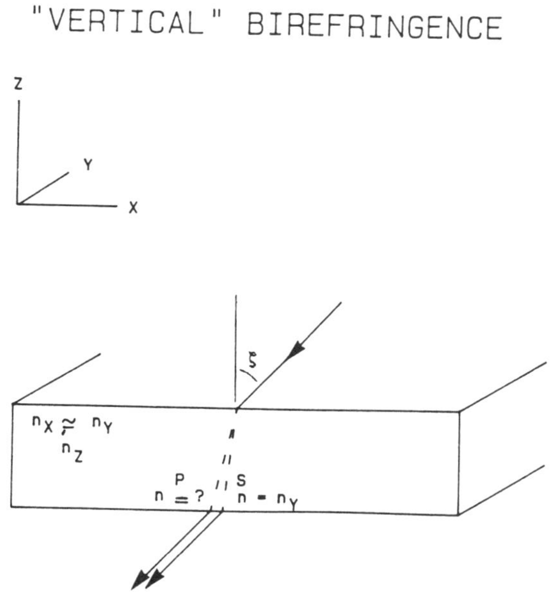

Figure 4

Before analyzing the effects of substrate retardation on M-O readout, we should consider just how the large retardation at non-normal incidence arises. In a previous article on the optics of cover sheets for DRAW optical disks, this author described how birefringence can have a large "vertical" component.[4] The general optical description of the substrate includes three refractive indices (n_x, n_y, and n_z) for light polarized along the axes of a particular coordinate system. The values of n_x, n_y, and n_z can change somewhat with position. The symmetry of the injection-molding process is such that we can identify x and y with the radial and azimuthal directions at a given point in the substrate disk, while the z axis is aligned normal to the disk. A light beam which is normally incident always has its E-field aligned in the x-y plane, so any retardation depends only on the ordinary birefringence, n_x-n_y, which is effectively minimized in modern injection-molding processes. On the other hand, light which passes obliquely through the substrate may sense n_z to some extent. And n_z differs from n_x and n_y by a huge value - nearly 10^{-3}. This vertical birefringence must be due in some way to the strong vertical shear conditions which are frozen in during the molding cycle.

The role of n_z in the propagation of obliquely incident light can be understood with the aid of Figure 4. Suppose that $n_x=n_y$ (or that the plane of incidence happens to include the x-axis). S-polarized light incident at any angle ζ will sense only n_y and will refract and propagate through the substrate accordingly. But for p-polarized light the E-field points in a direction intermediate between $\hat{x}$ and $\hat{z}$ so the effective index of refraction should be intermediate between n_x and n_z. Fresnal's equation of wave normals describes light propagation in the substrate:

$$\frac{s_x^2}{v^2-v_x^2}+\frac{s_y^2}{v^2-v_y^2}+\frac{s_z^2}{v^2-v_z^2}=0 \tag{6}$$

where $\hat{s}$ is the wave normal direction, v is the phase velocity (c/n, where n is the effective refractive index), $v_x \equiv c/n_x$, etc. For the case we are considering, $s_y = 0$ and two solutions are possible, corresponding to s and p polarization:

$$n_s=n_y,\ \text{and}$$

and

$$n_p \simeq n_x + (n_z-n_x)\ \sin^2\zeta/n_x2 \tag{7}$$

to first order in (n_z-n_x). Also to first order, the single-pass retardation is,

$$(n_x-n_y + (n_z-n_x)\ \sin^2\zeta/n_x2).t/\cos\zeta' \tag{8}$$

where t is the substrate thickness and ζ' is the propagation angle within the substrate, $\sin\zeta'=\sin\zeta/n_x$. The abscissa in Figure 3 is proportional to $\sin^2\zeta$; the linearity of the experimental curves adequately support this model.

The Jones matrix representation for the E-field at the detector in a general M-O readout system is

$$(1\ ,\ 0)\begin{pmatrix}\cos\omega & \sin\omega\\ -\sin\omega & \cos\omega\end{pmatrix}\begin{pmatrix}e^{i\delta/2} & 0\\ 0 & e^{-i\delta/2}\end{pmatrix}\begin{pmatrix}1 & \pm ke^{i\kappa}\\ \mp ke^{i\kappa} & 1\end{pmatrix}\begin{pmatrix}e^{i\delta/2} & 0\\ 0 & e^{-i\delta/2}\end{pmatrix}$$

$$\times\begin{pmatrix}\cos\omega & \sin\omega\\ \sin\omega & \cos\omega\end{pmatrix}\begin{pmatrix}f & 0\\ 0 & e^{i\phi}\end{pmatrix}\begin{pmatrix}\cos\theta & \sin\theta\\ -\sin\theta & \cos\theta\end{pmatrix}\begin{pmatrix}0 & 0\\ 0 & 1\end{pmatrix} \tag{9}$$

$$= (0\ ,\ \sin\theta\ f\ \ (\cos\ \delta +\ \ i\cos2\omega\sin\delta\ )$$

$$+ \cos\theta\ \ e^{i\phi}\ (\pm ke^{i\kappa}\ -\ i\sin2\omega\ \sin\delta\ )\ )$$

where δ is the single-pass retardation through the substrate (multiply nm of retardation by $2\pi/\lambda$), ω is the orientation of the principal retardation axes, ϕ is the system retardation as discussed in the previous section, and the overall multiplicative terms such as reflectivity and transmissivity have been omitted. Light rays coming through the center of the objective lens aperture encounter only the normal-incidence birefringence for which $\omega= 0^0$ or 90^0 because of the molding symmetry. But for light obliquely incident through the rest of the aperture vertical birefringence dominates and ω is determined by the orientation of the s- and p-polarizations; that is, ω simply corresponds to the angular position of the beam in the aperture.

Equation 9 is a good starting point for any realistic model of M-O readout. For the sake of a simplified discussion, consider the special case $\kappa=0$ and $\phi=0$ so that birefringence is the only source of retardation. Further, subsume f into θ via the definition $\tan\theta' = f\ \tan\theta$. The resultant E-field and detector currents in this case are, to first order in k,

$$E = (\ 0,\ \sin\theta'\cdot\cos\delta +\ i\ \sin\ (\theta' -2\omega)\cdot\sin\delta \pm\ k\ \cos\theta\ ') \tag{10}$$

$$DC = \sin^2\theta'\cdot\cos^2\delta\ +\ \sin^2(\theta\ '-2\omega)\cdot\sin^2\delta$$

$$S\ =\ \pm\ k\ \sin\ 2\theta'\ \cdot\ \cos\delta\ .$$

For normal incidence (ω=o), comparison with equation 2 shows that the only affect is a slight reduction in signal (by $\cos\delta$). But because of vertical birefringence, ω takes on all values around the aperture and the overall situation is much different.

Figure 5 illustrates a simple technique for visualizing the effects of vertical birefringence. Pass a strongly divergent light beam (NA comparable to that of the objective lens) through a polarizer, substrate sample, and second polarizer and look at the intensity distribution on a screen. When the two polarizers are crossed, the light is extinguished at the center of the aperture and along the arms of a "Zernicke cross". If the substrate is removed, or replaced with glass, the entire pattern is essentially extinguished. In single detection, θ is kept small so that the analyzer blocks out most of the DC term. But the effect of vertical birefringence, evident in Figure 5 and equation 10, is that the DC level and accompanying noise are greatly increased.

POLARIZED, DIVERGENT SOURCE

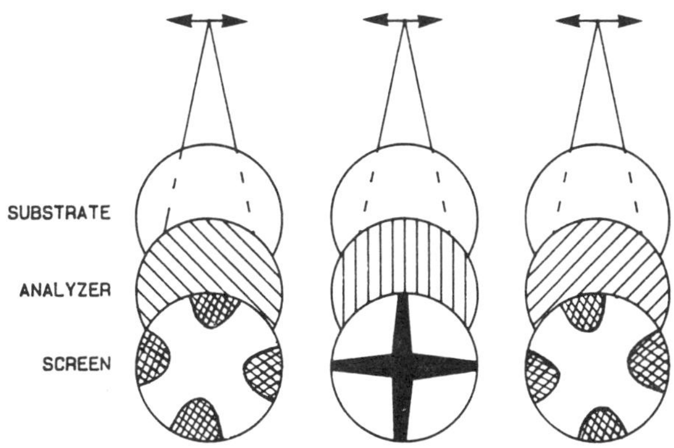

VISUALIZING "VERTICAL" BIREFRINGENCE

Figure 5

The influence of vertical birefringence in a differential detection system is more subtle, but no less serious. Figure 5 illustrates the light intensity on detectors with positive and negative analyzer angles. The total amount of light is ideally equal in the two cases so differencing the photocurrents should remove any offset. However, note that the intensity patterns are different in the two cases: the detectors pick up light from different parts of the aperture. Ideally, any defect in the recording layer or substrate affects the two detectors equally and is cancelled out in the differenced signal. But since the illumination patterns are now unequal, this type of "noise" is no longer cancelled out cleanly. For a well-filled aperture ($\sim$ 10% beam intensity at the edge) the offset between the two detector currents divided by the sum can approach $0.1 \sin^2\delta_{max}$ for the worst-case disturbance of the aperture, where δ_{max} is the retardation at the edge of the aperture. Recall from Figure 3 that for polycarbonate $\delta_{max} \sim 80$nm or $\sin\delta_{max} > 0.5$.

Summary

This paper is not a comprehensive treatment of retardation effects in M-O readout. For instance, substrate birefringence may make it impossible to compensate fully for a phase-shifted Kerr effect (κ). Nevertheless, several serious problems have been explained. Kerr effect phase shifts, which are particularly prevalent with multilayer "enhancement" structures, can significantly degrade the readout signal. Phase-shifts due to optical elements in the readout chain can cause similar degradation - or the two effects can be engineered to compensate for each other.

Substrate birefringence may cause the most serious retardation effects. Vertical birefringence from injection-molded polycarbonate substrates increases the background light (and noise) level for single-detection and it causes a serious unbalancing of the detector signals in differential detection. It is hoped that a clearer understanding of birefringence effects will motivate the development of techniques to overcome them and/or practical substrate materials to replace polycarbonate.

Acknowlegement

I wish to acknowledge helpful discussions with M. Mansuripur especially regarding the use of a QWP for retardation compensation.

REFERENCES

1 M. Mansuripur and G. Connell, J. Appl. Phys., 53, 4485 (1982).
2 Y. Yorozu, Sumitomo Metal Mining Co., private communication.
3 T. Deguchi, et al., Applied Optics, 23, 3972 (1984).
4 A. Marchant, SPIE Proceedings, 421, 43 (1983).
5 M. Born and E. Wolf, Principles of Optics, 5th ed., Pergamon, New York Section 14.2.2, 1975.

Reprinted with permission from *Journal of the Optical Society of America A*, Vol. 3(12), pp. 2086-2093 (December 1986). ©1986 Optical Society of America.

Distribution of light at and near the focus of high-numerical-aperture objectives

M. Mansuripur

College of Engineering, Boston University, 110 Cummington Street, Boston, Massachusetts 02215

Received December 20, 1985; accepted July 23, 1986

Classical diffraction theory is used to investigate the effects of high numerical aperture on the focusing of coherent light. By expanding the diffracted beam in plane waves, we show that the lens action can be expressed as a succession of three Fourier transforms. Furthermore, polarization effects are included in the model in a natural way. Some numerical results of the theory are also presented.

1. INTRODUCTION

To achieve high resolution in many optical systems of practical interest, it is necessary to use high-numerical-aperture objectives. For instance, in optical data storage one must focus the laser beam to extremely small spots for recording and readout of information. For a well-corrected lens, the limit of resolution is the radius of the Airy disk, which is equal to $0.61\lambda/\mathrm{NA}$. Here λ is the wavelength of light and $\mathrm{NA} = \sin(\alpha/2)$ is the numerical aperture, with α being the angle subtended by the exit pupil at the focal point. Standard approximations of the classical diffraction theory can be used to describe the field pattern around the focal plane, but, strictly speaking, these approximations are valid only for $\mathrm{NA} \lesssim \sqrt[4]{\lambda/f}$. When the focal length is $f = 4000\lambda$, for example, the upper limit of validity of the standard approximations is $\mathrm{NA} = 0.125$.

Hopkins[1] has shown that a more rigorous treatment yields almost the same results as the approximate method, provided that the aperture radius is taken to be $f\mathrm{NA}$ {as opposed to $f\tan[\arcsin(\mathrm{NA})]$}. The Hopkins method, however, is relatively complicated, and some of its assumptions are open to question. Moreover, it is not easy to generalize his treatment to situations in which the incident distribution is nonuniform. Other authors have studied the focusing problem in various regimes with different approximations, but, to our knowledge, there does not exist a comprehensive method of computing diffraction patterns that does not require approximations of one sort or another to the fundamental diffraction integral. To be sure, the fundamental integral of the classical theory is itself an approximation to the physical reality since it does not include the effect of the screen on the aperture field distribution. This approximation, however, turns out to be quite acceptable as long as the dimensions of interest are not comparable to or smaller than a wavelength.

In this paper we present a formulation of the Fresnel–Kirchhoff diffraction theory that allows polarization effects to be incorporated in a natural way. We show that the diffraction integral can be exactly evaluated in the near-field regime with two successive Fourier transforms; in the far-field regime, where the exact computations become impractical, the steepest-descent approximation applies to the integral, yielding the Fraunhofer formula that requires only one Fourier transform. Next we consider diffraction in the presence of a lens and show that for small numerical apertures, where the curvatures introduced by the lens are slight, the propagation formulas of Fresnel or Fraunhofer are still applicable. For large numerical apertures, however, these methods become excessively time consuming; in this regime, we factor out the fast-oscillating terms of the curvature function and obtain a practical formula for computing diffraction patterns using three successive Fourier transformations.

The organization of this paper is as follows: Section 2 describes the formulation of diffraction theory, including polarization, using plane-wave expansion of the incident distribution. Sections 3 and 4 are devoted to propagation in the far-field and near-field regimes, respectively. In Section 5 the effect of focusing on the distribution around the focal plane is discussed; also, some of the results are compared with the experimental or theoretical works of other authors. Results of numerical computations are presented in Section 6.

2. THEORY OF DIFFRACTION

In this section we introduce the framework for the analyses presented in the rest of the paper. Our approach to diffraction theory is only slightly different from the classical approach based on the Fresnel–Kirchhoff formulation,[2] and the results can be shown to be identical with those of the classical theory. The advantages of the adopted approach are that the range of validity of various approximations is easily identified and that extensions to situations in which these approximations are no longer applicable become straightforward; moreover, polarization effects can be included in the theory in a natural way. The approach is based on the plane-wave expansion of the incident distribution and the eventual recombination of the propagated waves.

Consider a plane, monochromatic wave of wavelength λ, traveling in the direction of the unit vector $\hat{\sigma} = (\sigma_x, \sigma_y, \sigma_z)$. The space–time dependence of this wave is given by

$$A(x, y, z, t) = A_0 \exp[i(2\pi/\lambda)(x\sigma_x + y\sigma_y + z\sigma_z)]\exp[-i\omega t]. \tag{1}$$

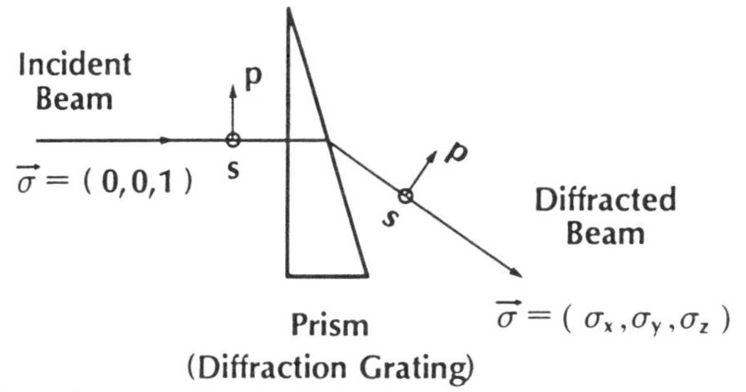

Fig. 1. Prism as a diffraction grating: A plane, linearly polarized incident beam is diffracted into a plane wave propagating in the direction $\hat{\sigma} = (\sigma_x, \sigma_y, \sigma_z)$. The polarization of the outgoing beam is related to the polarization of the incoming beam and the orientation of the prism as described in the text.

The polarization is in the plane perpendicular to $\hat{\sigma}$ but is otherwise arbitrary. Ignoring the time dependence for the time being and confining our attention to the plane $Z = 0$, we find that the amplitude distribution is given by

$$A(x, y) = A_0 \exp[i(2\pi/\lambda)(x\sigma_x + y\sigma_y)], \tag{2}$$

where $\sigma_x^2 + \sigma_y^2 \leq 1$.

Now, assuming an arbitrary distribution $t(x, y)$ in the plane $Z = 0$, define the Fourier transform of $t(x, y)$ as

$$T(S_x, S_y) = \iint_{-\infty}^{+\infty} t(x,y)\exp[-i2\pi(xS_x + yS_y)]dxdy. \tag{3}$$

The inverse Fourier-transform relation then yields

$$t(x, y) = \lambda^{-2} \iint_{-\infty}^{+\infty} T(\sigma_x/\lambda, \sigma_y/\lambda) \times \exp[i(2\pi/\lambda)(x\sigma_x + y\sigma_y)]d\sigma_x d\sigma_y. \tag{4}$$

Comparing Eqs. (2) and (4), one concludes that $t(x, y)$ is the superpositon of plane waves propagating in various directions. The amplitudes of these plane waves are determined by the Fourier transform of $t(x, y)$. Of course, there is the restriction on (σ_x, σ_y) that $\sigma_x^2 + \sigma_y^2$ must be less than unity, but the components of $t(x, y)$ that correspond to (σ_x, σ_y) beyond the unit circle are, in fact, evanescent waves that do not carry energy and do not travel more than a few wavelengths.[2] Consequently, except in the immediate neighborhood of the plane $Z = 0$, the representation of $t(x, y)$ as the superposition of plane waves is valid.

Turning now to the effects of polarization, let us assume that the incident beam is a linearly polarized plane wave, traveling in the Z direction, with a polarization vector along X. To determine the polarization of the diffracted beam along the vector $\hat{\sigma} = [\sigma_x, \sigma_y, (1 - \sigma_x^2 - \sigma_y^2)^{1/2}]$, imagine a prism (which happens to be the simplest diffraction grating) inserted between the incoming and outgoing plane waves, as shown in Fig. 1. Now consider the plane formed by the vector $\hat{\sigma}$ and the Z axis. The component of the incident polarization vector perpendicular to this plane retains its direction and becomes the s component of the deflected beam. The projection of the incident polarization onto the above plane, however, is reoriented as the beam goes through the prism and becomes the p component of the deflected beam. These two components are given by

$$s \text{ component} = |\sigma_y|(\sigma_x^2 + \sigma_y^2)^{-1/2},$$

$$p \text{ component} = |\sigma_x|(\sigma_x^2 + \sigma_y^2)^{-1/2}.$$

When the deflected beam arrives at the plane $Z = z_0$ it will have components of polarization along the X, Y, and X axes. These components will be denoted Ψ_x, Ψ_y, and Ψ_z and are given by

$$\Psi_x = \frac{[\sigma_x^2 (1 - \sigma_x^2 - \sigma_y^2)^{1/2} + \sigma_y^2]}{\sigma_x^2 + \sigma_y^2}, \tag{5a}$$

$$\Psi_y = \frac{-\sigma_x\sigma_y[1 - (1 - \sigma_x^2 - \sigma_y^2)^{1/2}]}{\sigma_x^2 + \sigma_y^2}, \tag{5b}$$

$$\Psi_z = -\sigma_x, \tag{5c}$$

where it can be verified that $\Psi_x^2 + \Psi_y^2 + \Psi_z^2 = 1$.

Combining Eqs. (1), (4), and (5), one arrives at the following expression for the distribution of light at $Z = z_0$:

$$t_\alpha(x, y, z_0) = \lambda^{-2} \iint_{\sigma_x^2+\sigma_y^2 \leq 1} \Psi_\alpha(\sigma_x, \sigma_y) T(\sigma_x/\lambda, \sigma_y/\lambda) \times \exp\{i(2\pi/\lambda)[x\sigma_x + y\sigma_y + z_0(1 - \sigma_x^2 - \sigma_y^2)^{1/2}]\} \times d\sigma_x d\sigma_y, \tag{6}$$

The subscript α in Eq. (6) indicates the polarization component of the final distribution and could be x, y, or z.

Equation (6) is our basic diffraction equation and, as shown in the following sections, is equivalent to the Fresnel–Kirchhoff equation in various limits. The restriction imposed on the domain of integration in Eq. (6) results in a power-transmission coefficient η less than unity. This coefficient is defined by

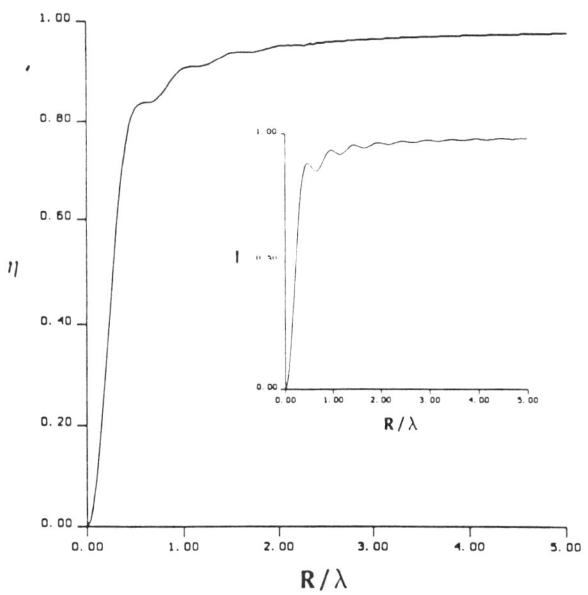

Fig. 2. Power-transmission coefficient for a circular aperture versus the normalized aperture radius. The inset shows the transmission coefficient obtained from the vector Smythe–Kirchhoff approximation according to the following equation: $\eta = 1 - (\lambda/4\pi R) \int_0^{4\pi R} J_0(x)dx$.

$$\eta = \frac{\displaystyle\iint_{\sigma_x^2+\sigma_y^2\leq 1} |T(\sigma_x/\lambda, \sigma_y/\lambda)|^2 d\sigma_x d\sigma_y}{\displaystyle\iint_{-\infty}^{+\infty} |T(\sigma_x/\lambda, \sigma_y/\lambda)|^2 d\sigma_x d\sigma_y}. \tag{7}$$

For a circular aperture of radius R, illuminated with a normally incident plane wave, the transmission coefficient calculated from Eq. (7) is

$$\eta = \int_0^1 (2/x) J_1^2(2\pi Rx/\lambda) dx = 1 - J_0^2(2\pi R/\lambda) - J_1^2(2\pi R/\lambda). \tag{8}$$

J_0 and J_1 are Bessel functions of the first kind. Figure 2 is a plot of η versus R/λ. The small-aperture limit in this figure is not expected to be correct, since in deriving Eq. (8) we have ignored the effect of the screen on the field distribution within the aperture. The inset in Fig. 2 shows the corresponding curve obtained from the vector Smythe–Kirchhoff approximation.[3] The two curves are in good agreement as far as general behavior is concerned. Small differences are due to the fact that the latter curve accounts for the (conducting) screen and, in part, to the approximate nature of this curve.

3. FRAUNHOFER OR FAR-FIELD DIFFRACTION

To derive the far-field diffraction formula from the results of the preceding section, consider the function

$$W(\sigma_x, \sigma_y) = x\sigma_x + y\sigma_y + z_o(1 - \sigma_x^2 - \sigma_y^2)^{1/2},$$

which appears in the exponent of the integrand in Eq. (6). This function has a saddle point at $(\sigma_{x0}, \sigma_{y0})$, where

$$\sigma_{x0} = x/(x^2 + y^2 + z_0^2)^{1/2},$$

$$\sigma_{y0} = y/(x^2 + y^2 + z_0^2)^{1/2}.$$

These are the angular coordinates of the point (x, y) in the observation plane as observed from the center of the aperture. Around this saddle point, the function W can be approximated as

$$W(\sigma_x, \sigma_y)/(x^2 + y^2 + z_0^2)^{1/2} = 1 - {}^1\!/_2[1 + (x/z_0)^2](\sigma_x - \sigma_{xo})^2 - (xy/z_0^2)(\sigma_x - \sigma_{xo})(\sigma_y - \sigma_{yo}) - {}^1\!/_2[1 + (y/z_0)^2](\sigma_y - \sigma_{yo})^2. \tag{9}$$

Using the method of steepest descent (see Ref. 2, App. III) and replacing for the exponent of the integrand in Eq. (6) from Eq. (9), one obtains

$$t_\alpha(x, y, z_0) = -(i/\lambda)\frac{\exp[i(2\pi/\lambda)(x^2 + y^2 + z_0^2)^{1/2}]}{z_0[1 + (x^2 + y^2)/z_0^2]} \times \Psi_\alpha(\sigma_{x0}, \sigma_{y0}) T(\sigma_{x0}/\lambda, \sigma_{y0}/\lambda). \tag{10}$$

Equation (10) is the well-known Fraunhofer diffraction formula[4,5] (except for the polarization factor Ψ_α). Note that this result is obtained directly from Eq. (6) by using the steepest-descent approximation without the need to go through the more complicated Kirchhoff theory.

The validity of the steepest-descent approximation is, in general, guaranteed if z_0 is much larger than the linear dimensions of the initial distribution and if the observation point (x, y) is not far form the origin in the observation plane. For the purpose of comparing the Fraunhofer formula with the Fresnel patterns obtained in the next section, we rewrite Eq. (10) below, using dimensionless parameters x', y', z_0' to represent coordinates in units of wavelength:

$$\begin{aligned} t_\alpha(\lambda x', \lambda y', \lambda z_0') = -i\,&\frac{\exp(i2\pi z_0')}{z_0'} \\ &\times \exp(i2\pi z_0'\{[1 + (x'^2 + y'^2)/z_0'^2]^{1/2} - 1\}) \\ &\times (1 - \sigma_{xo}^2 - \sigma_{y0}^2)\Psi_\alpha(\sigma_{xo}, \sigma_{y0}) \\ &\times [\lambda^{-2} T(\sigma_{xo}/\lambda, \sigma_{y0}/\lambda)], \end{aligned} \tag{11a}$$

where

$$\lambda^{-2} T(\sigma_x/\lambda, \sigma_y/\lambda) = \begin{cases} \mathcal{F}\{t(\lambda x', \lambda y')\}, & \sigma_x^2 + \sigma_y^2 \leq 1 \\ 0, & \text{otherwise} \end{cases}. \tag{11b}$$

Note that the second exponential term in Eq. (11a) is a curvature phase factor with radius of curvature z_0.

4. FRESNEL OR NEAR-FIELD DIFFRACTION

In general, Eq. (6) can be written as

$$\begin{aligned} t_\alpha(\lambda x', \lambda y', \lambda z_0') = \exp(i2\pi z_0') &\times \mathcal{F}^{-1}\{\lambda^{-2} T(\sigma_x/\lambda, \sigma_y/\lambda)\Psi_\alpha(\sigma_x, \sigma_y) \\ &\times \exp\{i2\pi z_0'[(1 - \sigma_x^2 - \sigma_y^2)^{1/2} - 1]\}\}, \end{aligned} \tag{12}$$

with $T(\sigma_x/\lambda, \sigma_y/\lambda)$ given by Eq. (11b). This is the general statement of the Fresnel diffraction.

As an example, consider a circular aperture of radius $R = 5\lambda$, illuminated with a plane, linearly polarized beam propagating in the Z direction with polarization vector along the X axis. The Fresnel number N of this aperture as observed from a point on the optical axis at $Z = z_0$ is given by[6]

$$z_0/\lambda = (1/N)(R/\lambda)^2 - (N/4). \tag{13}$$

At $N = 1$ (corresponding to $z_0 = 25\lambda$) the intensity distribution is as shown in Fig. 3. Figure 3(a) shows the intensity for the X component of polarization. The incident power on the aperture has been set to unity, and the calculated power for the distribution in Fig. 3(a) is 0.97; the peak intensity in this figure is $I_{max} = 0.05\lambda^{-2}$. Figure 3b shows the intensity for the Z component of polarization; the power in this component is 0.01, and the peak intensity is $0.18 \times 10^{-3}\,\lambda^{-2}$. The Y component (not shown here) is an order of magnitude below the Z component, and its computation is somewhat complicated owing to the existence of truncation and round-off errors that are of the same order of magnitude as the signal itself. In any event, the polarization effects in this example are small and can safely be ignored. In the absence of the polarization factor Ψ_α, the two-dimensional Fourier transforms in Eq. (12) can be reduced to one-dimensional Bessel transforms for circularly symmetric apertures. Figure 4 shows the radial distribution of intensity for the above aperture at $z_0 = 12\lambda$ (corresponding to $N = 2$) as computed with two successive Bessel transforms. The intensity here is normalized by the intensity at the aperture. Similar results can be obtained for other values of R and z_0, and all the well-

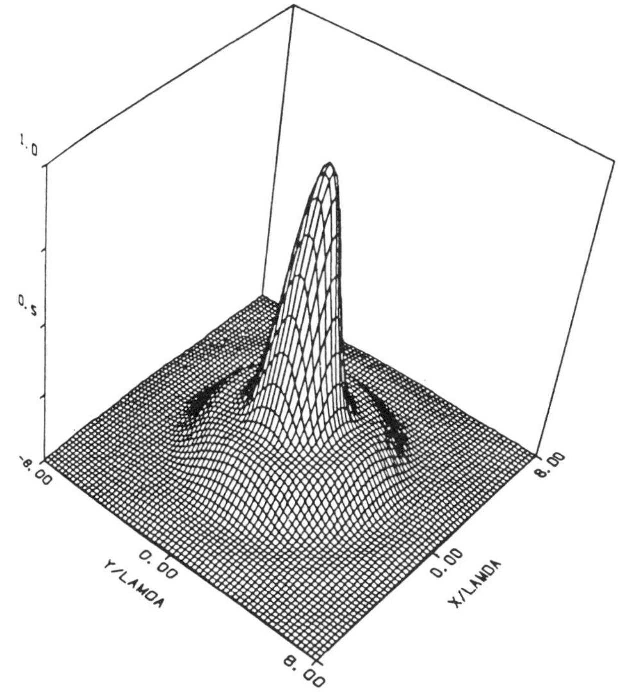

(a)

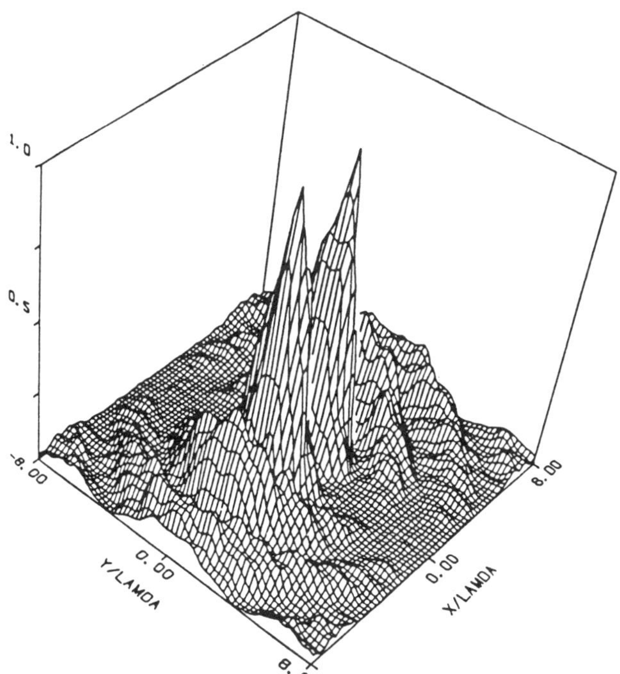

(b)

Fig. 3. Fresnel diffraction from a circular aperture of radius $R = 5\lambda$ at a distance $z_0 = 25\lambda$. The X and Y axes are normalized by the wavelength λ, whereas the vertical axis has units of maximum intensity, I_{max}. The incident beam is plane, propagating in the Z direction with unit incident power on the aperture. The incident polarization is linear in the X direction. (a) Intensity distribution for the X component of polarization, $I_{max} = 0.50 \times 10^{-1}$. (b) Intensity distribution for the Z component of polarization, $I_{max} = 0.18 \times 10^{-3}$.

known patterns of Fresnel diffraction can be computed from Eq. (12) by using two Fourier (or Bessel) transforms.

For values of N less than unity, computation of Eq. (12) becomes slow owing to the large number of oscillations of the phase factor. Fortunately, this is the Fraunhofer regime, where the steepest-descent approximation is applicable. With reference to Eqs. (11), consider a circular aperture of radius R. The first dark ring in the Fraunhofer pattern of this aperture is at the angular position $(\sigma_{x0}^2 + \sigma_{y0}^2)^{1/2} = 0.61/(R/\lambda)$, independent of z_0. Ignoring the term $N/4$ in Eq. (13) and approximating sine with tangent, we obtain for the radius of the first ring in the observation plane

$$r' = (x'^2 + y'^2)^{1/2} = 0.61\,(R/\lambda)/N.$$

Notice that the Fresnel pattern with $N = 1$ in Fig. 3(a) is already close to the Fraunhofer pattern with a hint of the first dark ring around $r' = 0.61 \times 5 = 3$. The steepest-descent approximation is usually acceptable for distances z_0 corresponding to $N \lesssim 1/2$.

5. DIFFRACTION IN THE PRESENCE OF A LENS

A perfect (aberration-free) spherical lens converts an incident plane wave to a spherical wave converging toward the focal point. Ideally, therefore, a lens is a phase object with the following amplitude-transmission function:

$$t(x, y) = \tau_0(x, y)\exp\{i(2\pi/\lambda)[f - (f^2 + x^2 + y^2)^{1/2}]\}. \quad (14)$$

$\tau_0(x, y)$ includes the aperture function, possible aberrations, and the incident amplitude distribution. As before, the incident beam will be assumed to be linearly polarized in the X direction. If the polarization is not linear, its X and Y components should be treated separately and the final results superimposed.

When the oscillations of the complex exponential term in Eq. (14) are not prohibitive [i.e., when $(f/\lambda)\mathrm{NA}^2$ is relatively small], it is possible to treat Eq. (14) as an aperture function and use the Fresnel or Fraunhofer propagation formulas of the preceding sections. For example, in Farnell's experi-

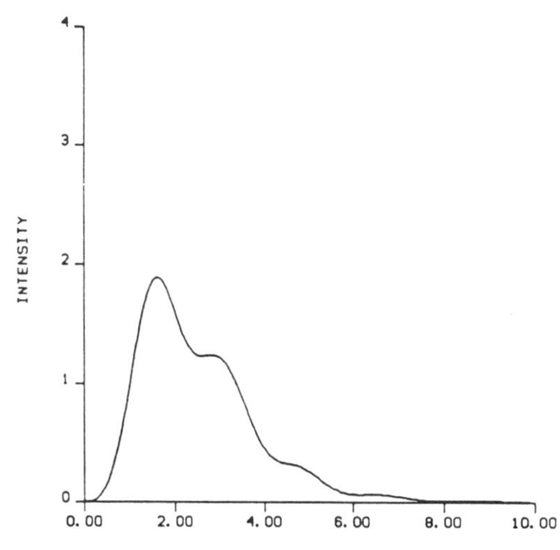

Fig. 4. Fresnel diffraction from a circular aperture of radius $R = 5\lambda$ at a distance $z_0 = 12\lambda$. The incident beam is plane with unit intensity at the aperture. The polarization effects are ignored. The intensity at the observation plane has circular symmetry; thus only the radial distribution is shown. Notice that the central spot is dark and the peak intensity is twice the intensity of the incident beam at the aperture.

ments in the microwave-frequency range ($\lambda = 3.22$ cm), the focal length and the numerical aperture were $f = 20\lambda$ and NA $= 0.36$, respectively.[7,8] To compute the diffraction patterns for this case we used Eq. (12) in conjunction with Eq. (14) and set $\tau_0(x, y) = 1$ in the aperture; the necessary array size was 256×256. We found that the maximum intensity on the optical axis occurs not at the focal point but at $z_0 = 18\lambda$, in agreement with Farnell's experimental findings. The computations also revealed that the maximum intensity is about 10% higher than the intensity at the focal point. As another example, we compared our results with theoretical results of Mahajan,[9] who calculated the axial intensity distribution for small numerical apertures. We used a 128×128 array for numerical computations corresponding to NA $= 0.85 \times 10^{-4}$ and $f = 139 \times 10^6\ \lambda$ and reconstructed the curve in Fig. 7 of Mahajan's paper.

For large values of $(f/\lambda)\text{NA}^2$, the rapid oscillations of the lens-transmission function in Eq. (14) reduce the efficiency of computations to the extent that time and memory requirements make such computations impractical. Fortunately, however, it is possible to factor out the fast-oscillating term and apply the Fourier transformations to the remaining function. To show this, we rewrite Eq. (14) as follows:

$$t(f\mathbf{x}, f\mathbf{y}) = \tau(f\mathbf{x}, f\mathbf{y})\exp[-i\pi f'(\mathbf{x}^2 + \mathbf{y}^2)]. \qquad (15a)$$

Here $\mathbf{x} = x/f$, $\mathbf{y} = y/f$, $f' = f/\lambda$, and

$$\begin{aligned}\tau(f\mathbf{x}, f\mathbf{y}) &= \tau_0(f\mathbf{x}, f\mathbf{y})\exp\{-i2\pi f'[(1 + \mathbf{x}^2 + \mathbf{y}^2)^{1/2} \\ &\quad - 1 - {}^1/_2(\mathbf{x}^2 + \mathbf{y}^2)]\} \\ &= \tau_0(f\mathbf{x}, f\mathbf{y})\exp\left\{i2\pi f'\sum_{n=2}^{\infty}(-1)^n\frac{(2n-3)!!}{2^n n!}(\mathbf{x}^2 + \mathbf{y}^2)^n\right\}.\end{aligned} \qquad (15b)$$

Since $t(f\mathbf{x}, f\mathbf{y})$ in Eq. (15a) is the product of two functions, its Fourier transform can be written as a convolution, namely,

$$\mathcal{F}\{t(f\mathbf{x}, f\mathbf{y})\} = \mathcal{F}\{\tau(f\mathbf{x}, f\mathbf{y})\} * (-i/f')\exp\{i(\pi/f')(u^2 + v^2)\}, \qquad (16)$$

where u and v are variables of the transform domain. Equation (16) can equivalently be written as

$$\begin{aligned}f^{-2}T(u/f, v/f) &= (-i/f')\exp[i(\pi/f')(u^2 + v^2)] \\ &\quad \times \iint_{-\infty}^{+\infty} h(u', v')\exp[-i(2\pi/f')(uu' + vv')]du'dv',\end{aligned} \qquad (17a)$$

where, by definition,

$$h(u, v) = \exp[i(\pi/f')(u^2 + v^2)]\mathcal{F}\{\tau(f\mathbf{x}, f\mathbf{y})\}. \qquad (17b)$$

Using dimensionless quantities $\sigma_x = u/f'$ and $\sigma_y = v/f'$, Eq. (17a) becomes

$$\lambda^{-2}T(\sigma_x/\lambda, \sigma_y/\lambda) = -if'\exp[i\pi f'(\sigma_x^2 + \sigma_y^2)]\mathcal{F}\{h(u, v)\}. \qquad (18)$$

Equation (18) is now used in the diffraction equation (12), yielding

$$\begin{aligned}t_\alpha(\lambda x', \lambda y', \lambda z_0') &= -if'\exp(i2\pi z_0')\mathcal{F}^{-1}\{\Psi_\alpha(\sigma_x, \sigma_y)\mathcal{F}\{h(u, v)\} \\ &\quad \times \exp\{-i2\pi z_0'[1 - {}^1/_2(f'/z_0')(\sigma_x^2 + \sigma_y^2) \\ &\quad - (1 - \sigma_x^2 - \sigma_y^2)^{1/2}]\}\}.\end{aligned} \qquad (19)$$

Equation (19) can be written in the following compact form:

$$t_\alpha(\lambda x', \lambda y', \lambda z_0') = -if'\exp(i2\pi z_0')\mathcal{F}^{-1}\{\Psi_\alpha(\sigma_x, \sigma_y)g(\sigma_x, \sigma_y)\}, \qquad (20a)$$

where

$$g(\sigma_x, \sigma_y) = \begin{cases}\exp\left\{i\pi\left[(f' - z_0')(\sigma_x^2 + \sigma_y^2) - 2z_0'\displaystyle\sum_{n=2}^{\infty}\frac{(2n-3)!!}{2^n n!}(\sigma_x^2 + \sigma_y^2)^n\right]\right\}\mathcal{F}\{h(u, v)\} & \\ 0, & \sigma_x^2 + \sigma_y^2 > 1\end{cases}. \qquad (20b)$$

The results can now be summarized as follows: To compute the diffraction pattern at $Z = z_0$ follow these steps:

(1) Calculate $\tau(f\mathbf{x}, f\mathbf{y})$ according to Eq. (15b). {Note that $\tau_0(f\mathbf{x}, f\mathbf{y}) = 0$ when $(\mathbf{x}^2 + \mathbf{y}^2)^{1/2} > \tan[\arcsin(\text{NA})]$}.

(2) Calculate $h(u, v)$ from Eq. (17b).

(3) Calculate $g(\sigma_x, \sigma_y)$ from Eq. (20b).

(4) Calculate $t_\alpha(\lambda x', \lambda y', \lambda z_0')$ from Eq. (20a). (This step should be performed separately for the three components of polarization.)

We have thus shown that a sequence of three Fourier transforms yields the diffraction pattern around the focal plane of a lens. It is instructive to derive the standard single-Fourier-transform lens formula from these results. In the limit when NA $\lesssim \sqrt[4]{\lambda/f}$, the infinite sums in Eqs. (15b) and (20b) are close to zero and can be ignored. Thus $\tau(x, y) = \tau_0(x, y)$ and $g(\sigma_x, \sigma_y) = \mathcal{F}\{h(u, v)\}$ at $z_0 = f$. If polarization effects are ignored, Eq. (20a) would yield

$$\begin{aligned}t(\lambda x', \lambda y', \lambda f') &= -if'\exp(i2\pi f')\mathcal{F}^{-1}\{\mathcal{F}\{h(u, v)\}\} \\ &= -i(f/\lambda)\exp(i2\pi f/\lambda)\exp[i(\pi\lambda/f) \\ &\quad \times (x'^2 + y'^2)]\mathcal{F}\{\tau_0(f\mathbf{x}, f\mathbf{y})\}.\end{aligned} \qquad (21)$$

Equation (21) is the well-known result of diffraction theory that is usually obtained from the Fresnel–Kirchhoff integral under standard approximations.[4]

5. RESULTS AND DISCUSSION

In this section we present results of numerical computations for a lens with $f = 4000\lambda$ and NA $= 0.3$. Figure 5(a) shows the function $\tau(x, y)$ versus distance from the lens center (only the real part of the function is shown); it has been assumed that $\tau_0(x, y) = 1$ within the aperture. The oscillations are due to the complex exponential function in Eq. (15b). After two Fourier transformations and multiplications by various phase factors in the process, the function $g(\sigma_x, \sigma_y)$ is obtained whose real and imaginary parts at $z_0 = f$ are plotted in Fig. 5(b) versus $(\sigma_x^2 + \sigma_y^2)^{1/2}$. Here, as in Fig.

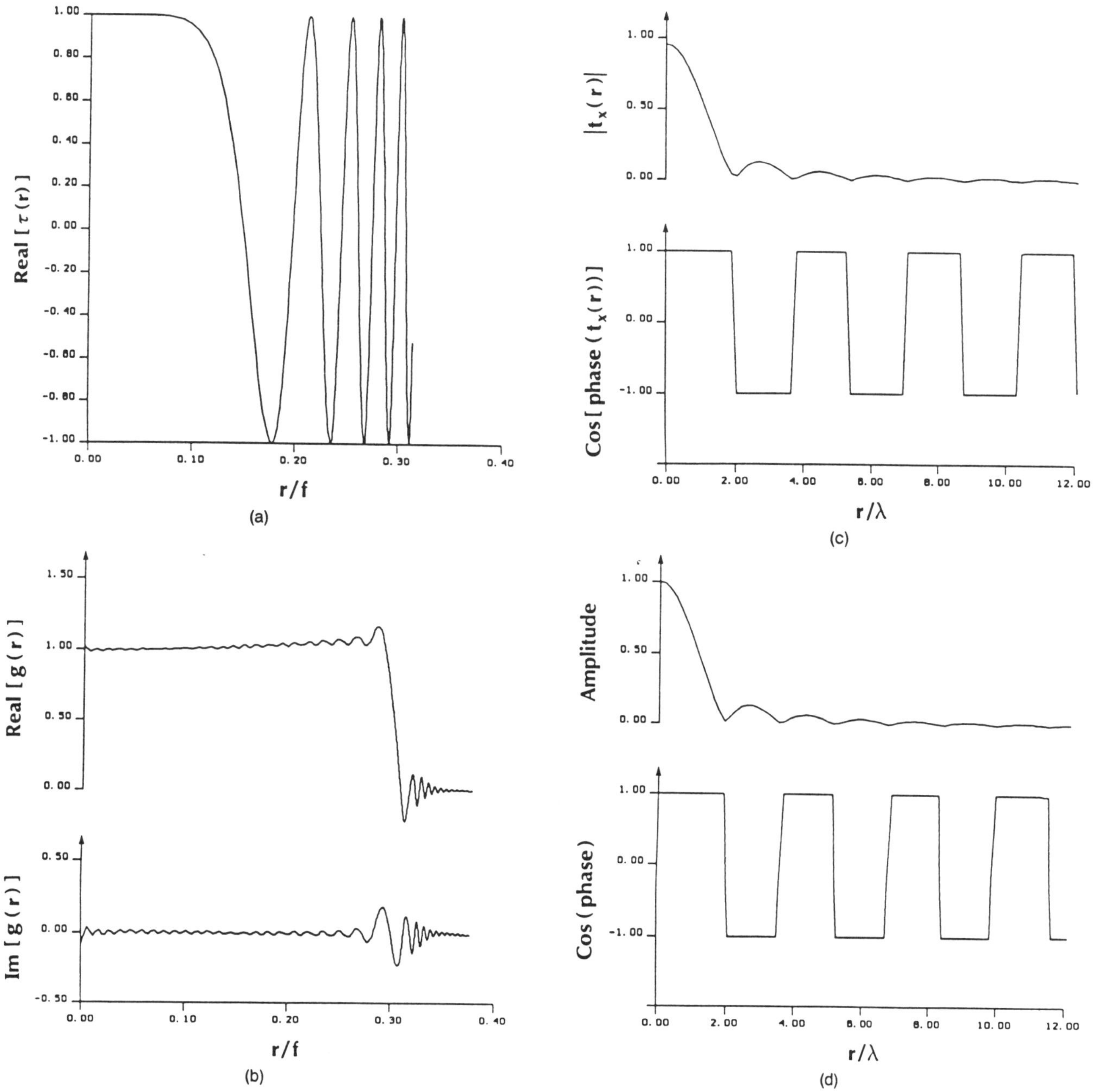

Fig. 5. Diffraction from aberration-free, spherical lens with $f = 4000\lambda$ and NA = 0.3. The incident beam is plane with unit amplitude, and the polarization effects are ignored. (a) The real part of the complex function $\tau(x, y)$ versus the normalized distance from the lens center. (b) The real and imaginary parts of the complex function $g(\sigma_x, \sigma_y)$ versus $(\sigma_x{}^2 + \sigma_y{}^2)^{1/2}$ at $z_0 = f$. (c) Amplitude and phase of the distribution at the focal plane. The amplitude is normalized by the aperture area of the lens. (d) Amplitude and phase of the distribution at the focal plane obtained with standard approximations.

5(a), the horizontal axis can be interpreted as the normalized distance from the lens center. Finally, a third Fourier transform as described by Eq. (20a) yields the distribution at the focal plane. We have ignored the polarization effects at this stage by setting $\Psi_\alpha = 1$ and computed $t(\lambda x', \lambda y', \lambda f')$, which is then plotted versus radial distance from the focal point as shown in Fig. 5(c). Both amplitude and phase of the distribution are shown here, and the amplitude is normalized by the aperture area of the lens. To compare this result with the standard approximate distribution, otherwise known as the Airy pattern, we have plotted the corresponding function, as described by Eq. (21), in Fig. 5(d). Notice that the peak amplitude in Fig. 5(c) is slightly less than that of Fig. 5(d) and that the zeros of the exact distribution are slightly shifted to the right of the corresponding zeros of the Airy pattern. These differences are expected to be more pronounced at larger numerical apertures.

The effects of polarization are shown in the two-dimensional plots of intensity at the focal plane in Fig. 6; a plane incident beam has been assumed with unit power in the aperture and linear polarization in the X direction. Figure 6(a) corresponds to the X component of polarization in the focal plane; the total power in this component is 0.976, and the peak intensity is $0.276\lambda^{-2}$. Figure 6(b) shows the inten-

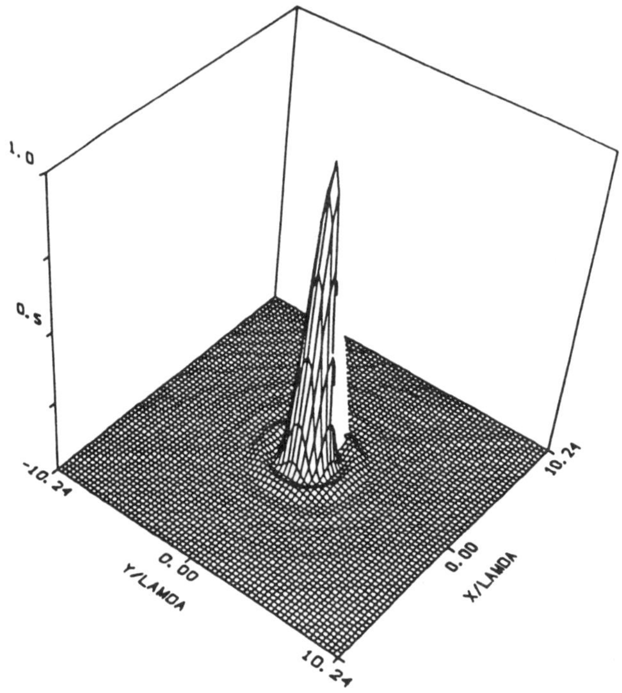

(a)

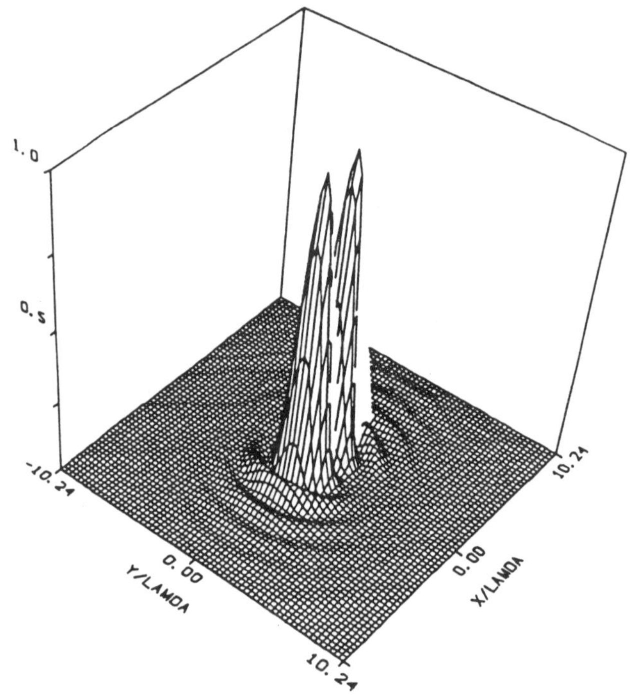

(c)

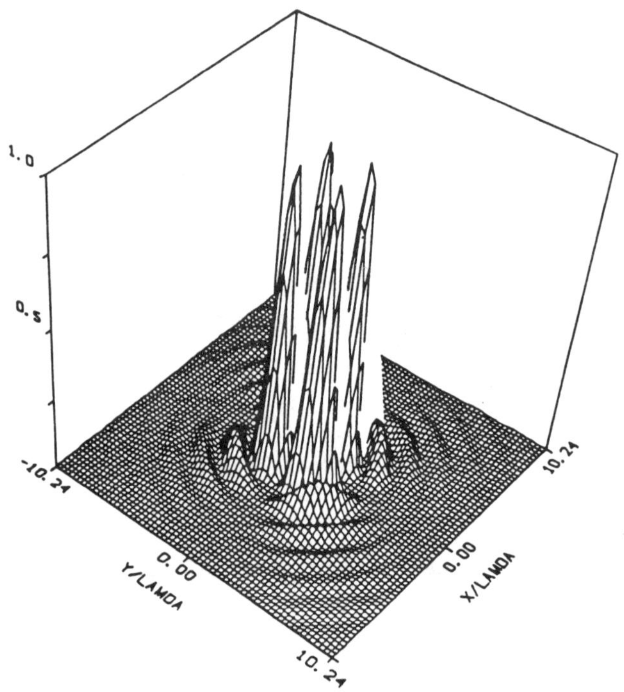

(b)

Fig. 6. Diffraction from aberration-free, spherical lens with $f = 4000\lambda$ and NA = 0.3. The incident beam is plane, is linearly polarized in the X direction, and has unit power in the aperture. (a) Intensity distribution for the X component of polarization at the focal plane. $I_{\max} = 0.276\lambda^{-2}$. (b) Intensity distribution for the Y component of polarization in the focal plane. $I_{\max} = 0.740 \times 10^{-5}\ \lambda^{-2}$. (c) Intensity distribution for the Z component of polarization in the focal plane. $I_{\max} = 0.335 \times 10^{-2}\ \lambda^{-2}$.

sity distribution for the Y component of polarization with a total power of 0.92×10^{-4} and a peak intensity of $0.74 \times 10^{-5}\ \lambda^{-2}$. Figure 6(c) corresponds to the Z component of polarization with a total power of 0.023 and a peak intensity of $0.335 \times 10^{-2}\ \lambda^{-2}$. Notice that the Y component has four peaks located in the four quadrants of the XY plane, while the Z component has only two peaks, which are separated in the direction of the incident polarization.

Finally, the intensity distribution at $z_0 = f + 15\lambda$ is shown in Fig. 7. (Only the X component of polarization is shown.) The total power in this component is again 0.976, but the peak intensity is only $0.04\lambda^{-2}$. The same distribution was found for $z_0 = f - 15\lambda$. Note that the depth of focus (λ/NA^2) for the lens studied here is about 10λ.

For the computations leading to Figs. 6 and 7 we used a 512×512 array and a standard two-dimensional fast-Fouri-

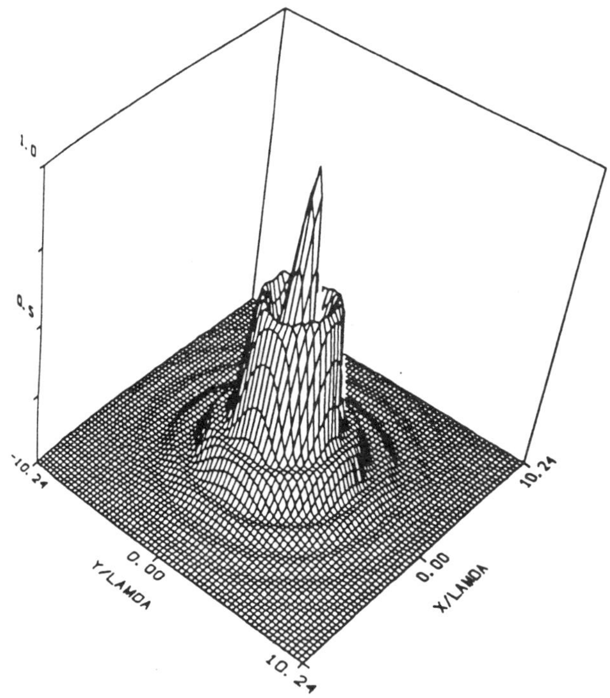

Fig. 7. Intensity distribution for the X component of polarization at $z_0 = f \pm 15\lambda$ (same lens as Fig. 6). $I_{max} = 0.4 \times 10^{-1}\ \lambda^{-2}$.

er-transform algorithm; on a VAX11/780 computer, the entire process took less than 10 min. For larger numerical apertures and/or focal lengths the array size increases {the linear dimensions of the array are proportional to the focal length and to the fourth power of tan[arcsin(NA)]}, but the computational requirements remain within the reach of present-day computers.

REFERENCES

1. H. H. Hopkins, "The Airy disk formula for systems of high relative aperture," Proc. Phys. Soc. London **55**, 116–128 (1943).
2. M. Born and E. Wolf, *Principles of Optics*, 6th ed. (Pergamon, Oxford, 1980).
3. J. D. Jackson, *Classical Electrodynamics*, 2nd ed. (Wiley, New York, 1975).
4. J. W. Goodman, *Introduction to Fourier Optics* (McGraw-Hill, New York, 1968).
5. M. V. Klein, *Optics* (Wiley, New York, 1970).
6. F. Jenkins and H. White, *Fundamentals of Physical Optics*, 1st ed. (McGraw-Hill, New York, 1937).
7. G. W. Farnell, "Measured phase distribution in the image space of a microwave lens," Can. J. Phys. **36**, 935–943 (1958).
8. Y. Li and E. Wolf, "Three-dimensional intensity distribution near the focus in systems of different Fresnel numbers," J. Opt. Soc. Am. A **1**, 801–808 (1984).
9. V. N. Mahajan, "Axial irradiance and optimum focusing of laser beams," Appl. Opt. **22**, 3042–3053 (1983).

Reprinted with permission from *Applied Optics,* Vol. 26(18), pp. 3974-3980
(September 15, 1987). ©1987 Optical Society of America.

Jones matrix analysis of magnetooptical media and read-back systems

William A. Challener and Thomas A. Rinehart

A Jones matrix approach is used to study a magnetooptic read-back system which uses a wave plate for maximizing the signal. The effects and interactions of Kerr rotation, Kerr ellipticity, birefringence in the substrate and in the optical system, and misalignment of the optical components are considered. Several criteria are developed for maximizing the signal and minimizing the noise.

I. Introduction

One of the most promising techniques for erasable optical recording is based on the polar magnetooptic (MO) Kerr effect. The MO Kerr effect can be understood in terms of circular birefringence and dichroism of the MO material. For normally incident light which is linearly polarized, the circular birefringence causes a small rotation θ of the plane of polarization of the reflected light. The circular dichroism introduces a small amount of ellipticity ε into the reflected light. When the state of magnetization of the MO material is reversed, the signs of both θ and ε reverse. In practice, the magnitude of the effect is enhanced by incorporating the MO material into a thin-film interference structure.

Since both θ and ε carry information about the state of magnetization of the material, the amplitude of the signal can be increased by designing the read-back system to make use of both θ and ε rather than θ alone.[1] Two techniques have been suggested for doing this. A variable phase retardation plate can be inserted in the read path to convert the elliptically polarized light which is reflected from the disk back into linearly polarized light.[2–4] Alternatively, a quarterwave plate can be inserted in the read path,[5] in which case the light is converted to nearly circular polarization. For both cases, the signal amplitude is increased by the factor $\sqrt{1 + (\varepsilon/\theta)^2}$ over the signal measured by a read-back system which does not make use of the ellipticity.

These are just two limiting cases, however, of a wide range of possibilities for using phase retarders to enhance the signal.

This paper is divided into six sections. Section II gives a description of the optical path of a typical read-back system and defines the appropriate quantities. Section III provides an analysis of this system when perfectly aligned for media which are free from substrate birefringence (e.g., glass disks). Section IV considers the effect of birefringence in the substrate (e.g., plastic disks), and Sec. V considers the effect of birefringence in the optics of the read-back system. The last section analyzes the optical system when it is slightly misaligned or mistuned.

II. Description of the Optics

The optical path for a typical MO read-back system is shown in Fig. 1. The source of linearly polarized collimated light is generally a semiconductor laser and appropriate optics. The axis of polarization is defined as the x axis. The light is incident on a polarizing beam splitter (PBS). The PBS is designed so that it transmits most of the light which is polarized along the x axis. A small fraction p_x of the light which is polarized along the x axis and a large fraction p_y of the light which is polarized along the y axis is reflected.

An objective lens focuses the light transmitted by the PBS onto the thin film-layer in the disk. The disk is characterized by a Kerr rotation θ, a Kerr ellipticity ε, and a substrate birefringence δ. The angle between the optical axes of the substrate and the optical axes of the read-back system is β. Part of the reflected light is split off by the PBS toward the detectors. It passes through a wave plate with a phase retardation of γ and an orientation of ξ with respect to the optical axes of the read-back system. It is then analyzed by another PBS, which is oriented at an angle α with respect to the optical axes of the read-back system. The analyzer

The authors are with 3M Company, Optical Recording Project, 1185 Wolters Boulevard, Vadnais Heights, Minnesota 55110.

Received 18 March 1987.
0003-6935/87/183974-07$02.00/0.

transmits all the light of one polarization to detector A and all the light of the other polarization to detector B. The signal amplitude is proportional to the difference in the light intensities on the two detectors. The Jones matrices which describe these optical elements are given in the Appendix.

III. Aligned Optics, No Birefringence

The signal amplitude is given by

$$S = (I_A^+ - I_B^+) - (I_A^- - I_B^-). \tag{1}$$

The plus and minus superscripts refer to the magnetic polarity of the written bit. The orientation of the analyzer which maximizes the signal can be found for an arbitrary lightwave incident on the analyzer defined by the electric field

$$E = \begin{bmatrix} A\pm \\ B\pm \end{bmatrix}.$$

In this case, the signal as defined by Eq. (1) is

$$S = X \cos 2\alpha + Y \sin 2\alpha, \tag{2}$$

where

$$X = (|A^+|^2 - |B^+|^2) - (|A^-|^2 - |B^-|^2), \tag{3}$$

$$Y = 2[\mathrm{Re}(A^+B^{+*}) - \mathrm{Re}(A^-B^{-*})]. \tag{4}$$

The criterion for mixing the signal with respect to the orientation of the analyzer α is then found by differentiation of Eq. (2). It is

$$\tan 2\alpha_0 = Y/X, \tag{5}$$

$$S_{\max} = \sqrt{X^2 + Y^2}. \tag{6}$$

With the analyzer oriented according to Eq. (5) and for an arbitrary light wave incident on the wave plate, defined by the electric field,

$$E = \begin{bmatrix} C\pm \\ D\pm \end{bmatrix},$$

Eqs. (3) and (4) become

$$X = M(\cos^2 2\xi + \sin^2 2\xi \cos\gamma) - N \sin 2\xi \cos 2\xi (1 - \cos\gamma) + P \sin 2\xi \sin\gamma, \tag{7}$$

$$Y = -M \sin 2\xi \cos 2\xi (1 - \cos\gamma) + N(\sin^2 2\xi + \cos^2 2\xi \cos\gamma) + P \cos 2\xi \sin\gamma, \tag{8}$$

where

$$M = (|C^+|^2 - |D^+|^2) - (|C^-|^2 - |D^-|^2), \tag{9}$$

$$N = 2[\mathrm{Re}(C^+D^{+*}) - \mathrm{Re}(C^-D^{-*})], \tag{10}$$

$$P = 2[\mathrm{Im}(C^+D^{+*}) - \mathrm{Im}(C^-D^{-*})]. \tag{11}$$

The criterion for maximizing the signal with respect to the orientation of the wave plate ξ is found by differentiation of Eq. (2) again. The result is

$$\cos 2\xi_0 = (P \cot\gamma)/N, \tag{12}$$

which implies the further restriction

$$\cos^2\gamma_0 \le N^2/(N^2 + P^2). \tag{13}$$

The maximum signal is now $S_{\max} = \sqrt{N^2 + P^2}$.

When there is no substrate birefringence, Eqs. (9)–

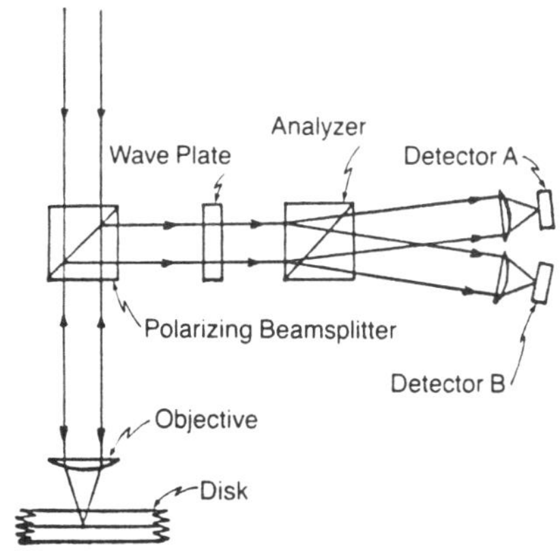

Fig. 1. Optical path of the MO recorder analyzed in the paper.

(11), expressed in terms of the Kerr rotation and Kerr ellipticity of the media, reduce to

$$M = 0, \tag{14}$$

$$N = 4p_xp_y\theta, \tag{15}$$

$$P = 4p_xp_y\varepsilon. \tag{16}$$

The three criteria for maximizing the signal [Eqs. (5), (12), and (13)] also reduce to

$$\cos^2\gamma_0 \le \theta^2/(\theta^2 + \varepsilon^2), \tag{17}$$

$$\cos 2\xi_0 = -\varepsilon \cot\gamma_0/\theta, \tag{18}$$

$$\tan 2\alpha_0 = -\frac{\theta^2 + \theta^2 \cos\gamma_0 + \varepsilon^2 \cos\gamma_0}{\varepsilon\sqrt{\theta^2 \sin^2\gamma_0 - \varepsilon^2 \cos^2\gamma_0}}. \tag{19}$$

The amplitude of the signal is

$$S_{\max} = 4p_xp_y\sqrt{\theta^2 + \varepsilon^2}. \tag{20}$$

A wave plate with a variable phase retardation (e.g., a Babinet-Soleil compensator) can be used to tune the read-back system for any value of θ and ε.[2] The phase plate must be oriented along the optical axes of the recorder so that $\xi_0 = 0$. According to Eq. (18), the amount of phase retardation must be adjusted so that

$$\tan\gamma_0 = -\varepsilon/\theta. \tag{21}$$

According to Eq. (19), the analyzer must be oriented at 45°. The light emerging from the wave plate is linearly polarized.

A second alternative is to use a quarterwave plate[5] oriented at 45°. In this case the analyzer must be tuned for the specific values of θ and ε by orienting it according to the relation

$$\tan 2\alpha_0 = -\theta/\varepsilon. \tag{22}$$

The electric field emerging from the wave plate is nearly circular in polarization in contrast to the previous case. This is illustrated in Fig. 2 for the particular example of $\varepsilon/\theta = 0.6$. In this figure, the electric fields emerging from the quarterwave plate for each magnetic polarity of the media are nearly circular ellipses. The orientation of the axes of the analyzer which maximizes the signal is also shown by two perpendicular

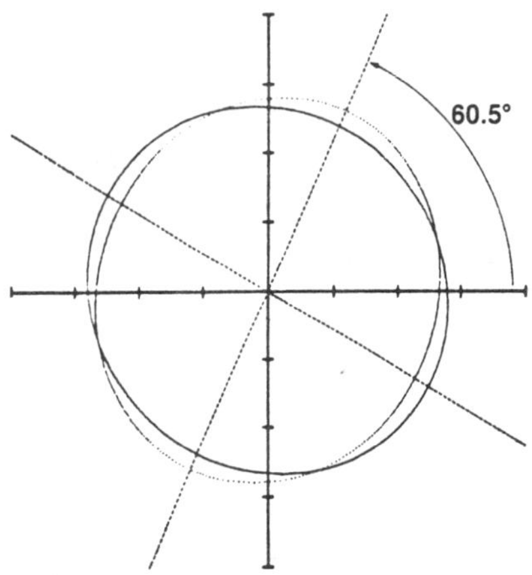

Fig. 2. Elliptically polarized electric field which emerges from the quarterwave plate is shown for both magnetic polarities of the MO layer. The orientation of the two axes of the analyzer which gives the maximum signal is also shown by orthogonal lines at an angle to the x-y coordinate system of the optics. This particular example is for $\varepsilon'/\theta' = 0.6$.

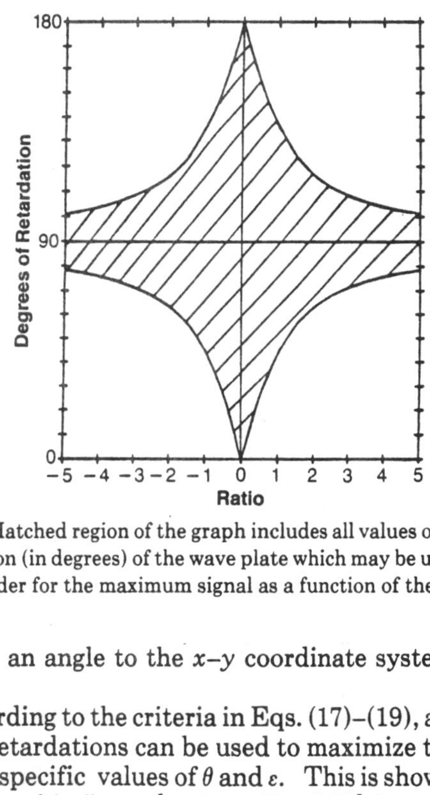

Fig. 3. Hatched region of the graph includes all values of the phase retardation (in degrees) of the wave plate which may be used to tune the recorder for the maximum signal as a function of the ratio ε'/θ'.

lines at an angle to the x–y coordinate system of the optics.

According to the criteria in Eqs. (17)–(19), a range of phase retardations can be used to maximize the signal for any specific values of θ and ε. This is shown in Fig. 3. From this figure [or from Eq. (17)] it can be seen that if the phase retardation of the wave plate is fixed, then the only value of retardation which can always be used to maximize the signal for any values of θ and ε is that of a quarterwave plate. For values of the retardation intermediate between the quarterwave plate and that defined by Eq. (21), the state of polarization of the light emerging from the wave plate will also be intermediate between nearly circular and linear.

IV. Birefringence in the Disk Substrate

When birefringence is present in the disk substrate, the amplitude of the signal is reduced. The average

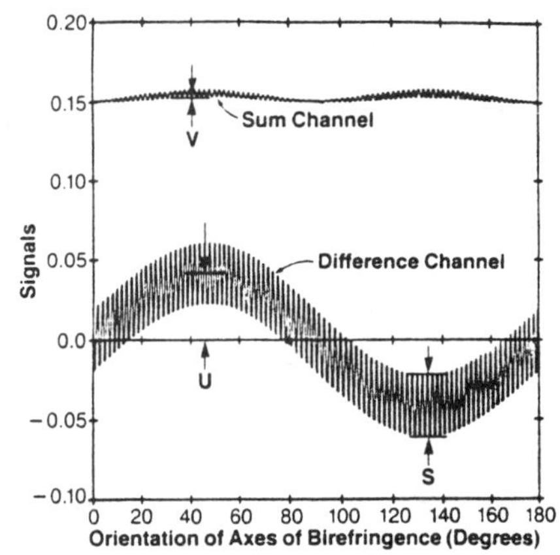

Fig. 4. Sum and difference channel signals for a written square wave are shown. The peak-to-peak signal amplitude is S, the unbalance is U, and the average value of the sum channel is V.

intensity of the light at each detector is no longer equal but varies as δ and β vary with the rotation of the disk. Because the detectors are unbalanced, the common mode noise increases. It is important to calculate the size of this effect to place reasonable limits on the sizes of δ and β which are allowable in the substrates.

When birefringence is present in the substrate of the disk, Eq. (9)–(11) become

$$M = 4\varepsilon \sin\delta \sin 2\beta p_y^2, \tag{23}$$

$$N = 4p_x p_y(\theta \cos\delta - \varepsilon \sin\delta \cos 2\beta), \tag{24}$$

$$P = -4p_x p_y(\varepsilon \cos\delta + \theta \sin\delta \cos 2\beta). \tag{25}$$

When the read-back system is aligned for media with no birefringence in the substrate, the light intensity at detectors A and B for media with birefringence in the substrate is

$$I_A^{\pm} = G^{\pm} + H^{\pm}, \tag{26}$$

$$I_B^{\pm} = G^{\pm} - H^{\pm}, \tag{27}$$

where

$$G^{\pm} = (\tfrac{1}{2})p_x^2(\cos^2\delta + \sin^2\delta \cos^2 2\beta) + (\tfrac{1}{2})p_y^2(\pm 2\varepsilon \sin\delta \sin 2\beta + \sin^2\delta \sin^2 2\beta), \tag{28}$$

$$H^{\pm} = p_x p_y[\pm(\theta^2 + \varepsilon^2)\cos\delta + \varepsilon \sin\delta \cos\delta \sin 2\beta - \theta \sin\delta \sin 2\beta \cos 2\beta]/\sqrt{\theta^2 + \varepsilon^2}. \tag{29}$$

It should be noted that these equations are independent of the phase retardation γ of the wave plate.

It is interesting to determine the effect of the birefringence in the substrate on three quantities; the peak-to-peak signal amplitude; the average value of the difference channel; and the average value of the sum channel. An illustration of these quantities for a specific case is shown in Fig. 4. The values chosen for this figure were $\theta = 1°$, $\varepsilon = 1°$, $\delta = 10$ nm at 830 nm, $p_x^2 = 0.15$, and $p_y^2 = 1.0$. The orientations of the wave plate and analyzer were chosen for perfect alignment in the absence of substrate birefringence. Signals similar to these can be observed for disks with extruded substrates as they are rotated by 180° because of the nearly perpendicular orientation of the axes of birefringence in the plane of the substrates.

A. Peak-to-Peak Signal Amplitude

The peak-to-peak signal amplitude is shown in Fig. 4 as S. It is independent of β and dependent on δ only in second order. It is given by

$$S = S_{max} \cos\delta. \tag{30}$$

B. Average Value of the Difference Channel

The average value of the difference channel is a measure of the balance of the detectors. When the detectors are unbalanced, the common mode noise increases. In Fig. 4, the unbalance is labeled U. From Eqs. (26)–(29), it is found to be

$$U = (\tfrac{1}{2}) [(I_A^+ - I_B^+) + (I_A^- - I_B^-)] \tag{31}$$

$$= p_x p_y (\varepsilon \sin 2\delta \sin 2\beta - \theta \sin^2\delta \sin 4\beta)/\sqrt{\theta^2 + \varepsilon^2}. \tag{32}$$

If the disk has a finite Kerr ellipticity, the birefringence in the substrate will contribute in first order to the unbalance of the detectors. This can be a sizable effect, as shown in Fig. 4. For an extruded substrate, the unbalance will modulate the signal amplitude at a frequency which is twice the rotational frequency of the disk. As a result, if the signal is observed for one complete revolution of the disk, the modulation of the signal from the unbalance will exhibit two complete periods. For the specific case of $\varepsilon = \theta$, Eq. (32) reduces to

$$U = \frac{1}{\sqrt{2}} p_x p_y (\sin 2\delta \sin 2\beta - \sin^2\delta \sin 4\beta). \tag{33}$$

For small δ, the first term in Eq. (33) dominates, and, therefore, the maximum unbalance occurs for $\beta = 45°$. The quantity $|U(\varepsilon = \theta)/(p_x p_y)|$ is graphed in Fig. 5 as a function of the birefringence δ for several different values of the orientation of the axes of birefringence β.

When the Kerr ellipticity of the disk is zero, the birefringence of the substrate only contributes in second order to the unbalance of the detectors. In this case

$$U(\varepsilon = 0) = -p_x p_y \sin^2\delta \sin 4\beta. \tag{34}$$

If the signal from a disk with an extruded substrate is measured, the unbalance causes a modulation of the signal at a frequency which is four times the rotational frequency of the disk. Therefore, if the signal is observed for one complete revolution of the disk, there will be four complete periods of the modulation of the signal amplitude due to the unbalance. The maximum unbalance occurs for $\beta = 22.5°$. The quantity $|U(\varepsilon = 0)/(p_x p_y)|$ is graphed in Fig. 6 as a function of the birefringence δ for several different values of the orientation of the axes of birefringence β.

It can be seen from Eq. (32) that the birefringence and the orientation of the axes of birefringence do not contribute independently to the unbalance. If β can be reduced, larger values of δ can be allowed, while

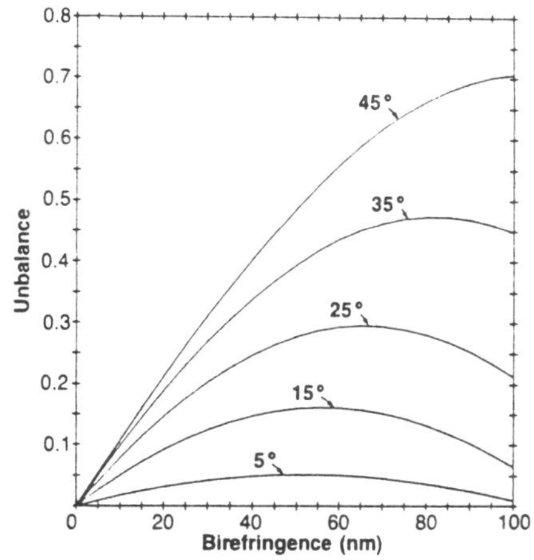

Fig. 5. Unbalance is plotted as a function of birefringence in the substrate at a laser wavelength of 830 nm for various orientations of the axes of birefringence when $\varepsilon = \theta$.

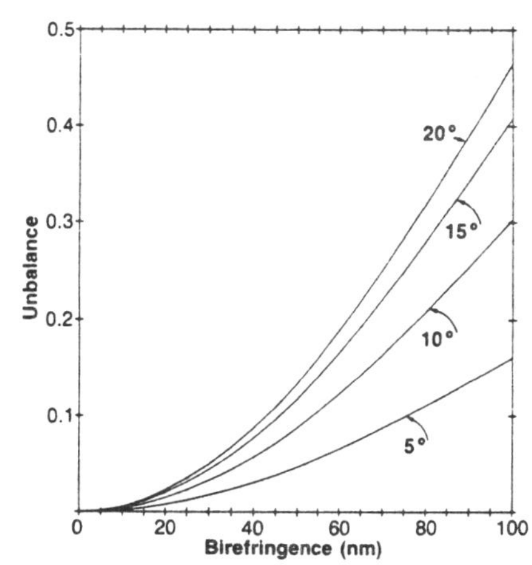

Fig. 6. Unbalance is plotted as a function of birefringence in the substrate at a laser wavelength of 830 nm for various orientations of the axes of birefringence when $\varepsilon = 0$.

keeping the same amount of unbalance. The stress lines are nearly radial in injection-molded substrates. As a result, β is much smaller in disks which are injection-molded than in those which are extruded, while the birefringence is usually comparable. Therefore, disks with injection-molded substrates typically show much less unbalance than disks with extruded substrates.

C. Average Value of the Sum Channel

The average value of the sum channel is a measure of the total energy falling on the detectors. In the absence of birefringence in the substrate of the disk, this quantity is

$$V_0 = (\tfrac{1}{2}) [(I_A^+ + I_B^+) + (I_A^- + I_B^-)] = p_x^2. \tag{35}$$

The average value of the sum channel when there is birefringence in the substrate of the disk is

$$V = V_0 + (p_y^2 - p_x^2) \sin^2\delta \sin^2 2\beta. \tag{36}$$

This is also shown in Fig. 4. The quantity is independent of θ and ε and depends on the birefringence δ only in second order. It is a non-negative number. Hence the presence of birefringence in the substrate of the media always acts to slightly increase the total energy falling on the detectors. The quantity $(V - V_0)/(p_y^2 - p_x^2)$ is graphed in Fig. 7 as a function of the birefringence δ for several values of the orientation of the axes of birefringence β.

D. Quality Factor

The ratio of the signal amplitude to the unbalance can be used as an approximate quality factor for disk substrates. This ratio is found to be

$$Q = 2(\theta^2 + \varepsilon^2)\cot\delta/[\varepsilon\cos\delta - \theta\sin\delta\cos2\beta)\sin2\beta]. \qquad (37)$$

In general, media with less Kerr ellipticity can tolerate more substrate birefringence. However, a detailed analysis of the noise sources in the read-back system must be made to determine the amount of ellipticity which maximizes the carrier-to-noise ratio (CNR). For example, if laser noise is the dominant noise source, the quality factor will be approximately proportional to the CNR. In this case, the CNR is minimized when the Kerr ellipticity is approximately equal to the Kerr rotation. On the other hand, if the read-back system is limited by detector shot noise, increasing the Kerr ellipticity increases the signal amplitude without significantly affecting the noise level, and so the CNR increases.

V. Birefringence in the Read-back System

Birefringence in the optics of the read-back system occurs primarily from the reflections at all mirror surfaces and beam splitters. This birefringence can be represented by a relative phase retardation of ϕ between the x-polarized light and y-polarized light reflected by the PBS.

A. Read-back System with no Wave Plate

For a read-back system without a wave plate, the signal is maximized when the analyzer is oriented at 45° to the optical axes of the system. The signal amplitude is

$$S = 4p_xp_y(\theta'\cos\delta - \varepsilon'\sin\delta\cos2\beta), \qquad (38)$$

where an effective θ' and ε' have been defined according to the relations

$$\theta' = \theta\cos\phi - \varepsilon\sin\phi, \qquad (39)$$

$$\varepsilon' = \theta\sin\phi + \varepsilon\cos\phi. \qquad (40)$$

The effect of birefringence in the substrate of the disk is to reduce the amplitude of the signal by a small amount proportional to cosδ. It will also cause a small modulation of the amplitude of the signal proportional to ε' as δ and β vary with the rotation of the disk. The only effect of birefringence in the optics of the read-back system on the amplitude of the signal is to convert the Kerr rotation and Kerr ellipticity in the disk into new effective quantities. Because $0 \le \theta' \le \sqrt{\theta^2 + \varepsilon^2}$, the signal amplitude may be either larger or smaller than the corresonding signal amplitude when

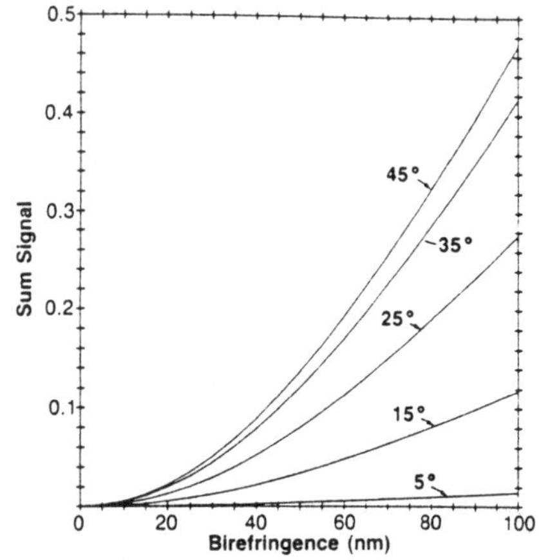

Fig. 7. Average value of the sum signal is plotted as a function of birefringence in the substrate at a laser wavelength of 830 nm for various orientations of the axes of birefringence.

there is no birefringence in the optics of the read-back system.

The unbalance is given by

$$U = -2p_xp_y[(\sin\phi\sin\delta\cos\delta)\sin2\beta + (\tfrac{1}{2})(\cos\phi\sin^2\delta)\sin4\beta]. \qquad (41)$$

It can be seen from this expression that there will be a first-order contribution to the unbalance from the birefringence in the substrate of the disk if there is any birefringence present in the optics of the read-back system. On the other hand, if the birefringence in the optics of the read-back system is zero, birefringence in the substrate of the disk contributes to the unbalance only in second-order.

It is also clear from Eq. (41) that the frequency dependence on β of these two possibilities is different. The former case is proportional to sin2β, while the latter case is proportional to sin4β. This observation suggests a simple test to detect the presence of birefringence in this type of read-back system. An extruded substrate with orthogonal optical axes in the plane of the substrate is coated with a reflecting layer. As this disk is spinning in the read-back system, the laser diode is modulated at a frequency which simulates written data on the disk. The envelop of the detected signal is observed for one complete revolution of the disk. If the observed envelop exhibits four complete periods for one revolution of the disk, the birefringence in the optics of the read-back system is negligible. On the other hand, if only two complete periods are observed, there is a significant amount of birefringence in the optics of the read-back system.

The amplitude of the sum channel is given by

$$V\pm = p_x^2(1 - \sin^2\delta\sin^2 2\beta) + p_y^2(\pm2\varepsilon\sin\delta\sin2\beta + \sin^2\delta\sin^2 2\beta). \qquad (42)$$

If the Kerr ellipticity of the media is zero, the amplitude of the sum signal will be affected only in second-order by birefringence in the substrate of the disk. If the Kerr ellipticity of the media is not zero, there will be a modulation of the sum signal which is first-order in the substrate birefringence. The Kerr ellipticity in this equation is the real ellipticity of the media, not the effective ellipticity defined in Eq. (40). Birefringence in the optics of the read-back system has no effect on the amplitude of the sum signal.

B. Read-back System with a Wave Plate

In a read-back system with a wave plate, Eqs. (17)–(19) still describe the correct criteria for maximizing the signal amplitude, if θ and ε are replaced everywhere by θ' and ε' according to Eqs. (39) and (40). The amplitude of the signal is given by

$$S = 4p_x p_y \sqrt{\theta^2 + \varepsilon^2} \cos\delta. \tag{43}$$

By making use of a wave plate there is no longer an undesirable modulation of the amplitude of the signal from the Kerr ellipticity of the media and the birefringence in the substrate of the disk as the disk rotates. However, the birefringence in the substrate still causes a small reduction in the amplitude of the signal.

The amount of unbalance is given by

$$U = p_x p_y (\varepsilon \sin 2\delta \sin 2\beta - \theta \sin^2\delta \sin 4\beta)/\sqrt{\theta^2 + \varepsilon^2}. \tag{44}$$

In contrast to the previous case, the unbalance is found to be independent of the birefringence in the optics of the read-back system. If the Kerr ellipticity of the media is zero, the birefringence in the substrate only contributes to the unbalance in second order.

In this case, the periodicity of the modulation of the read envelop depends on the characteristics of the media, not the read-back system. If the media makes use of an extruded substrate with orthogonal axes of birefringence in the plane of the substrate, two complete periods in the modulation of the read envelop for one revolution of the disk indicate that there is some Kerr ellipticity present in the media. Four complete periods in one revolution of the disk indicate that there is no Kerr ellipticity in the media.

The amplitude of the sum signal is given by

$$V = p_x^2(\cos^2\delta + \sin^2\delta \cos^2 2\beta) + p_y^2 \sin^2\delta \sin^2 2\beta. \tag{45}$$

In contrast to the previous case, there is no longer a modulation of the amplitude of the sum signal which is first order in the substrate birefringence, even in the presence of Kerr ellipticity in the media.

VI. Misalignments of the Optics

It is also important to determine the effects of a misalignment of either the wave plate or analyzer. If $\Delta\alpha$ is the angular deviation of the analyzer from its correct orientation α_0, the signal amplitude is

$$S = S_{max} \cos 2\Delta\alpha. \tag{46}$$

The signal amplitude as a function of angular deviation Δ_ξ of the wave plate from its correct orientation ξ_0 is

$$\begin{aligned} S = S_{max}[&\theta^2 \cos^2 2\Delta_\xi + \varepsilon^2 \cos 2\Delta_\xi \\ &+ \varepsilon\sqrt{\theta^2 \sin^2\gamma_0 - \varepsilon^2 \cos^2\gamma_0}\, \sin 2\Delta_\xi (1 - \cos 2\Delta_\xi) \\ &+ (\varepsilon^2 + \theta^2) \cos\gamma_0 \sin 2\Delta_\xi]/(\theta^2 + \varepsilon^2). \end{aligned} \tag{47}$$

For the special case of a quarterwave plate, this reduces to

$$\begin{aligned} S = S_{max}[&\theta^2 \cos^2 2\Delta_\xi + \varepsilon^2 \cos 2\Delta_\xi \\ &+ \varepsilon\theta \sin 2\Delta_\xi (1 - \cos 2\Delta_\xi)]/(\theta^2 + \varepsilon^2). \end{aligned} \tag{48}$$

The signal amplitude as a function of the deviation $\Delta\gamma$ of the phase retardation of the wave plate from its correct value γ_0 is

$$S = S_{max}\{1 - \varepsilon^2(1 - \cos\Delta_\xi)/[(\theta^2 + \varepsilon^2) \sin^2\gamma_0]\}. \tag{49}$$

For the special case of a quarterwave plate, this reduces to

$$S = S_{max}[1 - \varepsilon^2(1 - \cos\Delta\gamma)/(\theta^2 + \varepsilon^2)]. \tag{50}$$

In each of these cases, small deviations from perfect alignment only cause a reduction in the signal amplitude which is second order in the deviation. For example, if $\theta = 0.9°$ and $\varepsilon = 0.4°$, an error in the retardation of the wave plate of 20 nm at 830 nm causes a signal reduction of only 0.2%.

It is also interesting to calculate the signal amplitude for a read-back system which is optimized for media with a Kerr rotation of θ_0 and a Kerr ellipticity of ε_0 but which is actually using media with a rotation of θ and an ellipticity of ε. In this case, the amplitude of the signal is

$$S = S_{max}(\theta \cdot \theta_0 + \varepsilon \cdot \varepsilon_0)/(\theta_0^2 + \varepsilon_0^2). \tag{51}$$

The significance of this equation can be illustrated by an extreme example. If the read-back system is tuned for media with a rotation of 1° and an ellipticity of 1°, but media with a rotation of 1° and an ellipticity of −1° are used, the signal amplitude will be zero, even though the new media have potentially the same signal amplitude as the old media for a properly tuned read-back system.

VII. Conclusion

A Jones matrix analysis of a typical configuration for a MO read-back system demonstrates that a wave plate can be used to increase the signal amplitude by making use of both the Kerr rotation and Kerr ellipticity of the media. The phase retardation of the wave plate and the orientation of the wave plate and the analyzer must be chosen in accordance with the specific MO characteristics of the media, although small deviations from perfect alignment reduce the amplitude of the signal only slightly. Substrate birefringence can affect the balance of the difference channel in first-order in the presence of ellipticity in the media, although it is only a second-order effect in the amplitude of the signal and in the average value of the sum channel. Minimizing the angle between the axes of birefringence of the substrate and the optical axes of the read-back system (by choosing injection-molded substrates rather than extruded substrates) greatly reduces the unbalance of the difference channel. The presence of a wave plate in the optical system has the further advantage of eliminating the effects of birefringence in the optics of the read-back system.

Appendix

The Jones matrix which describes the disk is derived as follows. To first order, the circular birefringence of the MO layer introduces a small differential phase shift of $\pm\theta$, and the circular dichroism introduces a small differential absorption of $\pm\varepsilon$ into right and left circularly polarized light. This can be represented in circular coordinates by the Jones matrix

$$\tilde{D}_0 = \begin{bmatrix} 1 - \varepsilon + i\theta & 0 \\ 0 & 1 + \varepsilon - i\theta \end{bmatrix}, \quad \text{(A1)}$$

where the signs of θ and ε have been chosen for convenience. This matrix is transformed into Cartesian coordinates by the unitary transformation

$$\tilde{D}_1 = \sqrt{\frac{1}{2}} \cdot \begin{bmatrix} 1 & 1 \\ -i & i \end{bmatrix} \cdot \tilde{D}_0 \cdot \sqrt{\frac{1}{2}} \cdot \begin{bmatrix} 1 & i \\ 1 & -i \end{bmatrix} \quad \text{(A2)}$$

$$= \begin{bmatrix} 1 & -\theta - i\varepsilon \\ \theta + i\varepsilon & 1 \end{bmatrix}. \quad \text{(A3)}$$

The substrate birefringence in Cartesian coordinates is described by

$$\tilde{B} = \begin{bmatrix} \exp(-i\delta/2) & 0 \\ 0 & \exp(i\delta/2) \end{bmatrix}. \quad \text{(A4)}$$

This matrix, of course, does not take into account the effects of the index of refraction of the substrate along the axis perpendicular to the plane of the substrate. Because the beam is focused through the substrate, it has a small electric field component along this axis, but a Jones matrix approach is inadequate to deal with this complication.

The entire disk is then

$$\tilde{D}_2 = \tilde{B} \cdot \tilde{D}_1 \cdot \tilde{B} = \begin{bmatrix} \exp(-i\delta) & -\theta - i\varepsilon \\ \theta + i\varepsilon & \exp(i\delta) \end{bmatrix}. \quad \text{(A5)}$$

Finally, if the optical axes of the substrate are at an angle β with respect to the optical axes of the read-back system, another unitary transformation gives

$$\tilde{D} = \begin{bmatrix} \cos\beta & -\sin\beta \\ \sin\beta & \cos\beta \end{bmatrix} \cdot \tilde{D}_2 \cdot \begin{bmatrix} \cos\beta & \sin\beta \\ -\sin\beta & \cos\beta \end{bmatrix} \quad \text{(A6)}$$

$$= \begin{bmatrix} \cos\delta - i\sin\delta\cos2\beta & -\theta - i\varepsilon + i\sin\delta\sin2\beta \\ \theta + i\varepsilon + i\sin\delta\sin2\beta & \cos\delta + i\sin\delta\cos2\beta \end{bmatrix}. \quad \text{(A7)}$$

This is the Jones matrix of the disk in the coordinate system of the read-back system to first order in θ and ε.

The Jones matrix of the PBS is

$$\tilde{L} = \begin{bmatrix} p_x & 0 \\ 0 & P_y\exp(i\phi) \end{bmatrix}, \quad \text{(A8)}$$

where p_x is the fraction of the x-polarized electric field reflected by the beam splitter, p_y is the fraction of the y-polarized electric field reflected by the beam splitter, and ϕ is a relative phase shift between the two components which simulates the effect of birefringence in the optics of the read-back system.

The wave plate with a phase retardation of γ and an orientation of angle ξ with respect to the optical axes of the read-back system is described by

$$\tilde{W} = \begin{Bmatrix} [1 + \exp(i\gamma)] + [1 - \exp(i\gamma)]\cos2\xi & -[1 - \exp(i\gamma)]\sin2\xi \\ -[1 - \exp(i\gamma)]\sin2\xi & [1 + \exp(i\gamma)] - [1 - \exp(i\gamma)]\cos2\xi \end{Bmatrix}. \quad \text{(A9)}$$

Finally, the Jones matrix describing the analyzer (which is diagrammed as a Wollaston prism in Fig. 1) for light transmitted to detector A is

$$\tilde{P}_A = \begin{bmatrix} 1 + \cos2\alpha & \sin2\alpha \\ \sin2\alpha & 1 - \cos2\alpha \end{bmatrix} \quad \text{(A10)}$$

and to detector B is

$$\tilde{P}_B = \begin{bmatrix} 1 - \cos2\alpha & -\sin2\alpha \\ -\sin2\alpha & 1 + \cos2\alpha \end{bmatrix}, \quad \text{(A11)}$$

where the analyzer is oriented at an angle α with respect to the read-back system.

The electric field at the two detectors is then simply

$$\mathbf{E}_A = \tilde{P}_A \cdot \tilde{W} \cdot \tilde{L} \cdot \tilde{D} \cdot \begin{bmatrix} 1 \\ 0 \end{bmatrix}, \quad \text{(A12)}$$

$$\mathbf{E}_B = \tilde{P}_B \cdot \tilde{W} \cdot \tilde{L} \cdot \tilde{D} \cdot \begin{bmatrix} 1 \\ 0 \end{bmatrix}, \quad \text{(A13)}$$

and the light intensities are $I_A = |\mathbf{E}_A|^2$ and $I_B = |\mathbf{E}_B|^2$.

References

1. P. G. van Engen, K. H. J. Buschow, R. Jongebreur, and M. Erman, "PtMnSb, a Material with a Very High Magneto-optical Kerr Effect," Appl. Phys. Lett. **42**, 202 (1983).
2. M. Mansuripur, G. A. N. Connell, and J. W. Goodman, "Signal and Noise in Magneto-optical Readout," J. Appl. Phys. **53**, 4485 (1982).
3. M. Mansuripur, G. A. N. Connell, and J. W. Goodman, "Signal-to-Noise in Magneto-optic Storage," Proc. Soc. Photo-Opt. Instrum. Eng. **329**, 215 (1982).
4. D. Cheng, D. Treves, and T. Chen, "Static Tests of TbFe Films for Magneto-optical Recordings," Proc. Soc. Photo-Opt. Instrum. Eng. **329**, 223 (1982).
5. K. Sato and Y. Togami, "Magneto-optical Spectra of RF-Sputtered Amorphous Gd-Co and Gd-Fe Films," J. Magn. Magn. Mat. **35**, 181 (1983).

Reprinted with permission from *Applied Optics*, Vol. 28(2), pp. 335-339 (January 15, 1989). ©1989 Optical Society of America.

Jones matrix analysis of incident angle effects in magnetooptic storage media

Merritt N. Deeter, Jeffery T. Ingle, and Dror Sarid

A ray-tracing technique employing Jones matrices was used to model the polarization state of the reflected wavefront as a function of numerical aperture for the various types of magnetooptic media. The average reflectance and MO signal for differential readout were calculated for different numerical apertures by integrating over the wavefront. Quadrilayer structures which are designed to maximize the readout SNR are most affected by incident angle effects.

I. Introduction

Recently, we reported on calculations of the complex Kerr and Fresnel reflectivities as functions of incident angle and polarization for various magnetooptic media structures.[1] The calculations show that it may not always be accurate to use normal-incidence plane-wave values of reflectance and polar Kerr rotation to characterize a given media structure when in fact the structure is illuminated by the spectrum of plane waves that compose a focused beam. These incident angle effects become more pronounced as the numerical aperture (N.A.) increases, since more of the incident power strikes the disk at higher angles of incidence. The main conclusion of Ref. 1 was that quadrilayer structures (i.e., magnetic film structures composed of an active magnetic layer sandwiched between a dielectric layer on one side and a dielectric and a reflective layer on the other side) which are designed to be highly antireflective are most affected by obliquely incident rays. This is particularly significant because, for quadrilayer structures, the shot-noise-limited SNR as calculated from normal-incidence values is maximized when the reflectance is minimized.[1,2]

Jones matrix analysis is an important tool for modeling magnetooptical systems.[3] It has not, however, previously been used to investigate incident angle effects in magnetooptical media. The purpose of this paper is to use a simple polarization ray-tracing model to determine quantitatively the effects of incident angle on the net reflectance and shot-noise-limited SNR for different media structures and values of N.A. First, the theory which allows tracing the polarization state of each ray in the aperture is reviewed in some detail. Results for various structures and values of N.A. are then presented in the form of both 3-D plots and tabulated values of the net (integrated) reflectance and MO signal. Finally, the results are discussed in the context of actual magnetooptic readout systems.

II. Theory

The approach taken can be broken into four steps: First, the focusing aperture geometry is defined. Second, the Kerr and Fresnel reflectivities vs incident angle for the given media structure are calculated. Third, the reflectance and MO signal for each ray in a polar grid representing the focusing aperture are calculated. Finally, the reflectance and MO signal are integrated over the grid to produce a measure of the degradation due to incident angle effects.

The geometry of the focusing aperture is illustrated in Fig. 1. The incident wave is assumed to uniformly and completely fill the aperture and be linearly polarized along the x-direction. The approach of this paper is to divide the incident beam into a 2-D grid of discrete rays and trace the polarization state of each ray in the grid through the optical system. Similar analyses have previously been used to study the effects of substrate birefringence.[4,5] Thus rays which originate along the x axis in the aperture begin purely p-polarized while rays originating from the y axis begin purely s-polarized. Rays which originate from arbitrary points in the aperture contain components of both s- and p-polarization. These components must be treated separately because the conventional and Kerr re-

---

When this work was done all authors were with University of Arizona, Optical Sciences Center, Tucson, Arizona 85721; M. N. Detter is now with National Institute of Standards Technology, Electromagnetic Technology Division, Boulder, Colorado, 80303.

Received 27 May 1988.
0003-6935/89/020335-05$02.00/0.

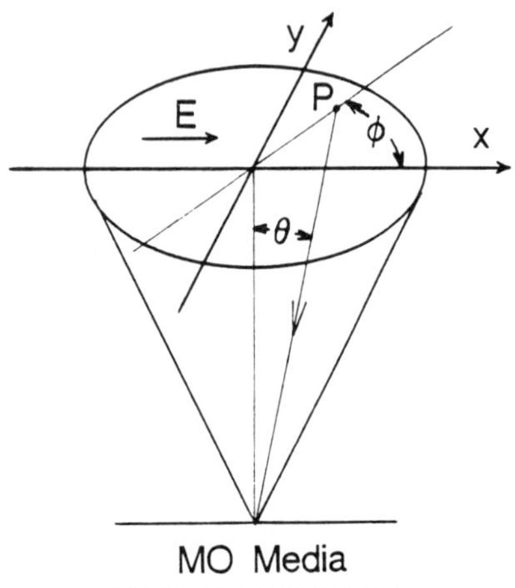

Fig. 1. Focusing aperture geometry. For the ray originating from point P, θ is the incident angle and ϕ is the angle between the x axis and the projection of the ray in the x–y plane.

flectivities are defined only for purely s- and p-polarized incident plane waves. In the Jones matrix formalism, this is accomplished by a rotation matrix which rotates from x,y coordinates to s,p coordinates.

The first component in the optical path is the media. The polarization state of each incident ray is modified through the Kerr reflectivities k_p and k_s as well as the standard Fresnel reflectivities r_p and r_s. As described previously,[1,6] Smith's matrix method of calculating the optical properties of magnetic film structures using 4×4 matrices is used. This theory applies for any incident angle, number of layers of films, incident light polarization, and orientation of magnetization. For a specific incident angle, the matrix representing the media is

$$A_{\text{med}} = \begin{bmatrix} \cos\phi & -\sin\phi \\ \sin\phi & \cos\phi \end{bmatrix} \begin{bmatrix} r_p & k_s \\ k_p & r_s \end{bmatrix} \begin{bmatrix} \cos\phi & \sin\phi \\ -\sin\phi & \cos\phi \end{bmatrix}, \tag{1}$$

where ϕ refers to the angle between the intercept of the plane of incidence for the given ray in the x–y plane and the x axis (see Fig. 1).

After reflecting from the media, the wavefront is recollimated by the focusing objective. The remaining components in the optical path are just those of the differential readout system. First, the wavefront passes through a phase plate which is designed to remove ellipticity from the beam. The matrix representing the phase plate is

$$A_{\text{ph}} = \frac{1}{2}\begin{bmatrix} 1 & 0 \\ 0 & 2\exp(i\delta) \end{bmatrix}. \tag{2}$$

The exact retardation δ is chosen to account for the ellipticity only for the normally incident ray, and thus other rays will in general retain some amount of ellipticity.

The final element of the readout system is the polarizing beam splitter. This component divides the beam into two orthogonally polarized beams, each of which is sent to a photodetector. The Jones matrices representing the polarizing beam splitter for each of the transmitted beams are simply those corresponding to polarizers oriented 90° to each other, i.e.,

$$A_1 = \begin{bmatrix} 1 & 0 \\ 0 & 0 \end{bmatrix} \quad \text{and} \quad A_2 = \begin{bmatrix} 0 & 0 \\ 0 & 1 \end{bmatrix}. \tag{3}$$

In differential detection, the intensities of the two orthogonally polarized beams should be made equal in the absence of magnetooptical effects. This may be accomplished physically by rotating the beam splitter 45° relative to the original x,y axes and mathematically by including another rotation matrix before the beam splitter matrices:

$$A_{\text{rot}} = \frac{1}{2}\begin{bmatrix} 1 & 1 \\ 1 & -1 \end{bmatrix}. \tag{4}$$

The Jones matrix product representing the electric fields at each detector can now be written as

$$\mathbf{E}_1 = [A_1][A_{\text{rot}}][A_{\text{ph}}][A_{\text{med}}]\begin{bmatrix} 1 \\ 0 \end{bmatrix}, \tag{5a}$$

$$\mathbf{E}_2 = [A_2][A_{\text{rot}}][A_{\text{ph}}][A_{\text{med}}]\begin{bmatrix} 1 \\ 0 \end{bmatrix}, \tag{5b}$$

where the input Jones vector at the right-hand side of each equation represents the original wavefront linearly polarized along the x direction.

At this point, the electric fields at each detector are multiplied by their complex conjugates to produce a value proportional to the actual intensity (and electrical signal) at each detector. The MO signal is derived from the difference of the two signals, i.e., $\Delta I = I_1 - I_2$. To separate the part of the differential signal due solely to magnetooptic effects, it is necessary to evaluate the differential signal for the case of $k_p = 0$ and $k_s = 0$ and subtract this value from the original differential signal:

$$I_{\text{MO}} = \Delta I - \Delta I(k_p = 0; k_s = 0). \tag{6}$$

The reflectance for each ray is also calculated. This is simply the sum of the two intensities I_1 and I_2:

$$I_{\text{Ref}} = I_1 + I_2. \tag{7}$$

After values of the MO signal and reflectance are calculated across the aperture, an integrated and normalized value is desired as a measure of the degradation due to incident angle effects. First, to obtain an average value of the function, the values are numerically integrated across the aperture and divided by the area of the aperture A:

$$\langle I_{\text{MO}} \rangle = \frac{\int I_{\text{MO}} dA}{A}; \tag{8}$$

$$\langle I_{\text{Ref}} \rangle = \frac{\int I_{\text{Ref}} dA}{A}. \tag{9}$$

These values are then normalized to values of the respective function for the normally incident ray to obtain a measure of the relative change in the function when all rays are considered:

$$I_{\text{MO,norm}} = \frac{\langle I_{\text{MO}} \rangle}{I_{\text{MO},\theta=0}}; \tag{10}$$

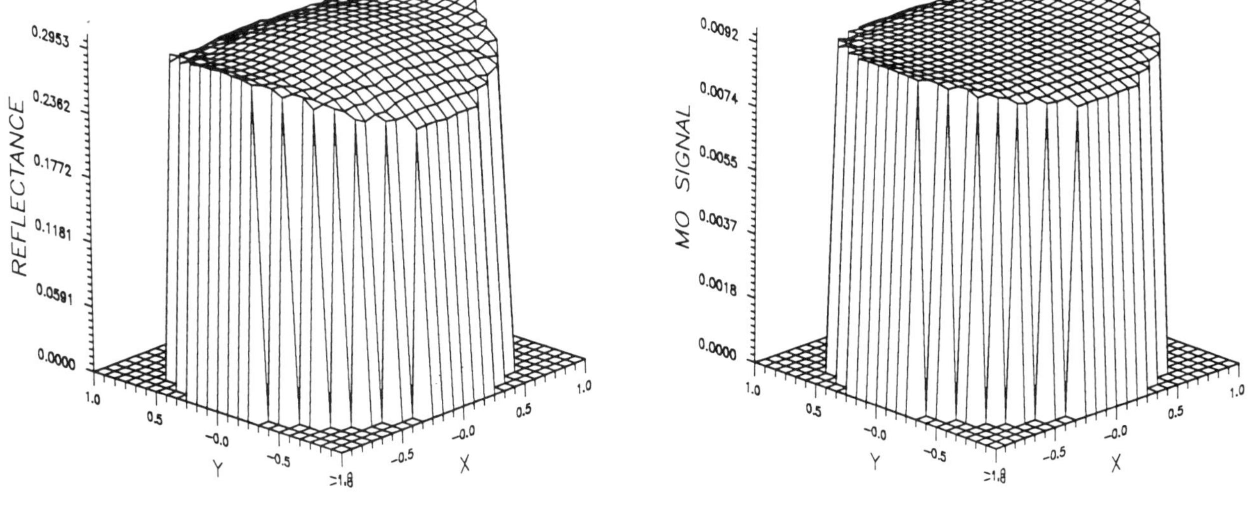

(a) (b)

Fig. 2. Results for the bilayer structure for an N.A. of 0.65. For this structure, neither the reflectance (a) nor the MO signal (b) are strong functions of incident angle.

$$I_{\mathrm{Ref,norm}} = \frac{\langle I_{\mathrm{Ref}} \rangle}{I_{\mathrm{Ref},\theta=0}} . \tag{11}$$

Thus, when the values of these two functions are near unity, incident angle effects are negligible. The shot-noise-limited SNR is proportional to the MO signal divided by the square root of the reflectance. Thus we define the SNR figure of merit F_{SNR} as

$$F_{\mathrm{SNR}} = \frac{I_{\mathrm{MO,norm}}}{\sqrt{I_{\mathrm{Ref,norm}}}} . \tag{12}$$

III. Results

To define the parameters for which incident angle effects become significant, three specific media structures were tested for numerical apertures of 0.50, 0.65, and 0.80. The structures investigated were a bilayer (antireflection dielectric coating over magnetic substrate) and two types of quadrilayer structure (AR coating, thin MO film, AR coating, and reflective substrate). Two types of quadrilayer were included to compare the effects of a tuned system for which the theoretical maximum in SNR occurs (and also the minimum in reflectance) and an off-resonance case for which the reflectance has been designed to be 10% and the theoretical SNR has not decreased significantly. This detuned structure is important because a disk reflectance value of at least 10% is generally considered necessary for tracking and focusing requirements.[7]

The materials used in the simulations were TbFe as the active MO material with a dielectric constant $\epsilon = -2.6 + i20.0$ and a magnetooptic constant $q = -0.016 + i0.00001$. The dielectric AR layers were SiO_2 with an index of $n = 1.5$. The reflective substrate was chosen to be aluminum with an index of $n = 1.3 + i7.0$. The incident medium was air, and the laser wavelength used was $\lambda = 840$ nm. Thus the AR layers were 140 nm thick.

The optical and magnetooptical responses of the bilayer for a numerical aperture of 0.65 are illustrated in 3-D plots in Figs. 2(a) and (b). These figures show that, for the bilayer case, neither the reflectance nor the MO signal vary appreciably for rays originating from different points in the aperture. Thus, for this case, characterizing the structure simply by its normal incidence performance appears to be a very good approximation. This is shown quantitatively in Table I, where it is seen that $I_{\mathrm{MO,norm}} = 0.990$ and $I_{\mathrm{Ref,norm}} = 0.990$.

Table I. Numerical Results for the Bilayer, Tuned Quadrilayer, and Detuned Quadrilayer Structures

		Numerical aperture		
	($\theta = 0$ Values)	0.50	0.65	0.80
Bilayer:				
$I_{\mathrm{MO},\theta=0}$	0.00931			
$I_{\mathrm{Ref},\theta=0}$	0.308			
$F_{\mathrm{SNR},\theta=0}$	0.0168			
$I_{\mathrm{MO,norm}}$		0.999	0.998	0.994
$I_{\mathrm{Ref,norm}}$		0.970	0.956	0.946
F_{SNR}		1.014	1.021	1.022
Tuned quadrilayer:				
$I_{\mathrm{MO},\theta=0}$	0.00111			
$I_{\mathrm{Ref},\theta=0}$	0.00126			
$F_{\mathrm{SNR},\theta=0}$	0.0313			
$I_{\mathrm{MO,norm}}$		0.980	0.943	0.879
$I_{\mathrm{Ref,norm}}$		1.020	1.138	2.564
F_{SNR}		0.971	0.884	0.549
Detuned quadrilayer:				
$I_{\mathrm{MO},\theta=0}$	0.00906			
$I_{\mathrm{Ref},\theta=0}$	0.110			
$F_{\mathrm{SNR},\theta=0}$	0.0273			
$I_{\mathrm{MO,norm}}$		0.988	0.988	0.985
$I_{\mathrm{Ref,norm}}$		0.965	0.957	0.978
F_{SNR}		1.006	1.007	0.996

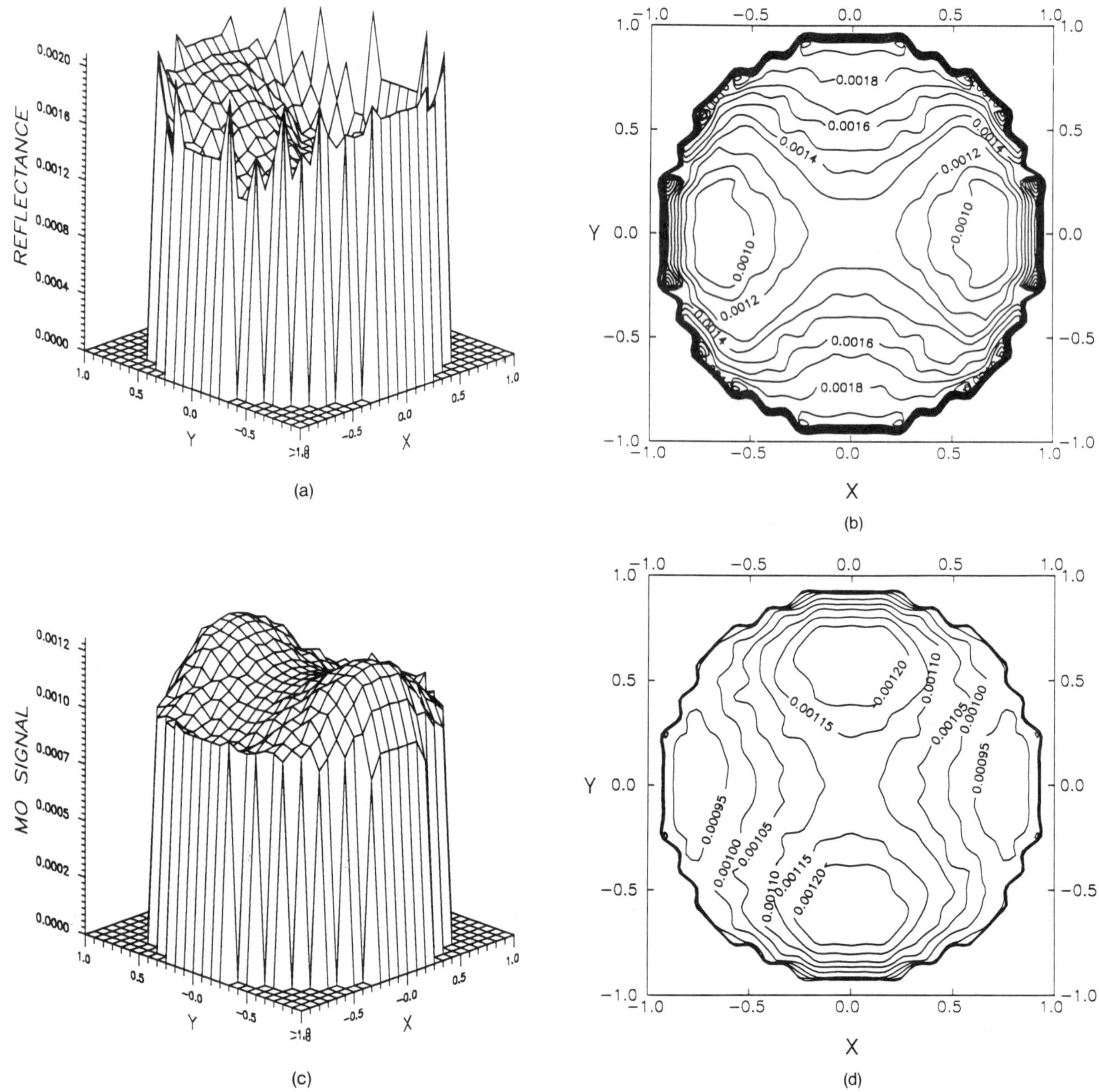

Fig. 3. Results for the tuned quadrilayer structure for an N.A. of 0.65. The reflectance (a), (b) and MO signal (c), (d) for this structure exhibit much stronger dependences on incident angle than for the bilayer. The contour plots (b), (d) were included to display the portions of the aperture obscured in the 3-D plots.

For the tuned quadrilayer case illustrated in Fig. 3, both the reflectance and MO signal show much larger features than for the bilayer. The sharp spikes which occur around the perimeter of the aperture in Fig. 3 are artifacts due to the process of conversion from a polar grid to a rectangular grid and are not important for the present discussion. The topographical (contour) plots of Figs. 3(b) and (d) were included since large areas of the surfaces of Figs. 3(a) and (c) are obscured. The stronger variations observed for this case relative to the previous case are confirmed numerically in Table I, which shows that $I_{\mathrm{Ref,norm}}$ rapidly increases and $I_{\mathrm{MO,norm}}$ and F_{SNR} rapidly decrease as the numerical aperture is increased.

Finally, we consider what happens when the MO film thickness is increased from the tuned quadrilayer thickness value, which might be done to increase the reflectance. An increase in the MO film thickness from 14 to 28 nm was found to increase the reflectance to the desired 10% (at normal incidence) with only an ~13% loss in theoretical SNR (at normal incidence). As seen in both Fig. 4 and Table I, the reflectance and MO signal for this structure are both much more uniform than for the previous quadrilayer resulting in a net SNR, which is within 1% of the normal incidence value.

It is important to analyze the results here in terms of their relevance to actual magnetooptic readout sys-

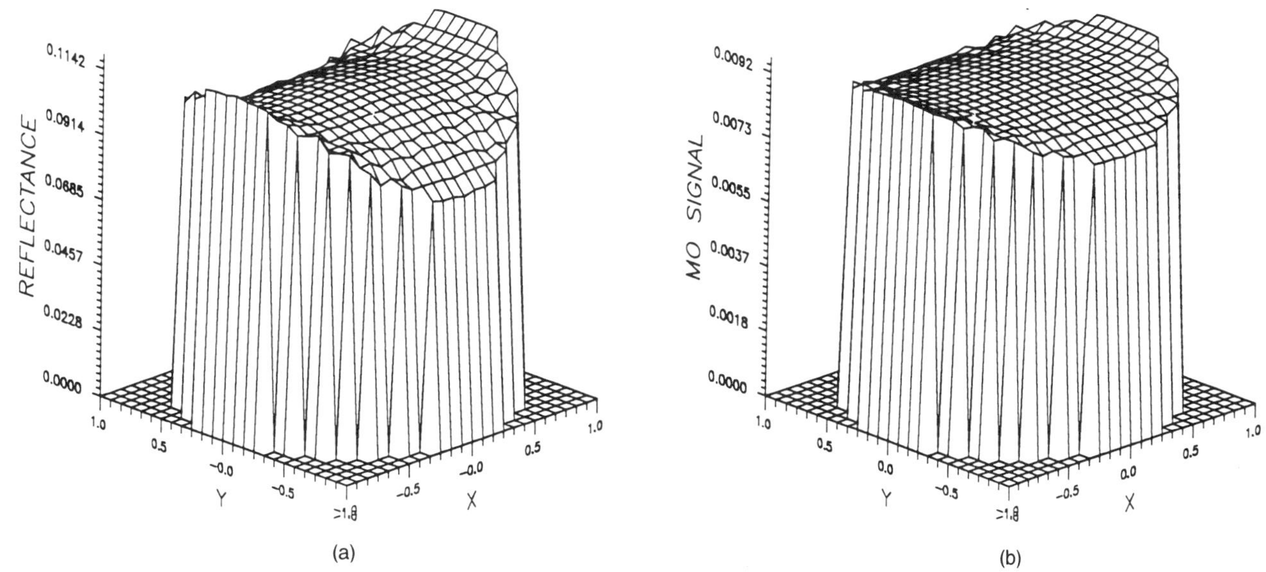

Fig. 4. Results for the detuned quadrilayer structure for an N.A. of 0.65. The reflectance (a) and MO signal (b) for this structure are much more uniform than for the tuned quadrilayer.

tems. In particular, two assumptions were implicitly made which might make direct comparison of the theoretical results presented here with experimental results from actual readout systems somewhat difficult. The first of these assumptions was that the original wavefront uniformly fills the focusing aperture. In reality, the incident wavefront is typically a truncated Gaussian beam. Since this type of wavefront has a larger percentage of its power at smaller angles of incidence, it should be expected that effects due to incident angle in this case would be less pronounced than for the uniformly illuminated aperture assumed here.

The other assumption was that each media structure was composed of isotropic materials, i.e., materials with no birefringence. Substrate birefringence is a recognized problem and has been treated theoretically by several authors.[3–5] Vertical birefringence, which is typically larger than in-plane birefringence, is particularly troublesome in high N.A. systems, because its effect increases with the component of the electric field in the vertical direction and thus increases with incident angle. Therefore, realistic media which exhibit birefringence would be more sensitive to effects of incident angle than the media investigated here.

IV. Conclusion

A Jones matrix model has been used to analyze the effects of incident angle on readout in magnetooptic storage media. The model traces the polarization state of each ray in a polar grid as the rays in the grid reflect from the MO media and then travel through a differential detection system. For purposes of comparison, a simple bilayer and two quadrilayer structures were analyzed. The quadrilayer structure which theoretically maximizes the readout SNR was found to be most sensitive to incident angle effects. Numerical results of average reflectance, MO signal, and shot-noise-limited SNR were presented which apply for a uniformly illuminated aperture and media with no birefringence.

This work was supported by the Optical Data Storage Center at the University of Arizona.

References

1. M. N. Deeter and D. Sarid, "Effects of Incident Angle on Readout in Magnetooptic Storage Media," Appl. Opt. **27,** 713 (1988).
2. M. Mansuripur, G. A. N. Connell, and J. W. Goodman, "Signal and Noise in Magneto-optical Readout," J. Appl. Phys. **53,** 4485 (1982).
3. W. A. Challener and T. A. Rinehart, "Jones Matrix Analysis of Magnetooptical Media and Read-Back Systems," Appl. Opt. **26,** 3974 (1987).
4. T. Toda *et al.*, "Analysis of Signal to Noise Ratio in Magneto-Optical Disk Using a Polarization Simulator," in *Technical Digest, Topical Meeting on Optical Data Storage* (Optical Society of America, Washington, DC, 1987), paper WB4.
5. A. B. Marchant, "Retardation Effects in Magneto-Optic Readout," Proc. Soc. Photo-Opt. Instrum. Eng. **695,** 40 (1986).
6. D. Sarid, M. N. Deeter, and V. Kahwaty, "Optical Probing of Magnetic Inert Layers," Appl. Opt. **26,** 3153 (1987).
7. A. E. Bell, "Antireflection Structures for Magneto-Optic Recording," in *Technical Digest, Topical Meeting on Optical Data Storage* (Optical Society of America, Washington, DC, 1987), paper ThD2.

Reprinted from *Optical Data Storage Topical Meeting,* Proc. SPIE Vol. 1078, pp. 279-286 (1989).

Optimizing the noise performance of a magneto-optic read channel

Anthony G. Dewey

IBM Research Division, Almaden Research Center
650 Harry Road, San Jose, CA 95120

Abstract

In a magneto-optic head with differential detection, the light level on the data detectors is determined by the ratio of a semi-polarizing beam splitter. This paper presents a way of measuring the various noise sources with a common set of values and uses this data to choose the beam splitter ratio that optimizes the SNR.

Introduction

One of the problems in the design a magneto-optic read-write head is determining the light level needed at the data detectors. Figure 1 is a schematic of a typical M-O head in which PBS1 and PBS2 are actually leaky beamsplitters that divert from the main optical path (laser to disk

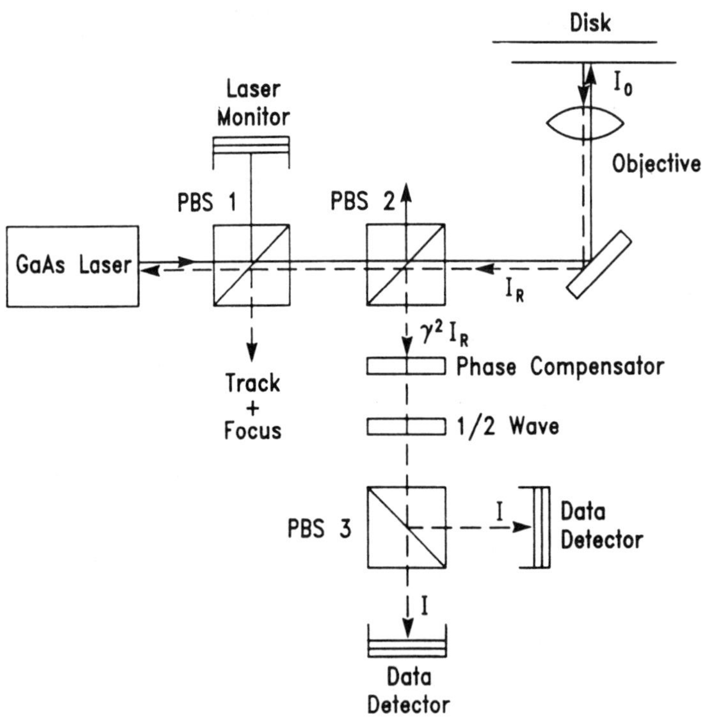

Fig. 1 Optical system of an M-O head

and back) some fraction of the p-polarized (unrotated) light. On the outward path this diverted light reduces the efficiency of the head and, especially from the point of view of writing on the disk, the goal would be to minimize this loss. On the return path the light diverted by PBS1 goes to the servo detectors to provide focus and track-error signals, and that by PBS2 goes to the data detectors.

If PBS2 were not leaky, the only light reaching the data detectors would be the Kerr-rotated signal which, for a typical M-O disk and a read level I_o of 1 mW, is about 10 nW. In this case the SNR would be limited by the electronic noise P_{ne} (dark noise) of the detectors (photo-diodes + amplifiers). By diverting a fraction γ^2 of I_R the light level I on each detector can be at least several μW and the SNR is improved.

Since the Kerr-rotated light is now a small fraction of the total, I is proportional to $\gamma^2 I_o$ and as shown in [1], the signal P_s is $\propto \gamma^2 I_o^2$. The noise level at each detector is the sum of the electronic noise P_{ne}; the shot noise P_{ns} and disk polarization noise P_{np}, both of which are $\propto$ I; and the laser noise P_{nl} and disk reflectivity noise P_{nd}, both of which are $\propto I^2$. The outputs of the two detectors go to a differential amplifier with a finite common-mode rejection ratio (CMRR). This differential detection scheme reduces, but does not eliminate, the common-mode noise (laser and disk reflectivity noise) particularly since the two data detectors can be balanced only for one polarization state of the disk. The effect of writing on the disk can be modeled as a reduction in the CMRR.

At low values of I the electronic noise will limit the SNR whereas at high values of I the residual $P_{nl} + P_{nd}$, being $\propto$ to I^2, will dominate. Between these extremes is an optimum value of γ that will give the best SNR. This paper shows how to measure the various noise components with a common set of values and determine the optimum light level I on the detectors for an optical head used in a disk tester.

Electronic Noise

The electronic noise is often specified by giving a plot of the noise spectrum ($nV/\sqrt{Hz}$) or by giving the noise voltage V_n (μV) in a given bandwidth B (MHz). Also specified is the sensitivity of the detector, R_D (V/W). One value that is not always given is the responsivity R (A/W) of the photo-diode at the laser wavelength. As mentioned in the introduction, we need enough light on the detectors to keep the electronic noise well below the one fundamental noise in the system, the shot noise. For this reason a convenient figure of merit for the detector is I_S, the light intensity at which the shot + electronic noise is 3 dB above the electronic noise ($P_{ns} = P_{ne}$). The shot noise voltage V_{ns} and current i_{ns} are given by:

$$V_{ns} = \frac{R_D}{R} i_{ns} = \frac{R_D}{R} \sqrt{2\, e\, R\, I\, B \times 10^6} \quad (V)$$

$$= R_D \sqrt{0.32\, I\, B / R} \quad (\mu V)$$

So I_S is given by:

$$V_{ns} = V_n = R_D\sqrt{0.32\, I_S\, B \,/\, R}$$

$$I_S = \frac{V_n^2\, R}{0.32\, R_D^2\, B} \quad (W)$$

The detectors used in our optical head are Hamamatsu S2858-01 hybrid PIN diode / preamps having a V_n of 92 μV in B = 10 MHz, and an R_D of 1.65×10^4. Assuming a responsivity R = 0.5 at the 830 nm wavelength of our laser, we can calculate an I_S of 4.9 μW.

Laser Noise

Laser noise is usually specified as RIN (relative intensity noise) defined as:

$$RIN_L = 10 \log\, (\Delta I^2 \,/\, I^2\, B) \quad (dB)$$

where I is the average laser power and ΔI is the rms power fluctuation in a bandwidth B. The noise depends strongly on the type of laser, the power level and particularly on feedback into the laser from an external reflector such as the disk.

In our optical system the laser is monitored by a low-noise PIN detector having an I_S of 7 μW. Figure 2 shows the laser noise spectrum when 2 mW was focused on an M-O disk and

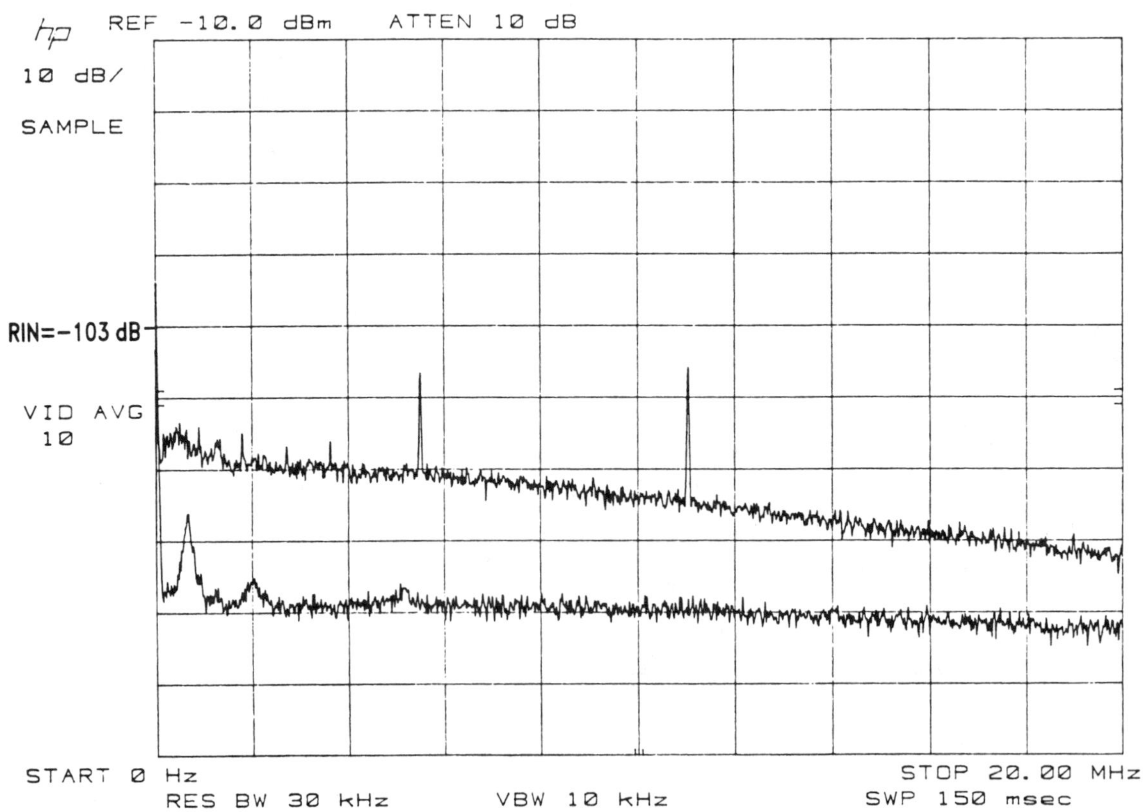

Fig. 2 Laser noise spectrum

the light level at the detector was 39 μW. The lower trace shows the electronic noise level, and the shot noise level is about 8 dB higher. The average detector output (into 50 Ω) was 0.71 V = + 10 dBm and the calculated RIN_L (B = 10 MHz) is -123 dB. This level was achieved by modulating the laser at ~ 200 MHz which significantly reduced the effect of feedback.

Shot Noise

If we know the responsivity R (A/W) of a detector we can also specify the shot noise as an RIN value.

$$RIN_S = 10 \log \left(i_{ns}^2 / i^2 B \right) \quad (dB)$$

$$= 10 \log \left(\frac{2 e R I B}{R^2 I^2 B} \right) \quad (dB)$$

With I in μW, this reduces to RIN_S = -125 - 10 log (RI) (dB)

Disk Reflectivity Noise

This noise component is measured using write-once optics, i.e., a PBS and a quarter-wave plate. With this system, laser feedback and hence laser noise is minimized. With sufficient light

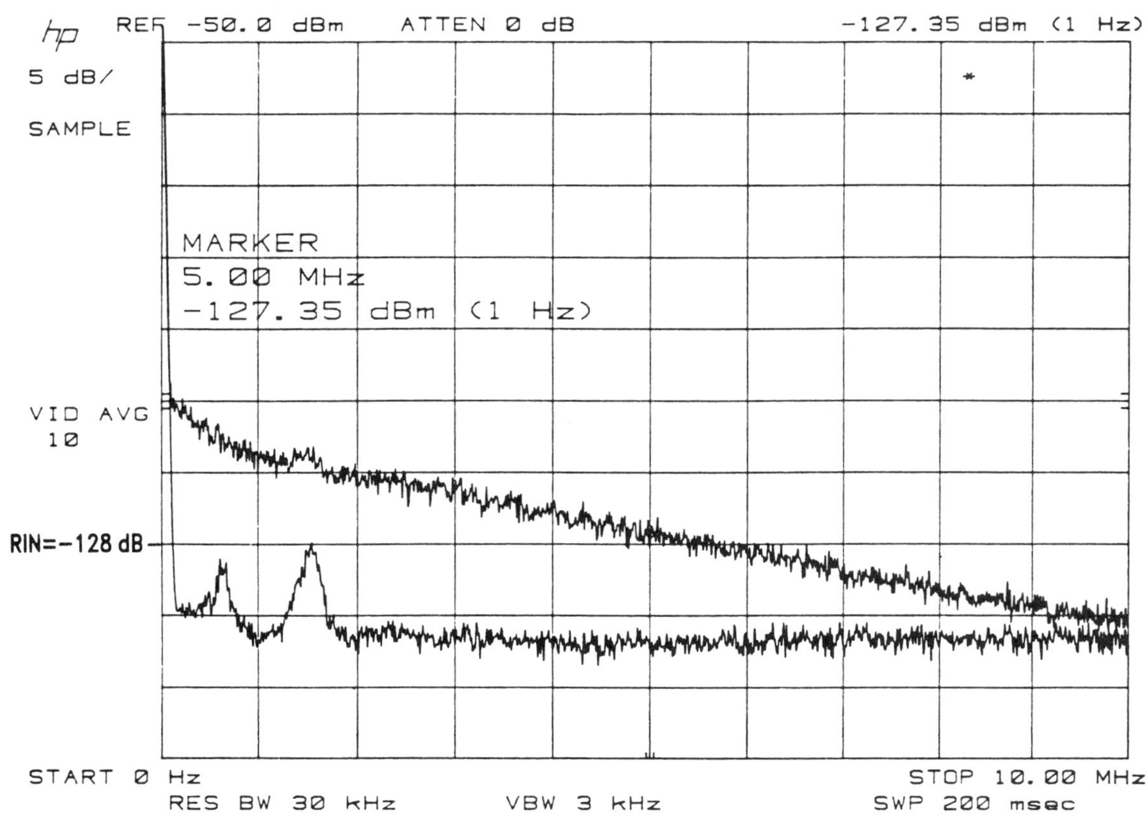

Fig. 3 Disk reflectivity noise spectrum

on a low-noise detector, the measured noise will be dominated by the disk reflectivity noise, and disk polarization noise will be negligible. Figure 3 shows such a noise measurement with an erased M-O disk spinning at 10 m/s. The average detector output into 50 Ω was 224 mV = 0 dBm. The lower trace shows the noise level with the disk stationary, from which the laser noise RIN_L is estimated to be -135 dB. By integrating the disk noise spectrum over the 10 MHz bandwidth the RIN_D is calculated to be -125 dB.

Residual Common-mode Noise

The sum of the laser noise RIN_L = -123 dB and the disk reflectivity noise RIN_D = -125 dB is a total noise RIN_T = -121 dB. This is reduced by the CMRR of the differential amplifier which, for the conditions of this test was approximately 27 dB. We can now define a light level at the detectors I_D at which the residual common-mode noise equals the shot noise:

$$RIN_T - CMRR = -125 - 10 \log (RI_D)$$

With R = 0.5 A/W and the above values we get I_D = 400 μW.

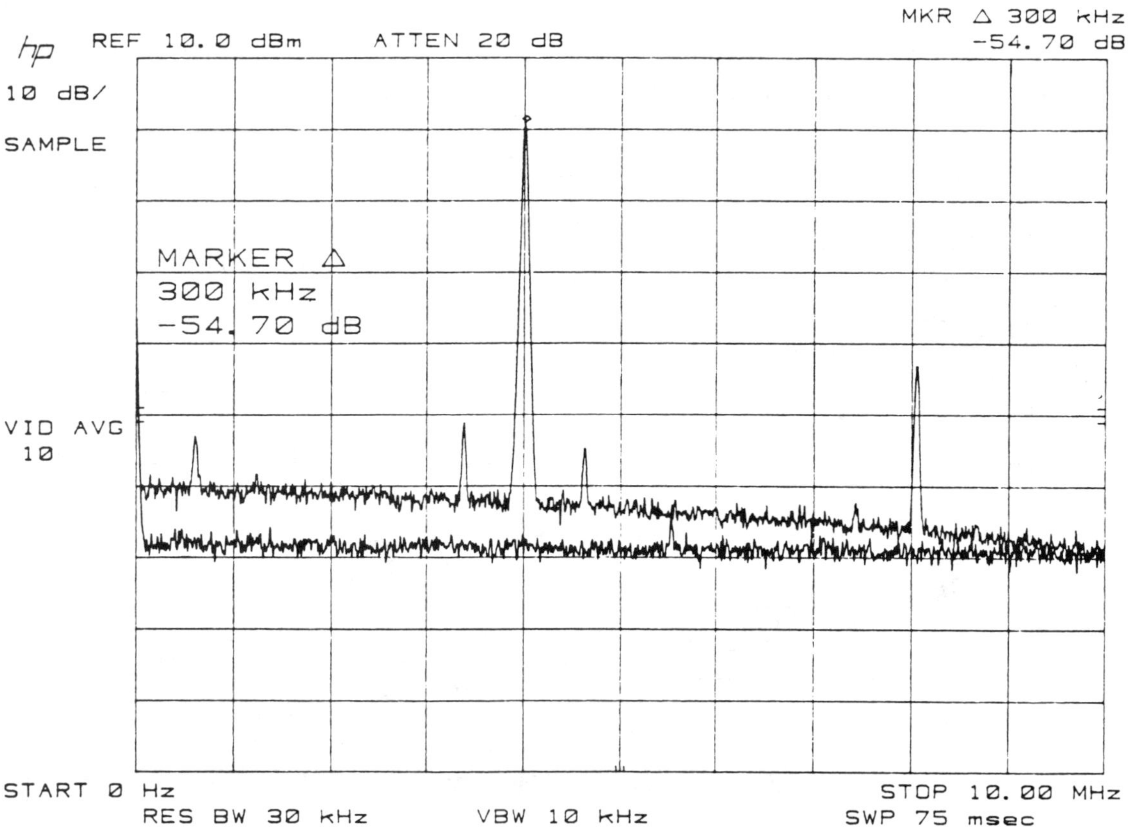

Fig. 4 Spectrum of the M-O read signal

Signal and Disk Polarization Noise

Figure 4 shows the result of writing a 4 MHz carrier on the M-O disk at a velocity of 10 m/s, a read level of 2 mW, and a detector light level I of 11 μW. The SNR in the 10 MHz bandwidth is calculated to be 28 dB. The lower trace, with the disk stationary, is the electronic + shot noise level, and the difference is the disk polarization noise, P_{np}. Actually P_{np} includes the writing noise caused when the 4 MHz signal is written on the disk: the noise on an erased track is lower. From the figure, the difference is calculated to be 6 dB and by blocking the light to the detectors we find the total noise is 11 dB above the electronic noise. Using these two figures we can calculate that:

$$P_{np} / P_{ns} = 4.4 = 6.4 \text{ dB}$$

Noise Chart

Knowing the SNR and all the noise components we can construct a chart as in Figure 5, showing their variation with I, as γ^2 is changed (with a fixed I_o). P_{ne} is constant and is chosen as the 0 dB level. P_{ns} increases at 10 dB / decade from its intersection with P_{ne} at $I = I_S$ and the residual common-mode noise increases at 20 dB / decade from its intersection with P_{ns} at $I = I_D$. P_{np} is 6.4 dB above P_{ns} and the total noise is found by summing the four components.

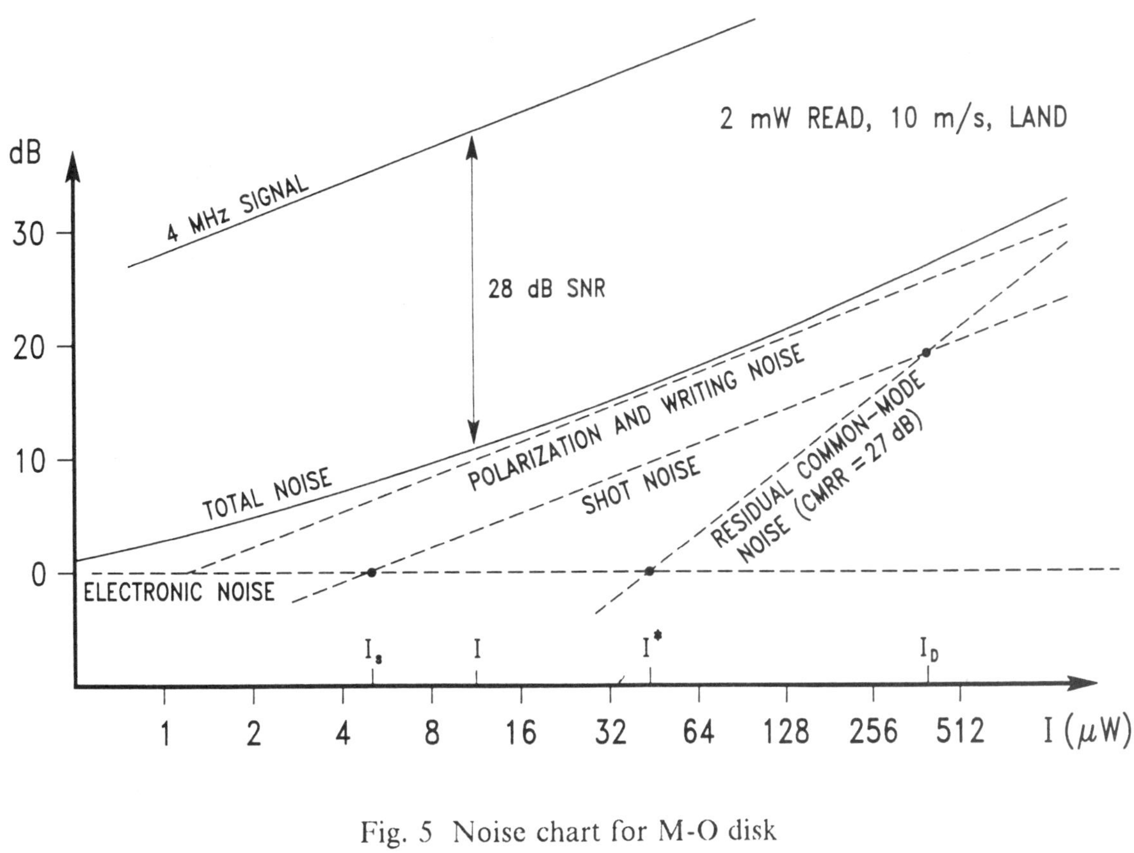

Fig. 5 Noise chart for M-O disk

The signal is 28 dB above the noise at $I = 11\ \mu W$. It can easily be shown that the maximum SNR is at $I = I^*$, the geometric mean of I_S and I_D which is also the intersection of the common-mode noise line with P_{ne}. In this example $I^* = \sqrt{4.9 \times 400} = 44\ \mu W$.

Measuring disk reflectivity noise on a separate drive with W-O optics is subject to errors caused by differences in the optical spot size, etc. An alternative method that is satisfactory for the present purpose of determining RIN_T (but can not be used to quantify RIN_D if $RIN_L \gg RIN_D$) is to block one of the M-O detectors. The result is shown in Figure 6 where, comparing with Figure 4, the signal is 6 dB lower, as expected, and the total integrated noise is about 2 dB higher. The lower trace is with both detectors blocked and shows the electronic noise level. Using these measurements and the CMRR of 27 dB we can calculate an RIN_T of -120.5 dB, which is approximately the same as the value quoted earlier.

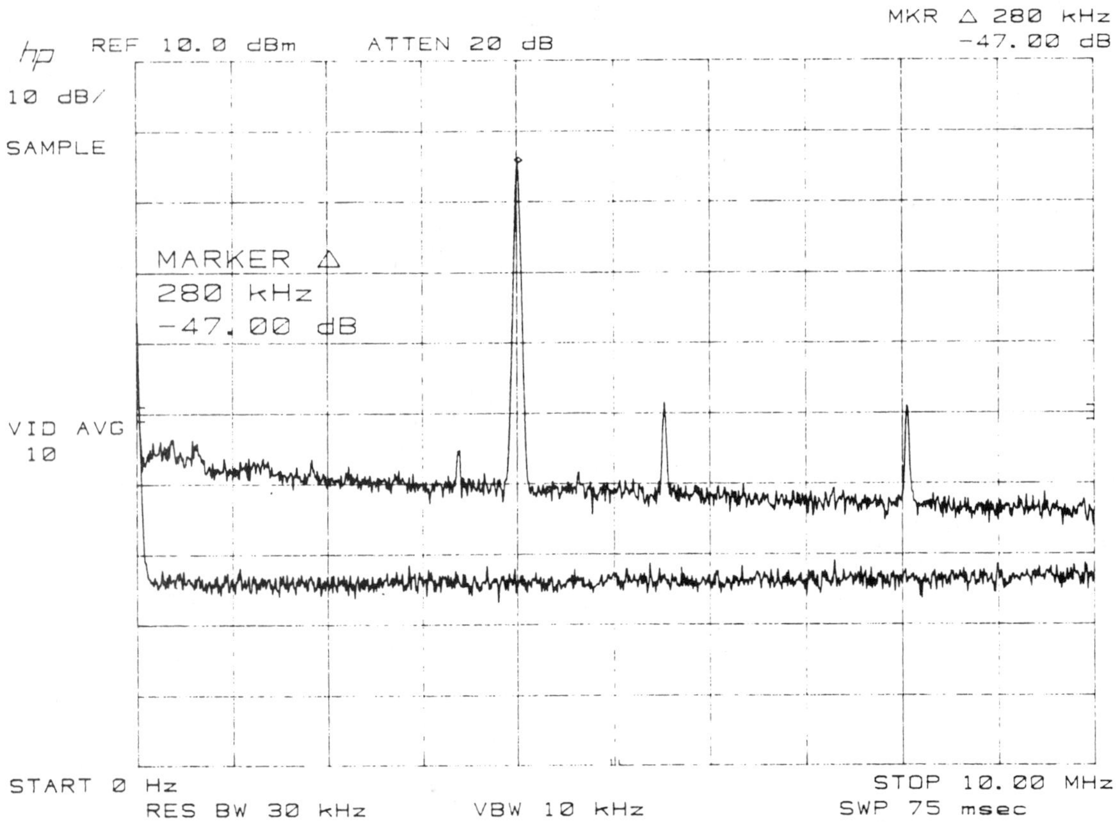

Fig. 6 Spectrum of M-O signal with one detector blocked

Conclusion

We have shown how to analyze the noise components in a M-O readout signal, have suggested a useful figure of merit I_S for photodetectors, and that disk reflectivity noise be quantified by an RIN value. Using these measurements we have shown how to design the head (vary γ^2) in order to optimize the SNR for the given set of conditions. In the example shown, the noise is dominated by shot noise + polarization and writing noise, and hence the SNR is fairly insensitive to I over a wide range. With any of the following conditions, however: higher P_{ne}, lower P_{np}, higher P_{nl}, higher P_{nd}, or lower CMRR, the optimum would be more well defined. In any case this technique allows one to evaluate the effect of higher detector, laser or disk noise, and lower CMRR, all of which will have some bearing on the cost of the M-O drive.

References

[1] Dewey, A. G., "Measurement and Modeling of Optical Disk Noise," Proc. SPIE, Vol. 695 - Optical Mass Data Storage II (1986), 72 - 78.

[2] Mansuripur, M., Connell, G. A. N. and Goodman, J. W., "Signal and Noise in Magneto- Optic Read-Out," J. Appl. Phys. 53 (6), June 1982, 4485 - 94.

[3] Treves, D. and Bloomberg, D. S., "Signal, Noise and Codes in Optical Memories," Optical Engineering, Vol. 25 No. 7, July 1986, 881 - 91.

Reprinted with permission from *Journal of the Optical Society of America A*, Vol. 6(5), pp. 786-805 (June 1989).

Certain computational aspects of vector diffraction problems

M. Mansuripur

Optical Sciences Center, University of Arizona, Tucson, Arizona 85721

Received July 6, 1988; accepted December 20, 1988

Fourier decomposition of a given amplitude distribution into plane waves and the subsequent superposition of these waves after propagation is a powerful yet simple approach to diffraction problems. Many vector diffraction problems can be formulated in this way, and the classical results are usually the consequence of a stationary-phase approximation to the resulting integrals. For situations in which the approximation does not apply, a factorization technique is developed that substantially reduces the required computational resources. Numerical computations are based on the fast-Fourier-transform algorithm, and the practicality of this method is shown with several examples.

1. INTRODUCTION

A. Background

The essence of diffraction theory is Huygens's principle, which states that every point on a given wave front can be treated as a source of radiation, emanating spherical wavelets that can be superimposed later to yield the light-amplitude distribution. A slight variation on this theme is the view that the wave front can be decomposed into a set of plane waves that propagate independently through space and that their superposition is the field at any given point.[1–4] The two views can be shown to be identical and to lead to the same results, although the latter approach is mathematically more convenient. The Fourier-transform technique is the natural vehicle for decomposition into and superposition of plane waves, and, in fact, the majority of problems in the classical theory of diffraction can be reduced to a pair of Fourier- and inverse Fourier-transform integrals. The trouble, from a computational point of view, is that usually the functions involved in these transformations are highly oscillatory and consequently require a large number of samples for proper representation. It turns out, however, that in many problems of practical interest one or the other of the Fourier integrals can be approximated quite accurately with the stationary-phase technique.[5,6] For instance, in the problem of far-field (Fraunhofer) diffraction from an aperture, the stationary-phase approximation applies to the second (superposition) integral, whereas in the case of diffraction from a lens the approximation is used to eliminate the first (decomposition) integral. In both cases the remaining integral can then be computed efficiently with the fast-Fourier-transform (FFT) algorithm.

Like any approximation, the stationary-phase technique is valid only over a certain range of the parameters involved. When the parameter set happens to be outside the proper range, one goes back to evaluate the original Fourier integrals more accurately. Near-field (Fresnel) diffraction from an aperture is but one representative of problems in which the stationary-phase approximation fails. Another example is diffraction from a lens with a small numerical aperture. It is well-known that at small numerical apertures the distribution becomes asymmetric with respect to the focal point and the maximum intensity occurs at a point that is somewhat closer to the lens than the point of geometric focus.[7-9] This contrasts sharply with the prediction of symmetry by theories that are based on the stationary-phase approximation.[2]

In this paper a factorization technique is presented that enables one to compute the Fourier integrals of diffraction theory more efficiently. The price of the reduction in the required number of samples (made possible by factorization) is an extra Fourier transform: instead of transforming a large array twice, one must now transform a smaller array three times. The obvious advantage of factorization is therefore the reduced level of computational resources (memory and CPU time) that a given problem requires. It is important to emphasize that factorization does not eliminate the need for the stationary-phase approximation (wherever it legitimately applies), nor does it further simplify problems in which the original Fourier and inverse Fourier integrals are readily computable. Simply stated, the factorization technique covers the gray boundary area between these two regions, where approximation is inaccurate and direct computation is costly.

The organization of the paper is as follows. In the remainder of this section the notation is introduced, and the background is provided for the problem formulation. In Section 2, we discuss the basic equations of vector diffraction theory and describe the factorization technique in detail. In Section 3 factorization is applied to problems of diffraction in the presence of a lens. The important case of a spherical lens is treated in detail in Subsections 3.A and 3.B, and the (rather exotic) cases of an astigmatic lens and a ring lens are treated in Subsections 3.C–3.E. In Section 4 the results are given for numerical computations based on the various methods described in this paper. The results are intended to give an appreciation for the range of applicability of each technique.

B. Preliminaries

The notation and some elementary aspects of the formalism used throughout the paper are as follows[10]:

1. $t(x, y)$ is a two-dimensional function defined on the xy

plane. When the x axis is normalized by a and the y axis is normalized by b, we define new coordinates $\mathbf{x}$ and $\mathbf{y}$, such that $x = a\mathbf{x}$ and $y = b\mathbf{y}$, and represent the function in this new coordinate system as $t(a\mathbf{x}, b\mathbf{y})$.

2. The Fourier transform of $t(x, y)$ is $T(u, v)$ and is written

$$T(u, v) = \mathcal{F}\{t(x, y)\} = \iint_{-\infty}^{\infty} t(x, y)\exp[-i2\pi(xu + yv)]dxdy. \tag{1.1}$$

If x and y are normalized by a and b, respectively, the Fourier transform of the scaled function is

$$\frac{1}{ab} T\left(\frac{u}{a}, \frac{v}{b}\right) = \mathcal{F}\{t(a\mathbf{x}, b\mathbf{y})\} = \iint_{-\infty}^{\infty} t(a\mathbf{x}, b\mathbf{y}) \times \exp[-i2\pi(\mathbf{x}u + \mathbf{y}v)]d\mathbf{x}d\mathbf{y}. \tag{1.2}$$

3. The inverse transform of $T(u, v)$ is $t(x, y)$ as follows:

$$t(x, y) = \mathcal{F}^{-1}\{T(u, v)\} = \iint_{-\infty}^{\infty} T(u, v)\exp[i2\pi(xu + yv)]dudv. \tag{1.3}$$

4. The discrete array used for numerical computations is initially in the $\mathbf{xy}$ plane, where $\mathbf{x} = x/\lambda$ and $\mathbf{y} = y/\lambda$ are dimensionless coordinates; i.e., $\mathbf{x}$ and $\mathbf{y}$ are in units of the wavelength λ. The length of the array in the $\mathbf{x}$ direction is L_{max}, and the number of pixels in this dimension is N_{max}. The corresponding parameters along the $\mathbf{y}$ axis are L_{may} and N_{may}. Because the arrays are transformed by the FFT algorithm, it is preferable that N_{max} and N_{may} be powers of 2. The array generally is centered at the origin, and the (I, J) pixel represents the point $\mathbf{x} = (I - N_{midx} - 1)L_{max}/N_{max}$, $\mathbf{y} = (J - N_{midy} - 1)L_{may}/N_{may}$, where $N_{midx} = N_{max}/2$ and $N_{midy} = N_{may}/2$. The spacing between nearest-neighbor pixels is $\Delta_x = L_{max}/N_{max}$ and $\Delta_y = L_{may}/N_{may}$. For the most part in this paper we have square lattices with $L_{may} = L_{max}$, $N_{may} = N_{max}$, and $\Delta_y = \Delta_x$. Thus, when there is no danger of confusion, we shall express the relationships in terms of L_{max}, N_{max}, and Δ, with the understanding that similar relationships exist for the $\mathbf{y}$ direction.

5. To normalize the x axis by a we divide L_{max} by a. Thus, if $t(x, y)$ is represented by an $N_{max} \times N_{may}$ array on an $L_{max} \times L_{may}$ rectangle, its scaled version $t(a\mathbf{x}, b\mathbf{y})$ has the same pixels but is on a rectangle of dimensions $(L_{max}/a) \times (L_{may}/b)$.

6. The discrete Fourier transform of an $L_{max} \times L_{may}$ array with $N_{max} \times N_{may}$ pixels is another array with the same number of pixels but with new dimensions $\tilde{L}_{max} \times \tilde{L}_{may}$. In general, $\tilde{L}_{max} = N_{max}/L_{max}$ and $\tilde{L}_{may} = N_{may}/L_{may}$.

7. When a linearly polarized plane wave traveling along the z axis encounters a prism, as shown in Fig. 1, its direction of propagation changes to $\hat{\sigma} = (\sigma_x, \sigma_y, \sigma_z)$. The effect of refraction on the polarization vector is to keep the S component unchanged but to reorient the P component such that it becomes perpendicular to $\hat{\sigma}$. The new polarization components Ψ_x, Ψ_y, and Ψ_z can thus be calculated for any incident polarization by simple geometric considerations. Table 1 shows the results for x-polarized (top three rows) and y-polarized (bottom three rows) incident beams. Notice that $\hat{\sigma}$ is a unit vector and thus that $\sigma_z = (1 - \sigma_x^2 - \sigma_y^2)^{1/2}$.

8. A coherent point source located at the origin gives rise to a spherical wave front. The amplitude distribution in a

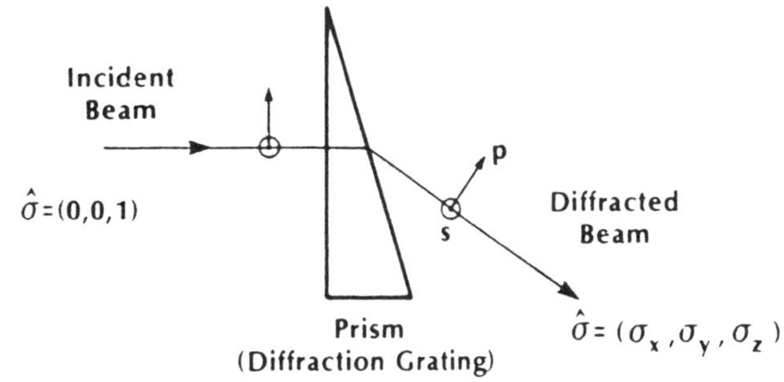

Fig. 1. Prism as a diffraction grating. A plane, linearly polarized incident beam is diffracted into a plane wave propagating along $\hat{\sigma} = (\sigma_x, \sigma_y, \sigma_z)$. The relationship between the incident and the diffracted polarizations is summarized in Table 1.

Table 1. Components of the Polarization Vector of a Plane Wave[a] Propagating along the Unit Vector $\hat{\sigma} = (\sigma_x, \sigma_y, \sigma_z)$

Component	Expression
Ψ_{yx}	$-\dfrac{\sigma_x\sigma_y}{1 + \sigma_z}$
Ψ_{yy}	$1 - \dfrac{\sigma_y^2}{1 + \sigma_z}$
Ψ_{yz}	$-\sigma_y$
Ψ_{xx}	$1 - \dfrac{\sigma_x^2}{1 + \sigma_z}$
Ψ_{xy}	$-\dfrac{\sigma_x\sigma_y}{1 + \sigma_z}$
Ψ_{xz}	$-\sigma_x$

[a] The beam is obtained by the refraction (through a properly oriented prism) of a linearly polarized wave propagating along $\hat{\sigma}_0 = (0, 0, 1)$. $\Psi_{\alpha\beta}$ is the β component of the refracted beam when the (unit-amplitude) incident beam is polarized along α.

plane perpendicular to the positive z axis is said to have a curvature $C = 1/z$. The curvature phase factor is given by

$$A(x, y) = \exp\left\{i \frac{2\pi}{\lambda C} [1 + C^2(x^2 + y^2)]^{1/2}\right\}. \tag{1.4}$$

If the coordinates are normalized by the wavelength such that $\mathbf{x} = x/\lambda$, $\mathbf{y} = y/\lambda$, and $\mathbf{z} = z/\lambda$, then we define the normalized curvature $\mathbf{C} = \lambda C$ in order to eliminate λ from Eq. (1.4). In this notation a positive curvature represents a diverging beam centered to the left of the xy plane, and a negative curvature corresponds to a converging beam toward a point to the right of the xy plane.

2. PROPAGATION IN FREE SPACE

A. Direct Application of Fresnel's Formula

It is well known that an amplitude distribution $t(x, y)$ in the xy plane can be considered the superposition of plane waves propagating along the unit vectors $\hat{\sigma} = [\sigma_x, \sigma_y, (1 - \sigma_x^2 - \sigma_y^2)^{1/2}]$. The complex amplitude of each plane wave is related to the Fourier transform of $t(x, y)$ as follows[1]:

$$\lambda^{-2}T\left(\frac{\sigma_x}{\lambda}, \frac{\sigma_y}{\lambda}\right) = \begin{cases} \mathcal{F}\{t(\lambda\mathbf{x}, \lambda\mathbf{y})\} & \sigma_x^2 + \sigma_y^2 \le 1 \\ 0 & \text{otherwise} \end{cases}. \tag{2.1}$$

Here $x = \lambda\mathbf{x}$, $y = \lambda\mathbf{y}$, and $\mathcal{F}\{\cdot\}$ is the two-dimensional Fourier-transform operator.

At a distance z from the origi the distribution is the superposition of the above plane waves, each multiplied by a proper phase factor in order to represent different propagation distances. Let us assume that each plane wave originates at the $z = 0$ plane by going through a properly oriented prism; then, if the incident beam is polarized along the α axis (α is either x or y), the contribution of the wave traveling along $\hat{\sigma}$ to the final polarization along β (β is x or y or z) will be $\Psi_{\alpha\beta}(\sigma_x, \sigma_y)$ as given in Table 1. Thus the amplitude distribution of β-polarized light at $z = \lambda\mathbf{z}$ is[10]

$$\tilde{t}_{\alpha\beta}(\lambda\mathbf{x}, \lambda\mathbf{y}, \lambda\mathbf{z}) = \mathcal{F}^{-1}\left\{\lambda^{-2}T\left(\frac{\sigma_x}{\lambda}, \frac{\sigma_y}{\lambda}\right)\Psi_{\alpha\beta}(\sigma_x, \sigma_y) \times \exp[i2\pi\mathbf{z}(1 - \sigma_x^2 - \sigma_y^2)^{1/2}]\right\}. \quad (2.2)$$

The Fourier-transform pair of Eqs. (2.1) and (2.2) constitutes the basic formula for propagation in the Fresnel regime. For numerical computations one needs a two-dimensional array to represent the initial and final distributions, as well as the intermediate functions. Let the linear dimension of this array (after normalization by λ) be L_{max}, and let N_{max} represent the number of samples (pixels) in this dimension. The sample spacing Δ must be small enough in order to represent properly the fine features of $t(\lambda\mathbf{x}, \lambda\mathbf{y})$. For instance, if the incident beam is uniform within a circular aperture of radius $R = \lambda\mathbf{R}$, then $\Delta \le \mathbf{R}/10$ is an appropriate choice. On the other hand, if the incident beam has a curvature $\mathbf{C}$, as in Eq. (1.4), then the fastest oscillations of the phase factor occur in the vicinity of the aperture edge, where the radial spacing between successive peaks (or valleys) is $\sim(1 + \mathbf{C}^2\mathbf{R}^2)^{1/2}/|\mathbf{C}|\mathbf{R}$. Consequently, Δ must satisfy the restriction

$$\Delta \le \min\left\{\frac{\mathbf{R}}{10}, \frac{(1 + \mathbf{C}^2\mathbf{R}^2)^{1/2}}{2|\mathbf{C}|\mathbf{R}}\right\}. \quad (2.3)$$

After the first transformation, the spacing between nearest-neighbor elements becomes $\tilde{\Delta} = 1/L_{max}$. The proper choice of $\tilde{\Delta}$ now depends on the functions that appear on the right-hand side of Eq. (2.2). The smallest features of $\lambda^{-2}T(\sigma_x/\lambda, \sigma_y/\lambda)$ in the $\sigma_x\sigma_y$ plane occur on a scale of $(R/\lambda)^{-1}$, where as before, R is the largest linear dimension of the aperture distribution $t(x, y)$. Thus $\tilde{\Delta} \le (10\mathbf{R})^{-1}$ is a good choice as far as $T(\sigma_x/\lambda, \sigma_y/\lambda)$ is concerned. The second function, $\Psi_{\alpha\beta}(\sigma_x, \sigma_y)$, is relatively smooth, and we need not be concerned about it at this point. The third function is an exponential phase factor that oscillates with increasing frequency as one moves away from the origin in the $\sigma_x\sigma_y$ plane. At a distance σ_{max} from the origin, the radial separation between successive peaks of this function is $\sim(1 - \sigma_{max}^2)^{1/2}/(\mathbf{z}\sigma_{max})$. Consequently, $\tilde{\Delta}$ must be small enough to sample these oscillations properly. σ_{max} itself depends on the spatial variations of $t(x, y)$; in fact, it is the largest spatial frequency contained in $t(\lambda\mathbf{x}, \lambda\mathbf{y})$. For the preceding example,

$$\sigma_{max} = \frac{|\mathbf{C}|\mathbf{R}}{(1 + \mathbf{C}^2\mathbf{R}^2)^{1/2}} + \frac{1}{2\mathbf{R}}. \quad (2.4)$$

Since $\tilde{\Delta}$ is the inverse of L_{max}, these considerations yield

$$L_{max} \ge \max\left\{10\mathbf{R}, \frac{2\mathbf{z}\sigma_{max}}{(1 - \sigma_{max}^2)^{1/2}}\right\}. \quad (2.5)$$

Relations (2.3) and (2.5) and Eq. (2.4) may be used for selecting L_{max} and N_{max} for numerical computations. Relation (2.5) is general and applies to all cases of Fresnel propagation, whereas relation (2.3) and Eq. (2.4) correspond to apertures illuminated with spherical waves only.

B. Extending the Range of Fresnel's Formula

As can be seen from relation (2.5), when the spatial variations of $t(\lambda\mathbf{x}, \lambda\mathbf{y})$ are rapid, (i.e., when σ_{max} is large) and/or when $\mathbf{z}$ is large, numerical computations may become impractical. One way around the problem is to separate the oscillating terms from Eq. (2.2). To this end, we define $G(\sigma_x, \sigma_y)$ as

$$G(\sigma_x, \sigma_y) = \lambda^{-2}T(\sigma_x/\lambda, \sigma_y/\lambda)\Psi_{\alpha\beta}(\sigma_x, \sigma_y) \times \exp\{i2\pi\mathbf{z}[(1 - \sigma_x^2 - \sigma_y^2)^{1/2} + \tfrac{1}{2}\eta(\sigma_x^2 + \sigma_y^2)]\} \quad (2.6)$$

and rewrite Eq. (2.2) as

$$\tilde{t}_{\alpha\beta}(\lambda\mathbf{x}, \lambda\mathbf{y}, \lambda\mathbf{z}) = \mathcal{F}^{-1}\{G(\sigma_x, \sigma_y)\exp[-i\pi\eta\mathbf{z}(\sigma_x^2 + \sigma_y^2)]\}. \quad (2.7)$$

Here η is a parameter that can be optimized to reduce the oscillations of the phase factor in Eq. (2.6). A method of finding the optimum η is given in Appendix A. Since the inverse transform of the product of two functions is the convolution of individual inverse transforms, Eq. (2.7) is written

$$\tilde{t}_{\alpha\beta}(\lambda\mathbf{x}, \lambda\mathbf{y}, \lambda\mathbf{z}) = \int\int_{-\infty}^{\infty} \mathcal{F}^{-1}\{G(\sigma_x, \sigma_y)\}\left(\frac{-i}{\eta\mathbf{z}}\right) \times \exp\left\{\frac{i\pi}{\eta\mathbf{z}}[(\mathbf{x} - \mathbf{x}')^2 + (\mathbf{y} - \mathbf{y}')^2]\right\}d\mathbf{x}'d\mathbf{y}'. \quad (2.8)$$

Let us define $H(\mathbf{x}, \mathbf{y})$ as follows:

$$H(\mathbf{x}, \mathbf{y}) = -i\eta\mathbf{z}\exp\left[\frac{i\pi}{\eta\mathbf{z}}(\mathbf{x}^2 + \mathbf{y}^2)\right]\mathcal{F}^{-1}\{G(\sigma_x, \sigma_y)\}; \quad (2.9)$$

then, when the normalized variables $X = \mathbf{x}/\eta\mathbf{z}$ and $Y = \mathbf{y}/\eta\mathbf{z}$ are used, Eq. (2.8) becomes

$$\tilde{t}_{\alpha\beta}(\lambda\mathbf{x}, \lambda\mathbf{y}, \lambda\mathbf{z}) = \exp\left[\frac{i\pi}{\eta\mathbf{z}}(\mathbf{x}^2 + \mathbf{y}^2)\right]\mathcal{F}\{H(\eta\mathbf{z}X, \eta\mathbf{z}Y)\}. \quad (2.10)$$

We summarize the results by outlining the steps in the computation of Fresnel propagation with an extended range:

1. Calculate $\lambda^{-2}T(\sigma_x/\lambda, \sigma_y/\lambda)$ from Eq. (2.1). As before, the sample spacing Δ depends on the spatial variations of $t(\lambda\mathbf{x}, \lambda\mathbf{y})$, and, for the simple example of an aperture with a spherical incident wave front, it is given by relation (2.3).
2. Calculate $G(\sigma_x, \sigma_y)$ from Eq. (2.6). The spacing $\tilde{\Delta}$ in the $\sigma_x\sigma_y$ plane is the inverse of L_{max}. To choose the correct L_{max}, one needs to know the maximum frequency σ_{max} of the input distribution. [σ_{max} for the case of an aperture with a spherical incident wave front is given in Eq. (2.4).] Once σ_{max} is determined, one calculates η_{opt} and $\tilde{\omega}_{max}$ by using the procedure outlined in Appendix A. The same considerations that led to relation (2.5) now yield

$$L_{max} \geq \max\{10\mathbf{R}, 2\mathbf{z}\hat{\omega}_{max}\}. \tag{2.11}$$

3. Calculate $H(\mathbf{x}, \mathbf{y})$ from Eq. (2.9). The largest distance from the origin of the $\mathbf{xy}$ plane that is now significant is $\mathbf{R} + \mathbf{z}\hat{\omega}_{max}$. At this distance the separation between successive peaks of the exponential phase factor in Eq. (2.9) poses another restriction on Δ as follows:

$$\Delta \leq \frac{\eta\mathbf{z}}{2(\mathbf{R} + \mathbf{z}\hat{\omega}_{max})}. \tag{2.12}$$

Note that Eq. (2.12) must be satisfied in addition to Eq. (2.3).

4. Normalize the axes $\mathbf{xy}$ of the domain of $H(\mathbf{x}, \mathbf{y})$ in order to obtain $H(\eta\mathbf{z}X, \eta\mathbf{z}Y)$. This is done simply by dividing L_{max} and L_{max} by $\eta\mathbf{z}$.

5. Calculate the final distribution according to Eq. (2.10). The separation between neighboring elements in the final array is $\tilde{\Delta} = \eta\mathbf{z}/L_{max}$ because of the normalization in the preceding step. The length of the final array is thus

$$\tilde{L}_{max} = \eta\mathbf{z}N_{max}/L_{max}. \tag{2.13}$$

The phase factor in Eq. (2.10) may be treated as a curvature $\mathbf{C} = (\eta\mathbf{z})^{-1}$, provided that $\mathbf{C}$ is sufficiently small.

C. Fresnel's Formula and the Stationary-Phase Approximation

When $\mathbf{z}$ is sufficiently large, the inverse transform in Eq. (2.2) can be evaluated with the stationary-phase approximation. The exponent of the integrand is

$$W(\sigma_x, \sigma_y) = \mathbf{x}\sigma_x + \mathbf{y}\sigma_y + \mathbf{z}(1 - \sigma_x^2 - \sigma_y^2)^{1/2}, \tag{2.14}$$

which has a stationary point at

$$\sigma_{x0} = \mathbf{x}/(\mathbf{x}^2 + \mathbf{y}^2 + \mathbf{z}^2)^{1/2}, \tag{2.15a}$$

$$\sigma_{y0} = \mathbf{y}/(\mathbf{x}^2 + \mathbf{y}^2 + \mathbf{z}^2)^{1/2}. \tag{2.15b}$$

The Taylor-series expansion of $W(\sigma_x, \sigma_y)$ around the stationary point is

$$\begin{aligned} W(\sigma_x, \sigma_y) = (\mathbf{x}^2 + \mathbf{y}^2 + \mathbf{z}^2)^{1/2}\{1 &- \tfrac{1}{2}[1 + (\mathbf{x}/\mathbf{z})^2](\sigma_x - \sigma_{x0})^2 \\ &- (\mathbf{xy}/\mathbf{z}^2)(\sigma_x - \sigma_{x0})(\sigma_y - \sigma_{y0}) \\ &- \tfrac{1}{2}[1 + (\mathbf{y}/\mathbf{z})^2](\sigma_y - \sigma_{y0})^2\} + \dots . \end{aligned} \tag{2.16}$$

After replacing $W(\sigma_x, \sigma_y)$ in Eq. (2.2) and completing the stationary-phase method, one obtains

$$\tilde{t}_{\alpha\beta}(\lambda\mathbf{x}, \lambda\mathbf{y}, \lambda\mathbf{z}) = -(i/\mathbf{z})\exp\left[i2\pi\mathbf{z}\left(1 + \frac{\mathbf{x}^2 + \mathbf{y}^2}{\mathbf{z}^2}\right)^{1/2}\right] \times \Psi_{\alpha\beta}(\sigma_{x0}, \sigma_{y0})(1 - \sigma_{x0}^2 - \sigma_{y0}^2)[\lambda^{-2}T(\sigma_{x0}/\lambda, \sigma_{y0}/\lambda)], \tag{2.17}$$

which is the standard expression for the far-field (Fraunhofer) diffraction.[11] Note that the exponential term is a curvature phase factor with a radius of curvature $\mathbf{z}$.

For numerical computations the sampling interval must be sufficiently small to represent the fine features of the incident waveform. The restriction on Δ is thus similar to that for the direct Fresnel method, and, when the aperture radius is $\mathbf{R}$ and the incident beam has a curvature $\mathbf{C}$, the upper limit of Δ is given by relation (2.3). The lower bound on L_{max}, however, is different here; the only requirement is for sufficient resolution in the output plane, namely,

$$L_{max} \geq 10\mathbf{R}. \tag{2.18}$$

Since σ_{x0} and σ_{y0} are the angular coordinates of the observation point $(\mathbf{x}, \mathbf{y}, \mathbf{z})$, a nonlinear scaling must be applied to the function in Eq. (2.17). The final array dimension is

$$\tilde{L}_{max} = \frac{\mathbf{z} \cdot N_{max}}{L_{max}}. \tag{2.19}$$

3. DIFFRACTION IN THE PRESENCE OF A LENS

A. Spherical Lens

Consider a spherical lens with a focal length f and a numerical aperture NA. Let the lens center be located at (x_c, y_c), and let the amplitude distribution in the plane of the entrance pupil be $\tau_0(x, y)$. The distribution at the exit pupil is then

$$t(x, y) = \tilde{\tau}_0(x, y)\exp\left\{-i\frac{2\pi}{\lambda}[f^2 + (x - x_c)^2 + (y - y_c)^2]^{1/2}\right\}. \tag{3.1a}$$

$\tilde{\tau}_0(x, y)$ is a scaled version of $\tau_0(x, y)$ as follows:

$$\tilde{\tau}_0(x, y) = \tau_0\left\{\frac{x - x_c}{\left[1 + \left(\frac{x - x_c}{f}\right)^2 + \left(\frac{y - y_c}{f}\right)^2\right]^{1/2}} + x_c, \frac{y - y_c}{\left[1 + \left(\frac{x - x_c}{f}\right)^2 + \left(\frac{y - y_c}{f}\right)^2\right]^{1/2}} + y_c\right\} \times \left[1 + \left(\frac{x - x_c}{f}\right)^2 + \left(\frac{y - y_c}{f}\right)^2\right]^{-1}. \tag{3.1b}$$

The scaling is in keeping with Abbe's sine condition for an aplanatic lens.[5] In addition, $\tilde{\tau}_0(x, y)$ may include any aberrations introduced into the beam while it is traveling between the entrance and the exit pupils.

Although the distribution in Eq. (3.1) can be propagated by using Fresnel's formula, the oscillatory nature of the phase factor at high numerical apertures will require a large number of samples. To avoid this problem, a factorization similar to that used in Section 2 is proposed. Here we define the function $\tau(x, y)$ as follows:

$$\tau(x,y)=\begin{cases}\bar{\tau}_0(x,y)\exp\left(-i\frac{2\pi f}{\lambda}\left\{\left[1+\frac{(x-x_c)^2+(y-y_c)^2}{f^2}\right]^{1/2}-\frac{1}{2}\eta\frac{(x-x_c)^2+(y-y_c)^2}{f^2}\right\}\right). \\ 0 \quad [(x-x_c)^2+(y-y_c)^2]^{1/2}>f\tan[\arcsin(\mathrm{NA})]\end{cases} \tag{3.2}$$

In this equation η is a parameter that can be optimized for a minimum number of samples in the xy plane. Equation (3.1a) then becomes

$$t(x,y)=\tau(x,y)\exp\left\{-i\frac{\pi\eta}{\lambda f}[(x-x_c)^2+(y-y_c)^2]\right\}. \tag{3.3}$$

Let us now define $\mathbf{f}=f/\lambda$, $X_c=\eta x_c/f$, $Y_c=\eta y_c/f$, $X=\eta x/f$, and $Y=\eta y/f$; then Eqs. (3.2) and (3.3) are written

$$\tau\left(\frac{f}{\eta}X,\frac{f}{\eta}Y\right)=\bar{\tau}_0\left(\frac{f}{\eta}X,\frac{f}{\eta}Y\right)\exp\left(-i2\pi\mathbf{f}\left\{\left[1+\frac{(X-X_c)^2+(Y-Y_c)^2}{\eta^2}\right]^{1/2}-\frac{1}{2}\frac{(X-X_c)^2+(Y-Y_c)^2}{\eta}\right\}\right), \tag{3.4}$$

$$t\left(\frac{f}{\eta}X,\frac{f}{\eta}Y\right)=\tau\left(\frac{f}{\eta}X,\frac{f}{\eta}Y\right)\times\exp\left\{\frac{-i\pi\mathbf{f}}{\eta}[(X-X_c)^2+(Y-Y_c)^2]\right\}. \tag{3.5}$$

The first step in propagating the distribution in Eq. (3.5) beyond the lens is finding its Fourier transform according to Eq. (2.1). However, Eq. (3.5) is the product of two functions, and its Fourier transform is consequently a convolution as follows:

$$\left(\frac{\eta}{f}\right)^2T\left(\frac{\eta u}{f},\frac{\eta v}{f}\right)=-\frac{i\eta}{\mathbf{f}}\iint_{-\infty}^{\infty}\mathcal{F}\left\{\tau\left(\frac{f}{\eta}X,\frac{f}{\eta}Y\right)\right\}\times\exp\{-i2\pi[X_c(u-u')+Y_c(v-v')]\}\times\exp\left\{\frac{i\pi\eta}{\mathbf{f}}[(u-u')^2+(v-v')^2]\right\}du'dv'. \tag{3.6}$$

Let $\sigma_x=\eta u/\mathbf{f}$ and $\sigma_y=\eta v/\mathbf{f}$ and rewrite Eq. (3.6) as follows:

$$\lambda^{-2}T\left(\frac{\sigma_x}{\lambda},\frac{\sigma_y}{\lambda}\right)=-\frac{i\mathbf{f}}{\eta}\exp\left[\frac{i\pi\mathbf{f}}{\eta}(\sigma_x^2+\sigma_y^2)-i\frac{2\pi}{\lambda}(x_c\sigma_x+y_c\sigma_y)\right]\times\iint_{-\infty}^{\infty}\mathcal{F}\left\{\tau\left(\frac{f}{\eta}X,\frac{f}{\eta}Y\right)\right\}\times\exp\left\{\frac{i\pi\eta}{\mathbf{f}}\left[\left(u'+\frac{x_c}{\lambda}\right)^2+\left(v'+\frac{y_c}{\lambda}\right)^2\right]\right\}\times\exp[-i2\pi(u'\sigma_x+v'\sigma_y)]du'dv'. \tag{3.7}$$

Again, let $\mathbf{x}_c=x_c/\lambda$ and $\mathbf{y}_c=y_c/\lambda$, and define $h(u,v)$ as

$$h(u,v)=-\frac{i\mathbf{f}}{\eta}\exp\left\{\frac{i\pi\eta}{\mathbf{f}}[(u+\mathbf{x}_c)^2+(v+\mathbf{y}_c)^2]\right\}\times\mathcal{F}\left\{\tau\left(\frac{\lambda\mathbf{f}}{\eta}X,\frac{\lambda\mathbf{f}}{\eta}Y\right)\right\}; \tag{3.8}$$

then Eq. (3.7) becomes

$$\lambda^{-2}T\left(\frac{\sigma_x}{\lambda},\frac{\sigma_y}{\lambda}\right)=\exp\left\{i2\pi\left[\frac{\mathbf{f}}{2\eta}(\sigma_x^2+\sigma_y^2)-\mathbf{x}_c\sigma_x-\mathbf{y}_c\sigma_y\right]\right\}\times\mathcal{F}\{h(u,v)\}. \tag{3.9}$$

When Eq. (3.9) is substituted into Eq. (2.2) and $g(\sigma_x,\sigma_y)$ is defined as

$$g(\sigma_x,\sigma_y)=\exp[i2\pi(\mathbf{z}-\mathbf{f})(1-\sigma_x^2-\sigma_y^2)^{1/2}\times\exp\left\{i2\pi\mathbf{f}\left[(1-\sigma_x^2-\sigma_y^2)^{1/2}+\frac{1}{2\eta}(\sigma_x^2+\sigma_y^2)\right]\right\}\times\exp[-i2\pi(\mathbf{x}_c\sigma_x+\mathbf{y}_c\sigma_y)]\mathcal{F}\{h(u,v)\}, \tag{3.10}$$

the final distribution can be written as

$$\bar{t}_{\alpha\beta}(\lambda\mathbf{x},\lambda\mathbf{y},\lambda\mathbf{z})=\mathcal{F}^{-1}\{\Psi_{\alpha\beta}(\sigma_x,\sigma_y)g(\sigma_x,\sigma_y)\}. \tag{3.11}$$

We summarize the results of this section in the following step-by-step method of calculating diffraction patterns for a lens:

1. Compute $\tau(x,y)$ from Eq. (3.2). The optimum η is obtained by the method described in Appendix B with $\theta_1=0$ and $\theta_2=\arcsin(\mathrm{NA})$. Since the maximum spatial frequency of the exponential phase factor in Eq. (3.2) is ω_{max}, the restriction on the sample spacing Δ at the aperture is

$$\Delta\le\min\left\{\frac{\mathbf{R}}{10},\frac{1}{2\omega_{max}}\right\}. \tag{3.12}$$

Here $\mathbf{R}=\mathbf{f}\tan[\arcsin(\mathrm{NA})]$, and ω_{max}, a function of NA, is given in Fig. 18 of Appendix B.

2. Normalize the coordinates $\mathbf{xy}$ by dividing L_{max} and L_{may} by $\mathbf{f}/\eta$; then calculate the Fourier transform of $\tau(x,y)$ in the new (normalized) coordinates XY. Multiply the result by the appropriate coefficient as in Eq. (3.8) to obtain $h(u,v)$. The extent of the function $h(u,v)$ in the uv plane is

$$\sigma_{max}=\frac{\mathbf{f}}{\eta}\left(\omega_{max}+\frac{1}{2\mathbf{R}}\right). \tag{3.13}$$

The phase factor in Eq. (3.8) has a maximum frequency of $(\sigma_{max}+\mathbf{r}_c)\eta/\mathbf{f}$, where $r_c=(x_c^2+y_c^2)^{1/2}$ is the radial distance of the lens center from the optical axis. Thus $\eta L_{max}/\mathbf{f}$ (the normalized L_{max} of the XY domain) must be greater than twice this frequency, namely,

$$L_{max}\ge\max\left\{10\mathbf{R},\frac{2\mathbf{f}}{\eta}\left(\omega_{max}+\frac{1}{2\mathbf{R}}\right)+2\mathbf{r}_c\right\}. \tag{3.14}$$

3. Calculate $g(\sigma_x, \sigma_y)$ from Eq. (3.10). It turns out that $\dot{\omega}_{max}$ in Appendixes A and B at optimum η are related by $\dot{\omega}_{max}^{(A)} = \dot{\omega}_{max}^{(B)}/\eta^{(B)}$, while $\eta^{(A)}\eta^{(B)} = 1$. The maximum frequency in the phase factor of Eq. (3.10) near the focus is therefore $(\mathbf{f}\dot{\omega}_{max}/\eta) + \mathbf{r}_c$. However, the coordinates have been normalized by $\mathbf{f}/\eta$; therefore

$$\Delta \le \frac{1}{2\left(\dot{\omega}_{max} + \dfrac{\eta \mathbf{r}_c}{\mathbf{f}}\right)}. \qquad (3.15)$$

Thus, when $\mathbf{z} \approx \mathbf{f}$, relations (3.12) and (3.15) combine to yield

$$\frac{L_{max}}{N_{max}} \le \min\left\{\frac{\mathbf{R}}{10}, \frac{1}{2\left(\dot{\omega}_{max} + \dfrac{\eta \mathbf{r}_c}{\mathbf{f}}\right)}\right\}. \qquad (3.16)$$

4. Calculate the final distribution from Eq. (3.11). The final array length will be

$$\tilde{L}_{max} = \frac{\mathbf{f}N_{max}}{\eta L_{max}}. \qquad (3.17)$$

B. Spherical Lens and the Stationary-Phase Approximation

With increasing numerical aperture and/or focal length, the method described in Subsection 3.A becomes impractical in terms of computer time and memory requirements. In this regime, however, the stationary-phase approximation applies, and we can directly calculate the Fourier transform of $t(x, y)$ in Eq. (3.1a). Defining $X = x/f$, $Y = y/f$, $X_c = x_c/f$, and $Y_c = y_c/f$, we write

$$\lambda^{-2}T\left(\frac{\sigma_x}{\lambda}, \frac{\sigma_y}{\lambda}\right) = \mathbf{f}^2 \int\int_{-\infty}^{\infty} \tilde{\tau}_0(fX, fY) \times \exp[-i2\pi \mathbf{f} W(X, Y)]\mathrm{d}X\mathrm{d}Y, \qquad (3.18a)$$

where

$$W(X, Y) = [1 + (X - X_c)^2 + (Y - Y_c)^2]^{1/2} + \sigma_x X + \sigma_y Y. \qquad (3.18b)$$

$W(X, Y)$ has a stationary point at

$$X_0 = X_c - \frac{\sigma_x}{(1 - \sigma_x^2 - \sigma_y^2)^{1/2}}, \qquad (3.19a)$$

$$Y_0 = Y_c - \frac{\sigma_y}{(1 - \sigma_x^2 - \sigma_y^2)^{1/2}}. \qquad (3.19b)$$

The Taylor-series expansion of $W(X, Y)$ around the stationary point is

$$\begin{aligned} W(X, Y) &= \sigma_x X_c + \sigma_y Y_c + (1 - \sigma_x^2 - \sigma_y^2)^{1/2} \\ &\times [1 + \tfrac{1}{2}(1 - \sigma_x^2)(X - X_0)^2 \\ &- \sigma_x\sigma_y(X - X_0)(Y - Y_0) \\ &+ \tfrac{1}{2}(1 - \sigma_y^2)(Y - Y_0)^2] + \ldots. \end{aligned} \qquad (3.20)$$

Replacing $W(X, Y)$ in Eq. (3.18a) and completing the stationary-phase approximation yields

$$\lambda^{-2}T\left(\frac{\sigma_x}{\lambda}, \frac{\sigma_y}{\lambda}\right) = \bar{\bar{\tau}}_0(\sigma_x, \sigma_y) \times \exp\{-i2\pi \mathbf{f}[(1 - \sigma_x^2 - \sigma_y^2)^{1/2} + \sigma_x X_c + \sigma_y Y_c]\}, \qquad (3.21a)$$

where

$$\bar{\bar{\tau}}_0(\sigma_x, \sigma_y) = \left(\frac{-if}{1 - \sigma_x^2 - \sigma_y^2}\right) \times \tilde{\tau}_0\left[x_c - \frac{f\sigma_x}{(1 - \sigma_x^2 - \sigma_y^2)^{1/2}}, y_c - \frac{f\sigma_y}{(1 - \sigma_x^2 - \sigma_y^2)^{1/2}}\right]. \qquad (3.21b)$$

$\bar{\bar{\tau}}_0(\sigma_x, \sigma_y)$ is a shifted, inverted, and scaled version of $\tilde{\tau}_0(x, y)$. The main attributes of this transformation are as follows:

1. $\tilde{\tau}_0$ is centered at (x_c, y_c), whereas $\bar{\bar{\tau}}_0$ is centered at the origin.
2. The radius of the exit pupil is f tan[arcsin(NA)], whereas $\bar{\bar{\tau}}_0(\sigma_x, \sigma_y)$ has a radius NA.
3. The total incident power is conserved; i.e., $|\tilde{\tau}_0(\lambda\mathbf{x}, \lambda\mathbf{y})|^2$ and $|\bar{\bar{\tau}}_0(\sigma_x, \sigma_y)|^2$ have the same integral in their domains.

The scaling of $\tilde{\tau}_0(x, y)$ in Eq. (3.21b) is equivalent to replacing the distribution at the exit pupil of an aplanatic lens with that at the entrance pupil. Thus, given the input distribution to such a lens, no further scaling is required. (The shift to the center and the inversion are still necessary, however.)

Equation (3.21a) is now substituted into Eq. (2.2) to yield

$$\tilde{t}_{\alpha\beta}(\lambda\mathbf{x}, \lambda\mathbf{y}, \lambda\mathbf{z}) = \mathcal{F}^{-1}\{\exp\{i2\pi[(\mathbf{z} - \mathbf{f})(1 - \sigma_x^2 - \sigma_y^2)^{1/2} - \mathbf{x}_c\sigma_x - \mathbf{y}_c\sigma_y]\}\Psi_{\alpha\beta}(\sigma_x, \sigma_y)\bar{\bar{\tau}}_0(\sigma_x, \sigma_y)\}. \qquad (3.22)$$

Equation (3.22) provides a simple expression for vector diffraction in the neighborhood of the focal plane. This equation is identical to the one derived by Wolf[1] and is valid only, when the stationary-phase approximation applies. Note that, in the absence of aberrations, $\tilde{t}(x, y, z) = \tilde{t}^*(2x_c - x, 2y_c - y, 2f - z)$. (* denotes a complex conjugate.) This symmetry with respect to the focal point is a consequence of the stationary-phase approximation and its absence in a physical system is a manifestation of the breakdown of the approximation.

The extent of $\bar{\bar{\tau}}_0(\sigma_x, \sigma_y)$ in the $\sigma_x\sigma_y$ plane is σ_{max} = NA, which means that the sample spacing that ensures proper sampling of the exponential phase factor in Eq. (3.22) is

$$\frac{\Delta}{\mathbf{f}} \le \min\left\{\frac{\text{NA}}{10}, \frac{1}{2}\left\{\mathbf{r}_c + \frac{|\mathbf{z} - \mathbf{f}|\text{NA}}{[1 - (\text{NA})^2]^{1/2}}\right\}^{-1}\right\}. \qquad (3.23)$$

For proper resolution in the **xy** plane we require that

$$L_{max} \ge 10\mathbf{f}\text{NA}. \qquad (3.24)$$

Of course, if the incident distribution has fine features or if better resolution is required, the values of N_{max} and L_{max} must be adjusted accordingly. The final length of the array will be

$$\tilde{L}_{max} = \frac{\mathbf{f}N_{max}}{L_{max}}. \qquad (3.25)$$

C. Astigmatic Lens

An astigmatic lens with focal lengths f_x and f_y, a numerical aperture NA (to be defined below), and center coordinates (x_c, y_c), is represented by the following amplitude distribution at the exit pupil:

$$t(x, y) = \begin{cases} \tau_0(x, y)\exp\left(-i\dfrac{2\pi}{\lambda}\left(\dfrac{2f_xf_y}{f_x+f_y}\right)\left\{1+\left(\dfrac{f_x+f_y}{2f_xf_y}\right)\left[\dfrac{(x-x_c)^2}{f_x}+\dfrac{(y-y_c)^2}{f_y}\right]\right\}^{1/2}\right) \\ 0 \qquad \left\{\dfrac{f_x+f_y}{2f_xf_y}\left[\dfrac{(x-x_c)^2}{f_x}+\dfrac{(y-y_c)^2}{f_y}\right]\right\}^{1/2} > \tan[\arcsin(\mathrm{NA})] \end{cases}. \quad (3.26)$$

This is not necessarily the simplest function to represent an astigmatic lens. Other functions may be suggested, and the following treatment would apply equally to them. The above function, however, has three appealing properties. The first is that when $f_x = f_y$ it reduces to the spherical lens of Eq. (3.1). The second is that in the paraxial approximation it becomes

$$t(x, y) = \tau_0(x, y)\exp\left\{-i\frac{\pi}{\lambda}\left[\frac{(x-x_c)^2}{f_x}+\frac{(y-y_c)^2}{f_y}\right]\right\}, \quad (3.27)$$

which clearly shows that f_x and f_y correspond to the curvature of the wave front along the x and y directions. The third property is that the exit pupil is an ellipse rather than a circle; this allows a single value of η to minimize the oscillations along both x and y axes. The optimum η is obtained from Appendix B with $\theta_1 = 0$ and $\theta_2 = \arcsin(\mathrm{NA})$. θ_2 is the half-angle subtended by the lens axis parallel to x at a distance of $[2f_x^2f_y/(f_x + f_y)]^{1/2}$. Equivalently, θ_2 is the half-angle subtended by the axis parallel to y at a distance of $[2f_xf_y^2/(f_x + f_y)]^{1/2}$.

The treatment of the astigmatic lens parallels that of the spherical lens in Subsection 3.A. The main steps are outlined below.

1. Calculate $\tau(x, y)$ from the following relation:

$$\tau(x, y) = \tau_0(x, y)\exp\left[-i2\pi\left(\frac{2\mathbf{f}_x\mathbf{f}_y}{\mathbf{f}_x+\mathbf{f}_y}\right) \times \left(\left\{1+\left(\frac{\mathbf{f}_x+\mathbf{f}_y}{2\mathbf{f}_x\mathbf{f}_y}\right)\left[\frac{(\mathbf{x}-\mathbf{x}_c)^2}{\mathbf{f}_x}+\frac{(\mathbf{y}-\mathbf{y}_c)^2}{\mathbf{f}_y}\right]\right\}^{1/2} - \frac{1}{2}\eta\left(\frac{\mathbf{f}_x+\mathbf{f}_y}{2\mathbf{f}_x\mathbf{f}_y}\right)\left[\frac{(\mathbf{x}-\mathbf{x}_c)^2}{\mathbf{f}_x}+\frac{(\mathbf{y}-\mathbf{y}_c)^2}{\mathbf{f}_y}\right]\right)\right]. \quad (3.28)$$

The sample spacings Δ_x and Δ_y must satisfy

$$\Delta_x \le \min\left\{\frac{\mathbf{R}_x}{10}, \frac{\left(\dfrac{1}{2}+\dfrac{f_x}{2f_y}\right)^{1/2}}{2\hat{\omega}_{\max}}\right\}, \quad (3.29a)$$

$$\Delta_y \le \min\left\{\frac{\mathbf{R}_y}{10}, \frac{\left(\dfrac{1}{2}+\dfrac{f_y}{2f_x}\right)^{1/2}}{2\hat{\omega}_{\max}}\right\}. \quad (3.29b)$$

Here R_x and R_y are the half axes of the (elliptical) exit pupil, and $\hat{\omega}_{\max}$, a function of NA, is given in Fig. 18 of Appendix B.

2. Calculate $h(u, v)$ as follows:

$$h(u, v) = -i\frac{(\mathbf{f}_x\mathbf{f}_y)^{1/2}}{\eta}\exp\left\{i\pi\eta\left[\frac{(u+\mathbf{x}_c)^2}{\mathbf{f}_x}+\frac{(v+\mathbf{y}_c)^2}{\mathbf{f}_y}\right]\right\} \times \mathcal{F}\left\{\tau\left(\frac{\lambda\mathbf{f}_x}{\eta}X, \frac{\lambda\mathbf{f}_y}{\eta}Y\right)\right\}. \quad (3.30)$$

Note that the x and y axes are normalized differently in Eq. (3.30). The limits imposed on $L_{\max}$ and L_{may} can be obtained by considerations similar to those that led to relation (3.14). Thus we have

$$L_{\max} \ge \max\left\{10\mathbf{R}_x, \frac{2\mathbf{f}_x}{\eta}\left[\frac{\hat{\omega}_{\max}}{\left(\dfrac{1}{2}+\dfrac{f_x}{2f_y}\right)^{1/2}}+\frac{1}{2\mathbf{R}_x}\right]+2|\mathbf{x}_c|\right\}. \quad (3.31)$$

A similar relation is obtained for L_{may}.

3. Compute $g(\sigma_x, \sigma_y)$ as follows:

$$g(\sigma_x, \sigma_y) = \exp\left\{i2\pi\mathbf{z}\left[(1-\sigma_x^2-\sigma_y^2)^{1/2}+\frac{\mathbf{f}_x}{2\eta\mathbf{z}}\sigma_x^2+\frac{\mathbf{f}_y}{2\eta\mathbf{z}}\sigma_y^2\right]\right\} \times \exp[-i2\pi(\mathbf{x}_c\sigma_x+\mathbf{y}_c\sigma_y)]\mathcal{F}\{h(u, v)\}. \quad (3.32)$$

If $\mathbf{f}_x$ and $\mathbf{f}_y$ are not too far apart, then in the neighborhood of the focus one can use the same approximations that led to relation (3.16) and conclude that in addition to relation (3.31) the following inequality must be satisfied:

$$\frac{L_{\max}}{N_{\max}} \le \min\left\{\frac{\mathbf{R}_x}{10}, \frac{\left(\dfrac{1}{2}+\dfrac{f_x}{2f_y}\right)^{1/2}}{2\left(\hat{\omega}_{\max}+\dfrac{\eta|\mathbf{x}_c|}{\mathbf{f}_x}\right)}\right\}. \quad (3.33)$$

A similar relationship exists for $L_{\mathrm{may}}/N_{\mathrm{may}}$.

4. Calculate the final distribution from

$$\tilde{t}_{\alpha\beta}(\lambda\mathbf{x}, \lambda\mathbf{y}, \lambda\mathbf{z}) = \mathcal{F}^{-1}\{\Psi_{\alpha\beta}(\sigma_x, \sigma_y)g(\sigma_x, \sigma_y)\}. \quad (3.34)$$

The final array dimensions will be

$$\tilde{L}_{\max} = \frac{\mathbf{f}_xN_{\max}}{\eta L_{\max}}, \quad (3.35a)$$

$$\tilde{L}_{\mathrm{may}} = \frac{\mathbf{f}_yN_{\mathrm{may}}}{\eta L_{\mathrm{may}}}. \quad (3.35b)$$

D. Astigmatic Lens and the Stationary-Phase Approximation

When the focal lengths are sufficiently large, the distribution of relation (3.26) can be Fourier transformed directly, provided that one invokes the stationary-phase approximation. Under these conditions,

$$\lambda^{-2}T\left(\frac{\sigma_x}{\lambda}, \frac{\sigma_y}{\lambda}\right) = \bar{\tau}_0(\sigma_x, \sigma_y)\exp\left(-i2\pi\left\{\left(\frac{2\mathbf{f}_x\mathbf{f}_y}{\mathbf{f}_x + \mathbf{f}_y}\right)\right.\right.$$

$$\left.\left.\times\left[1 - \left(\frac{1}{2} + \frac{\mathbf{f}_x}{2\mathbf{f}_y}\right)\sigma_x^2 - \left(\frac{1}{2} + \frac{\mathbf{f}_y}{2\mathbf{f}_x}\right)\sigma_y^2\right]^{1/2} + \mathbf{x}_c\sigma_x + \mathbf{y}_c\sigma_y\right\}\right), \quad (3.36a)$$

where

$$\bar{\tau}_0(\sigma_x, \sigma_y) = \frac{-i(\mathbf{f}_x\mathbf{f}_y)^{1/2}}{1 - \left(\frac{1}{2} + \frac{\mathbf{f}_x}{2\mathbf{f}_y}\right)\sigma_x^2 - \left(\frac{1}{2} + \frac{\mathbf{f}_y}{2\mathbf{f}_x}\right)\sigma_y^2}$$

$$\times\,\tau_0\left\{x_c - \frac{f_x\sigma_x}{\left[1 - \left(\frac{1}{2} + \frac{\mathbf{f}_x}{2\mathbf{f}_y}\right)\sigma_x^2 - \left(\frac{1}{2} + \frac{\mathbf{f}_y}{2\mathbf{f}_x}\right)\sigma_y^2\right]^{1/2}},\ y_c - \frac{f_y\sigma_y}{\left[1 - \left(\frac{1}{2} + \frac{\mathbf{f}_x}{2\mathbf{f}_y}\right)\sigma_x^2 - \left(\frac{1}{2} + \frac{\mathbf{f}_y}{2\mathbf{f}_x}\right)\sigma_y^2\right]^{1/2}}\right\}. \quad (3.36b)$$

Substituting Eq. (3.36a) into Eq. (2.2) yields

$$\bar{t}_{\alpha\beta}(\lambda\mathbf{x}, \lambda\mathbf{y}, \lambda\mathbf{z}) = \mathcal{F}^{-1}\left\{\exp\left(i2\pi\left\{\mathbf{z}(1 - \sigma_x^2 - \sigma_y^2)^{1/2} - \left(\frac{2\mathbf{f}_x\mathbf{f}_y}{\mathbf{f}_x + \mathbf{f}_y}\right)\right.\right.\right.$$

$$\times\left[1 - \left(\frac{1}{2} + \frac{\mathbf{f}_x}{2\mathbf{f}_y}\right)\sigma_x^2 - \left(\frac{1}{2} + \frac{\mathbf{f}_y}{2\mathbf{f}_x}\right)\sigma_y^2\right]^{1/2}$$

$$\left.\left.\left. - \mathbf{x}_c\sigma_x - \mathbf{y}_c\sigma_y\right\}\right)\Psi_{\alpha\beta}(\sigma_x, \sigma_y)\bar{\tau}_0(\sigma_x, \sigma_y)\right\}. \quad (3.37)$$

The proper sample spacing in the x direction is

$$\frac{\Delta_x}{\mathbf{f}_x} \le \min\left\{\frac{\mathrm{NA}}{10\left(\frac{1}{2} + \frac{\mathbf{f}_x}{2\mathbf{f}_y}\right)^{1/2}}, \frac{1}{2}\left(|\mathbf{x}_c| + \left|\frac{\mathbf{f}_x\mathrm{NA}}{\left\{\left(\frac{1}{2} + \frac{\mathbf{f}_x}{2\mathbf{f}_y}\right)[1 - (\mathrm{NA})^2]\right\}^{1/2}} - \frac{\mathbf{z}\mathrm{NA}}{\left[\frac{1}{2} + \frac{\mathbf{f}_x}{2\mathbf{f}_y} - (\mathrm{NA})^2\right]^{1/2}}\right|\right)^{-1}\right\}. \quad (3.38)$$

A similar expression exists for Δ_y. The length of the array is restricted as follows:

$$L_{\max} \ge \frac{10\mathbf{f}_x\mathrm{NA}}{\left(\frac{1}{2} + \frac{\mathbf{f}_x}{2\mathbf{f}_y}\right)^{1/2}}, \quad (3.39)$$

with a similar relationship for $L_{\max y}$. The final array dimensions are

$$\bar{L}_{\max} = \frac{\mathbf{f}_x N_{\max}}{L_{\max}}, \quad (3.40a)$$

$$\bar{L}_{\max y} = \frac{\mathbf{f}_y N_{\max y}}{L_{\max y}}. \quad (3.40b)$$

E. Ring Lens

The ring lens is a circularly symmetric lens with a cross section as shown in Fig. 2. A potential application of the ring lens is focus-error detection in optical disk systems.[12] A collimated beam is brought to focus by this lens on a ring of radius R_0 at the focal plane. The central region of the lens ($r < R_1$) is flat, and it is assumed that the light in this region is unaffected by the lens. The aperture radius is R_2. For lens-center coordinates (x_c, y_c), the exit pupil distribution is written as

$$t(x, y) = \tau_0(x, y)\exp\left[-i\frac{2\pi}{\lambda}\left(f^2 + \{[(x - x_c)^2 + (y - y_c)^2]^{1/2} - R_0\}^2\right)^{1/2}\right]. \quad (3.41)$$

Again, as in the case of a spherical lens, a quadratic term is factored out and treated analytically. The necessary steps are summarized below.

1. Calculate $\tau(x, y)$ with optimum η (see Appendix C) from the relation

$$\tau(x, y) = \begin{cases} \tau_0(x, y)\exp\left\{-\frac{i2\pi f}{\lambda}\left[\left(1 + \left\{\left[\frac{(x - x_c)^2 + (y - y_c)^2}{f^2}\right]^{1/2} - \frac{R_0}{f}\right\}^2\right)^{1/2} - \frac{1}{2}\eta\frac{(x - x_c)^2 + (y - y_c)^2}{f^2}\right]\right\} \\ 1 \quad [(x - x_c)^2 + (y - y_c)^2]^{1/2} \le R_1 \\ 0 \quad [(x - x_c)^2 + (y - y_c)^2]^{1/2} \ge R_2 \end{cases}. \quad (3.42)$$

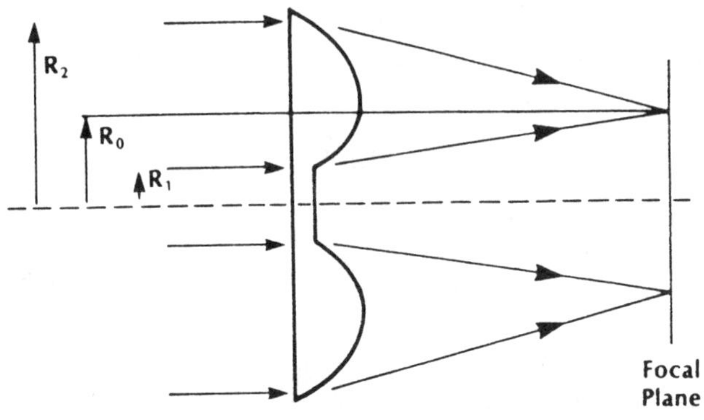

Fig. 2. Cross section of a ring lens. The central region ($r < R_1$) is flat and does not affect the incident beam. The curved surface peaks at $r = R_0$, and the aperture radius of the lens is R_2. A collimated beam is brought to focus by this lens on a sharp ring of radius R_0 at the focal plane. The amplitude distribution at the exit pupil is given by Eq. (3.41), and, in the absence of aberrations, the ring width is diffraction limited.

The spacing between the samples in the **xy** plane must satisfy relation (3.16).

2. Determine $h(u, v)$ from Eq. (3.8). The restriction on L_{max} is given by relation (3.14).

3. Calculate $g(\sigma_x, \sigma_y)$ from Eq. (3.10) and the final distribution from Eq. (3.11).

4. RESULTS AND DISCUSSION

Example 1

We study the diffraction pattern from a circular aperture of radius **R** = 500, illuminated with a plane uniform wave (**C** = 0). The integrated intensity within the aperture is set to unity. We choose $N_{max} = N_{may} = 256$, since it is a convenient array dimension for performing FFT's, and let $L_{max} = L_{may} = 5000$. (On a VAX 11/780 computer each 256 × 256 FFT takes only ~10 sec.) Now, for distances z that are not too far from the aperture, both direct and extended Fresnel techniques can be applied. A good guide for selecting distances in this

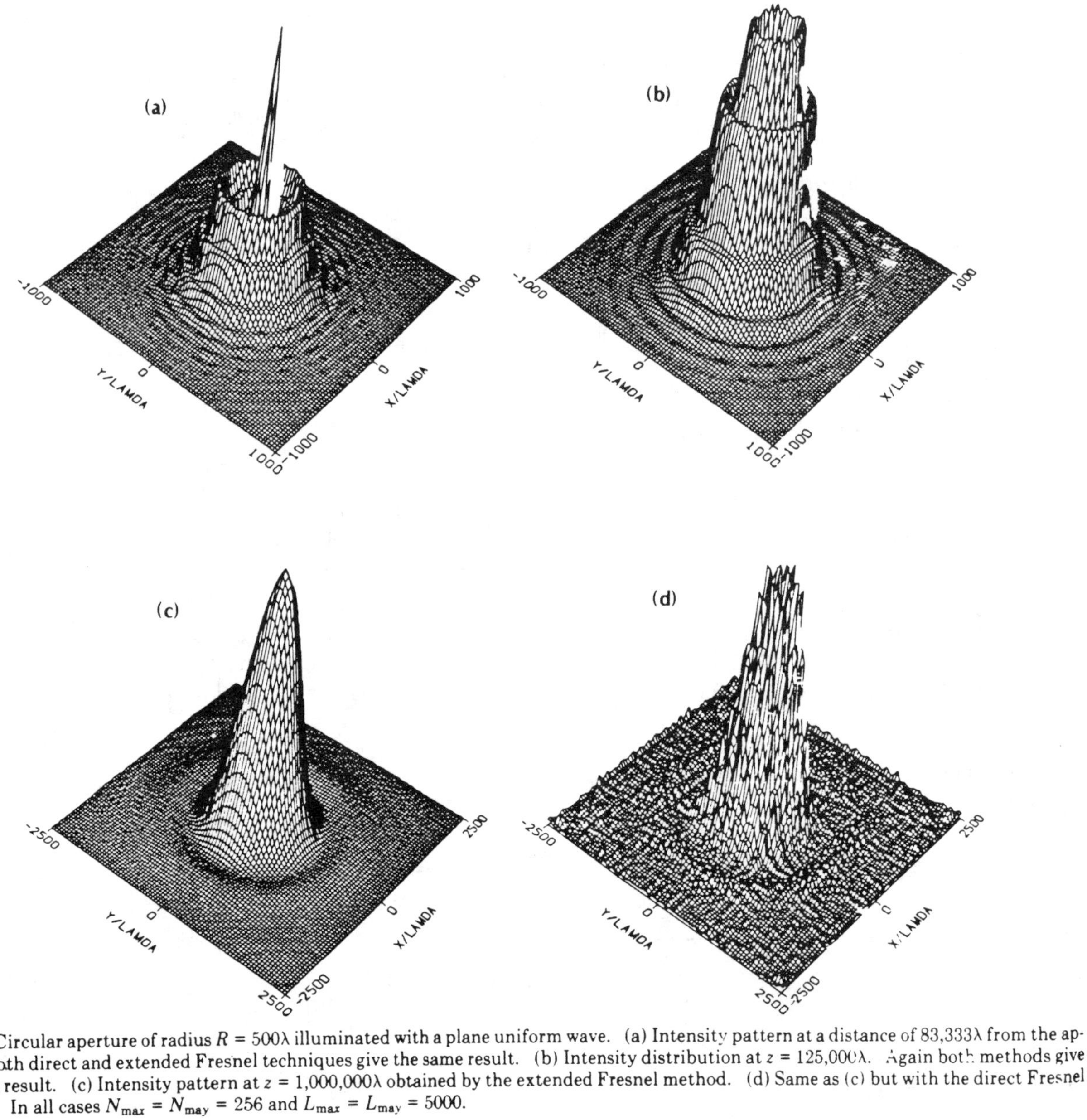

Fig. 3. Circular aperture of radius $R = 500\lambda$ illuminated with a plane uniform wave. (a) Intensity pattern at a distance of 83,333λ from the aperture; both direct and extended Fresnel techniques give the same result. (b) Intensity distribution at $z = 125{,}000\lambda$. Again both methods give the same result. (c) Intensity pattern at $z = 1{,}000{,}000\lambda$ obtained by the extended Fresnel method. (d) Same as (c) but with the direct Fresnel method. In all cases $N_{max} = N_{may} = 256$ and $L_{max} = L_{may} = 5000$.

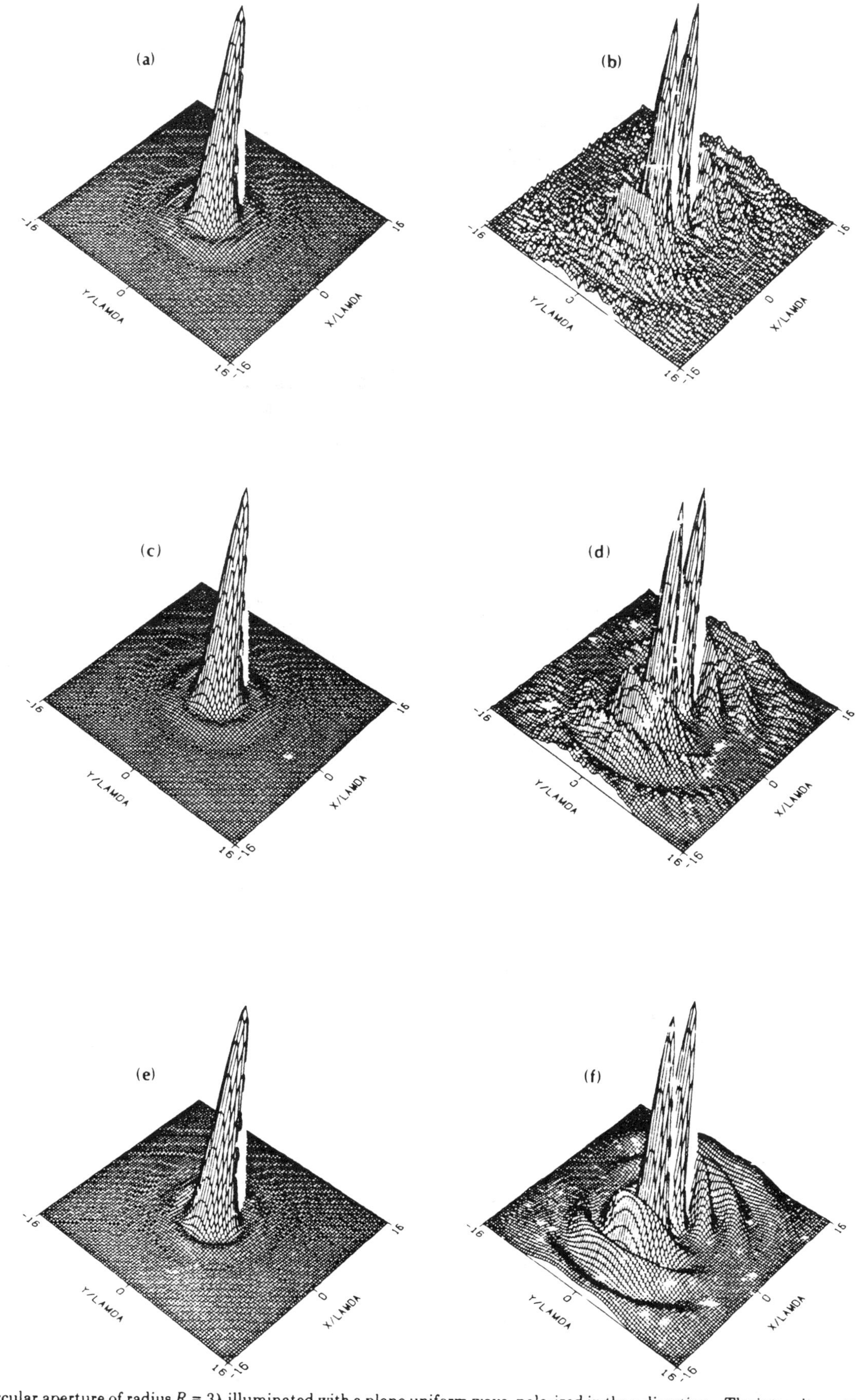

Fig. 4. Circular aperture of radius $R = 3\lambda$ illuminated with a plane uniform wave, polarized in the x direction. The intensity patterns are calculated at a distance of 20λ from the aperture. (a) and (b) show, respectively, the x and the z components of polarization obtained by the direct Fresnel method. (c) and (d) show the results of extended Fresnel calculations, and (e) and (f) show the results obtained from the Fraunhofer diffraction formula.

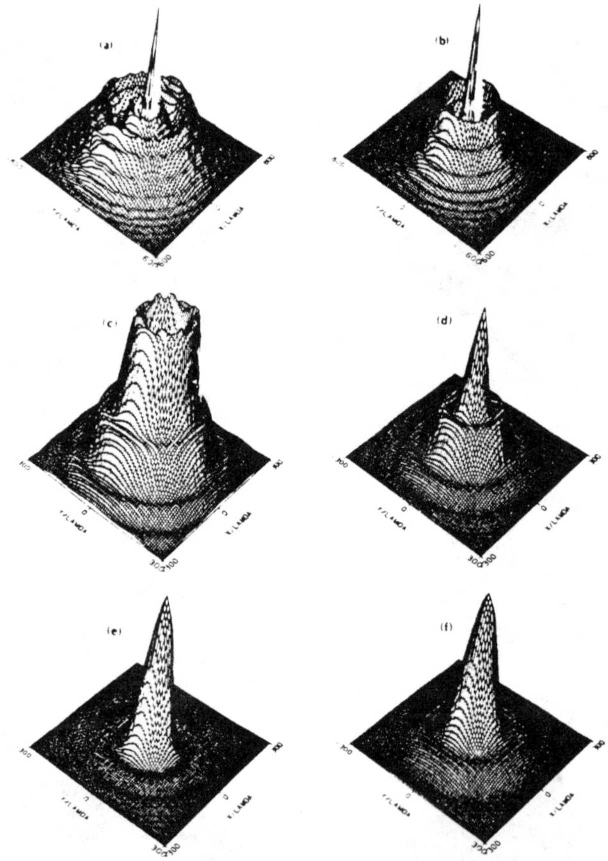

Fig. 5. Circular aperture of radius R = 1000λ illuminated with a uniform wave of curvature $\mathbf{C} = 5 \times 10^{-6}$ corresponding to a positive spherical lens of $\mathbf{f}$ = 200,000λ and NA = 0.005. The intensity patterns are calculated with the extended Fresnel technique, using $N_{max} = N_{may} = 512$ and $L_{max} = L_{may}$ = 10,000. (a) $\mathbf{z}$ = 100,000, (b) $\mathbf{z}$ = 125,000, (c) $\mathbf{z}$ = 150,000, (d) $\mathbf{z}$ = 160,000, (e) $\mathbf{z}$ = 200,000, and (f) $\mathbf{z}$ = 240,000.

not-too-far regime is the Fresnel number N defined by the relation[13]

$$\mathbf{z} = \frac{1}{N}\mathbf{R}^2 - \frac{N}{4}. \tag{4.1}$$

For instance, N = 3 corresponds to $\mathbf{z}$ = 83,333. The two methods give identical results in this case, and the resulting intensity pattern is shown in Fig. 3(a). A similar situation occurs when N = 2 and $\mathbf{z}$ = 125,000. Both direct and extended Fresnel methods give the same result, which is shown in Fig. 3(b).

As $\mathbf{z}$ increases, however, the advantages of the extended method become apparent. Figure 3(c) shows the intensity pattern at $\mathbf{z}$ = 1,000,000, calculated with the extended method, whereas Fig. 3(d) shows the result of direct computation. It is obvious that, for the direct method to work at large distances, the array size must increase, but the extended method can perform reliably even at distances that are deep in the far-field region. To check the validity of the extended method, we calculated the Fraunhofer pattern at $\mathbf{z}$ = 1,000,000; the resulting pattern was identical to that shown in Fig. 3(c).

Example 2

Consider a circular aperture of radius $\mathbf{R}$ = 3, illuminated with a plane uniform beam polarized along $\mathbf{x}$. The integrated intensity within the aperture is unity. Let $N_{max} = N_{may} = 256$ and $L_{max} = L_{may} = 30$. Figure 4 shows the intensity patterns

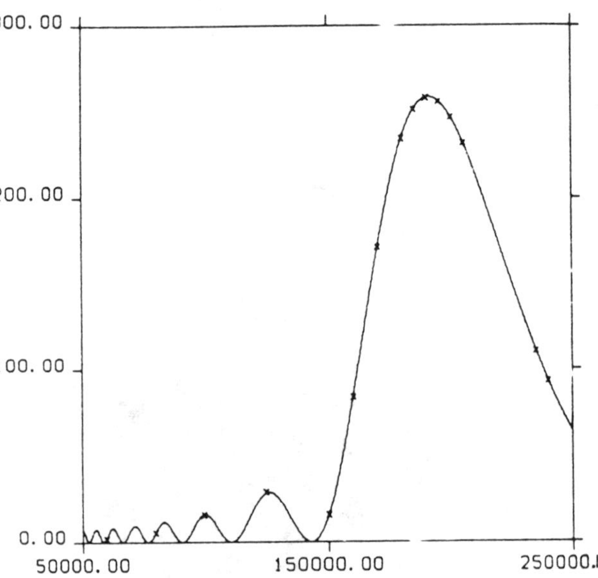

Fig. 6. Circular aperture of radius R = 1000λ illuminated with a uniform wave of curvature $\mathbf{C} = -5 \times 10^{-6}$. Intensity on the optical axis is shown here as a function of distance from the aperture. The solid curve is calculated from the analytic expression derived in Ref. 9. The crosses are obtained by numerical computations using direct (or extended) Fresnel method.

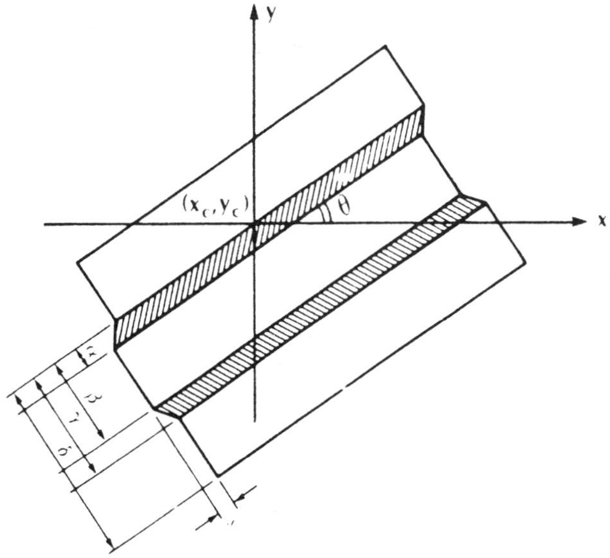

Fig. 7. Geometry of groove structure for a transmission phase diffraction grating. All parameters are in units of the wavelength λ.

for the x and z components of polarization at a distance of $\mathbf{z}$ = 20 from the aperture. (The y component was smaller than the numerical errors and could not be computed.) For comparison the results obtained with direct Fresnel, extended Fresnel, and Fraunhofer techniques are all shown in this figure.

Example 3

A circular aperture of radius $\mathbf{R}$ = 1000 is illuminated with a uniform wave front of curvature $\mathbf{C} = -5 \times 10^{-6}$. (The situation is similar to that of an incident plane wave on a positive spherical lens of $\mathbf{f}$ = 200,000 and NA = 0.005.) The intensity distributions at various distances from the aperture are shown in Fig. 5. Note the asymmetry with respect to the focal plane by comparing Fig. 5(d) with Fig. 5(f). This asymmetry is well known for lenses of small numerical aperture, and, in fact, an analytic expression for the axial intensity of such systems was derived by Mahajan.[9] We have reproduced Mahajan's results in Fig. 6 (solid curve) and shown several instances of our own direct (or extended) Fresnel computations (crosses). The agreement is extreme-

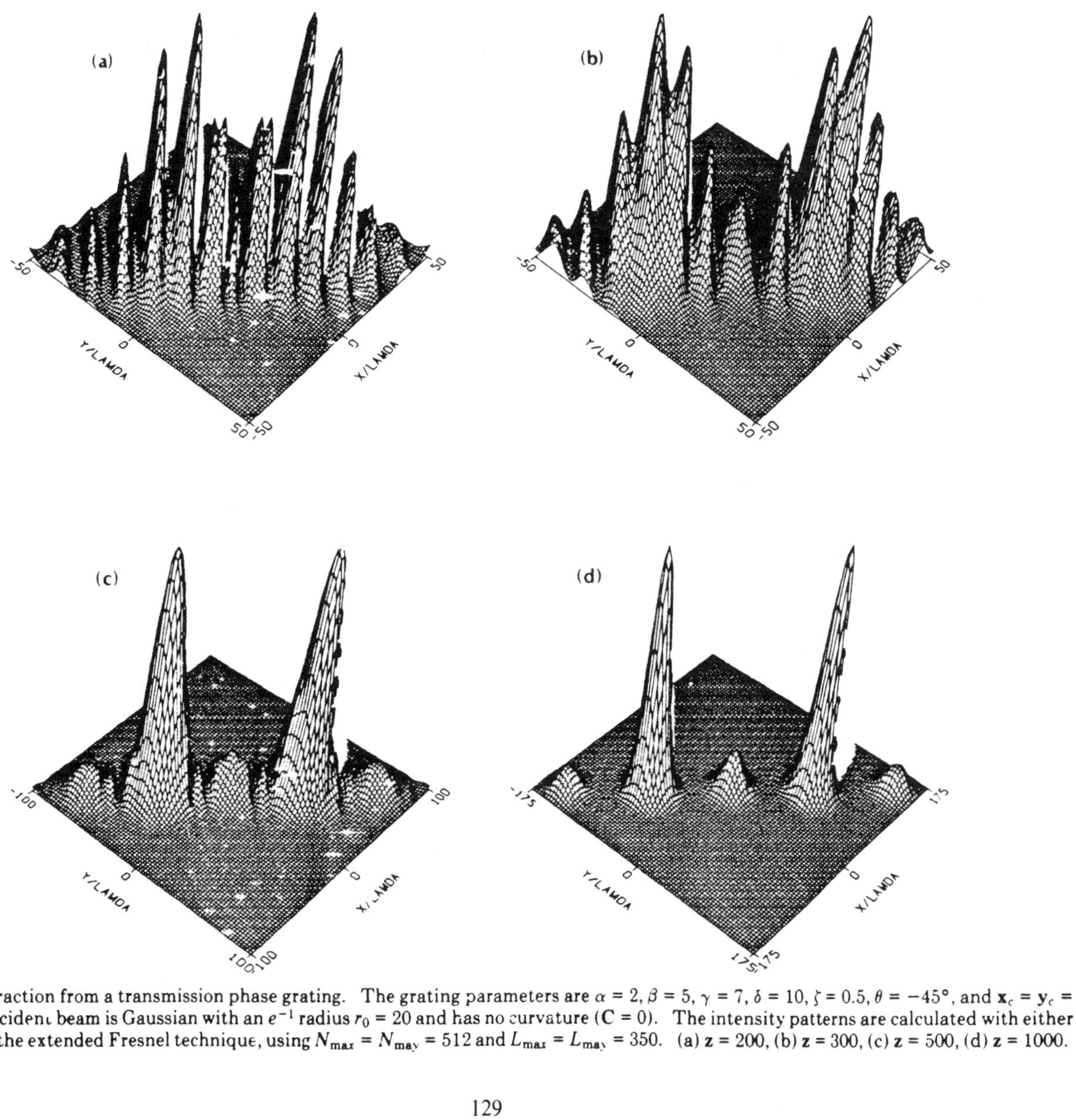

Fig. 8. Diffraction from a transmission phase grating. The grating parameters are $\alpha = 2$, $\beta = 5$, $\gamma = 7$, $\delta = 10$, $\zeta = 0.5$, $\theta = -45°$, and $\mathbf{x}_c = \mathbf{y}_c = 1.06$. The incident beam is Gaussian with an e^{-1} radius $r_0 = 20$ and has no curvature ($\mathbf{C} = 0$). The intensity patterns are calculated with either the direct or the extended Fresnel technique, using $N_{max} = N_{may} = 512$ and $L_{max} = L_{may} = 350$. (a) $\mathbf{z}$ = 200, (b) $\mathbf{z}$ = 300, (c) $\mathbf{z}$ = 500, (d) $\mathbf{z}$ = 1000.

ly good and shows the power and accuracy of the numerical method.

Example 4

With reference to Fig. 7, a transmission phase diffraction grating is specified with the parameters α, β, γ, δ, and ζ. The groove width is $\beta - \alpha$, and the land width is $\delta - \gamma$. The regions between land and groove have constant slope as determined by α and $\gamma - \beta$. If sharp boundaries are desired, one should set $\alpha = 0$ and $\beta = \gamma$. For further flexibility we have allowed the grating center (x_c, y_c) and orientation angle θ to be parameters as well. The aperture radius is R. If we assume that 10 samples per period of the grating are sufficient (this may not always be a good assumption) and that the incident beam is uniform with curvature **C**, the restriction on sample spacing will be

$$\Delta < \left[\frac{10}{\delta} + \frac{2|\mathbf{C}|\mathbf{R}}{(1 + \mathbf{C}^2\mathbf{R}^2)^{1/2}}\right]^{-1}. \tag{4.2}$$

The maximum spatial frequency is

$$\sigma_{\max} = \frac{|\mathbf{C}|\mathbf{R}}{(1 + \mathbf{C}^2\mathbf{R}^2)^{1/2}} + \frac{1}{\delta}, \tag{4.3}$$

which is used in relation (2.5) to determine L_{max} for direct Fresnel calculations. For Fraunhofer calculations L_{max} = 10**R** should be sufficient, and for extended Fresnel computations relation (2.11) applies. In the latter case, Δ must also satisfy relation (2.12).

Figure 8 shows the intensity patterns at various distances from a grating with parameters $\alpha = 2$, $\beta = 5$, $\gamma = 7$, $\delta = 10$, $\zeta = 0.5$, $\theta = -45°$, and $\mathbf{x}_c = \mathbf{y}_c = 1.06$. The incident beam is Gaussian with an e^{-1} radius $r_0 = 20$ and no curvature ($\mathbf{C} = 0$). Both direct and extended Fresnel methods give the near-field results in Figs. 8(a)–8(c), but the distribution in Fig. 8(d) must be calculated with the extended technique.

Example 5

Consider a cylindrical lens of focal length f and aperture radius R. The cylinder axis has an angle θ with the x axis, and the lens-center coordinates are (x_c, y_c). The aperture function is thus

$$t(x, y) = \tau_0(x, y)\exp\left(-i\frac{2\pi}{\lambda}\{f^2 + [(x - x_c)\sin\theta - (y - y_c)\cos\theta]^2\}^{1/2}\right)$$
$$= 0 \qquad [(x - x_c)^2 + (y - y_c)^2] > R^2. \tag{4.4}$$

If we assume that the incident beam has a curvature **C**, the spacing between samples must satisfy

$$\Delta \le \min\left\{\frac{\mathbf{R}}{10}, \left[2\text{NA} + \frac{2|\mathbf{C}|\mathbf{R}}{(1 + \mathbf{C}^2\mathbf{R}^2)^{1/2}}\right]^{-1}\right\}. \tag{4.5}$$

The maximum spatial frequency is

$$\sigma_{\max} = \text{NA} + \frac{|\mathbf{C}|\mathbf{R}}{(1 + \mathbf{C}^2\mathbf{R}^2)^{1/2}} + \frac{1}{2\mathbf{R}}, \tag{4.6}$$

which is used in relation (2.5) to determine L_{max} for direct Fresnel calculations. For Fraunhofer calculations L_{max} = 10**R** should be sufficient, and for extended Fresnel computations relation (2.11) applies. In the latter case Δ must also satisfy relation (2.12).

Figure 9 shows the intensity pattern in the neighborhood of the focus for a cylindrical lens with $\mathbf{R} = 2500$, $\mathbf{f} = 10^6$, $\theta = 45°$, and $\mathbf{x}_c = \mathbf{y}_c = 0$. The incident beam is a plane wave with a curvature $\mathbf{C} = 0$. For these computations $N_{max} = 512$ and $L_{max} = 40{,}000$.

Example 6

Consider a spherical lens with a NA = 0.5 and a focal length $f = 3500\lambda$. Let the incident beam be plane, linearly polarized along the x axis, and with a total power of unity at the entrance pupil. Also assume that the lens is aberration free and satisfies Abbe's sine condition. Figure 10 shows the intensity patterns for the three components of polarization at the focal plane. The x component has 93% of the incident power and is similar to the classical Airy pattern. The y component has less than 0.1% of the power but shows the four peaks expected from geometrical considerations. The remaining 7% of the power is in the z component shown in Fig. 10(c). This time we observe two peaks, separated along the direction of incident polarization (x), also as expected from geometrical optics. Figure 11 shows the intensity patterns for the x component at 5λ and 10λ away from the focal plane. All the results are obtained by the exact method, using L_{max} = 20,000 and $N_{max} = 980$.

Since the stationary-phase approximation also applies here, the results are in good agreement with the classical ones.[2] The particular set of parameters in this example were chosen, however, to show the practicality of the exact method for relatively large numerical apertures; the focal length and the numerical aperture are typical of the microlenses now being used in optical disk systems.

Example 7

We analyze the case of an astigmatic lens with $\mathbf{f}_x = 20{,}000$, $\mathbf{f}_y = 20{,}200$, and NA = 0.15. The incident beam is plane and uniform, and the effects of polarization are ignored. Figure 12 shows plots of intensity at the two focal planes and at a plane halfway between them. Both the exact method and the stationary-phase approximation yield the same results.

Example 8

Consider a ring lens with $\mathbf{R}_1 = 50$, $\mathbf{R}_0 = 100$, $\mathbf{R}_2 = 2500$, and a focal length $\mathbf{f} = 25{,}000$. The incident beam is plane and uniform, and the polarization effects are ignored. Figure 13 shows the intensity patterns at and near the focal plane, calculated with $N_{max} = N_{may} = 512$ and $L_{max} = L_{may} = 25{,}000$.

APPENDIX A

Consider the function

$$\omega_\eta(r) = (1 - r^2)^{1/2} + \tfrac{1}{2}\eta r^2, \qquad r_1 \le r \le r_2, \tag{A1}$$

where $0 \le r_1 \le r_2 \le 1$ and η is a real parameter. The objective is to find the value of η that minimizes the maximum of $|\dot{\omega}_\eta(r)|$ in the interval $[r_1, r_2]$. $\dot{\omega}_\eta(r)$ is the derivative of $\omega_\eta(r)$ with respect to r and is given by

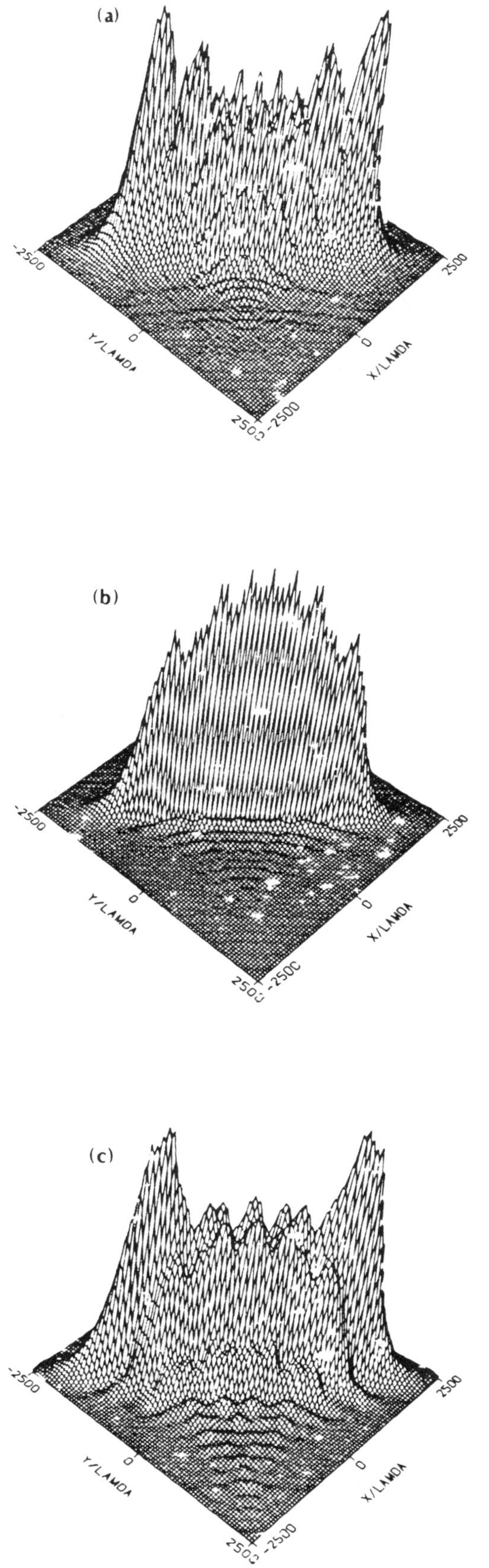

Fig. 9. Cylindrical lens with a focal length $f = 10^6\lambda$ and the aperture radius $R = 2500\lambda$. The incident beam is plane and uniform. The intensity patterns are obtained with the direct Fresnel method, using $N_{max} = N_{may} = 512$ and $L_{max} = L_{may} = 40{,}000$ (a) $\mathbf{z} = 0.75 \times 10^6$; (b) $\mathbf{z} = 10^6$; (c) $\mathbf{z} = 1.25 \times 10^6$.

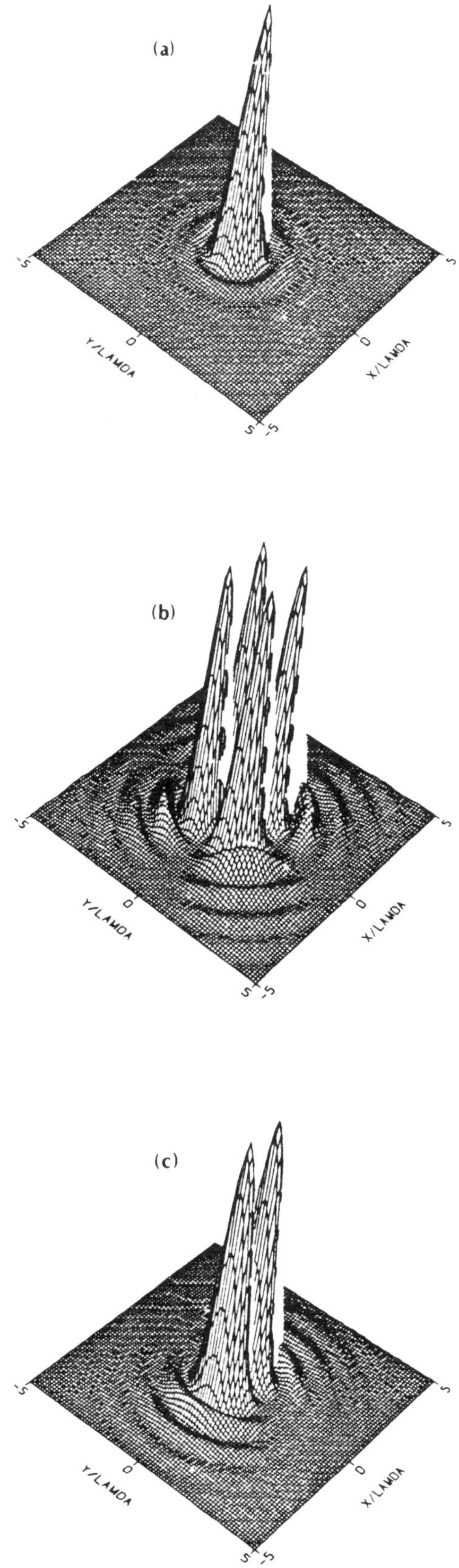

Fig. 10. Spherical lens with a numerical aperture NA = 0.5 and a focal length $f = 3500\lambda$, illuminated with a plane uniform wave, linearly polarized along the x axis. (a) Intensity pattern for the x component at the focus; (b) intensity pattern for the y component; (c) intensity distribution for the z component. For these computations, $N_{max} = N_{may} = 980$ and $L_{max} = L_{may} = 20{,}000$.

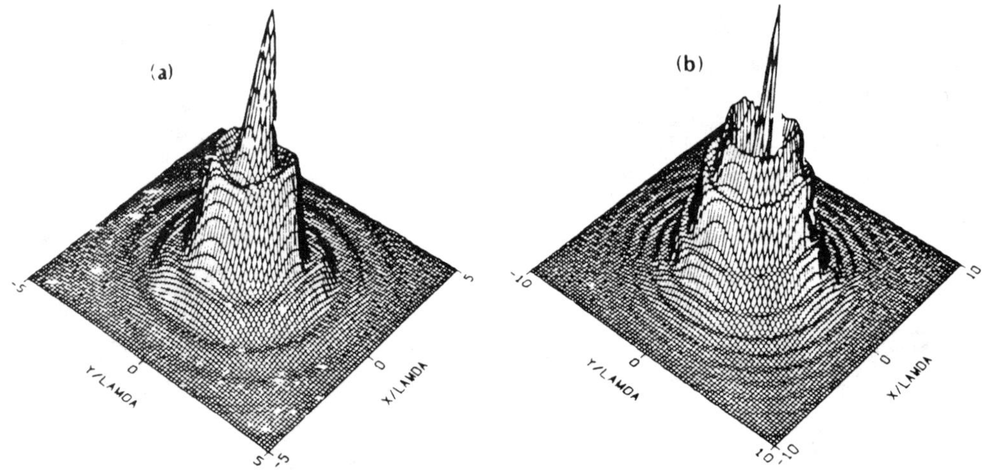

Fig. 11. Spherical lens with a numerical aperture NA = 0.5 and a focal length f = 3500λ, illuminated with a plane uniform wave, linearly polarized along the x axis. (a) Intensity pattern for the x component at a distance of 5λ from the focal plane; (b) intensity pattern for the x component at a distance of 10λ from the focal plane.

$$\dot{\omega}_\eta(r) = -(r^{-2} - 1)^{-1/2} + \eta r, \qquad r_1 \le r \le r_2 \tag{A2}$$

The function $\dot{\omega}_\eta(r)$ itself has the following derivative:

$$\ddot{\omega}_\eta(r) = -(1 - r^2)^{-3/2} + \eta, \qquad r_1 \le r \le r_2, \tag{A3}$$

which vanishes at

$$r_0 = (1 - \eta^{-2/3})^{1/2}. \tag{A4}$$

r_0 falls between r_1 and r_2, provided that

$$(1 - r_1^2)^{-3/2} \le \eta \le (1 - r_2^2)^{-3/2}. \tag{A5}$$

The value of $\dot{\omega}_\eta(r)$ at r_0 is

$$\dot{\omega}_\eta(r_0) = (\eta^{2/3} - 1)^{3/2}. \tag{A6}$$

Let us define θ_1 and θ_2 as follows.

$$\sin\theta_1 = r_1,$$

$$\sin\theta_2 = r_2; \tag{A7}$$

then the values of $\dot{\omega}_\eta(r)$ at r_1 and r_2 are

$$\dot{\omega}_\eta(r_1) = \eta \sin\theta_1 - \tan\theta_1, \tag{A8}$$

$$\dot{\omega}_\eta(r_2) = \eta \sin\theta_2 - \tan\theta_2. \tag{A9}$$

The maximum of $|\dot{\omega}_\eta(r)|$ over $[r_1, r_2]$, which will be referred to as $\dot{\omega}_{max}$, is at r_1 or r_2 or, in case r_0 happens to be within the interval, at r_0. Thus, among Eqs. (A6), (A8), and (A9), one has the maximum absolute value. Figure 14 shows a plot of these three functions versus η. Note that, according to relation (A5), the range of η over which r_0 falls between r_1 and r_2 is

$$\frac{1}{\cos^3\theta_1} \le \eta \le \frac{1}{\cos^3\theta_2}. \tag{A10}$$

The value of η at which $\dot{\omega}_\eta(r_1) = -\dot{\omega}_\eta(r_2)$ is

$$\eta_0 = \frac{\tan\theta_1 + \tan\theta_2}{\sin\theta_1 + \sin\theta_2}. \tag{A11}$$

If $1/\cos^3\theta_1 \ge \eta_0$, then $\eta_{opt} = \eta_0$; otherwise η_{opt} is between $1/\cos^3\theta_1$ and η_0 and corresponds to the crossing point of the curves $|\dot{\omega}_\eta(r_2)| = \tan\theta_2 - \eta\sin\theta_2$ and $\dot{\omega}_\eta(r_0) = (\eta^{2/3} - 1)^{3/2}$. The ordinate of the crossing point is $\dot{\omega}_{max}$.

Figure 15 shows plots of η_{opt} and $\dot{\omega}_{max}$ versus θ_2 for the special case in which $\theta_1 = 0$. Notice that with increasing θ_2 the optimum η rises above 1, first slowly and then rapidly. $\dot{\omega}_{max}$ also increases from zero at $\theta_2 = 0$ to 10^{-2} at $\theta_2 = 25°$ and to 1 at $\theta_2 = 75°$.

Figure 16 shows plots of $\cos\{2\pi \mathbf{f}[(1 - r^2)^{1/2} + \tfrac{1}{2}\eta r^2]\}$ versus r in the interval (0, 0.5), corresponding to $\theta_1 = 0$ and $\theta_2 = 30°$. The optimum value $\eta = 1.11395$ is used in Fig. 16(a), and the value $\eta = 1$ is used in Fig. 16(b). The value of $\mathbf{f}$ is 3500 in both plots. Notice how a nonoptimum value of η can increase the oscillations and thereby require a large number of samples for numerical computations.

APPENDIX B

Consider the function

$$\omega_\eta(r) = (1 + r^2)^{1/2} - \tfrac{1}{2}\eta r^2, \qquad r_1 \le r \le r_2, \tag{B1}$$

where $0 \le r_1 \le r_2$ and η is a real parameter. The objective is to find the value of η that minimizes the maximum value of $|\dot{\omega}_\eta(r)|$ in the interval $[r_1, r_2]$. $\dot{\omega}_\eta(r)$ is the derivative of $\omega_\eta(r)$ with respect to r and is given by

$$\dot{\omega}_\eta(r) = (1 + r^{-2})^{-1/2} - \eta r, \qquad r_1 \le r \le r_2. \tag{B2}$$

The second derivative of $\omega_\eta(r)$ is

$$\ddot{\omega}_\eta(r) = (1 + r^2)^{-3/2} - \eta, \qquad r_1 \le r \le r_2, \tag{B3}$$

which vanishes at

$$r_0 = (\eta^{-2/3} - 1)^{1/2}. \tag{B4}$$

r_0 belongs to $[r_1, r_2]$ provided that

$$(1 + r_2^2)^{-3/2} \le \eta \le (1 + r_1^2)^{-3/2}. \tag{B5}$$

The value of $\dot{\omega}_\eta(r)$ at r_0 is

$$\dot{\omega}_\eta(r_0) = (1 - \eta^{2/3})^{3/2}. \tag{B6}$$

Let us define θ_1 and θ_2 as follows:

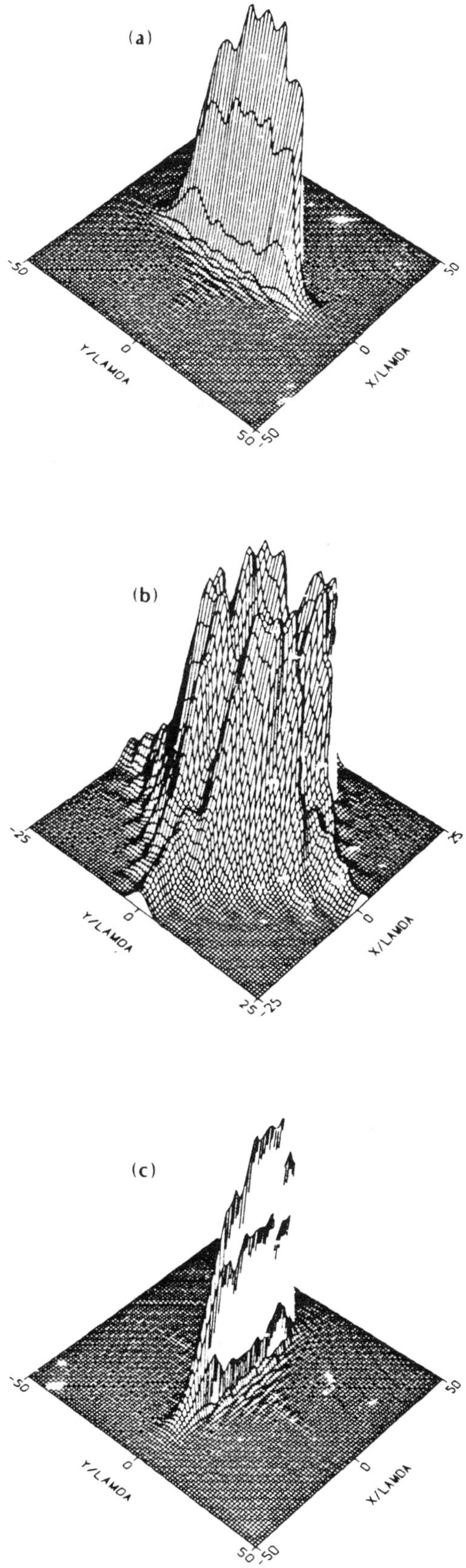

Fig. 12. Astigmatic lens with $f_x = 20{,}000\lambda$, $f_y = 20{,}200\lambda$, and NA = 0.15. The incident beam is plane and uniform. For these calculations, $N_{max} = N_{may} = 512$ and $L_{max} = L_{may} = 40{,}000$. The intensity pattern is shown at various distances from the lens: (a) $\mathbf{z} = 20{,}000$, (b) $\mathbf{z} = 20{,}100$, and (c) $\mathbf{z} = 20{,}200$.

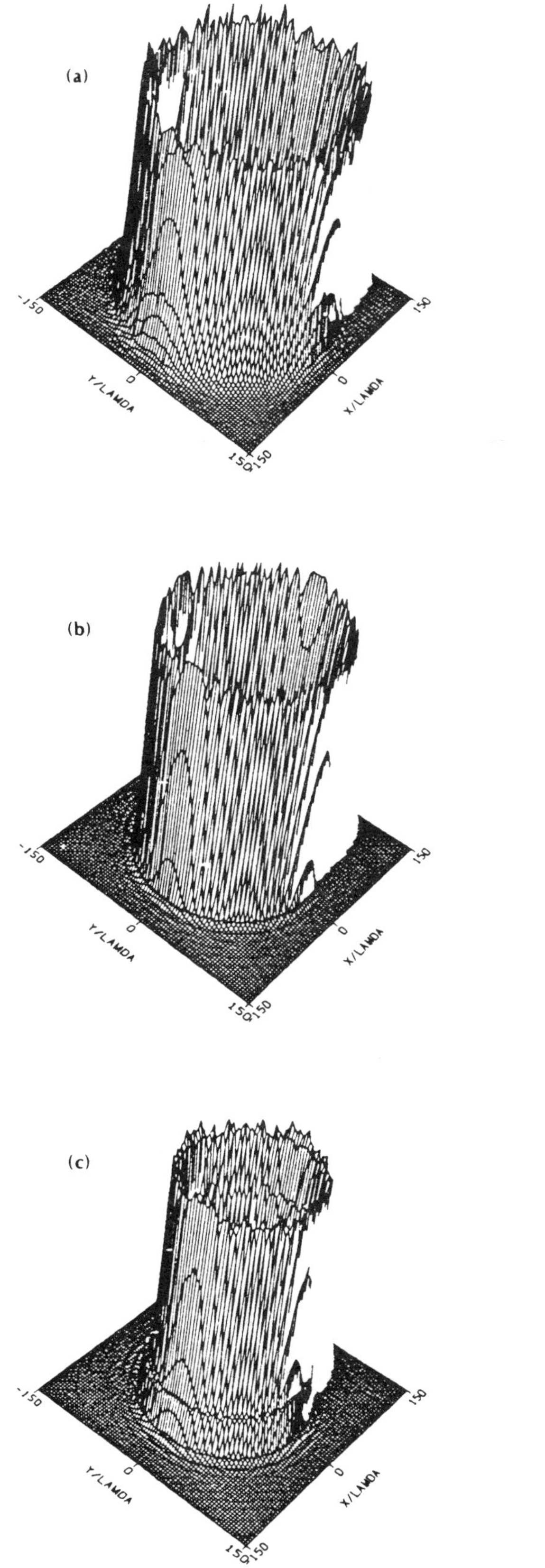

Fig. 13. Ring lens with $R_1 = 50\lambda$, $R_0 = 100\lambda$, $R_2 = 2500\lambda$, and $f = 25{,}000\lambda$. The incident beam is plane and uniform. The intensity patterns shown are at (a) $z = 24{,}750\lambda$, (b) $z = 25{,}000\lambda$, and (c) $z = 25{,}400\lambda$. For these computations $N_{max} = N_{may} = 512$ and $L_{max} = L_{may} = 25{,}000$.

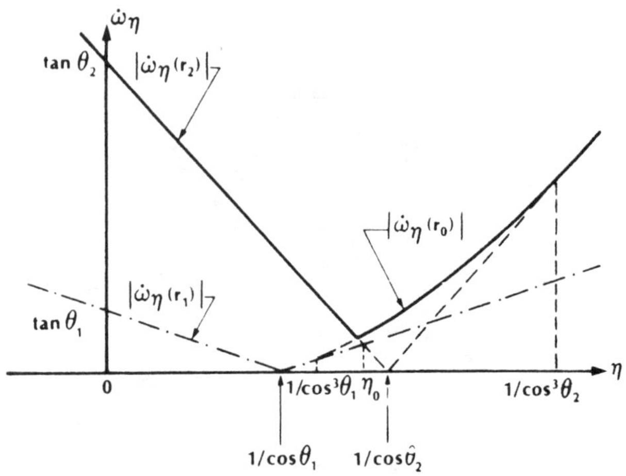

Fig. 14. Various peaks of $|\dot{\omega}_\eta(r)|$ in the interval $[r_1, r_2]$, plotted here as functions of η; the bold curve represents the largest peak. η_{opt} and $\dot{\omega}_{max}$ are the coordinates of the minimum point of the bold curve.

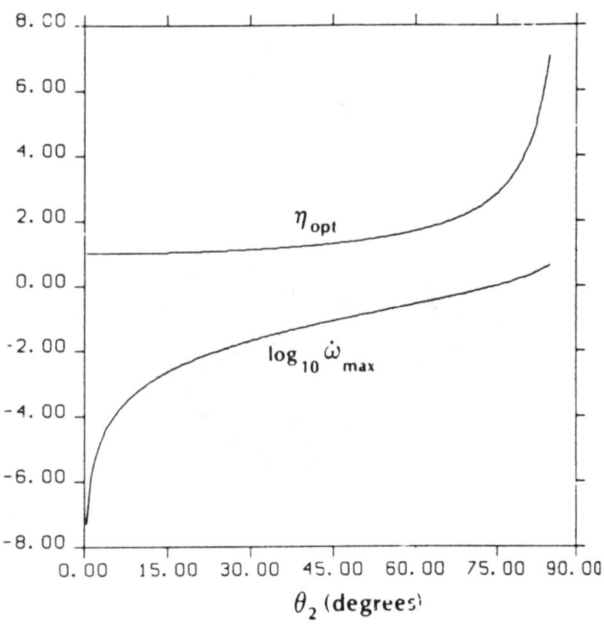

Fig. 15. η_{opt} and $\dot{\omega}_{max}$, as functions of θ_2 for $\theta_1 = 0$. Note that the scale for $\dot{\omega}_{max}$ is logarithmic. The particular value of θ_1 chosen here corresponds to $r_1 = 0$, while $r_2 = \sin\theta_2$.

$$\tan\theta_1 = r_1,$$

$$\tan\theta_2 = r_2; \tag{B7}$$

then the values of $\dot{\omega}_\eta(r)$ at r_1 and r_2 are

$$\dot{\omega}_\eta(r_1) = -\eta\tan\theta_1 + \sin\theta_1, \tag{B8}$$

$$\dot{\omega}_\eta(r_2) = -\eta\tan\theta_2 + \sin\theta_2. \tag{B9}$$

The maximum of $|\dot{\omega}_\eta(r)|$ over $[r_1, r_2]$, which will be referred to as $\dot{\omega}_{max}$, is at r_1 or at r_2 or, in case r_0 happens to be within the interval, at r_0. Thus, among Eqs. (B6), (B8), and (B9), one equation represents the maximum absolute value of the function. Figure 17 shows a plot of these three functions versus η. Note that according to relation (B5) the range of η over which r_0 lies between r_1 and r_2 is

$$\cos^3\theta_2 \leq \eta \leq \cos^3\theta_1. \tag{B10}$$

The functions $\dot{\omega}_\eta(r_1)$ and $-\dot{\omega}_\eta(r_2)$ cross at η_0, where

$$\eta_0 = \frac{\sin\theta_1 + \sin\theta_2}{\tan\theta_1 + \tan\theta_2}. \tag{B11}$$

If $\cos^3\theta_1 \leq \eta_0$, then $\eta_{opt} = \eta_0$; otherwise η_{opt} is between η_0 and $\cos^3\theta_1$ and corresponds to the crossing point of the curves $|\dot{\omega}_\eta(r_2)| = \eta\tan\theta_2 - \sin\theta_2$ and $\dot{\omega}_\eta(r_0) = (1 - \eta^{2/3})^{3/2}$. The ordinate of the crossing point is $\dot{\omega}_{max}$.

Notice that η_{opt} as obtained in Appendix A is the inverse of η_{opt} in this Appendix, provided that the corresponding θ's are identical. Of course, when θ_1 (or θ_2) in one problem is equal to

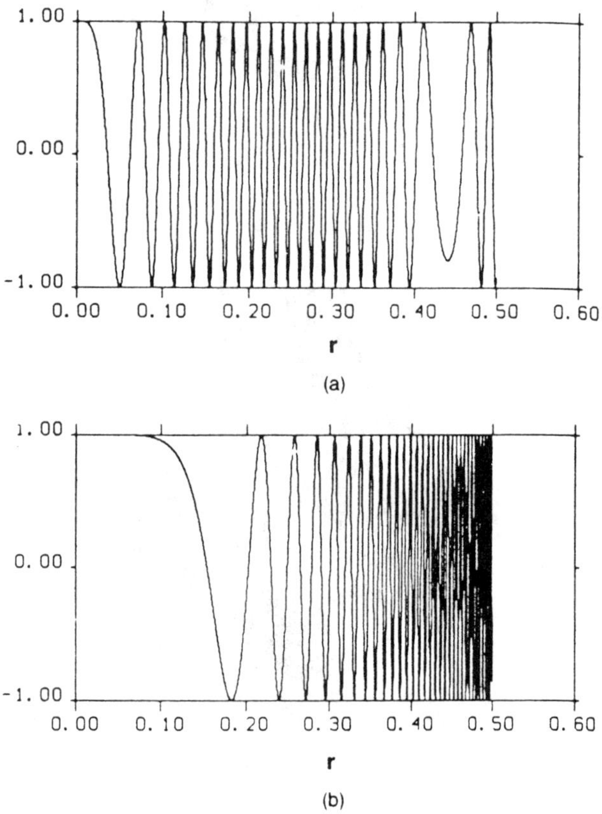

Fig. 16. Two plots of $\cos\{2\pi\mathbf{f}[(1 - r^2)^{1/2} + \frac{1}{2}\eta r^2]\}$ versus r. In (a) the optimum value of $\eta = 1.11395$ is used, whereas in (b) $\eta = 1$ is used. In both cases $\mathbf{f} = 3500$, $r_1 = 0$, and $r_2 = 0.5$.

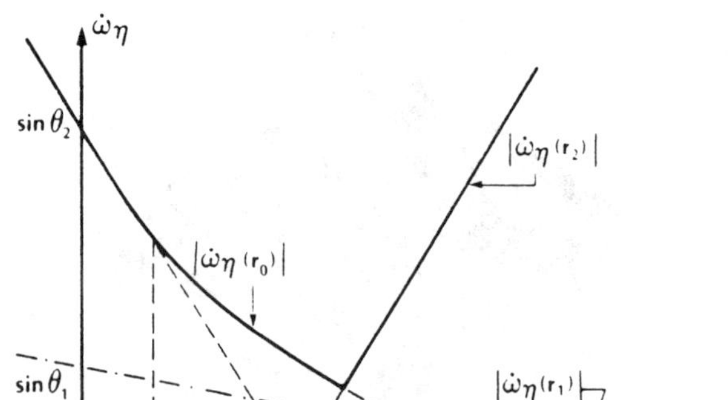

Fig. 17. Various peaks of $|\dot{\omega}_\eta(r)|$ in the interval $[r_1, r_2]$, plotted here as functions of η; the bold curve represents the largest peak. η_{opt} and $\dot{\omega}_{max}$ are the coordinates of the minimum point of the bold curve.

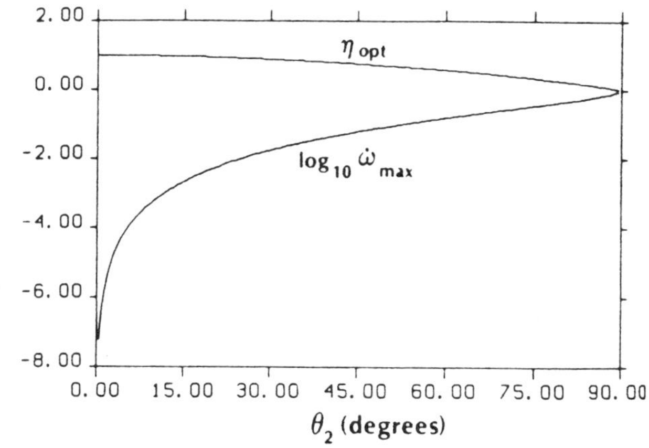

Fig. 18. η_{opt} and $\dot\omega_{max}$ as functions of θ_2 for $\theta_1 = 0$. Note that the scale for $\dot\omega_{max}$ is logarithmic. The particular value of θ_1 chosen here corresponds to $r_1 = 0$, while $r_2 = \tan\theta_2$.

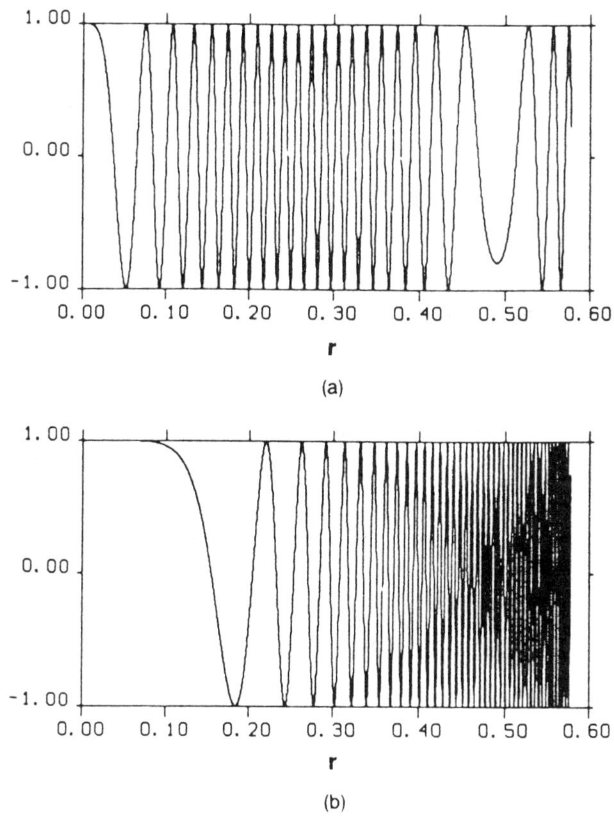

Fig. 19. Two plots of $\cos\{2\pi\mathbf{f}[(1+r^2)^{1/2} - \frac{1}{2}\eta r^2]\}$ versus r. In (a) the optimum value of $\eta = 0.8977$ is used, whereas in (b) $\eta = 1$ is used. In both cases $\mathbf{f} = 3500$, $r_1 = 0$, and $r_2 = 0.577$.

θ_1 (or θ_2) in the other, the r_1 (or r_2) values for the two problems are not equal, as can be observed by comparing Eqs. (A7) and (B7). When the corresponding θ's are identical, we have

$$\dot\omega_{max}{}^{(A)} = \tan\theta_2 - \eta_{opt}{}^{(A)}\sin\theta_2, \tag{B12}$$

$$\dot\omega_{max}{}^{(B)} = \eta_{opt}{}^{(B)}\tan\theta_2 - \sin\theta_2. \tag{B13}$$

Also, $\eta_{opt}{}^{(A)} = 1/\eta_{opt}{}^{(B)}$; therefore

$$\dot\omega_{max}{}^{(A)}/\eta_{opt}{}^{(A)} = \dot\omega_{max}{}^{(B)}, \tag{B14}$$

$$\dot\omega_{max}{}^{(B)}/\eta_{opt}{}^{(B)} = \dot\omega_{max}{}^{(A)}. \tag{B15}$$

Figure 18 shows plots of η_{opt} and $\dot\omega_{max}$ versus θ_2 for the special case in which $\theta_1 = 0$. Notice that with increasing θ_2 the optimum η drops below 1, while $\dot\omega_{max}$ rises rapidly from 0 at the origin to 10^{-2} at $\theta_2 = 25°$ and to 1 at $\theta_2 = 90°$.

Figure 19 shows plots of $\cos\{2\pi\mathbf{f}[(1+r^2)^{1/2} - \frac{1}{2}\eta r^2]\}$ versus r in the interval $[0, 0.577]$ corresponding to $\theta_1 = 0$ and $\theta_2 = 30°$. The value $\mathbf{f} = 3500$ is used in these plots. In Fig. 19(a) the optimum value $\eta = 0.8977$ is used, and the value $\eta = 1$ is used in Fig. 19(b). Notice how a nonoptimum η increases the oscillations and thereby requires a large number of samples for numerical computations.

APPENDIX C

Consider the function

$$\omega_\eta(r) = [1 + (r-R)^2]^{1/2} - \tfrac{1}{2}\eta r^2, \tag{C1}$$

where r is in the interval $[r_1, r_2]$ with $0 \le r_1 \le R \le r_2$. R is a constant, and η is a real parameter. The objective is to find the value of η that minimizes the maximum value of $|\dot\omega_\eta(r)|$ in the interval $[r_1, r_2]$. $\dot\omega_\eta(r)$ is the derivative of $\omega_\eta(r)$ with respect to r and is given by

$$\dot\omega_\eta(r) = (r-R)[1 + (r-R)^2]^{-1/2} - \eta r, \qquad r_1 \le r \le r_2. \tag{C2}$$

The second derivative of $\omega_\eta(r)$ is

$$\ddot\omega_\eta(r) = [1 + (r-R)^2]^{-3/2} - \eta, \qquad r_1 \le r \le r_2, \tag{C3}$$

which vanishes at the following two points:

$$r_0^\pm = R \pm (\eta^{-2/3} - 1)^{1/2}. \tag{C4}$$

These points will be in $[r_1, r_2]$, provided that

$$[1 + (r_2 - R)^2]^{-3/2} \le \eta \le 1, \tag{C5a}$$

$$[1 + (R - r_1)^2]^{-3/2} \le \eta \le 1. \tag{C5b}$$

The values of $\dot\omega_\eta(r)$ at $r_0^\pm$ are

$$\dot\omega_\eta(r_0^\pm) = \pm(1 - \eta^{2/3})^{3/2} - \eta R. \tag{C6}$$

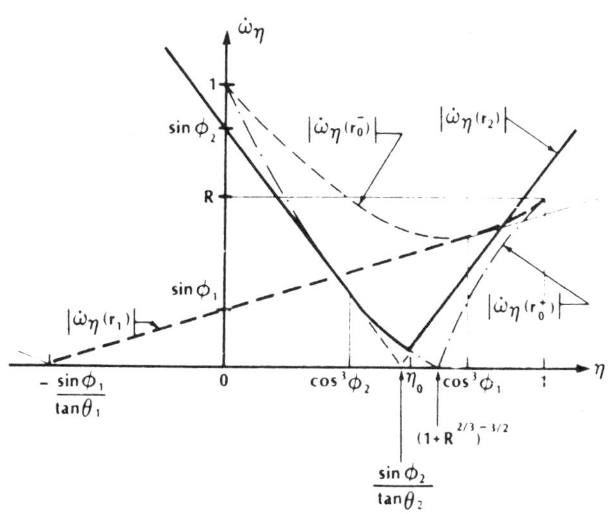

Fig. 20. Various peaks of $|\dot\omega_\eta(r)|$ in the interval $[r_1, r_2]$, plotted here as functions of η. The bold solid curve represents the largest peak to the right of R, whereas the bold dashed curve corresponds to the largest peak to the left of R. The larger of the two curves at any η thus gives the largest peak, and the optimum η is the point at which the largest peak is minimum.

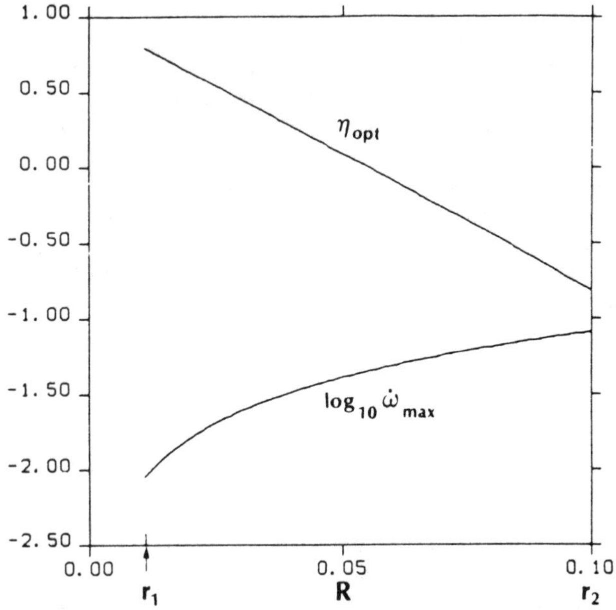

Fig. 21. η_{opt} and $\dot{\omega}_{max}$ as functions of R for $r_1 = 0.01$ and $r_2 = 0.1$. Note that the scale for $\dot{\omega}_{max}$ is logarithmic.

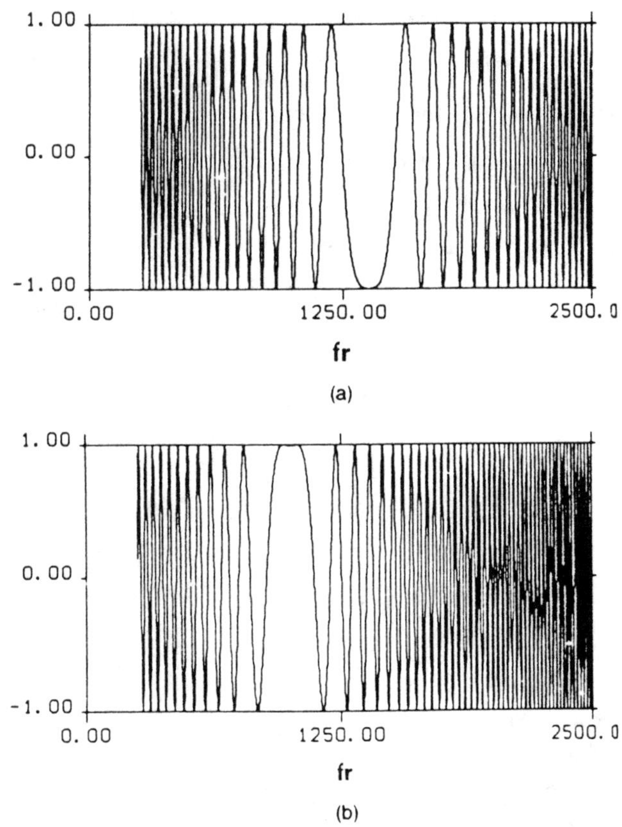

Fig. 22. Two plots of $\cos(2\pi \mathbf{f}\{[1 + (r - R)^2]^{1/2} - \frac{1}{2}\eta r^2\})$ versus $\mathbf{f}r$. In (a) the optimum value of $\eta = 0.272$ is used, whereas in (b) $\eta = 0$ is used. In both cases $\mathbf{f} = 25000$, $r_1 = 0.01$, and $r_2 = 0.1$.

Let us define θ_1, ϕ_1, θ_2, and ϕ_2 as follows:

$$\tan\theta_1 = r_1, \tag{C7a}$$

$$\tan\theta_2 = r_2, \tag{C7b}$$

$$\tan\phi_1 = R - r_1, \tag{C7c}$$

$$\tan\phi_2 = r_2 - R; \tag{C7d}$$

then the values of $\dot{\omega}_\eta(r)$ at r_1 and r_2 are

$$\dot{\omega}_\eta(r_1) = -\sin\phi_1 - \eta\tan\theta_1, \tag{C8}$$

$$\dot{\omega}_\eta(r_2) = \sin\phi_2 - \eta\tan\theta_2. \tag{C9}$$

The maximum of $|\dot{\omega}_\eta(r)|$ over $[r_1, r_2]$, which will be referred to as $\dot{\omega}_{max}$, is at r_1 or r_2 or, in case r_0^+ or r_0^- happens to be within the interval, at r_0^+ or r_0^-. Thus, among Eqs. (C6), (C8), and (C9), one equation represents the maximum absolute value of the function. Figure 20 shows a plot of these four functions versus η. Note that according to relations (C5) the ranges of η over which $r_0^\pm$ fall between r_1 and r_2 are

$$\cos^3\phi_2 \le \eta \le 1 \qquad (\text{for } r_0^+), \tag{C10a}$$

$$\cos^3\phi_1 \le \eta \le 1 \qquad (\text{for } r_0^-). \tag{C10b}$$

In Fig. 20 we have constructed two functions. The first, shown as a dark solid curve, represents the maximum of $|\dot{\omega}_\eta(r)|$ to the right of R and is the larger of the two functions $|\dot{\omega}_\eta(r_2)|$ and $|\dot{\omega}_\eta(r_0^+)|$ when the latter is in the region of relation (C10a). Evidently, there is no contribution to this curve from $|\dot{\omega}_\eta(r_0^+)|$ unless

$$\cos^3\phi_2 < \frac{\sin\phi_2}{\tan\theta_2}. \tag{C11}$$

The section of $|\dot{\omega}_\eta(r_0^+)|$ that is used here is between $\cos^3\phi_2$ and η_0, where η_0 belongs to the interval $[\sin\phi_2/\tan\theta_2, (1 + R^{2/3})^{-3/2}]$ and represents the crossing point of the curves for $|\dot{\omega}_\eta(r_2)| = \eta\tan\theta_2 - \sin\phi_2$ and $\dot{\omega}_\eta(r_0^+) = (1 - \eta^{2/3})^{3/2} - \eta R$. The second function, shown as a bold dashed curve, consists of $|\dot{\omega}_\eta(r_1)|$ when $\eta \le \cos^3\phi_1$ and of $|\dot{\omega}_\eta(r_0^-)|$ when $\eta \ge \cos^3\phi_1$. This function represents the largest value of $|\dot{\omega}_\eta(r)|$ when r is to the left of R. Thus we need to investigate the larger of the above two functions. Two possibilities exist:

1. $\sin\phi_1 > \sin\phi_2$. In this case η_{opt} is negative and is given by

$$\eta_{opt} = \frac{\sin\phi_2 - \sin\phi_1}{\tan\theta_2 + \tan\theta_1}. \tag{C12}$$

2. $\sin\phi_1 < \sin\phi_2$. In this case the optimum η is at the crossing of the two dark curves. The ordinate of the crossing point is $\dot{\omega}_{max}$.

Figure 21 shows plots of η_{opt} and $\dot{\omega}_{max}$ versus R for $r_1 = 0.01$ and $r_2 = 0.1$. When R is halfway between r_1 and r_2, $\eta_{opt} = 0$ and the high-frequency terms cannot be factored out. As R moves closer to r_1, however, $\dot{\omega}_{max}$ drops rapidly and η_{opt} increases toward 1.

Figure 22 shows plots of $\cos(2\pi \mathbf{f}\{[1 + (r - R)^2]^{1/2} - \frac{1}{2}\eta r^2\})$ versus $\mathbf{f}r$ for $\mathbf{f} = 25{,}000$, $\mathbf{f}r_1 = 250$, $\mathbf{f}r_2 = 2500$, and $\mathbf{f}R = 1000$. In Fig. 22(a) the optimum value $\eta = 0.272$ is used, and in Fig. 22(b), $\eta = 0$ is used. Note how optimum η can balance the oscillations at the two extremes of the aperture and consequently can minimize the required number of samples for numerical computations.

ACKNOWLEDGMENT

This study has been supported by a grant from the IBM Corporation.

REFERENCES

1. E. Wolf, "Electromagnetic diffraction in optical systems: an integral representation of the image field," Proc. R. Soc. Ser. A **253**, 349–357 (1959).
2. B. Richards and E. Wolf, "Electromagnetic diffraction in optical systems: structure of the image field in an aplanatic system," Proc. R. Soc. Ser. A **253**, 358–379 (1959).
3. D. R. Rhodes, "On the stored energy of planar apertures," IEEE Trans. Antennas Propag. **AP-14**, 676–683 (1966).
4. G. Borgiotti, "Fourier transforms method of aperture antennas problem," Alta Frequenza **32**, 196–204 (1963).
5. M. Born and E. Wolf, *Principles of Optics*, 6th ed. (Pergamon, Oxford, 1980).
6. K. Miyamoto and E. Wolf, "Generalization of the Maggi-Rubinowitz theory of the boundary diffraction wave—part I," J. Opt. Soc. Am. **52**, 615–625 (1962).
7. G. W. Farnell, "Measured phase distribution in the image space of a microwave lens," Can. J. Phys. **36**, 935–943 (1958).
8. Y. Li and E. Wolf, "Three-dimensional intensity distribution near the focus in systems of different Fresnel numbers," J. Opt. Soc. Am. A **1**, 801–808 (1984).
9. V. N. Mahajan, "Axial irradiance and optimum focusing of laser beams," Appl. Opt. **22**, 3042–3053 (1983).
10. M. Mansuripur, "Distribution of light at and near the focus of high-numerical-aperture objectives," J. Opt. Soc. Am. A **3**, 2086–2093 (1986).
11. J. W. Goodman, *Introduction to Fourier Optics* (McGraw-Hill, New York, 1968).
12. M. Mansuripur and C. Pons, "Diffraction modeling of optical path for magneto-optical disk systems," in *Optical Storage Technology and Applications*, D. Carlin, Y. Tsunoda, and A. Jamberdino, eds., Proc. Soc. Photo-Opt. Instrum. Eng. **899**, 56–60 (1988).
13. F. Jenkins and H. White, *Fundamentals of Physical Optics*, 1st ed. (McGraw-Hill, New York, 1937).

Reprinted from *Optical Data Storage,* Proc. SPIE Vol. 1316, pp. 70-80 (1990).

Computer modeling of optical storage media and systems

M. Mansuripur

Optical Sciences Center, University of Arizona, Tucson, AZ 85721

ABSTRACT

The status of theoretical analysis and numerical modeling efforts for optical data storage systems and media is reviewed. An attempt is made to provide guidelines for future research.

1. INTRODUCTION

With the recent introduction of magneto-optical disk systems in the market, optical storage products have completed a life cycle of some sort : all possible modes of optical storage (read-only, archival and erasable) are now commercially available. It is no longer far fetched to speak of single, universal drives that read mass-produced software and data-bases off the CD-ROMs, use high capacity erasable and removable magneto-optical disks for temporary storage and processing of information, and record archival files for permanent storage on WORM media. To be sure, there is still a multitude of problems to overcome : access times need to be reduced, the limits of reliability of the media must be firmly established, standards must be agreed upon and, last but not least, prices for drives and media should drastically decline. None of these problems, however, is being perceived as a fundamental roadblock and, with the current pace of research and development around the world, many of the technical problems and shortcomings can be expected to disappear in the near future. It is perhaps a good time now to sit back and take a long term look at the future of optical storage. We must begin to focus on questions which are less concerned with the short term viability of optical storage, but more related to the rate of growth of the technology in the steady state. Ultimately, one needs to understand the fundamental limits of the media and systems in order to approach those limits as closely as possible.

If the history of magnetic recording is any clue at all, it should be obvious that the three factors that need constant improvement are the access time, the data rate, and the storage density. While it is impossible to predict where new research and development leads us, it is not difficult to venture a few forecasts by extrapolating from certain trends in present research activities. One may observe, for instance, that both manufacturability and access times can benefit from integration and/or the use of holographic optical elements (HOEs), inasmuch as these techniques will enable us to design lighter heads which also require fewer adjustments [1-3]. Novel head and servo concepts such as autofocusing due to wavelength shift of the diode laser [4], magneto-optic signal detection using external cavity laser diode [5], and fast track acquisition by acousto-optic or electro-optic modulators [6] can also improve the access times. Data rates will be increased through the use of multiple heads, multiple lasers within a single head [7], increased linear densities, and faster-spinning disks. Higher storage densities will come, in part, from short wavelength lasers, but perhaps the main avenue for achieving higher density will remain modulation coding. Better run-length-limited (RLL) codes can be implemented only when the readout signal is of high quality and the noise level is low [8]. That is why there will be attempts to improve the signal to noise ratio either by improving the current media or by switching to entirely new classes of materials [9].

In-depth understanding of the properties of the media and systems of optical storage is a prerequisite to achieving the above goals. While no pretensions are being made here as to the knowledge of all the relevant results (or even all the relevant questions), the author will give a brief account of the ongoing theoretical analysis and numerical modeling efforts. It is hoped that this overview will help put the modeling efforts in perspective and encourage closer ties between theory and experiment. The following sections describe progress in the areas of record process modeling and analysis of the readout signal.

2. MODELING THE RECORDING PROCESS

All recordable media in use today rely on the rise in temperature to achieve local change in some physical property of the material. Therefore the first step in analyzing the write and/or erase processes is the computation of temperature profiles. Except in very simple situations, these calculations are done numerically using either the finite difference method [10-12] or the finite element technique [13]. An example of the results obtained with the finite difference method is given in Figure 1. (On a microVAX it takes about 10 minutes to generate these curves.) Generally speaking, the finite element technique produces similar results in comparable times. One shortcoming of the existing programs is that they assume flat surfaces for the disk and circular symmetry for the beam, which then allows them to proceed with solving the heat diffusion equation in two spatial dimensions (r and z in cylindrical coordinates). For more realistic calculations when the disk surface is grooved and/or when there is preformat information on the substrate, or when the beam has asymmetry due to aberrations or otherwise, one must solve the heat absorption and diffusion equations in 3-dimensional space.

Another problem of the existing models is that the optical and thermal parameters of the media are assumed to be time-independent. Usually, as the temperature varies, these parameters change (some of them appreciably). While it is not difficult to incorporate these changes in the numerical models, at the time of this writing it is almost impossible to obtain reliable data on the values of thermal parameters at a fixed temperature, let alone their temperature dependencies. It is hoped and expected that some effort in the future will be directed towards accurate thermal characterization of thin film media used in optical data storage. The situation is perhaps most critical for phase change media where the rates of heating and/or cooling in the vicinity of the melting point and in the neighborhood of the amorphous-to-crystalline phase transition determine the success or failure of the write/erase operations [14].

A major difficulty encountered in some thermal modeling efforts is that the geometry of the media changes during the heating cycle. This means that the thermal process cannot be completely decoupled from the mark formation process. (No such problems exist in magneto-optical recording where the physical structure of the media remains unchanged while the magnetic properties are affected.) Elementary methods for analyzing the thermal cycle in ablative and bubble forming media are already available [15], but more refinement and sophistication is still desired.

The process of thermomagnetic recording in erasable magneto-optic media is by far the most complex and, in some respects, the least understood mechanism of optical recording. Recent developments in exchange-coupled multilayer magnetic films with direct overwrite capability [16], and layered structures such as Co/Pt and Co/Pd with superior characteristics compared to the traditional RE-TM alloys [9], have only compounded the problems by introducing new levels of complexity in the structure of the media. Traditionally, the thermomagnetic recording process has been approached by the Huth model [17] which assumes a circular magnetic domain wall subjected to various expanding and contracting forces originating from magnetization and domain wall energy, with resistance to expansion and/or collapse provided by the coercivity [18, 19]. The important mechanism of initial domain nucleation is all but ignored. The temperature dependencies of the various parameters in the Huth model are obtained either by direct measurement (when possible at all) or derived from the mean-field theory [20]. Coercivity is always treated phenomenologically and its measured *macroscopic* values are used in a model that in every other respect is *micro*magnetic. Despite these shortcomings, the model has been successful in explaining the basic features of thermomagnetic recording and the observed stability of the recorded domains [21].

A different quasi-static model has recently been proposed which allows domains of arbitrary shape to be created/annihilated under thermomagnetic recoding/erasure conditions [22]. Instead of coercivity, this model incorporates the spatial variations of the magnetic parameters, which are believed to be the source of coercivity. Thus coercivity is no longer an input to the model but one of its by-products. This model has been used in computing the recorded domain shape in moving media, as well as estimating the effects of wall nonuniformity and jaggedness on readout noise and jitter [23, 24].

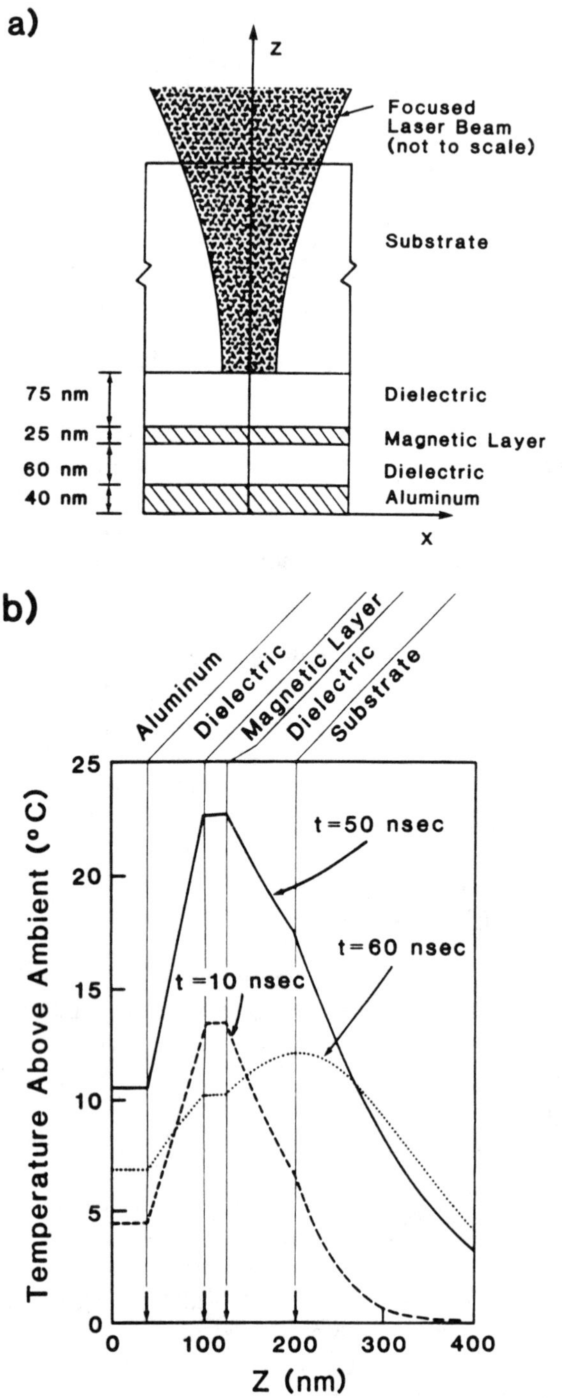

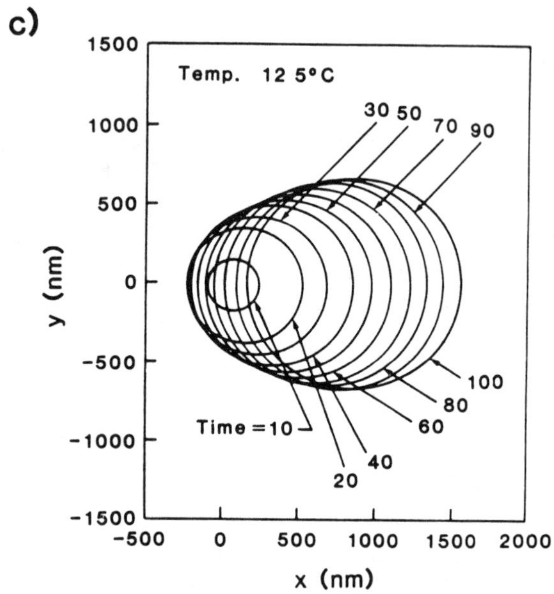

Figure 1. Temperature profiles in a multilayer structure as computed by the numerical solution of the light absorption and heat diffusion equations [11]. (a) Quadrilayer magneto-optic device. The focused Gaussian beam has wavelength λ = 830 nm, 1/e (intensity) radius r_0 = 560 nm, and is incident from the substrate side. (b) Temperature profiles at the beam center along the Z axis for three instants of time. The laser pulse is rectangular with 10 mW power and 50 nsec duration, and the disk is stationary. (c) Isotherms in the magnetic layer at the temperature of T = 125°C above the ambient. The laser pulse is rectangular with 10 mW power and 100 nsec duration, and the disk velocity is 10 m/sec along the negative X axis. The isotherms are shown at 10 nsec intervals and the maximum temperature reached in the film during the pulse is 250°C above the ambient.

More recently, large scale dynamic simulations for amorphous magnetic films have been performed on the massively parallel Connection Machine [25-27]. These simulations have brought about a fundamental understanding of domain nucleation and the mechanisms of coercivity in the media of optical recording. Figure 2 is an example of the results obtained in these simulations. It is now possible to investigate and assess the relative significance of the various sources of coercivity, be they small nonuniformities inherent in the amorphous structure of the material, or defects of varying origins. Dynamic simulations of the process of domain formation under conditions of thermomagnetic recording incorporate all the known physics of magnetization reversal without unjustifiable simplifications or approximations. They are, therefore, an invaluable tool in the analysis and design of future magneto-optical media with optimum characteristics.

In the future it will become necessary to investigate the interaction of domains and domain walls with the microstructural features that, intentionally or not, may exist on the substrate. Grooves, preformat marks, and pinholes have deteriorating or at least undesirable effects on the micromagnetics of the MO media. On the other hand, precision-etched marks at regular intervals along the track might help create stable, repeatable, and jitter-free magnetic domains. At present systematic studies of the interaction between magnetization and structural features in the above sense are rare. However, theoretical and experimental results in this area are highly desirable.

3. THE READOUT PROCESS

The diffraction theory of optical disk readout has been developed over the past several years and many excellent articles on this subject can now be found in the literature [28-30]. A good example of the type of computations presently achievable is given in Figure 3.

Ideally, one would like to have a computer program for vector diffraction calculations that allows the read beam to propagate through the various optical elements in the readout system, interact with the storage medium, and propagate back to the detector(s) for final analysis. One important ingredient of such a program is obviously the subprogram that provides accurate solutions to Maxwell's equations at the disk surface, where the light interacts with the small features inscribed on the storage medium. It is fair to say that at the present time this particular subprogram is far less developed than the rest of the program. Certain preliminary results regarding the interaction of polarized light with pits and grooves in simple disk structures exist [28,30,35] but much more needs to be done in this area. Figure 4 which shows the amplitude and phase patterns of a narrow strip of polarized light after reflection from the edge of a shallow groove is an example of the type of computations that can be performed today.

As the wavelengths become shorter and the numerical apertures increase, the noise arising from the interaction between light and the surface roughness increases, polarization dependent effects (specially those for the marginal rays) gain significance, and aberrations and misalignments become particularly bothersome. Under these circumstances it becomes crucial to better understand and thereby control the sources of media/system noise [36-38]. It is also important to design the system with appropriate tolerances. In this respect the diffraction modeling of the optical path is specially helpful.

Multilayer stacks containing dielectrics, metals, multiple magnetic layers, and birefringent substrates are quite well modelled at this point [39,40]. There exist comprehensive program packages that can compute reflection, absorption, and transmission for plane waves of arbitrary polarization, incident at arbitrary angles on multilayers of practically any degree of complexity. An instance of these calculations for a quadrilayer magneto-optic structure containing a single magnetic layer is shown in Figure 5. In the context of multilayer media, a problem which has heretofore received no attention is the electromagnetic interaction between light and the embedded marks within the multilayer. Theoretical investigations in this area will be very useful indeed.

Figure 2 (see next page). Nuclei of reverse magnetization in a thin film of amorphous rare earth-transition metal alloy, as simulated on the Connection Machine [25]. The simulated hexagonal lattice is 256 × 256, with magnetic parameters typical of TbFeCo alloys near the compensation composition. Each lattice site is assigned a magnetic dipole which interacts with its nearest neighbor dipoles via quantum mechanical exchange, and with all other dipoles in the lattice via the classical dipolar field (i.e., demagnetization). Also each dipole has a local easy axis with a small, randomly assigned deviation from the perpendicular to the plane of the lattice. The dipoles follow the dynamic equation of Landau, Lifshitz and Gilbert, while the effective magnetic fields under which individual dipoles gyrate are continuously updated. The nuclei in the picture were produced when a uniformly saturated lattice was subjected to a reverse external field whose strength was similar to the coercivity of the simulated material. The entire process of nucleation and growth has been monitored and recorded on video tape. What is shown here is merely one frame, corresponding to a particular instant of time during the reversal process. White regions represent areas of the lattice with magnetization perpendicular to the plane of the lattice and pointing "up". The magnetization in black regions is perpendicular and "down" while colorful areas represent domain walls. Each color corresponds to a particular direction of magnetization in the plane of the lattice.

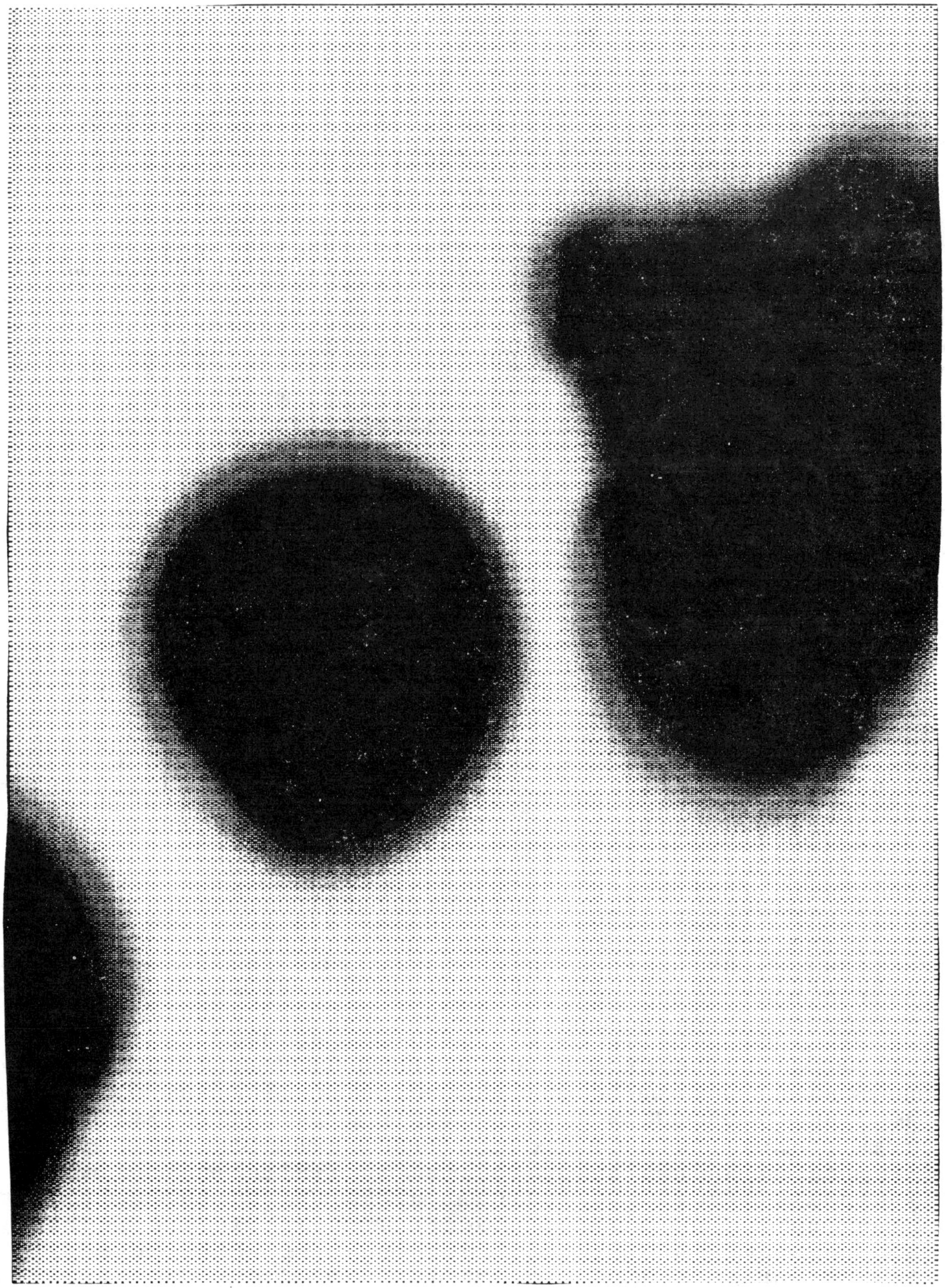

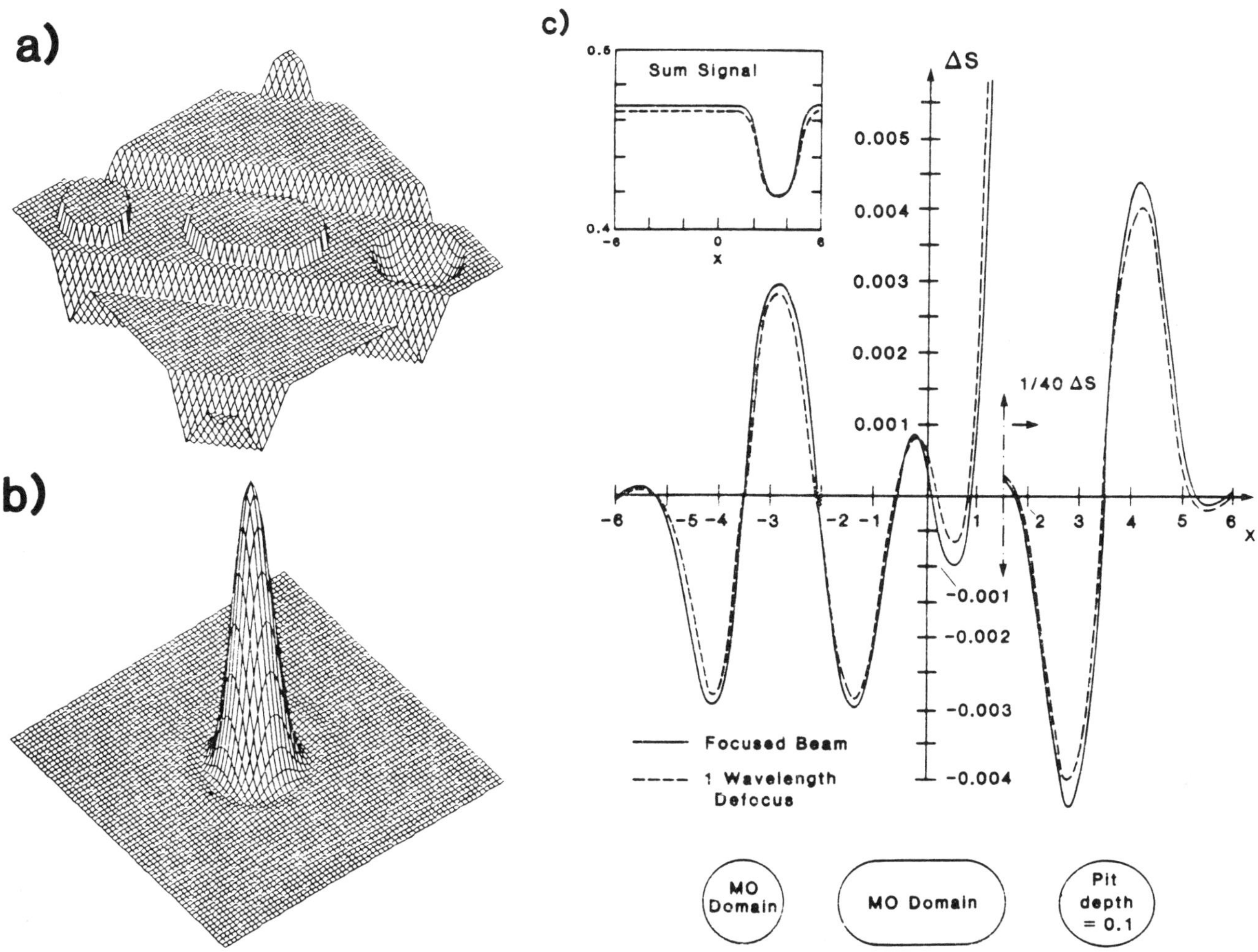

Figure 3. Calculation of the readout signal using diffraction theory. (a) Visualization of a small section from the disk surface. The V-shaped grooves are one wavelength (λ) wide and 0.16λ deep, and the width of the land region is 2λ. The disk surface has a reflectivity of 49% and a magneto-optic Kerr angle of 0.4 degree. Of the three marks shown on the middle track, the one on the right is a sphero-cylindrical pit with length = 2λ, width = 1.5λ, and depth = 0.1λ. The other two marks are magnetic domains which, for better visualization, are shown slightly raised above the land surface. (b) Intensity profile of the focused beam at the plane of the disk surface. The optical system consists of an objective lens with numerical aperture NA = 0.55 and focal length f = 4700λ. The objective focuses a circularly polarized, truncated Gaussian beam on the track and collects the reflected light, which will then continue to propagate a distance of 25000λ to an astigmatic lens. The astigmat has focal lengths $f_1 = 24700\lambda$ and $f_2 = 25300\lambda$. A quad detector located halfway between the two focal planes of the astigmat collects the light and generates the focus error and track error servo signals as well as the edge-detecting readout signal from the data [31-33]. (c) The read signal versus position of the beam along the track. The solid curve was calculated with the disk in perfect focus, while the dashed curve corresponds to a defocus of one wavelength. In both cases the beam was centered on the track. Note that the signal from the pit has been scaled down by a factor of 40 in order to fit on the same plot with the magneto-optic signals. The tail of the signal from the left edge of the pit interferes with the signal from the right edge of the large MO domain, rendering the latter undetectable. However, The left edge of the large MO domain as well as both edges of the small MO domain produce good signals and can easily be detected. The sum signal from the quad detector is shown in the inset, indicating a small drop in the amount of light refelected from the disk while the pit is being scanned. Vector diffraction theory [34] was used in these calculations throughout the entire optical path with the exception of the disk surface, where marks were treated as simple phase objects.

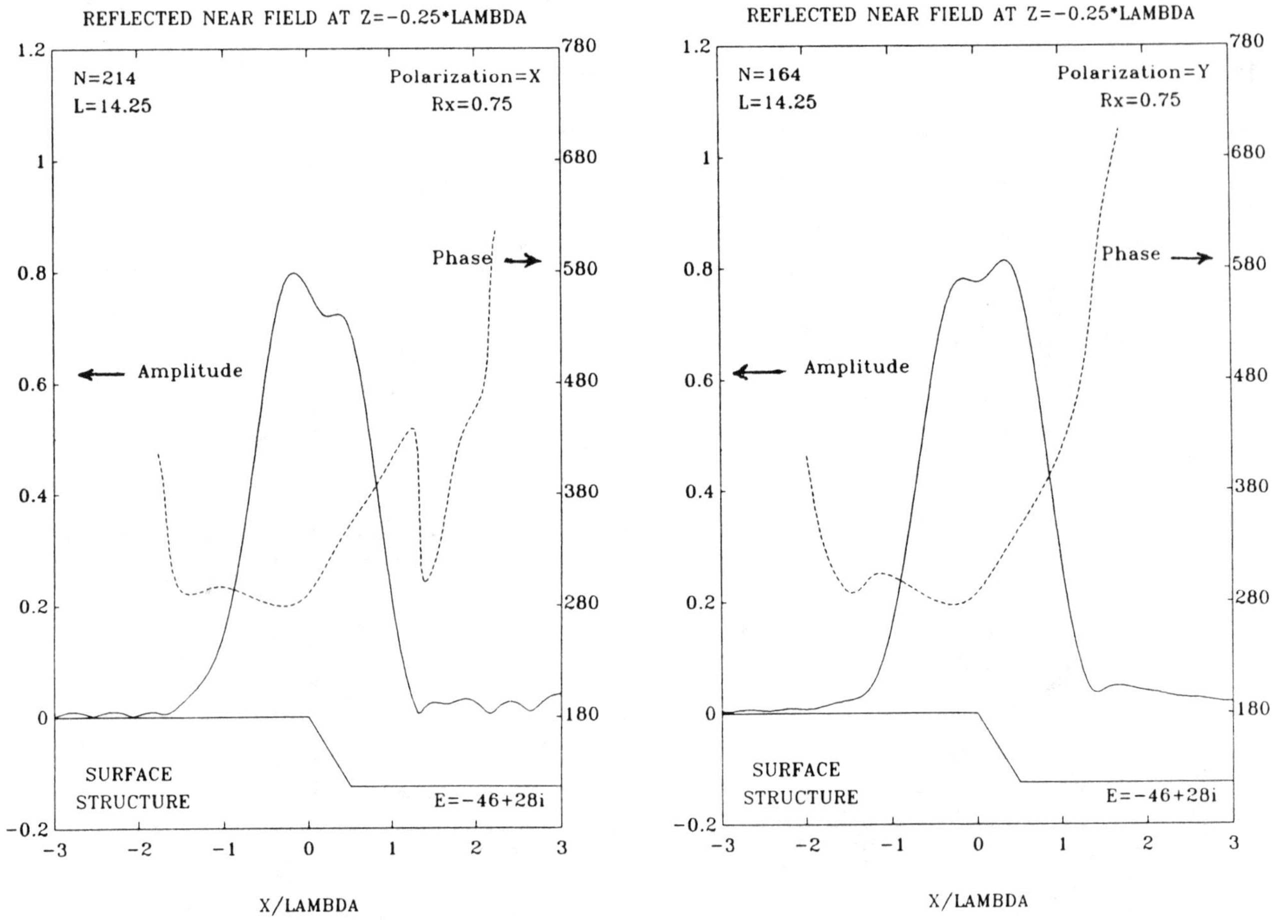

Figure 4. Distribution of near field amplitude and phase in reflection, corresponding to a strip of incident light focused on one edge of a wide, shallow groove. The incident beam is a strip parallel to the Y axis with uniform distribution along Y but has a Gaussian profile along X with 1/e radius in this direction given by $r_x = 0.75\lambda$. The medium of incidence is air and the groove is on the surface of aluminum whose assumed dielectric constant is $\epsilon = -46 + 28i$. The groove, which runs parallel to the Y axis and is much wider than the width of the incident beam along X, is trapezoidal in cross-section with a depth equal to 0.125λ. The groove edge under consideration extends from $X = 0$ to $X = 0.5\lambda$, as shown in the lower part of the figure. The two cases shown here correspond to incident polarization linear along X and linear along Y, respectively. For these computations the incident, reflected, and transmitted beams were assumed to be superpositions of N plane waves each, and the unknown amplitudes of the reflected and transmitted waves were obtained by matching the boundary conditions at N points on the interface.

Finally, Let us mention that another new and fruitful area of research in optical head modeling is the area of holographic optical elements and integrated waveguide-type devices [1-3]. Both areas need further progress in understanding, specially where high efficiency, polarization sensitive devices are concerned.

4. CONCLUDING REMARKS

In this article we have reviewed the state of the art in theoretical and computer modeling of the read/write/erase processes in optical data storage systems. Examples have been presented of the types of feasible calculations with the present levels of understanding and with the existing computers. Also a modest attempt has been made to provide guidelines for future research in this area.

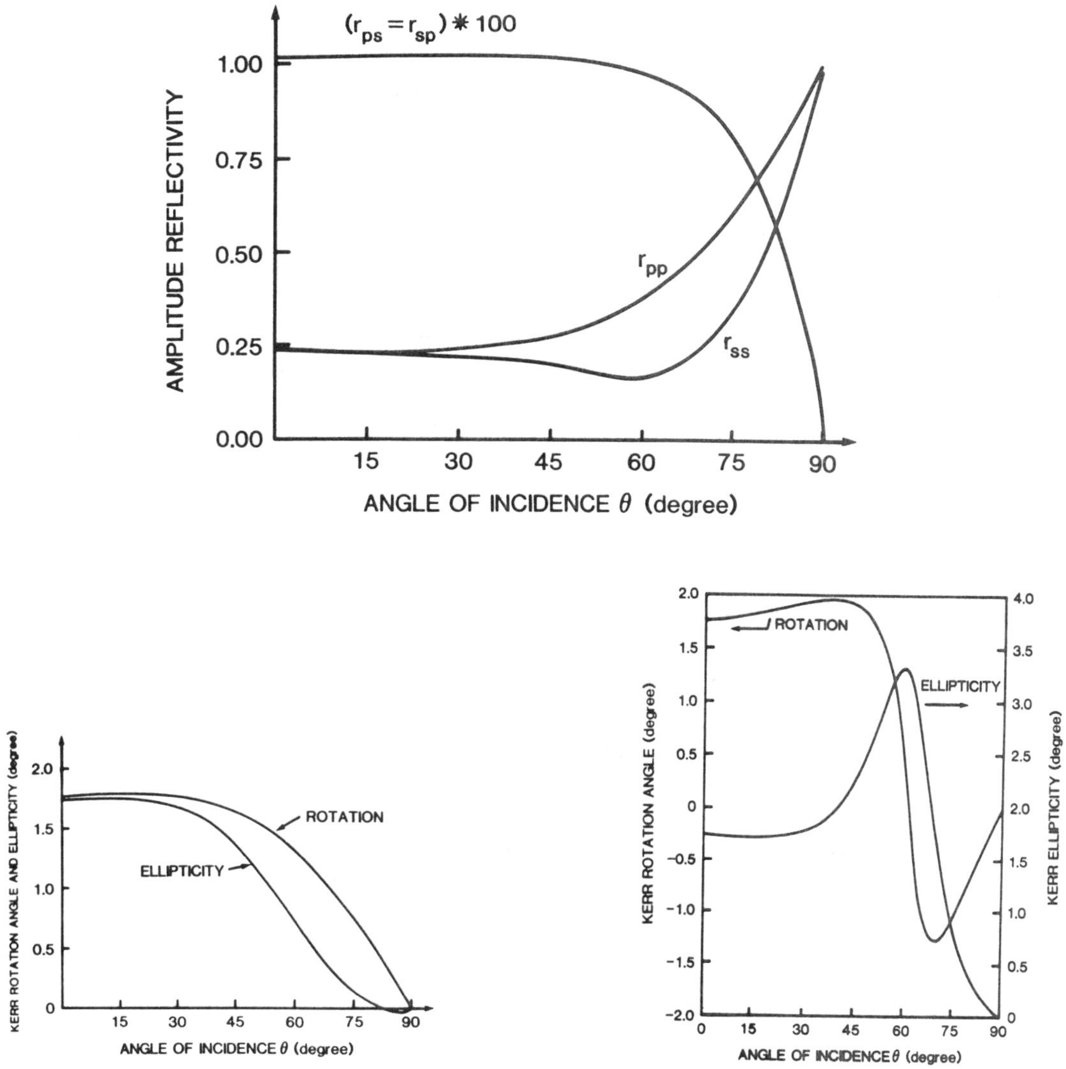

Figure 5 Reflection of linearly polarized plane wave from a quadrilayer magneto-optic device at oblique incidence. The quadrilayer consists of a glass substarte with refractive index $n_{sub} = 1.5$, coated with a reflecting layer of thickness 500 nm and complex refractive index (n, k) = (2.75, 8.31). A quarter-wave-thick layer of SiO_x (thickness = 143.2 nm) with (n, k) = (1.449, 0) separates the reflecting layer from the magneto-optic film which is 20 nm thick. The nonzero elements of the dielectric tensor of the MO layer are $\epsilon_{xx} = \epsilon_{yy} = \epsilon_{zz} = (-4.8984 + i\ 19.415)$ and $\epsilon_{xy} = -\epsilon_{yx} = (0.4322 + i\ 0.0058)$. The magnetic film is coated with another quarter-wave-thick layer of SiO_x and the medium of incidence is air ($n_{inc} = 1$). The incident beam is plane with wavelength $\lambda = 830$ nm.

(a) Various reflectivities as functions of the angle of incidence θ. Only magnitudes of the complex reflectivities are shown. r_{pp} and r_{ps} are the P and S components of the reflected beam when the incident polarization is P. Similarly, r_{sp} and r_{ss} correspond to S polarized incident light.

(b) Kerr rotation angle and ellipticity as functions of θ for P polarized incident light.

(c) Kerr rotation angle and ellipticity as functions of θ for S polarized incident light.

5. ACKNOWLEDGMENTS

This work has been made possible by support from the Optical Data Storage Center (ODSC) at the University of Arizona and, in part, by a grant from the Digital Equipment Corporation.

6. REFERENCES

1. A. Ohba, Y. Kimura, S. Sugama, Y. Urino and Y. Ono, "Holographic optical element with analyzer function for magneto-optical disk head", Paper 27D-24, presented at the International Symposium on Optical Memory (ISOM), Kobe, Japan, September 1989.
2. R. Linnebach, K. Gillessen, J. Frohlich and S. Greenard, "Application of holographic optical elements in optical memory devices", Paper 27D-20, presented at the International Symposium on Optical Memory (ISOM), Kobe, Japan, September 1989.
3. T. Suhara and H. Nishihara, "Possibility of super-resolution in integrated-optic disk pickup", Paper 27D-16, presented at the International Symposium on Optical Memory (ISOM), Kobe, Japan, September 1989.
4. S. Nakamura, T. Maeda and Y. Tsunoda, "Autofocusing effect due to wavelength change of diode lasers in an optical pickup", Applied Optics **26**, pp2549-2553 (1987).
5. N. Fukushima, K. Miura and I. Sawaki, "Detection of magneto-optic signals using an external cavity laser diode", SPIE Vol. **1078**, Proceedings of the 1989 Optical Data Storage Meeting in Los Angeles, California, pp90-93.
6. K. Matsumoto and T. Maeda, "Acousto-optic accessing in optical disks", Paper 28D-4, presented at the International Symposium on Optical Memory (ISOM), Kobe, Japan, September 1989.
7. R. Katayama, K. Yoshihara, Y. Yamanaka, M. Tsunekane, K. Yoshida, and K. Kubota, "Multi-beam magneto-optical disk drive for parallel read/write operation". SPIE Vol. **1078**, Proceedings of the 1989 Optical Data Storage Meeting in Los Angeles, California, pp98-104.
8. A. B. Marchant, *Optical Recording*, Addison-Wesley, Reading, Massachusetts, 1990.
9. W. Zeper and F. Greidanus, "Evaporated Co/Pt layered structures for magneto-optical recording", IEEE Trans. Magnet. 25, pp3764-3766 (1989).
10. P. Kivits, R. deBont, and P. Zalm, "Superheating of thin films for optical recording", Appl. Phys. **24**, pp273-278 (1981).
11. M. Mansuripur and G. A. N. Connell, "Laser-induced local heating of moving multilayer media", Appl. Optics 22, pp666-670 (1983).
12. S. C. Shin, "Thermal analysis of magneto-optical thin films under laser irradiation", J. Mag. Mag. Mater. **61**, pp301-306 (1986).
13. M. Bhattacharyya and Z. Cendes, "Finite element modeling of laser beam heating of magnetic films", J. Appl. Phys. **57**, pp3894-3896 (1985).
14. M. Chen and K. Rubin, "Progress of erasable phase-change materials", SPIE Vol. **1078**, Optical Data Storage Topical Meeting, pp150-156 (1989).
15. J. Halter and N. Iwamoto, "Thermal-mechanical modeling of reversible dye-polymer media", Proceedings of SPIE Vol. **899**, Optical Storage Technology and Applications, 1988.
16. K. Aratani, M. Kaneko, Y. Mutoh, K. Watanabe and H. Makino, "Overwriting on a magneto-optical disk with magnetic triple layers by means of the light intensity modulation method", SPIE Vol. **1078**, Proceedings of the 1989 Optical Data Storage Meeting in Los Angeles, California, pp258-264.
17. B. G. Huth, "Calculations of stable domain radii produced by thermomagnetic writing", IBM J. Res. Dev. **18**, pp100-109 (1974).
18. M. Mansuripur and G. A. N. Connell, "Energetics of domain formation in thermomagnetic recording", J. Appl. Phys. **55**, pp3049-3055 (1984).
19. P. Hansen, "Direct overwrite in amorphous rare earth-transition metal alloys", Appl. Phys. Lett. **50**, pp356-358 (1987).
20. M. Mansuripur and M. F. Ruane, "Mean field analysis of amorphous rare earth-transition metal alloys for thermomagnetic recording", IEEE Trans. Magnet. 22, pp33-43 (1986).
21. J. C. Suits, D. Rugar and C. J. Lin, "Thermomagnetic writing in TbFe : modeling and comparison with experiment", J. Appl. Phys. **64**, 252 (1988).
22. M. Mansuripur, "Magnetization reversal, coercivity, and the process of thermomagnetic recording in thin films of amorphous rare earth-transition metal alloys", J. Appl. Phys. **61**, pp1580-1587 (1987).

23. T. W. McDaniel and M. Mansuripur, "Numerical simulation of thermomagnetic writing in RE-TM films", IEEE Trans. Magnet. 23, pp2943-2945 (1987).
24. T. W. McDaniel, "Simulation of bit jitter in magneto-optic recording", J. Appl. Phys. **63**, pp3859-3861 (1988).
25. M. Mansuripur and R. Giles, "Simulation of the magnetization reversal dynamics on the Connection Machine", to appear in *Computers in Physics*, 1990.
26. M. Mansuripur, "Magnetization reversal dynamics in the media of magneto-optical recording", J. Appl. Phys. **63**, pp5809-5823 (1988).
27. M. Mansuripur and R. Giles, "Demagnetizing field computation for dynamic simulation of the magnetization reversal process", IEEE Trans. Magnet. **24**, pp2326-2328 (1988).
28. P. Sheng, "Theoretical considerations of optical diffraction for RCA video disk signals", RCA Review **39**, pp512-555 (1978).
29. H. H. Hopkins, "Diffraction theory of laser readout systems for optical video disks", J. Opt. Soc. Am. **69**, pp4-24 (1979).
30. G. Bouwhuis, J. Braat, A. Huijser, J. Pasman, G. van Rosmalen, and K. S. Immink, **Principles of Optical Disk Systems**, Adam Hilger Publishing Co., Bristol, 1985.
31. M. Mansuripur, "Analysis of astigmatic focusing and push-pull tracking error signals in magneto-optical disk systems", Appl. Optics **26**, pp3981-3986 (1987).
32. M. Mansuripur, "Detecting transition regions in magneto-optic disk systems", Appl. Phys. Lett. **55**, pp716-717 (1989).
33. J. S. Hartman, M. Lind and M. Mansuripur, "Read channel optical modeling for a bump-forming dye-polymer optical data storage medium", SPIE Vol. **1078**, Proceedings of the 1989 Optical Data Storage Meeting in Los Angeles, California, pp308-323.
34. M. Mansuripur, "Certain computational aspects of vector diffraction problems", J. Opt. Soc. Am. A **6**, pp786-805 (1989).
35. J. G. Dil and B. A. J. Jacobs, "Apparent size of reflecting polygonal obstacles of the order of one wavelength", J. Opt. Soc. Am. **69**, pp950-960 (1979).
36. M. Mansuripur, G. A. N. Connell and J. W. Goodman, "Signal and noise in magneto-optical readout", J. Appl. Phys. **53**, pp4485-4494 (1982).
37. A. G. Dewey, "Optimizing the noise performance of a magneto-optic read channel", SPIE Vol.1078, Proceedings of the 1989 Optical Data Storage Meeting in Los Angeles, California, pp279-286.
38. G. E. Acket, D. Lenstra, A. J. Den Boef and B. Verbeek, "Influence of feedback intensity on longitudinal mode properties and optical noise in index-guided semiconductor lasers", IEEE J. Quant. Elec. **20**, pp1163-1169 (1984).
39. K. Balasubramanian, A. Marathay, and H. A. Macleod, "Modeling magneto-optical thin film media for optical data storage", Thin Solid Films **164**, pp391-403 (1988).
40. M. Mansuripur, "Analysis of multilayer thin film structures containing magneto-optic and anisotropic media at oblique incidence using 2×2 matrices", to appear in J. Appl. Phys. (1990).

Section Two
Novel Detection Techniques

Reprinted with permission from *Applied Physics Letters*, Vol. 55(8), pp. 716-717 (August 21, 1989).

Detecting transition regions in magneto-optical disk systems

Masud Mansuripur
Optical Sciences Center, University of Arizona, Tucson, Arizona 85721

(Received 17 April 1989; accepted for publication 12 June 1989)

Several combinations of incident polarization and detection schemes suitable for direct detection of transitions in magneto-optical disk systems are described. One such combination uses a circularly polarized light with a single split detector in the far field. Another scheme uses linear polarization in conjunction with differential detection. In the first scheme the medium must be optimized for Kerr rotation while the second method requires maximum ellipticity.

The standard differential detection scheme for magneto-optical disk storage systems utilizes the polar Kerr effect.[1-3] In this scheme a linearly polarized beam of light is reflected from a perpendicularly magnetized medium where, upon reflection, the state of polarization becomes elliptical and the major axis of the ellipse rotates away from the incident direction of polarization. The senses of both rotation and ellipticity depend on the state of magnetization and constitute the magneto-optical signal. The differential detector converts this signal into a waveform which tends to be a good reproduction of the pattern of magnetization of the disk.

The purpose of this letter is to introduce a new readout scheme for the magneto-optical disk systems with certain potential advantages. One implementation of the new scheme, depicted in Fig. 1, utilizes circularly polarized incident light and relies on diffraction from domain boundaries in order to detect transitions of the magnetization state. One advantage of the proposed method is that, as in compact disk and write-once systems, the reflected beam can be directed away from the laser. In this way the laser feedback noise can be substantially reduced. Other advantages include compatibility with read-only and write-once pickup systems, total utilization of the laser power during recording and erasure, simpler optics, and direct detection of the transition region between oppositely magnetized neighboring domains (i.e., no need for differentiation of the readout signal). A disadvantage of the new technique is its smaller signal (by about 6.5 dB) as compared with differential detection. However, the advantages outlined above may outweigh this disadvantage in practice. In the remainder of this letter we derive the magnitude of the readout signal for the proposed scheme.

1. Preliminaries. A magnetic medium with perpendicular magnetization reflects right circularly polarized (RCP) light differently from left circularly polarized (LCP) light. If r_1 and r_2 (complex numbers) are the corresponding reflectivities in the "up" magnetized state of the medium, their roles will be reversed in the "down" magnetized state. We define r^+, r^-, ρ, and ϕ as follows:

$$r^+ = \tfrac{1}{2}(r_1 + r_2), \tag{1a}$$

$$r^- = \tfrac{1}{2}(r_1 - r_2), \tag{1b}$$

$$r^-/r^+ = \rho e^{i\phi}. \tag{1c}$$

The incident beam may be polarized in a variety of ways, but any state of polarization can be represented as a mixture of RCP and LCP, namely,

$$A_0 = a(\hat{x} + i\hat{y}) + b(\hat{x} - i\hat{y}). \tag{2}$$

Here a and b are complex numbers that represent the amplitudes of the two circularly polarized components and $\hat{x}$ and $\hat{y}$ are unit vectors in the Cartesian coordinate system. The incident beam propagates along $\hat{z}$.

When the incident beam is linearly polarized along the x axis, a and b must be equal. The reflected amplitude from a uniformly magnetized region will then be

$$\hat{A}_0 = a(r_1 + r_2)\hat{x} + ia(r_1 - r_2)\hat{y}. \tag{3a}$$

With reference to Eq. (1), Eq. (3a) is rewritten as

$$\hat{A}_0 = 2ar^+(\hat{x} + \rho e^{i(\phi + 90)}\hat{y}). \tag{3b}$$

Thus $\phi = \pm 90°$ corresponds to a medium which exhibits pure Kerr rotation, whereas $\phi = 0°$ or $180°$ represents a medium with pure ellipticity. In practice, ϕ can be adjusted by either multilayering or external phase compensation. The *intrinsic* Kerr angle θ_k is related to ρ as follows:

$$\rho = \tan \theta_k. \tag{4}$$

Finally, the intensity of the beam with amplitude A_0 as given by Eq. (2) is

$$I_0 = \tfrac{1}{2}(|A_x|^2 + |A_y|^2) = |a|^2 + |b|^2. \tag{5}$$

2. Reflection of a focused beam from the transition region between two domains. For simplicity we restrict this analysis to a one-dimensional case. Let the medium be up magnetized in the region $x > 0$ and down magnetized in the region $x < 0$. The transition is thus along the y axis. For circularly polarized incident light, the amplitude reflectivity of the medium will be

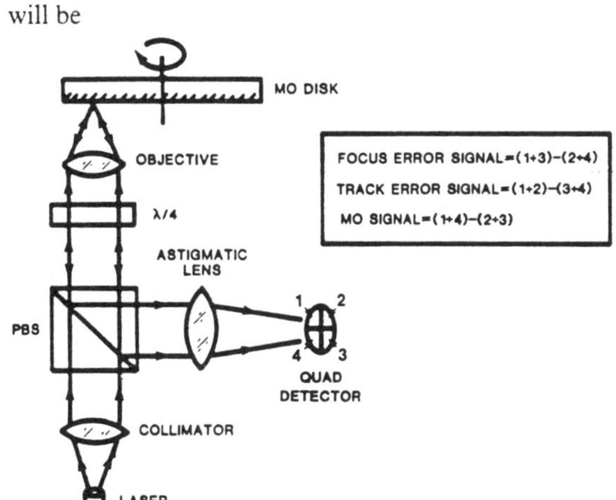

FIG. 1. Schematic diagram of the readout system utilizing circularly polarized light. In principle the servo signals for focus and track error can be obtained from the same quad detector used for detecting the magneto-optic signal.

$$r(x) = r^+ \pm r^- \operatorname{sgn}(x), \tag{6}$$

where the upper sign corresponds to RCP and the lower sign to LCP. The function sgn(x) is $+1$ for $x>0$ and -1 for $x<0$.

The focusing lens has focal length f and numerical aperture NA. In this one-dimensional analysis the lens is cylindrical with the cylinder axis parallel to the y axis. Let the distribution at the entrance pupil be circularly polarized with amplitude

$$A_1(x) = \frac{1}{\sqrt{2fNA}} \operatorname{Rect}\left(\frac{x}{2fNA}\right), \tag{7}$$

where Rect(x) is the rectangular function of x which is equal to 1 when $|x| < 1/2$ and 0 otherwise. The amplitude distribution at the focal plane is

$$A_2(x) = \frac{\sqrt{\lambda}}{\sqrt{2NA}} \left(\frac{\sin(2\pi NAx/\lambda)}{\pi x}\right). \tag{8}$$

The reflected distribution is the product of Eqs. (6) and (8). The Fourier transforms of $r(x)$ and $A_2(x)$ (after scaling by λ) are

$$\hat{r}(s) = r^+ \delta(s) \mp ir^-/\pi s, \tag{9}$$

$$\hat{A}_2(s) = \frac{1}{\sqrt{2NA\lambda}} \operatorname{Rect}\left(\frac{s}{2NA}\right). \tag{10}$$

Here s is the independent variable of the Fourier domain and $\delta(s)$ is Dirac's delta function. Convolving Eq. (9) with Eq. (10) yields the Fourier transform of the reflected distribution as

$$\mathscr{F}[r(\lambda x)A_2(\lambda x)] = \frac{1}{\sqrt{2NA\lambda}} \left[r^+ \operatorname{Rect}\left(\frac{s}{2NA}\right) \pm \frac{ir^-}{\pi} \ln\left|\frac{NA - s}{NA + s}\right| \right]. \tag{11}$$

With proper scaling and normalization the above equation represents the far-field pattern of the reflected beam. Noting that the diffracted rays beyond the lens aperture are blocked, we obtain the final distribution as follows:

$$A(x) = \frac{1}{\sqrt{2fNA}} \left(r^+ \pm \frac{ir^-}{\pi} \ln\left|\frac{fNA - x}{fNA + x}\right|\right) \operatorname{Rect}\left(\frac{x}{2fNA}\right). \tag{12}$$

The upper and lower signs correspond to RCP and LCP incident light, respectively. When the incident polarization is given by Eq. (1), the reflected beam must be resolved into its x and y components. In the far field we have

$$A_x(x) = \frac{1}{\sqrt{2fNA}} \left((a+b)r^+ + \frac{i}{\pi}(a-b)r^- \ln\left|\frac{fNA - x}{fNA + x}\right|\right) \times \operatorname{Rect}\left(\frac{x}{2fNA}\right) \tag{13a}$$

$$A_y(x) = \frac{i}{\sqrt{2fNA}} \left((a-b)r^+ + \frac{i}{\pi}(a+b)r^- \ln\left|\frac{fNA - x}{fNA + x}\right|\right) \times \operatorname{Rect}\left(\frac{x-}{2fNA}\right). \tag{13b}$$

The total reflected power is

$$\begin{aligned} P_r &= \int_{-\infty}^{\infty} \frac{1}{2}(|A_x|^2 + |A_y|^2)dx \\ &= (|a|^2 + |b|^2)\left[|r^+|^2 + \frac{|r^-|^2}{\pi^2} \int_0^1 \ln^2\left(\frac{1-x}{1+x}\right)dx\right] \\ &= I_0(|r^+|^2 + \tfrac{1}{3}|r^-|^2). \end{aligned} \tag{14}$$

3. The case of circularly polarized incident beam. Here $a = 1$ and $b = 0$. The readout system is identical to the read channel for the compact disk as shown in Fig. 1. The signal is the difference between the two detector halves and is obtained from Eq. (12) as follows:

$$\begin{aligned} \Delta S &= \int_{-\infty}^{0} |A(x)|^2\, dx - \int_0^{\infty} |A(x)|^2 dx \\ &= \frac{4 \ln 2}{\pi}(|r^+|^2 \tan\theta_k)\sin\phi. \end{aligned} \tag{15}$$

The signal is maximized for $\phi = \pm 90°$, i.e., the medium must be designed for maximum Kerr rotation and no ellipticity.

4. The case of the linearly polarized incident beam. Here $a = b = 1/\sqrt{2}$. The readout system uses standard differential detection, but each detector is split in two halves and the read signal is the difference between the two halves of each detector. The signals obtained from the individual arms of the system are identical and can be added in order to double the signal level. The polarizing beamsplitter in the differential detection scheme effectively superimposes the x and y components of the reflected beam and the detectors sense the sum (or the difference) of the fields in Eqs. (13a) and (13b). The signal is thus given by

$$\begin{aligned} \Delta S &= \int_{-\infty}^{0} \frac{1}{2}|A_x(x) + A_y(x)|^2\, dx \\ &\quad - \int_0^{\infty} \frac{1}{2}|A_x(x) + A_y(x)|^2\, dx \\ &= \frac{4 \ln 2}{\pi}(|r^+|^2 \tan\theta_k)\cos\phi. \end{aligned} \tag{16}$$

The maximum signal is obtained for $\phi = 0$, i.e., when the medium is optimized for maximum ellipticity and no Kerr rotation.

There are other ways to mix RCP and LCP in the incident beam in order to detect the transition regions. It seems, however, that all these techniques, when optimized, yield the same signal. The selection of one system over the other is, therefore, dictated by practical constraints of implementation, cost, and complexity. Finally, let us mention that the peak-to-peak signal of the standard differential detection is $8|r^+|^2 \tan\theta_k$. This result does not include the losses incurred at the beamsplitter where the incident and reflected beams are separated.

[1]R. L. Aagard, IEEE Trans. Magn. **MAG-9**, 705 (1973).

[2]M. H. Kryder, J. Appl. Phys. **57**, 3913 (1985).

[3]M. Mansuripur, G. A. N. Connell, and J. W. Goodman, J. Appl. Phys. **53**, 4485 (1982).

Reprinted with permission from *Japanese Journal of Applied Physics,* Vol. 28 (Suppl. 28-3), pp. 103-108 (1989).

Optics of Recording and Read-out in Optical Disk Systems

Joseph BRAAT

Philips Research Laboratories, P.O. Box 80000 5600 JA Eindhoven, The Netherlands

(Received September 4, 1989)

Successful read-out of digital optical disks largely depends on the system tolerances, *e.g.*, with respect to defocusing. Different read-out schemes (integral detection and confocal detection) are treated and the corresponding tolerance range is discussed.

§1. Introduction

The read-out of optical discs[1-3] is currently done by a scanning system which is shown schematically in Fig. 1(a) (the transmissive mode has been chosen for reasons of clarity). The information track on the optical disc is moved with respect to the read-out objective and the flux transmitted (reflected) by the disc is partially captured by a detector D. This system is identical with what is generally called nowadays a Type I scanning microscope.[4] The special way in which the information is arranged on an optical disc was exploited by Hopkins[5] to evaluate expressions for the harmonic components of the detected signal as a function of the relative position of the scanning spot in the radial direction. With some extensions of this model, it is possible to calculate cross-talk between tracks[6] and intermodulation effects and to estimate the magnitude of intersymbol interference in a digitally recorded signal.[7]

In Fig. 1(b), the Type II scanning microscope (or confocal microscope) is shown, which was first described by Minsky.[8] The light flux transmitted by the object is now collected by a second lens and focussed onto a very small detector (preferably much smaller than the extent of the ideal diffraction image of a point source by the collector lens). For a reflective optical disc system, this scanning method was proposed in conjunction with a diode laser. The exit window of the laser cavity plays the role of the small size detector and the current or laser power variations induced by feedback effects yield a kind of confocal signal.[9]

In the second section of this paper we describe the type of disc pattern we have adopted for the calculation of eye-patterns of digital signals and the corresponding eye jitter. In the third section we establish the analogy between the expressions for the harmonic components of the read-out signal in the Type I and Type II optical disc system. In the fourth section we give the results of calculations and measurements of eye jitter for the standard and for the confocal read-out method. The influence of defocusing on the eye pattern is evaluated. A comparison is made between the extended model used in this article and a linearized read-out model which has been treated by Kubota.[10]

§2. The Reflection Pattern of a Digital Disc

The configuration of the digital optical disc is shown in Fig. 2. A sequence of pits with varying lengths and varying intermediate distances is arranged along a track. This sequence is repeated every distance p where p is large (typically 10 to 15 μm) with respect to the extent of the scanning spot (typically 10 to 15 μm). The distance between tracks is q. Adjacent tracks are shifted forward and backward over a distance s which can be an arbitrary portion of the period p.

For the calculation of the detector signal one needs to evaluate the (periodic) reflection coefficient of the disc. In accordance with the notation in ref. 5, the reflection function of the reference (or central) track is given by

$$R_0(u, v)=\sum_{m=-\infty}^{+\infty} R_{00}(u-mp, v), \tag{1}$$

where the function R_{00} is only different from zero within the elementary centered cell with dimensions $p\times q$ of the periodic disc.

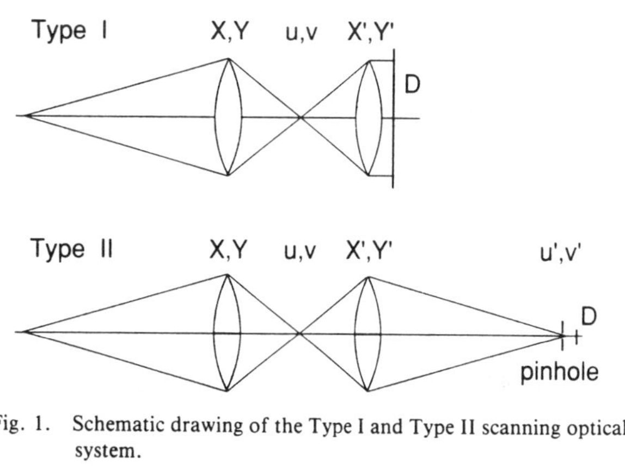

Fig. 1. Schematic drawing of the Type I and Type II scanning optical system.

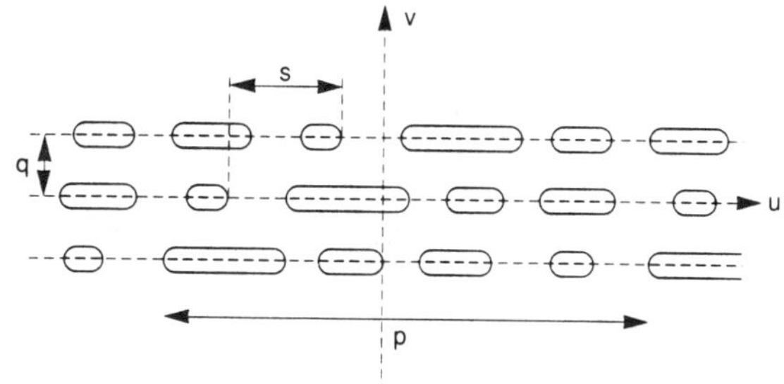

Fig. 2. The layout of the disk pattern. The enlarged period in the tangential direction is p, the radial period is q. Neighbouring tracks are shifted over a distance s.

The reflection function of the n-th track is given by

$$R_n(u, v)=\sum_{m=-\infty}^{+\infty} R_{00}(u-mp-ns, v-nq) \tag{2}$$

and the reflection of the entire disc is now represented by

$$R(u, v)=\sum_{m=-\infty}^{+\infty}\sum_{n=-\infty}^{+\infty} R_{00}(u-mp-ns, v-nq). \tag{3}$$

The reflection function of the elementary cell is written as a summation over basic cells corresponding to the possible pit lengths on a digital disc. When there are k different kinds of pits, the elementary reflection function is written

$$R_{00}(u, v)=l+\sum_{k=1}^{k_0}\sum_{l=1}^{l_0(k)}[R_k(u-d_{l,k}, v)-1], \tag{4}$$

where the index k denotes the type of pit, $l_0(k)$ the number of pits of type k and $d_{l,k}$ the shift of the center of a pit with respect to the center of the elementary cell. The function between brackets in eq. (4) is only different from zero at the location of a pit of type k.

The distances $d_{l,k}$ and the type of pits should be chosen such that the modulation scheme of the encoded digital signal is respected. Once the reflection functions $R_k(u, v)$ and their Fourier transforms have been established, the reflection function of the entire disc and its Fourier transform are easily calculated. The Fourier transform of an arbitrary sequence is obtained by applying the shift theorem to the Fourier transforms of the individual functions $R_k(u, v)$.

§3. The Harmonic Components of the Detector Signal

For a full treatment of the calculation of the harmonic components of the tangential signal we refer to Hopkins.[5] The complete expressions for the harmonic components in the case of a disc pattern with neighboring tracks shifted over a distance s turn out to be given by

$$A_0=P_1(0, 0)+2\sum_{\rho=1}^{\rho_{\max}}\left\{P_1(0, \rho)\cos\left(2\pi\frac{\rho}{q}v\right)+P_2(0, \rho)\sin\left(2\pi\frac{\rho}{q}v\right)\right\} \tag{5}$$

$$\begin{aligned}A_\mu=&2\left\{P_1(\mu, 0)\cos\left(2\pi\frac{s}{p}\frac{v}{q}\mu\right)+P_2(\mu, 0)\sin\left(2\pi\frac{s}{p}\frac{v}{q}\mu\right)\right\}+2\sum_{\rho=1}^{\rho_{\max}}\left\{\left[\{P_1(\mu, \rho)+Q_1(\mu, \rho)\}\cos\left(2\pi\frac{s}{p}\frac{v}{q}\mu\right)\right.\right.\\&\left.+\{P_2(\mu, \rho)+Q_2(\mu, \rho)\}\sin\left(2\pi\frac{s}{p}\frac{v}{q}\mu\right)\right]\cos\left(2\pi\frac{\rho}{q}v\right)+\left[\{P_2(\mu, \rho)-Q_2(\mu, \rho)\}\cos\left(2\pi\frac{s}{p}\frac{v}{q}\mu\right)\right.\\&\left.\left.+\{-P_1(\mu, \rho)+Q_1(\mu, \rho)\}\sin\left(2\pi\frac{s}{p}\frac{v}{q}\mu\right)\right]\sin\left(2\pi\frac{\rho}{q}v\right)\right\}\end{aligned} \tag{6}$$

$$\begin{aligned}B_\mu=&2\left\{-P_2(\mu, 0)\cos\left(2\pi\frac{s}{p}\frac{v}{q}\mu\right)+P_1(\mu, 0)\sin\left(2\pi\frac{s}{p}\frac{v}{q}\mu\right)\right\}+2\sum_{\rho=1}^{\rho_{\max}}\left\{\left[\{-P_2(\mu, \rho)-Q_2(\mu, \rho)\}\cos\left(2\pi\frac{s}{p}\frac{v}{q}\mu\right)\right.\right.\\&\left.+\{P_1(\mu, \rho)+Q_1(\mu, \rho)\}\sin\left(2\pi\frac{s}{p}\frac{v}{q}\mu\right)\right]\cos\left(2\pi\frac{\rho}{q}v\right)+\left[\{P_1(\mu, \rho)-Q_1(\mu, \rho)\}\cos\left(2\pi\frac{s}{p}\frac{v}{q}\mu\right)\right.\\&\left.\left.+\{P_2(\mu, \rho)-Q_2(\mu, \rho)\}\sin\left(2\pi\frac{s}{p}\frac{v}{q}\mu\right)\right]\sin\left(2\pi\frac{\rho}{q}v\right)\right\}.\end{aligned} \tag{7}$$

Here A_0 is the DC-component of the tangential signal and A_μ and B_μ are the amplitude of the cosine and sine components of the μ-th harmonic of the signal. The radial excursion off-track is denoted by v. The $P(\mu, \rho)$ and $Q(\mu, \rho)$-coefficients carry all possible contributions to the μ-th harmonic of the tangential signal and the ρ-th harmonic of the radial signal. Each contribution $\Delta P(\mu, \rho; m, n)$ consists of the product of the complex amplitudes of two diffraction orders R of the disc pattern and the corresponding cross-correlation coefficient C according to

$$\begin{aligned}\Delta P(\mu, \rho; m, n)=&R(\mu+m, \rho+n)R^\times(m, n)\\&\times C(\mu+m, \rho+n; m, n).\end{aligned} \tag{8}$$

The contributions to the Q-coefficients are obtained from

$$\begin{aligned}\Delta Q(\mu, \rho; m, n)=&R(\mu+m, n)R^\times(m, \rho+n)\\&\times C(\mu+m, n; m, \rho+n).\end{aligned} \tag{8a}$$

The indices 1 and 2 in the expressions (5)–(7) denote the real and the imaginary part of the P- and Q-coefficients.

In the case of the Type I read-out, the value of C is obtained numerically from the evaluation of the autocorrelation function of the pupil function of the scanning lens over a limited area (three circle overlap).

In the case of Type II read-out, the cross-correlation coefficient is written

$$C(\mu+m, \rho+n; m, n)=T(\mu+m, \rho+n)T^\times(m, n) \tag{9}$$

where $T(m, n)$ is represented by the expression

$$T(m, n)=\iint f_1(X, Y)f_2(-X+m/p, -Y+n/q)\,\mathrm{d}X\,\mathrm{d}Y \tag{10}$$

and where $f_1(X, Y)$ and $f_2(X, Y)$ are the complex pupil functions of the scanning and the collector lens. The modulus of these pupil functions represents the amplitude transmission of the lens and the phase part accounts for the wavefront aberration of the imaging pencils.

The numerical calculation of the harmonic components of the detector signal becomes time consuming when the fundamental periods p and q are large with respect to the size of the scanning spot. However, the cross-correlation coefficients $C(\mu+m, p+n; m, n)$ are calculated once; different digital sequences are obtained by introducing the modified values of the diffraction orders R in the expressions for the P- and Q-coefficients. In this way the amount of numerical work is drastically reduced. Note that in the general case of a disc with tracks shifted over a distance s, the value of n in each combination (m, n) should be replaced by $n-sm/p$.

§4. Results of Calculations and Experiments

In this section we present some results of calculations and experiments with a Type I optical system and calculated data of a Type II system. The disc format is consistent with the Compact Disc digital audio standard. The quantity of interest is the root mean square value of the signal jitter and the relative height of the digital eye pattern. These values are numerically derived from a certain number of digital sequences (typically 100).

Parameters which have been kept constant are

λ	0.79 μm
depth of the pits	0.12 μm
width of the pits	0.60 μm
pit wall angle	40°
track spacing	1.50 μm
scanning speed	1.25 m/s.

A nonuniform amplitude distribution in the entrance pupil of the scanning lens is taken in accordance with the far field pattern of a diode laser. The rim intensity in the tangential direction equals 0.75 and in the radial direction 0.15 with respect to the central intensity.

Parameters which have been varied are the detector geometry in the Type I system, the numerical aperture NA of the scanning and the collector lens and the focus setting of the lenses. In the Type II mode we have to discriminate between the reflective system where the defocussing of the scanning and the collector lens is identical while in a transmissive system the defocussing terms have opposite sign when the object is moved axially.

4.1 *Type I system: influence of detector geometry*

We have both calculated and measured the rms eye jitter Δ and the relative eye height I_e for the standard read-out system. The rms eye jitter is defined as

$$\Delta^2=\frac{1}{N}\left\{\sum_{j=1}^{N}[x_j^d-x_j^0]^2-\left\{\sum_{j=1}^{N}[x_j^d-x_j^0]\right\}^2\right\} \tag{11}$$

where x_j^d is the tangential position where the detected signal equals a chosen decision level. By varying the decision level one can monitor the change in Δ and establish the range I_e in decision levels which still yields zero error detection. An error is revealed when a detected position x_j^d is shifted over more than half the eye width w (231 ns or 0.29 μm) with respect to the desired position. When the scanning spot has circular symmetry, the desired positions are given by x_j^0; when the scanning shows asymmetry, the desired positions are corrected by an average shift which is given by the second term in eq. (11).

In order to increase the eye height I_e one can try to improve the signal transfer for the most sensitive sequences

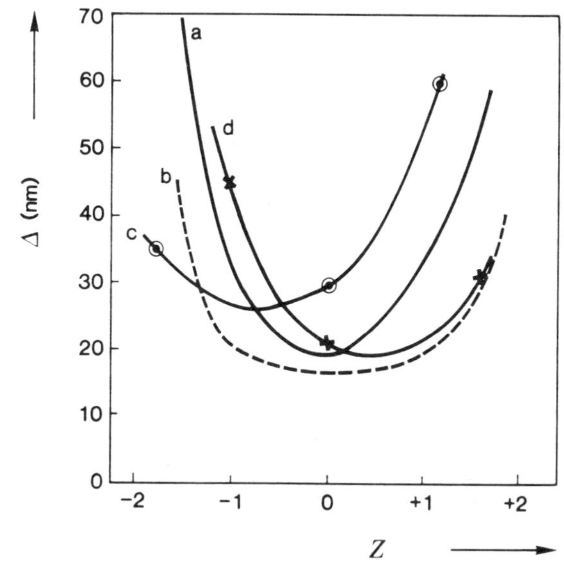

Fig. 3. Measured rms jitter values Δ in nm as a function of the defocusing z in the case of a Type I system. The defocusing is expressed in units of the focal depth and equals 1.75 μm ($NA=0.46$, $\lambda=0.79$ μm). Curve a) applies to the standard optical detection while curve b) has been obtained by adding an electronic compensation filter. Curves c) and d) are obtained by taking respectively a ring detector ($0.50\leq\gamma\leq0.85$) and a reduced area detector ($\gamma\leq0.50$).

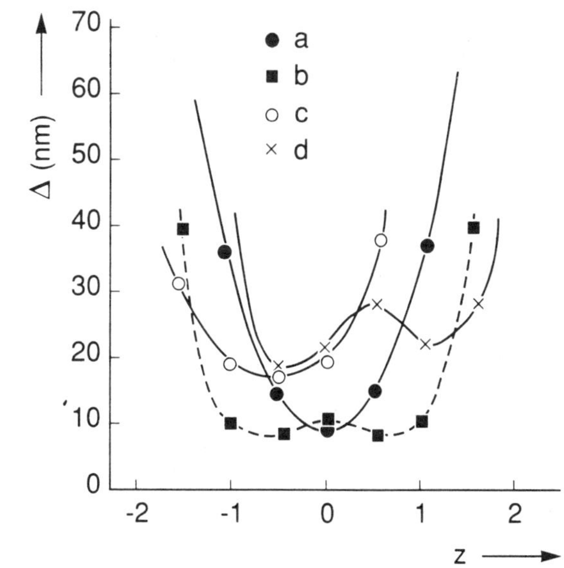

Fig. 4. Calculated rms jitter values Δ (the same legend as Fig. 3).

with the shortest pits and lands. Such a sequence approximately has a spatial frequency of NA/λ when the standard NA value of 0.46 is chosen. An obvious improvement is obtained by adding an electronic high-pass filter.[7] In Fig. 3 we show the measured values of Δ as a function of the defocusing z. This quantity is expressed in units of the focal depth (1.75 μm) of the standard system ($\lambda=0.79$ μm and $NA=0.46$). Curve a) represents the variation of Δ for the standard read-out system, while the extension of the detection circuit with a simple electronic filter yields curve b). The permitted value of Δ which still yields error-free read-out is approximately 45 nm (36 nsec at the scanning speed of 1.25 m/s). At this level the focal range for correct read-out has been extended by 30% for curve b) and approaches the value of 6 μm. The lowest measured values of Δ are 18 to 20 nm.

In Fig. 4 we show calculated values of Δ in a similar situation; curve a) gives the results for standard read-out, while in curve b) the influence of the electronic filter has been included. The difference in focal range between curve a) and b) is comparable in Figs. 3 and 4. The main difference stems from the minimum value of Δ which is 9 nm according to the calculations. The intrinsic jitter of the disc pattern as introduced in the mastering, copying and moulding operations and the noise at read-out are responsible for this discrepancy; the importance of the different contributions is unknown at this moment. Note the small increase in jitter in the optimum focus position for curve b of Fig. 4. This is due to the overshoot effects which are introduced by the filter on scanning a pit edge. A slight defocusing is sufficient to cancel this effect. In the real situation (Fig. 3), the steepness of the response is weakened by residual aberrations of the scanning spot and the intrinsic optical jitter is drowned by the extra sources of jitter mentioned above.

Another means to improve the transfer at the middle frequencies is a more coherent detection; this is easily done in a Type I microscope by reducing the extent of the detector. The well-known influence on the modulation transfer function MTF is shown in Fig. 5. Curve a) represents the standard response of an incoherent system (*NA* of the detector $\geq NA$ of the objective) while curve b) has been obtained by reducing the detector size to half of the *NA* of the objective (coherence ratio $\gamma=0.50$). When using the outer side of the detector ($0.50 \leq \gamma \leq 0.85$) one obtains curve c with an increased response close to the cut-off frequency $2NA/\lambda$.

The measured effects are shown in Fig. 3; curve c) corresponds to a ring detector ($0.50 \leq \gamma \leq 0.85$) while curve d) is obtained with the coherent detection scheme according to $\gamma \leq 0.50$. A slight increase in focal range is observed for both curves in the sense that their minima are broader. However, the optimum focal planes have drifted away from the plane corresponding to the best diffraction focus over a distance of at least half a focal depth. The same phenomenon is observed in Fig. 4 where the detector modifications have been included in the calculations. The focus drift is due to the presence of a phase pattern on the disc with an optical depth such that the phase between zeroth and *e.g.* first diffracted orders is

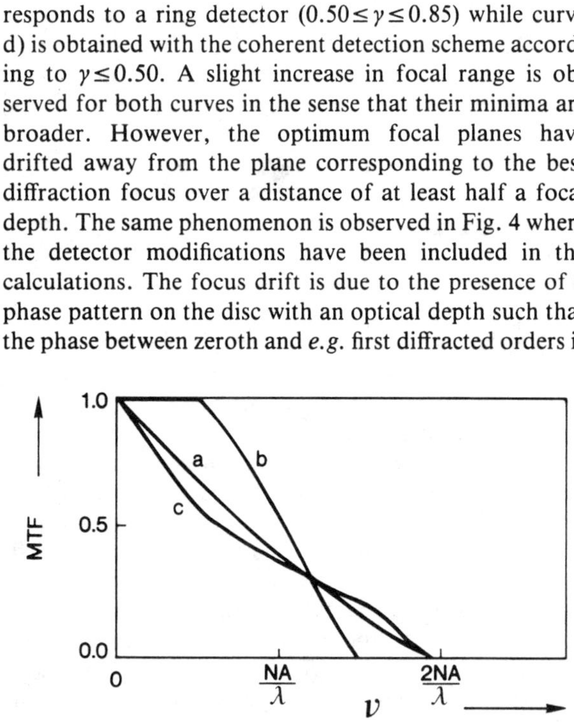

Fig. 5. The modulation transfer function *MTF* of a Type I system in the case of standard detection (curve a), partially coherent detection ($\gamma \leq 0.50$) and detection with a ring-shaped detector ($0.50 \leq \gamma \leq 0.85$).

not 0 (or π) but approximately $3\pi/4$ or $(-\pi/4)$. By a defocusing of the disc, the phase difference between undiffracted and diffracted light is changed. When considering a disc with a single frequency pattern, the phase difference between zeroth and first diffracted orders is not uniform over the entire detector.[11] By a proper defocusing, the phase difference can be adjusted close to 0 or π over a restricted detection area. The interferences between positive and negative diffraction orders then are in phase and the modulation depth of the signal originating from this restricted detector area shows a maximum depth of modulation despite the slight defocusing. When we try to maximize the most critical frequencies of the CD signal (close to NA/λ), we should introduce a defocusing of plus or minus half a focal depth ($\pi/4$) to optimize the middle frequencies in case of a ring detector or a coherent detector.

In Fig. 4, the calculated focal drift of curves c) and d) equals the measured values. However, curve d) shows a jitter increase at its center. This is caused by the overshoot of the edge response in the partially coherent detection mode. In this respect, the combination of the smooth edge response of the incoherent detection together with the electronic steepening (curve b) is a more versatile operation. Optically, one could simulate this by a more gradual decrease of the detector sensitivity towards the rim.

It is possible to combine curves c) and d) by splitting the standard detector into an inner and an outer ring and adding the two signals with proper weights as a function of the focal position. In this way, together with electronic correction, the focal range for error-free detection could still be extended.

4.2 *Type II system*

We have calculated the eye patterns and eye jitter for a Type II scanning system with an infinitely small detector. This is not a realistic option; however, it has been shown that up to a diameter of the detector of $0.4\lambda/NA$ the relative shape of the detected signal is quite comparable to the theoretical point detector response.[12]

A commonly mentioned advantage of a Type II scanning microscope is its axial slicing capability. For an optical disc system, this would mean a loss of signal in the defocused situation and, at a first glance, this should be considered more as a drawback. A second advantage of the Type II system is its sharpened response to an amplitude point object. This is due to the heavily nonlinear detection which takes place in a confocal system and makes the response proportional to the fourth power of the complex amplitude of the scanning spot. For a low level of intersymbol interference at the read-out stage of an optical disc, a large nonlinearity in the detection process could be a disadvantage.

In practice we have to reduce the harmful slicing effect of the confocal system. In order to achieve an acceptable focal range, it is necessary to compensate for the strongly varying signal strength by means of *e.g.* an automatic gain control. The correct centering of the small detector is also essential; in the case of a reflective system with feedback detection (SCOOP), this alignment is automatically obtained.

The calculated rms eye jitter Δ, the average signal level I_{av} and the absolute eye height I_e are shown in Fig. 6 for a confocal system with a correctly positioned point detector. The maximum frequency which is generated by a confocal system is $4NA/\lambda$ which is twice the classical cut-off frequency. This is due to nonlinear effects in the optical detection and should be considered as spurious resolution. The appearance of these frequencies above $2NA/\lambda$ is caused by the fact that certain diffraction orders (*e.g.* the orders +1 and −1) are brought to interference in the detector (image) plane even when they do not overlap in the far field (pupil) region. When the object is scanned, the intensity variations of the Young's fringe pattern associated with this interference are detected by the point detector. When the detector size is increased, this interference pattern is gradually averaged out. In our analysis, we have fully kept these extra nonlinear contributions in a confocal system with an infinitely small point detector.

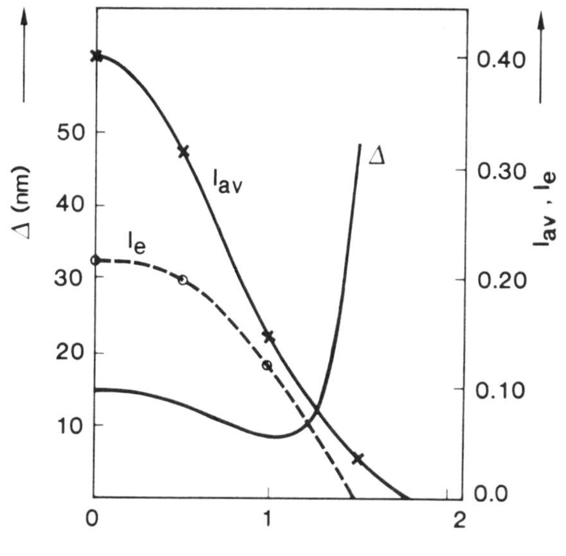

Fig. 6. The calculated rms jitter value Δ, the average signal value I_{av} and the (absolute) eye height I_e as a function of the defocusing z for confocal detection.

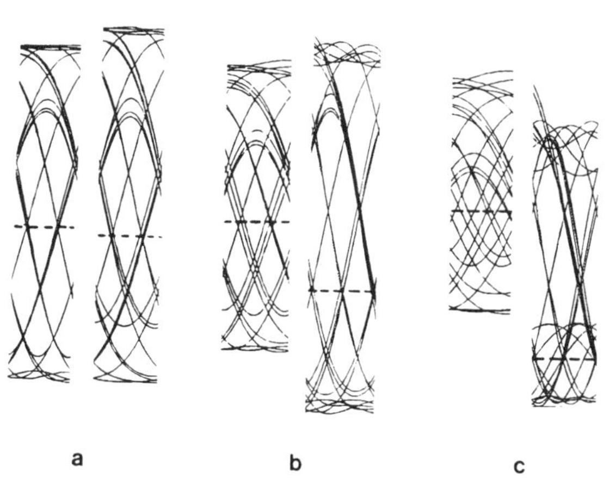

Fig. 7. The appearance of calculated digital eye patterns. The eye patterns sub a) are obtained in the correctly focused situation; the patterns sub b) correspond to a defocusing of one focal depth and those sub c) to one and a half focal depth. Each left eye is obtained by standard detection, the right eyes are the result of confocal detection (after normalization to unity). In each figure, the decision level which yields minimum jitter has been indicated.

The calculations show that the intrinsic jitter at optimum focus is slightly larger than with standard detection. This value decreases when going out of focus and reaches a minimum at a distance of approximately 1.25 focal depth and then increases rapidly at still larger defocusing values. The optimum in a defocused situation is due to the important reduction in average level (frequency zero) while the signal content at *e.g.* the critical middle frequencies is less reduced. This can easily be deduced from the behaviour in a defocused situation of the most important cross correlation coefficients (eq. (9)) which determine the signal. An improvement of the response to an edge of a scanning microscope in the defocused situation was also mentioned by Benshop.[13] At still larger defocusing values, the signal becomes heavily nonlinear because the spurious second harmonic content which is inherent to the confocal detection is only very slightly affected by defocusing. This causes the sharp increase of Δ beyond the value $|z| = 1.25$.

In Fig. 7 we have shown calculated eye patterns in the case of Type I and Type II detection with different amounts of defocusing. The confocal eye patterns have been normalized to unity. The larger focal range in the case of confocal detection is evident from this figure.

Instead of detection in the reflective mode, one can also consider confocal detection in transmission. The resulting jitter values as a function of defocusing are much worse than in reflection. The average level now is not affected. The middle frequencies are reduced like in the Type I read-out (see eq. (9)); however, the nonlinear contributions proper to confocal detection are less reduced and a larger jitter is observed. A defocusing of one focal depth already yields a Δ-value of 40 nm.

We finally have studied the eye jitter variation as a function of the relative spatial density, in this case by varying the numerical aperture of the scanning objective. The tolerance with respect to defocusing and the minimum achievable eye jitter are important features of a disc system. In Fig. 8 we show the calculated eye jitter values as a function of defocusing (the z-unit is one focal depth at a NA value of 0.46 and thus equals 1.75 μm) for the standard detection (a) and the confocal detection (b). From the figures it appears that the intrinsic jitter is higher in the confocal mode, but the focal range with error-free detection tends to be larger. A reduction in NA value of 15 to 20% at equal spatial density is possible in the confocal mode. Figure 8(a) shows that for the standard CD-system the actual NA-value of 0.46 is a good compromise between low intrinsic jitter and an acceptable focal range.

To conclude this section, we mention the difference in results between the full treatment we have adopted here and the linearized model.[10] By suppressing all contributions to ΔP and ΔQ of eq. (8) which do not contain a zero diffraction order, we conform to the linearized read-out model. In the case of the Type I system, the values of

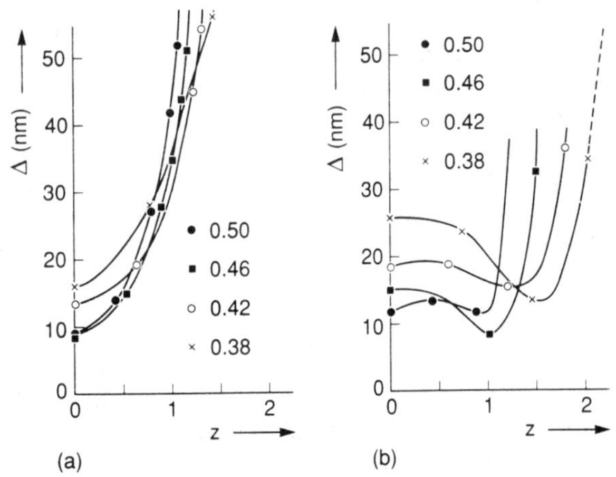

Fig. 8. The variation of the rms jitter Δ as a function of the amount of defocussing z for various values of the numerical aperture of the scanning (and collector) objective. One horizontal unit equals 1.75 μm. Figure a) applies to standard detection and Fig. b) to the confocal detection mode.

Δ tend to be 15 to 20% smaller for the linearized model. The linearization of the confocal read-out was not carried out, because this would suppress an essential part of the read-out process when the object is strong.

§5. Conclusions

We have analyzed the shape of the eye pattern when reading a digital disc. In the standard detection mode, an enlargment of the acceptable defocusing by optical means is difficult to achieve; electronic optimization of the optical transfer is more easily done. The confocal detection mode, although experimentally more difficult to implement, has a potentially larger tolerance range for defocusing and yields lower eye jitter values over this range. The spatial density could be increased by 30% with respect to the standard read-out mode.

Acknowledgement

Fruitful discussions with J. Benschop and C. van Uijen of this laboratory are gratefully acknowledged.

References

1) Ph. Rice *et al.*: "An Experimental Television Recording and Playback System Using Photographic Discs", Journal of the SMPTE **79** (1970) 997.
2) K. Compaan and P. Kramer: "The Philips VLP System", Philips Tech. Rev. **33** (1973) 178.
3) G. Bouwhuis and P. Burgstede: "The Optical Scanning System of the Philips 'VLP' Record Player", Philips Tech. Rev. **33** (1973) 186.
4) T. Wilson, C. Sheppard: "Theory and Practice of Scanning Optical Microscopy" (Academic Press, London) (1984) p. 3.
5) H. H. Hopkins: "Diffraction Theory of Laser Read-Out Systems for Optical Video Disks", Appl. Opt. **22** (1983) 4.
6) J. Braat: "Optically Read Disks with Increased Information Density", Appl. Opt. **22** (1983) 2196.
7) J. Braat: in "Principles of Optical Disc Systems" (Adam Hilger, Bristol/New York), (1985) p. 60.
8) M. Minsky: "Microscopy Apparatus", U.S. Patent 3.013.467 (1961).
9) Y. Mitsuhashi *et al.*: "Self-Coupled Optical Phickup", Opt Comm. **17** (1976) 95.
10) S. Kubota: "Aplanatic Condition Required to Reproduce Jitter-Free Signals in an Optical Digital Disk System", Appl. Opt. **26** (1987) 3961.
11) Reference 7, p. 42.
12) J. Benschop: "Signal Detection and Interpretation in Scanning Optical Microscopy", Thesis, University of Twente (The Netherlands; Dfl 9.95), (1989) p. 30.
13) Reference 12, p. 114.

Reprinted from *Optical Data Storage*, Proc. SPIE Vol. 1316, pp. 35-39 (1990).

High track density magneto-optical recording using a crosstalk canceler

K.Kayanuma* T.Iwanaga* H.Inada*
K.Okanoue* R.Katayama** K.Yoshihara**
Y.Yamanaka** M.Tsunekane** O.Okada***

* C&C Systems Research Laboratories, NEC Corp.
** Opto-Electronics Research Laboratories, NEC Corp.
*** Functional Devices Research Laboratories, NEC Corp.
1-1, Miyazaki 4-chome, Miyamae-ku, Kawasaki 213, JAPAN

ABSTRACT

High track density magneto-optical recording, using a crosstalk canceler which eliminates crosstalk from two adjacent tracks, has been developed. By this recording method, both land area and groove area can be used for data storage tracks, and twice the track density, compared with the conventional method, has been achieved. Applying this crosstalk canceler also realizes intersymbol interference compensation.

1 INTRODUCTION

In recent years, larger capacity and higher speed data transfer rate have been increasingly required of magneto-optical (MO) disk systems. To satisfy these requirements, high density recording has been found to be effective, and several methods, described below, have been proposed. For example, modulation code optimization as pit-edge recording, development of shorter wavelength lasers and spot size reduction by using a superresolution MO head, are well known[1-4]. From the data transfer rate improvement viewpoint, parallel read/write operation by multi-beam MO drive is known[5].

For recording density improvement, not only recording linear density but also track density increasing is important. When the data is recorded under high track density conditions, crosstalk between adjacent tracks usually disturbs the read signal from the MO disk. This paper describes a new method, which reads signals from three adjacent tracks by three-beam MO head, and eliminates any interference component by using a crosstalk canceler, that consists of adaptive digital filters. By applying this method, both land and groove area could be used for data storage track within a sufficiently low error rate. As a result, twice the track density, compared with conventional systems, has been obtained. In Section 2, drive component design and its basic operation are described. In Section 3, operational characteristics, achieved by applying the crosstalk canceler to the MO disk drive, are described.

2 CROSSTALK CANCELER

2.1 Crosstalk canceler overview

Spot size for a readout beam, focused on a disk using the conventional method, couldn't be smaller than the diffraction limit, which is proportional to the laser wavelength and inversely proportional to the objective lens numerical aperture (NA). Thus, as track pitch narrows, the crosstalk signal resulting from other pits, recorded on adjacent tracks, increases. This crosstalk influence has inhibited track density improvement. The readout signal contains crosstalk component and thermal noise. Due to crosstalk influence, utilization for recording information from the adjacent tracks has became important for obtaining correct readout data. The crosstalk canceler design is shown in Fig. 1. By using a three-beam MO head, sub-signals from two adjacent tracks can be read out simultaneously with the main signal, which is read out from the objective track. An adaptive digital filter eliminates crosstalk in the main signal, by generating a crosstalk component replica, having an opposite polarity, of the crosstalk component from sub-signals and adding them. Thus, a crosstalk canceled main signal is obtained as an adaptive digital filter output.

2.2 Recording media

High track density recording is implemented by recording data on both the land and groove areas. Both of these areas have the same width, 0.8μm, so that both can obtain the same readout signal levels. By using them both as recording tracks, a track density, twice that for conventional 1.6μm track pitch systems, has been obtained. Additionally, the media is highly sensitive and has a high carrier to noise (C/N) ratio.

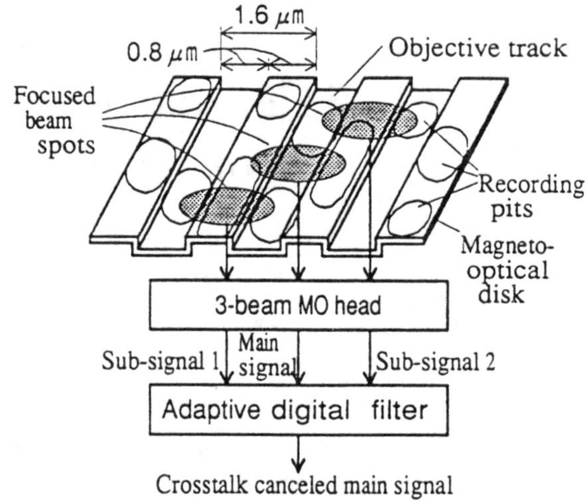

Figure 1. New high track density recording method

2.3 Three-beam MO head

The three-beam MO head, shown in Fig. 2, is constructed by arrayed lasers which could be independently driven, and which have a 0.83μm source wavelength. Individual lasers have a high output power, 50mW. Readout signals are detected by a two-segment photodiode array. The error signals for focusing and tracking are detected by a six-segment photodiode array.

The relationship between focused beams and tracks, when the signals are read out from three adjacent tracks, are shown in Fig. 1. The focused beam intervals are wider than the track pitch, more than 10μm. Therefore, laser beams are obliquely focused on the tracks by an objective lens, with a 0.55 numerical aperture. Accurate three-beam tracking is accomplished by beam rotation, using a Dove prism[5].

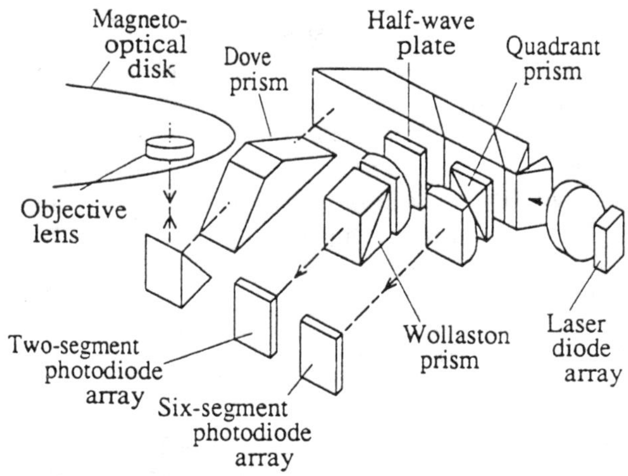

Figure 2. Three-beam MO head

2.4 Adaptive digital filter

The adaptive digital filter configuration is shown in Fig. 3. It consists of clock extract block, timing control block, equalizer and coefficient control block. The main signal and the sub-signals, read out simultaneously by the three-beam MO head, are respectively quantized into 8bit amplitude data. Recorded information on the objective track is picked up by eliminating the interference component by digital processing.

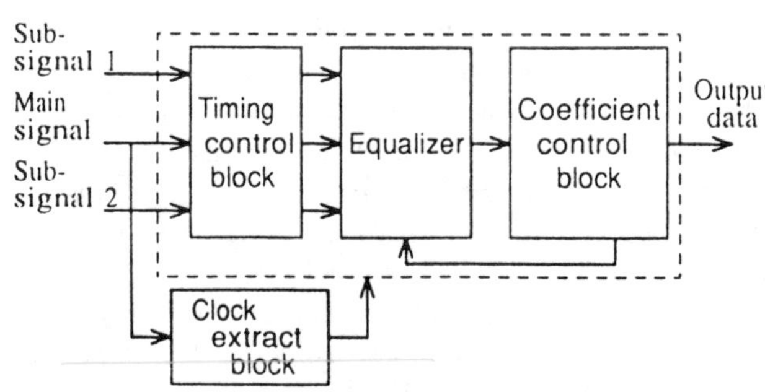

Figure 3. Adaptive digital filter configuration

2.4.1 Equalizer

There is a time lag between readout signals, proportional to the focused beam intervals. By controlling the disk rotation speed in constant linear velocity (CLV) mode, the time lag is fixed, and adjusted by the timing control block.

An equalizer, shown in Fig. 4, consists of three independent transversal filters and an adder, which adds their individual output signals. Each of the transversal filters has 9 taps. All of the coefficients are controlled independently by coefficient control block[6].

When the tap coefficients have converged into an ideal value, the main transversal filter acts to eliminate intersymbol interference caused by increasing linear density. The other two transversal filters generate a main signal crosstalk replica, having an opposite polarity, and cancel the crosstalk by adding individual outputs. Although crosstalk from outer tracks is contained in the sub-signals, their influence is negligible during the main signal equalization.

2.4.2 Coefficient control block

A coefficient control block, shown in Fig. 5, consists of a decision circuit, which carries out binary decision on the equalizer output data, and correlators, which generate tap coefficients for the transversal filters.

All tap coefficients, except for that for the main signal transversal filter center tap, have an initial value of 0, and the equalizer output is equal to the main signal. The decision circuit gives a binary decision on the equalizer output signal, and simultaneously implements the difference between the equalizer output and the decided value as an error ε. Each of the correlators, provided for individual variable coefficients, controls the coefficient in compliance with error ε and tap data, with a modified zero forcing (MZF) method[7]. Tap coefficients converge automatically to reduce the equalizer output error power, considering noise influence by the MZF method.

2.5 Special features

Theoretically, any crosstalk component and any intersymbol interference can be eliminated by the crosstalk canceler. Using a readout waveform simulator, -10dB crosstalk canceling was confirmed in experiments for a 20dB readout signal to noise ratio. If the adaptive digital filter is not mechanically adjusted, then the filter characteristics will reach optimum convergence automatically, since the read condition is the best possible.

This method can be used with other density improvement methods. By applying it to a shorter wavelength MO drive, higher recording density can be obtained, compared to the density realized by single shorter wavelength MO drives. It is also adaptable for other recording modulation codes. For example, convergence under (2,7) pit edge recording code modulation was also confirmed by simulator, with no problems. Furthermore, by applying the partial response multilevel detection method, an even higher recording density can be realized.

Since this crosstalk canceling method has a high recording density performance and a high speed transfer rate, due to increased linear recording density, it is suitable for image data file applications.

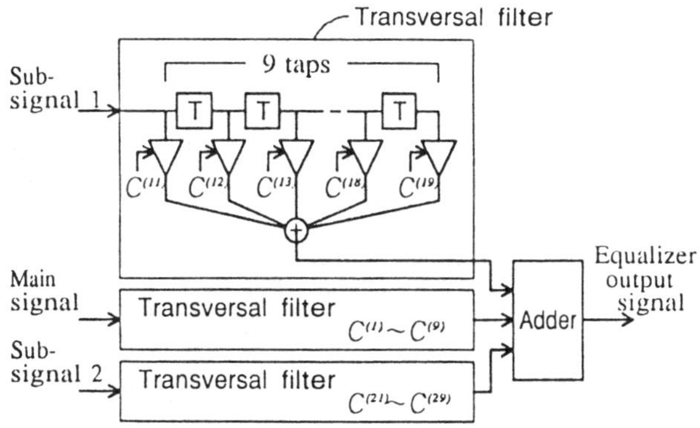

Figure 4. Equalizer

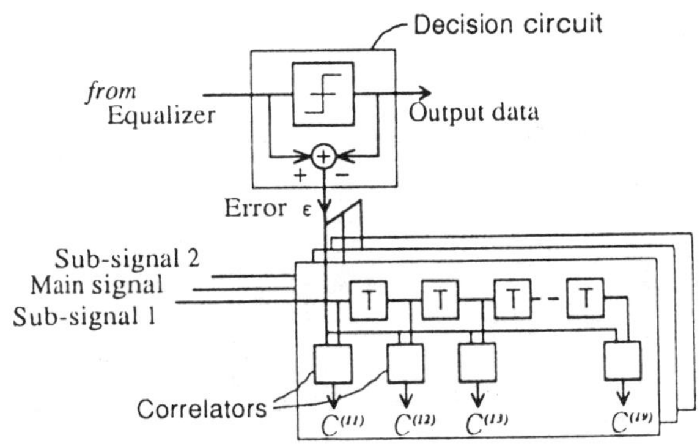

Figure 5. Coefficient control block

Track density	0.80μm/track
Linear density	0.83μm/bit
Laser wavelength	0.83μm
Objective lens NA	0.55
Recording code	NRZ
Data sequence	20bit M-sequence pseudorandom data
Crosstalk	-20.8dB
Error rate	5.1×10^{-5}

Figure 6. Recording conditions and appraised crosstalk and error rate

3 OPERATIONAL CHARACTERISTICS

3.1 Recording condition

Arrayed lasers in the three-beam MO head could be driven independently. Thus, recording under the same conditions as those for an ordinary MO head, could be realized by using any one of these lasers. Applied recording conditions are shown in Fig. 6. The recorded data

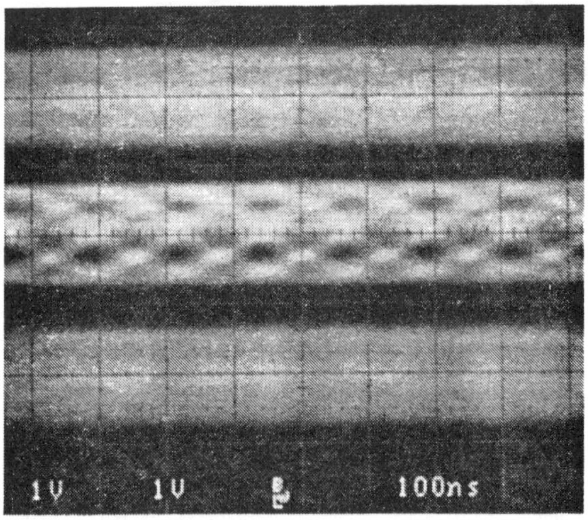

Figure 7. Read out signals

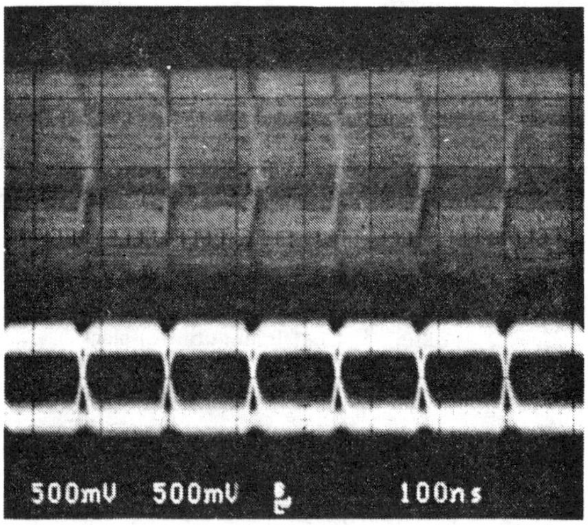

Figure 8. Before and after signal equalization (D/A converted data is shown)

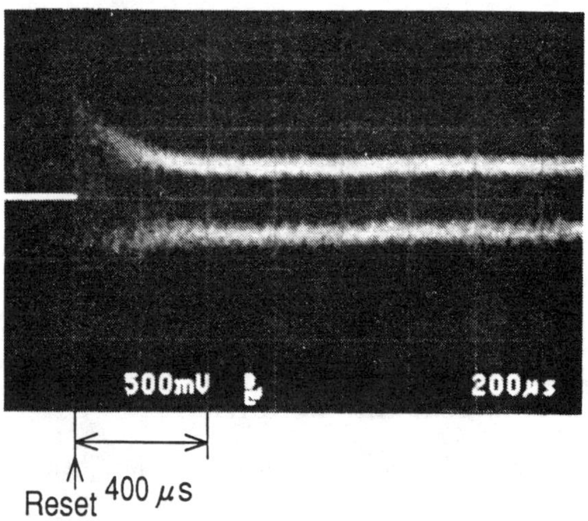

Figure 9. Equalizer output signal convergence

is 20bit M-sequence pseudorandom data, which is generated by characteristic equation $D^{20} + D^{17} + 1 = 0$. The data is recorded in non-return to zero (NRZ) code at bit rate 8Mbps. The recording linear density was $0.83\mu m$/bit at 6.6m/s linear velocity.

Figure 7 shows the signals read out by the three-beam MO head. The center waveform is the main signal, read out from the land area. The others are sub-signals, read out from adjacent grooves. These are amplified and sent through an anti-aliasing filter for A/D conversion. Due to extracting a clock signal for synchronization from the main signal, the sub-signals are not synchronized. The eye of the main signal is completely closed for binary decision. Crosstalk between adjacent tracks was -20.8dB.

3.2 Appraised results

Figure 8 shows main signal waves, before and after equalization. These wave forms are shown after D/A conversion for understanding ease. The equalizer output signal has a clearly opened eye for binary decision. A 5.1×10^{-5} error rate was obtained for a $0.8\mu m$ track pitch and a $0.83\mu m$/bit linear recording density.

Figure 9 shows the equalizer output signal, achieved by converging the tap coefficients. Output error gradually reduces, after the tap coefficients are reset, and the convergence is complete in about $400\mu s$.

4 CONCLUSION

High track density recording, using a crosstalk canceler, which consists of three adaptive digital filters, has been developed. By using both land area and groove area for data storage track, the applicability of this method has been ensured at high track density, twice the track density compared with conventional methods, and high recording linear density. By applying this method, a 5.1×10^{-5} error rate has been obtained at $0.8\mu m$ track pitch, $0.83\mu m$/bit recording linear density with NRZ recorded 20bit M-sequence pseudorandom data.

5 ACKNOWLEDGEMENT

The authors wish to thank Y. Yoshida, J. Namiki and H. Fukuchi for many useful suggestions.

6 REFERENCES

1) H. Sukeda et al., "High-Density Magneto-Optic Disk Using Highly Controlled Pit-Edge Recording", Proc. Int. Symp. on Optical Memory, p243, 1987.

2) Y. Yamanaka et al., "High Density Magneto-Optical Recording Using 0.67 μm Band High Power Laser Diode", IEEE Trans. Mag., vol.24, No.6, p2300, 1988.

3) Y. Yamanaka et al., "High Density Optical Recording by Superresolution", Tech. Digest, Int. Symp. on Optical Memory, p99, 1989.

4) T. Sekiguchi et al., "High Bit Rate Digital Video Recording Using Magneto-Optical Disks", Tech. Digest, Int. Symp. on Optical Memory, 1987.

5) R. Katayama et al., "Multi-Beam Magneto-Optical Disk Drive for Parallel Read/Write Operation", SPIE Proc., Topical Meeting on Optical Data Storage, vol.1078, p98, 1989.

6) J. Namiki et al., "Adaptive Receiver for Cross-Polarized Digital Transmission", IEEE 17th Int. Conf. Comm., p46.3.1, 1981.

7) D. Hirsch et al., "A Simple Adaptive Equalizer for Efficient Data Transmission", IEEE Trans., COM-18, 1, p15, 1970.

Reprinted with permission from *Applied Optics*, Vol. 30(2), 232-252 (January 10, 1991). ©1991 Optical Society of America.

Edge detection for magnetooptical data storage

Marc D. Levenson, Robert T. Lynch, Jr., and S. M. Tan

The polarization sensitive optical heterodyne methods used to detect polarization rotations in conventional magnetooptic data storage systems can be altered to produce nonzero electronic signals only when the polarization rotation changes. Such systems detect the edges of the magnetic domains optically and produce signals similar to conventional inductive magnetic recording channels. Both theoretical and experimental results indicate that optimized edge detection systems can have a performance comparable with present level detection systems but with less complexity. The maximum carrier-to-noise ratio (CNR) occurs for domains 1.3 μm in length when edge detection is used, and the CNR found experimentally at this maximum is 47 dB, comparable to level detection for this size domain.

I. Introduction

In magnetooptical (MO) data storage, information is encoded as a pattern of magnetized domains in a reflecting medium.[1] Linearly polarized light incident on a uniformly magnetized medium of this sort will be reflected with its plane of polarization rotated by a small angle. The sign of this Kerr rotation angle depends on the sense of magnetization; domains with the magnetization vector pointing upward toward the incident light produce one sign of rotation with downward magnetization producing the opposite rotation.

The basic challenge in this technology is the small magnitude of the Kerr rotation angle—typically 0.5°.[2-5] Today, the systems which read these domains detect changes in light level due to polarization rotation. This paper reports a new kind of detector which senses transitions between up and down magnetization. We present theoretical analysis, computer modeling, and experimental results which show carrier-to-noise ratio (CNR) levels above 47 dB at 5 MHz and 10 m/s, comparable to the high frequency performance of conventional level detectors.

The fundamental quantity in all optical phenomena is the amplitude of the light wave written as

$$\mathbf{E}(x,y,z,t) = \mathbf{E}(x,y,z)\exp[i(kz - \omega t)] \tag{1}$$

in the usual slowly varying complex amplitude notation. The total electric field amplitude here is a complex 2-D vector valued quantity, which depends on three spatial coordinates. However, only the values in certain planes are relevant to the discussion. It is convenient to define planes of constant $z = z_\alpha$ as in Fig. 1 and to express the amplitudes in those planes as

$$\mathbf{E}(x,y,z_\alpha) = \mathbf{E}(x_\alpha,y_\alpha) = \mathbf{E}\{\alpha\}. \tag{2}$$

Fourier optics and other operator and matrix techniques can be used to calculate the effect of propagation from one plane to another. We express the field at plane $\{\gamma\}$ in terms of that at plane $\{\alpha\}$ as

$$\mathbf{E}\{\gamma\} = \tilde{U}_{\{\gamma\}\{\alpha\}}\mathbf{E}\{\alpha\}, \tag{3}$$

where $\tilde{U}_{\{\gamma\}\{\alpha\}}$ is the propagation operator. Goodman[6] describes the conditions in which $\tilde{U}_{\{\gamma\}\{\alpha\}}$ can be approximated by a 2-D Fourier transform.

It is convenient to decompose the total electric field amplitude into a component $\mathbf{E}_0(x,y,z_\alpha) = \mathcal{E}_0\mathbf{a}_0(x_\alpha,y_\alpha)$, which would be present at plane z_α if there were no MO interaction and a signal amplitude $\mathbf{E}_s(x,y,z_\alpha) = \mathbf{E}_s(x_\alpha,y_\alpha;\Xi)$, which varies explicitly due to the data. The coefficient $\mathcal{E}_0$ parametrizes the output power of the laser collimator system. All amplitudes are linear in $\mathcal{E}_0$, while all currents and intensities scale as $\mathcal{E}_0^2$. The dimensionless vector amplitude $\mathbf{a}_0$ describes the spatial properties of the light in the absence of data. The data pattern is defined through a magnetization function $m(x_2,y_2;\Xi)$. The parameter Ξ measures the displacement of the pattern from the center line of the optical system.

The incident light will be assumed linearly polarized along the y-axis. The main effect of the MO data will be to produce an x-polarized component of the field. Thus the signal amplitude can be written

$$\mathbf{E}_s(x_\alpha,y_\alpha;\Xi) = \eta R E_x(x_\alpha,y_\alpha;\Xi)\hat{\mathbf{x}}, \tag{4}$$

where η is the (possibly complex) Kerr rotation angle

S. M. Tan is with University of Auckland, Auckland, New Zealand; the other authors are with IBM Research Division, Almaden Research Center, 650 Harry Road, San Jose, California 95120-6099.

Received 2 October 1989.
0003-6935/91/020232-21$05.00/0.

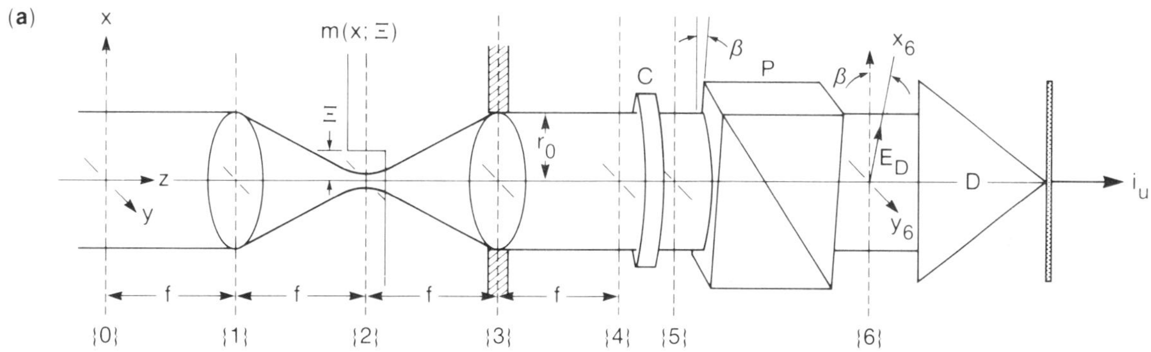

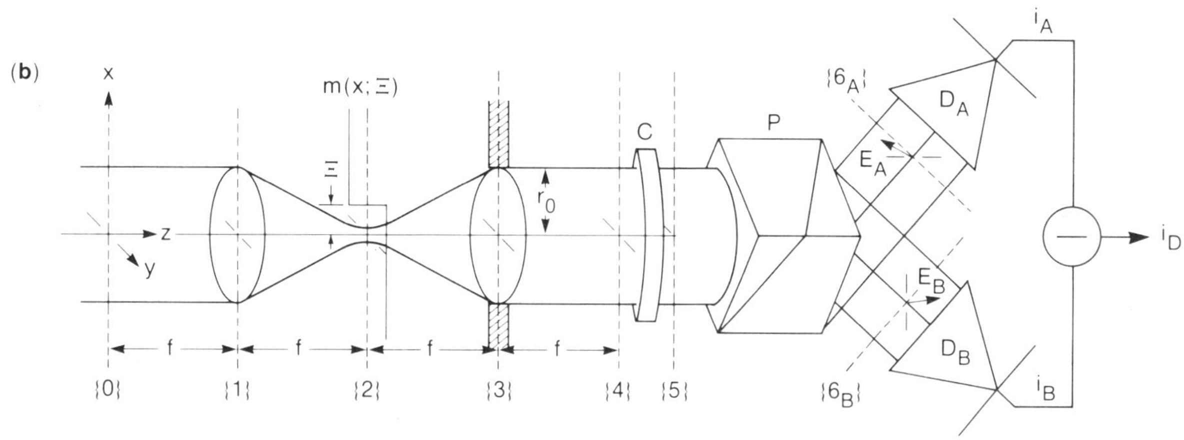

Fig. 1. Unfolded optical trains for conventional MO level detection systems. The y-polarized read beam is incident from the left. Significant planes are labeled with numbers in braces. The xyz coordinate system at plane {0} defines the general orientation. The storage medium lies at plane {2} between the focusing and recollimating objective planes {1} and {3}, respectively. The objective radius is r_0 with the fields vignetted to that radius at the input and output apertures at planes {0} and {4}. (a) Single ended level detector system where the polarization analyzer P is rotated by the small angle β. The polarization transmitted into the detector D deviates from the x-axis by β as shown in plane {6}. (b) Differential level detector system with two photodiodes. The polarization analyzer P directs light polarized at $-45°$ from the x-axis into detector D_A and light polarized at $+45°$ into detector D_B. Polarization compensator plates described in the text are labeled C.

and R is the amplitude reflectivity of the medium. The signal field $E_x(x_\alpha,y_\alpha;\Xi)$ depends on the input beam, magnetization pattern, and optical propagation to plane z_α.

The field present at the detector plane in the absence of a MO rotation $\mathbf{E}_{LO}$ acts as a local oscillator to facilitate photodetection. This component is simply that portion of $\mathbf{E}_0$ which reaches the detector plane.

It is convenient to choose a system of units where the intensity (generally in watts per square cm) is equal to the square of the electric field amplitude without prefactors. Thus we have $I = |E|^2$. The photoelectric current i produced by a detector must have the dimensions of amperes, and we define a dimensional quantum efficiency $q = i/\int IdA$ with this property.[7]

Many of the results in this paper involve integrals over products of electric field amplitude distributions. Such integrals can be represented compactly as

$$\langle E_j^*\{\alpha\}|E_\ell\{\alpha\}\rangle = \iint_{-\infty}^{\infty} dx_\alpha dy_\alpha E_j^*(x_\alpha,y_\alpha)E_\ell(x_\alpha,y_\alpha). \tag{5}$$

II. Analysis of Conventional Head Operation

The two sorts of detection scheme conventionally employed[8] are diagrammed in Fig. 1. The input and output beam paths have been unfolded for clarity in the figure. Elements such as leaky polarization beam splitters and halfwave plates may be necessary in the actual head but have been omitted here.

A. Single Polarization MO Detection

Figure 1(a) shows a single ended polarization selective optical heterodyne detection scheme. The incident read light is polarized along the y-axis. The polarization analyzer P is rotated from the direction orthogonal to the incident polarization direction by the bias angle β. Light transmitted through this polarizer is incident on the detector D, and the changes in the intensity at this detector reflect the changes in magnetization of the medium. The input light has vector amplitude $E_0(x_0,y_0)\hat{y}$ at plane {0}. Typically this is a Gaussian laser beam truncated at the radius of the objective lens. Fourier optics techniques[6] are used to transform the input field pattern at plane {0} to that at the medium plane {2}. The light reflected from the medium has the vector amplitude

$$\mathbf{E}_r(x_2,y_2;\Xi) = RE_0(x_2,y_2)\hat{\mathbf{y}} + E_s(x_2,y_2;\Xi)\hat{\mathbf{x}}, \tag{6}$$

where

$$E_s(x_2,y_2;\Xi) = \eta Rm(x_2,y_2;\Xi)E_0(x_2,y_2) = \eta_\ell RE_x(x_2,y_2;\Xi) \tag{7}$$

is the signal amplitude, η is the Kerr rotation angle (which may be complex to account for Kerr ellipticity), R is the amplitude reflection coefficient (assumed real), and $m(x_2,y_2;\Xi)$ is the magnetization. This function represents the pattern of up and down magnetization on the disk surface. The parameter Ξ measures the displacement of the pattern as the disk moves. In a 1-D approximation we have

$$m(x;\Xi) = \mathrm{sgn}_\Xi(x) = \begin{cases} -1 & x < \Xi, \\ 1 & x \geq \Xi, \end{cases} \tag{8}$$

where now the parameter Ξ represents the position of a transition between up and down magnetization with respect to the center line of the optical system.

Fourier optics techniques are then again used to transform the $\mathbf{E}_r(x_2,y_2;\Xi)$ field distribution at the medium to that at plane {4}. The x- and y-polarization components are treated independently. If the radius of the objective lens is r_O, the field distribution in plane {4}—the output aperture plane—is

$$\mathbf{E}_r(x_4,y_4;\Xi) = \mathbf{E}_r(\{4\};\Xi) = \mathrm{rect}_{r_O}(x_4,y_4)\tilde{U}_{|4||2|}\mathbf{E}_r(x_2,y_2;\Xi), \tag{9}$$

where the vignetting function in plane $\{\beta\}$ is given by

$$\mathrm{rect}_{r_O}\{\beta\} = \begin{cases} 1 & x_\beta^2 + y_\beta^2 < r_O^2, \\ 0 & \text{otherwise.} \end{cases} \tag{10}$$

Many systems include a compensator plate C to correct for ellipticity introduced by stray retardance and circular dichroism. The compensator retards the phase of the y-polarized light with respect to the x-polarized signal:

$$\mathbf{E}_r(x_5,y_5;\Xi) = \begin{Bmatrix} 1 & 0 \\ 0 & \exp(-i\Phi) \end{Bmatrix} \mathbf{E}_r(x_4,y_4;\Xi)$$
$$= \exp(-i\Phi)RE_0(x_4,y_4)\hat{\mathbf{y}} + E_s(x_4,y_4;\Xi)\hat{\mathbf{x}}. \tag{11}$$

Note that $E_0(x_4,y_4)$ is the field which would exist at plane {4} if there were no effect due to the magnetization and the disk reflectivity were unity. The 2×2 matrix in Eq. (11) is the Jones matrix[9] for the compensator C. The field transmitted through the polarizer is $\mathbf{E}_D(x_6,y_6;\Xi) = \hat{\boldsymbol{\beta}}\hat{\boldsymbol{\beta}} \cdot \mathbf{E}_r(x_5,y_5;\Xi)$, where $\hat{\boldsymbol{\beta}} = \sin\beta\hat{\mathbf{y}} + \cos\beta\hat{\mathbf{x}}$. The intensity at the detector is then $I_D(x_6,y_6;\Xi) = |E_D(x_6,y_6;\Xi)|^2$. For signal-to-noise reasons, the optimal value of β is small, typically 5°. For small angles such as this, a linear approximation suffices:

$$\mathbf{E}_D(x_6,y_6;\Xi) = \{\mathbf{E}_{LO}(x_5,y_5) + \mathbf{E}_s(x_5,y_5;\Xi)\} \cdot (\beta\hat{\mathbf{y}} + \hat{\mathbf{x}})(\beta\hat{\mathbf{y}} + \hat{\mathbf{x}}), \tag{12}$$

where the local oscillator amplitude at {5} is $\mathbf{E}_{LO}(x_5,y_5) = \exp(-i\Phi)R\mathbf{E}_0(x_4,y_4)$. To the lowest nonvanishing order in the small quantity η, the detector intensity is

$$I_D(x_6,y_6;\Xi) = |\mathbf{E}_{LO}(x_6,y_6)|^2 + 2\,\mathrm{Re}\{\mathbf{E}^*_{LO}(x_6,y_6)\mathbf{E}_s(x_6,y_6;\Xi)\}$$
$$= I_{LO}(x_6,y_6) + 2\beta R^2$$
$$\times \mathrm{Re}\{\eta \exp(i\Phi)E^*_0(x_4,y_4)E_x(x_4,y_4;\Xi)\}, \tag{13}$$

where $I_{LO}(x_6,y_6) = \beta^2R^2|\mathbf{E}_0(x_4,y_4)|^2$, the local oscillator intensity at the detector, does not depend on the position of the transition Ξ. The polarization of the light at the analyzer P and the amplitude transmitted to the detector are shown in Fig. 2(a) for plus and minus magnetization.

The current from the detector is obtained by integrating the two terms in Eq. (13) over the detector surface[7]:

$$i_D(\Xi) = i_{LO} + i_u(\Xi), \tag{14}$$

where

$$i_{LO} = q \int I_{LO}(x_6,y_6)dx_6dy_6 = q\beta^2R^2\langle E^*_0\{4\}|E_0\{4\}\rangle, \tag{15}$$

$$i_u(\Xi) = 2q\beta R^2 \int \mathrm{Re}\{\eta \exp(i\Phi)E^*_0(x_4,y_4)E_x(x_4,y_4;\Xi)\}dx_4dy_4$$
$$= 2q\beta R^2\,\mathrm{Re}\{\eta \exp(i\Phi)\langle E_0{}^*\{4\}|E_x(\{4\};\Xi)\rangle\}. \tag{16}$$

The number in braces indicates the plane where the projections are taken. Note that the detector currents are most conveniently expressed in terms of the fields at the objective output pupil plane {4}.

Only $i_u(\Xi)$ contributes to the desired MO signal, but i_{LO} is potentially a source of noise. The optimization of single ended polarization selective optical heterodyne detector systems of this sort has been discussed extensively in the literature.[10-12] At optimum, the bias angle β is such that noise due to i_{LO} is equal to the sum of the noise that does not depend on β. This generally requires $\beta \geq \eta$, with the equal sign only in the absence of technical noise.

B. Dual Polarization MO Detection

Dual polarization MO detection systems are identical to single ended systems up through plane {5}. The polarization analyzer P, however, has two exit ports (labeled A and B), and there are two corresponding detectors [Fig. 1(b)]. The polarization analyzer is oriented at $\beta = 45°$, so that the A output corresponds to the projection of the field $\mathbf{E}_r(x_5,y_5;\Xi)$ on the axis $\hat{\mathbf{A}} = (\hat{\mathbf{x}} + \hat{\mathbf{y}})/\sqrt{2}$, while the B output corresponds to the projection on $\hat{\mathbf{B}} = (\hat{\mathbf{x}} - \hat{\mathbf{y}})/\sqrt{2}$. Mathematically, the fields at the detectors are then

$$\mathbf{E}_A(x_6,y_6;\Xi) = \frac{\hat{\mathbf{x}} + \hat{\mathbf{y}}}{\sqrt{2}}\{\exp(-i\Phi)RE_0(x_4,y_4) + E_s(x_4,y_4;\Xi)\}, \tag{17a}$$

$$\mathbf{E}_B(x_6,y_6;\Xi) = \frac{\hat{\mathbf{x}} + \hat{\mathbf{y}}}{\sqrt{2}}\{\exp(-i\Phi)RE_0(x_4,y_4) - E_s(x_4,y_4;\Xi)\}, \tag{17b}$$

As in the former case, the intensities at the detectors are proportional to the squares of the fields

$$I_{\substack{A\\B}}(x_6,y_6;\Xi) = I_{LO}(x_y,y_6)$$
$$\pm R^2\mathrm{Re}\{\eta \exp(i\Phi)E^*_0(x_4,y_4)E_x(x_4,y_4;\Xi)\}, \tag{18}$$

where $I_{LO}(x_6,y_6) = 1/2\ R^2|E_0(x_5,y_5)|^2$, and only the second term contains information about the magnetization. Figures 1(b) and 2(b) illustrate the polarization

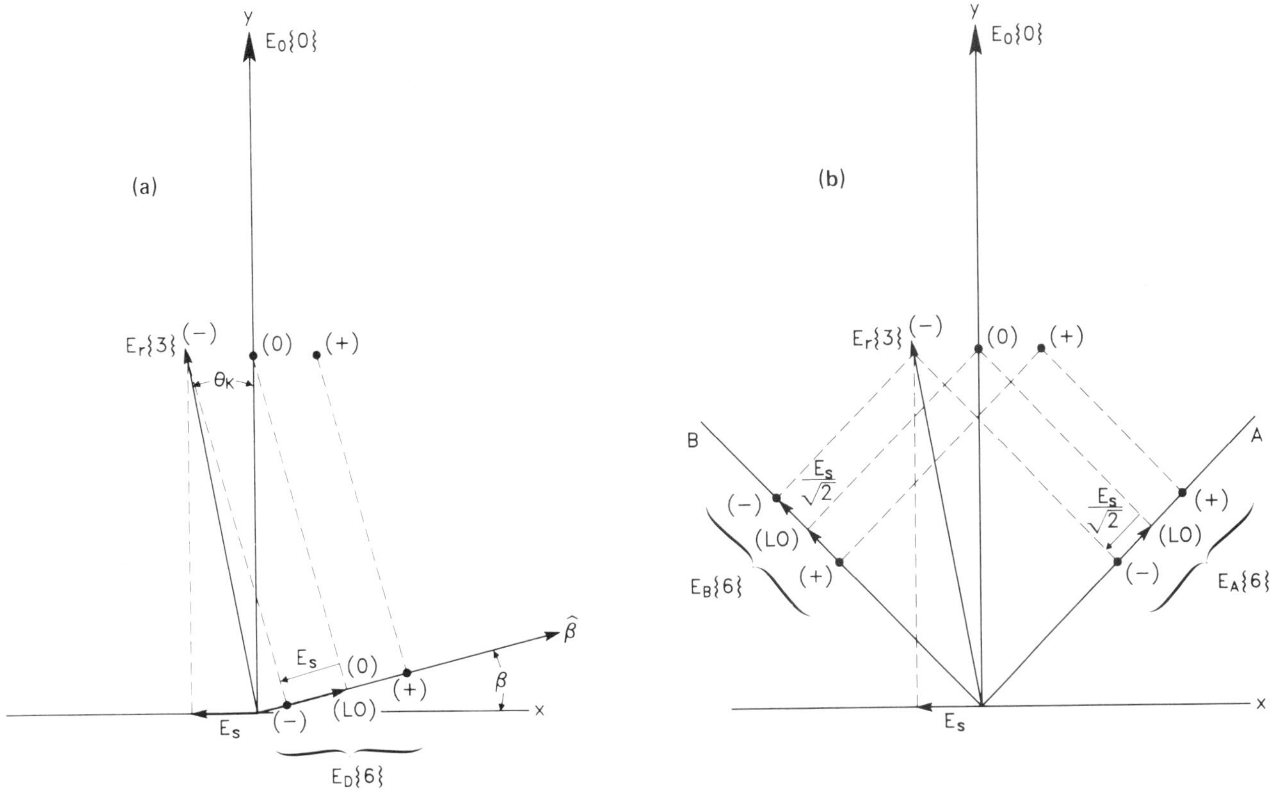

Fig. 2. Polarization changes and detector field amplitudes due to MO data. The input field $E_0\{0\}$ is polarized along the y-axis, and the reflected field in the absence of a MO rotation is denoted (0). Positive and negative magnetizations rotate the polarization of the reflected light in plane {3} as shown. The signal amplitude E_s also appears for negative magnetization. (a) The reflected fields are projected on an axis biased at the small angle β, producing the detector amplitudes $E_D\{6\}$. Positive magnetization produces a larger amplitude than negative with the local oscillator amplitude (LO) intermediate. The uniform detector current i_u is proportional to the square of E_D. The operation of a dual polarization differential detector appears in (b). The reflected polarization is projected onto two axes, each at 45° to the incident polarization. The Kerr rotation increases the amplitude at one detector and decreases it at the other. In the absence of rotation, the two amplitudes are equal. The detector signal i_D is proportional to the difference in the squares of the amplitudes.

at the analyzer as well as the projections onto the $\hat{\boldsymbol{A}}$ and $\hat{\boldsymbol{B}}$ directions. The term independent of Ξ again contributes nothing except noise. However, the actual signal from such a dual detection system is the difference in the currents from the two detectors:

$$\begin{aligned} i_D &= i_A(\Xi) - i_B(\Xi) \\ &= 2qR^2 \int \mathrm{Re}\{\eta \exp(i\Phi)E_0^*(x_4,y_4)E_x(x_4,y_4;\Xi)\}dx_4dy_4 \\ &= 2qR^2\, \mathrm{Re}\{\eta \exp(i\Phi)\langle E_0^*\{4\}|E_x(\{4\};\Xi)\rangle\}. \end{aligned} \tag{19}$$

If the differencing circuit works properly, all noise resulting from fluctuations of $I_{\mathrm{LO}}(x_6,y_6)$ cancels out. The quantum fluctuations (shot noise) on the local oscillator do not cancel and constitute the major noise component of a well engineered system.[13–15]

III. Fundamental Principles of MO Edge Detection

As an isolated MO transition at Ξ moves through the focused laser beam, the character of the reflected field $E_s(x_2,y_2;\Xi)$ changes. When $\Xi \gg w_0$, where w_0 is the beam radius at focus, the laser samples a region of uniform magnetization and $|E_x(x_2,y_2;\infty)| = E_0(x_2,y_2)$. This is a symmetric function of the displacement x_2, and this symmetric character is propagated through the axially symmetric system to the detector plane. Thus uniform magnetization produces the maximum possible overlap in Eqs. (16) and (19) and the maximum level detector signal. In contrast, a transition centered at the beam axis produces an x-polarized field distribution antisymmetric with respect to the dimension along the track: $E_x(x_2,y_2;0) = -E_x(-x_2,y_2;0)$. The antisymmetric character of the field is also propagated through the optical system to the detector plane. The overlap functions and signal currents in Eqs. (16) and (19) are thus zero when a transition is centered on the laser axis.

A. Single Polarization MO Edge Detection

In MO edge detection, the detector system is altered, or additional optical elements are inserted in the beam to produce a maximum overlap and a maximum signal when the x-polarized field is antisymmetric. The simplest system of this type is a modification of the single ended head of Fig. 1(a) and is shown in Fig. 3(a). A split detector has been substituted for the uniform detector of the previous figure. If the outputs of the

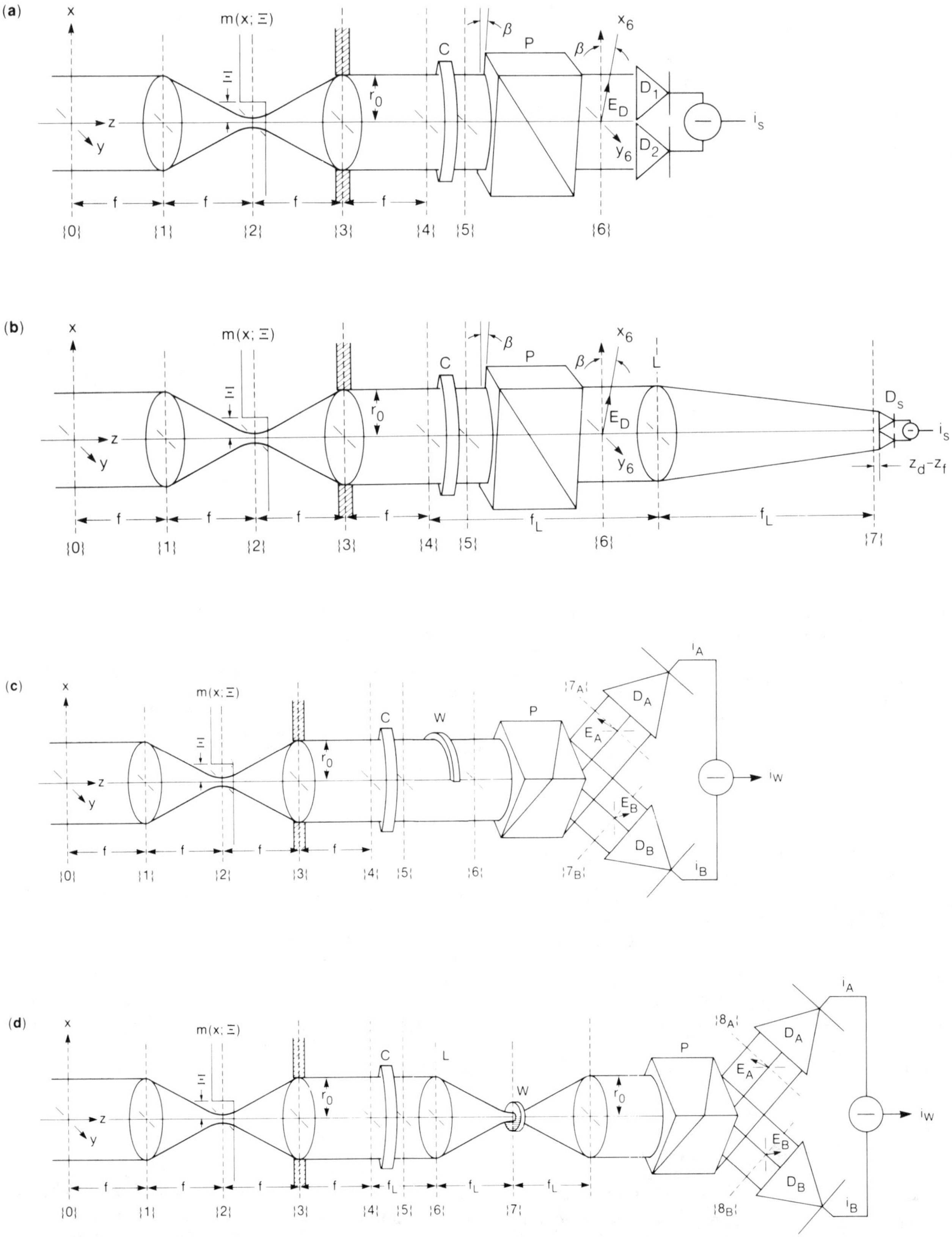

Fig. 3. Unfolded optical trains for MO edge detection: (a) Far field single polarization split detector system. The general scheme is similar to Fig. 1(a). The output signal is i_s. (b) Reimaging head design with a single split detector. The detector lies at axial distance z_d, which is near the focal plane of lens L at $z_f = z_7$. (c) Far field edge detection system with a semicircular halfwave plate matched filter W and separate photodiodes sensitive to orthogonal polarizations, as in Fig. 1(b). (d) Reimaging dual polarization detector system with a semicircular halfwave plate filter W at the focal plane of lens L. In this case, it is convenient to immerse the birefringent element in index matching fluid so that no phase shift occurs at $x_7 = 0$ for one polarization.

two sides of the split detector are subtracted from one another, the signal current is

$$i_s(\Xi) = 2q\beta R^2 \,\mathrm{Re}\left\{\eta \exp(i\Phi) \int \left[\int_0^{\infty} E_0^*(x_4,y_4)E_x(x_4,y_4;\Xi)dx_4 - \int_{-\infty}^{0} E_0^*(x_4,y_4)E_x(x_4,y_4\Xi)dx_4 \right] dy_4 \right\}$$
$$= 2q\beta R^2 \,\mathrm{Re}\{\eta \exp(i\Phi)\langle E_0^*\{4\}|\mathrm{sgn}_0\{4\}|E_x(\{4\};\Xi)\rangle\} \quad (20)$$

Again, the overlaps have been expressed in terms of the fields at the output pupil plane. Similar push–pull detection systems have been considered for read-only and write-once systems.[1] Such a detection system clearly produces no signal when the laser beam illuminates a uniformly magnetized region. Moreover, noise resulting from reflectivity fluctuations and laser power fluctuations cancels out. Ideally, only the quantum noise in the local oscillator remains. An optimized system of this sort allows a larger local oscillator intensity than the simple single polarization design. Note that the laws of optics produce a net $\pi/2$ phase shift between symmetric and antisymmetric fields which diverge from a common focus.[6,16] This implies that the phase shift Φ must be $\pi/2$ for optimum edge detection in this system when the Kerr rotation angle η is real. Such a phase shift can be easily obtained using the compensator C, which then becomes a quarterwave plate. An alternative explanation of this $\pi/2$ phase shift is given in Sec. IV. In conventional MO level detection (and polarization independent write-once systems), $\Phi = 0$ for optimum performance.

The Appendix shows how the x-polarized field at plane {2} can be decomposed into even and odd symmetry functions as

$$E_x(x_2,y_2;\Xi) = \mathcal{A}_0\{u(\Xi)\psi_+(x_2,y_2) + v(\Xi)\psi_-(x_2,y_2)\}, \quad (21)$$

where $\psi_+(x_2,y_2)$ and $\psi_-(x_2,y_2)$ are properly chosen, normalized even, and odd symmetry electric field profiles at the plane of the storage medium, respectively. The coefficients $u(\Xi)$ and $v(\Xi)$ are

$$u(\Xi) = \langle\psi_+^*\{2\}|\mathrm{sgn}_\Xi|\psi_+\{2\}\rangle, \quad (22)$$

$$v(\Xi) = \langle\psi_-^*\{2\}|\mathrm{sgn}_\Xi|\psi_+\{2\}\rangle. \quad (23)$$

By normalization $u(\infty) = 1$ and $u(0) = 0$. Similarly, $v(\infty) = 0$ and $v(0) = v_0 < 1$. Both $u(\Xi)$ and $v(\Xi)$ are real valued functions when $\psi_+\{2\}$ and $\psi_-\{2\}$ are real. The details depend only on the field distribution $E_0(x_2,y_2)$.

Typical MO head designs fill the aperture of the objective with the symmetric field E_0. Such head designs necessarily vignette the antisymmetric fields produced by a centered transition and essentially fail to transmit higher-order field patterns. Thus we are justified in a two-state analysis of such systems and need only consider the appropriate overlap coefficients of the two functions ψ_+ and ψ_- at relevant planes in the system. In terms of these functions, the signal currents in Eqs. (16), (19), and (20) can be written in terms of $u(\Xi)$ and $v(\Xi)$:

$$i_u(\Xi) = 2q\beta R^2 \mathcal{A}_0^2 \,\mathrm{Re}\{\eta \exp(i\Phi)\langle\psi_+^*\{4\}|\mathrm{rect}_{r_D}\{4\}|\psi_+\{4\}\rangle\}u(\Xi); \quad (24)$$

$$i_D(\Xi) = 2qR^2\mathcal{A}_0^2|\eta \exp(i\Phi)\langle\psi_+^*\{4\}|\mathrm{rect}_{r_D}\{4\}|\psi_+\{4\}\rangle|u(\Xi); \quad (25)$$

$$i_s(\Xi) = 2q\beta R^2\mathcal{A}_0^2 \,\mathrm{Re}\{\eta \exp(i\Phi)(\langle\psi_+^*\{4\} \times |\mathrm{rect}_{r_D}\{4\}\,\mathrm{sgn}_0\{4\}|\psi_-\{4\}\rangle\}v(\Xi), \quad (26)$$

where all dimensional factors are grouped to the left and $\mathcal{A}_0^2$ is the laser–collimator output power at plane {0}. In writing Eqs. (24)–(26) we have used the fact that planes {3}–{6} are very close to one another. The effect of the finite radius of the detector $r_D \leq r_O$ has been included explicitly, and the fields have been referred to plane {4}. Clearly, the Ξ dependence of the electronic signals reflects those of $u(\Xi)$ and $v(\Xi)$, and no other details are truly relevant. The Appendix gives an example of the use of this convenient method.

It is also possible to reimage the MO medium onto a split detector as shown in Fig. 3(b). In this optical system, the antisymmetric field accumulates a π phase shift at plane {7} with respect to the symmetric field.[6,16] The compensator plate C then must be set for $\Phi = 0$ when η is real.

B. Dual Polarization MO Edge Detection

One can easily imagine dual split photodiode edge detection schemes similar to the level detection head in Fig. 1(b). Such systems require careful alignment of the center line of both detectors and may not justify the additional complexity. The analysis of such systems follows directly from that above as the overlap integrals are entirely analogous to the level detector case.

More practical dual detector designs employ uniform photodiode surfaces and an additional optical element to project out the odd-symmetry portion of the field. The simplest such element is a halfwave plate, which covers half of the beam between planes {5} and {6} in Fig. 3(c). This matched filter is aligned with the ordinary axis of the birefringent material along $\hat{y}$ and the extraordinary axis along $\hat{x}$. The plate has no effect on the half of the beam where $x_6 > 0$. However, the phase of the x-polarized light is retarded by π for $x_6 < 0$. The transformation produced by this element is

$$\mathbf{E}_r\{6\} = \tilde{W}\mathbf{E}_r\{5\}, \quad (27)$$

where

$$\tilde{W}(x_5,y_5) = \begin{bmatrix} \mathrm{sgn}_0(x_5) & 0 \\ 0 & 1 \end{bmatrix} \quad (28)$$

is the Jones matrix for this nonsymmetric element. The rest of the detection system is identical to the conventional level detector in Fig. 1(b).

When the laser spot illuminates a uniformly magnetized region, the phase of the x-component of the field at plane {5} is uniform. The edge detector plate W reverses the sign of this field for $x_6 < 0$ without altering the field in the $x_6 > 0$ region. The result is a field pattern that changes sign in the center and produces no net contribution to the differential detector signal in Eqs. (19) and (25). However, when a magnetic transition lies at the center of the beam, the sign change in $\mathbf{E}_x$ produced by the medium is undone by the edge detector plate. The result is an increase in the current from one detector in Fig. 3(c) and a decrease in the current from the other detector. In the far field with $\Phi = \pi/2$, the net phase shift for $x_6 < 0$ is $3\pi/2$, while that for $x_6 > 0$ remains $\pi/2$. It is again convenient to refer all overlaps to the far field plane {4}. Mathematically, the difference signal produced by this dual polarization edge detector system is then

$$i_{w'}(\Xi) = 2qR^2\mathcal{A}_0^2 \operatorname{Re}\{\eta \exp(i\Phi)\langle\psi_+^*\{4\} \times |\operatorname{rect}_{r_W}\{4\} \operatorname{sgn}_0\{4\}|\psi_-\{4\}\rangle\}\upsilon(\Xi), \quad (29)$$

where $z_6 \approx z_4$. This is clearly analogous to that in Eq. (26) but with a larger local oscillator level, since the small factor β does not appear. The radius of the matched filter is $r_W \leq r_O$. The $\pi/2$ phase shift between symmetric and antisymmetric fields beyond focus also occurs in this case and requires $\Phi = \pi/2$ for a maximum signal. Such an edge detector system has less sensitivity to certain forms of optical noise than the single ended scheme of Fig. 3(a). Thus the additional complexity may be worthwhile, although it is important to note that both schemes produce signals proportional to the same function, $\upsilon(\Xi)$, and both require a $\pi/2$ phase shift between local oscillator and signal.

Figure 3(d) shows a reimaging dual polarization MO edge detection system. In this case, the partial half-wave plate filter W lies on the plane of the image of the medium formed by lens L. The beam is then polarization analyzed and directed into two uniform photodiodes. The difference signal has the same form as Eq. (29), but the reimaging of the medium eliminates the need for a $\pi/2$ phase shift and allows edge detection with $\Phi = 0$. The noise from this system, however, has subtle differences from those of the other edge detector systems. This case is treated more fully in Sec. VI.

IV. Two-Dimensional Fourier Optics Analysis

The Kerr rotation η for linear polarized light corresponds to a change in phase shift on reflection[2] of $+\eta$ for left-hand circular (LHC) polarized light and $-\eta$ for right-hand circular (RHC) polarization. Reversal of the magnetization reverses these phase shifts and thus also the Kerr rotation angle. It is, therefore appropriate to analyze the MO detection schemes in terms of a circularly polarized basis set.

When a transition between plus and minus magnetization is centered in the read spot, the phase of LHC polarized light is advanced on one side of the beam and retarded on the other. This phase shift causes a displacement of the reflected beam in the far field just as would a depression in the surface such as a tracking groove or a read-only disk data pit.[1] The phase shifts due to Kerr rotation are small; a rotation of $\eta = 0.5°$ corresponds to 0.0014 wavelengths or a 2.3-nm height difference between land and pit. Thus a push–pull detector designed to sense the far field shift of a circularly polarized read beam due to MO data would be overwhelmed by noise due to disk roughness (see Sec. V).

However, if two push–pull detectors are used, one sensitive to RHC polarized light and the other sensitive to LHC polarized light, the polarization independent noise due to disk roughness can be rejected as a common mode. This is exactly the strategy employed in the system of Fig. 3(c), except that the polarization filter W performs some of the electronic functions of the dual split detector system that would otherwise be needed. The $\Phi = \pi/2$ phase shift necessary for far field edge detection is thus explained here by the need to make the detector system sensitive to the relative profile displacements of circularly polarized modes.

The various level and edge detection methods discussed in this paper have been analyzed using a set of computer programs to perform numerically the 2-D Fourier transforms and spatial overlap integrals described in the preceding. The Fourier optics calculations involve 2-D fast Fourier transforms, which are done in FORTRAN.[17] The magnetization pattern on the MO film and the processing performed by the optical elements from the collection lens to the detectors are modeled using the APL2 programming language.

Scalar Fourier optics diffraction modeling assumes that the two normal polarization modes of the MO system (i.e., left- and right-hand circular polarizations) propagate through the optical system independently.[6] This simplification ignores some effects such as polarization aberrations due to focusing and any angular dependence of the Kerr effect and disk reflectivity function but is a reasonable approximation for our purposes. Thus the input light is decomposed into LHC and RHC components, and these two modes are considered separately using two invocations of the scalar model. We define our usual input beam of y-polarized light as being composed of equal amounts of LHC and RHC polarized light with a zero phase shift between them.

The effect of the MO layer is to impose a positive or negative phase shift on LHC polarized light, depending on whether the beam passes through the medium parallel or antiparallel to the magnetization. At the same time the MO layer imposes the opposite phase shift on RHC polarized light. The complex electric field amplitudes corresponding to each polarization state after interaction with the disk are then propagated to the output side of the collection lens by another Fourier transform. In the far field (i.e., plane {4} and beyond) the complex amplitudes for LHC and RHC polarized light at a particular plane can be used to calculate the amplitudes of any other pair of normal polarization modes according to the standard Jones vector formalism.[9] The polarization state, however, is in general not uniform across the beam, and so the two-element Jones polarization vector is a function of the transverse coordinates.

As an example of this 2-D analysis, consider the complex amplitudes for LHC and RHC polarized light for a series of beam positions relative to the racetrack domain shown in the inset in Fig. 6. Figure 4 shows the field amplitude and retardation in plane {4} for beam locations (a) well before and (b) directly on the edge of

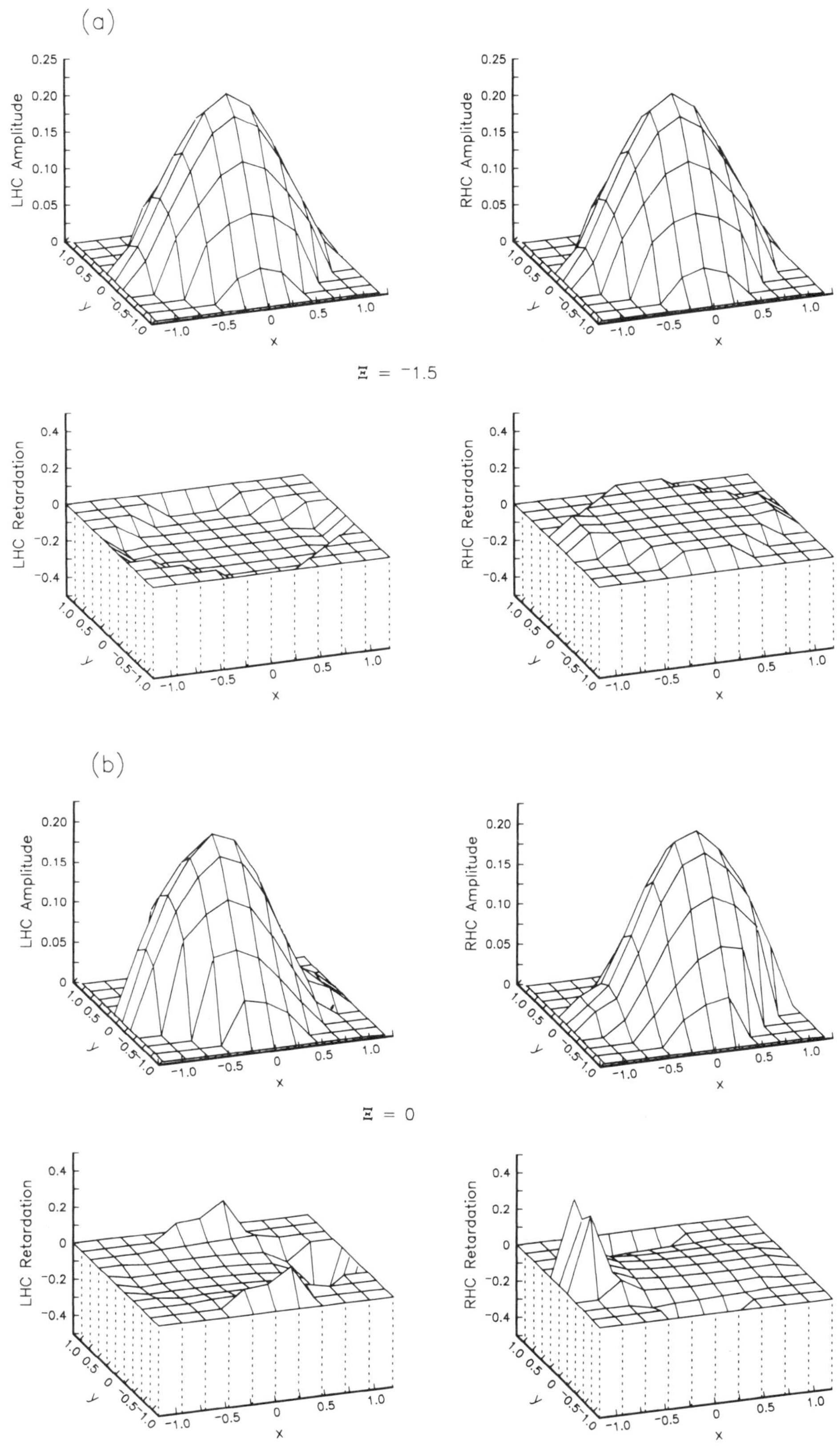

Fig. 4. Far field amplitude and retardation profiles for the circular polarized components of the read beam. The retardation is given in units of wavelength, and the phase shift is 2π times this quantity. These profiles correspond to plane {4} in Figs. 1 and 3. The parameter Ξ is the displacement of the center of the laser spot from the effective edge of the domain. (a) The laser spot is in a region of uniform magnetization, which produces equal and opposite retardations for the two components, but the beam amplitudes are centered on the detectors. (b) The laser beam is at the leading edge of the domain in Fig. 6. The centroids of the RHC and LHC beams are deviated in opposite directions from the center line, and the retardations are nearly unchanged.

the domain. In these plots the retardations are given in units of waves, and the phase shifts are equal to the retardations multiplied by 2π. In Fig. 4(a) the amplitude profiles of LHC and RHC polarized light are the same after interacting with the uniformly magnetized region of the disk as they were before the interaction, since the disk reflectivity is not affected by the magnetization, but the retardations for the two polarizations

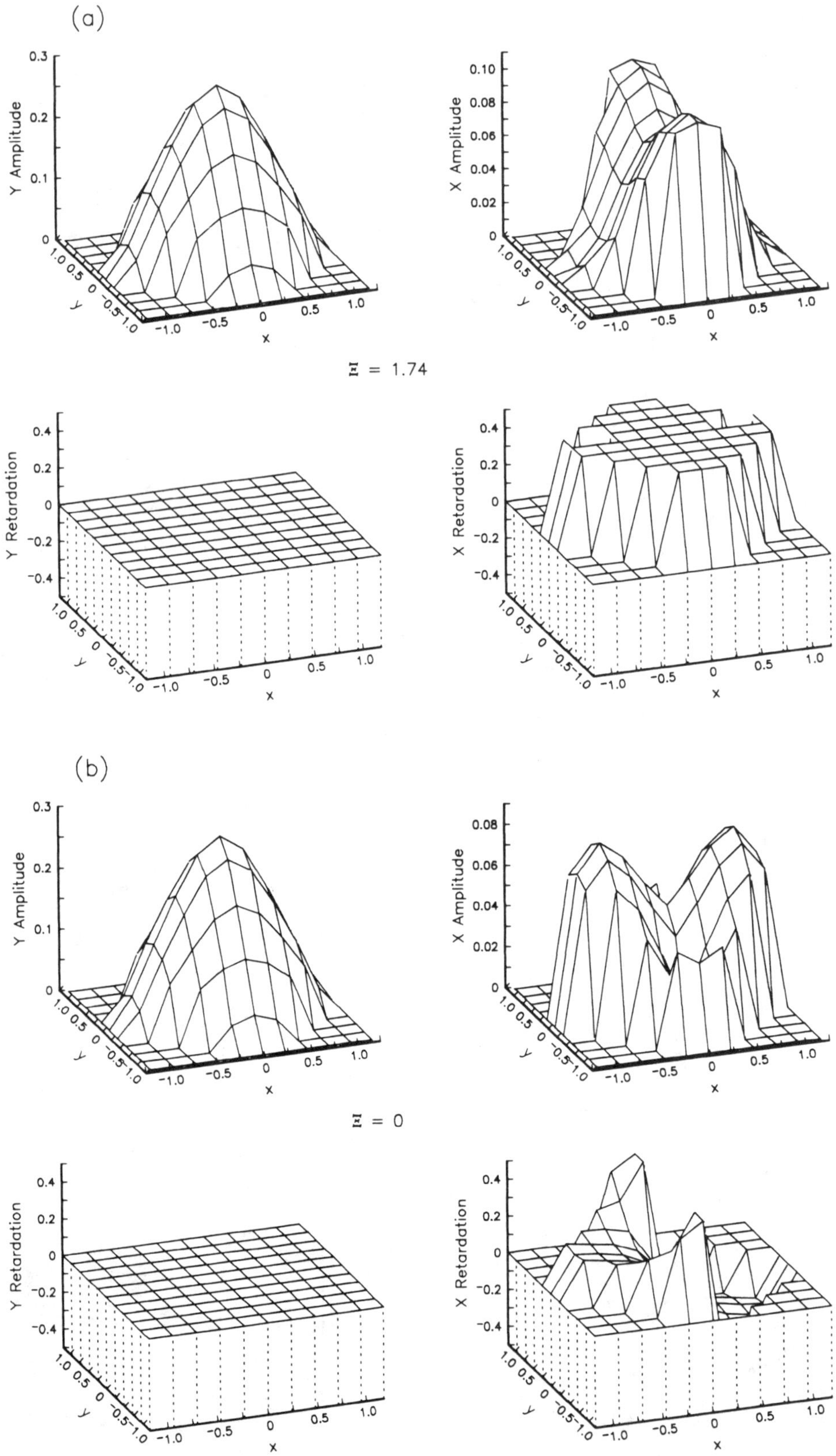

Fig. 5. Linearly polarized components of the far field read beam. The y-polarized component (on the left) is the local oscillator with a constant profile and zero retardation (in units of wavelength). The signal amplitude is x-polarized (on the right). (a) The laser spot is at the center of a written domain ($\Xi = 1.74\lambda$), which has a width smaller than the laser spot (see the inset in Fig. 6). The x-polarized amplitude pattern is broadened in the y-direction, and the retardation is $\lambda/2$, reflecting the rotation of the polarization in the opposite direction to the uniformly magnetized land. (b) The laser beam is exactly at the leading edge of the domain ($\Xi = 0$). The amplitude pattern shows the double humped structure of $|\psi_-|$, and the retardation reverses from $\lambda/4$ to $-\lambda/4$ at $x = 0$.

are shifted in opposite directions. The Kerr rotation was magnified approximately tenfold from its typical value to make the effects more noticeable in these amplitude-phase maps. When the beam is at the leading edge of the domain, as in Fig. 4(b), the magnetization pattern acts as a phase step of opposite sign for the two circular polarization states. The MO-induced phase step in the light at the disk surface results in a shift in the centroids of the amplitude profiles, causing them to lean in opposite directions at the exit plane of the collection lens. A pair of split detectors in this plane, sensitive selectively to LHC and RHC polarized light, would detect these centroid displacements and register a transition. When the beam is centered on the long domain (not shown), the amplitude profiles are again symmetric, while the retardations are opposite to what they are in Fig. 4(a).

By taking the appropriate linear combinations of the LHC and RHC complex amplitudes, the maps for linear polarizations $\hat{x}$ and $\hat{y}$, or $\hat{A}$ and $\hat{B}$, can be generated. Figure 5 shows the complex amplitudes for the input light $\hat{y}$ and the signal polarization $\hat{x}$ as described in Sec. II. When the magnetization is uniform (not shown) a fraction of the y input is transformed to the x-polarization, while the phases and amplitude profiles are not affected. Figure 5(a) shows what happens when the beam is centered on the domain. The effect of the finite width of the written domain is seen as a distortion of the x-amplitude profile in this 2-D analysis. The x-retardation, however, is uniform and equal to a ½ wave, indicating that the polarization rotation is reversed compared to the case of uniform magnetization. Figure 5(b) is the most interesting. Here the beam is centered on the leading edge of the domain. The retardation of the x-polarization is a ¼ wave for $x < 0$ and a −¼ wave for $x > 0$. The amplitude profile shows the expected double humped form of the antisymmetric function ψ_- described in the Appendix.

The complex amplitude maps can be processed by the Jones matrices associated with the various optical elements used,[9] and the optical detection can be performed numerically. In this way we can calculate the read signals for the various level and edge detectors described so far. In particular, we compare in Fig. 6 a dual polarization edge detector [Fig. 3(c)] and a dual polarization MO level detector [Fig. 1(b)]. The P–P signals are exactly in a 1:2 ratio for this isolated edge condition (that is, when the domain is sufficiently long compared to the beam size that the signals corresponding to the leading and trailing edges do not interfere). This 1:2 ratio between the edge and level detection signals is approximated by every correct comparison of schemes of this sort (see the Appendix).

The dual polarization level detector (solid curve in Fig. 6) operates as expected, and we will not describe this further. The edge detector operates as follows. The semicircular halfwave plate W reverses the polarization states of the $x < 0$ regions from LHC to RHC and vice versa. Thus, for example, the larger amplitude halves of the profiles in Fig. 4 would both be RHC, and the lower amplitude halves would be LHC after plate W. The compensator plate C simply turns LHC and RHC into $\hat{A}$ and $\hat{B}$ polarizations, which are separated by the polarizing beam splitter and sent to the two photodiodes. The difference signal between these two photodiodes is a maximum when the beam straddles an edge, as in Fig. 4(b), but is zero when the beam sees a uniform magnetization along $\hat{x}$, as in Fig. 4(a). The signal pulse generated by the edge detector is

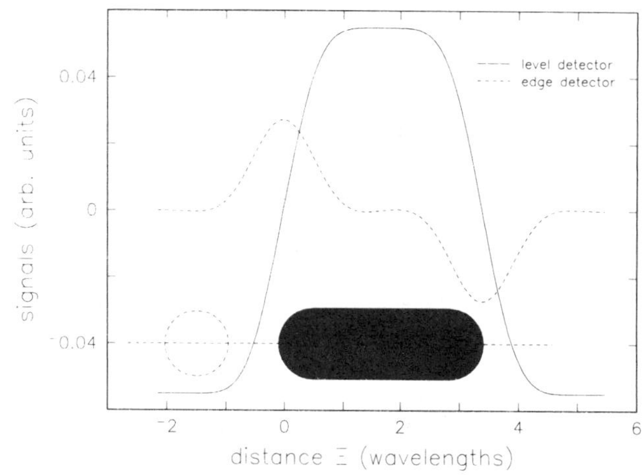

Fig. 6. Signals for the edge and level detectors calculated by 2-D Fourier optics. The edge detector peak is half of the height of the level detector maximum for this low frequency pattern. The level detector 10–90% distance is 0.97 μm for a 830-nm wavelength, which also happens to be the same as the FWHM for the edge detector. If the edge detector signal were perfectly Gaussian and the level detector were precisely the integral of this, the 10–90% distance would be 1.09 times the edge detector FWHM. The domain pattern and laser spot FWHM are shown in the inset.

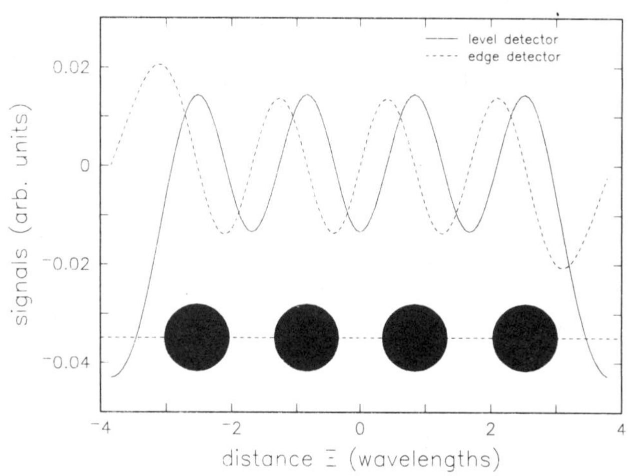

Fig. 7. Edge and level detector signals for the closely spaced four-domain pattern shown at the bottom. The central portion shows that the two detectors have equal signal strengths at high frequency.

calculated to be 0.97 μm in length (FWHM). The level detector gives a 10–90% transition length also equal to 0.97 μm in the same conditions.

The dual polarization level detector signal and the time integral of the edge detection signal, scaled to the same peak value, show very small differences. This indicates that the linear density potential for level and edge detection is the same.

To examine the performance of level and edge detection at higher spatial frequencies, the 2-D model was run using the domain patterns shown in the inset of Fig. 7. This figure compares the resulting read signals for dual polarization level detection and dual polarization edge detection. If we limit our attention to the central high frequency region of this plot, we see that the P–P signals are about equal.

The edge detector discussed above uses the far field light. In the reimaged case [Figs. 3(b),(d)] the fields at the detector reflect the 2-D shape of the domain on the disk. Note that if the shape of the detector (or filter) matches that of the domain, the high density performance can be improved. While this approach may not be valuable for racetrack shaped domains as in Fig. 6, the crescent domains produced by direct overwrite systems[18] yield larger signals at high density with matched filter edge detection than with level detection. Reimaging detector systems can also be designed to reject offtrack signals and noise and thus possibly allow improved areal density.

V. Additional Noise Sensitivities

All MO memories encounter noise from the electronics, storage medium, and laser.[10-15] Edge detection systems are more sensitive to certain optical noise sources and less sensitive to electronic and shot noise than differentiating level detection systems. The signal-to-noise analysis of level detection has been dealt with previously; the treatment here is very schematic.

All systems encounter electronic noise, which can be minimized by proper design of the preamplifier.[12,14] This noise does not depend on the signal or local oscillator levels. It is enhanced by electronic differentiation of the preamplifier signal.

Quantum noise (shot noise) and polarization fluctuations (resulting mostly from irregularities in the written data) produce a noise power linear in the local oscillator light level and proportional to the MO signal. This type of noise constitutes the ultimate noise floor in a well engineered system.[11,13,15]

Laser power fluctuations and reflectivity fluctuations in the medium produce a noise power that scales as the square of the local oscillator level.[19,20] These noise phenomena limit the magnitude of the optimum local oscillator and signal levels. Dual detector systems reject a large proportion of this last sort of noise as a common mode allowing larger MO signals without excessive noise.[15]

Edge detection systems are subject to all of these noise phenomena and a few additional ones. However, compared to single ended MO level detectors, edge detection systems are much less sensitive to laser power fluctuations, otherwise a very large effect.[20] All edge detectors have two equally illuminated photodiode segments which reject local oscillator fluctuations as common mode noise. In addition, the optical differentiating action of these systems eliminates the need for electronic differentiation, which enhances the electronic preamplifier and shot noise.

The additional noise sources are related to the motion of the beam on the detector surface. The centroids of semiconductor diode laser beams fluctuate in every plane due to spontaneous emission into higher-order modes of the laser cavity.[21] Even without feedback, this effect may produce a noise level 20 dB above the shot noise on a split detector, comparable with the power fluctuation noise.

There are actually two forms of this stochastic position noise: the centroid of the laser beam at the diode facet fluctuates, as does the direction of the beam. These two effects are uncorrelated; the former adds noise to single split detector reimaging edge detector systems, while the latter affects far field systems. Both effects can be eliminated by filtering the spatial mode of the laser diode with a polarization preserving optical fiber. Edge detection systems with two separated photodiodes sensitive to two orthogonal polarizations such as that in Figs. 3(c) and (d) are insensitive to laser position noise so long as no light is lost in the matched filter W. The laser position noise cannot be rejected electronically as a common mode; its level scales as the square of the local oscillator power and thus may require less optical power at the detector and less MO signal in the optimized system. The lessening of the optimum MO signal in turn makes even lower noise preamplifiers desirable.

Small scale roughness of the storage medium can also cause the position of the reflected spot to fluctuate at the detector plane. Such roughness cannot be entirely eliminated, but the magnitude of the resulting noise at signal frequencies can be made acceptable. This noise appears on the local oscillator field and can be minimized by reducing the power in that component. Unfortunately, such a strategy also reduces the MO signal.

We use the symmetric–antisymmetric fields defined in the Appendix to describe the substrate roughness-induced noise for single detector far field systems as in Figs. 1(a) and 3(a). The calculation for reimaging single detector systems in similar. We define the fluctuating height of the reflecting substrate surface to be

$$z_2 = \langle z_2 \rangle + \delta(x_2,y_2;\Xi), \tag{30}$$

where $\langle \delta \rangle = 0$ and $\langle |\delta|^2 \rangle \neq 0$. The incident field is $E_0\{2\} = \mathcal{A}_0 \psi_+\{2\}$. The medium introduces a phase shift on the reflected beam:

$$\begin{aligned} E_r\{2\} &= R\mathcal{A}_0 \exp[2ik\delta(x_2,y_2;\Xi)]\psi_+\{2\} \\ &\approx R\mathcal{A}_0\{1 + 2ik\delta(x_2,y_2;\Xi)\}\psi_+\{2\}. \end{aligned} \tag{31}$$

Expanding as in Eq. (21) we find that the local oscillator amplitude at the detector plane $\{6\}$ is

$$E_{LO}\{6\} = \beta R\mathcal{A}_0\psi_+\{6\} + 2i\beta R\mathcal{A}_0 k\Delta(\Xi)\psi_-\{6\}, \tag{32}$$

where

$$\Delta(\Xi) = \langle \psi_-^*\{2\}|\delta(\{2\};\Xi)|\psi_+\{2\}\rangle \tag{33}$$

is a stochastic mode conversion factor due to the fluctuating substrate surface. It is clear that the second term in Eq. (32) is a noise amplitude due to the roughness of the substrate. By symmetry, this amplitude does not contribute to the noise measured by a conventional level detector. However, it does contribute to the output of a split photodiode edge detector such as in Fig. 3(a). The arguments leading to Eqs. (20) and (26) result in a noise current of

$$i_n(\Xi) = 4q\beta^2R^2\mathcal{A}_0^2\,\mathrm{Re}\{i\langle\psi_+^*\{4\}|\mathrm{rect}_{r_0}\{4\}\,\mathrm{sgn}_0\{4\}|\psi_-\{4\}\rangle\Delta(\Xi)\}, \quad (34)$$

which has much the same structure as Eq. (26). Note that $i_n(\Xi)$ scales as β^2, not $\beta\eta$, because the surface roughness primarily affects the y-polarized light.

The importance of this noise is best estimated from the autocorrelation $\langle i_n^*(\Xi)i_n(0)\rangle$, and the Wiener-Khinchin theorem[22,23] can be used to calculate the spectrum of noise resulting from the stochastic slope fluctuations:

$$i_n^2(\nu) = \frac{1}{V}\int\langle i_n^*(\Xi)i_n(0)\rangle\exp(-2\pi i\nu\Xi/V)d\Xi, \quad (35)$$

where V is the linear velocity of the medium. It is this noise which appears in the output of a spectrum analyzer. Its exact form depends on the autocorrelation function of the medium roughness as averaged over the laser spot $\langle\Delta^*(\Xi)\Delta(0)\rangle$, and the noise spectrum depends on the Fourier transform of this quantity. Note that if both $\psi_+\{2\}$ and $\psi_-\{2\}$ are real, the product $\psi_+^*\{4\}\psi_-\{4\}$ is imaginary. Also $\Delta(\Xi)$ is real if it describes phase fluctuation but would be imaginary if it parameterized amplitude shifts [which would affect designs such as that in Fig. 3(b)]. At high frequencies, the noise spectral density is limited by the finite spatial bandwidth of the optical system. At low frequencies, the zero average displacement of the substrates implies that the noise scales as ν^2. Thus the most important length scale for substrate roughness is approximately the size of the laser spot at the medium.

If we assume that the optical system has the same effect on the noise as on the signal in Eq. (23), the Ξ dependence of the autocorrelation function becomes

$$\langle\Delta^*(\Xi)\Delta(0)\rangle = \langle|\Delta(0)|^2\rangle v(\Xi). \quad (36)$$

We can estimate the substrate roughness-limited SNR from Eqs. (26) and (34):

$$\frac{i_s(0)}{\langle i_n^2\rangle^{1/2}} = \frac{\eta}{4\beta k\langle|\Delta(0)|^2\rangle^{1/2}}. \quad (37)$$

It is clear that this SNR is increased as β is reduced. For typical values of $\beta = 0.1$, Kerr rotation $\eta = 0.01$, and $k = 2\pi/\lambda = 7.5\ \mu\mathrm{m}^{-1}$, the requirement for a current SNR >20 implies a maximum rms substrate roughness of $\langle|\Delta(0)|^2\rangle^{1/2} = 0.16$ nm. While this requirement may seem unduly restrictive, we have identified commercially available MO storage media which can meet it, even for this head design, which is most sensitive to roughness.

Dual polarization far field detector systems are less sensitive to this medium noise than single split detector systems, and reimaging designs are insensitive to angular deviations of the beam. However, the reimaging systems are sensitive to other effects which cause the position of the spot to move. In particular, split detector reimaging designs can be unbalanced by vibrations in the structure of the head, which physically translate the detector. Small scale reflectivity fluctuations of the storage medium deviate the centroid of the beam in reimaging designs. Al these factors imply that the quality of the storage medium and of the substrate must be very high for the adequate performance of MO edge detection schemes.

VI. Experimental Implementation

Six different MO edge detector systems were assembled to verify their characteristics. Quantitative data on signals and noise read from disks previously written using precision media testers were obtained and compared with data taken with a conventional level detector employing much of the same hardware. To avoid laser beam stochastic position noise, the output of the Sharp LT015MDO semiconductor diode laser was coupled into 1 m of Newport F-SPF polarization preserving optical fiber as shown in Fig. 8. The beam emerging from the fiber was collimated by a 10× microscope objective. The beam quality thus produced was not verified. Feedback from the input facet of the fiber lowered the oscillation threshold and raised the laser quantum efficiency near threshold. The laser was typically operated at a 60-mA current and produced 3-mW output from the fiber. Typically, the laser power at the storage medium was 1.5 mW. Laser power noise was suppressed by rf modulating the drive current at 195 mHz.[20]

The focus servo detected a portion of the return beam focused onto a knife-edge. A far field split detector provided the track-following servo signal. The aspherical objective lens had a numerical aperture of 0.5 and a focal length of 4.4 mm. The input beam diameter was chosen so that the objective vignetted the beam at the $1/e^2$ diameter. This lens was moved axially and transversely with a conventional dual voice coil actuator.

Most of the data reported here were obtained using sample Mitsubishi Chemical Co. disks. This storage medium employs a polycarbonate substrate with concentric 0.6-μm grooves and 1.0-μm lands. The nominal Kerr rotation angle is 0.5°, and the reflectivity was 20%.

A. Far Field Single Polarization Split Detector System

The basic difficulty in experimental implementation of single polarization split detector systems is that it is impossible both to reflect y-polarized read light and to transmit light polarized at the arbitrary angle β as in Fig. 1(a) or 3(a) with the same polarization beam splitter. The actual experimental system is shown in Fig. 9. Light from the laser/fiber/collimator system shown in Fig. 8 passes through a 70% transmitting beam splitter S and polarization analyzer P.

The y-polarized beam transmitted through the analyzer is then incident on a birefringent compensator plate C, which produces a small x-polarized component out of phase with the y-component. The operation of the head does not depend critically on the method of producing this x-component, which will act as the phase shifted local oscillator necessary for far field edge detection. We found it convenient to employ a quarterwave plate oriented with its axes at angle

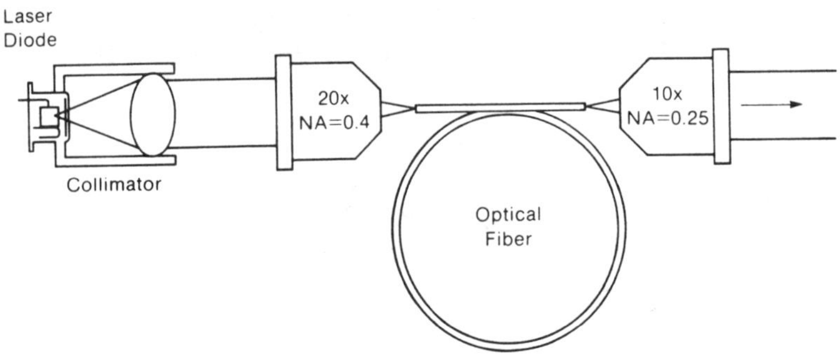

Fig. 8. Laser–collimator optical fiber system used to suppress stochastic position noise. The optical fiber must preserve polariztion.

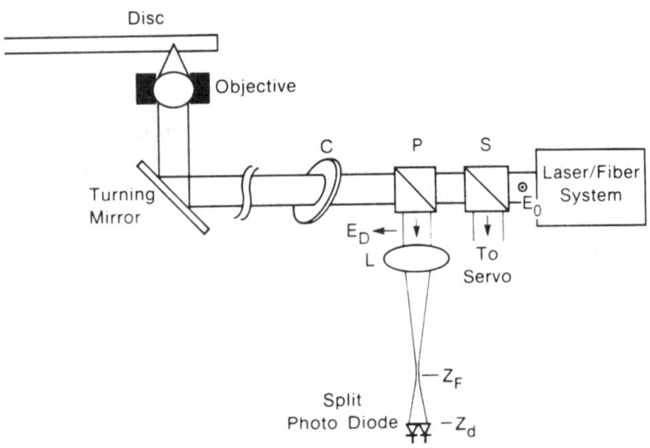

Fig. 9. Schematic of a single polarization split photodiode MO edge detector system. The portion to the left of the break in the beam has been rotated by 90° for clarity. The system shown here is a far field detector with a birefringent compensator plate C. The output of the laser collimator system E_0 is linearly polarized out of the page and propagates through the beam splitter S and the polarization analyzer P and into the compensator plate C. This device can be replaced by a Brewster plate in the reimaging systems. After the compensator (or Brewster) plate, the beam is turned and focused onto the medium by the objective lens. The reflected wave propagates back through the compensator (or Brewster) plate, and the component orthogonal to the initial polarization is directed by the polarization analyzer P into the lens L. That lens focuses the beam at axial position z_f. The split photodiode is at axial position z_d, which in this drawing is shown well behind focus.

β as the birefringent element. The Jones matrix for such a device is

$$\tilde{C} = \frac{1}{\sqrt{2}}\begin{bmatrix} 1 - i\cos 2\beta & -i\sin 2\beta \\ -i\sin 2\beta & 1 + i\cos 2\beta \end{bmatrix} = \frac{1}{\sqrt{2}}\begin{bmatrix} 1 - i & -2i\beta \\ -2i\beta & 1 + i \end{bmatrix}, \tag{38}$$

where the second equal sign applies when β is small. The turning mirror and objective do not alter the polarization. Thus the beam incident on the disk has the Jones vector representation[9]:

$$\mathbf{E}_0\{1\} = \frac{\mathcal{E}_0}{\sqrt{2}}\begin{bmatrix} -2i\beta \\ 1 + i \end{bmatrix} a_0\{1\}. \tag{39}$$

The medium at plane $\{2\}$ attenuates the beam by a scalar uniform factor R and alters the polarization of the read beam in a spatially nonuniform way as given by the Jones matrix:

$$\tilde{M} = \begin{bmatrix} 1 & \eta m(x_2;\Xi) \\ -\eta m(x_2;\Xi) & 1 \end{bmatrix}; \tag{40}$$

$$\mathbf{E}_r\{2\} = R\tilde{M}\mathbf{E}_0\{2\} = R\mathbf{E}_0\{2\} + \mathbf{E}_s(\{2\};\Xi). \tag{41}$$

The beam returning from the medium contains two amplitudes, $E_s\{2\}$, which depends on the magnetization $m(x_2;\Xi)$, and $\mathrm{RE}_0\{2\}$, which does not:

$$R\mathbf{E}_0\{2\} = \frac{R\mathcal{E}_0}{\sqrt{2}}\{-2i\beta\hat{\mathbf{x}} + (1 + i)\hat{\mathbf{y}}\}a_0\{2\}; \tag{42}$$

$$\begin{aligned} \mathbf{E}_s(\{2\};\Xi) &= [\tilde{M} - \tilde{I}]R\mathbf{E}_0\{2\} \\ &= \frac{1}{\sqrt{2}}\eta m(x_2;\Xi)R\mathcal{E}_0(1 + i)a_0\{2\}\hat{\mathbf{x}}. \end{aligned} \tag{43}$$

In Eqs. (41)–(43), terms beyond the lowest nonvanishing order in the small quantities β and η have been dropped.

This beam then propagates through the birefringent plate, and the x-polarized component is directed onto the split photodetector by the polarization analyzer. Mathematically, the field directed toward the detector is

$$\mathbf{E}_D(\{5\};\Xi) = \hat{\mathbf{x}}\hat{\mathbf{x}} \cdot \mathbf{E}_r\{4\} = R\mathcal{E}_0\{-2i\beta a_0\{4\} + \eta E_x(\{4\};\Xi)\}\hat{\mathbf{x}}. \tag{44}$$

Part of the y-polarized component is directed into the servo system by beam splitter S. The rest returns to the laser.

In the data detector beam line, the x-polarized beam is partially focused by a 10-cm focal length lens L. The detector can be moved through the resulting beam waist to fine tune the net phase shift between even and odd spatial distributions as indicated in Eq. (A12). The intensity distribution at the detector plane has a form analogous to Eq. (13):

$$I_D(\{6\};\Xi) = I_{\mathrm{LO}}\{6\} + 4R^2\beta\,\mathrm{Re}\{\eta\exp(i\Phi)E_0^*\{4\}E_x(\{4\};\Xi)\}, \tag{45}$$

where $I_{\mathrm{LO}}\{6\} = 4R^2\beta^2|E_0\{4\}|^2$ and the phase shift

$$\Phi = \frac{\pi}{2} - \cot^{-1}\left(\frac{z_d - z_f}{z_r}\right).$$

The split detector signal then has the same form as Eq. (20) and is optimized when $\arg\eta = \cot^{-1}[(z_d - z_f)/z_r]$, where z_d is the axial position of the detector, z_f is the position of the focus of the lens L and z_r is the Rayleigh range for that focus.

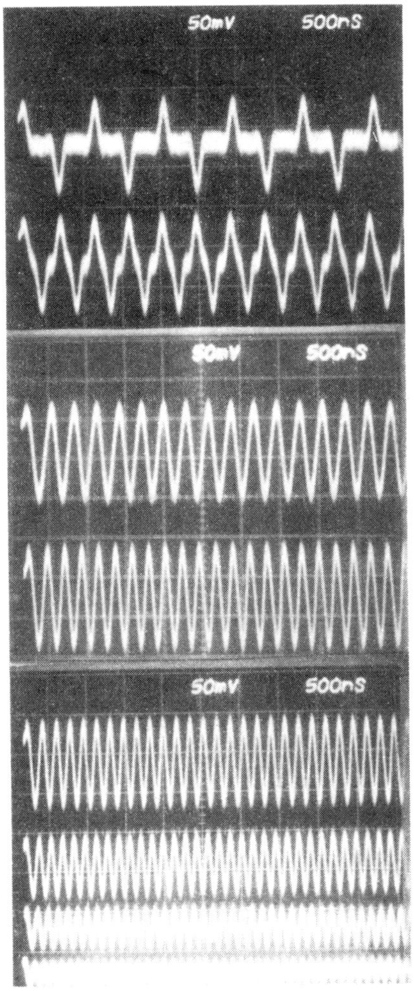

Fig. 10. Electronic signals from a far field edge detector. The experimental arrangement is shown in Fig. 9. The trace at the top corresponds to marks and gaps 5 μm in length, which produce a 1-MHz fundamental at 10 m/s. The succeeding traces are at 2, 3, 4, 5, 6, 7, and 8 MHz. All signals below 6 MHz ($\sim$0.8-μm mark and gap) are $\sim$130 mV P–P. Beyond 5 MHz the signals decrease in amplitude.

The optimum power level at the photodetector was 9 μW. The preamplifier system produced electronic noise that was equal to the quantum noise when there was 14 μW of light at the detectors. In practice, the noise level of i_s was well above the sum of electronic and quantum noise, although the common mode rejection of the differential amplifier system was more than sufficient to eliminate residual laser noise. We attributed the observed excess noise to the roughness of the disk substrate. This interpretation was reinforced by the observation that the spectral distribution of the noise scaled as expected with variations in the linear velocity of the track.

Figure 10 shows typical signals produced by patterns of regular reversals of magnetization along the track. The edge detector pulse shape approximates that predicted in Sec. IV. The pulse height was 65 mV, with an rms noise voltage of 4.2 mV. The pulse FWHM was 120 ns at a track velocity of $V = 11$ m/s. The maximum CNR in a 30-kHz bandwidth was 47 dB at 4.6 MHz and 11 m/s.

B. Reimaging Single Polarization Split Detector System

When the split detector is placed at (or near) the focal plane of lens L in Fig. 9, the phase shift required for the local oscillator is zero (or quite small). Such a local oscillator can be obtained by replacing the birefringent plate C in Fig. 9 with a diattenuating element. One such device is a Brewster plate oriented at 45° with respect to the y-axis. The Jones matrix for such a device is

$$\tilde{B} = \frac{1}{2}\begin{bmatrix} 1+p & 1-p \\ 1-p & 1+p \end{bmatrix}, \tag{46}$$

where p is the amplitude transmission coefficient for p-polarized light—typically 0.95. The transmission for s-polarization is unity. The y-polarized light transmitted by the beam splitter becomes

$$\mathbf{E}_0\{1\} = \frac{1}{2}\mathcal{E}_0[(1-p)\hat{\mathbf{x}} + (1+p)\hat{\mathbf{y}}]a_0\{1\}, \tag{47}$$

which is the input polarization. It is possible (but not necessary) to reorient the representation axes to make the analysis identical to that of Sec. II. In the current representation, the medium alters the input beam polarization in a nonuniform way:

$$\mathbf{E}_r\{2\} = R\tilde{M}\mathbf{E}_0\{2\} = R\mathbf{E}_0\{2\} + \mathbf{E}_s(\{2\};\Xi), \tag{48}$$

where

$$R\mathbf{E}_0\{2\} = \frac{R}{2}[(1-p)\hat{\mathbf{x}} + (1+p)\hat{\mathbf{y}}]\mathcal{E}_0 a_0\{2\}, \tag{49}$$

$$\mathbf{E}_s(\{2\};\Xi) = \frac{R}{2}\eta m(x_2;\Xi)\mathcal{E}_0[(1+p)\hat{\mathbf{x}} - (1-p)\hat{\mathbf{y}}]a_0\{2\}. \tag{50}$$

The light then propagates back through the Brewster plate, and the x-polarized component

$$\begin{aligned} \mathbf{E}_D(\{5\};\Xi) &= \hat{\mathbf{x}}\hat{\mathbf{x}} \cdot \tilde{B}\mathbf{E}_r\{4\} \\ &= \frac{R}{2}[\mathcal{E}_0(1-p^2)a_0\{4\} + \eta p E_x(\{4\};\Xi)]\hat{\mathbf{x}} \end{aligned} \tag{51}$$

is picked off by the polarization beam splitter and directed into the detector channel. The remaining y-polarized light goes to the servo channel or back to the laser.

The 10-cm focal length lens L then reimages the plane of the medium onto (or very near) the detector. The net phase shift between even and odd symmetry fields is $\zeta = \tan^{-1}(z_d - z_f/z_r) \approx 0$ and is included in the definition of the complex amplitudes at plane $\{7\}$. The intensity at the detector plane is

$$I_D = I_{\mathrm{LO}}\{7\} + R^2(1-p^2)p\,\mathrm{Re}\{\eta E_0^*\{7\}E_x(\{7\};\Xi)\}, \tag{52}$$

where $I_{\mathrm{LO}}\{7\} = (R^2/4)\,(1-p^2)^2\mathcal{E}_0^2 a_0\{7\}$; and the uniform and split detector signals can be calculated from Eqs. (16) and (19) for level detection and from Eq. (20) [or Eqs. (24)–(26)] for edge detection:

$$i_u = qpR^2(1-p^2)\mathcal{E}_0^2\,\mathrm{Re}\{\eta\langle\psi_+^*\{7\}|\mathrm{rect}_{r_D}\{7\}|\psi_+\{7\}\rangle\}u(\Xi); \tag{53}$$

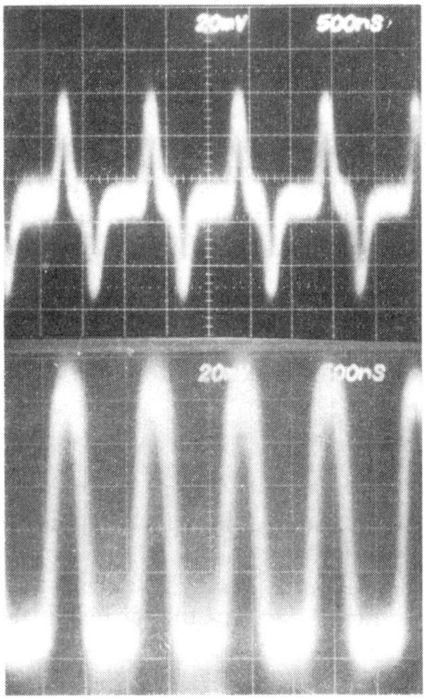

Fig. 11. Oscilloscope traces corresponding to reimaging edge detector output and level detector output produced by the same single polarization split detector system. The edge detector signal at the top is 95 mV P–P, while the level detector signal at this low frequency is 120 mV P–P.

The parameters of the operational system were very similar to that described in the preceding section. The power at the medium was 1.5 mW, with the power at the detector being 8.6 μW. The same storage medium and procedures were used. Obtaining clean edge detection signals required carefully centering the detector in the laser beam. If the center line of the split detector was offset, level detector signals appeared.

Typical traces of i_s and i_u signals appear in Fig. 11. The pulses produced by long domains were 45 mV P–P with a noise level of 4–5 mV (rms) in a 9-MHz bandwidth. On the same patterns, the single ended level detector signal i_u produced 120-mV P–P signals with 1.5-mV rms noise. This low SNR is typical for single ended level detectors which cannot reject laser noise as a common mode.[19,20] The ratio of the edge detector signal to the level detector signal was somewhat larger than predicted in Sec. IV.

The maximum CNR was 47 dB at 3.4 MHz and 11 m/s. In contrast, the single polarization edge detector gave a maximum CNR of 46 dB at 1.5 MHz. The CNRs of both detector systems are shown in Fig. 12 as a function of frequency along with the CNR of a fully optimized prototype drive employing a differential level detector. Above 4 MHz, both the signal level and CNR of the edge detector were superior to that of the single polarization level detector but were inferior to the optimized dual polarization level detector system.

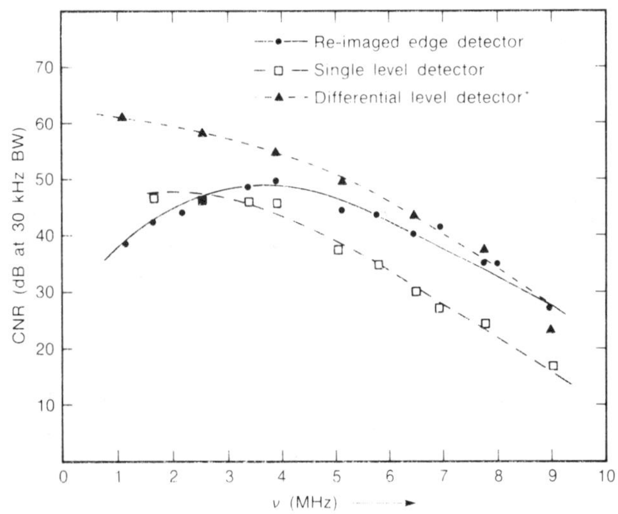

Fig. 12. Carrier-to-noise ratios for various detector systems. The linear velocity was 11 m/s, and the bandwidth was 30 kHz. The solid circles correspond to spectrum analyzer traces for the reimaging edge detector. The open squares are the single polarization level detector outputs from the same system, and part of the reduction in CNR results from unrejected laser noise. The triangles were produced by a fully optimized dual polarization differential level detector system reading the same optical disk.

$$i_s = qR^2p(1-p^2)\mathcal{E}_0^2\,\mathrm{Re}\{\eta\langle\psi_+^*\{7\}|\mathrm{rect}_{r_p}\{7\}\,\mathrm{sgn}_0\{7\}|\psi_-\{7\}\rangle v(\Xi). \quad (54)$$

In this case, level detector and edge detector signals are produced simultaneously.

C. Systems with Separate Detectors

Figure 13 shows the general scheme of our experiments with separated photodetectors and dual polarization. In these systems, a leaky polarization beam splitter at T reflects 100% of the x-polarized light and ~25% of the y-polarized light from the medium into the detector channel. Ideally, there is no phase shift between x- and y-polarization on reflection. There also is a polarization insensitive beam splitter at S, which separates the servo channel beams from the signal channel.

In the signal channel, the light passes through a compensator plate C, which imposes a phase shift Φ between x- and y-polarizations. A halfwave plate and polarization beam splitter then direct beams of nearly equal intensity and orthogonal polarization into the two separated photodiodes. In edge detection schemes, an optical filter, which includes a halfwave plate occluding half of the beam (see inset in Fig. 13), precedes the halfwave plate. In this system, the local oscillator level at the detector is larger than can constructively be used in single detector systems. In our case, that local oscillator power was 20 μW at each photodiode.

The amplitude reflected from the medium is given in Eq. (6). The beam splitter at T operates on this amplitude as

$$\mathbf{E}_T = \tilde{T}\mathbf{E}_r = \frac{1}{2}R\mathcal{E}_0 a_0\{4\}\hat{\mathbf{y}} + R\eta E_x(\{4\};\Xi)\hat{\mathbf{x}}, \quad (55)$$

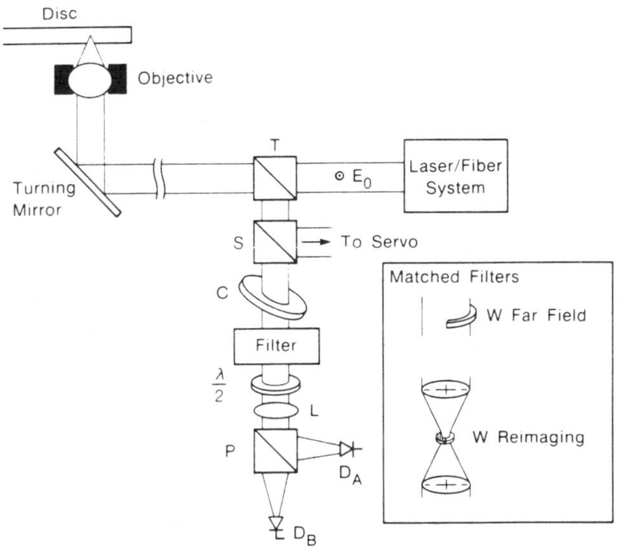

Fig. 13. Schematic of the dual polarization separated photodiode edge detector. Again the part to the left of the split has been rotated by 90° for clarity. The output of the laser collimator system is linearly polarized out of the page, and 75% of that power is transmitted though the leaky polarization beam splitter T and focused onto the disk. The orthogonal component of the reflected light along with 25% of the component in the incident polarization is reflected by the leaky beam splitter T into the detector and servo channels. The servo signals are split off at S. The compensator plate C and polarization dependent matched filter (inset) process the signal and local oscillator polarizations. A halfwave plate and polarization analyzer P direct equal average intensities corresponding to orthogonal polarizations into photodiodes D_A and D_B. The lens L merely insures that all light reaches the photodetectors, which remain in the far field region.

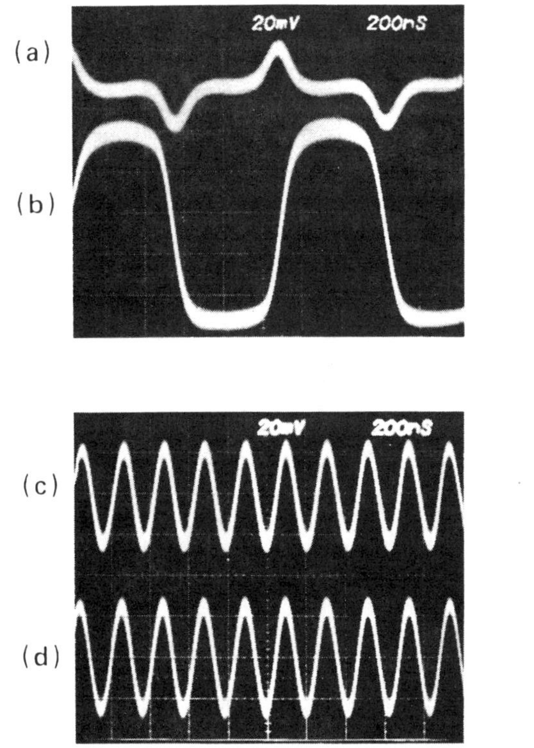

Fig. 14. Edge and level detector signals produced by a far field dual polarization separated detector system. The edge detector signals are in (a) and (c) with corresponding level detector signals in (b) and (d). While the level detector signal is higher at low frequency, the two become equal at high frequency. The linear velocity of the medium was 10 m/s.

where

$$T = \begin{bmatrix} 1 & 0 \\ 0 & 0.5 \end{bmatrix}$$

and $E_x(\{4\};\Xi) = \tilde{U}_{\{4\}\{2\}}m(\{2\};\Xi)a_0\{2\}$ is the signal field in plane {4} produced by the MO rotation in plane {2}. Beyond this plane, these systems act just as outlined in the theory of Eqs. (19) and (29) but with $E_{LO} = \frac{1}{2}R\mathcal{E}_0$ because of the effect of the leaky polarization beam splitter.

In the initial implementation, a halfwave plate oriented with the ordinary axis aligned along $\hat{y}$ intercepted half of the beam in the far field of the objective. This plate could be removed, but when it was in place it transformed the detection system into an edge detector as described by Eqs. (27) and (28). The optimum phase shift for edge detection was $\Phi = 90°$, and this was obtained by adjusting the compensator plate C. The resulting edge detection signals appear in Fig. 14(a). When the halfwave plate was removed, the system could be made to work as a level detector by readjusting the compensator plate to $\Phi = 0°$. The resulting signals appear in Fig. 14(b). The ratio of the optimum signal heights for lower frequency patterns was 0.44, comparable to expectations. The 4.5-MHz carrier shown in Figs. 14(c) and (d) gives the same signal level for each detector, again as predicted in Sec. IV.

The CNR in the 30-kHz bandwidth was measured for a variety of patterns with both the optimum level and edge detector systems. Results appear in Fig. 15. While the dual polarization systems are in principle less sensitive to media noise than the single polarization designs, optimized edge detection devices of each type give a similar performance. In particular, the maximum CNR for edge detection occurred at 3.4 MHz with V = 10 m/s. In these conditions the dual polarization edge detection system gave a 46.8-dB CNR, while the level detector gave 48.6 dB. At 5 MHz and beyond both systems gave the same CNR.

Very similar results were obtained with two split detectors and no partial halfwave plate. In this design, the signals from opposite sides of the two far field detectors were summed and the difference in the two sums taken electronically. The positioning of the two detectors was only moderately critical, and optimum performance occurred when the average currents on each side of each detector were made equal.

A reimaging separated detector system was arranged by focusing the reflected beam onto the partial halfwave plate with the 10-cm focal length lens in the scheme shown in the inset of Fig. 13. We found it useful to immerse the wave plate in index matching fluid to facilitate collection of the light downstream.

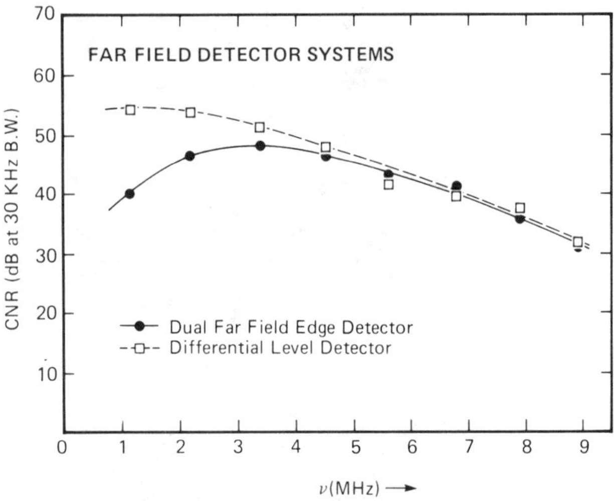

Fig. 15. Carrier-to-noise ratios for dual polarization separated detector systems. Both graphs were produced with the same hardware at 11 m/s and 30-kHz bandwidth. While the edge detector seems to give a lower CNR at low frequency, the actual peak height does not vary. Both systems give the same CNR above 5 MHz, corresponding to marks and gaps of ≤1.0 μm. The electronic noise for this system was only 3 dB below the observed noise floor. With ideal preamplifiers, the CNR values would be 3 dB higher.

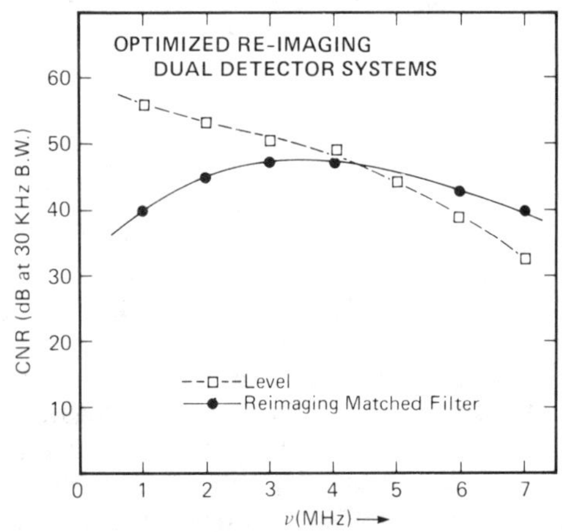

Fig. 16. Carrier-to-noise ratios for the optimized reimaging matched filter dual polarization edge detector and a corresponding level detector. The edge detector produces a better CNR above 4 MHz, which at the 10-m/s linear velocity corresponds to a 1.3-μm mark size. The level detector falloff at high frequency is partially due to a slight defocus condition to which the reimaging edge detector is relatively immune. The electronic noise level was <2 dB below the noise floor beyond 3 MHz. An ideal preamplifier would increase the CNR values by ~4 dB.

This system was very sensitive to misalignment and vibration but generally behaved as predicted. Peak and level detector signals for the same pattern were obtained at the same compensator plate settings by simply translating the partial wave plate transversely when the proper focus condition was found. The signal ratio was 0.6 at low frequency. The maximum CNR for the edge detector was 48 dB at 4 MHz and $V =$

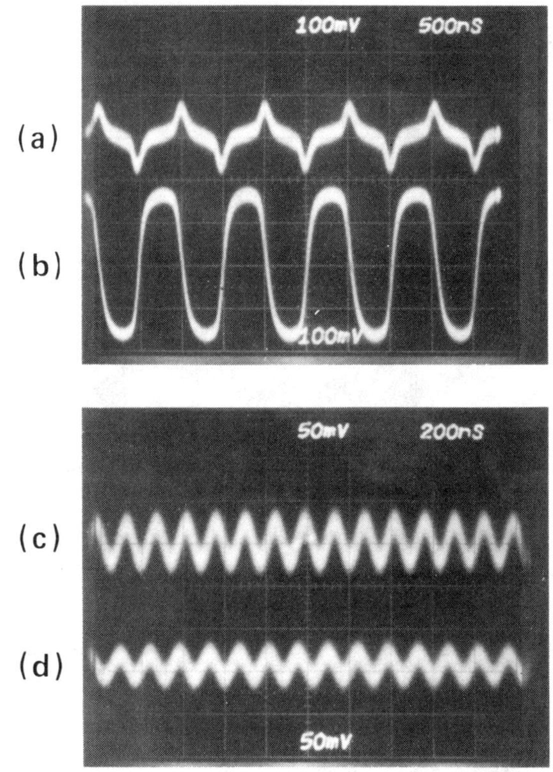

Fig. 17. Edge and level detector signals recorded simultaneously on a dual beam oscilloscope from a reimaging dual polarization split detector system. Rising edges produce negative edge detection peaks. At 1 MHz the edge detector signal in (a) is about half of the level detector in signal (b), but the 7-MHz edge signal in (c) is clearly larger than the level detector signal in (d). Such a system might be used in an all-optical conditional level detector channel. Linear velocity was 10 m/s.

10 m/s. In the same conditions the level detector gave 49 dB, as seen in Fig. 16.

When vibration and alignment problems can be controlled, a reimaging separated split detector system produces edge detection and level detection signals simultaneously, eliminating the need for electronic differentiation. Such a system would be relatively insensitive to disk substrate roughness but might respond unfavorably to other optical anomalies. Figure 17 shows oscilloscope traces produced by such a system. The CNR performance was essentially identical to Fig. 16.

VII. Promise for the Future

Optical edge detection of the transitions in MO memories has been shown to be possible. However, at present there is a penalty in both signal magnitude and noise sensitivity for these systems. The excess noise due to media roughness, nonuniform reflectivity, etc. can be made insignificant by reducing the local oscillator power level at the detectors. Present day detector–preamplifier systems require more than 7 μW of optical power to overcome the detector noise. When integrated GaAs detector amplifier systems with a power requirement of 0.5 μW or less become available,[24] the excess media noise should be easily eliminated. While such a detector amplifier system will still produce a larger SNR when used in a conventional level detector system, the performance of edge detectors should then become adequate.

The conventional dual differential level detector systems are complicated and sensitive to the phase shifts between x- and y-polarizaton. The single ended edge detector designs discussed in Sec. VI.A and B are simpler and sensitive only to the phase shift between even and odd symmetry fields in the same polarization. Such systems may be more feasible to implement in integrated optics than conventional detectors. Minimum signal light is rejected by such systems, which suggests that the SNR might indeed be higher with edge detection.

Optimized reimaging edge detection systems should show a performance comparable to conventional level detectors for domain sizes about equal to the laser spot size at the medium. Larger domains and lower frequencies will always produce bigger signals in level detection systems. When direct overwrite technologies are implemented, the importance of small domains and high frequencies will be further enhanced. Additionally, storage media with a CNR above 66 dB at low frequency (as measured with a level detector) will produce CNRs above 50 dB over a considerable bandwidth even with a poorly optimized edge detector. Edge detection might then become the technology of choice.

VIII. Conclusions

We have shown that optical methods of detecting the transitions between up and down magnetized domains exist. The spatial resolution of simple devices of this type has been shown equivalent to that of a conventional level detector followed by electronic differentiation. The SNR is better than that of a single ended MO level detector but not as good as differential MO level detection at low frequency. The dominant noise source is either substrate roughness or head vibrations.

Magnetooptic edge detection produces a smaller signal than level detection at low spatial frequency but equivalent signal levels at high frequency. While more sensitive to certain kinds of optical noise, the edge detectors reject most laser noise and yield a CNR comparable with differential level detectors. Further improvement in heads, electronics, and substrates will be needed to obtain superior signal-to-noise performance for these systems. However, the relative simplicity of the single polarization split detector designs may prove valuable for low end systems even with current technology.

The authors acknowledge valuable insights by James Zavislan, technical contributions of Pat Coleman and Charles Davis, and the loan of key apparatus by Frank Luecke.

Appendix: Reduced Basis Set Approximation for MO Systems

While Fourier optics methods are adequate to describe the phenomena encountered in MO data storage, they are mathematically cumbersome and provide little insight. The highly compact notation introduced in this paper facilitates writing the appropriate expressions, but evaluating enough of them to characterize a system can still be burdensome. For example, the propagation operator introduced in Eq. (3) has the Fourier integral representation

$$\tilde{U}_{\{\gamma\}\{\alpha\}}\mathbf{E}\{\alpha\} = \frac{-1}{\lambda(z_\gamma - z_\alpha)} \iint_{-\infty}^{\infty} dx_\alpha dy_\alpha \times \exp\left\{\frac{ik[(x_\gamma - x_\alpha)^2 + (y_\gamma - y_\alpha)^2]}{2(z_\gamma - z_\alpha)}\right\} \mathbf{E}(x_\alpha y_\alpha). \quad \text{(A1)}$$

For free space, note that as $\{\gamma\} \rightarrow \{\alpha\}$,

$$\tilde{U}_{\{\gamma\}\{\alpha\}} \rightarrow \iint_{-\infty}^{\infty} dx_\alpha dy_\alpha \delta(x_\gamma - x_\alpha)\delta(y_\gamma - y_\alpha). \quad \text{(A2)}$$

Many such expressions need to be evaluated to describe the systems of Figs. 1 and 3.

The full power of Fourier optics techniques is not really needed to describe systems that can transmit only a few spatial modes. We present here a simplified reduced basis set approximation that expresses all relevant quantities in terms of a few overlap integrals of at most four spatial functions.[16] Only two Fourier transforms need be calculated. The expressions obtained provide insight into the operation of the system and speed estimates of relevant quantities.

As an example this method is used here to estimate the dependences of the level and edge detection signals on the displacement of the transition Ξ from the center of the laser beam. Two spatial functions are necessary, one symmetric (ψ_+) and one antisymmetric (ψ_-). It is convenient to choose these functions to be normalized to unity. The symmetric function describes the distribution of the input beam at focus (plane {2}):

$$\psi_+(x_2,y_2) = E_0(x_2,y_2)/\langle E_0^*\{2\}|E_0\{2\}\rangle^{1/2}. \quad \text{(A3)}$$

The field distribution produced by simple reflection at the medium is then $E_0(x_2,y_2) = R\mathcal{A}_0\psi_+(x_2,y_2)$, where $|\mathcal{A}_0|^2$ is the laser–collimator output power at plane {0}.

The antisymmetric function should have the same phase fronts as ψ_+ and the minimum divergence consistent with the laws of optics. One method of defining ψ_- is to use Fourier optics techniques to propagate $E_x(x_5,y_5;0)$ back to plane {2}:

$$\psi_-(x_2,y_2) = \mathcal{D}\tilde{U}_{\{2\}\{5\}}E_x(\{5\};0), \quad \text{(A4)}$$

where $\tilde{U}_{\{2\}\{5\}}$ is the backward Fourier propagation operator,[6] and the normalization constant $\mathcal{D}$ is chosen so that

$$\langle \psi_-^*\{2\}|\psi_-\{2\}\rangle = 1. \quad \text{(A5)}$$

This gives a rigorously correct form for ψ_- but one which is not particularly convenient. The function $\psi_-\{2\}$ thus defined differs from $E_x(\{2\};0)$ as defined in

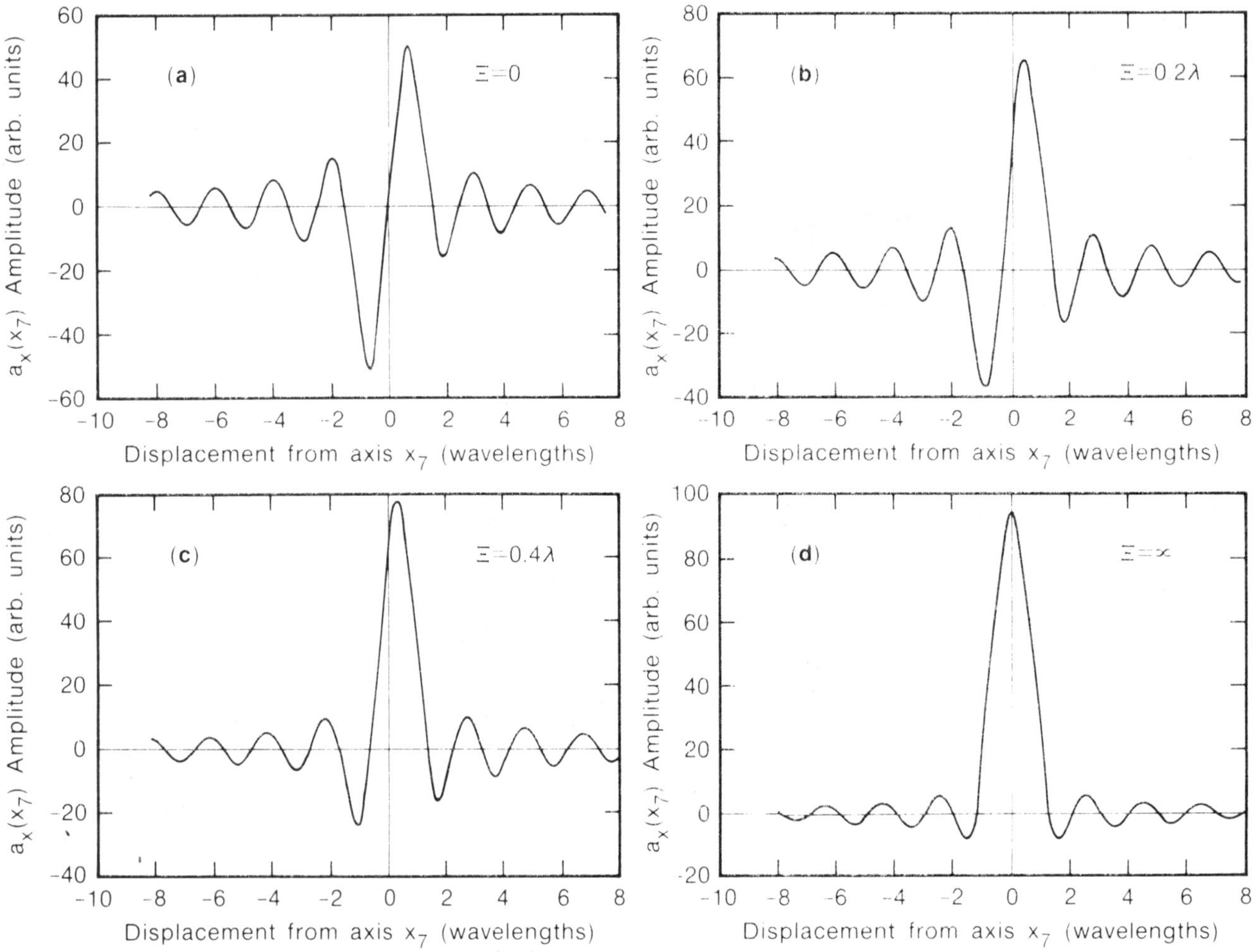

Fig. 18. Signal (x-polarized) field distributions for the reimaging detectors as a function of the displacement of the transition from the beam axis: (a) shows the purely antisymmetric function ψ_-, while (d) shows the symmetric profile of ψ_+ (i.e., the local oscillator field). As the transition moves away from the beam axis, the field distribution goes from antisymmetric to symmetric, remaining purely real for all displacements Ξ.

Eq. (7) in that the infinitely sharp step in $E_x(\{2\};0)$ has been appropriately smoothed by the finite apertures of the system. A typical $\psi_-\{2\}$ function appears in Fig. 18(a).

An alternative and more convenient approximation is

$$\psi_-(x_2,y_2) = \frac{d}{dx_2}\psi_+(x_2,y_2)\left\langle \frac{d}{dx_2}\psi_+^*(x_2,y_2)\middle| \frac{d}{dx_2}\psi_+(x_2,y_2)\right\rangle^{-1/2}, \tag{A6}$$

and this will be used later. In using Eq. (A6), one must deal properly with divergences that result from discontinuities in ψ_+. Such divergences can be eliminated when the output aperture at plane $\{4\}$ is slightly smaller than the input aperture in plane $\{0\}$. Alternatively, the detector radius r_D or spatial filter aperture r_W can be chosen to avoid nonphysical results. The final answer does not depend very sensitively on the choice of ψ_-.

Fourier optics techniques may be used to relate $\psi_+\{2\}$ and $\psi_-\{2\}$ to the corresponding functions in any other plane. The total electric field in that plane is the corresponding superposition. In particular, the formulas for the signal currents in Eqs. (24)–(26) result from using ψ_+ and ψ_- to project out the symmetric and antisymmetric portions of the field $E_x(x_4,y_4;\Xi)$ at plane $\{4\}$, as implied by Eqs. (16), (19), (20), and (29).

As an example of the use of a reduced state model, we apply it to estimate the signals which correspond to the head designs in Figs. 1(a) and 3(a). In the model, we assume that the input beam is a focused Gaussian, with a beam waist at the recording medium. We ignore vignetting of the input beam by the edges of the objective lens. However, the beam divergence may be chosen to match that of a realistic system where the edge of the lens lies near the $1/e^2$ point of the input beam intensity.

The input field is best described by the function

$$\psi_+(x,y,z) = \sqrt{\frac{2}{\pi}}\,\frac{1}{w(z)}\exp\left[-\frac{x^2+y^2}{w^2(z)}\right], \tag{A7}$$

where the beam waist is at plane $\{2\}$. The beam radius is then

$$w(z) = w_0\left\{1+\left[\frac{\lambda(z-z_2)}{\pi w_0^2}\right]^2\right\}^{1/2} \tag{A8}$$

The phase fronts are described by the function

$$H(x,y,z) = \exp\left\{-ik\left[\frac{x^2+y^2}{2R(z)} + z - z_2\right]\right\}, \qquad (A9)$$

where

$$R(z) = (z - z_2)\left[1 + \frac{\pi w_0^2}{\lambda(z - z_2)}\right] \qquad (A10)$$

and $k = 2\pi/\lambda$.

In the calculations that follow in this section, the function $H(x,y,z)$ is chosen as the phase reference at every plane, and the physical 3-D field pattern is proportional to $\psi_\pm(x,y,z)H(x,y,z)$. The wavelength can be chosen as 830 nm and the spot size at the medium as 920 nm to match the numerical calculations of Sec. IV. These choices imply that the beam waist $1/e^2$ radius is $w_0 = (2 \ln 2)^{-1/2} \times 920$ nm $= 780$ nm and that the convergence angle $\theta = \sin^{-1}(\lambda/\pi w_0) = 19.8°$.

The other field profile necessary in the analysis is proportional to the TEM_{01} mode profile with the same beam waist:

$$\psi_-(x,y,z) = 2\sqrt{\frac{2}{\pi}}\frac{x}{w^2(z)}\exp\left[-\frac{x^2+y^2}{w^2(z)}\right]\exp(-i\zeta), \qquad (A11)$$

where the phase shift

$$\zeta = \tan^{-1}\left[\frac{\lambda(z - z_2)}{\pi w_0^2}\right], \qquad (A12)$$

and z_2 is the position of the beam waist.

The overlap integrals in Eqs. (22) and (23) which describe the interaction with the edge of a domain are simply evaluated:

$$u(\Xi) = \sqrt{\frac{2}{\pi}}\frac{1}{w_0}\left[\int_{-\Xi}^{\infty}\exp(-2x^2/w_0^2)dx - \int_{-\infty}^{-\Xi}\exp(-2x^2/w_0^2)dx\right] = \mathrm{erf}\left(\frac{\Xi\sqrt{2}}{w_0}\right), \qquad (A13)$$

$$v(\Xi) = \frac{2}{w_0^2}\sqrt{\frac{2}{\pi}}\left[\int_{-\Xi}^{\infty}x\exp(-2x^2/w_0^2)dx - \int_{-\infty}^{-\Xi}x\exp(-2x^2/w_0^2)dx\right] = \sqrt{\frac{2}{\pi}}\exp(-2\Xi^2/w_0^2), \qquad (A14)$$

where $\mathrm{erf}(u)$ is the error function

$$\frac{2}{\sqrt{\pi}}\int_0^u \exp(-t^2)dt.$$

Since no other factor in Eqs. (24)–(26) depends on Ξ, these results are sufficient to determine the temporal profile of the electronic signal produced by a transition that moves past the focused spot at a linear velocity $d/dt = V$. Note that

$$v(\Xi) = \frac{w_0}{2}\frac{du}{d\Xi},$$

and thus a level detector followed by an electronic differentiator will produce the same sort of signal as a dc channel used with an edge detector. Noise issues aside, there is no intrinsic density advantage to be gained from one head or the other.

Equations (24)–(26) can be used to calculate the relative magnitudes of the level and edge detector signals. If we ignore the vignetting of the reflected beam by the objective and detector, the overlap integrals between the local oscillator and signal polarization are

$$\langle\psi_+^*\{4\}|\psi_+\{4\}\rangle = \frac{2}{\pi}\frac{1}{w^2(z_3)}\int\int_{-\infty}^{\infty}\exp\left[-2\frac{x^2+y^2}{w(z_3)}\right]dxdy = 1, \qquad (A15)$$

$$\langle\psi_+^*\{4\}|\mathrm{sgn}_0\{4\}|\psi_-\{4\}\rangle = \frac{2}{w^3(z_3)}\frac{4}{\pi}\int_{-\infty}^{\infty}\int_0^{\infty}x\exp\left[-2\frac{x^2+y^2}{w^2(z_3)}\right]dxdy = \sqrt{\frac{2}{\pi}}, \qquad (A16)$$

where we have used $z_6 \approx z_4 \approx z_3$. Thus, according to Eqs. (24) and (26), for optimum phase shift Φ the split and uniform detector signals have the following ratio:

$$i_s(0)/i_u(\infty) = \frac{2}{\pi} = 0.64. \qquad (A17)$$

The signal level for the edge detector suffers more from the vignetting of the reflected beam by the objective. Assuming that $r_O = w(z_4)$ the overlap integrals with vignetting are

$$\langle\psi_+^*\{4\}|\mathrm{rect}_{r_O}\{4\}|\psi_+\{4\}\rangle = 1 - \exp(-2) = 0.86, \qquad (A18)$$

$$\langle\psi_+^*\{4\}|\mathrm{rect}_{r_O}\{4\}|\psi_-\{4\}\rangle = \sqrt{\frac{2}{\pi}}\mathrm{erf}(\sqrt{2}) - \frac{4}{\pi}\exp(-2) = 0.59, \qquad (A19)$$

Implying a ratio of the maximum signal levels of

$$i_s(0)/i_u(\infty) = 0.55. \qquad (A20)$$

Note that this vignetting of the reflected beam does not affect the temporal profile, only the signal level.

This sort of analysis can be carried forward to the reimaged detector configuration described in the text or the dual detector schemes of Figs. 3(c) and (d). However, except for the value of the optimum phase shift, the results for signal profiles and ratio would be the same as in this simple case. Thus, with Gaussian input beams, all MO edge detector schemes produce signals with the temporal profile of the derivative of the level detector signal and peak signals (for isolated transitions) smaller by a factor of ~0.5.

References

1. G. Bouwhuis, J. Braat, A. Huijser, J. Pasman, G. van Rosmalen, and K. Schouhamer-Immink, *Principles of Optical Disk Systems* (Adam Hilger, Bristol, 1985).
2. M. J. Freiser, "A Survey of Magnetooptic Effects," IEEE Trans. Magn. **MAG-4**, 152–161 (1968).
3. A. E. Bell and F. W. Spong, "Antireflection Structures for Optical Recording," IEEE J. Quantum Electron. **QE-14**, 487–495 (1978).
4. M. Mansuripur and G. A. N. Connell, "Signal and Noise in Magnetooptical Readout," J. Appl. Phys. **53**, 4485–4494 (1982).
5. A. E. Bell, "Antireflection Structures for Magnetooptic Recording," in *Technical Digest SPIE Topical Meeting on Optical Data Storage, Vol. 10* (1987), pp. 138–139.
6. J. W. Goodman, *Introduction to Fourier Optics* (McGaw Hill, New York, 1968).
7. M. D. Levenson and S. S. Kano, *Introduction to Nonlinear Laser Spectroscopy* (Academic, Boston, 1988).

8. T. Deguchi, H. Katayama, A. Takahashi, K. Ohta, S. Kobayashi, and T. Okamoto, "Digital Magnetooptical Disk Drive," Appl. Opt. **23,** 3972–3978 (1984).
9. E. Hecht and A. Zajac, *Optics* (Addison-Wesley, Reading, MA, 1974), p. 268.
10. G. L. Eesley, M. D. Levenson, and W. M. Toiles, "Optically Heterodyned Coherent Raman Spectroscopy," IEEE J. Quantum Electron. **QE-14,** 44–49 (1978).
11. M. D. Levenson and G. L. Eesley, "Polarization Selective Optical Heterodyne Detection for Dramatically Improved Sensitivity in Laser Spectroscopy," Appl. Phys. **19,** 1–17 (1979).
12. A. G. Dewey, "Measurement and Modelling of Optical Disk Noise," Proc. Soc. Photo-Opt. Instrum. Eng. **695,** 72–78 (1986).
13. P. Grangier, R. E. Slusher, B. Yurke, and A. LaPorta, "Squeezed-Light Enhanced Polarization Interferometer," Phys. Rev. Lett. **59,** 2153–2156 (1987).
14. B. I. Finkelstein and W. C. Williams, "Noise Sources in Magnetooptic Recording," Appl. Opt. **27,** 703–709 (1988).
15. A. G. Dewey, "Optimizing the Noise Performance of a Magneto-Optical Read Channel," Proc. Soc. Photo-Opt. Instrum. Eng. **1078,** 279–286 (1989).
16. A. Yariv, *Quantum Electronics,* (Wiley, New York, 1975), Chap. 7.
17. V. B. Jipson and C. C. Williams, "Two-Dimensional Modeling of an Optical Disk Readout," Appl. Opt. **22,** 2202–2209 (1983).
18. D. Rugar, "Magnetooptic Direct Overwrite Using a Resonant Bias Coil," IEEE Trans. Magn. **MAG-24** 666–669 (1988).
19. M. Mansuripur, G. A. N. Connell, and J. W. Goodman, "Signal-to-Noise in Magnetooptic Storage," Proc. Soc. Photo-Opt. Instrum. Eng. **329,** 215–222 (1982).
20. B. J. Finkelstein and D. E. Call, "Laser Feedback Noise in Magnetooptical Recording," Proc. Soc. Photo-Opt. Instrum. Eng. **899,** 69–76 (1988).
21. M. D. Levenson, W. H. Richardson, and S. H. Perlumutter, "Stochastic Noise in TEM_{00} Laser Beam Position," Opt. Lett. **14,** 779–781 (1989).
22. N. Wiener, "Generalized Harmonic Analysis," Acta Math. **55,** 117–258 (1930).
23. A. Khinchin, "Korrelations Theorie der Stationëren Stochastischen Prozesse," Math. Ann. **104,** 604–615 (1934).
24. D. Rogers, T. J. Watson Research Center, I.B.M.; private communication (1989).

Reprinted with permission from *Applied Optics*, Vol. 29(20), pp. 3046-3051 (July 10, 1990). ©1990 Optical Society of America.

High density recording by superresolution in an optical disk memory system

Yutaka Yamanaka, Yutaka Hirose, Hiroaki Fujii, and Keiichi Kubota

High density optical disk recording using an optical superresolution technique is proposed. By setting a shading band in the center of the collimating beam, the focused spot size on the recording medium is reduced to 80% compared with that of a conventional optical recording. The readout signal degradation affected by the focused spot sidelobe on the recording medium is suppressed by spatially extracting the main lobe region from the reflected back beam using a slit on a reimaged plane. The linear recording density has been improved to 1.2 times as high as that for a conventional optical recording using a normal resolution optical head.

I. Introduction

Optical disk systems have a large data storage capacity compared with magnetic disk recording systems. For this reason, optical disk systems are being employed in document filing systems and image data filing systems. Large data capacity, however, is demanded in many applications, such as a high definition TV video recording system. Higher recording density than that of the present optical disk is essential for the development of such a larger capacity optical disk system without sacrificing its compactness.

To meet these demands, several approaches have so far been made, for example, improvements in the modulation code such as pit width modulation,[1] a duobinary code,[2] an offset track geometry,[3] or employment of high resolution optics using a higher numerical aperture objective lens. However, these methods sacrifice either the SNR (signal-to-noise ratio) in the readout signal or optics alignment tolerances of the optical head assembly. Shortening the light source wavelength of the recording optical head is another approach. At the present time, the available laser diode wavelength, however, is limited to ~0.6 μm.[4] Gas laser sources, such as a He–Cd^+ laser or an Ar^+ laser, are difficult to introduce into a practical compact filing system because of their larger size.

Keiichi Kubota is with NEC Corporation, Peripherals Division, 10, Nisshincho-Icchome, Fuchu City, Tokyo 183, Japan; the other authors are with NEC Optoelectronic Research Laboratories, 1-1, Miyazaki 4-chrome, Miyamae-ku, Kawasaki 213, Japan.

Received 10 October 1989.
0003-6935/90/203046-06$02.00/0.

Bearing in mind that we need a smaller spot using the currently available laser diode as a light source and a normal N.A. objective lens, the authors propose introducing superresolving optics into an optical disk head. Superresolution was studied for improving an optical system's resolving power.[5,6] A superresolved spot, however, has sidelobes, which are inherent in superresolution and may degrade the read/write characteristics in the optical head. This paper suggests a new optical head which enables use of the superresolution technique in high density optical recording.

In the following sections, the design of the optical head and numerical analysis are described. In Sec. IV, the experimental results of recording characteristics in a write-once optical head are shown.

II. Optical Head Design

Figure 1 shows an optical system for superresolution. The focused spot profile is determined by the pupil-plane filter set in the collimating beam prior to its incidence on the objective lens. To minimize the sidelobe intensity, the pupil-plane filters having an optical density distribution, such as $(1 - r^2)^n$, in the transverse direction were proposed.[5,6] However, such optical filters are difficult to fabricate for practical use. Considering the ease of manufacturing and the light power efficiency, a simple shading band, as shown in Fig. 1, was employed in our optical head. By introducing the shading band at the center of the collimating beam, the focused main spot size is reduced to less than the diffraction-limited size. The appearance of the sidelobe, however, is not negligible for optical recording application. To reduce the sidelobe's influence on the superresolving spot and to achieve high density read/write characteristics, a new optical head has been designed taking account of the following two points.

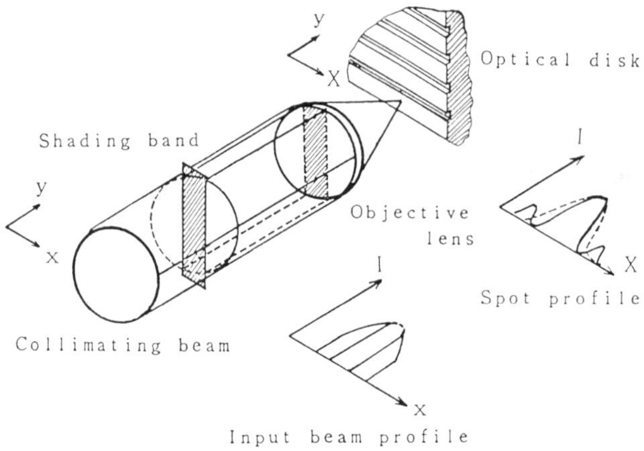

Fig. 1. Optical system for superresolution.

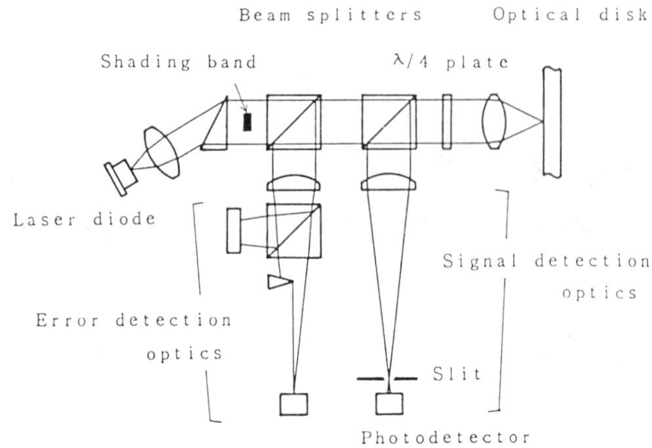

Fig. 3. Optical head design using superresolution.

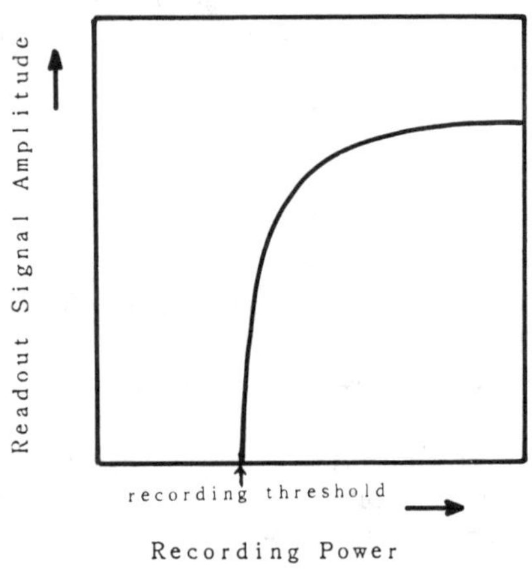

Fig. 2. Typical readout signal amplitude dependence on the recording power for a typical write-once recording medium.

The sidelobe may cause error pit writing in recording. In the optical recording, however, a recording medium has the threshold characteristics for recording power as shown in Fig. 2. The recording power for a practical medium is usually set ~2–3 times as high as the threshold power to assure stable recording. Therefore, to avoid recording error, the allowable sidelobe intensity must be ≲30% of the main spot intensity.

On the other hand, the readout signal may degrade due to sidelobe influence. In reading, the readout signal intensity is nearly proportional to the focused spot intensity. To suppress any readout signal degradation caused by sidelobe influence by rejection of the readout signal from the sidelobe region, new signal detection optics have been developed.

The optical design of the write-once optical head is shown in Fig. 3. In this head, the superresolution is made in one direction which is parallel to the track direction, because the residual distortion in the readout signal is easy to equalize by an electric filter circuit. Compared with a conventional optical head, a shading band and signal detection optics composed of a focusing lens and a slit are now added. In the detection optics, light reflected from a disk is focused and again forms a superresolving spot on the slit plane. The main lobe region of the focused light spot is spatially extracted using a slit and detected by a photodiode. The focusing and tracking errors on the disk are detected by the conventional knife-edge and push–pull methods, respectively, employing the error detection optics shown in Fig. 3. The error detection characteristics are almost the same as those of a conventional head.

II. Numerical Analysis

First, the superresolving optics in the optical head are optimized considering the recording condition. Next, the readout signal detection optics which minimize the sidelobe influence are analyzed.

To optimize the shading bandwidth, the superresolving spot profile has been numerically analyzed using a 1-D FFT. Figure 4 shows parameter values used for the calculation and the calculated results. These parameter values are those usually employed in a conventional head. The main spot size, main spot power, and sidelobe intensity are plotted as functions of the shading bandwidth. The shading bandwidth is normalized with an objective lens aperture. As the bandwidth increases, the main spot size decreases. On the other hand, the main spot power decreases and the sidelobe intensity increases.

From Fig. 4, the main spot size is reduced to 80% when the sidelobe intensity is 30%, which is the maximum permissible value to avoid pit writing error in recording as discussed above. The calculated shading bandwidth at this point is 26%.

The permissible value for the sidelobe intensity was confirmed by analyzing the temperature distribution of recording on the disk medium. No remarkable difference was noted at the 30% sidelobe intensity.

The effectiveness of the slit detection method was estimated by the 1-D FFT analysis. Figure 5 shows the calculated frequency dependence of the optical disk readout signal amplitude. Without slit detection, the signal amplitude for the superresolving spot (dash–dot line) is degraded at the low frequency region owing to the sidelobe interaction. Using the slit detec-

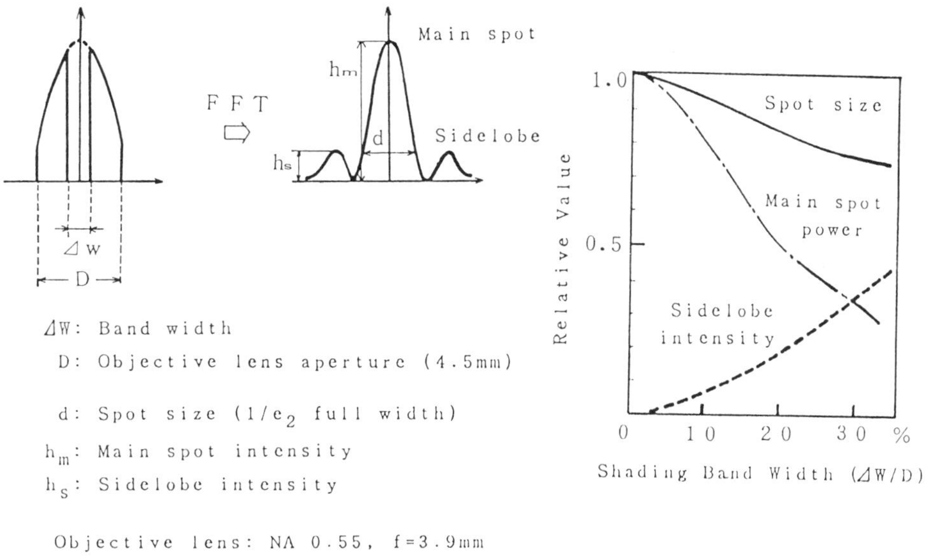

Fig. 4. Parameter values used for calculation and calculated spot size, main spot power, and sidelobe intensity dependence on the shading bandwidth.

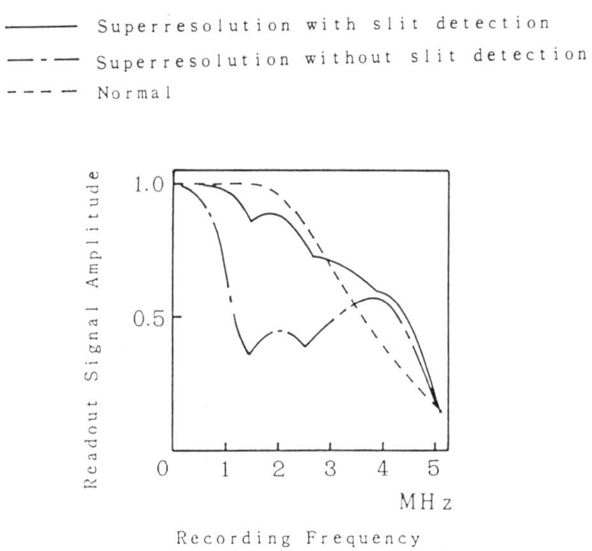

Fig. 5. Calculated readout signal amplitude with respect to recording frequency.

tion method (solid line), the degradation in the low frequency region has been drastically improved. The cutoff frequency becomes higher compared with that for a normal spot (dotted line). As is obvious from this result, it is possible to achieve high density recording by employing superresolution and slit detection optics. The change in the cutoff frequency is less than a few percent when the slit width changes ~10%; this means that the slit positioning tolerance is not severe. The residual ripple seen in the result shown by the solid line is caused by the objective lens aperture limit; that is, this is due to the fact the the reflected light pattern at the disk, which is a product of the focused spot profile by recording the pit reflection profile, is not perfectly reconstructed at the slit plane.

III. Experimental Results and Discussion

A. Spot Profile

An example of a cross-sectional view of the superresolving spot, observed with a magnifying lens and CCD camera, is shown in Fig. 6(a). It is clear that dark lines exist between the main spot and the sidelobes and that the superresolving profile is formed only in one direction. The spot profile was measured by the scanning knife-edge method. Figure 6(b) shows the measured spot profiles with respect to the normalized shading bandwidth. At any bandwidth, the spot size in the direction orthogonal to the superresolving direction is not changed. This and the above results support our use of 1-D analysis to evaluate the superresolving effect.

Figure 7 shows the measured spot size and sidelobe intensity to the main spot intensity ratio vs shading bandwidth. An ~80% main spot size and 30% sidelobe intensity, as theoretically predicted, were obtained at a 20% bandwidth. The main spot size was ~1.1 μm ($1/e^2$ full width) for a 0.55-N.A. objective lens at a 0.83-μm wavelength. The slight difference in bandwidth between the calculated (Fig. 4) and experimental results is mainly caused by the difference in parameter values between calculation and experiment.

Figures 8(a) and (b) show the spot photographs on the front and back planes of the slit, respectively. In this observation, a flat plate without pregrooves was used at the disk position to make clear images. It is clear that only the main spot region light transmits through the slit. In this experiment, a 10-μm width slit was used according to the image magnification (~10) between the disk plane and slit plane.

B. Recording Characteristics

Figure 9 shows the recording power dependence of the carrier-to-noise ratio (CNR) and second harmon-

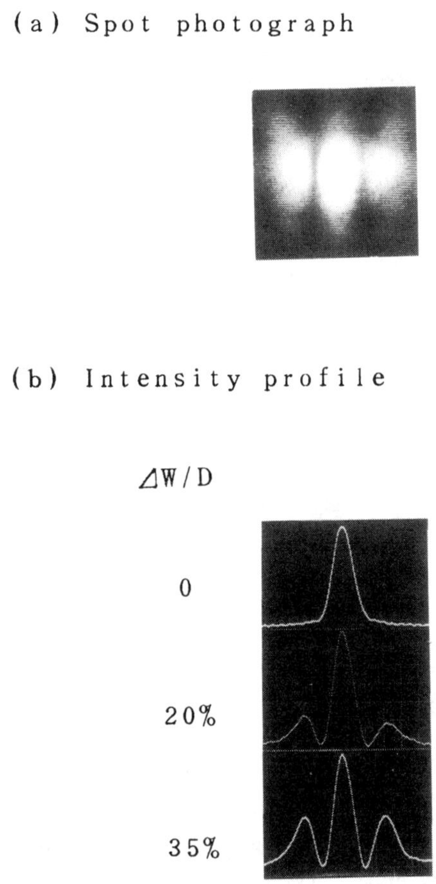

Fig. 6. (a) Cross-sectional photograph of the superresolving spot; (b) spot profiles measured for various shading bandwidths.

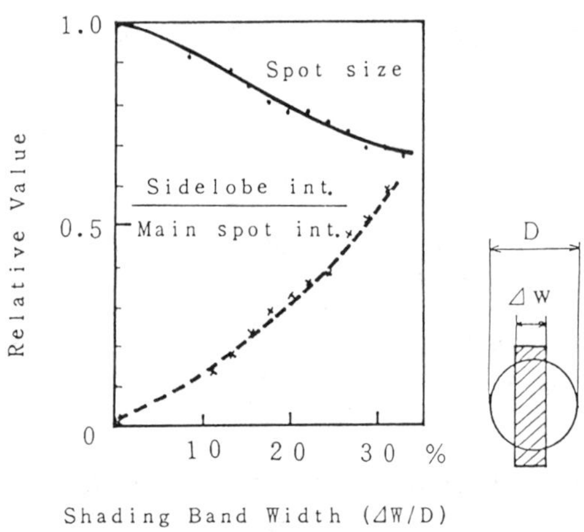

Fig. 7. Measured spot size and sidelobe intensity to main spot intensity ratio dependences on the shading width.

ics-to-carrier ratio (2nd/C) in the readout signal using a 5.25-inch Te alloy write-once optical disk. The 2nd/C is measured to decide a suitable recording power. Recording was done by a superresolving spot using various shading bandwidths, and the image was read by a normal spot to estimate only the effect of the recording spot profile difference. The recording power indicates the total power focused on the disk. Therefore, the recording power threshold becomes high as the bandwidth increases, owing to the increase

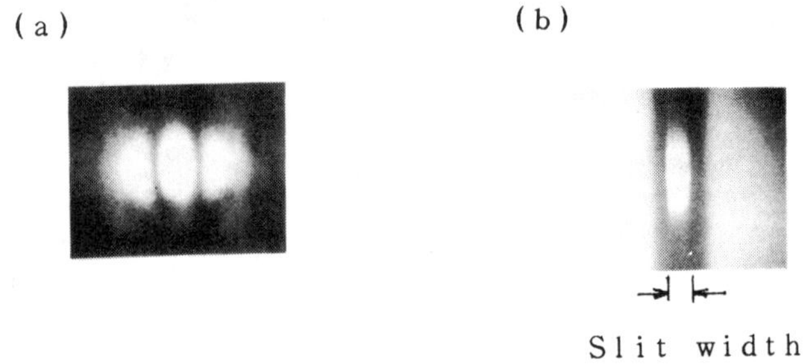

Fig. 8. Observed beam images on (a) the front plane of the slit and (b) the back plane of the slit.

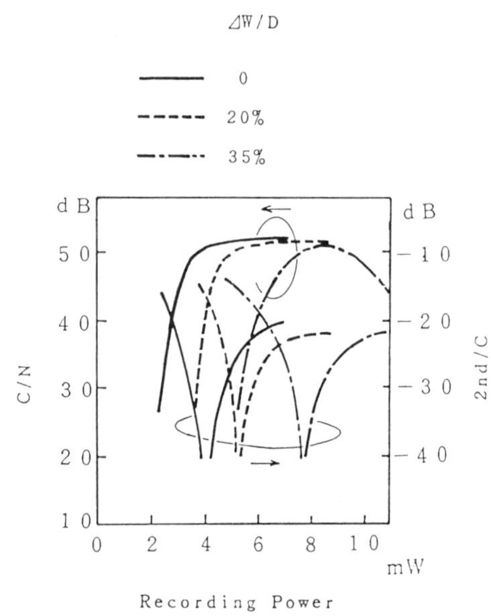

Fig. 9. Measured CNR and 2nd/C dependences on recording power and shading and width. Reading was done using a normal spot.

in the sidelobe power share. Except for the threshold difference, the result for the 20% bandwidth (dotted line) has the same power dependence as that for the no shading band (solid line). However, when the bandwidth is 35%, the readout CNR begins to fall at the high power region (dash–dot line). This experimental result indicates that the sidelobe intensity exceeded the recording threshold, and failure pits were formed when the shading bandwidth is 35%. Therefore, a 20% bandwidth is optimum for practical use.

C. Read–Write Frequency Characteristics

The recording frequency characteristics were measured using the same disk as used above. The recording power was set at a value where the second harmonic spectrum in a readout signal took the minimum value.

Figure 10 shows recording frequency characteristics without the slit detection method. The readout signal amplitude, using the superresolving spot for both recording and reading (solid line), was degraded at low frequency, as predicted in the calculation, while the readout signal amplitude using a normal spot to read the pits recorded by a superresolving spot (dash–dot line) was not degraded. On the other hand, when the reading spot was normal (dash–dot and dotted lines),

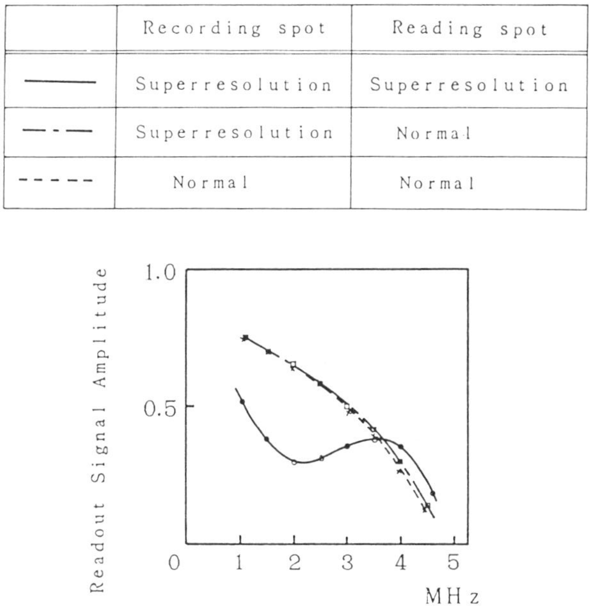

Fig. 10. Measured recording frequency characteristics without slit detection.

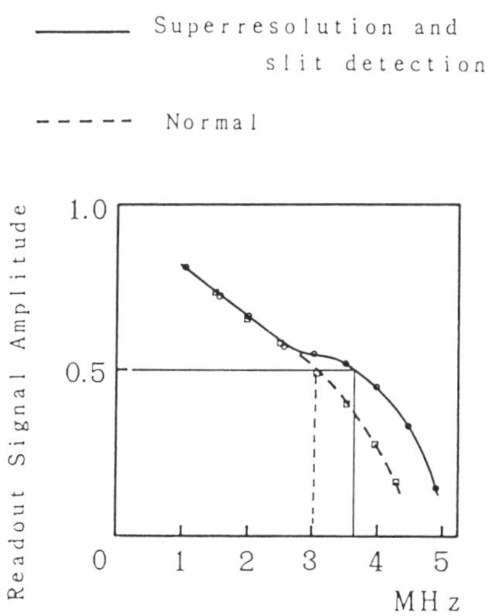

Fig. 11. Measured recording frequency characteristics using slit detection.

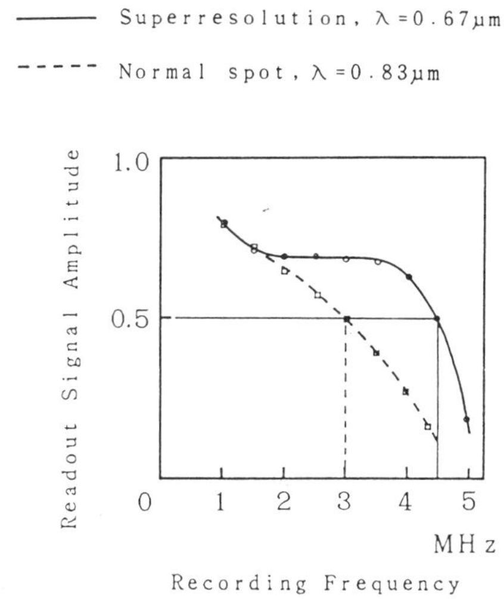

Fig. 13. Measured recording frequency characteristics using a visible light laser diode.

the readout signal amplitude was not degraded, and nearly the same characteristics were obtained regardless of the recording spot. These results mean that the degradation is caused by the sidelobe interaction in reading.

By using a low power normal spot, high density pits can be recorded. However, the pit size fluctuation increases compared with that using a superresolving spot.

Figure 11 shows the recording frequency dependence for the readout signal using the slit detection method. The amplitude degradation at low frequency without slit detection disappeared. The recording frequency for the superresolving spot, at which the signal amplitude becomes half of the maximum, is 1.2 times as high as that for a normal spot, indicating that linear density for this head can be increased ~1.2 times as high as that for a conventional head.

Figure 12 shows readout signal examples at 1 MHz. Many ripples were observed without slit detection as shown in Fig. 12(b). These ripples were suppressed by using the slit detection method shown in Fig. 12(a). This signal waveform is close to the signal waveform using a normal spot, as shown in Fig. 12(c).

Figure 13 shows the recording frequency characteristics measured for the optical head using the superresolution technique and a 0.67-μm wavelength visible

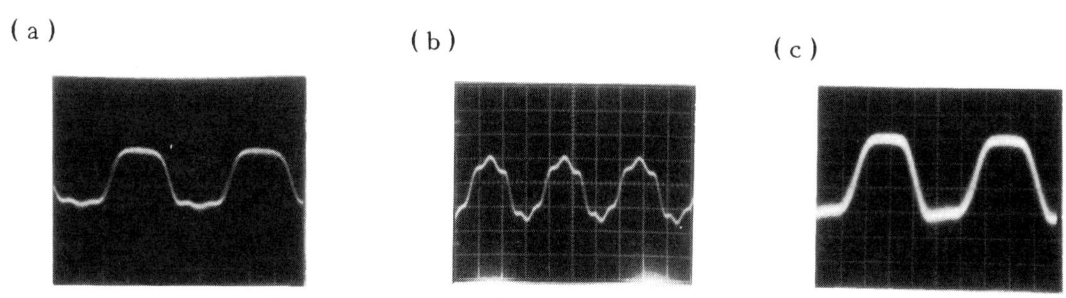

Fig. 12. Observed readout signal examples at 1 MHz using (a) superresolution spot with slit detection, (b) superresolution spot without slit detection, and a (c) normal spot.

light laser diode. Owing to the product effect of the superresolution and short wavelength, the cutoff frequency is increased to 1.5 times as high as that for a conventional head using a normal spot and a 0.83-μm wavelength laser diode.

IV. Conclusion

High density recording was accomplished using the superresolution technique with a slit for readout signal detection. The focused spot size was reduced to 80% of that for a conventional optical head. The linear recording density was increased to 1.2 times as high as that for a conventional head. Moreover, by employing the 0.67-μm wavelength visible light laser diode, the linear density was increased to 1.5 times as high as that for a conventional recording.

In this paper, the optical head for write-once media was described. As a matter of course, the superresolving effect can be applied to any other kind of optical recording, including read-only disks and magnetooptical disks.

The author would like to thank M. Sakaguchi, N. Nishida, T. Suzuki, and Y. Ono for their encouragement and advice.

References

1. H. Sukeda, M. Ojima, M. Takahashi, and T. Maeda, "High Density Magneto-Optical Disk Using Highly Controlled Pit-Edge Recording," in *Postdeadline Digest, ISOM'87* (1987), paper FC1.
2. T. Sekiguchi *et al.*, "High Bit Rate Digital Video Recording Using Magneto-Optical Disks," in *Technical Digest, ISOM'87* (1987), paper FB4.
3. G. Moore, "High Density Format For Sperry/Information Storage Inc. Second Generation Optical Drives," Proc. Soc. Photo-Opt. Instrum. Eng. **695,** 230–238 (1986).
4. Y. Yamanaka, K. Kubota, H. Fujii, K. Kobayashi, T. Suzuki, and H. Gokan, "High Density Magneto-Optical Recording Using 0.67 μm Band High Power Laser Diode," IEEE Trans. Magn. **MAG-24,** 2300–2304 (1988).
5. H. Osterberg and J. E. Wilkins, Jr., "The Resolving Power of a Coated Objective," J. Opt. Soc. Am **39,** 553–557 (1949).
6. W. Lukosz, "Optical System with Resolving Powers Exceeding the Classical Limit," J. Opt. Soc. Am. **56,** 1463–1472 (1966).

Section Three
Lasers and Laser Arrays

Reprinted with permission from *IEEE Transactions on Consumer Electronics,* Vol. CE-23(4), pp. 479-495 (November 1977). ©1977 IEEE.

SEMICONDUCTOR LASER PICKUP FOR OPTICAL VIDEO DISK PLAYER

Yoshito Tsunoda, Susumu Sawano, Hisashi Nakamura,
Kazutoshi Saito, Toshihisa Tsukada, and Yasutsugu Takeda
Central Research Laboratory, Hitachi Ltd.
Kokubunji, Tokyo, Japan

INTRODUCTION

Almost five years have passed since the concept of the optical video disk system was first presented, and several kinds of systems have been developed these few years.[1,2,3,4] Recently, the standardization of these confused optical video disks was made and optical video disk player systems have already reached the stage of commercial development.[5] Optical video disk systems offer several advantages to mechanical and capacitive video disk systems. High image quality, still and slow motion picture capability are some of its chief attractions. Unfortunately, large, complicated, and expensive optical pickups still remain a problem. Part of the difficulties is the use of the He-Ne laser as a light source. Although the He-Ne laser is one of the most stable and compact gas laser among several commercial available lasers, its size, over 200mm length and $30mm^{\phi}$ in diameter, is still too large for a compact optical pickup. Moreover, high voltage power supply of over 1.5KV are not desirable for home use systems. In addition, the optical system used in the pickup is quite complex because separate laser beams are used to detect the auto-focusing signal, the tracking signal, and the video signal. These factors make for a very expensive overall cost in labor for manufacturing and maintenance.

In an effort to develop a simple and small optical pickup for a low cost video disk player system, we have developed an optical pickup which employs a semiconductor laser as a light source. Semiconductor lasers have primarily been considered as a light source of optical communication. Their small size and ease in fabrication have prompted studies as to their applicability as light sources for video disk players. One of the difficulties is that the conventional Double Heterostructure (DH) semiconductor lasers have narrow stripe-like active regions (for example 1μm x 10μm) which make it difficult to focus their laser beams into circular spots with high efficiency on video disks. This causes difficulty in the reconstruction process. We approach this problem by using the Buried Heterostructure (BH) semiconductor lasers of which active regions are small and symmetric. This results in a more manageable beam and ease in the reconstruction of the video disk.

In this paper, we first describe the optical requirements of a light source for a video disk player system. In addition, the optical characteristics of the BH lasers are described and the possibility of their application to the light source of the optical video disk player is discussed. We then present an implementation of such a semiconductor laser pickup along with the optics needed for detection of servo control and video signals. Finally, experimental results of the reconstruction of the video disks by the newly developed optical pickup are presented.

OPTICAL REQUIREMENTS FOR A LIGHT SOURCE OF A VIDEO DISK PLAYER

The standard video disk designed for He-Ne laser read-out is made of a transparent polyvinyl chloride (PVC) of 1.1m thickness.[6] Frequency modulated video and audio signals are spirally recorded as a train of pits each of which is 0.6-0.8μm in width, 1-3μm in length, 0.11 μm in depth (corresponding to the 1/4 wavelength depth of a He-Ne laser in PVC), and a track pitch of 1.7-2.0μm. The recorded plane is metalized (usually by aluminum) with a protective layer applied on top of this reflective coating. To read the disk, a laser beam is focused on the recorded plane through the transparent disk and the reflected beam from the recorded plane is directed to a photo-detector through a focusing lens.

The spot size required to read the information pits is determined by the following conditions. The focused spot size must resolve the length which corresponds to the bandwidth required for a signal at the innermost radius (55mm) of the recorded area.[7] That is, the numerical aperture (NA) of the focusing lens should be selected so that the cut-off frequency f_c of the Modulation Transfer Function (MTF) of the lens at the innermost radius is equal to or more than the bandwidth f_m of the encoded signal as shown in the following equation.

$$NA \geq \frac{\lambda f_m}{4\pi f_R R_i}$$

where, λ is a wavelength of the laser, f_R is a rotation frequency of the disk (30Hz) and R_i is the innermost radius of the recorded area. In Fig. 1, the relations between the cut-off frequency f_c at the innermost radius and the numerical aperture NA of the lens are shown with the wavelength of the He-Ne laser (632.8nm) and the BH laser (830.0nm). When the NTSC directly encoding system[8] is used as a signal encoding system, f_m is about 13MHz (f_{md}) while for NTSC crossband system,[1] it is 10MHz(f_{mc}). Consequently, from Fig. 1 the required numerical apertures of the lens are 0.4 and 0.52 for the NTSC directly encoding system for He-Ne laser and BH laser, respectively. As for the NTSC crossband system, the required numerical apertures of the lens are 0.3 and 0.4 for He-Ne laser and BH laser, respectively. Spot sizes

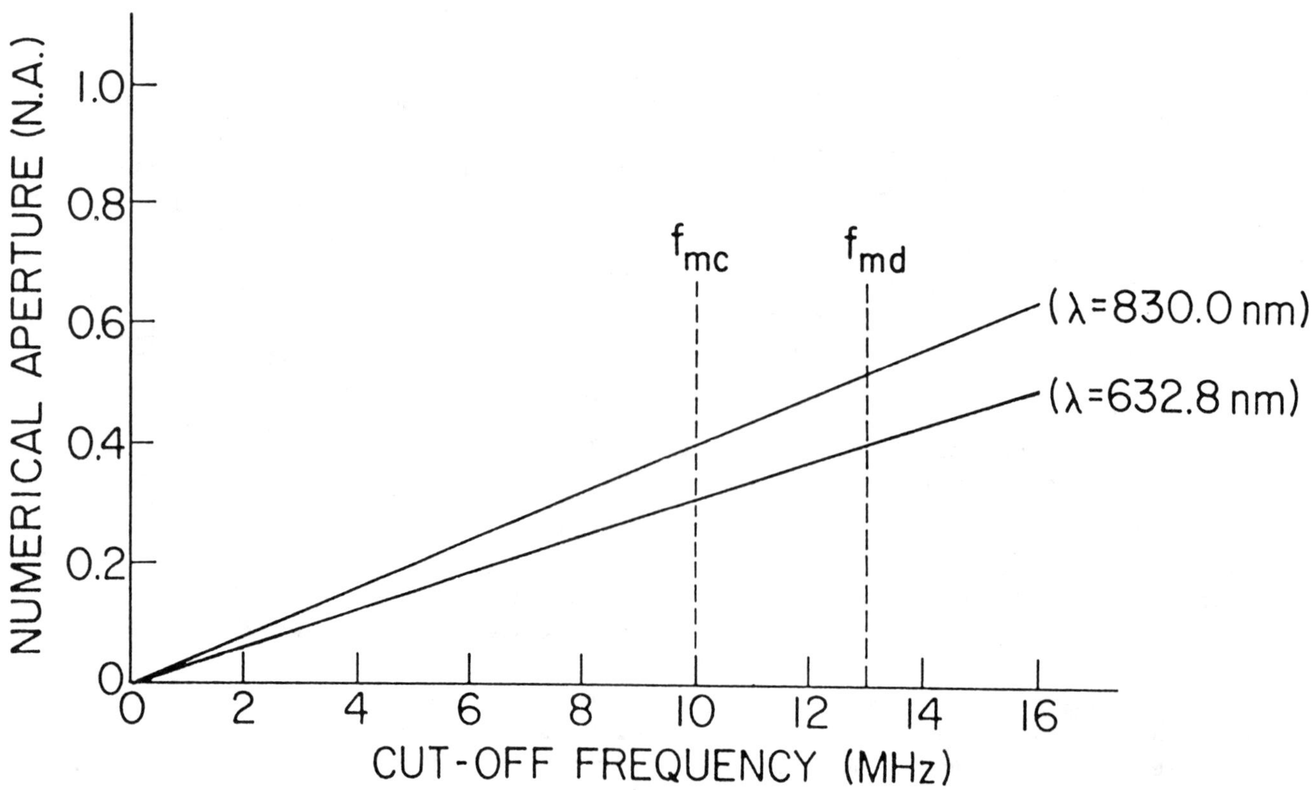

Fig. 1 Relation between the cut-off frequency and the numerical aperture of the lens.

on the disk are determined by the numerical apertures of the lens. Thus the corresponding spot sizes are about 1.2-1.3μm$^{\phi}$ and 1.6-1.7μm$^{\phi}$ (at $1/e^2$ intensity) for the NTSC directly encoding system and the NTSC crossband system respectively. These spot sizes are required at the innermost radius of the recorded area on the disk, while at the outer radius larger spot sizes are allowable. In addition to above conditions, the spot size in the radial direction is determined by a cross talk between adjacent tracks. For a track pitck of 1.7-2.0μm, the spot size in the radial direction should be about the same size as in the tangential direction in order to keep the cross talk to be less than -35dB. These requirements on the read spot size for the video disk implies that the radiating pattern of the semiconductor laser should be circular and small in order to enable it to be focused into a tiny spot efficiently with simple optics, such as conventional microscope objective lenses.

The required output power of the laser is determined by the following equation.

$$I_L = \frac{i_m}{M \eta \epsilon}$$

where I_L is the required minimum laser power, i_m is the minimum current which is required to get a sufficient signal-to-noise (S/N) ratio of a video preamplifier, η is a photo-sensitivity of the detector, M is the depth of the optical modulation of the video signal from the disk, and ϵ is a total optical efficiency of the pickup which is measured as a ratio of the laser power on the detector to the output power of the laser. For example, if we let $i_m = 4\mu A$, $\eta = 0.2\mu A/\mu W$, $M = 0.4$, and $\epsilon = 0.1$, then I_L is 500μW. In this case, the required output power of the laser should be about 1mW.

The wavelength of the semiconductor laser should be as close to that of the He-Ne laser as possible because a higher numerical aperture lens is required for a longer wavelength to obtain the same spot size on the disk. Moreover, the difference of the wavelength causes a decrease of the depth of modulation of the reconstructed video signal from the disk.

Polarization characteristic of the laser is utilized to increase the optical efficiency and to prevent the back talk effect in the He-Ne laser pickup. However, in case of the semiconductor laser, this characteristic is not critical since no back talk effect has been observed in the video frequency region experimentally.

Noise characteristic of the semiconductor laser should be the same as that of the He-Ne laser in order to obtain a high quality reconstructed image with a signal-to-noise ratio of over 40dB.

OPTICAL CHARACTERISTICS OF BURIED HETEROSTRUCTURE LASERS

The basic construction and the appearance of the BH laser are shown in Fig. 2. This laser has a filamentary GaAs active region completely surrounded by GaAlAs,[9] whereas the conventional DH laser has a thin and wide active region. This special structure is made by using a mesa-etching technique to leave a filamentary active region on the double heterostructure crystal grown by the continuous epitaxial method, and a liquid growth technique of GaAlAs to bury that filamentary active region. In general, the BH lasers have the following features:

1. The single lowest oscillation mode which is easy to be focused into a tiny spot is obtainable.
2. The oscillation mode is stable relative to changes in the exciting current.
3. The beam divergence is symmetric and thus easier to handle optically.
4. Working current and voltage are quite low and a small heat sink is sufficient.

The measured optical characteristics

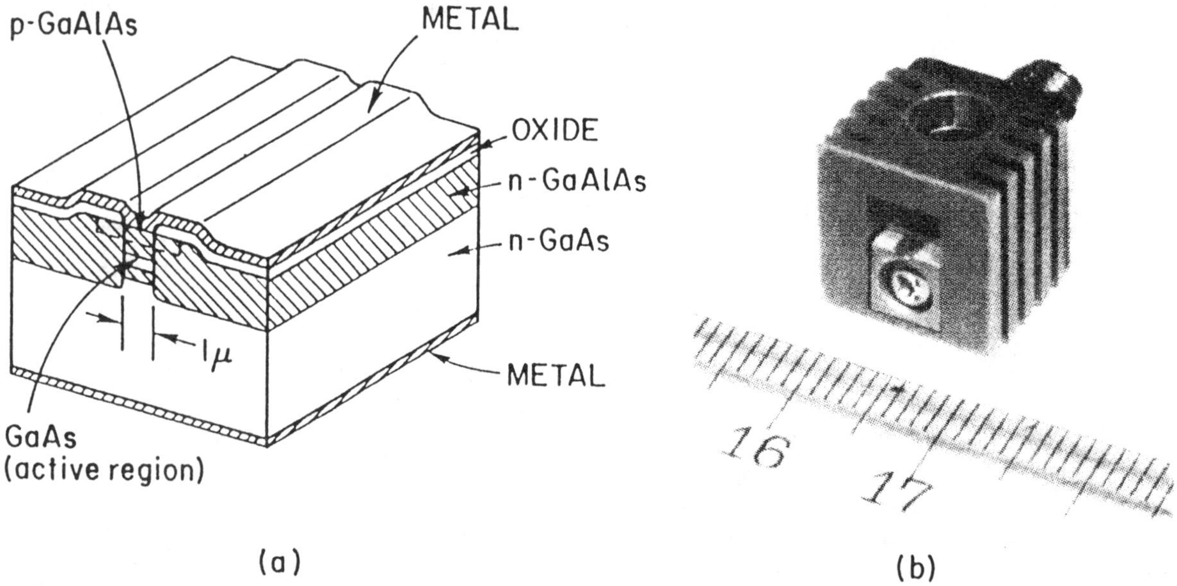

Fig. 2 (a) Basic construction and (b) the appearance of the BH laser.

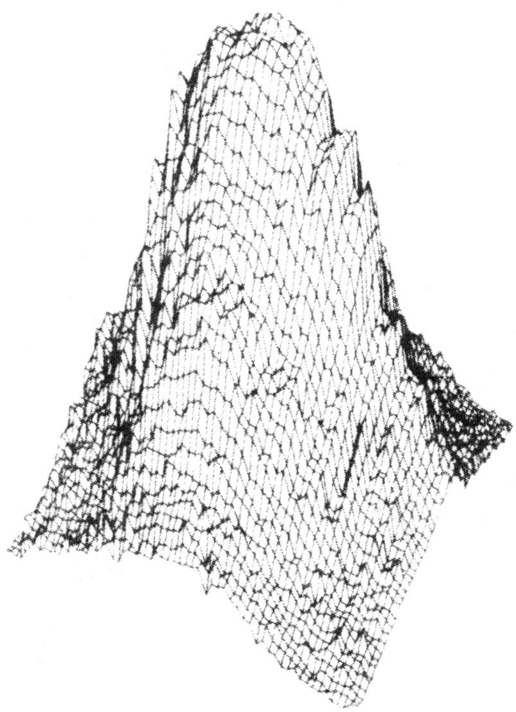

Fig. 3 3-dimensional plot of the far field pattern of the BH laser.

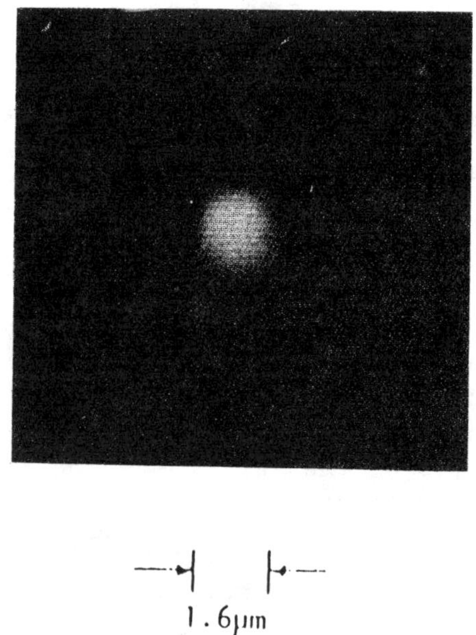

Fig. 4 Near field pattern shape of the BH laser (displayed on the TV monitor).

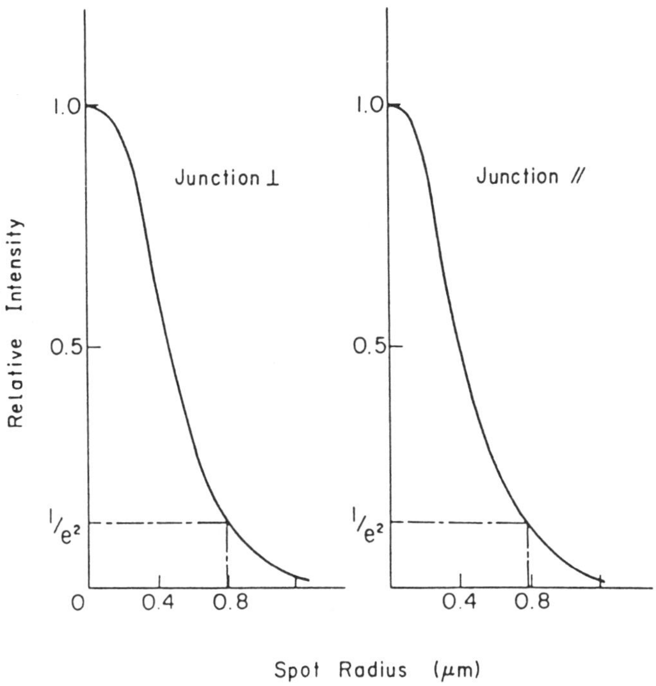

Fig. 5 Detailed cross section of the near field pattern of the BH laser.

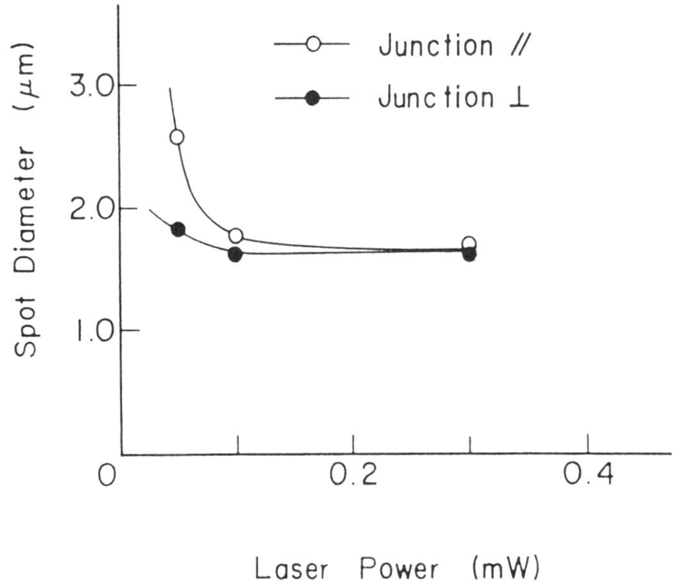

Fig. 6 Dependence of the spot size of the BH laser on the output laser power.

of the BH lasers are as follows.

Radiation Pattern of the Laser Beam

The far field pattern of the BH laser was measured. The BH laser was rotated with its active region at the center of the rotation at various angles of tilt while the radiation intensity was measured by a photo-multiplier through a pinhole which was set at points 10cm from the laser. The measured signals are processed by computer and a 3-dimensional plot of the intensity distribution is obtained as shown in Fig. 3. The intensity distribution is shown to be a rotational symmetric Gaussian-like distribution. The observed near field pattern is shown in Fig. 4. Figure 5 shows the detailed measurement of the cross section of the intensity. The cross sections in both parallel and perpendicular directions to the junction of the laser are shown to be Gaussian distributions with the same spot size of 1.6μm at $1/e^2$ intensity point. The dependence of the spot size on the output laser power is shown in Fig. 6. The spot size in the parallel direction to the junction is larger than that of the perpendicular direction in the output power region of below and around threshold (corresponding to an output power of 200μW). However, above threshold output power, both spot sizes become the same. This behavior in the range of low output power is due to the mixed radiation of the induced and the spontaneous emissions. Thus for a symmetric spot, the BH laser should be used in the output power region of more than 200μW.

Output Power Characteristic

Typical output laser power dependence on the excitation current is shown in Fig. 7. Threshold current is about 10mA which is more than one order lower than that of the conventional DH lasers (about 200mA). The output laser power of more than 1mW can constantly be maintained with 14-15mA input current with a high reliability.

Wavelength Characteristic

The typical lasing spectrum of the BH

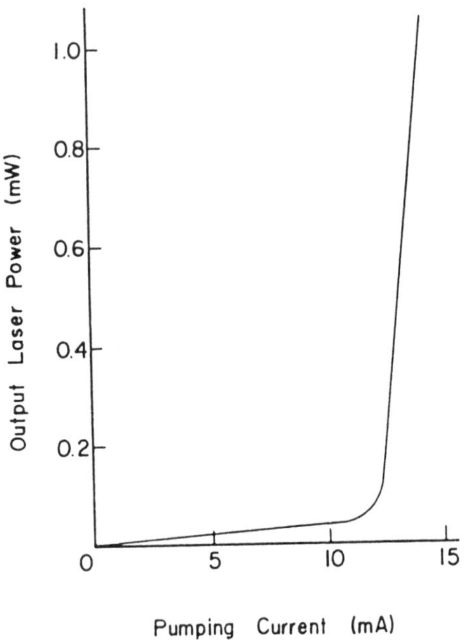

Fig. 7 Dependence of the output laser power of the BH laser on the excitation current.

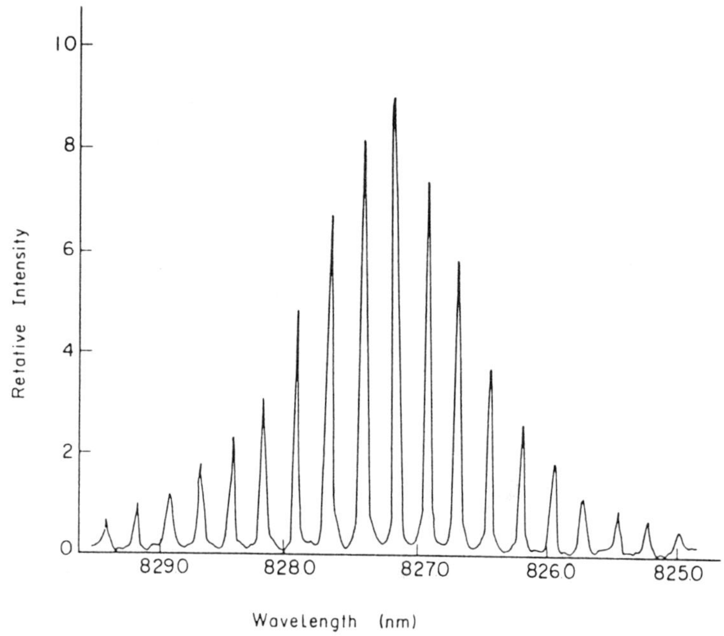

Fig. 8 Typical lasing spectrum of the BH laser.

laser is shown in Fig. 8. Several lasing lines can be seen around the wavelength of 830.0nm. This wavelength of the BH laser causes about 15% decrease of the depth of the modulation to that of the He-Ne laser. However, this decrease is allowable to obtain a high quality reconstructed image with a signal-to-noise ratio of over 40dB. This wavelength difference can be positively utilized to detect the tracking signal of the pickup in a technique described later.

Polarization Characteristic

The output laser beam of the BH laser is linear polarized in the perpendicular direction to the junction. The extinction ratio depends on the output laser power as is shown in Fig. 9. The low extinction ratio seen below and around the output power corresponding to the threshold current is due to the mixed radiation of the polarized induced emission and the non-polarized spontaneous emission. Above the output power of 200μW the extinction ratio becomes constant to more than 20:1. This ratio is lower than that of the He-Ne laser (more than 100:1). If the polarization characteristic is only used to increase the optical efficiency of the pickup then this ratio is sufficient. However, to eliminate the back talk effect, this 20:1 ratio would not be sufficient. Fortunately, no back talk effect can be observed in the video frequency region (0-15MHz) when the BH laser is used in the pickup.

Noise Characteristic

The allowable relative noise power of the laser to obtain the reproduced image with a signal-to-noise ratio of more than 40dB is about 0.01% in the video frequency region. The noise characteristic of the BH laser is shown in Fig. 10. It can be seen that the noise power in the video frequency region of the BH laser shows a keen peak at the threshold current (I_{th})

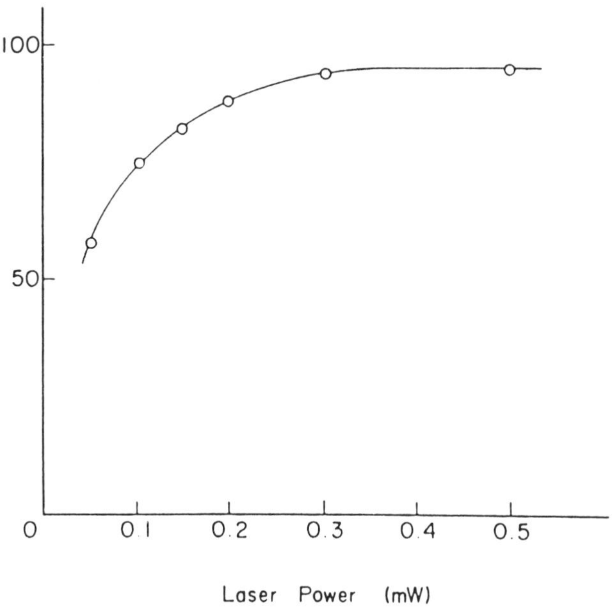

Fig. 9 Dependence of the excitation ratio of the BH laser on the output laser power.

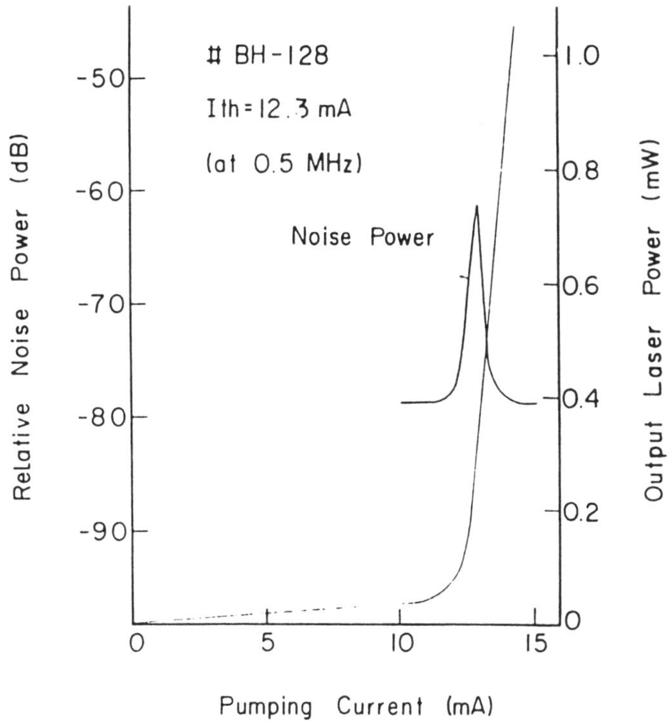

Fig. 10 Dependence of the relative noise power of the BH laser on the excitation current.

region in the same manner as DH lasers. However, noise power in the current region of more than $1.2I_{th}$ is shown to be about 0.01% (corresponding to -80dB in the figure) or the same as that of the He-Ne laser. This region is equivalent to an output power range of more than 500-600μW. Fortunately, this output power is consistent with the results of the required output power specification discussed previously.

The characteristics of the BH laser are summarized in Table 1. Through above measurements and discussion, the optical characteristics of the BH lasers are shown to satisfy the conditions required of the light source for an optical pickup.

A PRACTICAL PICKUP STRUCTURE

The basic configuration of an optical system used in the semiconductor laser pickup is shown in Fig. 11. The pickup consists of two main parts. One portion is the semiconductor laser. While the other portion is the optical system needed to direct a laser beam to the disk and to detect the reflected signal. Optical system consists of the focusing optics section which is used to project the laser beam as a tiny spot on the disk, and the servo signal detecting section which is used to detect the focusing and the tracking error signals from the rotating disk. For simplicity, these two optics sections are combined into one optical system. In addition to these optical system, there are electronics for processing the video and servo signals.

The radiating laser beam from the BH laser symmetrically diverges at about 60 degrees angle beam because of the diffraction of the nearly 1μm$^{\phi}$ aperture. This diverging beam is captured and imaged on the conjugate point by using a coupling lens. This conjugate image is focused on

TABLE 1

Summarized characteristics of the BH laser

	B-H Injection Laser	He-Ne Laser
Wavelength	830.0 nm	632.8 nm
Transverse Mode	TE_{oo}	TEM_{oo}
Output Power	> 1 mW	> 1 mW
Noise (video range)	≦ 0.01 %	≦ 0.01 %
Beam Diameter	< 2 μm^{ϕ}	0.75 ± 0.05 mm^{ϕ}
Beam Divergence	1 ~ 2 rad	1.2 mrad
Driving Voltage	2 V	1.6 KV (Trigger, 7KV)
Current	10 mA	5 mA
Dimension	Less Than 10 mm cubic (with Heat Sink)	30 mm^{ϕ} x 220mm

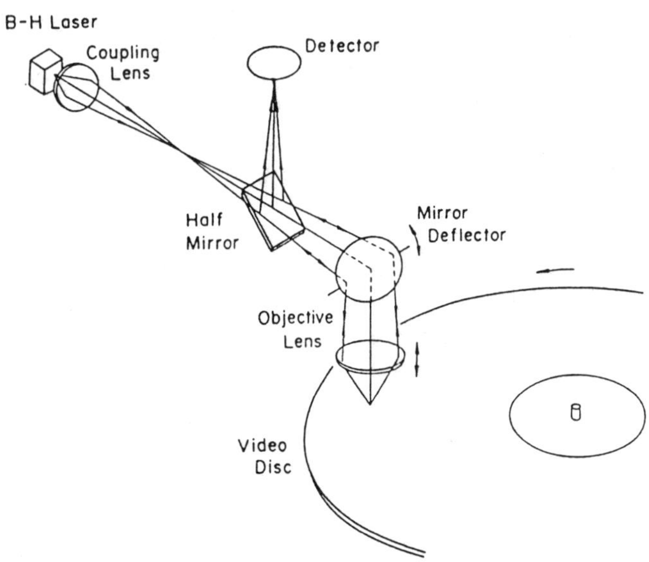

Fig. 11 Basic configuration of the semiconductor laser pickup.

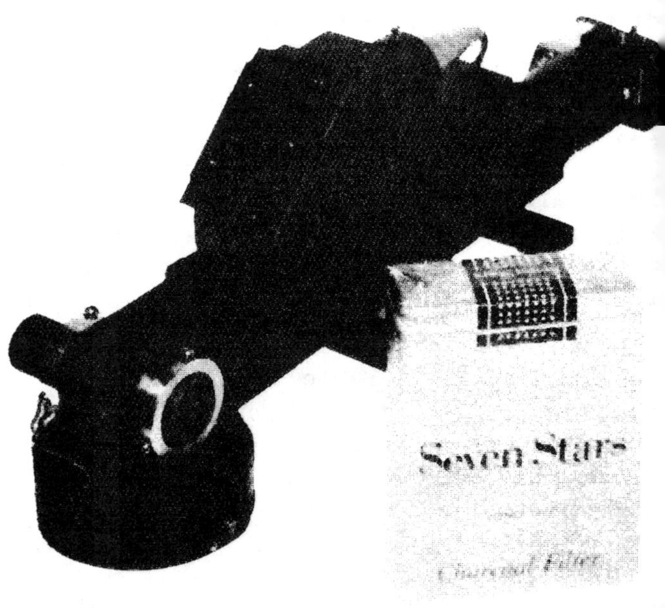

Fig. 12 Appearance of the semiconductor laser pickup.

the disk by using another focusing lens. The spot size on the disk is determined by the numerical aperture of the lenses and the intensity distribution of the laser beam itself. From the results of the far field and the near field patterns measurements of the BH lasers, it can be seen that the intensity distribution is Gaussian-like. Taking this distribution into consideration, the numerical aperture of the coupling lens is selected to be 0.4. As for the focusing lens, both NA 0.4 and 0.5 lenses can be used for the NTSC crossband system and the NTSC directly encoding system video disks, respectively.

The recorded surface of the disk moves several hundreds micronmeters vertically during its rotation because of the distortion and inclination of the disk. Auto-focusing servo control is required to keep a tiny focused spot on the tiny pits on the disk. The required accuracy of the focusing servo control is roughly determined by the focusing depth of the lens which is proportional to $\lambda/(NA)^2$. For an He-Ne laser system with a λ of 632.8nm and a lens of NA 0.4, the focusing depth is about $\pm 2\mu m$. In the case of the BH laser, λ is 830.0nm and NA are 0.4 and 0.5, and the focusing depths are $\pm 2.6\mu m$ and $\pm 1.7\mu m$, respectively. These values are almost equal to that of the He-Ne laser. Several simple optical methods to detect the focusing error signals have been developed.[10] By using these methods the moving lens is driven to control the focusing on the disk. It was found that the control range of the auto-focusing servo system needs to be about $\pm 100\mu m$ with an accuracy of $\pm 1.5\mu m$.

The information track moves in the radial direction 100-200μm during the rotation of the disk because of possible eccentricity and the distortion of the disk. To detect the tracking error signals, the diffraction effect of the laser beam by the information pits is utilized. The diffraction pattern of the laser beam by the pit is calculated on the plane of the focusing lens. The depth of the pit is 0.16μm (1/4 wavelength of the He-Ne laser) and the calculations are done both for the He-Ne laser (λ = 632.8nm) and the BH laser (λ = 830.0nm). In the case of the He-Ne laser, the symmetry of the pattern distribution in the direction of the spot movement is maintained. The pattern distribution of the BH laser shows an asymmetry due to the mismatching of the wavelength of the BH laser and the depth of the pit. That is to say, the laser beam diffracted by the pit is not only modulated in intensity (this effect is used to detect the video signal), but also contains the information as to the tracking error in its asymmetry distribution. It was found that the control range of the tracking servo system needs to be about $\pm 150\mu m$ with an accuracy of $\pm 0.2\mu m$.

No special optics are used for the detection of the time base error (jitter) signal. The phase shift of the horizontal sync frequency (f_H = 15734.264Hz) is detected by using an AFC circuitry. The mirror deflector (not shown in Fig. 11) for the jitter compensation is driven in the tangential direction by this time base error signal.

The video signal is obtained by detecting the total modulation of the reflected laser beam from the disk.

By using this simple optical pickup, the video and all the servo control signals can be detected. There are however some interactions between the servo control signals which would decrease the overall accuracy of the pickup servo control. These interactions can generally be eliminated by proper electronic compensation.

EXPERIMENTAL RESULTS

An engineering model of the semiconductor laser pickup based on the structure and the principle described in the previous section was designed and constructed. The appearance of the pickup is shown in Fig. 12. The pickup is shown with the BH laser mounted on a $1cm^3$ heat sink, optical components, and the

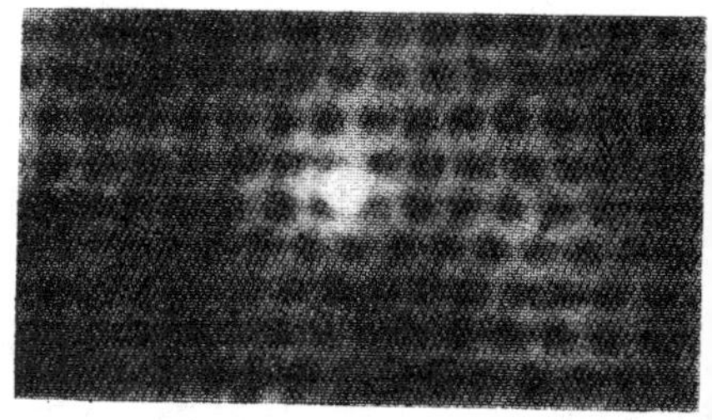

Fig. 14 Focused spot pattern on the information pits.

Fig. 13 Appearance of the experimental model video disk player.

detector. The size of the whole pickup is less than 200mm long, 50 mm high, and 20mm wide. The total weight is less than 300 grams. The total number of parts is seven including the laser and the detector. The singular optical axis considerably simplifies the alignment of the detector and the remainder of the optical system. The total optical efficiency is about 15%. The maintenance is greatly reduced due to the stability associated with the short optical path. The pickup was mounted on the experimental model video disk player as is shown in Fig. 13. In Fig. 14, the focused laser spot pattern on the information pits is shown. The laser output power is 500-600μW. An example of the reconstructed image is shown in Fig. 15. The NTSC directly encoding system and the NTSC crossband system video signals were recorded on the disk, with pits 0.8μm in width, 0.16μm in depth and tracking pitch of 2μm. The final signal-to-noise ratio of the demodulated images is more than 40dB which is comparable to conventional video signal sources. The measured characteristics of the auto-focusing servo control and the tracking servo control are as follows: Auto-focusing has a control range of ±100μm with an accuracy of ±1.5μm; and tracking has a control range of ±150μm with an accuracy of ±0.2μm. These values satisfy the conditions required to reconstruct the image of the average video disks.

CONCLUSION

The possibility of the utilization of the Buried Heterostructure semiconductor laser as a light source of the optical pickup for the optical video disks was demonstrated. Using this semiconductor laser pickup the video disks with the NTSC directly encoding system and the NTSC crossband system signals were reconstructed with a signal-to-noise ratio of more than 40dB. This pickup has the

Fig. 15 Reconstructed image from the optical video disk by using the semiconductor laser pickup.

following features when compared with the conventional He-Ne laser pickups.

1. The total volume of the pickup is quite small since the size of the BH laser is less than $1cm^3$ even with a heat sink. Moreover, the power requirements (DC 2V, 10-15mA) are simple.

2. The single axis optics for the detections of the video, the auto-focusing and the tracking signals considerably simplify the structure of the pickup. This simplicity and compactness of the pickup significantly contribute to the commercial viability of an optical video disk player system.

3. The semiconductor laser pickup is shown to be available with the video disks made for the He-Ne laser read-out.

The lifetime of the BH lasers in the CW operation is one of the most interesting and important matters for their applications. The results of the accelerated life test of the BH laser shows a promising and encouraging data of 1500 hours at 70°C. This value corresponds to about 10^5 hours at 30°C.

The authors are grateful to S. Yonezawa, K. Tatsuno, T. Maeda, M. Izumita, and T. Miyauchi for the discussion and their help in experiments.

REFERENCES

1. K. Compaan and P. Kramer, Phillips Tech. Rev., 33, p. 178, 1973.

2. G.W. Hrbek, J. of SMPTE 83, p. 580, 1974.

3. G. Broussaud, E. Spitz, et. al., Trans. BTR, p. 332, Nov., 1974.

4. Y. Tsunoda, K. Tatsuno, K. Kataoka and Y. Takeda, Appl. Opt., 15, p. 1398, 1976.

5. R. Adler, IEEE Trans. Consumer Elect., p. 302, Nov., 1976.

6. J.S. Winslow, IEEE Trans. Consumer Elect., p. 318, Nov., 1976.

7. L.J. Laub, IEEE Trans. Consumer Elect., p. 258, August, 1976.

8. P.W. Bogels, IEEE Trans. Consumer Elect., p. 309, Nov., 1976.

9. T. Tsukada, J. Appl. Phys., 45, p. 4899, 1974.

10. C. Bricot, J.C. Lehureau, C. Puech and F. Le Carvennec, IEEE Trans. Consumer Elect., p. 304, Nov., 1976.

BIOGRAPHIES

Yoshito Tsunoda

Yoshito Tsunoda was born in Tokyo, Japan, in 1944. He received the B.A., M.S. and Ph.D. degrees, all from the University of Tokyo, Tokyo, Japan, in 1968, 1971, and 1976, respectively.

In 1971, he joined the Central Research Laboratory, Hitachi Ltd., Tokyo, Japan, where he is now a research engineer. He has been engaged in the research of high storage density holography memory, three dimensional displays and optical video disks. During 1976-1977, he is staying at Stanford University, Stanford, CA as a visiting researcher.

Dr. Tsunoda is a member of the Optical Society of America, the Institute of Electronics and Communications Engineers of Japan and the Japan Society of Applied Physics.

Hisashi Nakamura

Hisashi Nakamura was born in Tokyo, Japan, in 1935. He received the B.S. degree in Electrical Engineering from Tokyo Institute of Technology in 1960.

He has been with the Central Research Laboratory, Hitachi Ltd., since 1961. From 1962 to 1971 he was engaged in work on Pulse-Code-Modulation communication equipments for sound signal and television signal transmission. Since 1972, as a senior researcher of the laboratory, he has been responsible for the development of the video recording systems.

Mr. Nakamura is a member of the Institute of Television Engineering of Japan and the Institute of Electronics and Communications Engineers of Japan.

Susumu Sawano

Susumu Sawano, senior researcher, has been with Central Research Laboratory, Hitachi Ltd., Tokyo, Japan, since 1960. He received his BSME from Yokohama National University, Yokohama, Japan, in 1954. His principal field of interest at CRL, Hitachi Ltd., has been research and development of automatic control systems.

He was formerly a control engineer, Tokyo Kiki Engineering Co., Kawasaki, and a senior researcher of Guidance and Control Group, National Space Development Agency of Japan, Tokyo.

Mr. Sawano holds several patents and is the author of six papers on control theory and servomechanisms. Mr. Sawano is a member of the editorial board of the Society of Instrument and Control Engineers of Japan since 1976, and is a member of the Japan Society of Mechanical Engineers.

Kazutoshi Saito

Kazutoshi Saito was born in Akita, Japan, in 1939. He received the B.S. degree in polymer science from the University of Hokkaido, Sapporo, Japan, in 1963.

In 1963, he joined the Central Research Laboratory Hitachi Ltd., Tokyo, Japan. He has been engaged in research and development of amorphous photo-sensors, light-emitting diodes, and semiconductor lasers.

Mr. Saito is a member of the Japan Society of Applied Physics.

Toshihisa Tsukada

Toshihisa Tsukada was born in Toyama, Japan, in 1939. He received the B.S., M.S., and Ph.D. degrees, all from the University of Tokyo, Tokyo, Japan, in 1963, 1965, and 1968, respectively.

In 1968, he joined the Central Research Laboratory, Hitachi Ltd., Tokyo, Japan, where he is now a senior researcher. He has been engaged in research and development of light-emitting diodes, semiconductor lasers, and amorphous thin film devices. During 1975-1976, he stayed at Cornell University, Ithaca, N.Y., as a faculty member.

Dr. Tsukada is a member of the Institute of Electronics and Communications Engineers of Japan and the Japan Society of Applied Physics.

Yasutsugu Takeda

Yasutsugu Takeda was born in Tokyo, Japan, in 1935. He received the B.S. degree in Electronics from the University of Tokyo in 1958.

In 1958, he joined the Central Research Laboratory, Hitachi Ltd., Tokyo, Japan, where he has been engaged in the research of linear electron accelerators, linacs, spatial modulators, holographic storages, three dimensional displays and optical video disks. He received the Ph.D. degree from the University of Tokyo in 1973. His present position is the manager of the first department of the Central Research Laboratory, Hitachi, Ltd.

Dr. Takeda is a member of the Institute of Electronics and Communications Engineers of Japan, Institute of Electrical Engineers of Japan, Japan Society of Physics, and the Japan Society of Applied Physics.

Reprinted with permission from *Optical Engineering*, Vol. 22(4), pp. 464-472
(July/August 1983). ©1983 Society of Photo-Optical Instrumentation Engineers.

Multichannel digital optical disk memory system

D. C. Kowalski
D. J. Curry
L. T. Klinger
G. Knight*
Xerox Palo Alto Research Center
3333 Coyote Hill Road
Palo Alto, California 94304

Abstract. This paper describes a prototype computer-controlled digital optical disk memory system that was designed and built for an office-of-the-future application and also to track the technology, assess issues, and provide answers to a host of application-oriented requirements. The system uses gas lasers for writing and reading. Unique features of the system are simultaneous eight-channel parallel readout, write channel addressing, and a readout channel incorporating a single custom photodetector array for extracting data as well as focus and tracking error signals. The problems of media sensitivity and archivability are discussed, and sensitivity measurement techniques are described with typical examples shown.

Keywords: digital optical disk; write-once memory system; parallel readout; integrated photodetector.

Optical Engineering 22(4), 464-472 (July/August 1983).

CONTENTS

1. INTRODUCTION

Digital optical disk memory systems employing the technology base developed for the video disk are beginning to emerge from research into development. These laser operated optical-media storage and retrieval systems promise to be a viable solution to mass storage problems associated with magnetic storage systems. The optical disk has the capability of providing three major important improvements to storage technology: (1) presently around 40 times higher recording density on removable media,[1] (2) faster access to large amounts of data, and (3) a promise of archivability. Presently, the only drawback is that the optical disk is a write-once memory. This is not an apparent limitation in many specialized data storage applications. Fast access to large amounts of data, as well as archivability and removability of media, are the predominant considerations for use of the optical disk.

Here we describe a system that uses yet another property of the optical disk technology that cannot be implemented or used as effectively by its magnetic counterpart. This property is the multibeam or channel capacity a lens possesses by the very nature of the physical optics. Thus, a lens can support a number of write and read channels in parallel—a property that can increase the data rate of an optical disk system significantly over that of the magnetic disk. RCA already has reported on a high data rate system that uses the input and output multichannel technique.[2,3] The system described here is slightly less ambitious in design goals. It uses a single addressable (scanning) write beam and a multichannel read beam. A multibeam array is passed through the lens to achieve a read spot array that allows the system to read eight parallel data channels simultaneously and provide servo control signals. The spot array on the disk is two-dimensional, whereas the one employed in the very high data rate system reported by RCA uses a one-dimensional spot array. The following sections describe the overall system with emphasis placed on the optics, servos, digital software designs, and the media measurement process.

2. DESIGN CONSIDERATION

The primary motive for developing a research prototype optical disk system was to explore its use as a storage device for the office of the future. The high storage density and fast random access capability of the optical disk make it a suitable choice for storing images for future-generation duplicators. Advanced duplicators will easily operate at data rates in excess of 60 Mbits/s.[4] For the optical disk to be useful in this application, it will have to output at such high data rates. On the other hand, the input rate to the disk can be significantly lower since it will be loaded by a computer operating near a data rate of 10 Mbits/s.

To explore this type of duplicator application, a design goal of 10 and 80 Mbits/s was selected for the input and output data rates of the experimental optical disk system. To achieve this different write and read capability, a unique optical head was developed. The head uses a single track addressable write channel and nine parallel readout channels. To illustrate the way the optical head is used in writing and

*Currently at Optimem, Unit of Shugart Associates, 350 Potrero, Sunnyvale, CA 94086.

Paper 1884 received May 17, 1982; revised manuscript received Nov. 15, 1982; accepted for publication Nov. 23, 1982; received by Managing Editor Jan. 5, 1983. This paper is a revision of Paper 329-02 which was presented at the SPIE conference on Optical Disk Technology, Jan. 26-28, 1982, Los Angeles, California. The paper presented there appears (unrefereed) in SPIE Proceedings Vol. 329.

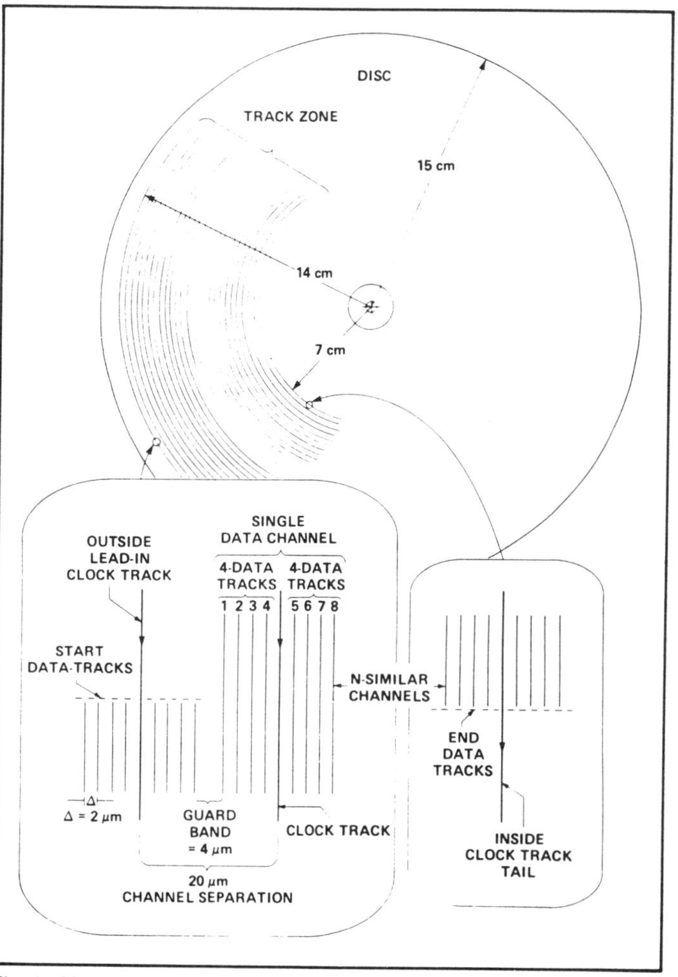

ɪig. 1. **Multichannel optical disk format.**

eading data, consider the drawing in Fig. 1. Shown is a 30 cm iameter disk with a track zone in the outer annulus. An expanded iew of the data format is shown below the disk at three different ositions along the spiral track. Concentric circular tracks have the ame format. The disk is preformatted initially with a clock track at a itch of 20 μm. It takes a little over two minutes to preformat the isk. Spaced at 2 μm intervals on each side of the clock track are four ata tracks. Each data track is written individually at a 10 Mbits/s ate by means of the addressable single-write channel. The clock rack has two important functions: it is used to derive accurate teering signals to control the position of the write and read spots, nd to derive consistent bit and word clocks during reading and riting. The eight data tracks together with the clock track define a ingle data channel since they can be read out simultaneously in arallel to achieve an 80 Mbits/s output data rate. Thus, the specified esign goals for a low input and high output rate can be met by an ptical head using an addressable single write channel and nine arallel readout channels. The optical head will be described in more etail in a following section.

. OPTICAL DISK SYSTEM

he optical disk system is shown in the drawing in Fig. 2. The main arts of the system are the write/read station, the read channel and rvo electronics, and the computer and controller. The optics, ectromechanical drives, and lasers are all mounted on a stationary atform below the upper baseplate of the write/read station. The nal output focusing lens is mounted beneath the disk. The station as two gas lasers: a high power argon laser for writing and a low ower helium-neon (HeNe) laser for readout. In this system the disk moved to access the data. The disk is positioned with a precision

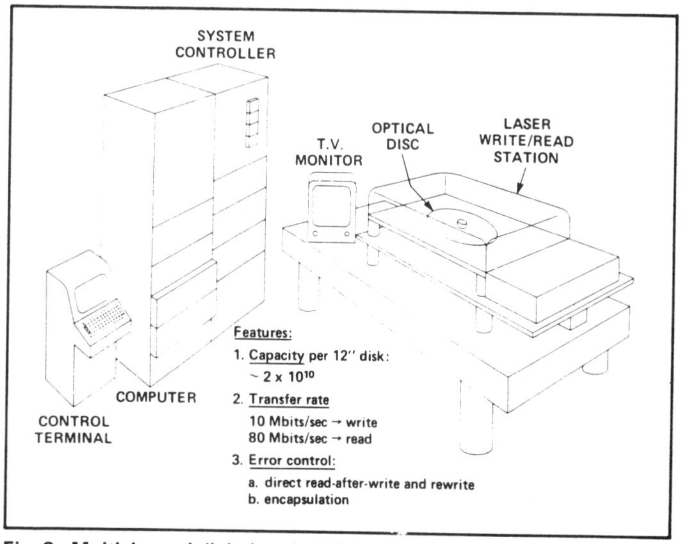

Fig. 2. Multichannel digital optical disk memory system.

ball-screw drive dc servo system. Accurate position information is obtained by the controller using a linear encoder attached to the disk carriage. The readout and servo electronics are mounted on the station for operational convenience and signal bandwidth and noise considerations. The write and read station is self-contained and it can be operated manually from switches and controls on the electronic cards and manual control panel. The baseplate just below the disk can be used as a Bernoulli air bearing surface for stabilizing thin floppy optical disks when necessary. The baseplate also has small air holes for generating an air cushion below the disks when they are installed. This feature protects floppy disks from scratches and contact damage during routine installation. The air cushion is automatically turned off when final spin velocity is reached.

The optics package below the top plate contains a TV camera and imaging optics which display a 30 μm diameter area of the disk on a monitor next to the station. The TV display turned out to be a very valuable tool for alignment and analysis of the system performance. The writing, reading, focus, and tracking performances were readily observable.

The block diagram in Fig. 3 illustrates in some detail what the system contains and the way the overall system is interconnected. The NOVA computer provides the general interactive control of the system through software developed for this specific application. On command from the computer, the controller writes the clock track, provides synchronization of data, and writes and reads sectored data or blocks of data at arbitrary locations on the disk surface. In general, most of the control inputs to the station pass through the read and servo electronics from either the manual or digital control panel. Most of the system's basic modes of operation can be controlled from the manual panel. This capability offers a significant advantage when system maintenance and calibration are required.

4. WRITE AND READ OPTICS

The major components in the write and read optical head are shown in the drawing in Fig. 4. The system has two gas lasers: a high power argon for writing at 458 nm wavelength and a lower power HeNe laser for reading data. In future systems gallium arsenide laser diodes will replace such gas lasers. The two beams from the lasers are transmitted through and operated on by the various components in the optical paths before they are combined and focused on the disk by the output lens. The write (blue) beam is modulated at a high rate to thermally mark the media on the spinning disk. The read (red) beam is reflected and modulated by the thermally marked media. The reflected light is imaged on a small PIN photodiode array mounted on a standard 16 pin DIP (dual in-line package) ceramic integrated circuit (IC) header called the VTF detector (video-track-

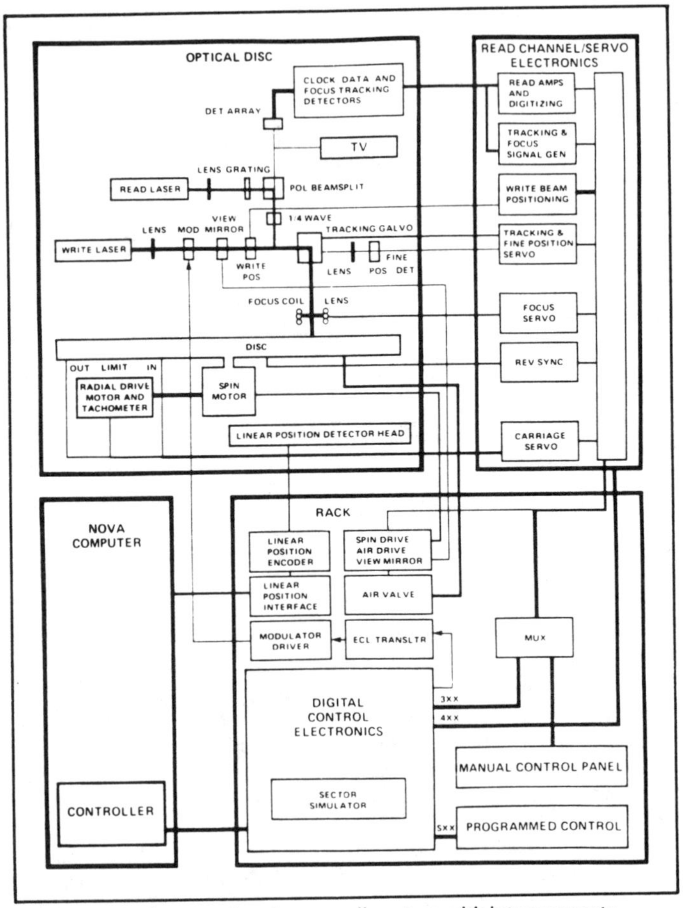

Fig. 3. Block diagram showing overall system with interconnects.

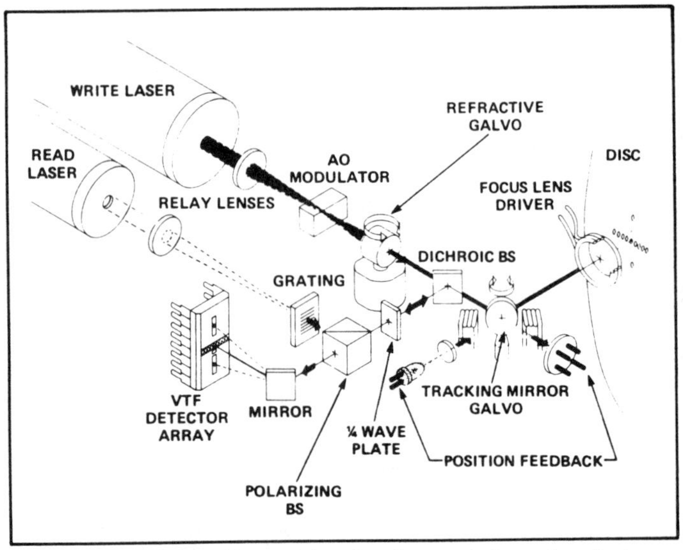

Fig. 4. Optical disk and beam-related optics, controls, and readout.

ing-focus). The output detector array together with the two-dimensional phase grating that is located at the input side of the read channel are the components that give the optical head multichannel capability. From them are derived all the data, tracking, and focus control signals. Not shown on the drawing, so as to minimize its complexity, are three reflecting elements used for vignetting correction between the mirror galvanometer and the focus lens.

5. WRITE CHANNEL

The write channel can deliver as much as 30 mW of laser power to the disk in the approximate form of a Gaussian spot 0.8 μm in diameter at its half-power points. The first element along the optical path from the argon laser is the spot-forming lens. This lens focuses the raw beam down to an appropriate size spot inside a fast acousto-optic (AO) modulator. Binary data from the computer at a 10 Mbits/s rate drives the modulator to transfer the data information to the laser beam. This beam is imaged and minified on the disk by the output focus lens. Between the modulator and the output lens is the refractive galvanometer, a dichroic beam combiner, and a tracking mirror galvanometer.

The refractive galvanometer moves the spot by tilting a thick parallel glass plate. In the tilted plate the beam is refracted and effectively displaced at the output. The refractive principle allowed the use of a moving iron galvanometer open-ended without feedback to accurately position the write spot within 0.1 μm of the specified track position. A standard mirror galvanometer would require some form of position feedback control to achieve the same pointing accuracy.

The next element in the path is the dichroic mirror. It is used as a high efficiency bichromatic beam combiner that transmits the write beam and reflects the read beam towards the mirror galvanometer. The mirror galvanometer is used to move both the write and read beams together for tracking the data. The tracking galvanometer is used for overcoming dynamic mechanical errors and for fast access to data within the lens field of view. The output focus lens is the final element in the optical train and a critical component. The numerical aperture (NA) of the lens defines the final output spot size on the disk. To hold the spot size accurately, the output lens must be automatically controlled to maintain a constant distance between the lens and disk surface. This is done by means of a electromechanical drive similar to an acoustic speaker with a hole through the center for the light and lens. This is a standard technique for focus control and is used in most present optical disk systems.[5] A notable exception is the focus control technique using a Bernoulli stabilizer—a floating air bearing shoe containing the output lens.[6]

6. MULTIBEAM READOUT CHANNEL

An optical read channel must deliver light to the disk and return it to a detector for electronic detection of the recorded data bits, and it must do this efficiently so that the light power reaching the disk will not alter the written data. In addition, the read channel must provide some optical technique for generating servo control signals for automatic control of the spot focus and tracking of the data bits. The read channel described here accomplishes all of these functions efficiently with the minimum amount of components. The main components in the system are the two-dimensional phase grating and the detector array that give the system its multichannel read and control capability.

Referring again to Fig. 4, the raw laser beam from the HeNe laser is focused to a spot in front of the phase grating. The diverging beam from the spot passes through the phase grating, which generates a two-dimensional array of light beams. The beams are reflected by the diagonal in the polarizing beam splitter and pass through a quarter-wave phase plate and on to the dichroic beam splitter. The combination of polarizing beam splitter and quarter-wave phase plate is an optical technique for increasing the optical efficiency of a read channel. They allow almost all the reflected light from the disk to reach the output detector array. The combination also eliminates feedback into the read laser which, if not removed, will cause extraneous laser noise in the read channel. At the dichroic mirror the read beams are reflected and combined with the write beam and pass on to the mirror galvanometer and through the output lens. The function of these last two components has been described in the previous section. The appearance of the final output spot pattern is depicted on the disk in the drawing, where it is shown locked on to a single data track. The reflected light from the disk is returned along the same optical path until it reaches the polarizing beam, where now, due to the quarter wave plate, it is transmitted to the output detector array. The output

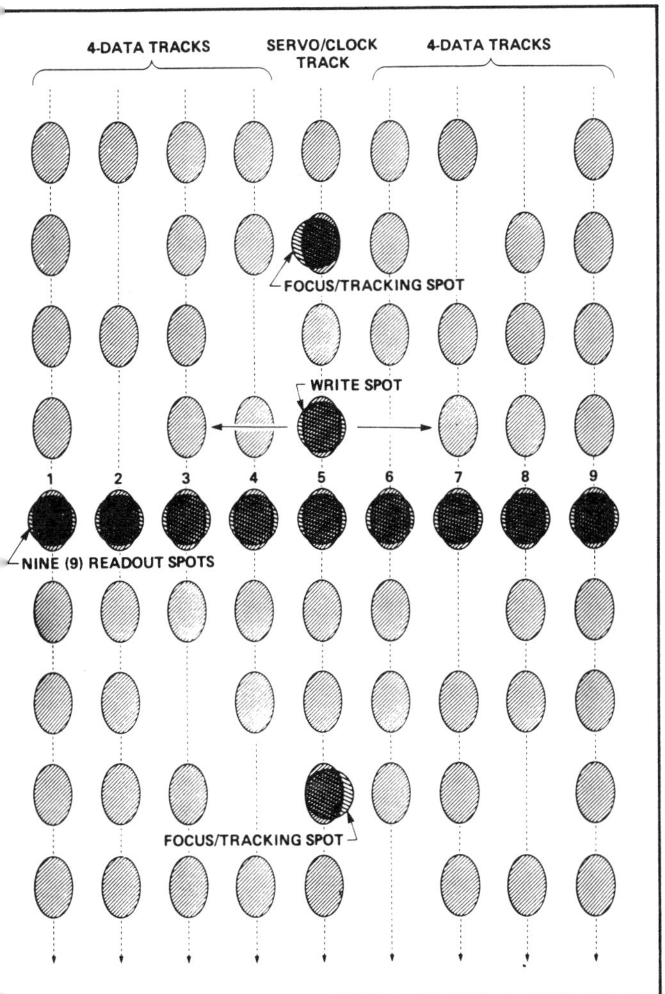

g. 5. Read spot array on written disk.

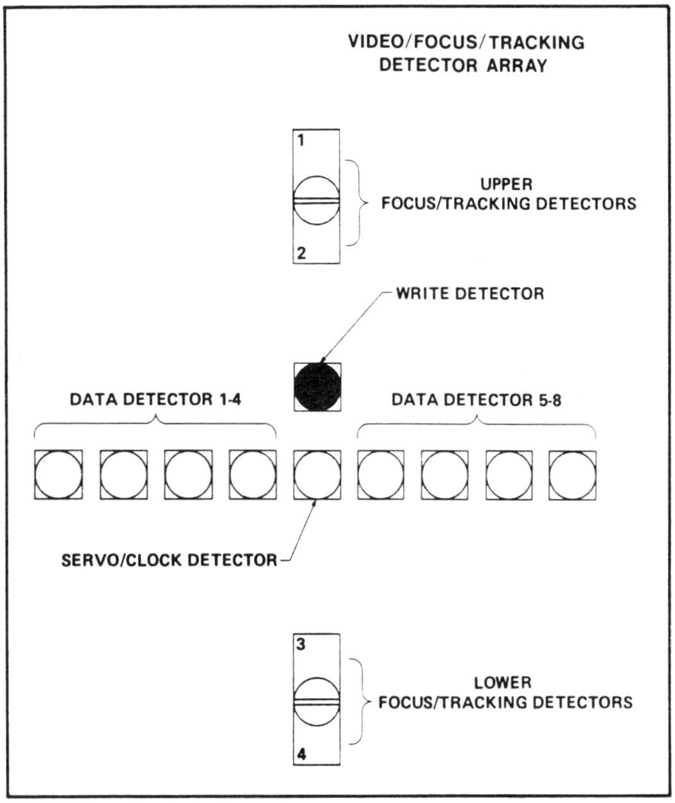

Fig. 6. Multichannel detector array.

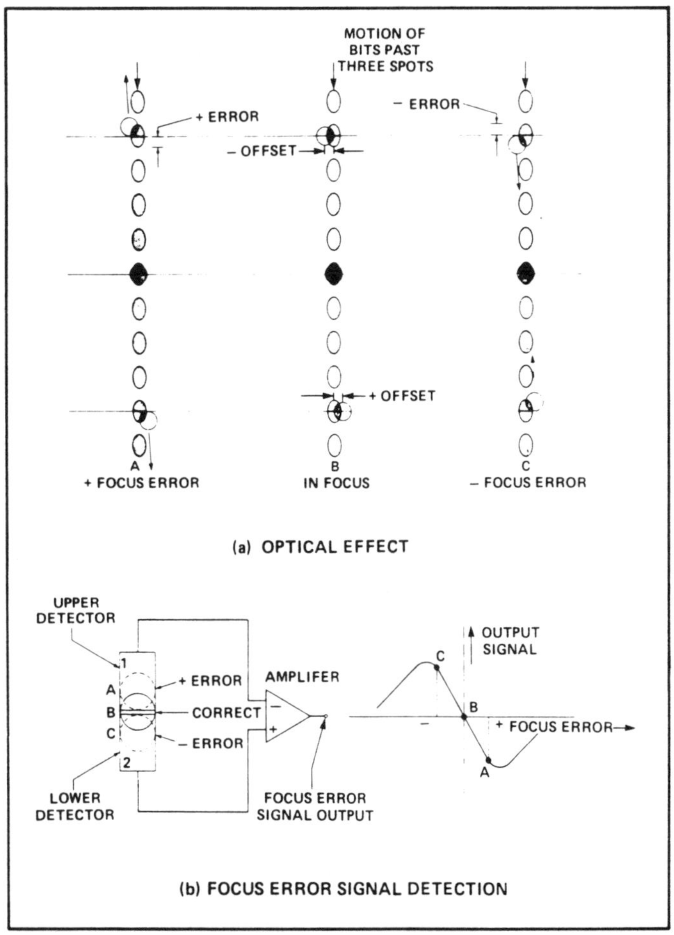

Fig. 7. Focus error signal generation. (a) optical effect; (b) focus error signal detection.

ns reimages the spots from the disk on the individual elements on ıe detector array.

The drawings in Figs. 5 and 6 show the relationship between the pots on the disk and the reflected spots on the detector array. The :rvo clock and data bits are shown as they appear moving past the ine readout and two focus and tracking control spots. The position f the write spot is shown in its center position above the clock :adout spot. The write spot can be positioned accurately on either de just above the other readout spots. In this configuration the read pot lags the write spot and allows for direct reading after write, a :chnique named DRAW by Philips.[7,12] The DRAW technique pro-ides the optical disk user with a means of observing the write errors nd the potential of fast error control. The layout of the detector rray is shown in Fig. 6 with the returned reflected read spots imaged n the appropriate elements. The function of the servo clock and ata detector elements is obvious. The write detector element is used) verify the center position of the refractive write galvanometer efore writing of the data tracks begins. The performance of the pper and power pair of detector elements for focus and tracking rror detection is not so obvious and will be described in the follow-ig sections.

The focus error signal is generated by the motion of the upper and wer outrigger spots across the spacing between the split detectors s the disk goes in and out of focus. The method is well known and is alled the skew-beam focus error signal generation technique. The ptical effect and the signal detection electronics are depicted in Figs. (a) and 7(b). With the disk surface in focus, the upper and lower ontrol spots are shown centered across a track of data. Now con-der only the upper spot and its reflected image on the upper detector pair in the array. When the disk is in focus, the reimaged spot is centered across the split detector pair and the electrical signal from the differential amplifier is equal to zero. As the disk moves in and out of focus, the outer spots move away or towards the center depending on the direction of the disk error. This is depicted in the drawings, which also show the motion across the detectors and the resulting focus error signal. The error signal is linear for a significant range of focus error and then drops off to zero each side of focus as the spots get large due to the defocusing effect and drop in intensity.

The tracking error signal is generated by the same two reflected outrigger spots and detector elements that are used in the focus technique. The optical effect and the signal detection electronics are depicted in Figs. 8(a) and 8(b), which are very similar to those in the previous figure. The relative positions of the three spots and the data track are shown in Fig. 8(a) for three conditions: the correct on-track and the shift right and left conditions. The on-track situation is shown in the center drawing with the two spots symmetrically placed with respect to the track. The reflected spots are centered on the upper and lower detectors as before; however, for track error detection the two elements in each pair are now treated as single elements. The average signal of these detectors is subtracted by the differential amplifier to give the tracking error signal. As the track moves to the right or left of the stationary spots, it affects the ratio of the light reaching the detectors and generates a tracking error signal of the type shown in Fig. 8(b). The error signal has a linear range that is nearly as wide as the spot size on the disk. This method of generating the tracking error signal is usually referred to as the three-spot tracking technique[8] and is presently used in Magnavox and Pioneer video disk players.

7. MULTICHANNEL PHASE GRATING

The purpose of the grating is to produce a two-dimensional array of eleven spots on the disk: nine spots for reading data tracks and two for focus and tracking control. It is desirable to have the nine data spots as uniform in intensity as possible, whereas the two focus and tracking spots must be perfectly balanced in intensity and tilted slightly to straddle the servo clock track. The grating was designed using the pulse-width modulation technique.[9] The fabrication process involved computer generation of a large two-dimensional binary density pattern, photoreduction, contact printing on photoresist, and final etching of the phase pattern on a half-inch diameter glass disk. The photographs in Fig. 9 show the computer-generated binary grating pattern and the resulting laser spot array for both a four and eight data channel system. It can be seen that the intensity of the data channel readout spots is uniform. The tilt of the vertical servo control spots A is about 1.4°; this corresponds to 0.5 μm of offset on the disk between the upper and lower spots. The ratio of the intensities of the control spots A to the data spots can be adjusted by varying the modulation of the grating pattern in either direction. For instance, the diffraction intensity in vertical direction, spots A, can be reduced by cutting the amount of modulation of the grating in the vertical direction. The uniformity of the data channel spots is limited by the plotting resolution of the graphics plotter used to generate the binary grating pattern. The grating is highly efficient and places 70% of the transmitted light into the 11 precisely positioned diffraction-limited spots with accurately controlled intensities.

8. SERVO CONTROLS

The accuracy to which the data on the disk must be tracked to generate acceptable output signals is in the order of ±0.1 μm in both the axial (focus) and radial directions of the disk. Focus control is accomplished by displacing the output lens with an electromagnetic coil similar to ones used in loudspeakers. The required frequency bandpass of the focus servo for this system is, in general, close to that reported for the video disk players.[10] Radial tracking control is achieved with a moving-magnet-type mirror galvanometer which also uses an electromagnetic drive coil. In the optical disk memory system there is an added requirement for fast access of data within the field of view of the output lens. This requires an auxiliary fine

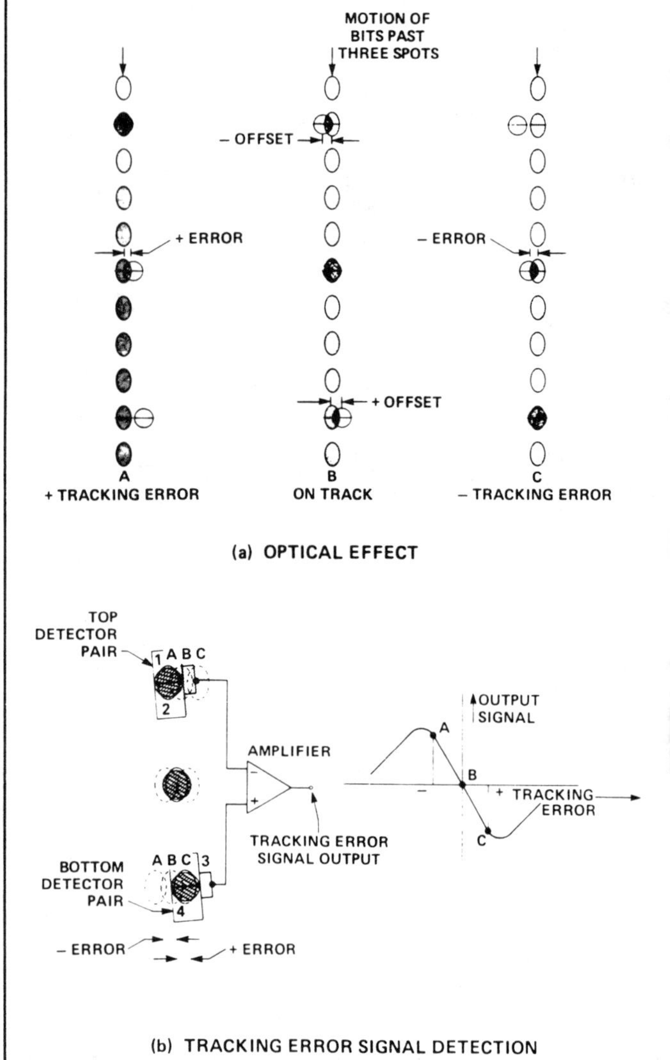

Fig. 8. Radial tracking error signal generation. (a) optical effect; (b) tracking error signal detection.

position loop which shares the radial tracking galvanometer as its output transducer.

9. FOCUS CONTROL

The focus control shown in Fig. 10 has an open-loop unity gain frequency (f_1) of 4.0 kHz and, thereby, a closed-loop response of approximately 2.5 kHz. This is more than sufficient to achieve the desired ±0.1 μm accuracy with an 1800 rpm rotation rate. Since the load at times consisted of rather heavy output lenses modified from microscope objectives, a dual-diaphragm construction was used to achieve good off-axis stability. The resulting spring strength makes it useful to employ integral feedback in the loop to achieve a broad operating range with zero error resulting from the spring. Mathematically this yields type-3 loop operation. A simple open-loop scan function is used in conjunction with a validation signal from the data channel to close the loop when the media target passes through focus. With this approach, a valid focus linear range of 4 μm has proven to be quite sufficient.

10. RADIAL TRACKING SERVO

The radial tracking servo shown in Fig. 11 consists of three interactive loops. The tracking loop achieves high speed data track follow-

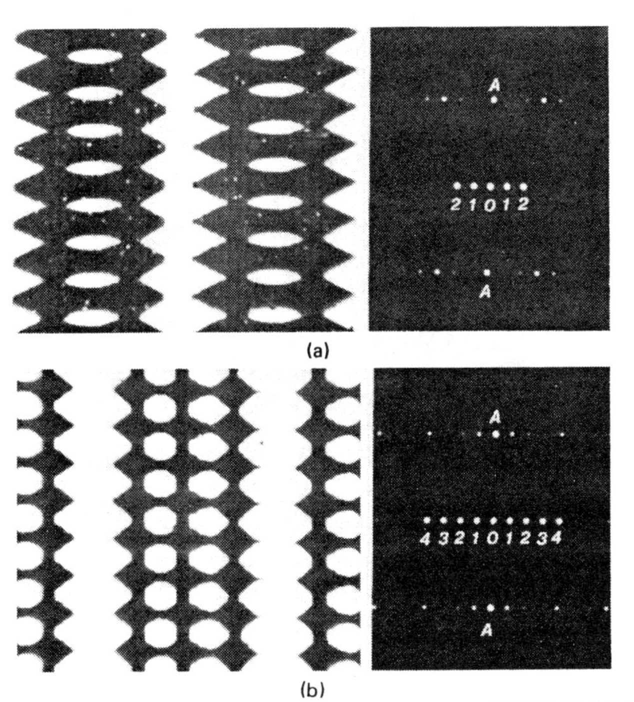

Fig. 9. Two-dimensional phase grating and read spot array. (a) four data channels; (b) eight data channels.

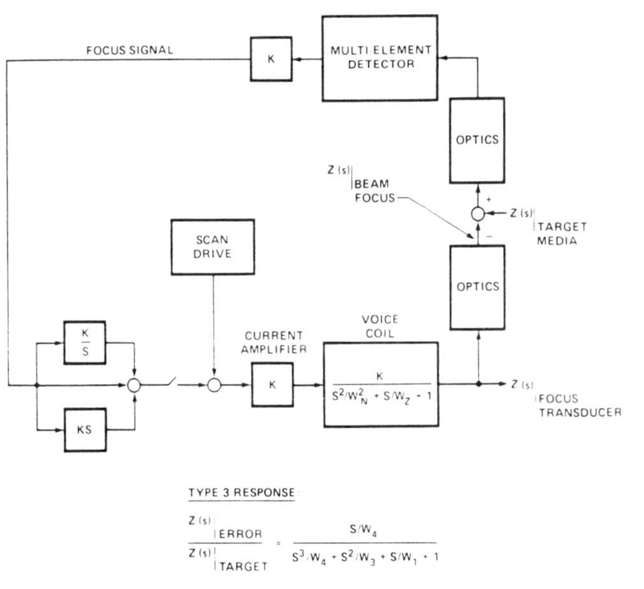

Fig. 10. Block diagram of the focus servo.

ing, while the fine position loop is used to command high speed seeks or command the coarse carriage loop to center the track in the optical field of view. This arrangement has operated equally well for both spiral and concentric tracks with no functional change in operation. The fine position detection is performed with a light-emitting diode (LED) which reflects off the galvanometer mirror (at an angle different from the read and write beams) into a split detector. For experimental convenience, the coarse radial positioning is accomplished by moving the disk spindle with a motor/tachometer attached to a precision lead screw drive assembly. A linear position detector sends a digital carriage position signal to the controller, but is not included in the closed-loop operation.

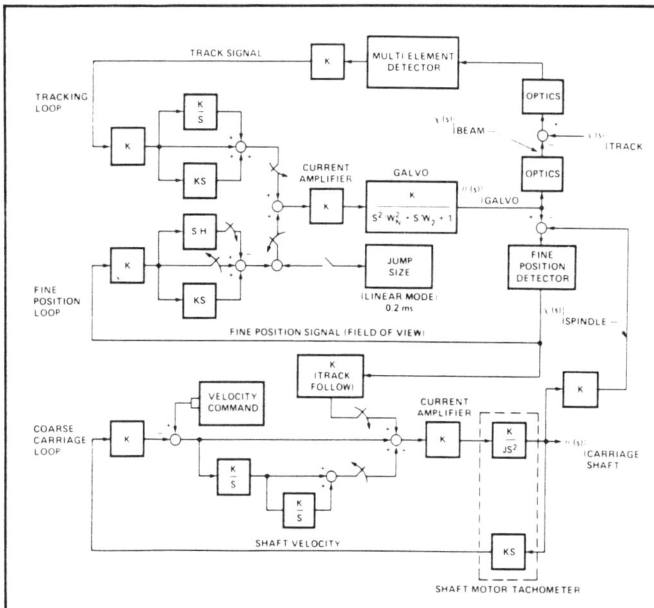

Fig. 11. Block diagram of composite radial servo.

The tracking loop has an f_1 of 2.0 kHz and includes integral feedback for type-3 operation. The fine position loop also has an f_1 of 2.0 kHz and operates as a type-2 system for holding position and as a type-1 system for executing fast seeks in its linear mode. Seek/capture times of 0.5 ms for distances less than 16 μm are typical. The sample and hold function shown allows the fast seeks to be referenced to the track being followed.

The coarse carriage loop has an f_1 of 80 Hz and operates in either of two modes. In the track follow mode it centers the track in the optical field of view by using the fine position detector output and the shaft velocity signal in type-1 operation, while the tracking loop controls the galvanometer. The servo/clock track generation mode is used to drive the carriage at a precise velocity for writing spiral tracks (or concentric tracks with a velocity equals zero command). The precision lead screw velocity, with its first and second integrals, is fed back to successfully eliminate effects of the lead screw friction on track pitch. This results in type-3 operation. Net inaccuracy of pitch is then due only to the machanical compliance between the shaft and spindle, precision lead screw error, or tachometer nonlinearity. In all cases, for both the focus and radical axis loops, integral feedback is restored to zero if the servos are not operating in their linear mode.

11. OPTICAL DISK

The central problem in an optical disk memory system is the optical disk. Here by the optical disk we mean the recording media, the substrate, and, finally, the disk package. The media requirements have two conflicting demands: high sensitivity to the write laser, and long term stability or archivability. Presently there are a number of companies with sizable development efforts aimed at producing the ideal media. Since the media are still in the developmental state and in most cases proprietary, the question of media will not be addressed. Here we will discuss and describe the substrate and the disk package as they relate to supplying a single machine with suitable optical disks for simultaneous development of both the machine and media. An excellent survey of reported state-of-the-art media for optical storage can be found in Ref. 11.

11.1. Substrates

The selection of substrates for the final optical disk package will require stringent specifications on optical quality and thickness uniformity; however, when disks are considered for a single developmental machine, the optical and thickness tolerances can be lowered such that suitable disks can be hand selected and cut from commercially available materials. Most of the substrates for this program were procured in this manner. Since the machine was designed to work with both floppy and rigid disks, and in both first and second surface record and playback mode, a number of techniques were tried for selecting and fabricating the disks. Here first surface implies direct contact of the recording media by the write and read laser spots without an intermediate optical surface between the media and the final output lens; in contrast, second surface refers to the use of an intermediate layer for the protection of the media from dust and scratches. With the maturing of the program, the floppy disks were completely replaced by single rigid disks and the sealed disk pair commonly referred to as the air sandwich disk.[12] The single rigid disk could be used in both the first and second surface mode, while the sandwich-type disk was designed to be used only in the second surface mode.

The dimensions of the presently used substrate are those of the commercial video disks. The diameter is 30 cm and the thickness is 1.1 mm. The substrate material that appeared to be the most suitable optically and in terms of material media interactions is polymethyl methacrylate, also called PMMA and Plexiglass.[13] The disks were cut from nominal 40 mil thick stock of large 3×4 foot sheets of PMMA. The first step in the selection was to measure the exact thickness of the sheet in 3 in. increments with a linear array of precision dial gages. Contours of equal thickness were then plotted on the protective paper surface, and a 30 cm diameter disk was used to select areas of appropriate thickness and tolerance. The yield averaged approximately three disks per sheet. The selected areas were cut into 32 cm squares and then stacked in units of 20 to be cut on a lathe into disks with accurate center holes. The next step in the process required performing a static balance measurement of the disk. This involved the use of a side-mounted hub with a friction-free bearing. The disk was treated as a pendulum, and the period of the disk swing gave a measure of the imbalance. The selected disks were then sent to the coating labs, where, in a clean room, the protective paper was removed and the disks were run through a thorough cleaning process before the recording media was applied.

11.2 Sealed disk package

The air sandwich disk is an attractive structure for sealing and protecting the media. Even in a laboratory environment it became apparent that such a structure was needed since a large number of disks were handled. After trying a number of chemical adhesives and solvents, it was clear that sealing the upper and lower disks to the separators was a difficult task that required further study and development. For the sake of expedience, a simple alternative was tried using double-sided adhesive paper of the type used to mount photographic pictures. The double-sided adhesive separator turned out to be ideally suited for construction of the air sandwich disk with a simple assembler used in the disk coating facility.

A simple disk assembler was designed and built. Figure 12 shows a drawing of the assembler together with the separate parts of the disk. The structure of the disk, with its annular outside ring and inside disk, is evident from the exploded view of the separated part on the guide rod. The assembler consists of two vacuum hold-down plates which are guided together by an accurate center rod having the exact diameter of the center hole in the disk. These two plates have identical vacuum channels that hold the parts as they are assembled. The rod serves to align the center holes of the upper and lower disk and provides a means of applying pressure to the completed assembly for positive sealing of the disks. The vacuum holds the different parts as they are laminated together in a step-by-step assembly process. A convenient aspect of the mounting material was that it had a paper protective coating on each side; this made cutting of the outer and inner separators a simple process. The cover paper offered complete protection to the material until used in the final stages of fabrication of the disk package.

Besides protecting the media both before and after writing, the sandwich-type structure provides two other benefits: (1) the disk

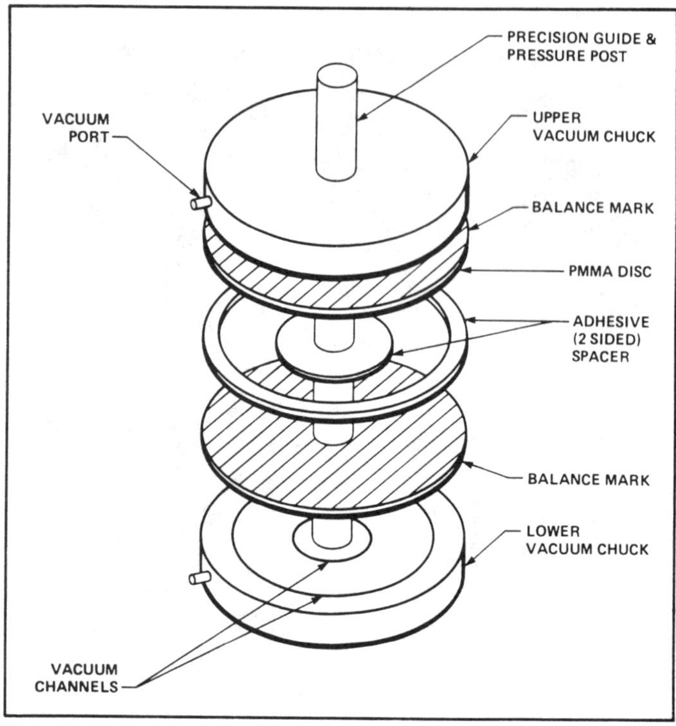

Fig. 12. Disk assembler.

package is more rigid and therefore has better spin characteristics, and (2) the disk package can be balanced statistically extremely well since the two disks can be joined together so that they cancel out each other's imbalance. The air sandwich design and assembly technique described here was strictly a stopgap measure to allow the immediate use of environmentally protected disks; however, due to its simplicity, the process is still being used in the laboratory.

12. DIGITAL CONTROLS

The system has four levels of control. The lowest level is implemented as various toggle switches located on the servo control subsystems themselves. These allow the subsystems to be checked out individually and manually in the system and also allow alignment functions to be carried out nearly independently of higher levels of control. Functions at this level include such things as focus enable, manual write channel select, tracking galvanometer lock, and radial slew speed select.

The next level of control is referred to as the manual panel. This array of switches allows the control signals on the interface cables to be manipulated conveniently, although statically. The functions available at this level are similar to those at the subsystem level. This control level is used heavily during system alignment.

The third level of control, referred to as the drive control panel, provides status monitoring and manual access to the dynamic control modes implemented by a programmed sequencer inside the drive electronics. These modes are identical to those with which the computer interacts with the drive. In addition to the programmed sequencer, the drive electronics include the read bit and word clock generation phase lock loop, write clock pattern generators, serializer/deserializer circuitry, and two programmable counters. These electronics also include boards for simulating various-sized data sectors, and interfaces for a linear position encoder and high speed defect measurement counters. Control modes at this level range in complexity from simple radial motion commands to fairly involved tracking, clock/data track differentiation, and clock lockup procedures.

The highest level of control is provided by a NOVA 800 minicomputer using a control board which implements direct memory access, automatic header/sector detection, programmable sector formatting, and first-in, first-out (FIFO) buffering of data to and from the drive electronics. Using the language BCPL, a large body of controlling software has been written. Besides a myriad of special test functions, some of the higher level functions implemented include

- preformatting single and multiple concentric or spiral tracks;
- performing defect scans of various areas of the disk;
- sampling and displaying arbitrary areas of the disk;
- reading and writing data files.

This last function involves successful interaction of all the various subsystems as well as a viable defect protection scheme.

13. SYSTEM PERFORMANCE

The complete operation of the multichannel optical disk system was achieved, and all aspects of the parallel channel readout and single addressable write channel were proven feasible. The subsystems carried out digital commands from local and computer control. A major accomplishment that was central to the parallel channel operation was writing data headers in synchronization with the prerecorded servo clock track. The importance of this operation was that it included nearly all internal controller and drive electronic processes, such as the phase lock loop, write and read delay adjustment, sector memory formatting, read detect logic, and, finally, data channel transfer and write software.

As a demonstration of the success of this device, print files were received off the Ethernet local network, stored onto the disk, recovered error free from the disk, and sent out over the Ethernet to be printed at a remote printing station.

Simultaneous with the building and testing of the system, a media development program was started to supply the optical write/read system with disks. During this period it became apparent to us, as it has to others in this field, that the major unresolved problem in the developing technology was the media. Therefore, after the multichannel concept was proven feasible, the optical disk memory system was devoted almost exclusively to media testing and evaluation. The following sections describe the media evaluation and testing effort, specifically the measurement technique, with some typical performance results.

14. MEDIA TESTS

The media sensitivity was measured by conducting tests in which the write laser power was changed from a high to a low level in small increments to the point where the media stopped responding and no output data signal was observed from the preamplifier. At each write power level, the output signal was monitored on an oscilloscope and measured with a power spectrum analyzer. The oscilloscope displayed two traces of the output signal: a high-sweep display of approximately 10 cycles of the bit pattern, and a low-sweep display of one complete revolution of the bit pattern showing mainly the envelope of the output signal. The high-sweep display was used to measure the signal contrast ratio, the ratio of the difference to the sum of the read signal excursions, and to note the write power at which the 50% duty cycle in the playback signal occurred. The 50% duty cycle write power was used as one important measure for comparing media. The low frequency display showing the readout signal envelope always gave a good indication of the gross condition of the disk substrate and media.

A Tektronics 7L13 spectrum analyzer was used to measure the signal-to-noise (SNR) power spectrum of the output signal. For this measurement a 30 kHz bandwidth was used, together with a sweep display of 2 MHz per cm on the horizontal display of the oscilloscope. The SNR of the full bandwidth of the readout channel can be calculated from the 30 kHz data if the noise spectrum is flat within the response of the red channel bandwidth. At this stage of the tests if the media warranted any further consideration, optical microscope ×1000 pictures were taken of the written tracks, and then scanning electron microscope pictures were taken of the individual bits. A typical set of data is shown in Fig. 13. This data gave sufficient information to categorize the media. One further step was taken

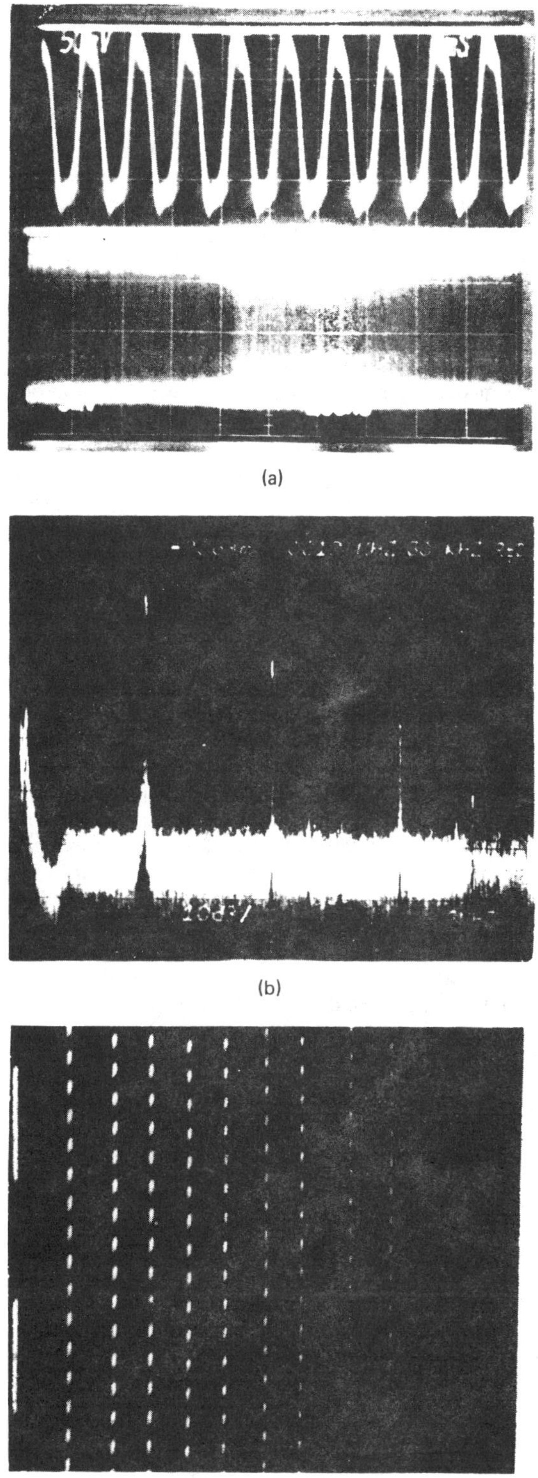

(a) (b) (c)

Fig. 13. Typical oscilloscope and microscope pictures of write power test. (a) high- and low-sweep; (b) power spectrum display; (c) tracks on disk.

when the media showed much promise: surface defect statistics were measured on sealed disks of the specific media. This test measures the ratio of the total scanned area to the defect area and also gives the size of the average defect. More elaborate statistical defect measurements and studies are planned. Typically, a raw BER (bit error rate) of 10^{-4} has been measured, with some media approaching 10^{-5}. Once a BER of 10^{-5} is approached, our present optical quality of the substrate will have to be improved. Presently, there appears to be a one-to-one correlation between measured substrate surface quality and BER.[14]

15. MEDIA PERFORMANCE

The general results of some dynamic recording and playback tests are shown in Fig. 14, where the SNR of the media is plotted as a function of the laser power and energy density. The contrast ratio has been omitted to simplify the presentation. All data shown refer to a disk spin rate of 30 rps and a radius of 12 cm. Typical response curves are shown for monolayer tellurium, monolayer tellurium alloy, inverted (second surface) tellurium, and DRM (deformable recording media) of the type reported at the June 1980 meeting of CLEO.[15,16] For comparison the monolayer tellurium shown will be used as the reference media. This media was also selected as our benchmark material for comparison purposes and system calibration.

The curves show that both the monolayer and trilayer DRM material have only slightly better sensitivity than the 0.3 nJ/$(\mu m)^2$ of the monolayer tellurium. However, the monolayer DRM is about 10 dB better in SNR performance and the trilayer DRM only about 7 dB.[17] The inverted tellurium trilayer is of interest because it forms a natural encapsulating structure with the media between the substrate and the relatively thick aluminum reflector. Although the SNR of this media is good, >50 dB, it requires about twice the write power of the DRM and tellurium media. We are currently investigating tellurium alloys that have been reported to give high SNR performance and a promise of archivability.[18] The sensitivity curve shows the test results for one of our early samples. This sample requires 0.5 nJ/$(\mu m)^2$ of write energy as compared to 0.3 nJ/$(\mu m)^2$ for plain tellurium; however, as reported, it has about 5 dB better SNR. The curves shown are not to be construed as having final significance, but as representative of our typical measurements for the specific purpose of describing our measurement technique. Besides the media shown in the curves, iron dispersion, silver halide, and dye type of media have been investigated. The effect of overcoating protective layers such as SiO_2 and PMMA was also studied with the same conclusions reached as other investigators, i.e., that a protective layer can mechanically constrain the media marking process and thereby cause degradation of the write sensitivity.[19,20]

16. CONCLUSION

An experimental optical disk system was built and tested that operates at a 10 Mbits/s input and 80 Mbits/s output data rate. The design concept of multichannel parallel operation for eventual laser printer use was proven feasible: files were stored on the optical disk and reproduced by a laser printer duplicator.

17. ACKNOWLEDGMENTS

The authors gratefully acknowledge the assistance of R. I. Johnson for coating the optical disks and C. DePuy for operating and maintaining the system and also for collecting data. For their designs of various media and coating facilities we thank G. A. N. Connell and C. C. Tsai and also T. W. Smith and T. M. Ward of Xerox Webster Research Center.

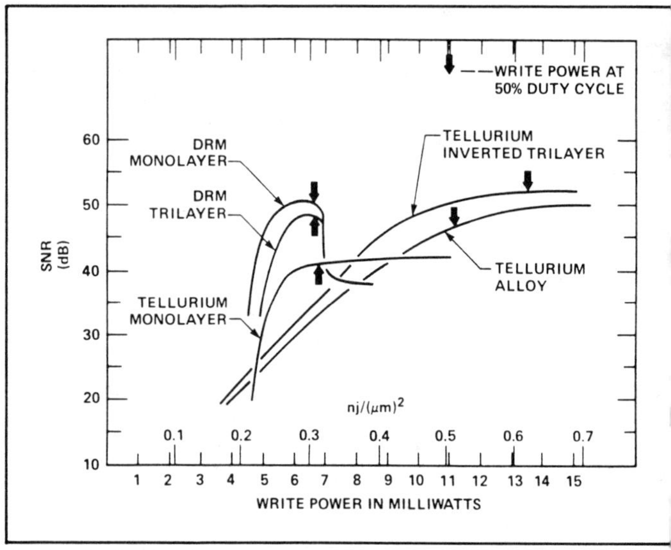

Fig. 14. Typical media sensitivity curves.

18. REFERENCES

1. P. Snigier, Digital Design 11(10), 44(1981).
2. G. J. Ammon and G. W. Reno, Laser Focus, p. 22 (December, 1981).
3. G. J. Ammon, Proc. SPIE 200, 64(1979).
4. G. K. Starkweather, *Laser Applications*, Vol. 4, Academic Press, New York (1980).
5. P. J. M. Janssen and P. E. Day, Philips Tech. Rev. 33(7), 190(1973).
6. P. J. Rice, H. F. Frohbach, N. A. Peppers, J. R. Young, L. F. Schaefer, and G. A. Pierce, J. Appl. Photo. Eng. 6, 62(1980).
7. R. Pepperl and I. Sander, Opt. Acta 24(4), 427(1977).
8. C. H. F. Velzel, Appl. Opt. 17(13), 2029(1978).
9. Wai-Hon Lee, Appl. Opt. 18(13), 2152(1979).
10. P. W. Bogels, IEEE Trans. Cons. Elec. 22, 309(1976).
11. H. Brody, Laser Focus, p. 47 (August, 1981).
12. G. C. Kenney, D. Y. K. Lou, R. McFarlane, A. Y. Chan, J. S. Nadan, T. R. Kohler, J. G. Wagner, and F. Zernike, IEEE Spectrum 10(2), 33(1979).
13. T. W. Smith, J. Vac. Sci. Tech. 18(1), 100(1981).
14. D. Y. Lou, G. M. Blom, and G. C. Kenney, J. Vac. Sci. Tech. 18(1), 78(1981).
15. J. Cornet, C. Lehureau, and F. LeCarvennec, CLEOS Conf. June (1981).
16. W. Robbins, R. Freese, T. Smith, and R. Wilson, CLEOS Conf. June (1981).
17. G. A. N. Connell, R. I. Johnson, D. C. Kowalski, and C. W. DePuy, Proc. SPIE 329, 174(1982).
18. M. Terao, K. Shigematsu, M. Ojima, Y. Taniguchi, S. Horigome, and S. J. Yonezawa, J. Appl. Phys. 50(11), 6881(1979).
19. M. Chen and V. Marrello, J. Vac. Sci. Tech. 18(1), 75(1981).
20. R. A. Bartolini, J. Vac. Sci. Tech. 18(11), 70(1981).

Reprinted from *Optical Data Storage Topical Meeting*, Proc. SPIE Vol. 1078, pp. 98-104 (1989). ©1989 Society of Photo-Optical Instrumentation Engineers.

Multi-beam magneto-optical disk drive for parallel read/write operation

Ryuichi Katayama, Kazuhiro Yoshihara, Yutaka Yamanaka,
Masaki Tsunekane, Kiminori Yoshida* and Keiichi Kubota

Opto-Electronics Research Laboratories, NEC Corporation
*R&D Planning and Technical Service Division, NEC Corporation
1-1, Miyazaki 4-chome, Miyamae-ku, Kawasaki 213, Japan

ABSTRACT

A multi-beam optical disk drive, which improves the data transfer rate for the magneto-optical disk systems, has been developed. It employs a newly developed image rotating actuator for tracing multi-tracks simultaneously, and photodiode arrays for simplifying signal detection circuits. Four-channel parallel read/write operation has been accomplished. The carrier-to-noise ratio is greater than 55dB, and the crosstalk between channels is lower than -36dB. A data transfer rate, four times that for an ordinary single-beam optical disk drive, has been realized.

1. INTRODUCTION

Optical disk systems, using a laser beam for data recording and reading, are foreseen for use as high density information storage systems. Recently, erasable magneto-optical disk systems have become commercially available. However, there is a strong demand for higher data transfer rate systems, such as digital video image applications. A multi-beam optical disk drive[1,2] can meet this demand, for the following reasons.
(1) Parallel recording and reading on neighboring multi-tracks can be accomplished.
(2) Real time error detections, before and after recording, are possible.

A parallel tracking is necessary for multi-beam optical disk drive, in order to form individual beam spots on more than one track simultaneously. However, the necessity for achieving this tracking was not taken into account in the conventional multi-beam optical disk drive. Moreover, the size for the conventional multi-beam optical head was fairly large. In order to put the multi-beam optical disk drive into practical use, its overall size must be as small as that of the ordinary single-beam optical head.

This paper presents a multi-beam optical disk drive, which improves the data transfer rate for the magneto-optical disk systems by parallel read/write operation. It employs a newly developed image rotating actuator for tracing multi-tracks simultaneously, and photodiode arrays for simplifying signal detection circuits.

The optical head is described in Sect. 2, and parallel read/write characteristics for the optical disk drive are described in Sect. 3.

2. OPTICAL HEAD

In order to form individual beam spots, irradiated from the optical head, on more than one track simultaneously, the conventional tracking method, which moves beam spots laterally across a track, is insufficient. The reason is that, when the optical head moves between the inner track and the outer track of the disk, the track pitch varies in appearance, owing to the carriage mechanism inaccuracy. So as to compensate for this track pitch variation, rotating the beam spots is necessary in addition to moving the beam spots in parallel, as shown in Fig. 1. Rotating the beam spots is also effective in compensating for other kinds of track pitch variation, caused by the track eccentricity and the track winding of the disk.
Additionally, in order to achieve the same overall size as the ordinary single-beam optical head, both the error signal detection circuit and the data signal detection circuit must be simplified.

The structure of the newly developed multi-beam optical head for magneto-optical disks is shown in Fig. 2. The Dove prism, inserted in the optical circuit, is used for the image rotating actuator. An individually addressable four-channel monolithic laser diode array is used as the light source. The quadrant prism[3] and the six-segment photodiode array are used for error signal detection. The Wollaston prism and the two-segment photodiode array are used for data signal detection.

Four channels are used for this optical head, considering the output power from the laser diode array and aberration in the optical circuit. The numerical aperture for the collimating lens in a multi-beam optical head must be lower than that in an ordinary single-beam optical head, in order to reduce the aberration for off-axis beams. However, the use of low numerical aperture lens decreases the light power efficiency for the optical head. In this optical head, the numerical apertures for the collimating lens and the objective lens are optimized to 0.3 and 0.5, respectively. As a result, a focused beam spot size, smaller than 1.6μm (full width at $1/e^2$ maximum), and a light power efficiency, more than 25%, are obtained for each channel. Therefore, the use of the laser diode array with 40mW output power makes it possible to obtain more than 10mW output power on the optical disk for each channel, which is sufficient for practical use.

Figure 3 shows the image rotation principle. The Dove prism is inserted into the optical circuit. When the incident beam is totally reflected at the bottom surface of the prism, its X component is inverted and its Y component is not inverted. Therefore, the beam spots images are inverted versus the Y-Z plane. Rotating the Dove prism by θ degrees around the optical axis makes it possible to rotate beam spots by 2θ degrees.

The image rotating actuator structure is shown in Fig. 4. The Dove prism is hung from a plate spring. A coil is attached to each side of the prism, and two magnets are placed outside the coils. These coils and magnets generate a force which rotates the prism around the plate spring. The Dove prism is dynamically servo-controlled by the electrical feedback loop in the same way as the conventional focusing and tracking servo systems, with a frequency range wider than 1kHz. This method only needs Dove prism rotation, which is easier than rotating the entire optical head.

Figure 5 shows an external view of the fabricated image rotating actuator. Its overall size is 23mmx15mmx15mm.

Figure 6 shows a cross-sectional view of the laser diode array, which was newly developed for this optical head. The array is grown by the metal-organic vapor phase epitaxy technique. Laser diodes are spaced at 100μm intervals. 20μm wide grooves between the laser diodes are formed for electrical isolation by etching. The array is soldered to patterned electrodes on a submount in a junction-down configuration, so as to reduce both the thermal resistance of each channel and the thermal crosstalk between channels. The lasing wavelength is 0.78μm, and a maximum output power exceeding 60mW is obtained for each channel.

Output power for each channel can be individually monitored at the rear of the laser diode array for automatic power control operation, as shown in Fig. 7. In order to reduce the optical crosstalk between channels, optical waveguides are formed between the laser diode array and the photodiode array by using the heatsink and the silicon substrate with grooves on it. Each output beam from the rear facet of the laser diode is guided to the corresponding photodiode through the optical waveguide, with less than -40dB crosstalk. Both the optical waveguides and the photodiodes are spaced at 100μm intervals, like the laser diodes.

Figure 8 shows the layout for the quadrant prism error signal detection method. The reflected beam from the optical disk is divided into four sections. These beams are refracted in four different directions by the quadrant prism. Beams through sections A and C of the prism are focused on the vertical split line in the central quadrant detecting section of the six-segment photodiode. On the other hand, beams through sections B and D of the prism are focused on the two detecting sections at both ends of the six-segment photodiode.

Beam spots on the six-segment photodiode array are shown in Fig. 9. The focusing error signal is obtained from the central quadrant detecting sections for the six-segment photodiode in Channel 2, by adding the two pairs of diagonal section's outputs and subtracting each summation, according to the Foucault method principle. The tracking error signal is obtained from the two detecting sections for both ends of the six-segment photodiode for Channel 1, by subtracting each section's output, according to the push-pull method principle. The image rotating error signal is obtained by using the difference between the tracking error signals for Channel 1 and Channel 4.

Figure 10 shows the layout for the Wollaston prism data signal detection method. The polarization plane for the incident beam is rotated by 45 degrees by the half-wave plate. This beam is divided into two orthogonal polarization components by the Wollaston prism. They are focused on different detecting sections of the two-segment photodiode. The data signal for each channel is individually obtained, using the difference in the outputs between the two detecting sections.

Figure 11 shows an external view of the fabricated multi-beam optical head. Its overall size is 75mmx56mmx36mm. The use of photodiode arrays in the signal detection circuits makes it possible to achieve almost the same overall size as for an ordinary single-beam optical head.

3. PARALLEL READ/WRITE CHARACTERISTICS

The four-channel parallel read/write characteristics for the optical disk drive were measured, using a 5.25 inch TbFeCo magneto-optical disk.

Figure 12 shows the readout signals and the beam positions at 5.65m/s linear velocity and 1MHz recording frequency. Image rotating servo power is on in Fig. 12(a), and power is off in Fig. 12(b). When the image rotating servo power is on, all beams are on the track. Therefore, the data signals are correctly reproduced. On the other hand, when the image rotating servo power is off, only one beam (Channel 1) is on the track. The other beams (Channels 2 to 4) are out of the track. Therefore, signal amplitudes for Channels 2 to 4 decrease, and the data signals are incorrectly reproduced.

Figure 13 shows the four-channel parallel readout signals at a 11.3m/s linear velocity. The recording frequency is about 1MHz, and the recording pulse width is about 100ns. The recording pulse phases are varied between channels. Good parallel readout signals were obtained.

The carrier-to-noise ratio for each channel was measured at a 11.3m/s linear velocity. Figure 14 shows typical readout signal spectra for Channel 2. The carrier-to-noise ratio was greater than 55dB for 1MHz recording frequency and greater than 52dB for 3.7MHz recording frequency, as shown in Fig. 14(a) and Fig. 14(b), respectively. When the recording frequency is 3.7MHz, the data transfer rate becomes 5.5Mbps for each channel. Therefore, the 22Mbps data transfer rate is achieved by four-channel parallel read/write operation.

Crosstalk between the channels was also measured. Figure 15 shows the beam spots configuration used for crosstalk measurement. First, in order to measure the crosstalk originating in the optical circuit (the laser diode array and the photodiode array) separately from that originating in the neighboring tracks of the optical disk, the beam spots were placed not on neighboring tracks, but on those adjacent to neighboring tracks, as shown in Fig. 15(a). Then, in order to measure the total crosstalk originating in the optical circuit and the neighboring tracks of the optical disk, the beam spots were placed on neighboring tracks, as shown in Fig. 15(b). The 2MHz signal was recorded and reproduced by Channel 2, and the 1MHz signals were recorded and reproduced by Channel 1 and Channel 3. The crosstalk was defined as the ratio of the carrier level at 2MHz to that at 1MHz for Channel 2.

Figure 16 shows the readout signal spectra for Channel 2 for crosstalk measurement. The crosstalk originating in the optical circuit was lower than -36dB, as shown in Fig. 16(a). On the other hand, the total crosstalk originating in the optical circuit and the neighboring tracks of the optical disk was lower than -31dB, as shown in Fig. 16(b). These are sufficiently low values for practical use.

The block diagram for the coded data input/output experiment using this multi-beam magneto-optical disk drive is shown in Fig. 17. The write data and the read data were first stored in the write memory and the read memory, respectively. Then, they were tramsmitted between the personal computer and the magneto-optical disk drive. The data signals were encoded and decoded according to the 2-7 modulation code. The clock signal was detected only from Channel 1, and it was used to simultaneously reproduce the data signals for all channels.

The color graphic image data were recorded and reproduced. One graphic image was recorded on four neighboring tracks of the disk. The data signals were divided into red, green, blue and intensity signal components. These signal components were assigned to Channels 1 to 4, respectively. Therefore, only one disk rotation is necessary for reproducing one graphic image, using this multi-beam optical disk drive, while four disk rotations are necessary, using the ordinary single-beam optical disk drive. Figure 18 shows an example of the reproduced color graphic image.

4. CONCLUSION

A four-channel multi-beam optical disk drive was developed, for improving the data transfer rate for the magneto-optical disk systems. A new tracking method, using an image rotating actuator, was developed in addition to the conventional tracking method. The overall size for this optical head was as small as that for the ordinary single-beam optical head, by employing the photodiode arrays in the signal detection circuits. The carrier-to-noise ratio was greater than 55dB, and the crosstalk between channels was lower than -36dB. As a result, a data transfer rate, four times that for an ordinary single-beam magneto-optical disk drive, was realized.

The development of a higher power laser diode array or a lower aberration optical circuit will make it possible to increase the number of channels.

5. ACKNOWLEDGMENT

The authors would like to thank Mitsuhito Sakaguchi, Nobuo Nishida, Kazuaki Hotta, Toru Suzuki, Tonao Yuasa, Kenji Endo, Toshio Ohkubo, Isao Kagaya and Tadashi Yoshiura for their encouragement and advice. The authors would also like to thank Hiroshi Yoshida and Kazuo Yamanaka for their assistance in preparing the monitor of the laser diode array.

6. REFERENCES

1. D. B. Carlin, J. P. Bednarz, C. J. Kaiser, J. C. Connolly and M. G. Harvey, "Multichannel optical recording using monolithic arrays of diode lasers: errata," Appl. Opt. Vol. 23, No. 24, pp. 4613-4619 (1984)

2. M. Itoh, H. Kawano, S. Shimonou and T. Yoshiura, "Real-time error detection using a laser diode array optical head," Technical Digest of CLEO'85, WM48, pp. 130-131 (1985)

3. Y. Yamanaka and Y. Ohta, "New optical head focusing and tracking error detection method with quadrant prism," Technical Digest of OEC'86, D5-8, pp. 226-227 (1986)

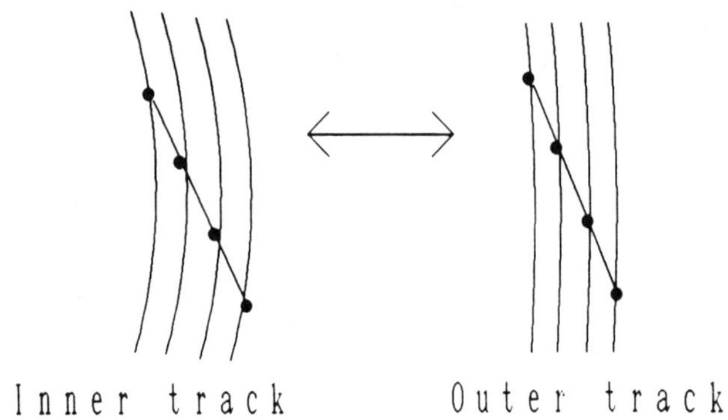

Figure 1. Necessity for rotating beam spots.

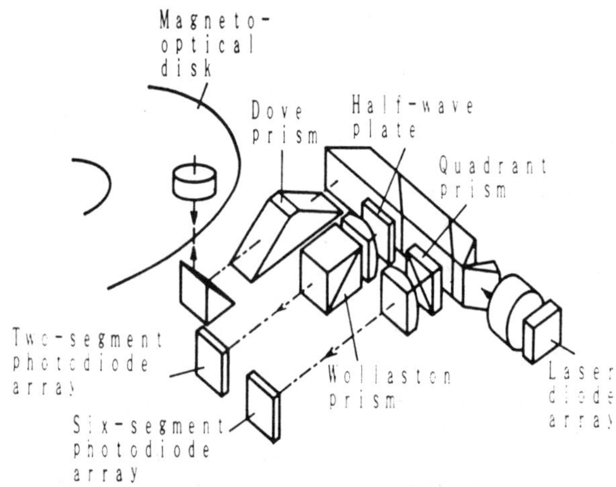

Figure 2. Multi-beam optical head for magneto-optical disk.

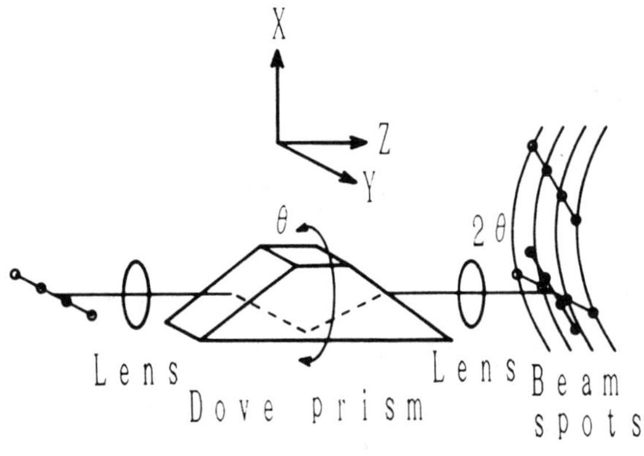

Figure 3. Image rotation principle.

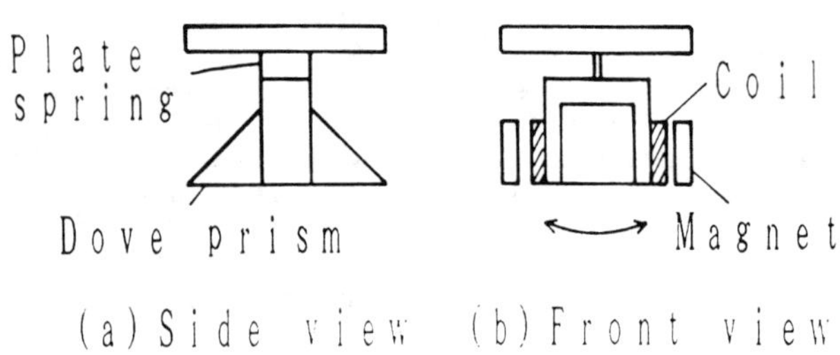

Figure 4. Image rotating actuator structure.

23mmX15mmX15mm

Figure 5. Fabricated image rotating actuator external view.

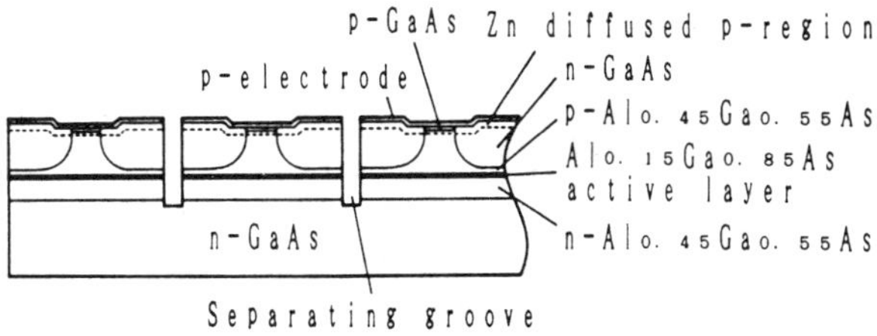

Figure 6. Laser diode array cross-sectional view.

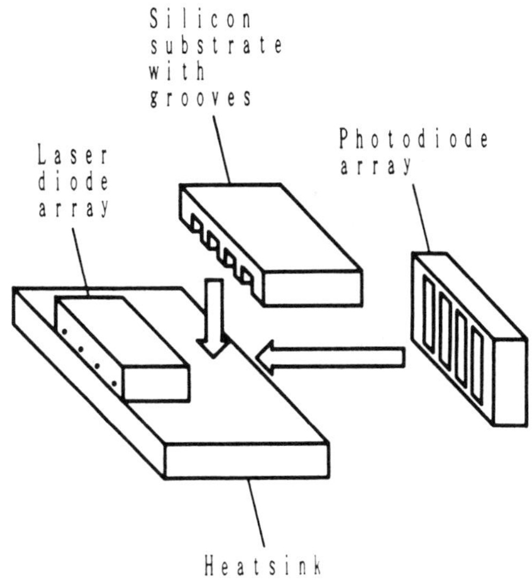

Figure 7. Method for monitoring output power from each channel.

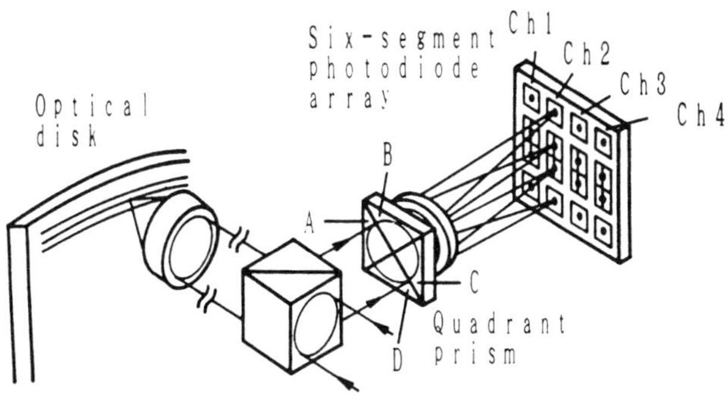

Figure 8. Quadrant prism error signal detection method layout.

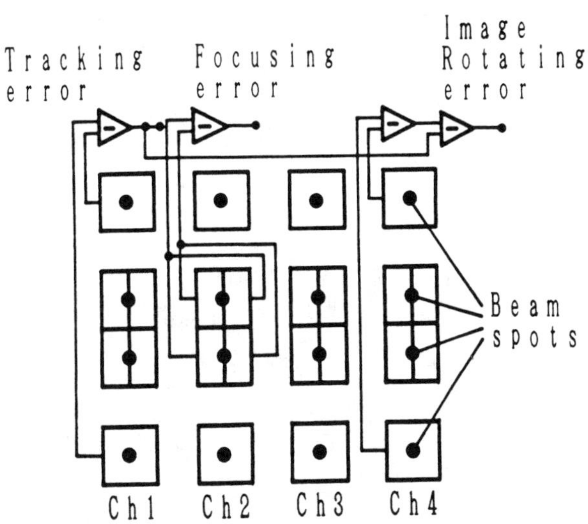

Figure 9. Beam spots on the six-segment photodiode array.

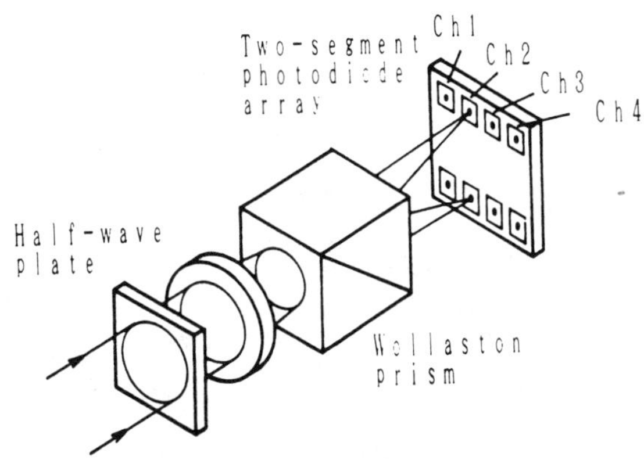

Figure 10. Wollaston prism data signal detection method layout.

Figure 11. Fabricated optical head external view.

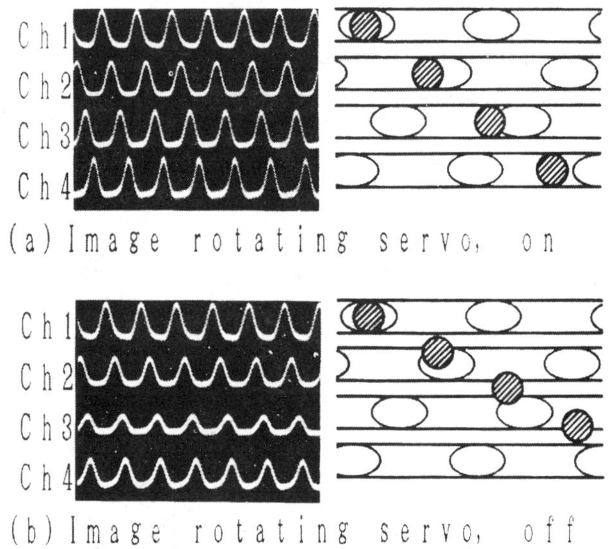

Figure 12. Readout signals and beam positions.

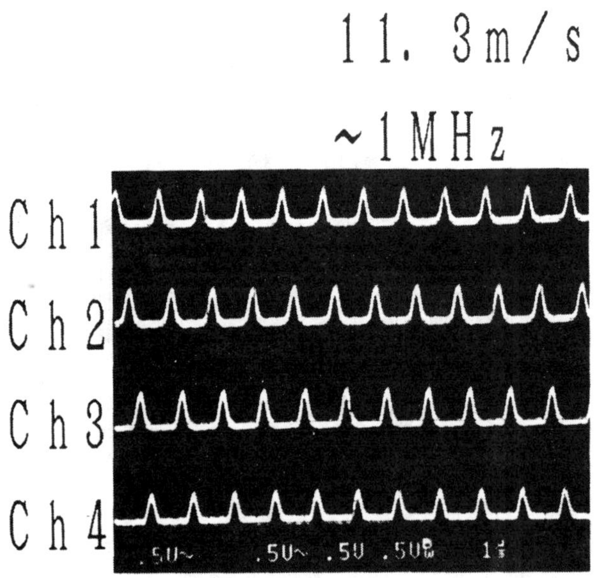

Figure 13. Parallel readout signals.

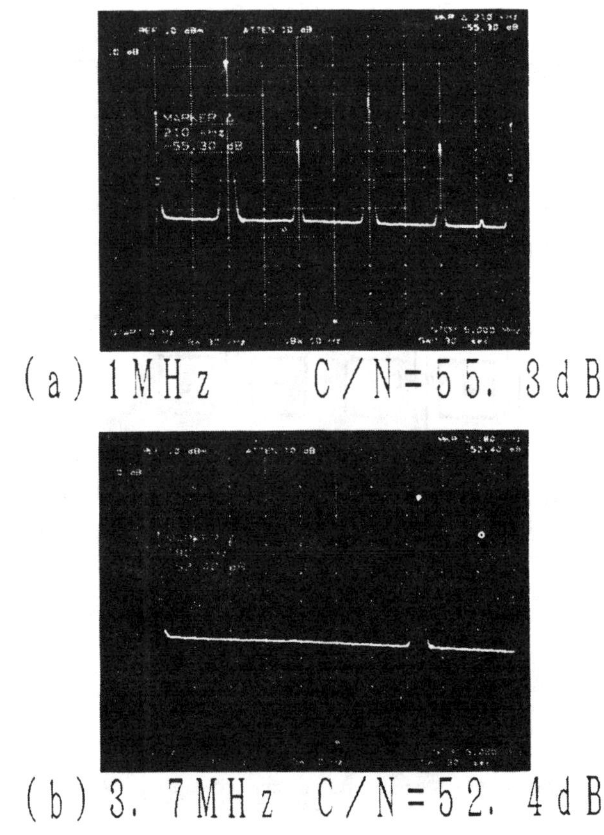

Figure 14. Readout signal spectra.

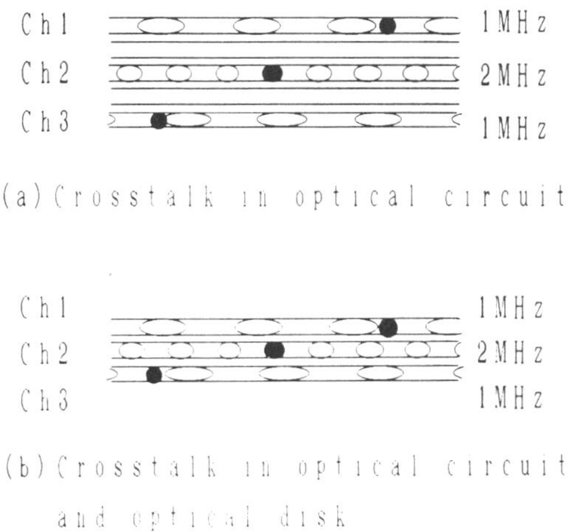

Figure 15. Beam spots configuration for crosstalk measurement.

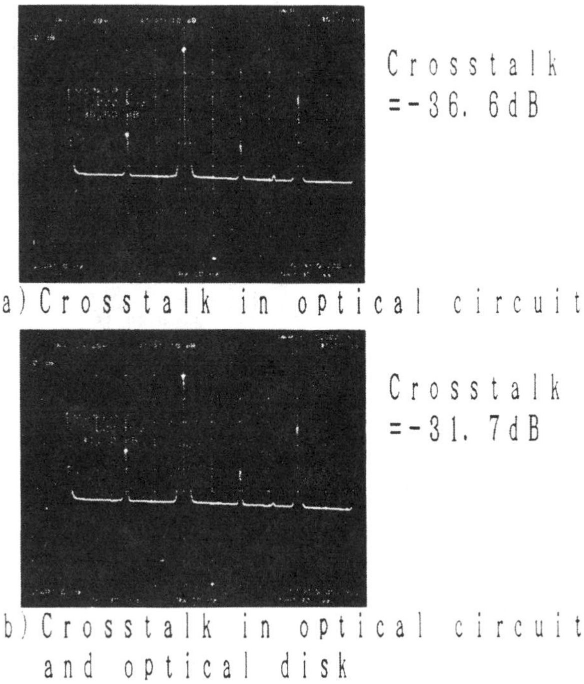

Figure 16. Readout signal spectra for crosstalk measurement.

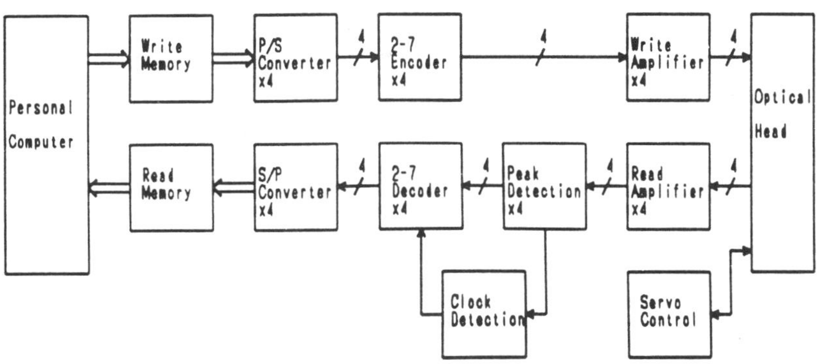

Figure 17. Coded data input/output experiment block diagram.

Figure 18. Reproduced graphic image.

Reprinted with permission from *IEEE Transactions on Magnetics*, Vol. 24(6), pp. 2300-2304 (November 1988). ©1988 IEEE.

HIGH DENSITY MAGNETO-OPTICAL RECORDING USING 0.67μm BAND HIGH POWER LASER DIODE

Y.Yamanaka, K.Kubota, H.Fujii, K.Kobayashi, T.Suzuki, H.Gokan*
Opto-Electronics Res. Labs., *Microelectronics Res. Labs.,
NEC Corp., 1-1,Miyazaki-Yonchome,Miyamae-ku,Kawasaki 213, Japan

ABSTRACT

High density optical recording on a magneto-optical disk has been achieved by a compact optical head, which employs a 0.67μm band high power visible-light laser diode. The recording density has been improved to 1.5 times that for a conventional optical recording, using a 0.83μm laser diode, as a result of the increase in linear density and track density. The C/N ratio for the readout signal is greater than 50dB.

INTRODUCTION

Optical disk systems have a large data storage capacity, compared with magnetic disk recording systems. Recently, erasable magneto-optical disk systems have appeared in practical use. These systems will replace both floppy disk and rigid magnetic disk systems in office automation or information storage systems. However, there are increasing demands for larger data capacity or higher recording density. For example, high definition TV video recording requires a larger storage capacity, compared with normal TV video recording. A 3.5 inch disk, which has been developed so as to reduce disk driver size, requires higher recording density, compared with a 5 inch disk.

To meet these demands, several approaches have been proposed. The density of pit width modulation recordings are higher than pit position modulation recordings.[1] A duobinary code makes it possible to increase the linear density.[2] An offset track geometry makes it possible to increase the track density.[3] However, these methods sacrifice recording characteristics. The recording density in the conventional recording system has been determined by the focused light spot size on the disk. This spot size can be reduced by using a higher numerical aperture objective lens. However, optical alignment tolerances between an optical head and a disk become small.

One of the most essential methods to increase the recording density is to reduce the light source wavelength of the recording optical head. In a conventional optical head, the source wavelength was limited to about 0.8μm,[4] which is the wavelength for the AlGaAs laser diode. The calculated wavelength dependence of recording density, normalized at 0.83μm, is shown in Fig.1. It was assumed that the density is proportional to the square of focused spot size. According to Fig.1, about 1.5 times higher recording density is expected at 0.67μm wavelength.

This paper describes high density magneto-optical recording, using a high quality magneto-optical head containing a 0.67μm band high power visible-light laser diode. The structures and characteristics of the laser diode, optical head and magneto-optical disk are described. The recording characteristics are also described.

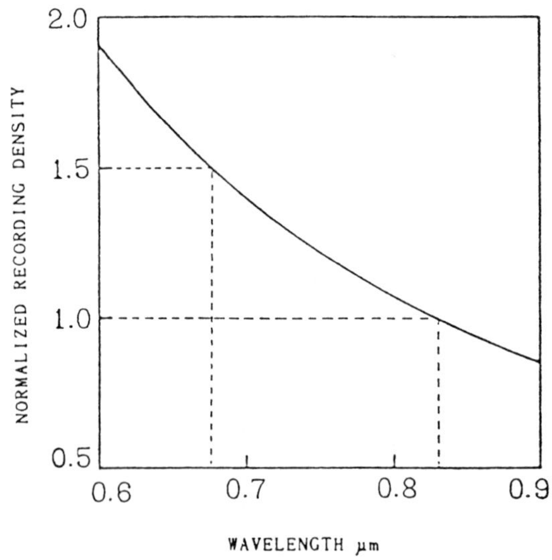

Fig.1 Recording density wavelength dependence normalized at 0.83μm

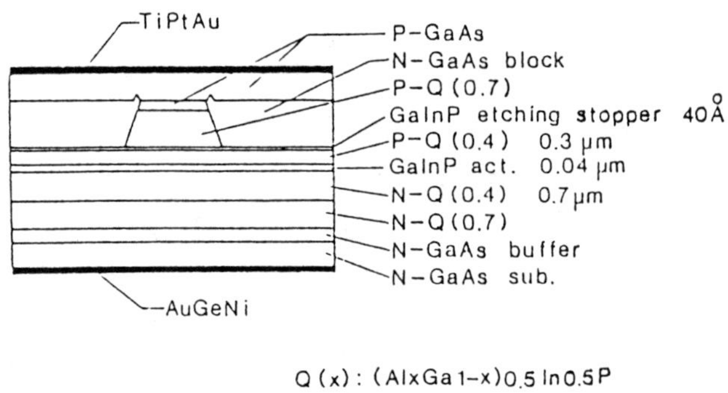

Fig.2 AlGaInP laser diode cross-sectional view.

VISIBLE-LIGHT LASER DIODE

It is extremely difficult to achieve shorter (<0.78μm) wavelength high power laser diodes using the AlGaAs alloy semiconductor, due to its energy band gap. Fortunately recent progress in semiconductor material technology has made it possible to use a completely new semiconductor alloy AlGaInP. Room-temperature CW operation of AlGaInP visible-light laser diodes has been achieved[5-7] and long-life characteristics have been reported for the first time.[8] In order to obtain a high power visible emission, a transverse-mode stabilized laser diode with highly reflective and anti-reflective mirrors was developed.[8] A cross-sectional view of this

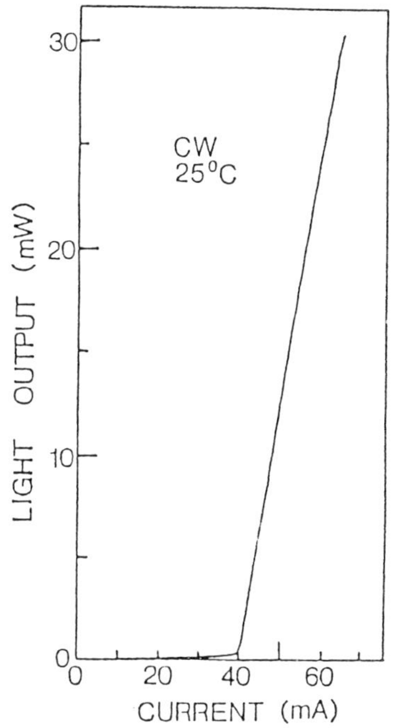

Fig.3 CW light output vs. injected current.

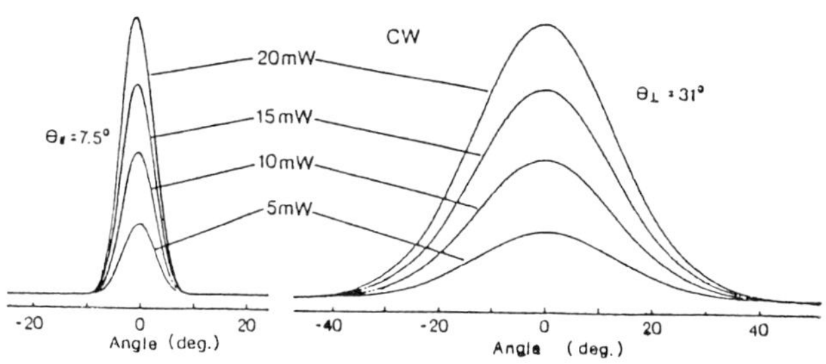

Fig.4 Far-field patterns for different output powers.

laser diode is shown in Fig.2. Epitaxial growth was carried out by low-pressure MOVPE.

Figure 3 shows light output vs. injected-current characteristics. The CW threshold current is about 40mA, and maximum available light output power is greater than 30mW. The optical output power is sufficient for use in a magneto-optical head. Typical emission wavelength for the AlGaInP laser diode employed in the optical head was 667nm. A typical far-field pattern is shown in Fig.4. Stable fundamental-mode operation has been achieved, at least up to 20mW output power. The FWHMs (full widths at half-maximum) values for the beam divergence angles, parallel and vertical to the junction plane, are 7.5 degrees and 37 degrees, respectively.

OPTICAL HEAD

In order to realize a stable servo control, a high SNR readout signal and a simple optical design,[10] the quadrant prism method for servo error detection[10] and the birefringence wedge prism method for data signal detection were developed.

The optical design of the magneto-optical head is shown in Fig.5. A cross section of the collimated laser output is transformed from an ellipsoid to a circle by a beam shaping prism. This light travels through the beam splitters, and is focused into a circular spot by an objective lens (NA=0.55). The focused spot diameter on a medium's surface was 0.7μm (FWHM). The reflected light from the medium is fed to both a signal detection circuit and a servo error detection circuit.

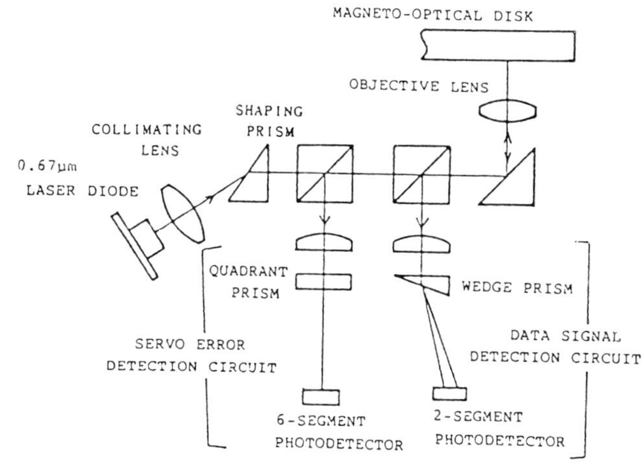

Fig.5 Magneto-optical head optics.

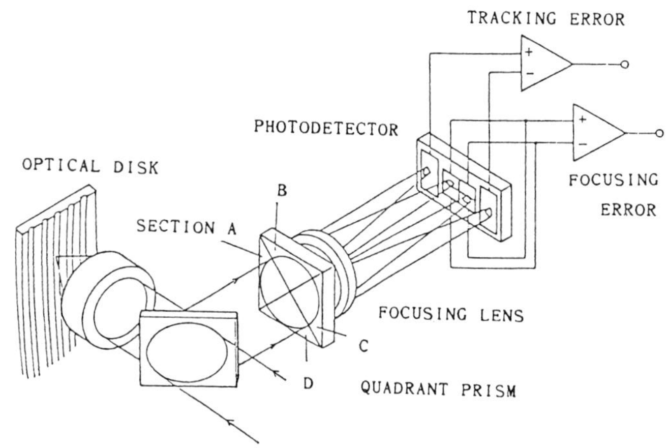

Fig.6 Quadrant prism method layout.

Figure 6 shows the fundamental layout for the quadrant prism method. The light beam reflected from the disk is divided into four sections. These beams are then refracted in four different directions by the quadrant prism. Beams through sections A and C of the prism are focused on separated detecting sections of the photodetector. The tracking error signal is obtained from these two sections, according to the push-pull method. On the other hand, beams through sections B and D are focused on the vertical split line of the quadrant detecting section. Summing the two pairs of the diagonal sections and subtracting each summation produces the focus error signal, according to the double knife-edge method or the Foucault method principle. This quadrant prism method has high sensitivity and low signal offset characteristics. The focus error signal offset is 1/10 times that for a conventional knife-edge method.[10] Therefore, it was possible to divide circuits between the servo error detection and the data signal detection so as to obtain a high SNR data signal.

Figure 7 shows the layout for the birefringence wedge prism method. The rutile wedge prism has a principal axes at 45 degrees to incident P-polarized light. Owing to its birefringence, the signal light is divided into parallel and perpendicular components, with reference to the principal axes. They are refracted at different angles from each other, and focused on different sections of the 2-segment photodetector.

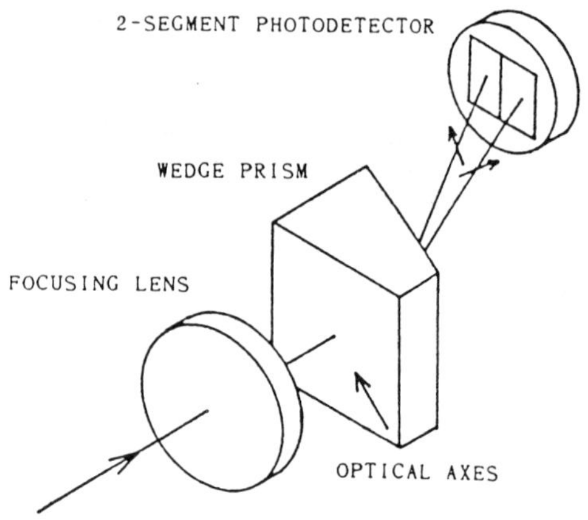

Fig.7 Birefringence wedge prism method layout.

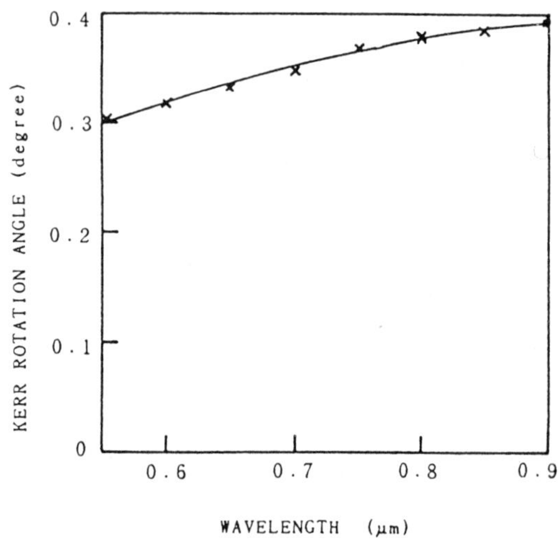

Fig.8 Kerr rotation angle wavelength dependence.

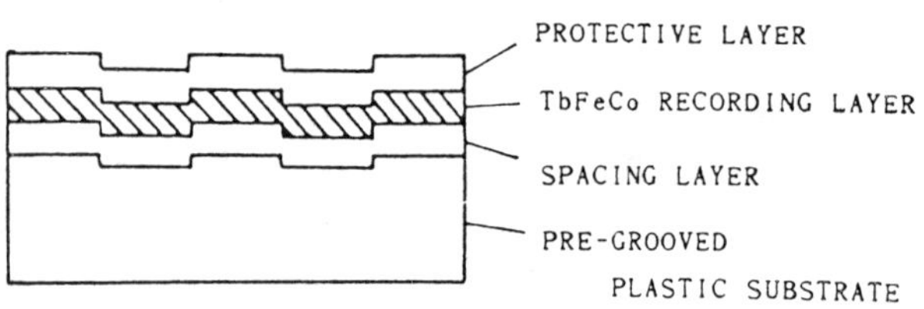

Fig.9 Magneto-optical disk construction.

Because no phase plate was used, low phase difference (< 5 degree) between P-polarized and S-polarized light could be achieved. By causing this difference in detector outputs, a high SNR data signal could be obtained.

MAGNETO-OPTICAL DISK

A TbFeCo magneto-optical disk was used to evaluate the recording characteristics. The Kerr rotation angle wavelength dependence for TbFeCo is shown in Fig.8. The Kerr rotation angle at 0.67μm is about 90% of that at 0.83μm. Estimated readout carrier level degradation is less than 1dB. So, a TbFeCo disk is useful at this short wavelength. Figure 9 shows the disk construction.[11]

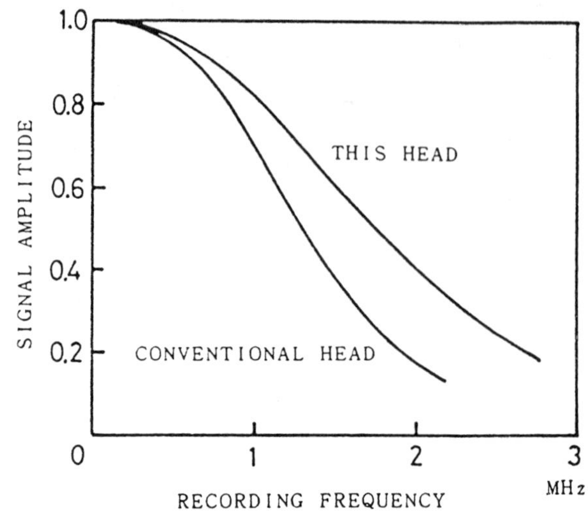

Fig.10 Readout signal amplitude vs. recording frequency.

Injection molded 130mm diameter pregrooved PC substrates were used. Groove pitch was changed from 1.0μm to 1.8μm, so as to measure the track density characteristics. Over the substrate, a SiN spacing layer, a TbFeCo recording layer and a SiN protective layer were successively deposited, using the RF magnetron sputtering method. The recording layer was 800Å thick. Spacing layer thickness, which was determined to enhance Kerr rotation, was 700Å.

RECORDING CHARACTERISTICS

The recording characteristics for this optical head were measured and compared with that for a conventional head employing a 0.83μm laser diode. The readout signal amplitude dependence on the recording frequency is shown in Fig.10. The recording pulse has 50% duty cycle, and the linear velocity was 2.8m/s. The recording frequency for this head, at which the signal amplitude becomes half of the maximum, is 1.3 times as high as that for a conventional head. This means that linear density for this head is 1.3 times as high as that for a conventional head.

Crosstalk from adjacent tracks was measured. Carrier level for the readout track (1.5MHz) was compared with that for sidetracks (1MHz), as shown in Fig.11. The crosstalk dependences on the track pitch are shown in Fig.12. Track pitch can be reduced from 1.6μm to 1.3μm by using this head, without any degradation in crosstalk characteristics.

Typical examples of a readout signal and a signal spectrum at 1MHz frequency are shown in Fig.13. Recording was carried out at 7mW recording power, 400Oe magnetic field and 5.6m/s linear velocity. A 53dB C/N ratio was obtained at 1mW reading power. This value is a few dB lower than that for a conventional head because of feedback induced noise in the visible-light laser diode (RIN∼-110dB/Hz). This noise will be suppressed by optimizing the laser structure. A polarization microscope photograph of the recording pits is shown in Fig.14. Groove pitch is 1.3μm. No degradation, caused by using the short wavelength, was observed.

As a result of these measurements, a total recording density, as great as 1.5 times that reached previously, has been achieved without any signal degradation. Therefore, over 1GB data capacity will be realized in a 5.25 inch 2 sided disk without changing the coding format.

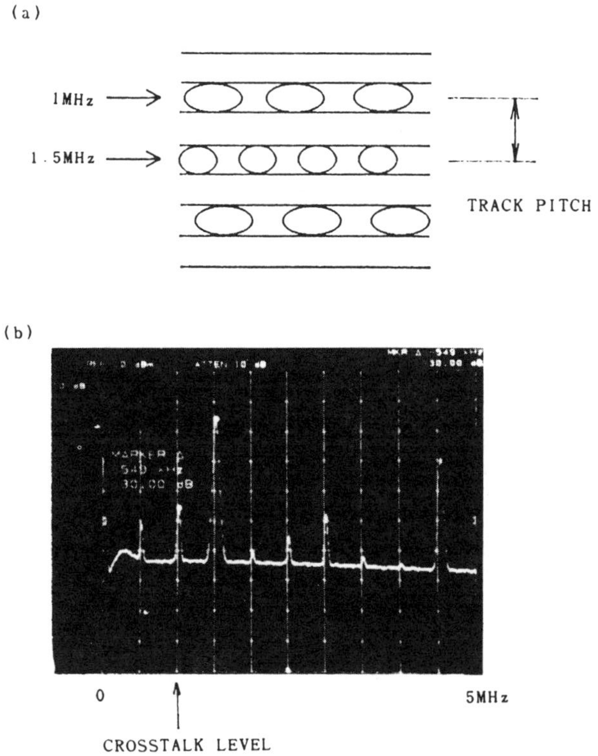

Fig.11 Crosstalk measurements:
(a)Recording formation (b)Readout spectrum.

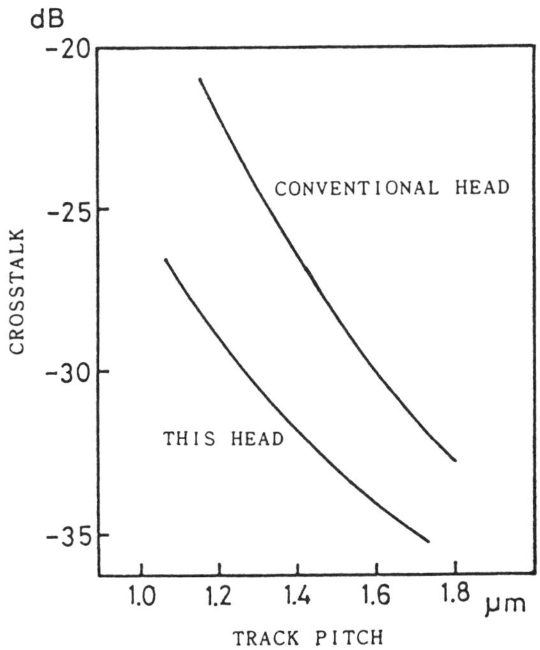

Fig.12 Crosstalk vs. track pitch.

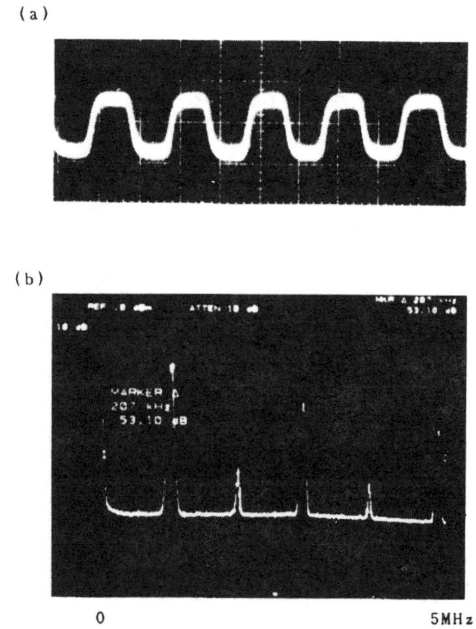

Fig.13 Recording and reading characteristics:
(a)Readout signal (b)Signal spectrum.

Fig.14 Polarization microscope photograph of recording pits on 1.3μm pitch pregrooves.

CONCLUSION

High density magneto-optical recording was accomplished using a high quality optical head with a 0.67μm band high power laser diode. It has been confirmed that the TbFeCo disk has sufficient recording characteristics at 0.67μm. Over 1GB data capacity, in a practical 5.25 inch 2 sided disk recording system, will be achieved by future improvement in AlGaInP laser diode characteristics.

ACKNOWLEDGMENTS

The authors would like to thank M.Sakaguchi, N.Nishida, K.Kobayashi and S.Esho for their encouragement and advice, and K.Toki for preparing the disks.

REFERENCES

(1) H.Sukeda, M.Ojima, M.Takahashi and T.Maeda, " High Density Magneto-Optical Disk Using Highly Controlled Pit-Edge Recording," Post Deadline Digest of ISOM'87, FC1,pp.1-4,(1987).

(2) T.Sekiguchi, H.Kato, Y.Sasaki, T.Arimura, H.Inada, T.Iwanaga, K.Toki and M.Okada," High Bit Rate Digital Video Recording Using Magneto-Optical Disks," Technical Digest of ISOM'87,FB4,pp.269-272,(1987).

(3) G.Moore," High Density Format For Sperry/Information Storage Inc. Second Generation Optical Drives," Proceedings of SPIE Conference on Optical Mass Data Storage,vol.695,pp.230-238,(1986).

(4) S.Nakamura, M.Ojima, T.Nakao and K.Mizuishi,"Compact Two-Beam Head With A Hybrid Two-Wavelength Laser Array For Magnet-Optical Recording,"Technical Digest of ISOM'87,TB5,pp.183-186,(1987).

(5) K.Kobayashi, S.Kawata, A.Gomyo, I.Hino and T.Suzuki, "Room-Temperature CW Operation of AlGaInP Double Heterostructure visible Laser," Electronics Letters, vol.21,pp.931-932,(1985).

(6) M.Ikeda, K.Nakano, Y.Mori, K.Kaneko and N.Watanabe, "Room-Temperature Continuous-Wave Operation of an AlGaInP Mesa Stripe Laser,"Applied Physics Letters, vol.48,pp.89-91,(1986).

(7) M.Ishikawa, Y.Ohba, H.Sugawara, M.Yamamoto and T. Nakanishi, "Room-Temperature CW Operation of InGaP/InGaAlP Visible Light Laser Diodes on GaAs Substrates Grown by Metalorganic Chemical Vapour Deposition,"ibid.,vol.48,pp.207-208,(1986).

(8) A.Gomyo, K.Kobayashi, S.Kawata,I.Hino and T.Suzuki, "Aging Characteristics of AlGaInP/GaInP Visible-Light Laser (λ_L=687nm)," Electronics Letters, vol.23, p.85, (1987).

(9) H.Fujii,K.Kobayashi,S.Kawata,A.Gomyo, I.Hino, H.Hotta and T.Suzuki,"High-Power Operation of a Transverse-Mode Stabilized AlGaInP Visible Light (λ_L=683nm) Semiconductor Laser," Electronics Letters, vol.23, pp.938-939,(1987).

(10) Y.Yamanaka and Y.Ohta,"New Optical Head Focusing And Tracking Error Detection Method With Quadrant Prism," Technical Digest of OEC'86,D5-8,pp.226-227,(1986).

(11) M.Okada, M.Nakada,K.Toki, H.Gokan, T.Iwanaga and H. Inada, "Magneto-Optical Disks Using PC Substrates," Technical Digest of Optical Data Storage Topical Meeting,ThC2-1,pp.123-126,(1987).

Reprinted from *Optical Storage and Scanning Technology*, Proc. SPIE Vol. 1139, pp. 149-154 (1989).

A Triple-Density Optical Disk System Using A Frequency-Doubled Nd:YAG Laser

Michio Oka, Toshiyuki Kashiwagi and Shigeo Kubota

Sony Corporation, Corporate Research Laboratories
1-7-4 Konan, Minato-ku, Tokyo 108, Japan

ABSTRACT

A diode-laser-pumped intracavity frequency-doubled Nd:YAG laser which radiates 532nm green light is an excellent candidate for a higher-density optical disk system. Using a stabilized green laser and an aspheric plastic lens with a numerical aperture of 0.6, we report the playback of a triple-density video disk. An NTSC-formatted analog video signal and optical-performance test signal were recorded on 0.9μm-pitch tracks with a linear velocity of 6.25m/sec. The disk was replicated on a Compact-Disk-size glass substrate using the photopolymerization technique. The reproduced signal had crosstalk less than -40dB and a carrier-to-noise ratio of 58dB at a bandwidth of 30kHz with a pit size of 0.39μm. The measured test signal indicated that the laser spot quality was nearly diffraction limited.

1. INTRODUCTION

After 15 years, the storage density of an optical disk system using a He-Ne laser (wavelength λ=633nm) and an objective lens with a numerical aperture (NA) of 0.4 is still not obsolete.[1] The track interval of a disk should be comparable to the spot diameter, approximately given by λ/NA, and is equal to 1.6μm. The success of the Compact-Disk system can be attributed to the use of a compact laser diode (LD) and a precision-molded plastic lens [2], but the spot diameter unavoidably become larger when a GaAlAs LD with λ=0.78μm and a molded aspheric lens with NA=0.45 were employed, that is λ/NA=1.73μm. As the acceptable level of crosstalk from adjacent tracks in a digital data reproduction system is less stringent than that in a video disk system, essentially the same track interval - 1.6μm - was used.

It would seem that a shorter wavelength light source and higher NA objective lens are required to increase the density. The recently available 680nm LD [3] will make possible an increase in track density of only 15%. Although a lithium niobate nonlinear waveguide device [4] has been proposed to halve the LD wavelength, its spot quality may have to be improved before the device can be applied to an optical disk system.

A diode-laser-pumped intracavity frequency-doubled Nd:YAG laser using potassium titanyl phosphate ($KTiOPO_4$: KTP), which radiates 532nm green light, has two main advantages over other short wavelength light sources: higher energy-conversion efficiency and diffraction-limited wavefront quality. Although the advantages, the instability of the laser output has until now prevented its practical use.[5] In a previous paper, we reported a method to stabilize the output which provides for independent oscillation of two polarization modes inside a laser cavity without mode coupling.[6] An optical pickup was fabricated using the stabilized green laser and an aspheric plastic lens with NA=0.6. The calculated spot size in Airy disk diameter is 1.08μm and is 1/1.76 of a conventional pickup with λ=780nm and NA=0.5.

In this paper, a high-density optical disk system using the low-noise solid state green laser and a quarter-micron-length-pit-embossed optical disk is described. A disk mastering and replication process is also shown.

2. FREQUENCY DOUBLED Nd:YAG LASER

2.1. Requirements for the light source in optical disk systems

Requirements for the light source in optical disk systems are considered as follows.

(1) Spatially coherent, TEM_{00} and diffraction limited:
rms wavefront error < 0.07λ (Marechal criterion),
(2) low noise: S/N >80dB with bandwidth 30kHz, particularly for analogue signal playback,
(3) sufficient power: > 1mW, for a read only disk, > 20mW, for a writable disk,
(4) efficient
(5) reliable

(6)compact
(7)mass-producible (inexpensive).

Performance of frequency-doubled Nd:YAG laser:

(1) Frequency-doubled Nd:YAG laser is spatially coherent, fundamental transverse mode and diffraction limited.
(2) low noise: S/N >80dB.
(3) Sufficient power for read only disk is available with 200mW pumping, and applicable to writable disk system with 1W pumping and an external modulation.
(5) Maximum efficiency of 6% is obtained at 1W pumping.
(5) Reliability of the laser mainly depends on an LD. MTTF of the LD is 12000h at room temperature.
(6) Cavity length is 4cm and the size can still be reduced by future cavity refinement.
(7) High power LDs and nonlinear crystal are now too expensive to built in the consumer electronics products.

From the above, a frequency doubled Nd:YAG laser is a potential candidate for a future light source in optical disk systems.

2.2. Structure of frequency-doubled Nd:YAG LASER

Figure 1 shows the 1W-class broad-area, gain guided Sony SLD304 [7] laser diode used to pump a 7mm-long Nd:YAG rod with an Nd^{3+} concentration of 1.1 at.%. To correct divergence and astigmatism of the pumping beam, two cylindrical lenses are employed, one with a focal length of 60mm to correct the divergence perpendicular to the junction plane of the LD and the other with a focal length of 10 mm to correct the divergence parallel to the junction plane. One end of the Nd:YAG rod and the output mirror with 40mm radius of curvature are highly reflective coated at wavelength 1.06μm to form a laser cavity. A 3mm-thick, type-II cut KTP crystal is inserted in the cavity. With a cavity length of 40mm, the calculated radius of the Gaussian spot in the KTP is about 70μm. A quarter wave plate (QWP) at fundamental laser wavelength λ=1.064 μm is inserted with its fast axis making 45° to the E axis of the KTP in order to reduce laser output noise.

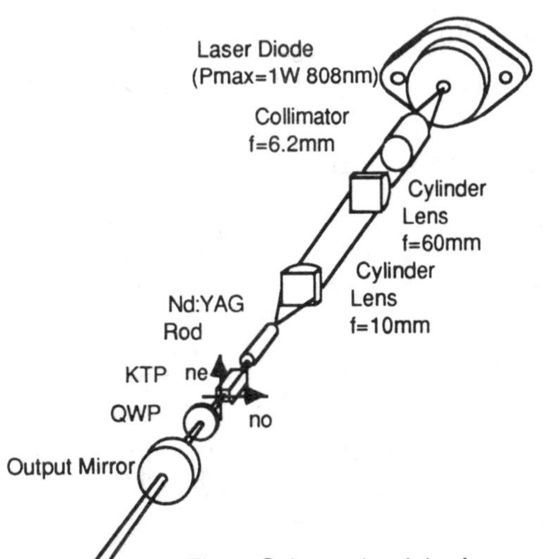

Fig. 1 Schematic of the frequency doubled Nd:YAG laser. The fast axis of the QWP makes an angle 45° with respect to the E axis of the KTP.

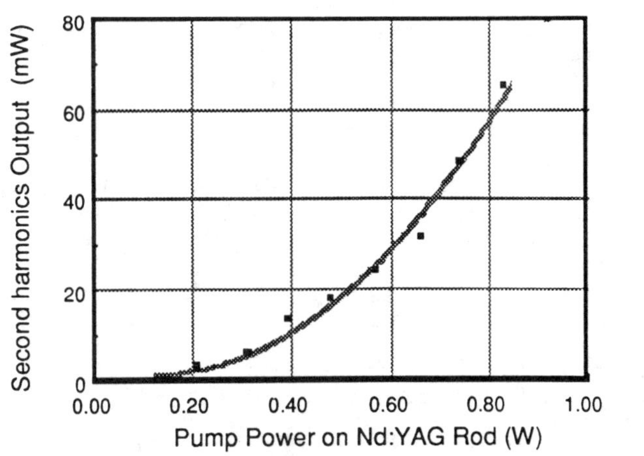

Fig. 2 Second harmonics output characteristics of the frequency doubled Nd:YAG laser.

2.3. Output Characteristics

Typical output characteristics for up to 1W pumping using the SLD304 are given in Fig.2. Maximum conversion efficiency from LD power on the rod to second harmonics output power is 6%.[8] We found that instability is effectively suppressed by inserting the QWP in the cavity. The QWP is arranged so that the azimuthal angle of the crystal axis of the KTP is 45° with respect to the fast axis of the QWP. The theoretical explanation is that the coupled loss term between the two polarization modes in the rate equation, describing oscillation modes in the laser cavity, is eliminated by this positioning of the QWP. Figure 3 shows the oscilloscope trace of second harmonics and its spectrum. S/N ratio of 80dB is obtained at 5MHz (bandwidth 30kHz). Pumping power of less than 200mW is sufficient for the about 2mW green output required for reading out information from an optical disk.

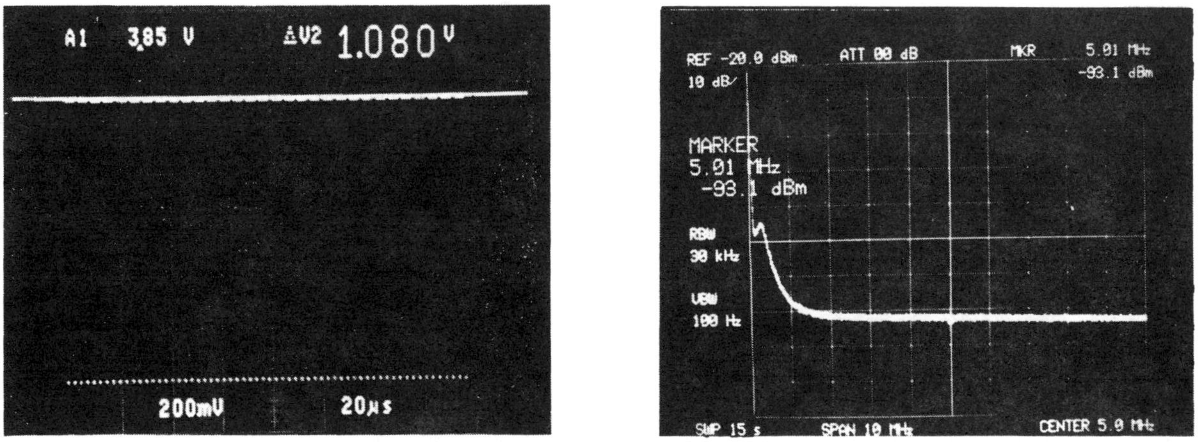

(a) (b)

Fig. 3. (a) Oscilloscope traces of the laser output. The dashed line shows the ground level. (b) Noise spectrum measured by a spectrum analyzer. Resolution bandwidth, 30kHz; video bandwidth, 30kHz.

3. OPTICAL HEAD

3.1. Structure of Green Laser Optical Head

Figure 4 shows the structure of the optical head and the 2mW green laser employed. A dichroic plate filters out the fundamental light component (1.06μm) and transmits nearly 100% green light. A TEM_{oo} second harmonics beam with a Gaussian beam waist (W) must be expanded to fill the objective lens aperture (A). When A/W is approximately 0.6, the optimum aperture-filling condition is obtained. The expander is built with a plano concave lens and a positive doublet with focal lengths of -3.3mm and 25mm, respectively. A grating which generates side spots for so-called 3-spot tracking error signal detection and a polarization beam splitter (PBS) are located inside the expander. A newly developed aspheric

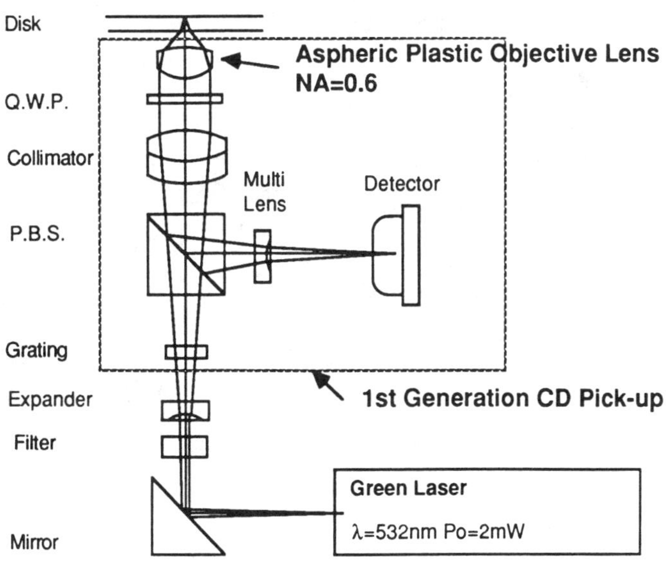

Objective Lens: f=3.6, NA=0.6
Gausian beam condition: A/W~0.6
Focus Servo Method: Astigmatic Method
Tracking Servo Method: 3-spot Method

Fig. 4. Structure of green laser optical pick-up

plastic objective lens with NA=0.6, EFL=3.6mm, working distance=1.2mm is employed. The transmitted wavefront error of the lens measured by a digital interferometer was less than 0.04 rmsλ on axis, which indicates a diffraction limited spot. Using the green laser and the objective lens, the spot size is 1.1μm in Airy disk diameter. Transmitted power through the objective lens is 0.8mW, and the couping efficiency is 40%. The readout signal is reflected back to a PIN photodetector by the QWP and the PBS. A multi-lens with cylindrical and concave surfaces is used for so-called astigmatic focusing error detection. It should be noted that this combination of techniques for servo error sensing is often used in conventional CD pickups.

4.DISK FABRICATION

The master disk of a triple density optical disk is fabricated using the 442nm wavelength of a He-Cd laser and an objective lens with NA=0.9. The setup for master-disk cutting is shown in Fig. 5. The setup is the same as a conventional system except for the NA=0.9 objective lens to make the cutting spot diameter as small as 0.6μm. With careful adjustment of resist sensitivity, baking temperature and development condition, even smaller recorded pits were obtained, as is shown in Fig.6. An NTSC-formatted analog video signal and test signal for evaluating optical characteristics such as MTF, crosstalk and C/N are recorded along spiral tracks with 0.9μm pitch at a linear velocity of 6.25m/sec. The disk is replicated on a CD-size glass substrate using photopolymerization. Table I compares the conventional CDV disk and the triple-density disk.

Table I. Specifications of conventional CDV disk and triple-density optical disk

	CDV Disk	Triple Density Disk
diameter	120mm	120mm
substrate	polycarbonate	glass
thickness	1.2mm	1.2mm
track pitch	1.7μm	0.9μm
pit depth	135nm	105nm
linear velocity	11m/s	6.25m/s
relative areal density	1	3
play back time	5min	15min (R39-58 mm) 25min (R25-58 mm)
format	NTSC	NTSC

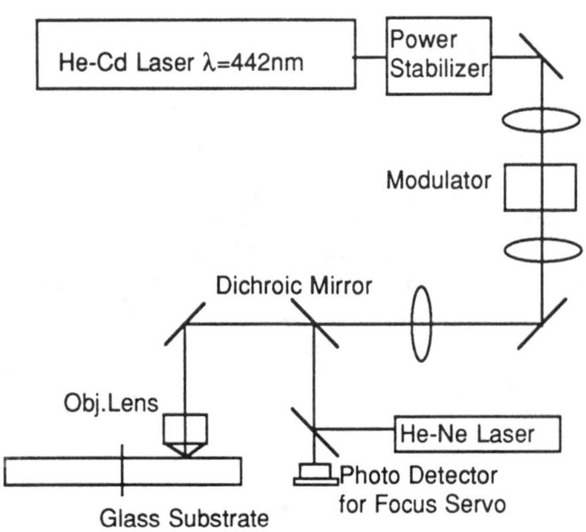

Fig. 5. Optical layout of a master-cutting machine. A He-Cd laser (λ=442nm) and an objective lens with NA=0.9 are used.

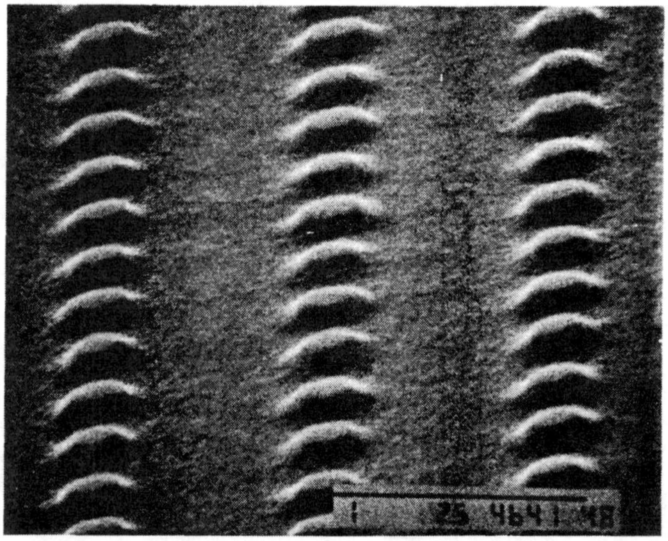

Fig. 6. A SEM photograph (x50,000) of the triple density optical disk; track pitch=0.9μm,. pit length 0.4μm, and pit width 0.3μm.

5.READOUT SIGNAL EVALUATION

5.1.MTF

One way to evaluate the spot quality is to measure the frequency response of the readout signal. Several spatial frequency components (75 to 2100 lines pair/mm) are recorded on a disk as a test signal. The plotted curve of the signal level at each frequency, measured by a spectrum analyzer, gives the modulation transfer function (MTF). As shown in Fig.7, the measured and calculated MTF's indicate that the laser spot quality is nearly diffraction limited and resolution up to 2000 lines/mm is obtained.

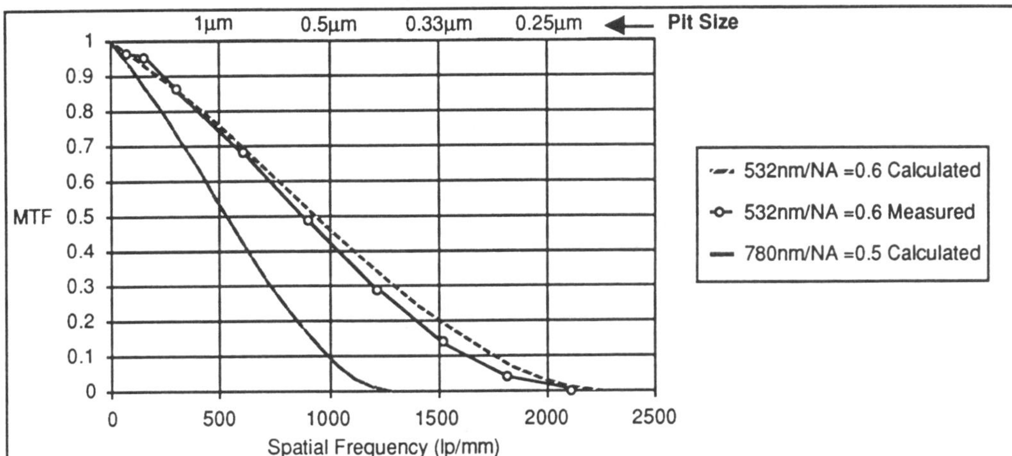

Fig. 7. Measured and Calculated Modulation Transfer Function (MTF)

5.2. Crosstalk

Crosstalk from the adjacent tracks is measured by reading a test disk on which three signals of different frequencies have been recorded on three different tracks-ABC,ABC.... As shown in Fig.8, the crosstalk level is -42dB at a track pitch of 0.9μm, which is sufficiently low for NTSC video playback.

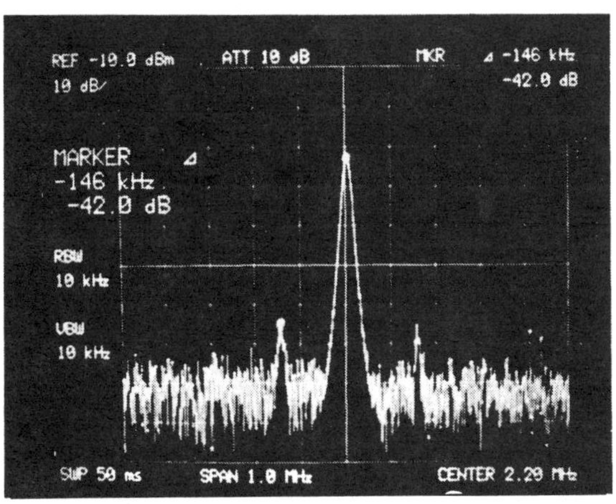

Fig. 8. Crosstalk from adjacent track, measured by a spectrum analyser. Crosstalk is -42dB when track pitch is 0.9μm.

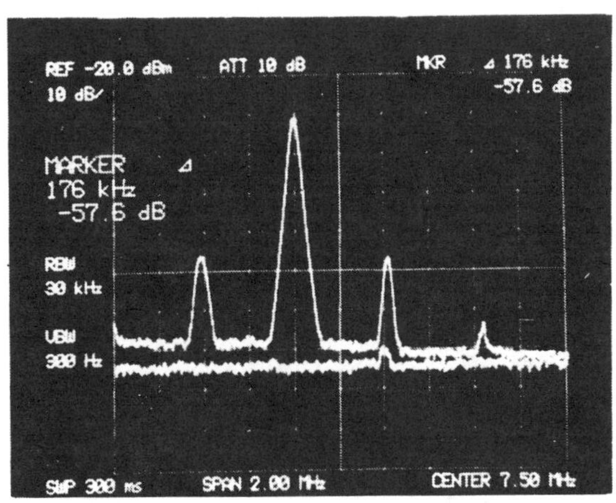

Fig. 9. Measured C/N ratio. C/N is 58dB at bandwidth of 30kHz, when track pitch is 0.9μm and pit size 0.39μm.

5.3. Carrier to noise ratio

A carrier-to-noise ratio (C/N) of 58dB is measured at f=7.16MHz with a sampling bandwidth of 30kHz, where the pit size is 0.39μm. Figure 9 shows spectrum analyzer traces. The bottom trace shows the noise level from the laser source and photo-detector, which indicates that the C/N of the readout signal is limited by disk noise. For an NTSC analog video signal, a C/N of 60 dB is acceptable. The measured C/N shows that 60dB is nearly achieved with the green optical head. An improvement in C/N is possible with more precise tuning in a disk replication-process condition and optimization of pit shape condition.

5.4. Video Playback

Finally, playback of the triple-density video disk is carried out using the test disk. An NTSC video signal recorded on the disk is decoded by a conventional CDV-player circuit. The obtained picture was of good quality and in good agreement with previously mentioned MTF, crosstalk and C/N.

6.CONCLUSIONS

Playback of a triple density video disk using a laser-diode-pumped intracavity frequency-doubled Nd:YAG laser was carried out for the first time. Although the high power LD for pumping and the nonlinear crystal are still too expensive for use in consumer products, the laser proved to be applicable as the light source of a very high-density optical disk system. We also confirmed that the disk can be mastered and replicated.

ACKNOWLEDGEMNET

Authors would like to thank Dr.S.Miyaoka for his helpful advices and encouragement.

REFERENCES

1. K.Compaan and P.Kramer, Philips Tech. Rev. **33**, 178 (1973).
2. S.Kubota, Proc. OSA '84, TuS2 (1984).
3. M.Ikeda, K.Nakano, A.Toda, Y. Mori and C. Kojima, Jpn. J. Appl.Phys., **26** Suppl.26-4, 101 (1987).
4. T.Taniuchi and K.Yamamoto, Proc. ECOC '86, TuC5 (1986).
5. T.Baer, J.Opt.Soc.Am.B. **3**, 1175 (1986).
6. M.Oka and S.Kubota, Opt.Lett. **10**, 805 (1988).
7. T.Mamine and Y.Fujii, J.Electron Eng. **24**, 58 (1987).
8. M.Oka and S.Kubota, Microoptics News, **7**, 1 (1988). (in Japanese)

Reprinted from *Optical Data Storage,* Proc. SPIE Vol. 1316, pp. 194-198 (1990).

Compact Blue Lasers For Optical Recording Applications

W. J. Kozlovsky, W. Lenth and W. P. Risk
IBM Research Division, Almaden Research Center, 650 Harry Road, San Jose, CA 95120

ABSTRACT

Compact blue laser sources with output powers of 10 and 40 mW at 473 and 428 nm have been developed using non-linear frequency upconversion of GaAlAs diode lasers. These devices are attractive for application to high–density optical data storage.

1. INTRODUCTION

Compact blue laser sources offer the potential for optical recording at areal storage densities of several gigabit/sq. inch.[1] The development of wide–bandgap semiconductor injection lasers is hampered by severe material fabrication problems. However, nonlinear optical processes can be utilized to convert the near–infrared output of the presently available GaAlAs diode lasers to wavelengths in the blue spectral region. We have used such techniques for the development of two blue laser sources with output powers that are attractive for optical recording applications. The first device is based on intracavity frequency doubling of a diode–pumped 946–nm Nd:YAG laser. Another device uses direct frequency doubling of a GaAlAs diode laser by coupling its output to a resonator containing a nonlinear crystal.

2. FREQUENCY DOUBLING OF A DIODE–PUMPED 946–nm Nd:YAG LASER

Diode–laser pumping of miniature Nd:YAG lasers offers several advantages for the generation of short–wavelength radiation by nonlinear frequency upconversion. High–power laser diode arrays can be used to pump the Nd:YAG laser, and the relatively poor spatial and spectral mode properties of these laser arrays are thus converted to the TEM_{00} spatial mode and narrow spectrum of the solid–state laser. In addition, the nonlinear crystal can be placed inside the solid–state laser cavity to take advantage of the high circulating intensity of the internal laser field for efficient intracavity second harmonic generation (SHG).

In order to generate blue radiation we have developed a diode–pumped Nd:YAG laser that operates on the 946–nm laser transition.[2,3,4] This transition has a lower gain cross section than the standard 1064–nm laser line, and in addition, exhibits a reabsorption loss. A comprehensive modeling analysis was used for designing the cavity and pump geometries and optimizing the length of the Nd:YAG crystal.[4,5] The experimental configuration is shown in Figure 1. The collimated output of a 0.5–W broad–area 809–nm laser diode was passed through cylindrical optics which were used to circularize the pump beam and reduce its astigmatism so that it could be focused to a tight waist in the Nd:YAG crystal. A dielectric input mirror coating was deposited directly on one side of the Nd:YAG crystal. The length of the cavity (~5 cm) was adjusted to be near–hemispherical. The crystal chosen for frequency doubling was potassium niobate ($KNbO_3$). The crystal was cut in a special orientation to permit phasematched SHG at room temperature. Our experiments showed that this crystal orientation also results in Nd:YAG laser oscillation with the correct linear polarization for efficient SHG.[4] As a consequence of this design, no additional polarization elements had to be placed inside the laser cavity.

The length of the Nd:YAG laser crystal must be carefully chosen. The reabsorption loss characteristic of the 946–nm laser transition increases with crystal length; on the other hand, efficient absorption of the 809–nm diode–laser pump

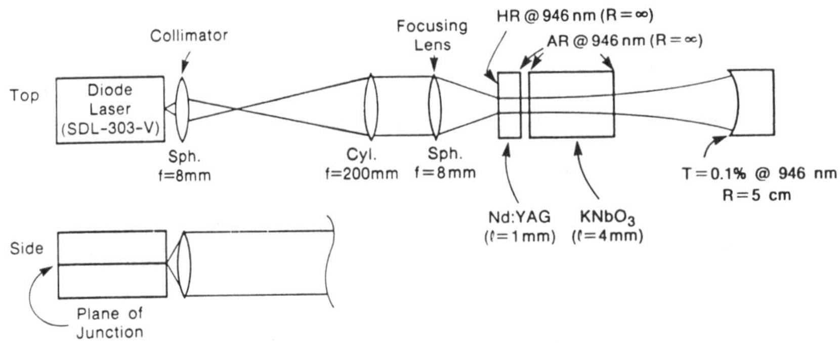

Figure 1. Experimental configuration for generation of 473–nm light by intracavity frequency doubling of a diode–pumped 946–nm Nd:YAG laser.

light requires a reasonably long Nd:YAG rod length. Supported by modelling calculations we have used a 1.5–mm long Nd:YAG crystal as a comprise between these two effects. Using a 4–mm long $KNbO_3$ crystal placed in close proximity to the cavity beam waist of the Nd:YAG laser, a blue 473–nm output power of 9.5 mW was produced with 420 mW of diode laser pump power. The blue output beam was slightly elliptical and had Gaussian intensity distributions in the x and y directions.

3. FREQUENCY DOUBLING OF A DIODE–LASER USING AN EXTERNAL CAVITY

Direct frequency doubling of GaAlAs diode lasers potentially allows simple designs for blue laser devices. Due to the relatively low power available from GaAlAs diode lasers, field enhancement techniques are required for efficient blue generation. By coupling the diode–laser radiation to an external cavity that contains the $KNbO_3$ crystal, the diode–laser radiation can build up to a high circulating intracavity intensity which results in efficient second harmonic generation.[6,7,8] The resonant frequency of the external cavity and the output frequency of the diode laser must be locked together for the build–up to occur, since small changes in the resonance frequency of the external cavity due to self–heating effects in the crystal[7,9] or small changes in the laser frequency caused by current (~2 GHz/mA) and temperature (~40 GHz/°C) variations can result in substantial fluctuations of the infrared power inside the resonator and thus in the blue output. We have used a simple, effective frequency locking technique and have achieved stable operation of a frequency–doubled semiconductor diode laser with 41 mW of blue output power.

The experimental set–up is shown in Figure 2. A monolithic ring cavity for frequency doubling was fabricated by polishing 2 spherical faces and a TIR (total internal reflection) surface onto a 7–mm long $KNbO_3$ crystal; dielectric coatings were applied to the spherical faces.[9] The 856–nm GaAlAs diode laser used was a ridge–type, single quantum–well, graded–index double heterostructure laser.[10] In order to lock the laser frequency to the $KNbO_3$ resonator of ~200–MHz linewidth, we modified the Pound–Drever locking technique[11] often used for stabilizing dye and argon lasers by exploiting the convenient modulation and tuning properties of the semiconductor laser. For this method, weak FM–sidebands were produced around the ~5 MHz–wide laser emission line by superimposing a weak 292–MHz current on the dc injection current using capacitive coupling (see Figure 2). By using a double–balanced mixer for phase–sensitive detection of the heterodyne spectroscopy signal[12] contained in the light reflected from the input surface of the resonator, a strong servo signal indicative of the match of the laser and resonator frequencies was generated.[13] This servo signal was amplified and used to control the dc–current of the diode laser so that frequency stabilization was obtained by tuning of the laser frequency so that it tracked the cavity resonance.

With the servo loop closed the GaAlAs laser frequency remained locked to the nonlinear cavity resonance. Stable 428–nm output for several hours was observed; the measured relative intensity noise (RIN) at 5 MHz was –120 dB. As shown in Fig. 3, an output power of ~41 mW at 428 nm was obtained with 105 mW of near–infrared power incident on the $KNbO_3$ resonator, for an nonlinear conversion efficiency of 39%. The diode laser output was ~160 mW and the transmission of the diode–laser optics was approximately 65%. The electrical input to the diode laser was ~400 mW, resulting in an overall electrical–to–blue conversion efficiency of ~10%.

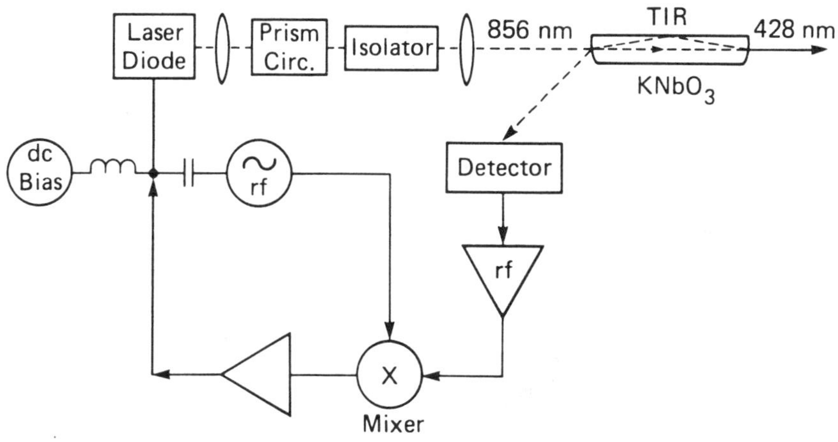

Figure 2. Experimental configuration for generation of 428–nm light by resonantly enhanced SHG of a GaAlAs laser diode.

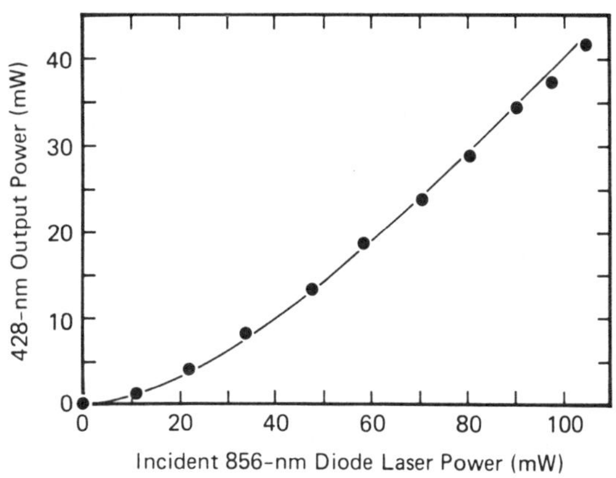

Figure 3. Generated blue power at 428 nm versus diode laser input power at 856 nm.

As expected for our resonator design, the 428 nm output beam had a slightly elliptical Gaussian intensity distribution (1.1:1). Using a 0.65 N.A. lens, the blue output beam was focused to an essentially diffraction limited spot size of 0.4 μm (FWHM).

4. SUMMARY

These experiments demonstrate that nonlinear frequency upconversion techniques can permit the design of compact, efficient blue laser devices based on the presently available GaAlAs diode lasers. These blue sources have attractive output powers and performance characteristics that are compatible with the requirements of high–density optical recording.

5. ACKNOWLEDGMENT

The authors acknowledge the fabrication of high–power 856–nm GRIN–SCH GaAlAs diode lasers by the staff of the Laser Science and Technology Department at the IBM Zurich Research Laboratory.

6. REFERENCES

1. D. Rugar, C.–J. Lin and R. Geiss, "Submicron Domains for High Density Magneto–Optic Data Storage," *IEEE Trans. Magn.*, vol. MAG–23, pp. 2263–2265, 1987.

2. T.Y. Fan and R.L. Byer, "Continuous–wave operation of a room–temperature, diode–laser–pumped, 946–nm Nd:YAG laser," *Opt. Lett.*, vol. 12, pp. 809–811, 1987.

3. W.P. Risk and W. Lenth, "Room–temperature, continuous–wave, 946–nm Nd:YAG laser pumped by laser–diode arrays and intracavity frequency doubling to 473 nm," *Opt. Lett.*, vol. 12, pp. 993–995, 1987.

4. W.P. Risk and W. Lenth, "Efficient Frequency Doubling of a Diode–Pumped 946–nm Nd:YAG Laser," in Tunable Solid State Lasers, Vol.5 of the OSA Proceeding Series, M. L. Shand and H. P. Jenssen, eds. (Optical Society of America, Washington, D.C., 1989) pp. 329–331.

5. W.P. Risk, "Modeling of longitudinally pumped solid–state lasers exhibiting reabsorption losses," *J. Opt. Soc. Am.–B*, vol. 5, pp. 1412–1423, 1988.

6. G. J. Dixon, Carrol E. Tanner, and Carl E. Wieman, "432–nm source based on efficient second–harmonic generation of GaAlAs diode–laser radiation in a self–locking external resonant cavity," *Opt. Lett.*, vol. 14, pp. 731–733, 1989.

7. Lew Goldberg and Myuang Chun, "Efficient generation at 421 nm by resonantly enhanced doubling of GaAlAs laser diode array emission," *Appl. Phys. Lett.*, vol. 55, pp. 218–220, 1989.

8. T. Baer, M. S. Keirstead, and D. F. Welch, "Efficient frequency doubling of a diode laser," Conference on Lasers and Electro–Optics, 1989 Technical Digest Series, Vol. 11 (Optical Society of America, Washington, D. C. 1989) p. 322.

9. W. J. Kozlovsky, C. D. Nabors, and R. L. Byer, "Efficient Second Harmonic Generation of a Diode–Laser–Pumped CW Nd:YAG Laser Using Monolithic MgO:$LiNbO_3$ External Resonant Cavities," *IEEE J. Quant. Electron.*, vol. QE–24, pp. 913–919, 1988.

10. C. Harder, P. Buchmann, and H.P. Meier, "High–Power Ridge–Waveguide AlGaAs GRIN–SCH Laser Diode," *Electron. Lett.*, vol. 22, pp. 1081–1082, 1986.

11. R. W. P. Drever, J. L. Hall, F. V. Kowalski, J. Hough, G. M. Ford, A. J. Munley, and H. Ward, "Laser Phase and Frequency Stabilization Using an Optical Resonator," *Appl. Phys. B*, vol. 31, pp. 97–105, 1983.

12. W. Lenth, "High Frequency Heterodyne Spectroscopy with Current–Modulated Diode Lasers, " *IEEE J. Quant. Electron.*, vol. QE–20, pp. 1045–1050, 1984.

13. Armin Sollberger, Arja Heinamaki, and Hans Melchior, "Frequency Stabilization of Semiconductor Lasers for Applications in Coherent Communication Systems," *J. Lightwave Tech.*, vol. LT–5, pp. 485–491, 1987.

Section Four
Lenses

Reprinted with permission from *Optics Letters*, Vol. 4(2), pp. 70-72 (February 1979).

Realization of a bi-aspherical objective lens for the Philips Video Long Play system

J. Haisma

Philips Research Laboratories, Eindhoven, The Netherlands

E. Hugues and C. Babolat

Centre de Recherches et de Calculs Optiques, 10 boulevard de Verdun, Courbevoie 92400, France

Received October 27, 1978

An objective lens for focusing monochromatic laser light on an optical video disk has been machined in polymethyl methacrylate on a lathe numerically controlled by a computer. The objective is a single-element lens and has two steeply aspherical surfaces. It has a numerical aperture of 0.43, and its field has a diffraction-limited quality over a radius of 300 μm. Some details of the computer-numerically controlled hydraulic lathe are given.

Introduction

Optical video and audio disk systems have drawn much attention in the last 10 years.[1–3] For the optical readout unit, the essential components nowadays are a He–Ne laser (λ = 6328 Å), a beam-splitting prism, radial and tangential mirrors, and a diffraction-limited objective with a numerical aperture typically of 0.4. The introduction of an AlGaAs-diode laser in the optical pickup makes the light path so small that the whole optics can be incorporated in a single *xyz* actuator,[4] which results in a substantial decrease of the total number of optical components needed. Moreover, the feedback effect in the laser can then be applied for high-frequency signal detection. In this setup the objective lens system and the diode laser are the most essential components.

The introduction of aspherical surfaces into an optical system has the advantage of reducing the number of its elements. In the Video Long Play (VLP) system,[1] we have applied an objective lens consisting of a single element whose two surfaces have an aspherical shape, the "bi-asphere."[11] This single element is used instead of the objective lens of spherical components consisting of three or four elements, depending on the design. Moreover, the bi-asphere provides more freedom in the free space between the disk and the objective, i.e., it has a larger back focal length and it diminishes the total weight of the readout element. This is of importance for automatic focusing through a protective, transparent layer onto the information layer of the disk.

This machined plastic lens is only a first step in the technological procedure of converting spherics to aspherics. It implies, for instance, that optical, aspherical lenses with a higher numerical aperture can be made easily by using the capabilities of the novel lathe. The process still has to be evaluated in more detail to permit an improved technology of producing bi-aspheres in a cost-effective way compared with the manufacture of conventional spherical objectives.

The Bi-asphere

The lens used in the VLP player should have a diffraction-limited optical quality with a numerical aperture of at least 0.40 in order to resolve adequately the information stored in the successive grooves on the disk; the groove pitch is 1.67 μm. The field diameter required to cover the eccentricity effect on the disk rotation is 600 μm. The lens material in this first experiment is a well-known optical, machinable plastic, polymethyl methacrylate, (Perspex or Plexiglas) with a refractive index of n = 1.4908 for λ = 6328 Å at room temperature. In a design[5–8] of an aspherical lens with the field performance mentioned above, it is necessary to fulfill Abbe's sine condition to diminish the coma to a tolerable amount up to the border of the field. Besides, the condition for stigmatism has to be respected for the axial image point. This is the reason that both surfaces have to be asphérized.[7] Finally, the astigmatism and field curvature can be minimized by adapting the thickness of the lens.

Having fixed the distance between the conjugate planes and the magnification of the lens, we are left with an infinite number of solutions depending on the specific paraxial conditions one wants to start from. As a final design[11] we have chosen a lens that is an acceptable compromise between field quality, i.e., the magnitude of the residual astigmatism and field curvature, and the application tolerances to be specified for realizing the lens, i.e., the amount of decentering, the tolerable tilt of one lens surface with respect to the other, and deviations of the ideal lens thickness. In the final design the reduction of aberrations in the field to a tolerable level has been considered as the most important requirement. Under these conditions a bi-aspherical lens could be prescribed that has a numerical aperture of 0.43 and a useful flat-field diameter of 600 μm. Figure 1 is a diagram of this VLP bi-asphere.

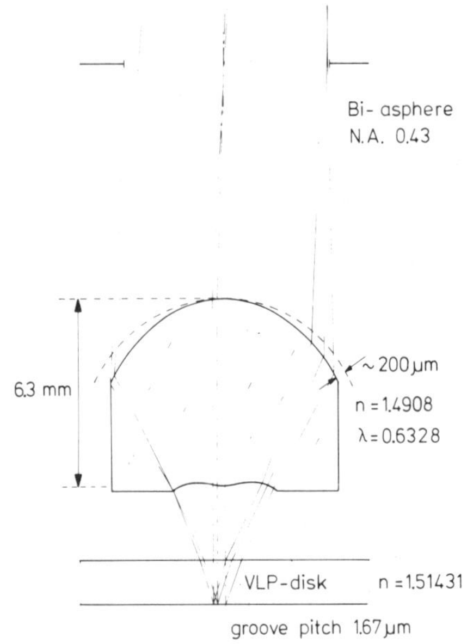

Fig. 1. Diagram of the bi-asphere and its focusing on the VLP disk.

Lathe and Machining Details

The first series of these bi-aspherical, objective lenses has been produced on a computer-numerically controlled (cnc) lathe, the so-called COLATH, designed and constructed at Philips Research Laboratories. This cnc lathe is provided with hydrostatic bearings and hydraulic drives. (This lathe is an advanced successor of the one described in Ref. 9.) As measuring elements in the closed-loop servo system of the machine, laser interferometers are used. The stability of the lathe is very high. Typical values are tilt stiffness, ~27 kg/μm and axial stiffness, ~10 kg/μm.

The cutting tool, a diamond single cutter, is brazed on a metal holder and ground to a high degree of roundness, i.e., the deviation from an ideal circle over the periphery of the diamond is <0.1 μm over 60° of arc. The roughness of a surface thus machined is less than $R_t \sim 0.04$ μm.

The prescribed curves of the aspherical surfaces are approximated by linear interpolations to be kept within certain limits. This tolerance means maximum permissible deviations of 0.05 μm at both sides of the prescribed, ideal curve.

The Plexiglas used, 245, Röhm, is a material that can be machined to an optical-surface quality. However, plastic is generally not so suitable for use as the material for diffraction-limited optical elements. Nevertheless, the results obtained with it indicate that, as a first tentative material, under reasonable but not exaggerated temperature conditions (room temperature) it may be applicable. However, as far as the compatibility of lens design and manufacturing tolerances is concerned, we were able to draw important conclusions (see below).

The machining time of an element was about 20 min.; 25 elements were machined in a one-day run; 20 lenses were qualified between very good and good (probably because of the misalignment of the axes of the two surfaces); the rest showed some axial astigmatism. This means that each surface of the elements that were qualified very good deviates from the ideal shape by about 0.2 μm or less, the good ones between 0.2 and 0.4 μm. Figure 2 shows a Twyman–Green interferogram[10] of an objective lens qualified as very good (satisfying the Maréchal criteria).

Finally, the objectives were tested at room temperature in a VLP player, and they showed a good picture performance as well as a good still picture, not different from those made with a well-designed four-element spherical objective. These bi-aspheres were tested at more elevated temperatures, and this showed that a definite change in quality occurs as soon as plastic deformations of the Plexiglas appear; an upper limit of temperature to be allowed will be 40 °C.

Conclusions

From all these results it can be concluded that

1. Diffraction-limited optical requirements can nowadays be met with advanced mechanical means differing from the conventional method applied in optics.
2. The linear approximation of a prescribed nonlinear curve is not at all detrimental to the optical performance in itself, although it gives rise to a slight extra light scattering.
3. The realization of these steeply aspherical optical elements by the described mechanical means implies another way of tackling the problem of the optics of aspherical surfaces with a high degree of asphericity (aspherics), and the flexibility to achieve elements with a higher numerical aperture is relatively unlimited.
4. This experiment has revealed that diffraction-limited, optical, aspherical components can be realized (at least in plastics) and applied.
5. The performance at room temperature of the

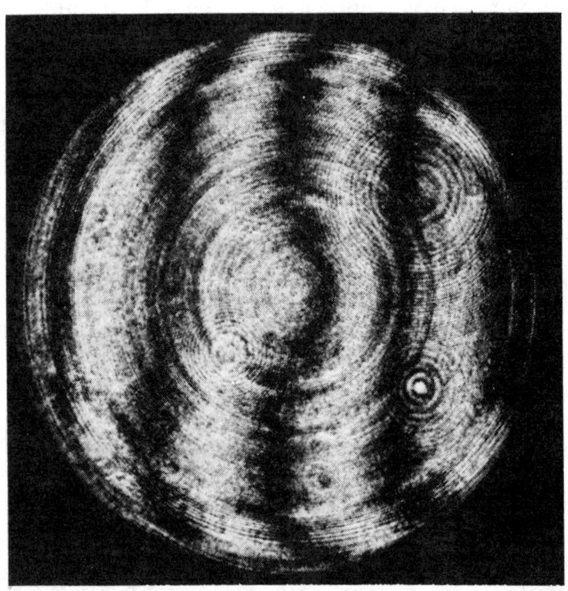

Fig. 2. Twyman–Green interferogram of a bi-asphere.

bi-aspheres satisfies the specifications for a video player.

6. In this specific application deviations from the ideal curve up to a few tenths of a micrometer are still acceptable.

The successful production of this high-quality lens stimulates the development of aspherics for mass production in harder materials. This first realization shows that the design of the aspherical lenses and the machining of them into the relevant shape have been mastered.

This result of our project on aspherics springs from the outstanding attainments of many of the cooperating members of the Philips Research Laboratories in The Netherlands and the Centre de Recherches et de Calculs Optiques in France. The authors should like to express their gratitude to all of them for their valuable contributions.

References

1. K. Compaan and P. Kramer, Philips Tech. Rev. **33,** 178 (1973).
2. G. Bouwhuis and J. J. M. Braat, Appl. Opt. **17,** 1993 (1978).
3. P. W. Bögels, IEEE Trans. Commun. Electron. **CE-22,** 309 (1976).
4. K. Bulthuis, in *Conference Proceedings Laser 77, Opto-Electronics,* W. Waidelich, ed. (IPC Science and Technology Press, Guildford, England, 1977), pp. 637–642.
5. R. Straubel, Z. Instrumentenkd. **61,** 240 (1941).
6. R. Straubel, Z. Instrumentenkd. **63,** 153 (1943).
7. G. D. Wassermann and E. Wolf, Proc. Phys. Soc. **62B,** 2 (1949).
8. B. Jurek, Opt. Acta **6,** 222 (1959).
9. H. J. J. Kraakman and J. G. C. de Gast, Philips Tech. Rev. **30,** 117 (1969).
10. J. G. Dil, P. F. Greve, and W. Mesman, Appl. Opt. **17,** 553 (1978).
11. E. A. Hugues, "Single Optical Lens with Large Aperture and Large Field," U.S. Patent 4,027,952 (June 7, 1977).

Reprinted with permission from *Optical Engineering*, Vol. 22(4), pp. 473-478 (July/August 1983). ©1983 Society of Photo-Optical Instrumentation Engineers.

Design and testing of lenses for optical disk technology

P. Kuttner
Optische Werke G. Rodenstock
Isartalstrasse 43
D 8000 München 5, West Germany

Abstract. Optical disk technology today finds widespread application in las vision, compact disk, and data storage equipment. The required specificatio of lenses which meet these requirements are discussed. Their image quality given by rms of wave aberration, by point spread function, by encircled ener or by modulation transfer function (MTF). Due to the automatic servo focusi technique, a certain amount of field curvature can be tolerated. Measur results of image quality are compared with theoretical values.

Keywords: lens design; optical disk technology; optical data storage; image quality.

Optical Engineering 22(4), 473-478 (July/August 1983).

CONTENTS

1. INTRODUCTION

The digital storage and retrieval of information on the optical disk is presently being discussed in the form of a number of practical applications, tested in laboratory models, and even introduced on the market in versions such as the Laser Vision (LV), former Video Long-Playing Disk (VLP), and as the Compact Disk (CD). The optical disk technology is generally considered to be very promising for information storage. The advantages of this technique over other storage techniques are its higher storage capacity, rapid access, and a considerable reduction of manufacturing costs.[1] The technique generally consists in burning small holes in a thin film deposited on a rotating disk. Tellurium deposited on a carrier is an especially suitable material. Plastic disks are likewise being used as carrier materials. Holes of less than 1 μm diameter can be burned into a layer 0.06 μm thick.[2] This results in a storage density of about 10^8 bits/cm^2. Storage capacities of 2×10^{10} bits are quite feasible. For the future even considerably higher storage capacities are in discussion.

A laser is controlled by a light modulator and focused onto the film by a high-aperture objective. The holes are formed in the film with an illuminance of over 10^6 W/cm^2, which heats up the material, partly melts the metal, and opens the hole. The required illuminance is attained with lasers of 10 mW output. The irradiation time used for burning the holes is 0.5 μs.[2,3]

With a diffraction-limited uniformly illuminated objective, the intensity distribution in the image plane is given by the Airy pattern. The first dark ring has a diameter d_A given by

$$d_A = 1.22 \frac{\lambda}{NA} , \qquad (1)$$

where λ is the used wavelength and NA the numerical aperture of the objective. Some 84% of the total incident energy is concentrated the central Airy disk.

The intensity distribution in the image plane for a Gaussian la beam is likewise Gaussian.[4,5] Truncation effects of the incide Gaussian beam will influence the image pattern. For laser light w transverse electromagnetic (TEM_{00}) mode output, the beam dia ter D_b is usually specified for the $1/e^2$ intensity level. If the specified beam diameter is 1/2 of the lens diameter D_l, the $1/e^2$ sp size diameter is given by the relationship

$$d_{1/e^2} = \frac{4}{\pi}\frac{\lambda f}{D_b} = \frac{4}{\pi}\frac{\lambda}{NA} ,$$

where f is the focal length of the objective.[6] Some 86% of the to incident energy is concentrated in this spot size. Only for laser be diameters

$$D_b \leq 0.5\, D_l ,$$

no essential power loss is caused by truncation.

In addition to numerical aperture and wavelength, the attaina diameter of formed holes is a function of the correction of objective, the type of illumination, and the threshold value, melti and evaporation properties of the storage layer. However, it a depends on the irradiation time and thus on energy density. Dur readout of the information, the material characteristics of the stor layer and the energetic characteristics of the laser beam do not contr ute to the resolution so that the optical characteristics of the object alone will determine the readout quality. If the resolving power of objective is not sufficient, the readout signal may be disturbed b signal from the neighboring trace. On the other hand, the cross-t effect is of only minor importance during writing of information du the threshold nature of the hole-forming process.

2. REQUIREMENTS MADE OF OBJECTIVES FOR DI TECHNOLOGY

Optical disk technology is suitable for three different types of ap cations: LV, CD, and data storage (DS). The following characteris specifications can be given for these three applications, althou there are certain variations among different manufacturers. Tab gives the general specifications for these systems. For semiconduc lasers, radiation emitted is extended to the bandwidth of nearl nm.[7] The variation of wavelength location at a constant temperat of 30°C for some semiconductor lasers can be ±10 nm, but includi a temperature range of 0°C to 60°C, the variation will be ±20 n Therefore, in Table I is given the spectral region needed for semic ductor lasers.

Paper 1886 received June 2, 1982; revised manuscript received Dec. 21, 1982; accepted for publication Jan. 11, 1983; received by Managing Editor Jan. 17, 1983. This paper is a revision of Paper 329-22 which was presented at the SPIE conference on Optical Disk Technology, Jan. 26-28, 1982, Los Angeles, CA. The paper presented there appears (unrefereed) in SPIE Proceedings Vol. 329.

TABLE I. System Data for LV, CD, and DS Systems

	LV	CD	DS Write	DS Read
Light source	Gas laser: HeNe, Semiconductor laser	Semiconductor laser	Gas laser: HeNe, Ar, Semiconductor laser	Gas laser: HeNe, Semiconductor laser
Nominal wavelength	632, 780 nm	780 nm	488, 632, 825, 870 nm	632, 825 nm
Spectral region	632 760–800 nm	780–820 nm	780–820 nm 800–850 nm	632, 780–860 nm 800–850 nm
Trace interval	1.6 μm	1.6 μm	1.6 μm–2.0 μm	
Trace width	0.6 μm	0.6 μm	0.6 μm–0.8 μm	
Trace depth	0.1 μm	0.1 μm	0.06 μm–0.1 μm	
Plate material	Plastic	Plastic	Glass, Plastic, Aluminum	

TABLE II. Specifications for Objectives Used in Disk Technology

	LV	CD	DS
NA	0.40–0.65	0.40–0.45	0.45–0.60
Focal length	4.5, 5, 8, 9 mm	4.5, 5 mm	4, 4.5, 6 mm
Field diameter	0.2, 0.4, 0.8 mm	0.04–0.08, 0.2 mm	0.04, 0.2, 0.4 mm
Cover glass thickness	1.3 mm	1.2 mm	1.2–1.5 mm
Free working distance	0.8–3.5 mm	~2.3 mm	0.75–2 mm
Vignetting	0%	0%	0%
Transmission	>97%	>95%	>95%
Weight	1.6 g–3.7 g	0.6 g, 1.6 g	0.6 g, 4.3 g
Number of elements in actual objectives	4	3, 4	3, 4, 6, 7, or 8

Table II gives the specifications for the required objectives. In the early stage of optical disk technology, lenses with field diameters of 0.2 or 0.4 mm were needed. Today a tendency of a reduced field size of 0.04 mm or even less could be required. A continuous drive of the optical head in the radial disk direction is required for small field size lenses. Due to refocusing in the actuator, field curvature up to radii of $r \sim f/2$ is generally admissible, f standing for focal length. In an objective of 5 mm focal length with an admissible field curvature of 2.5 mm, the amount of defocusing for an image height of 0.1 mm is 2 μm. A control circuit will automatically refocus the system by this amount. The admission of this type of field curvature makes it possible to correct astigmatism and to reduce coma to a large extent. For a reduced field size of about 0.08 mm, the weight of 0.6 g is realized with a three-element version; however, for a fully color-corrected lens, seven or eight elements are needed and a weight of 4.3 g is attained.

Due to diffraction, every objective has a depth of field $2\Delta z$ within which there is only a minimum loss of image quality. For a uniformly illuminated objective, defocusing by the amount of

$$\Delta z = \frac{\lambda}{4\,NA^2} \tag{4}$$

will reduce the intensity on the axis to about 95%.[8] Table III gives the depth of field for systems of NA 0.45, 0.5, 0.55, and 0.6, for wavelengths of 632 nm and 800 nm. For a Gaussian laser beam the depth of field depends on the amount of truncation. With an increase of truncation, which means with a reduction of the effective NA value of the beam, the depth of field increases.

TABLE III. Depth of Field of Optical Disk Objectives

NA	$2\Delta z$ in μm for λ = 632 nm	$2\Delta z$ in μm for λ = 800 nm
0.45	1.5	1.9
0.5	1.2	1.6
0.55	1	1.3
0.6	0.9	1.1

No chromatic correction is required for objectives using a given monochromatic wavelength. These are all systems based on gas lasers, which are generally LV systems apart from some developmental systems in the DS field. In Fig. 1 the allowed longitudinal chromatic aberration Δz of LV type objectives used with HeNe lasers is shown as function of wavelength. These objectives have NA values of 0.4, focal lengths of 8.3 mm, and image heights of 0.4 mm.

If semiconductor lasers are used, the bandwidth of the individual unit has to be taken into account in the lens design. Hence, objectives used with semiconductor lasers require chromatic correction over a wavelength region of 5 nm.

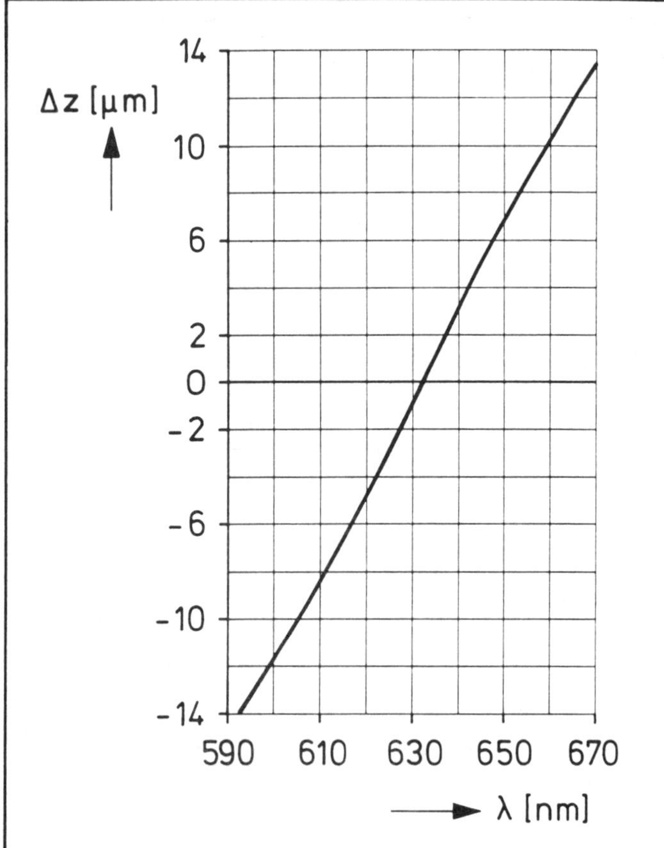

Fig. 1. Longitudinal chromatic aberration Δz of LV type objectives used with HeNe laser: NA = 0.4, f = 8.3 mm.

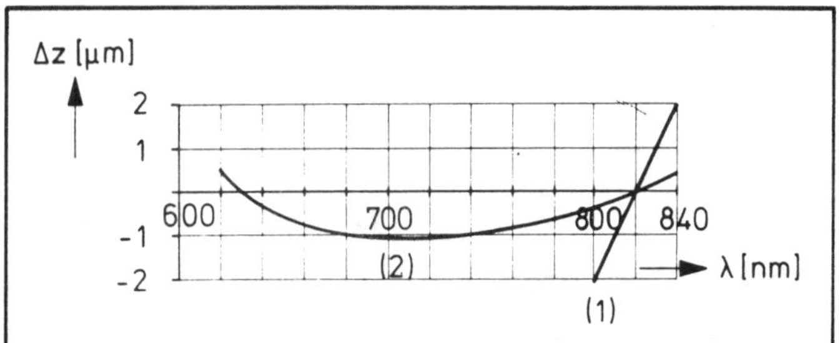

Fig. 2. Longitudinal chromatic aberration Δz of (1) CD type objective used with semiconductor laser; (2) DS type objective, achromatic correction for 632 and 820 nm.

If the objective is used independently from writing, for reading only, a reduction of the allowed longitudinal chromatic aberration to about one-third of the value of non-color-corrected systems is necessary. Automatic focusing in the actuator allows precise adjustment to the individual wavelength of the laser unit used. In sophisticated systems, the secondary chromatic spectrum in the used wavelength range should be less than the depth of field given in Table III. In combination with automatic focusing, this will ensure appropriate image quality for each individual unit. CD objectives and some DS lenses belong in this category.

If writing and reading is simultaneous and at different wavelengths, the objective has to be achromatic for the two wavelengths. For this purpose, however, the chromatic lateral aberration also has to be reduced. In the whole image field lateral color aberrations between writing and reading wavelengths smaller than 0.05 to 0.1 μm can be tolerated. DS objectives especially belong in this category. In Fig. 2 the allowed longitudinal chromatic aberration of objectives used with semiconductor lasers is shown, as well as that of a full

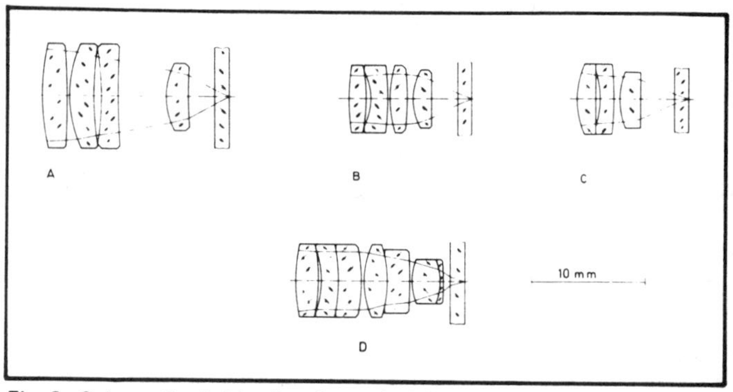

Fig. 3. Schematic of objectives for optical disk technology (a) LV objective; (b) CD objective; (c) CD, DS objective with reduced image field; (d) DS objective with full color correction.

achromatic version. The central wavelength of the semiconductor laser is assumed to be 820 nm in both cases; for the DS type lens generally the other wavelength is 632 nm. The digitalized information in LV and CD disks and also in some DS applications is protected under a 1 to 1.5 mm thick cover glass. Normally polymethyl methacrylate (PMMA) or polycarbonate with refractive indexes of 1.49 or 1.55, respectively, are used as the cover glass. Therefore, the planoparallel cover glass is part of the image-forming system, and allowance has to be made already in the lens design. Generally optical disk objectives are used for an infinite conjugate.

A schematic representation of the types of objectives for optical disk technology is given in Fig. 3. The element shown on the right is the cover glass plate that is part of the optical formula. Table IV gives the characteristic data of these objectives, f standing for focal length and y′ for half-field diameter. Also added are the main application and the weight of these lenses. Objectives of the type (A) for LV applications were among the first ones to be designed for optical disk technology. They are used with monochromatic laser light. Especially the needed large image field is the dominant factor for the lens design. For CD and some DS applications the image field is very much reduced. Therefore, for that purpose smaller and lighter lenses can be designed. Objectives of type (B) belong in this category. For image fields smaller than 0.08 mm, a light three-element lens of the type (C) was realized. The spectral range of the semiconductor lasers used calls for some attention in the color correction. However, the achromatization for two wavelengths and the reduction of lateral color aberration to values smaller than 0.1 μm result in objectives of type (D).

3. PERFORMANCE OF OBJECTIVES FOR OPTICAL DISK TECHNOLOGY

The image quality of lenses is preferably indicated by one of the following functions: wave aberration, point image, and modulation transfer function (MTF). These three functions are interrelated mathematically. Each of them can be computed and measured. The type of application will determine the most appropriate form of application. Determination of the wave aberration seems to be particularly well suited for computing an objective for use with the optical disk technology because it is a very sensitive function for determining the image quality of high speed, nearly diffraction-limited

TABLE IV. Data of Objectives for Optical Disk Technology

Objective	Application	NA	f[mm]	λ[nm]	y′[mm]	weight
A	LV	0.4	8.3	632	0.4	3.6 g
B	CD	0.45	4.6	800–840	0.12	1.6 g
C	CD, DS	0.5	5	800–840	0.04	0.6 g
D	DS	0.6	4	632, 800–840	0.05	4.3 g

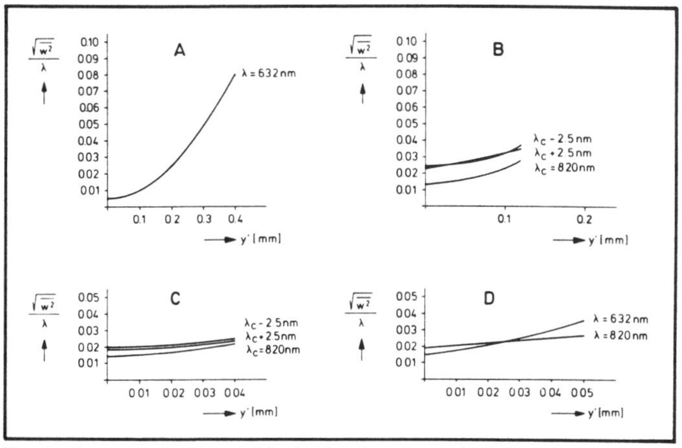

Fig. 4. Computed values 1/N of objectives for optical disk technology: (a) LV objective; (b) CD objective; (c) CD, DS objective with reduced image field; (d) DS objective with full color correction.

lenses. The root mean square (rms) of wave aberration can be expressed in parts of the wavelength λ by the formula

$$\sqrt{\overline{w^2}} = \frac{\lambda}{N}, \tag{5}$$

where the mean is taken over the pupil, and w is the wave aberration and N a positive rational number. The square of Eq. (5),

$$\overline{w^2} = \frac{\lambda^2}{N^2}, \tag{6}$$

is also used to describe lens performance. This gives the relationship between λ^2 and the mean square wave aberration.

Figure 4 shows the rms of the computed wave aberration divided by the used wavelength λ, or by interpretation of Eq. (5) the values $1/N = \sqrt{\overline{w^2}}/\lambda$ as a function of the half-field diameter y′ for the four objectives depicted in Fig. 3. The LV objective (A) has been computed for rigorously monochromatic light. It has an rms of wave aberration of less than λ/100 on-axis, which in the field drops to λ/12 at 0.4 mm half-field diameter.

Objective (B) has a reduced longitudinal chromatic aberration. The rms of wave aberration is given for the central wavelength λ_c of 820 nm and for the wavelengths

$$\lambda = \lambda_c \pm 2.5 \text{ nm}$$

up to a half-field diameter of 0.12 mm. Thereby the image quality for an individual band emitted by a semiconductor laser is described. Focused for the central wavelength λ_c, the objective has an rms of wave aberration of λ/77 on-axis, which drops to λ/42 at wavelengths shifted by ±2.5 nm. The quality on-axis is slightly less than in objective (A), but this is made up for by a more uniform distribution of image quality over the required field.

Objective (C) has a reduced longitudinal chromatic aberration. Due to the reduced half-field diameter of 0.04 mm, the design resulted with a three-element lens. Focused for the central wavelength λ_c, the objective has an rms of wave aberration of λ/70 on-axis, which drops to λ/50 at wavelengths shifted by ±2.5 nm. Approximately this image quality is realized over the whole image field.

Objective (D) is achromatic for the two wavelengths. The rms of wave aberration for 820 nm is ~λ/50 in the whole field. In the same focal plane the image quality for 632 nm is somewhat better on-axis and somewhat less in the field.

The wave aberration determines the shape of the resulting point image. The size and symmetry of the point image obtained is particularly important for reading out the information stored by the optical disk technique. In the disk technique the information is stored in the form of a successive series of holes. It therefore appears feasible to interpret the signal modulation produced on readout of two successive holes approximately as the MTF value of the spatial frequency corresponding to the reciprocal of the distance between the centers of the holes. The MTF of the objective can be determined from the point image or from the pupil function.

4. MEASURING PERFORMANCE OF OBJECTIVES FOR OPTICAL DISK TECHNOLOGY

Objectives intended for optical disk technology should be manufactured with the utmost care. It should be remembered that we are dealing with nearly diffraction-limited objectives with a resolving power of about 1200 c/mm that must be very light in weight. During automatic focusing and trace control they are subject to oscillation frequencies of up to 4 kHz and still have to give full performance. Optical performance can be measured by interferometric measurement of wavefront, by assessment of point image, or by measurement of MTF.

4.1. Interferometric measurement of wavefront with automatic processing of interference fringes

A suitable commercial instrument for the purpose is the Zygo interferometer and a ZAPP processor.[9] The interferometer used is of the Fizeau type. First, the laser light is expanded by a beam expander that forms an emerging plane wave. This plane wave is partially reflected and partially transmitted by the reference surface, forming the reference wave and the measurement wave, respectively. The plane measurement wave passes the objective. For diffraction-limited objectives the wave emerging is spherical. A spherical mirror with the center of curvature in the focus of the lens reflects the spherical wave forming a convergent spherical wave. On its way back this wave passes again through the lens and emerges as a plane wave, which after the reference surface will interfere with the reference wave. Wave aberrations of the objective will form first a nonspherical emergent wave that on the way back results in a nonplane measuring wave that interferes with the plane reference wave. The ZAPP processor evaluates automatically the interference fringes and results with rms of wave aberration. Figure 5 shows the design of this interferometer. This allows the wavefront to be determined with an accuracy up to λ/100.

4.2. Assessing performance on the basis of the point image

Generally optical disk objectives are corrected for an infinite conjugate. During testing the objective is illuminated by an expanded laser beam with a constant amplitude cross section over its clear aperture. The degree of collimation can be adjusted to accommodate a finite conjugate objective. The point image formed by the objective is projected onto a ground glass screen by a fairly symmetrical 100× microscope objective and a 16× eyepiece. The projected point image may be judged visually, scanned, or photographed. Point images affected by various aberrations are compiled in the *Atlas of Optical Phenomena*.[10]

Figure 6 shows typical aberrations in the point image. These photos were taken with type (A) objectives according to Table IV, with an NA of 0.4, a focal length of 8.3 mm, and with λ = 632 nm. Figure 6(a) shows the point image of a nearly symmetrical objective. Figure 6(b) shows a point image with astigmatism. Figure 6(c) shows the point image of an objective with coma as a predominant aberration. Figure 6(d), finally, shows a more pronounced coma error.

These aberrations are essentially caused by tilted elements in the objective, but also by errors of curvature and strain due to pressure on the elements. The tilt angles of the surface can be measured. Once these angles are known, the wave aberration for the corresponding objective can be computed. From the result, the rms of wave aberration and the corresponding intensity distribution in the point image can be calculated. The point images computed in this manner are shown in Fig. 7. Figure 7(a) shows the computed point image of the nearly symmetrical objective, Fig. 7(b) the point image for the objective with astigmatism,

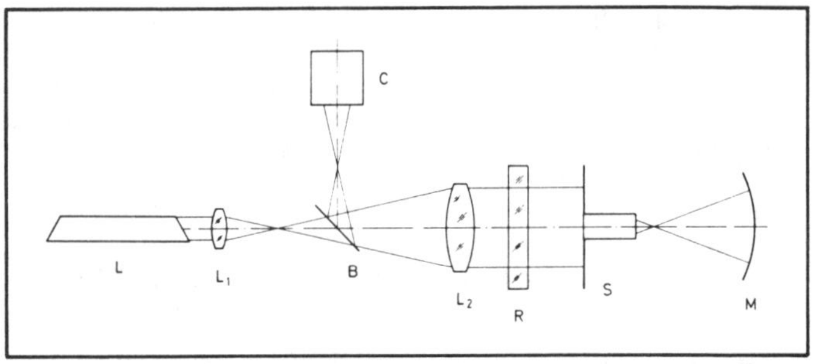

Fig. 5. Interferometer for wavefront measurement with automatic processing of interference fringes. L: laser, L_1; L_2: beam expander; B: beam splitter; R: planoparallel reference surface; S: objective under test; M: spherical mirror; C: TV viewing system.

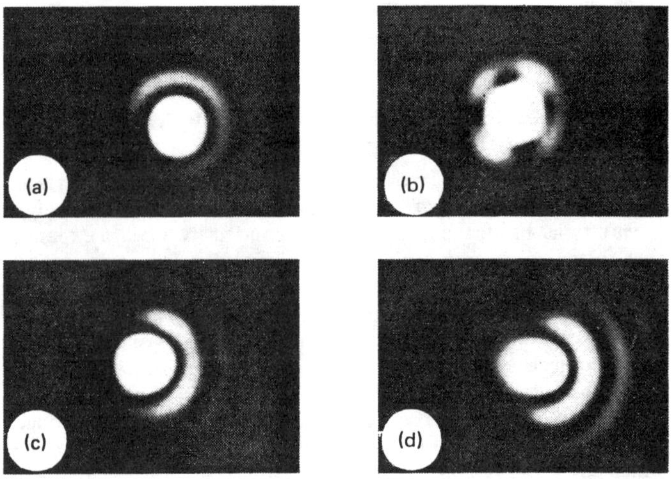

Fig. 6. Point image of objective of NA 0.40, focal length 8.3 mm, λ = 632 nm: 6(a) nearly symmetrical objective; 6(b) objective with astigmatism; 6(c) and 6(d) objective with coma.

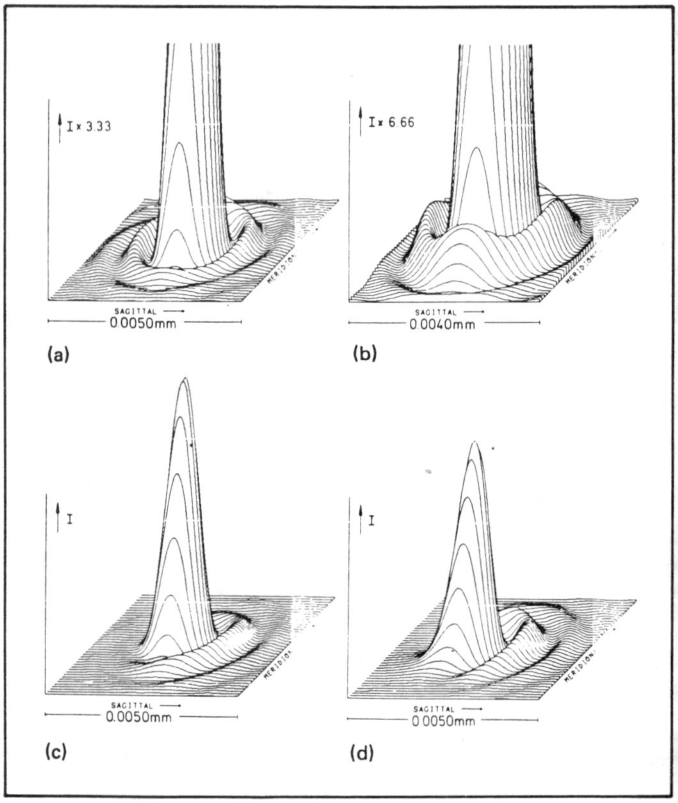

Fig. 7. Computed point images of objective of NA 0.40, focal length 8.3 mm, λ = 632 nm: 7(a) nearly symmetrical objective, ordinate scale exaggerated 3.33×; 7(b) objective with astigmatism, ordinate scale exaggerated 6.66×; 7(c) and 7(d) objective with coma, unexaggerated ordinate scale.

and Figs. 7(c) and 7(d) for the objectives with coma.

It should be mentioned that the wave aberrations and point images thus obtained only contain the effects of measured tilt angles. They do not contain the effect of curvature errors and lens strain. The point image of a nearly aberration-free objective as per Fig. 6(a) and the point images of the two objectives with coma as per Figs. 6(c) and 6(d) can be explained with the tilt angles found. The objective with astigmatism as per Fig. 6(b) not only has surface tilt angles but is affected also by additional strain and errors of curvature. Including measured tilt angles for the elements, the calculations gave the result that the first bright Airy ring shows a variation in illuminance by the factor 1.4 when changing the azimuth by 90°. However, direct measurement of this illuminance in the projected point image gave a three times difference. This means that about half of the actual astigmatism is caused by surface tilt angles in the objective, while the other half is caused by errors of curvature and strain.

The energy distribution in the point image can be measured by scanning with a pinhole. For precise results, a small pinhole is required. However, the smaller the pinhole the less is the energy to be detected. Therefore, very often in practice the scan is done with a small slit. Then, instead of the point image the fine-line image is registered. For a diffraction-limited lens the energy distribution in the point image I(x) is given by the Airy formula.[11]

In normalized form it can be written as

$$\frac{I(x)}{I(o)} = \left(\frac{2\,J_1(x)}{x}\right)^2 , \tag{7}$$

where

$$x = \frac{2\pi}{\lambda}\,NA\,y' , \tag{8}$$

J_1 is the Bessel function of first order, and y′ is the radial coordinate in the image plane.

In normalized form the energy distribution in the fine-line image L(x) is given by[12]

$$\frac{L(x)}{L(o)} = \frac{3\pi}{8x^2}\,H_1(2x) , \tag{9}$$

where H_1 is the Struve function of first order.[13]

The first dark ring in the Airy formula is given by Eq. (1); it encircles 84% of the total energy. The first diffraction minima in the fine-line image are found at

$$y' = \pm 1.16\,\frac{\lambda}{2NA} . \tag{10}$$

Thus, the central part of the fine-line image is concentrated in the distance

$$d = 1.16\,\frac{\lambda}{NA} \tag{11}$$

and contains 88.6% of the total energy. Keeping in mind these differences, for energy reasons the point image is allowed to be scanned with a small slit.

The point images scanned with a slit 1 μm wide are shown in Fig. 8. Figure 8(a) gives the resulting fine-line image of the nearly symmetrical objective. Figure 8(b) shows the fine-line images for azimuth angles of 0° and 45° for the objective with astigmatism. Figure 8(c) gives the fine-line image with low coma and Fig. 8(d) the one with pronounced coma. It can be seen from these that for the two objectives with coma an agreement better than 10% was obtained for measured and calcu-

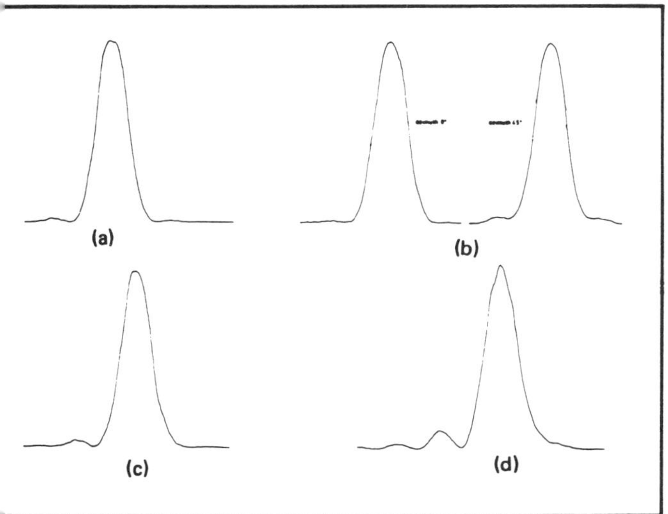

ig. 8. Measured fine-line images of an objective of NA 0.40, focal length .3 mm, λ = 632 nm. (a) Nearly symmetrical objective; (b) objective with stigmatism; (c) and (d) objective with coma.

ABLE V. rms of Wave Aberration for Objectives of NA 0.40, Focal ength 8.3 mm, λ = 632 nm (computed with allowance made for neasured surface tilt angles)

bjective	Major aberration	$\sqrt{\overline{w^2}}$
a	nearly diffraction-limited	~ λ/41
b	astigmatism	~{λ/40*, λ/28†}
c	coma	~ λ/20
d	coma	~ λ/10

with surface tilt errors only
with surface tilt errors and errors of curvature as well as strain
based on measurement observation)

'ABLE VI. Relationship between rms of Wave Aberration and Reduction of MTF

$\sqrt{\overline{w^2}}$	ΔMTF
λ/50	~−0.01
λ/20	~−0.05
λ/10	~−0.15

ited illuminance ratios between the first fringe maximum and the enter. Therefore, for these two objectives the surface tilt angles may e considered a major cause of the coma encountered. Table V gives e calculated rms of wave aberration for these objectives.

3. Measuring of MTF

he incoherent MTF is measured in a machine in which a fine-line nage is first scanned and the MTF then determined from this by ourier transforming.[12] This set-up is shown in Fig. 9.

In nearly diffraction-limited optical disk objectives, the admissi-le levels of image degradation result in the greatest reduction of ITF values in the medium frequency range, while the reduction is less at higher frequencies. This effect is slightly dependent on the type of aberration encountered. The approximate relationship between rms of wave aberration and a reduction of MTF at half-limit frequency is given in Table VI. The effect of various image errors on the MTF is given in Ref. 14. In particular, G. Black and E. F. Linfoot have studied the effect of spherical aberration[15] and M. De the effect of astigmatism.[16]

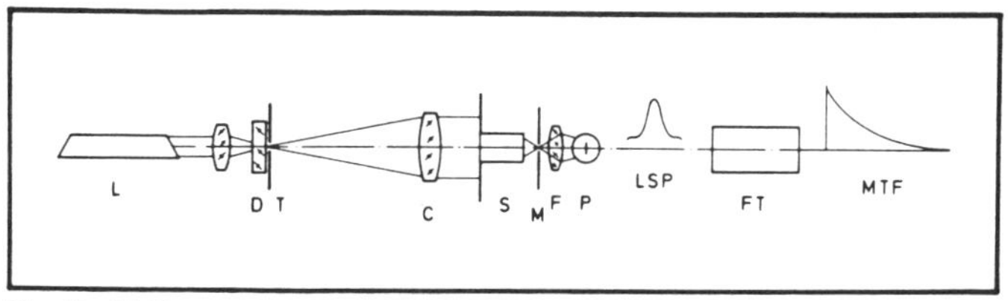

Fig. 9. MTF meter. L: HeNe laser; D: ground glass; T: test slit; C: collimator; S: test piece; M: scanning slit; F: field lens; P: photomultiplier; LSP: line-spread function; FT: Fourier transform; MTF: modulation transfer function.

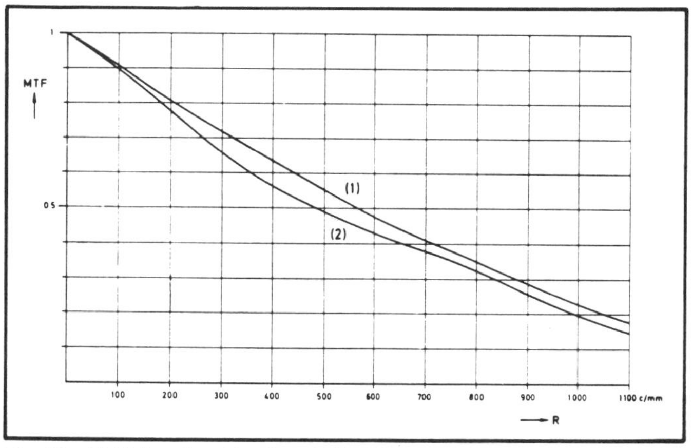

Fig. 10. MTF data for objective of NA 0.50, focal length 8.4 mm, λ = 632 nm: (1) MTF for azimuth 0°; (2) MTF for azimuth 120°.

Figure 10 shows the measured incoherent MTFs of a uniformly illuminated optical disk objective with an NA of 0.5 and a focal length of 8.4 mm at the wavelength of 632 nm as a function of spatial frequency R up to 1000 c/mm. The MTF measurements are given for the best and the worst azimuth. The variation of MTF points to an rms of wave aberration of about λ/20.

5. REFERENCES

1. M. Marshall, Electronics, 97 (May 1981).
2. R. Pepperl and I. Sander, Opt. Acta 24, 427(1977).
3. Electro Optical Systems Design, 6 (Jan. 1979).
4. H. Kogelnig, Bell Syst. Tech. J. 44, 455(1965).
5. J. D. Zock and T. C. Lee, Appl. Opt. 11, 2140(1972).
6. R. E. Hopkins and B. J. Buzawa, Opt. Eng. 15(2), 90(1976).
7. J. C. J. Finck, H. J. M. van der Laak, and J. T. Schrama, Philips Techn. Rundschau. 39, Nr. 4, 101(1981/82).
8. M. Born and E. Wolf, *Principles of Optics*, p. 440, Pergamon Press, London (1959).
9. P. Forman, Proc. SPIE 192, 41(1979).
10. M. Cagnet, M. Françon, and J. C. Thierr, *Atlas of Optical Phenomena*, p. 25, Springer Verlag, Berlin (1962).
11. M. Born and E. Wolf, *Principles of Optics*, p. 395, Pergamon Press, London (1959).
12. P. Kuttner, Appl. Opt. 7, 1029(1968).
13. E. Jahnke and F. Emde, *Tables of Functions*, p. 211, Dover Publ., New York (1945).
14. M. Born and E. Wolf, *Principles of Optics*, p. 485, Pergamon Press, London (1959).
15. G. Black and E. Linfoot, Proc. Roy. Soc. London, Ser. A 239, 522(1957).
16. M. De, Proc. Roy. Soc. London, Ser. A 233, 96(1955).

Reprinted with permission from *Applied Optics,* Vol. 25(19), pp. 3340-3344 (October 1, 1986). ©1986 Optical Society of America.

Gradient-index objective lens for the compact disk system

Hisami Nishi, Hiroyuki Ichikawa, Minoru Toyama and Ichiro Kitano

A radial gradient-index aplanatic singlet pickup objective lens for the compact disk (CD) system has been designed, manufactured, and evaluated. A plano-convex aplanatic singlet was designed by optimizing the refractive-index coefficients of the gradient-index medium, lens thickness, and one curvature. The necessary ion exchange technology and surface fabrication technology have been developed to successfully fabricate the objective lens. The gradient-index singlet was mounted in a CD player, and tests confirmed that the lens meets the system requirements.

I. Introduction

The objective lens for the compact disk (CD) system requires diffraction-limited performance to focus the laser beam into a small spot on the optical disk. The optical performance of commercial gradient-index rod lenses for copy machines is relatively low compared to CD system requirements. Ordinarily, a CD objective is composed of three homogeneous glass elements, but, recently, examples of molded aspheric singlet objective lenses have been reported.[1-4]

While aspheric lenses provide one alternative solution, studies on the high-performance capabilities of radial gradient-index lenses with spherical surfaces have also been investigated.[5-8] Now a plano–convex radial gradient singlet lens, meeting the design requirements of the CD objective lens, has been developed. In this paper, the lens design is discussed, and the results of the gradient-index CD objective lens fabrication are presented.

II. Design

A. Ray-tracing

For the design of diffraction-limited gradient-index (GI) systems, a lens design program which can trace real rays through the GI medium is indispensable. The general ray equation is written as

$$\frac{d}{dS}\left[n(\mathbf{r})\,\frac{d\mathbf{r}}{dS}\right] = \nabla n(\mathbf{r}), \tag{1}$$

where $\mathbf{r}$ is the position vector of a point on the ray, $n(\mathbf{r})$ is the refractive index, and S is the length along the ray. The radial GI distribution is expressed by

$$n^2(r) = n_0^2[1 - (gr)^2 + h_4(gr)^4 + h_6(gr)^6 + \ldots], \tag{2}$$

where $n(r)$ is the refractive index at the distance r from the optical axis, n_0 is the axial refractive index, g is the quadratic gradient-index constant, and h_4,h_6 are higher-order gradient-index coefficients. Paraxial ray-tracing can be easily done by solving the equation on condition that the higher-order coefficients (h_4,h_6,etc.) are negligible. The cardinal points of a lens, such as its focal and principal points, are thus obtained. Real rays can be traced by numerical evaluation of Eq. (1) with the refractive-index distribution given by Eq. (2) using the Runge-Kutta method.[9]

A lens design program for both real and paraxial ray tracing in GI media was developed for a minicomputer. In this program, damped least-squares (DLS) optimization is available.

B. CD Objective Lens

The objective lens system is illustrated in Fig. 1. The general lens specifications require a numerical aperture (N.A.) of 0.45 and a working distance (WD) of 1.5–2.5 mm. Furthermore, the lens must be diffraction-limited, and the rms wavefront error (OPD rms) must not exceed the Marechal criterion, that is, OPD rms $\leq$ 0.07 λ rms over an image field of 0.075-mm radius. Therefore, the lens must be corrected for both spherical aberration (SA) and coma.

Before optimizing the GI design, the following design restrictions were imposed:

The authors are with Nippon Sheet Glass Company, Central Research Laboratory, 1, Kaidoshita, Konoike, Itami City, Hyogo Prefecture 664, Japan.

Received 28 January 1986.
0003-6935/86/193340-05$02.00/0.

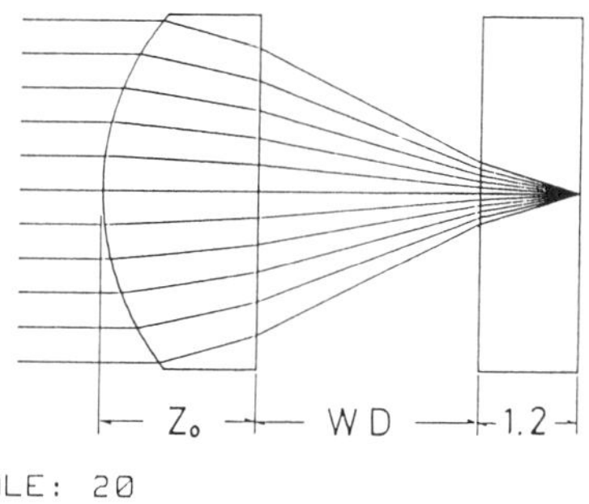

Fig. 1. Plano–convex radial GI lens for the CD system (type B).

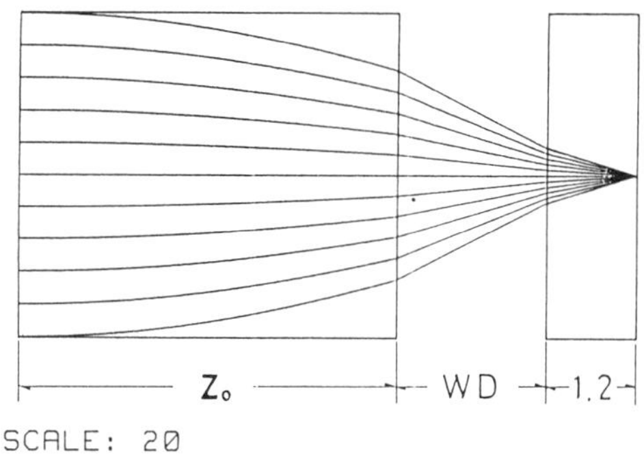

Fig. 2. Plano–plano radial GI lens for the CD system (type A).

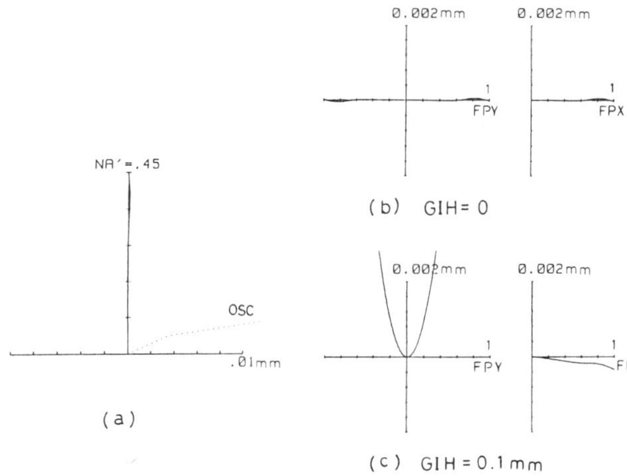

Fig. 3. Type A lens: (a) longitudinal spherical aberration and (b), (c) ray intercept vs normalized entrance pupil coordinate.

Table I. Optimized Design of Type B Lens[a]

$R_1 =$ 3.254	$R_2 =$ ∞
$Z_0 =$ 1.884	$n_0 =$ 1.5800
$g =$ 0.1273	$h_4 =$ −0.863
$h_6 =$ −0.653	$h_8 =$ −66.9
EFL = 4.53	N.A. = 0.45
WD = 2.67	
$n_{disk} =$ 1.57	$t_{disk} =$ 1.2

[a] Lens units: millimeters.

(1) no more than one spherical surface;
(2) a radial GI distribution that is manufacturable;
(3) infinite conjugate system; and
(4) entrance pupil located at the first lens surface.

In the following subsections, a plano–plano GI lens and a plano–convex GI lens are considered. Design and tolerance results are given for the optimized radial GI objective lens.

C. Plano–Plano GI Lens (Type A Lens)

The plano–plano-type lens shown in Fig. 2 has parallel plano surfaces and is the simplest singlet. At the first stage of design, the feasibility of this lens type was considered. The SA correction is easily accomplished by optimizing the higher-order refractive-index coefficients h_4,h_6,h_8,etc. For coma correction, there is no effective degree of freedom. During optimization, no aplanatic solution was obtained even though all the lens parameters were used as variables. Figure 3 shows one result of calculations where the effective focal length (EFL) is 4.5 mm. The coma makes its field of view too small for use in the CD objective lens.

D. Plano–Convex GI Lens (Type B Lens)

To add more degrees of freedom, one spherical refracting surface was allowed with the radial GI medium. Earlier researchers had considered GI objective lenses with two spherical surfaces,[7,8] but this increases the surface polishing costs. For the plano–convex-type lens, there are two possible configurations: one configuration has the spherical surface on the infinite conjugate side, and the other is the reverse.[6] Aplanatic solutions can be obtained in either configuration, but the former type is preferred because it has a shorter lens thickness and a longer WD. The optimized lens parameters for a typical GI objective lens are given in Table I. In this design, the major characteristic is that the spherical surface contributes more than 80% of the total lens power. Figure 4 shows the longitudinal SA and ray intercept curves at two image heights. During the design it was found that steeper curvatures introduce higher orders of spherical aberration. Higher-order SA could then be corrected by optimizing the refractive-index coefficients. For coma correction, the type B lens has one surface power contribution and the medium's power contribution to the overall lens power so that the lens has a degree of freedom equivalent to homogeneous lens bending. In the merit function, ray displacement, offense against the sine condition (OSC), and rms wavefront error (OPD rms) were used. In Fig. 4(a), the OSC curve of type B lens (dashed line) shows satisfactory correction of OSC compared with the result of the type A lens (Fig. 3).

Zernike polynominal fitting to the wavefront error was carried out to understand the limiting characteristics of the type B design. In Table II, the Zernike coefficients of the type B lens at a Gaussian image height (GIH) = 0.1 mm are listed. The SA terms (*W*40, *W*60, and *W*80) and the third-order coma term (*W*31) are well corrected. Third-order astigmatism (*W*22) contributes 94% of the wavefront error. From this analysis, the type B lens is shown to be aplanatic as required for the CD system.

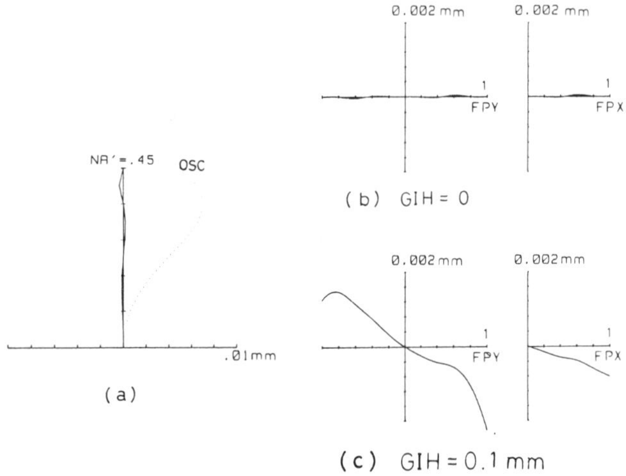

Fig. 4. Type B lens: (a) longitudinal spherical aberration and (b), (c) ray intercept vs normalized entrance pupil coordinate.

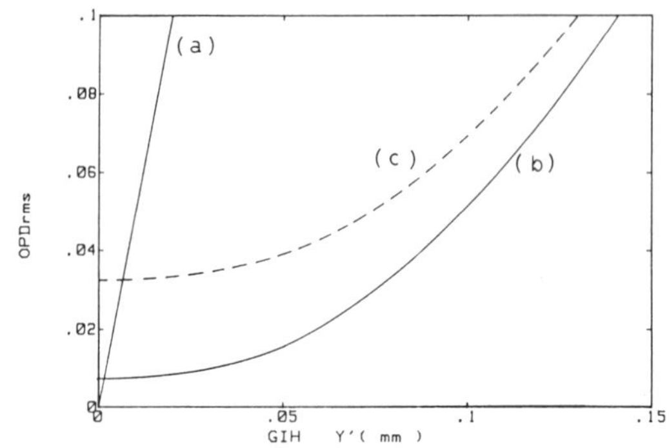

Fig. 5. Wavefront error: (a) type A lens: (b) type B lens; and (c) homogeneous triplet.

Table II. Zernike Coefficients of Plano–Convex Radial GI Singlet (Type B) at GIH = 0.1 mm[a]

*W*nm	Aberration	Coefficients
*W*40	SA3	0.01618
*W*60	SA5	0.02964
*W*80	SA7	0.00983
*W*31	CM3	0.00000
*W*51	CM5	0.02480
*W*22	AS3	0.11856

[a] SAn = *n*th-order spherical aberration.
CMn = *n*th-order coma.
ASn = *n*th order astigmatism.

Table III. Tolerance of Refractive-Index Coefficients Expressed by the Index Change at the Periphery[a]

Lens type	Plano–plano (type A)	Plano–convex (type B)
Δ	6.5%	3.6%
(Δn_0)	±0.0234	±0.0043
$\Delta nr\ (\Delta g)$	±0.0036	±0.0006
(Δh_4)	±0.0001	±0.0007

[a] $\Delta = (n_0 - nr)/n_0 \times 100$ (%);
Δ = relative refractive-index difference;
n_0 = refractive index at the center;
nr = refractive index at the periphery;
$\Delta nr = nr(n_0,g,h_4,\ldots) - nr(n_0 - \Delta n_0,g,h_4,\ldots)$ etc.

The relationship between (GIH) and OPD rms is plotted in Fig. 5 where the type B GI objective is compared with the type A objective and a homogeneous triplet lens. In a type A lens the field of view is limited by third-order coma. On the other hand, the type B lens has a wider image field than the homogeneous triplet. The type B objective satisfies all the required specifications of the CD objective lens.

E. Tolerance

Tolerancing of radial GI lenses must include tolerancing of the index profile and the tolerancing of the lens surface tilts with respect to the gradient axis. Realization of a designed lens depends on allowing a reasonable tolerance for the refractive-index profile. In Table III, tolerances of the refractive-index profiles are compared for type A and type B lenses. All values were converted from GI coefficient tolerances to refractive-index change tolerances for ease of comparison. Tolerance limits are defined as the change from the design value that increases the OPD rms up to 0.03 waves. From these results, the h_4 tolerance of the type A lens is the most severe and only one-seventh of the type B lens tolerance. At the current level of ion exchange technology, the desired value of Δ and the refractive-index tolerances of the type B lens are consistently obtained. On the contrary, the tolerance on surface tilts is very severe, because the surfaces must

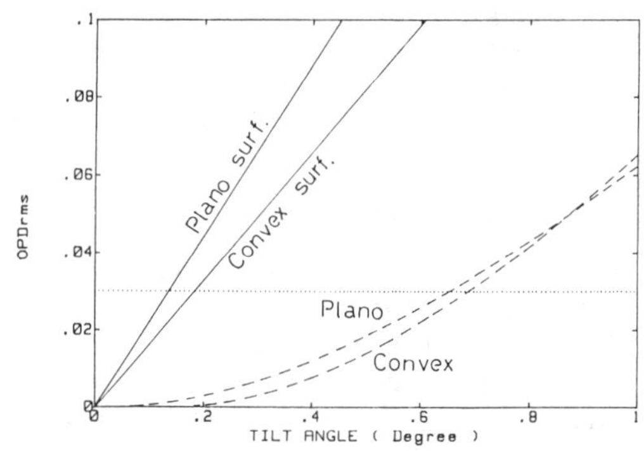

Fig. 6. Tolerance of surface tilt. Solid curve shows before recentering, and dashed curve shows after recentering.

be generated in alignment with the gradient axis of symmetry. The relationship between a single surface tilt and the resulting OPD rms increase is plotted in Fig. 6. From these results, the tolerance of surface tilt should be within 11 and 8 min of arc for the spherical and plano surfaces, respectively. To achieve these alignment requirements, a special grinding technology was developed. Additionally, it is possible to relax the tilt tolerances by adjusting the tilt of the total lens during mounting. The dashed lines in Fig. 6 represent the OPD rms values after this mounting adjustment. By this adjustment, tolerance of the surface tilt is enlarged, but adjustment for each lens is needed.

Table IV. Specifications of Plano-Convex Radial GI Singlet (Type B)

Wavelength	780 nm
EFL	3.56 mm
N.A.	0.45
WD	1.85 mm
Conjugate length	Infinity
GIH max	0.075 mm
OPD rms	0.07 Waves
Weight	0.05 g
Dimensions ($D \times Z$)	4×1.6 mm

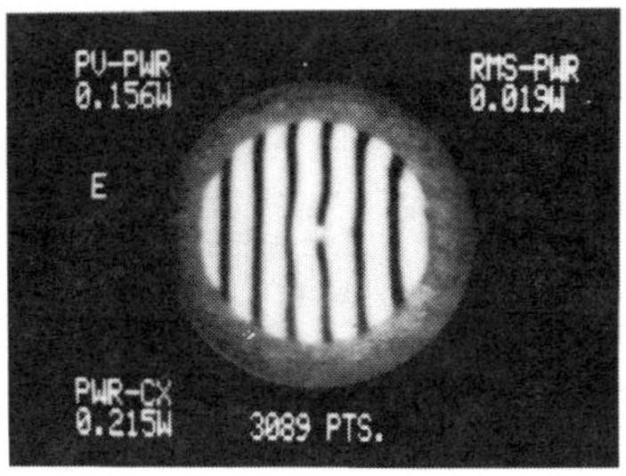

(a) GIH=0 OPDrms=0.019 waves

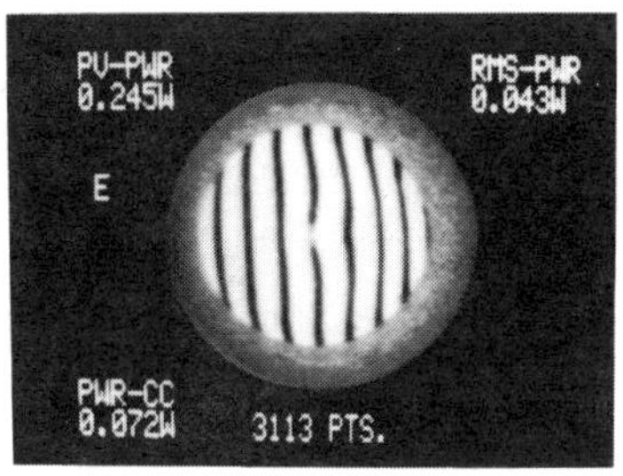

(b) GIH=0.07mm OPDrms=0.043 waves

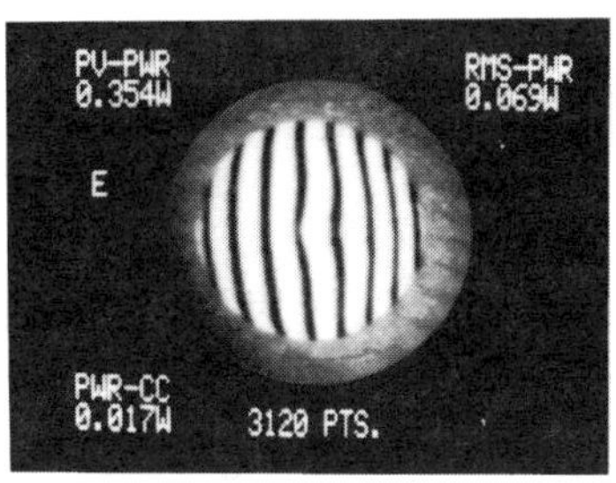

(c) GIH=0.10mm OPDrms=0.069 waves

Fig. 7. Wavefront interferograms by Fizeau interferometry.

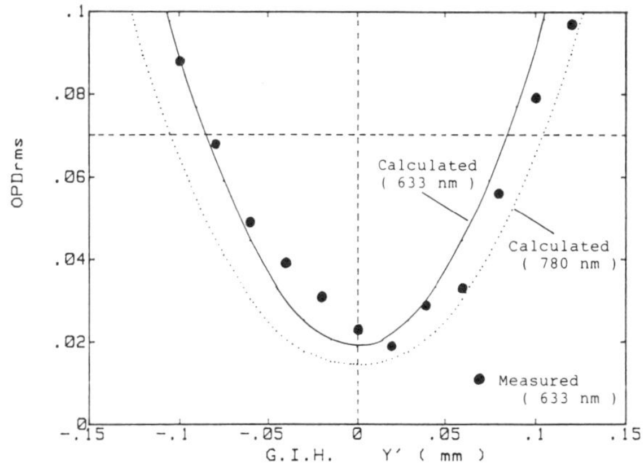

Fig. 8. Comparison between calculated and measured values of wavefront error.

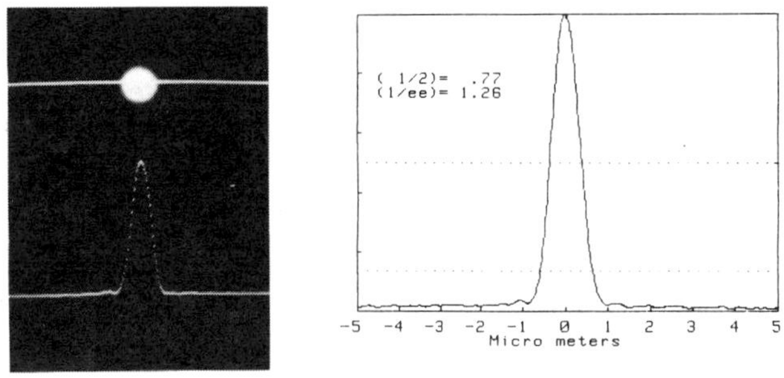

(a) (b)

Fig. 9. Point image and its intensity profile at λ = 633 nm: (a) point image and (b) normalized intensity profile.

III. Fabrication and Evaluation

A. Fabrication

The necessary technologies to fabricate successfully the type B GI objective lens have been developed. As the starting material, a water and chemical resistant multicomponent glass was developed for a modified ion exchange technique. By using a large 4-mm diam glass rod, the ion exchange time is long enough to precisely control the ion exchange conditions and repeatedly obtain the desired refractive-index profile. The spherical lens surface is generated on a modified curve-generating machine specifically designed to suppress surface tilts. As a result, diffraction-limited radial gradient-index objective lenses are now routinely fabricated to meet the CD system specifications given in Table IV.

B. Evaluation

First, a ZYGO MK-III Fizeau-type interferometer was used to test the transmitted wavefront errors of the GI objective lens. In Fig. 7, typical on-axis and off-axis interferograms show the diffraction-limited performance of the lens, OPD rms $\leq 0.07\ \lambda$ rms. In Fig. 8, calculated OPD rms values and the measured values are shown for several image heights. In these curves, solid and dashed curves correspond to the calculated values at 633- and 780-nm wavelengths, respectively. The calculated and measured OPD rms values are in good agreement. Note that the design was optimized for use at the 780-nm wavelength, although testing was done at 633 nm. Figure 9 shows the on-axis point image and its intensity profile at 633 nm. The intensity profile is symmetric, and the $1/e^2$ spot size is 1.26 μm in diameter. The image spot is symmetric and nearly diffraction-limited.

As a practical test, a GI objective lens was mounted into a commercially available CD player. A clean electrical signal from the CD optical pickup head was obtained, and the resulting audio output was of high quality.

IV. Conclusion

A gradient-index objective lens for the compact disk system has been developed. By optimizing the lens configuration, aplanatic solutions were obtained while maintaining the plano–convex shape. Choosing the design configuration with the spherical surface toward the infinite conjugate made it possible to enlarge the refractive-index tolerance and reduce the lens thickness, thereby reducing the lens weight. The necessary ion exchange technology and surface fabrication technology have also been developed to fabricate successfully the objective lens. The major advantages of the gradient-index glass CD objective lenses are its good thermal and chemical durability, mechanical stability, and light weight (0.05 g). These advantages make the lenses appropriate for use in compact disk systems that require high performance over a wide range of environmental conditions such as in a car audio player.

We would like to acknowledge N. Araki and P. McLaughlin for their kind support and advice and A. Akazawa and M. Kobayashi for making the gradient-index rods by the ion exchange technique.

References

1. K. Shintani and S. Kubota, "Evaluation of a Diffraction-Limited Plastic Biaspheric Objective Lens," in *Technical Digest, Conference on Lasers and Electro-Optics* (Optical Society of America, Washingon, DC, 1984), paper WB2.
2. T. Kiriki, N. Izumiya, K. Sakurai, and T. Kojima, "Plastic Aspheric Lens for the Compact Disc System," in *Technical Digest, Conference on Lasers and Electro-Optics* (Optical Society of America, Washington, DC, 1984), paper WB3.
3. R. O. Maschmeyer, R. M. Hujar, L. L. Carpenter, B. W. Nicholson, and E. F. Vozenilek, "Optical Performance of a Diffraction-Limited Molded-Glass Biaspheric Lens," Appl. Opt. **22,** 2413 (1983).
4. J. J. M. Braat, A. Smid, and M. M. B. Wignakker, "Design and Production Technology of Replicated Aspheric Objective Lenses for Optical Disk Systems," Appl. Opt. **24,** 1853 (1985).
5. K. Fujii, H. Ueno, and T. Yamagishi, "New Gradient-Index Rod Lenses with Low Aberration," in *Technical Digest, Topical Meeting on Gradient-Index Optical Imaging Systems*, Kobe, Japan (1983), paper E4.
6. H. Nishi, M. Toyama, T. Yamagishi, K. Fujii, and H. Ueno, "Plano–Convex Gradient-Index Singlet," in *Technical Digest, Topical Meeting on Gradient-Index Optical Imaging Systems*, Kobe, Japan (1983), paper C2.
7. Y. Aono, "Optical Design of Pick-Up Objectives with Gradient Index Elements," Micro Opt. News **1,** 126 (1984).
8. K. Kikuchi, S. Ishihara, H. Shimizu, and J. Shimada, "Design of Gradient-Index Spherical Lenses for Optical Pickup System," Appl. Opt. **19,** 1076 (1980).
9. A. Sharma, D. V. Kumar, and A. K. Ghatak, "Tracing Rays Through Gradient-Index Media: a New Method," Appl. Opt. **21,** 984 (1982).
10. M. Toyama, H. Nishi, and I. Kitano, "Gradient Index Objective for the Compact Disk," in *Technical Digest, Topical Meeting on Gradient-Index Optical Imaging Systems*, Palermo, Italy (1985), paper F1.
11. H. Nishi and M. Toyama, "Design of an Aplanatic GRIN Singlet for the Compact Disk System," in *Technical Digest International, Lens Design Conference* (1985), paper TuB5.

Reprinted from *Current Developments in Optical Engineering and Diffraction Phenomena,* Proc. SPIE Vol. 679, pp. 49-56 (1986).

Optical performance of molded-glass lenses for optical memory applications

Richard O. Maschmeyer

Research and Development Division, Corning Glass Works
SP-PR-10, Corning, New York 14831

Abstract

Optical performance results are reported for four different molded-glass lenses. All four lenses are intended for use as focus objectives in optical memory applications, including Compact Disc, Laser Disc, and optical storage pickups. All lenses have measured performance below 0.04 waves RMS on-axis at 0.780 micron. The lenses differ in the different performance features they emphasize. These features include excellent off-axis performance, longer working distance, a finite conjugate, and a higher numerical aperture.

Introduction

Directly molding optical elements without subsequent finishing operations has numerous potential advantages. Perhaps the most significant advantage is the capability of molding aspheric surfaces. For example, properly designed aspheric lenses can minimize spherical aberration and off-axis coma whose correction with conventional spherical optics would require the use of additional elements. It is not unusual for an aspheric element to be capable of performance equivalent to three or more spherical elements in an appropriate design. Reducing the number of elements can lead to simplified mounting, lower mass, lower light losses, more compact designs, and lower costs. These advantages are well realized in molded-glass lenses for focus objectives for optical pickups.

This article reports the design parameters and measured optical performance of four molded-glass biaspheric lenses intended for use as focus objectives in optical memory applications, including Compact Disc, Laser Disc, and optical storage players. The lenses are similar to one reported in a previous publication[1] but have better optical performance. All four lenses are single element, biaspheric designs intended to operate at 0.780 microns wavelength. All four lenses have on-axis optical performance less than 0.04 waves RMS. The lenses differ, however, in that they emphasize different performance features. The first design has excellent off-axis performance. The second design has a longer working distance at the expense of off-axis performance. The third design has a finite conjugate which incorporates the function of both focus objective and the laser collimator into a single element. The fourth design has a higher numerical aperture of 0.50 suitable for Laser Disc or optical storage pickups. It is clear that one of the advantages of the molded-glass-lens process is that lens designs can be tailored to match the needs of specific applications.

Lens designs

The molded-glass lenses described in this paper are all biaspheric singlets. The lenses are shown schematically in Figure 1. Representative samples of the four lenses are shown in Figure 2. Note that they are similar in basic shape but differ in diameter, mass, thickness, the size of mounting reference surfaces, and the details of the optical surfaces. The optical design and performance parameters of the lenses are listed in Table 1. The optical designs are well optimized as can be seen from their "as-designed" image quality. The designs are also well optimized for the fabrication tolerances of the molding process. Fabrication tolerances for these designs are similar to those discussed in a previous publication.[2] The expected optical performance of the lenses was calculated using ray tracing/tolerancing techniques of Optical Research Associates' Code V adapted to the peculiarities of molded-glass aspheric lenses. As noted previously[1], the predictive capability of these design techniques is excellent. Because the glass-molding process allows excellent relative replication of the mold[2], no process iterations are required to achieve the necessary level of surface figure accuracy. Thus, the time to develop new lens designs is very short.

The four lenses described in this article differ in that they each emphasize different performance features. Each feature might be of importance in a different pickup design. For example, the expected optical performance of lens 350010 across the field of view is exceptionally constant, changing only 0.001 wave RMS at an image height of 0.075 millimeters. Such constant performance might be an advantage because pickup focus objectives spend most of their time working off-axis. The penalty for such uniform performance in this lens is a shorter working distance of 1.45 mm. Lens 350022 has a longer working distance of 1.85 mm but at the expense of less constant off-axis optical performance. Lens

Table 1. Lens Design and Performance Parameters

Optical design specifications

Parameter	Lens: 350010	350022	350030	350040
Number of elements	1	1	1	1
Wavelength (nm)	790	780	780	780
Numerical Aperture	0.47	0.47	0.45/0.09	0.50
Entrance Pupil Diameter (mm)	4.2	4.2	4.45	4.50
Effective Focal Length (mm)	4.47	4.47	4.1	4.5
Working Distance (mm)	1.51	1.85	2.57	1.87
Magnification	Infinite	Infinite	5	Infinite
Disc Thickness (mm)	1.2	1.2	1.2	1.25
Disc Refractive Index	1.55	1.55	1.55	1.49
As-designed RMS Image Quality (waves) 0 microns image height	≤0.003	≤0.002	≤0.003	≤0.004

Fabricated performance specifications

Parameter	Lens: 350010	350022	350030	350040
As-fabricated RMS Image Quality (waves) 0 microns image height	<0.030	<0.030	<0.035	<0.040
75 microns image height	≤0.031	≤0.050	≤0.050	≤0.050
Birefringence (nanometers)	<2.	<2.	<2.	<2.
Mass (gm)	0.54	0.24	0.24	0.32
Axial Thickness (mm)	4.5	3.4	3.0	3.37
Physical Diameter (mm)	7.2	5.4	6.33	6.7

350030 is a finite conjugate design. It incorporates the function of both the focus objective and the laser collimator into a single element, offering the advantage of fewer components in the pickup. Lens 350040 has a higher numerical aperture of 0.50 for Laser Disc or optical storage application.

Lens measurements

A series of lenses of the four designs described above were molded from a noncommercial specialty optical glass melted by Corning Glass Works. This glass, coded C 05-50, has a He d-line refractive index of 1.605 and an Abbe number of 50.4.

Optical performance of the molded lenses was measured with two different heterodyne interferometers. The interferometers were a Spectra-Physics DIAD model 550 in a Twyman-Green configuration and a Zygo Mark III in a Fizeau configuration. Both instruments operate at a wavelength of 0.633 microns. Both instruments allow an absolute phase measurement of the transmitted wavefront and digital reduction and analysis of data, including Zernike analysis. The Zernike analysis was performed both by commercially-available software called WISP and by computer programs written at Corning. Both instruments and both computer programs gave comparable results.

Figure 3 shows a schematic of a typical test configuration for on-axis testing. Off-axis testing was accomplished by displacing the lens a known distance. The glass plate is the optical equivalent of the transparent portion of the memory disc. Its presence in the optical path is compensated for in the optical design of the lenses. The design of lens 350030, the finite-conjugate design, also compensates for some optics in the long conjugate

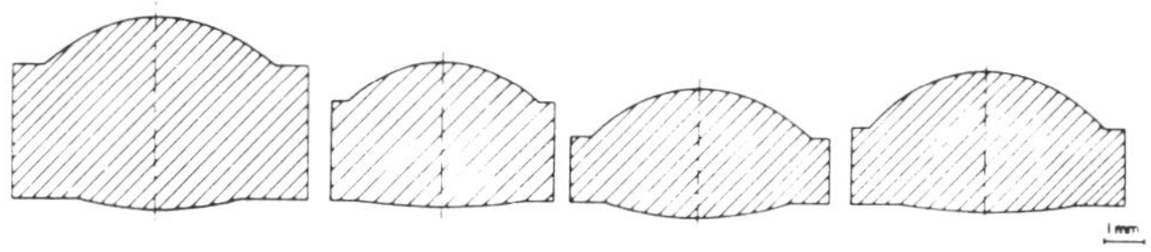

Figure 1. Schematic shapes of the four designs in cross-section. From left to right, the designs are 350010, 350022, 350030, and 350040.

Figure 2. Photographs of lenses of the four designs. From left to right, the lenses are designs 350010, 350022, 350030, and 350040.

side. The test configuration for this lens requires a spherical mirror in place of the retro-reflecting flat. Alignment of the optical components is critical to accurate measurements. In some designs, the conjugate length and/or plate thickness under test must be adjusted to compensate for spherochromatism arising from not testing the lens at its design wavelength. Under these adjusted test conditions, data measured at 0.633 microns may be converted to 0.780 microns by multiplying by the ratio of the wavelengths.

Optical performance results

The measured RMS wavefront error of the molded lenses typically ranged from 0.01 to 0.03 waves at the application wavelength of 0.780 microns. Since lens designs 350022, 350030, and 350040 appear to be of greater commercial interest to the pickup industry, more data will be presented on their performance. Figure 4 shows the distributions of RMS wavefront error for one hundred samples of lenses 350022, 350030, and 350040.

Wavefront maps for representative lenses are shown in Figure 5. The numeric information in these figures is generally self-explanatory: "PV-PWR" is the peak-to-valley difference of the map; "RMS-PWR" is the RMS average of the map, and "PTS" are the number of data points in the measurement. Note that the data in these maps are reported at the measurement wavelength of 0.633 microns. It is clear from the maps that lens 350022 has more high order spherical aberration associated with mold fabrication.

Although RMS wavefront error is a convenient single number for evaluating focus objective performance, it discards much useful data. In pickup applications, for example, the size and shape of the focused spot of the objective lens is critical to performance. Figure 6 shows the point spread functions of the lenses calculated from the wavefront data in Figure 5. These calculations were performed by the commercially-available program WISP. The

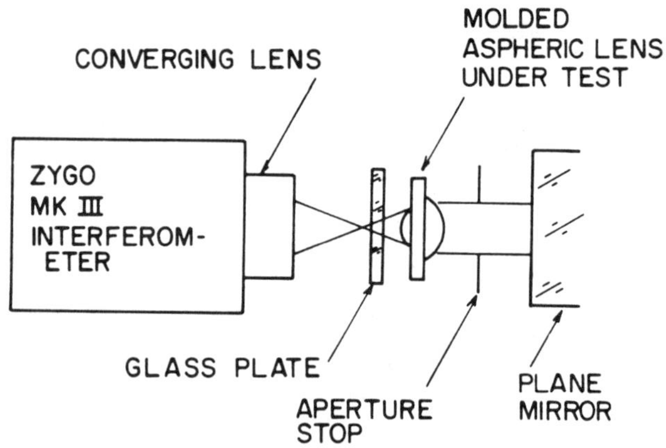

Figure 3. Schematic of the interferometer configuration used to measure the molded lenses.

Strehl number is the ratio of the light intensity at the peak of the diffraction pattern of the actual lens to that at the peak of a perfect lens. Strehl numbers of 0.96, 0.97, and 0.98 indicate a very high concentration of energy in the central spot. Note also the excellent rotational symmetry of the spots in these figures.

Another useful characterization of focus-objective performance is the distribution of aberrations in the wavefront. For the representative lenses in Figures 7, 8, and 9, eighth order Zernike analysis of the data performed by WISP shows the components in Table 2. The large residual in lens 350022 appears to be high order spherical aberration.

Table 2. Aberration distributions in typical lenses.

	Waves RMS (0.780 microns)		
Component	Lens 350022	Lens 350030	Lens 350040
Spherical	0.006	0.013	0.007
Coma	0.009	0.008	0.009
Astigmatism	0.004	0.010	0.006
Residual	0.023	0.011	0.016

Other important lens properties

Operating performance of the lens with changes in temperature are important in pickup applications. The thermal expansion of the glass used to mold the lenses is 1.6×10^{-5} C^{-1}. Its change in refractive index with temperature is -1.1×10^{-5} C^{-1}. The opposite sign of these coefficients tend to compensate each other, making the wavefront error of the lens at best focus relatively insensitive to temperature. The finite conjugate design, lens 350030, is a good example, as its performance is particularly sensitive to temperature. Its RMS wavefront error at best focus changes only 0.001 waves over the range of -5 C to +55 C. The focal lengths of all the lenses change significantly with temperature, but this change is unimportant in an autofocus application such as a pickup.

All four lens designs have flats at the edge of the optical area which serve as mounting and/or alignment surfaces for either optical or mechanical alignment. Because the flat is created in the molding process, it provides a very accurate reference for the optical axis of the lens. Furthermore, the monolithic, all-glass design makes the lens insensitive to thermal cycling and mechanical vibration.

The glass composition has weathering characteristics which are comparable to commercial optical glasses such as Corning BCD C 20-60 or Schott SK16. Conventional magnesium fluoride coatings improve durability and yield transmission in excess of 95%. Multilayer coatings may be used where the application requires higher transmission. The coatings pass standard specifications for abrasion and adhesion. Static electricity is not traditionally a problem with glass optics.

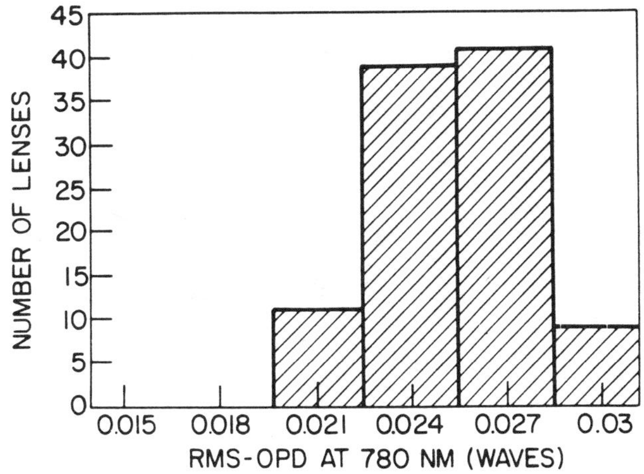

Lens 350022

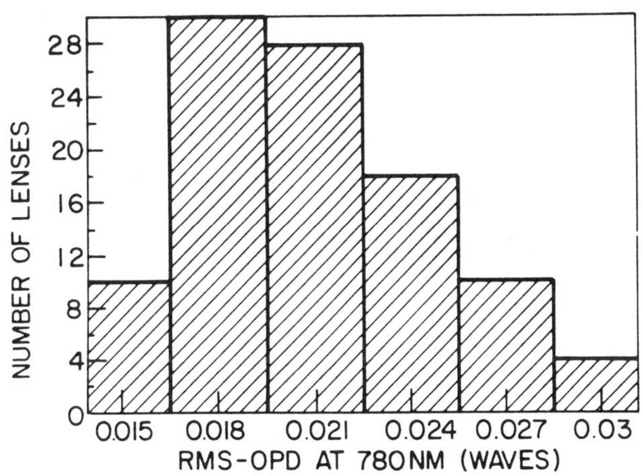

Lens 350030

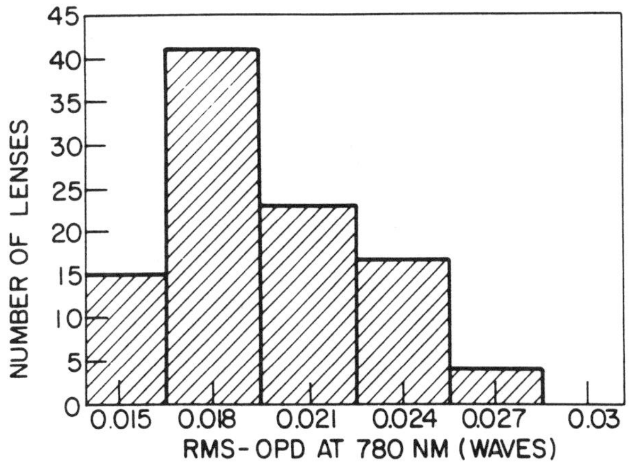

Lens 350040

Figure 4. Distribution of RMS wavefront errors for one hundred samples of lenses 350022, 350030, and 350040.

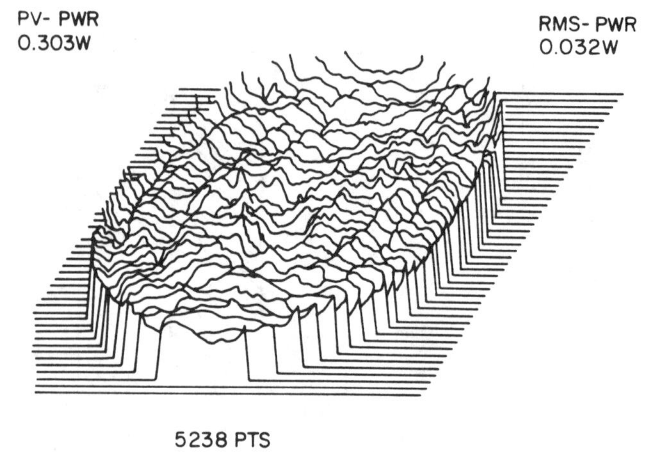

Lens 350022

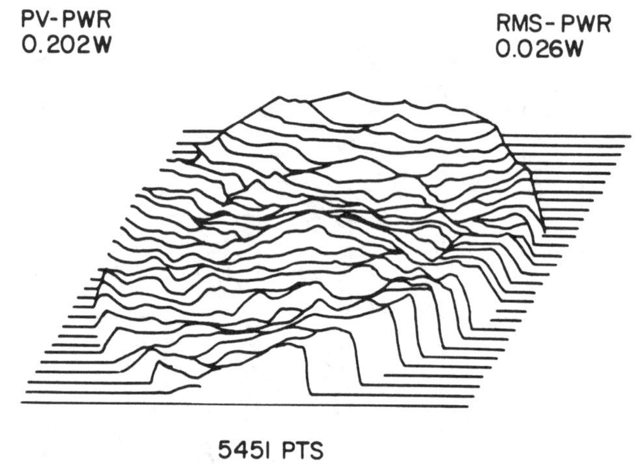

Lens 350030

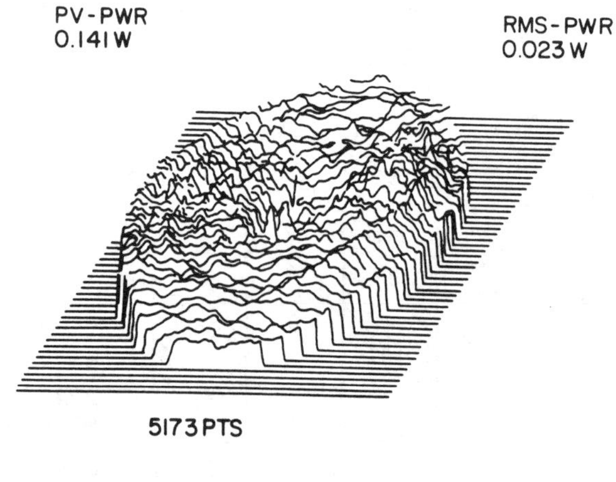

Lens 350040

Figure 5. Wavefront maps of typical lenses 350022, 350030, and 350040. Note that the P-P and RMS wavefront error are reported in this map at the measurement wavelength of 0.633 microns. Performance at 0.780 microns would be 0.026, 0.021, and 0.020 waves RMS respectively.

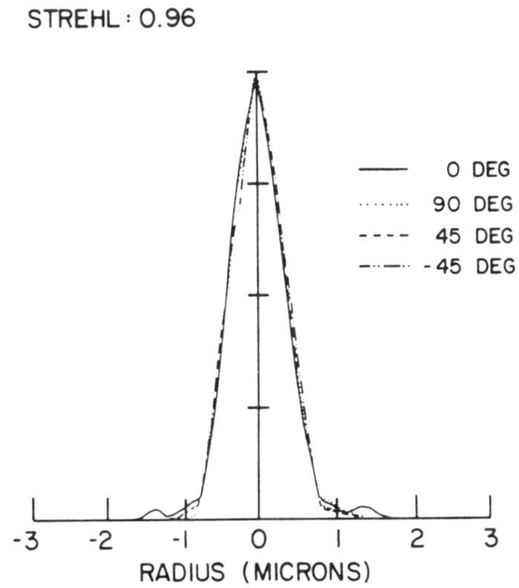

Lens 350022

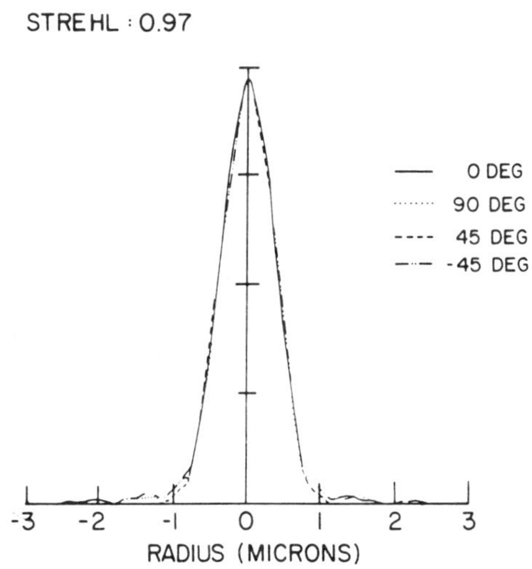

Lens 350030

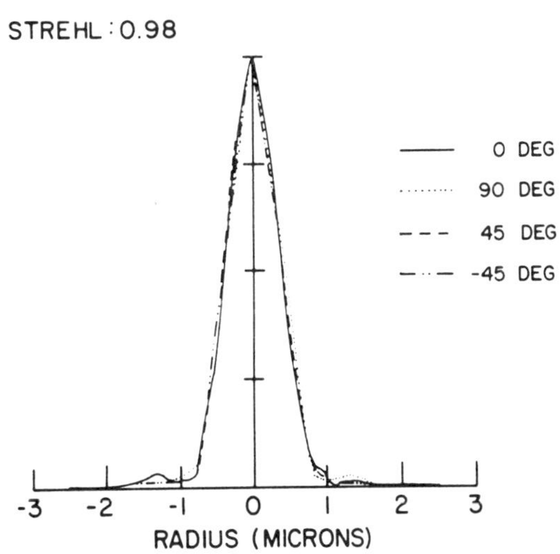

Lens 350040

Figure 6. Point spread functions for lenses 350022, 350030, and 350040 calculated from the wavefront data in Figure 5.

Summary

Design specifications and optical performance measurements of four single-element, high-precision, molded-glass biaspheric lenses are discussed. All four lens designs are intended for use as focus objectives for optical memory applications, including Compact Disc, Laser Disc, and optical storage pickups. All four operate at 780 nm and have wavefront errors below 0.040 waves RMS on-axis. All designs have fabrication tolerances similar to those previously published.[2] The designs differ in that they emphasize different performance features, such as good off-axis performance, longer working distance, a finite conjugate, and a higher numerical aperture. Transmitted wavefront aberrations of molded lenses were measured using heterodyne interferometers. Zernike and point spread function analysis are also presented. The monolithic structure and the thermal properties of the glass make the lenses relatively insensitive to temperature change.

Acknowledgment

The author would like to thank the following people who contributed to this work: R. L. Childs, W. R. Christoff, G. M. Cunningham, M. A. Fitch, J. M. Grottenthaler, G. Guillevic, R. J. Hagerty, R. M. Hujar, P. I. Kingsbury, D. Leroy, P. J. Majestic, D. G. Pellman, D. J. Pucci, K. R. Salisbury, H. G. Shafer, M. E. Vance, and T. W. Webb.

References

1. R. O. Maschmeyer, R. M. Hujar, L. L. Carpenter, B. W. Nicholson, and E. F. Vozenilik, Appl. Opt. 22, 2413 (1983).

2. R. O. Maschmeyer, C. A. Andrysick, T. W. Geyer, H. E. Meissner, C. J. Parker, and L. M. Sanford, Appl. Opt. 22, 2410 (1983).

Reprinted with permission from *Japanese Journal of Applied Physics*, Vol. 26, Suppl. 26-4, pp. 121-126 (1987). ©1987 Japanese Journal of Applied Physics.

Lens and Optics for Optical Disk System

YASUHIRO TANAKA, Yoshitomi NAGAOKA and Masaaki UEDA

Wireless Research Laboratory, Matsushita Electric Industrial Co., Ltd.
1006, Kadoma-shi, Osaka 571, Japan

Various types of pick-up lenses and optical systems for pick-ups are discussed. We evaluated optical performances of objective lenses. Aspherical lenses for CD player pick-ups have wider field angle compared to the conventional triplet objective lens. We also evaluated reliability of both plastic and glass molded lenses which are designed in the same way. Plastic lens caused spherical aberration due to temperature change. Optical pick-ups for CD players are remarkably simplified by means of aspherical lens or holographic element. On the other hand, pick-ups for write once or erasable optical disk system consist of many optical components. For these pick-ups, we developed a single bi-aspheric molded glass lens which has high NA (0.55) and remarkable light weight (85mg).

1. INTRODUCTION

Optical disk technology today finds widespread application in compact disk, write once and erasable optical disk system. More than 8 million CD (Compact Disk) players were produced in 1986. Owing to their high memory density, DRAW (direct read after write) disks and erasable optical disks are expected to expand their market. The optical pick-up is a key device in the optical disk systems. For these optical pick-ups, high accuracy optical components are required; aspherical lenses and other optical components have been proposed to meet these requirements. In this paper, optical systems for optical pick-ups and the current state of optical lenses, as well as a molded glass lens we developed for use in optical disk systems, will be discussed.

2. OPTICAL PICK-UP LENS

Among the various components constituting an optical pick-up, an objective lens is the most important, in terms of both technology and cost. Various types of objective lenses have been developed to date. There are basically two types of optical systems used for optical pick-ups. One type uses a collimator lens and an objective lens, as shown in Fig. 1 (a); the other type uses a lens sytem which combines a function of a collimator lens and an objective lens, as shown in Fig. 1 (b).[1] In the following sections, various types of lenses for use in CD players will be discussed.

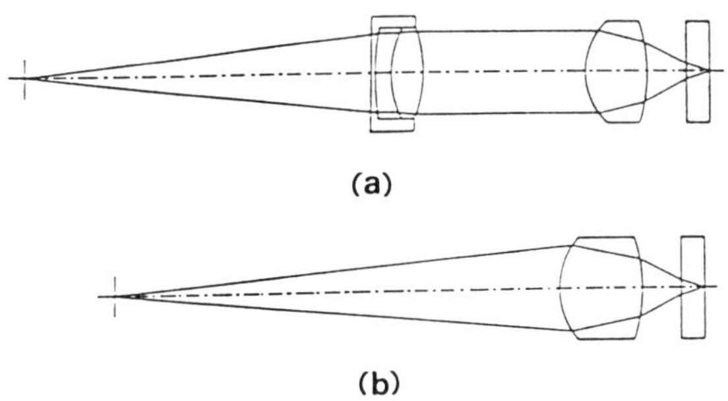

Fig. 1. Comparison of optical pick-up systems. (a) Collimator and objective lens system. (b) Single lens system.

2.1 Lenses for CD player pick-ups

Tables I and II list, respectively, those objective lenses and collimator lenses under development or in practical use at present. A conventional objective lens consists of 3 or 4 shperical lenses.[2] In this construction, it is very difficult to build up these lenses without decentering. Further, a high accuracy lens barrel is needed.

The molded glass lens is highly reliable and costs less.[3-5] Molded plastic lenses, lightweight and mass-producible, are widely used in various types of CD players.[6-8] The optical

Table II. The optical design parameters of collimator lenses for CD players.

Material		Glass	Plastic
Construction		Aspheric	Grating
Numerical aperture		0.14	0.11
Focal Length	(mm)	17	24
Thickness	(mm)	2	2
Diameter	(mm)	6	5.3
Weight	(mg)	220	—

Table I. The optical design parameters of objective lenses for CD players.

Type		Infinite Conjugate				Finite Conjugate
Material		Plastic	Glass	Glass	—	Glass
Construction		Bi-aspheric	Bi-aspheric	G R I N	Micro-Fresnel	Bi-aspheric
Numerical aperture		0.47	0.47	0.45	0.45	0.45
Focal length	(mm)	4.5	4.47	3.56	2.7	3.9
Working distance	(mm)	1.9	1.85	1.85	—	2.0
Thickness	(mm)	3.5	3.4	1.6	1.5	4.2
Diameter	(mm)	7.4	5.4	4	4	5
Weight	(mg)	120	240	50	—	300

performances of plastic lenses vary with temperature. Thanks to the autofocusing mechanism, however, focus shift of the objective lens does not affect the optical disk system. In collimator lenses, on the other hand, focus shift cannot be compensated because the lenses are set in a fixed position. Therefore there is no practical plastic collimator lens.

Replicated objective lens comprises a spherical glass lens substrate and a layer of plastic having aspherical outer profiles.[9,10] Since most of the lens power resides in the glass lens, this type of lens is free from focus shift due to temperature change.

A graded index objective lens which is made by polishing one side of the lens to a spherical surface have sufficient performance for use in CD players.[11]

2.2 Optical performance of lens

Fig. 2 shows the optical performance of each lens in relation to field angle. Optical performance is generally represented by RMS wave-front aberration. As compared to the conventional spherical lens system, the aspherical plastic lens shows better diffraction limit performance in wider field angle.[7]

The micro-fresnel lens [12], which is formed on a flat base plate, cannot satisfy the sine condition; therefore, a large coma aberration appears at the off-axis. Its field angle is only about 0.1°, as shown in Fig. 2. In the case of lower numerical aperture collimator lenses, however, about 0.03 λ is obtained for the field angle of 0.5°. This off-axis performance is satisfactory for normal use. Accordingly, this type of collimator lens is used for CD player pick-up.[13]

We developed the molded glass finite conjugate objective lens for portable CD player. For infinite conjugate type objective lenses, wider field angle merely permits lower assembling accuracy. For finite conjugate type objective lenses on the other hand, wider field angle is required. Because there are several factors that cause the positions of the laser and the lens to deviate relative to each other. Firstly, there is lens movement due to tracking and focusing. Secondly, there are the alignment errors of the laser, mirror and lens. Thirdly, there is a problem that affects portable CD players, because the lens is mounted on springs. The lens position moves according to the angle of the pick-up. All these factors cause discrepancies in the positions of the laser and lens relative to each other. This means that with a single lens pick-up system where the laser and the lens are separate, the lens needs sufficient off-axis performance.

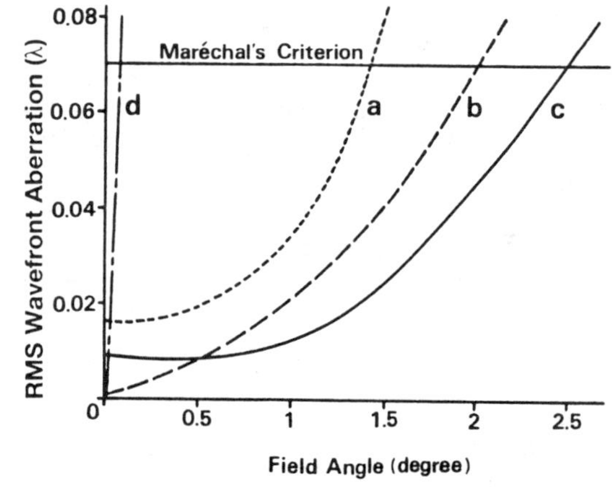

Fig. 2. Optical performances of the objective lens (NA 0.45) for CD players. (a) Conventional lens system. (b) Plastic aspheric lens. (c) Finite conjugate molded glass lens. (d) Micro fresnel lens.

As shown in Fig. 2, the off-axis wave-front aberration of the finite conjugate objective lens which we developed for CD player application is 0.05 λ for 2.2° of field angle. This shows that the lens has remarkably higher performance than the plastic aspherical lens, because refractive index can be selected relatively easily for molded glass lenses, whereas this is difficult for plastic lenses. The relation between refractive index and off-axis performance is shown in Fig. 3. The figure shows that a higher refractive index gives better off-axis performance.

2.3 Reliability

The properties of plastics change with temperature and humidity.[14] PMMA and PC are the chief materials of plastic optical components. PMMA is little sensitive to birefringence, on the other hand, PC has a high sensitivity to it. So PMMA is widely used as optical pick-up lenses, even

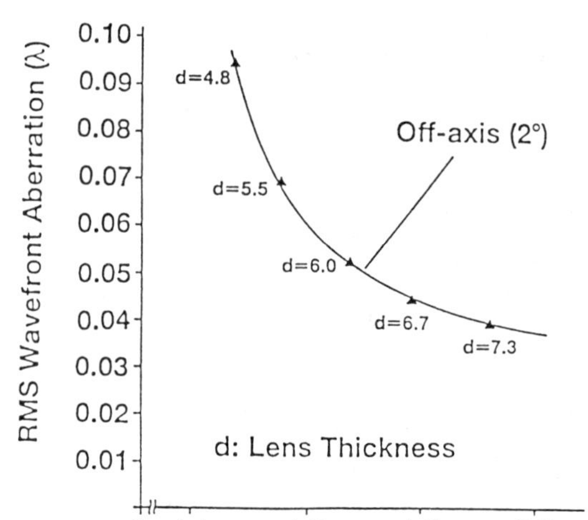

Fig. 3. Off-axis wavefront aberration vs refractive index. The lens thickness was chosen so that the smallest off-axis wavefront aberration was produced for each refractive index.

Table III. Comparison of the characteristics of materials for lenses.

Material	Plastic(PMMA)	Glass(SF8)
Temperature Coefficient for Refractive Index	-1.2×10^{-4} 1/°C	5×10^{-6} 1/°C
Coefficient of Linear Expansion (at 20°C)	7×10^{-5} 1/°C	7×10^{-6} 1/°C
Water absorption (be saturated)	2.0%	——

though it has disadvantage of absorbing water up to 2%. Table III shows example properties of PMMA and glass. PMMA has a temperature dependency of refractive index over ten times that of glass; its linear expansion coefficients is also ten times greater.

We manufactured prototypes of a glass molded lens of the finite conjugate type and a plastic lens with the same specifications, focal length: 3.9 mm, numerical aperture: 0.45, magnification: -0.2. Measurements were carried out to determine how optical performance changes with temperature. We measured the wave-front aberration with Fizeau interferometer. The result is shown in Fig. 4. Generally, lenses have aberration even at normal temperature, due to finishing and process errors. However, this sort of higher order aberration is almost independent of temperature change. On the other hand, the 4th order spherical aberration of plastic lens increases with temperature. By contrast, the spherical aberration of molded glass lens remains unchanged. This is due to the difference in the temperature coefficient of refractive index. Fig. 5 shows the wave-front aberrations at 70°C for two types of lenses. It can be seen that the plastic lens has greater speherical aberration.

2.4 Evaluation of aspherical lens

There are a variety of methods for evaluating the lens properties: point image observation, transmitted wave-front aberration measurement by the interferometer, measurement of electric properties as an optical pick-up such as jitter, and so forth. Of these, the measurement of the wave-front aberration is considered an excellent method in that it is quantitative and capable of the evaluation of optical performance of the lens itself and the feedback of the results of machining and molding. Wave-front aberration can be analyzed into each aberration using a Zernike polynominal fitting. An example wave-front aberration and its decomposition into each aberrations are shown in Fig. 6. Most of the aberration caused by slight change in each parameter, such as lens thickness, radius of curvature, refraction index and lens surface tilt, are classified as 4th order aberrations. Thus, if spherical aberration is observed, the cause may derive from deviations in lens thickness, radius of curvature or refractive index resulting from the original design. Asymmetric aberration, such as coma aberration, is usually caused by a tilted or decentered lens surface.

When the lens was designed into no aberration, higher order aberration and some portion of spherical aberration directly represent deviations from the designed aspheric surfaces. Fig. 7 shows the molds for glass molding and the lens configuration, both of which are measured by the ultra-high accuracy 3-dimensional profilometer [15], based on laser heterodyne interferometer. The profilometer can measure aspheric surfaces to the order of 0.5 μm accuracy. Fig. 7 (a) and (b) respectively show the deviations of the 1st and 2nd surfaces of the molds from the original design. Both surfaces have an approximately 0.1 μm deep dent at their center. Fig. 7 (c) shows the interferogram of a lens molded by these molds. The interferogram is also disturbed at the center.

Let us suppose the configuration error of a lens to be d, the refractive index of the lens to be n. The change in optical path distance P can then be expressed by the following equation:

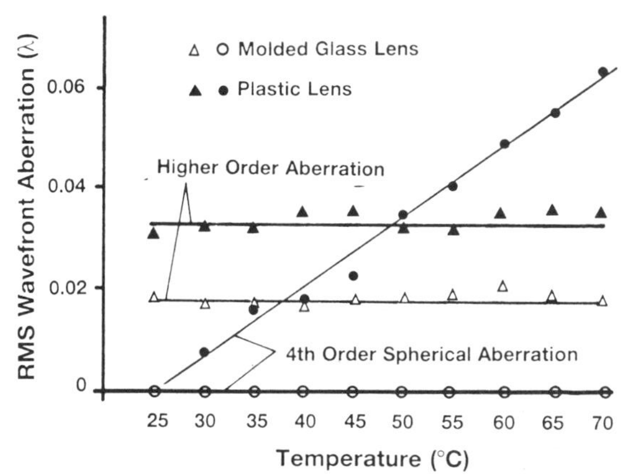

Fig. 4. Experimental results of wavefront aberration caused by temperature.

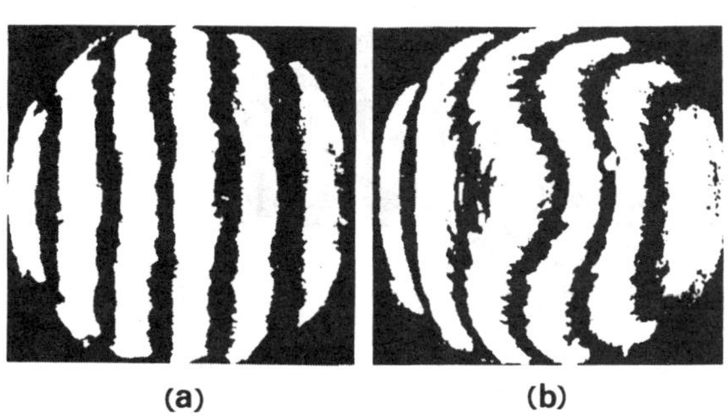

Fig. 5. The interferogram in temperature of 70°C (a) for molded glass lens and (b) for plastic lens.

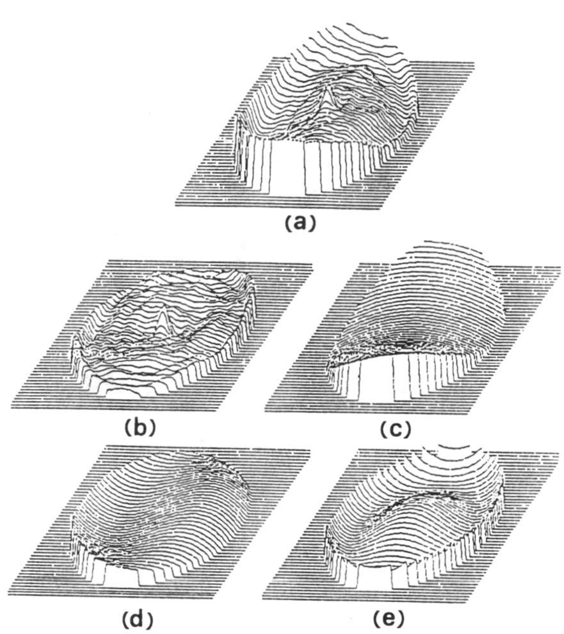

Fig. 6. (a) Total wavefront aberration and its Zernike polynomial fitting. (b) Higher order aberration. (c) Astigmatism. (d) Coma aberration. (e) Spherical aberration.

$$P = (n - 1)d \qquad (1)$$

This indicates that the displacement on a lens surface can be calculated if the wave-front aberration and the refractive index of the lens can be measured. Fig. 7 (d) illustrates displacement from the designed data which was calculated using the phase data of the interferogram. The dent at the center was calculated at about 0.2 μm deep. The displacement agrees quite well with the sum of the displacements of two molds. This means the mold configuration is very accurately transferred to the molded glass lens.

3. OPTICAL PICK-UP FOR CD PLAYER

CD players are produced in large quantities, and correspondingly the mass-productivity and lower cost of the pick-ups are required. The optical pick-up system for a CD player in the early days is shown in Fig. 8 (a).[16] The system consisted of about 20 optical components, including 11 glass components.

A high accuracy lens barrel is required to assemble two or more conventional spherical lenses. Some of recently developed pick-ups consist of fewer components by eliminating the polarizer and

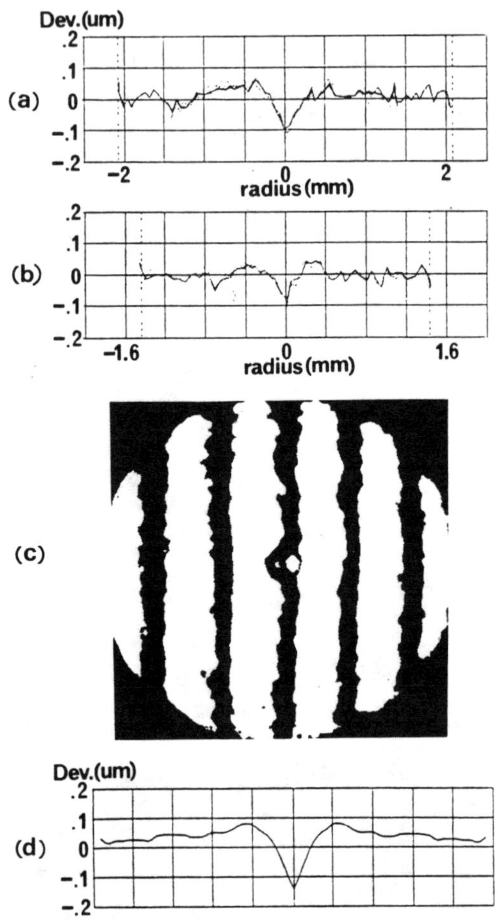

Fig. 7. Measurement of an aspheric mold and a bi-aspheric lens. (a) The first surface of the mold. (b) The second surface of the mold. (a) and (b) plot differences from the design data. (c) The interferogram of the bi-aspheric lens. (d) The cross section of the interferogram.

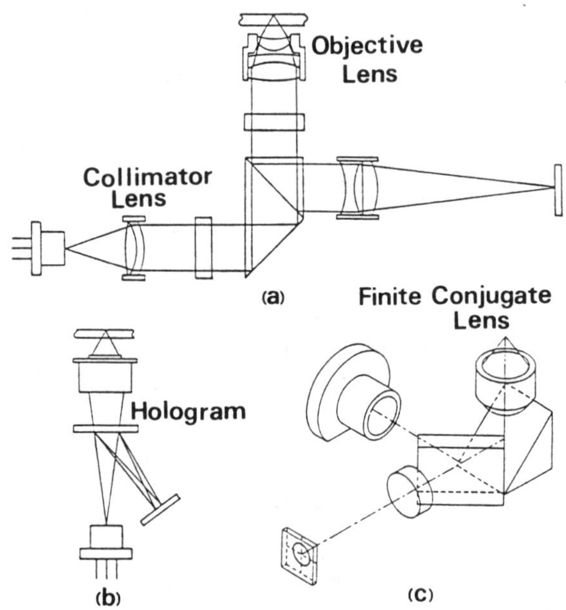

Fig. 8. Optical pick-ups for CD players.

by simplifying the error signal detection optical system. Fig. 8 (b) presents a pick-up that uses a hologram element.[17] A single hologram element not only separates beams but also functions as an error signal detection optical system. Basically, it is composed of only two optical elements.

Fig. 8 (c) shows a remarkably small pick-up for CD players.[18,19] Optical pick-up size was reduced by the following improvements:

1) Collimator and objective lens replacement with a high performance finite conjugate type molded glass lens
2) Optical system to generate astigmatism with the parallel glass plate tilted at 45 degrees to detect error signals.
3) Building a concave lens into the optical detection system, making it a so-called "telephoto type", thus reducing the distance from the photo detector

Fig. 9 gives an external view of the pick-up for CD player. Only 13 mm high, it consists of just four optical components.

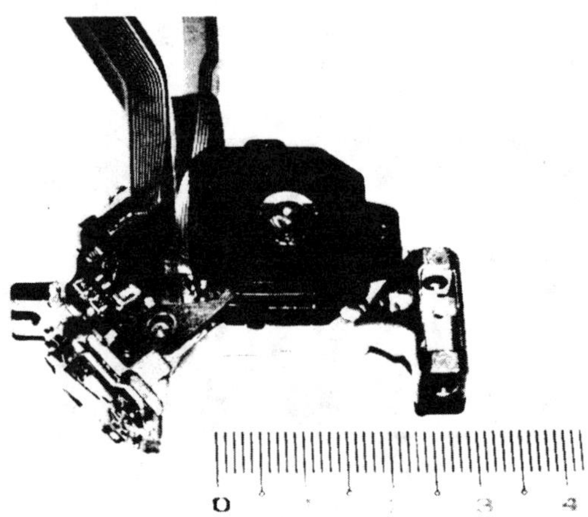

Fig. 9. The portable CD player pick-up with the molded glass lens.

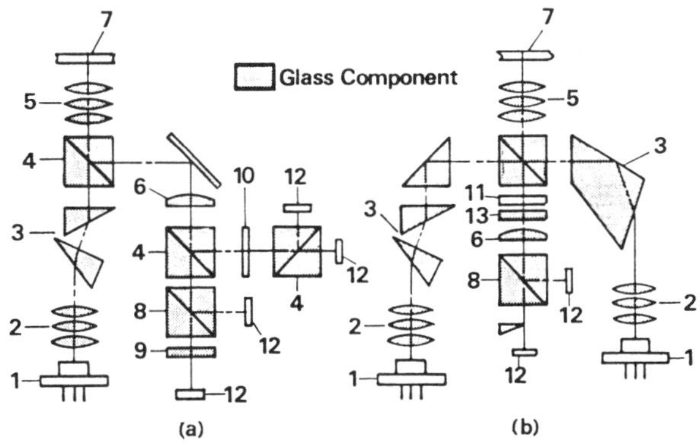

Fig. 10. Optical pick-ups for (a) magneto optical disk and (b) phase change type disk. 1. Laser; 2. Collimator lens; 3. Beam shaper; 4. P.B.S.; 5. Objective lens; 6. Convex lens; 7. Disk; 8. Half mirror; 9. Cylindrical lens; 10. λ/2 plate; 11. λ/4 plate; 12. Photo detector; 13. Filter.

4. OPTICAL PICK-UP FOR DRAW AND ERASABLE OPTICAL DISK SYSTEM

4.1 Optical system for pick-up

Not so many optical disk systems are currently produced for document filing, data storing, etc. Under such circumstances, the optical pick-ups for this type of systems consist mostly of conventional optical components. Fig. 10 shows the optical pick-up systems for phase change type [20] and magneto-optical disks.[21] It is understood that the optical systems are complicated when compared to the pick-up systems for CD players. This is due to the fact that

1) an optical system is additionally required to improve the laser light efficiency and to correct the beam shape,
2) two beams having different wave lengths must be combined and separated in phase change type disks,
3) an optical system is needed to detect Kerr rotation angle in magneto-optical disks, and so on.

However, these optical pick-up systems will also be simplified in the near future.

4.2 High numerical aperture pick-up lens

In general a high numerical aperture lens requires high precision. Fig. 11 shows shift in wave-front aberration when the lens parameters have changed from the design data. As compared to NA of 0.45 for CD players, the change in performance is 3 or 4 times with NA of 0.55.

We have developed molded glass, high numerical aperture lenses suitable for use in DRAW and erasable optical disk systems. Table IV shows the lens specifications.

The objective lens, with NA of 0.55 (type D), is remarkably light in weight. A weight of as little as 85 mg and a working distance as long as 2.1 mm were achieved by using glass with a specific gravity of 3.12 and by limiting lens thickness to 1.8 mm. The weight is less than that of a conventional plastic objective lens. An objective lens moves for focusing and tracking, so heat emission from the coils sometimes affects the performance of the lens. A single molded glass objective lens, however, is hardly affected by heat, therefore the outside diameter can be reduced to the limit. This also realized a super-light objective lens.

Fig. 12 shows the designed optical performances of type C and type D. Fig. 13 shows the wave-front aberration of the objective lens with NA of 0.55 (type D). Fig. 14 shows the cross-sectional

Table IV. The optical design parameters of newly developed molded glass lenses for write once and erasable optical disk systems.

Lens type		A	B	C	D
Numerical aperture		0.30	0.50	0.53	0.55
Focal length	(mm)	9.0	4.49	4.25	3.9
Working distance	(mm)	7.43	2.14	2.28	2.07
Thickness	(mm)	2.3	2.8	2.0	1.8
Diameter	(mm)	6.4	6.4	6.0	6.0
Weight	(mg)	250	270	140	84

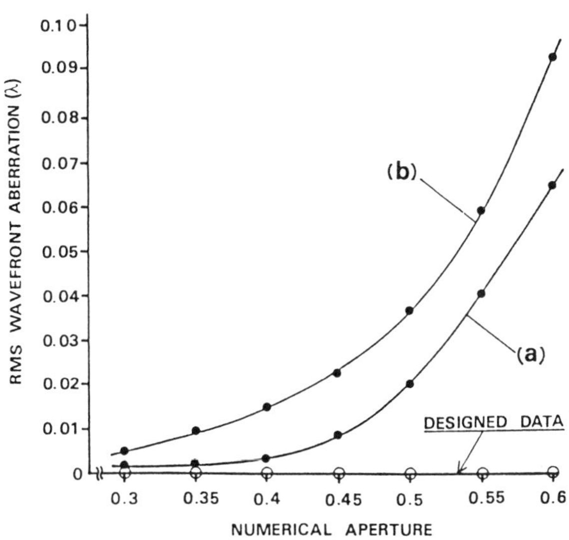

Fig. 11. Optical performance changes when (a) the lens thickness has changed by 50 μm and (b) the refractive index has changed by 0.005 from the design data.

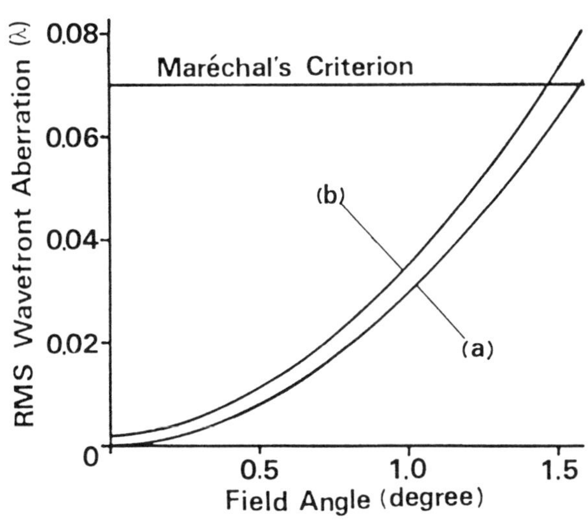

Fig. 12. Designed Optical performances of (a) the collimator lens (type A) and (b) the objective lens (type D).

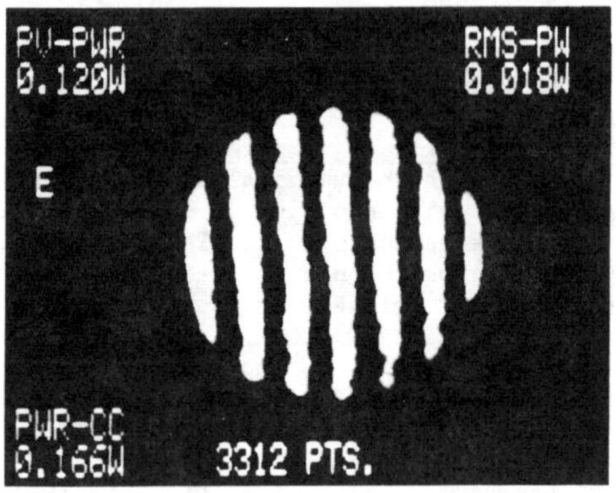

Fig. 13. The interferogram of the molded glass lens (type D).

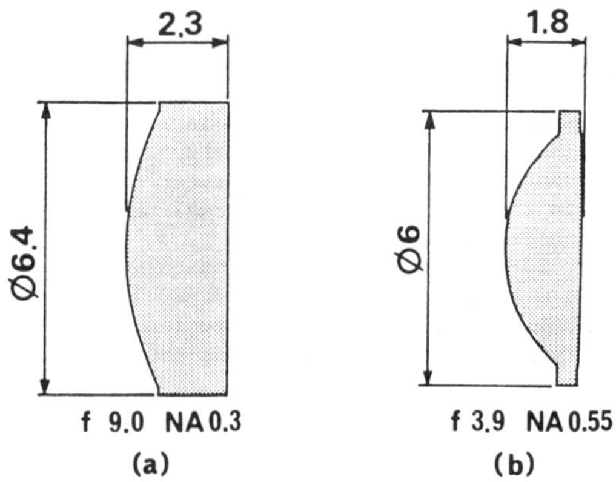

Fig. 14. Schematic shapes of (a) the collimator lens (type A) and (b) the objective lens (type D) in cross section.

views of collimator and objective lenses (type A and D). The collimator lens has a flat surface and an aspherical surface.

ACKNOWLEDGEMENT

We are deeply grateful to Mr. M. Nagasawa and all the members committed to this project for their guidance in our study and to the staff of Sumita Optical Glass Manufacturing Co. and of Hi-Fi Audio Division and Disk Systems Division of Matsushita Electric Industrial Company for their kind cooperation in this development and Mr. Yoshizumi and Mr. Murao for their cooperation in measuring the aspherical configurations.

REFERENCES

1) G. Bouwhuis et al.: Principles of Optical Disc Systems (Adam Hilger, 1985) Chap. 2, p.81.
2) P. Kuttner: Opt. Eng. 22 (1983) 473.
3) Y. Tanaka, Y. Nagaoka, and M. Ueda: Workshop on OF & T Tech. Dig., WB5 (1986).
4) R. O. Macshmeyer, R. M. Hujar, L. L. Carpenter, B. W. Nicholson, and E. F. Vozenilek: Appl. Opt., 22 (1983) 2413.
5) P. L. Ruben: Appl. Opt., 24 (1985) 1682.
6) J. Haisma, E. Hugues, and C. Babolat: Opt. Lett., 4 (1979) 70.
7) T. Kiriki, N. Izumiya, K. Sakurai, and T. Kojima: CLEO '84 Tech. Dig. (1984) WB3.
8) K. Shintani, and S. Kubota: CLEO '84 Tech. Dig. (1984) WB2.
9) D. Visser, T. G. Gijsbers, and R. A. M. Jorna: Appl. Opt. 24 (1985) 1848.
10) J. J. M. Braat, A. Smid, and M. M. B. Wijinakker: Appl. Opt. 24 (1985) 1853.
11) H. Nishi, H. Ichikawa, M. Toyama, and I. Kitano: Appl. Opt. 25 (1986) 3340.
12) K. Takusagawa, T. Niriki, T. Suemitsu, T. Nishino, and S. Suzuki: Micro Optics News 4 (1986) 98.
13) Y. Higuchi, et al.: Proc. of the 47th Spring Meeting of the Jap. Soc. of Appl. Phys. (1986) 3a-X-9.
14) T. Kiriki: Japanese J. Opt. 12 (1983) 460.
15) K. Yoshizumi, T. Murao, J. Masui, R. Imanaka, and Y. Okino: Appl. Opt. 26 (1987) 1647.
16) M. Itoh, H. Nakamura, T. Abe, N. Hirose, and S. Tanaka: National Technical Report 29 (1983) 746.
17) Y. Kimura, Y. Ono, S. Sugama, and Y. Ohta: Proc. of the 33th Autumn Meeting of the Jap. Soc. of Appl. Phys. (1986) 30p-ZE-1.
18) H. Nakamura et al.: National Technical Report 32 (1986) 490.
19) Y. Tanaka, Y. Nagaoka, and M. Ueda: Proc. of Optical Memory Symposium '86 (1986) p.99.
20) T. Ishida, K. Ishibashi, T. Satoh, S. Ohhara, and T. Yoshida: Proc. of Optical Memory Symposium '86 (1986) p.121.
21) H. Ishibashi, R. Shimizu, M. Imura, M. Kuwamoto, H. Yamamoto, S. Tanaka, and T. Shimamoto: ISOM '87 Tech. Dig. (1987) WA5.

Reprinted with permission from *Japanese Journal of Applied Physics,* Vol. 26, Suppl. 26-4, pp. 135-140 (1987). ©1987 Japanese Journal of Applied Physics.

Spherical Grating Objective Lenses for Optical Disk Pick-ups

Kenya Goto, Kazushige Mori, Genichi Hatakoshi*, Shunsuke Takahashi

Electron Tube & Device Group, and R&D Center*, Toshiba Corporation
72, Horikawa-cho, Saiwai-ku, Kawasaki 210, JAPAN

Plane grating collimator lenses [GCLs] and spherical grating objective lenses [SGLs] have been developed and installed in different optical pick-ups for CD players. The relief profile for the grating lenses have been designed to obtain high diffraction power efficiency of about 80% for the GCLs and about 65% for the SGLs. The blazed grating profile has been engraved by diamond turning lathes. GCLs were mass-produced by the plastic injection molding method and installed in more than 80 thousand optical pick-ups for portable and general audio use. SGLs have a much higher NA and can be used as objectives for the optical pick-ups. A thin plastic relief grating profile for the SGLs is formed by the photo-polymerization method on a flat face of a plano-spherical glass substrate. The field angle for the SGLs is more than ±0.5 degree, within the Marechel Criterion.

1. INTRODUCTION

More than 7 million plastic injection molded aspheric objective lenses were mounted in various optical pick-ups for CD players last year. One reason is the lower unit price and another is the light weight. However, they have a weak point in that each mounting surface must be adjusted in a pick-up head by slightly inclining the lens with regard to the disc surface in order to decrease the coma aberration of the lens. This aberration is generated from the gate effect inherent in the injection molding process. Canceling the coma aberration angle direction and the mount adjusting angle direction decreases coma aberration. Moreover, besides this fault, plastic lenses are apt to cause other troubles such as a warping problem due to binding tension between the lens and the lens holder, expansion and refractive index change due to temperature change and high humidity, and birefringence problem in the magnetooptical memory use. Plane grating collimators (GCL) had been developed [1] and mounted on pickups,TOPH7820, for portable and general audio uses. More than 80 thousand sets were mass-produced last year [2] and still work very well in both The United States and The United Kingdom.

Using grating technology similar to that in GCL spherical grating lenses (SGL) have been developed not for collimators but for objectives. They are free from coma and astigmatic wavefront aberration and can eliminate the lens inclination adjusting process, which makes the pick-up manufacturing process high. Also, the SGL can avoid other problems listed above. The SGLs, mounted in the optical pickups for CD players, work fine, if they are used in conjunction with laser diodes with narrower dispersion wavelength than about 10 nm in 780 nm wavelength region. The newly developed SGL is not much higher in production cost than an aspheric plastic lens, and can avoid its demerits.

2. DESIGN

A conceptional illustration and specifications designed for the SGL are shown in Fig.1 and Table I, respectively. High NA (numerical aperture=0.5) plane grating lenses (PGL) had already been developed [3-5], but their diffraction efficiencies were not high (30-40%), owing to the conventional EB lithographic fabrication technique. What is still worse, PGLs must be used while keeping the lens and the disc face parallel within 5′, since an inclined laser beam incidenct into the PGL causes coma aberration and jitter value increment in a CD pick-up and in a CD player, respectively.

By means of the CNC machining technique, where blazed grating dies are generated using diamond turning lathes as shown in Fig.2 the diffraction efficiency improvement of the plane grating lens had been greatly attained upto 82 % in the case of smaller NA (=0.1) in order to apply it for the collimator (GCL) in the pick-up. Figure 3 is a cross sectional view of a GCL, where Rm and Hm are

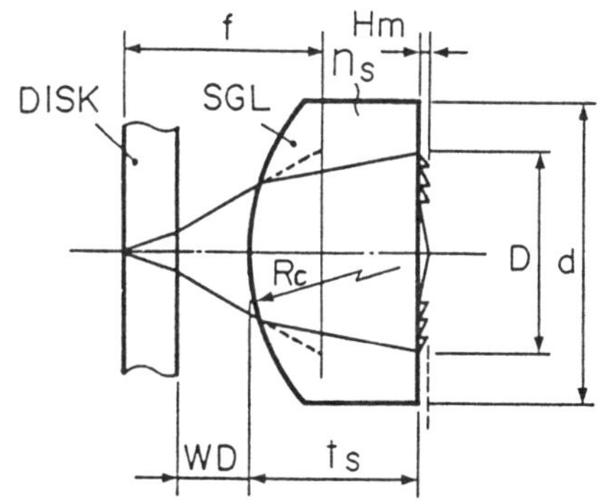

Fig.1 A conceptional illustration of a spherical grating objective lens of NA=0.47 for a pick-up

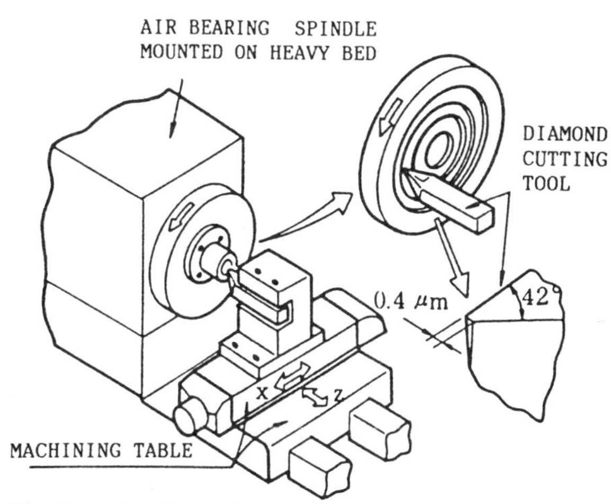

Fig.2 A diamond turning lathe illustration. This permits machining axially symmetric blazed grating metal dies for both GCLs and SGLs.

Table I. Design specifications for high NA Spherical Grating Lenses for optical disc objectives

DIE TYPE FOR THE SPHERICAL GRATING LENSES FOR OPTICAL DISC HEAD	Symbol	Infinite SGLs				Finite Conjugate	
nominal number of SGL stampers	#	# 1	# 2	# 3	# 4	# 5	
Numerical aperture (NA=sin θ)	NA	0.47	0.45	0.45	0.45	a=0.45	b=0.06
Spherical radius: substrate curvature	Rc (mm)	4.5	4.0	4.25	4.5	4.0 (±0.01)	
Focal length	f (mm)	4.3	4.5				
Effective diameter of the SGL	D (mm)	4.05				3.8	
Lens outer diameter	d (mm)	6.0					
Lens thickness	ts (mm)	3.5					
Working distance	WD (mm)	1.7	2.2	2.1	2.0	1.8	30.8
Maximum grating pitch in the center	Λ_{max} (μm)	95.7	106.1	102.8	100.3	90.3	
Minimum grating pitch at outermost	Λ_{min} (μm)	2.9	4.3	3.7	3.3	2.8	
Blazed grating height	Rm (μm)	1.53	1.53	1.54	1.54	1.54	
Total number of grating rings	ν(rings)	417	303	335	361	398	
SGL substrate refractive index	ns	1.511					
Designed center wavelength	λ (nm)	785(±5nm)					

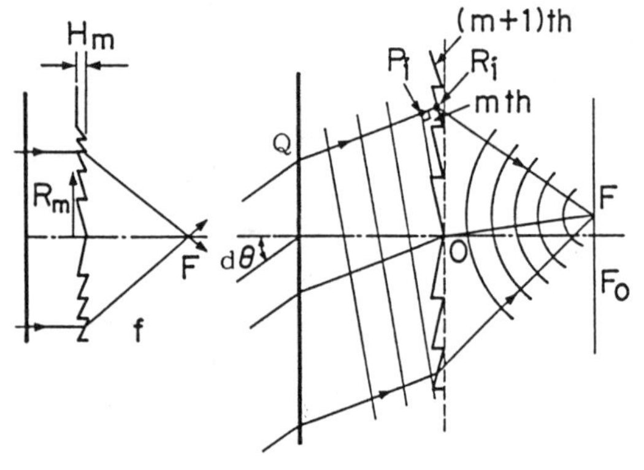

Fig.3. A cross sectional view of a plane parallel plastic Grating Collimator Lens for designing ring radius and height for the mth blazed grating. They can be represented by eqs.(1)-(2).

$$R_m = \{2m\lambda f + (m\lambda)^2\}^{1/2} \quad (1)$$

$$H_m = \lambda / \{n_s - f/(f^2 + R_m^2)^{1/2}\} \quad (2)$$

$$dLi = n_s \cdot \overline{Pi\,Ri} + \overline{Ri\,F} - (\overline{OF} + m\lambda) \quad (3)$$

$$dW = \{\langle dLi^2\rangle - \langle dLi\rangle^2\}^{1/2} \quad (4)$$

where $\langle Xi\rangle$ is the average value for Xi,"i" and "m" are the integers which also mean the ring order started from the center of the grating. "n_s" is the refractive index for the glass substrate.

Optical path difference between $\overline{OF}$ and $\overline{PiRiF}$ is represented as dLi and can be defined by eq.(3). Wavefront aberration dW is given by eq.(4), in which focus point F is defined as a point of the smallest dW. Effective incident angle dθ, where dW is given by the value below than 0.033 λ (30/λ) shown in r.m.s. value, is ±0.5 degree for infinite conjugate high NA SGLs and is ±1.0 degree for finite conjugate high NA SGLs, whereas plane grating collimators (GCLs) are limited to only smaller NAs for the same inclination angle dθ as high numerical aperture SGL objectives. Figure 4 shows degradation in the wavefront aberration versus incident angle increment.

Wavelength dependency for the grating lens focal length is derived from eq.(1) and given by;

$$df = -f \cdot d\lambda / \lambda \quad (5)$$

A result of the calculation for the equation is shown in Fig.5 , when NA and the focal length are 0.13 and 20 mm respectively. It is shown that the dW increment is only 0.017 for 20 nm wavelength change at the 780 nm laser wavelength.

Grating pitch $\Lambda(R_m)$, which is a function of grating radii R_m, is given by :

$$\Lambda(R_m) = \lambda\{1 + (f/R_m)^2\}^{1/2} \quad (6)$$

Minimum pitch in maximum radius in a grating lens is then given by :

$$\Lambda_{min} = \lambda\{1 + (f/R_{m\,max})^2\}^{1/2} \quad (7)$$

Minimum pitch and height for a GCL, designed for 780 ±5 nm wavelength, are calculated from eqs.(7) and (2) respectively. They are 6 μm and 1.6 μm for a GCL of f=20.3 mm and D(effective diameter) =5.2 mm.

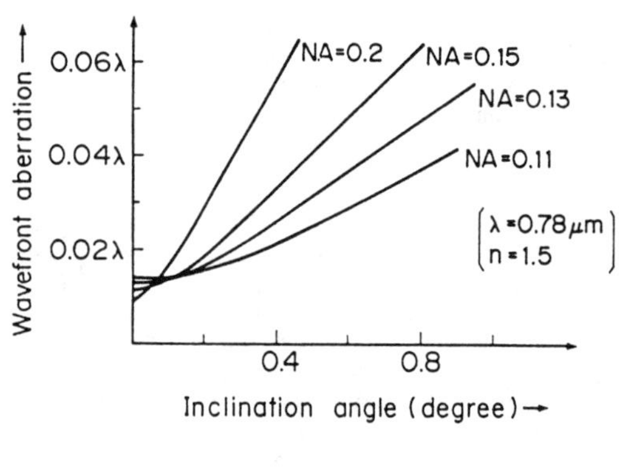

Fig.4 Wave aberration degradation versus incident angle change for GCLs with NA variation

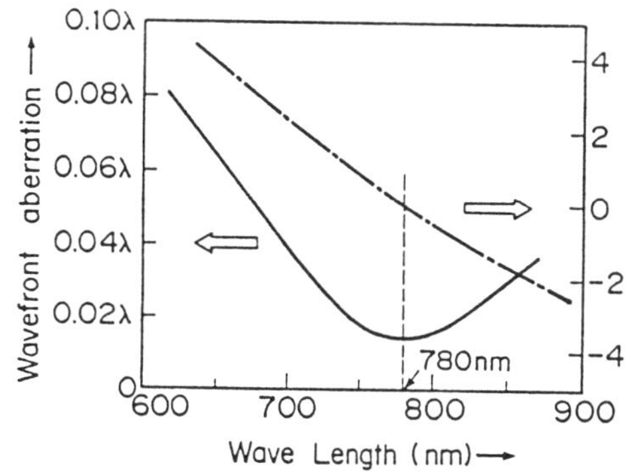

Fig.5 Wave aberration degradation and focal length shift in conjunction with the change in wavelength used for a 0.13 NA fixed GCL

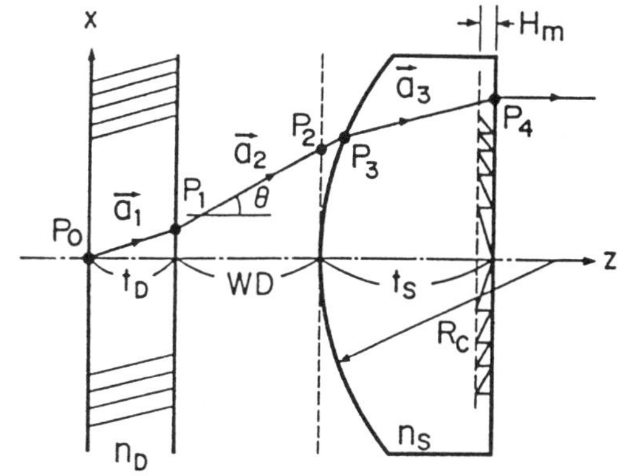

Fig.6 Design and grating pitch calculation for a high NA spherical grating lens (SGL) for the optical pick-up objective

Design and pitch calculation for a SGL as an objective lens for a disc pick-up, are described in Fig.6, where point (P) is expressed by the X-Z coordinates :

$$P_0(X_0, Z_0),\quad P_1(X_1, Z_2),\quad P_2(X_2, Z_2),$$

$$P_3(X_3, Z_3),\quad P_4(X_4, Z_4)$$

Light wave vectors and individual parameters are given by:

$$\vec{a}_1 = (L_1, N_1),\quad \vec{a}_2 = (L_2, N_2),\quad \vec{a}_3 = (L_3, N_3) \qquad (8)$$

n_S : refractive index of SGL, n_D : refractive index of disc

t_S : SGL thickness , t_D : disc thickness, λ : wavelength

WD : working distance, NA : numerical aperture

Rc : radius of spherical curvature (substrate radius of SGL)

Calculations on each grating radius (Rm) were carried out, using a grating formula (eq.9) which means that path difference, between the axial ray and any out of axial ray, is equal to the integer (m) times one wavelength of the laser beam. That is:

$$\overline{P_0 P_4} - (n_D \cdot t_D + WD + n_S \cdot t_S) = m\lambda \qquad (9)$$

$$\overline{P_0 P_4} = \overline{L_1} + \overline{L_2} + \overline{L_3} \qquad (10)$$

Where $\overline{L_1}$, $\overline{L_2}$, and $\overline{L_3}$ are the lengths between the points P_0, P_1, P_2, P_3, P_4 in Fig.6. we can solve eq.(9) and calculate the grating radius (Rm) for a SGL by the ray tracing method, using those given parameters such as n_S, Rc, n_D, t_S, t_D, WD, etc. in the order of P_0, P_1, P_2, P_3, P_4.

Other known quantities are :

$$Z_1 = t_D,\quad Z_2 = t_D + WD,\quad Z_4 = t_D + WD + t_S \qquad (11)$$

Calculation of $\vec{a}_1$:

$$L_1 = X_1/(X_1^2 + Z_1^2)^{1/2},\quad N_1 = Z_1/(X_1^2 + Z_1^2)^{1/2} \qquad (12)$$

$$X_{1\,max} = NA \cdot (t_D/n_D)\{1-(NA/n_D)\}^{1/2} \quad \because L_{1\,max} = NA/n_D \qquad (13)$$

From Snell's law :

$$L_2 = n_D \cdot L_1 \qquad (14)$$

and using eq.(12) x_2 is given by:

$$X_2 = X_1 + n_D \cdot L_1 \cdot (Z_2 - Z_1)/\{1 - n_D^2 \cdot L_1^2\}^{1/2} \qquad (15)$$

$\vec{a}_3$ is obtained from the refraction at spherical surface at P_3.

As point P_4 is on the vector $\vec{a}_3$, x_4 and z_4 are given by x_3, L_3 and z_3, N_3. From eq.(8), we obtain

$$X_4 = X_3 + L_3\,(Z_4 - Z_3)/N_3 \qquad (16)$$

Using eqs.(8)-(14) L_1, L_2, and L_3 are given by :

$$L_1 = n_D \cdot (X_1^2 + Z_1^2)^{1/2} \qquad (17)$$

$$L_2 = \{(X_3 - X_1)^2 + (Z_3 - Z_1)^2\}^{1/2} \qquad (18)$$

$$L_3 = n_S \cdot \{(X_4 - X_3)^2 + (Z_4 - Z_3)^2\}^{1/2} \qquad (19)$$

Those computer calculated data for each grating ring radius Rm and Hm were input to the diamond turning machine for generating SGL stampers.

3. FABRICATION

The SGL consists of a thin plastic film and a spherical glass substrate, as shown in Fig.1. The plastic film, whose thickness is about 10 μm, is formed by a stamper engraved with a large number of chirped and rotationally symmetric inverted blazed grating rings, and bonded to plane surface of a plano-convex spherical substrate using the photo-polymerization (2p) method shown in Fig.7.

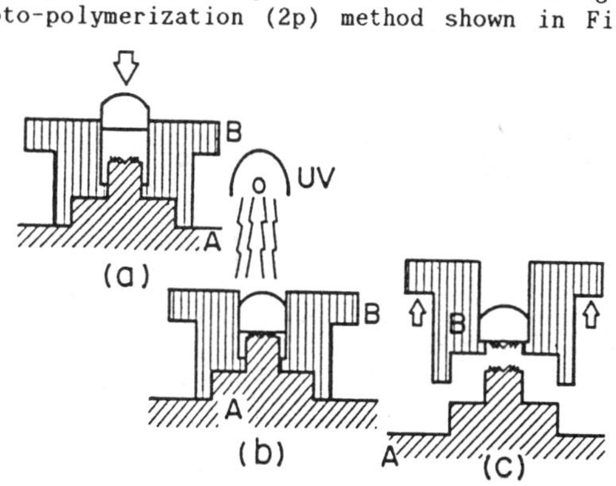

Fig.7 Photopolymerization (2p) film forming and bonding process to the glass substrate in SGL fabrication. 7(a) is a layer forming cavity. 7(b) is ultra-violet irradiation process. 7(c) is the SGL removing process from the stamper.

The inverted grating stamper had been machined by a computer numerical controlled diamond turning lathe which has a single point diamond tool with 55 degree included angle. Figure 8 shows a cross sectional blazed grating shape near the center for a SGL taken by a surface profile meter. Interferometric microphotographs of the stamper, both near the center and near the outermost edge are shown in Fig.9. They show that the micro-machining by this method has been carried out very well.

Five kinds of chirped grating lens stamper were designed and machined for the SGL fabrication. Four infinite conjugate categories, which are #1, #2, #3, and #4, and one finite conjugate category which is #5, were fabricated. Their specifications are shown in Table I. NA values for each kind of SGL are 0.47 for #1 and 0.45 for the rest. Each spherical substrate has the same outer diameter, the same thickness, and the same refractive index but has different spherical radius Rc values, 4.0, 4.25, 4.5mm. Maximum and minimum pitch for the chirped grating in the SGLs are 106.1 and 2.8 μm, respectively, as shown in Table I. Total number of

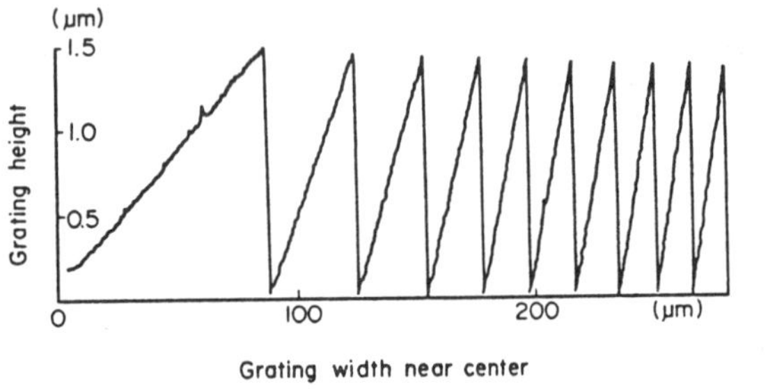

Fig.8 Blazed grating shape around the SGL center The vertical scale is enlarged, compared with the horizontal scale.

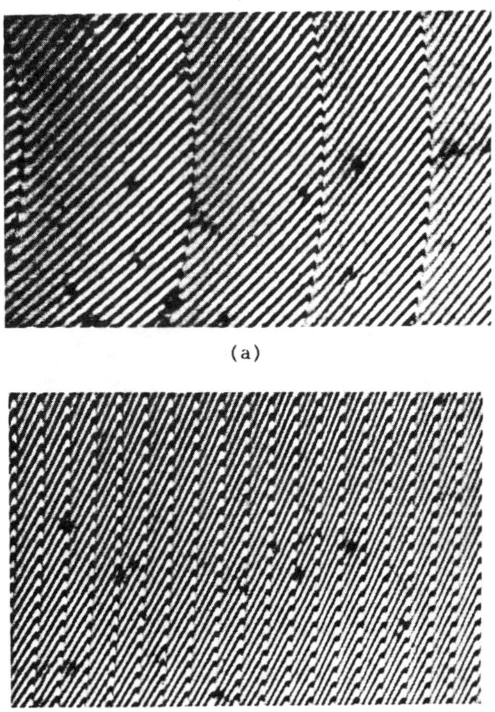

(a)

(b)

Fig.9 Interferometric microscopic photographs of a spherical grating objective. (a) Blazed shapes near the center. (b) Blazed shapes at outermost

grating rings in the SGL for the finite conjugate SGL is 398, which is more than that for the infinite SGLs with the same numerical aperture.

It was found that the spherical substrate not only functions to widen the allowable incident angle and the grating pitch, but also functions to decrease the number of grating rings. The higher NA in the SGL requires a larger number of grating rings. However, even the infinite conjugate SGL, whose NA is as large as 0.47, requires only 417 rings, in comparison with 728 rings in the PGL for the same NA [3-6]. They are shown in Table I for the SGL and eq.(20) below for the GCL:

$$\nu\,(\text{number of grating rings}) = (f/\lambda)\{(1-NA^2)^{-1/2}-1\} \qquad (20)$$

4. EXPERIMENTAL

Figure 10 shows the experimental diffraction efficiency measurement for the #5 finite conjugate SGL which has no anti-reflection (AR) coated. The experimental data for the infinite conjugate SGL #2 is shown in Fig.11. The diffraction efficiency for the grating lens is defined in this paper as the focussed central beam intensity percentage through an experimental lens versus through an AR coated perfect conventional glass triplet lens using the same semi-conductor laser beam. If the SGL is AR coated, the intensity of the SGL in the Figs.10 and 11 will be raised at least 4%. The experimental maximum values for diffraction efficiency for both the finite conjugate SGL and the infinite conjugate SGL are 78% and 65% respectively with 4% Fresnel loss on one surface.

The theoretical maximum diffraction efficiency, that will be discussed in the next section, is 85% without 4% Fresnel reflection loss on a spherical surface of the SGL. Then the theoretical limit will be 81% for no AR coating . The difference between theoretical and measured values are 3% for the #5 finite conjugate SGL and 16% for the #2 SGL. Those data show that almost ideal blazed grating had been formed in our experiment for finite SGLs.

Figure 12 shows a calcurated plot for the wave aberration change for various kinds of radius (Rc) for SGLs versus radius difference between designed and finished spherical substrates. Figures 11 and 12 show how important the radius difference Rc is in SGLs. It was found that a smaller radius Rc causes a larger wavefront aberration increment with a same radius error value.

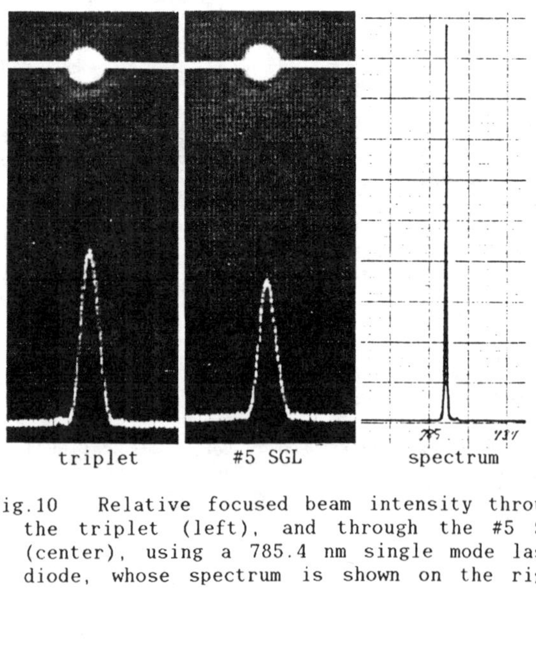

Fig.10 Relative focused beam intensity through the triplet (left), and through the #5 SGL (center), using a 785.4 nm single mode laser diode, whose spectrum is shown on the right

Figure 13 shows increments of wave aberration versus the inclination angle with various SGLs. The allowable angle for the infinite conjugate SGL is as small as one third of the triplet and also the biaspheric objective, but the finite conjugate SGL has a wider angle, which is about twice that for the infinite SGL. It also shows fairly better characteristics in the relative intensity change versus the lens displacement test in Fig.14, where the SGL moves parallel to the disc surface and the intensity decrement is evaluated.

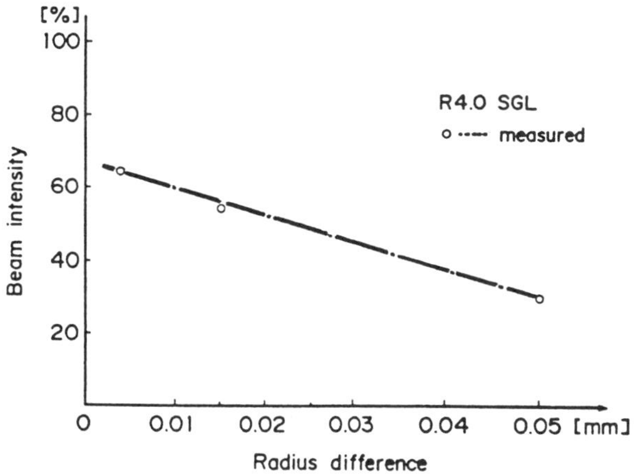

Fig.11 Experimental data for measuring diffraction efficiency. Beam central intensity perentage versus an ideal glass triplet. Results depend on the SGLs radius differences.

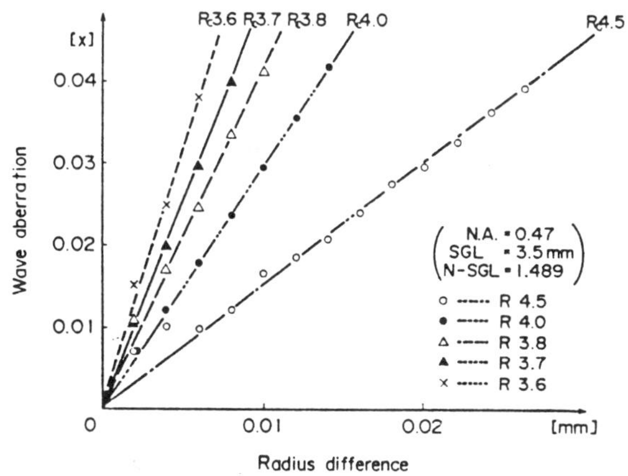

Fig.12 Calculated plot for wave aberration dW versus radius differences, with parameter Rc

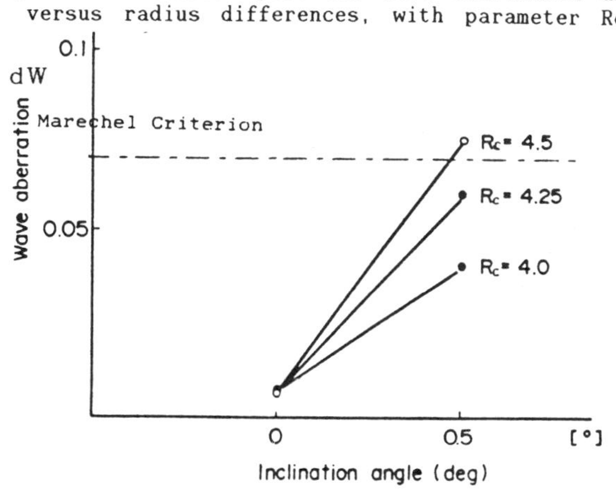

Fig.13 Wave aberration versus inclination angle

Many samples were fabricated. It was found that the stamper life is very long (bearable more than 100 thousand times in both SGLs and GCLs). The #1 SGLs, whose weight is 0.2 g, were mounted on TOPH7810 [7], as shown in Fig.15, and on TOPH7812, which are TOSHIBA commercial CD pick-ups. They were set in the portable CD player, XR-P9 and in the component player, XR-Z40, that were originally mounted with the 0.47 NA triplet objective.

There was no practical problem involved in playing the usual CD discs, though the evaluation data shown in Table II are slightly worse than in the glass triplet case.

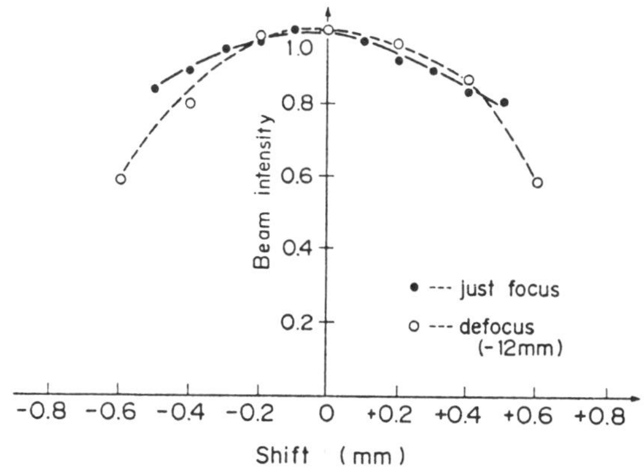

Fig.14 Intensity change in the focused beam through the SGL (#5 objective) versus the lens shift in the radial direction of the CD disc

Fig.15 SGLs (left) and optical pick-up for TOPH7810, on which the SGL is mounted (right)

Table II Evaluation data for an SGL mounted on CD players, compared with the triplet

Objective	Spherical Grating #1		Triplet
Mounted PUH	TOPH7810	TOPH7812	TOPH7812
J(3t)	25.0 ns	26.5 ns	19.0 ns
I(top)	33.5 μA	13.1 μA	13.5 μA
I(11t)	21.9 μA	8.3 μA	9.6 μA
I(3t)	12.0 μA	4.5 μA	6.5 μA
Modulation	65.3 %	69.3 %	71.0 %
MTF	58.1 %	54.1 %	68.0 %

5. RESULTS

It was found that the SGLs might be used in the optical pick-up with less time involved in the mounting process and could be produced using a longer life stamper than the plastic lenses, due to their having no coma aberration and no forced pressure imposed during replication. Infinite SGLs have ±0.5 degree allowable inclined angle to the disc surface, that is very wide, compared with the plane grating lenses (PGLs). The finite-conjugate SGL has more wider angle, ±1.0 degree. The SGLs may be used in the Magneto-Optical disc system, because they have no birefringence in themselves.

Diffraction power efficiency for the relief grating drawn in Fig.16 is solved from the electro magnetic wave equation. Results are as follow.

Diffraction efficiency η_m for the mth order diffracted and transmitted TE wave is given by:

$$\eta_m = \mathrm{Re}\ (\beta_m / \beta_0)\ |\tilde{T}_m|^2 \qquad (21)$$

where

$\beta_m = (k_0^2 n_s^2 - K_m^2)^{1/2}$ = Z component of wave number for the mth order of diffracted wave in region III

$\beta_0 = k_0 n_0 \cos\theta$ = Z component of wave number for the mth order of incident wave in region I

$\tilde{T}_m$ = expansion coeffecient of electric field for the mth order diffracted wave, obtained by solving wave equation in region II which is given by equ.(22)

$$\sum_m \left(\frac{\delta^2 \tilde{A}_m}{\delta Z^2} - K_m^2 \tilde{A}_m \right) \exp(-i K_m x) +$$

$$k_0^2 \sum_{m'} \sum_{l} \tilde{a}_{m'} \tilde{A}_l \exp(-i\,[k_0 n_0 \sin\theta + (l - m')K]\,x) = 0 \qquad (22)$$

$K_m = k_0 n_0 \sin\theta + mK$, m, l, m' = integer numbers

K = grating vector = $2\pi/\Lambda$, Λ = grating pitch

k_0 = wave number = $2\pi/\lambda$, λ = wavelength

θ = incidence angle for the grating

n_0 = refractive index of air

a_m, $a_{m'}$ = expansion coeffecients of permittivity ε

$\tilde{A}_m$, $\tilde{A}_l$ = expansion coeffecients of electric field

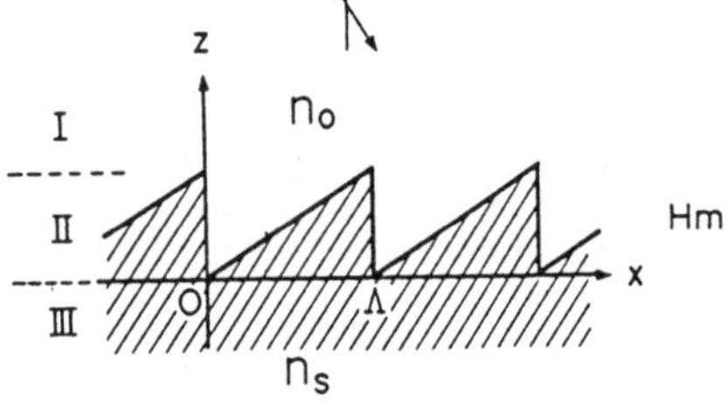

Fig.16 Relief grating shape and symbols used for solving the electromagnetic wave equations to find theoretical diffraction power efficiency

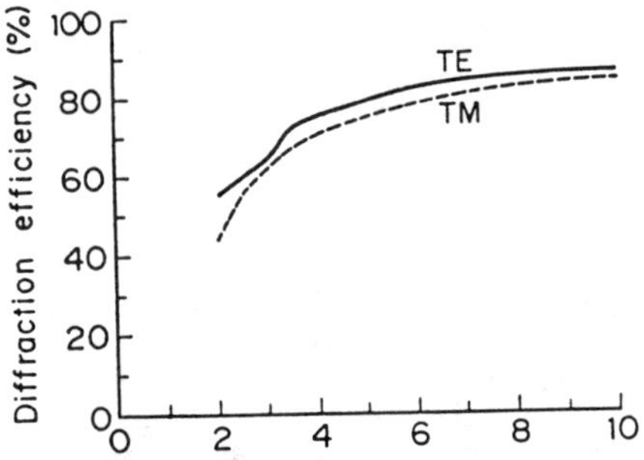

Fig.17 Diffraction efficiencies for fundamental diffracted transmission waves versus normalized grating period obtained by computer analysis of electromagnetic TE and TM wave eqations

The result, shown in Fig.17 with the computer analysis of eqs.(21-22) and similar eqations for TM wave, indicates diffraction power efficiencies for the 1st order diffracted light transmitted from the normalized grating period Λ/λ. The value Λ/λ is 5.5 at the outermost edge and 136 at the center for the #2 SGL. Since it is 7-8 on an average, with Gussian beam profile in this work, theoretical efficiency can be estimated at 85% from Fig.17. SGL efficiency is greatly affected with the radius error value of the finished glass substrate, which should be within ±30μm of designed spherical radius size. Achieving this is not difficult, as compared with the aspherical lens accuracy of less than ±0.5μm . Rc can be finished within ±10μm accuracy with conventional simple and low cost polishing methods. From this point of view, and the experimental data shown in Figs.10 and 11 the average efficiency for SGLs in a mass-production of 4.5mm Rc will be more than 75% and 60% each for the finite and the infinite conjugate categories. A SGL fault is wavelength selectivity, owing to the nature of the grating, which requires narrower spectrum linewidth. According to experiment, it was found that the diviation should be within ±5nm at the center wavelength of 780 nm.

6. CONCLUSION

Besides the first mass-production and the first shipment for high diffraction efficiency GCLs to so many pick-ups, high NA SGLs had also been developed for objectives having features as follow;

(1) Less mounting process for mounting onto pick-up. It is free from lens inclination adjusting mechanisms, because there is no coma aberration, when it is mounted upon a flat level of pick-ups
(2) High efficiency close to theoretical limit of 85% without AR coating, because of a wider grating pitch(about twice that for PGL),which enable a low cost diamond turning machine to be used.
(3) Simple designing and programing, that keep the machining from being difficult, with more fidelity than the biaspheric lens cases.
(4) Less effect from temperature and humidity change in both refractive index and expansion coefficient, because the plastic film thickness is at most 10μm and flat with thin 1.5μm grating
(5) No birefringence in the lens itself
(6) Long and durable stamper life time compared with the aspheric plastic injection molds.

It was found that the finite conjugate SGL is more recomendable than the infinite conjugate SGL for both fabrication and utilization to the objective lens in the optical disc use.

7. ACKNOWLEDGEMENTS

Authors greatfully acknowledge Mr. Katsunobu Ueda in the Toshiba manufacturing engineering lab. for the SGL stamper machining and Mr. Koichi Takahashi in solid state device engineering first department of Toshiba electron tube and device division for his encouragement and helpful advice.

8. REFERENCES

1)K.Goto & K.Mori; O plus E,No.76,(March,1986)p.75
2)S.Ohkoshi et.al.; Toshiba Review 41,(1986)p.555
3)G.Hatakoshi & K.Goto;IOOC'83,29A2-2,(June 1983)
4)K.Tatsumi,K.Nukui et al;Appl.Opt.23,(1984)p.1742
5)G.Hatakoshi & K.Goto; Appl. Opt.24, (1985)p.4307
6)T.Fujita,H.Nishihara; Opt.Lett.,7,12,(1982)p.578

Reprinted with permission from *IEEE Transactions on Consumer Electronics*, Vol. CE-33(4), pp. 520-530 (November 1987).

SINGLE LENS CD PLAYER PICKUP SYSTEM USING A BI-ASPHERIC MOLDED GLASS LENS

M. Sunohara, Y. Tanaka,
Y. Nagaoka, M. Ueda
Matsushita Electric
Kadoma, Osaka, Japan

K. Azuma
Matsushita Electric
Moriguchi, Osaka, Japan

1. Introduction

Since compact disc (CD) players were first introduced in 1982, they have been prevailing at a remarkable speed. The number of shipped CD players is assumedly eight to nine millions in 1986. Following this first generation CD player of fixed installation type, portable or radio cassette CD players and CD players for car use are supposed to prevail as a second generation. For these CD players, mass production, lower cost and performance reliability in hot and humid environment are demanded.

The pickup system is the key component of a CD player and is expected to satisfy various requirements of new CD players. Various methods have been researched to simplify the pickup system. Among them, lenses are as important pickup component as laser diodes to promise better performance and lower cost CD players.

Conventionally, a spherical lens dominates as an objective lens in the CD player pickup system. In recent years, however, various lenses such as a plastic aspheric lens[1], a molded glass lens[2], a gradient index lens[3] and a replicated lens[4] are reported to be suitable as the objective lens. When one of these lenses is used as an objective lens, the pickup system may contain only one objective lens. Since the system requires collimator lenses as well, however, it needs at least two lenses if the number of required collimator lenses could be reduced to one.

The typical conventional pickup system contains four or five lenses including collimator lenses and objective lenses. We succeeded in replacing these four or five lenses with a single bi-aspheric lens. We have also developed a high precision glass molding technique allowing mass production of reliable bi-aspheric lenses.

The optical pickup system using the bi-aspheric lens of our development has been simplified substantially for application in a portable CD player so that it is superior in performance, size and cost to any conventional system.

In this paper, we describe the optical design, its evaluation, characteristic of the mass-produced lens, the construction of the pickup system and its reliability in temperature and humidity fluctuation.

2. Lens Design and Evaluation

2.1 Aim of lens design

There are basically two types of CD player optical pickup system. One uses collimator and objective lenses. Laser beam is made parallel by the collimator lens and focused on the disc by the objective lens. We call this type two-unit system. The typical two-unit system comprises two spherical glass lenses as collimator lens and three spherical glass lenses as objective lens. Recently, bi-aspheric plastic lenses are often used as objective lens. Fig. 1 (a) shows a two-unit optical system using plastic objective lens. The other type uses a lens system which integrates the functions of the collimator and objective lenses. We call this type one-unit system. An example of the one-unit optical system is shown in Fig. 1 (b). We have employed the one-unit system since the summer 1983. Fig. 2 shows our optical pickup system.

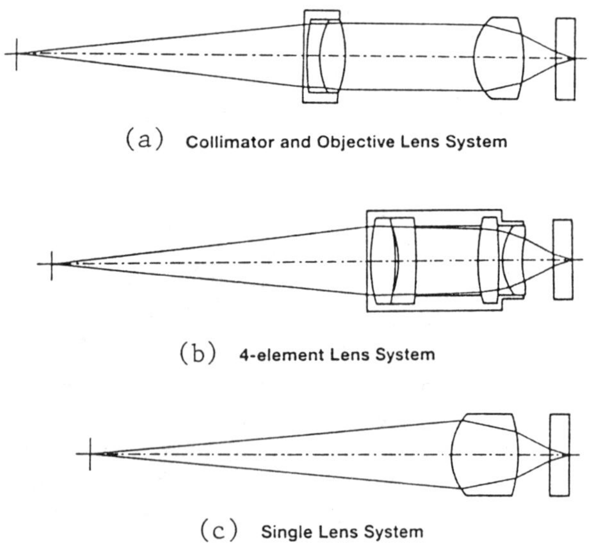

Fig. 1 Comparison of Optical Pickup Systems

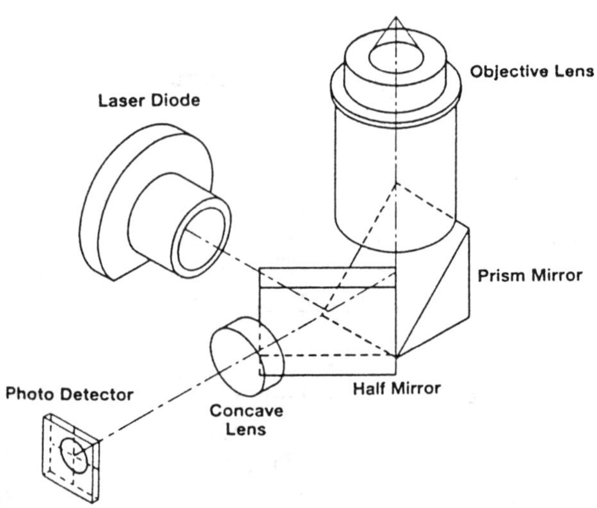

Fig.2 Optical Pickup System Configuration

In comparison with the two-unit optical system, the one-unit system has the following features.

1) Since the system does not involve collimator lens, its construction is simple and the distance between the laser diode and the disc is substantially reduced (See Fig. 1), realizing a compact, low cost pickup system.
2) The one-unit system does not require alignment process which the two-unit system requires to allow the collimator lens to make laser beam parallel.
3) The one-unit system needs apparently higher numerical aperture (NA) for the objective lens than the two-unit type because laser beam is not parallel at the laser diode (LD) side.
4) Aberration compensation is necessary because the LD position relative to the objective lens changes as the lens moves for tracking and other reasons.

Thus, the one-unit system requires higher performance objective lens. In other words, if a high performance objective lens can be realized, the one-unit system is by far superior to the two-unit system. As shown in Fig. 2, the one-unit system that does not contain collimator lens has enabled us to produce a very small pickup system.

However, the conventional one-unit system of four-element lens type has the total length of about 10mm, taking the substantial portion of the entire pickup height.

When an infinite conjugate plastic lens is used as an objective lens in the two-unit system, the focal point shift with temperature fluctuation does not provide any problems because the pickup has an auto focusing function. However, since the refractive index of a plastic lens changes with temperature[5], the temperature fluctuation causes spherical aberration as well as the focal point shift. Therefore, although the two-unit optical system of a CD player generally uses plastic objective lens, reliability of plastic must be studied carefully if it is to be used for lenses of more strict specifications. Considering its poor heat resistivity, it seems difficult to use plastic lenses in a CD player pickup for car use. The refractive index of a glass lens also changes with temperature and humidity fluctuation, although its change is negligibly small compared with that of a plastic lens.

With the above points taken into consideration, we have developed a molded glass lens system, replacing the four-element lens of our one-unit system with a single bi-aspheric lens without sacrificing the performance of the four-element lens required to realize small, low cost and temperature and humidity fluctuation-independent CD players such as portable or radio cassette CD players and CD players for car use. We call this new type integral system. Fig. 1 (c) shows the integral optical system. This system makes it possible to realize a thinner and more compact optical pickup.

2.2 Lens design

2.2.1 Numerical Aperture (NA)

For a CD player, the NA of the objective lens at the disc side is typically 0.45 to 0.47. Since the objective lens of the one-unit system has a finite conjugate, NA_0 converted for parallel beam is calculated as follows:

$$NA_0 = (1 + |m|)\ NA \qquad (1)$$

Since NA at the disc side is fixed, the magnification "m" depends on the LD side NA which in turn depends on the laser beam availability. The smaller "$|m|$" makes the lens design simpler, closing to that of the two-unit system, but simultaneously causes lower laser beam availability, making it necessary to raise the laser power. When NA at the disc side is 0.45 and that at the LD side 0.1, for instance, the magnification "m" is -0.22. Therefore, NA_0 converted for parallel beam is 0.55. In general, lenses with higher NA is more difficult to design and manufacture and provides higher performance dependence on temperature and humidity.

2.2.2 Off-axis performance

What must be considered in designing the one-unit lens system is the off-axis performance of the lens. The on-axis performance can be made small easily by using an aspheric surface. The required off-axis performance is closely connected with the specifications of the pickup itself. Fig. 3 shows the principle of our optical pickup system. The off-axis performance is required when the position of the objective lens displaces relative to the LD. The displacement occurs to the following three reasons. Firstly, since the lens is installed separately from the fixed optical system including a laser diode, a mirror, etc., the lens alone is moved when tracking is performed. However, a small actuator is sufficient to move the small, light lens alone, so that the entire volume of the pickup can be made compact.

Secondly, a catoptric system such as prism mirror is incorporated in the fixed optical system as an essential requirement for a thinner pickup system. However, possible installation position or angle error of the mirror can result in the displacement of the lens relative to the LD. To make the assembling of the pickup system easier, the off-axis performance of the lens is vital.

Thirdly, our pickup system holds the lens by four-wired suspension which deflects by the tare of the lens. The deflection varies depending upon the posture of the CD player, that is, whether it is vertical or lateral, and causes the lens to move with respect to the LD particularly when the pickup is built in a portable or a radio cassette player.

If displacement of the relative position between the LD and the lens occurs due to any of the above reasons, off-axis light enters the lens. This is why the lens requires appropriate off-axis performance.

In designing our lens, we anticipated

1) alignment error, for laser, mirror and lens as ±0.3mm, and

2) lens movement due to tracking as ±0.15mm and
3) lens movement by the CD player posture as ±0.25mm.

Therefore, the total object height is 0.7mm which is equivalent to the angle of incidence of about 1.8 degree and the image height of 0.14mm.

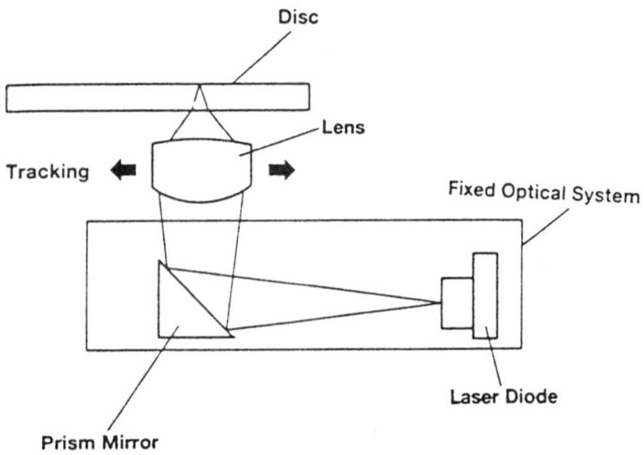

Fig. 3 Optical Pickup System

2.2.3 Lens design

In order to attain higher performance, we designed our integral lens, taking the thickness as a parameter. Fig. 4 shows lens thickness plotted against RMS wavefront aberration on the assumption that refractive index is 1.78, focal length 4.8mm and magnification -0.23. The aim of our design was to satisfy the spherical aberration and sine condition requirements. The on-axis wavefront aberration is virtually stable when the lens thickness is changed. In contrast, the off-axis RMS wavefront aberration reaches the minimum at a particular lens thickness. With our design, the off-axis wavefront aberration is the minimum at the lens thickness of 6.7mm.

The second design parameter was the refractive index. The lens thickness was selected so that the smallest off-axis wavefront aberration was produced for each refractive index. The specifications of the lens are the same as described above. As shown in Fig. 5, the higher refractive index produces better off-axis wavefront aberration. At the refractive index of 1.5 or smaller, the off-axis performance drops suddenly. It is apparent that the plastic integral lens with low refractive index is not suitable for optical pickup use. The higher refractive index is accompanied by the thicker lens, naturally resulting in the weight increase of the lens. Therefore, it is essential to select an appropriate refractive index, balancing the required off-axis performance with the weight.

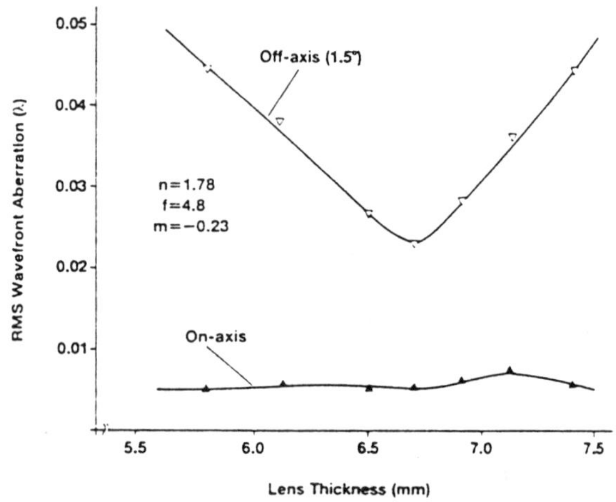

Fig. 4 Wavefront Aberration vs Lens Thickness

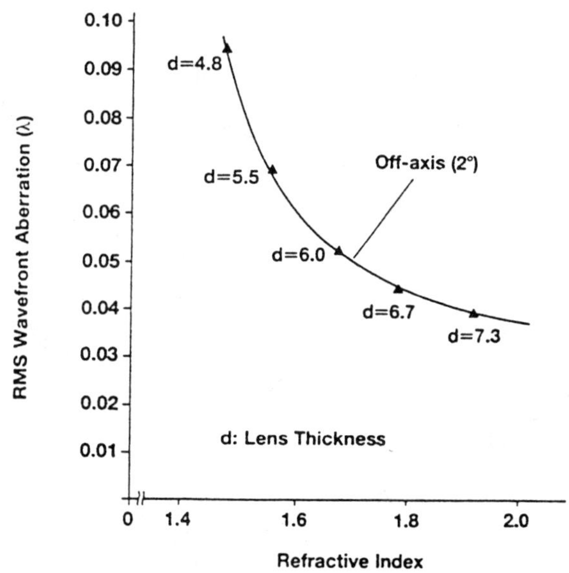

Fig. 5 Off-axis Wavefront Aberration vs Refractive Index

2.2.4 Design specifications

All the above points were taken into account in designing our integral lens. Table 1 compares the specification of the integral lens with those of the conventional four-element lens system.

Table 1 Comparison of Specifications

	New molded glass lens	Current lens system
Focal Length	3.9 mm	5.0 mm
NA	0.45	0.47
Working Distance	2.0 mm	1.7 mm
Lens Length	4.2 mm	10.7 mm
Diameter	5.0 mm	6.8 mm
Weight	0.30 g	1.02 g
The Number of Lenses	1	4

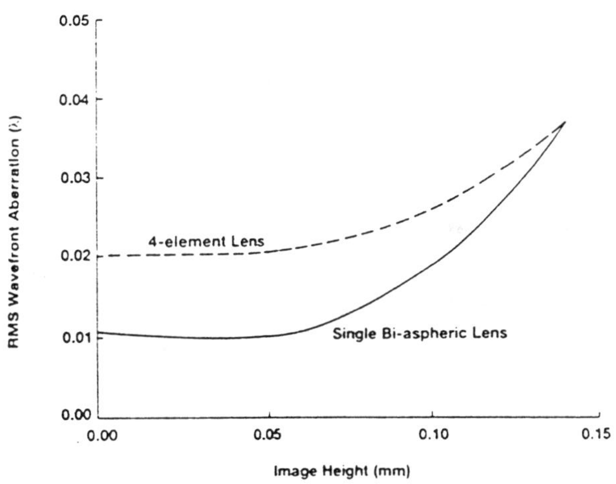

Fig. 6 Comparison Wavefront Aberration

Fig. 6 shows the wavefront aberration of our integral lens. At the required image height of 0.14mm, the wavefront aberration is 0.035 lambda, well within the diffraction limit. In addition, the integral lens presents higher performance than the conventional four-element lens system at every image height within 0.14mm. Since the new lens is made up of only one lens, it does not need a lens barrel and alignment process.

2.3 Evaluation of lens

The transmitted wavefront aberration of the molded lens was measured with a Fizeau interferometer to determine the optical performance.

It is extremely difficult to maintain the optical axis of the lens perpendicular to the disc during the measurement. The lens inclination with respect to the disc causes coma aberration. Fig. 7 plots the actual measurements of the fourth order coma aberration caused by the tilting disc. As shown, the fourth order coma aberration changes linearly whereas higher order aberration attributable to the lens itself shows no change. This means that if the lens itself has coma aberration, it can be canceled by tilting the lens relative to the disc. Our pickup is provided with a skew adjustment mechanism for rotating the lens with the principal point of the lens as the rotation axis as shown in Fig. 8. The original purpose of this mechanism is to correct the inclination of the lens mounted in the actuator. Coma aberration of the lens itself is canceled by rotating the lens with respect to the disc.

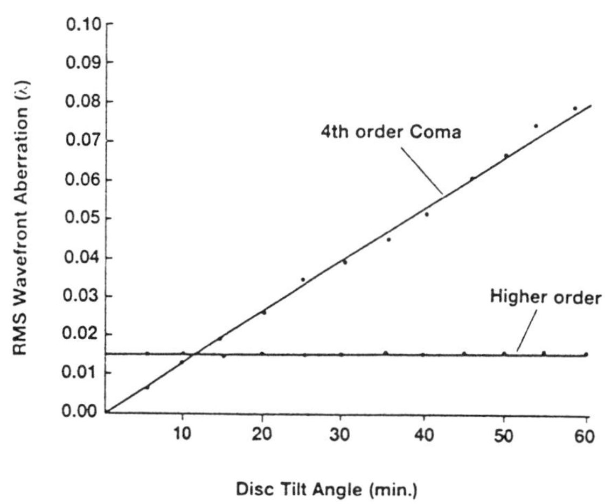

Fig. 7 Wavefront Aberration Caused by Disc Tilt

Jitter is the typical characteristics of a CD player pickup. The RMS wavefront aberration of the lens, especially coma aberration has significant influence on the jitter. Since fourth order coma aberration can be canceled in our pickup, it was excluded from our wavefront aberration measurement. Meanwhile, astigmatism, if it is small, does not cause jitter to change[6]. Therefore, what affects the jitter most is the measured RMS wavefront aberration from which coma aberration and astigmatism are subtracted.

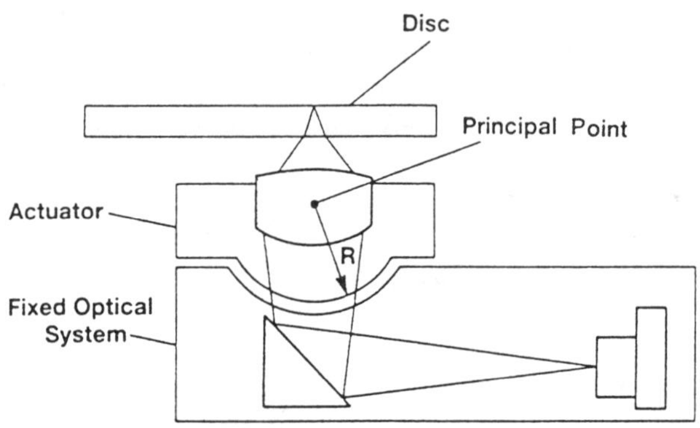

Fig. 8 Skew Adjustment Mechanism

3. Lens Performance

Fig. 9 shows the wavefront aberration of 16 different lenses molded with the same die. The measured values widely vary depending upon the angle of the lens. In contrast, the values subtracting coma aberration and astigmatism are stable, indicating that the molding process is stable.

All the lenses mass-produced by us have presented wavefront aberration of 0.03 lambda or less.

4. Pickup

We have manufactured a portable CD player pickup using this newly developed bi-aspheric molded lens. The height of the pickup is 13.3mm, 26% thinner than the conventional 18mm tall pickup. The light, small lens has enabled us to realize a small, thin pickup with improved oscillation resistance and smaller power consumption. Fig. 10 shows photographs of the portable CD player pickup using our integral lens and the one using the four-element lens system. The thickness of the portable CD player using the former pickup is 22.9mm, about 30% thinner than the conventional portable CD player which is 31.9mm thick.

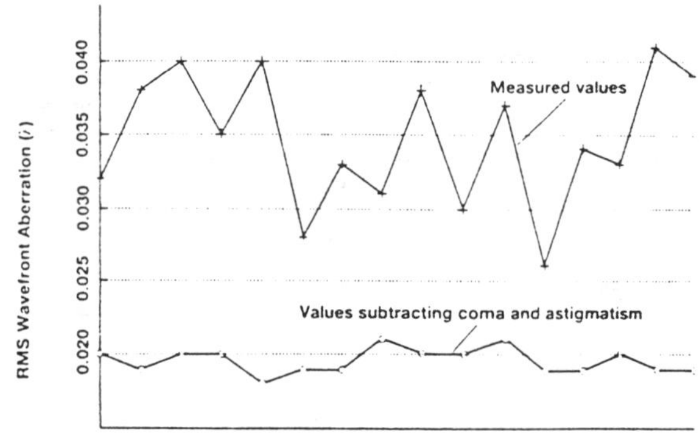

Fig. 9 Wavefront Aberration of the Lenses Molded by the Same Die

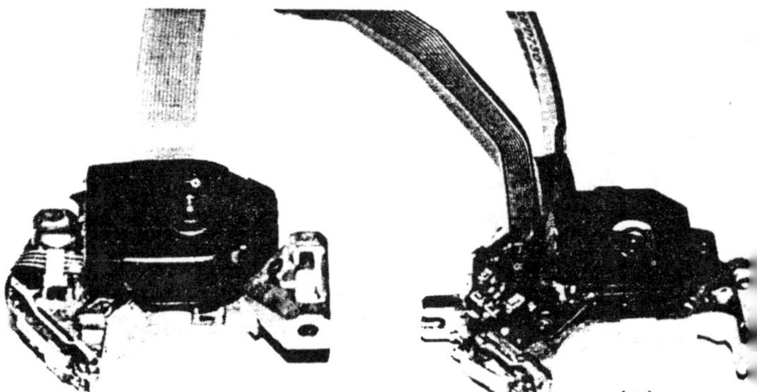

(a) (b)

Fig. 10 The Portable CD Player Pickups with (a) the Old Lens System and (b) the Molded Glass Lens

5. Pickup Reliability

5.1 Temperature and humidity characteristics of wavefront aberration

A portable or radio cassette CD player and a CD player for car use are supposed to be used in a severe outdoor environment. It is anticipated that temperature and humidity fluctuation will change the refractive index and the surface figure of the lens, increasing the wavefront aberration and resulting in larger jitter. We measured the wavefront aberration of an integral glass lens, raising the lens temperature from 25°C to 70°C. For comparison sake, the wavefront aberration of an integral plastic lens was also measured under the same temperature condition for the plastic water absorptions of 0% and 1.5%. Fig. 11 shows our experimental results of the temperature dependence of the RMS wavefront aberration of the glass and plastic lenses. Apparently, the glass lens is independent of temperature and humidity, whereas the RMS wavefront aberration of the plastic lens increases with the temperature. The wavefront aberration also increases with water absorption. The RMS wavefront aberration at water absorption of 1.5% is about 0.015 lambda higher than that as water

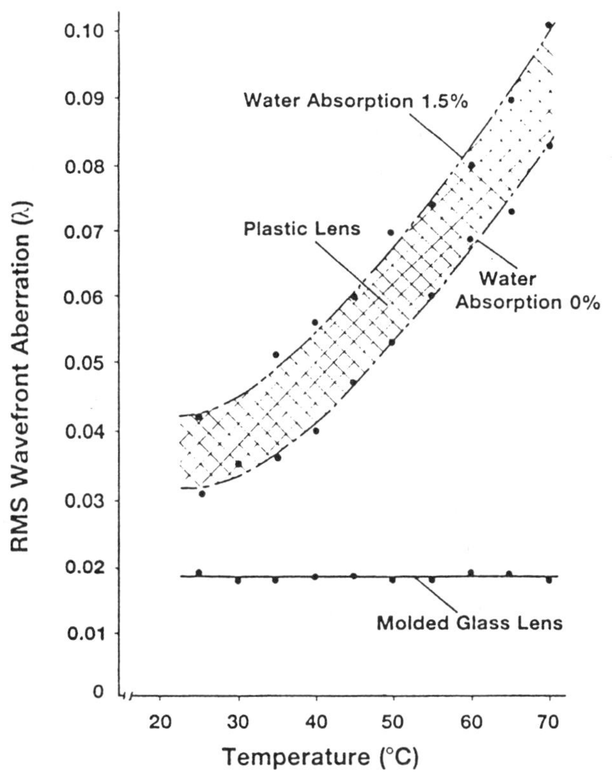

Fig. 11 Experimental Results of RMS Wavefront Aberration Caused by Temperature and Humidity

absorption of 0%. This is attributable to the dimensional and refractive index changes caused by the almost saturating water absorption of 1.5%. Temperature characteristics of various aberrations are shown in Figs. 12 through 14. Fig. 12 shows measured temperature characteristic of the fourth order spherical aberration at the water absorption of 0%. For convenience in comparison, the fourth order spherical aberrations of the molded glass lens and of the molded plastic lens at 25°C are equally set to "0". This measurement has revealed that the fourth order spherical aberration of the molded glass lens is stable while that of the plastic lens increases linearly by 0.0635 lambda. The measurement thus obtained for the fourth order spherical aberration of the plastic lens is somewhat higher than but virtually conforms to the calculated result obtained by simulation and indicated by dot line in Fig. 12. Fig. 13 shows the measured temperature characteristic of the higher order aberration of the molded glass and plastic lenses for water absorption of 1.5%. The measured values are stable at 0.018 lambda and 0.033 lambda respectively. Fig. 14 shows the measured temperature characteristics of the fourth

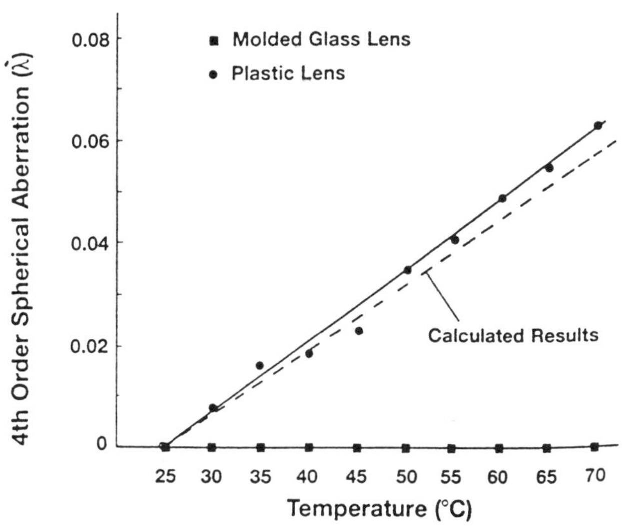

Fig. 12 Experimental Results and Calculated Results of Spherical Aberration Caused by Temperature

order spherical aberration and the higher order aberration of the plastic lens for water absorption of 0% and 1.5%. As the temperature rises from 25°C to 70°C, the fourth order spherical aberration and higher order aberration increase about 0.02 lambda and about 0.08 lambda respectively.

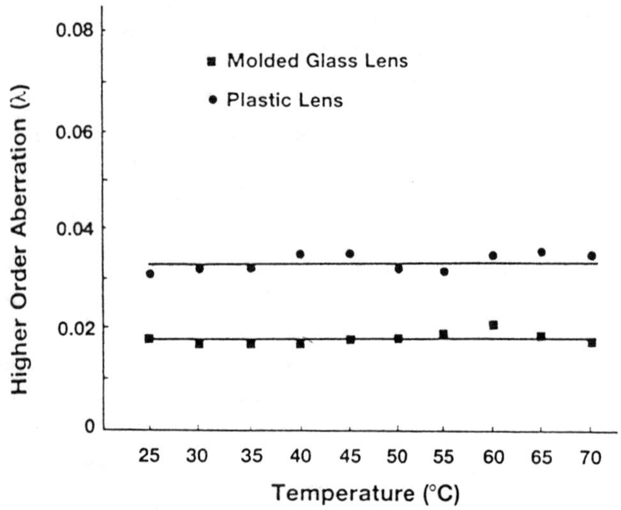

Fig. 13 Experimental Results of Higher Order Aberration Caused by Temperature

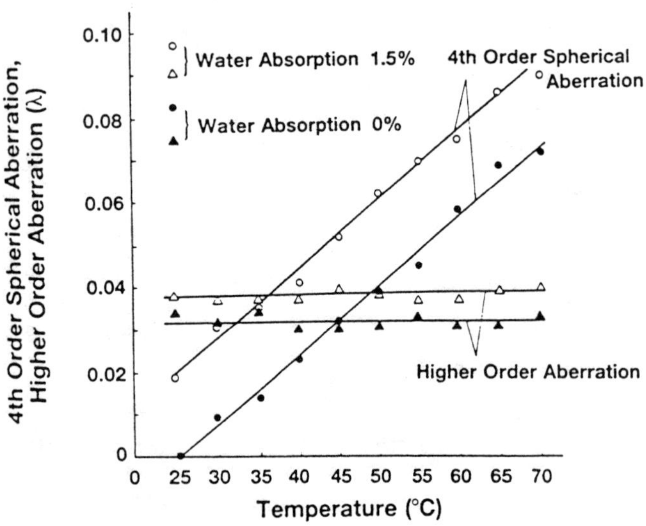

Fig. 14 Experimental Results of Spherical Aberration and Higher Order Aberration Caused by Temperature and Humidity in Plastic Lens

5.2 Wavefront aberration and jitter

The optical pickup of a CD player is always expected to provide a good reproduction performance. In this regard, it is vital to suppress optical aberration as low as possible. The allowable aberration is 0.05 lambda. The quality of reproduced signal is generally expressed by jitter. It is believed that satisfactory reproduction requires about 30 nsec jitter. In our pickup, as described earlier, the coma aberration is canceled by the skew adjustment mechanism shown in Fig. 8 so that it need not be taken into account. As long as it is at a relatively low level, astigmatism does not affect the jitter. To determine the influence of the fourth spherical aberration and higher order aberration on the jitter, we manufactured integral lenses whose surface figure differs significantly from the design figure by deviating the molding conditions intentionally from the optimum values. We measured the RMS wavefront aberration of these integral lenses with surface figure error. We also measured the jitter of these lens mounted in pickups. The results are shown in Figs. 15 through 18. It is understood from these results that astigmatism and higher order aberration do not affect the jitter but that RMS wavefront aberration and the fourth order spherical aberration do. Fig. 17 shows the relationship between the fourth order spherical aberration and the jitter. Fig. 19 shows this relationship in terms of the fourth order spherical aberration W_{SA} and the amount of change $\Delta\sigma$. The following equation (2) holds between W_{SA} and $\Delta\sigma$.

$$\Delta\sigma = 1300\, W_{SA}^2 \qquad (2)$$

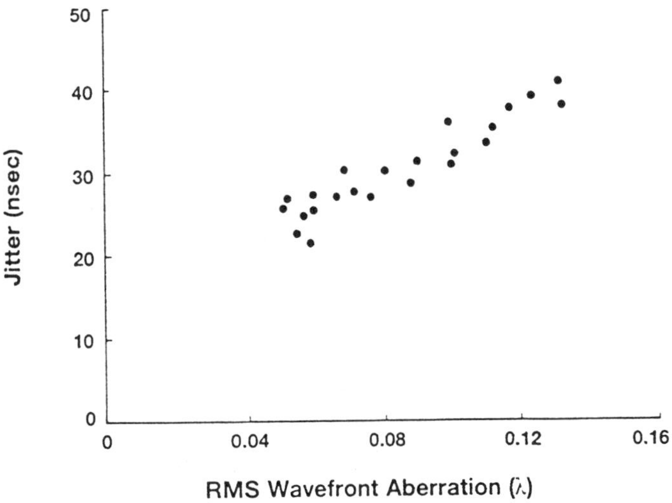

Fig. 15 Experimental Results of Jitter Caused by RMS Wavefront Aberration

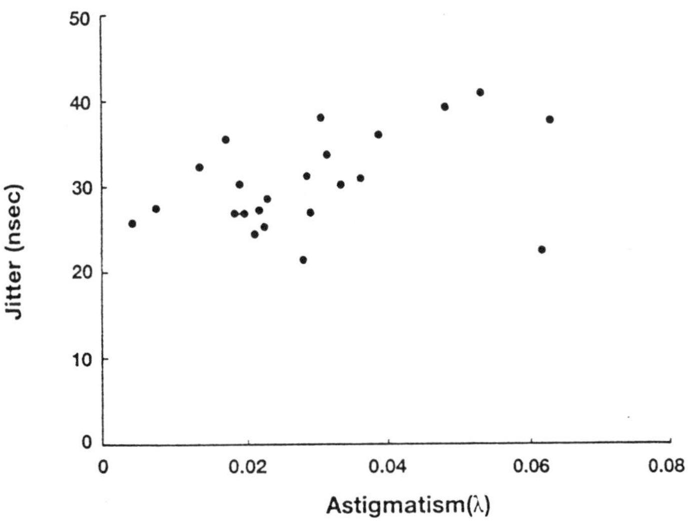

Fig. 16 Experimental Results of Jitter Caused by Astigmatism

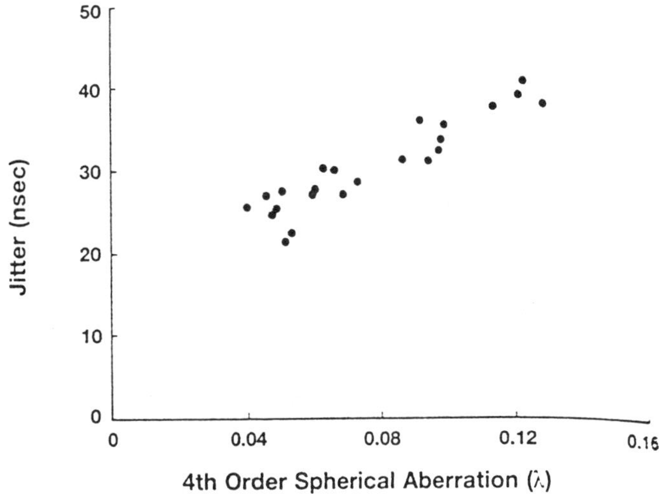

Fig. 17 Experimental Results of Jitter Caused by Spherical Aberration

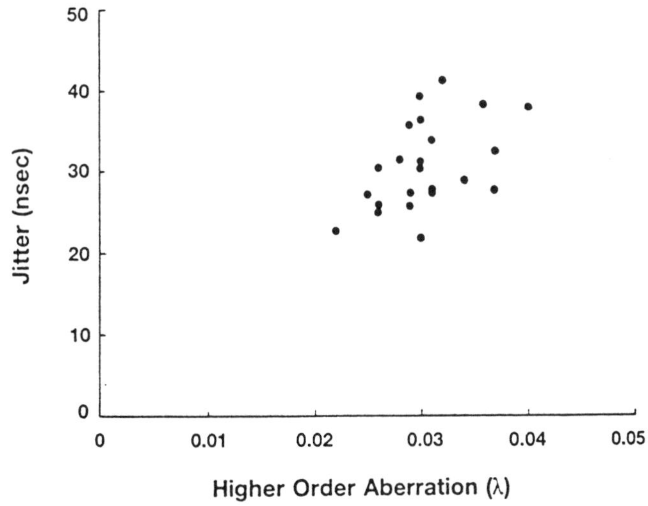

Fig. 18 Experimental Results of Jitter Caused by Higher Order Aberration

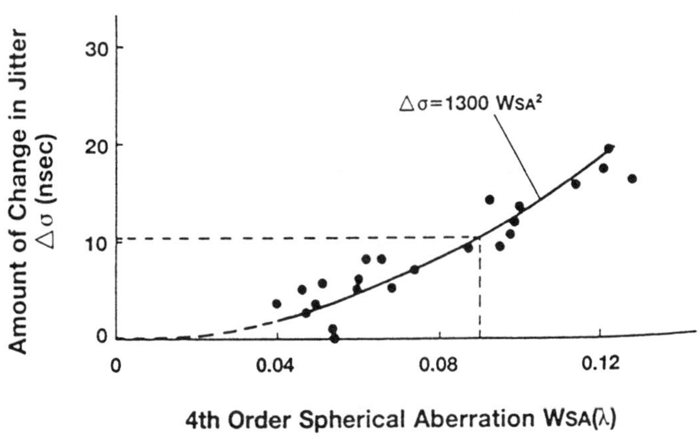

Fig. 19 Relationship between Amount of Change in Jitter and Spherical Aberration

5.3 Pickup reliability

The discussion presented in the sections 5.1 and 5.2 above shows that the RMS wavefront aberration of the integral molded glass lens is independent of temperature and humidity fluctuation. This means that jitter is independent of temperature and humidity. Therefore, the pickup using our integral molded glass lens is extremely reliable. In contrast, the RMS wavefront aberration of the integral molded plastic lens can be higher about 0.06 lambda than that shown in Fig. 18 and the fourth order spherical aberration can be higher about 0.09 lambda than that shown in Fig. 14. According to the calculation using the equation (2):

$$\Delta\sigma = 1300 \times 0.09^2 \simeq 11 \text{ (nsec)}$$

jitter deterioration of about 11 nsec is anticipated. Consequently, the RMS wavefront aberration exceeds 0.05 lambda and it is possible for jitter to exceed 30 nsec.

6. Conclusion

We developed a lens that satisfies compactness, low cost and high resistance to temperature and humidity fluctuation, the requirements by a portable or radio cassette CD player and a CD player for car use. As a result, we have realized a single lens CD player pickup system in which the conventional four or five spherical lenses are replaced by a single bi-aspheric lens without deviating from the specifications required for the CD player optical pickup system. The high precision glass molding technique also developed by us has made it possible to mass-produce the lens.

The portable CD player pickup using the bi-aspheric molded glass lens is superior in performance, size and cost to the conventional type. Thickness of the pickup is reduced 26% from 18.0mm for the conventional pickup using the four-element spherical lens system to 13.3mm for the pickup using the new lens.

The spherical aberration of the integral molded plastic lens increases with temperature and humidity, deteriorating the jitter characteristic of the pickup, whereas that of the integral molded glass lens we have developed is not affected by the temperature and humidity fluctuation. From this fact, it is clear that the integral molded glass lens is more reliable than the integral molded plastic lens.

We believe the pickup using the high precision molded glass lens will expand its application range not only to CD players of fixed installation type but also to CD-ROMs, optical discs, optical magnetic discs and other fields demanding higher performance and reliability.

7. Acknowledgement

We are deeply grateful to Mr. M. Nakajima and all the members committed to this project for their guidance in our study and to the staff of Sumita Optical Glass Manufacturing Co. and of Hi-Fi Audio Division of Matsushita Electric for their kind cooperation in this development.

8. References

1) T. Kiriki et al.: "Plastic aspheric lens for compact disc system," CLEO '84 Tech. Dig. (1984) WB3.
2) R.O. Maschmeyer et al.: "Optical performance of a diffraction-limited molded-glass bi-aspheric lens," Appl. Opt., 22 (1983) 2413 ∿ 2415.
3) H. Nishi et al.: "Gradient index objective lens for the optical disc system," Japanese J. Opt. 15 (1986) 303 ∿ 308.
4) Joseph J.M. Braat et al.: "Design and production technology of replicated aspheric objective lenses for optical disc systems," Appl. Opt., 24 (1985) 1853 ∿ 1855.
5) T. Kiriki: "Refractive Index of Plastic Lens," Japanese J. Opt. 12 (1983) 460 ∿ 463.

6) S. Kubota et al.: "An analysis of eye diagram phase jitter in optical disc system III," Japanese J. Opt. 14 (1985) 140 ~ 143.

Biographies

Masaaki Sunohara is a Senior Engineer of Wireless Research Laboratory at Matsushita Electric Industrial Co., Ltd., Osaka, Japan. He received his B.E. degree and M.S. degree from Kobe University in 1971 and 1973, respectively. In 1973 he joined Matsushita, where he has been engaged in the development of optical disc and optical devices.

Yasuhiro Tanaka is an Engineer of Wireless Research Laboratory at Matsushita Electric Industrial Co., Ltd., Osaka, Japan. He received his B.E. degree and M.S. degree from Osaka University in 1980 and 1982, respectively. In 1982 he joined Matsushita, where he has been engaged in the development of optical devices.

Yoshitome Nagaoka is a General Manager of Wireless Research Laboratory at Matsushita Electric Industrial Co., Ltd., Osaka, Japan. He received his B.E. degree and Ph. D. from Osaka University in 1964 and 1980, respectively. In 1964 he joined Matsushita, where he has been engaged in the development of video equipments and optical devices.

Masaaki Ueda is a Manager of Wireless Research Laboratory at Matsushita Electric Industrial Co., Ltd., Osaka, Japan. He received his B.E. degree from Tokyo University in 1968. In 1968 he joined Matsushita, where he has been engaged in developing various kind of sensors and electronic components.

Koichi Azuma is a Manager of disc development office of Audio Sector at Matsushita Electric Industrial Co., Ltd., Osaka, Japan. He received his B.E. degree and M.S. degree from Shizuoka University in 1971 and 1973, respectively. In 1973 he joined Matsushita, where he has been engaged in the development of audio equipments and optical disc.

Reprinted from *Optical Data Storage,* Proc. SPIE Vol. 1316, pp. 210-214 (1990).

Optical characteristics of micro Fresnel lens for high density memory system

Tetsuo MAEDA, Shiro OGATA, and Koichi IMANAKA

Central R&D Laboratory, OMRON Corporation,
Shimokaiinji, Nagaokakyo-city, Kyoto, 617 Japan

ABSTRACT

We have proposed a novel optical memory system, that is, multilayered optical storage system utilizing a micro Fresnel lens(MFL)[2,3]. The multiwavelength laser light through the MFL is focused on the point where the surface of each layer is laid, since the MFL focal length depends on the incident laser wavelength. We discussed and demonstrated the dependences of MFL's characteristics such as wavefront aberration, focal length, and focused beam spot size on the incident laser wavelength.

1. INTRODUCTION

In these years, the development of optical data storage systems has been advanced and become to be practical use. In these storage systems, however, the memory density is subject to the limitation with the focused optical spot size, and would not be met the demand for higher memory capacity in the near future. In order to overcome such a problem, we have proposed a novel high density and high signal processing speed optical data storage system, that is, multilayered optical storage system[1]. In the system, we have proposed to adopt the micro Fresnel lens(MFL) with unique optical characteristic of wavelength dependency, and multilayered memory media in which each layer has wavelength dependent optical activities. In the previous paper[1], we reported the basic concept for the system, but experimental data were not enough.

In the present paper, we begin with reviews of proposed multilayered memory system, MFL's optical characteristics and fabrication process, and report the detailed optical characteristics of the MFL, focussing and wavefront characteristics for several incident laser wavelengths. The results show that the proposed high density memory system will be promising when the memory media are developed.

2. MULTILAYERED OPTICAL STORAGE SYSTEM

The MFL has an unique optical characteristic that the focal length depends on incident light wavelength as shown in Fig.1, since the MFL works on the theory of diffraction. When an incident laser light with wavelengths of λ_1, λ_2 and λ_3 illuminates the MFL, focused spots of light, which correspond to the incident wavelengths are obtained on the same optical axis. So when memory layers are set on corresponding focal points, focused each light can read and write data on a corresponding memory layer.

The proposed system[1] is illustrated in Fig.2. It mainly consists of multiwavelength laser light source, an objective MFL, multilayered optical memory media, a grating lens for laser light de-multiplexing, and photodetectors. The multiwavelength light source includes a laser diode array, and each laser diode (λ_1, λ_2, λ_3, λ_4) has a different wavelength. These wavelengths are focused by the MFL as separated spots of light on the corresponding surface of the memory layer (L_1, L_2, L_3, L_4). The grating lens deflects the lights reflected from the surface of the memory layers to photodetectors (P_1, P_2, P_3, P_4). In this system, memory data density increases proportionally to the number of memory layers, and high signal processing speed is expected by parallel processing of each layer or wavelength.

3. MFL'S OPTICAL CHARACTERISTICS AND FABRICATION PROCESS

3.1. Theoretical analysis of MFL

Micro Fresnel lens(MFL) is suitable for compact optical devices, because of small lens diameter, short focal length, thin and planer structure[4], and easy production[5]. The MFL is constructed with stepped setback of many divided annular zones as shown in Fig.3. Because the zone periods of the MFL outer portion are close to the optical wavelength, the MFL works on the theory of diffraction.

The optical characteristics of an MFL include wavelength dependence of the focal length and wavelength dependence of the spot diameter at the focal plane. The deviation of focal length δf is approximately given by

$$\delta f= -\delta \lambda \times f_d / \lambda_d \qquad (1)$$

Where f_d is design focal length and $\delta \lambda$ is the wavelength deviation from designing wavelength λ_d. That is, focal length f is in inverse proportion to wavelength λ. The theoretical spot diameter 2ω at the focal plane is calculated by

$$2\omega= K \times \lambda / \tan\theta \qquad (2)$$

Where λ is the illuminating wavelength, and θ is one-half of the angular aperture. θ also depends on λ and the minimum grating pitch Λ_{min} of the MFL as in eq.(3).

$$\theta= \sin^{-1} (\lambda / \Lambda_{min}) \qquad (3)$$

From eq.(2),(3), the focal spot size remains fairly constant, even when the incident wavelength λ varys.

3.2. Fabrication process

The MFL fabrication process[5] is shown in Fig.4. First, a master MFL was fabricated on a glass substrate by using an electron-beam lithography technique(a)-(c). An indium tin oxide(ITO) layer was deposited on the glass substrate to prevent charging-up during an electron-beam writting beforehand. The blazed profile was formed by a suitably chosen electron-dose distribution because the etching rate of the resist depends on the electron dose during development. Next, a stamper was prepared by a conventional nickel electro-forming method(d),(e). The cross-sectional pattern of the stamper is the just inverse of that of the master. After potting a monomer of an UV-curable resin between the stamper and a base glass substrate with anti-reflection coating, photo-polymerization process which apply UV-light to the resin resulted in replica MFLs(f),(g). The cross-sectional profile of the replica MFL was same as that of the master MFL. Fig.5 shows a microphotograph of fabricated MFL which has 780nm design wavelength, 0.25 numerical aperture, ϕ0.5mm lens diameter, and 1 mm focal length. Fig.6 shows a cross-sectional scanning-electron-microscope(SEM) photograph of the MFL.

4. EXPERIMENTAL RESULTS

Fig.7 shows the theoretically calculated and measured focal lengths; these agree well with each other. Fig.8 shows the calculated and measured focused spot diameters. Here the spot diameter 2ω is determined where optical intensity drops to $1/e^2$ at the focal plane. Spot diameters are close to the diffraction-limited spot diameters and little depend on the wavelength.

We measured the optical properties of replicated MFL at room temparature. Fig.9 shows the wavefront pattern passing through the MFL measured by a wavefront analyzer (Zygo WA8100). The result which shows a wavefront aberration of 0.035λ RMS value was sufficiently lower than Marechal criterion $\lambda/14$. Therefore, the diffraction limited characteristics should be promised for the MFL. Fig.10 shows the dependence of the MFL wavefront aberration on the incident laser wavelength of 0.63,0.73,0.78 and 0.83μm: The RMS values of wavefront aberration as a function of incident laser wavelength are plotted. We use the replica MFL with parameters of λ_d=0.78μm, numerical aperture N.A.=0.25, lens diameter D=ϕ0.5mm. These RMS values little depend on the wavelength, and show enough low wavefront aberration even for laser wavelengths different from MFL's design wavelength λ_d. These results show that the multilayered optical memory system can be realized by using the MFL.

5. CONCLUSIONS

We proposed a new optical memory system by using unique optical characteristic of the micro Fresnel lens(MFL) that the focal length depends on the incident wavelength. This system would be able to achieve high memory data density and high signal processing speed. We discussed and demonstrated the dependences of MFL's characteristics of wavefront aberration, focal length, and focused beam spot size on the incident laser wavelength. The results made us sure that the MFL can be applicable to the multilayered optical memory system, when the multilayered optical memory medium is developed.

6. REFERENCES

1. S.Ogata,S.Aoyama and T.Yamashita,MOC/GRIN'89,PP304-307(1989).
2. T.Fujita,H.Nishihara and J.Koyama,Opt.Lett.12,P613(1981).
3. S.Aoyama,S.Ogata,T.Inoue and T.Yamashita,CLEO'88,ThM49(1988).
4. S.Ogata,H.Sekii,T.Maeda,H.Goto,T.Yamashita,and K.Imanaka,IEEE Photonics Tech.Lett.Vol.1(1989).
5. M.Tanigami,S.Ogata,S.Aoyama,T.Yamashita,and K.Imanaka,IEEE Photonics Tech. Lett.Vol.1(1989).

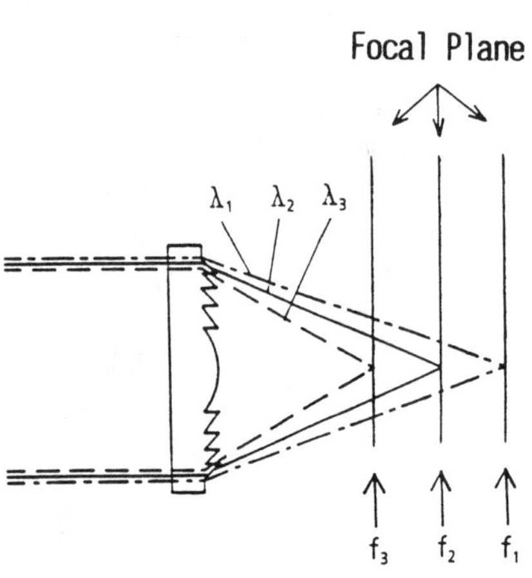

Fig.1. Focal length versus incident wavelength.

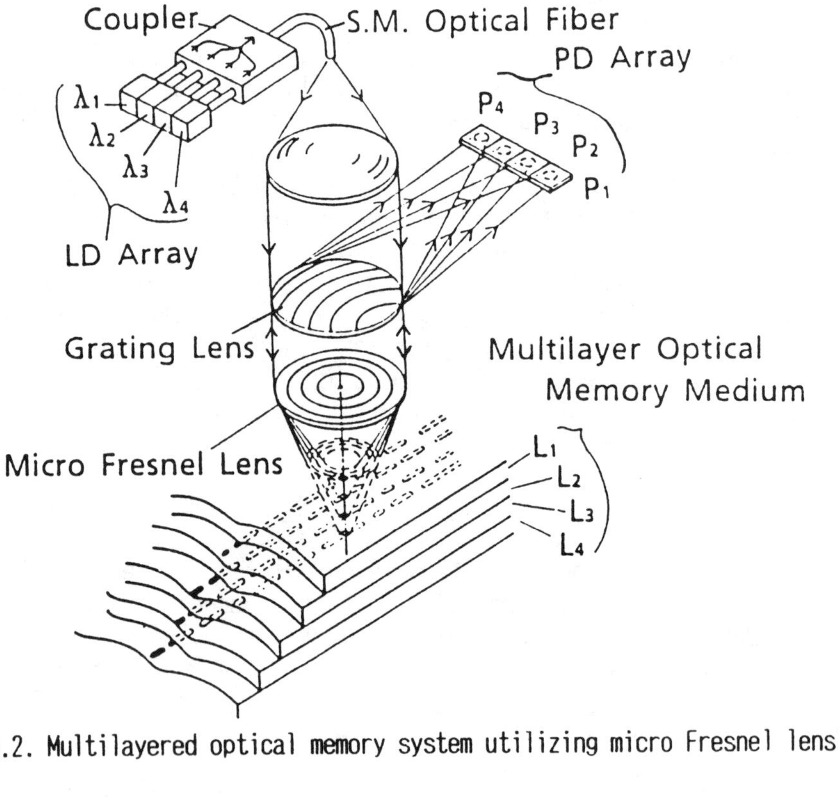

Fig.2. Multilayered optical memory system utilizing micro Fresnel lens.

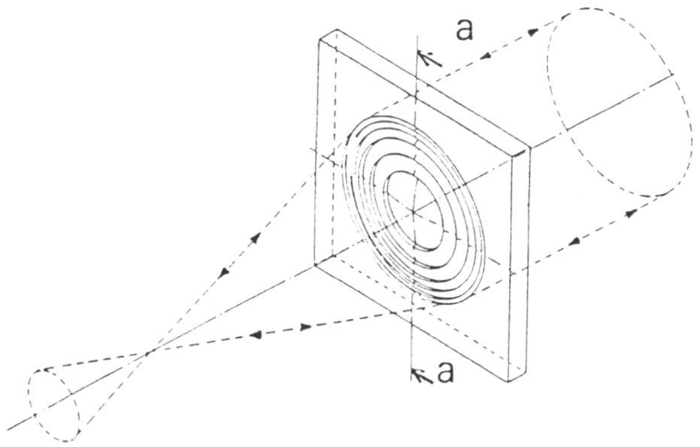

Fig.3(a). Illustration of a micro Fresnel lens.

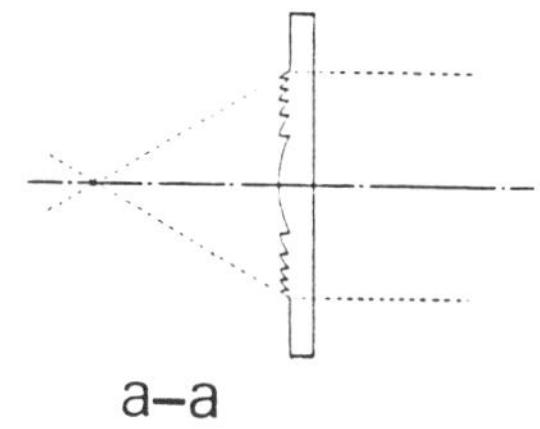

Fig.3(b). Cross-sectional view of a micro Fresnel lens.

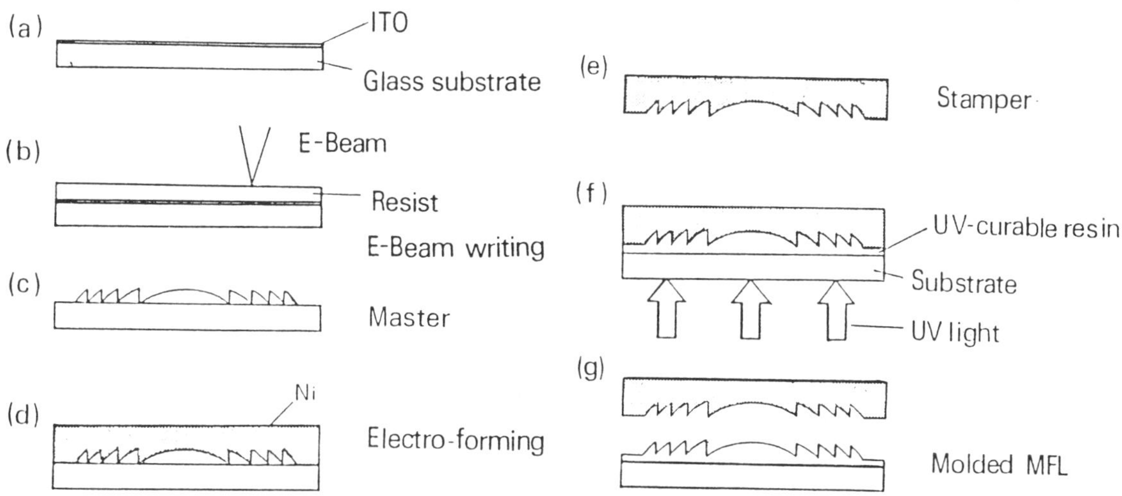

Fig.4. Fabrication process of a micro Fresnel lens.

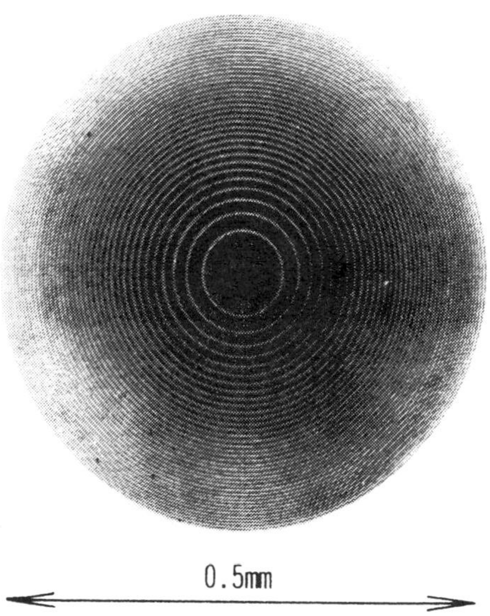

Fig.5. Microphotograph of fabricated micro Fresnel lens.

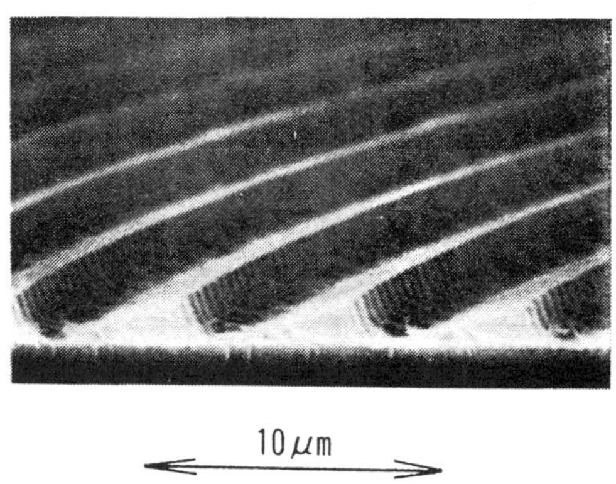

Fig.6. SEM cross-sectional photograph of micro Fresnel lens.

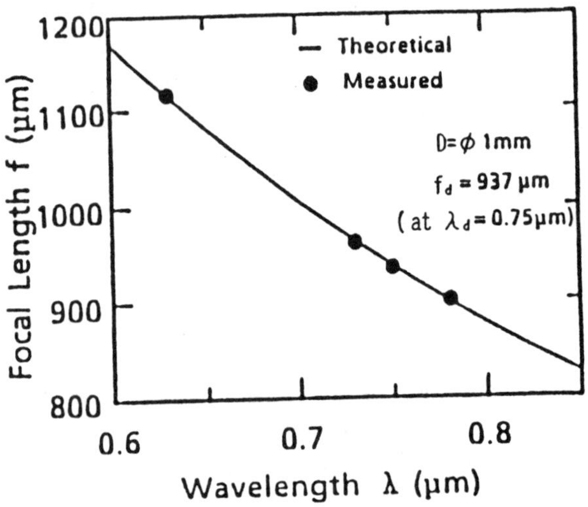

Fig.7. Wavelength dependence of focal length.

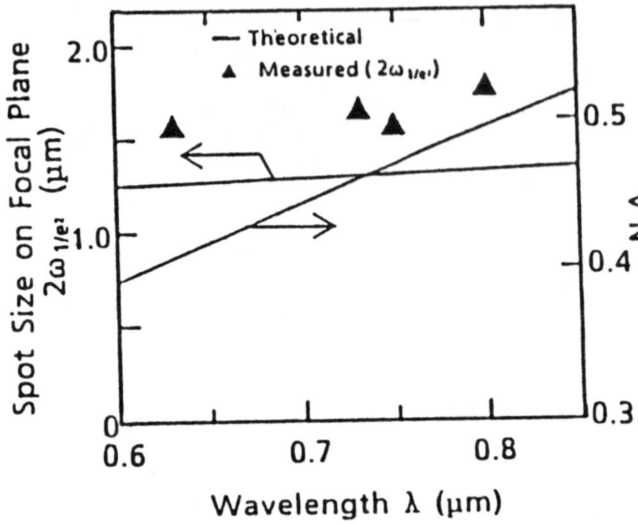

Fig.8. Spot size at focal plane versus wavelength.

Fig.9. Wavefront pattern passing through the lens. (Zygo WA8100)

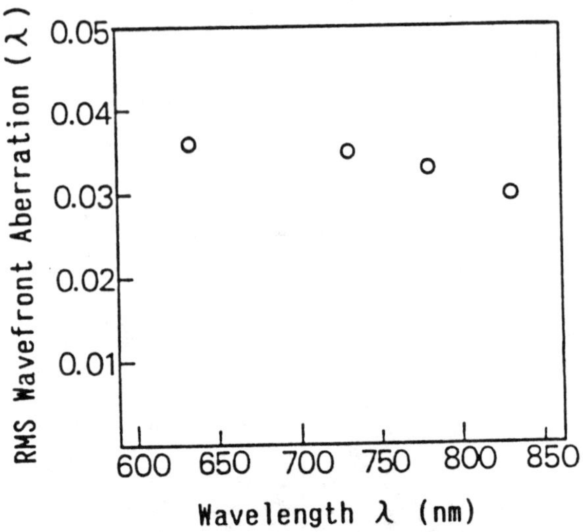

Fig.10. Wavefront aberration versus wavelength.

Section Five
Holographic Optical Elements

Reprinted with permission from *Optical Engineering*, Vol. 28(6), pp. 650-653 (June 1989). ©1989 Society of Photo-Optical Instrumentation Engineers.

Holographic optical head for compact disk applications

Wai-Hon Lee, MEMBER SPIE
Pencom International Corporation
776 Palomar Avenue
Sunnyvale, California 94086

Abstract. Considerations in using holographic optical elements and a laser/detector hybrid device for optical disk systems are discussed. The process for fabricating the holographic optical elements is briefly reviewed.

Subject terms: holography; optical data storage; holographic optical elements.

Optical Engineering 28(6), 650-653 (June 1989).

CONTENTS

1. INTRODUCTION

Since the advent of holography, there has been a strong interest in replacing the lenses used in optical systems by holographic optical elements (HOEs). Because HOEs are diffractive optical elements, it is difficult to use them for applications in which a coherent light source is not available. In the past decade, with the availability of semiconductor lasers, it has been much simpler to employ an HOE in many optical data storage and retrieval systems. For example, Ohki and Kojima[1] proposed the use of HOEs to replace the collimating lens, beamsplitter, and objective lens in compact disk optical heads. In late 1986 Kimura et al.[2] presented a paper on using an HOE to replace the beamsplitter, the cylindrical lens, and the three-beam grating in optical heads for compact disk players. The HOE allows a reduction of component count in the manufacturing of the compact disk optical head. At the same time, another holographic optical head was developed by Lee.[3,4] Unlike Kimura's optical head, Lee's optical head contains a laser/detector hybrid device. In late 1987 Tatsumi et al.[5] published a paper on the use of a reflection HOE in a compact disk optical head. Again, in Tatsumi's optical head the laser diode and the photodetector are not packaged into a hybrid device. Without using a laser/detector hybrid device, the alignments of Kimura and Tatsumi's optical head is similar to a conventional optical head, where the position of the photodetector is adjusted to the center of the returned beam. On the other hand, the photodetector in Lee's optical head is fixed in position. The alignment in Lee's optical head is achieved by the adjustment of the HOE.

Invited Paper AH-106 received Dec. 9, 1988; revised manuscript received Feb. 16, 1989; accepted for publication Feb. 21, 1989. This paper is a revision of Paper 884-14, presented at the SPIE conference Computer-Generated Holography II, Jan. 11-12, 1988, Los Angeles, Calif. The paper presented there appears (unrefereed) in SPIE Proceedings Vol. 884.

The importance of using an HOE is not only to simplify the optical system to achieve cost reduction but also to provide a better means to align the optical system to provide reliable signals for servo control of the optical storage systems. In this paper we discuss the design and alignment of a holographic optical head that uses a laser/detector hybrid device. The process for fabricating the holographic optical elements is also briefly reviewed.

2. HOLOGRAPHIC OPTICAL HEADS FOR CDs

Figure 1 shows an optical head based on the HOE and laser/detector hybrid technology. In operation, the laser beam from the diode laser in the hybrid device passes through a collimating lens. This beam is focused on the disk through an objective lens mounted on an actuator. In the forward path, the laser beam also passes through a diffraction grating (DG) and an HOE. The diffraction grating is used to produce the tracking error signals, and the HOE is used to produce the focusing error signal, as will be explained.

The diffraction grating splits the laser beam into three beams to produce three focused spots on the disk surface. The arrangement of the three spots is shown in Fig. 2. About 65% of the incident light remains in the center spot. The diffraction efficiencies at the two extreme spots are roughly 10%. The center spot is for retrieving information; the two extreme spots are for obtaining the tracking error signals. When the center spot is on track, the two extreme spots are symmetrically placed on each side of the track, as shown in Fig. 2(b). On the other hand, when the center spot is off track, as shown in Figs. 2(a) and 2(c), the light returned from one of the extreme spots will be more intense than from the other extreme spot. This difference in signal levels corresponds to the tracking error.

The HOE diffracts the outgoing beam into three beams. The optical efficiencies of the undiffracted and the diffracted beams are about 50% and 15% of the incident energy, respectively. Only the undiffracted beam is focused on the disk surface. The diffracted beams from the HOE are not used in the forward optical path. On the return path, the three beams from the diffraction grating are reflected from the disk and pass through the HOE again. One of the first-order diffracted beams from the HOE is now focused in conjunction with a collimating lens on a quadrant detector located near the laser diode chip. An example of a hybrid device[4] is shown in Fig. 3. As shown, the emitting surface of the laser diode is separated from the detector surface by about 1 mm and the lasing junc-

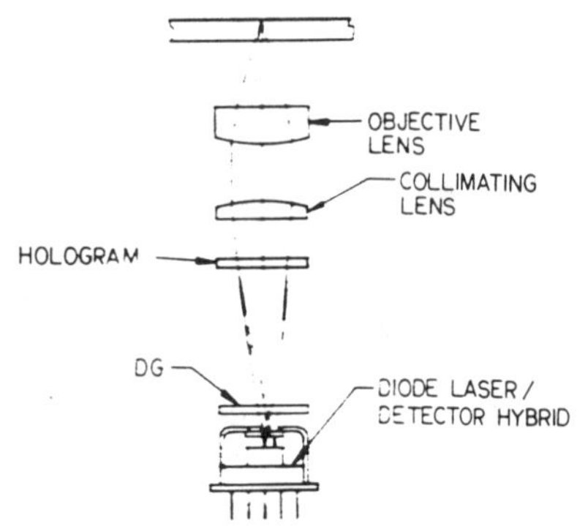

Fig. 1. Holographic optical head using three beams for tracking.

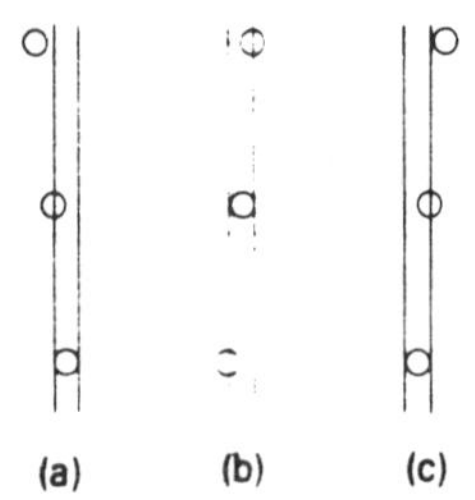

Fig. 2. Three beam trackings: (a) and (c) center beam off track; (b) center beam on track.

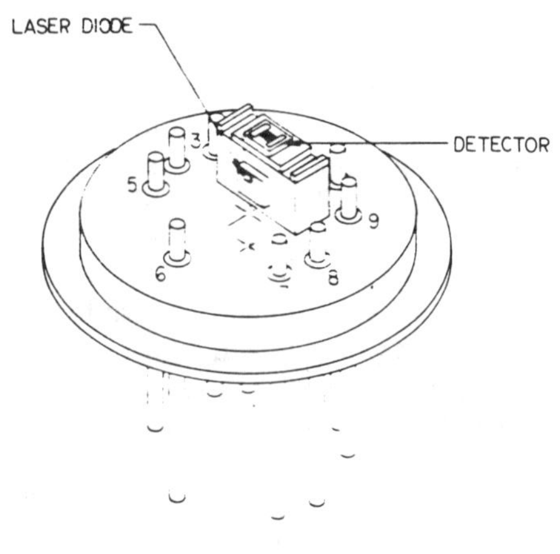

Fig. 3. Example of laser/detector hybrid device.

tion is separated from the center of the photodetector by about 1 mm. The spatial frequency of the HOE is designed to diffract the returned light to the center of the detector. The HOE also serves as a cylindrical lens to produce a focus error signal. This holographic cylindrical lens introduces astigmatism to the returned beam. The focusing condition of the laser beam on the disk surface can be detected from the signal generated by the astigmatic beam on the detector, as shown in Fig. 4. When the spot is in focus, the beam on the detector is round. If the point of focus is above or below the disk's surface, the beam on the detector is either a vertical elliptical beam or a horizontal elliptical beam. The focus error is then simply given by (D1 + D3) − (D2 + D4).

One of the advantages of using an HOE in optical heads is the flexibility and simplicity it allows in the optical system design. For example, in the optical head with a collimating

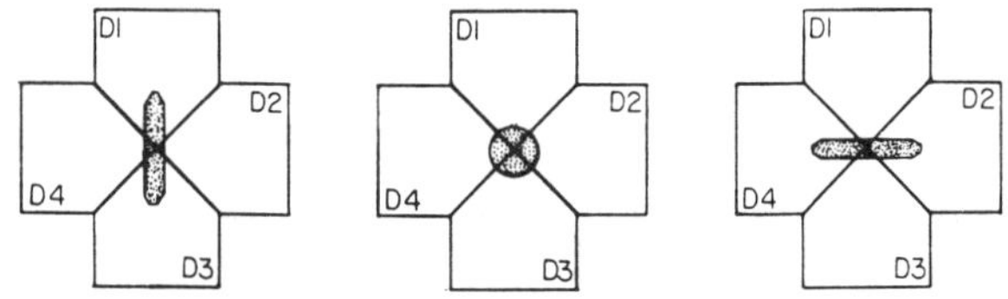

Fig. 4. Focus error detection using astigmatic beam.

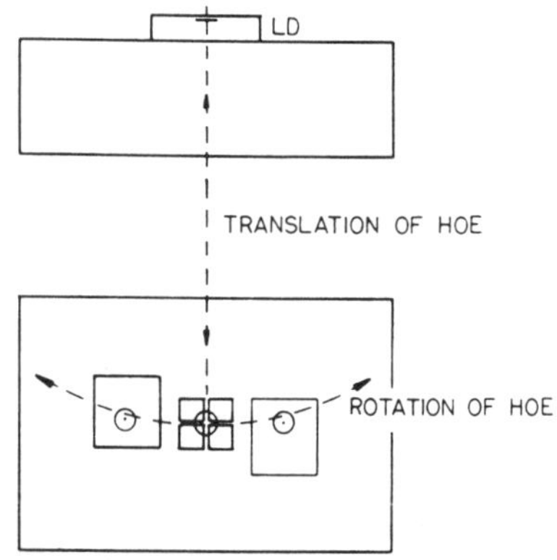

Fig. 5. Alignment of returned beam on photodetector.

lens, the HOE can be placed either outside or inside the collimating lens. With the HOE outside the collimating lens, it is possible to use a collimating lens with larger NA to collect the laser light. For a laser diode with divergence angle between 15° and 26°, the light collection efficiency can be as high as 70% or more with a collimating lens of NA = 0.25. The focal length of such a lens is about 9 mm and is too short for use in a conventional compact disk optical head, which requires space within the collimating lens for a half mirror or a beamsplitter. It is obvious in Fig. 1 that the collimating lens and the objective lens can be combined into a single aspherical lens. Moreover, the HOE can be of either the transmissive or the reflective type.

3. ALIGNMENT WITH HOEs

The astigmatic beam needed to generate the focus error signal for the compact disk player is formed by an HOE with the phase function

$$\phi(x,y) = \frac{\pi}{\lambda}\left(\frac{x^2}{F_x} + \frac{y^2}{F_y}\right) . \qquad (1)$$

This phase function represents a parabolic wavefront with different focal lengths along the x and y axes. When a laser beam passes through the HOE, the beam will have two astigmatic foci. The HOE with the phase function in Eq. (1) also has the property that when it is translated across a laser beam, it causes the laser beam to change its direction of propagation. This is one of the principles used in many holographic laser beam deflection systems.[6] Laser beam deflection can be the result of the rotation of the grating. These two properties of the HOE are used in the holographic head in Fig. 1 to align the returned beam precisely on the center of the detector. Figure 5 shows the relationship between the laser diode and the detector. The laser diode is along the optical axis. Hence, as the HOE is rotated, the diffracted beam is rotated about the laser diode. A linear translation of the HOE will move the diffracted beam in the radial direction. It is possible to find a position of the HOE to produce a precise placement of the return beam on the detector.

4. FABRICATION OF COMPUTER-GENERATED HOEs

Many HOEs are made by recording the interference pattern between two laser beams. This requires setting up a very stable optical table to produce such an optical component for laser beam deflection. An added difficulty in making an HOE by the interference method is that the object wavefront for the HOE sometimes cannot be derived from an existing optical component. The use of the computer-generated hologram technique[7] solves these two difficulties. The fringe pattern corresponding to a particular HOE is calculated by a computer following a mathematical description. These fringes, after being calculated, are plotted by an e-beam device or a photoplotter. In some cases even a large pen plotter can be used. After the pattern is created and photoreduced to the proper dimensions, it can be used as a mask for the production of the thin phase HOEs. In this section we briefly discuss how to control the diffraction efficiency of a thin phase HOE and a method for its fabrication.

4.1. Light efficiency

One of the deterrents to employing an HOE in many optical systems is the light efficiency of these components. Maximum diffraction efficiency in the +1 diffracted order of the diffractive optical elements ranges from 5% to 100%. It is well known that when amplitude holograms are made on a photographic plate, the maximum diffraction efficiency in the first diffracted order is about 6%. A thin phase hologram recorded on photoresist material or bleached photographic plate has a maximum diffraction efficiency of about 40%. Of course, a thick phase hologram recorded on a dichromated gelatin plate, or photopolymer, has a theoretical diffraction efficiency of 100% in the first of the diffracted orders.

The reason an HOE can be used in optical heads for the compact disk player discussed above is that more than one diffracted order from the HOE is utilized by the optical system. Therefore, the diffraction efficiency in one diffracted order is not as important. The main consideration in designing the HOE is obtaining a high total light utilization rather than trying to concentrate all of the incident light into one diffracted order. For example, in the case of the optical head in Fig. 1, the 0 order, +1 order, and −1 order from the three-beam grating are used to produce three spots on the disk surface. This three-beam grating has 65% of the incident light remaining in the 0 order and diffracts 10% of the incident light into each +1 and −1 diffracted order. The total light efficiency of the three-beam grating is, therefore, 85%. Similarly, the computer-generated HOE used in optical heads in Fig. 1 has about 50% of the incident light in the 0 order and diffracts 17% of the incident light into each of the first diffracted orders. Although this HOE has only 17% of the incident light in the first diffracted order, it fulfills the requriement of the optical head by providing enough beam intensity to the recording medium and

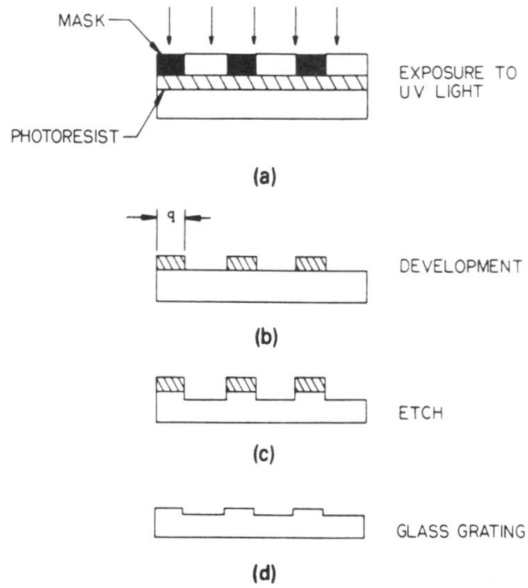

Fig. 6. HOE fabrication process.

to the detector. One common feature between the HOE and the three-beam grating in Fig. 1 is that there is more than 50% of the incident light remaining in the 0 order. This requirement of these two HOEs can be met by thin phase holograms, which can be replicated easily either by a photographic process or by an embossing process.

A simple formula[8] to calculate the normalized intensity of each diffracted order of a periodic thin phase grating is

$$A_0 = [(2q-1)\sin\theta]^2 + \cos^2\theta \ ,$$
$$A_l = \left[2\left(\frac{\sin m\pi q}{m\pi}\right)\sin\theta\right]^2 , \tag{2}$$

where m is the diffracted order and q is the duty cycle of the grating, as shown in Fig. 6, and is related to the optical thickness of the grating d by the formula

$$\frac{d(n-1)}{\lambda} = \frac{\theta}{180} , \tag{3}$$

where λ is the wavelength of the light source and n is the refractive index of the recording material. The diffraction efficiency in percent is obtained by multiplying A_0 or A_l by 100. For example, if we choose q to be ½ for the grating shown in Fig. 6, the phase angle to obtain 70% efficiency in the 0 order beam is 33.2°. For $\lambda = 0.8$ μm and $n = 1.5$, the optical thickness d is about 0.295 μm. The diffraction efficiency at the first order is about 12%.

In practice, it is difficult to maintain the duty cycle q at exactly ½. Hence, the obtainable efficiency could be slightly off the optimum value predicted by the above formulas. In any case, the total light efficiency of a thin phase grating after combining the contributions from the 0 order and other higher orders is high enough to be useful for many practical applications.

4.2. Fabrication process

Another feature of the HOE discussed above is that the space-bandwidth product of those components is small compared to the space-bandwidth products of most lenses. Space-bandwidth product is a measure of the complexity of the HOE. In many cases the space-bandwidth product is also equal to the number of fringes in the HOE. For example, the spatial frequency for the three-beam grating is about 40 line pairs/mm and the diameter of the grating is about 4 mm. The space-bandwidth product or the number of fringes of this three-beam grating is 160. The typical space-bandwidth product for the HOE used in the optical head ranges from 600 to 3000.

For moderate space-bandwidth products the mask for this type of HOE can be generated using an e-beam or a photoplotter. Once the mask is created, it can be replicated to glass plates overcoated with photoresist. Since the photoresist is a transparent material having a refractive index close to that of the substrate material, the diffraction efficiency of the grating is determined by the optical thickness of the photoresist layer as given by the above formulas. Two methods can be used to control the thickness of the photoresist. One method is to control the thickness during the spin coating process. The other is to use a thick layer of photoresist. The thickness variation of the relief structure for the grating is obtained by thinning out the photoresist layer in a controlled manner. The photoresist layer that has been exposed to light will be dissolved by the development solution. The amount of photoresist to be removed is determined by the exposure time and the development time during the replication process.

Figure 6 illustrates a process in which the grating structure is etched on a glass substrate. The mask is first copied to a photoresist layer. After being developed, the plate is immersed in hydrofluoric acid to remove the proper thickness of glass material. Once the etching rate and etching time are established, the fabrication process is very repeatable and reliable.

Recording the HOE on glass produces a very stable optical component. Unlike components recorded on photoresist, the glass component can be cleaned whenever required.

5. SUMMARY REMARKS

We have discussed some general principles of the holographic optical head. There are many ways to use HOEs in optical heads; however, to take full advantage of the HOE, we must also package the laser and the detector into a hybrid device. Only then can we significantly reduce the alignment complexity in optical heads for compact disk players and for optical memory systems.

The search for opportunities to replace glass lenses by HOEs continues. There is still a need for more efficient recording methods and materials; however, with the practical applications discussed, we have demonstrated that HOEs have their place in optical system designs whenever they can provide unique functions that cannot be performed in a convenient way by refractive optical components. In those situations, the total light utilization by the HOE can be very efficient compared to conventional optical components. Moreover, HOEs can be computer generated and can be recorded on thin phase material such as photoresist or even etched on glass.

6. REFERENCES

1. H. Ohki and C. Kojima, U.S. Patent 4,458,980 (July 1984).
2. Y. Kimura, Y. Ono, S. Sugama, and Y. Ohta, "An optical head using multifunctional hologram for CD players," Microoptics News 5(3), 228 (1986). In Japanese.
3. W.-H. Lee, U.S. Patent 4,731,772 (March 1988).
4. W.-H. Lee, U.S. Patent 4,757,197 (July 1988).
5. K. Tatsumi, T. Matsushita, and S. Ito, "A multi-functional reflection type grating lens for the CD optical head," Microoptics News 6(3), 151 (1988). In Japanese.
6. O. Bryngdahl and W.-H. Lee, Appl. Opt. 15, 183 (1976).
7. W.-H. Lee, "Computer-generated holograms: techniques and applications," in *Progress in Optics*, Vol. 16, E. Wolf, ed., North-Holland, Amsterdam, p. 121 (1978).
8. W.-H. Lee, "High efficiency multiple beam gratings," Appl. Opt. 18(13), 2152–2158 (1979).

Wai-Hon Lee earned his doctorate degree from the Massachusetts Institute of Technology and has accumulated more than 20 years' experience in the optical data storage field. Dr. Lee began his career with Harris Corp. working on holographic optical memory and optical processors. He then joined Xerox Palo Alto Research Center (PARC) as a member of the research staff in the area of holography and optical data storage. Later, as manager of laser recording for Xidex, he developed laser recording materials and low cost data storage systems. He was also a senior optoelectronic engineer at Optical Storage International, where he was responsible for enhancing the manufacturing process of optical subsystems in optical disk drives. Currently, Dr. Lee is the chief technical officer of Pencom International Corp.

Reprinted from *Holographic Optics: Optically and Computer Generated,* Proc. SPIE Vol. 1052, pp. 150-157 (1988). ©1988 Society of Photo-Optical Instrumentation Engineers.

Computer generated holographic optical elements for optical disk memory read write heads

Yuzo Ono, Yasuo Kimura and Nobuo Nishida

NEC Corporation, Opto-Electronics Research Laboratories
1-1, Miyazaki 4-chome, Miyamae-ku, Kawasaki 213, Japan

ABSTRACT

A new concept optical head, using a holographic optical element has been developed. This holographic element combines three optical functions, beam splitter, focusing error detection and tracking error detection optics functions. A polarizing beam splitter function is also combined for magneto-optical disk head use. These holographic optical elements are computer generated holograms fabricated using electron beam lithography technique. A new optical configuration has been designed for this optical head to remove the influence of wavelength variation in the laser diode used for a light source. The off-set signal in tracking error signal has been reduced by introduction of the holograms which are used only for detecting tacking error. The stable operation and high read out S/N ratio have been obtained by applying this holographic optical element.

1. INTRODUCTION

The success of high density recording in optical disk memories, such as compact disk (CD), laser video disk (VD), write once optical disk (WO) and magneto-optical disk (MO), is based on the highly accurate optical interconnection capability of coherent light between recording pit and read write optical head. To realize this optical interconnection, the optical head has three optical functions in addition to light beam focusing on a disk: splitting the back reflected beam on a disk, which carries the read out signals, from the outward optical path; focusing error detection optics; and tracking error detection optics. Conventional optical heads have employed several bulky polished optical elements to realize these three optical functions. For example, they are a beam splitter prism, a cylindrical lens and a spherical lens pair, and a grating, respectively. This complicated optics set up has prevented realizing small size and light weight optical heads.

In optical disk memories, however, light weight and small size optical heads are required for high speed track access. Further, low cost is required for optical heads (pickups) used in home consumer players, such as CD and VD. To satisfy these requirements, the authors have developed holographic optical elements (HOEs) which combine the above mentioned three optical functions and have applied them to CD, VD and MO optical disk heads.[1~7]

This paper describes principles of error signal detection and beam splitter functions for the optical heads using HOEs, fabrication method and characteristics for HOEs and operation characteristics for developed CD, VD and MO optical heads.

2. ERROR SIGNAL DETECTION PRINCIPLE

2.1 Focusing error signal detection principle

To explain the error signal detection principle for optical heads using HOEs, the optical configuration for a CD player head is shown in Fig. 1. This optical head consists of only four optical elements: a laser diode (LD) used for a light source; a holographic optical element (HOE) consisting of four different holograms, $H_1 \sim H_4$; an imaging lens; and a photo detector (PD) having six sensor segments, $S_1 \sim S_6$. The operating principle is as follows. The light emitted from the LD transmits through the HOE and is focused on the disk by the imaging lens. About a half of the light emitted from the LD is, of course, diffracted out of the optical axis by the HOE. The light reflected on the disk carrying the recorded data is transformed into a spherical wave, converging to the light emitting point on the LD by the imaging lens, and illuminates the HOE. In the individual hologram, $H_1 \sim H_4$ on the HOE, the interference fringes, generated by a divergent spherical wave, which diverges from the LD, and a divergent spherical wave, which diverges from point P_n $(n : 1 \sim 4)$ on the PD, are recorded, respectively, in accordance with the subscript numeral n. Point P_1 lies on the X-axis, which is a boundary between S_1 and S_3, while point P_2 lies on the X-axis, which is a boundary between S_2 and S_4. Points P_3 and P_4 lie on S_5 and S_6, respectively, as shown in Fig.1. Therefore, when the light beam is exactly focused on the disk, the incident light on the hologram, $H_1 \sim H_4$, is diffracted so as to focus on P_1 to P_4, respectively, as shown in Fig.1. The compact disk read out signal is obtained by the sum of the $S_1 \sim S_4$ output powers.

In this optical head, the focusing error signal is derived by using the double knife edge method, that is Foucault's method. The boundary line between holograms H_1 and H_2 acts effectively as a knife edge. Figure 2 shows the spot shapes on the PD. When the disk is in focus (Fig.2(b)), the two central light spots focus in the X-axis. When the disk moves too close to the head (Fig.2(a)), the two central defocused light spots lie on only S_2 and S_3. On the contrary, when the disk moves too far away (Fig.2(c)), the two defocused light spots lie on only S_1 and S_4. Therefore, focusing error signal S_{fe} is given by $S_{fe} = (V_1 + V_4) - (V_2 + V_3)$, where $V_1 \sim V_4$ are output powers for $S_1 \sim S_4$, respectively.

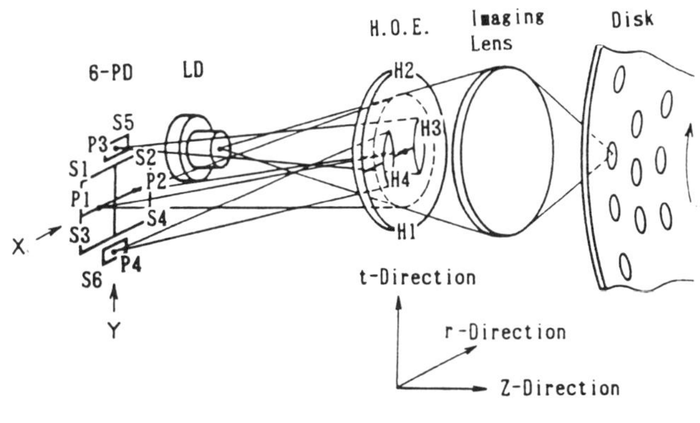

Fig.1 Optical configuration of holographic optical head for CD players.

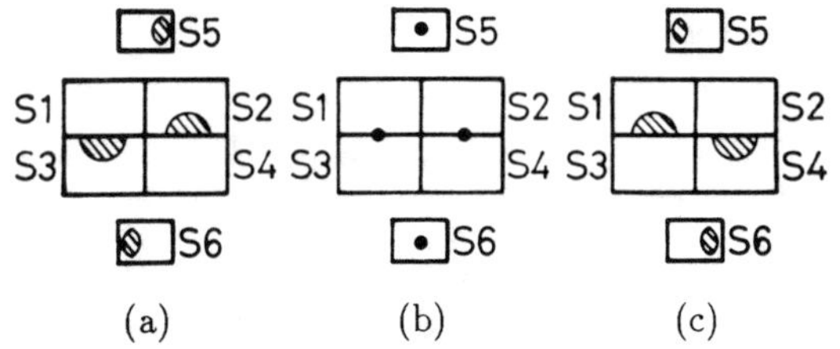

Fig.2 Spot shapes variation on PD segments $S_1 \sim S_6$ caused by defocusing. (a) When the disk is too close. (b) In focus position. (c) Too far away.

2.2 Reduction in wavelength fluctuation influence

Holographic optical elements have not been applied to an optical head, because of the sensitivity to wavelength fluctuation in the light source. In this optical head, the HOE is designed so that HOE lateral aberration of due to the wavelength variation in the light source, that is, variation in diffraction angle, arises along the X-axis on the PD. The size of each sensor segment for the PD is designed sufficiently larger than the fluctuation width of the focused spots on the PD, caused by wavelength variation, to avoid the influence. On the other hand, to remove the longitudinal aberration due to the wavelength variation, that is, the variation in the hologram focal power, holograms are designed to make the focal powers be zero by generating holograms using interference between two almost identical divergent spherical waves. Therefore, the focusing error signal is not affected by the wavelength variation..

The wavelength variation influence, that is, the appearance of the focus error signal off-set, has been numerically analyzed using the ray-tracing method[8]. Figure 3 shows the optical configuration and coordinate system. The hologram is on the $x - y$ plane. The light emitting point on the LD and the light beam converging point on the PD are arranged at $(0, 0, z_{LD})$ and $(0, y_f, z_f)$, respectively. In the numerical analysis, the following values are used; distance between LD and hologram is 18 mm, distance between LD and the PD center is 7.6 mm, longitudinal magnification for the imaging lens is 20.67, designed wavelength is 780 nm.

The numerical results shows that the focusing error signal off-set, caused by the wavelength deviation due to LD production lot difference, can be removed by the PD position adjustment in the manufacturing process. The focusing error signal off-set, due to the wavelength fluctuation during the head operation, was also analyzed. When the PD is located at a point where $z_{LD}^2 = y_f^2 + z_f^2$, the focusing error signal off-set due to wavelength fluctuation is minimized to $\Delta F/\Delta\lambda = 0.0025\,\mu m/nm$, as shown in Fig.4. This value is far smaller than an acceptable focusing error tolerance for the optical disk operation.

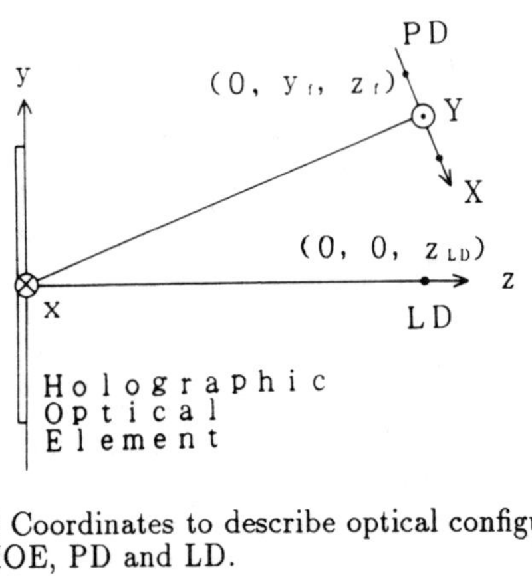

Fig.3 Coordinates to describe optical configuration for HOE, PD and LD.

2.3 Tracking error signal detection principle

Tracking error signal is derived by using the push-pull method. A usual push-pull tracking method, however, has signal off-set problems, as shown in Fig.5. With any movement of the tracking actuator, as shown in Fig.5(a), an optical axis discrepancy arises between the imaging optics and the error detection optics. In such a case, the cross-section of the reflected beam on PD_1 and PD_2 becomes different. Consequently, a signal off-set in the tracking error signal is caused. Any appreciable inclination of the disk will also cause this same type of signal off-set, as shown in Fig.5(b). In order to suppress this signal off-set, tracking holograms H_3 and H_4 are arranged symmetrically on both off-axis sides of the HOE. The tracking error signal S_{te} is derived by the incident light power difference between holograms H_3 and H_4. That is, S_{te} is given by $S_{te} = V_5 - V_6$, where V_5 and V_6 are output power values for S_5 and S_6 sensor segments, respectively.

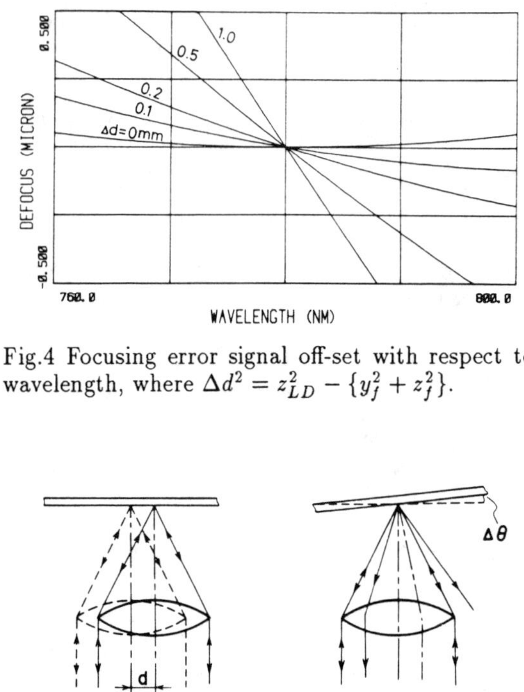

Fig.4 Focusing error signal off-set with respect to wavelength, where $\Delta d^2 = z_{LD}^2 - \{y_f^2 + z_f^2\}$.

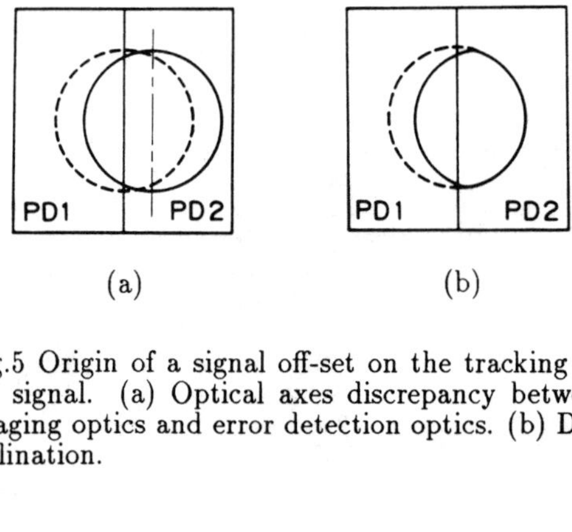

Fig.5 Origin of a signal off-set on the tracking error signal. (a) Optical axes discrepancy between imaging optics and error detection optics. (b) Disk inclination.

In this configuration, the shapes for tracking holograms H_3 and H_4 are optimized, in order to greatly reduce the tracking signal off-set. In the optimization, two factors are taken into account, the discrepancy value between the axis for imaging optics and error detection optics, generated by tracking actuator movement, and light power distribution in the reflected beam on the disk.

Figure 6 shows the relation between the axis for imaging optics and error detection optics. The light power distribution in the reflected beam has a split image, one crescent on each side of the ball shape, as shown in Fig.6. The dotted parts in Fig.6 show the most sensitive part for tracking error in the reflected beam. In these regions, the zero-th order diffracted light overlaps with the plus or minus first order light diffracted by pits on the disk. Tracking holograms H_3 and H_4 are arranged so that H_3 and H_4 are always in the right and left dotted parts, respectively, in spite of any tracking actuator movement, as shown in Figs.6 (a) and (c). In Fig.6, the s value shows the maximum tracking actuator movement permitted within the above mentioned condition. The light power distribution for each dotted part of the reflected beam is almost uniform in each dotted part, as the calculated distribution shows in Fig.7, so that there is no change in the incident light power for an individual tracking hologram with any tracking actuator movement. Therefore, signal off-set is suppressed. From the light power distribution analysis, the shape and configuration of the tracking holograms are given by,

$$\begin{cases} \{x + J(\frac{f\lambda}{p} + s)\}^2 + y^2 \leq r^2, \\ (x - Js)^2 + y^2 \leq r^2, \end{cases} \tag{1}$$

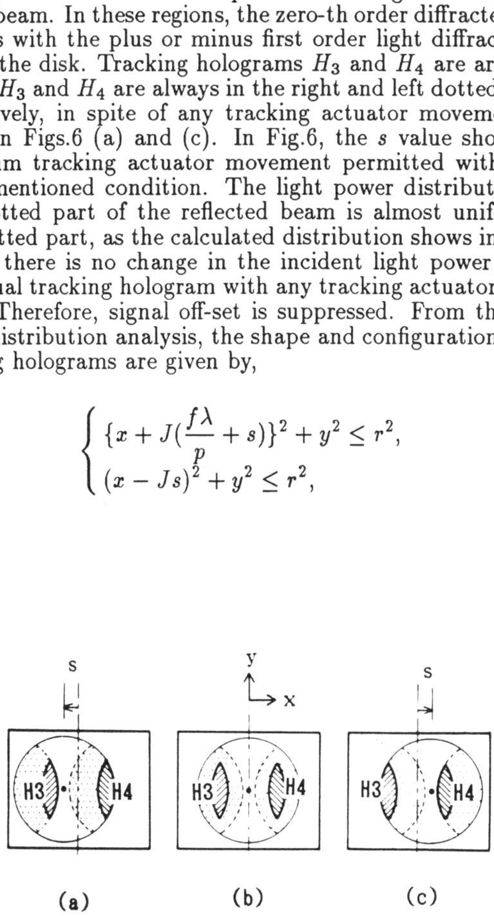

Fig.6 Relation between axes for imaging optics and error detection holograms.

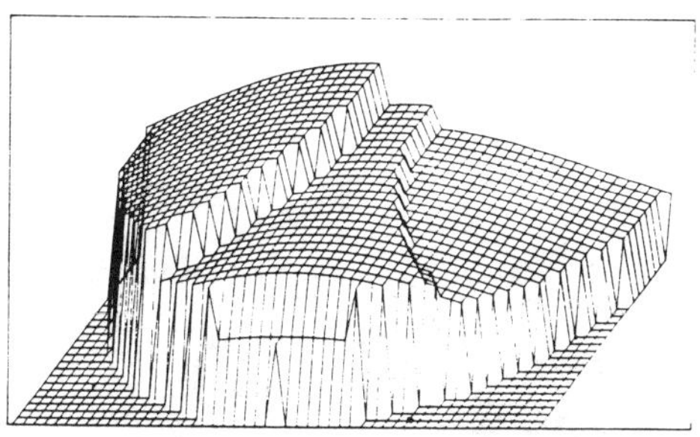

Fig.7 Light power distribution example, showing reflected beam cross section when focused beam is off-track position.

where,

J : +1 for H_3 and -1 for H_4,
x and y : coordinates shown in Fig.6 (b),
f : focal length of objective lens,
λ : wavelength,
p : track pitch of the disk,
r : objective lens radius.

To decide the s value, both actual tracking actuator movement and the light power needed to detect the tracking error signal must be taken into account. When the s value is 13% of the objective lens diameter, the total incident light power to both tracking holograms is about 18% of the total reflected light power. This s value corresponds to 0.5 mm for a 4 mm diameter objective lens. These values are acceptable for an optical head used in a compact disk system. Figure 8 shows calculated tracking off-set results for both a conventional push-pull method and for this method. In this figure, Tracking error signal off-set is defined as the signal level shift normalized by its own amplitude. Tracking actuator movement is also normalized by the imaging lens aperture. In the calculation, a divergent wave from a usual laser diode, which has Gaussian light power distribution, is assumed. Figure 9 shows the envelope variation in tracking error signal for this optical head, generated by the ±0.5 mm tracking actuator movement. In this figure, vertical and horizontal axes show the tracking error signal amplitude and tracking actuator movement, respectively. From these figures, it is clear that this method is very effective for suppressing tracking error off-set.

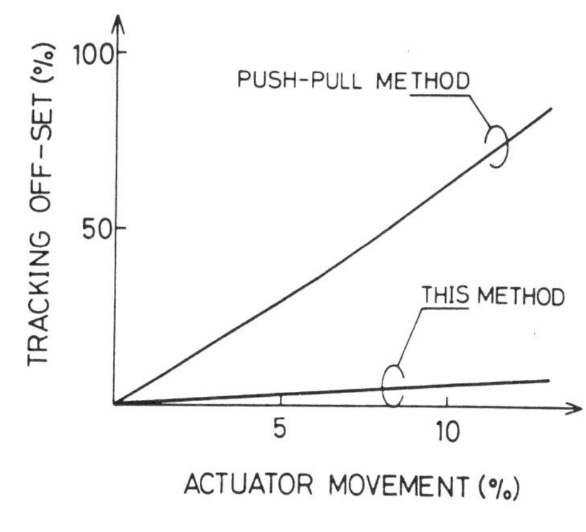

Fig.8 Calculated tracking error off-set results for both push-pull method and this method.

Fig.9 Envelope variation in tracking error signal for this method, generated by ±0.5 mm tracking actuator movement.

3. BEAM SPLITTER FUNCTION

3.1 Beam splitter for CD head

In the optical head for CD players, shown in Fig.1, both transmitted light and first order light diffracted by HOE are used. Therefore, the total light power efficiency is given by multiplying the transmission and first order diffraction efficiencies.

In the case of a thin surface relief holograms, m-th order diffraction efficiency η_m is give by,[9]

$$\eta_m = |b_m|^2, \tag{2}$$

where,

$$b_m = \frac{1}{T}\int_0^T exp\{j\phi(x)\}exp(-j\frac{2\pi m}{T}x)dx, \tag{2'}$$

where $\phi(x)$ is phase shift function for the grating and T is the grating pitch. For a rectangular relief profile grating, shown in Fig.10 (a), total light power efficiency is given by,

$$\eta = \frac{1}{\pi^2}(1 - cos^2\phi_0), \tag{3}$$

where ϕ_0 is phase difference between land and groove of the grating. The theoretical maximum value for total light power efficiency is $1/\pi^2$ (equal to 10.1%) when $\phi_0 = \pi/2$.

When the holograms have a sawtooth relief profile, as shown in Fig.10 (b), total light power efficiency η is given by,

$$\eta = sinc^2(\frac{\phi_0}{2})sinc^2(\frac{\phi_0}{2} - \pi d). \tag{4}$$

It has the theoretical maximum value, which is $16/\pi^4$ (equal to 16.4%), when $\phi_0 = \pi$. For an actual grating, however, the relief profile shape has some deviation from the sawtooth relief profile, as shown in Fig.10 (c). In this figure, d shows the degree of this deviation. For this grating, total light power efficiency η is given by,

$$\eta = \{\frac{\phi_0}{\phi_0 - 2\pi(d-1)}sinc(\frac{\phi_0}{2})sinc(\frac{\phi_0}{2} - \pi d)\}^2. \tag{5}$$

Figure 11 shows the calculated total light power efficiency versus deviation d for various ϕ_0 values. In a grating fabricated by the oblique incident angle ion beam etching (IBE) method,[10] d was about 0.7 ~ 0.9. Under this condition, high efficiency, which exceeds 14%, can be realized when the phase difference is about $0.8\pi \sim \pi$. Therefore, the fabrication condition is not so severe. According to this theoretical estimation, a holographic optical element, whose efficiency is 14.8%, was fabricated using the IBE method. This value includes reflection loss. Figure 12 shows an SEM photograph of a cross section of the fabricated HOE.

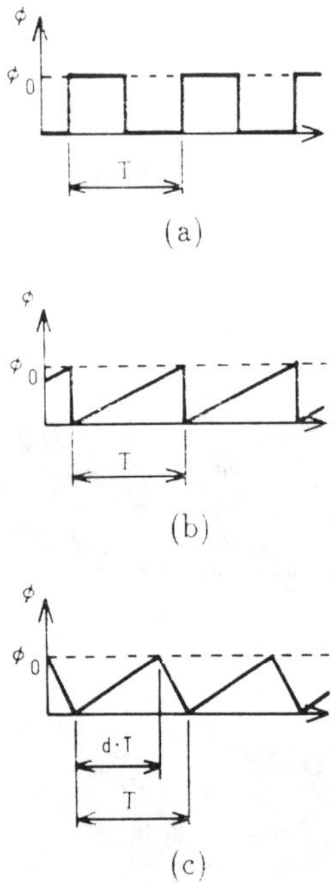

Fig.10 HOE relief profiles. (a) Rectangular. (b) Sawtooth. (c) Actual holograms.

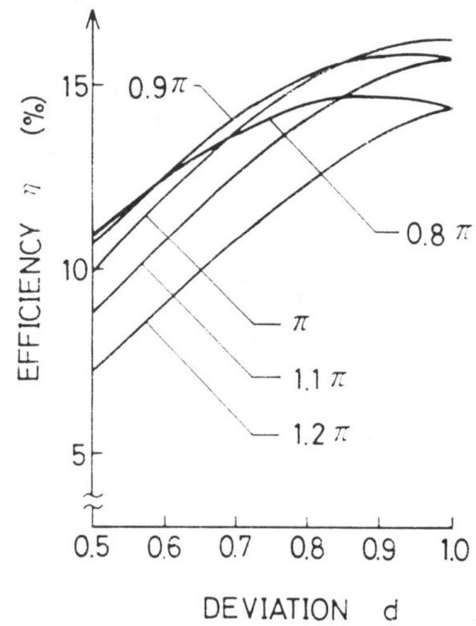

Fig.11 Relation between efficiency and deviation d.

Fig.12 Cross-sectional SEM photograph of a hologram fabricated by IBE method.

3.2 Polarizing beam splitter for MO disk head

In magneto-optical disk memories, a polarized light component involving a recorded signal, which is generated by Kerr effect, is very small, because polarization rotation generated by the Kerr effect is about 0.5 degrees. Therefore, to read out the recorded signal with a high C/N ratio, power distribution, depending on polarization is required for the polarizing beam splitters, PBS_1 and PBS_2, used in the conventional MO head, as shown in Fig.13. To realize a compact MO head, an HOE, which can be combined with beam splitter and error detecting optics functions, is applied for the PBS_2. The HOE must bring the polarized light component, which is generated by Kerr effect, to the read out signal detecting optics without appreciable light power loss. The HOE must also distribute about 20% of the polarized light component perpendicular to the polarized light component generated by the Kerr effect to the error detecting optics, in order to detect error signal and disk sector index signal. The HOE must possess these polarizing characteristics.

Since the high spatial frequency diffraction grating, whose pitch d is smaller than wavelength λ, generates only first order and zero-th order diffracted lights, it is suitable for use as a beam splitter. The diffraction efficiency for such gratings shows large polarization dependence.[11] A grating with $\lambda/d > 1.6$ has nearly zero diffraction efficiency for TM wave (polarization perpendicular to the grating grooves), but the diffraction efficiency for TE wave (polarization parallel to the grating grooves) is varied with differing grating groove depth h values. For example, when $\lambda/d = 1.6$ and $h/d = 0.5$ for Bragg diffraction geometry, diffraction efficiencies are $\eta_1(\mathrm{TM}) = 0\%$, $\eta_1(\mathrm{TE}) = 20\%$, $\eta_0(\mathrm{TM}) = 100\%$ and $\eta_0(\mathrm{TE}) = 80\%$. This distribution just matches the required PBS characteristics in the MO head shown in Fig.13.

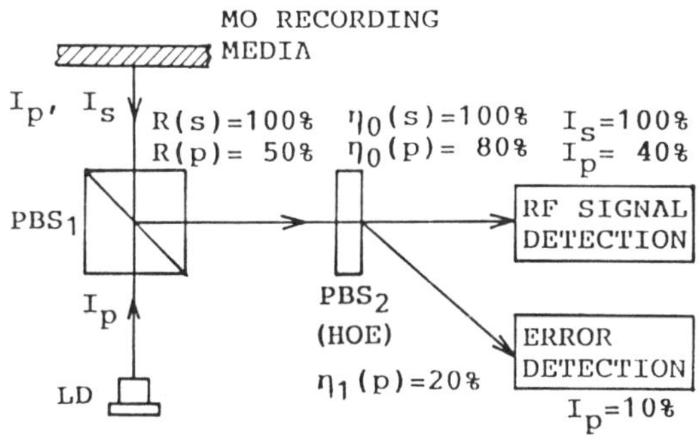

Fig.13 Polarizing characteristics required for PBSs used for MO head.

4. HOLOGRAM FABRICATION PROCESS

The interference fringe pattern for the holograms is a hyperbolic group, generated by the interference between two divergent spherical waves. In the coordinates shown in Fig.3, the equation for the interference fringe pattern is

$$\frac{(y+\frac{T}{2Q})^2}{\frac{1}{Q}(\frac{T^2}{4Q}-U)} - \frac{x^2}{\frac{1}{4P}(\frac{T^2}{4Q}-U)} = 1, \tag{6}$$

where P, Q, T and U are constants determined by the arrangements of two light emitting points and the wavelength, which are given by

$$\begin{cases} P = (\lambda n)^2, \\ Q = 4y_f^2 - 4(\lambda n)^2, \\ T = 4y_f^2\{z_{LD}^2 - (y_f^2 + z_f^2) + (\lambda n)^2\}, \\ U = y_f^2 + z_f^2 + z_{LD}^2 + (\lambda n)^2 - 4(\lambda n)^2 z_{LD}^2, \end{cases} \tag{7}$$

where, λ is the wavelength of the LD and n is an integer.

The fabrication process is shown in Fig.14. The original reticle pattern is fabricated, using electron beam (EB) lithography, to arrange the four holograms precisely. First, the pattern, magnified five times, is drawn by EB. A photomask, on which many holographic optical elements are arranged, is fabricated by reducing the magnified pattern to one fifth, using a photorepeater. This reduction process is effective to reduce stage error and the quantization error in the EB drawing. Holographic optical elements are easily duplicated by contact printing in a photoresist layer on the glass substrate, using this photomask. The fabricated hologram is 6 mm in diameter. Holograms H_3 and H_4 are formed 0.75 mm from the optical axis. An individual area occupies 7.2% in the cross-section of incident reflected beam from the disk. Average pitch for the hologram is about 2.5 μm. Figure 15 shows a photograph of the hologram fabricated for use in a CD player head. For the hologram etched using the IBE method, the total light power efficiency was 14.8% as mentioned before. For mass HOE production, plastic injection molding is applied, using an electroformed nickel mold.

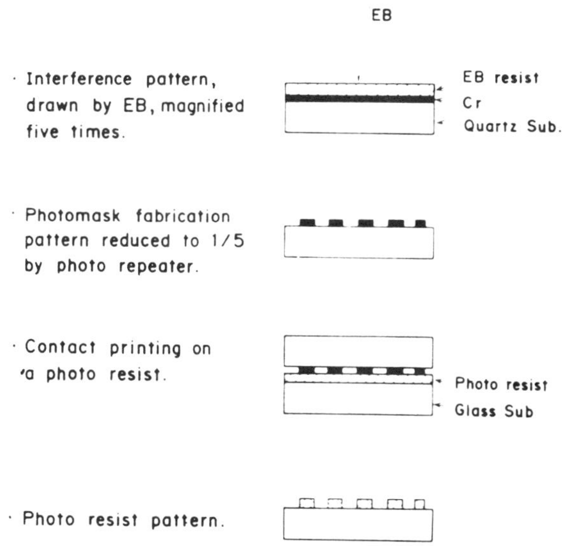

Fig.14 Holographic optical element fabrication process.

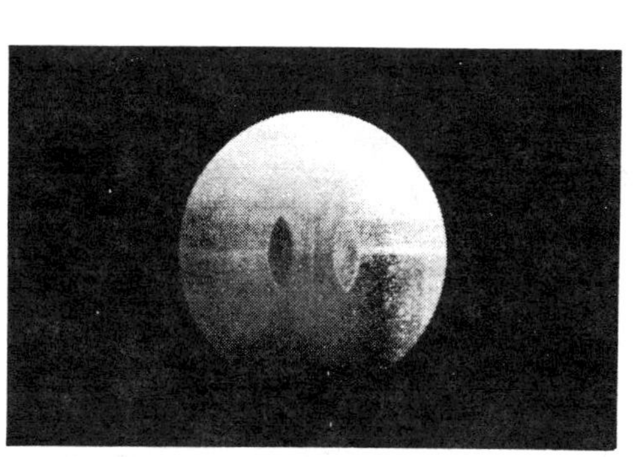

Fig.15 Fabricated HOE.

To fabricate the polarizing HOE for MO heads, the original photomask is fabricated, using direct electron beam (EB) drawing. In the practical fabrication process, to correct any astigmatism caused by an oblique incident angle substrate, the astigmatic wave is assumed instead of a spherical wave. The interference fringe pattern, therefore, is a complicated implicit function. The averaged pitch for the interference fringe is designed to be 0.52 μm, so that $\lambda/d = 1.6$ at λ=0.83 μm. Holographic optical elements are duplicated in a photoresist layer on the glass substrate, using this mask. The fabricated hologram has a 6 × 10 mm quasi-ellipse shape.

Figure 16 shows the diffraction efficiency distribution for the fabricated HOE, whose groove depth $h = 0.25$ μm. At the designed wavelength ($\lambda = 0.83$ μm, $\lambda/d = 1.6$), the polarizing characteristics required for the MO head are obtained.

5. APPLICATION FOR OPTICAL DISK HEADS

5.1 Application for CD head

The optical configuration and operating principle have already been explained, using Fig.1. The fabricated holographic optical head is shown in Fig.17. The focusing error detection characteristics are shown in Fig.18. The linear signal detection region for the focusing error detection characteristics is about 20 μm. Signal play back from the CD was realized, using this optical head employing a tracking and focusing error servo. The read out signal from this optical head typically had about 18 nsec jitter for a $3T$ signal. The read out signal from a CD is shown in Fig.19.

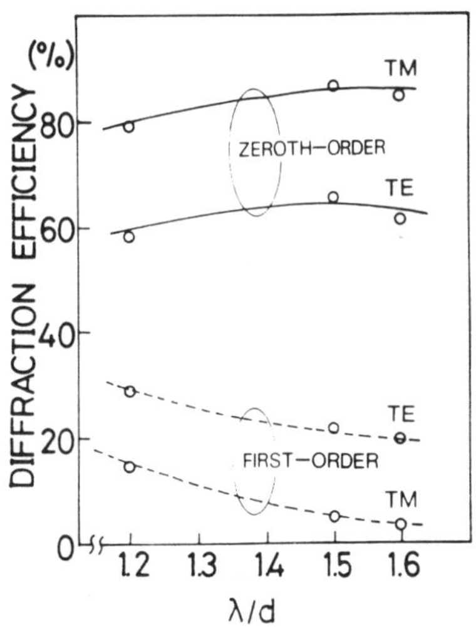

Fig.16 Polarization dependence of diffraction efficiency versus grating pitch.

Fig.17 External view of fabricated holographic optical head for CD players.

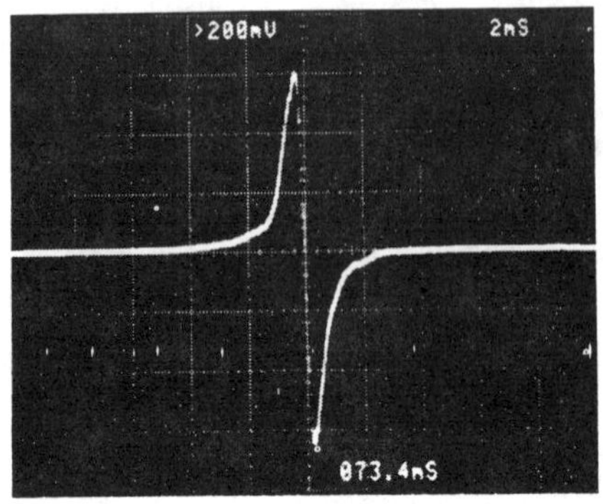

Fig.18 Focusing error detection characteristics.

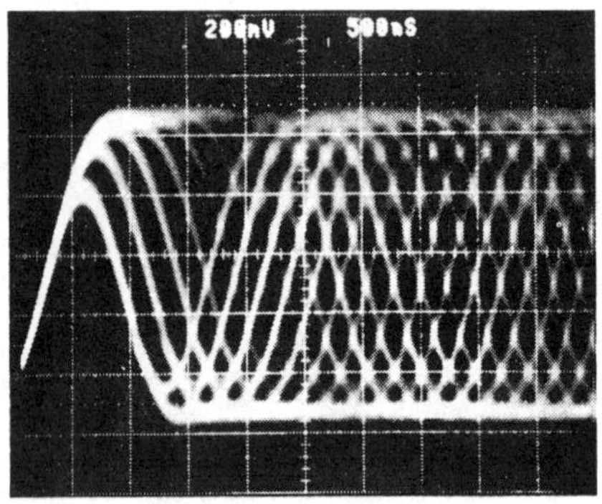

Fig.19 CD read out signal.

Use of the reflection HOE, instead of a transmission HOE is very advantageous to make the optical head thin. The HOE combines a mirror function in addition to the above mentioned three functions. NEC Home Electronics, Ltd. has already begun the mass production of optical heads using such a reflection HOE.

5.2. Application for laser video disk head

The optical isolator function is required for a video disk head to attain a read out with high S/N ratio by suppressing the laser diode noise due to reflected light incident on the LD, because the signal is recorded in analog form. In this optical head, an optical isolator, consisting of a polarizing beam splitter (PBS) and a quarter wave plate, is used. The HOE is used in the light path split by the PBS, as shown in Fig.20. The HOE acts as a beam splitter and the RF read out signal is derived from the zero-th order diffracted light. The focusing and tracking error signals are derived from the first-order diffracted light. For the light source, a visible light LD, composed of AlGaInP, whose wavelength range was 660 ~ 680 nm, was used.[12] The HOE was designed so that wavelength $\lambda = 670$ nm. The S/N ratio for the read out signal, measured by the video S/N meter, was about 42 dB. Figure 21 shows the read out signal spectrum.

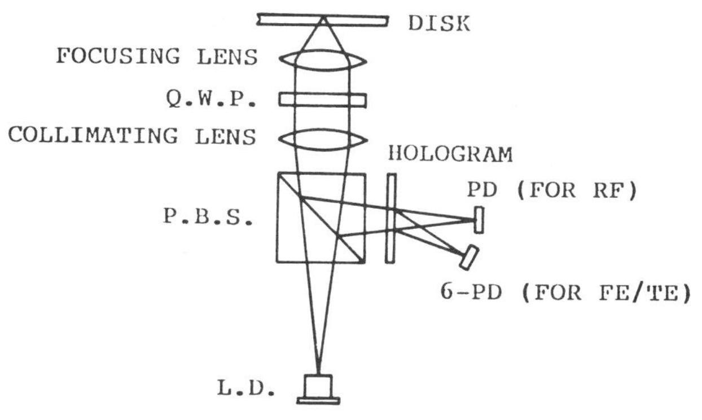

Fig.20 Optical configuration of holographic optical head for laser video disk players.

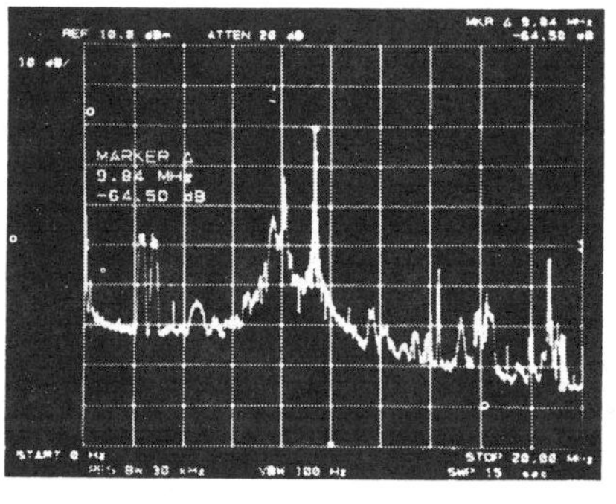

Fig.21 Read out signal spectrum for laser video disk.

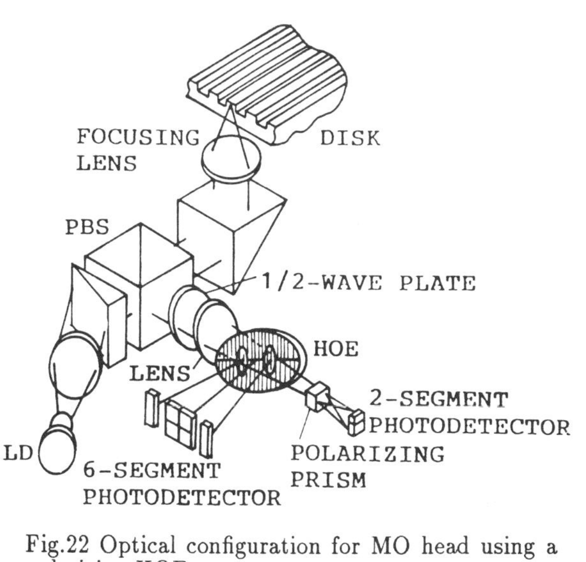

Fig.22 Optical configuration for MO head using a polarizing HOE.

5.3 Application for MO disk optical head

The optical configuration for an MO head, using the polarizing HOE, is shown in Fig.22. The light reflected back from the magneto-optical disk is split by the PBS prism. Its polarization direction is rotated by the half wave plate, so that the Kerr read out polarization component becomes TM polarized with regard to the HOE. Then, it becomes incident on the HOE at the Bragg angle. The first-order diffracted light, which is split into four beams, is used for error signal detection. The zero-th order diffracted light is used for signal read out through the polarizing prism.

The HOE consists of four different holograms, the same as that for CD and VD heads. Focusing and tracking error signals are derived using the above mentioned principles. Kerr read out signal is derived from the two-segment photodetector using a differential method.

In read and write experiments using the proposed MO head with the HOE, more than 57 dB C/N ratio was obtained for pit width recording at 1 MHz recording frequency, as shown in Fig.23.

A reflection grating shows strong polarization dependence in a large pitch region, in comparison with a transmission grating.[13] Besides, a reflection HOE can be used to construct a more compact head, in comparison with a transmission HOE.[7]

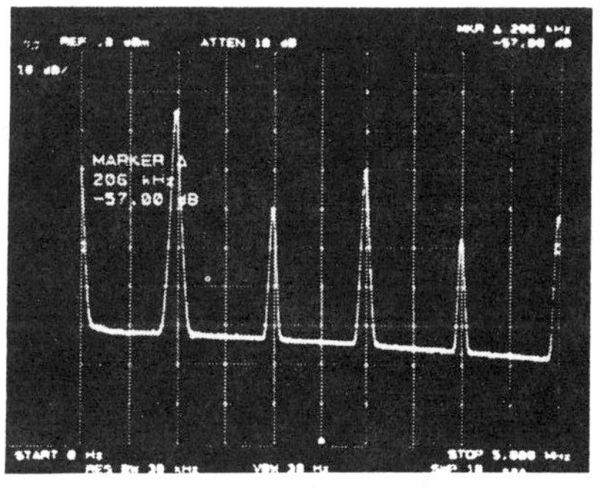

Fig.23 Read out signal spectrum for MO disk.

6. CONCLUSION

A new concept optical head for optical disk memories has been developed using the holographic optical element. The holographic optical element has three optical functions, required for an optical head; splitting reflected beam, focusing error detection and tracking error detection. In addition to these, a polarizing beam splitter function, using a high spatial-frequency hologram, is also combined for a magneto-optical disk head. This optical head is free from wavelength variation, in spite of the hologram use. The off-set signal in the tracking error signal has been reduced, using an optimized holographic optical element. These HOEs are computer generated holograms and have been fabricated by using electron beam lithography and ion beam etching techniques. In applications for CD, VD and MO disks, stable operations with high read out S/N ratio have been obtained.

7. ACKNOWLEDGEMENTS

The authors would like to appreciate to members of the Audio and Video Products Division and TV and New Media Products Division, NEC Home Electronics Ltd., and members of Memory Research Laboratory, NEC Corp., for their help with the experiments.

8. References

1. Y. Kimura, S. Sugama and Y. Ono, "A compact optical head using holographic optical element," *Tech. Digest of 4th Topical Meeting on Optical Data Storage*, 178 (1987).
2. Y. Kimura, S. Sugama and Y. Ono, "Compact optical head using a holographic optical element for CD players," Appl. Opt. **27**, 668 (1988).
3. Y. Kimura, S. Sugama and Y. Ono, "High performance optical head using optimized holographic optical element," *Tech. Digest of Int. Symp. on Optical Memory 1987*, 195 (1987).
4. Y. Kimura and Y. Ono, "Polarizing holographic optical element for magneto-optical disk heads," *Tech. Digest of 1st MICROOPTICS Conference*, 162 (1987).
5. Y. Kimura, S. Sugama and Y. Ono, "High performance optical head using optimized holographic optical element," Jpn. J. Appl. Phys. **26**, Suppl. **26-4**, 131 (1987).
6. Y. Ono, Y. Kimura and S. Sugama, "Holographic optical head for optical disk memories," *Proc. of First Int. Symp. on Industrial Uses for Holography*,(Oxford University Press), to be published.
7. A. Ohba, Y. Kimura, S. Sugama, R. Katayama, M. Okada, Y. Ono and N. Nishida, "Compact magneto-optical disk head using reflection polarizing holographic optical element," to be presented at 5th topical Meeting on Optical Data Storage.
8. W. H. Holloway and R. A. Ferrente, "Computer Analysis of Holographic Systems by Means of Vector Ray Tracing," Appl. Opt. **20**, 2081 (1981).
9. R. Magnusson and T. K. Gaylord, "Diffraction efficiencies of thin gratings with arbitrary grating shape," J. Opt. Soc. Am. **68**, 806 (1978).
10. K. Kosuge, S. Sugama, Y. Ono and N. Nishida, "Ion-etched blazed holographic zone plate," *Conference digest of 13th con-gress of the International commission for Optics*, 526 (1984).
11. K. Yokomori, "Dielectric surface-relief gratings with high diffraction efficiency," Appl. Opt. **23**, 2303 (1984).
12. K. Kobayashi, S. Kawata, A. Gomyo, I. Hino and T. Suzuki, "Room-temperature CW operation of AlGaInP double-heterostructure visible lasers," Electron. Lett. **21**, 931 (1985).
13. E. G. Loewen, M. Neviere and D. Maystre, "Efficiency optimization of rectangular groove gratings for use in the visible and IR regions," Appl. Opt., **18**, 2262 (1979).

Reprinted from *Optical Data Storage*, Proc. SPIE Vol. 1316, pp. 159-166 (1990)

Compact Magneto-optical Disk Head using a Holographic Optical Element with Analyzer Function

Akitomo Ohba, Seijin Sugama, Ryuichi Katayama, Yutaka Urino, and Yuzo Ono

Opto-Electronics Research Laboratories
NEC Corporation
1-1, Miyazaki 4-chome, Miyamae-ku, Kawasaki 213, Japan

Abstract

A compact magneto-optical disk head was constructed using a newly developed holographic optical element. The element has an analyzer function with more than 15dB extinction ratios for data signal detection, in addition to focusing and tracking error signal detection functions. The element was fabricated in a birefringent $LiNbO_3$ substrate using the proton exchange method, which is a well known waveguide fabrication technique. In Read/Write experiments using the head, sufficient carrier-to-noise ratio values for practical use were obtained. As a typical result, a 47dB carrier-to-noise ratio was obtained for 3.7MHz pulse recorded signal.

1. Introduction

In an optical data storage system, a small-size and light-weight optical disk head is required for high speed track access. Holographic optical elements have sufficiently high potential to realize a compact optical head, because it can embody both focusing and tracking error signal detection functions as well as beam splitter function. Compact optical heads for a compact disc (CD) and a magneto-optical(MO) disk have been developed by incorporating the elements[1,2]. In magneto-optical disk heads, an analyzer is required for data signal detection in comparison with the CD optical heads. The authors previously proposed a holographic optical element with an analyzer function in addition to error signal detection functions[3]. For realizing the analyzer function, this element was fabricated in a birefringent $LiNbO_3$ substrate, using the proton exchange method which is a well known optical waveguide fabrication technique. Using the element, the authors have developed a compact magneto-optical disk head which can be constructed with eight optical elements. This paper describes the principle of the element and optical head operation, the element fabrication process as well as head operational characteristics.

2. HOE analyzer function

For data signal detection in magneto-optical disk heads, the differential detection method is used for data signal detection to obtain a readout signal with high carrier-to-noise ratio, because, in the method, a common mode noise component in a detected signal can be eliminated. An analyzer such as a polarizing beam splitter or Wollaston prism is indispensable for this method. It splits the light reflected back from a disk into +45 and -45 degree polarized-light components. In addition to the analyzer, to set the polarization plane for the reflected light in a 45 degree direction with respect to the analyzer axis, a half wave plate is also necessary. To realize a compact magneto-optical disk head, the authors previously proposed a holographic optical element with the differential signal detection function[3] and error signal detection functions[1,2]. In the element, both the analyzer function and the half wave plate function, described above were realized by appropriately introducing a birefringent holographic grating[4].

Figure 1 shows the element cross-sectional structure. The element is fabricated in a birefringent $LiNbO_3$ substrate, using the proton exchange method[5], which is a well known optical waveguide fabrication technique. By the proton exchange, the $LiNbO_3$ refractive index changes in different ways for extraordinary and ordinary lights of the $LiNbO_3$ crystal. That is, the extraordinary refractive index for the proton-exchanged regions is higher than that for the substrate, while the ordinary refractive index is a little lower. Changes in refractive index values for extraordinary and ordinary light are considered to be 0.11 and -0.04 for 830nm wavelength light. Figure 2 shows the grating phase distributions for these lights. The dielectric films on the proton-exchanged regions cancel the phase shift in the proton-exchanged regions for ordinary light, as shown in Fig.2(a). That is, the element does not act as a grating for ordinary light. On the other hand, the proton-exchanged region depth is designed so that the phase shift, including the dielectric films, is π radian for extraordinary light, as shown in Fig.2(b). In a rectangular grating, when its phase shift is π radian, the 0th order diffraction efficiency is 0%. Therefore, the element completely diffracts extraordinary light, while it does not diffract ordinary light at all. The sum of +1st and -1st order diffraction efficiencies is 81%[4].

Extraordinary light indicates polarized-light, parallel to the $LiNbO_3$ optic axis. Ordinary light indicates polarized-light, perpendicular to the optic axis. That is, the $LiNbO_3$ optic axis corresponds to the analyzer axis. When the $LiNbO_3$ optic axis is set in an about 45 degree direction with respect to the polarization plane for the light reflected back from a disk, as shown in Fig.3, the element acts as both a half wave plate and an analyzer required for the differential detection.

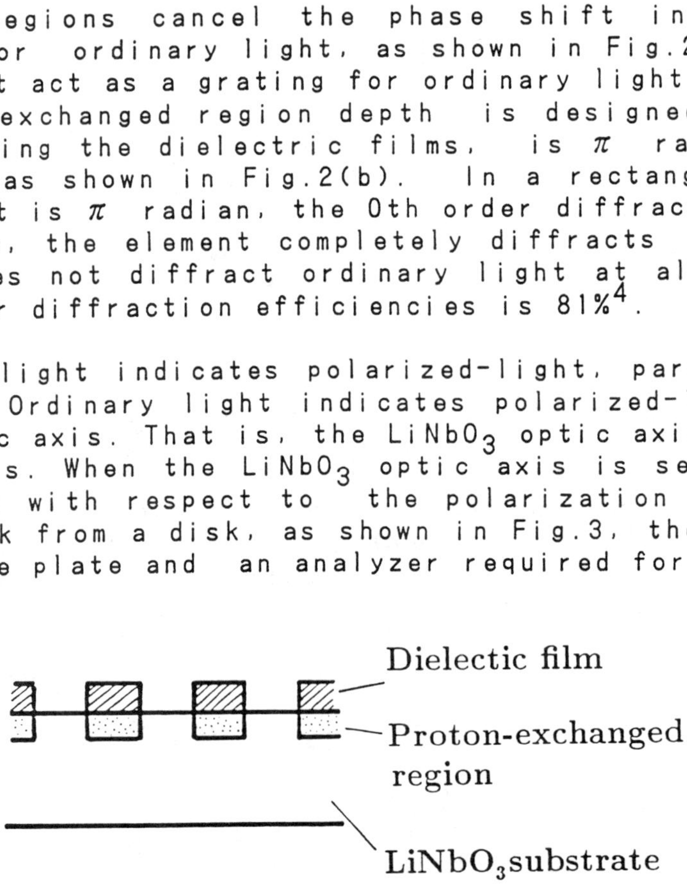

Figure 1. Holographic optical element cross sectional profile.

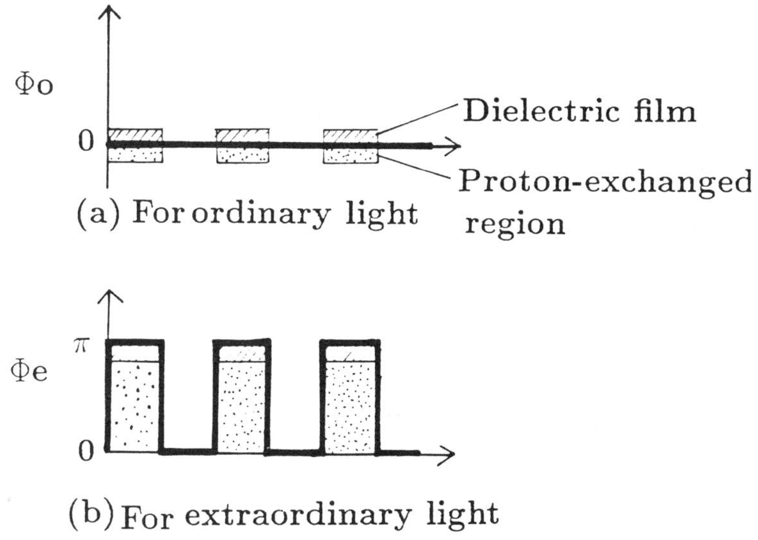

Figure 2. Holographic optical element phase distribution (bold lines).

3. Optical head operational principle

The optical configuration for the magneto-optical disk head, using the holographic optical element, is shown in Fig.4. The light reflected back from a magneto-optical disk is reflected by the polarizing beam splitter (PBS) and becomes incident on the holographic optical element through the lens.

This element consists of four different holograms, H1 to H4, for both focusing and tracking error signal detections, as shown in Fig.3. These hologram fringe patterns are calculated by a computer and fabricated using electron beam lithography. Individual holograms have different fringes generated by interference between a spherical wave diverging from point P0, to which the incident light converges, and individual spherical waves diverging from points Pn (n=1-4), to which the +1st order diffracted beams from individual holograms converge. Therefore, the +1st order diffracted beams from individual holograms H1 to H4 converge to points P1 to P4, respectively, on the photodiode with eight segments(8-PD).

A focusing error signal is detected from two beam spots at points P1 and P2, using the double knife edge method. The division line between holograms H1 and H2 acts as a knife edge. A tracking error signal is detected from two beam spots at points P3 and P4, using the push-pull method. A readout signal is detected from the difference between the sum of +1st and -1st order diffracted light powers and the undiffracted light power, using the differential detection method.

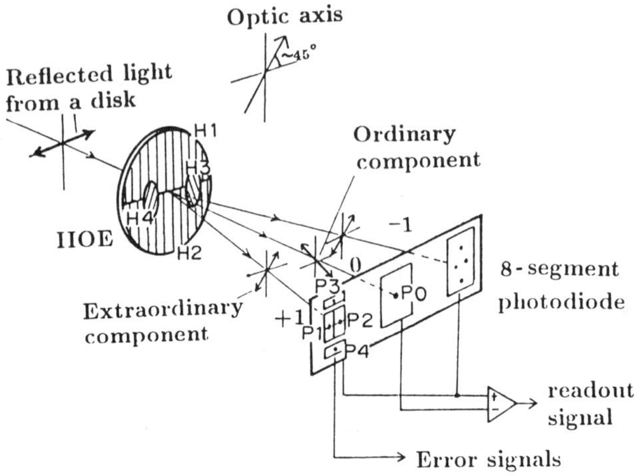

Figure 3. Holographic optical element (HOE) operational principle.

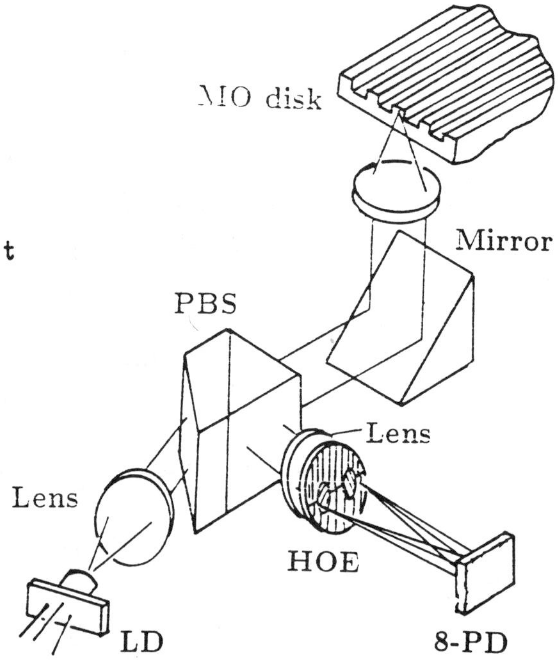

Figure 4. Optical head configuration using a holographic optical element (HOE) for a magneto-optical (MO) disk.

Table 1. Specifications for a fabricated holographic optical element.

Size	6 mm ϕ
Grating pitch	≈ 50 μm
Diffraction angle	≈ 1°
Substrate	X-cut $LiNbO_3$ 0.5mm t
Dielectric film thickness	1400 Å
Proton-exchanged region depth	3.1 μm

4. Experimental

4.1 HOE Fabrication

The HOE fabrication process is as follows. First, a photomask with four hologram patterns, shown in Fig.3, was fabricated, using electron beam lithography to arrange the holograms precisely. An X-cut $LiNbO_3$ crystal was used as an element substrate. A metallic film patterned on the substrate, using this photomask, was used as a protection mask during proton exchange process. Then, by soaking the substrate in molten benzoic acid, proton-exchanged regions were formed, where Li^+ in $LiNbO_3$ was replaced with H^+. Finally, after removing the metallic mask, a photoresist film was put on the proton-exchanged regions as the phase shifter dielectric film, using the photomask. Table 1 shows fabricated element specifications. The proton-exchanged region depth was 3.1μm and the dielectric film thickness was 1400Å. The fabricated element has a 6mm diameter circular shape, 0.5mm thickness and about 50μm grating pitch. Diffraction angle is about 1 degree.

In the fabricated element, the 0th order and the 1st order diffraction efficiencies for ordinary light were 81.3% and 1.7%, while the efficiencies for extraordinary light were 2.5% and 59.1%. The resultant extinction ratios were -15.1dB for the 0th order and -15.4dB for the 1st order.

4.2 Head characteristics

Figure 5 shows a photograph of the magneto-optical disk head constructed using the fabricated holographic optical element. Figure 6 shows the diffracted beam spots converging onto the eight segment photodiode. In the Read/Write experiments, focusing and tracking error signal were detected from the +1st diffracted light. For data signal detection, only -1st order and 0th order diffracted lights were used for the easiness of the amplifier circuits. Figures 7(a) and (b) show readout signals for pit width recording at 1MHz recording frequency and for pulse recording at 3.7MHz in 1800rpm rotation speed. Their recording radii in a magneto-optical disk were 60mm for the 1MHz signal and 30mm for the 3.7MHz signal. Linear velocities in their radii correspond to 11.3m/sec and 5.7m/sec, respectively. Readout light power was 1mW. Figures 8(a) and (b) show their readout signal spectra. For the 1MHz signal, a 54.7dB carrier-to-noise ratio was obtained and for the 3.7MHz signal, a 46.9dB ratio, which satisfies the ISO standard, was attained. These values are sufficient for practical use.

5. Summary

A compact magneto-optical disk head was constructed, using a new holographic optical element made from a birefringent $LiNbO_3$ crystal. The element includes the analyzer function required for data signal detection, in addition to both focusing and tracking error signal detection functions. Carrier-to-noise ratios with sufficient values for practical use were achieved, while simple optics were used.

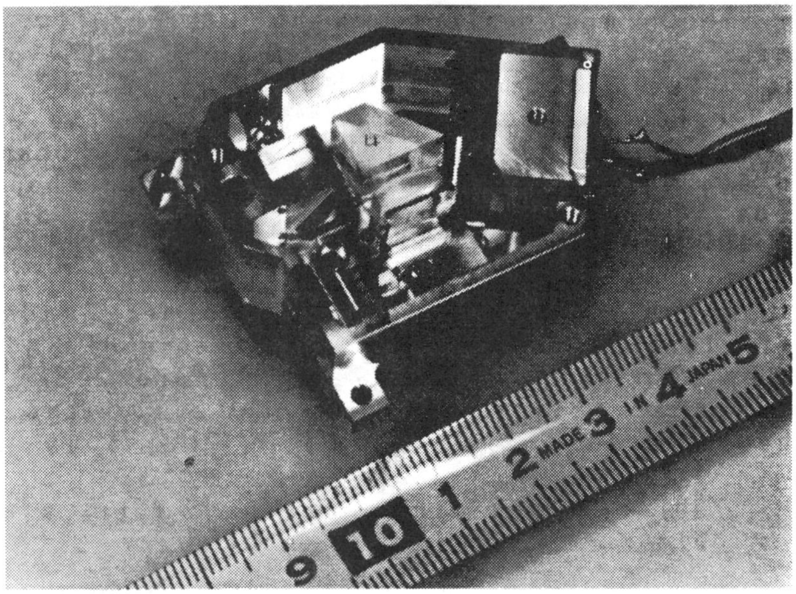

Figure 5. Magneto-optical disk head using a fabricated holographic optical element.

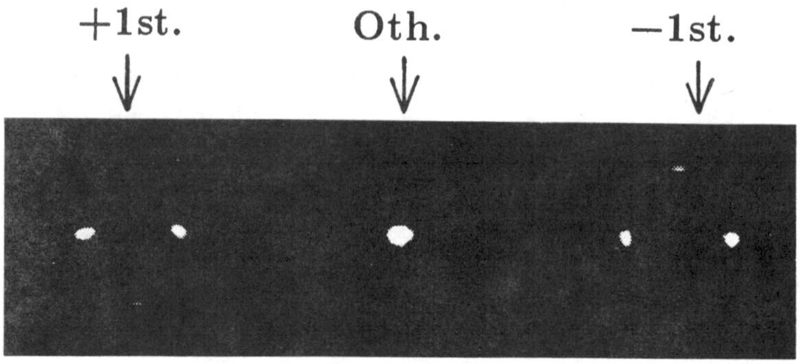

Figure 6. Diffracted beam spots converging onto the eight segment photodiode. A diffracted beam spot at P4 in Fig. 3 is brighter than that at P3, because the beam focusing onto a magneto-optical disk is not on track.

6. References

1.Y. Kimura, S. Sugama and Y. Ono, "Compact optical head using a holographic optical element for CD players," Appl. Opt. 27, pp.668-671, Feb. 1988.

2.A. Ohba, Y. Kimura, S. Sugama, R. Katayama, M. Okada, Y. Ono, and N. Nishida, "Compact magneto-optical disk head using reflection polarizing holographic optical element," Proceedings of Optical Data Storage Topical Meeting, SPIE Proc. 1078, pp.127-130, Los Angeles, Jan. 1989.

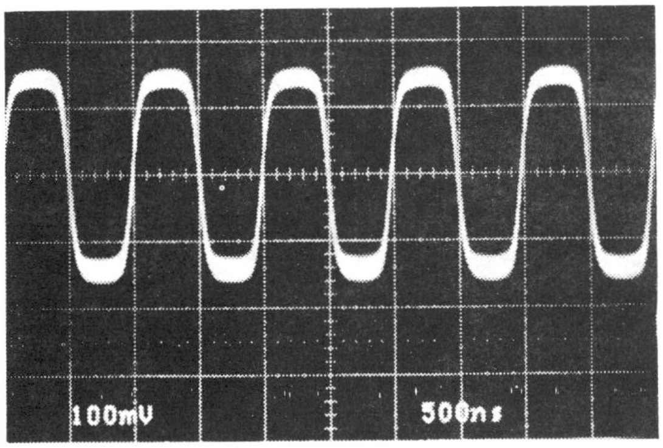

(a)For 1MHz pit width recording signal. (r=60mm)

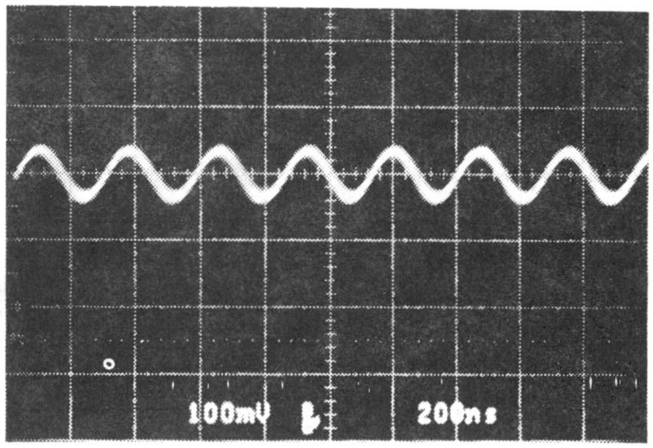

(b)For 3.7MHz pulse recording signal. (r=30mm)

Figure 7. Readout signal waveforms. Disk rotation speed was 1800rpm. Readout power was 1mW.

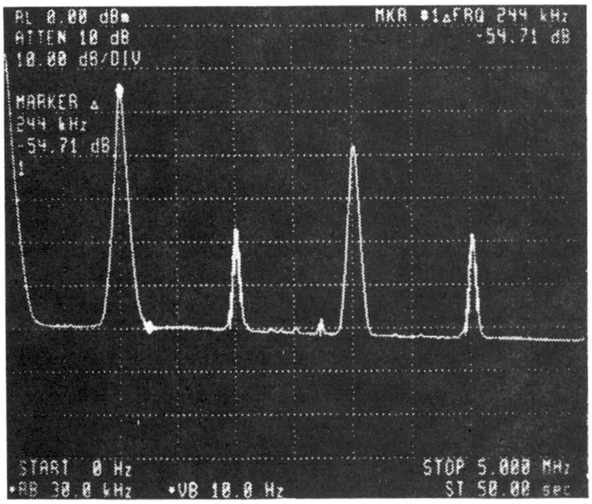

(a) For the 1MHz signal.

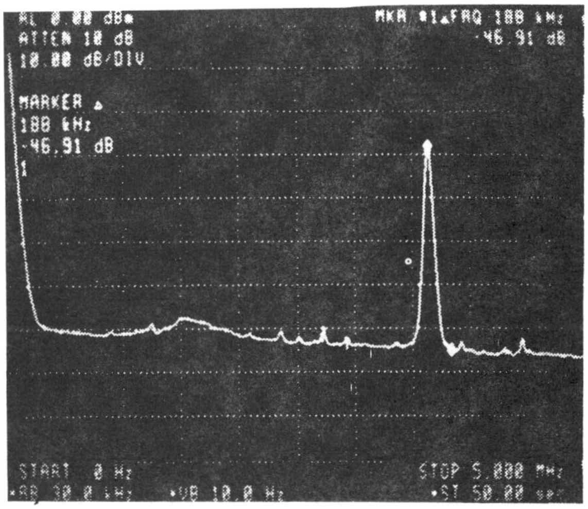

(b) For the 3.7MHz signal.

Figure 8. Readout signal spectra.

3.A. Ohba, Y. Kimura, S. Sugama, Y. Urino, and Y. Ono, "Holographic optical element with Analyzer Function for Magneto-optical Disk Head," Post-deadline Papers Technical Digest of International Symposium on Optical Memory '89, pp.7-8, Kobe, Sept. 1989.

4.Y. Urino, H. Nishimoto and Y. Ohta, "Birefringent grating polarizer," Technical Digest of Second Optoelectronics Conference, pp.167-169, Tokyo, 1988.

5.J. L. Jackel, C. E. Rice and J. J. Veselka, "Proton exchange for high-index waveguides in $LiNbO_3$," Appl. Phys. Lett. 41, pp.607-608, Oct. 1982.

Reprinted with permission from *Applied Optics,* Vol. 29(34), pp. 51275130 (December 1, 1990). ©1990 Optical Society of America.

Holographic optical element for an optical disk head with spot-size detection servo optics

Yoshiaki Komma, Shin-ichi Kadowaki, Yoshikazu Hori, and Makoto Kato

A CGH construction of a holographic optical element (HOE) for an optical disk head with spot-size detection by means of superimposing two Fresnel zone plates is studied. Optimum parameter values for the ratio of segment to grating pitch (duty) and grating height are clarified from the viewpoint of total light power efficiency in the round trip optical system. The influences of higher-order diffracted beams near the first-order diffracted beams used to get signals on signal detection are shown to be small. The HOEs with almost optimum parameter values are fabricated and the desired diffracted spots, the expected total light power efficiency, and the small influence of higher-order diffracted beams are observed.

I. Introduction

A number of holographic optical elements (HOEs) have been studied to simplify the servo signal detection optics in optical pickup heads.[1,2] Further development in integrating holographic heads has been of much interest.[3,4] The construction of such microoptics, however, requires mechanical alignment precision of the order of microns as well as allowance for wavelength shift of the laser diodes over several nanometers.

We previously presented a HOE scheme to use spot-size detection, which is not strongly affected by mechanical misalignment or the wavelength shift.[5] In this paper we report on the analysis and optimization of the HOE in detail.

II. Spot-Size Detection Method[6]

The schematic construction of the proposed spot-size detection (SSD) servo optics for detecting a focusing error (FE) signal is shown in Fig. 1. A light beam emitted from a laser diode (LD) passes through the HOE, and the zeroth-order beam is focused on the disk. The reflected beam from the disk illuminates the HOE again and then generates a pair of spots P_1 and P_2 on the photodetector (PD). The focal positions of the two spots are located on both sides of the PD plane. When the disk deviates from the focal position of the objective lens (L_1), the spot size of one of the diffracted beams on the PD plane is decreased, and that of the other is increased as shown in Fig. 1(b).

Thus the focusing error signal FE is obtained according to the size of those two spots by differential detection on the six-element detectors as shown in Fig. 1. The detector boundaries are almost parallel to the spatial frequency axis of the FZPs, so that the wavelength shift hardly influences the FE signal.

III. Optimization of the HOE Pattern

Considering the round trip path, total light power efficiency for servo signal detection which is given by multiplying the zeroth- and first-order diffraction efficiencies must be analyzed and optimized to have maximum value. To reduce noise in detecting the servo signal, it is necessary to make the total light power efficiency as large as possible.

We used a CGH pattern as shown in Fig. 2. It was obtained by superimposing two Fresnel zone plates with different optical axes.[5]

This pattern of the HOE causes unnecessary higher-order diffracted beams due to superimposing two patterns with binary profiles. Then we analyze the intensities of these unnecessary higher-order diffracted beams near the diffracted beams used to get signals, to determine whether they impose offset or noise to the servo signal.

To simplify the calculation, we assume that (1) each grating of the HOE is a linear line, (2) the pitch of each grating is uniform, and (3) the phase profile of the HOE is rectangular.

Because the focal points of spots P_1 and P_2 are close to the focal plane of collimator lens L_2 (Fig. 1), curva-

The authors are with Matsushita Electric Industrial Company, Ltd., Semiconductor Research Center, Optoelectronics Research Laboratory, 3-15 Yagumonakamachi, Moriguchi, Osaka 570, Japan.

Received 15 November 1989.
0003-6935/90/345127-04$02.00/0.

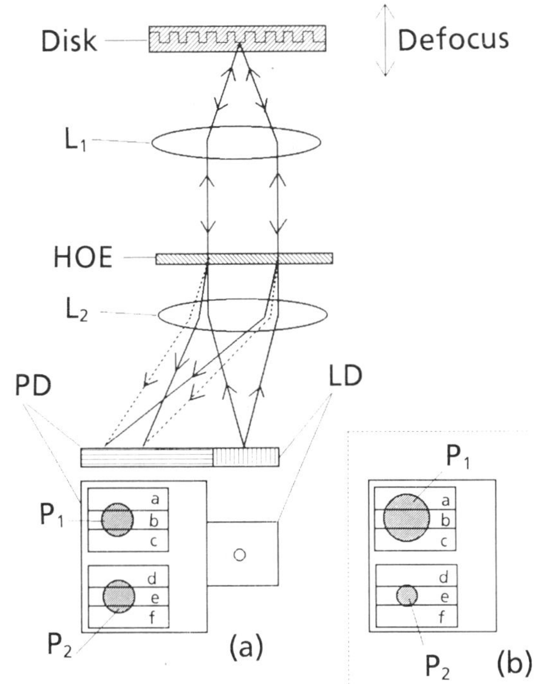

Fig. 1. Focusing servo optics using a HOE with six-element photodetectors: L_1, objective lens; L_2, collimating lens; *PD*, photodetector; *LD*, laser diode; P_1,P_2, first-order diffracted beams in situations of (a) focus and (b) defocus; $FE = (a + c - b) - (d + f - e)$.

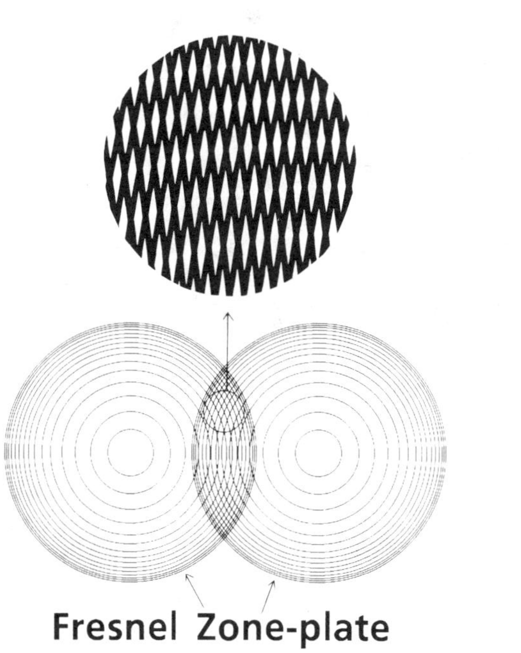

Fig. 2. Part of a pattern of the HOE showing the CGH construction of FZPs.

tures of gratings are small and pitches of gratings vary slightly. Therefore it is proper to adopt assumptions (1) and (2). Assumption (3) depends on the process of HOE fabrication. In this paper, however, we adopt it as a first-order approximation.

Diffraction efficiencies of the HOE as 2-D binary

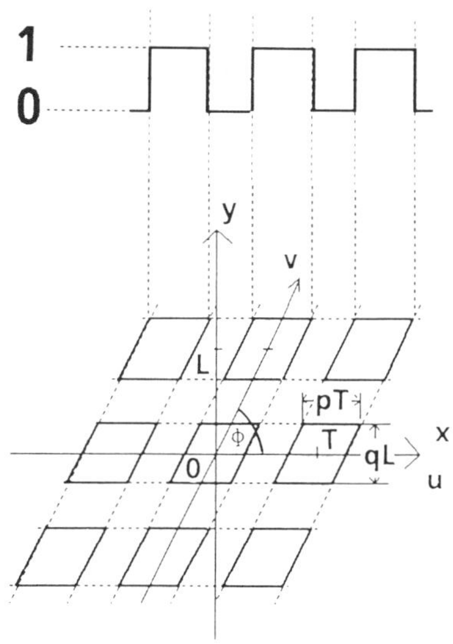

Fig. 3. Two-dimensional binary function $f(x,y)$ which periodically takes the value of 0 or 1.

phase gratings are analyzed[7] as follows. The function $f(x,y)$ shown in Fig. 3 periodically takes the value of 0 or 1, the ratios of the segment width to the grating pitch of T(x-axis) and L(y-axis) are p and q, respectively, and the crossing angle between the gratings is $\Phi(0 < \Phi < \pi)$.

With the function $f(x,y)$ the complex amplitude transmittance profile of the phase grating is given as

$$g(x,y) = [2f(x,y) - 1] \sin\theta + i \cos\theta, \tag{1}$$

where $i = \sqrt{-1}$, and the phase modulation amplitude of θ is

$$\theta = \frac{1}{2} \cdot \frac{2\pi(n_D - 1)D}{\lambda}.$$

Here D is the grating height, n_D is the refractive index of the HOE material, and λ is the wavelength of the light. Since $f(x,y)$ is periodic, it can be expanded into a Fourier series as follows:

$$f(x,y) = \sum_{m,n} C_{mn} \exp[2\pi i(mu + nv)],$$

where

$$u = x/T - y/T \cdot \cot\Phi, \quad v = y/L,$$

$$C_{mn} = \frac{1}{TL} \int\int_s f(x,y) \exp\{-2\pi i[m(x/T - y/T \cdot \cot\Phi) + n \cdot y/L]\} dxdy = \frac{\sin(m\pi p)}{m\pi} \cdot \frac{\sin(n\pi q)}{n\pi}, \tag{2}$$

where S is a unit segment of the grating.

From Eqs. (1) and (2) we find that diffraction efficiencies are independent of T, L, and Φ, and the amplitudes of the first-order diffracted waves $a_{0,1}$ and $a_{1,0}$ from the phase grating are

$$a_{0,1} = 2p \cdot \frac{\sin\pi q}{\pi} \cdot \sin\theta, \tag{3}$$

$$a_{1,0} = 2q \cdot \frac{\sin\pi p}{\pi} \cdot \sin\theta, \tag{4}$$

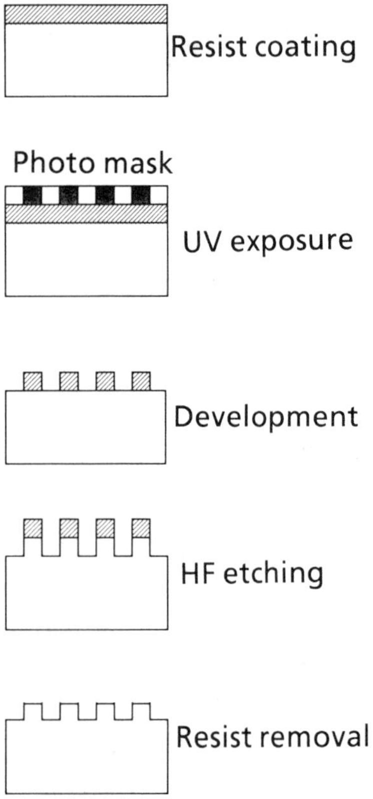

Fig. 4. Fabrication process of the HOE.

where the suffix 0,1 or 1,0 corresponds to $m = 0, n = 1$ or $m = 1, n = 0$, respectively. To get the FE signal without offset, values of $a_{0,1}$ must be equal to $a_{1,0}$; this is satisfied when $q = p$.

The total light power efficiency η is given as

$$\eta = \eta_{0,0} \times (\eta_{0,1} + \eta_{1,0}), \tag{5}$$

where $\eta_{0,0}$ is the diffraction efficiency of the zeroth-order beam and $\eta_{0,1}$ and $\eta_{1,0}$ are those of the first-order beams.

The total light power efficiency η when $q = p$ is thus derived:

$$\eta = [(2p^2 - 1)^2 \sin^2\theta + \cos^2\theta] \cdot \frac{8}{\pi^2} \cdot \sin^2\theta \cdot p^2 \cdot \sin^2\pi p. \tag{6}$$

From Eq. (6) we find the maximum value of $\eta = 7.17\%$ at $p = 0.589$, $\theta = 0.267\ \pi$. The higher-order diffracted beams closest to the first-order beams may possibly impose noise on the FE signal. The efficiency of these higher-order beams is derived, from Eq. (2), as

$$\eta_{-1,2} = \left(\frac{1}{2\pi^2} \cdot \sin\pi p \cdot \sin 2\pi q \cdot \sin\theta\right)^2. \tag{7}$$

If we assume the constraint that η be the maximum, $\eta_{-1,2}$ is estimated to be 0.15% and the ratio $\eta_{-1,2}/\eta_{0,1}$ is 2.1%.

IV. Experiment

Some HOEs were fabricated experimentally on a silica plate as a relief type phase hologram (Fig. 4) through some conventional lithographic process (Fig. 5), and the desired first-order beams were obtained as shown in Fig. 6 ($\eta_{0,1}$,$\eta_{1,0}$). The spot-size variation at defocused conditions was also observed to be the same as expected. The higher-order diffracted beams ($\eta_{-1,2}$,$\eta_{2,-1}$) closest to the first-order beams were very weak, although they were overexposed to be revealed in the picture.

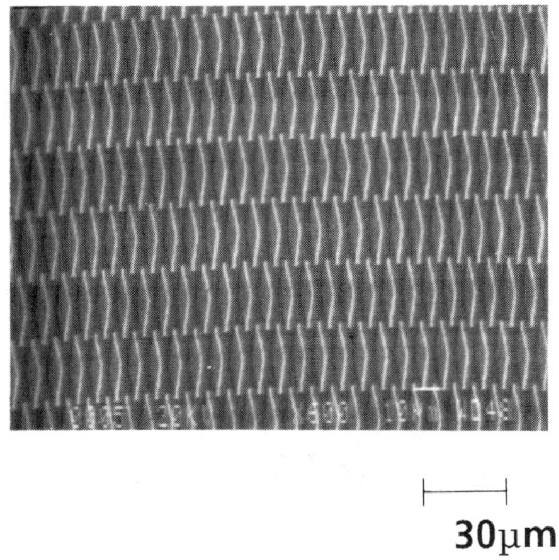

Fig. 5. SEM photograph of an example of a fabricated HOE on a silica plate as a relief type phase hologram through some conventional lithographic process.

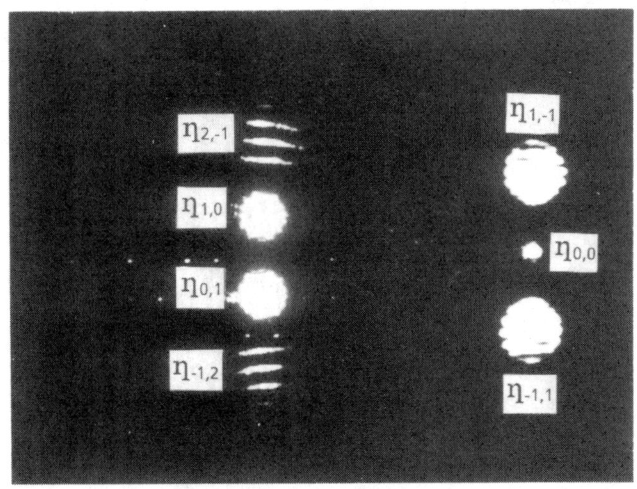

Fig. 6. Picture of diffracted beams from the HOE: η_0, the zeroth-order diffracted beam; $\eta_{0,1}$, $\eta_{1,0}$, the first-order diffracted beams; $\eta_{-1,2}$, $\eta_{2,-1}$, high-order diffracted beams near $\eta_{0,1}$, $\eta_{1,0}$. The two spots $\eta_{-1,1}$, and $\eta_{1,-1}$ adjacent to the zeroth-order diffracted beam may not make noise in the servo signal because they are not near $\eta_{0,1}$, or $\eta_{1,0}$.

Table I shows the measured characteristics of the fabricated HOEs and the calculated values for comparison. They were measured with the wavelength of 633 nm (sample 1) or 780 nm (samples 2 and 3). The pitch of sample 1 is ~30 μm while that of samples 2 and 3 is ~8 μm. In sample 1, in which duty p was much smaller than the optimum value of 0.59, the values of

Table I. Characteristics of the HOE

θ :phase modulation of gratings

p:ratio of segment width to pitch of gratings

η_1, η_B:diffraction efficiencies

$\eta_1=\eta_{0,1}+\eta_{1,0}$, $\eta_B=\eta_{-1,2}+\eta_{2,-1}$

η :total light power efficiency

No.		θ/π	p	η_B/η_1 (%)	η (%)
1	Measured	0.467	0.31	29	3.3
	Calculated			23	3.5
2	Measured	0.337	0.57	2.0	6.0
	Calculated			1.4	6.4
3	Measured	0.266	0.67	5.1	6.5
	Calculated			4.3	6.8
optimum condition		0.267	0.59	2.1	7.2

$\eta_{-1,2}$ and $\eta_{2,-1}$ were large. The maximum total light power efficiency of 6.5% was obtained for the HOE with almost optimum phase modulation and duty p, and intensities of the $\eta_{-1,2}$ and $\eta_{2,-1}$ were as small as predicted.

Some discrepancies between the measured and calculated values are considered to be derived mainly from the fact that the profile of the fabricated grating was not completely rectangular.

The maximum total light power efficiency was sufficient to obtain FE signals in our experiment.[8] The power ratio of the overlapping higher-order diffracted beams to the first-order beams for samples 3 and 4 was found to be acceptable in the SSD servo signal detection system.

V. Conclusion

A simplified CGH construction of an HOE for achieving spot-size servo signal detection (SSD) optics, which has a large allowance for mechanical misalignment of optical components and wavelength shift of the laser diode, was investigated by means of superimposing two Fresnel zone plates with different optical axes.

The optimum parameter values of the ratio of segment width to grating pitch (duty) and grating height in the round trip optical system were clarified; the maximum value of total light power efficiency η is 7.17%, when the segment width ratio p to the grating pitch is 0.589, and the phase modulation of the grating is 0.267π, the diffraction efficiency of the higher-order diffracted beams due to the superimposing is shown to be small.

The almost optimum condition HOEs were experimentally fabricated photolithographically, and the desired diffracted spots, the expected total light power efficiency, and the small influence of higher-order diffracted beams were observed.

The authors wish to express their thanks to T. Saimi for useful discussions throughout this work and also to H. Yamamoto and F. Sogawa for their support during the experiments.

References

1. Y. Kimura, S. Sugama, and Y. Ono, "High Performance Optical Head Using Optimized Holographic Optical Element," Jpn. J. Appl. Phys. Suppl. **26-4,** 131–134 (1987).
2. K. Tatsumi, T. Matsushita, and S. Ito, "A Multi-Functional Reflection Type Gratings for the CD Optical Head," Jpn. J. Appl. Phys. Suppl. **26-4,** 127–130 (1987).
3. W. H. Lee, "Optical Head Using Hologram Lens for Both Beam Splitting and Focus Error Detection Functions," U.S. Patent 4,731,772 (Mar. 1988).
4. H. Jacobus, "Apparatus for Optically Scanning an Information Plane Wherein a Diffraction Grating Splits the Beam into Subbeams," U.S. Patent 4,665,310 (May 1987).
5. M. Kato, T. Saimi, S. Kadowaki, and Y. Komma, "Holographic Servo-Signal Detection in Optical Heads," in *Technical Digest, Topical Meeting on Optical Data Storage* (Optical Society of America, Washington, DC, 1989), pp. 82–85; Proc. Soc. Photo-Opt. Instrum. Eng. **1078,** 122–126 (1989).
6. S. Ohara, T. Yoshida, and T. Ishida, "Optical Head," U.S. Patent 4,724,533 (June 1984).
7. W. H. Lee, "High Efficiency Multiple Beam Gratings," Appl. Opt. **18,** 2152–2158 (1979).
8. S. Kadowaki, Y. Komma, Y. Hori, and M. Kato, "Integrated Holographic Optical Pickup Head with Wide Allowance Range," Jpn. J. Appl. Phys. Suppl. **28-3,** 189–192 (1989).

Section Six
Head Integration/Miniaturization

Reprinted with permission from *IEEE Journal of Quantum Electronics,* Vol. QE-17(7), pp. 1257-1263 (July 1981).

Calculation and Experimental Verification of Two-Dimensional Focusing Grating Couplers

DETLEF HEITMANN AND CARMEN ORTIZ

Abstract—**A ray tracing method has been performed to calculate the focus spot size of two-dimensional focusing grating couplers that are produced by holographic interferometry of freely propagating wavefronts. The calculations have been used to optimize the constructing optical configuration. Diffraction limited spot diameters of about 2 μm can be achieved for couplers at wavelength 632.8 nm. These spot sizes have been experimentally verified for coupling and focusing TE modes which are guided in sputtered glass films on Si substrates.**

Introduction

LIGHT can be coupled out of a waveguide by the well-known grating coupling principle [1]. A linear grating with a uniform periodicity couples the light into a parallel beam. By using a chirped periodicity light is focused into a line [3]. Chirped and curved gratings in integrated optics have been reported in [3]. By a special chirped and curved grating structure, the light of a guided wave can be coupled out and focused into a small spot [4], [5].

In video disk and similar optical storage systems components are needed to focus light onto a small region of the optical storage material. We are interested in an integrated optical implementation of these focusing devices. The integrated optical chip might consist of light sources formed by an array of GaAs lasers [6] or precisely aligned and coupled fibers [7] with an external laser light source (He-Ne), beam forming waveguide lenses, the mentioned focusing grating coupler, and the detecting photodiode. Suitable substrates which allow integration of optical and electrical components can be Si substrates with sputtered glass waveguides [8], [9].

One can think of different ways to produce the special chirp and curvature of the focusing grating coupler for this purpose, e.g., performing a numerical calculation of the grating structure (see below) and using electron lithography, or applying holography with guided waves [10], in which case the wavelength λ_C for the construction of the hologram and the working coupler λ_R have to be identical. In this paper we investigate if this grating coupler can be fabricated by constructing a hologram at the substrate with a shorter wavelength $\lambda_C < \lambda_R$ by using a superposition of two expanded freely propagating light beams (Fig. 1). The wavefronts of these beams are modified by conventional analog optical elements (spherical and cylindrical lenses). The disadvantage of this method is that, due to the chromatic and spherical "aberrations" for the reconstruction with a longer wavelength, only a larger than diffraction limited spot size can be achieved. However, by a suitable combination of these lenses, the spot size can be minimized. Since high quality optical elements are used for the construction, high quality holograms with low scattering can be produced and the recording wavelength can be chosen to be suitable for exposing available photoresists.

Manuscript received October 3, 1980; revised March 16, 1981.

D. Heitmann is with the IBM Research Laboratory, San Jose, CA, on leave from the Institut für Angewandte Physik, Jungiusstrasse 11, 2 Hamburg 36, Germany.

C. Ortiz is with the IBM Research Laboratory, San Jose, CA, on leave from the Institut de Optica, Serrano 121, Madrid 6, Spain.

Calculations

We assume a planar waveguide structure covered with a thin layer of a holographic material to form a thin or surface hologram. We calculate first the chirp and curvature of a surface grating which is produced by a superposition of two wavefronts and perform, in a second step, a ray tracing if this grating is illuminated with a third wavefront.

We are using the following characterizations: the substrate plane is the x-y plane with the z-axis being the surface normal. $\vec{k} = (k_x, k_y, k_z)$ is the wavevector of a wavefront with $|\vec{k}| = 2\pi/\lambda$. $\vec{k} = (k_x, k_y)$ is the projection of this wavevector on the surface. λ_C or λ_R are wavelengths where the indexes C and R refer to the wavelength of construction and reconstruction of the hologram, respectively. We assume that the half-space on top of the substrate is air ($n = 1$). A guided wave can be characterized by its wavevector $\vec{k}_\perp = (k_x, k_y)$ with $|\vec{k}_\perp| = q \cdot 2\pi/\lambda$ where q is the effective index of refraction for the guided wave; $\vec{g} = (g_x, g_y)$ is the grating vector in a certain point of the hologram with $\Lambda = 2\pi/|g|$ being the periodicity of the grating.

An arbitrary wavefront can be described by the distribution of the wavevectors $\vec{k}_\perp(x, y)$ in the x-y plane. A superposition of two coherent wavefronts with indexes I and II, respectively, results in a grating with a locally varying grating vector

$$\vec{g}(x, y) = \vec{k}_\perp^{\,\mathrm{I}}(x, y) - \vec{k}_\perp^{\,\mathrm{II}}(x, y). \tag{1}$$

If this grating is illuminated with a wavefront III, then diffracted waves with a wavevector distribution $\vec{k}_\perp^{(n)}(x, y)$ are created [11]:

$$\vec{k}_\perp^{(n)}(x, y) = \vec{k}_\perp^{\,\mathrm{III}}(x, y) + n\vec{g}(x, y) \qquad n = \pm 1, 2, 3 \ldots \tag{2}$$

where n is the order of diffraction.

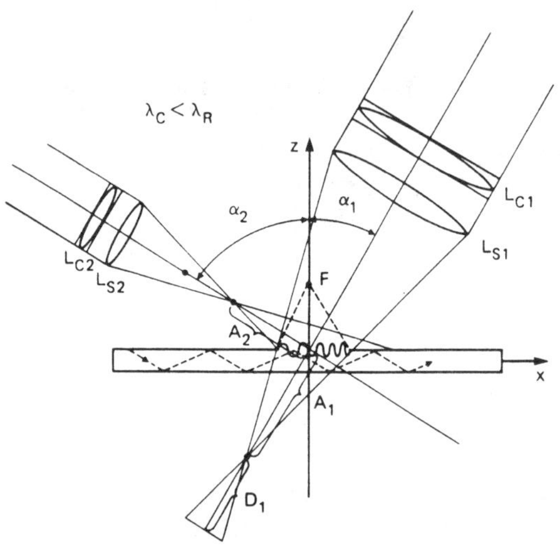

Fig. 1. Schematic for the construction of two-dimensional focusing grating couplers with freely propagating wavefronts. The wavefronts are modified by a combination of spherical lenses $L_{S1,2}$ and cylindrical lenses $L_{C1,2}$. The dashed lines show in a schematic way the focusing of a guided wave at wavelength $\lambda_R > \lambda_C$.

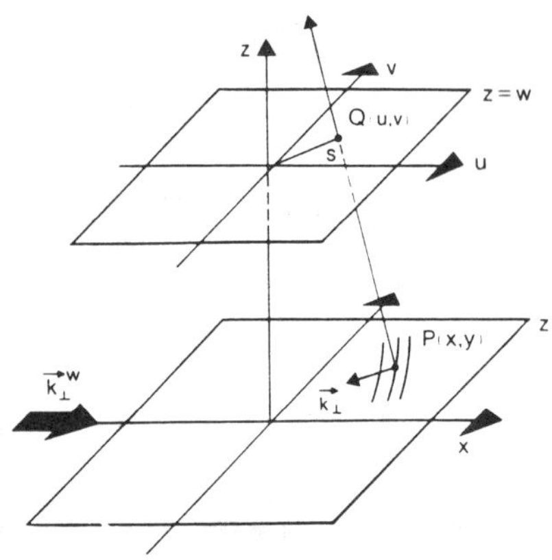

Fig. 2. Ray tracing for a grating coupler. Each point of the grating area $P(x, y)$ transmits a ray which traverses a parallel plane $z = w$ in a certain point $Q(u, v)$ of this plane.

For $|\vec{k}_\perp^{(n)}(x,y)| < 2\pi/\lambda_R$, a ray propagates from the point $P(x,y)$ into the upper half-space (see Fig. 2). This ray traverses a plane $z = w$, which is parallel to the x-y plane in a certain point $Q(u, v)$. The coordinates of this point are

$$\vec{u} = (u, v) = (x, y) + \frac{w}{\sqrt{\left(\frac{2\pi}{\lambda_R}\right)^2 - (k_x^{(n)})^2 - (k_y^{(n)})^2}} \cdot (k_x^{(n)}, k_y^{(n)}). \quad (3)$$

For $|\vec{k}_\perp^{(n)}(x,y)| < n_s \cdot 2\pi/\lambda_R$, waves are diffracted into the substrate (n_s = index of refraction for the substrate). If $k_\perp^n(x,y)$ is equal to $q_m \cdot 2\pi/\lambda_R$ with q_m the effective index of refraction of a guided wave with the same polarization, then Bragg reflection occurs. Diffraction into the substrate and Bragg reflection should be avoided or suppressed by the appropriate choice of substrate material, an additional reflecting sublayer or a thickness of the waveguiding film, respectively.

We now calculate the distribution $\vec{k}_\perp(x,y)$ for some special wavefronts. A beam, propagating parallel to the x-z plane with an angle of incidence α with respect to the substrate normal, is modified by lenses (see, e.g., beam I of Fig. 1). A combination of a spherical and a cylindrical lens focuses the light into two perpendicular lines. If the distance between these lines is D, and A is the distance of the focal line in the y-direction to the origin (see Fig. 1, beam I), then the calculation yields

The distribution for a guided wave propagating in the x-direction is

$$k_x = q \cdot \frac{2\pi}{\lambda}, \quad k_y = 0. \quad (5)$$

From a combination of these results we can calculate gratings and ray patterns: the difference of two freely propagating or guided wavefronts [(4) and (5)] gives, according to (1), the holographic grating structure (here $\lambda = \lambda_C$ is the wavelength of construction). If this grating is illuminated with one of the wavefronts of (4) and (5) at the wavelength $\lambda = \lambda_R$, we can calculate the ray pattern in a plane $z = w$ according to (2) and (3).

By an appropriate choice of A or D equal to infinity, simpler lens combinations are included. We have also calculated further lens combinations, nonparallel propagating guided waves (if beam forming waveguide lenses are used on the chip), and the loci of the grating rules, which are needed for electron lithography fabrication [see also Fig. 3(c)].

Discussion

In the following we discuss results from numerical calculations for a constructing wavelength $\lambda_C = 457.9$ nm and a wavelength of reconstruction $\lambda_R = 632.8$ nm (except for some results in Figs. 4 and 5 for which $\lambda_R = 830.0$ nm). We assume an effective index of refraction of $q = 1.5$, which fits for 7059 glass waveguides of about 0.5 μm thickness on Si-SiO_2 substrates. Geometrical distances are given in dimensionless units, since we see from (3)-(5) that distances in the u-v plane increase in proportion with corresponding distances in the x-y plane, if the ratio of A_1, A_2, D_1, D_2, x, y with respect to w remains constant. We first neglect the influence of the finite

$$k_x = \frac{2\pi}{\lambda} \frac{-(A \sin\alpha + x)}{\sqrt{A^2 \cos^2\alpha + (A \sin\alpha + x)^2 + \frac{y^2}{[1 + D/(A + x \sin\alpha)]^2}}} \quad (4a)$$

$$k_y = \frac{2\pi}{\lambda} \frac{-y}{\sqrt{y^2 + (A^2 \cos^2\alpha + (A \sin\alpha + x)^2)\,[1 + D/(A + \sin\alpha)]^2}}. \quad (4b)$$

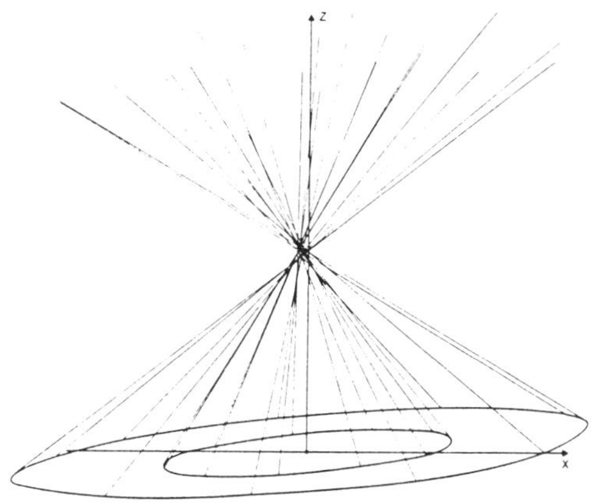

Fig. 3. (a) Three-dimensional plot of rays, emitted by a grating coupler. Rays from two circles of radius 0.5 and 1 in the grating area and 6 angular positions are shown to be focused in a distance $F = 3$ (the scale in the z-direction is compressed). The grating coupler operates at the wavelength $\lambda_R = 632.8$ nm for a guided wave with $q = 1.5$. This grating has been constructed with an optimized holographic setup at wavelength $\lambda_C = 457.9$ nm. Due to the optimization a very small focus spot can be achieved.

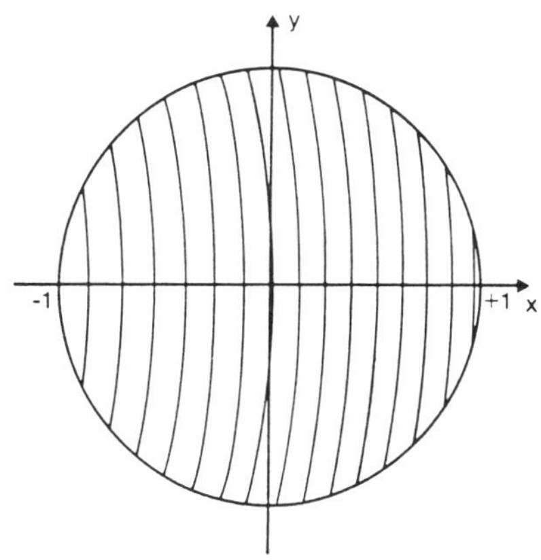

Fig. 3. (c) Curvature and chirp of the grating of Fig. 3(a). The distance between the lines is in proportion to the grating periodicity.

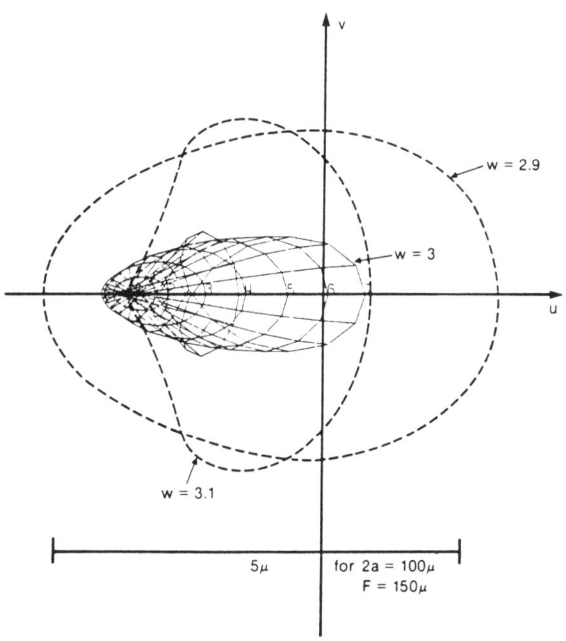

Fig. 3. (b) Ray pattern for the grating of Fig. 3(a). For points $P(x, y)$ on 7 circles with radii $r = 0$ (number 1), $r = 0.16$ (number 2), to $r = 1$ (number 7), and 24 angular positions of the grating plane, the corresponding points $Q(u, v)$ in the plane $w = 3$ are calculated and connected inside the fully drawn line. This means that the crossing point of two lines is a ray traversing point. The dashed lines are the envelope of the ray traversing points in the planes $w = 2.9$, 3.1, respectively. The scale in the lower part corresponds to a grating area diameter of $2a = 100$ μm.

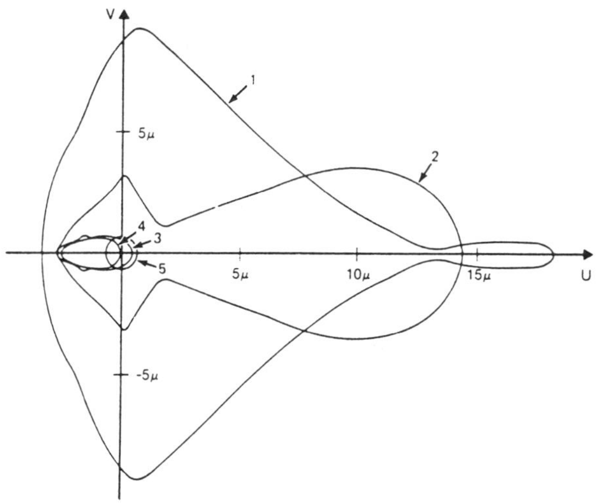

Fig. 4. Envelope of the ray traversing points in the focal plane for different construction setups. Numbers 1 and 2 are constructed with two beams of symmetrical angles of incidence, 3 and 4 with asymmetrical angles of incidence, and 5 is calculated for a grating coupler of the wavelength $\lambda_R = 830$ nm. (For more details see text.) The scale is related to a grating diameter $2a = 100$ μm. The focus distance is $F = 150$ μm.

grating size or, equivalently, the finite number of grating rules. This influence is discussed below.

Fig. 3(a)-(c) shows results for a grating which is produced by a superposition of a parallel beam with angle of incidence $\alpha_2 = -75^\circ$ and a convergent beam with angle of incidence $\alpha_1 = 7.5^\circ$ and focus distance $A_1 = 4.25$. Fig. 3(a) is a three-dimensional plot of the ray tracing. Some rays from two radii $a = 0.5$ and $a = 1$ for the grating area are shown. The grating couples out a parallel guided wave of $q = 1.5$ propagating in the x-direction. The light is concentrated in a distance F of about 3.

The ray pattern in different planes $z = w$ is shown in Fig. 3(b). We have chosen a calculation mesh of points $P(x, y)$ located on seven circles with radii $0, 0.16, \cdots 1$, and 24 radial positions $\lambda = 0^\circ, 15^\circ, \cdots$ to 345° in the grating plane. The corresponding points $Q(u, v)$ (see Fig. 2) in the plane $w = 3$ are shown and connected inside the fully drawn line of Fig. 3(b). The numbers refer to the number of the radius in the calculation mesh circles. The dashed lines in Fig. 3(b) are the envelope of the points $Q(u, v)$ for planes in a distance $w = 2.9$ and 3.1, respectively. We see that this grating focuses the light from a grating area of diameter $2a = 2$ into a spot with dimensions 0.06 and 0.03 in the u- and v-direction, respectively.

Fig. 3(c) is a calculation of the chirped and curved grating. The distance between the lines is in proportion to the grating periodicity. For this example $q = 1.5$, $F = 3$, $a = 1$, and the periodicity varies from about $\Lambda = 350$-535 nm. The largest difference between this grating, which has been fabricated with freely propagating waves at $\lambda_C = 458$ nm, and the "ideal" focusing grating, which can be calculated as a hologram of a parallel guided wave [(5) with $q = 1.5$] and a focused beam [(4) with $A = 3$, $D = \infty$, $\alpha = 0$] at $\lambda_C = 633$ nm, is very small (about 1 percent of the periodicity) and can hardly be shown in Fig. 3(c).

OPTIMIZATION

It is our goal to focus the light into the smallest possible spot. For a certain focus distance $W = F$, the spot size caused by aberrations increases with the growing radius of the grating area. However, taking into account the diffraction due to a finite number of grating lines, which is discussed in more detail below, a small F/a ratio is desirable. For a given F/a ratio and constructing and reconstructing wavelength, the spot size can be influenced by an appropriate choice of the components and arrangement for the constructing setup.

Fig. 4 shows the envelope of the ray traversing points Q as explained in Fig. 3(b) for gratings of diameter $2a = 100$ μm and focus distance $F = 150$ μm. The parameters ($A_1, A_2, \alpha_1, \alpha_2$) for the setup have been found in the following way: for nine points $P_i(x, y)$ equally distributed over the grating area, the distance $S_i = \sqrt{(u_i - u_f)^2 + (v_i - v_f)^2}$ of the ray traversing point from the assumed focus point $Q_f(u_f, v_f)$ has been calculated for various combinations of the parameters above. The optimum parameter combination corresponds to a minimum value of $S_M = \Sigma_i S_i$. The optimizations in Fig. 4 are performed for focus spots being close to the normal in the center of the grating center.

For the curves 1 and 2, a symmetrical setup with $\alpha_1 = -\alpha_2 = 32.8°$ is chosen. In curve 1, beam II is parallel and the focus distance A_1 of a spherical lens is varied. An optimum spot size is found for $A_1 = 190$ μm. (All values correspond to a grating radius $a = 50$ μm.) The spot size can be decreased by using an additional cylindrical lens in beam II. By varying A_1 and A_2, an optimum combination is found for $A_1 = 215$ μm and $A_2 = -1100$ μm. The envelope of the ray traversing points is shown in curve 2. A further improvement can be achieved by using asymmetrical angles of incidence, especially the nearly grazing incidence of one of the beams. For a parallel beam II with an angle of incidence $\alpha_2 = -75°$, which is suitable in an experimental setup, an optimization of A_1 and α_1 gives a smallest spot size for $A_1 = 212$ μm and $\alpha_1 = 7.5°$ (curve 3). This spot area can be slightly decreased by using a cylindrical lens in beam II with $A_1 = 215$ μm and $A_2 = -1250$ μm (curve 4). A similar spot size is also found for a superposition of a parallel beam II and a combination of a spherical lens and a cylindrical lens in beam I (variation of D_1).

If the constructing wavelength λ_C is smaller than λ_R/q, then a grating coupler can be constructed with a parallel beam II and a beam I incident in normal direction ($\alpha_1 = 0$) on the substrate. Best results are found using only one spherical lens. Curve 5 in Fig. 4 shows the spot size for a grating coupler at $\lambda_R = 830.0$ nm (GaAs laser) constructed with a parallel beam II ($\alpha_2 = -55.8°$) and a focused beam II ($\alpha_1 = 0, A_1 = 280$ μm).

We also calculated grating couplers with a nonnormal coupling direction [the focus spot $Q(u_f, v_f)$ is not normal over the grating center]. Even slightly smaller spot sizes are found for backward direction couplers (u_f is negative).

INTENSITY PROFILE

Our discussion of the spot size above is related to the envelope of all ray traversing points and does not take into account the intensity. If we look at the ray pattern inside the solid line of Fig. 3(b), we notice that most of the rays are concentrated in a smaller area than indicated by the envelope. The intensity profile of the spot can be calculated if the efficiency of the grating in each point of the grating area is known and the damping of the guided wave while traveling to this point is taken into account. The efficiency depends mainly on the modulation amplitude of the grating. Thus, by increasing the modulation amplitude in the direction of the traveling wave in an appropriate way, the entire grating area and numerical aperture of the grating coupler can be used. If it is not possible to vary the efficiency of the grating in an optimum way, then the actual efficiency of the grating can be taken into account for the optimization of the grating production parameters by weighting each ray according to the local efficiency and the damping.

F/a RATIO AND FINITE GRATING SIZE

In Fig. 5 we illustrate the dependence of the spot size as a function of the ratio F/a. We characterize the spot size by the maximum dimensions of the envelope of the ray pattern. The spot size increases with a smaller F/a ratio.

As mentioned before, the spot size caused by the grating aberrations decreases in proportion with a smaller grating diameter. However, for a small grating area the boundary diffraction or, equivalently, the finite number of grating lines comes into play. The curves labeled S_D in Fig. 5 show the diffraction limited spot diameter for a lens of diameter $2a$ and focus distance F:

$$S_D = \frac{1.22 \times \lambda}{\mathrm{NA}} = 1.22 \cdot \lambda \cdot \sqrt{1 + \left(\frac{F}{a}\right)^2}$$

where NA means numerical aperture.

For a certain grating radius a, an optimum spot size can be achieved by choosing the F/a ratio so that the diffraction spot size and aberration spot size have the same value. Fig. 5 shows that for a grating radius $a = 50$ μm, an F/a ratio of about three is an appropriate value leading to spot diameters of about 2-3 μm.

Slightly smaller spot sizes can be achieved by making the grating area and focus distance even smaller. For example, the optimized lineal dimension S_u for the grating of the curve labeled by full rectangles in Fig. 5 is for a grating area radius $a = 50$ μm: $F/a = 3$ with a corresponding value $S_u = 2.5$ μm and

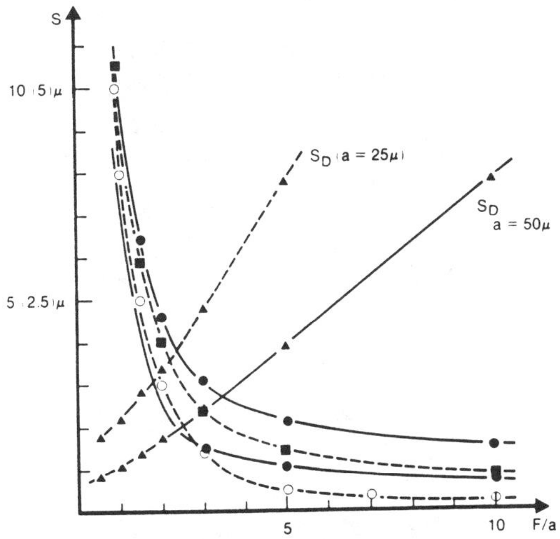

Fig. 5. Lineal dimensions of the focus spot envelope as a function of the F/a ratio. The full circles correspond to a grating coupler constructed with a parallel beam II with angle of incidence $\alpha_2 = -75°$ and a focused beam I with angle of incidence $\alpha_1 = 7.5°$ and optimized focus distance A_1. The upper curve is the lineal dimension S_u in the u-direction, the lower curve the dimension S_v in the v-direction. The curve characterized by full rectangles shows how the linear dimension S_u can be decreased by an additional cylindrical lens in beam II in an optimized position. The open circles show the diameter of the spot area for grating coupler at λ_R = 830 nm constructed with a parallel beam II ($\alpha_2 = -55.8°$) and an optimized focused beam I ($\alpha_1 = 0°$). The scale is related to a grating area radius a = 50 µm. (Values in brackets for a = 25 µm.) The curves labeled S_D are the diffraction limited spot diameters for λ_R = 632.8 nm.

a = 25 µm: F/a = 2.3 and S_u = 1.8 µm. However, a very high efficiency grating is needed to couple the light out for this small grating.

Grating Fabrication and Experimental Setup

A schematic of the waveguides is shown in the upper part of Fig. 6. A 0.5 µm thick layer of sputtered 7059 Corning glass was used as a waveguide [8], [9]. The waveguide is isolated from the silicon substrate by an SiO_2 film of 1 µm thickness. The 0.5 µm glass film supports one TE and one TM mode. The holograms are recorded as a surface grating in a film of 0.4 µm Shipley AZ 1350 B photoresist. For the experiments discussed here, we are using the photoresist grating itself as an output coupler. In the area to the right of the grating, the photoresist is removed by a suitable masking technique. This method has the advantage of fast production of the grating coupler and an immediate investigation of its optical properties. The disadvantage is that, in the region where the film is covered with the photoresist, the combined thickness of glass and resist allow two pairs of modes to propagate; thus, even light of one polarization is focused into two spots according to the conversion of intensity in the two modes at the photoresist edge. (This can be avoided by etching the grating profile into the surface of the waveguiding film. We demonstrated this, however, due to poor etching facilities only with very low efficiency.)

For the recording of the holograms we are using a superposition of two expanded and spatially filtered beams of an argon laser at wavelength λ_c = 457.9 nm. The substrate is tilted with respect to the symmetry axis of the beams. The optical element is a high quality microscope objective with a long working distance (Leitz UMK) with a power of 32 and with a maximum numerical aperture of 0.4. By varying the aperture of this objective, different grating areas and thus different F/a ratios can be achieved.

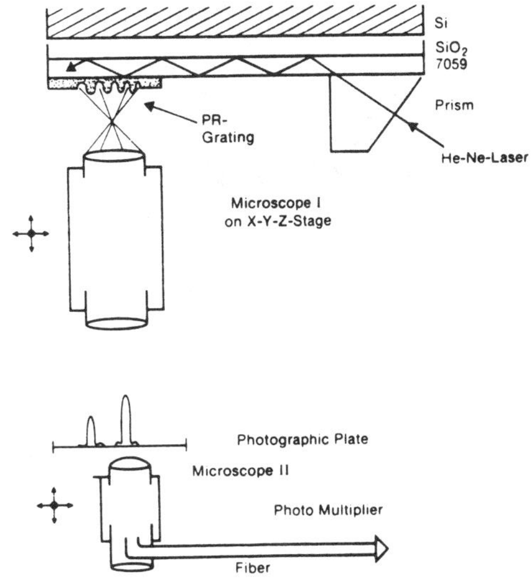

Fig. 6. Schematic of the waveguide with the grating coupler and the investigation arrangement. The grating plane and the focal plane can be observed with a microscope I. In addition, the image can be projected on a photographic plate or can be scanned by a seeond microscope with a light sensitive element.

The grating couplers are investigated with an arrangement shown in Fig. 6. The light of TE polarization is coupled into the waveguide with a prism. The out-coupled light is observed with a microscope I. Objectives with a high power (60X) and numerical aperture (0.85) have to be used to investigate all the rays of grating couplers with a low ratio F/a. If the object plane of the microscope lies in the grating plane the lineal dimensions of the grating can be measured and the light intensity distribution of the guided wave can be scanned by moving the microscope on a high resolution (2 µm) x-y-z stage. In the focal plane the intensity distribution of the focus spot can be similarly investigated. By using a projection ocular we record the image on a photographic plate. In addition, this projection can be investigated with a second microscope. With a fiber we couple the light of a certain area in the ocular plane of this microscope into a photomultiplier and measure the intensity profile of the projected and enlarged focus spot as a function of u and v in the focal plane.

Experimental Results

Fig. 7(a) shows a microscopic photograph of the spot in the focal plane of a grating coupler with focal length F = 1 mm. The grating area is about 0.5 mm in width and 0.6 mm in length. This grating has been constructed with a parallel beam II with an angle of incidence $\alpha_2 = 65.8°$ and a focused beam I with $\alpha_1 = 10°$ and a focal distance A_1 = 1.3 mm. Fig. 7(b) shows the calculated ray pattern in the focal plane using the

Fig. 7. (a) Photograph of the spot in the focal plane of an experimental grating coupler with a focal length $F = 1$ mm. The spot size of this coupler is determined by the aberration of the grating coupler.

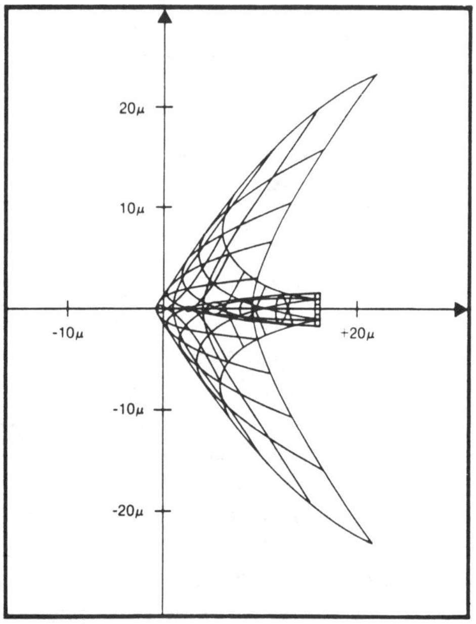

Fig. 7. (b) Calculated ray pattern for the grating of Fig. 7(a).

ray tracing method previously described. For the calculation we chose a rectangular mesh of 10 × 10 points $P(x, y)$ in the grating area of size 0.5 × 0.6 mm^2. The rays starting from these points traverse the focal plane in points $Q(u, v)$. The position of these points is indicated by the crossing point of two lines in Fig. 7(b). Within the accuracy for which the effective grating area and efficiency of the experimental grating can be taken into account, good agreement between theory and experiment exists and shows that the spot size of this grating is aberration limited.

It was shown before that the aberration limited spot size of holographic grating couplers of equal F/a ratio decreases in proportion with a smaller focal length and grating diameter. The decrease of spot size is limited for small grating areas by the diffraction of the finite grating size. For an optimal grating coupler the spot size determined by the aberrations and the diffraction limitation have the same values. These optimum conditions can be achieved for a grating coupler with focal length $F = 150$ μm and grating diameter $2a = 100$ μm.

Fig. 8 shows microscopic photographs of the grating plane (a) and the focal plane (b) for an experimental grating coupler

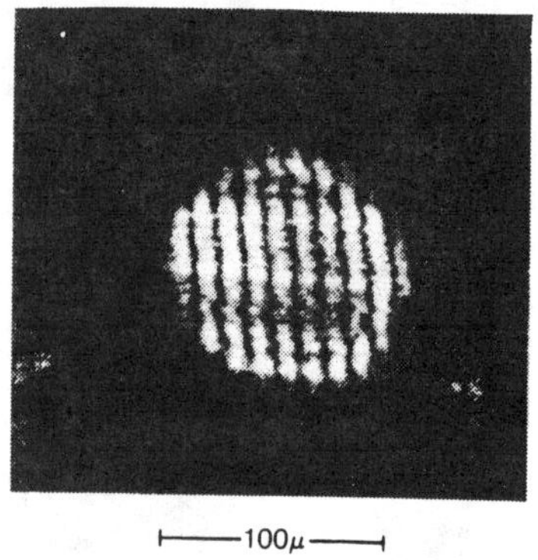

(a)

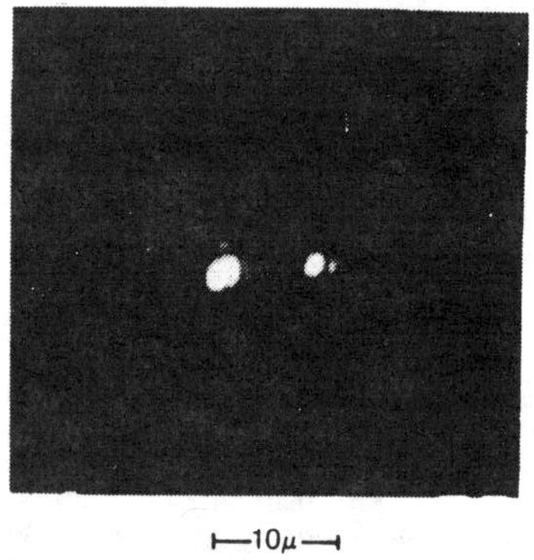

(b)

Fig. 8. Microscopic photograph of a grating coupler with focal length $F = 150$ μm and grating diameter $2a = 100$ μm. (a) Grating plane. (b) Focal plane.

which has been constructed with a parallel beam II with angle of incidence $\alpha_2 = 75°$ and a focused beam I with $\alpha_1 = 7.5°$ and distance of the focus $A_1 = 210$ μm. The additional structure in the light intensity of the grating area [Fig. 8(a)] is caused by the coherent superposition of the two TE modes propagating in the compound film of 7059 glass and photoresist. Fig. 8(b) shows the focal plane in a distance of 150 μm. The two spots occur due to the two modes. The left spot in Fig. 8(b) is over-exposed. A scan of the intensity profile of the spot with a microscope photometer (as described in Fig. 6) shows that the intensity is similar to an Airy pattern with the intensity decreasing to 10 percent of maximum at a diameter of 2.2 μm. Similar gratings with the same focal length of 150 μm and a smaller diameter have been investigated and show nearly rotationally symmetric focus spots of larger diameter limited by diffraction; gratings with a smaller F/a ratio also show larger spots with an asymmetric shape caused by the grating aberration. These results were predicted by the calculations. All these gratings with small grating area and focal length were found to have a very low level of scattered light.

Since our goal in this work is to investigate the optical quality of the grating couplers, we used relatively small modulation amplitudes and efficiencies for the gratings in order to achieve only slowly decreasing intensity over the grating area (about a 30 percent decrease over the entire grating area). For strongly modulated gratings we achieved output coupling efficiencies up to about 30 percent. However, gratings with higher efficiencies show larger spot sizes, which can be explained by the smaller effective grating area and hence stronger diffraction. This indicates that high efficiency grating couplers need a specially increasing profile of the modulation height to achieve optimum use of the entire grating area and numerical aperture of the grating coupler. The efficiency of the grating couplers discussed here is also reduced because light is diffracted into the substrate. This can be avoided by evaporating a highly reflecting material beneath the grating area. For our small grating dimensions of 100 μm the damping caused by such a coating would be expected to be small.

Conclusions

Two-dimensional focusing grating couplers for guided waves at He-Ne and GaAs laser wavelengths can be fabricated by using holographic interferometry at wavelength $\lambda = 457.9$ nm. By applying a ray tracing method, we calculated the focus spot size and optimized the constructing optical setup. With an appropriate setup grating couplers of focus distance $F = 150$ μm and grating area diameter $2a = 100$ μm achieve spot diameters of about 2 μm. We have demonstrated experimental two-dimensional focusing grating couplers for He-Ne laser light. The focus spot size of grating couplers with a large grating area is determined by the aberrations. For small grating areas and focal lengths diffraction limited spot sizes can be achieved. For an optimized construction setup spot sizes of 2.2 μm have been obtained experimentally.

Acknowledgment

The authors thank the late Dr. R. V. Pole and Prof. C. Imbert for valuable discussions, and Dr. F. Sequeda and his group, CSS of IBM Research, for the preparation of excellent sputtered glass films.

References

[1] M. L. Dakss, L. Kuhn, P. F. Heidrich, and B. A. Scott, "Grating coupler for efficient excitation of optical guided waves in thin films," *Appl. Phys. Lett.*, vol. 16, pp. 523-525, June 1970; H. Kogelnik and T. P. Sosnowski, "Holographic thin film couplers," *Bell Syst. Tech. J.*, vol. 49, pp. 1602-1608, Sept. 1970.

[2] A. C. Livanos, A. Katzir, A. Yariv, and C. S. Hong, "Chirped-grating demultiplexers in dielectric waveguides," *Appl. Phys. Lett.*, vol. 30, pp. 519-521, May 1977; A. Katzir, A. C. Livanos, J. B. Shellan, and A. Yariv, "Chirped gratings in integrated optics," *IEEE J. Quantum Electron.*, vol. QE-13, pp. 296-304, Apr. 1977; W. W. Ng, C. S. Hong, and A. Yariv, "Holographic interference lithography for integrated optics," *IEEE Trans. Electron Devices*, vol. ED-25, pp. 1193-1200, Oct. 1978.

[3] P. K. Tien, "Method of forming novel curved-line gratings and their use as reflectors and resonators in integrated optics," *Opt. Lett.*, vol. 1, pp. 64-66, Aug. 1977.

[4] M. Miler and M. Skalsky, "Stigmatically focusing grating coupler," *Electron. Lett.*, vol. 15, pp. 275-276, May 1979.

[5] D. Heitmann and R. V. Pole, "Two-dimensional focusing holographic grating coupler," *Appl. Phys. Lett.*, vol. 37, pp. 585-587, Oct. 1980.

[6] J. D. Crow, L. D. Comerford, J. S. Harper, M. J. Brady, and R. A. Laff, "Gallium arsenide laser-array-on-silicon package," *Appl. Opt.*, vol. 17, pp. 479-485, Feb. 1978.

[7] J. T. Boyd and S. Siram, "Optical coupling from fibers to channel waveguides formed on silicon," *Appl. Opt.*, vol. 17, pp. 895-898, Mar. 1978.

[8] J. E. Goell and R. D. Standley, "Sputtered glass waveguide for integrated optical circuits," *Bell Syst. Tech. J.*, vol. 48, pp. 3445-3448, Dec. 1969.

[9] D. B. Anderson, R. L. Davis, J. T. Boyd, and R. R. August, "Comparison of optical-waveguide lens technologies," *IEEE J. Quantum Electron.*, vol. QE-13, pp. 275-282, Apr. 1977; D. B. Anderson, "Integrated optical spectrum analyzer: An imminent chip," *IEEE Spectrum*, vol. 15, pp. 22-29, Dec. 1978.

[10] T. Suhara, H. Nishihara, and J. Koyama, "Waveguide holograms: A new approach to hologram integration," *Opt. Commun.*, vol. 19, pp. 353-358, Dec. 1976; A. Wüthrich and W. Lukosz, *Appl. Phys.*, vol. 21, pp. 55-64, Jan. 1980.

[11] This method is well known as the vector closure method. See, for example, M. R. Latta and R. V. Pole, "Design techniques for forming 488-nm holographic lenses with reconstruction at 633-nm," *Appl. Opt.*, vol. 18, pp. 2418-2421, July 1979.

Detlef Heitmann received the Ph.D. degree from the Institute of Applied Physics, Hamburg, Germany, while working on solid state spectroscopy with fast electrons and light, especially excitation and application of surface plasmons.

In 1979 and 1980 he held an IBM Post-Doctoral Fellowship at the IBM Research Lab, San Jose, CA, while working on integrated optics. He is currently with the Institute of Applied Physics, Hamburg, doing basic research on two-dimensional electron gases in MIS systems.

Carmen Ortiz was born in Madrid, Spain, on April 12, 1947. She received the M.S. degree in physics from the University of Madrid, Madrid, Spain, in 1970 and the Ph.D. degree from the University of Stuttgart, Stuttgart, Germany, in 1974.

From 1974 to 1976 she worked in the Max Planck Institute for Material Sciences under a contract of the German Society of Research (D.F.G.) on order–disorder phenomena in metallic alloys. In 1976 she joined the Institute of Optics (Spanish Scientific Council) in Madrid, Spain, where her research has centered on sensitive thin film materials. Since 1979 she has been a Visiting Scientist at the IBM Research Laboratory, San Jose, CA, where she has been studying thin films used in laser technology.

Reprinted with permission from *IEEE Journal of Lightwave Technology,* Vol. LT-4(7), pp. 913-918 (July 1986). ©1986 IEEE.

An Integrated-Optic Disk Pickup Device

SHOGO URA, TOSHIAKI SUHARA, HIROSHI NISHIHARA, MEMBER, IEEE, AND JIRO KOYAMA

Abstract—A monolithic-integrated optical-disk pickup device capable of detecting readout and focus/tracking error signals is proposed, designed, and fabricated. The integrated-optic disk pickup device (IODPU) is constructed by integrating a focusing grating coupler, a grating beam splitter, and photodiodes in a film waveguide on a Si substrate. The function of detecting readout and error signals is demonstrated by butt-coupling a laser diode to the fabricated IODPU.

I. Introduction

RECENTLY THERE HAS been much interest in optical disk memory systems, since they are a great advantage in mass data storage. An optical-disk pickup head is currently constructed with bulk micro-optics and needs complex and time-consuming fabrication processes, e.g., polishing and assembling of the micro-optics. However, if the pickup head is constructed with integrated optics using a waveguide, the planar fabrication technique can be employed and there would be great improvement in producibility, reduction of size, weight and cost, and enhancement of application flexibility.

In this paper, we will propose and demonstrate a monolithic integrated-optic disk pickup device (IODPU) capable of detecting readout and focus/tracking error signals. Such a device needs a component which focuses the guided wave into a point on a disk, a beam splitting component which deflects the reflected wave into photodetectors, and the photodetectors in the same waveguide. We employ a focusing grating coupler (FGC) as a focusing component of the device. A FGC is a kind of waveguide hologram [1]. It has chirped and curved grating pattern to couple a guided wave and a spherical wave focusing in free space [2]-[5]. The FGC's fabricated by the holographic technique [2], [3], however, exhibit a large aberration due to the difference in wavelength between the fabrication light and the coupling light. By the electron-beam direct writing technique, aberration-free FGC's, for use at an arbitrary wavelength, can be fabricated [4], [5]. We have studied the FGC's of the latter type to obtain good focusing characteristics and high coupling efficiency for the integrated-optics approach [4]. By integrating this type of FGC, a grating beam splitter and photodiodes in a film waveguide on a Si substrate, we constructed the IODPU [6]. Section II describes the design consideration of the IODPU. The device fabrication and experimental results are reported in Section III.

Manuscript received November 18, 1985; revised February 5, 1986.
The authors are with the Department of Electronics, Faculty of Engineering, Osaka University, Osaka 565, Japan.
IEEE Log Number 8608490.

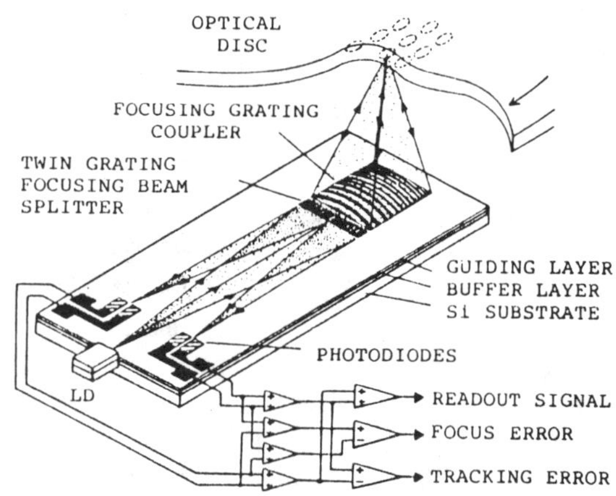

Fig. 1. Schematic view of the proposed IODPU.

II. Device Description and Design Consideration

A. Device Construction and Pickup Operation

Fig. 1 shows the schematic illustration of the proposed IODPU. The waveguide is formed with a glass guiding layer and a SiO_2 buffer layer on a Si substrate. A FGC, a twin grating focusing beam splitter (TGFBS), and a photodiode array are integrated in this film waveguide. The guided wave diverging from the butt-coupled laser diode (LD) is focused by the FGC into a point on a disk. The wave reflected by the disk is collected and coupled back into the waveguide by the same FGC. Then the TGFBS divides the reflected wavefront into halves, deflects (beam splitting), and simultaneously focuses into four photodiodes located at a distance slightly shorter than that of the input point (the LD position). This complex function of the TGFBS minimizes the number of the components. Thus the disk signal is read out. Although the forward wave diffracted by the TGFBS results in two additional spots on the disk, this beam can not be coupled back into the waveguide since it is different in wavefront and propagating direction from the undiffracted wave.

The focus/tracking error signals also are obtained by a simple processing of the photocurrents based on the Foucault/push-pull methods [7], respectively. The focusing and tracking can be achieved by actuators controlled by the error signals. The disk displacement from the focal plane, namely, the focus error, causes the focus displacement of the back-coupled guided wave along the y-axis according to the imaging relation of the FGC (see Section II-B). In the absence of the focus error, the IODPU is designed so that the wave deflected by the TGFBS focuses into the center gap of the neighboring photodiodes as

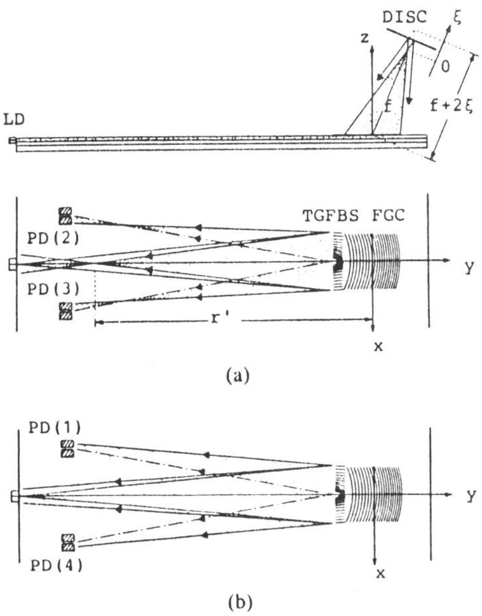

Fig. 2. Schematic diagram of the focus error detection. (a) The disk is too far. (b) The disk is too near.

shown in Fig. 1. When the disk goes away from the IODPU, the focus point of the back-coupled guided wave moves toward the FGC, and this wave is absorbed by the inner photodiodes PD(2) and PD(3) as shown in Fig. 2(a). When the disk comes near, on the other hand, it is absorbed by the outer ones PD(1) and PD(4) as shown in Fig. 2(b). Thus we can get the focus error signal by making the subtraction in photocurrent of the inner and outer photodiodes. The IODPU is to be set so that the disk track lies on the $y - z$ plane and the tracking error occurs in the x direction. The light intensity distribution is symmetric with respect to the y-z plane in the absence of the tracking error. However, it becomes asymmetric in the presence of the error. Accordingly, we can get the tracking error signal by subtracting in photocurrent a photodiode from the other side one with respect to y-axis, for example, PD(2)–PD(3).

B. Focusing Grating Coupler (FGC)

Grating Pattern: The grating pattern can be calculated from the phase difference between the input wave and the output wave coupled by the grating. The input wave is the guided wave diverging from the LD, and the output one is the focusing wave in the air, so the m-th line pattern equation of the FGC was obtained to be (see Fig. 3)

$$N\sqrt{x^2 + (y + r)^2} + \sqrt{x^2 + (y - f \cdot \sin\theta)^2 + (f \cdot \cos\theta)^2} = m\lambda + Nr + f \tag{1}$$

where λ is the wavelength in free space and N is the mode index of the waveguide.

Imaging Relation: When the wavefront of the reflected wave from the disk is the original one with which the grating pattern has been calculated, the wavefront of the guided wave coupled back by the FGC coincides with that

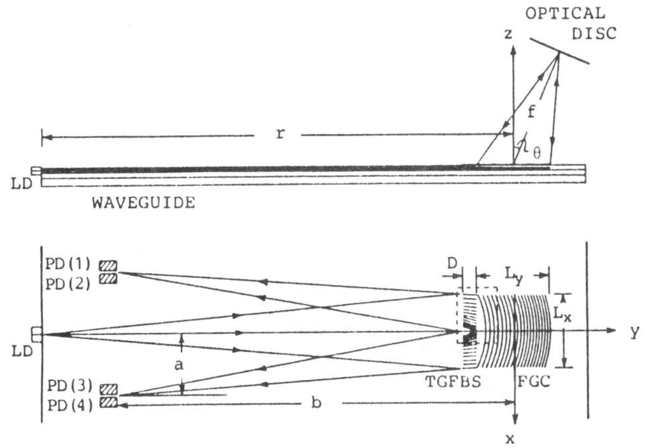

Fig. 3. Configuration of components and wavefronts on the waveguide plane and its cross section.

of the wave diverging from the LD, except for the propagation direction. The wavefront, however, varies with the disk displacement along the normal to the disk surface. By introducing a Gaussian reference sphere and comparing the series expansion coefficients with those of the actual wave, the imaging relation was obtained to be written as

$$\frac{N}{r'} + \frac{1}{f + 2\xi} = \frac{N}{r} + \frac{1}{f} = \text{const.} \tag{2}$$

where r' and ξ are defined in Fig. 2 and $f + 2\xi$ is a length from the FGC to a diverging center. This expression shows that the relation of the FGC coincides with that of conventional lenses except that N is not unity. This is an important relation indicating that the disk displacement affects the focus position of the back-coupled guided wave.

Focusing Characteristics: For simplicity of analysis, let us consider the case of $\theta = 0$. The light amplitude distribution on the focal plane is given by Fourier transform of the guided wave amplitude distribution over the FGC aperture on the waveguide. In the case that the amplitude over the aperture is uniform, the focus spot 3-dB full width is the diffraction-limited value and can be written as

$$2\sigma_x = 0.88 f\lambda/L_x$$
$$2\sigma_y = 0.88 f\lambda/L_y. \tag{3}$$

In our case, the amplitude on the y-axis is proportional to $\exp(-\alpha L_y)$ reflecting the guided wave radiation, and the spot width depends on the radiation decay factor α, too. To obtain the small focus spot, α should be small.

Coupling Efficiency: The output coupling efficiency of the FGC can be written as

$$\eta^o(L_y) = \eta^a\{1 - \exp(-2\alpha L_y)\} \tag{4}$$

where η^a is the fractional power for the first-order diffraction into the air. To obtain a high efficiency for a given L_y, α and η^a should be large.

The input coupling efficiency depends on the wavefront

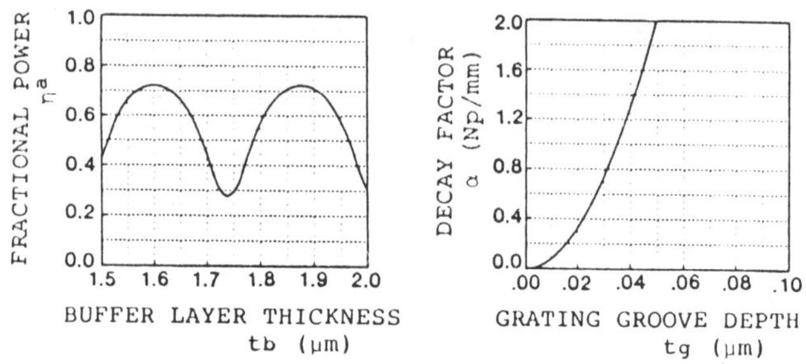

(a) (b)

Fig. 4. (a) Calculated dependence of η^a on t_b. (b) Calculated dependence of α on t_g in $t_b = 1.86$ μm. Refractive indices of the air, guiding layer, buffer layer, and substrate are 1.0, 1.55, 1.46, and 3.75, respectively. The guiding layer thickness is 0.95 μm, and $\lambda = 0.79$ μm.

configuration of the incident wave reflected by the disk. We consider here the efficiency including the influence of the disk displacement shown in Fig. 2. By applying the reciprocity theorem, the input coupling efficiency can be written theoretically as

$$\eta^i(L_y, \xi) = \eta^o(L_y) \cdot \frac{S^2(u) + C^2(u)}{u^2} \cdot \left\{ \frac{2\alpha L_y}{\exp(\alpha L_y) - \exp(-\alpha L_y)} \right\} \quad (5)$$

$$S(x) = \int_0^x \sin\left(\frac{\pi}{2} t^2\right) dt$$

$$C(x) = \int_0^x \cos\left(\frac{\pi}{2} t^2\right) dt, \; u = \frac{L_y \cos\theta}{f} \sqrt{\frac{\xi}{\lambda}}.$$

The first factor of the product on the right-hand side of (5) is the output coupling efficiency. The second one means the efficiency reduction arising from the difference in wavefront between the focusing wave and the reflected one, and is unity when $\xi = 0$. The third one shows the effect of the amplitude distribution over the FGC aperture and results in 0.72 when $\alpha L_y = 1$.

Parameter Optimization: The output angle θ was determined to be 15° so that any unwanted Bragg reflection and higher order radiation are eliminated; when $\theta = 0°$, which may be the most suitable for the actuation, the second-order coupling may result in the strong Bragg reflection. The factor α was determined to be $1/L_y$ to realize good focusing characteristics (3-percent increase of focus spot size compared to (3)) and high efficiency (nearly 90-percent diffraction efficiency) simultaneously.

By applying the analysis presented by T. Tamir *et al.* [8], we obtained the analytical expressions of α and η^a in terms of the waveguide and the grating parameters. Fig. 4 shows the calculated η^a and α. The dependence of η^a on the buffer layer thickness t_b is periodic due to the interference between the air radiation and the radiation reflected by the buffer-layer/substrate interface. The thickness t_b was determined to be 1.86 μm to have the maximum of η^a, and the grating groove depth t_g was to be 0.035 μm to have $\alpha L_y = 1$ for $L_y = 1$ mm. Then the theoretical

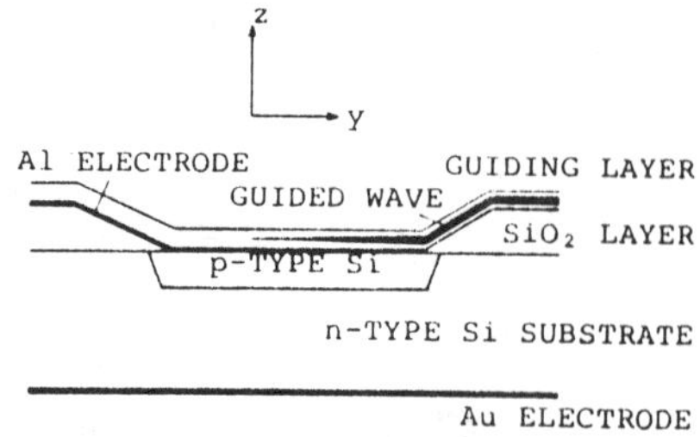

Fig. 5. Cross section of photodetecting area.

output coupling efficiency and the diffraction-limited focus spot size are calculated to be 60 percent and 1.4 μm according to (4) and (3), respectively.

C. Twin Grating Focusing Beam Splitter (TGFBS)

Grating Pattern: The TGFBS also is fabricated by the electron-beam direct writing technique. The grating pattern can be obtained in a similar fashion to the FGC. The input wave is the guided wave converging back into the input point (the LD position), and the output one is the focusing wave into the detecting points for each half, so the m-th line pattern equation can be written as

$$\sqrt{x^2 + (y + r)^2} - \sqrt{(x \pm a)^2 + (y + b)^2} = m\lambda/N + r - \sqrt{a^2 + b^2}. \quad (6)$$

Diffraction Efficiency: The forward wave diverging from the LD passes through the TGFBS, while the backward wave from the FGC is deflected by the same TGFBS. To obtain the highest efficiency of the device, the diffraction efficiency of the TGFBS should be 50 percent. The diffraction in the Bragg regime is utilized for this high efficiency. The diffraction efficiency is given by

$$\eta_{BS} = \sin^2(\kappa D/\sqrt{\cos\delta \cdot \cos\epsilon}) \quad (7)$$

where κ, D, δ, and ϵ are the coupling coefficient, the coupling length, the angle between the incident wave normal and the y-axis, and the angle between the diffraction wave normal and the y-axis, respectively. Therefore, the condition of 50-percent efficiency can be expressed by

$$\kappa D/\sqrt{\cos\delta \cdot \cos\epsilon} = \pi/4. \quad (8)$$

Parameter Optimization: The diffraction angle ϵ was determined to be 7.8 degrees to satisfy the condition of the Bragg operation. For the grating of the same structure as the FGC, the coefficient κ was calculated to be 8.2 mm^{-1} and then the length D was determined to be 95 μm from (8).

D. Photodiode Array

Fig. 5 illustrates a cross section of the photodetecting area on n-type Si substrate along the y-direction. The oxidized SiO_2 buffer layer is tapered from guiding area to detecting area, and the guided wave is led into a photodiode without serious scattering. The element size was determined to be 150 × 50 μm^2 to get large photocurrents and high S/N ratio.

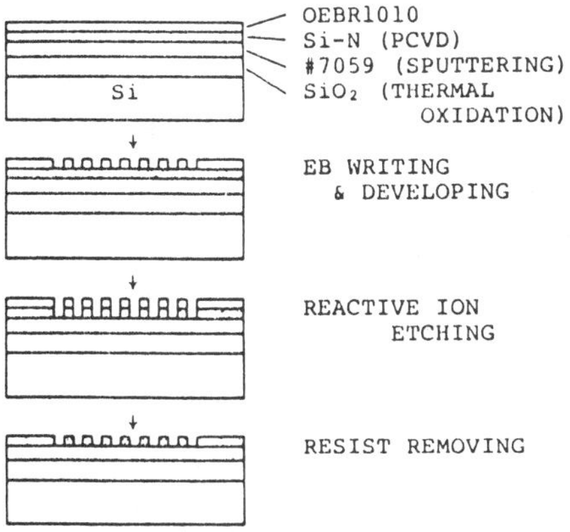

Fig. 6. Fabrication process of FGC/TGFBS.

TABLE I
SPECIFICATIONS OF THE FABRICATED IODPU

Light Source GaAlAs LD	Wavelength	λ=0.79 μm
Waveguide Single Mode Glass/SiO_2/Si	Guiding Layer (#7059 glass) Buffer Layer (SiO_2) Mode Index (TE_0)	0.95 μm t_b=1.86 μm 1.52
Grating Components FGC	Cladding Layer (Si-N) Focal Lengths Output Angle Aperture Period Theoretical Coupling Efficiency Diffraction-limited Focus Spot Width	t_g=0.035 μm f=2.0 mm r=10.5 mm θ=15.0 Deg 1.0 x 1.0 mm^2 0.52 - 0.75 μm 60 % 1.4 μm
TGFBS	Detecting Point Size Period Theoretical Diffraction Efficiency	a=1.165 mm b=9.00 mm 1.0 x 0.095 mm^2 3.80 - 4.04 μm 50 %
Photodiode Array	Element Size	150 x 50 μm^2 (10 μm Gap)

III. Fabrication and Experimental Results

A. Device Fabrication

The specifications of the fabricated IODPU are summarized in Table I. At first Si (n-type, (100)) substrate was thermally oxidized to grow the SiO_2 buffer layer, and a p-n photodiode array was fabricated by the planar process with photolithography. Then the Corning #7059 glass wave-guiding layer and the Si-N cladding layer were deposited by sputtering and PCVD methods, respectively. Fig. 6 shows the fabrication process of the FGC/TGFBS. The grating pattern was directly written by an electron beam in positive resist coated on the waveguide. The writing time including the computing time was approximately 30 min. The pattern was transferred to the Si-N layer by reactive ion etching to form the grating grooves. Fig. 7 shows a microphotograph of the fabricated FGC/TGFBS pattern. Then the photodiode contact windows were opened by chemical etching, and the waveguide/substrate was cut and cleaved for LD butt-coupling into chips of 5 × 12 mm^2 area. Finally the chip was mounted on an IC package, wire bonded, and a LD was coupled. Fig. 8 shows a photograph of the fabricated IODPU.

B. Experimental Results

Focusing Characteristics: The guided wave of TE_0 mode excited by a LD was diffracted by the FGC and focused in the air. This focus spot was projected through a ×40 objective lens on a CCD image sensor and observed by a CRT and a synchroscope. Fig. 9 shows the focus spot and the CCD video signal trace indicating the intensity distribution on the focal plane. The 3-dB spot width was measured to be 3.5 μm, while the diffraction-

Fig. 7. Microphotograph of the fabricated FGC/TGFBS pattern. (The portion indicated by broken line in Fig. 3).

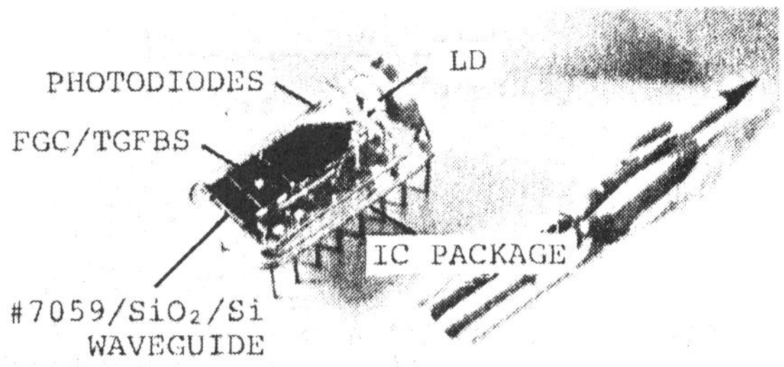

Fig. 8. Photograph of the fabricated IODPU.

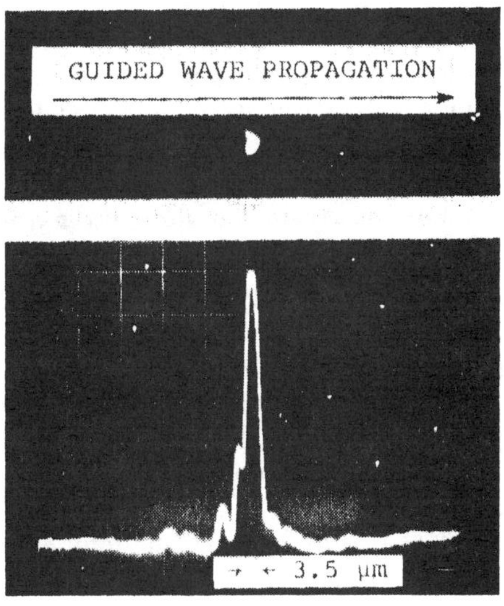

Fig. 9. Light intensity distribution on the focal plane.

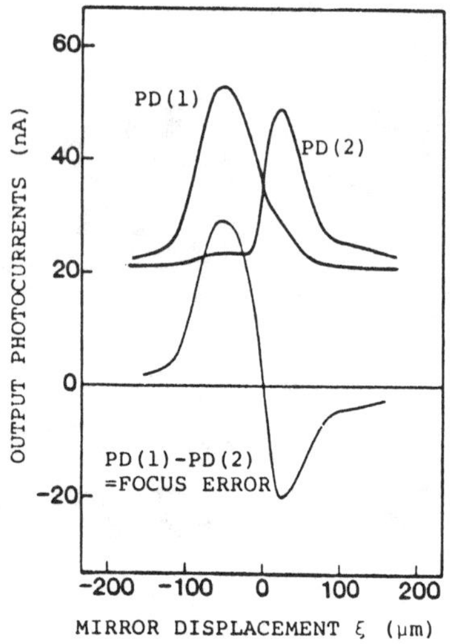

Fig. 10. Measured output photocurrents of the fabricated IODPU.

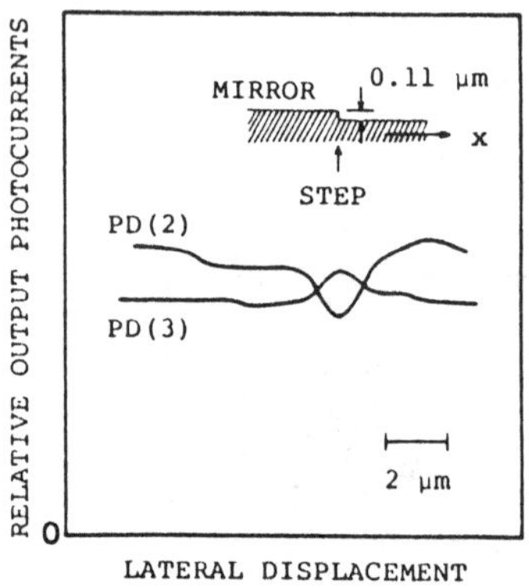

Fig. 11. Measured output photocurrents of the fabricated IODPU.

limited value is 1.4 μm. The broadening of the spot would be due to the fabrication error in the FGC pattern. It is necessary for achieving submicron spot size to fabricate the FGC having more accurate pattern and larger aperture.

Pickup Operation: A mirror was used to simulate the disk and was able to be actuated along the normal of the focal plane of the FGC. The output photocurrents of the integrated photodiodes were measured and recorded while the mirror was actuated. The measured photocurrents are plotted against the mirror displacement in Fig. 10. The LD power was 3 mW and the air gap length between the LD and the IODPU chip was 30 μm. The maximum photocurrent was 50 nA per photodiode while the estimated value is about 300 nA (the fraction of the butt-coupled power in the FGC aperture ~0.03, the total efficiency of the grating components ~0.03, and the efficiency of the photodiode ~0.12 A/W).

The photocurrent of the outer-side photodiode PD(1) had the peak at a negative value of ξ corresponding to too near position of the mirror, and the inner-side PD(2) had the peak at a positive ξ on the other hand. Thus it was confirmed that the focus error signal can be obtained by the subtraction of these photocurrents shown by the thinner line PD(1)-PD(2) in Fig. 10.

To confirm the function of detecting the tracking error, a mirror having a step configuration simulating the edge of a disk pit was set on the focal plane and actuated along the x-direction. Fig. 11 plots the output photocurrents of PD(2) and PD(3) against the displacement of a mirror having 0.11-μm step. These curves show the rise and fall at the step, indicating that the tracking error signal can be obtained by the subtraction of these photocurrents.

IV. Conclusion

An integrated-optic disk pickup device (IODPU) was fabricated for the first time. Although the performance obtained so far is not enough to the practical application, the fundamental operation was demonstrated. The experimental works are being continued to improve the focus spot size, photocurrent level, and error detection sensitivity, and the effect of the spectrum variation of the LD is also under study.

Acknowledgment

The authors wish to thank LSI R&D Lab. and Consumers' Electronics Development Lab., Mitsubishi Electric Co., for their cooperation in preparing the photodiodes.

References

[1] T. Suhara, H. Nishihara, and J. Koyama, "Waveguide holograms: A new approach to hologram integration," *Opt. Commun.*, vol. 19, pp. 353–358, 1976.

[2] M. Miler and M. Skalsky, "Chirped and curved grating coupler focusing both outgoing beam and guided wave," *Opt. Commun.*, vol. 33, pp. 13–16, 1980.

[3] D. Heitmann and C. Ortiz, "Calculation and experimental verification of two-dimensional focusing grating couplers," *IEEE J. Quantum Electron.*, vol. QE-17, pp. 1257–1263, 1981.

[4] T. Suhara, H. Nishihara, and J. Koyama, "High-performance focusing grating coupler fabricated by electron-beam writing," presented at Topical Meet. Integrated Guided-Wave Optics, Kissimmee, FL, Apr. 24–27, 1984, paper ThD4.

[5] G. Hatakoshi, H. Fujima, and K. Goto, "Waveguide grating lenses for optical couplers," *Appl. Opt.*, vol. 23, pp. 1749–1753, 1984.

[6] T. Suhara, S. Ura, H. Nishihara, and J. Koyama, "An integrated-optic disc pickup device," presented at 5th IOOC—11th ECOC '85, Venezia, Italy, Oct. 1–4, 1985.

[7] G. Bouwhuis and J. J. M. Braat, "Video disk player optics," *Appl. Opt.*, vol. 17, pp. 1993–2000, 1978.

[8] T. Tamir and S. T. Peng, "Analysis and design of grating couplers," *Appl. Phys.*, vol. 14, pp. 235–254, 1977

*

Shogo Ura was born in Osaka, Japan, on January 9, 1960. He received the B.S. and M.S. degrees in electrical engineering from Osaka University, Osaka, Japan, in 1982 and 1984, respectively.

He is a graduate student of the Department of Electronics, Faculty of Engineering, Osaka University, and is engaged in research on integrated optics.

Mr. Ura is a member of the Institute of Electronics and Communication Engineers of Japan.

Toshiaki Suhara was born in Wakayama, Japan, on November 13, 1950. He received the B.S., M.S., and Ph.D. degrees in electronics engineering from Osaka Universigy, Osaka, Japan, in 1973, 1975, and 1978, respectively.

He joined the Department of Electronics, Faculty of Engineering, Osaka University, in 1978 as a Research Associate. He is currently engaged in research on integrated optics, holography, optical memory, and applications of electron beams. He received the Paper Award in 1978 and the Young Engineer Award in 1980 both from the Institute of Electronics and Communication Engineers (IECE) of Japan. He is a co-author of a volume "Optical Integrated Circuits" (in Japanese).

Dr. Suhara is a member of the IECE of Japan, the Japan Society of Applied Physics, and the Laser Society of Japan.

*

Hiroshi Nishihara (S'59–M'68) was born in Osaka, Japan, on August 4, 1937. He received the B.S., M.S., and Ph.D. degrees in communication engineering from Osaka University, Osaka, Japan, in 1960, 1962, and 1965, respectively.

He then joined the Department of Electronic Engineering, Faculty of Engineering, Osaka University, and has been a Professor since 1984. His areas of research interest include holography, integrated optics, and optical-fiber sensor techniques. He received the Paper Awards in 1978 and 1983 and the Book Award for "Optical Wave Electronics" in 1980 all from the Institute of Electronics and Communication Engineers (IECE) of Japan, and the Research Award in 1982 from the Laser Society of Japan.

Dr. Nishihara is a member of IECE Japan, the Japan Society of Applied Physics, and SPIE.

*

Jiro Koyama was born in Tokyo, Japan, on November 29, 1924. He received the B.S. and Ph.D. degrees in electrical engineering from Tokyo Institute of Technology, Tokyo, Japan, in 1946 and 1958, respectively.

He was first associated with the Agency of Aeronavigation, working on radio controls of air traffic. In 1950, he joined the Electrical Communication Laboratory, Nippon Telegraph and Telephone Public Corporation, Tokyo, where he was initially engaged in research on microwave and millimeter-wave tubes. From 1967 to 1972, he worked in the fields of traveling-wave-type amplifiers in GaAs bulk semiconductor and optical information storages. Since 1973, he has been a Professor at Osaka University, where he and his associates are now engaged in research on optical wave electronics.

Dr. Koyama is a member of the Institute of Electronics and Communication Engineers of Japan and the Optical Society of America.

Reprinted with permission from *Applied Optics*, Vol. 28(20), pp. 4360-4365 (October 15, 1988). ©1989 Optical Society of America.

Supersmall flying optical head for phase change recording media

Hiroo Ukita, Yoshitada Katagiri, and Susumu Fujimori

A supersmall optically switched laser (OSL) head is proposed. A laser diode attached to an air bearing slider forms a complex cavity together with a recording medium, and its light output is detected by a photodetector placed at the other end of the laser facet. Data signals and track error signals read from the sampled servo marks are successfully detected. The signal amplitude variation caused by the flying height change is much reduced, and the SNR is increased to 36 dB (40 kHz to 20 MHz for a phase change medium) by decreasing the reflectivity of the laser facet facing the medium to <5%.

I. Introduction

Current optical disk systems are limited in their use for data storage because of their low data transfer rate and long access time. To realize a high performance optical disk, we have proposed a microoptical head[1] based on laser switching by light reflected back from a phase change medium several micrometers from the laser diode.

Fukushima and Sawaki[2] have proposed another method of detecting magnetooptic signals with a long (several tens of millimeters) external cavity laser diode.

This paper describes a very short cavity optical head for phase change recording media, that is, a flying head structure, data signal and track error signal detection mechanism in a OSL mode, a method of suppressing signal amplitude variation due to flying height change and SNR improvement.

II. OSL Head Structure

Figure 1(a) shows an OSL head used with a rotating optical disk. The head operates in a sealed dust-free environment. Fig. 1(b) shows a detailed OSL head consisting of a laser diode and a photodetector, both attached to an air bearing slider. Fig. 1(c) shows the schematic structure of a beam-converging laser diode.[3] The laser is a planar stripe type with a ridged waveguide. The electrode pattern is 2 μm wide at each end and 4 μm wide in the center region. The full widths at half-maximum of the emitted light are 0.8 and 1 μm.

Figure 2 shows a photograph of an experimental apparatus consisting of an OSL head and a 130-mm diam glass substrate disk. The head is mounted on the same type of compliant spring as used in magnetic disk drives. An OSL head is used just like a magnetic head. However, it has higher data reliability than a magnetic head because of its 10 times greater flying height.

III. Data Signal Detection Mechanism

A. Readout Principle

OSL head signal detection is based on complex cavity laser light output characteristics. Figure 3 shows the OSL head configuration. Here R_1 and R_2 are the power reflectivities of the laser diode (LD), and R_3 is that for the external recording medium. The flying height h between the laser facet and recording medium surface is just a few micrometers. A fraction of the light is fed back into the active region, and it strongly affects the light output, which is detected by the photodetector (PD) located at the other end of the laser facet.

The laser light output is either strong stimulated emission light P^h corresponding to a high reflection part (R_3^h) or weak spontaneous emission light P^l corresponding to a low reflection part (R_3^l), as shown in Fig. 4. That is, the laser is switched according to the light fed back from the recording medium. The threshold current I_{th}^h for the high reflectivity R_3^h of the medium is 42 mA and I_{th}^l for the low reflectivity R_3^l is 60 mA. Signals are detected as the light output difference between the two states under proper bias current. In the present experiment, the reflectivities of the laser facets are $R_1 = 0.32$ (facing the photodetector) and $R_2 = 0.06$ (facing the medium). The medium reflectivities in the two states are $R_3^h = 0.35$ and $R_3^l = 0.04$.

The authors are with NTT Applied Electronics Laboratories, 9-11 Midori-cho 3-chome, Musashino-shi, Tokyo 180, Japan.

Received 15 March 1989.
0003-6935/89/204360-06$02.00/0.

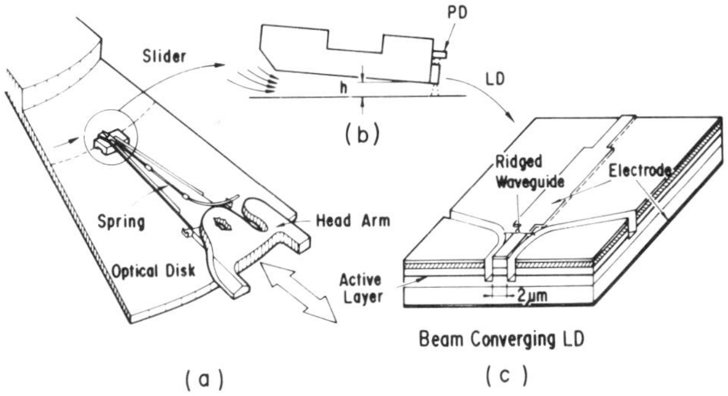

Fig. 1. (a) Schematic representation of a new type optical disk, (b) OSL head on a slider, and (c) a beam-converging laser diode with a taper ridged waveguide.

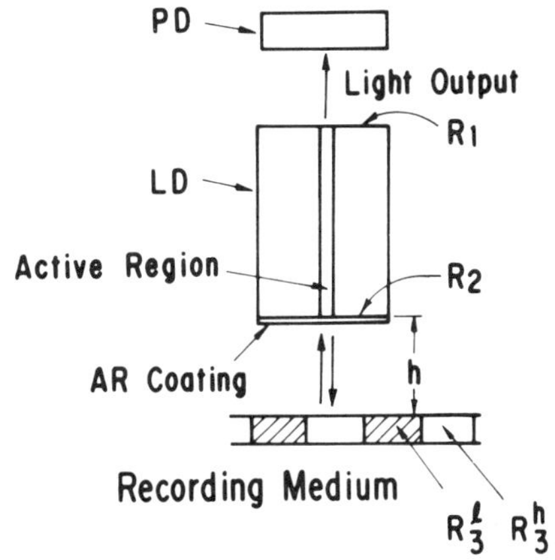

Fig. 3. Configuration of OSL head.

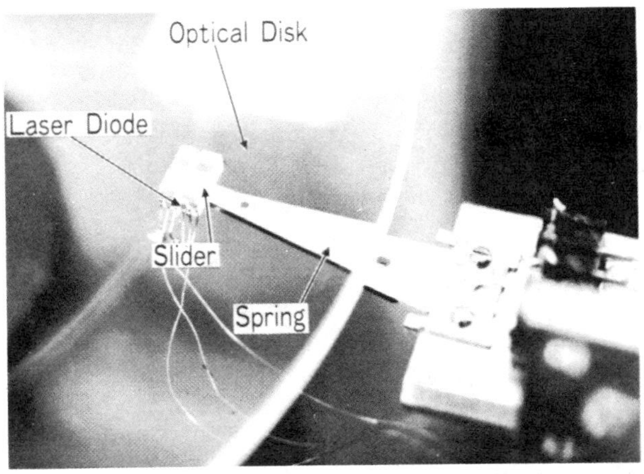

Fig. 2. Experimental setup consisting of an OSL head on a slider and a 130-mm diam rotating optical disk.

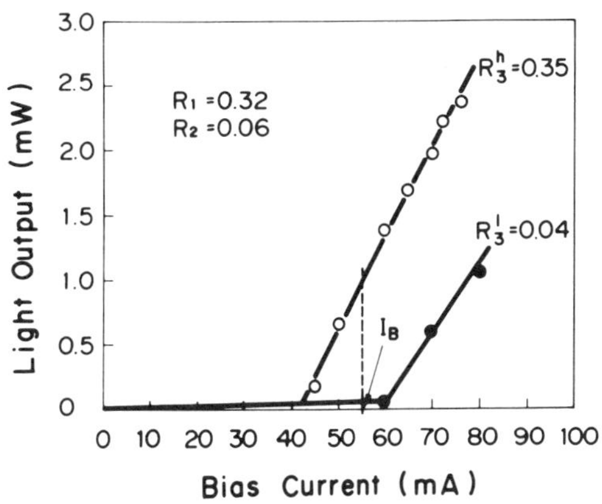

Fig. 4. Relation between the light output of an OSL head and bias current. The medium reflectivities in the two states are $R_3^h = 0.35$ and $R_3^l = 0.04$. Data signals are detected as light output difference between the two states under proper bias current.

Figure 5 shows the bias current dependence on normalized signal amplitude S_{pp} (measured from base to peak) and normalized rise time T_r (measured from 10 to 90% of the leading edge). They are defined as

$$S_{pp} = S_{pp}(I)/S_{pp}(60 \text{ mA}),$$

$$T_r = T_r(I)V/d,$$

where V is the disk velocity and d is the spot diameter on the medium. Both S_{pp} and T_r increase with bias current and nearly saturate at around the threshold current of 60 mA for a medium low reflectivity.

The effects of bias current, flying height, and reflectivities R_1, R_2, and R_3 on the readout signal have been analyzed using rate equations with optical feedback. The analysis introduces coupling coefficient ϵ (the ratio of the fraction of coupled-back light power from the external cavity to the light power within the diode cavity).[1]

B. Light Output Ratio in the Complex Cavity Laser

To increase the data signal S_{pp}, or threshold difference $I_{th}^l - I_{th}^h$, the laser facet facing the medium is antireflection (AR) coated with a $(Si_3N_4)_x(SiO_2)_{1-x}$ film prepared by sputtering a silicon target in an oxygen–nitrogen gas environment.

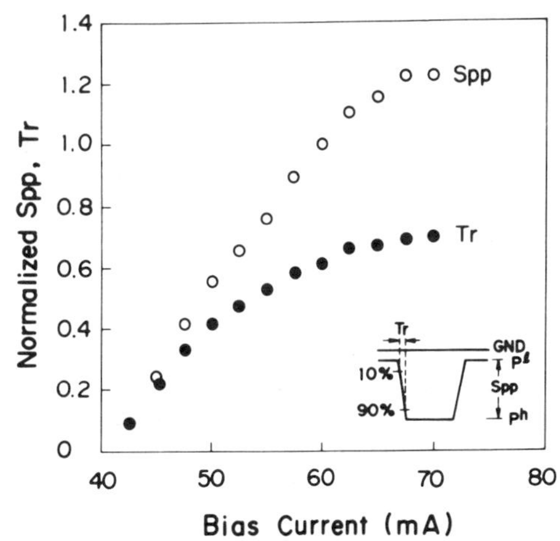

Fig. 5. Signal amplitude and leading edge response to bias current. The white circles are S_{pp} (measured from base to peak), and the black circles are T_r (measured from 10 to 90% of the leading edge).

An antireflection (AR) coating applied to one facet substantially increases the threshold current and light output power ratio of the two facets.[4] Figure 6 shows the relationship between the threshold current ratio of

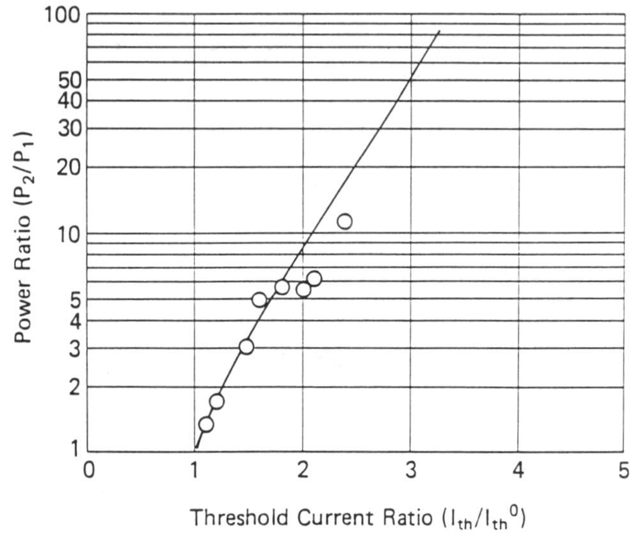

Fig. 6. Threshold current ratio vs power ratio for CSP laser diodes with various facet reflectivities.

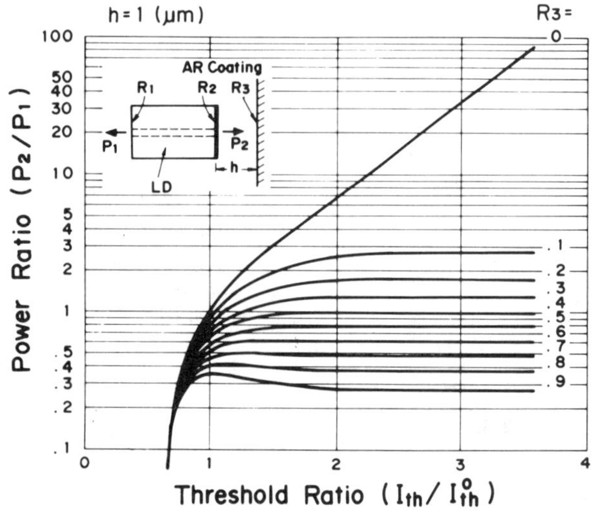

Fig. 7. Calculated power ratio of a complex cavity laser for an external cavity length $h = 1$ μm with external medium reflectivity R_3 as a parameter. The threshold currents of an AR coated and an uncoated laser are I_{th} and I_{th}^0, respectively.

an AR coated I_{th} laser to an uncoated I_{th}^0 laser and the power ratio of the two facets. The circles indicate the experimental values at room temperature, while the solid line shows the theoretical result.[5]

The power ratio for a complex cavity laser is calculated using effective reflectivity $R_2^{eff.}$ (Ref. 1) instead of laser facet reflectivity R_2.[4] Figure 7 shows the complex cavity power ratio for $h = 1$ μm with the medium reflectivity R_3 as a parameter. The figure clarifies that the power ratio of the complex cavity laser is determined mainly by the reflectivity R_3 for the AR coated laser, whose threshold current ratio is >2. [AR coating reflectivity is <1% (Ref. 5).] When $R_1 = 0.32$ and $R_3 = 0.35$, the power ratio P_2/P_1 is nearly equal to unity.

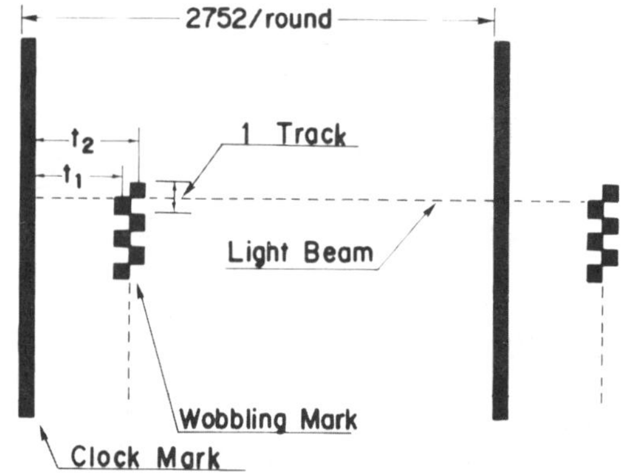

Fig. 8. Sampled servo marks consisting of clock marks and wobbling marks.

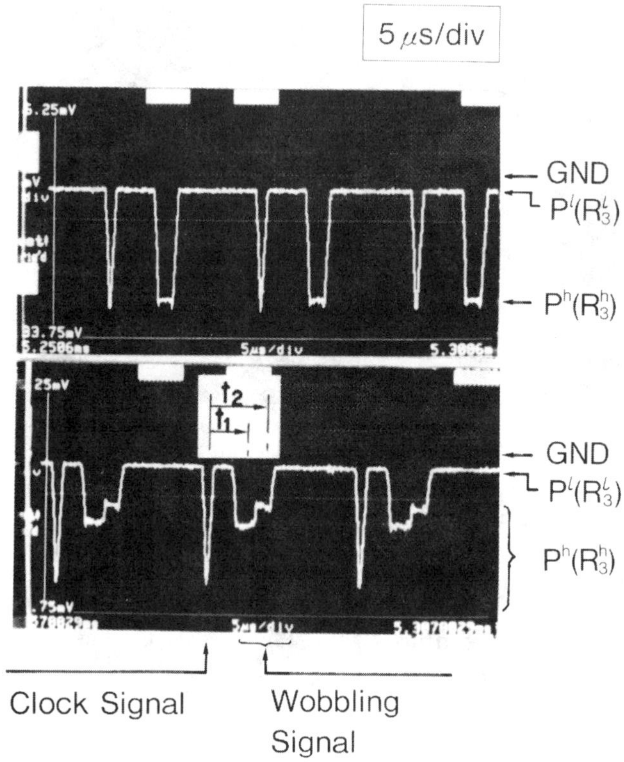

Fig. 9. Examples of sampled servo signal waveforms. Track error signals are detected as the light output difference between the two wobbling signals at sampled times t_1 and t_2.

IV. Track Error Signal Characteristics

Since the OSL head has only one detector, the track error signal is obtained as the light output amplitude difference between two successive times instead of the spacial light power difference refected back from the tracking groove. Figure 8 shows pseudosampled marks consisting of clock marks and wobbling marks, both of chromium patterns. The clock marks are long and rectangular, and the wobbling marks are quasisquare. The lengths between two clock marks at the experimental radius were 4.3 and 13 μm. There are 2752 sampling marks per revolution (twice the standard number) for the various experiments.

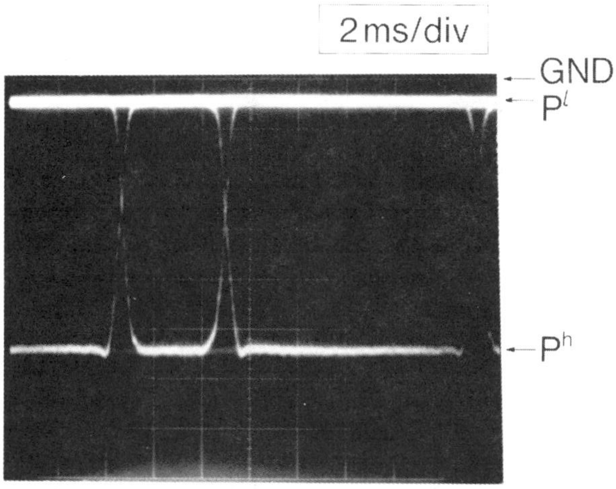

Fig. 10. Cross marks (X) formed by the envelope of wobbling signals. The marks are found to be linearly dependent on the light beam deviation from the track center.

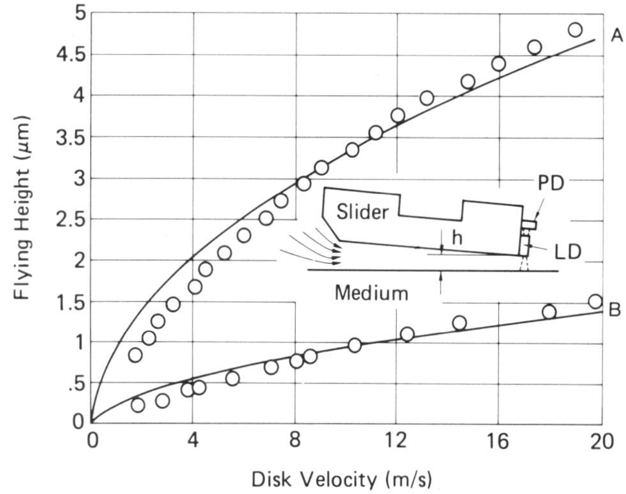

Fig. 11. Relationship between flying heights of two kinds of Whitney type slider (*A* and *B*) and disk velocity. The circles indicate experimental measurements, while the solid lines indicate numerical results.

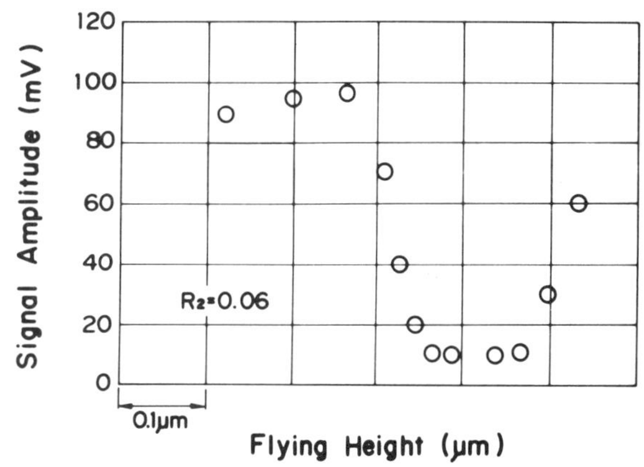

Fig. 12. Variation of the readout OSL head signal amplitude against slider flying height. The optical disk consists of a SbTe phase-change medium sandwiched between dielectric films.

Sampled servo signals corresponding to on-track and off-track light beams are shown in Fig. 9. The laser bias current is 60 mA. The upper figure shows an off-track signal, for which a laser beam scans only one time t_2 of the wobbling marks, while the lower figure shows a nearly on-track signal, for which the laser beam scans two times (t_1 and t_2) of the wobbling marks. Actual track error signals can be detected[6] as the light output difference between the two wobbling signals at the times t_1 and t_2, as shown in Fig. 9.

Figure 10 shows the wobbling signal envelope. Three cross marks (X) can be seen in the figure. These marks are formed by the envelope of the wobbling signals such as those shown in Fig. 9. The wobbling signal peaks at every t_1 and t_2 form the two straight lines of the cross mark. The straightness of the cross marks indicates that the amplitude of the wobbling marks corresponds linearly to the light beam deviation from the track center and promises exact on-track operation of the light beam.

V. Signal Amplitude Fluctuation Suppression by AR Coating

A. Flying Height Dependence on Signal Amplitude

The relationship between the trailing edge flying height for two kinds of Whitney slider (*A* and *B*) and the disk velocity is shown in Fig. 11. The flying height was measured by a composite cavity interferometry method, each count from the peak signal representing 0.42 μm ($\lambda/2$). Data were taken at a radius $r = 51$ mm, while the disk rotation speed was varied from 400 to 4000 rpm. The circles indicate the measured flying height, while the solid lines indicate the square root of disk velocity.

The head signal amplitude variation against the flying height for laser facet reflectivity $R_2 = 0.06$ is shown in Fig. 12. The optical disk is a phase-change medium (SbTe) on a glass substrate.[7] Strong signal amplitude modulation in the head signal is a result of the flying height (external cavity length) variation due to the disk velocity shown in Fig. 11. Maximum peaks are experienced every half wavelength of flying height. In an actual constant angular velocity optical disk system, the flying height increases as the radius increases.

B. Effect of AR Coating on Laser Facet

The flying height dependence curve for the complex cavity laser light output is shown in Fig. 13 with R_2 reflectivity for the medium facing laser facet as a parameter. The interference between the reflected light from the laser facet and that from the medium causes both light outputs *A* and *B* to have peaks every half-wavelength. Since the coupling-back light decreases with increasing spacing, the light ampliude also decays exponentially with increasing flying height.

On the other hand, the smaller facet reflectivity laser *A* ($R_2 = 0.02$) has smaller light output modulation and higher sensitivity than the larger facet reflectivity laser *B* ($R_2 = 0.07$). This is because the interference effect decreases and the coupling efficiency increases

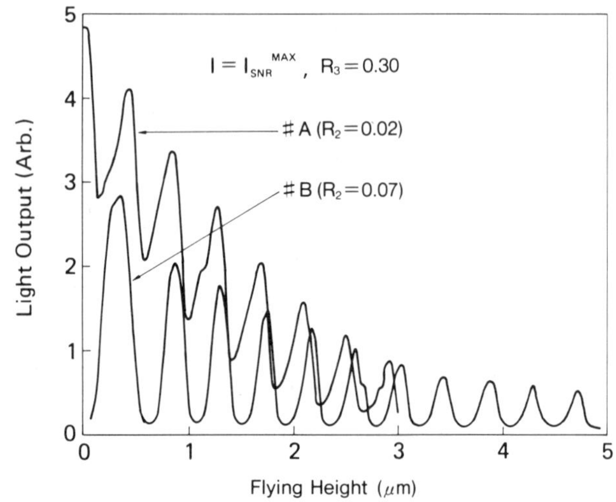

Fig. 13. Flying height dependence on complex cavity laser light output with laser facet reflectivity R_2 facing the medium as a parameter.

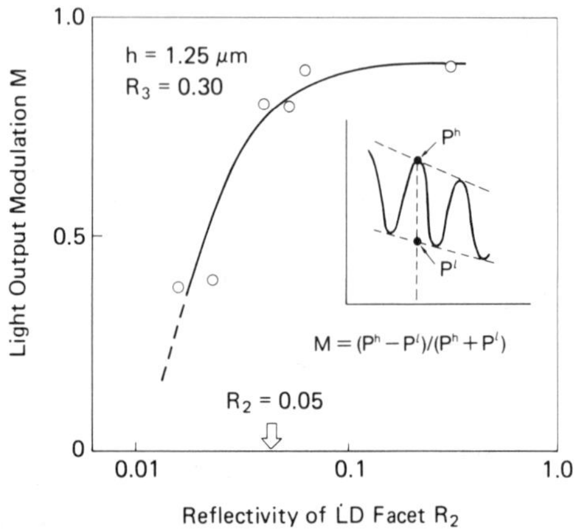

Fig. 14. Relationship between the decrease in complex cavity laser light output amplitude modulation and the reflectivity of the laser facet facing the medium, where flying height h = 1.25 μm and the medium reflectivity R_3 = 0.30.

when facet reflectivity is reduced. It has been found that facet reflectivity reduction is useful not only in head signal fluctuation suppression but also in increasing head signal sensitivity.

The dependence of the head signal on the flying height, shown in Fig. 12, corresponds to the light output dependence on B (large facet reflectivity) in Fig. 13. The relationship between the decrease in light output amplitude modulation and the reflectivity of the laser facet facing the medium is shown in Fig. 14. It should be noted that an abrupt modulation drop occurred at a facet reflectivity of 5%.

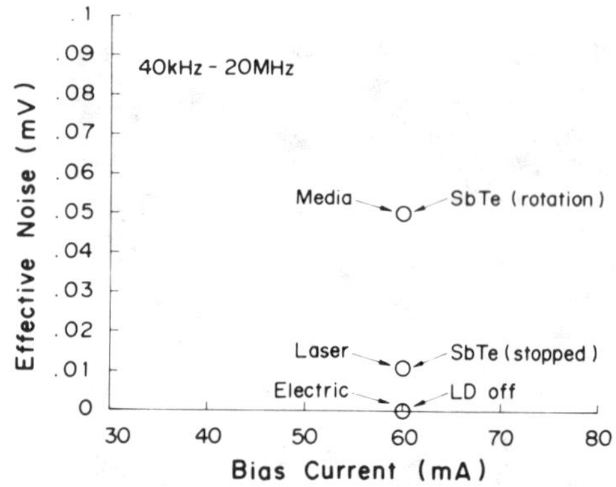

Fig. 16. Effective noise levels for an OSL head.

This result suggests that the signal amplitude variation due to the flying height change will decrease if we use a laser with a facet reflectivity of <5%.

VI. Dynamic Signal-to-Noise Ratio Improvement

We investigated the feedback noise$_{rms}$ and data signal amplitude for various head drive conditions using a phase-change medium. The phase-change medium structure consists of a recording layer SbTe (100 nm) sandwiched between dielectric films. An example of an OSL head noise spectrum from 0 to 20 MHz is shown in Fig. 15. The light output power was P = 1 mW when the disk (R_3 = 0.35) was not rotating. The figure shows that laser noise, stimulated by an external cavity only a few micrometers long, exists mainly at low frequencies.

To evaluate the laser noise quantitatively, we defined the SNR as

$$\mathrm{SNR} = 20 \log S_{pp}/2\sqrt{2}N_{\mathrm{rms}},$$

where S_{pp} is the data signal amplitude and N_{rms} is the effective noise, defined as the rms of the noise spectrum integral (40 kHz to 20 MHz).

The major noises in the OSL head are electronic, laser, and media noises. There are two kinds of laser

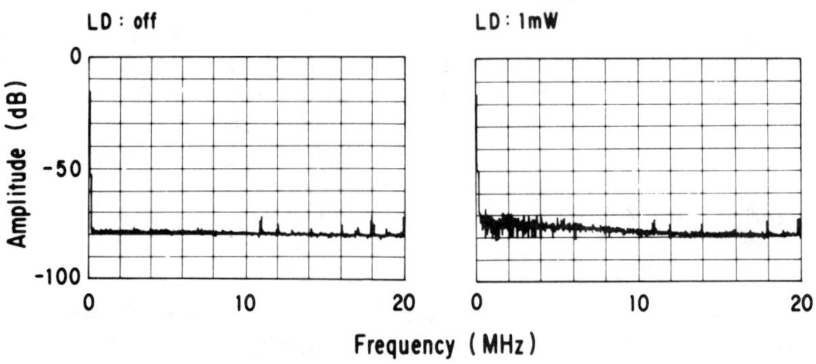

Fig. 15. Frequency spectrum example of intensity noise for an OSL head.

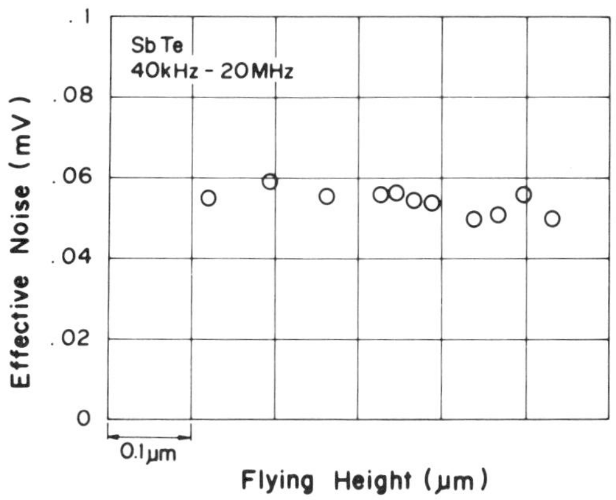

Fig. 17. Dependence of effective noise N_{rms} on flying height.

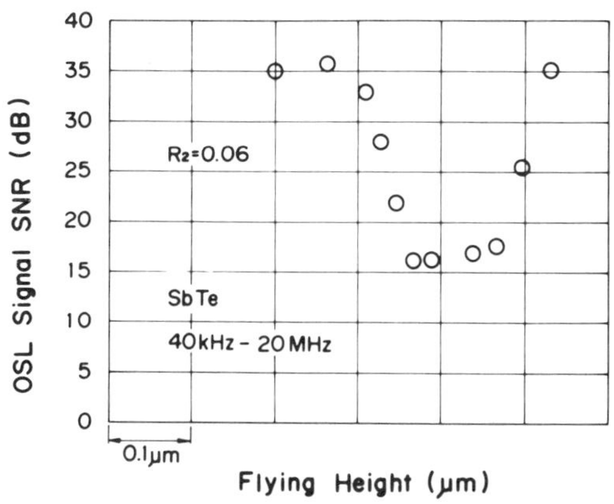

Fig. 18. SNR vs flying height for an OSL head whose laser facet reflectivity facing the medium is 0.06. The flying height between an OSL head and recording medium surface varies with disk velocity.

noise: (1) the very low frequency light variation due to flying height change (variation in external cavity length) and temperature fluctuation (variation in threshold current) and (2) high frequency mode hopping and feedback noises. Figure 16 shows the electronic noise level (with the laser power off), the laser noise level (with the disk stopped), and the media noise level (with the disk rotating). In the media noise, the disk motion noise[8] is added. This is caused by variation in reflectivity for the medium and fluctuation in the head-to-medium flying height.

Figure 17 shows the effective noise N_{rms} dependence on flying height. N_{rms} does not change much when the flying height changes. However, data signal amplitude S_{pp} varies with the flying height change through disk rotation time (Fig. 12). As a result, the SNR varies with flying height, as shown in Fig. 18. The peak SNR occurred at $h = N\lambda/2$ (N, natural number; λ, wavelength), and the valley SNR occurred at $h = (2N + 1)\lambda/4$. A SNR value of 36 dB was achieved for a phase-change medium ($R_3^h = 0.25$, $R_3^l = 0.04$) in a dynamic readout condition. It is expected that a higher SNR and a smaller difference between peak and valley of SNR values can be obtained by decreasing the reflectivity of the laser facet facing the recording medium, as shown in Fig. 13.

VII. Conclusion

A supersmall flying optical head for phase-change recording media was proposed. The head consists of a beam-converging laser diode and photodetector, both on an air bearing slider.

The data signal detection mechanism and track error signal characteristics were analyzed experimentally. Signal amplitude variation due to the flying height change can be suppressed, and the SNR can be improved by reducing the reflectivity of the laser facet facing the recording medium to <5%.

Except for writing noise, the SNR value of 36 dB using an SbTe recording medium sputtered on a glass disk satisfies the requirements of an optical disk system. The OSL head track following will be accomplished using the sampled servo method.

It is expected that the supersmall OSL head will be used in next generation optical disk systems for on-line files.

The authors would like to thank Hisao Takata, Kiyoshi Itao, and Tomoyuki Toshima of the NTT Applied Electronics Laboratories for their encouragement.

They would also like to acknowledge Yoshihiro Isomura, Naoyuki Tamaru, Akira Morinaka, Yuji Uenishi, Renshi Sawada, and Hidenao Tanaka of the NTT Applied Electronics Laboratories for their help with experiments and helpful discussions on the LD structure and Katsuhiko Kurumada and Nobuyori Tsuzuki of the NTT Opto Electronics Laboratories for offering laser chips.

References

1. H. Ukita, Y. Katagiri, and Y. Uenishi, "Readout Characteristics of Micro-Optical Head Operated in Bistable Mode," Jpn. J. Appl. Phys. **26**, Suppl. 26-4, 111–116 (1987).
2. N. Fukushima and I. Sawaki, "Detection of Magneto-Optic Signal Using External Cavity Laser Diode," in *Technical Digest, Topical Meeting on Optical Data Storage* (Optical Society of America, Washington, DC, 1989), pp. 55–58.
3. Y. Uenishi, Y. Isomura, R. Sawada, H. Ukita, and T. Toshima, "Beam Converging Laser Diode by Taper Ridged Waveguide," Electron. Lett. **24**, 623–624 (1988).
4. M. Ettenberg, H. S. Sommers, Jr., H. Kressel, and H. F. Lockwood, "Control of Facet Damage in GaAs Laser Diodes," Appl. Phys. Lett. **18**, 571–573 (1971).
5. H. Ukita, K. Mise, and Y. Katagiri, "Simple Measurement of the Reflectivity of Antireflection-Coated Laser Diode Facet," Jpn. J. Appl. Phys. **27**, L1128–L1130 (1988).
6. N. Tamaru, H. Hosaka, S. Hara, N. Funakoshi, and K. Kogure, "A Study of a Flexible Optical Disk Drive," in *Technical Digest, Topical Meeting on Optical Data Storage* (Optical Society of America, Washington, DC, 1989), pp. 59–62.
7. S. Fujimori, S. Yagi, H. Yamazaki, and N. Funakoshi, "Crystallization Process of Sb-Te Alloy Films for Optical Storage," J. Appl. Phys. **64**, 1000–1004 (1988).
8. B. I. Finkelstein and W. C. Williams, "Noise Sources in Magnetooptic Recording," Appl. Opt. **27**, 703–709 (1988).

Reprinted from *Holographic Optics II: Principles and Applications,* Proc. SPIE Vol. 1136, pp. 92-99 (1989). ©1989 Society of Photo-Optical Instrumentation Engineers.

Integrated-optic disc pickup devices using waveguide holographic components

Toshiaki Suhara and Hiroshi Nishihara

Osaka University, Department of Electronics, Faculty of Engineering
2-1 Yamada-Oka, Suita, Osaka, 565 Japan

ABSTRACT

The authors' work is reviewed on implementation of integrated-optic disc pickup devices using waveguides and holographic components. The integrated-optic disc pickup (IODPU) is constructed by integrating a focusing grating coupler (FGC), a grating beam splitter and photodiodes in a waveguide on Si, and is capable of detecting readout signal and tracking/ focusing error signals. The gratings are fabricated by computer-controlled electron-beam writing and reactive ion etching. Integration of detection optics for magneto-optical disc pickup is also presented. The device uses a trifocal-FGC, which serves as a polarization splitter based on waveguide mode dispersion, and performs differential detection of MO read-out signal. The design considerations, fabrication processes and experimental works are presented including the most recent results. Although the performances obtained so far are not sufficient for practical application, the elementary functions have been successfully demonstrated.

1. INTRODUCTION

There has been much research and development interest in optical disc memory because of the high memory density and capacity and the many applications. Although many systems have already been commercialized, there are increasing demands for higher performances, function and production efficiency. Pickup heads for disc data readout are currently constructed with microoptics; they need complex and time-consuming fabrication processes, e.g., polishing and assembling of bulk components, and further miniaturization seems to be diffucult.

If a pickup head is constructed with integrated optics[1] using a thin-film waveguide, the planar process can be employed for the fabrication. The use of holographic components in the integrated device is very advantageous because complex functions can be realized with minimized component number. Such implementation of disc pickup would lead to great improvements in producibility, reduction in size, weight and cost, and enhancement of applications.

The authors are conducting a research project on the integrated-optic implementation of optical disk pickup, based on the works on integrated optics and grating (holographic) components.[2] The authors have proposed a series of device constructions and have continued theoretical and experimental works, since they proposed and demonstrated a specific implementation of integrated-optic disc pickup device (IODPU) for the first time.[3] Although the performances obtained so far are not sufficient for practical application, the elementary functions have been successfully demonstrated. The work collected interest of Japanese and foreign research groups, and stimulated the applications of holographic components to optical disc pickup heads.

In this paper, the authors would like to review their work on the integrated-optic disc pickup devices and report the current status of technology. The design considerations, fabrication processes and experimental work are presented including the most recent results. The feasibility and problems in achieving the operational performances are also discussed.

2. FOCUSING GRATING COUPLERS

The integrated-optic implementation of pickup head requires an appropriate waveguide-type component for focusing a guided wave in a planar waveguide into a diffraction-limited spot on a disc in free space, or a component for coupling a guided wave to a beam in free space. The component must have structure suitable for fabrication by planar process. We have adopted focusing grating couplers (FGC) as a waveguide component satisfying these requirements.

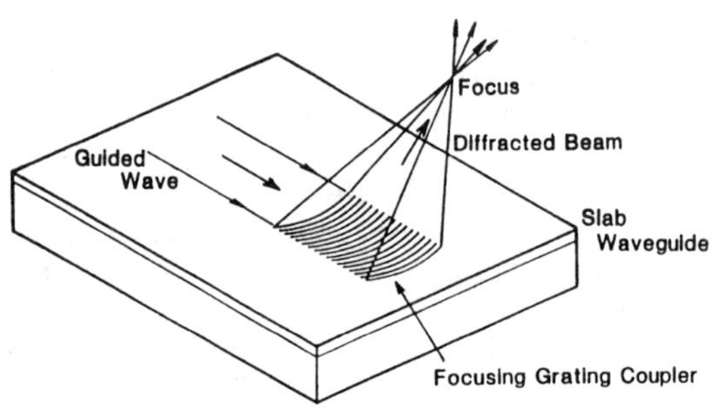

Fig. 1 Focusing grating coupler.

Figure 1 illustrates the prototype configuration of FGC. A FGC, which is a curved and chirped grating evolving from a grating coupler for integrated optics, incorporates the input/output coupling and the focusing by the wavefront conversion based on the principle of holography. The possibility of adding a complex wavefront conversion to a grating coupler was first demonstrated in waveguide holograms which aimed at an approach to hologram integration.[4] Apllication of point-image waveguide hologram (FGC) to disc pickup was considered. The early FGC fabricated by the holographic interference recording[5], however, suffered from the aberration problem caused by the wavelength difference between the recording laser and the laser diode, and therefore the performances suitable for pickup application were not obtained. The problem motivated the FGC fabrication by the electron-beam (EB) writing technique.[6,7] The grating pattern is calculated from mathematical formulae describing the phase difference between input and output wavefronts, and is then written by a scanning electron beam. It is thus possible to design and fabricate a FGC which is free of aberration at an arbitrary wavelength. Although the FGC pattern has submicron period and the EB writing involves the drawback of rather low throughput, the holographic contact printing technique[8] can be employed for efficient mass production.

A FGC exhibits optical properties different largely from those of conventional lenses, since a FGC performs focusing function based on the wavefront conversion through diffraction and one of the input/output waves is a discrete guided mode. The important features of a FGC are as follows:[6,9]

1. The imaging relation between object and image points is strongly affected by the properties of guided wave with reduced spatial dimension.
2. The profile of light amplitude within the FGC aperture (pupil function) is not uniform but exhibits exponential decay associated with the guided-wave attenuation due to the radiation.
3. The diffraction (coupling) efficiency and the focus spot size can not be designed independently, and therefore, optimization is required to find the compromise between them.
4. The input and output coupling efficiencies are not equal; the former is reduced from the latter for any input with mismatch to designed wavefront.

Designing of a FGC involves 1) designing a single-mode planar waveguide by the standard technique, 2) deriving from the input/output wavefront expressions the mathematical formulae to describe the FGC pattern for EB control, and 3) determining the FGC parameters optimized for the maximum efficiency and the diffraction-limited spot size. The analytical expressions, which are derived[1] based on the the coupled-wave theory or the perturbation theory, for uniform-period grating couplers, can be used for the FGC design step 3) to have a reasonable accuracy.[2,3,9] The output coupling efficiency of a FGC can be written as

$$\eta_{fgc} = p^a [1 - \exp (-\alpha_r L)], \tag{1}$$

where α_r is the radiation decay factor, p^a the fractional power for the first-order air radiation, and L the length of FGC along the guided-wave propagation. To obtain a high efficiency, $\alpha_r L$ should be large. The spot width, however, becomes larger than the diffraction-limited value for L, when α_r is too large (the effective aperture $1/\alpha_r$ smaller than L). Thus the spot size and the efficiency must be traded off. The condition for obtaining a high efficiency without substantial spot broadening is given by $\alpha_r L=1$, which can be realized by appropriate choice of the grating material and parameters. Figure 2 shows the result calculated for Si-N grating couplers on a Glass/SiO_2/Si waveguide. In the structure using a reflective (Si) substrate, the fractional air power p^a can be maximized, without blazing of grating, by choosing the SiO_2 buffer layer thickness so that the reflected substrate radiation interferes constructively with the direct air radiation to give the enhanced total air radiation. It was found that the maximum output coupling efficiency obtainable simultaneously with a diffraction-limited spot size is approximately 60 % ($p^a_{max}=0.72$, $\alpha_r L=1$).

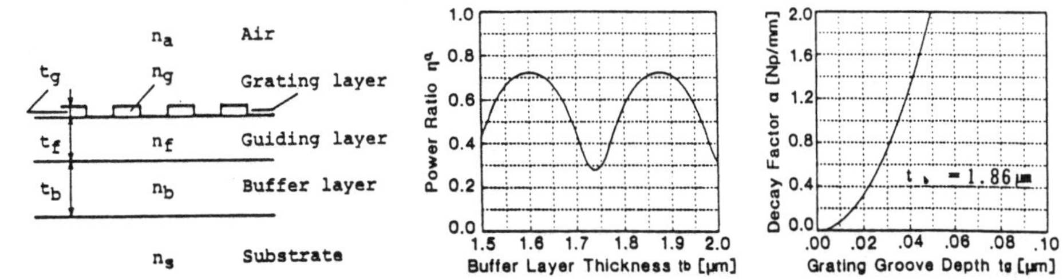

Fig. 2 Cross section and design diagrams of grating coupler in #7059/SiO_2/Si waveguide.

It is in theory possible to design and fabricate an aberration-free FGC, as mentioned above. The possible fabrication error, however, causes aberrations. A deviation of the optical wavelength from the design value also leads to aberration problem, since a FGC is a dispersion element. Theoretical analysis[10] showed that the major aberrations are astigmatism and coma. The fabrication accuracy required to obtain the diffraction-limited focus spot size becomes stringent rapidly with increasing numerical aperture (NA). The maximum tolerable error in FGC linear dimension and in wavelength deviation are estimated to be approximately ±0.6% and ±0.9%, respectively, for NA=0.24, and ±0.09% and ±0.1% for NA=0.45.

3. INTEGRATED-OPTIC DISC PICKUP DEVICE FOR READ-ONLY DISC MEMORY

3.1 Device Construction and Working Principle

Figure 3 shows the schematic illustration of a monolithic integrated-optic disc pickup device (IODPU) capable of data readout and focusing/tracking error detection.[2,3,11] The error signals are required to drive actuators for the automatic servo operation. The focusing/trucking error detections are based upon the Foucault/push-pull methods,[12] respectively, which are modified for implementation by grating components in waveguide.

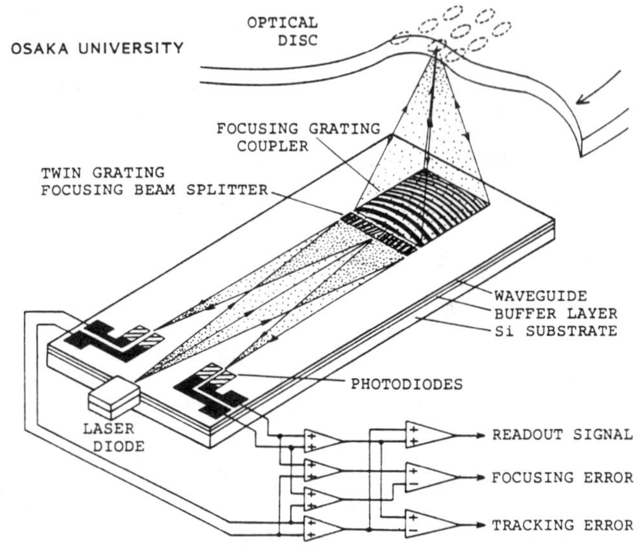

Fig. 3 Integrated-optic disc pickup (IODPU).

The device is constructed by integrating a focusing grating coupler (FGC), a twin grating focusing beam splitter (TGFBS), and photodiodes in a film waveguide on a Si substrate. The guided wave diverging from a butt-coupled laser diode (LD) is focused by the FGC into a point on a disc. The wave reflected by the disc is collected and coupled back into the waveguide by the same FGC. The TGFBS divides the reflected wavefront into halves, deflects (beam splitting), and simultaneously focuses into two detecting points at a distance slightly shorter than that of the input point (LD). It should be noted that the combined function of the TGFBS leads to the minimization of component number, which is very imoprtant in integrated-optics devices. At the detecting points, the guided waves are transferred to four PD elements fabricated on the surface of the Si substrate. Thus the disc data is read out and the fucusing/trucking error signals are also obtained by simple processing of the photocurrents.

The coplanar transmission-type Bragg grating,[1] which is fabricated in the same layer and by the same process as the FGC, is adopted for TGFBS. The grating pattern is calculated from the phase difference between the guided wave converging back into the input point (LD) and the waves focusing into the detecting points. The diffraction efficiency (power splitting ratio) of a Bragg grating is given by

$$\eta_{gbs} = \sin^2 (\kappa D), \tag{2}$$

where D is the coupling length and κ the coupling coefficient, which is determined by the grating groove depth and can be calculated in the coupled-wave analysis.[1] The optimum efficiency, η_{gbs} = 50 %, which maximizes the round-trip throughput, is readily obtained by tailoring the coupling length D.

A series of modifications of IODPU has also been proposed as integrated-optics devices for disc pickup. A pickup can be constructed by combining a waveguide FGC/TGFBS device and an external objective lens as shown in Fig. 4.[13] The FGC for such device is designed so that it couples a diverging guided wave to a collimated beam. The fabrication accuracy requirements are less stringent for the small-NA collimater FGC. The use of an external lens releaves the load to the actuator, although the construction involves reduced integration advantages. Figure 5 shows a simpler waveguide device for pickup.[14] A bifocal focusing grating coupler (BFGC), which couples an external beam to guided waves converging into two focuses (detection points), incorporates beamsplitter, prism and lens functions.

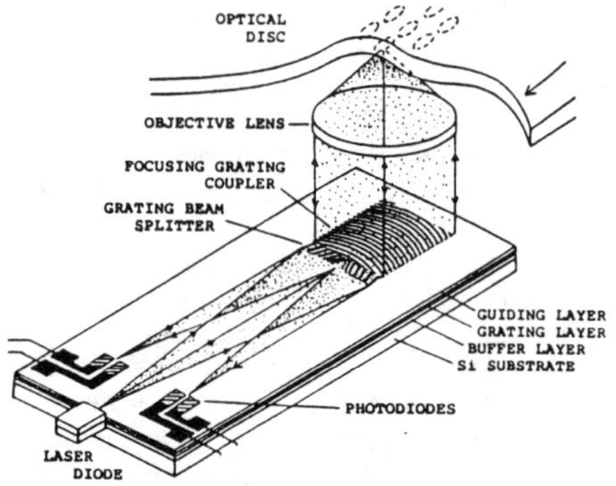

Fig. 4 Pickup using an external lens.

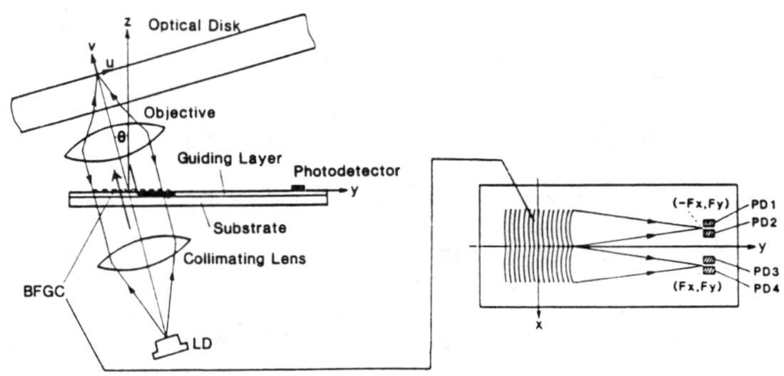

Fig. 5 Pickup using a bifocal FGC.

3.2 Theoretical Readout Response Analysis

The overall optical performances of a pickup are described and evaluated in terms of the readout response.[12] The well-known theory of response for the conventional lens pickup by Hopkins,[15] however, does not apply for an IODPU which uses a FGC having the singular properties pointed out in 2. A theoretical analysis of IODPU readout response was made based upon the scalar diffraction theory, to clarify the unique properties and facilitate the optimum design of IODPU. It was found that the IODPU response for the resolution in direction along disc pit length, which is affected by a filtering effect resulting from the discreteness of a guided mode, is equivalent to that of a confocal detection optics (i.e., a scanning microscope[17] of Type II). Accordingly, an IODPU exhibits a resolution higher than that of a conventional lens pickup (SLDPU; Square-aperture Lens DPU). Figure 6 compares the calculated eye patterns of an IODPU and a SLDPU having the same aperture (1.2x1.5 mm^2, focal length 2 mm). It can be seen that the IODPU gives better eye pattern with smaller jitters and intersymbol interferences indicationg the improved resolution. This result implies that there is a possibility of obtaining required readout performances in an IODPU of a NA smaller than that requied in conventional pickups.

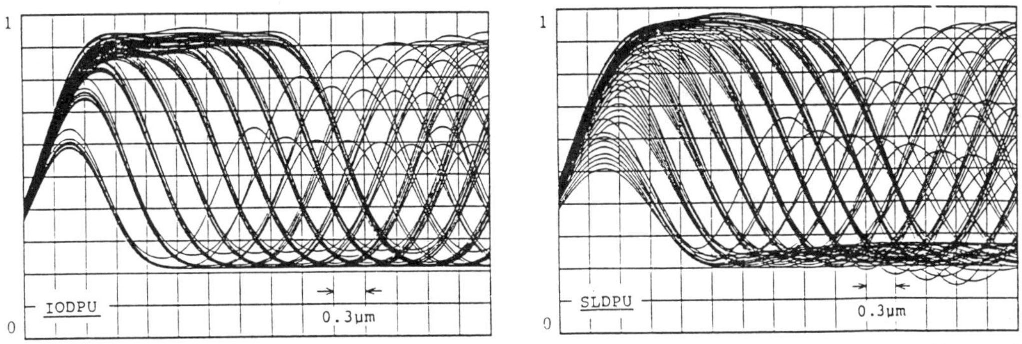

Fig. 6 Calculated eye patterns of IODPU and pickup using square-aperture lens (SLDPU).

3.3 Device Fabrication

An example of the specifications of the fabricated prototype IODPUs are summarized in Tab. 1. In the EB writing of submicron FGC pattern, it is very difficult to have sufficient accuracy in the pattern connection by mechanical stage shift, and therefore the pattern is written in one frame by the electron beam scanning. Thus the writing area is rather limited. The 1x1 mm^2 aperture design of FGC in the IODPUs fabricated so far was adopted under this limitation.

At first Si substrate was thermally oxidized to grow the SiO_2 buffer layer, and PN photodiode array was fabricated by the standard semiconductor processes. The local oxidization (LOCOS) technique was used to prepare the tapered-SiO_2 PD coupling region. Then the Corning #7059 glass wave-guiding layer and the Si-N layer for gratings were deposited by sputtering and PCVD methods, respectively. Figure 7 shows the fabrication process of grating components.

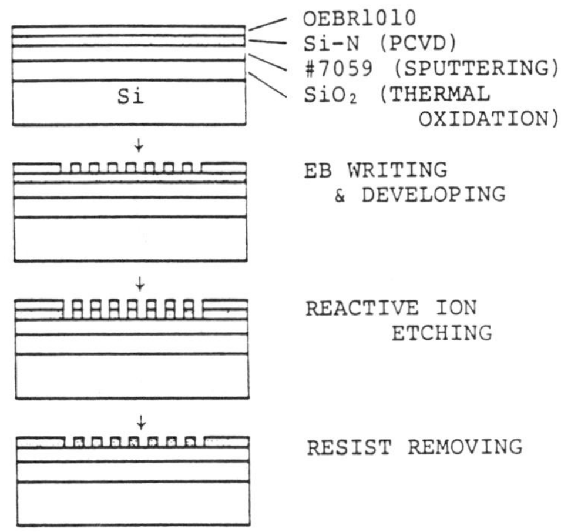

Fig. 7 Grating fabrication process.

Tab. 1 Specifications of the fabricated prototype IODPU.

Light source	GaAlAs LD	λ = 0.79 μm
Waveguide Single mode Glass/SiO_2/Si	Guiding layer Buffer layer Mode index	0.95 μm #7059 glass 1.86 μm SiO_2 n_e = 1.52 (TE_0 mode)
Grating components Focusing grating coupler	Cladding layer Focal lengths Output angle Aperture Period	0.035 μm Si-N f = 2.0 mm, r = 10.5 mm θ = 15.0° 1.0 × 1.0 mm^2 0.52 ∿ 0.75 μm
Twin grating focusing beam splitter	Detecting point Size Period	a = 1.165 mm, b = 8.50 mm 1.0 × 0.096 mm^2 3.80 ∿ 4.04 μm
Photodiodes	Element size	50 × 150 μm^2 (10 μm gap)

The FGC/TGFBS pattern was directly written by curved-line scanning of EB in positive resist coated on the waveguide. The writing time including the computing time was approximately 30 min. The pattern was then transferred to the Si-N layer by reactive ion etching (RIE) to form the grating grooves. Figure 8 shows a microphotograph of a part of FGC/TGFBS pattern. Then the photodiode contact windows were opened and the waveguide was cut and cleaved for butt-coupling into 5x12 mm^2 area chips. Finally the chip was mounted on an IC base with a LD (Mitsubishi ML4402, 3 mW). Figure 9 shows a view of the fabricated IODPU.

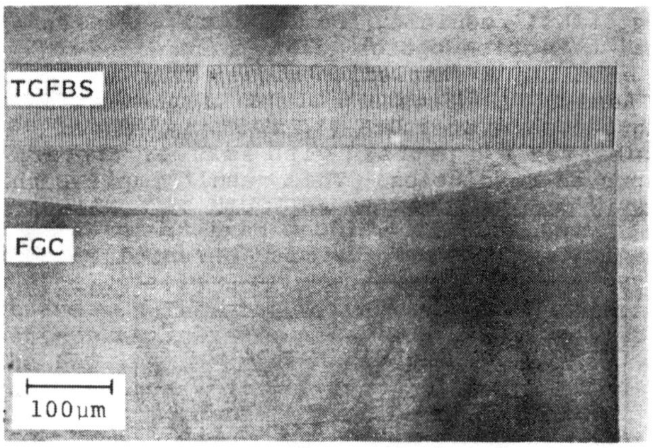

Fig. 8 Microphotograph of FGC/TGFBS gratings.

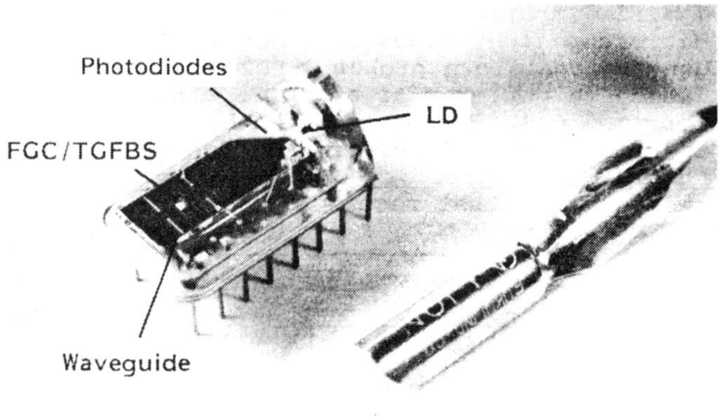

Fig. 9 Fabricated prototype IODPU.

3.4 Experimental Results

The 3-dB focus spot width of the fabricated IODPU was measured to be approximately 2 µm, while the theoretical diffraction-limited value is 1.4 µm. The FGC out-coupling efficiency was measured to be apploximately 40 %. The effect of the LD lasing spectrum on the IODPU output spot was experimentally examined, and it was confirmed that the stable single-mode lasing is essential for the IODPU operation.[18]

The pickup operation was examined by using a small mirror to simulate a disc and actuating it near the FGC focus.[3,11] Output photocurrents of around 50 nA were obtained with a LD power of 3 mW. Figure 10 shows the focusing error signal obtained when the mirror was actuated to give deviation from focus. The qualitative focusing servo operation was obtained in the experiment.[19] The tracking error detection function was also confirmed by actuating a mirror having a step simulating the disc pit edge. Figure 11 shows the readout signal obtained by moving a mirror with 1.5µm-width 0.15µm-step linear grooves (an enlarged simulation of disc pits) without servo operation.

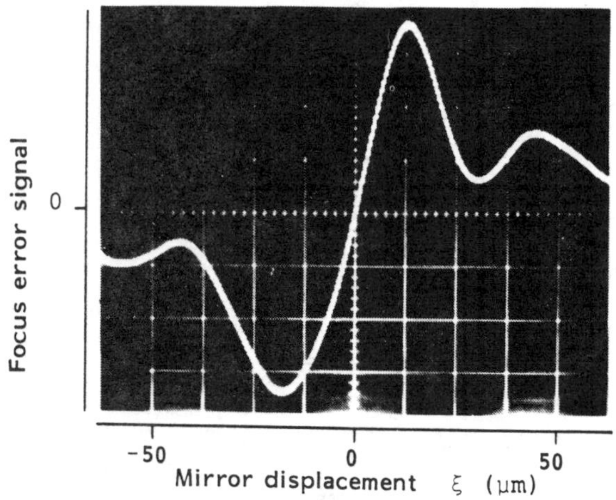

Fig.10 Obtained focusing error signal.

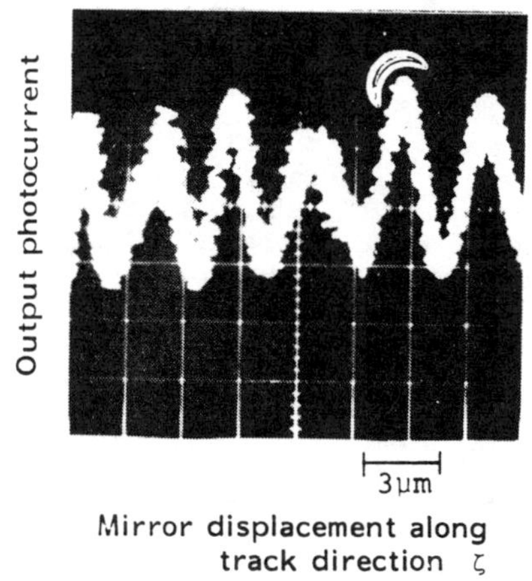

Fig.11 Readout signal obtained by using a stepped mirror simulating a disc.

4. INTEGRATION OF DETECTION OPTICS FOR MAGNETOOPTICAL DISC PICKUP

4.1 Device Construction and Working Principle

A pickup head for magnetooptical (MO) disc system, which is being developed as a high-density read/write memory, requires complicated optics for detecting the small polarization rotation due to the magnetic Kerr effect, and therefore the device simplification by integration is strongly desired. Use of a trifocal focusing grating coupler (TFGC) for this purpose was proposed and demonstrated.[20]

Figure 12 illustrates the proposed MO disk pickup configurations.[21] The pickup consists of a waveguide device, external lenses, a beam splitter and a laser diode (LD). The detection optics is constructed by integrating a TFGC, which consists of three segments, and five PD elements in a waveguide on Si substrate. In the waveguide, TE and TM modes can propagate and the propagation constants are slightly different from each other. The TFGC is designed so that, for an incidence angle, the central segment and the side segments excite TM and TE waves, respectively; the grating period is different for central and side segments.

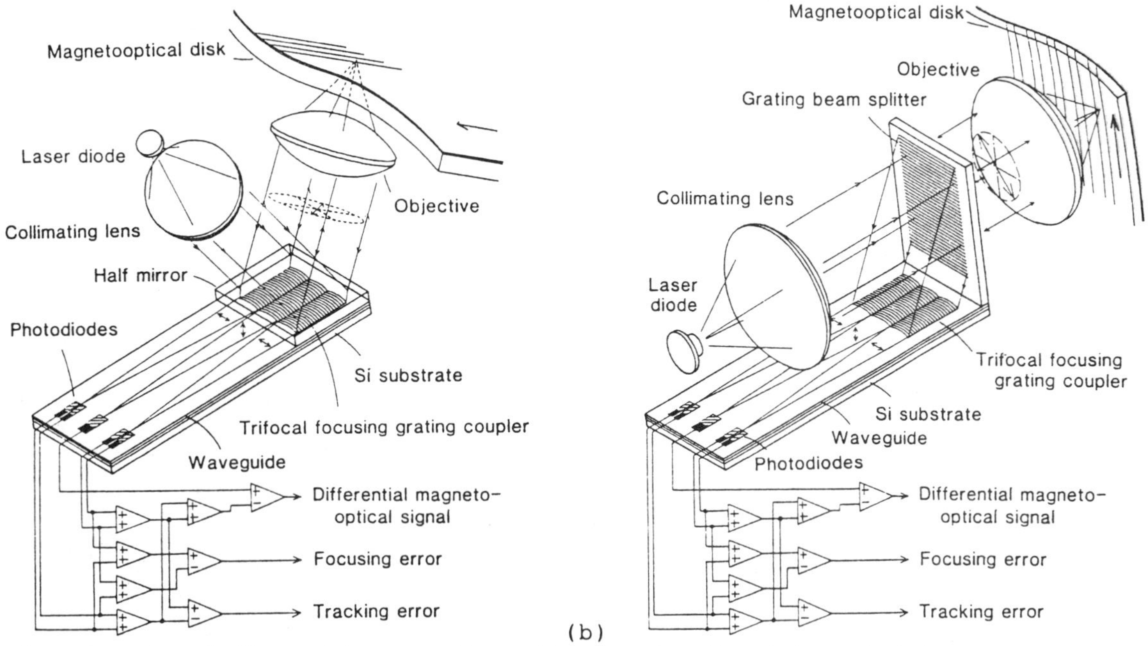

Fig.12 Magnetooptical disc pickup configurations using the integrated detection optics.

Figure 13 illustrates the principle of the polarization rotation detection by the TFGC, which is based on differential method.[20,22] Because of the above-described design of the TFGC, the p-polarization component of the beam from the disc is coupled selectively to the TM wave in waveguide, and the s- component to the TE wave. Intensities of the TM and TE waves vary with $\sin^2\theta$ and $\cos^2\theta$ dependences, respectively, on the polarization angle θ of the incident beam. Thus the TFGC serves as a polarization splitter based on the TE/TM mode dispersion. Intensities of the TM and TE waves vary with phases opposite to each other when a Kerr rotation is imposed on the beam polarized at around $\theta = 45°$. The MO readout signal is obtained by differential amplification of the TE and TM photocurrents. The focusing/tracking error signals are also obtained from the photocurrents of the side PD elements based on the Foucault/push-pull methods.

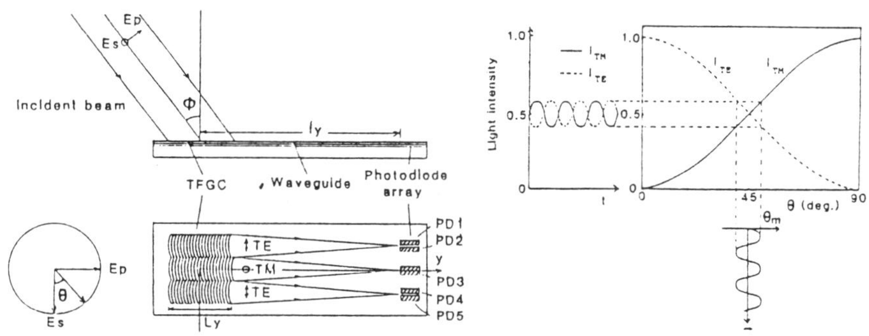

Fig.13 Differential detection of polarization rotation by a TFGC.

Theoretical analysis on the wavelength dependence of the TFGC coupling efficiency for constant incidence angle showed that the TFGC exhibits a narrow bandwidth resulting from the grating dispersion and the guided-mode discreteness. This implies that, for the pickup of Fig. 12 (a) using a half mirror (HM) as a beam splitter, stable single-mode operation of LD is essential since mode instabilities would disable the detection of small polarization rotation. The pickup construction shown in Fig. 12 (b) using a grating beam splitter (GBS) was proposed to eliminate this problem.[21] Theoretical analysis showed that it is possible to cancell out the dispersion of TFGC by that of GBS, by appropriate design of grating parameters; the combined GBS/TFGC efficiency is flat over a wide (> 10 nm) wavelength range. Thus the pickup operation is insensitive to the LD wavelength fluctuation, and even multimode operation by RF driving of LD can be employed for stabilization.

4.2 Fabrication and Experimental Results

The differential detection of small polarization rotation by TFGC was first demonstrated in a prototype divice using a #7059/Pyrex glass waveguide.[20] The TFGC was fabricated by the similar process as described in 3.3, and a separately prepared 5-element PD was butt coupled to the waveguide. The waveguide devices shown in Fig. 12 was fabricated in a #7059/SiO_2/Si waveguide by the processes similar to those for IODPU.[21] The HM for the device of Fig. 12 (a) was prepred separately by evaporating Cr on a glass substrate and was attached on the TFGC area. A photograph of the device is shown in Fig. 14. The GBS (period: 1.32μm, groove depth: 0.20μm) for the device of Fig. 12 (b) was prepared by EB writing and RIE of Si-N layer on a separate Pyrex substrate.

The polarization rotation detection function was examined, in the fabricated three types of the waveguide device, by using a MO modulator (Farady rotator) to give a sinusoidal polarization modulation (θ_m: ±0.4°pp) simulating the Kerr rotation by a MO disc.[20,21,22] The output TE and TM photocurrents exhibited modulation of opposite phases. Figure 15 shows the output signals from the device of Fig. 12 before attatching the beam splitter. The photocurrent revel of approximately 2μ A was obtained with a 30 mW LD (Mitsubishi ML6702A). It was confirmed that the MO signal was doubled and the common-mode noise was reduced in the read-out signal obtained by the differential I-V conversion of the photocurrents. The focusing error signal was obtained over a ±30 μ m mirror displacement range. The TFGC - GBS configuration (the device of Fig. 12 (b)) was examined also with a LD in multimode operation by RF driving, and the differential detection of polarization rotation was confirmed with the improved stability but without reduction of photocurrent revel.

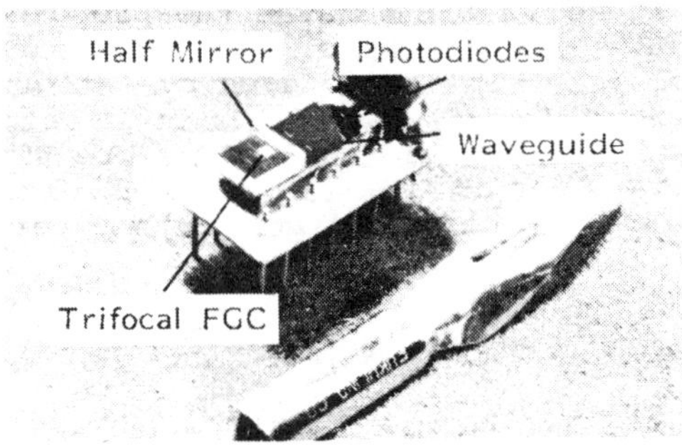

Fig.14 Fabricated detection device for integrated MO disc pickup.

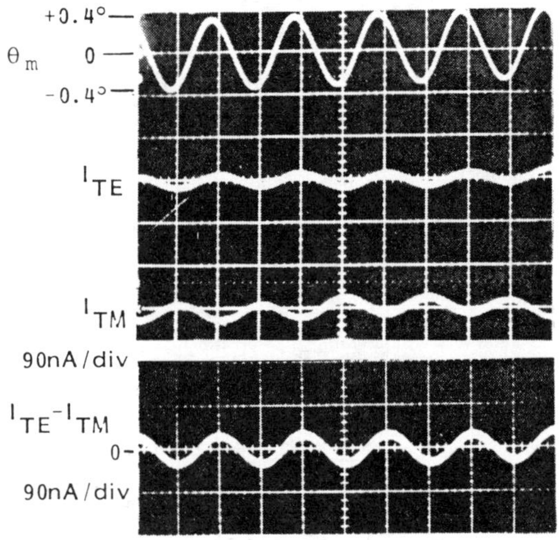

Fig.15 Obtained MO readout signals.

5. DISCUSSIONS AND CONCLUSIONS

The authors' work has been reviewed on integrated-optic implementation of disc pickup using waveguides and holographic components. A number of device constructions has been proposed and the feasibilities have been examined. Through these examples, efficient use of the advantages of waveguide holographic components based on the EB writing technique, i.e., the versatilities (multi- or combined functions) and the integration compatibilities, has been visualized. Although the performances obtained so far are not sufficient for practical application, the elementary functions have been successfully demonstrated in the simulation experiments. Most of the results are encouraging, but there are still some problems to be solved before the practical applications.

The most important subjects for the future work to achieve the operational performances in the proposed pickup devices are as follows:

1. assuring the stable single-mode lasing of LD under the working conditions, and overcoming the sample-to-sample wavelength deviation of LD,
2. attaining the FGC pattern fabrication accuracy required to obtain the diffraction-limited focus spot size.

As for the first subject, improvements in the pickup construction are possible but they tend to require modifications against the integration and simplification. Future developments of dynamic stabilized LDs, e.g., DFB LDs, interference-cavity LDs, QW LDs, etc., are therefore strongly desired for the solution of the problem. To discuss the second subject, it should be noted that all the results presented in this paper have been obtained with devices fabricated by using an EB writing system[2] using an ordinary scanning electron microscope (SEM). Most recently, a new high-precision EB writing system has been installed and started working in the authors' laboratory. Now we can write FGC patterns, which are corrected for the disc-clad aberration, of up to 3x3 mm^2 aperture with pattern accuracy of 0.05λ rms error. The IODPU optimum designing based on the read-out response analysis, as well as design improvements for waveguide scattering reduction and FGC efficiency enhancement, are also in progress. We are thus restarting the fabrication work toward attaining the operational performances, and the results will be reported in the near future.

To conclude this review, the authors would mention that the applications of the technology presented here are not limited to disc pickup devices. An integrated-optic interferometer position sensor[23] evolving from the IODPU, and an integrated acoustooptic printer head[24] and an integrated-optic RF spectrum analyser[25], which are FGC applications extended to $LiNbO_3$ waveguides, have been proposed and demonstrated by the authors.

6. ACKNOWLEDGMENTS

The authors wish to thank S. Ura, H. Sunagawa, M. Horita, M. Oyama and H. Ishimaru for their contributions to the project. A part of this work was supported by Scientific Research Grant-in-Aid from the Ministry of Education, Science and Culture. They gratefully acknowledge the support through the Fund for International Communication Research by KDD Engineering and Consulting Foundation. They also thank LSI R&D Lab. and Consumers' Electronics Development. Lab., Mitsubishi Electric Co., for the cooperation in preparing the photodiodes.

7. REFERENCES

1. H. Nishihara, M. Haruna and T. Suhara, Optical Integrated Circuits, McGraw-Hill, New York (1989).
2. T. Suhara and H. Nishihara, "Integrated-optics components and devices using periodic structures," IEEE J. Quantum Electron., QE-22, 845-867 (1986).
3. T. Suhara, S. Ura, H. Nishihara and J. Koyama, "An integrated-optic disc pickup device," Int. Conf. Integrated Optics and Optical Fiber Communications (IOOC'85), 117-120, Venezia, Oct. 1-4 (1985).
4. T. Suhara, H. Nishihara and J. Koyama, "Waveguide holograms: A new approach to hologram integration," Opt. Commun., 19, 353-358 (1976).
5. D. Heitmann and C. Ortiz, "Calculation and experimental verification of two-dimensional focusing grating couplers," IEEE J. Quantum Electron., QE-17, 1257-1263 (1981).
6. T. Suhara, H. Nishihara and J. Koyama, "High-performance focusing grating coupler fabricated by electron-beam writing," Topical Meeting on Integrated and Guided-Wave Optics, ThD-4, Kissimmee, Apr. 24-27 (1984).
7. G. Hatakoshi, H. Fujima and K. Goto, "Waveguide grating lenses for optical couplers," Appl. Opt., 23, 1749-1753 (1984).
8. T. Suhrara, H. Tanaka, S. Ura and H. Nishihara, "Holographic printing of EB-written submicron gratings for integrated optics (in Japanese)," Annual Meeting of Japan. Soc. Appl. Phys., 4p-L-13 (1986).
9. S. Ura, T. Suhara, H. Nishihara and J. Koyama, "Focusing grating for integrated optical-disc pickup device (in Japanese)," Trans. IECE of Japan, J68-C, 803-811 (1985).
10. S. Ura, T. Suhara and H. Nishihara, "Aberration characterizations of a focusing grating coupler in an integrated-optic disc pickup device," Appl. Opt., 26, 4777-4782 (1987).
11. S. Ura, T. Suhara, H. Nishihara and J. Koyama, "An integrated-optic disc pickup device," IEEE J. Lightwave Tech., LT-4, 913-918 (1986).
12. G. Bouwhuis, J. Braat, A. Huijser, J. Pasman, G. van Rosmalen and K. S. Immink, Principles of Optical Disc Systems, Adam Hilger, Bristol (1985).
13. M. Horita, S. Ura, T. Suhara and H. Nishihara, "An integrated-optic pickup using an external focusing lens (in Japanese)," Ann. Meet. Japan. Soc. Appl. Phys., 28pZE13 (1987).
14. H. Sunagawa, T. Suhara and H. Nishihara, "Bifocal gocusing grating coupler for optical disc pickup (in Japanese)," Annual Meeting of IECE of Japan, 451 (1986).
15. H. H. Hopkins, "Diffraction theory of laser read-out systems for optical video discs," J. Opt. Soc. Amer., 69, 4-24 (1979).
16. T. Suhara and H. Nishihara, "Analysis of read-out response in integrated-optic disc pickup device (in Japanese)," Japan. J. Opt., 18, 82-90 (1989).
17. T. Wilson and C. Sheppard, Theory and Practice of Scanning Optical Microscopy, Academic Press, London (1985).
18. S. Ura, T. Suhara and H. Nishihara, "Laser diode spectrum in integrated-optic disc pickup device," Optoelectronics Conf., A8-4, Tokyo, July 29-31 (1986).
19. M. Oyama, S. Ura, T. Suhara and H. Nishihara, unpublished work, Osaka Univ. (1988).
20. H. Sunagawa, S. Ura, T. Suhara and H. Nishihara, "An integrated optic detection device for magnetooptical disc pickup," Int. Symp. on Optical Memory, FA7, Tokyo, Sept. 16-18 (1987).
21. T. Suhara, H. Ishimaru, S. Ura and H. Nishihara, "Integration of detection optics for magnetooptical disc pickup," to be presented at Int. Conf. on Integrated Optics and Optical Fiber Communications, Kobe, July 18-21 (1989).
22. S. Ura, H. Sunagawa, T. Suhara and H. Nishihara, "Focusing grating coupler for polarization detection," IEEE J. Lightwave Tech., LT-6, 1028-1033 (1988).
23. T. Suhara, S. Ura and H. Nishihara, "An integrated-optic interferometer position sensor," Conf. on Laser and Electro-Optics, FQ3, Baltimore, April 26 - May 1 (1987).
24. T. Suhara, N. Nozaki and H. Nishihara, "An integrated acoustooptic printer head," Europ. Conf. on Integrated Optics, 119-122, Glasgow, May 11-13 (1987).
25. T. Suhara, K. Hara and H. Nishihara, "An integrated-optic spectrum analyzer using a focusing grating (in Japanese)," Rev. Laser Eng., 17, 22-31 (1989).

Reprinted with permission from *Applied Optics*, Vol. 29(15), pp. 2315-2319 (May 20, 1990). ©1990 Optical Society of America.

Rotationally symmetric construction optics for a waveguide focusing grating

George N. Lawrence, Kenneth E. Moore, and Patrick J. Cronkite

Waveguide grating couplers have been under study as a method of focusing laser diode radiation on the surface of an optical data storage device. In this letter, we describe a simple method of holographically constructing a waveguide grating coupler at visible wavelengths which may be used with collimated infrared laser radiation to form high numerical aperture, diffraction limited beams. The design method includes the necessary compensation aberrations to offset the effects of differences between construction and end-use wavelengths.

I. Introduction

The focusing grating coupler (FGC), developed for application to optical data storage,[1] is of considerable interest for low cost, highly stable optical readout heads. An optical schematic of the device is shown in Fig. 1. During construction of the grating, an interference pattern due to beams diverging from points C_1 and C_2 is recorded on the surface of the waveguide. During use, a laser diode is optically coupled to the slab laser waveguide. The light in the slab propagates to the FGC, where it leaves the slab and focuses to a point. After reflection from the surface of the optical recording medium the light returns to the FGC as a diverging beam. After a second diffraction from the FGC, the light again propagates in the waveguide, this time in the reverse direction. (The detection and construction optics are omitted from the drawing for clarity.)

The device is intended for operation in the near infrared, but it is convenient to form the FGC by holographic methods with a visible laser. E-beam techniques have been recently used to construct gratings with near diffraction-limited performance.[2] However because of stability problems, it is difficult to construct gratings larger than a few millimeters with existing *e*-beam equipment.[1] Construction at one wavelength and reconstruction at another results in aberrations of the optical beam, which must be compensated by the design of the construction optics. In the original research on holographic construction of these devices, a combination of spherical and cylindrical lenses was used to form the holographic construction optics.[3-6]

Rotationally symmetric construction optics for fabricating conventional gratings have been investigated.[7-9] In these studies, the application was to focus a collimated beam from a laser diode with a grating constructed by certain visible wavelength lasers.

Construction of waveguide gratings relies on principles similar to conventional gratings but is more complex because the waveguide mode is at grazing incidence and propagates in the waveguide with nonunity index of refraction. Additionally, the waveguide grating is more difficult to design because none of the standard lens design codes—without special modifications—model waveguide gratings.[6,7] In contrast, most major lens design codes model conventional holograms. For the particular method described in this paper, no special ray tracing procedures are required. In this paper, we present the theory for the method and a detailed discussion of the design procedure.

II. Theory

We can construct the gratings with simple, rotationally symmetric optics if the laser diode radiation is collimated in the waveguide. An integrated optic waveguide lens is a common device[10] and can readily provide the collimation required. With a collimated waveguide mode, we can assure exact rotational symmetry so that spherical aberration is the only aberration to be controlled in the holographic construction system.[11] Spherical aberration is easily controlled and we will present a simple system that accomplishes that task.

Shown schematically in Fig. 2 are a top and side view of a waveguide grating and the new construction design. The laser diode illumination is collimated by a waveguide collimating lens. The beam labeled C_1 is a collimated construction beam at angle θ from the nor-

The authors are with University of Arizona, Optical Sciences Center, Tucson, Arizona 85721.

Received 7 September 1989.

0003-6935/90/152315-05$02.00/0.

mal to the surface such that the phasefront exactly matches the phasefront of the waveguide mode. This collimated beam interferes with a second construction beam diverging from point C_2. Beam U_1 is the collimated beam in the waveguide and U_2 is the beam converging to a focus for end-use. If λ_c is the construction wavelength and λ_u the end-use wavelength, the waveguide grating equation is

$$\frac{1}{\lambda_u}\hat{n}\times(\hat{r}_{u_1}-n_0\hat{r}_{u_2})=\frac{1}{\lambda_c}\hat{n}\times(\hat{r}_{c_1}-\hat{r}_{c_2}), \tag{1}$$

where $\hat{r}$ is the unit vector and the subscript refers to the beam. The condition for formation of the waveguide grating is

$$\frac{y}{\lambda_c}\sin\theta-\Gamma(x,y)+\frac{R_c}{\lambda_c}\left[\sqrt{1+\frac{r^2}{R_c^2}}-1\right]=0, \tag{2}$$

where y is the longitudinal coordinate on the surface, x is the transverse coordinate, r is the square root of the sum of x^2 and y^2, R_c is the normal distance from the construction point C_2 to the surface, and $\Gamma(x,y)$ is the phase of the signal recorded on the surface. The coordinates x, y, and r are measured from point on the grating directly below C_2. The condition for diffraction of the end-use beam is

$$-\frac{yn_0}{\lambda_u}-m\Gamma(x,y)-\frac{R_u}{\lambda_u}\left[\sqrt{1+\frac{r^2}{R_u^2}}-1\right]=0, \tag{3}$$

where R_u is the normal distance from the end-use focus point to the surface, n_0 is the effective index of the waveguide (generally close to the index of refraction of the waveguide layer) and we have reversed the signs of the beam phase terms to account for the reversal of direction of the end-use beams from the contruction beams. The symbol m represents the integer value of the diffracted order number. Choosing $m = -1$ will allow both Eqs. (2) and (3) to be satisfied by the same $\Gamma(x,y)$, i.e., the same grating, if

$$\frac{1}{\lambda_c}\sin\theta=\frac{n_0}{\lambda_u}, \tag{4}$$

which sets the angle requirement for the collimated construction beam and

$$\frac{R_c}{\lambda_c}\left[\sqrt{1+\frac{r^2}{R_c^2}}-1\right]=\frac{R_u}{\lambda_u}\left[\sqrt{1+\frac{r^2}{R_u^2}}-1\right], \tag{5}$$

which matches the phasefronts for Beams C_2 and U_2 Matching the collimated beam angle is straightforward. For a typical case of $\lambda_u = 0.78\ \mu m$, $\lambda_c = 0.48\ \mu m$, and $n_0 = 1.5$; the angle is 67.38°. Note that for construction in air, the wavelengths cannot be made too close together, i.e.,

$$n_o<\frac{\lambda_u}{\lambda_c}. \tag{6}$$

Identical phasefronts for the collimated beams is relatively easily achieved by forming good collimated beams and setting the proper angle. The diverging beams may be made identical in phase by setting the conjugate distances correctly and by spherical aberration correction. A key feature is that the optical phase

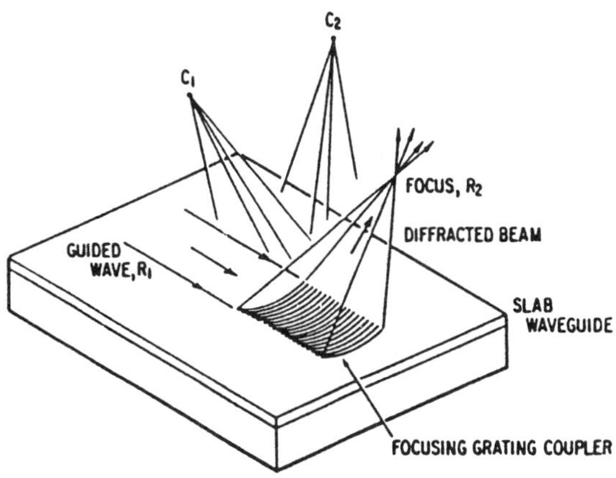

Fig. 1. Focusing grating coupler optical data head. A laser diode is butt coupled into a waveguide. The guided mode propagates in the waveguide until it encounters a grating consisting of a corrugation of the surface of the waveguide. The grating out-couples light into a focused beam if the grating period is properly chirped. In principal, the grating may be constructed from two points at visible wavelengths for use in the infrared. In practice severe aberration will result from the wavelength shift if two simple points are used for construction.

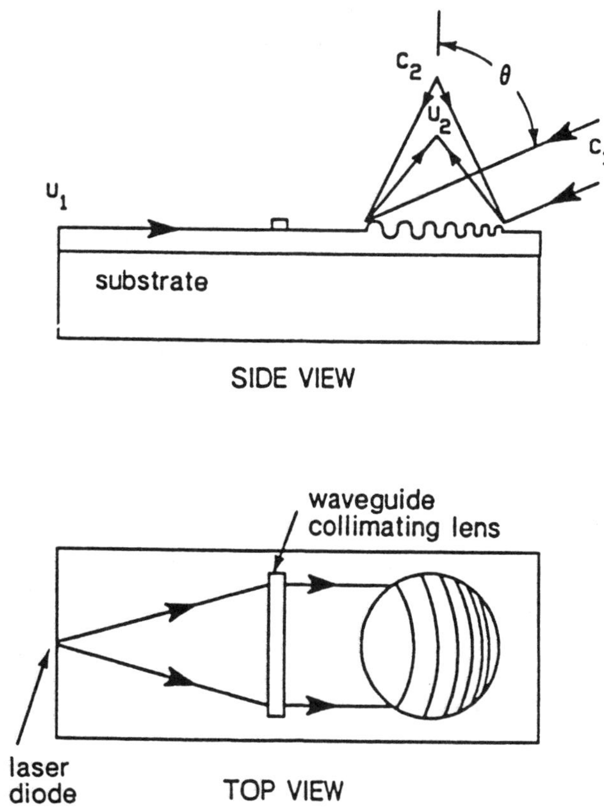

Fig. 2. Configuration for forming a waveguide hologram using rotationally symmetric optics. Light from a laser diode at U_1 is injected into the edge of the waveguide. A waveguide lens is used to collimate the waveguide mode. A grating consisting of corrugations in the waveguide diffracts some of the light into a beam focused at point U_2. The construction beams must create an interference pattern which exactly matches the frequency needed for the waveguide grating. One of the construction beams is collimated at a specific angle of incidence. The other is a rotationally symmetric beam.

error and the correction optics are strictly rotationally symmetrical. The phase difference between the diverging beams may be expanded in a power series:

$$\phi_u - \phi_c = 0 = 2\pi\left[\frac{r^2}{2}\left(\frac{1}{\lambda_u R_u} - \frac{1}{\lambda_c R_c}\right) - \frac{r^4}{8}\left(\frac{1}{\lambda_u R_u^3} - \frac{1}{\lambda_c R_c^3}\right) + \ldots\right]. \tag{7}$$

To second order, we can satisfy the phase condition by

$$\lambda_u R_u = \lambda_c R_c. \tag{8}$$

The r^4 and higher order terms represent spherical aberration and must be compensated. The fourth order term is

$$A = -\frac{1}{8\lambda_u R_u}\left(\frac{1}{R_u^2} - \frac{1}{R_c^2}\right), \tag{9}$$

where we have assumed Eq. (8), for equal optical power, is satisfied. Higher coefficients are similarly derived.

For a typical grating of 3-mm diameter and focal length of 1.5 mm, the construction beam would have a focus at 2.4375 mm from the surface. The end-use system has a numerical aperture of 0.7, the numerical aperture for the visible illumination is 0.52. Correction of the spherical aberration for such a high numerical aperture system is difficult because spherical aberration of very high order is present.

III. Design Technique

We may perform the design by calculating the spherical aberration terms for all significant orders and using these terms as target values in the construction process. A more convenient method has been found which allows the construction optics to be designed as a completely conventional system without consideration of holographic elements. The quality of the image formed by the diffracted end-use beam can be determined by analysis using the construction beam. The path lengths of rays in the diffracted beam can be matched to path lengths at the construction wavelength by introducing a fictitious medium whose index is n_f. The fictitious medium is placed at the surface of the waveguide during design of the construction optics, and the rays which are refracted at the waveguide surface are propagated backward to the virtual focus which corresponds to the end-use focus. The path length matching condition is

$$\frac{n_f R_u}{\lambda_c}\left[\sqrt{1 + \frac{r^2}{R_u^2}} - 1\right] = \frac{R_u}{\lambda_u}\left[\sqrt{1 + \frac{r^2}{R_u^2}} - 1\right], \tag{10}$$

where

$$n_f = \frac{\lambda_c}{\lambda_u}. \tag{11}$$

The fictitious index is $n_f = 0.6154$ for the wavelengths of our example. Using this fictitious glass, we can perform the calculation as a classical imaging problem. The configuration is shown schematically in Fig. 3. The source point at S is readily created by a focused laser beam with a clean-up pinhole aperture. The lens, shown schematically, focuses the beam and

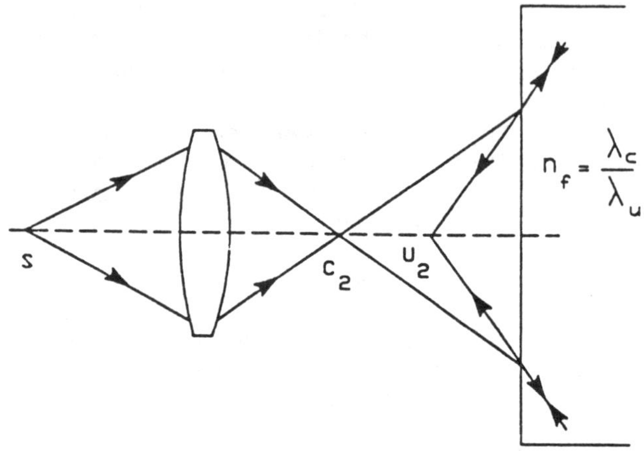

Fig. 3. Schematic of holographic construction design using a fictious glass. The source point of the laser is focused with special aberration construction optics to achieve the divergence and aberration compensation. At the surface of the grating, the diverging light enters a fictious medium of index <1. The diverging beam is analyzed at the virtual focus in the fictious glass at U_2. When good correction is achieved in the fictious glass at U_2, the system will create the correct hologram when used with a collimated beam at the proper angle.

adds compensating aberration. The beam passes nominally through the best fit point source location C_2 and then diverges. At the grating surface, a flat medium interface is applied with the light entering into a medium of index n_f and proceeding backward to a virtual focus. If the fictitious index is chosen to fit Eq. (11), then optics which result in a good focus in this classical analysis will be correct for the construction beam.

Use of the fictitious glass is a considerable aid to the design process and eliminates the need for a special holographic equation for waveguide gratings. Some lens programs will not allow specification of a glass with index less than unity. In that case, index scaling may be used to perform the analysis. The virtual focus at U_2 may be placed in air and all other indices scaled by λ_u/λ_c. All ray trajectories will be identical so the transverse ray errors will be identical to the unscaled model.

The design of the holographic construction optics for the divergent beam does not depend on the effective index of the waveguide—only on the two wavelengths. Consequently, we can use the same construction optics for a variety of waveguide propagation constants and accommodate the differences by changing the incident angle of the collimated construction beam.

Focusing grating couplers that form convergent beams which come off the substrate at an angle may also be constructed with the same optical design by using a decentered aperture of reduced size. This is shown schematically in Fig. 4. The maximum angle with respect to the surface is, of course, limited by the numerical aperture of the system. It is interesting that a good design for these skewed beams consists of a

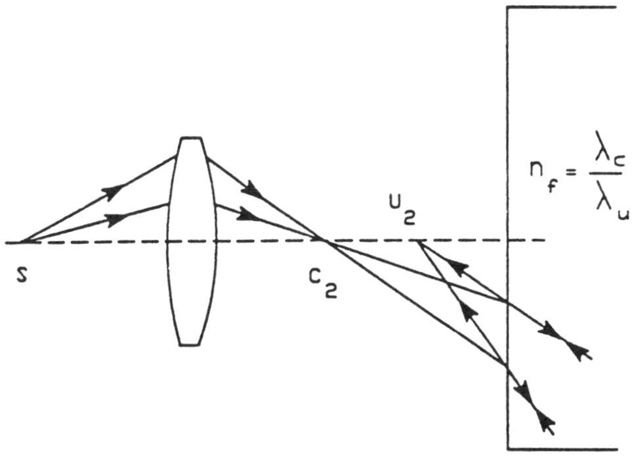

Fig. 4. Gratings that form skewed focusing beams may be constructed with the rotationally symmetric construction optics by using a decentered aperture to limit the beam. The origin of the construction beam (with suitable aberrations) is point C_2. The resulting grating will cause the laser diode to come to a focus at point U_2. The skewed section must fall inside the cone defined by the numerical aperture of the construction optics.

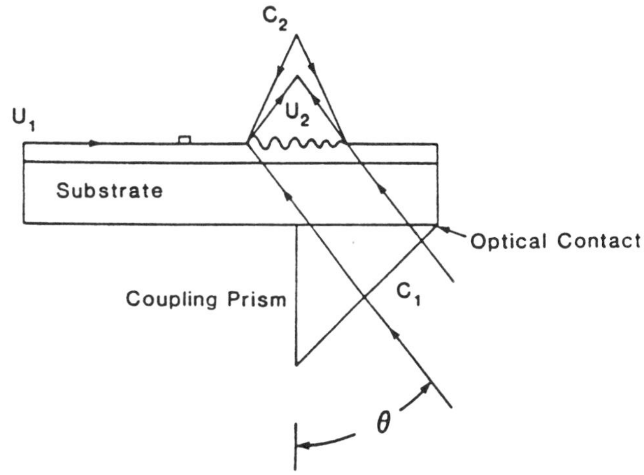

Fig. 5. Optical schematic for grating construction through substrate. By coupling the collimated beam C_1 in to the substrate through a 45° prism, incident angles can be achieved that are beyond the critical angle for the waveguide. The shortened construction wavelength in the substrate permits the use of shorter and end-use wavelengths than would be possible if the grating were constructed in air.

system with rotational symmetry and an optical axis normal to the substrate surface.

IV. Construction in Medium Other than Air

The restriction on the wavelengths specified by Eq. (6) prevents the use of some desirable wavelength combinations, such as $\lambda_c = 0.48\ \mu m$ and $\lambda_u = 0.63\ \mu m$. This combination yields, according to Eq. (6), a maximum n_0 of 1.3125, which is smaller than is readily achievable in current SOL-GEL waveguides. To alleviate this restriction, the construction of the grating can be performed through the waveguide substrate if the substrate is transparent. This changes the wavenumber matching requirement to

$$\frac{n_s}{\lambda_c} \sin\theta = \frac{n_0}{\lambda_u}, \tag{12}$$

and the condition on the wavelength ratio to

$$n_0 < \frac{\lambda_u}{\lambda_c} n_s, \tag{13}$$

where n_s is the index of the substrate. If the condition specified by Eq. (6) is not met, then the angle θ required is beyond the critical angle for the waveguide medium. To couple the collimated construction beam into the substrate, an angled face must be provided. This can be achieved by cutting the substrate at an angle, or more easily, by coupling into the medium via a prism. This principle is shown schematically in Fig. 5. A 45° prism will work well, and the exact θ of incidence on the waveguide can be externally controlled by the angle of incidence to the prism.

V. Design Example

We illustrate the theory with a design suitable for testing the concept. The parameters of the system are listed in Table I. For this example, we assume the laser diode is collimated. We have designed for an aerial image for simplicity rather than including the optical thickness of the optical data storage medium.

Figure 6 shows a layout of a design that achieves diffraction limited performance. The design is a derivative of the Offner null lens configuration, consisting of a field lens and a two-element objective lens. All elements are convex-plano for ease of manufacture.

A plot of the wavefront error as a function of aperture is shown in Fig. 7. The residual aberration is due to very high order spherical aberration. The design is sensitive to numerical aperture. With a slight reduction in numerical aperture a singlet may be used as the objective lens. This design is not claimed to be optimum but is intended, merely, to prove that reasonable designs may be achieved using the procedures described in this letter.

VI. Conclusions

A procedure has been developed which results in rotationally symmetric, holographic construction optics for IR waveguide grating couplers which allow fabrication at visible wavelengths. The use of a fictitious glass enables the designer to use classical methods and standard lens design programs. A sample design has been presented which is capable of a high numerical aperture with three plano-convex elements.

Table I. Parameters of Sample Design

End-use wavelength	0.78 μm
Construction wavelength	0.48 μm
End-use numerical aperture	0.7
Diameter of grating	3 mm
Working distance	1.5 mm
Theoretical spot size (0.78 μm)	1.3 μm
Fictious glass index	0.61538
Glass for construction optics	BK-7, Schott

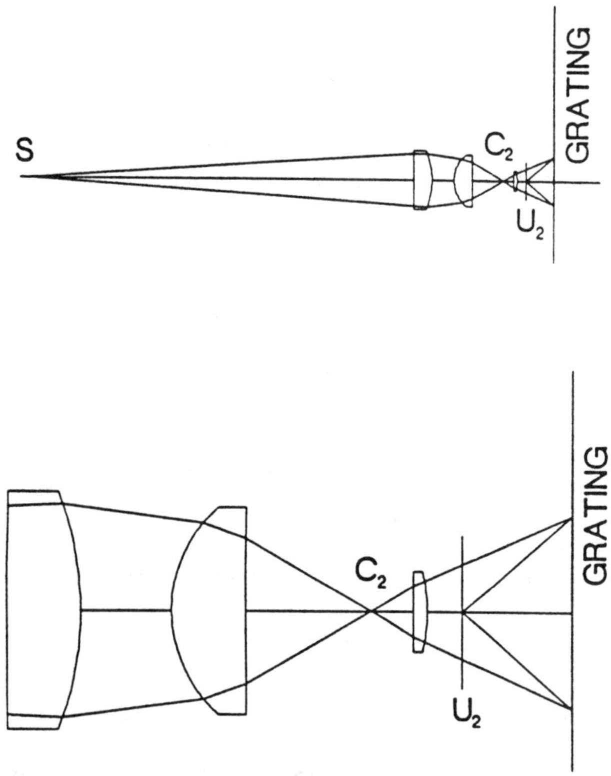

Fig. 6. Optical layout of a sample holographic construction lens. The source point is indicated by S. This should consist of a pinhole aperture illuminated by a focused laser beam of $\lambda = 0.48\ \mu m$. The two element objective lens adds the requisite aberration and reimages S to Point C_2. The light diverging from C_2 strikes a plane surface at the position of the grating substrate. For the purposes of design, the grating surface is modeled as the boundary between air and a fictious glass of index $n = \lambda_c/\lambda_u$. The design is optimized to form a well corrected virtual image in the fictious glass. The resulting design will work correctly for holographic construction. A full layout is shown as well as an enlarged view of the elements.

This research was funded by the Optical Data Storage Center, University of Arizona. Design work was done on the optical design codes SYNOPSYS[12] and Code V.[13]

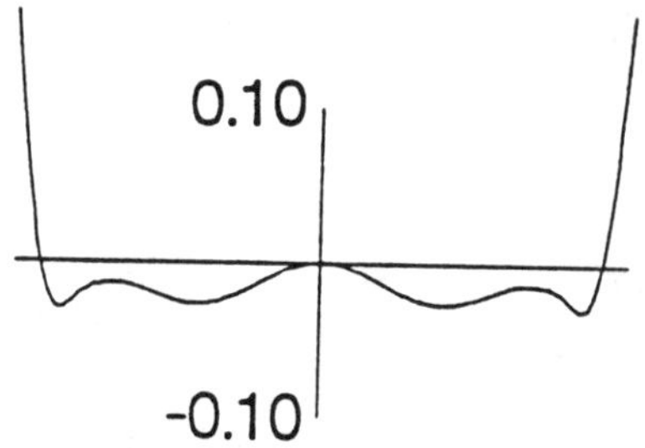

Fig. 7. Plot of wavefront aberration versus aperture position for the design of Fig. 5 at a maximum scale of 0.1 wavelengths. The aberration is measured in air at 0.48 microns. At 0.78 microns the aberration will be only 61% of the amount shown. The design exhibits high order spherical aberration which causes a small drop in the Strehl ratio.

References

1. S. Ura, T. Suhara, H. Nishihara, and J. Koyana, "An Integrated-Optic Disk Pickup Device," IEEE J. Lightwave Technol., LT-4, 913–918 (1986).
2. F. Sogawa, Y. Hori, M. Kato, "Fabrication of Almost Aberration Free Focusing Grating Coupler with Large Numerical Aperture," MOC/GRIN '89, Tokyo, pp. 126–129 (1989).
3. T. Suhara, H. Nishihara, and J. Koyama, "Waveguide Holograms: A New Approach to Hologram Integration," Opt. Commun. **19**, 353–000 (1976).
4. D. Heitmann and R. V. Pole, "Two-dimensional Focusing Holographic Grating Coupler," Appl. Phys. Lett. **37**, 585–000 (1980).
5. D. Heitmann and C. Ortiz, "Calculation and Experimental Verification of Two-Dimensional Focusing Grating Couplers," IEEE J. Quantum Electron. **QE-17**, 1257–1263 (1981).
6. P. Cronkite and G. Lawrence, "Aberration Minimization of Focusing Grating Coupler under Wavelength Shift," Appl. Opt. 27, 679–683 (1988).
7. T. Kuwayama, Y. Nakamura, N. Taniguchi, and S. Suda, "Aberration Corrected Off-Axis Holographic Lens," in *Conference Digest, ICO-13 Sapporo '84* **C6-6** (1984), p. 520.
8. S. Hasegawa, M. Kato, F. Yamagishi, H. Ikeda, and T. Inagaki, "Holographic Lens (3)—Aberration Correction Method," in *Extended Abstract, JSAP, 32nd Spring Meeting*, **31p-P-7** (1985), p. 103.
9. M. Kato, T. Iimura, S. Hasegawa, F. Yamagishi and H. Ikeda, "Holographic Lens (5)—Two-layer holographic lens," in *Extended Abstract, JSAP, 46th Autumn Meeting*, **2p-H-8**, (1985), p. 55.
10. T. Tamir, *Integrated Optics, Topics in Applied Physics* Series, Vol. 7, (Springer-Verlag, New York, 1975), p. 102.
11. G. Lawrence and P. Cronkite, "Practical Design Methods for Holographic Construction of Waveguide Focusing Grating Couplers," in *Technical Digest, Course on Opto Electronics & Laser Applications* (SPIE, Bellingham, WA, 1989); p. 1051-20.
12. SYNOPSYS is a proprietary product of Optical Systems Design, East Boothbay, Maine.
13. Code V is a proprietary product of Optical Research Associates, Pasadena, California.

Reprinted from *Storage and Retrieval Systems and Applications*, Proc. SPIE Vol. 1248, pp. 28-35 (1990).

Magneto-optical recording enhanced by magnetic recording techniques

Seiya Ogawa and Miyozo Maeda

FUJITSU LABORATORIES LTD.
10-1 Morinosato-Wakamiya, Atsugi 243-01, Japan

ABSTRACT

Front illumination is the fastest type of optical disk. Front illumination also allowed us to design a very small and light head with a beam size of 1 mm. It also allowed us to use an opaque aluminum substrate, which can rotate at high speeds. If optical disks are mounted in an enclosure like those for hard disks, they are protected from contamination and front illumination would be practical.

1. INTRODUCTION

Magneto-optical (MO) disks have high recording density, duplication capability, and media exchangeability. Hard disks, however, have high capacity and are fast and inexpensive.

MO disks are now too slow to be used in image recording. To speed up the data transfer, rotation must be faster, and to speed up access, the optical head must be lighter. Present use of back illumination makes it difficult to design a small head with a small lens, because minimum focal length is limited by substrate thickness.

The higher recording density of magnetic disks comes from the limitless reduction of head flying height(the gap between head and media). Small flying height enables a small magnetic head, which has a small magnetic induction, resulting in high speed switching of magnetic field.

If an optical head is located very near the MO media, as in the case in magnetic recording, MO disks could also have a small optical head and high speed switching. MO recording will be enhanced by magnetic recording techniques, and front illumination is very promising. This paper proposes front illumination and shows its advantages.

2. SPEED COMPARISON OF MO DISKS AND MAGNETIC DISKS

Figure 1 shows the configuration of the head and media in MO and magnetic disks. In the MO disk, the head is located far (2 mm) from the media and the beam passes through the substrate (called back illumination). This prevents head crash and contamination has little effect. Since the bias coil is far from recording layer, its induction becomes large and it can not switch at over 100 kHz.

In the hard disk, the head is small and close (less than 0.2 μm) to media that both access time and data transfer rate are very short, although sometimes the head crashes.

Figure 2 compares MO[1] and magnetic heads. The MO head is much larger and heavier. The height of the head is 20 mm for the MO head and 2 mm for the magnetic head. These numbers seriously affect the three dimensional recording density; capacity/volume.

Typical examples of 5.25 inch MO and magnetic disks are compared in Table I. There are eight disks in the hard disk drive, but only one MO disk. The MO disk has much slower data transfer, latency time, and access time, and the MO drive is more expensive. When media exchangeability is taken into account, bit cost of MO decreases drastically.

The MO disk must incorporate some of the advantages of the hard disk, especially speed. As the first step, we developed the FACOM 6443 Magneto-Optical Disk Sub-System[2], in which 8 disks are stacked on a spindle, media is not exchangeabe (Fig. 3). Eight 8-inch two-sided disks are stacked and the total capacity is 8.9 GB. They rotate at 1800 rpm. The assembly is sealed to avoid contamination by dust.

Tracks are accessed in two steps: media-to-media access and on-media access. Media-to-media access takes 3 s and the on-media access takes 200 ms, on average. The access time is much better than conventional mass storage devices, such as magnetic tape libraries or optical jukebox units, whose access times are more than 10 seconds.

As the next step, front illumination media and small optical head must be combined in a sealed drive to avoid contamination. Drastic improvement in speed is expected.

3. FASTER ACCESS

3.1 Lighter head

Access speed is determined mainly by head weight. To reduce weight, we have to reduce head size, which is determined by beam size. For back illumination, the focal distance must be larger than substrate thickness and, therefore, the beam size cannot be reduced. Figure 4 shows the relation between the beam size **d** and focal distance **f**. Since **d** is given by NA × **2f**, a drastic reduction of **f** is necessary to reduce **d**. If we use back illumination, it is clear that we cannot reduce the focal distance, since focal distance is substrate thickness plus the working distance (Fig. 5). When substrate thickness is 1.2 mm, a beam size of over 2.5 mm is required for a working distance of 1 mm. However, if the substrate had no thickness, a 1 mm beam size is usable (Fig. 6). This means that front illumination is promising for reduction of size and weight.

As is mentioned before, the head size, especially the head height is important. The height can be reduced to 2 or 3 mm for a 1 mm beam (Fig. 7). This drastically increases the capacity/volume ratio.

Figure 8 shows that access time decreases as beam size and head weight becomes smaller. A beam size of 1 mm will lighten the head to ten grams or so and shorten the access time to 20 - 30 ms which is close to the access time of magnetic disks.

3.2 Front illumination

Front illumination media is illustrated in Fig.9. On the each side of the substrate, there is a pregrooved 2P layer, a recording layer and a thin protective. The head shown in the figure has a 1 mm beam, is 2 mm high, and only weighs 10 g.

With front illumination, the substrate does not have to be transparent. Table II compares various substrate materials. Aluminum substrates are good for front illumination. Advantages of aluminum are inexpensiveness and toughness, which enable rotation up to 10,000 rpm.

In back illumination, the tilt of media with respect to the beam axis seriously degrades focusing (Fig. 10). Only a tilt angle of 0.1 deg is allowed for a 1.2 mm thick disk. More than 5 deg is allowed for a 0.03 mm thick disk, that is, for front illumination. The intensity distribution of beam spots are compared in Fig. 10 for back and front illumination for a disk tilt of 1 deg. This tilt problem becomes very important because there is a large margin of substrate manufacturing in mass-production.

We can see from Fig. 10, that a lens with a NA as large as 0.75 can be used.

4. FASTER ROTATION

4.1 Media

The aluminum substrate is best suitable for front illumination. Our goal for the time being was a rotation of 5,400 rpm. Aluminum is durable for more than 5,400 rpm, but various factors limit speed. Another advantage of aluminum is its conductivity, preventing static build-up. Therefore, unlike plastic substrates, aluminum substrates are scarcely affected by dust.

Recording sensitivity and laser power also affect rotation speed. The linear velocity dependence of recording laser power is shown in Fig. 11 for a highly-sensitive amorphous rare-earth-and-transition metal alloy. The most powerful commercially available laser diode is about 50 mW. Since the laser power available on the media is roughly 1/3 of 50 mW, the linear velocity is limited to 35 m/s, or 5,400 rpm.

4.2 Beam spot servos

In high speeds, focusing and tracking servos are problems. If we suppose conventional mechanics, which have a servo gain of 60 dB up to 30 Hz, focusing error depends on rotation speed (Fig. 12). Media deflection must be reduced and resonant frequency must be raised. To reduce tracking error, mechanical deviation and microscopic concentricity of grooves must be improved.

5. DATA TRANSFER RATE AND OVERWRITE

When rotation is 5,400 rpm and minimum bit length is 0.75 μm, reading at 4 MB/s can be easily achieved. To write at 4 MB/s requires overwriting.

For overwriting, we assume a very small coil 200 μm above the recording layer (Fig. 13). For overwriting, 100 Oe is enough. On the contrary, the magnetic head flies about 0.2 μm over the media. The magnetic field leaks from the core gap, so the field is very large. In the magneto-optical head, the number of turns determines field and inductance (Fig. 14). The field is 100 Oe at several 100 nH, which is small enough to drive 1.5 MB/s overwriting.

With an optical head with small coil whose flying height is several μm (coil-to-recording layer distance is 10 μm including the protective layer), the coil diameter can be reduced much more and its inductance is drastically reduced, resulting in 4 MB/s overwriting. In hard disks, flying height is so small that heads sometimes crash. However, the flying optical head does not have such severe conditions.

Table III summarizes the performance of three different MO disks. The performance is very close to that of hard disks.

6. MAGNETO-OPTICAL DISK DRIVES

Two kinds of MO disk drives using 5.25 inch disks will be designed: a conventional media-exchangeable type and the media-fixed type (Fig. 15). The latter has several disks, each surface of which has at least one head. This will be made possible through a drastic cost reduction of the heads.

With five 5.25 inch disks stacked in a drive, capacity will be 10 GB, data rate 4 MB/s, and access time 20 ms.

Now MO disks are used because of their exchangeability, but in the near future, MO disks will begin to replace hard disks because of capacity, speed, reliability and, of course, price.

7. CONCLUSION

At present the magneto-optical head is large and heavy, causing slow access. Back illumination makes it difficult to design a small head with a small lens, because minimum focal length is limited by the substrate thickness.

We have shown that front illumination enabled a much lighter head (10 grams) with a beam size of 1 mm. Front illumination also allowed the use of an opaque aluminum substrate, which is both inexpensive and able to rotate at high speeds. Front illumination media is insensitive to small tilt angles. When bias-coil-to-recording layer distance is small, field modulation overwriting will be possible.

If optical disks are placed in a disk enclosure like those for hard disks, they will be protected from contamination and front illumination will be practical.

With five 5.25 inch disks stacked in an enclosure, capacity is 10 GB, data transfer rate 4 MB/s, access time 20 ms, and rotation speed 5,400 rpm.

8. REFERENCES

1. M. Moritsugu, S. Arai, A. Futamata, F. Abe, and K. Ogawa, "New optical head for magneto-optic library units", Proc. of SPIE, Vol. 1078 Optical Data Storage Topical Meeting , pp. 131-137, 1989.

2. Y. Inoue and S. Hatayama, "FACOM 6443 Magneto-Optical Disk Sub-System", Proc. of IEEE Supercomputing '88, pp. 266-271, 1988.

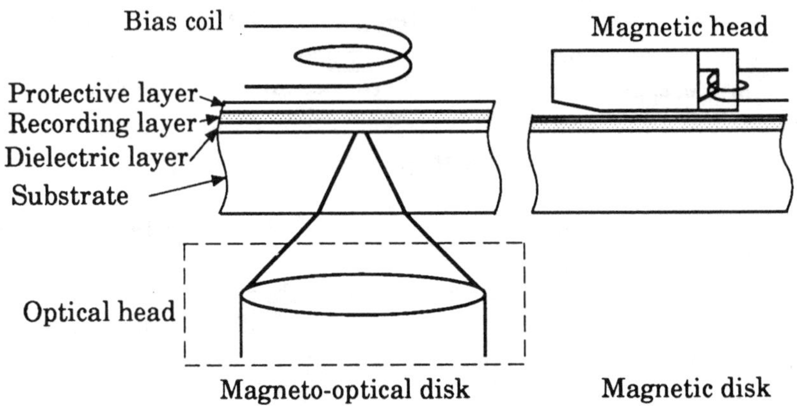

Fig. 1. Magneto-optical disk and magnetic disk.

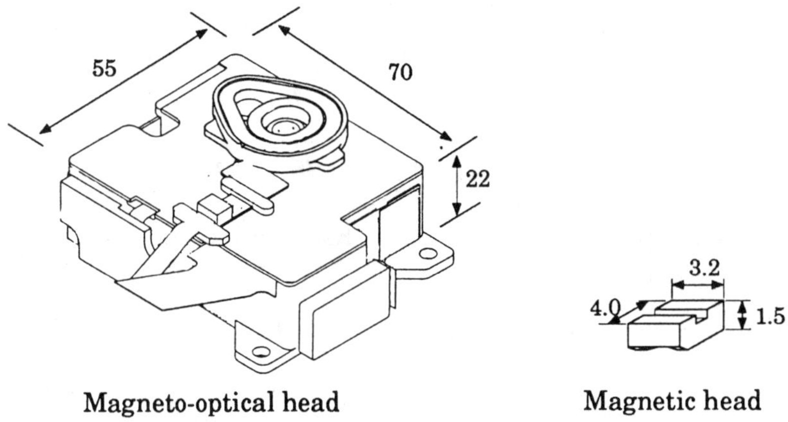

Fig. 2. Sizes of magneto-optical head and magnetic head.

Table I Characteristics of magneto-optical and magnetic disks.

	Magneto-optical disk	Magnetic disk
Total capcity (MB)	867	780
BPI	25 000	29 000
TPI	16 000	1 700
Area density (Mb / $inch^2$)	400	49
Capacity/Volume (GB / m^3)	72	300
Data transfer rate (MB / s)	1.0	1.8
Latency time (ms)	13	8.3
Access time (ms)	100	16
Price (drive) ($)	3 500	2 000
(media) ($)	300	-
Bit cost (drive) ($/MB)	0 - 5.5	2.6

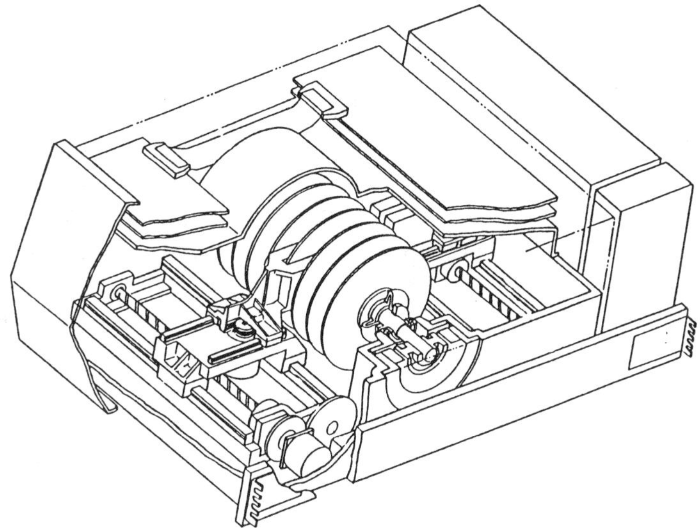

Fig. 3. FACOM6443 Magneto-Optical Disk Sub-System.

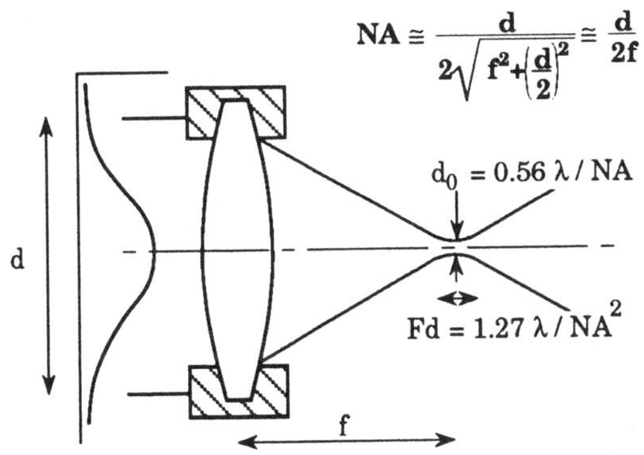

Fig. 4. Relationship between beam size d and focal length f.

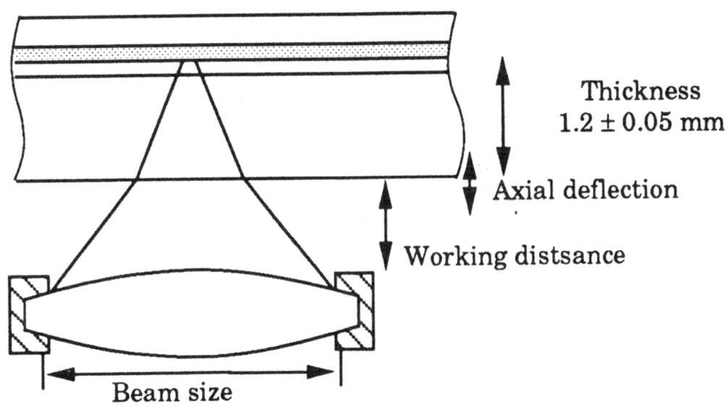

Fig. 5. Back illumination.

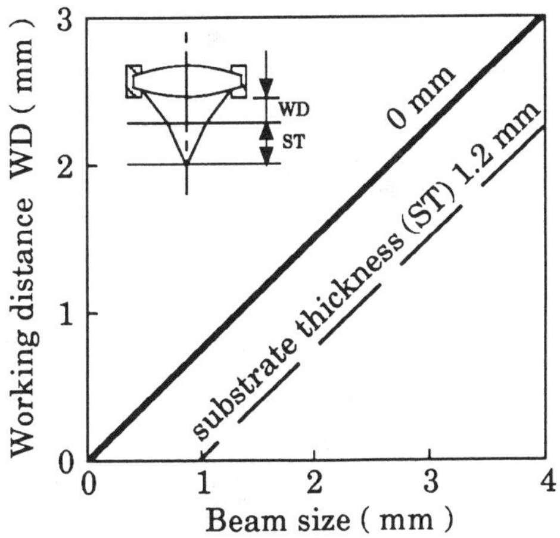

Fig. 6. Working distance vs. beam size.

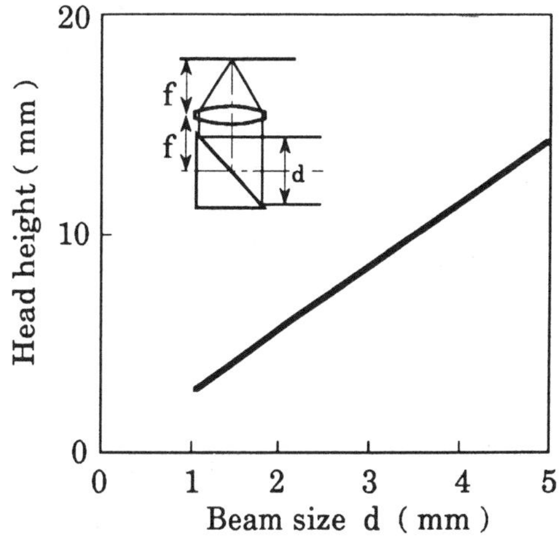

Fig. 7. Head height vs. beam size d.

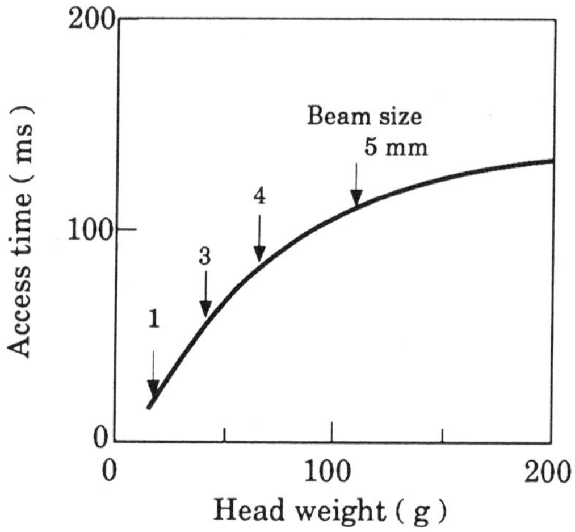

Fig. 8. Access time vs. head weight.

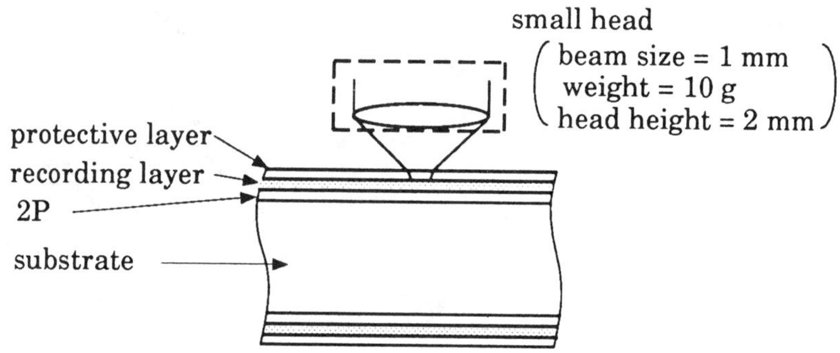

Fig. 9. Structure of front illumination.

Table II 5.25" substrates.

	Plastic	Glass	Aluminum
Max rotation (rpm)	3 000	5 000	10 000
Density (g/cm^3)	1.3	2.6	2.7
Price ($)	1	40	5
Advantage	Grooved, Inexpensive	Waterproof	Tough, Conductive, Inexpensive
Disadvantage	Shape changes	Fragile	Opaque

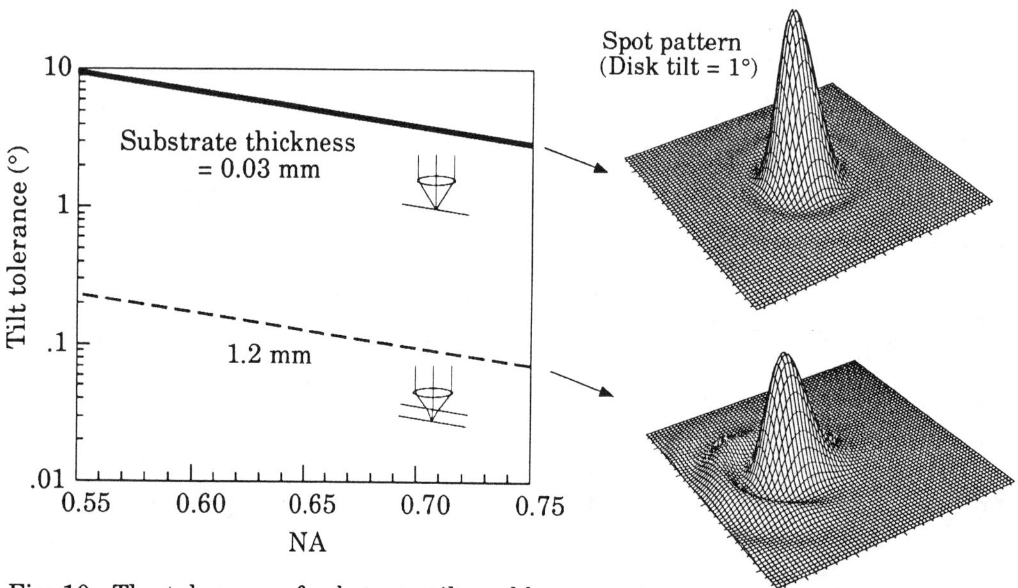

Fig. 10. The tolerance of substrate tilt and beam spot pattern.

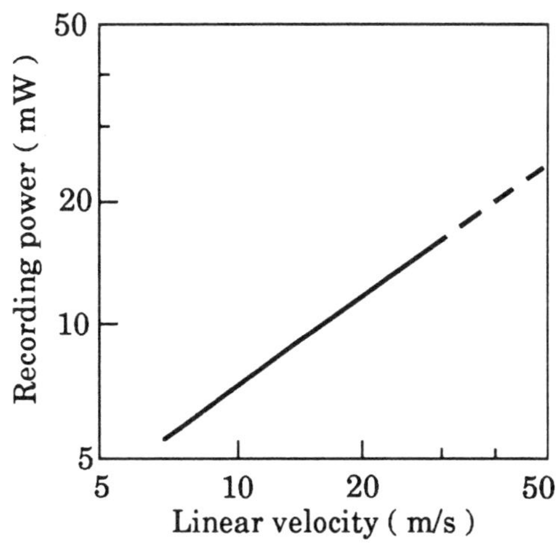

Fig. 11. Recording power dependence on linear velocity.

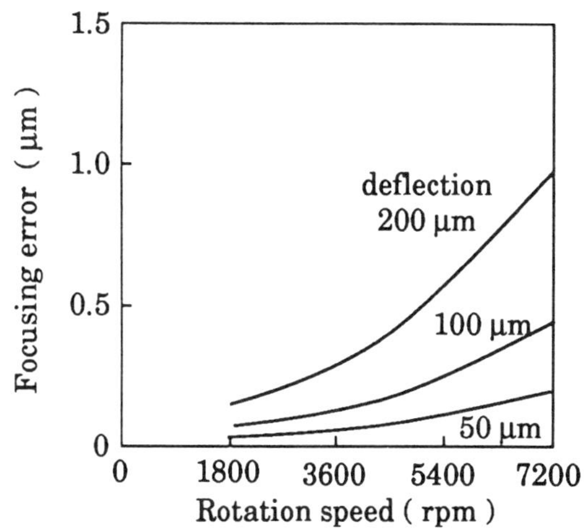

Fig. 12. Focusing error vs. rotation speed.

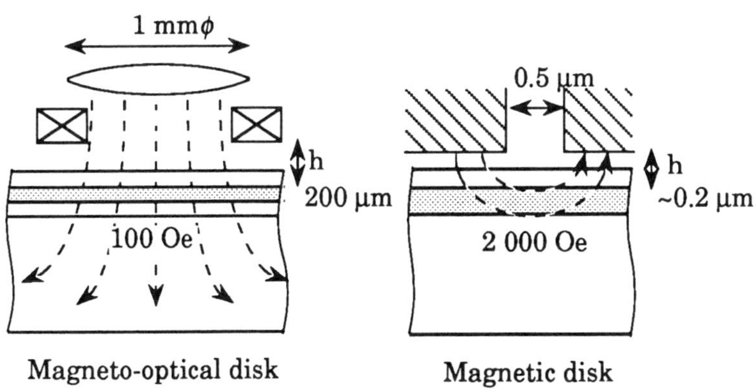

Fig. 13. Sizes of magneto-optical and magnetic disks.

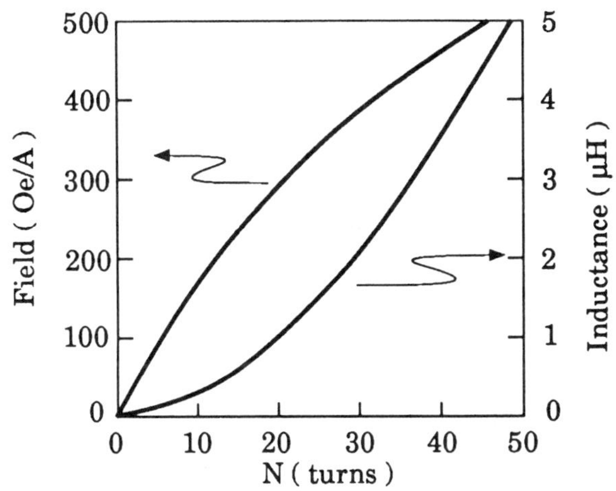

Fig. 14. Field strength from coil used in Fig. 13.

Table III Performance of 5.25" MO disks.

	MO	MO (with coil)	MO (with flying coil)
O/W	Non O/W	O/W	O/W
Bit length (μm)	0.75	2	0.75
Transfer rate (MB/s)	4 (read)	1.5	4
Access time (ms)	20	30	20
Rotation speed (rpm)	5 400	5 400	5 400

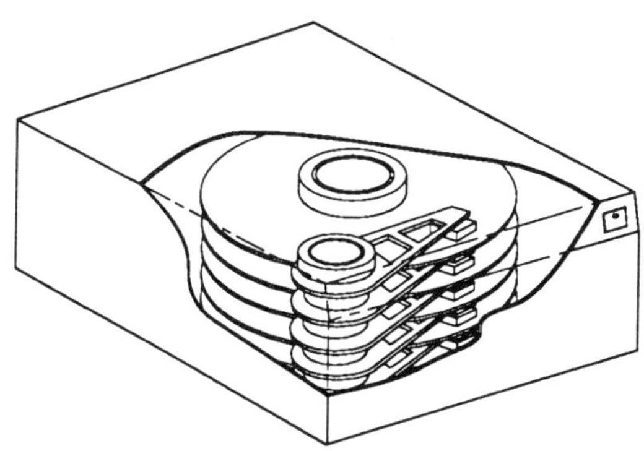

Fig. 15. Example design of MO disk drive.
Media are fixed and heads are placed on all media surfaces.

Reprinted from *Storage and Retrieval Systems and Applications,* Proc. SPIE Vol. 1248, pp. 88-95 (1990).

Recent studies of miniaturization of optical disk pickups in Japan

Hiroshi Nishihara

Osaka University, Department of Electronic Engineering
Faculty of Engineering, Suita, Osaka 565 Japan

ABSTRACT

Miniaturization of optical disk pickups is very important for higher data rate, faster access and lower cost, which are all required in future computer data storage systems. Various proposals that have been made along this direction in Japan for the last few years are reviewed. Proposed pickups for miniaturization are divided into the following four ones: (1) Conventional type pickups: these are miniaturized further by reducing the size and weight of each component as well as the number of components. (2) A laser/photodiode chip pickup: the laser is optically switched by relief of external recording medium, and the pickup is mounted on a flying slider, which eliminates the focus servo system. (3) Pickups with holographic optical elements: HOEs are used for replacement of discrete components. Since a single hologram can have several functions such as beam splitting and focusing, the number of those components can be reduced. Various types of complicated HOEs are proposed for tracking/focusing error detection. (4) A waveguide-type pickup: several optical components such as focusing grating, splitting grating and four photodiode are integrated on a single waveguide chip. These miniaturization research activities will become more important for future development of disk systems.

1. INTRODUCTION

Comparing with conventional magnetic disk and tape systems, data transfer rate and access time in optical disk systems are worse approximately by one order of magnitude, as shown in Fig. 1.[1] The development goal for access time is less than 20 ms as short as that of magnetic disk drivers. To realize this goal, miniaturization of optical disk pickup is very important.

The present pickups commonly consist of several discrete components such as a laser diode, a focusing lens, prisms, polarizers and four-segment photodiodes. One direction of miniaturization is to reduce the size and weight of each discrete components. (This makes assembling more difficult.) Much efforts are being made along this direction in many companies. The achieved values so far, for example, is 50 x 20 x 40 mm^3 and 20 g.[2] For breakthrough, therefore, it seems that we need other approaches.

Various proposals can be classified in four types: conventional type, laser chip type, HOE(holographic optical element)contained type and waveguide integration type.

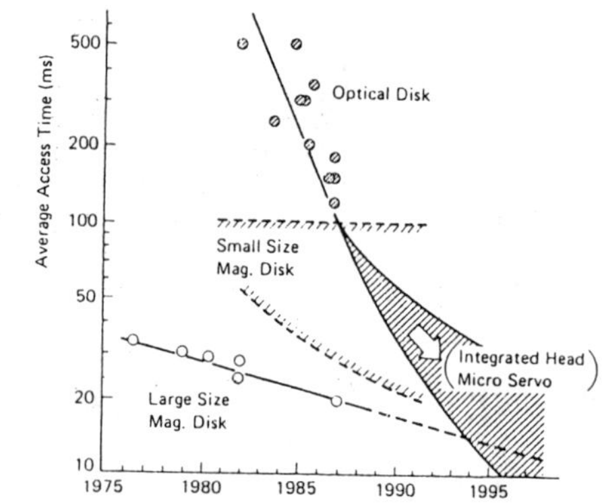

Fig. 1 Trends in access time of optical and magnetic disk systems[1]

2. CONVENTIONAL TYPE PICKUP

Miniaturization of conventional type pickups, which are composed of bulk components, are being carried on in many companies in Japan, such as Hitachi, Mitsubishi, Toshiba, Sony, NEC, Matsushita, Olympus, Pioneer and so on. As one example, the studies in Mitsubishi Electric Corporation [2] is introduced here. The first pickup for CD-ROM, as shown in Fig. 2, was produced in 1983. It was larger and heavier. By reducing the size and weight of each component as well as the number of components, new versions have been miniaturized. The configuration of the latest one is shown in Fig. 3. In Fig. 4, trends in miniaturization is illustrated with respect to weight, size(volume) and the number of components. This reduction trend will continue, as the structure and materials are optimized more properly. This reduction motivation at present, however, comes mainly from necessity of mass production or lower cost rather than shorter access time.

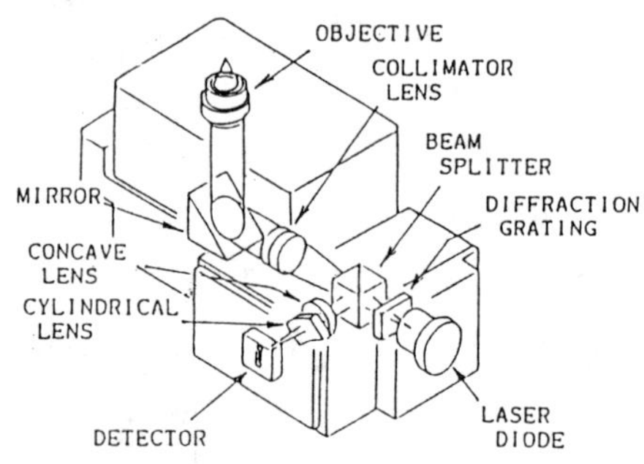

Fig. 2 The first version of conventional type CD pickup in Mitsubishi[2]

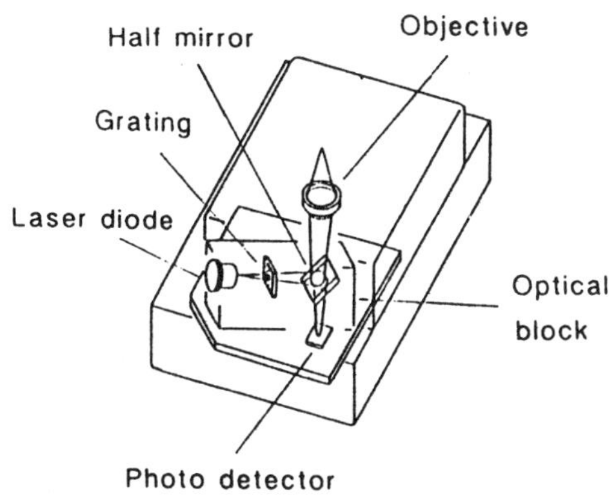

Fig. 3 The latest version in Mitsubishi[2]

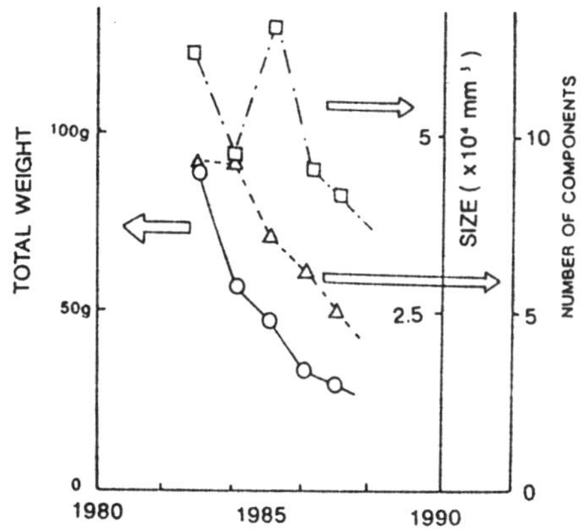

Fig. 4 Miniaturization trend of conventional type of CD-ROM pickups

3. OPTICALLY SWITCHED LASER PICKUP[3]

3.1. Configuration

This was proposed by Ukita's group. The configuration[3] is shown in Fig. 5 schematically. The pickup is based on an optically switched laser (OSL). They pointed out that it has the following advantages:

(1) air bearing technology determines the spacing and eliminates the focus servo system making the head much smaller and lighter than the self- coupled optical pickup.

(2) both spontaneous and stimulated emission states of the laser are used to detect the data signal, which makes the head simpler, smaller and cheaper.

Fig. 5 (b) shows a schematic illustration of the OSL pickup consisting of a laser and an external recording medium. R_0 and R_1 are the power reflectivities of the laser and R_f is that of the external medium. The laser operation is influenced by the feedback output power into the active region, and consequently the laser output varies.

The laser is at the spontaneous emission state at low reflectivity (0-bit) and is at the stimulated emission state at high reflectivity(1-bit), as shown in Fig. 6.

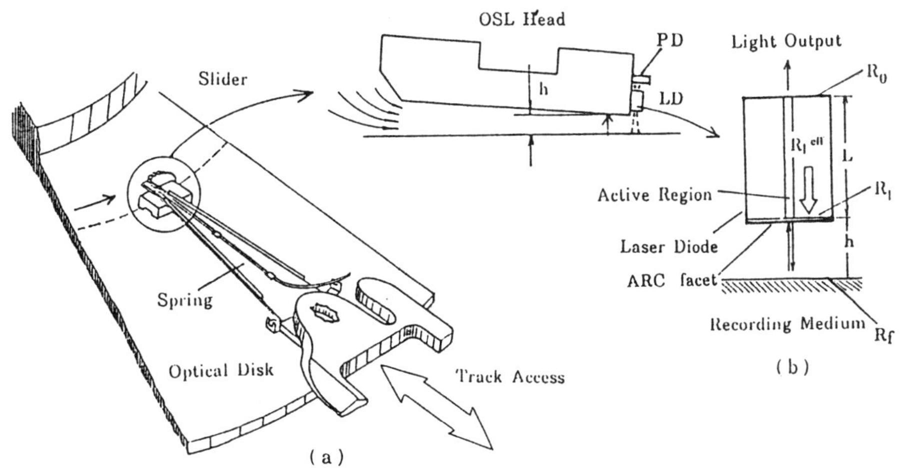

Fig. 5 Schematic illustration of an optically switched laser pickup[3]

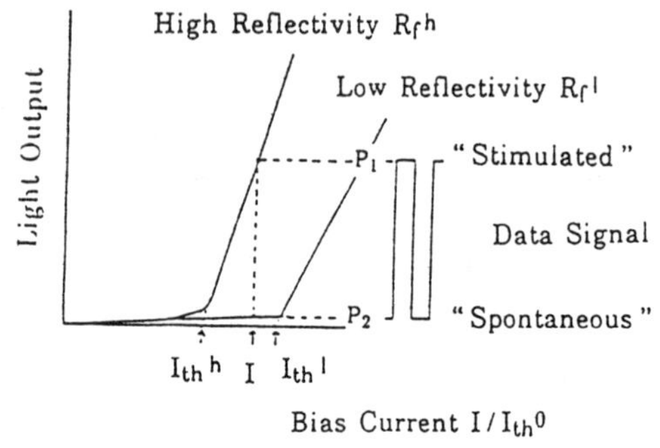

Fig. 6 Signal detection principle of OSL pickup[3]

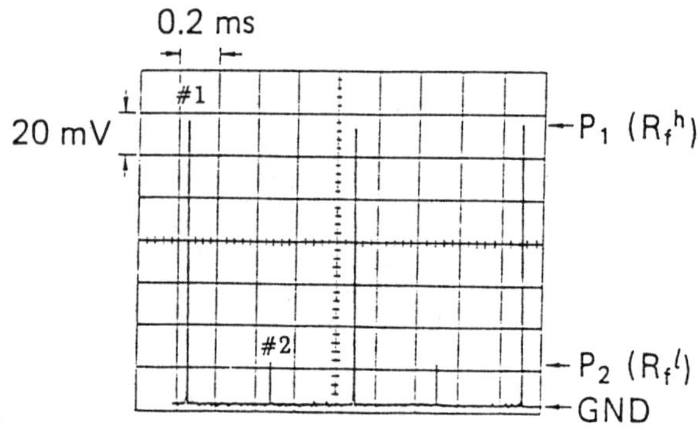

Fig. 7 Example of data signal[3]

3.2. Performance

The light output depends on the spacing between laser facet and recording medium and also the reflectivity of the two surfaces. The maximum peaks are experienced every half wavelength. The permissible spacing fluctuation is estimated to be less than 0.1 um, which was achieved experimentally for a 130 mm glass substrate disk. The switching ratio increases, as the spacing h and the reflectivity of the laser facet R_1 decrease. Therefore, anti-reflection coating (ARC) on laser facet is useful for increasing data signal. Reflectivity 0.05 of ARC facet was achieved, whereas that of cleaved facet is 0.32.

Fig. 7 shows an oscilloscope trace of the output of the PIN diode, attached close to the other end of the LD. Experiments were made using a disc with model pits. Marks #1 and #2 are the stimulated and spontaneous emission output, respectively. It is found that the readout SNR is high and that the signal amplitude fluctuation originated from the spacing between the LD facet and the glass substrate disk is very small.

The questions are whether or not the head can have enough resolution to high recording density disks, and how one picks up tracking error signals.

4. PICKUPS WITH HOLOGRAPHIC OPTICAL ELEMENT

4.1. Pickups with transmission HOE (four-beam splitter)

A transmission-HOE type pickup[4] is proposed by Ono's group (Nippon Electric Company). The pickup consists of only four optical elements, and is able to operate independently from wavelength variation in the light source, in spite of using an HOE.

The configuration is shown in Fig. 8. The pickup consists of four optical elements; laser diode (LD), photodetector(PD), imaging lens and the HOE. The HOE consists of four different holograms. In each hologram, H1 -H4, the interference fringe is generated by a spherical wave, which diverges from the laser diode, and another spherical wave, which diverges from point Pn on the photodetector, and is recorded one by one.

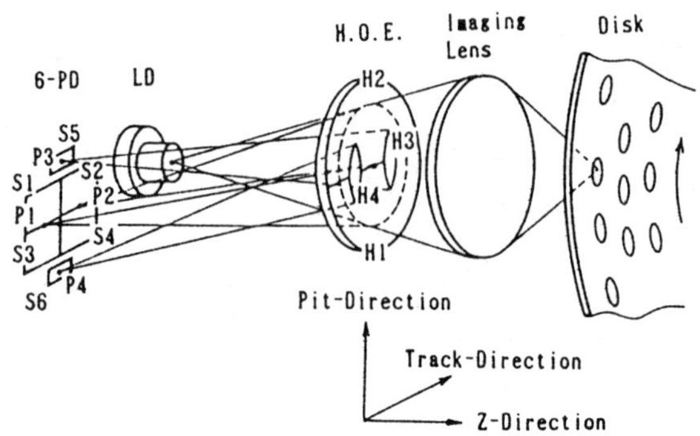

Fig. 8 Pickup with transmission type four-beam splitter HOE[4]

The operation principle is in the following. The laser light transmitted through the HOE is focused on the disk by the imaging lens. The light reflected from the disk is transformed into a spherical wave, converging to the emitting point of the laser diode by the imaging lens, which illuminates the HOE. The light incident on individual holograms, H1~H4, is diffracted so as to focus on P1 to P4, respectively. Focusing error signal and tracking error signal are derived using the principle of double knife-edge method and push-pull method, respectively.

Generally holograms are sensitive to wavelength variation in the light source. In this pickup, lateral aberration, that is, variation in diffraction angle due to the wavelength variation, arises along the lateral dividing line of the photodetector, on which there are two focus spots, P1 and P2, as shown in Fig. 8. Therefore, the focusing error signal is not affected. The longitudinal aberration is removed, since the holograms are generated by using interference between two almost identical divergent spherical waves in order to make the focal power zero.

Holograms are relief type, and the relief profile was optimized. Holograms were fabricated using the IBE(ion beam etching) method. Theoretical efficiency is estimated to be 15 %.

The tracking error signal is derived by the push-pull method. The signal is derived by the difference in incident light power for these tracking holograms. In this configuration, the tracking holograms H3 and H4 shapes are optimized in order to greatly reduce the tracking signal off-set. Fig.9 shows the relation between the axis for imaging optics and error detection optics. The tracking holograms H3 and H4 are arranged in the right and left dotted parts, respectively, in order that only the reflected beam in each dotted part illuminates H3 and H4, in spite of any tracking actuator movement, as shown in Figs. 9(a) and (c). It was experimentally confirmed that this type pickup works well.

Recently this group tried to apply the similar HOE made of birefringent material($LiNbO_3$ crystal) for a pickup of magneto-optical disks.[5]

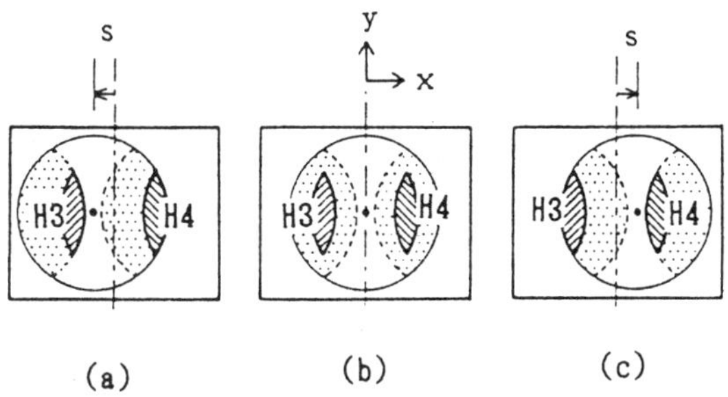

Fig. 9 Explanation of hologram functions[4]
(Relation between the axis for imaging optics and error detection optics, when the actuator is: (a)moved to the left, (b)at the neutral position, and (c)moved to the right.)

4.2. Pickup with reflection HOE

Tatsumi's group[6] (Mitsubishi Electric Corporation) has developed a pickup which has a reflection type grating lens (RGL). The lens has four functions: beam splitting, focusing tracking error detection and mirror.

Fig. 10 shows a schematic view of the optical head. The head is composed of the LD, the PD, a single aspherical lens and the RGL. The actuator drives only the single aspherical lens. With this arrangement, the beam emitted from the LD is incident upon the RGL, and then the zero-order diffraction beam is focused onto the disc by the single aspherical lens. The beam reflected by the disc is again converged by the single aspherical lens, and thereafter is incident upon the RGL. The first-order diffraction beam is converged onto the QD by the RGL. As shown in Fig.6, the optical axis of the diffraction beam is not coincident with the optical path of the beam from the LD. Therefore, only the beam having the information is detected by the QD. On the contrary, zero-order diffraction beam by the RGL is reflected and converged onto the LD. For focus error detection, the astigmatic method is used. The tracking error signal may be easily obtained by employing a push-pull method.

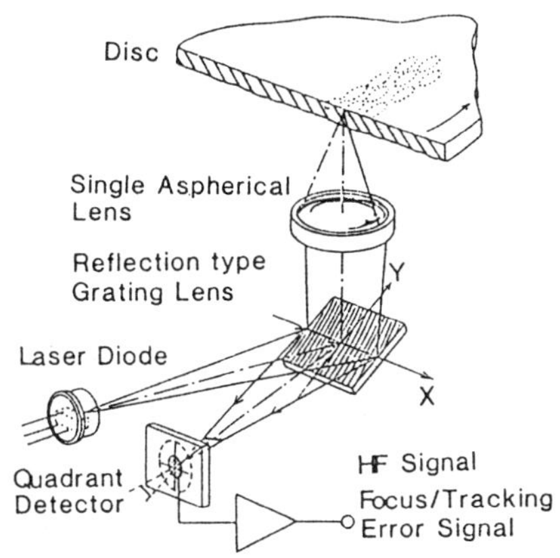

Fig. 10 Pickup with reflection HOE (grating lens)[6]

The design parameters are as follows: the distance between the laser diode and the RGL is 16 mm, the basic distance between the RGL and the QD is 20 mm, and the first-order diffraction angle is 17.5 degrees. The wavelength of the LD is 780 nm.

In order to make clear the performance of this optical head, an experiment was made. The eye pattern of read-out signals is shown in Fig. 11. The standard deviation of the jitter for the 3T signal is 22 ns. The tracking range was approximately 0.2 mm, as using the push-pull method for the tracking error signal detection. It was confirmed that the read-out signal quality was comparable to usual ones.

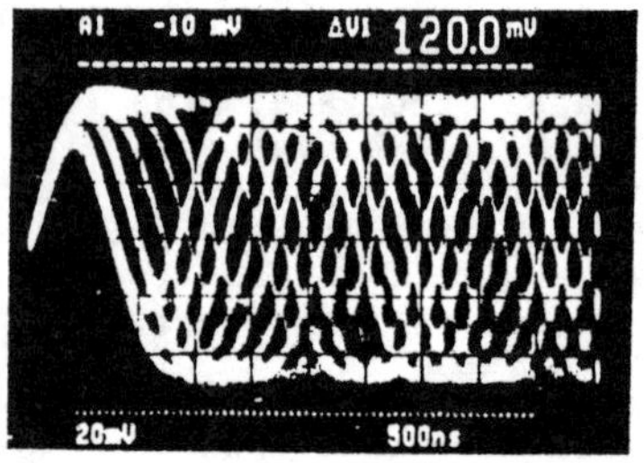

Fig. 11 Eye pattern of readout signal from CD[6]

4.3. Pickup with transmission HOE (two-beam splitter)

Kurata's group (Sharp Corporation) and Coops' group (Philips Netherland) have developed another type of HOE pickup.[7] The HOE is mounted on the top of the package with a laser and a photodiode. The servo signal offset, due to the wavelength variation of the laser, has been compensated by optimization of the geometry of the PD and by highly accurate mounting technique of the two components. The number of components of the pickup is reduced from 6 to 3. As a result, the size and weight are reduced, and the environmental stability is improved. At the same time the high quality performance of the pickup has been successfully achieved.

The configuration is shown in Fig. 12. Two functions of HOE are a beam splitter and and a wedge-prism. The HOE deflects the beam which has been reflected and modulated by the disc in the direction of the PD. Then, HOE splits the reflected beam into two sub-beams to generate servo signals.

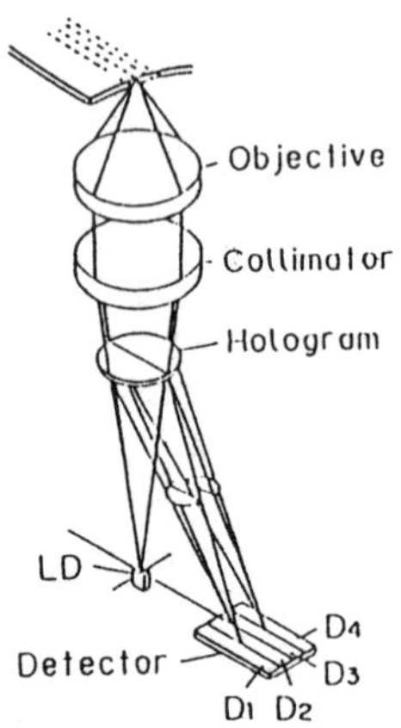

Fig. 12 Pickup with HOE[7]

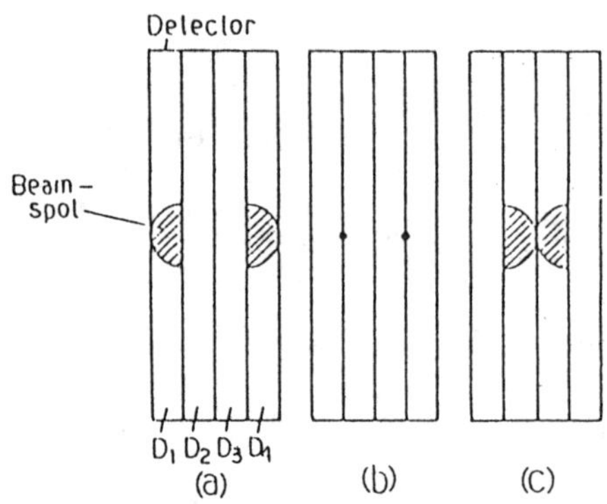

Fig. 13 Spot variation on four-segment PD[7]

Some portion of light emitted from a LD is transmitted through a HOE as the zero-order diffracted beam. Using a collimator, the diverging beam is collimated and is directed to an objective lens which focuses the beam into a small spot of light on the surface of the information layer of the disc. The reflected and diffracted light from the disc is converged by the objective lens and the collimator lens and is incident on the HOE. HOE is divided into two grating sections by a division line oriented along a tangential direction of the disc. The beam is split into two sub-beams and diffracted to the four-segment PD direction. The formation of the light spots on the PD is shown in Fig. 13.

To obtain a maximum read-out signal, the depth of the groove must be optimized. The groove depth was fixed at 425 nm. The total efficiency was 8 % with 45 % transmission and 18 % first order diffraction efficiencies.

In order to evaluate the HOE-LD-PD unit with respect to servo signals an evaluation and test equipment was constructed. The unit was also installed in a CD player in order to test its actual performance in the application. The typical FES signal and RF signal are evaluated. The standard deviation of the jitter for an I3-signal is less than 20 ns. Temperature characteristics of the read-out signal fluctuation is shown in Fig. 14. No degradation of the readout signal is observed, this fact means that the wavelength shift of the LD due to the temperature variation is completely compensated by the photodiode geometry.

Kato's group in Matsushita is also interested in HOE-contained pickups. They tried to sophisticate the shape of segmented photodiodes.[8]

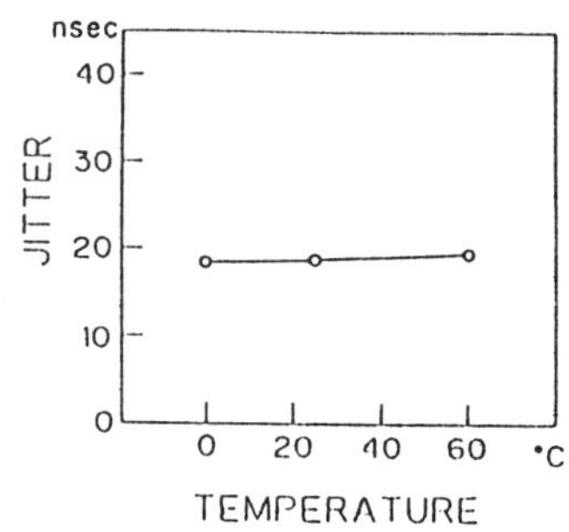

Fig. 14 Temperature dependency of RF signal fluctuation[7]

4.4. Subjects of HOE-contained pickups

1. Alignment of HOE in pickups is very critical just as same as conventional lenses. In another word, alignment redundancy, that is one of hologram features, is not utilized.
2. There are several proposals of holographic lenses including micro Fresnel lenses.[9] These holographic lenses could also be used in pickups, resulting in less weight, although the lenses must be designed to be insensitive to wavelength variation.

5. WAVEGUIDE INTEGRATION

If a pickup head is constructed with integrated optics using a thin-film waveguide, the planar process can be employed for the fabrication. The use of holographic components in the integrated device is very advantageous because complex functions can be realized with minimized component number. Such implementation of disc pickup would lead to great improvements in producibility, reduction in size, weight and cost, and enhancement of applications. This version was proposed by the author's group[10] in Osaka University.

The author's group are conducting a research project on the integrated-optic implementation of optical disk pickup, based on the works on integrated optics and grating (holographic)theoretical and experimental works, since they proposed and demonstrated a specific implementation of integrated-optic disc pickup device(IODPU) for the first time. Although the performances obtained so far are not sufficient for practical application, the elementary functions have been successfully demonstrated.

5.1 Waveguide type pickup

Fig. 15 shows the schematic illustration of a monolithic integrated-optics disc pickup for read-only disc memory. This device is capable of data readout and focusing/tracking error detection. The error signals are required to drive actuators for the automatic servo operation. The focusing/tracking error detections based upon the Foucault/push-pull methods, respectively, which are modified for implementation by grating components in waveguide.

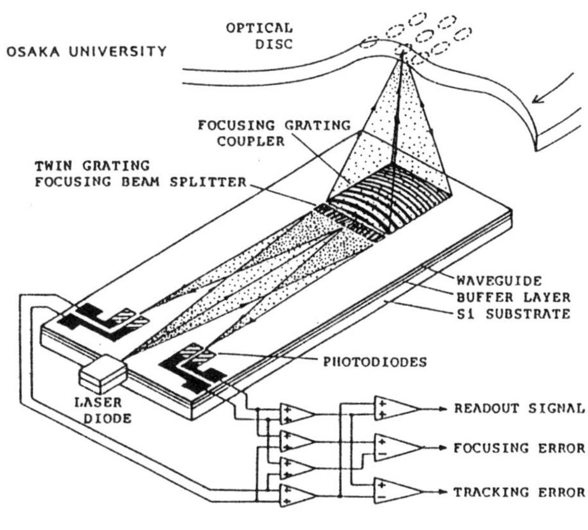

Fig. 15 Waveguide type (integrated optic) pickup[10]

The device is constructed by integrating a focusing grating coupler(FGC), a twin grating focusing beam splitter (TGFBS) and photodiodes in a film waveguide on a Si substrate. The guided wave diverging from a butt-coupled laser diode (LD) is focused by the FGC into a point on a disc. The wave reflected by the disc is collected and coupled back into the waveguide by the same FGC. The TGFBS divides the reflected wavefront into halves, deflects(beam splitting), and simultaneously focuses into two detecting points at a distance slightly shorter than that of the input point(LD). It should be noted that the combined function of the TGFBS leads to the minimization of component number, which is very important in integrated optics devices. At the detecting points, the guided waves are transferred to four PD elements fabricated on the surface of the Si substrate. Thus the disc data is read out and the focusing/tracking error signals are also obtained by simple processing of the photocurrents.

The grating pattern is calculated from mathematical formulae describing the phase difference between input and output wavefronts, and is then written by an electron beam scanning. It is thus possible to design and fabricate a FGC and a TGFBS which are free of aberration at an arbitrary wavelength. Although the FGC has submicron period and the EB writing involves the drawback of rather low throughput, the holographic contact printing technique can be employed for efficient mass production.

5.2 Waveguide type pickup for MO discs

A series of modifications of IODPU has also been proposed as magneto-optic disc pickup,[11] which is shown in Fig. 16.

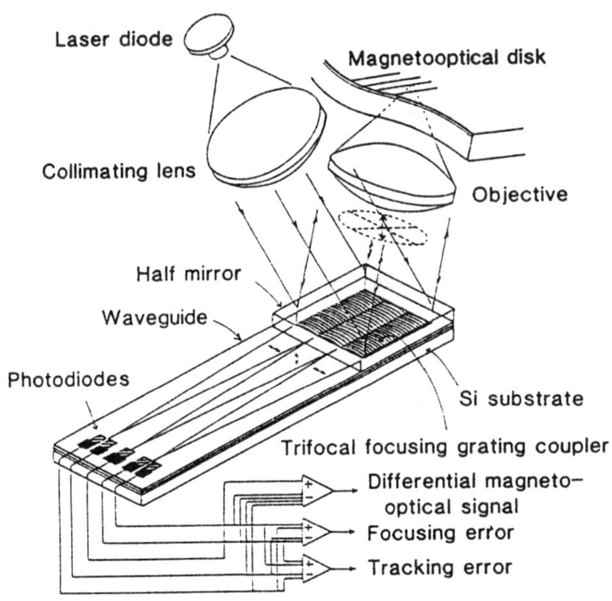

Fig.16 Magnetooptic disc pickup with polarization detection[11]

The pickup consists of a wageguide device, external lenses, a beam splitter and a laser diode (LD). The detection optics is constructed by integrating a TFGC, which consists of three segments, and five PD elements in a waveguide on Si substrate. In the waveguide, TE and TM modes can propagate and the propagation constants are slightly different from each other. The TFGC is designed so that, for an incidence angle, the central segment and the side segments excite TM and TE waves, respectively; the grating period is different for central and side segments.

The differential detection of small polarization rotation of +-0.4 degrees was first demonstrated in a proto-type device using a #7059-Pyrex glass waveguide/SiO2/Si-substrate.

5.3. Subjects of waveguide type pickups

Although the performances obtained so far are not sufficient for practical application, the elementary functions have been successfully demonstrated in the simulation experiments. Most of the results are encouraging, but there are still some problems to solved before the practical applications.

The most important subjects for the future work to achieve the operational performances in the proposed pickup devices are as follows:
1) assuring the stable single-mode lasing of LD under the working conditions, and overcoming the sample-to-sample wavelength deviation of LD,
2) attaining the FGC pattern fabrication accuracy required to obtain the diffraction-limited focus spot size.
These problems will be solved by the optimum design based on the read-out response analysis, as well as design improvements for waveguide scattering reduction and FGC efficiency enhancement, and an improved EB writing system that has been installed recently. The results will be reported in the near future.

6. CONCLUSIONS

Recent studies of miniaturization of optical disk pickups in Japan are reviewed. Pickups are "fruits" which are produced by combined technology of electronics, optics and mechanics.
Miniaturization is being carried on from the following points of view:
1) Simplification of configuration
2) Reduction of the number of components
3) High-grade optimized design (less redundancy)
4) Development of new functional components such as micro Fresnel lens and grating lens
5) Optical integration on waveguides
6) Configuration suitable for mass production
7) Lower cost

After all, pickups are required to satisfy the following functions: fast accessibility, high-density recordability, compatibility for various systems such as ROM, write once and erasable systems.

It is difficult to judge which direction is the best one for miniaturization. Various new types will be proposed, and pickups will be made rapid progress in next 10 years to compete magnetic storage systems.

7. REFERENCES

1. K. Itao and S. Hara, "High Performance Optical Disk Storage System," Proc. Int. Symp. on Optical Memory (ISOM), 1987, pp.177-182
2. Catalogues from Mitsubishi Electric Corporation
3. H. Ukita, Y. Isomura and H. Tanaka, "An Optically Switched Laser(OSL) Head for Optical Recording," Proc. SPIE, vol. 1139, 20, 1989.
4. Y. Kimura, S. Sugama and Y. Ono, "High Performance Optical Head using Optimized Holographic Optical Element," ibid,1987, pp.131-134.
5. A. Ohba, Y. Kimura, S. Sugama, Y. Urino and Y. Ono, "Holographic Optical Element with Analyzer Function for Magneto-Optical Disk Head," Proc. ISOM, 1989 (Kobe), pp. 359-361.
6. K. Tatsumi, T. Matsushita and S. Ito, "A Multi-functional Reflection Type Grating Lens for the CD Optical Head," ibid, 1987, pp. 127-130.
7. Y. Kurata, H. Yamaoka and T. Ishikawa, "CD Optical Pickup using a Computer Generated Holographic Optical Element," SPIE vol.1139, Optical Storage and Scanning Technology, 1989, pp.161-168.
8. S. Kadowaki, Y. Komma, Y. Hori and M. Kato, " Integrated Holographic Optical Pickup Head with Wide Allowance Range," ISOM '87, 27D-13, Tokyo.
9. H. Nishihara and T. Suhara, "Micro Fresnel Lenses," Progress in Optics vol. 24, pp. 3-37, 1987.
10. T. Suhara and H. Nishihara, "Integrated-Optic Disc Pick Devices Using Waveguide Holographic Components," Proc. SPIE, vol. 1136, 1989, pp.92-99.
11. S. Ura, H. Sunagawa, T. Suhara and H. Nishihara, "Focusing Grating Couplers for Polarization Detection," IEEE, J. of Lightwave Tech. vol. 6, 6, pp. 1028-1033, 1988.

Section Seven
Servos—Focus and Tracking

Reprinted with permission from *IEEE Transactions on Consumer Electronics*, pp. 304-308 (November 1976).

OPTICAL READOUT OF VIDEODISC

C. Bricot, J. C. Lehureau and C. Puech
presented by F. Le Carvennec
Thomson-CSF
Laboratoire Central de Recherches
B.P. 10 - 91401, Orsay, France

- INTRODUCTION -

At first sight optical readout of videosignal stored on a disc seems to be a difficult and critical task due to what is paradoxically one of its most interesting features : no mechanical contact between the readout "stylus" and the information carrying surface of the disc. Since the readout "stylus" used here is a light beam, it cannot be simply mechanically guided by a groove, the way it is accomplished, for example with conventional audio records.

This paper intends to demonstrate that the servoing of the optical readout stylus on the track is not a critical process. Various simple, reliable, low cost solutions are possible and give the opportunity to take benefit of the main features of optical readout : no wear, possibility of tricks.

Work done at Thomson concerning optical readout has dealt mostly with simplification of solutions without being prejudicial to performances. New developments concerning this topics will be emphasized. Results obtained from prototypes built according to some of these solutions will be illustrated by some data concerning picture quality.

II - GENERALITIES ON OPTICAL READOUT - SOME ORDERS OF MAGNITUDE -

Technical problems encountered in the realization of a videodisc storage system are mostly due to the necesseraly high density of information. In order to store a 30 minutes TV program on a 30cm diameter disc, an area of less than $1mm^2$ is allowed for each single frame. This information density is much higher than those used at this time in magnetic storage system (by a factor of 15 to 30). Several solutions have been proposed and investigated to solve the problems of writing and reading this information. At Thomson we chose the solution where the videosignal is stored in the form of "binary" microinformations arranged along à spiral (with a typical pitch of 1.6/um) (figure 1). The coding system is a frequency modulated signal leading to a modulation of the duty factor of the "binary" microinformations (modulation of the zero crossings of the electrical signal).

Optical readout of those microinformations is accomplished by means of a focused laser beam (of the order of 1/um). Transmitted or reflected light is modulated by these microinformations and detected with photocells. The original electrical signal is then restored (figure 2, 3).

Practically several ways of storage can be used : phase relief modulation, absorption modulation, reflection modulation. One interesting advantage of phase relief modulation is that it is well adapted to known replication processes (figure 4). It renders possible replication by embossing a thermoplastic material with a metallic stamper obtained from the master. This process is similar to the one used in audio domain.

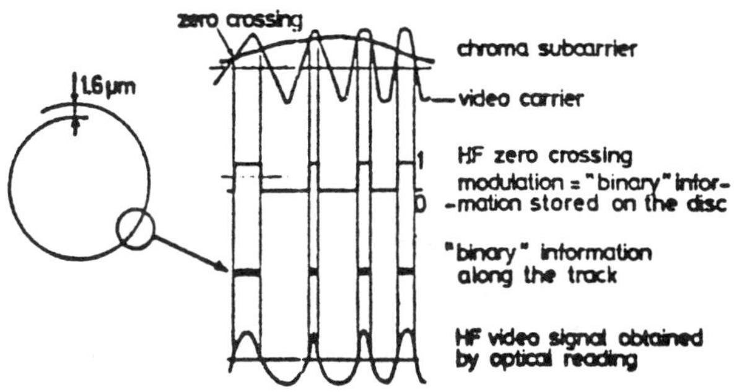

- Figure 1 -

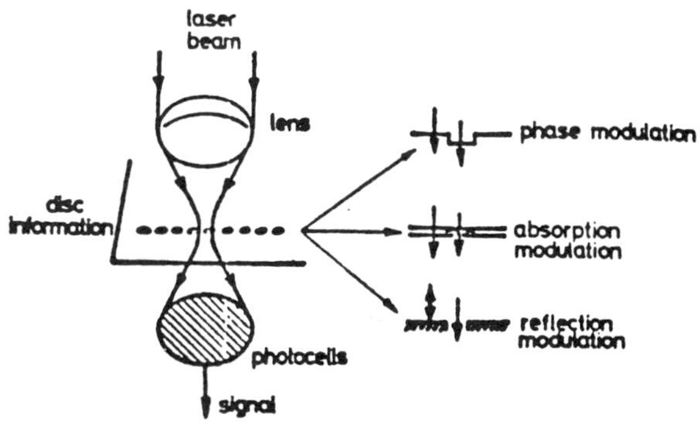

TRANSMISSIVE READOUT

- Figure 2 -

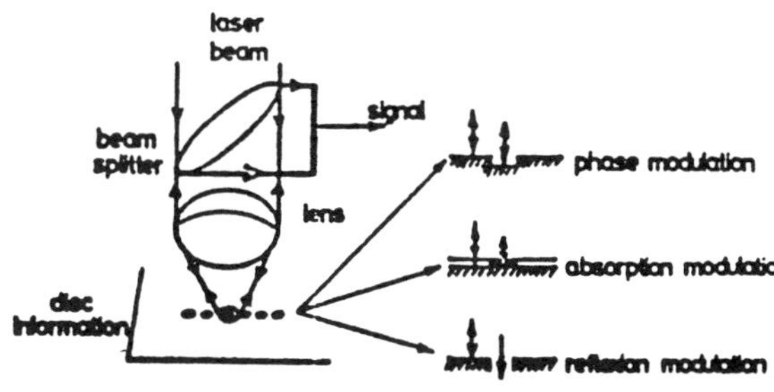

REFLECTIVE READOUT

- Figure 3 -

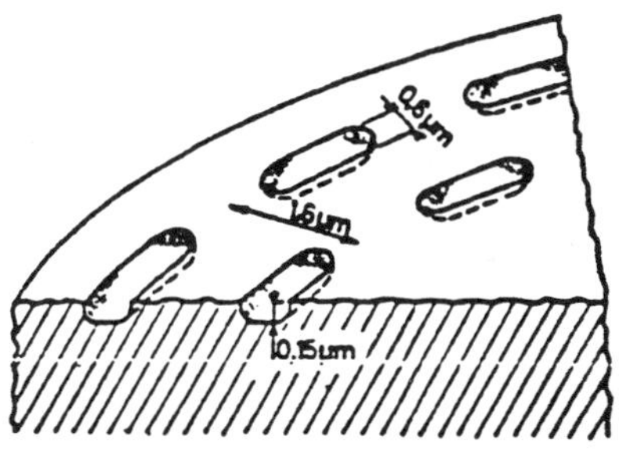

CUT VIEW OF A THOMSON VIDEODISC WITH RELIEF MODULATION

- Figure 4 -

Several demonstrations have been made by Thomson with laboratory prototypes utilizing the type of modulation described above (Cannes 1974 - Montreux 1975). Discs used were transparent flexible embossed discs. Since those demonstrations, studies have dealt with :

1 - Simplification of technology, improvement of performance, quality, and reliability.

2 - Adaptation and optimization of technical options in relation with various fields of application : consumer market, institutional market and professional market.

This paper gives results concerning work being done in the field of optical readout.

The highest temporal frequency to be stored depends on the chosen coding scheme. This is typically 8 to 9 MHz for European TV standards for additive coding. In the case of a synchronous disc (1500 rpm) this leads to a maximum spatial frequency of the order of 1400 cycles/mm on the inner track at radius 7cm. Such spatial frequencies can be written and read by optical means with a good signal to noise ratio when high numerical aperture optics is used (typically N.A. = 0.4 for readout). Only small depth of focus is attainable with high N.A optics. Therefore a precise positionning of the information with respect to the focal point of the light beam is necessary. This light beam acts as a "reading stylus" without any mechanical contact (figure 5), and its position has to be controlled very accurately in order to obtain a good readout (figure 6) :

$\pm$ 2 μm in the vertical direction

$\pm$ 0.1 μm in the radial direction

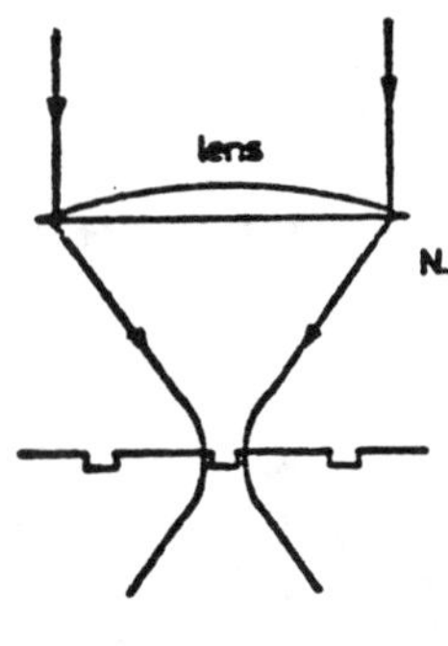

OPTICAL STYLUS

- Figure 5 -

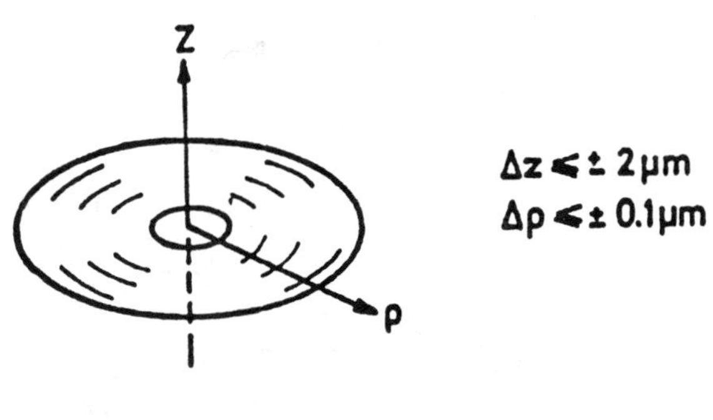

OPTICAL READOUT PRECISION

- Figure 6 -

As mentioned above, guiding of the optical beam with such a high accuracy may appear at first thought very delicate. However several solutions have been studied and experimented at Thomson and we intend to describe the most promising ones. Analysis of performances, simplicity, reliability of those solution shows that optical readout is well adapted to the various fields of videodisc applications. Among other advantages one can see the following ones :

- no wear of disc, because of no mechanical contact between the reading stylus and the disc.
- possibility of tricks such as stop frame, slow and fast forward and backward motion, due to the fact that it is possible to deflect quickly a light beam.

III - VERTICAL TRACKING

Two kinds of solutions appear to be suitable :

- Aerodynamic stabilization.
- Vertical servoing of the readout objective.

III.1 - Aerodynamic stabilization

Aerodynamic stabilization is well adapted to flexible discs. Two types of systems, passive and active stabilizers have been considered. Two solutions have been particularly studied.

-a- U-Shaped stabilizer (figure 7).

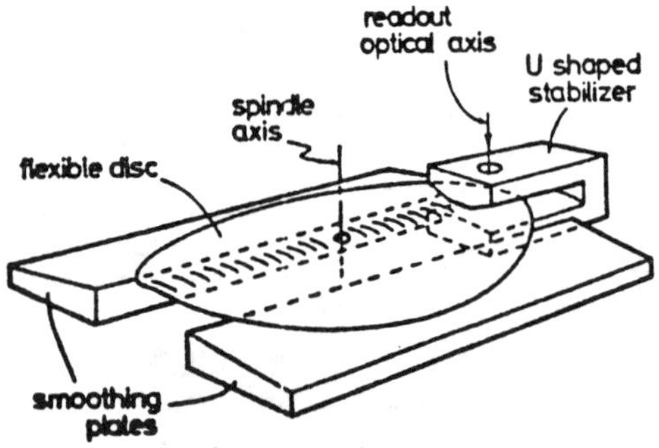

U SHAPE STABILIZER

- Figure 7 -

The first one to be used, consists of a U-Shaped guide in which the rotating disc is introduced. When the disc is rotating, air is forced between the upper and lower parts of the guide, creating a combination of aerodynamic forces. This rather simple and efficient solution has already been described and demonstrated several times. Its major drawback is inherent to the fact that it is necessary to use a narrow gap between the two guides to obtain the desired stabilization for all discs (typically 50 microns in addition to disc thickness). In dusty environment this may cause severe damages of disc surface when large dust particles are passing through the gap. For this reason the U-Shaped stabilizer has been given up and the following solutions have been proposed to overcome this drawback.

-b- Stabilization by Bernouilli effect (figure 8)

In place of a passive aerodynamic stabilizer an active aerodynamic one is being used. A properly distributed air flow is blown on the rotating disc through a nozzle placed under the readout objective. According to Bernouilli effect, the laminar flow created between the nozzle and the disc provides a local stabilization of the disc surface. This type of stabilization is more efficient than the one obtained with the U-Shaped stabilizer. Furthermore dust particles, which would otherwise pass under the readout lens and affect the quality of the videosignal, are blown away due to the air flow. This solution however demands the realization of a compact air pump.

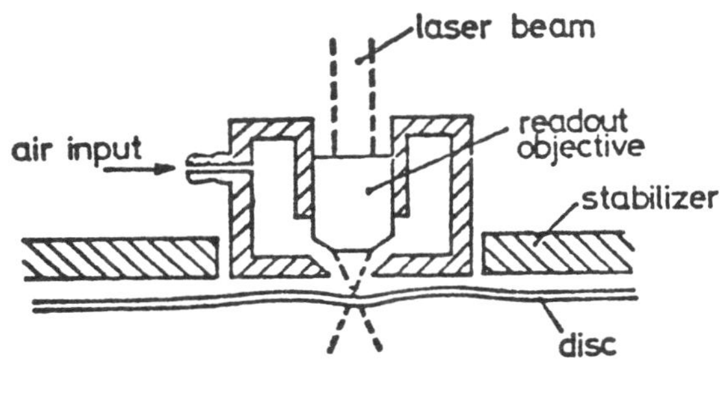

"BERNOUILLI" STABILIZER

- Figure 8 -

III.2 - Vertical servoing of the readout objective.

An other approach to vertical tracking consists in the servoing of the readout objective. The displacement of the objective can be simply obtained by the use of an electrodynamic coil similar to the ones utilized in loudspeakers (figure 9).

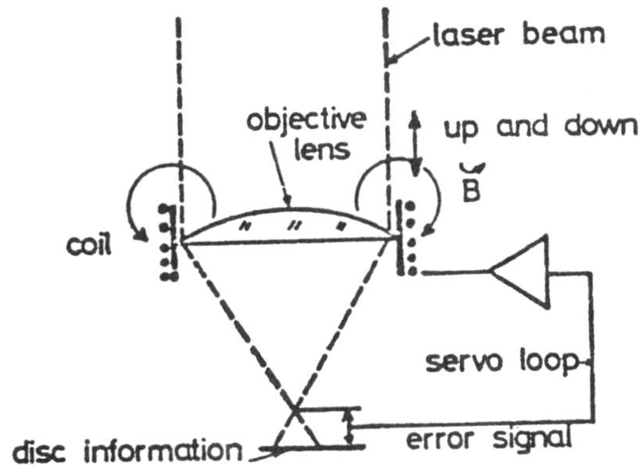

VERTICAL FOCUS SERVOING BY A COIL

- Figure 9 -

The driving signal of the servo loop can be obtained by several ways. Work has been emphasized, on the use of simple optical sensors. The most interesting ones appear to be those of the "single light spot" type (no need of extra light beam in addition to the readout beam) using only few optical and electronic components. Two sensors of this kind adapted to transparent or reflective discs are particularly suitable. These ones use energy reflected by the information carrying surface which is always sufficient to give a good signal to noise ratio for the error signal.

- Astigmatic sensor (figure 10a, b).

The error signal is derived by means of a system including a cylindrical lens (placed on the beam reflected by the disc) and 4 photocells (connected by pairs, oriented at 45° with respect to the axis of the lens and placed in the plane of least diffusion Po on the figure). When the disc surface is in focus, the light spot centered on the 4 photocells is circular, and the electrical signal obtained from the differential amplifier is equal to zero. When the disc surface is out of focus, the light spot becomes elliptical. Depending on whether the disc is positionned too low or too high, the long axis of the ellipse is in one direction or at 90°. The signal obtained from the differential amplifier is significant of the sign and of the amplitude of defocusing. The dynamic range depends on the focal length of the cylindrical lens. It can be adjusted at a typical value of ± 50 microns (figure 11).

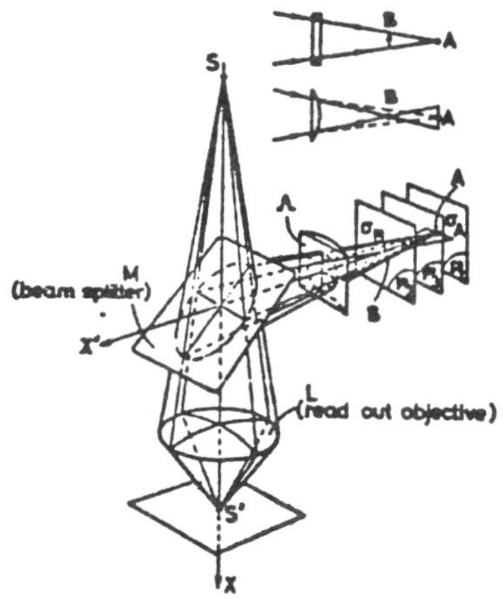

ASTIGMATIC SENSOR
(system schematic)

- Figure 10.a -

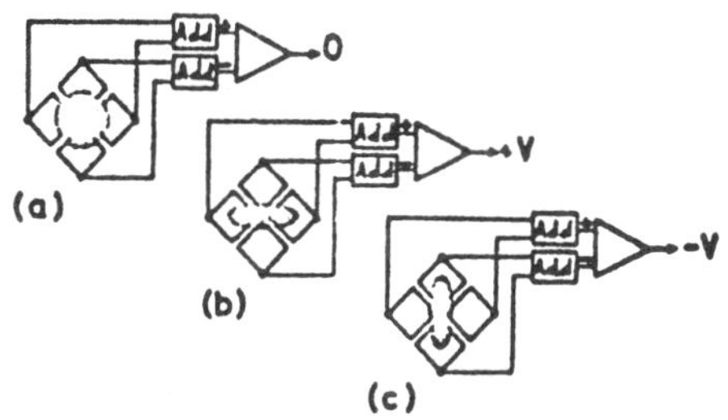

(a) disc in focus
(b) disc out of focus - too far -
(c) disc out of focus - too close -

ASTIGMATIC SENSOR
TYPICAL PHOTOCELLS ILLUMINATION

- Figure 10.b -

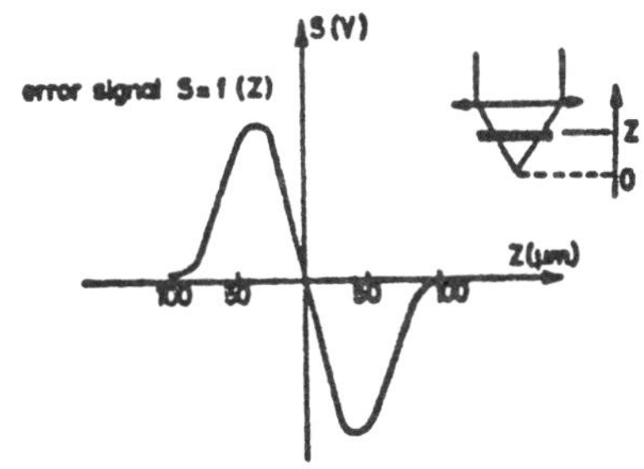

ASTIGMATIC SENSOR

- Figure 11 -

- Asymmetric sensor (figure 12)

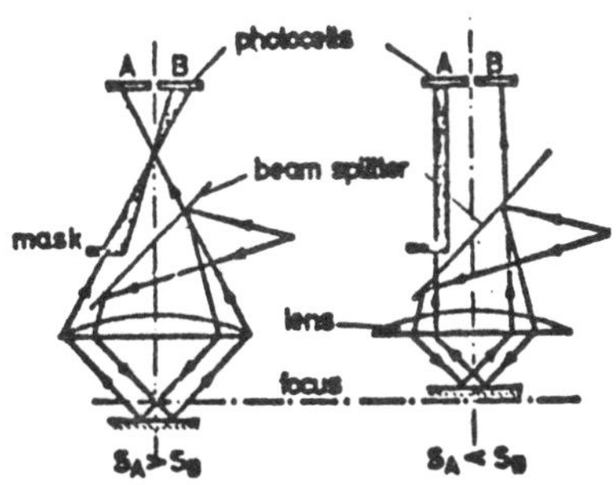

ASYMMETRIC SENSOR

- Figure 12 -

This new solution is even more simple than the preceeding one. The only additional part consists of a mask (screen) placed on the reflected beam. Two photocells connected to a differential amplifier provide the error signal. When the disc is out of focus, depending on whether it is too low or too high, the mask projects a shadow on one or the other of the two photocells. Again the signal obtained from the differential amplifier provides us with desired informations concerning the sign and the amplitude of defocusing (figure 13).

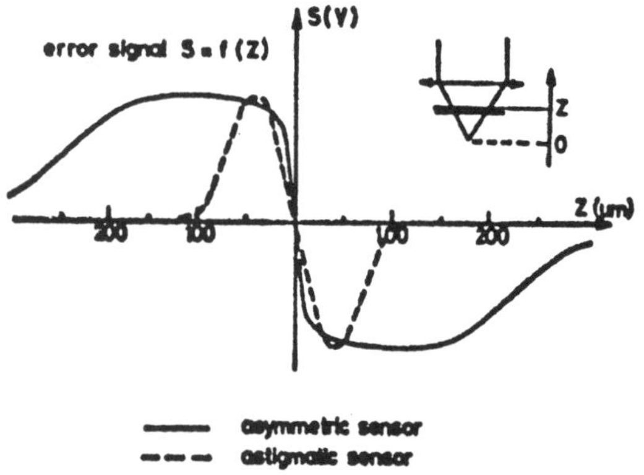

- Figure 13 -

This simple system needs few elements and is easy to adjust. Moreover due to its very large dynamic range (typical value : ± 200 um), it renders obsolete the logical sequence, otherwhise necessary to find the right focus.

IV - RADIAL TRACKING -

The radial tracking can be accomplished by deflection of the reading light beam driven by a significant error signal (figure 14).

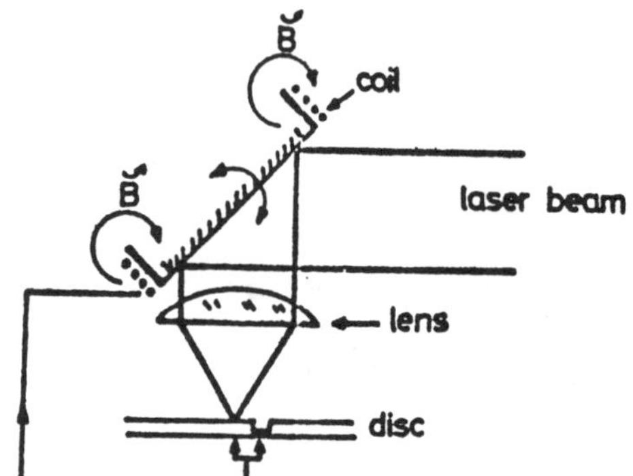

- Figure 14 -

Deflection of the light beam can be obtained by an electrodynamically actuated tilting mirror. The error signal is derived in a way similar to the one used to get the H.F. signal. For radial tracking work has been again emphasized on "single light spot" systems since they avoid the complexity inherent to the use of additional light spots. The choice of the system depends on the type of modulation used on the disc.

1)- In the case of "phase" discs (one particular type of discs with relief modulation) it is possible to make use of the asymmetry in the light beam intensity distribution which occurs when the light spot is not well centered on the track. This asymmetry is detected by two photocells (figure 15.a, b). This process, described on many occasions, is very simple. However it can not be used to track "amplitude" modulated discs.

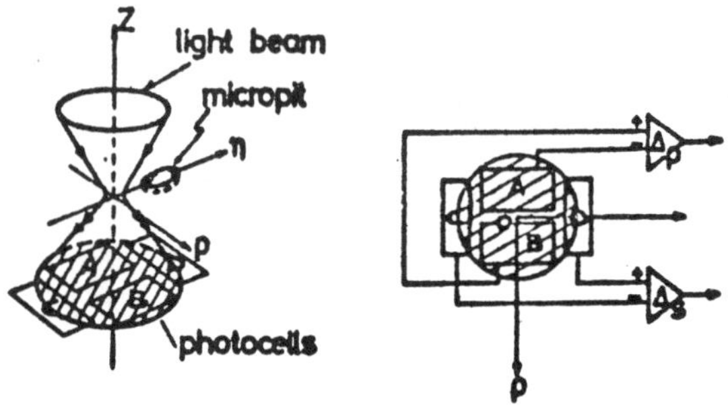

LIGHT SPOT ON TRACK AXIS

- Figure 15.a -

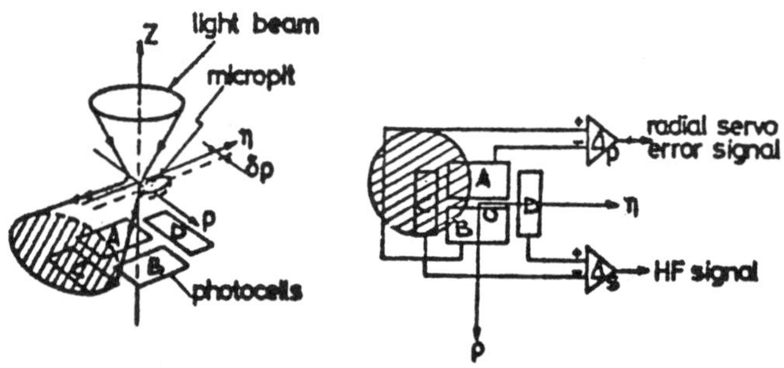

LIGHT SPOT OUT OF TRACK AXIS

- Figure 15.b -

2)- An other simple system (figure 16), well adapted to "amplitude" discs, detects the error of track position at the focal point of the reflected beam. When the track is not well centered there is an asymmetric light distribution detected by two photocells.

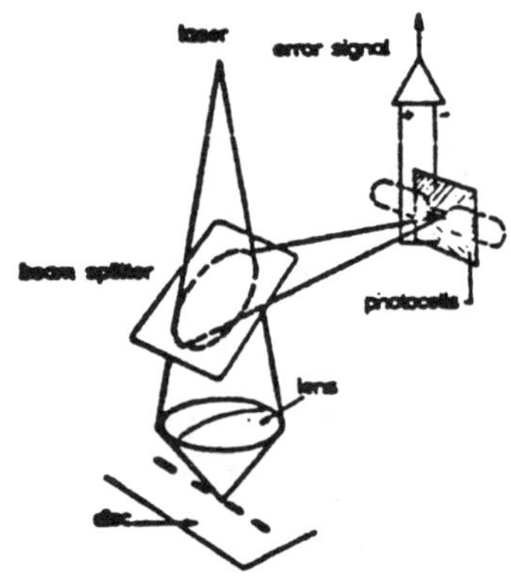

- Figure 16 -

3)- A very interesting solution (figure 17) well adapted for the two types of discs (amplitude and phase) and the two modes of readout (transmission and reflection) can also be used. The readout light spot is wobbled in the radial direction (at a typical frequency of 60KHz) inducing an amplitude modulation of the intensity of the transmitted or reflected light beam. This modulation may be heterodyned by the wobbling signal, providing the radial error signal. One can achieve simply light spot wobble of small amplitude, by actuating an additional mirror, place on the light path, with a piezoelectric element. Excellent results are obtained with such a system and makes it an interesting choice.

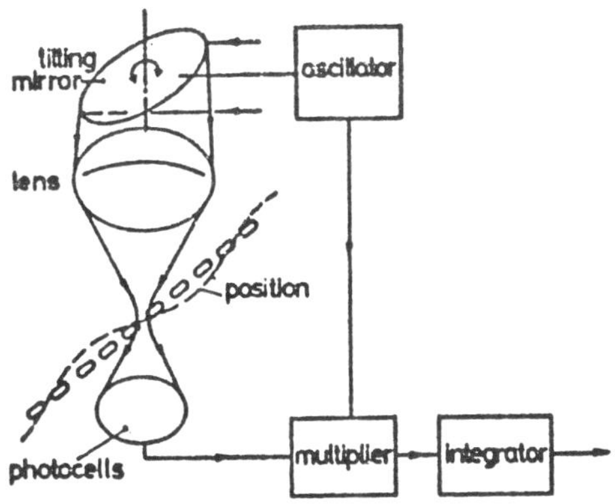

SPOTWOBBLE TRACKING

- Figure 17 -

CARRIER $\frac{\text{signal}}{\text{noise}}$ (30 KHz bandwith)		62 dB
INTERMODULATION $\left(\frac{\omega_0 \pm \omega_1}{\omega_0}\right)$		50 dB
LUMINANCE $\frac{\text{signal}}{\text{noise}}$ (3 MHz bandwith)		45 dB
$\frac{\text{signal}}{\text{moire}}$ (luminance)		42 dB

TYPICAL RESULTS

- Figure 19 -

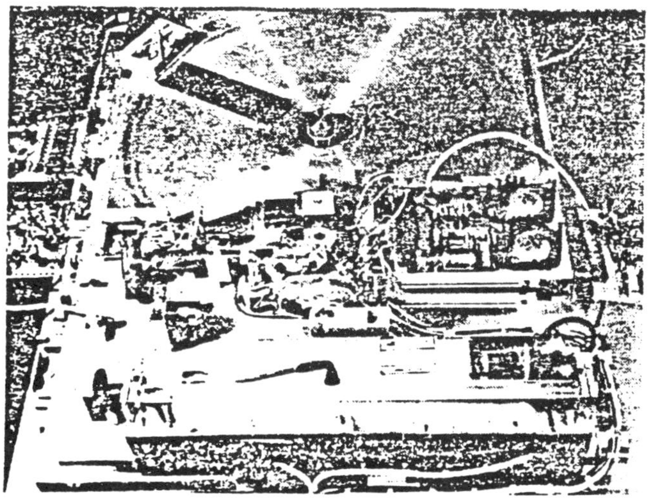

- Figure 18 -

V - MAIN FEATURES OF THOMSON PLAYERS - CONCLUSION -

Several prototypes have been built according to the various solutions described above (in transmissive mode as well as in reflective mode). It has been demonstrated that optical solutions are well qualified for various fields of applications (consumer market, institutional market and professionnal market). Choice of right solutions depends on the type of microelements used to store information ("phase" discs or "amplitude" discs) and on foreseen applications.

At the present time, Thomson, thinks that system utilizing transparent flexible discs (in jackets or cassettes) are well suited for mass production markets. They lead to simple players and low cost discs.

One of the prototype is shown on figure 18. Image quality obtained, for example, on a player using transmissive readout mode with flexible embossed discs is characteristized by a signal to noise ratio in the luminance (3 MHz bandwidth), and by an intermodulation, equal or better than 45 dB (figure 19).

Reprinted with permission from *Applied Optics*, Vol. 17(13), pp. 2013-2021 (July 1, 1978).

Position sensing in video disk readout

J. J. M. Braat and G. Bouwhuis

In video disk systems based on contactless optical readout, the position of the readout light spot is dynamically controlled with the aid of servo systems. The input of the radial or vertical servo systems in the readout unit is fed with a bipolar function of the position error that has to be corrected. This paper describes a method that generates a radial and a vertical error signal by making use of the phase relations between diffracted orders in the far field of the video disk. These phase relations are derived from the rf signals that appear at the different outputs of a composite photodetector. It is shown that in this way auxiliary beams or supplementary optical elements can be avoided in the readout system.

I. Introduction

The storage and readout of information on disks by optical means are currently the subject of a great deal of interest.[1-5] The information is stored mostly as a sequence of depressions, absorbers, or scatterers that forms a spiral or a set of concentric circles on the disk. Optically, the information structure may be seen as a regular, single-frequency grating in the radial direction. In the tangential direction, each track can be considered as a more complex grating containing a broad spectrum of spatial frequencies, which corresponds to the distribution of the depressions along the track. During readout of the disk the positions of the centers of the depressions and their geometries restore the time-dependent signal that originally has been recorded on the disk.

On optical video disks, the distance between neighboring tracks and between subsequent depressions can be very small (1 μm or 2 μm), and the readout beam ($\lambda = 0.63\ \mu$m) must have a rather high numerical aperture (e.g., 0.40) in order to give sufficient resolution. As a consequence, the focal depth is only a few microns, and the demands on the vertical positioning of the readout light spot are severe. Comparable tolerances hold for the radial positioning of the light spot because cross-talk from a neighboring track has to be kept below a certain level. Servo systems are therefore indispensable.

This paper describes a method that generates error signals for the focusing and the radial tracking servo systems without making use of auxiliary beams or supplementary optical elements.[6]

The authors are with Philips Research Laboratories, Eindhoven, The Netherlands.

Received 21 June 1977.
0003-6935/78/0701-2013$0.50/0.

In Sec. II the diffraction of light by a 1-D grating is treated. Special attention is paid to the phase relations between the diffracted orders as a function of focusing. In Sec. III the generation of the focus error signal is described. In Sec. IV the extension to 2-D gratings is made in order to illustrate the derivation of the radial error signal. In the last section the applicability of the method is discussed.

II. Diffraction by a 1-D Grating

Using the elementary grating law for an incident plane wave, we proceed to the diffraction of a spherical defocused wave. Next, the influence of a grating movement on the phase of the diffracted orders is treated. An expression is derived for the time dependent intensity of the diffracted light in the far field of the grating.

A. Diffraction of a Plane Wave

The diffraction of a plane wave by a grating structure is shown in Fig. 1. The grating consists of a sequence of equidistant rectangular troughs with a width w and a depth d. The period of the grating is denoted by p. After diffraction by this phase structure, the incident plane wave is split into a set of plane waves traveling in directions given by the angle η_J and which are determined by the grating law:

$$\sin\alpha + \sin\eta_J = J(\lambda/p), \qquad (1)$$

where α is the angle of incidence, λ the wavelength of the light, and J the order number of the diffracted wave. In the following, we will confine our attention to the zeroth ($J = 0$) and first diffracted orders ($J = \pm 1$).

The grating law only predicts the direction of propagation of the different orders but not their strength and mutual phase. In a simplified way, these phase relations can be deduced from Fig. 2, where the complex

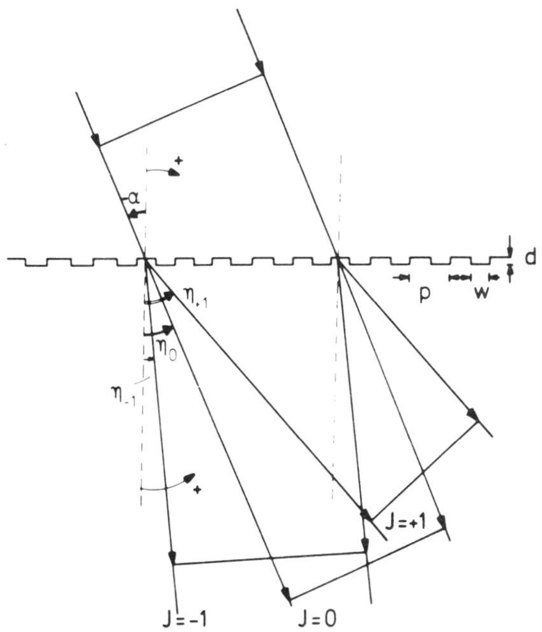

Fig. 1. The diffraction of a plane wave by a grating. The normal to the incident wavefront makes an angle α with the normal to the grating surface. The period of the grating equals p. The width of the grooves is denoted by w. Their depth is equal to d.

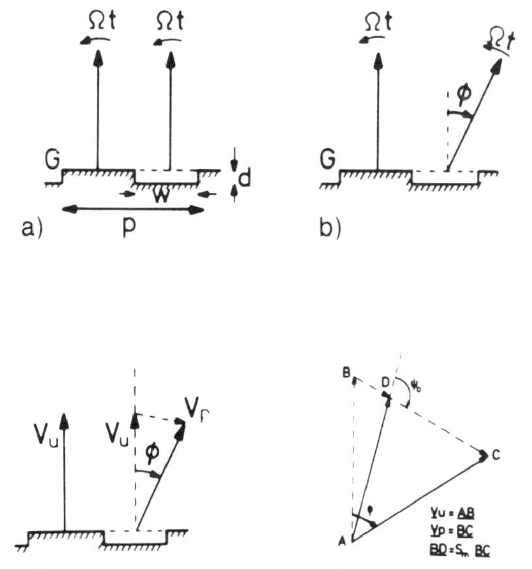

Fig. 2. The phase vectors of the incident and reflected wave. (a) The phase vectors of the incident wave have all equal phase and rotate at the light frequency Ω. (b) After reflection at the grating, the phase vector at the location of a groove has suffered a phase delay of ϕ. (c) The vector rotated by an angle ϕ can be decomposed in an undisturbed vector $\mathbf{V}_u$ and a periodic vector $\mathbf{V}_p$. (d) The zeroth order is represented by the vector **AD;** the first order is proportional to **BC.** The phase difference between the zeroth and first order is given by the angle ψ_0.

amplitudes of the waves are represented by phase vectors. The phase vectors of the incident plane wave are shown in Fig. 2(a). After reflection [Fig. 2(b)] one of the phase vectors has suffered a phase shift ϕ of

$$\phi = -2\pi \cdot [(2d)/\lambda]. \tag{2a}$$

This angle ϕ is called the phase depth of the grating.

The rotated phase vectors can be decomposed into an undisturbed vector $\mathbf{V}_u$ and a new vector $\mathbf{V}_p$ [Fig. 2(c)]. The amplitude A_0 of the zeroth order of the diffracted light is the sum of $\mathbf{V}_u$ and the average value of the periodic amplitude pattern $\mathbf{V}_p$. Denoting the relative modulated area of the grating by S_m the amplitude of the zeroth order equals $\mathbf{V}_u + S_m \mathbf{V}_p$ [Fig. 2(d)]. In case of a 1-D grating, the value of S_m is w/p.

The amplitude of the first diffracted orders $A_{\pm 1}$ is proportional to the vector $\mathbf{V}_p$. The phase difference ψ_0 between the zeroth and first diffracted orders is shown in the vector diagram of Fig. 2(d). From the diagram, the following equation for ψ_0 can be deduced:

$$\tan(\psi_0) = \frac{-\cot(\phi/2)}{1 - 2S_m}. \tag{2b}$$

In the following we consider the amplitude and phase of diffracted orders in three different cases that are of practical interest.

1. *Shallow Phase Grating [Fig. 2(d)]*

If d is much smaller than the wavelength of the light, the angle ϕ remains very small. It can easily be seen that the amplitudes of the zeroth and first diffracted orders equal

$$\left.\begin{aligned} A_0 &= 1 \\ A_{\pm 1} &\propto \frac{w}{p}\phi \end{aligned}\right\} d \ll \lambda; -1 \ll \phi < 0. \tag{3}$$

The phase difference ψ_0 between the zeroth and a first diffracted order is approximately equal to $-\pi/2$.

2. *Deeply Modulated Phase Grating ($\phi \approx -\pi$)*

The amplitudes of the diffracted orders can be written as [Fig. 3(a)]

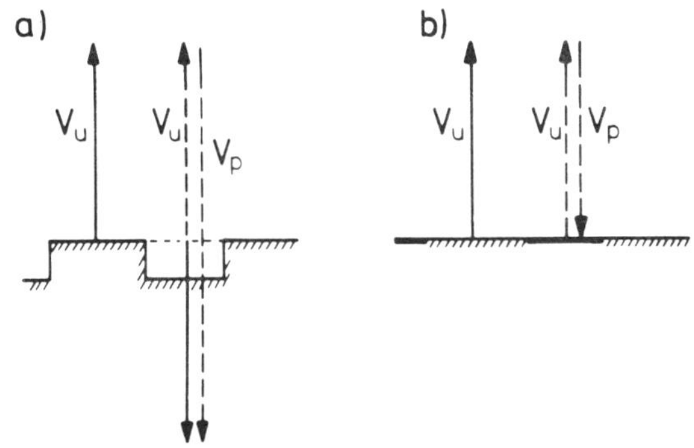

Fig. 3. The phase of the periodic vector $\mathbf{V}_p$ with respect to the undisturbed vector $\mathbf{V}_u$ in case of (a) a deep phase grating and (b) an amplitude grating.

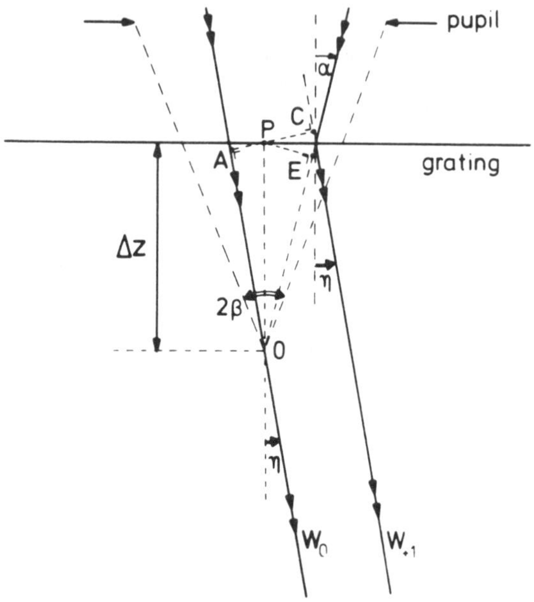

Fig. 4. Diffraction of a spherical wave by a grating. Two diffracted waves, W_0 and W_{+1}, that travel in the direction given by the angle η are shown. W_0 is the zero order of a plane wave with its normal parallel to AO. W_{+1} is the first order of an incident plane wave traveling in the direction OE given by the angle α. Δz is the amount of defocusing of the spherical wave with respect to the grating surface. The numerical aperture of the incident wave is determined by the angle β.

$$\left.\begin{aligned} A_0 &= 1 - 2\frac{w}{p} \\ A_{\pm 1} &= -\frac{2w}{p} \end{aligned}\right\} d = \frac{\lambda}{4};\ \phi = -\pi. \quad (4)$$

The phase difference ψ_0 between the zeroth and a first diffracted order is equal to $-\pi$; when $w/p = 0.5$, the zeroth order disappears.

For phase grating structures with $0 > \phi > -\pi$, the phase angle ψ_0 grows from the initial value $-\pi/2$ to the value $-\pi$.

3. *Amplitude Grating*

In the same way as for a phase grating, the light amplitude immediately after the grating can be decomposed into the sum of a uniform part $\mathbf{V}_u$ and a periodic part $\mathbf{V}_p$. $\mathbf{V}_p$ only exists behind the black sections of the grating and is in phase opposition with $\mathbf{V}_u$. The amplitude of the diffracted orders now becomes [Fig. 3(b)]

$$\left.\begin{aligned} A_0 &= 1 - \frac{w}{p} \\ A_{\pm 1} &= -\frac{w}{p} \end{aligned}\right\} \text{amplitude-structure.} \quad (5)$$

The phase difference ψ_0 in this case equals $-\pi$. With respect to ψ_0, a black-and-white grating behaves in a similar way as a phase grating with a phase depth ϕ equal to $-\pi$.

B. Diffraction of a Spherical Wave by a Grating

Optical video disks are usually read out with the aid of a scanning light spot. The light spot is generated by focusing a laser beam with a sufficiently large numerical aperture on the disk. We consider the phase relations between zeroth and first orders of the incident spherical wave; these relations vary as a function of the amount of defocusing Δz of the spherical wave with respect to the grating surface.

The wave converging to 0 and limited by an aperture can be represented by a set of plane waves, whose angles of incidence vary from $-\beta$ to $+\beta$ (Fig. 4). After diffraction by the grating several plane waves travel in the direction η: the zeroth order of a plane wave W_0 that was incident at an angle η and the first diffracted orders $W_{\pm 1}$ of the waves that were incident under an angle α, satisfying the grating equation

$$\sin\alpha + \sin\eta = \pm\frac{\lambda}{p}. \quad (6)$$

The contribution of W_0 and W_{+1} to the energy diffracted in the direction η depends upon the phase difference between W_0 and W_{+1}, thus on the mutual phase of A and C (Fig. 4). In this section the defocus contribution to the mutual phase is considered only. Consequently the wavefronts PE and PC of the incident wave W_0 and its diffracted wave W_{+1} may be considered to be in phase. Hence, the defocus contribution $\Delta\theta_{+1}$ to the phase difference between A and C equals that between A and E:

$$\Delta\theta_{+1} = \frac{2\pi}{\lambda}\{OA - OE\} = 2\pi\frac{\Delta z}{\lambda}(\cos\eta - \cos\alpha), \quad (7)$$

where $\Delta z = OP$. With the aid of Eq. (1), Eq. (7) can be written as

$$\Delta\theta_{+1} = \frac{2\pi}{\lambda}\cdot\frac{\Delta z}{2}\cdot\frac{\lambda}{p}\left(\frac{\lambda}{p} - 2\eta\right), \quad (8)$$

using the approximations

$$\sin\epsilon = \epsilon - \epsilon^3/6,$$
$$\cos\epsilon = 1 - \epsilon^2/2.$$

In Fig. 5(a), the far-field diffraction pattern of the grating illuminated by the spherical wave has been depicted. The zeroth order is represented by a circle centered in M_0. The midpoints $M_{\pm 1}$ of the first diffracted orders are situated at

$$M_{\pm 1} = \left(\pm\frac{\lambda}{p},0\right). \quad (9)$$

We are interested in the region where a zeroth and a first order overlap. In Fig. 5(b), the phase difference between the zeroth and first diffracted orders in the presence of a certain amount of defocusing Δz has been depicted, as expressed by Eq. (8). The value of $\Delta\theta_{-1}$ is obtained from Eq. (8) by putting $\eta = -\eta$. At exactly halfway the centers M_0 and $M_{\pm 1}$ of the orders, the phase difference $\Delta\theta_{\pm 1}$ is zero and independent of Δr. This can easily be deduced from Fig. 4, where η and α then have the same value.

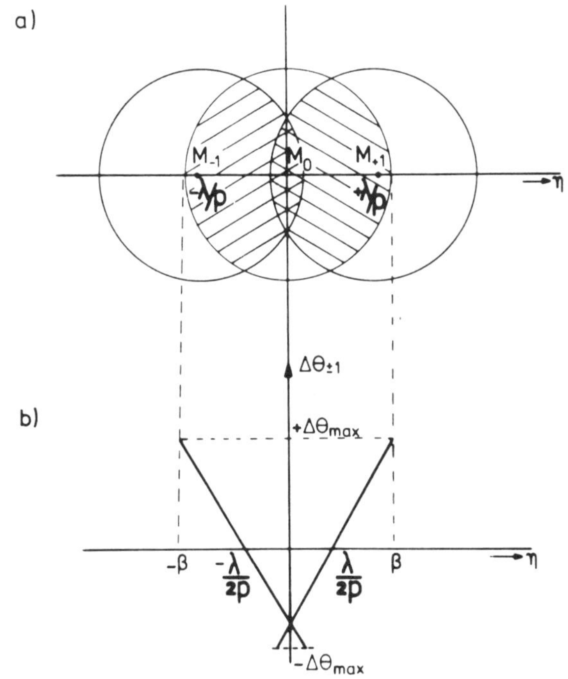

Fig. 5. (a) The far-field diffraction pattern of the grating when illuminated by a spherical wave. In the hatched regions the zeroth order and first diffracted orders interfere. (b) The phase difference between the zeroth and first diffracted orders in the regions of overlap in the presence of defocusing.

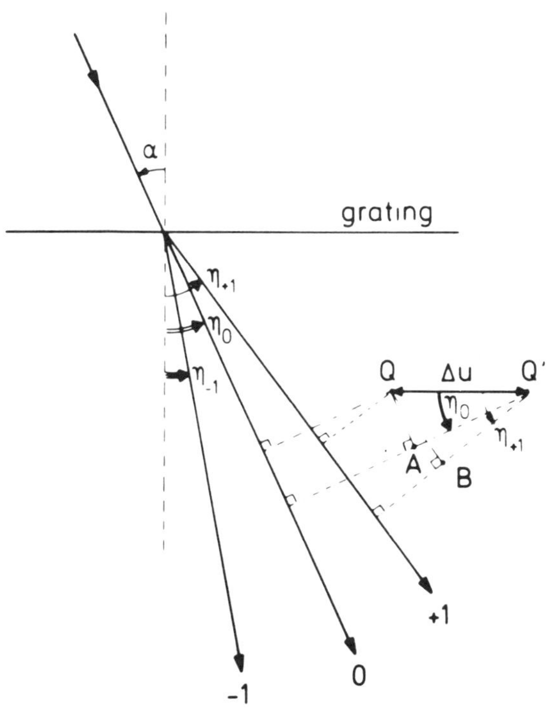

Fig. 6. The effect of a grating displacement Δu on the phase difference between diffracted orders. The increase in phase difference can be derived from the path difference $QA - QB$.

Because of interference between a zeroth and a first diffracted order, fringes appear in the region of overlap. The spatial frequency of the fringes is proportional to the focus error Δz.

C. Moving Grating

The movement of the video disk with respect to the light spot gives rise to phase changes between various orders. In Fig. 6 we consider the phase angle between the zeroth and the first order as a function of the position of the grating with respect to a reference point Q. A grating displacement Δu is equivalent to an identical displacement of the reference point from Q to Q'. In Q' the phase difference between the zeroth and the plus first-order wave has increased by an amount

$$\xi_{+1} = \frac{2\pi}{\lambda}\{QB - QA\} = \frac{2\pi}{\lambda}\Delta u\{\sin\eta_{+1} - \sin\eta_0\}. \qquad (10)$$

Substituting Eq. (1) we obtain

$$\xi_{\pm 1} = \pm(2\pi\Delta u)/p. \qquad (11)$$

When the grating is being translated at a uniform speed, the phase angle $\xi_{\pm 1}$ becomes a linear function of time

$$\xi_{\pm 1} = \pm\omega t. \qquad (12)$$

In case of a video signal, $\omega/2\pi$ is in the 10-MHz range.

D. Interference of the Diffracted Waves

For the total phase difference $\delta_{\pm 1}$ between the zeroth and first diffracted orders we now have three contributions, viz., the displacement phase change $\xi_{\pm 1}$, the phase angle ψ_0, depending on the grating geometry, and the angle $\Delta\theta_{\pm 1}$ due to defocusing. Substituting Eq. (12), $\delta_{\pm 1}$ equals

$$\delta_{\pm 1}(t) = \pm\omega t + \psi_0 + \Delta\theta_{\pm 1}. \qquad (13)$$

In the regions of overlap of the zeroth and first diffracted orders, the irradiance can be represented by

$$I_{\pm 1}(t) \propto 1 + \mu\cos[\delta_{\pm 1}(t)], \qquad (14)$$

where μ is a modulation index, smaller than unity, determined by the mutual strength of the orders. In the presence of a certain amount of defocusing, the phase angles $\delta_{\pm 1}(t)$ vary as a function of the far-field coordinate η in a region of overlap [Fig. 5(b)]. It is convenient to normalize the coordinate η with respect to the aperture angle β of the incident beam, yielding the coordinate x:

$$x = \eta/\beta. \qquad (15)$$

Substituting Eqs. (8), (13), and (15) into Eq. (14), we obtain

$$I_{\pm 1}(x,t) \propto 1 + \mu\cos\left[\pm\omega t + \psi_0 + \frac{\pi\beta^2\Delta z}{\lambda}x_0(x_0 \mp 2x)\right], \qquad (16)$$

where x_0 equals $\lambda/p\beta$.

From Eq. (16) it can be seen that in the presence of defocusing, $I_{+1}(x,t)$ and $I_{-1}(x,t)$ represent two fringe patterns that propagate in the same direction. The direction of propagation is reversed when Δz changes sign, and this phenomenon is exploited for the generation of a focus error signal.

III. Focus Error Signal

The direction of propagation of the defocusing fringes is detected by putting two small detectors in the overlapping region of the zeroth and a first order. The difference signal of the two detectors yields a derivative value of the intensity pattern that contains the sign of the focus error. The sum signal of the two detectors is used as a phase reference signal. Multiplication of both signals yields an appropriate focus error signal.

A. Difference Signal of the Detectors

In Fig. 7 the overlapping region of the zeroth and the plus first order is shown. Near the rim of the zeroth order two small rectangular detectors, DT_1 and DT_2, are present centered at the position $x = x_1$. Expressed in the normalized angular coordinate x, the width of each detector is denoted by Δx.

The difference signal of the two detectors can be represented by

$$I_{DF}(t) \propto -\Delta z \left[\frac{2\pi\mu x_0 \beta^2(\Delta x)}{\lambda}\right] \times \sin\left[\omega t + \psi_0 + \frac{\pi\beta^2 x_0 \Delta z(x_0 - 2x_1)}{\lambda}\right]. \quad (17)$$

In deriving Eq. (17), the detector width is supposed to be very small.

B. Generation of the Focus Error Signal

A change of sign of the focus error Δz induces a phase shift of π to the high frequency difference signal $I_{DF}(t)$ under the condition that $2x_1 = x_0$. This phase shift can be detected by comparing the phase of $I_{DF}(t)$ with another high frequency signal, whose phase is independent of the focus error.

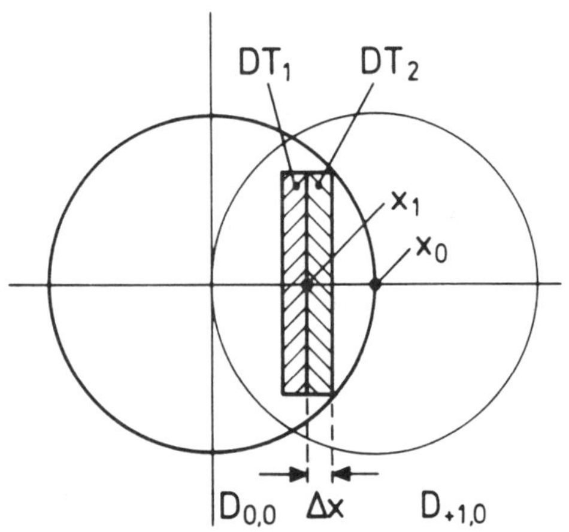

Fig. 7. The split detector (DT_1 and DT_2) in the readout pupil. The width of each detector is equal to Δx. The midpoint of the split detector is situated in $(x_1,0)$, the midpoint of the diffracted order $D_{+1,0}$ in $(x_0,0)$.

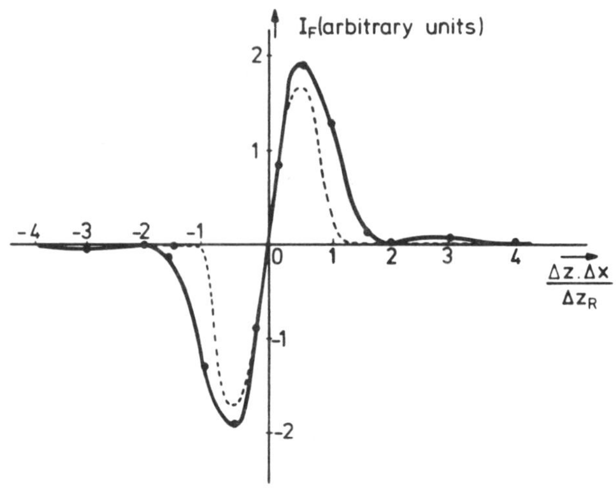

Fig. 8. The focus error signal I_F. The defocusing Δz is expressed in units, $\Delta z_R/\Delta x$. This quotient approximately equals the increased focal depth obtained by using a detector with a limited width Δx. The drawn line applies to a purely periodic disk. The dotted line shows how the error signal decreases in a real situation. This is caused by the different signal content of the neighboring tracks.

Taking together the signals from the two detectors, DT_1 and DT_2, we obtain for the high-frequency part of the sum signal

$$I_{SF}(t) \propto \cos\left[\omega t + \psi_0 + \frac{\pi\beta^2\Delta z}{\lambda} x_0(x_0 - 2x_1)\right]. \quad (18)$$

After a phase shift of $I_{SF}(t)$ over $+\pi/2$ rad in the frequency band around ω, the multiplication of $I_{SF}(t)$ and $I_{DF}(t)$ yields the signal

$$I_F(\Delta z,t) \propto \Delta z + \text{terms}(2\omega t), \quad (19)$$

where the condition $2x_1 = x_0$ does not apply anymore.

Filtering out the components at the frequency 2ω, a bipolar signal of Δz is obtained. Such a signal can be used directly as the input signal of a focus servo system that reduces the value of Δz to 0. Until now, the detector width Δx is taken to be very small. In practice, one should take into account the finite width of the detectors. For large values of Δz, the period of the defocusing fringes becomes comparable with the detector width, and apparent contrast reversals of the fringe pattern are observed. These reversals are falsely interpreted as changes in sign of the focus error Δz. They are avoided by giving a triangular shape to the detectors as the Fourier transform of a triangle is always positive. The detector signals are calculated taking into account the finite width Δx of each detector. The dc component of the product of the signals $I_{SF}(t)$ and $I_{DF}(t)$ then yields the focus-error signal.

This calculated focus error signal is shown in Fig. 8. The focus error has been normalized with respect to the quantity $\Delta z_R/\Delta x$. Δz_R denotes the depth of focus of the illuminating beam with full aperture:

$$\Delta z_R = \lambda/(2\beta^2) \quad \text{(Rayleigh criterion).} \quad (20)$$

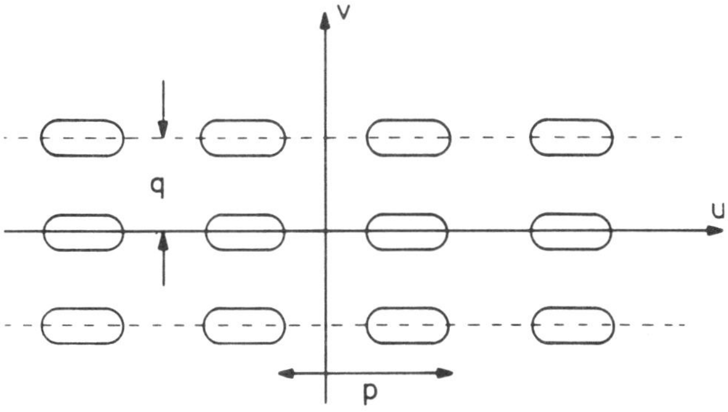

Fig. 9. The arrangement of diffracting obstacles on the disk. For the sake of simplicity the signal content is identical for all the tracks. The tangential and radial periods are denoted by p and q, respectively.

Figure 8 shows that the focus error signal I_F is well defined over a distance $\Delta z_R/\Delta x$ on each side of focus. This distance is commonly called the acquisition range of the servo system. Apparently, this method of generating a focus error signal yields an acquisition range that is $(1/\Delta x)$ times larger than the depth of focus of the full aperture beam. At still larger values of Δz, the focus error signal quickly drops off to zero. In practice, a factor $(1/\Delta x)$ equal to 5 can be attained, yielding an acquisition range of approximately 10 μm. The acquisition range determines how easily the initial focusing on the information layer of the video disk can be achieved. In practice, the value of 10 μm, mentioned above, has proven to be sufficiently large.

The focus error signal has been calculated supposing ideal aberration-free optics. The influence of small aberrations of the illuminating beam on the focus setting is treated in the Appendix.

IV. Radial Error Signal

The generation of the focus error signal could be done by a 1-D treatment of the diffraction problem. For the derivation of a radial error signal, an extension to two dimensions has to be made. Interferences between diffracted orders that yield information about the radial position of the readout spot are considered. The radial error signal is the result of a phase comparison between a reference signal and a second signal whose phase depends on the radial error.

A. Diffraction Pattern of a 2-D Grating

In Fig. 9, the arrangement of diffracting obstacles (absorbers, scatterers, or depressions) on an information-carrying disk is shown. Only a very small part of the spiral track is visible, and over this small distance the obstacles are essentially arranged along straight lines. A double periodicity is present. The local period p in the tangential direction u corresponds to the time signal that has been recorded on the disk; the period q in the radial direction v is the constant spacing between tracks. If the disk is illuminated by a spherical wave, the resulting diffraction pattern is shown in Fig. 10. Apart from the tangentially diffracted orders, $D_{+1,0}$ and $D_{-1,0}$, radially diffracted orders, $D_{0,+1}$ and $D_{0,-1}$, are also present. Moreover, diagonally diffracted orders $D_{\pm 1,\pm 1}$ can be observed.

Using the dimensionless coordinates (x,y), defined by Eq. (15), the midpoint $M_{\pm 1,\pm 1}$ of the diffracted order $D_{\pm 1,\pm 1}$ is located at

$$(x,y)_{M_{\pm 1,\pm 1}} = (\pm x_0, \pm y_0) \tag{21}$$

with

$$x_0 = \lambda/\beta p, \qquad y_0 = \lambda/\beta q. \tag{22}$$

The phase difference of the diagonally diffracted orders and the zeroth order depends on the position of the readout spot in both the tangential and the radial direction (Fig. 9); the tangential displacement introduces a phase shift that equals $2\pi\Delta u/p = \omega t$ according to Eqs. (11) and (12). In the same way a radial displacement v_0 introduces a phase shift $2\pi v_0/q$. The total phase difference between a diagonally diffracted order and the zeroth order thus equals

$$\delta_{\pm 1,\pm 1}(t,v_0) = \pm\omega t \pm 2\pi v_0/q + \psi_0. \tag{23}$$

B. Generation of the Radial Error Signal

In order to obtain a radial error signal, we put one detector in the region of overlap of the zeroth order and the order $D_{+1,+1}$ (midpoint $M_{+1,+1}$) and a second detector in the overlapping region of the zeroth order and the order $D_{+1,-1}$ (midpoint $M_{+1,-1}$). The detector signals, $I_{+1,+1}(t,v_0)$ and $I_{+1,-1}(t,v_0)$, are mainly due to the interference between the strong zeroth order and the orders $D_{+1,0}$ and $D_{+1,\pm 1}$:

$$I_{+1,\pm 1}(t,v_0) \propto \cos(\omega t + \psi_0) + \mu_1 \cos(\omega t \pm 2\pi v_0/q + \psi_0), \tag{24}$$

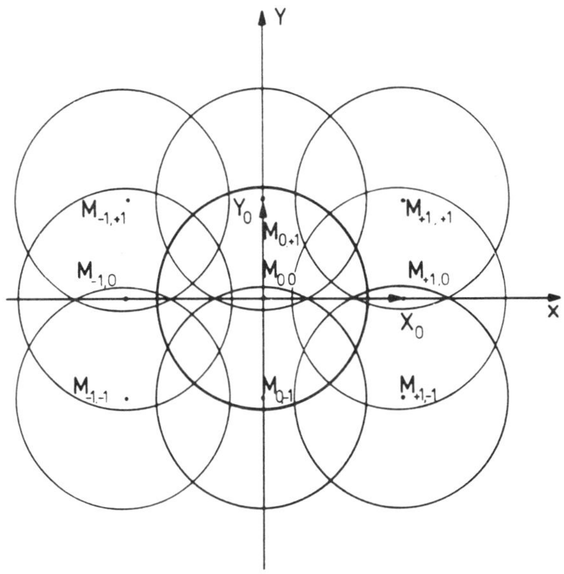

Fig. 10. The 2-D diffraction pattern of the video disk. Using the normalized angular coordinates, the rim of the zeroth order is represented by the unit circle.

where μ_1 is a constant close to unity. The difference signal of the two detectors is represented by

$$I_{DR}(t,v_0) \propto \sin(2\pi v_0/q)\sin(\omega t + \psi_0). \quad (25)$$

The sum signal of the detectors is proportional to

$$I_{SR}(t,v_0) \propto [1 + \mu_1 \cos(2\pi v_0/q)]\cos(\omega t + \psi_0). \quad (26)$$

From Eqs. (25) and (26), it can be seen that the phase of the high-frequency difference signal is shifted over an amount of π when v_0 changes sign. The sum signal serves as a reference signal whose phase is independent of the tracking error v_0.

After an electronic phase shift of I_{SR} over $\pi/2$ rad, the multiplication of the sum and the difference signals yields the signal

$$I_R(t,v_0) \propto [1 + \mu_1 \cos(2\pi v_0/q)]\sin(2\pi v_0/q) + \text{terms}(2\omega t). \quad (27)$$

Filtering out the frequency components at 2ω, a bipolar function of the tracking error results. This signal can be used as the input signal for a servo system that centers the readout light beam on the middle of the track.

V. Experimental Results

A. Experimental Arrangement

The experiment was performed by playing back a standard video disk with a broadband FM TV signal recorded on it. In the playback unit (Fig. 11), the four detectors needed for the generation of both error signals are placed in the far-field region of the video disk. The far-field region of the disk is the exit pupil plane P of the objective O. This exit pupil is imaged onto the detectors by means of the auxiliary lens L. When the radial tracking mirror M is placed close to the exit pupil of the objective, the movement of this mirror does not change the position of the pupil image in the detector plane. The detectors are integrated on one chip. A triangular shape is adopted for each detector in order to avoid contrast reversals at large defocusing values. The detectors are placed near the rim of the beam (Fig. 12) in order to assure the existence of error signals even at high spatial frequencies in the tangential direction. The high-frequency reference signal for deriving both the radial and vertical error signal is obtained by taking the sum of detectors 1, 2, 3, and 4:

$$I_S(t,\Delta z,v_0) = I_1 + I_2 + I_3 + I_4. \quad (28)$$

The focus difference signal equals

$$I_{DF}(t,\Delta z) = (I_1 + I_4) - (I_2 + I_3), \quad (29)$$

while the radial difference signal is expressed by

$$I_{DR}(t,v_0) = (I_1 + I_2) - (I_3 + I_4). \quad (30)$$

It should be noted that the amplitude of the reference signal, $I_S(t,\Delta z,v_0)$, depends on the defocusing Δz and the tracking error v_0. The quantity of interest, the phase of the reference signal, is not affected by variations of Δz and v_0. The phase shift of $\pi/2$ rad that has

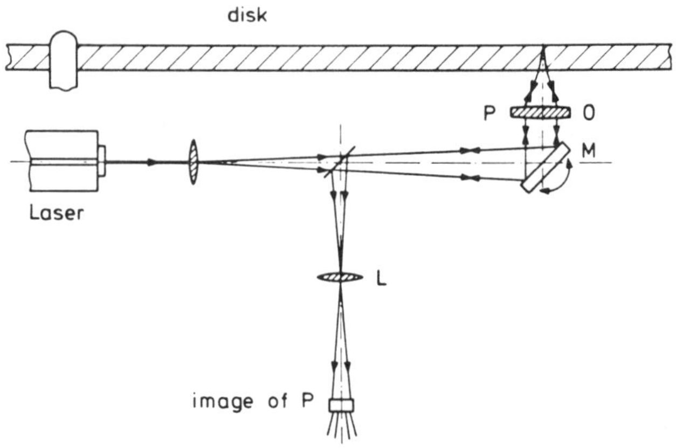

Fig. 11. Schematic view of the optical setup. Radial tracking and focusing of the information are achieved by tilting the mirror M and by moving vertically the objective O.

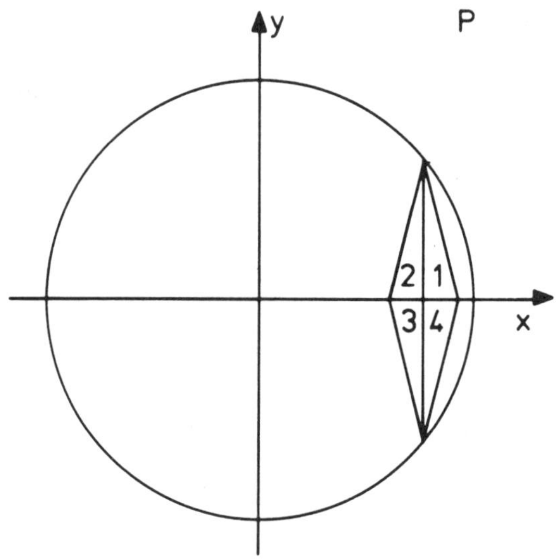

Fig. 12. Detector arrangement in the far-field region of the disk.

to be introduced electronically to the reference signal is obtained by means of a time delay that corresponds to a quarter of a period at the carrier frequency of the rf signal.

Figure 13 illustrates the focus error signal $I_F(\Delta z)$ derived according to the method of Sec. III. In the playback unit, the focus servo system was driven at very low power. Under this condition, the focused readout beam cannot follow all the vertical movements of the rotating disk. The envelope of the rf signal is affected by these out-of-focus movements as is shown by the middle trace of the photograph. The out-of-focus movement with values up to 6 μm means that the focus error signal is not equal to zero but shows residual variations around zero level.

The lower trace of Fig. 13 shows the residual focus error signal derived according to the method described in this paper. As a comparison we show in the upper trace the residual error signal according to the Foucault

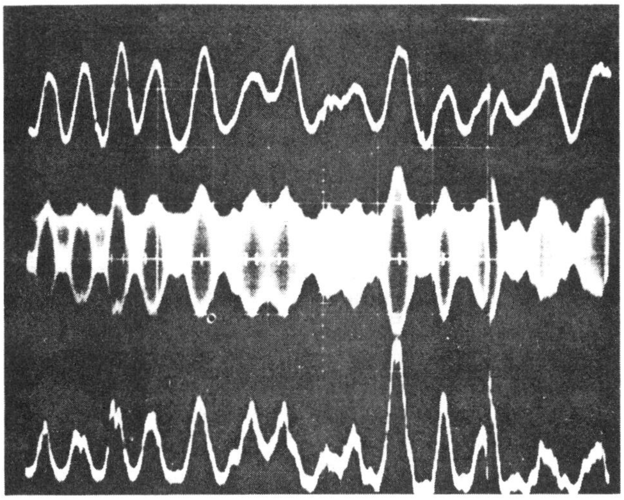

Fig. 13. The rf signal (middle trace) and the residual focus error signal (upper and lower trace) in case of bad focusing on the disk. The horizontal scale corresponds to the distance along the track circumference. The upper oscilloscope trace shows the focus error signal generated by the Foucault method; the lower trace is the focus error signal according to the method described in this paper.

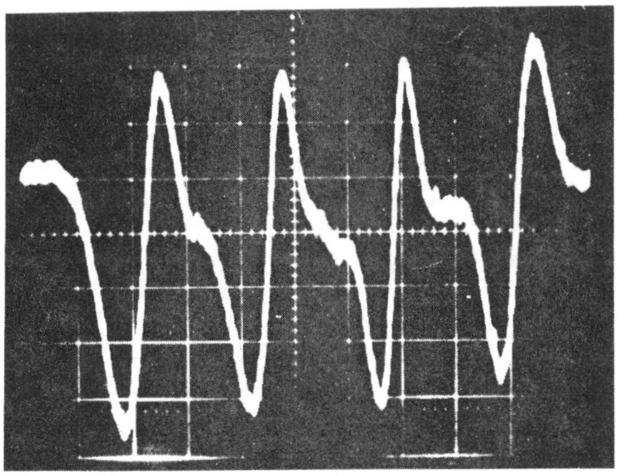

Fig. 14. An oscilloscope trace of the radial error signal $I_R(v_0)$. The radial mirror is not driven by the servo system, and the scanning spot is running freely through the tracks of the disk.

method,[6] which makes no use at all of the rf signal on the disk. Both signals approximately have the same shape and can freely be exchanged as an input signal for the focus servo system.

In Fig. 14 we show an oscilloscope trace of the radial error signal $I_R(v_0)$. The section of the signal that is used for tracking is situated in the neighborhood of $v_0 = 0, \pm q, \ldots$. Here the slope of the signal is steepest, and a stable tracking of the information is achieved. At values of v_0 in the neighborhood of $\pm q/2$, $\pm 3q/2, \ldots$, the slope of $I_R(v_0)$ is rather ill defined; the reference signal almost disappears halfway between tracks. In practice, the position on track is automatically found.

With respect to position tolerances of the detectors, only displacements in the y direction (see Fig. 12) are harmful. A position error in this direction introduces a radial offset of the tracking readout spot. The tolerance is approximately one tenth of the imaged pupil diameter in order to assure tracking with an offset less than 0.2 μm.

B. SNR of the Error Signal

The usual noise sources like shot noise of the photons, preamplifier noise, and noise due to the surface roughness of the disk have to be maintained at a low level in order to assure a good quality rf signal. A typical value for the SNR is 40 dB over the full 5-MHz bandwidth of a signal. The bandwidth required for the error signals (approximately 25 kHz) is rather small compared with the bandwidth of the high-frequency signal. In spite of the small detector area used for the extraction of the error signals, the SNR of these signals is better than 40 dB because of their small bandwidth.

Another source of noise is the amplitude modulation of the light source, especially if the light source is a laser with a large coherence length suffering from feedback into the cavity. These amplitude modulations can be canceled out easily by electronic means (ratio detection).

The feedback into the laser equally introduces a distortion of the light distribution across the beam causing unbalance of the detectors. In a similar way, weakly contrasted moving fringe patterns due to coherent stray light can disturb the equilibrium between the light fluxes on the two detectors that have to be subtracted. These unbalance effects finally determine the reliability of the error signals.

VI. Conclusion

Phase relations between high-frequency signals detected in different sections of the readout pupil depend on the radial and vertical position of the light spot with respect to the track that is scanned. The comparison of the phase of signals that are tracking and focusing dependent with the phase of a high-frequency reference signal yields a radial and a vertical error signal. The high-frequency signals are obtained with the aid of a composite detector that is placed in the far-field region of the video disk. The method is independent of the type of video disk (shallow or deep phase structures, amplitude structures).

The alignment of the detector is not critical. A focus offset of the readout beam is not possible with this method, for the information itself on the disk is used for the generation of a vertical error signal. The acquisition range of the vertical servo system is approximately ± 10 μm.

The authors thank H. H. Hopkins for a private communication in which he gave an extensive treatment of the diffraction of light by 2-D periodic structures like video disks. This paper will be published shortly.

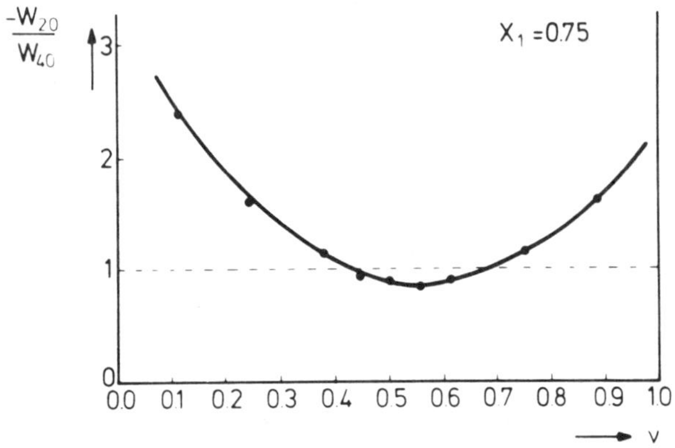

Fig. 15. The focus setting W_{20} in the presence of primary spherical aberration W_{40} as a function of the normalized spatial frequency ν in the tangential direction. The cutoff frequency ($\nu = 1$) equals $2N/\lambda$ mm^{-1}.

Appendix: Influence of Aberrations on the Focus Setting

In Sec. III.B, the variable phase difference between the zeroth and a first diffracted order was calculated for a general point (x,y) in the common area of these orders in the presence of a certain amount of defocusing Δz. If, in addition, the incident spherical wave suffers from some intrinsic aberrations, the phase difference at a point (x,y) of the pupil slightly changes and influences the focus error signal. It is useful to express the defocusing and the aberrations as a departure W_a from the ideal spherical waveform in the pupil:

$$W_a = W_{20}(x^2 + y^2) + W_{31}x(x^2 + y^2) + W_{40}(x^2 + y^2)^2, \quad \text{(A1)}$$

where W_{20} is the defocusing, W_{31} the primary coma, and W_{40} the primary spherical aberration coefficient, all expressed in wavelengths at the pupil rim ($x^2 + y^2 = 1$). The pupil aberration $W_{a,0}$ of the zeroth order is given by Eq. (A1); the pupil aberration, $W_{a,1}$ of the first order, can be obtained by substituting $x \rightarrow (x - x_0)$ into Eq. (A1). Each of the two focus error detectors, DT_1 and DT_2, measures an aberration difference W_d between the zeroth and first order:

$$W_d(x) = [W_a(x) - W_a(x - x_0)]. \quad \text{(A2)}$$

The change in W_d over a small distance Δx in the readout pupil can be represented near the x axis ($y = 0$) by

$$\Delta W = \left(\frac{\partial W_d}{\partial x}\right) \Delta x$$
$$= \Delta x \cdot x_0[2W_{20} + 3W_{31}(2x_1 - x_0) + 4W_{40}(3x_1^2 - 3x_1x_0 + x_0^2)]. \quad \text{(A3)}$$

In Eq. (A3), x_0 is proportional to the spatial frequency in the tangential direction ($0 < x_0 < 2$), and x_1 stands for the position of the detectors in the readout pupil ($0 < x_1 < 1$).

The focus-error detectors, DT_1 and DT_2, force the control loop to reduce the function $\Delta\dot{W}$ to zero value. From Eq. (A3), it is seen that without any aberrations the ideal focus setting $W_{20} = 0$ is obtained. In the presence of aberrations, a balancing of the aberration with a certain amount of defocusing occurs. This new focus setting does not correspond to the best aberration balance for all spatial frequencies simultaneously.

As an example, we consider the influence of a certain amount of spherical aberration W_{40} on the focus setting. The position x_1 of the midpoint of the split detectors, DT_1 and DT_2, is taken equal to 0.75. The ratio, $-W_{20}/W_{40}$, as a result of the balancing, is shown in Fig. 15 as a function of the normalized spatial frequency in the tangential direction. The dotted line indicates the optimum focus setting according to the Rayleigh criterion when spherical aberration is present: $W_{20} = -W_{40}$. The graph shows that this best focus is obtained at spatial frequencies close to half the cutoff frequency. For very low and very high frequencies, the defocusing shows a greater departure from best focus. Fortunately, the signal is much less sensitive to defocusing in these regions. However, using this focusing method, the permitted tolerance on spherical aberration is smaller than the tolerance according to Maréchal's criterion ($W_{40} < 1\lambda$). As a practical limit, W_{40} should not exceed half of a wavelength.

References

1. K. Compaan and P. Kramer, Philips Tech. Rev. **33,** 178 (1973).
2. G. Broussaud, E. Spitz, C. Tinet, and F. Le Carvenec, IEEE Trans. Broadcast Telev. Receivers **BTR-20,** 332 (1974).
3. G. W. Hrbek, J. Soc. Motion Pict. Telev. Eng. **83,** 580 (1974).
4. K. D. Broadbent, IEEE Trans. Broadcast Telev. Receivers **BTR-20,** 338 (1974).
5. Y. Tsunoda, K. Tatsuno, K. Kataoka, and Y. Takeda, Appl. Opt. **15,** 1398 (1976).
6. G. Bouwhuis and J. J. M. Braat, Appl. Opt. **17,** 1993 (1978).

Reprinted from *Optical Mass Data Storage II,* Proc. SPIE Vol. 695, pp. 141-146 (1986). ©1986 Society of Photo-Optical Instrumentation Engineers.

Interlinked Tracking Servo Technology

S.Arai, T.Iwamoto, K.Hamada and K.Ogawa

FUJITSU LABORATORIES LTD., KAWASAKI, JAPAN

Abstract

The well known push-pull method has a problem in tracking offset caused by deviation of the beam axis and disk inclination. The tracking offset degrades the tracking servo performance and the track following range is conventionally restricted to about ±50 μm. To solve this problem and enlarge the track following range, we have developed a new tracking servo technology which uses two actuators interlinked by servos. This method completely eliminates the deviation of the beam axis and enables a track following range of ±400 μm.

Introduction

The 2-stage positioning servo consisting of a fine tracking actuator and a coarse carriage (voice coil motor) is widely used in optical disk drive. This type of optical disk drive has a wide track following range, because the VCM follows most of the disk eccentricity at low frequency and the fine tracking actuator follows only at high frequency. However a stepping motor is more suitable for low end disk drives because of its low cost and stable accessing. But if a stepping motor is used as a coarse carriage, the fine tracking actuator must follow all components of disk eccentricity. In the push-pull method the axis of the beam reflected from disk deviates due to the movement of the tracking actuator as shown in Figure 1. This phenomenon causes tracking offset and restricts the following range to only about ±50 μm. This problem has been a big obstacle in designing optical disk drives using a stepping motor.

New concept (the ILT method) for tracking servo technology

To solve the tracking offset problem, the authors have developed a new tracking servo technology which we call interlinked tracking (ILT). Figure 2 shows the concept. This method uses two separate tracking actuators: the 2-dimensional actuator and the galvano mirror. These two actuators are driven simultaneusly, keeping the positional relation where the incident beam reflected on the galvano mirror always passes through the back focal point of the objective lens mounted in the 2-dimensional actuator. This relationship guarantees that the reflected beam from the disk returns on the same path and never causes the above mentioned deviation. This IlT method enlarges the track following range considerably.

Optical head used in experiment

The optical head used in our experiments is shown in Figure 3. The beam from the laser diode passes through the polarizing beam splitter, is reflected by the galvano mirror, passes through the objective lens mounted in the 2-dimensional actuator. The beam returns along the same optical path, is reflected by the polarizing beam splitter, goes to the quadrant detector which derives the servo signals and the data signal. Each actuator has a position sensor consisting of LED and photo diode.

Tracking servo with ILT

ILT works as follows. The tracking actuator is driven by the push-pull signal. The deviation from the relation (The beam reflected on the galvano mirror goes through the back focal point of the objective lens) is detected by comparing the two signals from the two position sensors. This compared output signal drives another actuator to follow the tracking actuator, keeping the mutual relation. In this way, ILT is performed as a double closed loop.

Figure 4 is the block diagram of the ILT system. G_t is the transfer function of the tracking actuator. G_l is that of the linked actuator. K-factor is a constant value which is determined by the spatial arrangement of the two actuators. K-factor is written as

$$K = \frac{E_l}{E_g} = \frac{l}{f}$$

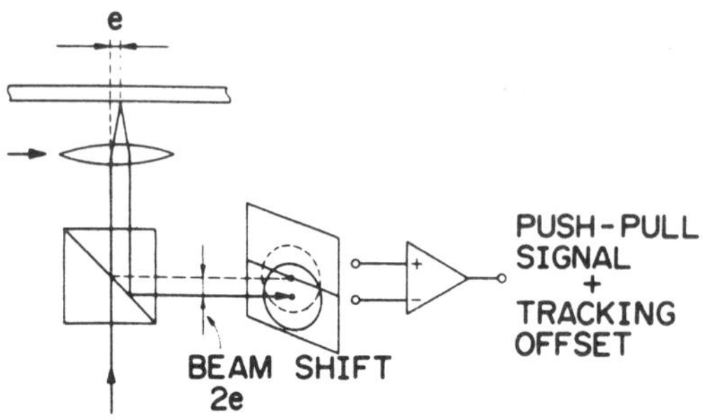

Figure 1. Beam shift in the push-pull method

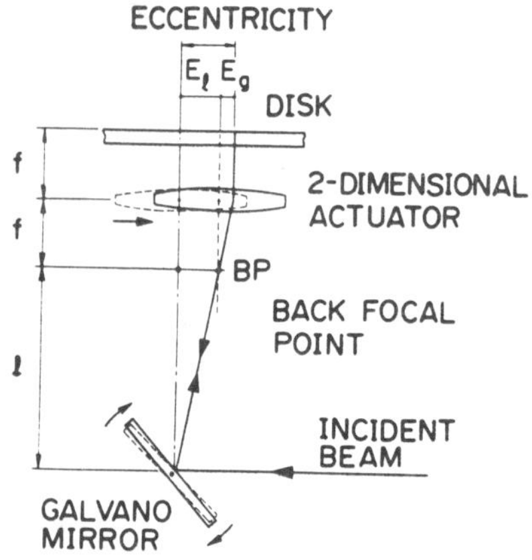

Figure 2. Concept of the ILT method

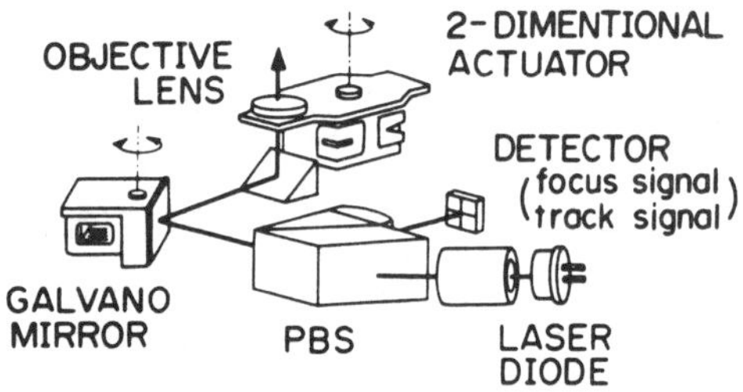

Figure 3. Optical head used in experiment

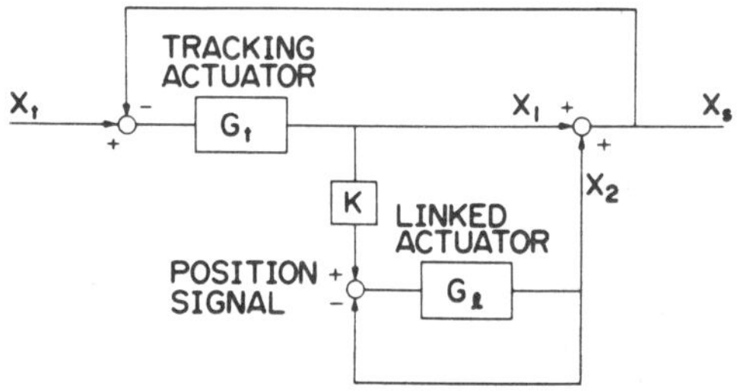

Figure 4. Block diagram of ILT serve loop

E_l : Beam position affected by the 2-dimentional actuator
E_g : Beam position affected by the galvano mirror
f : Focal length of the objective lens
l : Distance between the back focal point and the galvano mirror

The open loop transfer function G_O can be written as

$$G_0 = G_t \left(1 + K \cdot \frac{G_l}{1 + G_l}\right) \qquad K = k \text{ or } \frac{1}{k}$$

Beam position $X_1$from the tracking actuator and beam position X_2 from the linked actuator can be written as

$$X_1 = \frac{G_t}{1 + G_t(1+KG')} \cdot X_t$$

$$X_2 = \frac{1 + G'G_t}{1 + G_t(1+KG')} \cdot X_t$$

$$G' = \frac{G_l}{1 + G_l}$$

where X_t is the position of the destination track. For low frequency, X_1, X_2 and G' are written as

$$X_1 = \frac{1}{1 + K} \cdot X_t$$

$$X_2 = \frac{K}{1 + K} \cdot X_t$$

$$G' = 1$$

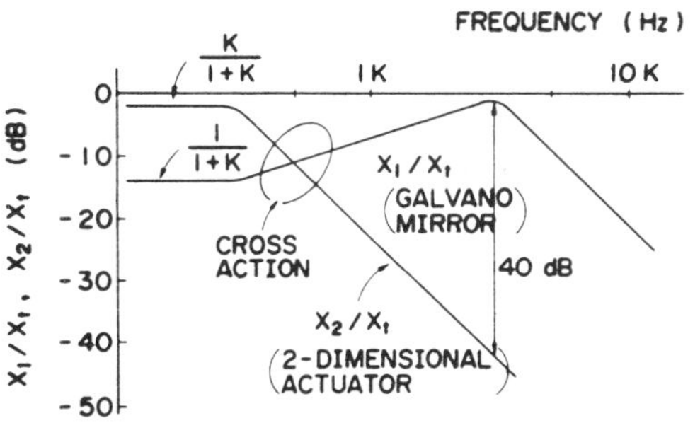

Figure 5. Movement of two actuators in the ILT method (TYPE A)

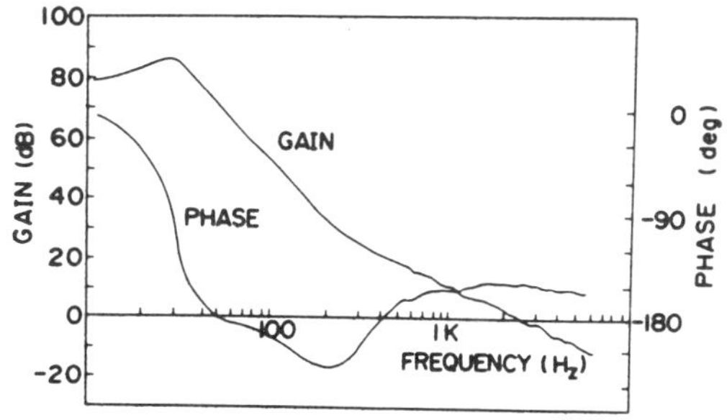

Figure 6. Open loop gain in the ILT method (TYPE A)

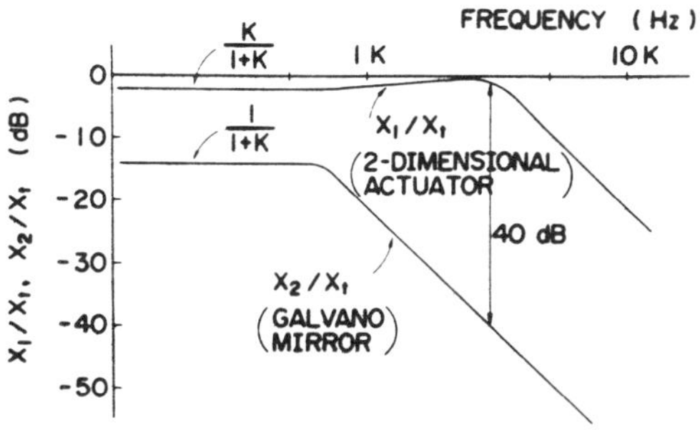

Figure 7. Movement of two actuators in the ILT method (TYPE B)

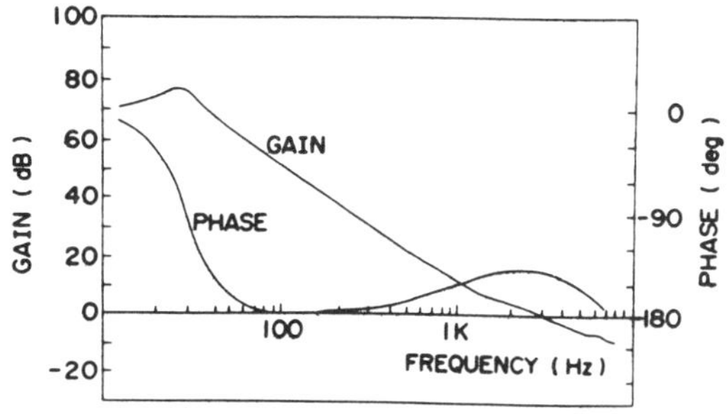

Figure 8. Open loop gain in the ILT method (TYPE B)

The mutual relation between the two actuators is maintained as long as $X_2/X_1=k$. In the ILT method, 2 types of servo control are possible.

TYPE A: The galvano mirror is chosen as the tracking actuator and the 2-dimensional actuator serves as the linked actuator. In this case the open loop gain at low frequency becomes G_t (1+k). This means the open loop gain is (1+k) times as large as in the case without linking. The value is 14 dB at k=4. However the linking response frequency is restricted to 1/10 of that of the main tracking loop if the linking loop gain is to be 40 dB lower at the zero cross frequency. Also the cross action between these two actuators occurs at an intermediate frequency as shown in Figure 5, 6.

TYPE B: The 2-dimensional actuator is chosen as the tracking actuator and the galvano mirror serves as the linked actuator. In this case the open loop gain at low frequency is almost the same as in the case without linking, because (1+k) becomes (1+1/k) at k=4. However, the response frequency can be 1/4 of that of the main tracking loop. It is 2.5 times as high as in TYPE A. Experimentally, 1 kHz of linking is possible where the zero cross frequency is 4 kHz. This linking servo loop is sufficiently tough and reliable, compared with TYPE A and the conventional 2-stage servo using VCM. Also no cross action is observed, because the movement of the galvano mirror is always smaller than that of the 2-dimensional actuator as shown in Figure 7, 8.

Performance of the ILT method

The linking open loop gain G_l can be 25 dB for TYPE A, and 40 dB for TYPE B. The conventional tracking method using only the 2-dimensional actuator causes 2e of beam shift (the deviation between the beam incident beam and the beam reflected from the disk), where e is disk eccentricity. In the ILT method the beam shift BS can be written as

$$BS = 2 \cdot e \cdot \frac{K}{1 + K} \cdot \frac{1}{G_l}$$

This value is negligibly small. For example, BS is only 3.2 μm for TYPE B when the disk eccentricity is 200 μm, where with the conventional method it is 400 μm. The tracking offset caused by this beam shift is respectively 0.003 μm and 0.4 μm as shown in Figure 13.

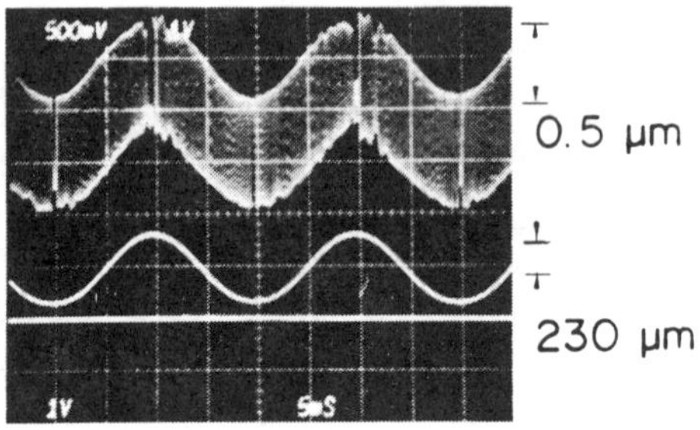

Figure 9. Tracking offset without the ILT method
Upper trace : Push pull signal
Middle trace: Position signal of 2-dimensional actuator
Lower trace : Position signal of galvano mirror

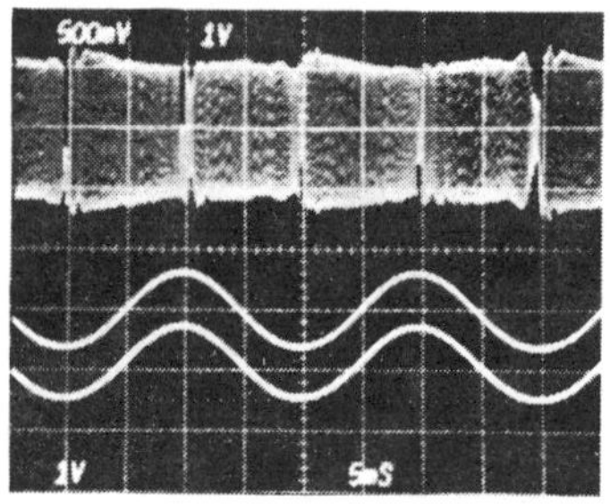

Figure 10. Disapearance of tracking offset while two actuators are interlinked
Upper trace : Push-pull signal
Middle trace: Position signal of 2-dimensional actuator
Lower trace : Position signal of galvano mirror

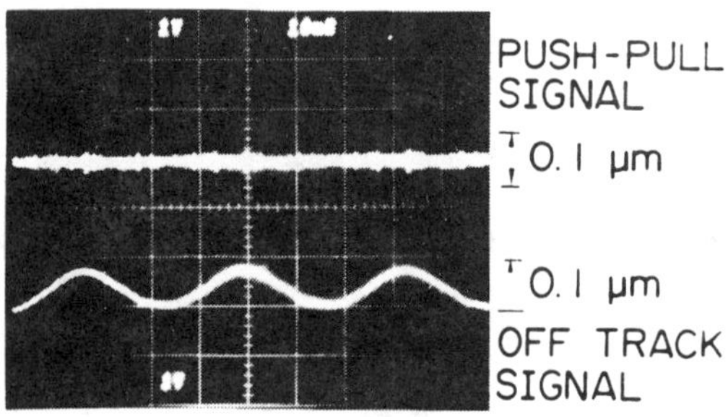

Figure 11. Off track signal sampled from ID region without the ILT method (Disk eccentricity : 35 μm)

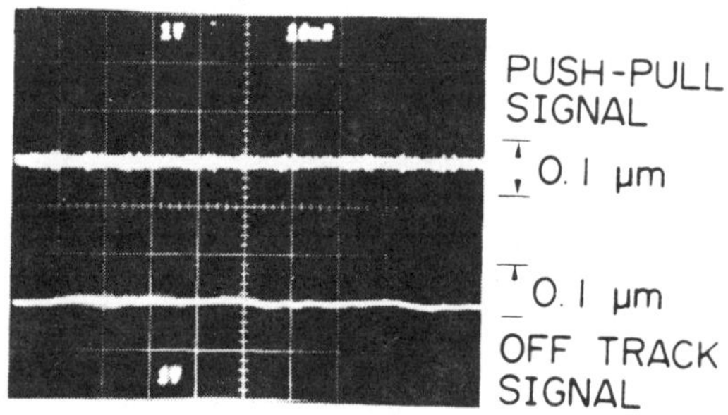

Figure 12. Off track signal sampled from ID region with the ILT method (Disk eccentricity : 35 μm)

Therefore the track following range of the ILT method is decided only by the residual value of the servo loop, while in the conventional method it is restricted by the beam shift.

Figure 9 shows the beam shift. In the photograph the 2-dimensional actuator is vibrated from outside at 50 Hz and 230 μm of amplitude. The galvano mirror is locked by its own position sensor signal. ±0.5 μm of the tracking offset is observed, which means the track following range is estimated to be 50 μm assuming an off track limitation of ±0.05 μm. Figure 10 shows the performance of the ILT method. The galvano mirror is linked so as to follow the 2-dimensional actuator. No track offset is observed in the push-pull signal.

Figure 11 shows the genuine off-track signal obtained by the sampling method that the off track signal is sampled on the ID region and smoothed with a low pass filter. The offtrack signal in Figure 11 and 12 are sampled at 480 Hz and smoothed with the 30 Hz low pass filter. In the conventional method, 0.1 μm off-track is observed at a disk eccentricity of 35 μm in figure 11. On the other hand, using the ILT method, no off-track is observed at the same disk eccentricity as shown in figure 12. Even if standardization valeue of disk eccentricity limit is about 100 μm, off track at the disk rotation frequency is never observed while the ILT method is working.

Fast access by multi-jumping

The beam on the disk can follow ±256 tracks under the ILT method when the stepping motor is stopped. This makes multi-jumping for fine access more reliable, because the linking response frequency is sufficiently high and tracking offset never occurs during multi-jumping. This is a big advantage from the view point of high speed random access.

	BEAM SHIFT	TRACKING OFFSET	TRACK FOLLOWING RANGE
LENS SHIFT	2e	0.4 μm	± 40
ILT	$2e \cdot \frac{K}{1+K} \cdot \frac{1}{G_\ell}$	0.003 μm	± 400 (Servo Limit)
		⇧ [e : 200 μm, K : 3, G_ℓ : 40 dB]	⇧ [± 0.05 μm of OFF TRACK LIMIT]

Figure 13. Beam Shift and track following range

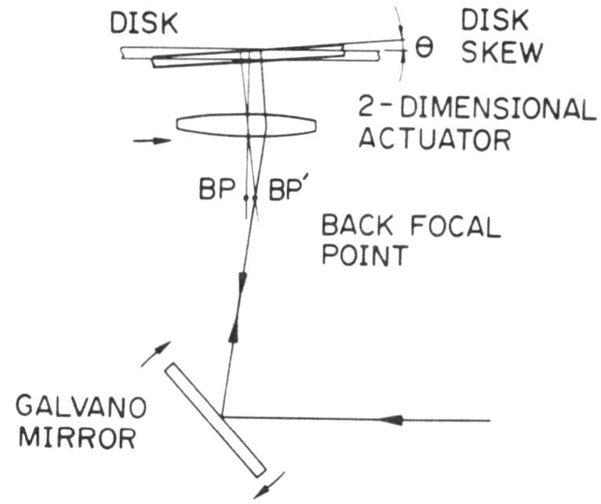

Figure 14. Compensation of disk inclination for disk inclination

Influence of focusing

The back focal point BP moves in the direction of focusing. If the back focal point moves from BP_0 to BP_1 for focusing, this influence is written as

$$BS = \frac{2 \cdot z \cdot e}{l + f}$$

where z is the distance between BP_0 and BP_1, e is disk eccentricity. This value is only 3.1 μm at z=200 μm and e=300 μm. This value is only 6 μm, corresponding to 0.01 μm of tracking offet. Therefore the ILT is not disturbed by movement of the back focal point during focusing.

Compensation for disk inclination

Disk inclination causes the beam shift. The ILT method can compensate for disk inclination by changing the mutual relation slightly so that the reflected beam on the galvano mirror follows the back focal point BP' which is displaced from BP by disk inclination as shown Figure 14. The authors have been studying the necessity for this compensation. It depends on the quality of the optical disk.

Conclusions

The new tracking technology (the ILT method) has many features as follows.

1) Track following range is ten times as large as conventional method with no aid from VCM.
2) Use of a stepping motor as a coarse carriage is feasible.
3) The lock mode stabibilizes coarse access by the head carriage.
4) The interlinked mode stabilizes multi-jumping over long distances.
5) Ability of compensation for disk inclination

Acknowledgement

The authors would like to thank Mr.T.Tatsuta and Mr.H.Tajiri of Fujitsu Limited, Mr.M.Hiyane and Mr.M.Fujino of Fujitsu Laboratories for their guidance and encouragement. The authors also would like to thank their collaborators for their help and valuable discussions.

References

1 S.Arai, I.Shibata, K.Ogawa and O.Ogawa, A new servo method which increases track following range of a one-beam optical head, Proceeding of SPIE, Optical Discs and Applications, Vol.421, pl43-147, 1983.
2 H.Inada, T.Nomura, S.Shimonou and Y.Ichiyama, Tracking servo system with new optical head locked objective lens, Proceeding of SPIE, Optical Discs and application, Vol.421,pl36-142, 1983.
3 S.Arai, K.Hamada and K.Ogawa, A new tracking servo system, Topical Meeting of Optical Disc Society, ThBB3-1, 1985.
4 S.Yonezawa, Composite Prewobbling Optical Disk Memory, Topical Meeting On Optical Data Storage, ThBB4-1, 1985.

Reprinted from *Optical Mass Data Storage II*, Proc. SPIE Vol. 695, pp. 172-180 (1986).

Scanning techniques for optical data storage

David K. Towner

Disc Memory Division, Hewlett-Packard Co.
Mail Stop 342, P.O. Box 39, Boise, Idaho 83707

Introduction

It seems almost paradoxical that beams of light can be moved and steered at very high speeds using a variety of scanning methods, yet the optical disk drives now being designed and marketed for data storage applications have comparatively long access times. Knowing that optical data storage has unrealized potential is of interest, but of more immediate concern is the recognition that poor access performance is a serious design issue. Magnetic disk drives offer average seek times in the 15-25ms range, compared to about 80-500ms (or more, for CD ROMS) for current optical drives. This performance disparity exists, in part, because the relatively massive "optical heads" in use today cannot be transported across the radius of a disk as quickly as a stack of much lighter magnetic heads. Any potential distance advantage that the optical drive might have, due to its substantially higher track density, is offset by the magnetic drive's use of a multi-disk stack. As a result, the drive must achieve similar radial accelerations during seeks if it is to have similar average access times. The inability of current optical drives to approach the access speeds of comparable magnetic drives significantly reduces the competitiveness of optical products in major segments of the very large data storage market. This shortcoming is especially disturbing when we know that opto-mechanical scanners typically operate in the 1-10ms range and that non-mechanical scanning techniques can be substantially faster than that.

This design study resulted from an attempt to find out if scanning techniques could be applied to the design of optical data storage systems. An obvious goal was to identify geometries which offer the possibility of very rapid access to large amounts of data. A secondary goal was to identify alternatives to well known approaches, in the hope of finding systems with different, and perhaps useful, sets of properties. Other than the expectation that a rotating recording medium would be used, no initial restrictions were placed on the geometry of the optics or mechanics. Instead, fundamental design issues were addressed directly, and the various system geometries resulted. A goal of this paper is to stimulate thinking and discussion and to invite consideration of non-conventional approaches to the design of optical data storage systems.

A basic requirement

Diffraction effects control the dimensions of the focused recording spot. A converging beam numerical aperture (NA) in the range of .47 to .55 is needed at the recording medium. A choice in this range will yield micrometer sized spots and sufficient irradiance for reading and writing with a GaAlAs laser diode source.

Scanners for optical recording

Given that an objective lens, in some form, is needed in order to focus a read/write beam onto the surface of the storage medium at about .5 NA, there are two fundamentally different ways of scanning the focused spot. A scanning element located in the converging beam between the lens and focus, as shown in Figure 1, is referred to as a post-objective scanner. If the scanning element is located in the collimated or low NA space on the other side of the objective, it is appropriately known as a pre-objective scanner; illustrated in Figure 2. As we will see, scanners of these two types have very different properties.

Post-objective scanning is very attractive, where it can be used, because the objective lens is only required to provide good image quality near the optical axis. As a result, the objective can often be compact, simple to design, and relatively inexpensive. Unfortunately, it is impractical to consider post-objective scanning at these numerical apertures for several reasons. First, the marginal rays at one side of the ray cone are incident on the scanning mirror at very large angles and the beam focus is only barely beyond the marginal rays at the other side. As a result, the mirror must be quite large compared to the scanned field and the possibility of scanning large fields at high speed is lost. Also, unless the objective lens and scan mirror are very large, the image surface will be tightly curved. In all cases it would be difficult to design a scanning mechanism to operate in the restricted space between the objective lens and the recording medium. One final concern is the large range of incidence angles at the scan mirror. This makes it potentially difficult to achieve uniform performance of the mirror coatings everywhere in the beam. Figure 1 illustrates the difficulties of using post-objective scanning at these numerical apertures.

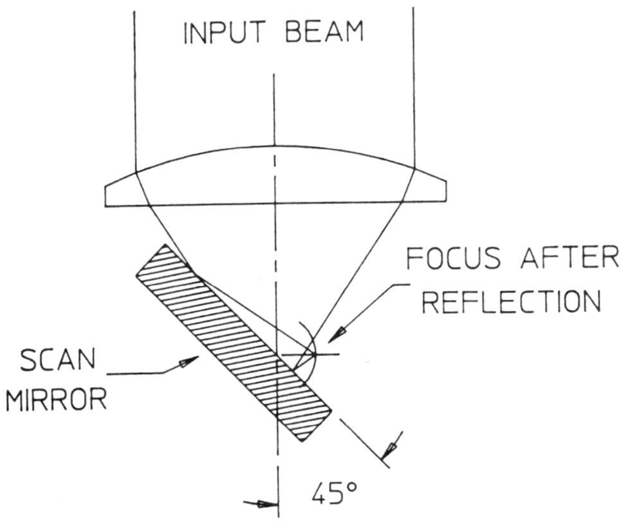

Figure 1. Post-objective scanning in a high NA beam.

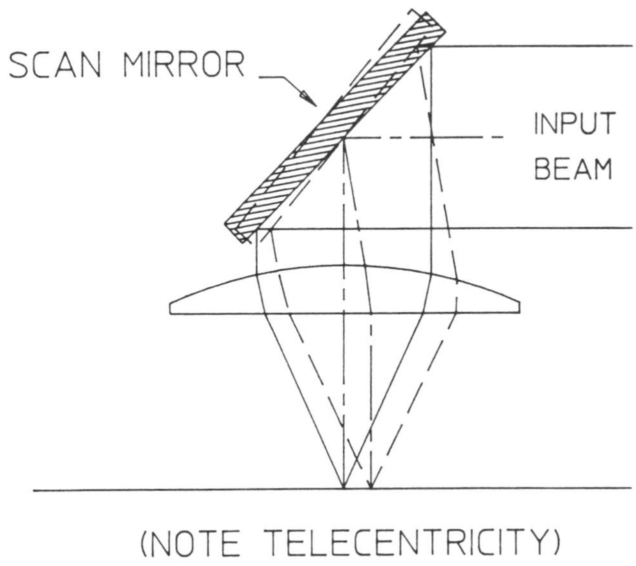

Figure 2. Pre-objective scanning.

Pre-objective scanners are free from the problems described above, but they must be used in conjunction with an objective lens that provides high quality imagery over the entire scanned distance. As a result, they are often considered impractical for accessing large numbers of tracks in optical recording systems, because designs for affordable high NA lenses with large image fields are simply not available. This is not strictly true, however, and pre-objective scanning should not be dismissed without more serious examination. Figure 2 schematically illustrates a pre-objective scanning system. This particular system is telecentric, causing the chief ray to remain perpendicular to the disk during scanning; a desirable condition for maximum light recovery during readout. Pre-objective scanning has been used in the form illustrated for tracking and high-speed short seeks in several optical disk drive designs, but the range of focused spot motion in these systems has been limited to about ±100um by the available image field of the objective lens.

Conventional wide-field high NA objective lenses

An obvious starting point in the search for moderate to high NA objective lenses that are well corrected over large image fields is the catalog of one of the major microscope manufacturers. It isn't surprising that the microscope objectives designed for visual use are closely matched to the aperture and field limitations of the eye, in combination with commonly used eyepieces.

The Lagrange (or parxial) invariant is, among other things, a measure of the number of resolvable points a lens can image. In our case, it is also a measure of the number of tracks a scanner could access using a given lens. It is defined as:

$$I = n\bar{u}y - nu\bar{y} \quad \text{or, for high NA systems as:} \quad I' = ny\sin(\bar{U}) - n\bar{y}\sin(U) \tag{1}$$

Note that u and y are the paraxial ray slope and height, and U is the real ray angle. Barred quantities refer the chief (or principal) ray, and the unbarred quantities refer to the marginal ray.

Figure 3 is a plot of NA versus image field diameter for an assortment of miscroscope objectives. Note how closely they cluster about the curve indicating a Lagrange invariant of Ø.25mm. Examples of several special purpose lenses are also shown, including designs for microlithography and scanning microscopy. For our purposes, objective lenses start to get interesting as their invariants approach about 2.5mm. If we superimpose "iso-element" contours on Figure 3a, and approximately connect points designating lenses with equal numbers of elements, we see that designs having both large image fields and high NA's also have the greatest number of elements (Figure 3b). These lenses are among the most difficult of all lenses to design and manufacture, and are invariably large and very expensive. Figure 4 shows the design of a microlithography lens capable of providing extremely high quality imagery at .42 NA over a 7.Ømm diameter field (I'=1.47). Despite the use of many elements and the finest design and manufacturing techniques available, lenses of this type still fall short of our goal. These lenses have a large number of elements as a consequence of the geometric increase in the magnitude of many aberration types as aperture and field are

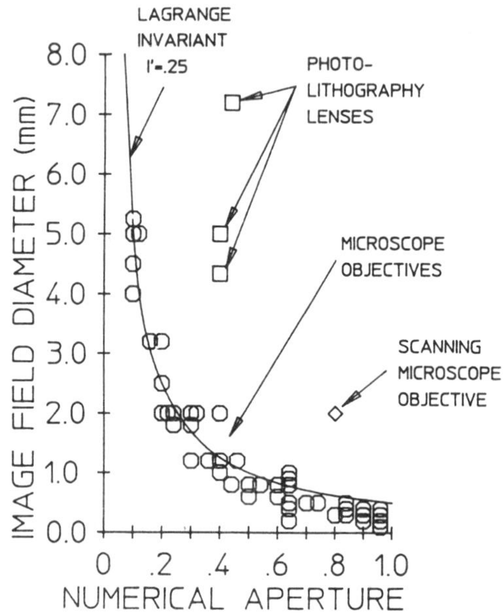

Figure 3a. Image field diameter vs. NA for objective lenses.

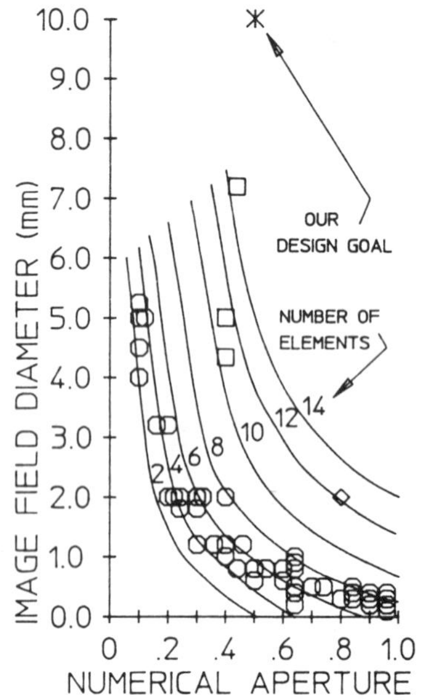

Figure 3b. Number of elements vs. NA and image field diameter.

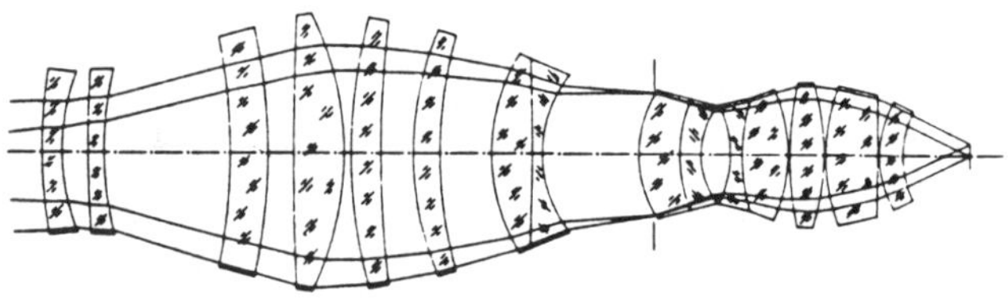

Figure 4. A high performance lens for microlithography.
(Drawing courtesy of Carl Zeiss, Inc., used by permission)

increased. Increasing numbers of elements are needed to give the designer the additional degrees of freedom he needs to maintain control of the image quality. If we take a slightly different point of view, we can attribute this predicament to the designer's choice of rotational symmetry about a unique "optical" axis. As the image field grows larger, the ray bundles from off-axis points traverse paths which are increasingly unlike the paths for the on-axis point. The task of achieving a uniformly high level of aberration correction across a wide image field in the face of this rapidly degenerating symmetry is challenging in the extreme.

Lenses with monocentric symmetry

We are now in the territory of lenses that are too complex and expensive for our purposes, but still do not give us the NA and image field size that we need. It is apparent that an entirely different approach is needed if we are to proceed. One way out of our dilemma is to design a lens based on a different class of symmetry.

If we choose point symmetry instead of rotational axial symmetry, and require all optical surfaces (including the image) to be concentric about a single point, a monocentric optical system is formed. The symmetry of such a system assures us that, if all chief rays pass through the center of symmetry, the image quality will be identical for all field points. As a result, only the field-independent aberrations must be corrected. These include the various orders of spherical aberration and longitudinal chromatic aberration, if the source happens to be broad-band or multi-wavelength. It might appear that this rule of symmetry is so restrictive that we would be unable to design useful lenses of this type; but as we shall see, this is not the case.

A simple two-element monocentric design is shown in Figure 5. Note that the entrance and exit pupils are both located at the center of symmetry and that any line drawn through this point is an identical optical axis. This particular cemented doublet design has slightly less than one wave P-P wavefront aberration at .5 NA, when focused for minimum RMS spot size. The image quality is, of course, constant over the entire 12.75mm unvignetted scan length, and the Lagrange invariant (I') is 3.18mm. Additional monocentric elements could be added to improve the image quality to any desired degree, but as we will see, this may not be the best way to correct the system. Since we have chosen hemispherical rather than full spherical symmetry, the angular field of this particular lens is limited by vignetting as the marginal rays approach the edges of the elements. In this example, the focal length of the lens is 20.0mm, the entrance pupil radius is 6.9mm, and the total lens length is 30mm. The full un-vignetted angular field of 53 degrees approaches 120 degrees if we allow about 30% vignetting. The radius of curvature of the image surface in this case is 13.8mm. (Note that a weakly converging input beam has been used for reasons that will become clear in the discussion of pupil relays.)

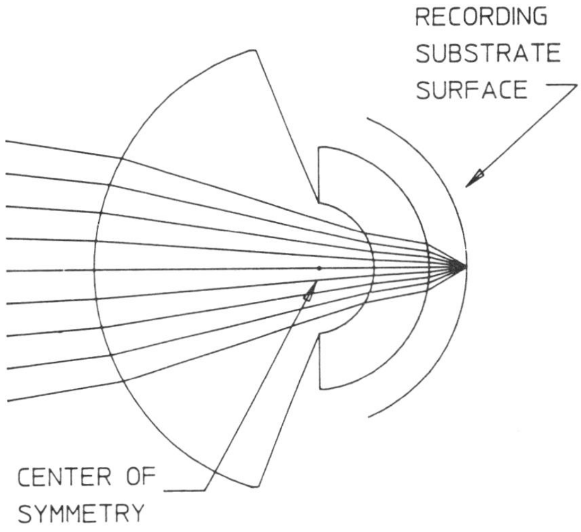

Figure 5. A simple monocentric doublet.

We have solved one problem by designing a simple, very wide field lens that operates at sufficiently high NA, and can provide the necessary image quality. One wave of aberration is obviously not yet good enough, but a practical means of correcting the residual will be described later. In the process of designing the lens, however, we have created several new problems:

1) Focus-error correction by moving the objective lens is not practical.
2) Since the entrance pupil is buried in glass and inaccessible, the scanning element must be imaged onto the pupil with a field element.
3) The image surface is curved and the recording substrate must conform to the scanned arc.

Before examining the the first of these, let us consider how much field could actually be accessed by a practical scanning element.

Limitations due to the scanning element

Much like lenses, scanners are able to access only a limited number of resolvable points. Once again, the Lagrange invariant is a convenient way to measure and describe this ability. If we elect to use a galvanometer mirror scanner, the Lagrange invariant, thus the number of accessible tracks, is determined by the dimensions of the scan mirror and the total output beam deflection angle. The General Scanning G-350[1] is an example of a galvo scanner which offers a large excursion angle in combination with reasonably high speed and adequate mirror dimensions. This scanner has a shaft rotation range of ±25 degrees. This provides a 1-theta full-field scan range of 50 degrees and a 2-theta range of 100 degrees, (1-theta and 2-theta scanning will be described in the next section[2]). In the following analysis, a beam diameter of 14mm, an objective lens NA of .5, and a track pitch of 1.6um have been assumed.

Calculation of the number of accessible tracks in a galvo-limited system

For 1-theta scanning:
Lagrange invariant at the galvo:
$I' = ny\sin(\bar{U})$ ($\bar{y}=0$ at a pupil)
$I' = \sin(25\text{deg}) \times 7\text{mm} = 2.96\text{mm}$

Lagrange invariant at the recording medium:
$I' = n\bar{y}\sin(U)$ ($y=0$ at an image)
$\sin(U)$ = objective NA = .5
$I' = \bar{y}NA = .5\bar{y}$

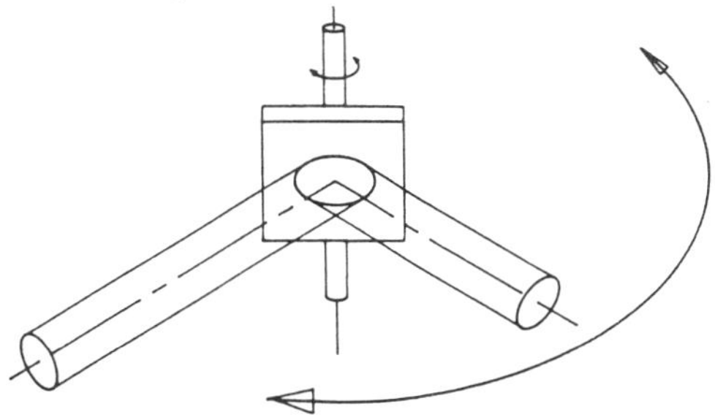

Figure 6. A 2-theta scanner.

Combining the two expressions for I':
I'= 2.96mm = .5ȳ
ȳ = 5.92mm
2ȳ = 11.84mm (full field for flat-field scan)
track pitch = 1.60um

No. tracks = 7400 (flat-field scan)
No. tracks = 7640 (circular field scan)

For 2-theta scanning:
ȳ = 10.72mm
2ȳ = 21.44mm
No. tracks = 13,405 (flat-field scan)
No. tracks = 15,280 (circular field scan)

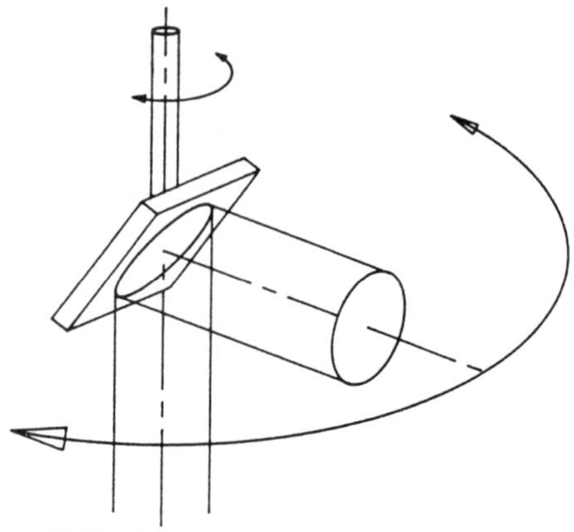

Figure 7. A 1-theta scanner.

If these tracks were arranged at the outside of a 130mm diameter disk and the in-track bit density used for the inside track is equivalent to that of a 16kByte track at a radius of 30mm (typical for current systems), then the following capacities would result for one surface:

Table 1. Accessible capacities of galvo-limited systems

	Flat-field lens	Monocentric lens
1-theta scan	190.1MB	194.6MB
2-theta scan	275.7MB	289.7MB

Note: A linear relationship between radius and track length was used in calculating these capacities. This is not strictly true for the circular scan but the comparison remains valid.

We can safely conclude from this that the limits on the scanned field imposed by the galvo scanner are generous enough to justify more careful study of the imaging optics. If we could design an optical system in which the field was limited by the galvo, it would indeed be interesting.

Scanning systems based on a monocentric objective lenses

The 2-theta geometry shown in Figure 6 is most commonly used and provides an output beam scan angle of twice the deflection angle of the mirror. In this case, the mirror is parallel to the axis of rotation and the input and output beams are perpendicular to the axis of rotation. The 1-theta scanner shown in Figure 7 provides an output beam which scans an angle equal to the mirror rotation angle. This is accomplished by inserting the input beam coaxially with the mirror's rotation axis. The mirror is mounted at 45 degrees to the axis of rotation and the output beam scans in a plane perpendicular to this axis. Both of these scan methods are of interest to us, but one may be far more useful in certain systems than the other.

Let's return now to the three problems created by a monocentric objective lens. First, because this lens would be large and heavy compared to the lenses typically used for optical recording, it seems impractical to correct focus errors by moving the objective lens directly. It is unreasonable to expect to achieve the necessary focus servo bandwidths with any practical focus motor design if this approach is used. The curved field further compounds the problem because the direction of the required motion varies across the field. A more practical approach is to correct focus errors in the input beam prior to both the scanner and the objective. This can be done by first forming an intermediate image in the stationary input beam with a pair of well-corrected lenses, as shown in Figure 8. One of these lenses is actuated directly, using a conventional voice-coil focus motor. The divergence angle of the exiting beam is slightly altered as is the final focus at the recording medium. The NA at the intermediate image is chosen to provide the desired longitudinal magnification between the focus motor and the final focus. Note that the focusing system must be included in the overall optical design to insure that the spherical aberration introduced by conjugate shifts during focus correction is well controlled.

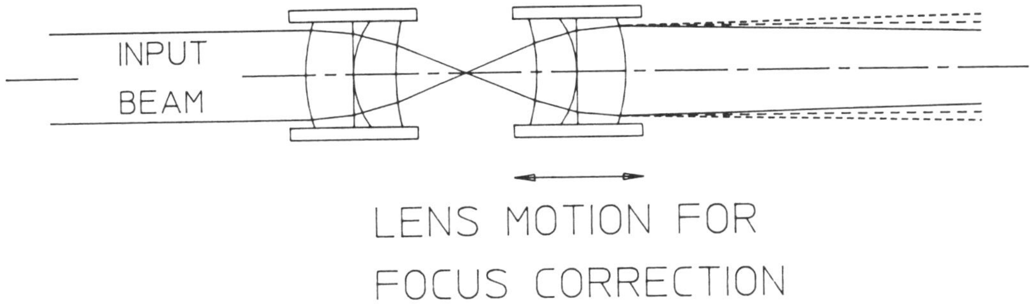

Figure 8. Focus correction prior to the scanning element and the objective lens.

The second problem we noted concerned the location of the entrance pupil of the monocentric lens. In order to maintain the symmetry we have worked to achieve, all chief rays must pass through the center of curvature of the objective lens. This is equivalent to saying that the scanning element must be located either at the entrance pupil of the objective or at a location which is image conjugate to it. Only the latter choice is available since this pupil is in glass. Although various types of pupil relays could be used, only those that approach monocentric symmetry are of interest here.

At this point, some departure from rigorous monocentric symmetry is required, otherwise, all pupils and their images will coincide at one point. A good design choice is a spherical mirror acting as a 1:1 pupil relay as shown in Figure 9. Off-axis aberration types (coma and astigmatism) will be introduced by this relay, but they will be constant over the entire scanned field if the axis of galvo rotation passes through the center of curvature of the relay mirror. This result is of great potential value. It may not be necessary to design a more elaborate relay or objective lens to achieve the desired level of aberration correction. Instead, since all aberrations of both the relay and the lens are constant for all scanned field positions, and since we have easy access to a non-scanning input beam, the final wavefront correction can be done with an aspheric (or holographic) corrector plate. This approach is most practical if the aberrations needed in the corrector plate are small.

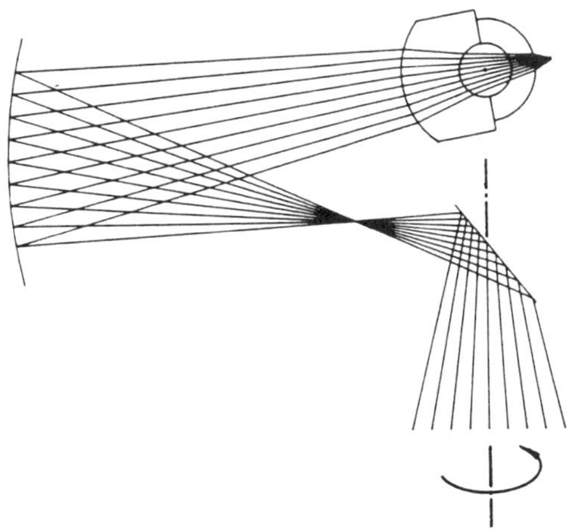

Figure 9. A single-element pupil relay used with a monocentric lens.

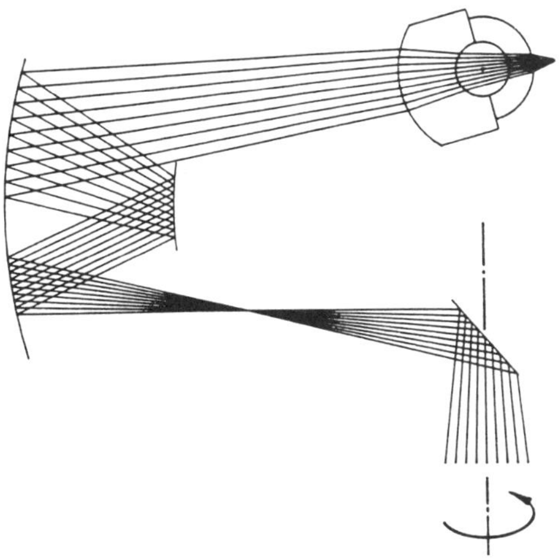

Figure 10. A 2-element pupil relay.

Figure 10 illustrates a two-element reflective relay which preserves the desired rotational symmetry and offers substantially better aberration correction than a single mirror. The two mirrors are spherical and concentric, and provide unit magnification pupil imaging. Note that in both figures above, the scanning takes place perpendicular to the plane of the drawing.

Let us now turn our attention to the issue of the curved image surface. The focus of the scanning beam traverses a circular arc that must coincide closely with the surface of the recording medium. Clearly, the media substrate cannot be a flat disk, but must be a surface of revolution generated by rotating the scanned arc about the line which becomes the axis of media rotation. Several such shapes are possible for a given scanned arc, including one sphere and a variety of torics. Two of these are shown in Figure 11.

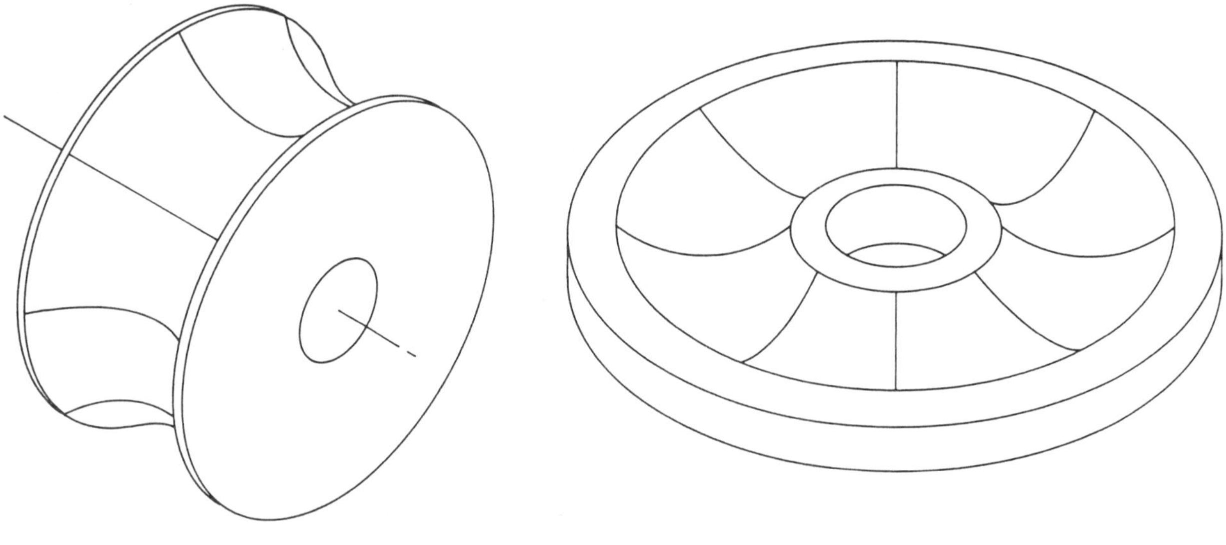

Figure 11. Two of the possible recording substrates for curved-field scanners.

An alternate way to configure the pupil relay isn't a relay at all. Instead of imaging the scan mirror onto the entrance pupil of the lens, two electronically coupled scanning mirrors are used. Figure 12 illustrates this approach. In effect, the two scan mirrors provide two degrees of freedom, allowing simultaneous control of both the angle and the transverse position of the beam entering the objective. This enables us to scan the beam angularly into the objective lens while keeping the chief ray centered in the pupil. Since this "relay" has no power, the input beam can be collimated, and path length variations during scanning will not cause focus shifts. For very large scan angles, it may be desirable to split the second mirror into multiple adjacent scan mirrors, otherwise this mirror becomes large and speed is compromised. This approach has the important advantage that by using plane mirrors and collimated beams, this relay does not introduce any aberration into the beam, or any variation of aberration with scan position at the final focus. Since there are no field aberrations, either a fully-optimized monocentric objective or a corrector plate in the non-scanning space can be used to achieve the required aberration control.

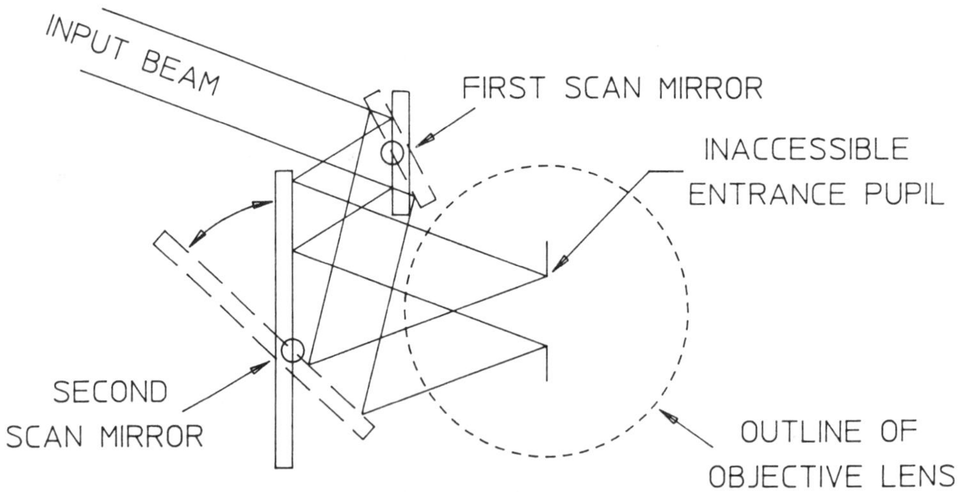

Figure 12. Two electronically coupled galvos acting as both a scanner and a relay.

Flat-field zonal scanners

An entirely different approach offers the advantage of using conventional flat disks and permits the design of an objective lens with unusually relaxed requirements. Instead of requiring the lens to perform well over its entire field, all we ask is that it be reasonably well corrected at one field position. We then take advantage of axial symmetry by scanning the focused spot along a circular zone in the plane of the disk. Figure 13 shows the scanned path of such a system on a disk.

A scanner of this type can be configured in a number of ways. The principal requirements are that the central ray of the output beam must form a constant angle with the lens axis and must always pass through the center of the entrance pupil of the objective. This scanning system also has the property that any uncorrected aberrations that exist do not vary over the scan path. Once again, this makes it possible to correct aberration residuals in the non-scanning input beam rather than fully optimizing the objective lens. In this case, however, the corrector must rotate with the scanning beam, since the aberrations contain field-dependent terms that arise in the coordinate system of the rotating meridional plane. This requirement is satisfied by attaching the corrector plate directly to a 1-theta scanner. Figures 14 and 15 illustrate two embodiments of this approach.

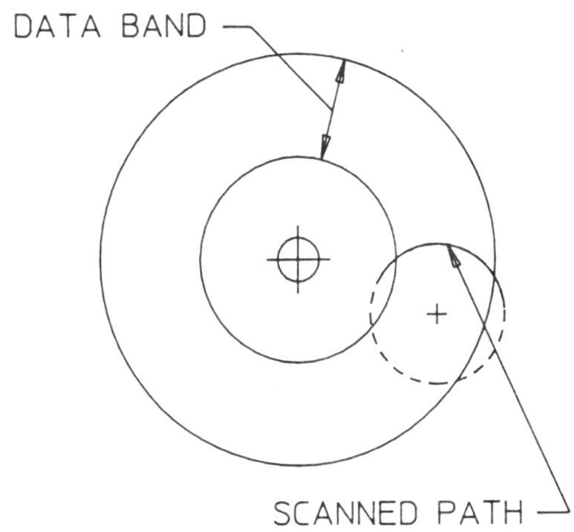

Figure 13. The scanned path of a flat-field zonal scanner.

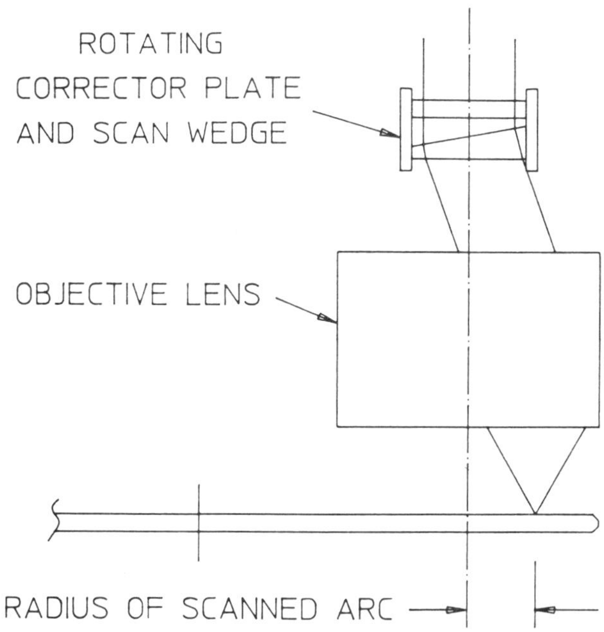

Figure 14. A refractive zonal scanner

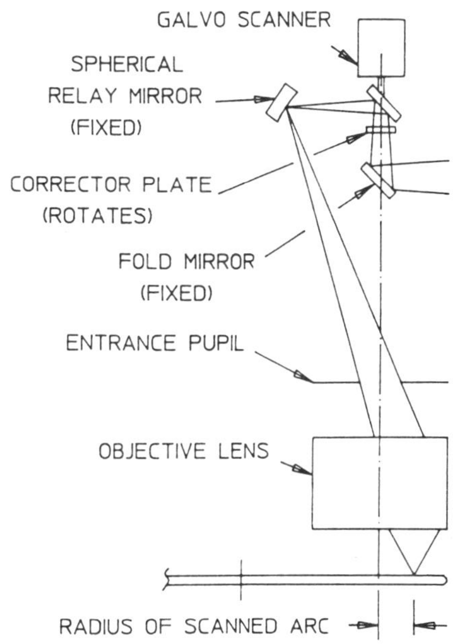

Figure 15. A reflective zonal scanner.

The relay mirror shown in Figure 15 is spherical with its center of curvature on the system axis. The scanning occurs perpendicular to the drawing and the relay mirror is long in that direction.

Figure 16 shows a multi-element objective lens designed to give good image quality at a substantial field radius. Note that this lens is telecentric; i.e., the chief ray is perpendicular to the disk. This allows efficient light recovery during read-out.

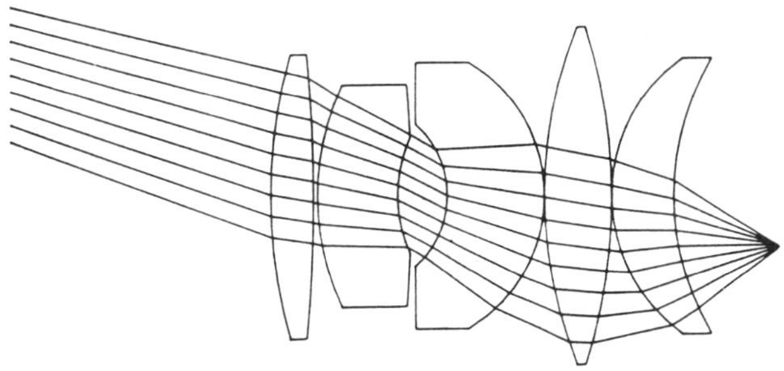

Figure 16. A telecentric objective lens for a zonal scanner.

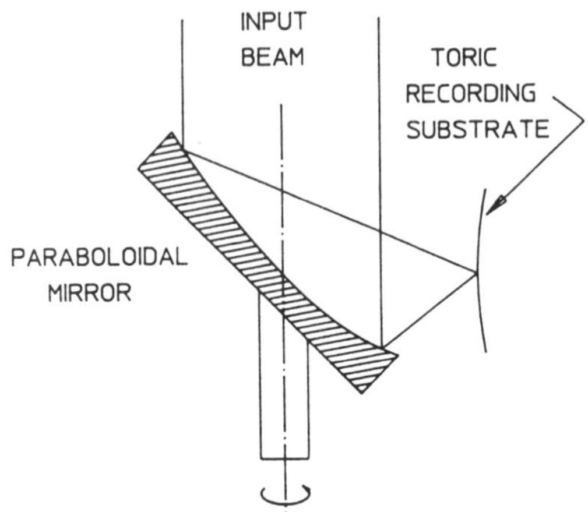

Figure 17. A 1-theta scanner with power.

Other geometries

The design approaches discussed here lend themselves to extension and modification in a variety of ways. Acousto-optic (AO) deflectors can be used in place of scanning mirrors in cases where the additional speed justifies the reduction of scan angle. An interesting variation would be an AO deflector buried in a monocentric lens. In this case, the scanning element is located at the pupil, and only the output side is scanned. Thus, the design of the input side need not be monocentric. We can also consider adding optical power to the scanning element itself, in ways that preserve symmetry during scanning. A simple example of this approach is the paraboloidal mirror scanner shown in Figure 17. The central ray of the input beam coincides with axis of rotation of the mirror in this example.

Some of the design concepts described here could be modified for two-dimensional scanning over large fields. This invites us to consider designs with smaller capacity that do not require rotating media.

Conclusions

A question I've heard asked many times is, "Why can't we use a small mirror to scan the focused beam across the entire radius of the disk at very high speed?" In a sense, this paper is an attempt to answer that question. In summary form, the answer is that something very much like that can be done, but only by departing in important ways from traditional design approaches. There are several design concepts implicit in these departures:

1) Correct focus errors ahead of the objective, in non-scanning space.
2) Utilize maximum symmetry in the design of objective lenses.
3) Take advantage of that symmetry by relaxing the image quality requirements of the objective lens, and correct the residual aberrations in the non-scanning space.
4) Make use of non-planar recording substrates.
5) Do not require the scanned path to be a straight line.
6) Use electronics instead of optics to control the beam path where possible.

Since this investigation was motivated by a quest for improved performance through reduced seek time, let's see if we've gained anything for our efforts. For a galvo-motor like the one referenced previously, we could expect a maximum angular acceleration of about 4.5 rad/sec ^{2} . A constant acceleration seek over an average angular extent of 16.6 degrees would require about 1.6ms. The servo bandwidth would be in the range of 500-1000hz, resulting in about 3ms additional time for settling after the seek. In reality, we can't drive the maximum rated current through the galvo instantaneously, due to the inductance of the coils. To account for this, several additional milliseconds are required (the exact number depends on the driver electronics). A reasonable estimate of the total average seek time would be 8-10ms. This is about an order of magnitude improvement over the current optical drives and a healthy factor of 2 better than most magnetic drives. This is certainly good enough to attract some attention if a practical embodiment were offered at a reasonable price.

References

1. A Product of General Scanning Inc., 500 Arsenal St., Watertown, MA.

2. Marshall, G.F., "Scanning Devices and Systems," Applied Optics and Optical Engineering, Vol. 6, Academic Press, 1980

Reprinted with permission from *Japanese Journal of Applied Physics,* Vol. 26, Suppl. 26-4, pp. 195-197 (1987).

A Floating-Lens Actuator

G.E. van Rosmalen

Philips Research Laboratories, P.O. Box 80.000, 5600JA Eindhoven, The Netherlands

An actuator for optical recorder applications is described in which the objective lens is electromagnetically suspended in a configuration of magnetic coils, free from mechanical contact. The coil configuration is such that the lens can be driven in five degrees of freedom. An optical strategy has been devised to generate the necessary error signals for servo control.

1. Introduction.

The positioning problem encountered in scanning a micron-sized spot over the structure in an optical disc system can be handled using a servo system. Such a servo system has two important aspects. One is the spot position detector which delivers position error signals (in optical disc systems known as radial, tangential and focus error signals), and the second is the actuator which is used to adjust the relative position of the objective lens with respect to the disc.

A common approach is to actuate two degrees of freedom of the lens for focussing and radial tracking. Tangential errors correspond to time-base deviations which can be corrected electronically in the signal processing of the data stream from the disc. However, in video applications, where the signal frequencies are high, an actuator incorporating tangential movements (a so-called 3D actuator) is of practical interest. Although designs of such an actuator are available using mechanical suspension, the limited bandwidth which can be achieved because of mechanical resonances has stimulated the current investigation of a purely electromagnetic suspension system which is described in this paper.

2. The 5D Actuator.

As for any rigid body, the objective lens of an optical recording system has six degrees of motional freedom, these being three orthogonal translations and three rotations. In a 3D actuator, the free objective lens translations are determined by the vertical (focus), radial and tangential spot position control loops. One of the three remaining degrees of freedom, rotation a round the optical axis of the lens, is optically irrelevant. The final two rotations correspond to a tilt of the optical axis of the lens with respect to the optical axis of the rest of the light path and therefore need to be controlled by independent tilt suspensions driven by optically-generated tilt error signals. The 3D actuator which is described consists basically of a 5D actuator in which the three translations are actively driven while the two tilt modes are actively confined to sufficiently small excursions.

3. Error Signal Detection.

To understand the generation of the error signals needed for the servo control systems, it is necessary to consider the design of an optical recording light path incorporating the 5D actuator. The light path shown in Fig. 1 consists of six parts: a laser, a semi-reflecting beam splitter cube incorporating a cylinder lens on one surface, a collimator lens to form a parallel beam, an objective lens with a diaphragm to define the lens pupil and a photodetector. The latter consists of a composite structure containing, in the center, a small quadrant cell detector which permits the normal generation of a focus error signal (using the astigmatic method), the generation of the push-pull radial error signal and the detection of the central aperture signal representing the information in the track of the disc.

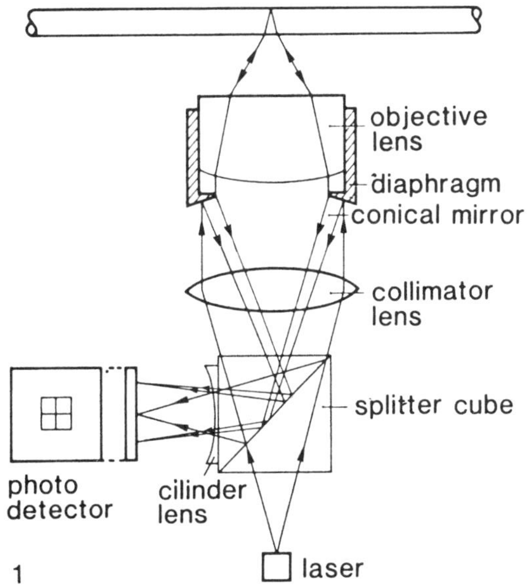

Fig. 1

The tilt error signal is generated by replacing the normal diaphragm stop by a conical ring mirror with its axis parallel to the optical axis of the lens. Light reflected from the parallel incident beam by the conical mirror forms a ring pattern on the photodetector. With an extra ring-shaped diode structure on the detector (see Fig. 1), the position of this light ring with respect to the center of the detector structure can be measured.

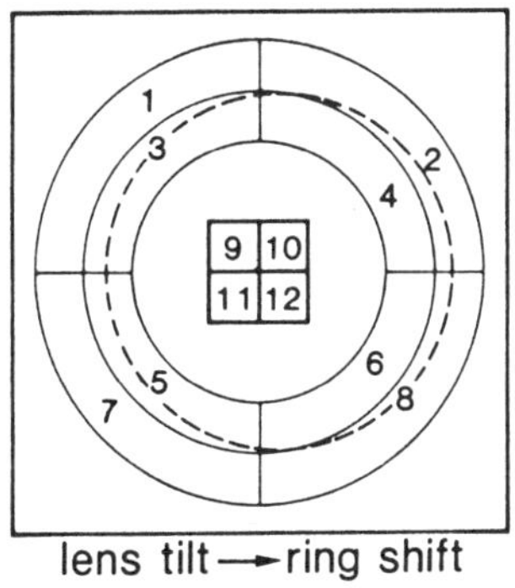

Fig. 2 lens tilt → ring shift

A tilt of the objective lens (including the conical mirror) results in a displacement of the light ring over the diode chip as illustrated by the dotted line in Fig. 2. The light-sensitive areas on the periphery of the detector consist of two concentric rings, each divided into four sectors (numbered from 1 to 8 in Fig. 2) with the dividing line (e.g., between sectors 1 and 3, 2 and 4, etc.) having the same radius as the median radius of the light ring. Considering a lens tilt giving a shift of the light ring to the position of the dotted line in Fig. 2 (a shift parallel to the horizontal axis of the figure), then it can be seen that this shift can be detected by the combination of diode signals:

$$(S_1 + S_7 + S_4 + S_6) - (S_3 + S_5 + S_2 + S_8).$$

A lens tilt around an axis perpendicular to the first is represented by:

$$(S_1 + S_2 + S_5 + S_6) - (S_3 + S_4 + S_7 + S_8).$$

To detect a lens shift corresponding to radial or tangential tracking, a second diaphragm or optical stop is used in the light path (see Fig. 1) which reduces the diameter of the parallel light beam such that, in the neutral position, only half of the conical mirror is illuminated. Considering Fig. 1, if the lens were to move to the left, the right-hand side of the mirror would move more deeply into the beam (and conversely for the left-hand side) leading to an asymmetry in the intensity distribution in the reflected ring. This asymmetry and hence the displacement of the lens can be detected via the diode signal combinations:

$$(S_1 + S_3 + S_5 + S_7) - (S_2 + S_4 + S_6 + S_8)$$

and, in the perpendicular direction:

$$(S_1 + S_2 + S_3 + S_4) - (S_5 + S_6 + S_7 + S_8).$$

For completeness, the focus error signal from the astigmatic method is derived via the combination:

$$(S_9 + S_{12}) - (S_{10} + S_{11}),$$

and the push-pull radial error signal from:

$$(S_9 + S_{11}) - (S_{10} + S_{12}).$$

4. The 5D Drive.

Starting from the availability of error signals representing five degrees of freedom of the objective lens, consideration can be given to the design of a 5D drive for the lens. The requirements on the drive include an insensitivity to rotation around the optical axis, and a sufficient stroke in the orthogonal translations to accommodate the desired focus, radial tracking and tangential tracking movements. The solution which was chosen is illustrated in Figs. 3 and 4. The objective lens was glued into a permanent magnet ring and the lens/magnet combination placed in a special configuration of six banana-shaped coils such that the lens can be magnetically levitated in this structure.

Fig. 3

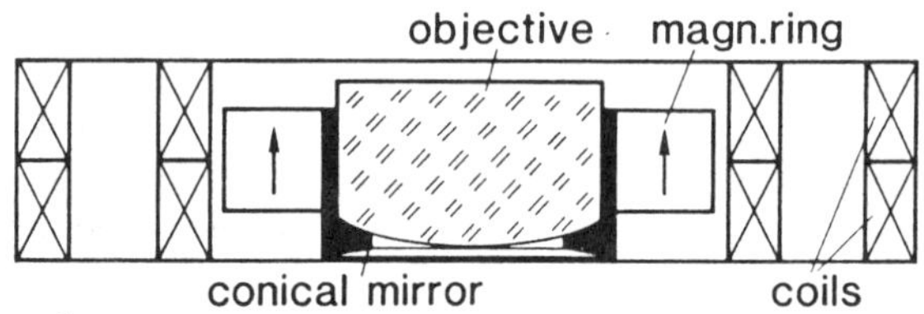

Fig. 4

To explain the operation, the current model can be used. A magnet ring, magnetized axially, can be represented by a ring of air with an electric current along its surfaces. In Figs. 5 and 6, this ribbon current (neglecting the contribution from the inside surface of the magnet ring) and the coil currents are indicated in a simplified form as concentrated currents with the dots and crosses indicating the current direction. As can be seen in Fig. 5, if the current directions in both coils are opposite, a vertical force is generated on the magnet ring.

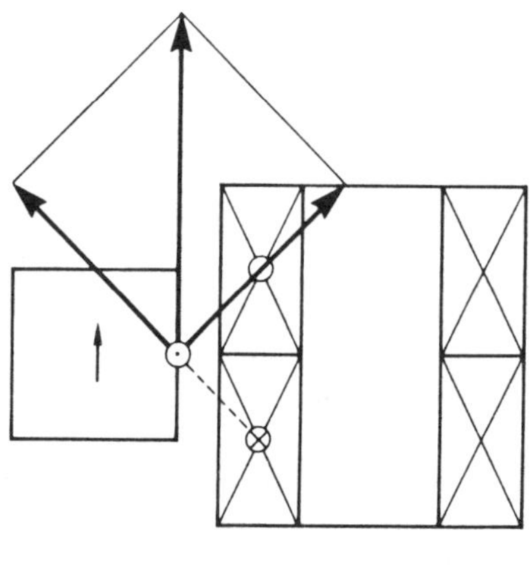

Fig. 5

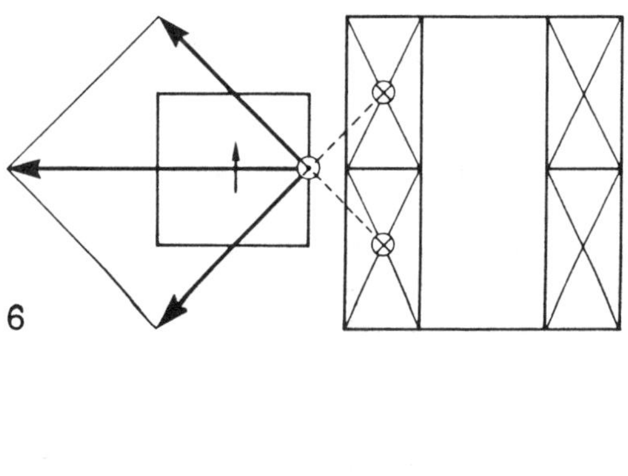

Fig. 6

Parallel currents generate a horizontal force (Fig. 6). Since there are three sets (every 120°) of two coils, each degree of the five degrees of freedom can be driven by an appropriate current combination.

It is convenient to define a coordinate system such that the z-axis corresponds to the focus direction, the y-axis to the tangential movement and the x-axis to the radial movement.

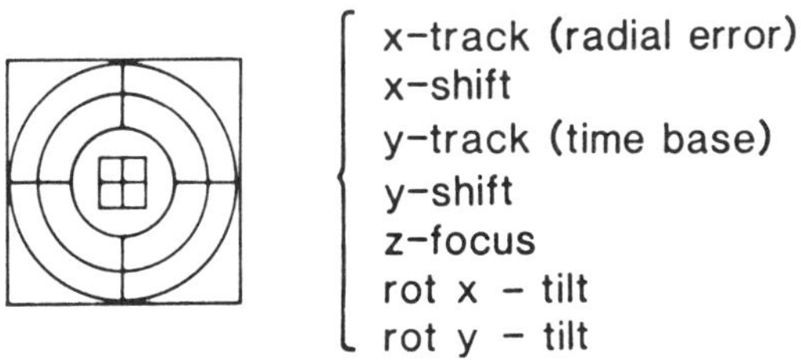

Fig. 7

The two relevant tilts are around the x and y axes. In Fig. 7, the seven position signals normally employed in optical recorders are listed, and, as explained above, are available from the photodetector chip. Fig. 8 illustrates schematically how the five drive signals can drive the relevant degrees of freedom via the configuration of six magnetic coils. The electronic coupling is illustrated in Fig. 9 and consists of a variable gain to provide sufficient reduction of the servo bandwidth and a lead network to give damping for a stable functioning of the servo.

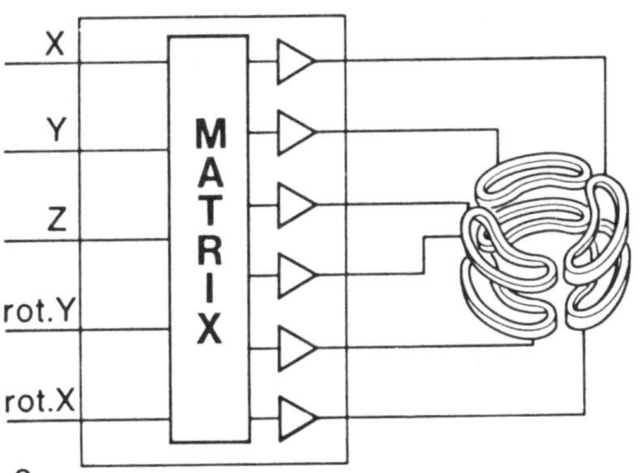

Fig. 8

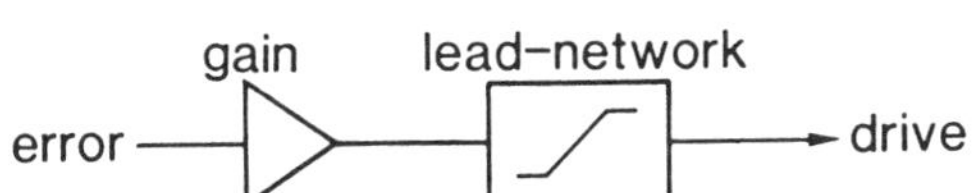

Fig. 9

5. Start-up and Operation.

Starting from the situation with the lens/magnet combination resting in an arbitrary position on the mounting base of the coil structure, the start-up procedure is the following. First, the servo loops corresponding to the x and y rotations (rot x and rot y) are connected. The lens/magnet remains in point contact with the mount but its z-axis is aligned with the incident light beam and the reflected light ring centers on the diode ring structure. If now the servo x and y loops are connected the lens will shift to the middle of the coil structure. In this situation, the 5D actuator has become a 1D actuator in that the lens can only be moved up and down in a straight line. If a disc is present above the lens, then a ramp signal is applied to the z drive to levitate the lens/magnet combination. When the lens arrives at a position near focus (judged from the focus error signal), the focus servo loop is connected and the ramp voltage disconnected. The lens now floats at the focus position, centered and without tilt.

Since large radial strokes have to be accommodated, this actuator is incorporated in a carriage system, in which the carriage has the function of keeping the coil structure nearly centered around the lens. Fine radial and tangential tracking is performed using the x-track and y-track capabilities of the 5D actuator, whereas the x-shift and y-shift signals are used as error signals for the carriage position and turntable motor speed control, respectively.

6. Results and Discussion.

The actuator shown in Fig. 3 has been built into an experimental PAL video disc player. The performance was more than adequate for this application. A noticeable positive feature of this concept is the low level of acoustic noise generation due to the small radiating surface area. The mass of the moving lens/magnet combination was 1 gram. The NdFe permanent magnet had a diameter of 10 mm and a height of 2 mm. The lens was a mono asphere with an NA of 0.5 and a diameter of 6 mm. The six coils were 2 mm in height with a cross-section of 1 mm. To compensate for the gravitational force, the coil power consumption is 40 mW per coil, leading to a total static dissipation of 240 mW. Because the average accelerations due to the tracking movements are limited to 1 g, the total coil power consumption in the video player application was limited to 1 W. The vertical stroke was 2 mm (peak-to-peak) and the radial and tangential strokes 1 mm (p-p).

Reprinted with permission from *Applied Optics,* Vol. 26(13), pp. 2549-2553 (July 1, 1987). ©1987 Optical Society of America.

Autofocusing effect due to wavelength change of diode lasers in an optical pickup

Shigeru Nakamura, Takeshi Maeda, and Yoshito Tsunoda

A novel autofocusing effect in an optical pickup using a diode laser has been discovered. This effect occurs when a diode laser beam is focused on a disk surface through a focusing lens with chromatic aberration, and a certain amount of the reflected light from the disk is fed back to the diode laser. The wavelength of the diode laser changes with the movement of the disk to keep the amount of optical feedback to the diode laser at a maximum. As a result, focusing of the laser beam onto the disk may be automatically controlled without any lens actuators.

I. Introduction

Diode lasers have been successfully employed as light sources in optical disk pickups. In an optical pickup, a light beam emitted from a diode laser is focused into a spot of ~1-μm diameter on the disk surface by a focusing lens, in order to write and/or read data points with a diameter of <1 μm. Since the disk can move in an optical axis direction during disk rotation, an autofocusing mechanism is necessary with an accuracy in the focal depth range of about ± 1 μm.[1] A laser beam reflected from the disk and passed through the focusing lens is transmitted to focus the error detection optics. According to the focusing error signal, the focusing lens is moved in an optical axis direction by a lens actuator so that the focusing position of the laser beam follows the disk movement.

When employing diode lasers in an optical pickup, one of the most serious problems is noise in the output light intensity.[2] It is well known that diode lasers are significantly affected by optical feedback,[3] because the optical gain of a diode laser is high and has a much wider spectrum than cavity mode spacing. In addition, the reflectivity of a laser resonator is low. One of the methods proposed to reduce the diode laser noise in an optical pickup is to purposely feed light back partially to the diode laser after reflection from the optical disk.[4] As a result, a single-mode diode laser is forced to oscillate in multiple longitudinal modes so that the noise level is relatively low and stable against optical feedback.

In our optical pickup where the diode laser noise is reduced by this intentional optical feedback, an extraordinary focus error signal has been discovered: the focus error signal is suppressed to near zero for a disk movement range of plus or minus several micrometers from the original focusing point, without actuating the focusing lens. This indicates that the focusing position of the laser beam automatically follows the disk movement.

This paper presents an experimental study on this novel autofocusing effect in an optical pickup using a diode laser.

II. Extraordinary Behavior of the Focusing Error Signal

Figure 1(a) shows the optical configuration of the optical disk pickup used in our experiments. A laser beam emitted from a diode laser is collimated by a collimating lens. The collimated laser beam passes through a polarization beam splitter and a quarter-wave plate and is then focused onto an optical disk by a focusing lens. The laser beam reflected from the disk passes through the focusing lens again and is partially reflected by the polarization beam splitter, which is used to obtain a focusing error signal. The focusing error detection optics consists of a spherical lens, a cylindrical lens, a knife-edge and a detector split into two parts.[1,5] The knife-edge is inserted between the two astigmatic foci formed by the cylindrical lens when the disk is at the focal plane of the focusing lens; the orientation of the knife-edge is at 45° with respect to the orientation of the astigmatic focal lens. Accordingly, a semicircular laser beam is rotated on the detector according to disk movement as shown in Fig. 1(b), and the focusing error signal can be obtained from the

The authors are with Hitachi, Ltd., Central Research Laboratory, 1-280, Higashi-Koigakubo, Kokubunji, Tokyo 185, Japan.

Received 5 December 1986.

0003-6935/87/132549-05$02.00/0.

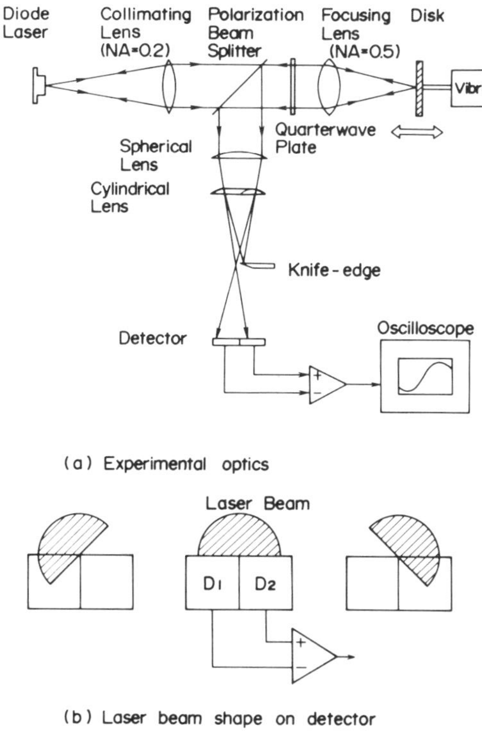

Fig. 1. Experimental setup for measuring focusing error signals in an optical pickup with a diode laser.

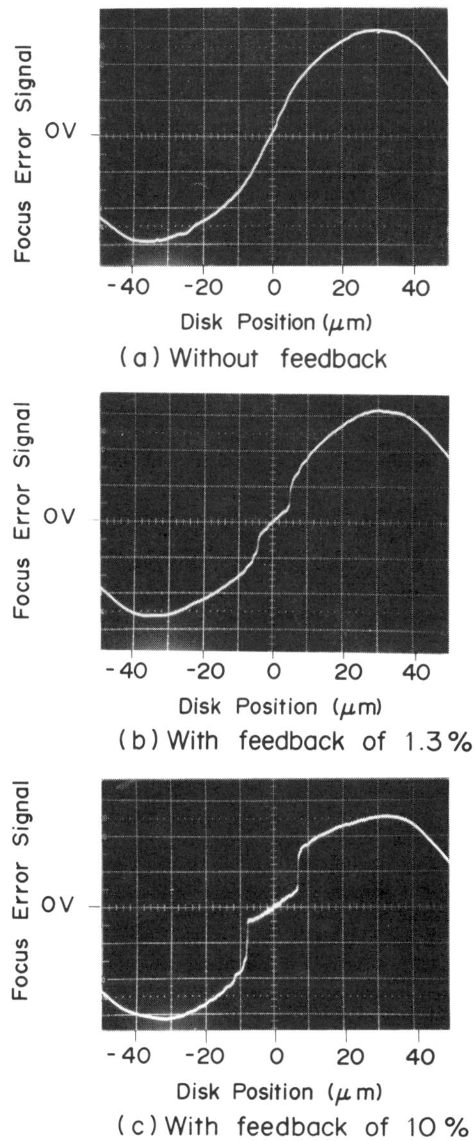

Fig. 2. Focusing error signals measured with various feedback ratios. Chromatic aberration of the focusing lens is 0.7 μm/nm.

difference between signals derived from detector segments D_1 and D_2.

The laser beam reflected from the disk partially passes through the polarization beam splitter and is returned to the diode laser for optical feedback. The feedback ratio, which is the ratio of the feedback light power to the laser output power, can be changed by rotating the quarterwave plate.

Diode lasers used in this experiment were the CSP (channeled-substrate-planar) type of GaAlAs double-heterostructure lasers, with a typical wavelength of ~830 nm. The numerical apertures (N.A.) of the collimating lens and the focusing lens are 0.2 and 0.5, respectively. Maximum value of the feedback ratio is ~10% in this setup. The optical disk is coated with aluminum and moved by a vibrator in the optical axis direction.

A typical focusing error signal measured without feedback is shown in Fig. 2(a). The zero of the disk position means that the disk surface is at the focus. Within a disk position range of about ±15 μm, the focusing error signal changes linearly in proportion to the disk movement. However, when the optical feedback is 1.3%, which is the best feedback ratio for diode laser noise suppression, the focusing error signal suddenly decreases to a half-level within a disk position range of ±4 μm, as shown in Fig. 2(b). When the optical feedback is increased to 10%, the focusing error signal decreases to a 30% level within a disk position range of ±8 μm, as shown in Fig. 2(c).

When diode lasers receive optical feedback, this behavior in the focusing error signal can also be observed with other types of CSP diode laser and focusing error detection optics. The large decrease of signal level in the focusing error signal indicates that the focusing position of a laser beam moves automatically toward the disk surface within the above ranges. This focusing position movement of the laser beam has no unfavorable influence but rather a useful effect (autofocusing) on the autofocusing servo system of the optical pickup. This is because the focusing position of the laser beam moves with the disk movement, without any lens actuator and error detection optics.

The mechanism of this autofocusing effect was investigated and is reported in the next section.

II. Experimental

To investigate the reasons for this focusing position movement, the spectrum of the diode laser was observed. For this purpose, the setup shown in Fig. 1 is

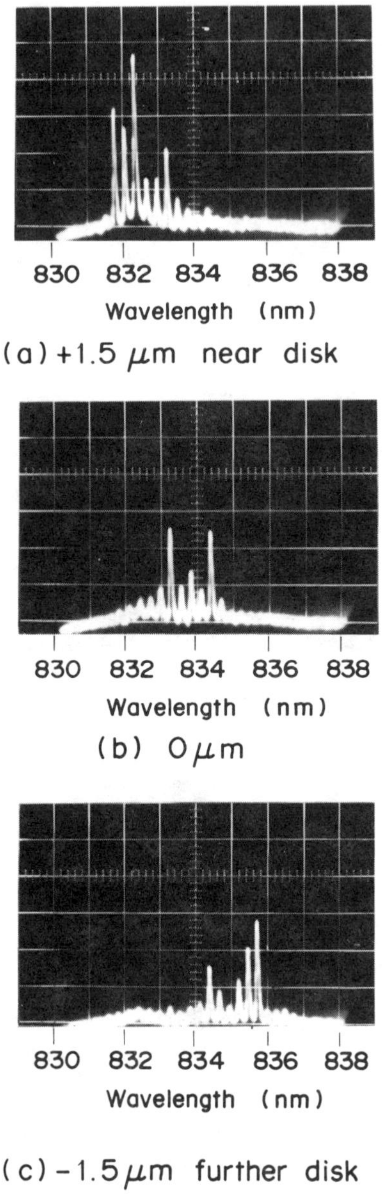

Fig. 3. Observed spectra of a diode laser at various disk positions, where optical feedback is kept at 10%.

modified to use a grating monochromator (plus a TV camera system) in place of the focusing error detection optics. The optical feedback to the diode laser is kept at 10%.

The spectra observed at various disk positions are shown in Fig. 3. Figure 3(b) shows the spectrum when the disk is at the standard position corresponding to the disk position of 0 μm in Fig. 2. The diode laser is oscillating in multiple modes centering around the wavelength of 834 nm. When the disk is moved toward the focusing lens from the standard position by 1.5 μm, the center wavelength changes to 832.3 nm, as shown in Fig. 3(a). The wavelength shift is −1.7 nm. When the disk is moved away from the focusing lens from the standard position by 1.5 μm, the center wavelength changes to 835.5 nm, as shown in Fig. 3(c). The wavelength shift is +1.5 nm.

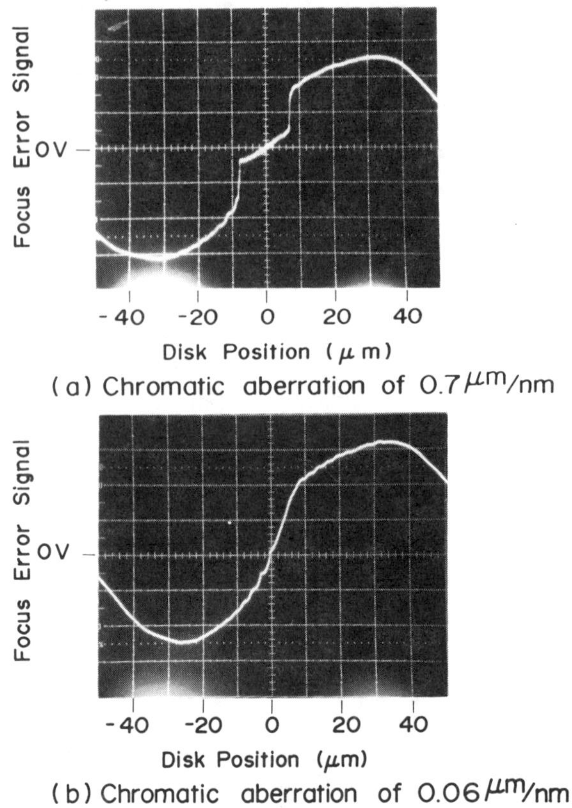

Fig. 4. Focusing error signals for two focusing lenses with different chromatic aberrations, where optical feedback is kept at 10%.

Based on the results of this experiment, we have assumed the following. Focusing lenses generally have more or less chromatic aberration. If the wavelength of the diode laser changes, the focusing position moves due to the chromatic aberration.

To confirm this assumption, we have performed the same measurements using two focusing lenses with different chromatic aberrations. Optical feedback is kept at 10%. The experimental results are shown in Fig. 4. Figure 4(a) shows a focusing lens with chromatic aberration of 0.7 μm/nm; in this case, which is exactly the same as Fig. 2(c), the autofocusing effect was observed. Figure 4(b) shows a focusing lens with chromatic aberration of 0.06 μm/nm; in this case, no autofocusing effect was observed. This experimental result shows that the focusing position movement is caused by the combined effect of the wavelength shift of the diode laser and the chromatic aberration of the focusing lens.

From the experimental results shown in Fig. 3, we quantitatively studied the focusing position movement of a laser beam. The chromatic aberration of the focusing lens used in the Fig. 3 experiment is +0.7 μm/nm. In the case shown in Fig. 3(a), the wavelength shift of the diode laser is approximately −1.7 nm so that the focusing position shift is estimated at 1.2 μm toward the focusing lens. This value nearly coincides with the actual disk movement of 1.5 μm minus the residual off-focus position of 0.4 μm observed in Fig. 2(c). In the case shown in Fig. 3(c), the wavelength shift is approximately +1.5 nm and the chromatic aberration is +0.7 μm/nm, so that the focusing position shift is estimated at 1.1 μm away from the focusing lens. This value also agrees with the actual disk movement of 1.5 μm minus the residual off-focus position of 0.4 μm.

As the observation in Fig. 2 shows, the focusing error is automatically reduced within a certain disk position range; however, it is not completely suppressed. Figure 5 shows the relationship between the focusing error suppression ratio and the optical feedback ratio. The focusing error suppression ratio is defined as the ratio of the focusing error signal change to the original focusing error signal and is calculated using the data of Fig. 2.

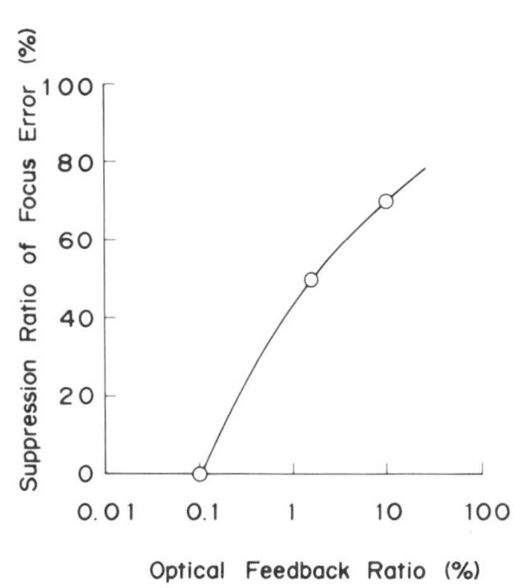

Fig. 5. Focusing error suppression ratio vs optical feedback ratio.

IV. Discussion

A single-mode diode laser usually oscillates in a single longitudinal mode. However, it is well known that a single-mode laser can oscillate in multiple longitudinal modes with optical feedback from external reflectors.[6] An optical disk plus a focusing lens with chromatic aberration serves as the external feedback reflector having wavelength selectivity.

Referencing Fig. 6(a), let us consider a case when the disk is placed at position P, where a laser beam with wavelength λ_P is just focused on the disk surface. The solid curve l_P in Fig. 6(b) shows the optical feedback efficiency in the laser cavity as a function of wavelength, which has a peak at wavelength λ_P. The diode laser cavity modes are also shown in Fig. 6(b). The wavelength separation $\Delta\lambda$ between the cavity modes is given as $\Delta\lambda = \lambda^2/2nL$, where n is the index of refraction and L is the cavity length. In our case, $L = 300\ \mu$m, $n = 3.5$, and $\lambda = 0.83\ \mu$m, so that $\Delta\lambda = 0.33$ nm. Laser oscillation occurs at the modes where the loss in the external cavity is small. Therefore, the oscillating spectrum of the diode laser takes on the form shown in Fig. 6(c). When the disk moves to position Q in Fig. 6(a), the laser beam of wavelength λ_P is defocused on the optical disk, while that of wavelength λ_Q is just focused on the optical disk surface (and mostly returned into the laser cavity), due to the chromatic aberration of the lens. Consequently, optical feedback efficiency into the laser cavity vs the wavelength of the diode laser changes to the dotted line, l_Q, shown in Fig. 6(b), which has a peak at wavelength λ_Q. Thus, the laser spectrum changes and now concentrates around the wavelength λ_Q mode, as shown in Fig. 6(d). This explains why the focusing position of the laser beam automatically follows the moving disk surface.

Usually, an optical disk surface moves vertically in a range of about ±0.5 mm during disk rotation of ~30 Hz, and small disk vibration of several micrometers peak-to-peak in the frequency range from several times to about ten times that of the disk rotation is added due to undulation of the disk surface. Consequently, a wide dynamic range of ~1 mm and a high-speed response of ~1-kHz range are needed for a focusing lens actuator in an optical pickup. To attain automatic focusing control with an accuracy in the focal depth range of about ±1 μm, a high dc gain of ~60 dB is necessary for a focusing servo system including the lens actuator.

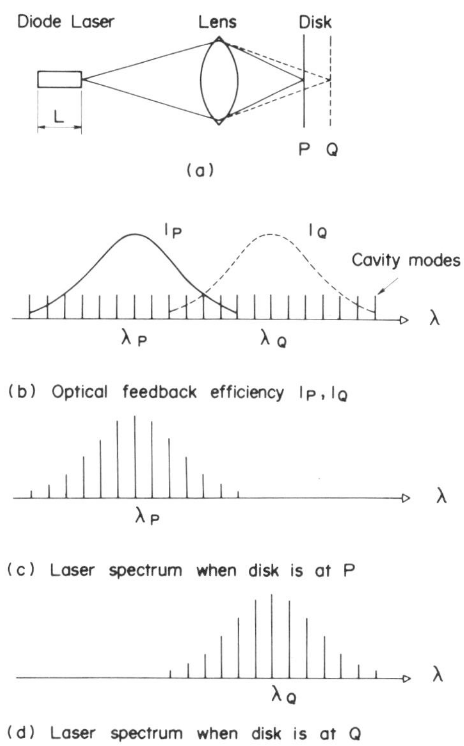

Fig. 6. Explanation of the autofocusing mechanism due to wavelength change of diode lasers.

The autofocusing effect is useful for the focusing servo system in terms of gain and response. In the case of the feedback ratio of 10%, the suppression ratio is 70% as shown in Fig. 5. This suppression provides a servo gain of 10 dB so that the required dc gain of the focus servo system can be decreased to ~50 dB. Furthermore, autofocusing occurs with very high speed. The response time of the wavelength change due to the optical feedback is of the order of a nanosecond for a laser-to-disk distance of ~10 cm. Therefore, the required response for the lens actuator can be decreased to the range of disk rotation frequency (~30 Hz). This can contribute to a simpler and smaller lens actuator.

To obtain a higher suppression ratio, the effective amount of optical feedback into the laser cavity must be increased by using a coupling lens of large numerical aperture and/or coating an antireflection layer on the incident side of the diode laser facet.

The disk movement range where the autofocusing effect occurs is determined by three factors: the spectral width of the diode laser gain for laser oscillation, the chromatic aberration of the lens system, and the level of optical feedback. The gain spectrum of a GaAlAs diode laser has about a 20-nm width.[7] In our experiments, shown in Fig. 2, the diode laser wavelength shift was ±11 nm, which almost covers the total gain spectral width. Therefore, larger chromatic aberration of a focusing lens is needed to extend the autofocusing range. Conventionally, the combination lens used for the focusing lens of an optical pickup consists of spherical convex lenses and concave lenses. Large chromatic aberration can be obtained by using convex lenses of large dispersion glass and concave lenses of small dispersion glass. Serial arrangement of such lenses results in a larger total chromatic aberration. Another approach for larger chromatic aberration is a grating lens. However, its diffraction efficiency is small and making a high N.A. lens is difficult.

Is it possible that an optical pickup with the autofocusing effect may not need a lens actuator? Drastic improvement in both the defocus suppression ratio and the autofocusing range are necessary for an optical disk exhibiting large vertical movement. However, for a disk moving in the range of ~10-μm p–p, a pickup with no lens actuator could be possible with the autofocusing effect.

V. Conclusions

An autofocusing effect without a lens actuator has been observed in a diode laser optical pickup. This novel phenomenon has been attributed to the combined effects of optical feedback by the reflection from the disk and chromatic aberration of a focusing lens, causing the laser wavelength to change according to the disk movement. An autofocusing range of ±8 μm and a focusing error suppression ratio of up to 70% were attained with an optical feedback of 10% and chromatic aberration of 0.7 μm/nm. This effect can be used in the focusing servo system of an optical pickup, decreasing the response of the lens actuator and the gain of the focusing servo system. A short pickup without a lens actuator may be possible for an optical disk of small vertical vibration.

The authors would like to thank N. Chinone, M. Ojima, and Z. Tsutsumi of Hitachi, Ltd. for their encouragement and valuable insight during this work.

References

1. Y. Tsunoda, S. Horigome, and Z. Tsutsumi, "Optical Digital Data Storage Technologies with Semiconductor Laser Head," Proc. Soc. Photo-Opt. Instrum. Eng. **382,** 252 (1983).
2. A. Arimoto and M. Ojima, "Diode Laser Noise at Control Frequencies in Optical Videodisc Players," Appl. Opt. **23,** 2913 (1984).
3. R. Lang and K. Kobayashi, "External Optical Feedback Effects on Semiconductor Injection Laser Properties," IEEE J. Quantum Electron. **QE-16,** 347 (1980).
4. T. Sugiyama and H. Suenahá, "Method and Apparatus for Reducing Semiconductor Laser Optical Noise," U.S. Patent 4,532,619 (1985).
5. S. Nakamurá, T. Maedá, T. Kakú, Y. Tsunodá, W. Takasugi, and T. Kanedá, "Focal Position Detecting Optical Apparatus," U.S. Patent 4,450,547 (1984).
6. T. Kanade and K. Nawata, "Injection Laser Characteristics Due to Reflected Optical Power," IEEE J. Quantum Electron. **QE-15,** 559 (1979).
7. B. W. Hakki and T. L. Paoli, "Gain Spectra in GaAs Double-Heterostructure Injection Lasers," J. Appl. Phys. **46,** 1299 (1975).

Reprinted with permission from *Applied Optics*, Vol. 26(18), pp. 3981-3986 (September 15, 1987). ©1987 Optical Society of America.

Analysis of astigmatic focusing and push–pull tracking error signals in magnetooptical disk systems

Masud Mansuripur

Methods of geometrical optics are used in deriving a simple expression for the focus error signal in terms of the parameters of an astigmatic system. The extent of validity of this result is then examined by comparison with diffraction calculations. Diffraction analysis also permits the study of push–pull tracking error signal from pregrooved disk surfaces.

I. Introduction

Focusing and tracking control subsystems are two essential components of any optical disk drive.[1–7] To achieve high-density recording and readout one must use an objective lens with a large numerical aperture (N.A.), since the spot diameter at focus is inversely proportional to N.A. This choice of N.A., however, results in a small depth of focus which is approximately given by $\lambda/\text{N.A.}^2$. At the typical wavelength of $\lambda = 0.8\ \mu$m and with N.A. = 0.5, the depth of focus is $\sim 3\ \mu$m, thus necessitating the stringent control of the lens position with respect to the storage layer of the optical disk. At the same time, a high-track-density requirement translates into the need for accurate positioning of the focused spot in the radial direction. In practice, with a typical track pitch of 2 μm, one cannot tolerate radial positioning errors of more than 0.1 μm.

A practical method of controlling the spot position, used in almost all optical disk systems in use today, is to mount the objective in a pair of voice coils that supports two independent perpendicular motions along the directions of interest. Using one of several possible techniques for position error detection, the deviation of the spot from the desired location is sensed, and the resulting bipolar signals are fed back to the voice coils. The closed loop system thus obtained can maintain focus and track errors at or below the acceptable levels at all times.

The author is with Boston University, College for Engineering, Boston, Massachusetts 02215.

Received 6 January 1987.
0003-6935/87/183981-06$02.00/0.

Among the many techniques suggested for the derivation of the focus error signal, the astigmatic technique seems to be the most versatile. As shown in Fig. 1, this method employs a mildly astigmatic lens L_2 in the path of the reflected light and a quad photodetector that is placed in the region between the two focal planes of L_2. Although the performance of this system is well understood there seems to be a lack of consensus as to the optimum choice of parameters. Therefore, in the next section, we derive a closed form for the focus error signal based on geometrical optics and discuss the choice of parameters. Section III is concerned with diffraction calculations of the beam pattern at various cross sections of the system; the extent of validity of the geometrical results is also examined in this section.

The track error signal is most easily derived from a pregrooved disk using the push–pull method. The required groove depth for this technique is $\lambda/8$, where λ is the optical wavelength; no additional optical elements are required, and the signals obtained from the quad detector of Fig. 1 contain all the necessary information for control of the tracking voice coil. In contrast to the other optical disk technologies, application of the push–pull method in magnetooptical systems is made simple by the lack of interference from the recorded data. While in the read-only and write-once devices the recorded data modulate the amplitude and/or the phase of the readout beam; in magnetooptics it is the polarization that is modulated, and, as such, both focus and track error signals are rendered independent of the recorded information pattern. Diffraction analysis of the track error signal on pregrooved magnetooptical disks is the subject of Sec. IV.

It is worthwhile mentioning at this point that in practice the beam may be offset from the optical axis by a varying amount that depends on the track eccentricities.[6] The resulting offset from the center of the

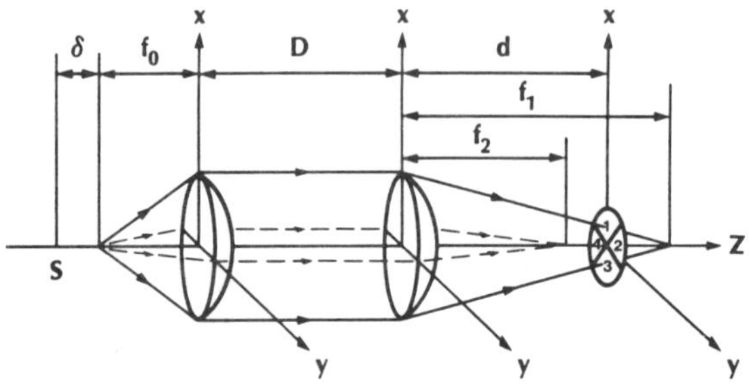

Fig. 1. Schematic diagram of the astigmatic focus-error-detection system. When the disk surface has grooves of depth λ/8, the same system yields the track error signal in the push–pull mode. Note that δ, the distance between S and the focal point of L_1, is twice the actual separation of the disk from the objective's focal plane.

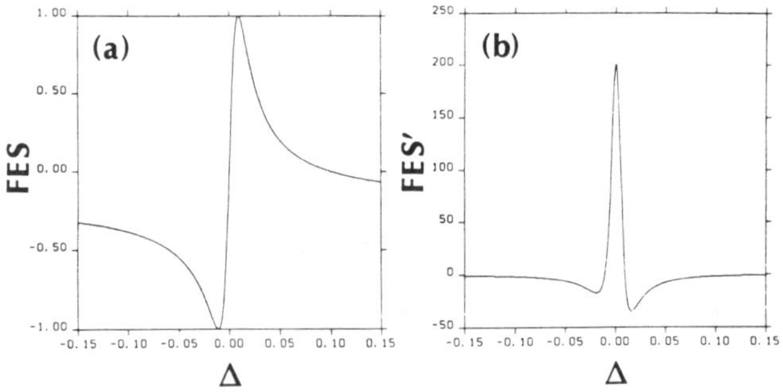

Fig. 2. (a) FES vs Δ as obtained from geometrical optics considerations leading to Eq. (3). Although the plot is obtained for the specific values of $\alpha = 100$ and $\beta = 10$ it is typical of all other values of α and β. The maximum of FES occurs at $\Delta_{max} = 1/(\alpha + \beta)$, while its minimum occurs at $\Delta_{min} = -1/(\alpha - \beta)$. The zero-crossings occur at $\Delta = 0$ and $\Delta = 1/\beta$. (b) Derivative of FES with respect to Δ for $\alpha = 100$ and $\beta = 10$. The extrema of this curve occur at Δ_0, Δ_1, and Δ_2 which are related to α and β through Eq. (5).

quad detector could be several tens of micrometers and, therefore, comparable to the spot size itself. Thus it is advisable to use a photodetector array in place of the quad detector and correct for this offset whose magnitude and direction are available to the system at all times. Our analysis, therefore, need not consider complications arising from the presence of this offset.

II. Analysis of the Astigmatic Focus Error Signal Using Geometrical Optics

An astigmatic focus-error-detection scheme is depicted in Fig. 1 where a point source S, separated along the optical axis by a distance δ from the objective's focal point, is shown to the left of the objective lens L_1. For an infinity-corrected objective, S is the image of the original light source formed by the objective and the reflecting mirror that is the disk surface. δ is thus twice the distance between the disk and the focal plane of L_1. We adopt the following sign convention: δ is positive when the source has moved away from L_1 and negative when it has moved toward L_1. The remaining system parameters are defined as follows:

f_0 = objective's focal length;
D = distance between the objective L_1 and the astigmatic lens L_2;
f_1, f_2 = focal lengths of the astigmatic lens ($f_1 > f_2$);
d = distance of the detector plane from the astigmatic lens.

The spot formed on the detector plane is in general an ellipse with horizontal and vertical diameters h and w, respectively. When S is at the focus of L_1 we require that h be equal to w. This requirement fixes the position of the detector plane at

$$d = \frac{2f_1f_2}{f_1 + f_2}, \tag{1}$$

without imposing restrictions on other parameters. Next we define the focus error signal as

$$\mathrm{FES} = \frac{h^2 - w^2}{h^2 + w^2} \tag{2}$$

and carry out straightforward, although somewhat tedious, calculations based on geometrical optics that yield

$$\mathrm{FES} = \frac{2\alpha(1 - \beta\Delta)\Delta}{1 - 2\beta\Delta + (\alpha^2 + \beta^2)\Delta^2}. \tag{3}$$

Here $\alpha = 2f_1f_2/(f_1 - f_2)f_0$ is a measure of astigmatism for L_2, normalized by the objective's focal length; $\beta = (D/f_0) - 1$ is the normalized distance between L_1 and L_2; and $\Delta = \delta/f_0$ is the normalized separation of S from the objective's focal point.

A typical plot of FES vs Δ for fixed values of α and β is shown in Fig. 2(a). Note that the maximum value of the function is +1 and is attained at $\Delta_{max} = 1/(\alpha + \beta)$, while its minimum is −1 and is achieved at $\Delta_{min} = -1/(\alpha - \beta)$. The first zero-crossing occurs at $\Delta = 0$, and the slope of the curve at this point is 2α. The second zero-crossing occurs at $\Delta = 1/\beta$ independent of α. In the limit when Δ goes to ±∞ the function approaches $-2\alpha\beta/(\alpha^2 + \beta^2)$. The derivative of FES with respect to Δ is given by

$$\frac{d}{d\Delta}\mathrm{FES} = \frac{2\alpha[1 - 2\beta\Delta - (\alpha^2 - \beta^2)\Delta^2]}{[1 - 2\beta\Delta + (\alpha^2 + \beta^2)\Delta^2]^2}. \tag{4}$$

Figure (2b) shows a plot of FES′(Δ) vs Δ for fixed values of α and β. The extrema of FES′(Δ), which are the inflection points of FES(Δ), occur at

$$\Delta_n = \frac{\beta + 2\alpha\cos(\{2n\pi + \arccos[2\alpha\beta/(\alpha^2 + \beta^2)]\}/3)}{\beta^2 - \alpha^2}; \quad n = 0,1,2. \tag{5}$$

Note, in particular, that the maximum slope of FES does not occur at $\Delta = 0$ but rather at $\Delta = \Delta_2$; however, as long as β is much less than α (and this is almost always the case in practice) one can verify that the two slopes are nearly the same.

To put the matters in perspective consider the typical values of system parameters shown in Table I. For these parameters $\alpha = 421$ and $\beta = 4.26$; thus the range of positions of the disk (with respect to the focal plane) over which the FES is suitable for use in closed loop feedback control is $(\Delta_{min}f_0/2, \Delta_{max}f_0/2) = (-4.56\ \mu\mathrm{m},$

Table I. Numerical Values of the System Parameters used in the Computations

Parameter	Definition	Numerical value
λ	Wavelength	0.8 μm
f_0	Objective's focal length	3.8 mm
D	Distance between L_1 and L_2	20.0 mm
f_1,f_2	Focal lengths of L_2	20.25 mm, 19.75 mm
d	Distance between L_2 and detectors	20.0 mm
N.A.$_0$	Objective's N.A.	0.50
N.A.$_1$	Astigmatic lens' N.A.	0.12
r_0	Incident Gaussian beam's $1/e$ radius at L_1	2.2 mm
Ω	Track pitch	2.4 μm
γ	Groove width	1.2 μm
ζ	Groove depth	0.1 μm

+4.47 μm). Note in comparison that the depth of focus (λ/N.A.^{2}) for this particular objective is 3.2 μm.

If the second zero-crossing of the FES curve in Fig. (2a) is considered as a measure of the acquisition range, $f_0/2\beta$ = 446 μm is the acquisition range of the preceding system. However, because of the SNR limitations and diffraction effects, the practical acquisition range is expected to be somewhat lower than this analysis indicates.

The focus error signal of Eq. (3) is reduced by the effects of diffraction. These effects are studied in the next section.

III. Effects of Diffraction on the Focus Error Signal

In this section we present the results of numerical calculations for the system of Fig. 1 with parameters as specified in Table I. For a detailed description of the numerical scheme used in these computations the reader is referred to Ref. 8. The assumed amplitude distribution of the incident beam on the objective is Gaussian with e^{-1} radius r_0 = 2.2 mm, which is equal to the objective's aperture radius. The pregrooved disk is parallel to and in the vicinity of the focal plane of L_1; its groove cross section is rectangular with 0.1-μm depth, 2.4-μm track pitch, and equal reflectivity over the land and groove. The land and groove have equal widths, and the tracks are oriented at 45° with respect to the axes of the astigmatic lens. In the following discussion all lengths are normalized by the wavelength λ and, therefore, expressed as dimensionless quantities.

Figure 3(a) shows the intensity distribution at the focal plane of the objective. Had the incident beam been plane, the standard Airy pattern would have been expected at the focal plane, while for a Gaussian beam we expect a Gaussian distribution. However, the incident beam here is a truncated Gaussian, and therefore, the focused spot of Fig. 3(a) shows a faint ring around a central bright spot. The radius of the bright spot is ~1.25λ, close to the expected value of 0.6λ/N.A. When the disk is exactly at the focal plane and the spot centered on the groove, the reflected beam has the intensity distribution of Fig. 3(b) at the plane of L_1, the deviation of this pattern from the original Gaussian

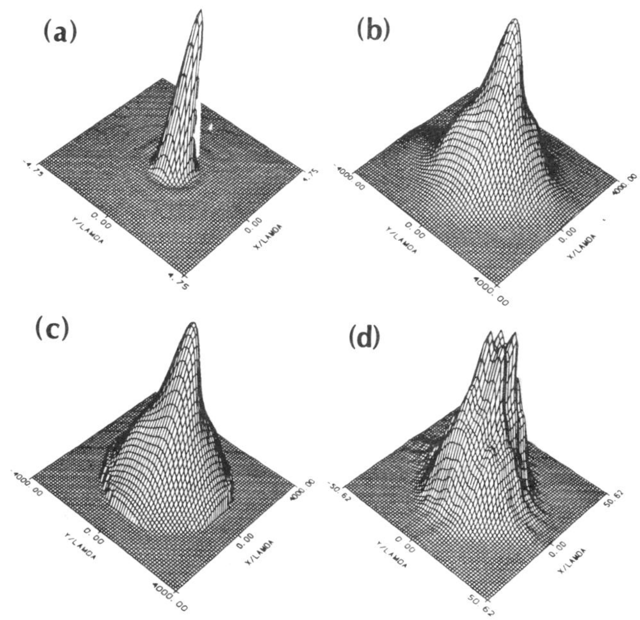

Fig. 3. Distribution of the reflected light intensity at various cross sections of the system of Fig. 1 with parameters specified in Table I. The assumed incident beam at L_1 has a Gaussian amplitude distribution with $1/e$ radius equal to 2.2 mm (same as the radius of L_1). The disk is in the focal plane of the objective, and the focused spot is ontrack: (a) intensity distribution at the disk surface; (b) intensity distribution at the plane of the objective lens L_1; (c) intensity distribution at the plane of the astigmatic lens L_2; (d) intensity distribution at the plane of the photodetectors.

being mainly due to the groove structure of the disk. The finite aperture of the objective blocks some of the light at this point before it reaches the astigmatic lens L_2. The size of the beam and the traveling distance are such that the propagation from L_1 to L_2 is best described by the Fresnel diffraction equations.[8] In fact, the propagation distance is so short that the only significant change in the intensity distribution is the peripheral blocking at the objective's rim as shown in Fig. 3(c), which is the intensity distribution at the plane of L_2. Finally, Fig. 3(d) shows the distribution at the plane of detectors where a quad detector receives equal amounts of light on its four sections. If these four signals are denoted by Q_1, Q_2, Q_3, and Q_4 and the focus error signal is defined as

$$\text{FES} = \frac{(Q_1 + Q_3) - (Q_2 + Q_4)}{Q_1 + Q_2 + Q_3 + Q_4}, \qquad (6)$$

FES is equal to zero for the distribution of Fig. 3(d).

Figures 4–6 are similar to Fig. 3 except that they correspond to different positions of the disk with respect to the objective's focal plane. In units of λ, these figures represent out-of-focus distances of +5, +3, and −3, respectively. The sign convention is the same as before, namely, positive distances are those that are farther from the lens. In all these figures the beam is centered on the groove, and thus the tracking error is zero. It is clear from (d) in each figure that two of the detectors receive more light than the other two, and,

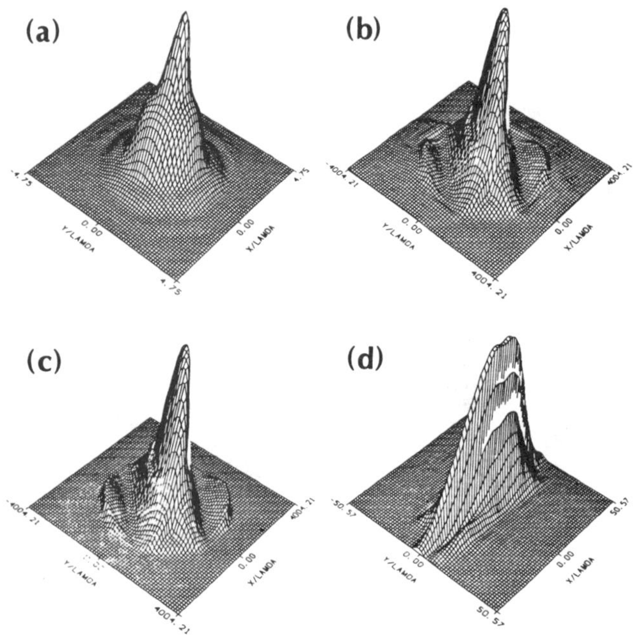

Fig. 4. Same as Fig. 3 but for the disk being out-of-focus by $+5\lambda$. The plus indicates that the disk is far from the lens.

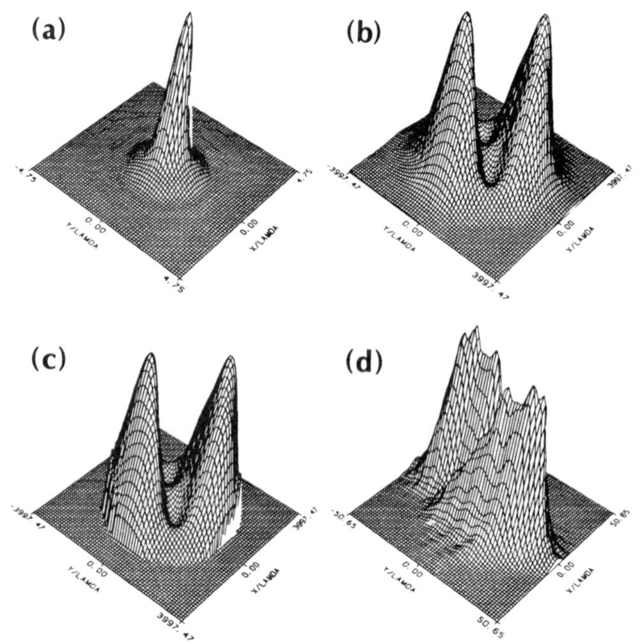

Fig. 6. Same as Fig. 3 but for the disk being out-of-focus by -3λ.

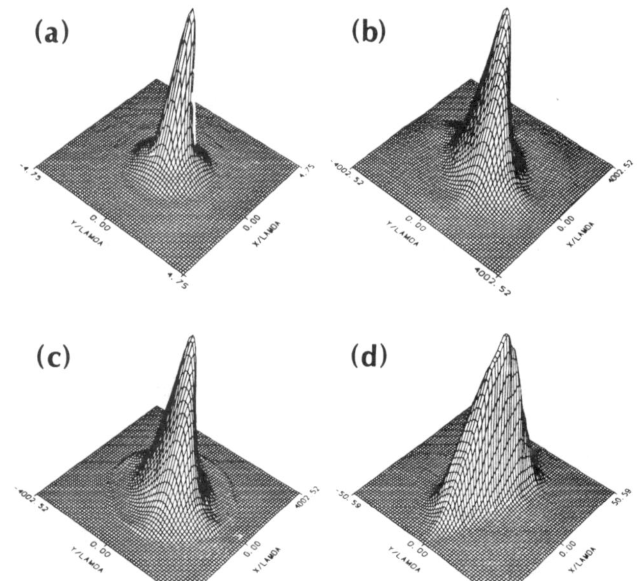

Fig. 5. Same as Fig. 3 but for the disk being out-of-focus by $+3\lambda$.

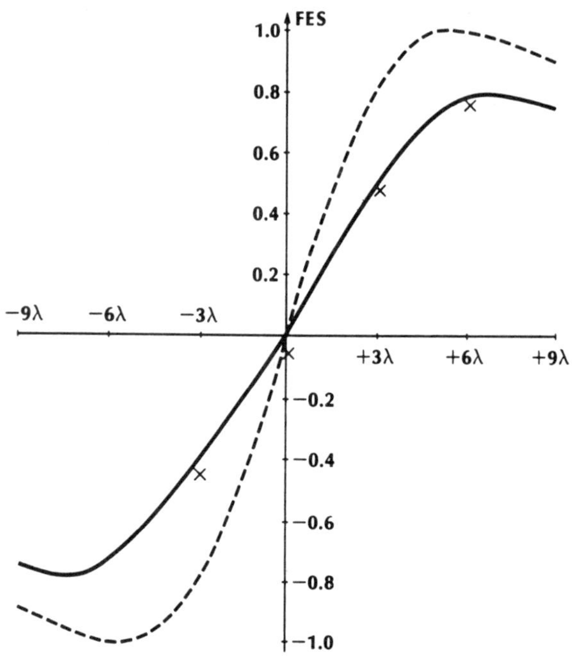

Fig. 7. Focus error signal vs focus error distance for the system of Fig. 1 with parameters of Table I. The broken curve is the geometric result of Eq. 3, while the solid curve is obtained from diffraction calculations. The × points represent the focus error signal when in the diffraction calculations the spot is assumed offtrack by various amounts.

therefore, the net focus error signal as defined by Eq. (6) is nonzero.

The results of these diffraction calculations are summarizd in Fig. 7 where the FES of Eq. (6) vs focus error distance is plotted (solid curve) and compared with the geometrical optics result of Eq. (3) (broken curve). These computations show quite clearly that the geometric calculation can accurately predict the linear range of the FES signal and that diffraction effects play a secondary role in evaluating the system performance. Finally, the points indicated by × in Fig. 7 correspond to the FES obtained when the beam is off-track by various amounts; apparently, the presence of tracking error has only a small effect on the focus error signal.

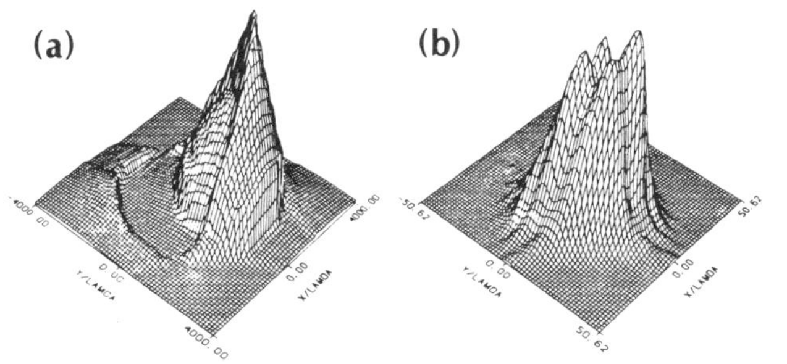

Fig. 8. Intensity distribution after reflection from the disk: (a) at the plane of the objective lens and (b) at the plane of the detectors. The disk is in the focal plane of L_1, but the focused spot is offtrack by -0.6 μm.

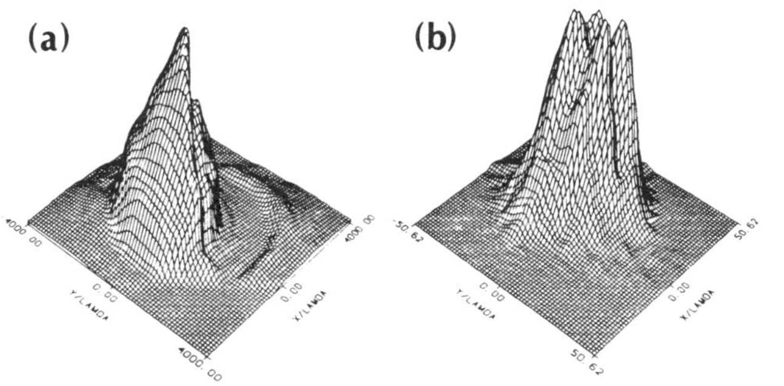

Fig. 9. Same as Fig. 8 but for an offtrack distance of $+0.45$ μm.

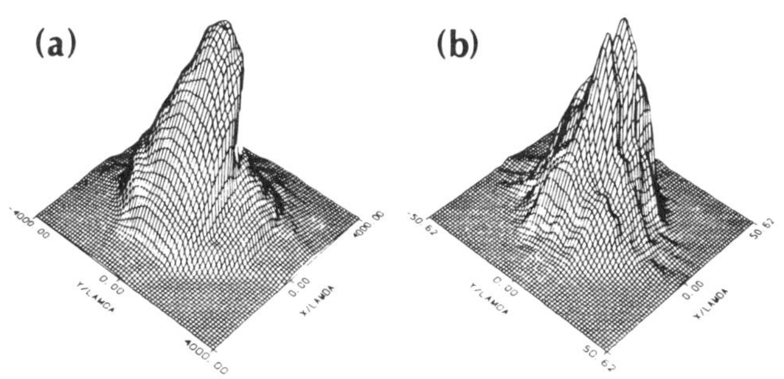

Fig. 10. Same as Fig. 8 but for an offtrack distance of $+0.15$ μm.

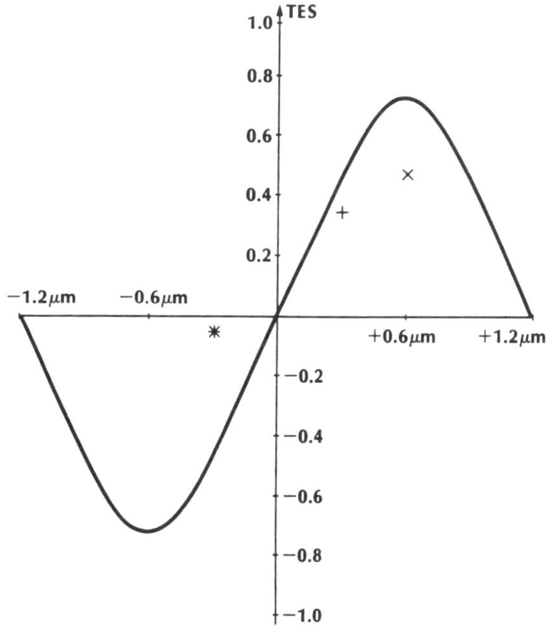

Fig. 11. Track error signal vs position of the spot as obtained from diffraction calculations. The solid curve corresponds to situations where the disk is in-focus. The points indicated by ×, +, and * represent TES when the disk is out-of-focus by -3λ, $+3\lambda$, and $+6\lambda$, respectively.

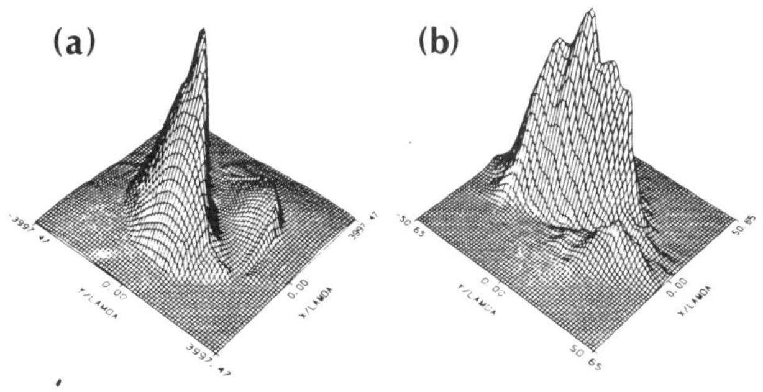

Fig. 12. Intensity distribution after reflection from the disk: (a) at the plane of the objective lens and (b) at the plane of the detectors. The disk is out-of-focus by -3λ, and the focused spot is offtrack by $+0.6$ μm.

IV. Track Error Detection

In terms of the signals Q_1, Q_2, Q_3, and Q_4 obtained from the quad detector, the push–pull track error signal is given by

$$\mathrm{TES} = \frac{(Q_1 + Q_2) - (Q_3 + Q_4)}{Q_1 + Q_2 + Q_3 + Q_4} \cdot \tag{7}$$

As long as the beam is centered on the track, this track error signal is equal to zero. Once the beam moves away from the track center the imbalance of $Q_1 + Q_2$ with respect to $Q_3 + Q_4$ generates a bipolar signal suitable for feedback control of the beam position. Figures 8–10 correspond to situations where the beam is focused but is offtrack by -0.60, $+0.45$, and $+0.15$ μm, respectively. In each figure, (a) represents the reflected intensity distribution at the objective L_1, and (b) is the pattern of intensity at the detectors. Note in all these cases that the intensity distribution on the detectors is similar to that of Fig. 3(d), when the spot is in-focus and ontrack, with the only major change being a shift of the distribution away from the center of the detector plane.

The shift of the intensity distribution on the detector plane, resulting from the shift of the focused spot from the groove center, generates a track error signal according to Eq. (7). The resulting TES vs position of the spot on the disk is plotted in Fig. 11. The points identified by ×, +, and * in Fig. 11 represent the TES when the disk is out of focus by -3λ, $+3\lambda$, and $+6\lambda$, respectively. Unlike FES, which is only slightly affected by the accuracy of tracking, it is clear from Fig. 11 that the inability to maintain focus can significantly reduce the TES.

Finally, in Fig. 12 we show the intensity patterns when the spot is both out-of-focus and offtrack. In this figure (a) and (b) correspond, respectively, to the intensity at the plane of L_1 and on the detectors when the out-of-focus distance is -3λ and the beam is off-track by $+0.6$ μm.

V. Concluding Remarks

We have derived an expression for the focus error signal in terms of the parameters of an astigmatic focus error detection system and shown that the performance depends on two lumped parameters which we have defined as α and β. In fact, for most practical purposes, β turns out to be a minor influence whereby the choice of α alone determines the performance of the focusing servo. Given that the track error signal (at least to a first approximation) is determined by the groove structure, we conclude that the only criterion for choosing the remaining system parameters is the compactness of the head assembly in face of the spatial constraints on the positioning of the elements.

References

1. G. Bouwhuis and J. J. M. Braat, "Recording and Reading of Information on Optical Disks," in *Applied Optics and Optical Engineering, Vol. 9*, R. Shannon and J. C. Wyant, Eds. (Academic, New York, 1983).
2. W. H. Meiklejohn, "Magneto-optics: A Thermomagnetic Recording Technology," IEEE Proc. **74,** 1570 (1986).
3. M. Yamamoto, A. Watabe, and H. Ukita, "Optical Pregroove Dimensions: Design Considerations," Appl. Opt. **25,** 4031 (1986).
4. T. Murakami, K. Taira, and M. Mori, "Magnetooptic Erasable Disk Memory with Two Optical Heads," Appl. Opt. **25,** 3986 (1986).
5. S. L. DeVore, "Radial Error Signal Simulation for Optical Disk Drives," Appl. Opt. **25,** 4001 (1986).
6. D. K. Cohen, W. H. Gee, M. Ludeke, and J. Lewkowicz, "Automatic Focus Control: The Astigmatic Lens Approach," Appl. Opt. **23,** 565 (1984).
7. J. J. M. Braat and G. Bouwhuis, "Position Sensing in Video Disk Readout," Appl. Opt. **17,** 2013 (1978).
8. M. Mansuripur, "Distribution of Light at and Near the Focus of High Numerical Aperture Objectives," J. Opt. Soc. Am. A **3,** 2086 (1986).

Reprinted with permission from *Japanese Journal of Applied Physics,* Vol. 26, Suppl. 26-4, pp. 187-190 (1987).

A New Focus Servo Method for Magneto-Optical Disk Systems

Michio Oka, Atushi Fukumoto, Kiyoshi Osato, and Shigeo Kubota

Audio-Video Technology Center, Sony Corporation
1-7-4, Konan, Minato-ku, Tokyo, 108 Japan
Telephone:(03) 458-7880 Facsimile:(03) 458-7343

The influence of birefringence of a polycarbonate disk upon focus error signals in a magneto-optical disk system is analyzed. We propose a useful focus servo method employing a micro-prism, which is well suited for sampling tracking method.

1. INTRODUCTION

Magneto-optical (MO) disk systems for data storage are quickly evolving for practical use. For this application, injection-molded polycarbonate (PC) disk substrates have attracted many manufacturers for its mass-producibility and reliability. Although efforts to reduce the birefringence have been made, PC substrates tend to have a significant amount of birefringence in the direction perpendicular to the surface of the substrate. Since the magneto-optical pick-up detects the small Kerr rotation angle, it is inevitably sensitive to birefringence. Recently the influence of birefringence on the S/N ratio of a magneto-optical signal has been reported. [1], [2], [3]

In this paper, we analyze the influence of birefringence upon focus error signals employing a model of the biaxial refractive index ellipsoid. Since the incident beam is focused through the substrate, the influence of birefringence greatly differs at each incident angle. Since the polarization state of the reflected beam depends on the incident angle, it is spatially in homogeneous. This results in an intensity distribution anomaly after the beam passes through the analyzer and it causes the offset of focus error signal in some configurations of optical pick-ups. The offset of focus error in an auto-focusing loop leads to defocusing, which may degrade data reliability.

Although a simple solution to avoid the offset is to add non-polarized optical components to the focusing servo, it is not allowed for a higher grade system which requires a simplified and miniaturized optical pick-up.

The astigmatism focus method and the knife-edge method are generally used to detect focus error in optical disk systems. The offset in focus error signal can be calculated by integrating the intensity distribution over the area of each photo detector pattern. Calculated results indicate that our newly developed Micro-Prism Detector (MPD) method has small offset.

The principle of the MPD method is to compensate the offset of the knife edge focus error signal by subtracting the complimental knife edge focus error . To embody this method, we devised the hybrid Micro-Prism Detector, a detector on which a micro prism with a dimension of 1mm X2mmX1mm is attached.

Combining this method with the sampling tracking method, we constructed a compact magneto-optical pick-up, which was satisfactorily used with a mass-produced PC disk substrate. Since the sampling tracking method does not require any special optical configulation to detect the tracking error signal, we can take advantage of one detector detection of magneto-optical signal and servo signal, which contributes to the simpler and lighter optical pick-up.

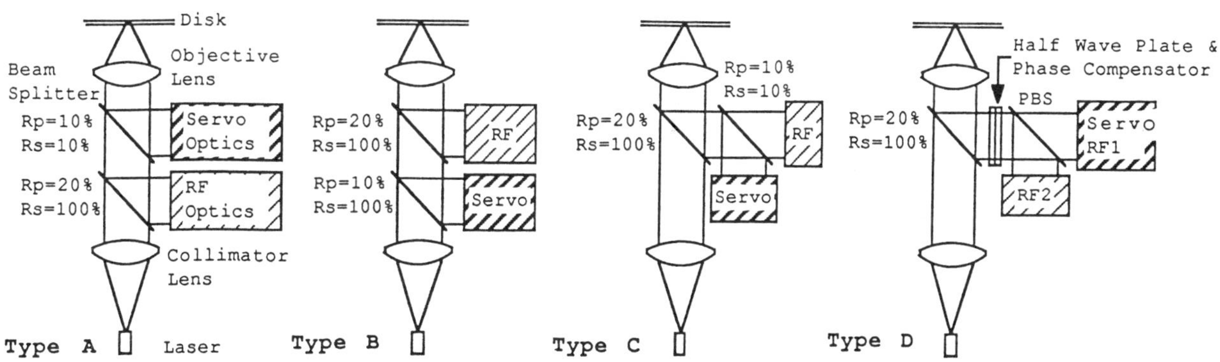

Fig.1 Various Magneto-Optical Pick-Ups

2. ANALYSIS OF BIREFRINGENCE

The influence of birefringence upon focus error signals depends on the configulation of optical pick-up. We analysed four types as shown in Fig. 1. The effect of birefringence can be well understood employing Jones Matrix Formalism. [4] For example, the wave vector on the servo detector of Type D optical pick-up can be written as follows.

$$E\begin{pmatrix} x \\ y \end{pmatrix} = R(45°)T(1,0)R(-45°)C(\delta)T(rp,1)\, R(\alpha)C(2\phi)R(-\alpha)\begin{pmatrix} 1 \\ 0 \end{pmatrix}, \qquad (2\text{-}1)$$

where

$$R(\phi)=\begin{pmatrix} \cos\phi & -\sin\phi \\ \sin\phi & \cos\phi \end{pmatrix}$$

$$C(\phi)=\begin{pmatrix} \exp(i\phi/2) & 0 \\ 0 & \exp(-i\phi/2) \end{pmatrix}$$

$$T(a,b)=\begin{pmatrix} a & 0 \\ 0 & b \end{pmatrix}$$

ϕ single pass birefringence of substrate
α azimuth of birefringnece
δ phase compensator
rp amplitide reflection coefficient of beam splitter.

In the analysis, we employ a model of the biaxial refractive index ellipsoid, of which the optical axis (Z axis) is perpendicular to the surface of the substrate and its principal optical axis may rotate around the Z axis as shown in Fig. 2.

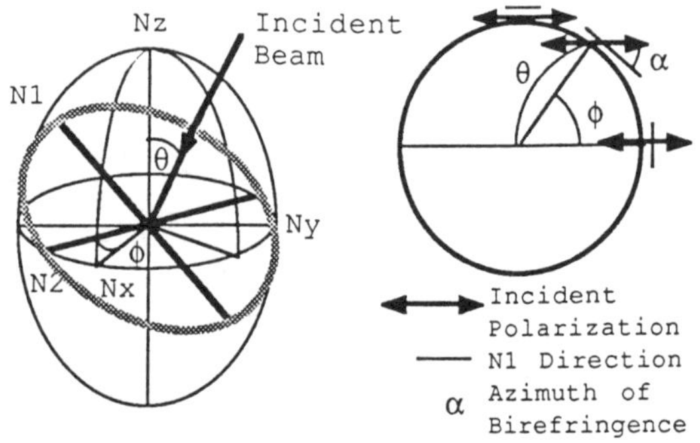

Retardation=$2\pi/\lambda\,(N2-N1)d/\cos\theta$

Fig.2 A Model of the Biaxial Refractive Index Ellipsoid

We calculated the intensity distribution anomaly for various magneto-optical pick-ups and for a disk with a typical amount of birefringence ($\Delta N=6\times10^{-4}$). As shown in Fig.3, calculated results indicate that the birefringence causes the offset of focus error signal in some configurations of optical pick-ups. The result for type A is not shown because type A is insensitive to the polarization change. Although type D suffers most, its simplicity of optical configulation is attractive.

The astigmatism focus method and the knife-edge method are generally used to detect focus error in optical disk systems.[5] The offset in focus error signal can be calculated by integrating the intensity distribution over the area of each photo detector pattern, as shown in Fig.4. Namely,

$$\text{Focus Error}=\Sigma_i\, P_i \iint_{DETi} |E(x,y)|^2 dxdy. \qquad (2\text{-}2)$$

i for each detector element
Pi Polarity of each detector
DETi Integration area of each detector.

Simple algebra is shown for type D optical pick-up.

$$I_{\pm}=|E(x,y)_{\pm}|^2=1/2\ [rp^2(\cos^2\phi +\sin^2\phi\cos^2 2\alpha)+\sin^2 2\alpha\sin^2\phi \pm rp(\sin^2\phi\sin 4\alpha\cos\delta +\sin 2\phi\sin 2\alpha\sin\delta)], \qquad (2\text{-}3)$$

($\pm$) indicates two polarities of analyser angle.

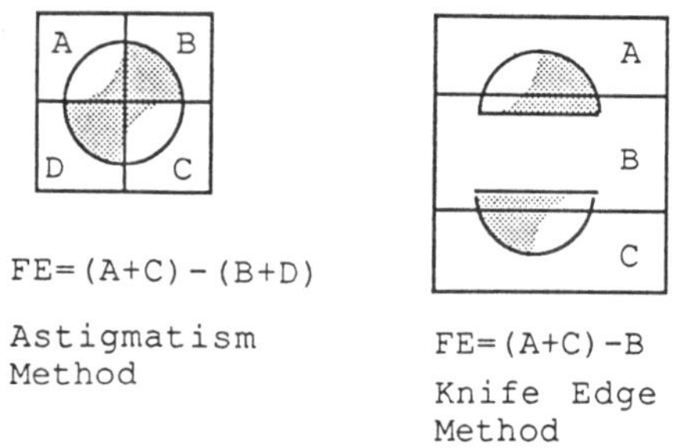

Fig.4 Focus Detecting Method

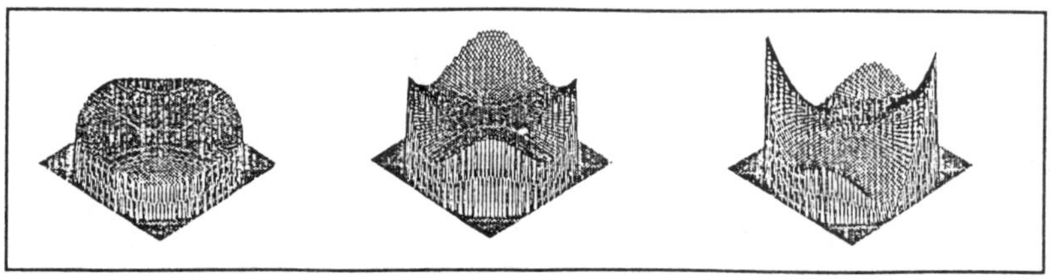

Type B Type C Type D

Δn=6E-4, Phase compensation=30°.

Fig.3 Intensity Distribution Anomaly due to Birefringence

The focus error signal which is compensated by complimental error signal is represented by

Compensated Focus Error=

$$\Sigma_i P_i \iint_{DETi} (I_+ - I_-)dxdy. \quad (2-4)$$

Using eq. (2-3),

$$I_+ - I_- = 2rp(\sin^2\phi \sin 4\alpha \cos\delta + \sin 2\phi \sin 2\alpha \sin\delta).$$

The single pass birefringence of substrate ϕ and its azimuth α are the functions of the position on the detector. Assuming the symmetry of the birefringence and the detector pattern with respect to y axis, the integration of eq.(2-4) goes zero because terms including $\sin 4\alpha$ and $\sin 2\alpha$ are odd function of y coordinate. This result suggests that the focus error offset of the knife-edge method which is symmetric with respect to the axis can be compensated, if the symmetry of the birefringence is assumed.

The birefringence of actual disk substrates may not be symmetric in rigorous meaning. The focus error offsets of various optical pick-ups were calculated numerically using Jones matrix and integrating it over the detector. Parameters used in the calculation are the vertical index difference of $\Delta n_z = 6 \times 10^{-4}$,the horizontal single pass birefringence of 30nm and its optical axis rotation of ±15°. The results of calculation are shown in Table 1, where the offset is normalized by the total intensity. The amount of defocus can be estimated by the normalized offset devided by the sensitivity, which is determined by the spot size and the magnification of focus servo optical system. Typical sensitivity chosen for the knife-edge focus method is around 10%/μm. Therefore, we can estimate the defocus of MPD method around 0.2μm which is well within the practical limit.

	B type	C type	D type	MPD
Astigma Method	4%	17%	48%	---
Knife Edge Method	3%	8%	12%	2%

$\Delta n = 6 \times 10^{-4}$,Phase Compensation=30°, Horizontal Birefringence=30nm, Direction= ±15°.

Table 1. Calculated Focus Offset of Various Pick-Ups

3.PRINCIPLE OF MICRO PRISM DETECTOR

The principle of the MPD method is to compensate the offset of the knife edge focus error signal by subtracting the complimental knife edge focus error.

To obtain complimental error signal, two detectors are located before and after the focal point. Each detector can detect the knife edge focus error simply by means of 3 elements detectors. Width of each element are determined so that photo current of outer elements and that of center element are balanced at the just focus position.

To embody this method, we deviced the hybrid Micro-Prism Detector, a detector on which a micro prism with a dimension of 1mmX2mmX1mm is attached. The schematic structure of the MPD is shown in Fig.5. The MPD assemmbled on a TO-5 package is shown in Photo. 1.

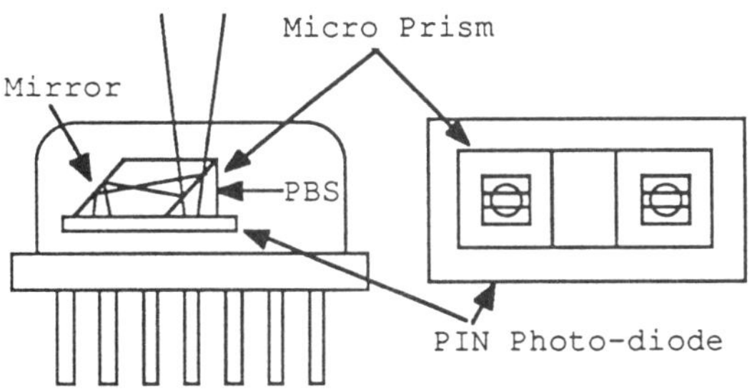

Fig.5 Structure of Micro Prism Detector

Photo.1 Photogragh of Micro Prism Detector

Combining MPD method with the sampling tracking method, all signals, namely the recorded signal, the embossed signal, the focusing servo signal and the tracking servo signal, can be detected by one device. An example of the magneto-optical pick-up utilizing the MPD is shown in Fig.6.

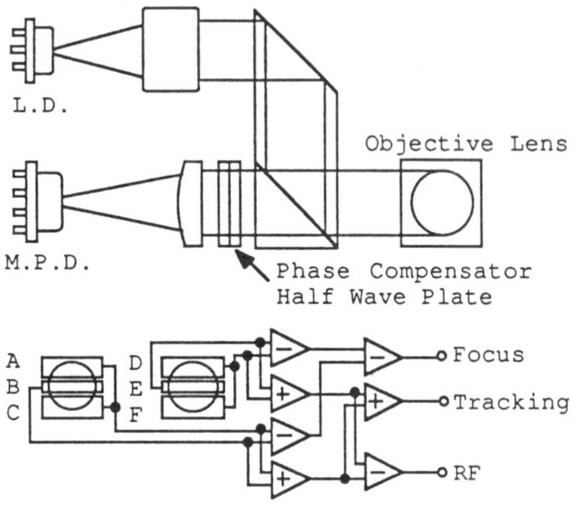

Fig.6 Example of Magneto-Optical Pick-Up employing Micro Prism Detector

4.EXPERIMENTAL RESULT

The measured focus error signal of the Micro-Prism Detector is shown in Fig.7. The pull-in length of about 25μm was obtained. The sensitivity normalized by the total intensity is about 10%/μm.

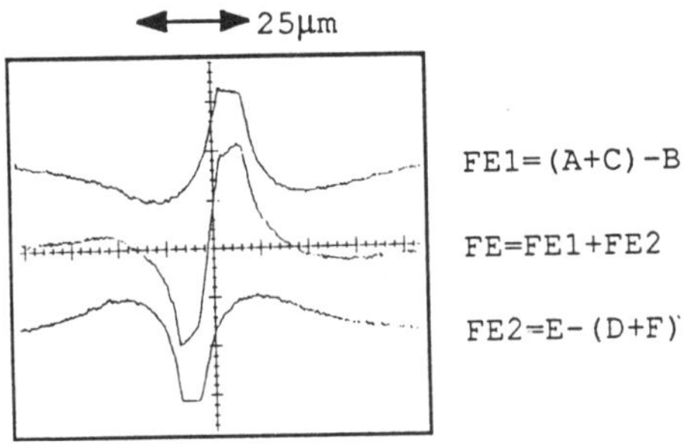

Fig.7 Measured Focus Error Signal of Micro Prism Detector

The MPD is applied to the sampling tracking format magneto-optical pick-up. The Photo.2 shows the pick-up without the objective lens and the actuator. The magneto-optical signal and the focusing servo signal of this pick-up is shown in Photo.3, which indicates that the MPD is satisfactorily used with a mass-produced PC disk substrate.

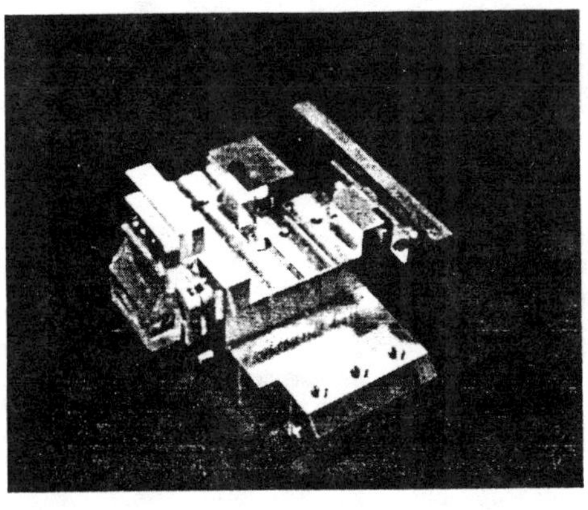

Photo.2 Photogragh of Optical Pick-Up

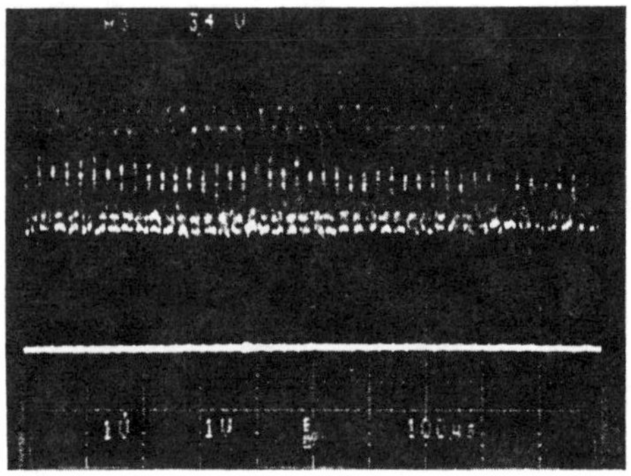

Photo.3 Magneto-Optical Signal and Focus Servo signal of Sampling Format Disk

5.CONCLUSION

We analysed theoretically the influence of birefringence upon focus error signals of various magnato-optical pick-ups. The simulated results indicates that the offset caused by the birefringence could deteriorate focusing quality seriously in some configurations of optical pick-ups.

To compensate the offset, we deviced the Micro-Prism Detector. Although the MPD could be used with various types of tracking methods, we combined it with the sampling tracking method for its simplicity to constract a compact magneto-optical pick-up.

Using the pick-up, we confirmed experimentally the effectiveness of the MPD method applying it to a PC disk substrate with a fairly large vertical birefringence.

The MPD method is, therefore, a nice solution to compensate the offset caused by birefringence and is believed to be suitable for further miniaturization and hybrid integration of magneto-optical pick-up.

ACKNOWLEDGEMENT

The authors would like to thank Dr. Miyaoka and Mr. Tsukamura for their encouragement. We also wish to thank Dr. Matsuda for fabricating the MPD.

REFERENCES

1.)A.Yoshizawa et al.,"Optical Anisotropy of PC Substrate for Magneto-Optical Disk and Its Effect on Carrier to Noise Ratio" Digest of Optical Memory Symposium '86 p33(in Japanese).

2.) T. Toda et al.,"Analysis of Signal to Noise Ratio in Magneto-Optical Disk Using a Polarization Simulator", Digest of Optical Memory Symposium '86, p19(in Japanese).

3.) D. Treves et al., "Effect of Birefringence on Optical Memory System", SPIE PROC. Vol.695 Optical Data Storage II '86, p262.

4.) A. Marchant, "Retardation effects in magneto-optic readout", SPIE PROC. Vol.695, Optical Data Storage II '86, p270.

5.) S. Etoh et al., "C-SSD Focus Servo Method", Digest of 31th Japan Applied Physics Conference '84, p87 (in Japanese)

Reprinted with permission from *Japanese Journal of Applied Physics*, Vol. 28, Suppl. 28-3, pp. 171-175 (1989).

Focusing Servo Method Using the Autofocusing Effect due to Wavelength Change of Diode Lasers

Hiroyuki TSUCHINAGA, Shigeru NAKAMURA, Masahiro OJIMA and Seiichi MITA

Central Research Laboratory, Hitachi Ltd. 1-280, Higashi-Koigakubo, Kokubunji, Tokyo 185

(Received August 31, 1989)

An autofocusing effect due to wavelength change of diode lasers has been investigated. The following range of an optical spot towards an optical disk widens to maximum of 35 μmp-p as the chromatic aberration of a lens increases to 2.7 μm/nm. This effect occurs at high laser power, and can be used in a focusing servo in an optical pickup to write and/or read data pits.

§1. Introduction

Diode lasers have been used recently as light sources in optical disks. In an optical disk, a light beam emitted from a diode laser is focused on a disk by a focusing lens into a spot with diameter of about 1 μm, to write and/or read data pits with diameter of less than 1 μm. Since the disk can move in a range of about 1 mmp-p in the direction of the optical axis during disk rotation, a focusing servo mechanism is necessary with an accuracy in a focal depth of about 1 μmp-p. In general, the focusing servo mechanism consists of focusing error detection optics and a mechanical lens actuator. A laser beam reflected from the disk, passes through the focusing lens, and is transmitted into the focusing error detection optics. The lens actuator moves the focusing lens mechanically so that the focusing spot can follow the disk. Optical disks having higher data recording density and higher data transfer rate are desirable. In order to record small data pits on a disk rotating with high speed, high gain and high speed response are necessary for the focusing servo mechanism. But a mechanical lens actuator with high gain and high speed response will become too large and too complex. Therefore, a simple and compact focusing mechanism is required.

We have previously reported an autofocusing effect due to a wavelenght change of diode lasers.[1] This effect occurs when a diode laser beam is focused on a disk through a focusing lens with chromatic aberration, and a certain amount of the reflected beam from the disk is fed back to the diode laser. The wavelength of the diode laser changes with the disk displacement to keep the amount of optical feedback to the diode laser at a maximum. As a result, focusing of the laser beam onto the disk is automatically controlled without mechanical lens actuators. In the previous experiments, focusing spots could follow the disk in the 16 μmp-p range of disk displacement.

To apply the autofocusing effect to the focusing servo in optical pickups, it is necessary to widen the following range of a focusing spot. The following range of a focusing spot may be determined by the wavelength change of a diode laser and chromatic aberration of the focusing lens. The wavelength change of a diode laser may be caused by optical feedback to the diode laser. Therefore, it is necessary to investigate the autofocusing behavior against optical feedback to the diode laser and chromatic aberration of the focusing lens.

This paper presents the successive experiments on widening the following range of the focusing spot by an autofocusing effect due to the wavelength change of diode lasers, according to the ideas stated above.

§2. Experimental

Figure 1 shows the configuration of an optical setup to measure focusing error signals, which is the same setup as in the previous report.[1] An emitted laser beam from a diode laser is collimated by a collimating lens with a numerical aperture of 0.3, passes through a polarization beam splitter(PBS) and a quarter wavelength plate($\lambda/4$), and is then focused on a mirror with a reflectivity of 70% (instead of using an optical disk) by a focusing lens with a numerical aperture of 0.5. The reflected beam from the mirror passes through the focusing lens again, and is divided into two beams by the PBS. One of these beams is fed back to the diode laser as optical feedback. The

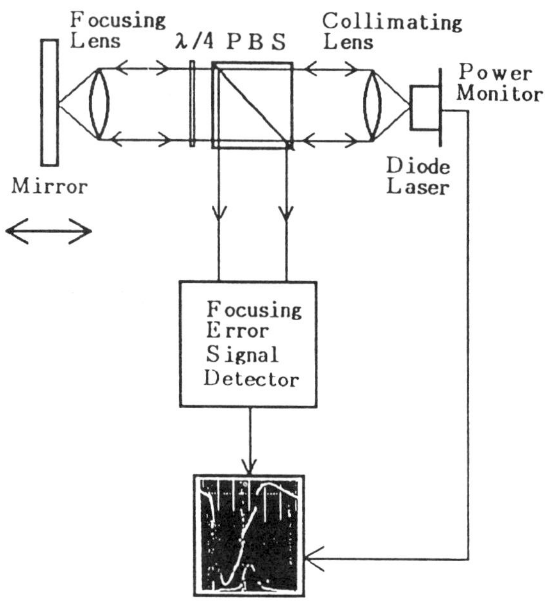

Fig. 1. Experimental setup for measuring focusing error signals.

other is used to obtain focusing error signals. We used the Rotational Half-Moon Method[1] for detecting focusing error signals, which is also the same as in the previous report.

In this setup, the optical feedback ratio to the diode laser can be changed from near 0% to about 42%, by rotating the $\lambda/4$ plate. The diode lasers used in this experiment were CSP (channeled-substrate planar) type, which usually oscillate in a single-longitudinal mode. Their typical wavelength is 830 nm and maximum light power is 30 mW. These diode lasers have power monitoring detectors behind them.

Figures 2(a) and 2(b) are typical focusing error signals and power monitoring signals, measured using a focusing lens with a chromatic aberration of 0.7 μm/nm. When no laser beam is fed back to the diode laser, the focusing error signal changes linearly within the mirror displacement range of 20 μmp-p from an original focusing point, and the power monitoring signal of the diode laser does not change, as shown in Fig. 2(a). However, when 10% of emitted laser beam is fed back to the diode laser by rotating the $\lambda/4$ plate, the focusing error signal decreases to 60% of the original focusing error level within the disk displacement range of 10 μmp-p, as shown in Fig. 2(b). This results in the focusing spot moving 40% of a disk displacement towards the disk by the autofocusing effect. Within the same disk displacement range, the power monitoring signal of the diode laser increases, which shows that an autofocusing effect is caused by returning the laser beam after reflection from the mirror to the diode laser.

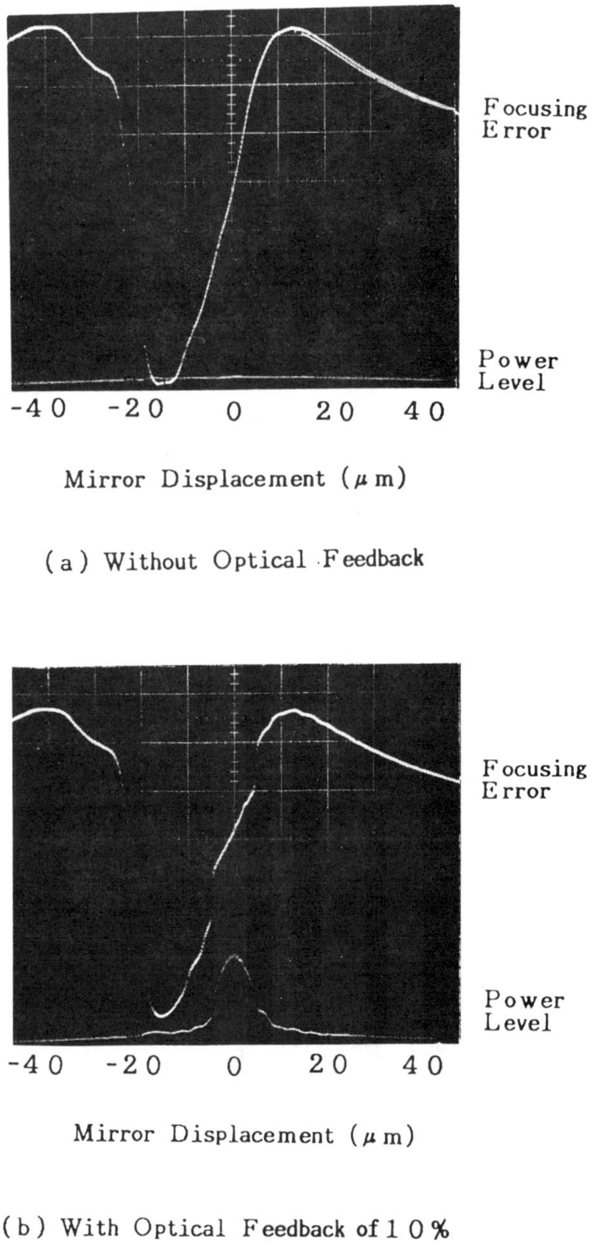

Fig. 2. Typical focusing error signals and power monitoring signals. Chromatic aberration of the focusing lens is 0.7 μm/nm. (a) Without optical feedback. (b) With an optical feedback of 10%.

§3. Experimental Results

At first, to investigate the autofocusing behavior, relationships between the autofocusing range and optical feedback to the diode laser with different chromatic aberration of the focusing lenses were measured. Autofocusing range is given by measuring range where the focusing error signal is suppressed. To obtain different chromatic aberration, focusing lenses were serially arranged along an optical axis. In the case of small chromatic aberration of 0.7 μm/nm, the autofocusing range widens a little to 14 μmp-p as the optical feedback ratio increases to an available value of 32%, as shown in Fig. 3. However, the autofocusing range expandes as chromatic aberration increases. In the case of an optical feedback ratio of 32%, the autofocusing effect can occur within a disk displacement range of 37 μmp-p using a focusing lens with a chromatic aberration of 2.7 μm/nm.

The autofocusing behavior of a focusing spot can be observed by the modulation of the reflected light beam due to tracking grooves of an optical disk. Figure 4(a) shows the modulated signal of the reflected light beam

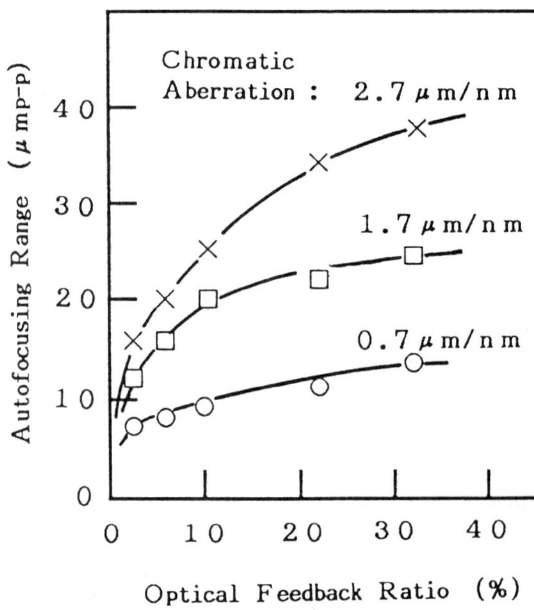

Fig. 3. Autofocusing range vs. optical feedback ratio.

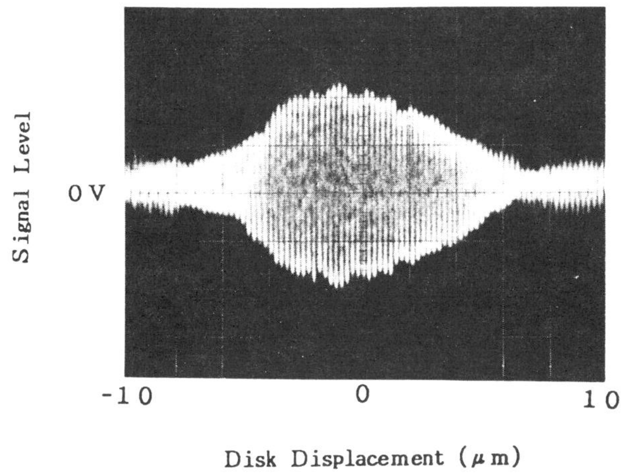

(a) Without Optical Feedback

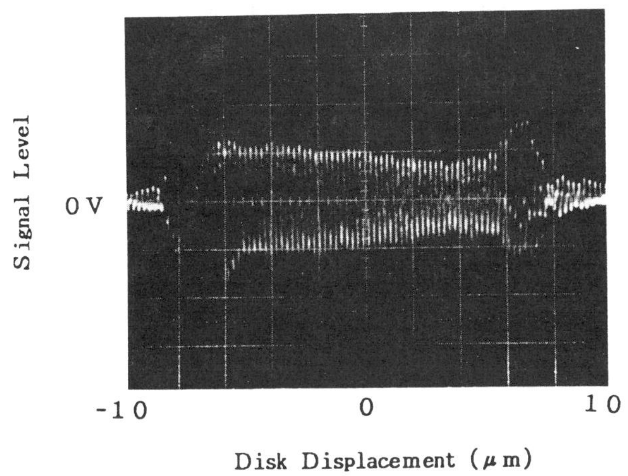

(b) With Optical Feedback of 42%

Fig. 4. The behavior of the modulation range of reflected ligth beam by tracking grooves on a disk against the disk displacement. Chromatic aberration is 0.7 μm/nm. (a) Without optical feedback. (b) With an optical feedback of 42%.

through a high pass filter without optical feedback to the diode laser, with chromatic aberration of 0.7 μm/nm. The amplitude of the modulated signal is suddenly decreased as the disk is displaced from an original focusing point. However, in the case of an optical feedback ratio of 42%, the range of signal modulation widens by about 10 μmp-p of disk displacement as shown in Fig. 4(b). This agrees with the experimental result of the autofocusing range shown in Fig. 3, and suggests that the focusing spot follows the disk displacement.

The focusing error is automatically reduced within the autofocusing range by the autofocusing effct, however, it is not completely suppressed. To write and/or read data pits on an optical disk, it is required that the focusing error is within the focal depth of the focusing spot. Therefore, we investigated the following range of the focusing spot by the autofocusing effect, where the original focusing error is suppressed within the focal

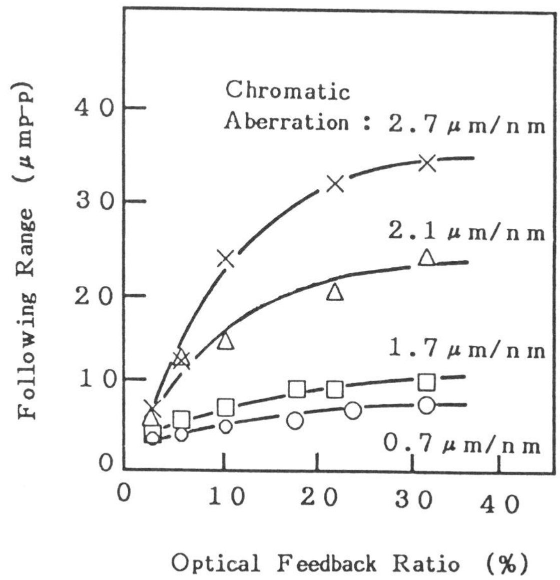

Fig. 5. Following range vs. optical feedback ratio.

depth. The focal depth is well known to be given as λ/NA^2, where λ is the wavelength of laser beam and NA is the numerical aperture of a focusing lens. In our setup, λ is 830 nm and NA is 0.5, giving a focal depth of 3.32 μmp-p. In the case of small chromatic aberration of 0.7 μm/nm, the following range widens a little to 10 μmp-p as the optical feedback ratio increases to an available value of 32%, as shown in Fig. 5. However, the following range expandes as chromatic aberration increases. In the case of an optical feedback ratio of 32%, the focusing spot can follow the disk displacement range of 35 μmp-p within the focal depth, using the focusing lens with chromatic aberration of 2.7 μm/nm.

§4. Discussion

In an optical pickup using focusing lens with chromatic aberration, when the disk moves towards the focusing spot from far away and just reaches at the autofocusing range, the wavelength of the diode laser changes and the focusing spot moves nearer the disk. The movement of the focusing spot can be estimated by the autofocusing range. Figure 6 shows the relationship between the autofocusing range and chromatic aberration using data from Fig. 3. The autofocusing range changes linearly with chromatic aberration in the case of the same optical feedback ratio. This linear relationship suggests that the maximum wavelength change is constant with the same optical feedback ratio for different chromatic aberration. Figure 7 shows the relationship between the maximum wavelength change and the optical feedback ratio. Maximum wavelength change increases as the optical feedback ratio increase, and reaches 15 nm with an optical feedback ratio of 32%. If the optical feedback ratio is increased more than 32%, maximum wavelength change may reach a value of 20 nm, which is a spectral gain width of GaAlAs diode laser.[2] The reflectivity of an optical disk is about 40%. When the numerical aperture of the collimating lens is 0.5, then 80% of the laser beam emitted from the diode laser can be focused onto the op-

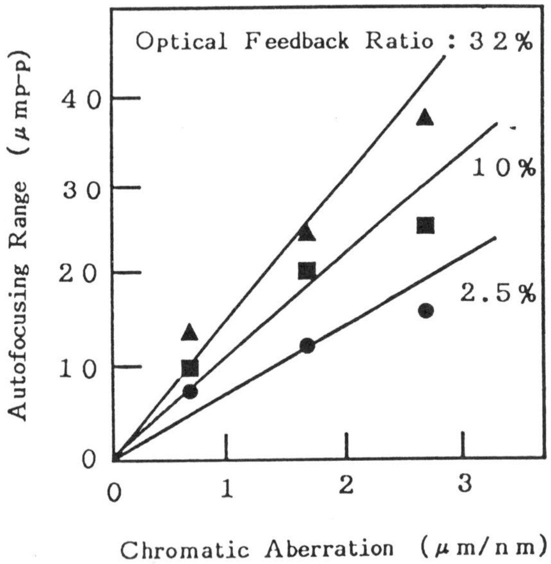

Fig. 6. Autofocusing range vs. chromatic aberration.

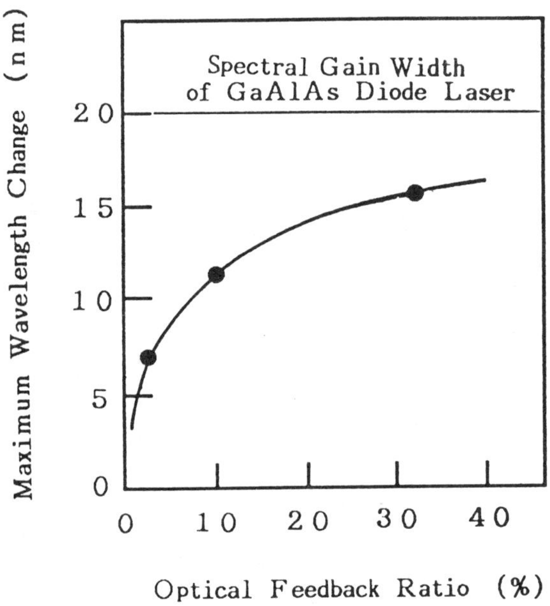

Fig. 7. Maximum wavelength change vs. optical feedback ratio.

tical disk. Thus it is possible to feed 32% of the emitted laser beam back to the diode laser.

Figure 8 shows the relationship between the following range and chromatic aberration with an optical feedback ratio of 32% using data from Fig. 5. The following range increases only a few micrometers for a chromatic aberration less than 2 μm/nm, but the following range suddenly increases for a chromatic aberration more than 2 μm/nm, and reaches to a maximum value of about 35 μmp-p for a chromatic aberration of 2.7 μm/nm. To investigate the sudden increase of the following range against chromatic aberration, we consider the relationship between them as follows.

The focusing spot is linear with displacement. We define the supprssion ratio R_s as the ratio of the focusing error change to the original focusing error, which can be calculated using the data as shown in Fig. 2. When the

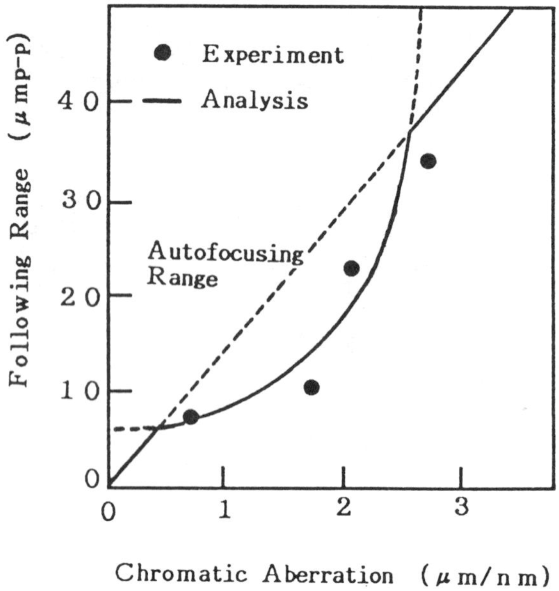

Fig. 8. Analytical results of the following range against chromatic aberration with the optical feedback ratio of 32%.

disk is displaced δ from an original focusing point, the focusing spot moves $\delta * R_s$ towards the disk by the autofocusing effect. Therefore, the amount of remaining focusing error is $\delta - \delta * R_s$. The following range is given by the disk position where the amount of remaining focusing error is less than a half of the focal depth of the diode laser beam. Therefore, the condition for the following range is given by

$$|\delta - \delta * R_s| \leqq \frac{\lambda}{2 * NA^2}. \tag{4.1}$$

From eq. (4.1), following range FR is obtained as

$$FR(R_s) = \frac{\lambda}{(1 - R_s) * NA^2}. \tag{4.2}$$

Since λ and NA are constant in this experimental setup, the following range FR is found to be dependent on only the suppression ratio R_s. Therefore, the behavior of following range against chromatic aberration as shown in Fig. 8, suggests that the suppression ratio changes by chromatic aberration. Figure 9 shows the relationship between suppression ratio and chromatic aberration. For chromatic aberration larger than about 0.7 μm/nm, the plotted data are on a line expressed by

$$R_s(CA) = 0.4 + 0.2 * CA, \tag{4.3}$$

where CA is chromatic aberration of the focusing lens. The solid line shown in Fig. 8, is the calculated result of the following range FR by using eq. (4.3) and eq. (4.2). In the region of chromatic aberration larger than about of 2.7 μm/nm, the analytical result of following range is larger than autofocusing range. But, since the focusing spot can not move over the autofocusing range, the following range is equal to the autofocusing range in this region of chromatic aberration.

From the results stated above, the focusing range of 20 ~ 35 μmp-p can be achieved by the autofocusing effect. Using an aerofloating mechanism, the disk displacement

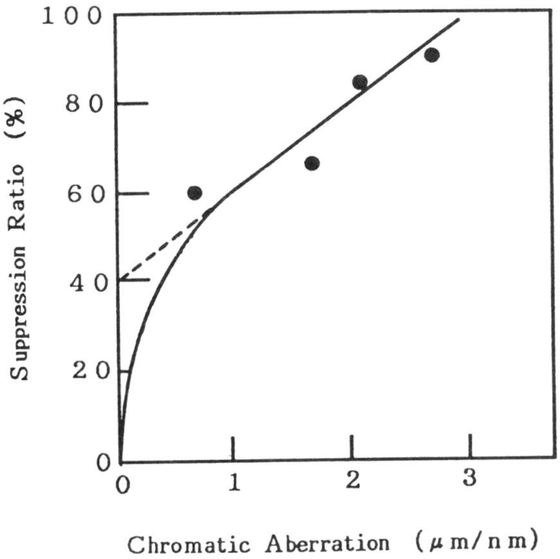

Fig. 9. Suppression ratio vs. chromatic aberration with the optical feedback ratio of 32%.

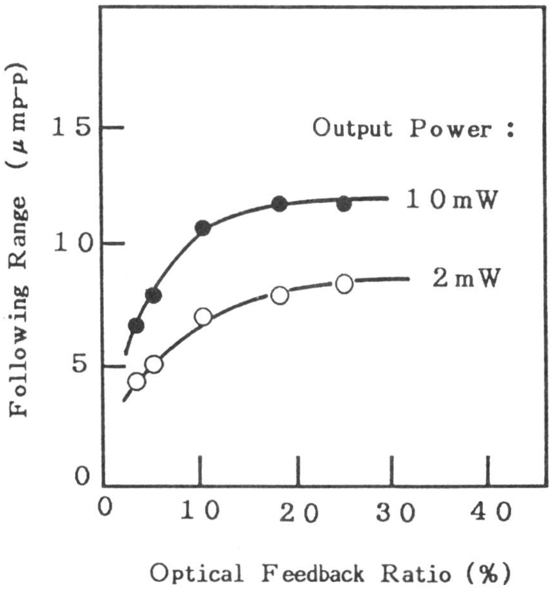

Fig. 10. The behavior of the following range against the optical output power of the diode laser with different optical feedback ratio.

against an optical head may decrease to within 5 μmp-p, and the disk thickness change may be 10～20 μmp-p, so a focusing servo using the autofocusing effect can be possible for an optical disk pickup.

To investigate the autofocusing behavior with output power of the diode lasers is very important, because for writable optical disk, a high power laser beam emitted from the diode laser is fed back to the diode laser. Figure 10 shows the relationship between the following range and optical feedback ratio with different output power of a diode laser. The following range widens by 3～4 μmp-p as output power increases from 2 mW to 10 mW. This is enough to convince us that the autofocusing effect is usefull when writing data pits with a high level of output power.

In this paper, the autofocusing phenomena of diode laser systems with optical feedback from optical disks, and chromatic aberration of the focusing lenses has been investigated. The autofocusing effect occurs when the reflected laser beam is fed back to the diode laser. On the other hand, increased noise has been observed in the light output of the diode laser with the optical feedback, but this is decreased when the diode laser oscillates in a multi-longitudinal mode with 1～2% of optical feedback. The behavior of the diode laser in the multi-longitudinal mode with optical feedback must be physically clarified before this effect can be used in a real optical disk, especially to read data pits on optical disks.

§5. Conclusion

An autofocusing effect due to a wavelength change of diode lasers has been investigated with respect to possibility of widening the following range where a focusing spot moves towards an optical disk displaced in the direction of an optical axis. From results of measuring the autofocusing behavior, by changing the optical feedback ratio to the diode laser with different chromatic aberration of a focusing lens, we confirmed that the following range is strongly determined by chromatic aberration of the focusing lens. The maximum value of the following range is 35 μmp-p with a chromatic aberration of 2.7 μm/nm and an optical feedback ratio of 32%. This effect occurs by increasing the output power of the diode laser, hence it is effective for writing data pits with high level output power as well as when reading them with low level. Combined with an aerodynamic floating mechanism, the autofocusing effect can be used in a focusing servo for an optical disk of vertical vibration of about 35 μmp-p.

Acknowledgment

The authors are much indebted to F. Hara, S. Yonezawa, Y. Tsunoda, Y. Eto and M. Nakamura of Hitachi Ltd. for their encouragements and valuable advices during this work.

References

1) S. Nakamura, T. Maeda and Y. Tsunoda: Appl. Opt. **26** (1987) 2549.
2) B. W. Hakki and T. L. Paoli: J. Appl. Phys. **46** (1975) 1299.

Reprinted with permission from *Japanese Journal of Applied Physics*, Vol. 28, Suppl. 28-3, pp. 205-208 (1989).

A New Flying Optical Head

K. YASUKAWA, S. YAMAGUCHI, A. MURAKAMI, H. GOTO, N. NISHIMURA and K. UEYANAGI

Electric Imaging & Devices Research Laboratory, Fuji Xerox Co, Ltd., 2274, Hongo, Ebina, Kanagawa, Japan

(Received November 17, 1989)

A new flying optical head was developed for attaining the high speed accessing. The weight of the moving part of the flying optical head is reduced to less than 10 gr by using the flying slider. The variation of distance between an objective lens and a disk is largely reduced by floating the objective lens on the disk surface. The autofocusing is done by axially moving one of the relay lenses, which is mounted on the fixed part of the head. The moving part is carried by a linear motor to do track-following as well as track-seeking. The mechanical resonant frequency for the linear motor is as high as more than 5 kHz, and it can sufficiently follow disk tracks under the condition of the radial runout of ±45 μm and the rotational speed of 1800 rpm. The new flying optical head is successfully applied for read/write of optical disk data.

§1. Introduction

High speed accessing[1] as well as high data transfer has been one of the most significant technical subjects in applying the optical disk for the computer file memories. A split type optical system with a flying optical head has been proposed for attaining high speed accessing. The slider of the conventional flying head[2] floats closely on the disk surface like that of a flying magnetic head. The spacing between the slider and the surface is maintained about 1 μm for taking away a focusing actuator from the head. In the optical head system with the flying head, special structured disks are required, of which media is closely placed on the disk surface to avoid defocusing caused by the variation of disk thickness. Also, optical disks cannot be exchanged in order to prevent crush of the flying slider to disks caused by dust particles.

A new flying optical head was successfully applied for read/write of optical disks data for the first time without sacrificing the disk exchangeability. The flying slider of the flying head floats more than 5 μm away from the disk surface. The weight of the moving part in the flying optical head is remarkably reduced by taking away lens actuators from the moving part and therefore both the track-following and the track-seeking operation are performed by one linear motor. The optical head would be the most appropriate for high speed accessing, because the weight of the moving part can be ultimately reduced and comparable to a magnetic flying head. In this paper, the characteristics of the flying optical head is presented.

§2. Configuration

The optical system of the flying optical head is a split type one that consists of two parts: a fixed part and a moving part as shown in Fig. 1. This system is arranged for write/read of a write/once disk or a phase/change-type disk. The fixed part consists of a laser diode, a beam splitter, a pair of relay lenses, focus/tracking error detecting optics, and read-out signal detecting optics. A pair of relay lenses is arranged to form a telescopic system and one of the relay lenses is axially moved by a focusing actuator mounted on the fixed part.

The moving part consists of an objective lens, prism, flying slider and moving coil of a linear motor. The objective lens and the reflecting prism are arranged on the flying slider and floats on a disk surface by the slider. The area of the air bearing is as large as 130 mm^2 in order that the spacing to a disk is maintained to be more than 5 μm to prevent crush caused by dust particles. The flying slider is also made of lubricious material to reduce disk surface damage caused by the crush.

The flying slider is attached to a moving coil by a spring suspension and is carried by the linear motor in the radial direction of a disk. The weight of the moving part including the moving coil is reduced to less than 10 gr by taking away all lens actuators from it and by fabricating the coil with a square shaped Al wire.

§3. Focusing Control

Optical pass length between an objective lens and a disk media varies with the axial runout of the disk and thickness fluctuation of the disk substrate (usually about 20 μm in glass substrates), which causes defocusing of a laser spot at the disk media. The pass length variation caused by the runout is largely reduced by making the flying slider float on the disk surface. But the slider cannot follow the disk runout thoroughly because its air bearing is made soft for maintaining the space far larger than that of a magnetic head.

The correction of the defocusing caused by the pass length fluctuation, that is, the autofocusing is performed by moving the one of the relay lenses in the direction of the optical axis.

3.1 *Characteristics of the flying slider*

Flying height of the slider, which corresponds to the spacing, varies with the disk velocity as well as the axial runout. These dependencies are shown in Fig. 2 and Fig. 3, respectively. The flying height almost linearly depends on the disk velocity when the axial runout is as small as 10 μm (Fig. 2). It takes minimum value of 5.5 μm at the velocity of 7 m/sec, which corresponds to that of the in-

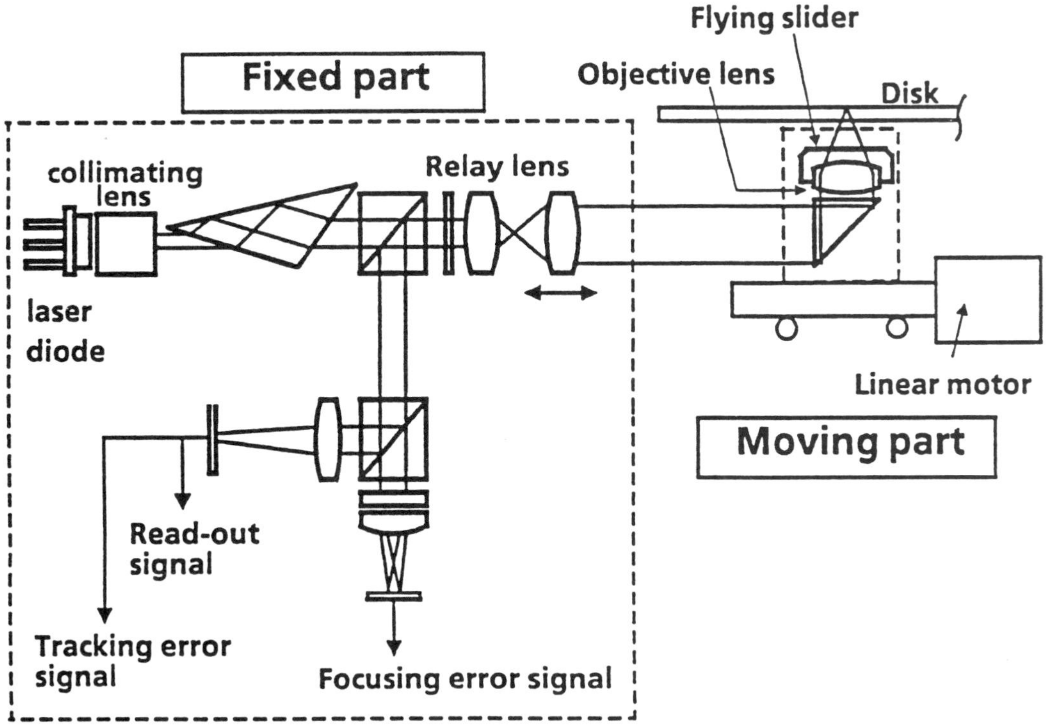

Fig. 1. The configuration of the split-type flying optical head.

nermost region of a 5.25 "disk at 1800 rpm and becomes 12 μm at the outermost. The tilt angle of the objective lens caused by the axial runout is less than 1.5 mrad, which is fairly smaller than the allowed tilt angle of disks (5 mrad).

The fluctuation is as small as less than 2 μm over the recording area when the axial runout is negligible (Fig. 2) and it increases rapidly with the axial runout as is shown Fig. 3. And at last the slider causes crush at the runout of 100 μm. Total flying height fluctuation brought out by the runout and disk velocity variation over the disk area is estimated to be about 30 μm from the data shown in Figs. 2 and 3.

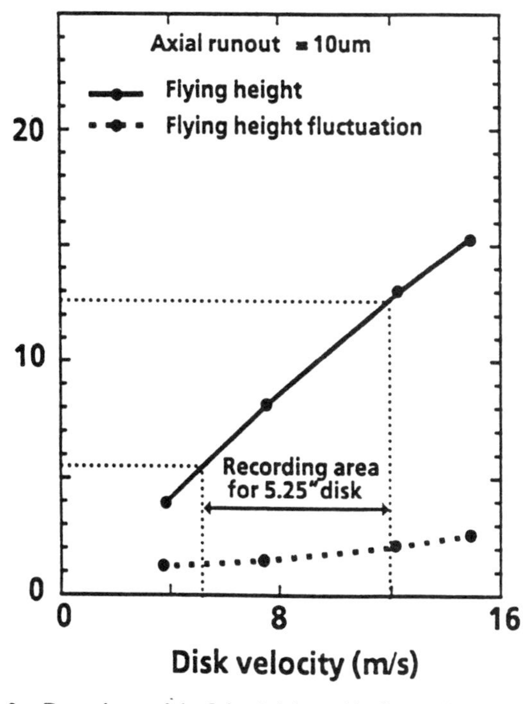

Fig. 2. Dependence of the flying height and its fluctuation on the disk velocity.

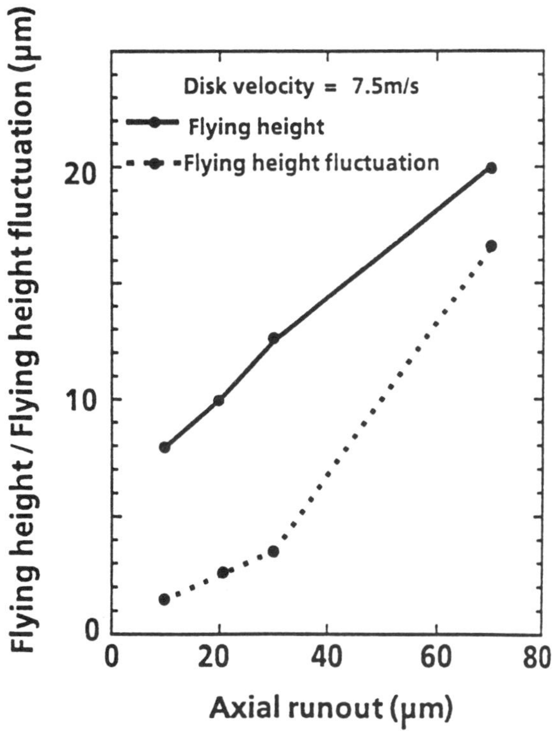

Fig. 3. Dependence of the flying height and its fluctuation on the axial runout.

3.2 *Characteristics of autofocusing*

The amount of the defocusing is the sum of the fluctuation and the substrate thickness and is estimated to be about 50 μm, which must be compensated by the autofocusing servo of the relay lens. The amount of the defocusing must be sufficiently smaller than the autofocusing range, which is limited by the wavefront aberration caused by the displacement of the lens. The relay lens is initially adjusted so that the laser beam is well focused to a diffraction limited spot at a disk media when the incident beam to the the objective lens is collimated at the axial runout of zero. As the axial runout becomes large, the wavefront aberration will increase with the displacement of the relay lens and the spot size of the laser beam at the displacement will spread.

The magnitude of the aberration is easily estimated with the intensity of the diffraction focus. In general, when the decreasing of the intensity is less than 20%, the aberration of an optical system can be regarded to be well corrected (Marechal's criterion[3])). According to the up-and-down motion of the disk, the relay lens moves back and forth in the direction of the optical axis to compensate the defocusing and the intensity of the laser varies, accordingly. The dependence of the peak and total intensity on the disk axial motion is shown in Fig. 4.

When the disk goes away from the objective lens (plus direction in the Fig. 4), the peak intensity increases at first and becomes maximum around the zero point and then decreases, on the other hand the total power gradually decreases. It is because the incident beam to the objective lens changes from convergent to divergent and the outskirts of the laser beam limited by the aperture of the objective lens will gradually increases. The decreases of the peak power is found to be less than 20% in the range of the disk displacement of 60 μm, that is, from -20 μm to 40 μm. The range is large enough to compensate the defocusing caused by the variation of flying height and so forth. The variation of the spot size is surely less than 10% in this range (Fig. 5), which does not affect the write/read characteristics of optical disks.

It is important for attaining controllable autofocusing servo that the moving distance of relay lens is mostly proportional to that of the disk. The proportionality is ascertained with the data shown in Fig. 6. The doted line and

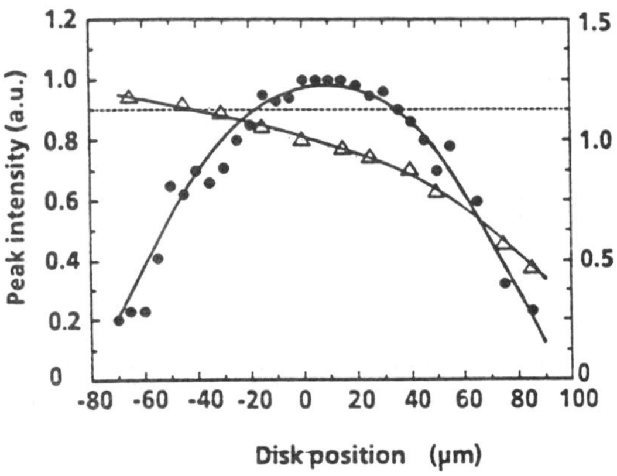

Fig. 4. Dependence of peak intensity and toal power on the disk position.

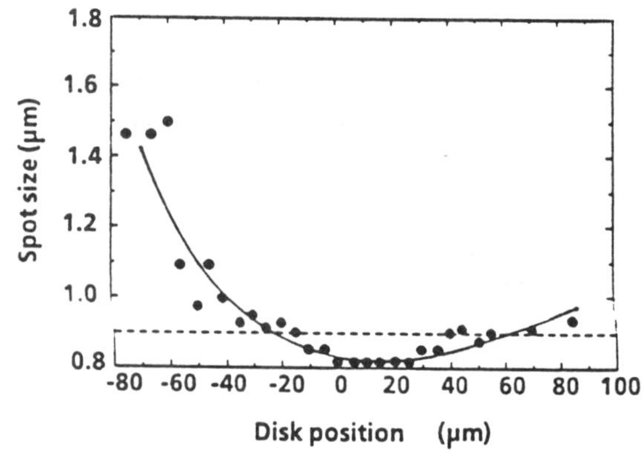

Fig. 5. Dependence of the spot size of focused spot on the disk position.

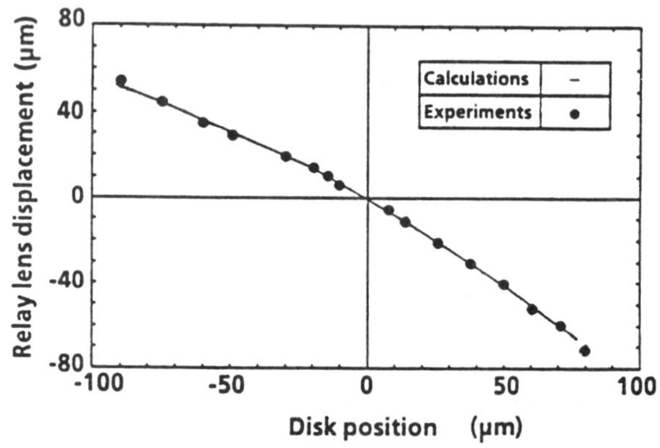

Fig. 6. Relationship of relay lens position and disk position Dots show experimental results and solid line is a theoretical calculation.

solid one represent experimental results and calculated one, respectively, which agree well with each other.

§4. Tracking Control

Servo configuration is largely simplified by introducing the flying optical head, because both tracking and seeking operation are made by one linear motor (Fig. 7). And switchover between the two operations is done automatically by adding the both signals to the driver of the linear motor. On the other hand, it is said the frequency response of a linear motor would be more than 10 kHz for attaining the stable tracking servo. The frequency response of the moving part, shown in Fig. 8, is restricted by the vibration of the spring suspension. Its first resonance frequency is about 5.5 kHz, which is much higher than that of a conventional linear motor used for track seeking, but is considered to be not enough for the tracking. The tracking servo bandwidth is to be estimated 1.5 kHz from the closed loop response curve (Fig. 9), which is narrower than a typical bandwidth of a conventional tracking servo system. But it can sufficiently follow disk tracks under the condition of the radial runout of ± 45 μm and the rotational speed of 1800 rpm. The frequency response will be improved by reducing the weight of the moving part to the half. The acceleration of as high as 15 G is obtained with the linear motor, by which seek time would be reduced to less than 25 msec.

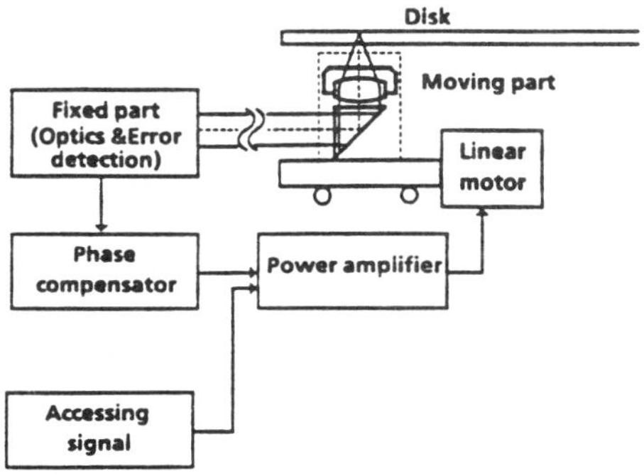

Fig. 7. Block diagram of the tracking servo system.

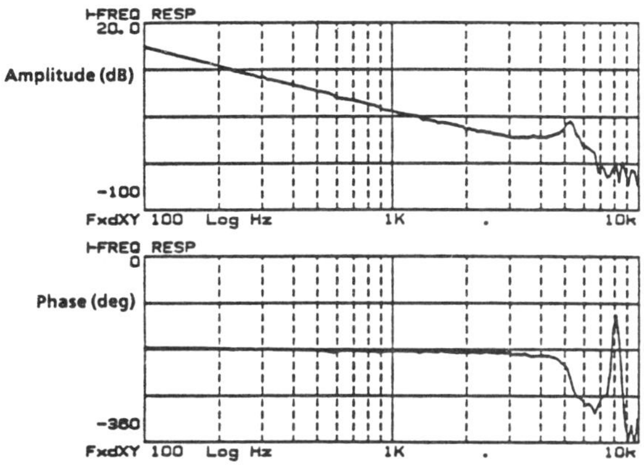

Fig. 8. Transfer function of the linear motor.

Write/read of write/once disks is performed with the optical head under the condition of the disk eccentricity of ±35 μm and the rotation speed of 1800 rpm (Fig. 10). Figure 10 (b) and Fig. 10(b) show the read-out signals and tracking signals with the tracking servo loop is open and closed, respectively. The tracking servo control and write/read of the disks are successfully attained and the envelope of the read-out signal is flat and the residual tracking error is estimated to be less than ±0.1 μm with the error signal as shown in Fig. 10(b).

§5. Conclusion

The flying optical head was successfully applied for read/write of optical disks for the first time. The weight of its moving part can be reduced to less than 10 gr, which enable one linear motor to make both track-following and track- seeking operation. The first resonant frequency of the moving part is more than 5 kHz and the tracking servo bandwidth is 1.5 kHz. The focusing servo control for the flying optical head was attained by axially moving one of relay lenses. Its focusing range is about 60 micron, which is adequate for absorbing the variation of the flying height.

Acknowledgements

The authors will express their gratitude to Mr. Shunsuke Tomiyama, Manager of Electric Imaging and Devices Laboratory, for his support and encouragement.

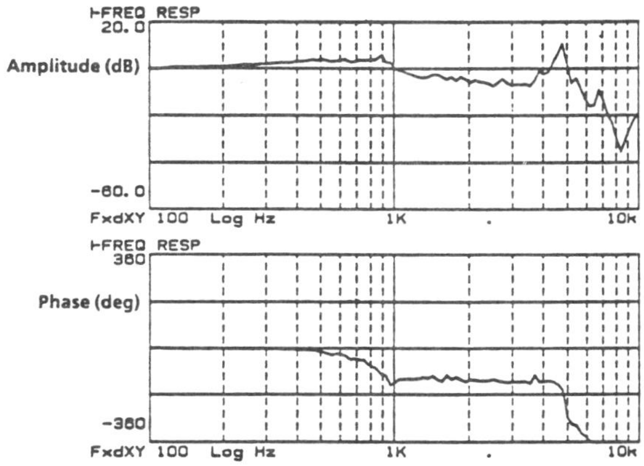

Fig. 9. Tracking servo closed loop curve.

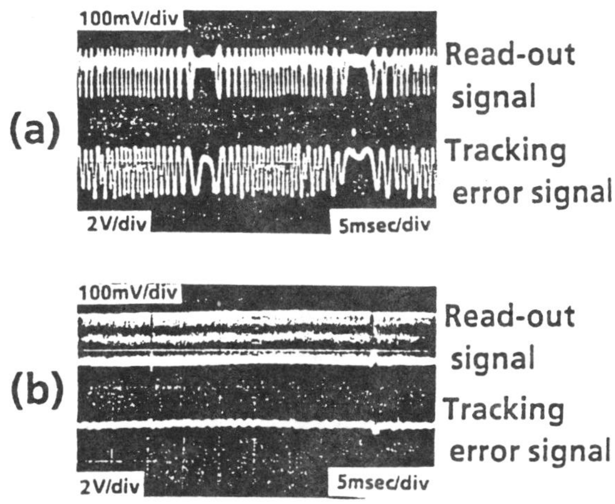

Fig. 10. Photographs of read-out signal and the tracking error signal under the condition of a disk eccentricity of ±35 μm and the rotational speed of 1800 rpm. (a) Focusing servo loop is closed and tracking servo loop is open. (b) Both servo loops are closed.

Table I. Characteristics of the linear actuator.

Weight (moving part)	<10 gr
Movable range	30 mm
Force constant	0.5 N/A
Maximum current	3 A

References

1) K. Koumura, F. Tanizawa, T. Iwanaga, H. Inada and K. Yoshida: SPIE Optical Data Storage Topical Meeting Vol. 1078 (1989) 239.
2) H. Ukita, Y. Katagiri and Y. Uenishi: Jpn. J. Appl. Phys. Suppl. 26-4 (1987) 111.
3) M. Born and E. Wolf: "PRINCIPLE OF OPTICS" (PERGAMON PRESS, 1970) 4th ed., p. 469.

Reprinted with permission from *Japanese Journal of Applied Physics,* Vol. 28, Suppl. 28-3, pp. 73-76 (1989).

High Speed Accessing by Direct Accessing

Kazuhisa KOUMURA, Fuminori TAKIZAWA, Takahiko TSUJISAWA and Hiroshi INADA

C&C Systems Research Laboratories, NEC Corp. Kawasaki 213

Accessing performance for magneto-optical disk drive, especially the address seeking error with the direct accessing method, was evaluated. The seeking error, due to incorrect track counting, which is caused by the tracking error signal disturbance, can be reduced to ±10 tracks by a seeking stroke modification. The 24 msec address seeking performance without disk revolution latency by direct accessing method is presented.

§1. Introduction

Recently, magneto-optical disk drives have become available for use in computer systems. In order to more effectively use their advantages; i.e. large capacity and removable media, a high-speed accessing technique is required and studies have been made on magneto-optical disk drives.[1-3]

The authors have studied the direct accessing method to realize high-speed track seeking. A 21 msec coarse track seeking time has been presented along the 130 mm diameter magneto-optical disk, using a split optical head.[1] Also, the direct accessing method was implemented in this system.

In this method, an optical beam position is estimated by track cross point counting during the track seeking period. The track cross points are detected using tracking error signal from the composite continuous servo (CCS) format disk. Using this estimated position, the optical beam can be positioned directly onto the desired target track. However, fine track seeking is required to counteract the address seeking error, caused by the disk revolution with a spiral track and incorrect track counting during the track seeking period.

In the track seeking experiments for International Standardization Organization (ISO) standard CCS format disk, large numbers of the address seeking errors due to the tracking error signal disturbance were observed. This address seeking error causes long fine seeking time, and accessing performance is reduced.

In this paper, the address seeking error with direct accessing method was considered as a result of some experiments. The experiments indicated that the address seeking error distribution was less than average address seeking error. High-speed accessibility with direct accessing is presented with the coarse seeking stroke modification by the average address seeking error.

§2. Accessing Performance by Direct Accessing Method

The direct accessing method is characterized by optical beam position estimation with track cross counting during the track seeking period. The track cross point is detected from periodic tracking error signal alternation. So, at the end of the track seeking period, the optical beam is positioned directly on the target track.

When ideal tracking error signal detection is assumed for ISO standard CCS format disk, the track seeking error by direct accessing method is considered as incorrect track counting, due to the disk revolution with a spiral track and missing tracking error signal at the mirror zone in a preformat area.

Considering this seeking error, accessing sequence for a magneto-optical disk is formed as follows.

a) Coarse track seeking: Position the optical beam onto or near the target track by direct accessing.
b) Fine track seeking: Move the optical beam accurately to the target track by track jumping.
c) Waiting: Waiting for the disk to revolve.

In the practical system, the tracking error signal is disturbed by the preformat patterns and defects on the disk, and an incorrect track cross point is detected. In another words, incorrect optical beam position is estimated during track seeking. Address seeking error increases the fine seeking time, and accessing performance reduces. In order to confirm the accessing performance, the address seeking error needs to be estimated.

§3. Address Seek Error Evaluation System and Results

3.1 *Evaluation system*

The experimental system configuration is illustrated in Fig. 1. A split optical head is implemented, and moving parts weight is reduced to 30 g. The track following mode for the tracking servo system is formed by cooperation between the Galvano-mirror and positioner actuator. Tracking servo band width is 3 kHz. Figure 2 shows a tracking servo system block diagram for the track seeking. The optical beam is carried by only the positioner actuator in the velocity control mode. The servo bandwidth for velocity control is 2 kHz. The direct accessing method is implemented, and the optical beam position is estimated by track cross counting, detected from tracking error signal. The optical beam velocity is controlled to follow the reference velocity, which is defined versus the remaining seeking distance and saved in the velocity table. The maximum reference velocity is 1 m/sec, maximum deceleration is 150 m/s^2. To make sure of the positioning stability at the end of the track seeking period, the deceleration is decreased smoothly from 150 m/s^2 to 0. The optical beam velocity is also detected from the

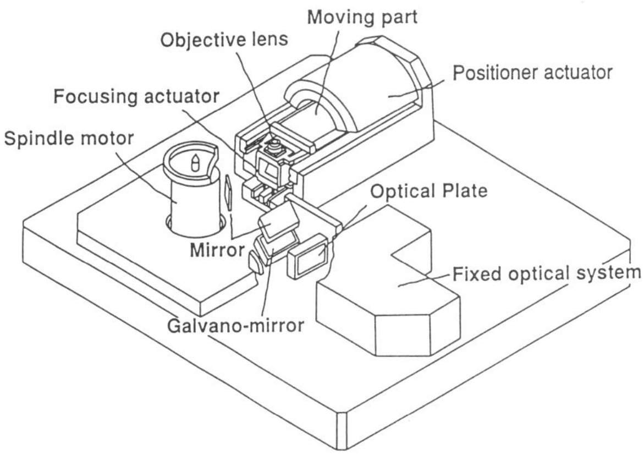

Fig. 1. 130 mm diameter magneto-optical disk drive configuration.

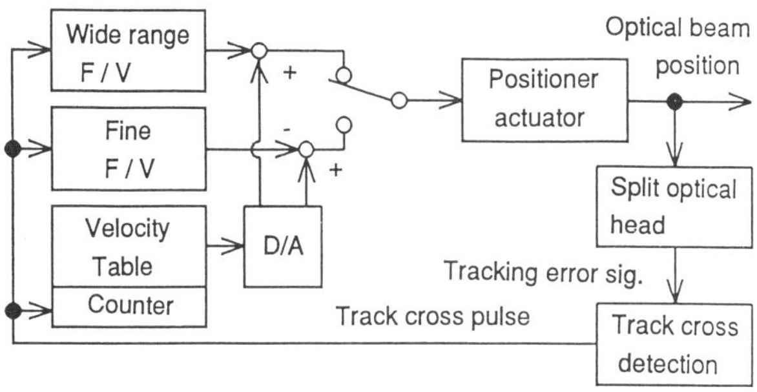

Fig. 2. Tracking servo blockdiagram (Velocity mode).

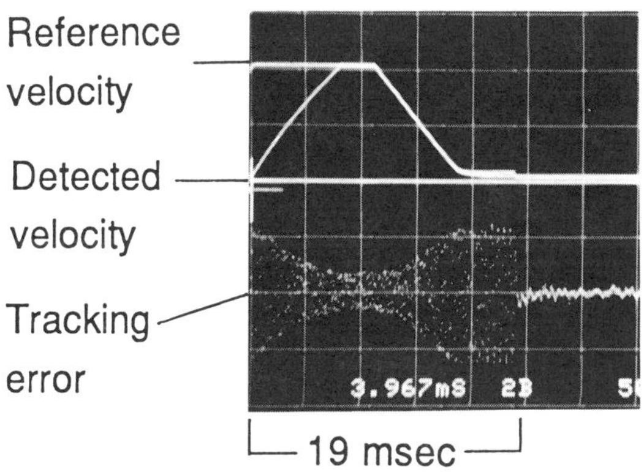

Fig. 3. Wave form in 6000 tracks coarse seeking.

tracking error signal by two charge-pump frequency-voltage converters. The first one detects its velocity for all of the velocity range. The second one detects only 1/10 of the velocity range for maximum speed, and is used to control velocity at low speed. In this system, the disk revolution speed is fixed to 1800 revolutions per minute (rpm). The velocity signals and tracking error signal during the 6000 tracks seeking period are shown in Fig. 3. The average coarse track seeking time for a 130 mm diameter optical disk is 19 msec.

3.2 *Address seeking error evaluation by direct accessing*

The address seeking error is estimated repeatedly as the difference between the desired seeking stroke and the actual seeking stroke in the experiments. The actual track seeking strokes are obtained from read-out address differences, recognized before and after track seeking. The evaluation results are shown in Fig. 4. In this figure, positive value means that the optical beam is positioned over the desired stroke. In Fig. 4, the center line shows the average address seeking error, and both side lines show the address seeking error distribution in 90 percent. When the 1 msec track jump time is assumed for the fine track seeking, total address seeking time is obtained as indicated in Fig. 5. The accessing performance by the direct accessing method is rapidly reduced by the address seeking error increase.

At this time, the disk revolution cannot be considered as a main reason for more than 2 tracks address seeking error, because the disk revolution speed is 1800 rpm, 33.3 msec per revolution, and maximum coarse track seeking time is 36 msec, as shown in Fig. 5. So, almost all of the address seeking error with long seeking stroke will occur due to incorrect track counting.

In order to determine the cause for the address seeking error, the tracking error signal and the track cross pulse series generated from track cross detecting block are observed during direct accessing, as shown in Figs. 6(a)–6(d). The incorrect track cross detections, due to the tracking error signal disturbance, are observed as follows.

Figure 6(a) shows the tracking error signal in low seeking velocity. The tracking error signal is disturbed at the sector mark in preformat area and the mirror zone. The illegal track cross point is detected at this point. In fast seeking velocity, the tracking error signal is lost at the mirror zone in the disk format. Incorrect counting is illustrated in Fig. 6(b). Figures 6(c) and 6(d) show the tracking error signal disturbance on the preformatted sector mark in fast seeking velocity. The tracking error signal detecting gain is decreased at preformat area, and

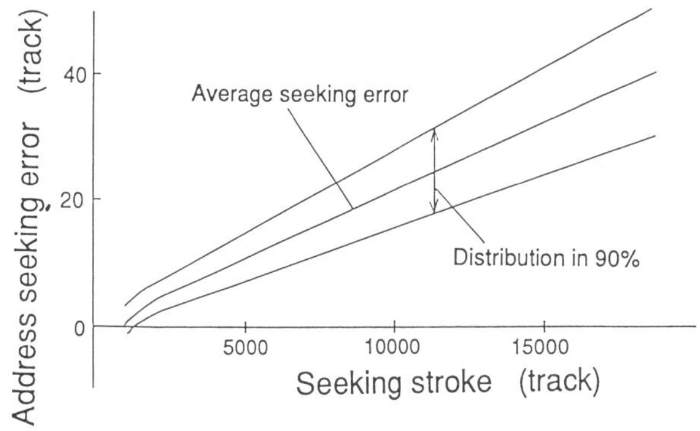

Fig. 4. Address seeking error by direct accessing.

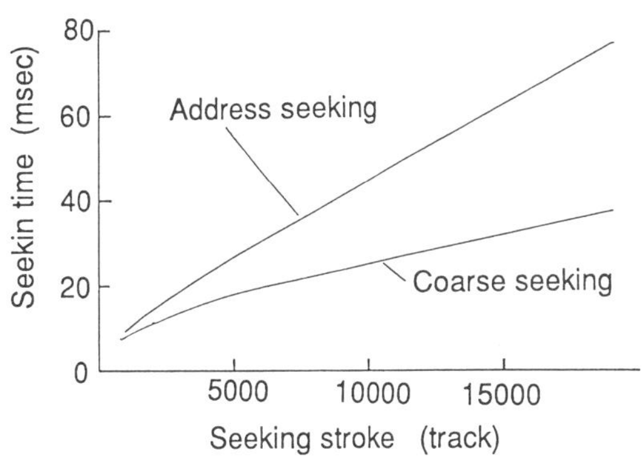

Fig. 5. Address seeking performance containing fine seeking.

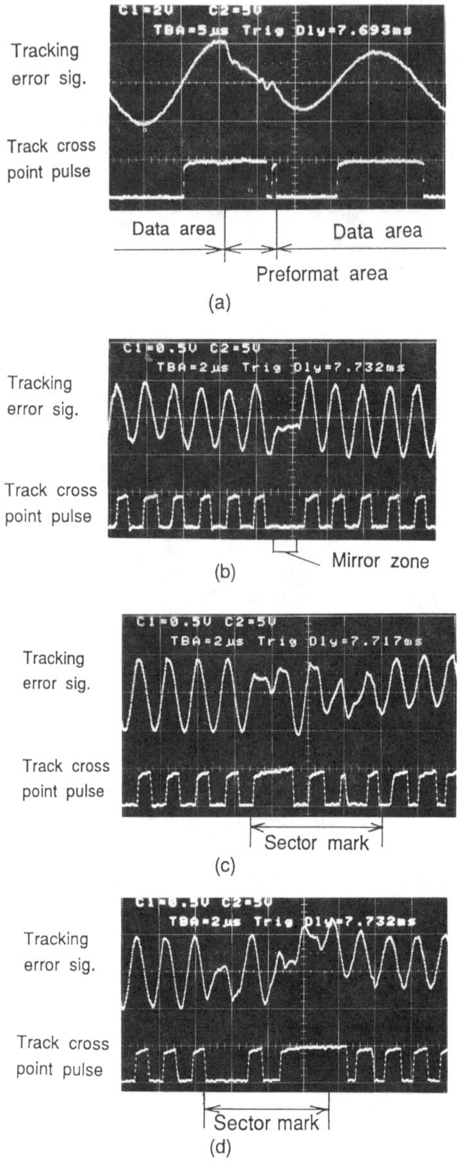

Fig. 6. Incorrect track cross point detection by tracking error signal disturbance. (a) At low seeking velocity. (b)-(c) At fast seeking velocity.

irregular track cross detection is observed.

From these results, the tracking error signal disturbance is determined as the main reason for the incorrect track counting at the preformat area and the mirror zone on the disk. Moreover, these incorrect countings, due to tracking error signal disturbance, appear irregularly. So, it is difficult to predict or compensate for the irregular counting in real time action.

§4. Examination to Reduce Address Seeking Error

The direct accessing method has large advantages for highspeed accessing, though it has a problem in regard to address seeking error used for the ISO standard CCS format disk. However, the address seeking error distribution relative to their average seeking error are less than or equal to the average seeking error, as shown in Fig. 4. Using this seeking error characteristics, the address seeking error reduction with coarse seeking stroke modification is examined.

In this approach, the seeking stroke is modified by the average address seeking error, corresponding to the desired seeking stroke, as follows.

a) Calculate the desired seeking stroke from target track address and the track address on which the optical beam is positioned.
b) Calculate the modified seeking stroke as a subtraction by average seeking error from desired seeking stroke.
c) Carry out the coarse track seeking with this modified stroke.

This seeking stroke modification reduces the average track seeking error to 0. The remaining address seeking error is caused by the seeking error distribution relative to the average seeking error.

The relative address seeking errors relation to the average seeking error for desired strokes of 1000,6000 and 15000 tracks are shown in Fig. 7. From more than 90 percent experiments, ±10 tracks address seeking error distribution, relative to their mean, is observed. The average address seeking error, weighted by the difference between the actual seeking error and their mean, is obtained as 7 tracks for maximum track seeking stroke. The weighted mean for less than 6000 tracks is obtained as less than 4 tracks. This result shows that the remaining average seeking error with the proposed seeking stroke modification, is reduced into ±4 tracks for 6000 tracks stroke. The seeking error weighted mean with seeking stroke modification is shown in Fig. 8.

§5. Conclusion

The address seeking error influence to an accessing performance with the direct accessing method is described. The seeking error is evaluated on the experimental system, and 40 tracks average address seeking error is obtained for maximum track seeking stroke. This error is

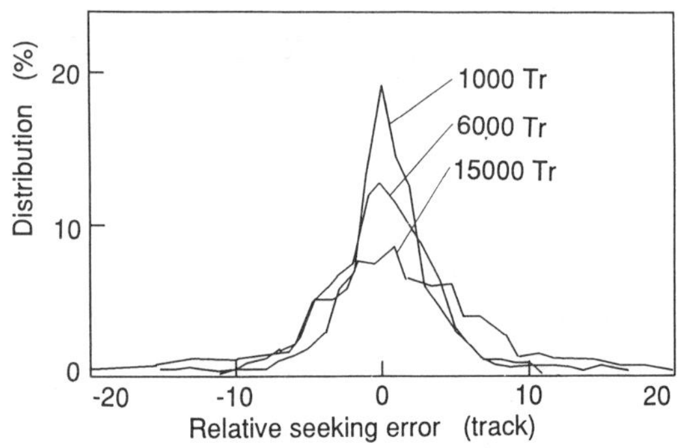

Fig. 7. Relative address seeking error distribution.

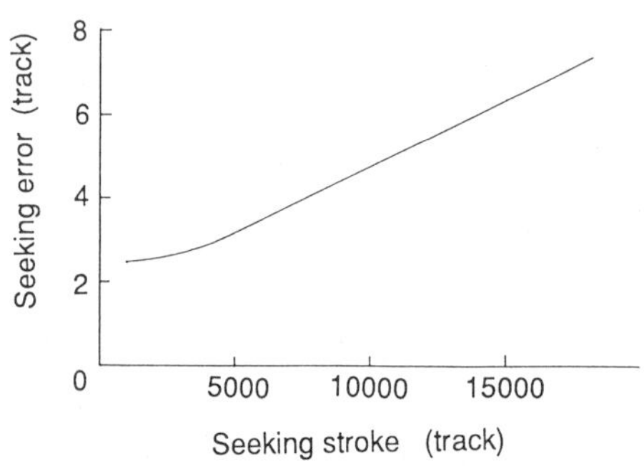

Fig. 8. Average seeking error by seeking stroke modification.

large enough to decreases the practical accessing performance.

By the tracking error signal observation during the track seeking period, the main cause for the address seeking error is confirmed as the tracking error signal disturbance at the sector mark and mirror zone.

The track seeking stroke modification is implemented into the coarse track seeking, because real time compensation is difficult in order to compensate for these errors. From more than 90 percent experiments, less than ± 10 tracks seeking error distribution relative to average seeking error was observed. Using the proposed seeking stroke modification, the average fine seeking stroke can be reduced to ± 7 tracks for maximum stroke and ± 4 tracks for less than 6000 tracks.

Implementing this technique to the magneto-optical disk drive developed in the laboratory, the accessing performance in address seeking is shown as follows,

$$\text{(Average coarse seeking time)} = 19 \text{ (msec)} \quad (1)$$

$$\text{(Average address readout time)} = 1 \text{ (msec)} \quad (2)$$

$$\text{(Average fine seeking time)} = 4 \text{ (msec)} \quad (3)$$

$$\text{(Average address seeking time)} = (1) + (2) + (3) = 24 \text{ (msec)} \quad (4)$$

This result shows the 24 msec average address seeking performance without disk revolution waiting time, by the direct accessing method with seeking stroke modification.

Acknowledgments

The authors wish to thank Mr. Fukuchi for his encouragement in this work. They extended their special thanks to Mr. Yamanaka and Mr. Yoshida for cooperation.

References

1) K. Koumura, F. Takizawa, T. Iwanaga, H. Inada, Y. Yamanaka and K. Yoshida: SPIE Proc. 1078 Optical Data Storage Topical Meeting, Los Angels 1989 (SPIE, Washington, 1989) p. 239.
2) H. Inada, T. Nomura, T. Iwanaga and K. Koumura: SPIE Proc. 695 Optical Mass Data Storage II, San Diego 1986 (SPIE, Washington, 1986) p. 130.
3) H. Ishibashi, R. Shimizu, M. Imura, M. Kuwamoto, H. Yamamoto, S. Tanaka and T. Shimamoto: Technical Digest Int. Symp. on Optical Memory, Tokyo 1987, WA5 p. 17.

Reprinted with permission from *Japanese Journal of Applied Physics,* Vol. 28, Suppl. 28-3, pp. 335-340 (1989).

Acousto-Optic Accessing in Optical Disks

Kiyoshi MATSUMOTO and Takeshi MAEDA

Central Research Laboratory, Hitachi Ltd.,
1-280, Higashi-Koigakubo, Kokubunji, Tokyo 185 Phone 0423-23-1111, Fax 0423-23-1900

(Received August 31, 1989)

An acousto-optic deflector has been used as a fine actuator to improve the ability to pull in a track. Combining this A/O system with split optics which make it possible to get high acceleration, we have achieved an access time without latency of 14 ms in a 12″ optical disk system.

§1. Introduction

Shortening the access time in optical disks is a high priority for a high performance large capacity optical file system. For magnetic disks, the average access time without latency is about 12–16 ms, while that of optical disks is longer by a factor of 10. Therefore, improving the access performance is essential, if optical disks are to be used as an external computer file.

There are two major problems in achieving high access speed.

a) The first is to get high acceleration, as stated in many papers, and split optical systems have been developed to reduce the weight of the moving part.[1-3] However there is a problem in this system of the optical axis displacement. A rotating-optical-plate actuator has been developed as one solution,[4] but further improvements are necessary for dynamic displacement.

b) The second problem, which we believe to be more important, is to improve the ability to pull in a track. In typical optical disk systems, it is impossible to pull in a track unless the relative velocity between the laser spot and the track is reduced to less than 3 ~ 4 mm/s. It takes about 5 ~ 10 ms for the velocity to settle to this limit. The higher the acceleration becomes, the more notable this overtime appears.

To improve the ability to pull in a track, an acousto-optic (A/O) deflector has been introduced to act as a fine actuator, and the overtime mentioned above can be negligibly small. By combining this A/O system with split optics, an access time of 14 ms has been achieved in a 12″ optical disk system.

Our system is designed for 12″ disks. But of course, the techniques described in this paper are applicable to smaller size disk systems. In the future, wave guide optics including an SAW deflector[5] will be used as fine actuators to achieve high speed access in small size disk systems.

Some of the problems and solutions in achieving high speed access will be described in §2. In §3 we will describe the characteristics of the experimental system, while in §4 experimental results are reported.

§2. Problems and Solutions to Achieve High Speed Access

The velocity profile of accessing a track in optical disk systems is shown in Fig. 1. The access time can be divided into four periods as follows:

(1) Time to move the optical head coarsely and for the spot velocity to be reduced to a value low enough to pull in the target track.

(2) Time to pull in the target track.

(3) Time to confirm the address of the track.

(4) Time to compensate the seek error.

Periods (1) and (2) are called "macro seek", and period (4) "micro seek" in this paper.

In period (1), the head velocity is controlled to follow a velocity curve which is designed to move the head to the target in the shortest time. Higher acceleration and shorter stroke are effective to shorten the time for this period. To get higher acceleration, split optical systems have been proposed as mentioned above, but there is another problem which arises.

In period (2), the tracking servo starts to pull in the target track. At this time, the velocity of the spot must be lower than a certain limit, which is determined mainly by the accuracy of the laser spot positioning, settling time and bandwidth of the tracking servo.

On the other hand, because of velocity deviation, the velocity of the spot is not zero, even if it is controlled to be zero. This deviation is caused by velocity control error, friction of the mechanism, tremor of the body and so on. Furthermore, it has a tendency to increase according to increases in the acceleration.

In typical electro-mechanical controlled optical disk systems, the allowable upper limit of the velocity is ap-

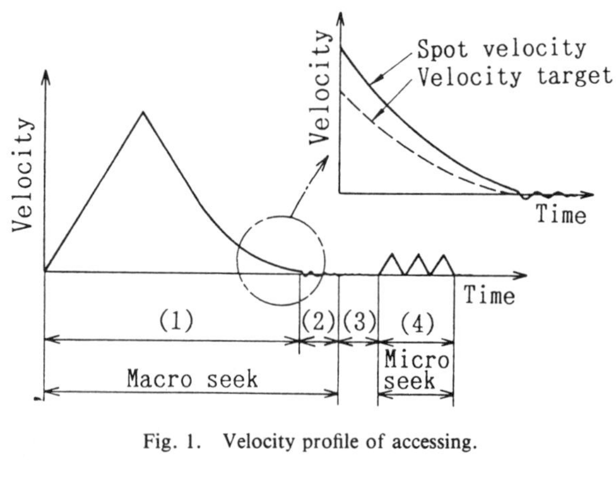

Fig. 1. Velocity profile of accessing.

proximately 3 ~ 4 mm/s, so deceleration must be decreased at the end of seek to reduce the velocity deviation to this limit. This is shown in Fig. 1. The overtime caused by this is estimated to be about 5 ~ 10 ms.

In magnetic disk systems, this overtime is not necessary because the track pitch is ten times wider and the allowable velocity limit is 20 ~ 30 mm/s. To make the access time in optical disks comparable to that of magnetic disks, the allowable limit, that is to say, the ability to pull in a track must be improved.

We have introduced an A/O deflector into a typical electro-mechanical system to act as a fine actuator. A/O deflectors have no harmful sub-resonances which are seen in mechanical actuators such as galvano mirrors, and therefore the bandwidth of the servo can be extended.

In period (3), the address of the track is confirmed and if the pulled-in track is not the target, micro seek is carried out in period (4).

To make macro seek more accurate, the method of full-cross-track count has been proposed.[6,7] We do not employ this method because the disks must not have preformatted data. This gives a restriction to the system, that is, disks must be formatted by MO signal.

We developed a Laser Scale, which has a resolution of 0.4 μm. This acts as an external head position and velocity sensor. Combining the Laser Scale and correction method of disk eccentricity, the accuracy of macro seek can be improved. Furthermore, using an A/O deflector as a fine actuator, micro seek can be carried out in a short period as long as the target is within its dynamic range.

§3. System Description and Characteristics

3.1 *Configuration of the system*

The system configuration is shown in Fig. 2.

Our system is designed for 12″ disks. A voice coil motor (VCM) is employed as a coarse actuator. To reduce the weight of the moving part, split optics are employed. And as a result a maximum acceleration of 30 G was obtained.

To shorten the stroke, two pairs of optics are employed, and each pair is assigned to inner and outer tracks. The maximum stroke is reduced to 35 mm and the average macro seek time is reduced by a factor of almost

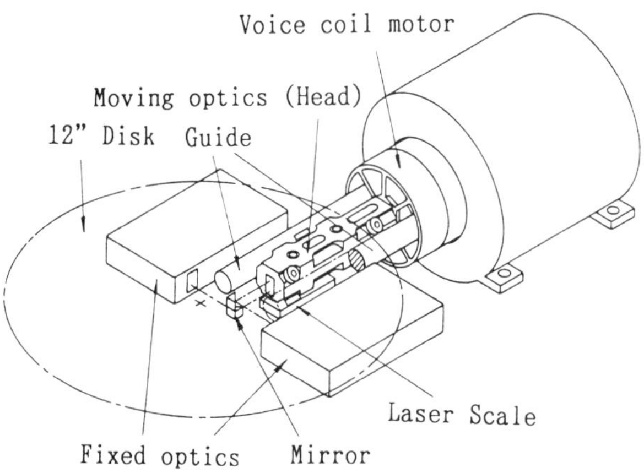

Fig. 2. System configuration.

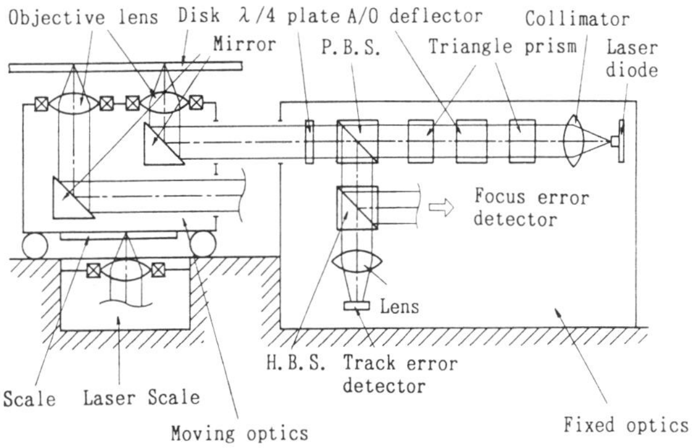

Fig. 3. Optical configuration.

$1/\sqrt{2}$.

The moving part carries only AF actuators and reflection mirrors, and the other optics including A/O deflectors are situated in the fixed parts.

The Laser Scale is fixed on the base, and on the bottom of the moving part is fixed a scale.

3.2 *Optics*

The optical configuration of the system is shown in Fig. 3. Two pairs of optics are employed to shorten the stroke. The optics are divided into two parts, the moving part and the fixed part.

The moving part consists of objective lenses and reflection mirrors, and the fixed part consists of an A/O deflector and other optics. Push-pull detection is applied for track error signals.

Triangle prisms are used to get a sheet-beam, which improves the deflection efficiency of the A/O deflector. The characteristics of the A/O deflector are described in next paragraph.

On the bottom of the moving part is fixed a scale which has a pitch of 1.6 μm. Onto this scale, a laser beam from the Laser Scale is focused to detect the head position and velocity. 3-spot method is applied for the signal detection. Two signals of 1.6 μm pitch, which have a phase difference of 90° between each other, are obtained. By operation with these signals, the resolution of 0.4 μm is obtained.

3.3 *Acousto-optic deflector as a fine actuator*

The acoustic wave traveling in the A/O material causes a periodic change in refractive index. This acts as a grating and deflects the laser beam.

The deflection angle is proportional to the driving frequency, and this frequency can be altered by a voltage-controlled oscillator (VCO). In other words, when acousto-optic deflector is used as a fine actuator, the laser spot movement by the fine actuator can be known without any sensors and observers, and can be controlled even without feedback.

Instead of wave guide optics including SAW deflector, a bulk A/O deflector is employed because coupling efficiency is superior for now. The frequency response of this device is shown in Fig. 4. The amplitude characteristic is flat to more than 100 kHz, and there is no harmful

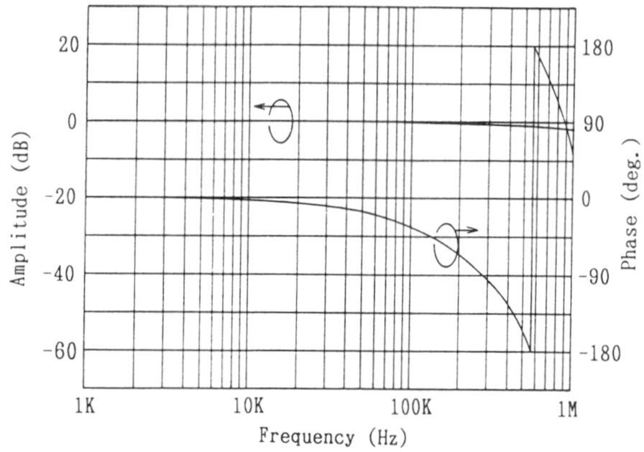

Fig. 4. Frequency response of A/O deflector.

subresonances which are seen in mechanical actuators such as galvano mirrors. These limit the bandwidth of the servo.

Phase lag increases according to the frequency, because there is a 2 μs delay between the change in driving frequency to the change in the deflection angle. This comes from the size of the deflector and the sonic speed of the material.

Driving frequency is limited to 85.0±11.0 MHz from a viewpoint of efficiency of the optics. The deflection efficiency of the A/O deflector is 90% at the maximum, and 70% at the minimum in this range. The change of the deflection angle is ±2.53 mrad, and the focal length of the objective lens is 4.35 mm, so the optical spot movement on the disk by the fine actuator is about ±11 μm.

3.4 *Offset-free track error detection using an A/O deflector*

In split optical systems, offset of track error signal is a serious problem when push-pull method is used for the detection. This offset is caused mainly by optical axis displacement due to

(1) movement of the moving part.
(2) movement of the fine actuator.

The offset caused by (1) can be reduced for example by an adjustment of the optical axis and moving part. However, the offset caused by (2) is inevitable as the distance between the fine actuator and the objective lens is large in split optical systems. Furthermore, in our system, the bandwidth of the fine actuator servo is wider than usual systems, and the fine actuator moves faster. This means that the mechanical canceling method is impossible.

We developed a new dynamic offset-elimination method which uses a wobbling technique based on an A/O deflector. This method eliminates the offset directly from the track error signal.

The track error signal E with offset can be described as

$$E=a\sin(2\pi x)+b \tag{3.1}$$

where x: the laser spot position normalized by the track pitch, a: amplitude of the track error signal, b: the offset.

By A/O deflector, the laser spot can be wobbled in both directions, and the track error signals are described as follows.

$$E_{+}=a\{\sin(2\pi x)\cos(2\pi w)+\cos(2\pi x)\sin(2\pi w)\}+b \tag{3.2}$$

$$E_{-}=a\{\sin(2\pi x)\cos(2\pi w)-\cos(2\pi x)\sin(2\pi w)\}+b \tag{3.3}$$

where w: wobbling movement normalized by the track pitch. Using the following operations, the offset can be eliminated.

$$E_1\equiv E-\frac{1}{2}(E_{+}+E_{-})=a\{1-\cos(2\pi w)\}\sin(2\pi x) \tag{3.4}$$

$$E_2\equiv E_{+}-E_{-}=2a\sin(2\pi w)\cos(2\pi x) \tag{3.5}$$

E_1 is the track error signal without the offset, and E_2 is the cross track signal which has a phase difference of 90° from E_1. In our system, E_2 is not used.

The diagram of the operation circuit and the timing chart for E_1 are shown in Fig. 5 and Fig. 6. The wobbling period T_w is 8 μs because of the time lag of the deflector and the settling time of the sample/hold. The wobbling interval T_i is variable under the limitation of $T_i \geqq T_w$.

T_i causes phase lag to the servo, so when the bandwidth is wide, T_i is set at as short interval as possible. If T_i is set at T_w, the phase lag is −90° at 62.5 kHz.

The effect of this elimination method can be seen in Fig. 7. An automatic gain controller (AGC) is used for amplitude compensation.

3.5 *Velocity control servo*

Figure 8 shows the block diagram of velocity servo which is used in macro seek period.

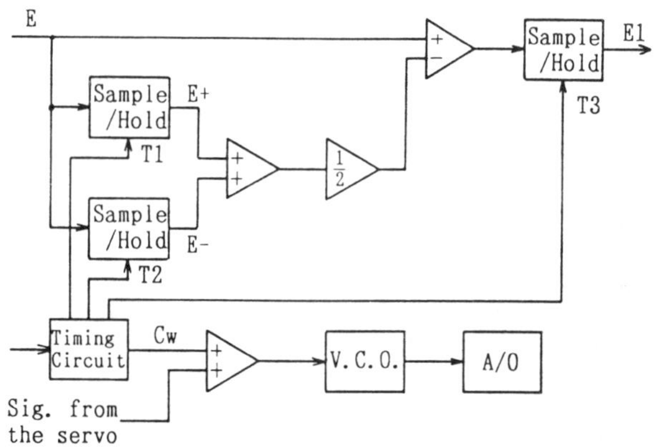

Fig. 5. Operation circuit for offset-free track error detection.

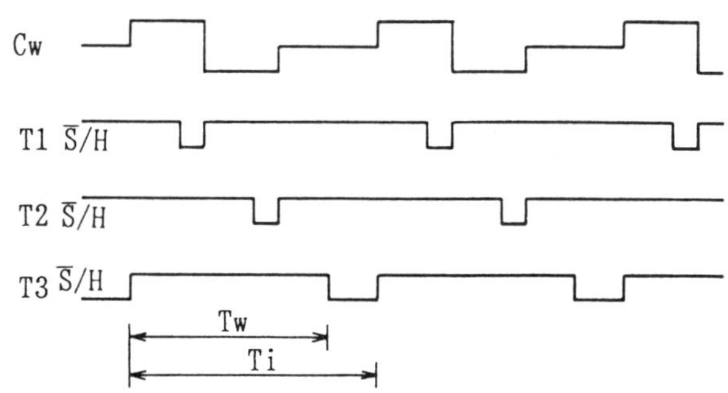

Fig. 6. Timing chart for wobbling.

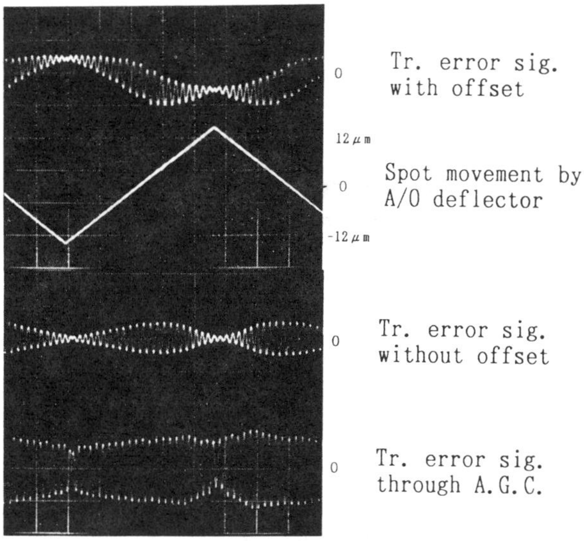

Fig. 7. Effect of offset elimination.

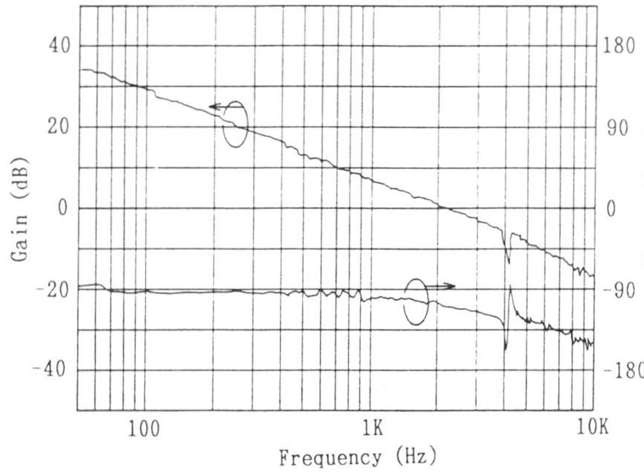

Fig. 9. Open loop characteristics of velocity servo.

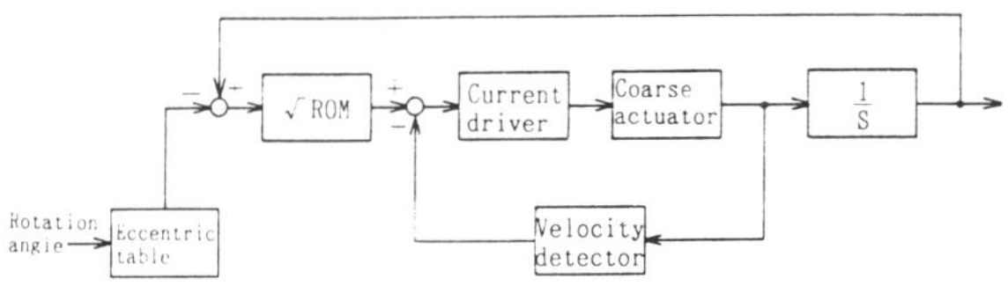

Fig. 8. Block diagram of velocity servo.

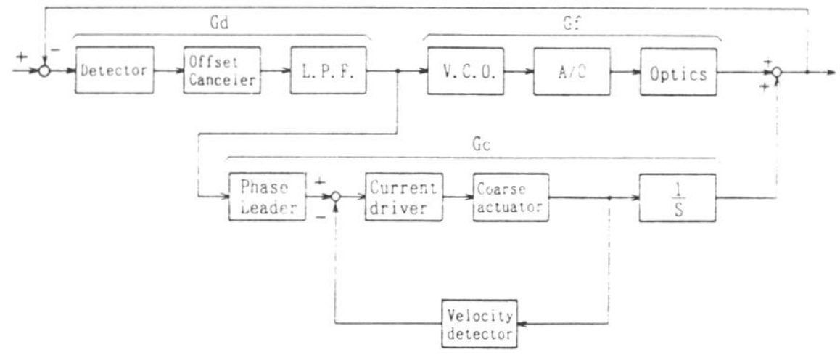

Fig. 10. Block diagram of 2-stage-servo.

The eccentric table has the data about the eccentric movement and velocity of the disk. These were measured when the disk rotation started. The target position and control velocity are compensated by this data according to the disk rotation angle, and thus the accuracy of the macro seek and the stability in pulling in a track are both improved.

The head velocity is detected by F/V converter, using signals from the Laser Scale. However it is necessary to use the VCM current to compensate for the frequency characteristic of the F/V. The resolution of the Laser Scale is 0.4 μm, so the time delay of the detected velocity matters little.

Figure 9 shows the open loop characteristic of the velocity servo. A resonance caused by stretching vibration of the VCM can be seen at 4.2 kHz, but it does not effect the velocity servo. The bandwidth is set at 2.3 kHz, thus makes it possible to reduce the velocity control error to 20 mm/s even when the deceleration is 30 G. This can be achieved without decreasing in the deceleration as shown in Fig. 1.

3.6 *Tracking servo*

For tracking, we used 2-stage-servo system[8] to expand the dynamic range of the A/O deflector. Figure 10 shows the block diagram of the servo.

A low pass filter (LPF) of first order lag is inserted to control the bandiwth of the fine actuator loop. The A/O deflector is driven by a voltage controlled oscillator (VCO). The control signal between LPF and VCO represents the movement of the fine actuator directly, so detectors and observers which are seen in typical systems are not necessary. The coarse actuator is driven by this signal, and acts to keep the fine actuator at a neutral position.

When the laser spot reaches the target track at the end of macro seek, the head has a certain velocity V_a caused by several factors as mentioned in §2. The allowable limit of V_a to pull in a track is determined mainly by the accuracy of the laser spot positioning and bandwidth of the tracking servo.

Figure 11(a) shows how the maximum laser spot velocity is related to the bandwidth and the damping factor of the coarse actuator servo, taking into account that overshoot is limited to 11 μm by the dynamic range of the laser spot movement by the A/O deflector. In our system, this bandwidth is set at 600 Hz, and the damping factor at 0.5.

Figure 11(b) shows the maximum laser spot velocity as a function of servo bandwidth and allowable following error. When the bandwidth is set at 40 kHz, it is possible to pull in a track of 1.2 μm pitch at the velocity of 40 mm/s.

Figure 12 shows an example of open loop characteristics of the fine actuator servo. Because of the delay time by the offset eliminator, maximum bandwidth of the fine actuator servo is about 50 kHz. In our access experiments, the bandwidth is set at 30 kHz out of consideration for the stability of the servo.

3.7 *Fine seek*

In the A/O deflector, the change of the deflection

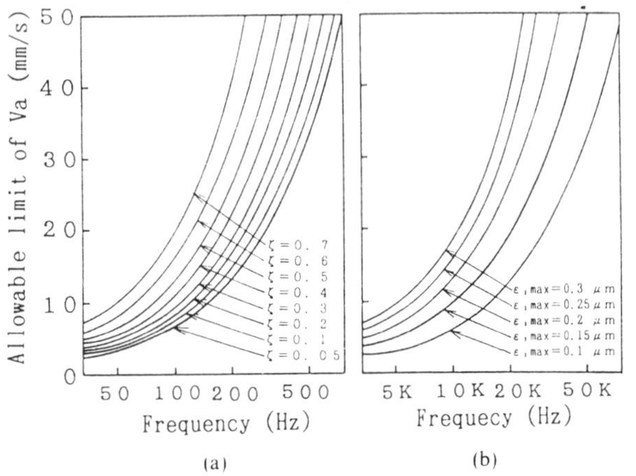

Fig. 11(a). Maximum laser spot velocity as a function of bandwidth of coarse actuator servo GdGc/(1+GdGf).

Fig. 11(b). Maximum laser spot velocity as a function of bandwidth of fine actuator servo GdGf/(1+GdGc).

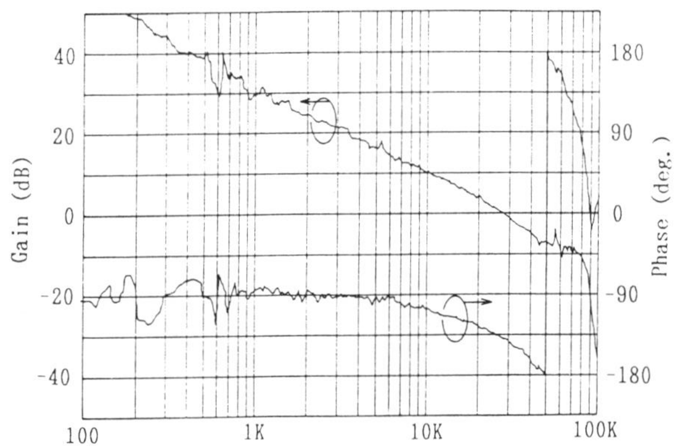

Fig. 12. Open loop characteristics of fine actuator servo (GdGf).

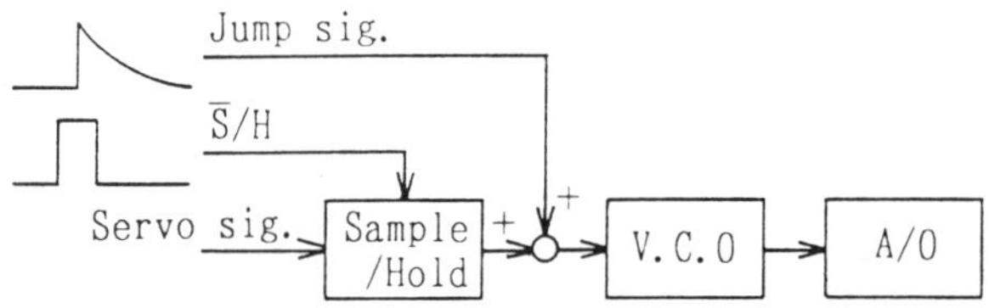

Fig. 13. Block diagram of the jump circuit.

angle is in proportion to the change of the driving frequency. If the target track is within the dynamic range of the A/O deflector, the laser spot can jump to the target directly. It takes about 2 μs for the laser spot to jump, and it takes 20-30 μs for the fine servo to settle.

Figure 13 shows the block diagram of the jump circuit. The jump control signal is added to the VCO, and is decreased slowly so that the fine and coarse servo can follow.

§4. Experimental Results

4.1 *Pulling in a track*

Figure 14 shows the track error signal and the laser spot movement by the A/O deflector when pulling in a track whose pitch is 1.2 μm at a velocity of about 30 mm/s. The bandwidth of the fine actuator servo is 30 kHz.

Settling time is estimated to be less than 100 μs. When the tracking servo is on, the laser spot movement by the A/O deflector represents the overshoot of the coarse actuator.

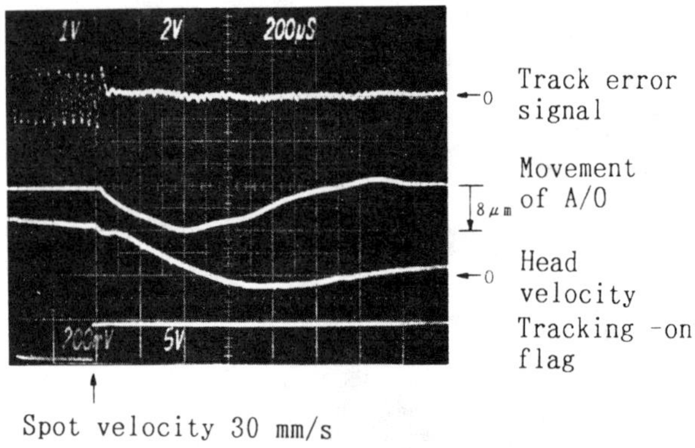

Fig. 14. Track error signal, spot movement by A/O deflector and head velocity when pulling in a track.

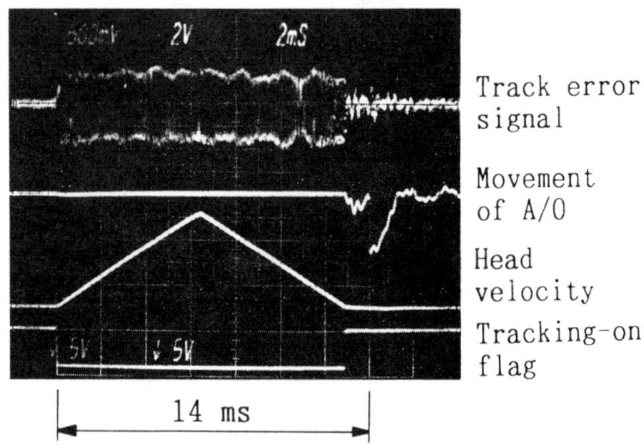

Fig. 15. Track error signal, spot movement by A/O deflector and head velocity when accessing 11.67 mm.

4.2 *Track accessing*

Figure 15 shows the track error signal, movement of the A/O deflector and detected head velocity for accessing 11.67 mm stroke, which is one third of full stroke.

The head is accelerated at 28 G, and decelerated at 26 G in macro seek. Maximum speed is about 1.8 m/s. The fine actuator is locked during the macro seek. The macro seek time is 13 ms.

The vibration of the objective lens caused by hasty change of the deceleration is surpressed by the fine actuator servo, so track address can be confirmed soon after the macro seek. Micro seek starts 1 ms after the macro seek, and in this case, the spot jumps 7 tracks directly to the target. It takes less than 100 μs for micro seek.

The average access time is estimated to be about 14 ms in our system.

§5. Conclusions

In achieving high speed access, the necessity to improve the ability to pull in a track has been pointed out.

This ability has been improved by introducing an A/O deflector as a fine actuator.

A new offset-free track error detection method using an A/O deflector has been developed.

By combining this A/O system with split optics, high speed access has been verified through experiment. An average access time without latency of 14 ms has been achieved for 12″ optical disk system.

Acknowledgments

The authors would like to thank Dr. Kameyama of Hitachi Odawara Works, Mr. Ichikawa of Mechanical Engineering Research Lab. and Mr. Kasai of Central Research Lab. Hitachi Ltd. for their advice and discussion.

References

1) T. Maeda, Y. Tsunoda and Z. Tsutsumi: IECEJ Tech. Report (Image Recording), IE83-71, (1983) [in Japanese].
2) J. Ichihara: Proc. of JSPE 1985 Spring, pp. 667, (1985) [in Japanese].
3) H. Ishibashi, S. Tanaka, R. Shimizu, M. Kuwamoto, M. Imura, H. Yamamoto and T. Shimamoto: "High Speed Accessing Magneto-Optical Disk Drive", SPIE Optical Storage Techinology and Applications, 899-02, Jan. (1988).
4) K. Koumura, F. Takizawa, H. Inada, Y. Yamanaka and K. Yoshida: "High Speed Accessing Using Split Optical Head", SPIE Vol. 1078 Optical Data Storage Topical Meeting p. 239, (1989).
5) A. Arimoto, K. Muraoka, T. Shimano, K. Senda and S. Ishikawa: "Wave Guide Deflector for Optical Disk Tracking Actuator Using Surface Acoustic Wave Device", Proc. of IGWO '89 (MBB5), pp. 37–40, (1989).
6) T. Maeda, M. Kasai and Y. Tsunoda: National Conf. Record 1983 IECE Japan, 1022, (1983) [in Japanese].
7) M. Moriya, A. Nakata and S. Kanemaru: National Conf. Record 1983 IECE Japan, 1210, (1983) [in Japanese].
8) T. Maeda, M. Uno and M. Takasago: "Two Stage Actuator Servo Method", Proc. of Optical Memory Symposium '85, pp. 189–194, (1984).

Reprinted with permission from *Applied Optics,* Vol. 29(2), pp. 247-250 (January 10, 1990). ©1990 Optical Society of America.

Waveguide optical deflector for an optical disk tracking actuator using a surface acoustic wave device

Akira Arimoto, Kouji Muraoka, Takeshi Shimano, Kazuto Senda, and Sachiko Ishikawa

Optical disks are widely used in data processing. However, one of the problems is slow access time compared with that of magnetic disks, because in the former dual mechanical actuators such as a galvanomirror or voice coils are used. To achieve higher speed access, it is necessary to replace mechanical components with an electronically controlled system. A surface acoustic device is introduced into the fine access actuator and its feasibility is tested. Less than 1-ms fine access can be expected.

I. Introduction

Optical disks are widely used in data processing systems as well as in compact disk and videodisk systems. Rapid progress is now being achieved in terms of data capacity and rewritability. Data capacity is expected to increase with shorter wavelength laser diodes[1] and rewritable optical disks with magnetooptic and phase change recording materials now available.[2] However, current major unsolved problems are slow data transfer and tracking access times compared with those of magnetic disk systems. This slowness is because optical disks employ a dual actuator system comprised of coarse and fine actuators, each one utilizing mechanical components such as galvanomirrors. These components have slow response times and take more than 20 ms to stabilize, because of the mechanical vibration. It is difficult to design a coarse actuator without mechanical components, but these components can be removed from the fine actuator. For faster access, waveguided and electronically controlled components are promising. In this study, ultrahigh speed tracking access is attained using a surface acoustic wave (SAW) deflector in the optical disk pickup.

II. Necessity for Reduced Access Time

Figure 1 compares the access times of optical and magnetic disks. In conventional 13.3-cm (5.25-in.) optical disk systems, the total access time is 60 ms. Of this 60 ms, coarse access occupies 40 ms and fine access takes 20 ms. On the other hand, the access time of the magnetic disk is only 20 ms, that is, one-third that of the optical disk. This difference is due to the track pitch difference between the two disk systems and the use of a dual mechanical actuator in the optical disk system. If an electronically controlled SAW deflector replaces the mechanical fine actuator, the fine access time can be expected to be <1 ms, which is negligible compared with the coarse access time. This reduction in the fine access time is investigated in detail.

III. Optical Pickup Using a SAW Deflector

The optical pickup configuration is shown in Fig. 2.[3] Ti-diffused $LiNbO_3$ is used as the waveguide. The laser beam emitted from the laser diode is sent to the waveguide through a rutile prism and a coupling lens. SAWs are excited in the waveguide by an Al interdigital transducer (IDT) located on the substrate. The center frequency is 250 MHz. The laser beam deflected by the SAW is collimated through a cylindrical lens placed behind the SAW deflector and is sent to the optical disk through the beam splitter and objective lens. The RF signal and the spot control signals are obtained by the photodetectors. A tracking servomechanism is utilized to reduce tracking error signals. A voltage controlled oscillator (VCO) circuit is used to change the tracking error voltage to correspond to the SAW frequency change. The grating placed between the cylindrical lens and the beam splitter is used to get the tracking signals.[4] Plus minus first-order beams are used for this.

IV. Fundamental Properties of the SAW Deflector

Before performing the tracking experiments, several basic properties of the SAW deflector were investigated.

The authors are with Hitachi, Ltd., Central Research Laboratory, 1-280 Higashi-Koigakubo, Kokubinji, Tokyo 185, Japan.

Received 14 July 1989.
0003-6935/90/020247-04$02.00/0.

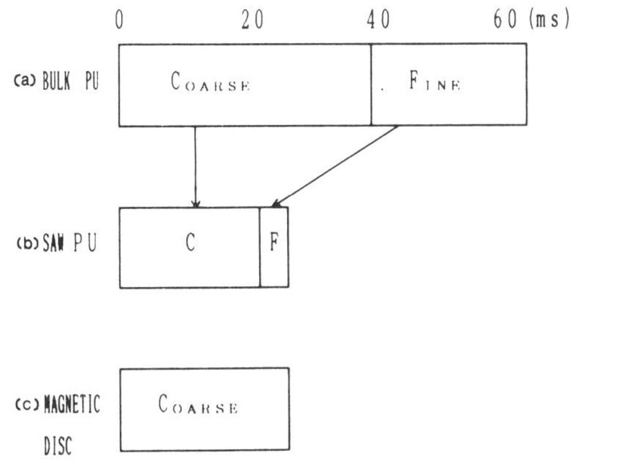

Fig. 1. Comparison of access time between optical and magnetic disk systems.

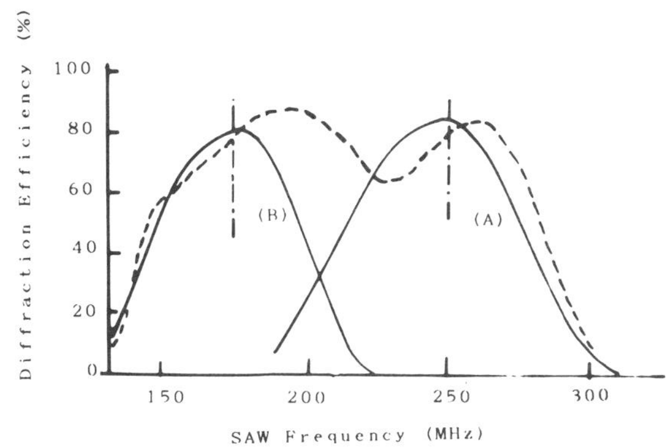

Fig. 3. Deflection bandwidth.

A. Deflection Bandwidth

The diffraction efficiency vs SAW deflector frequency properties are shown in Fig. 3. When an IDT (A) with 250-MHz center frequency is used, over 40% efficiency is obtained between 210 and 290 MHz. The bandwidth is 80 MHz. This value corresponds to a 40 disk track bandwidth when using an objective lens with 4.5-mm focal length. The deflector can cover more than 100 disk tracks when two IDTs, one with a center frequency of 175 MHz (B) and the other of 250 MHz, are utilized simultaneously. This is represented by the dashed line. These numbers are quite sufficient for fine access in optical disk systems.

B. Deflection Speed

The deflection speed between two positions is checked by switching back and forth between frequencies. This speed is detected with a streak camera. The measured switching time is ~10 μs, as shown in Fig. 4; this time ensures that the fine access time is <100 μs—much faster than that obtained when using a mechanical deflector.

C. Wavefront Aberration

A 0.349-λ wavefront aberration was obtained, as can be seen in Fig. 5, when the collimated output beam from the SAW deflector was sent to a Mach-Zehnder interferometer. Even though this wavefront is not aberration free, a focused laser spot of ~2 μm can be expected. The aberration is mainly caused by the error in displacement of the cylindrical lens placed behind the deflector. In this experiment, the data track pitch is 3 μm.

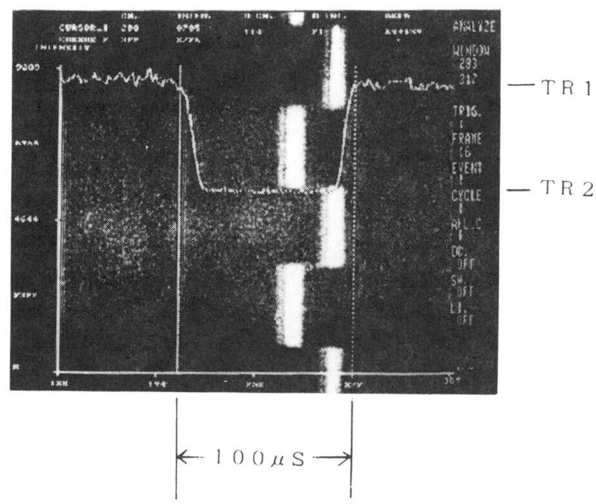

Fig. 4. Deflection speed properties.

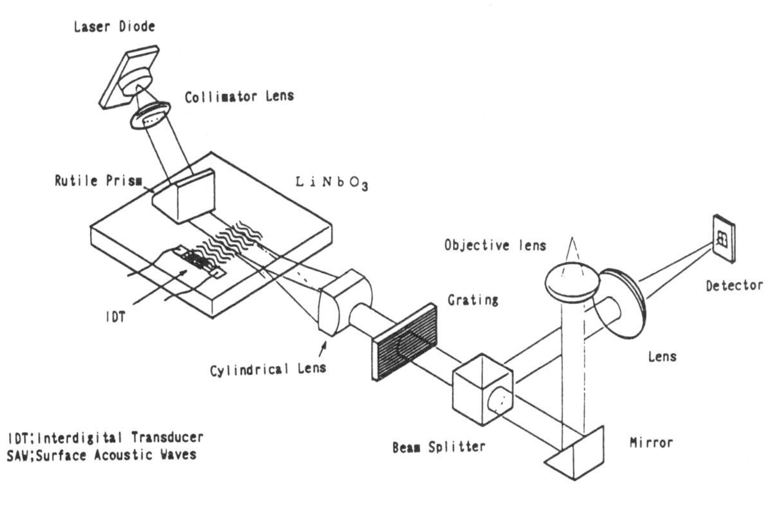

Fig. 2. Optical pickup using the SAW deflector.

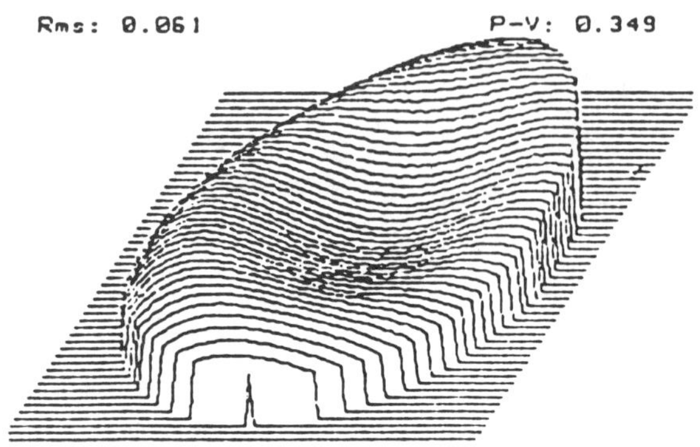

Fig. 5. Wavefront aberration.

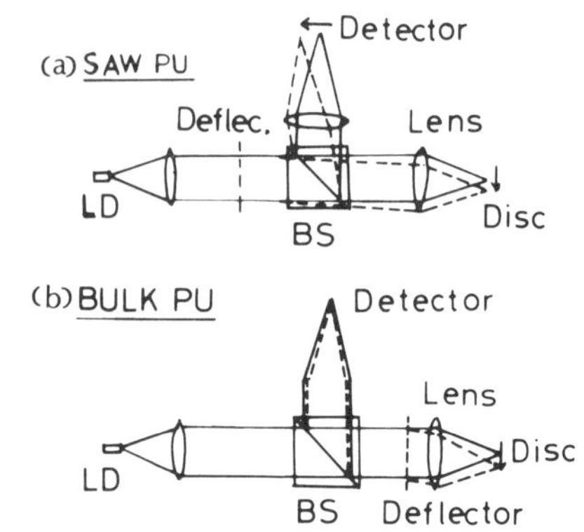

Fig. 6. Comparison of the optical pickup configuration: (a) SAW pickup and (b) conventional pickup.

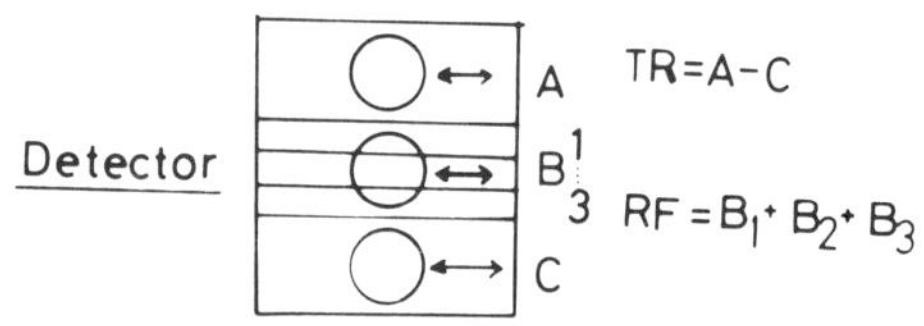

Fig. 7. Five-segmented detectors in this experiment.

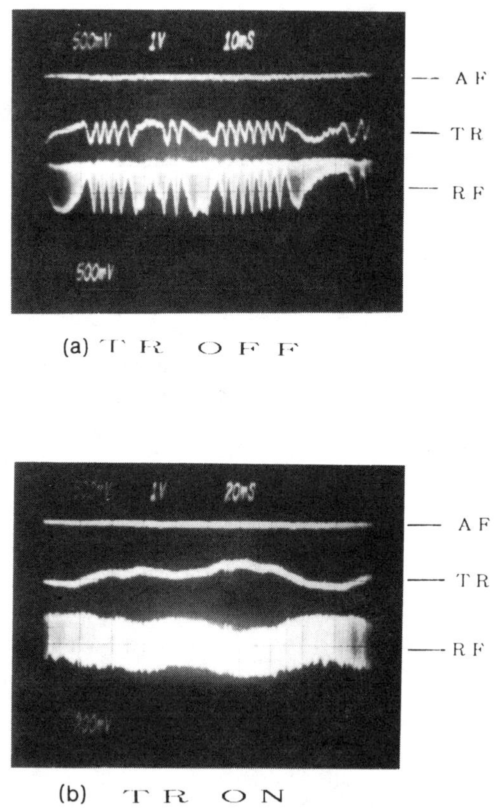

Fig. 8. Tracking experiment: (a) TR off and (b) TR on.

V. Configuration Difference Between Conventional Bulk Pickup and SAW Waveguide Pickup

The optical configurations of conventional and SAW pickups are shown in Fig. 6.

The main difference between these two pickups is their deflector position. In the conventional bulk pickup, the deflector for tracking is placed between the beam splitter and the objective lens. In this case, the laser spot on the detector moves only slightly, even when the laser beam is deflected by the mechanical deflector. This is because the laser beam passes through the same path on the way to and from the optical disks, thus going through the deflector twice. In the SAW waveguide pickup, the deflector is placed between the collimating lens and the beam splitter, because it is difficult to insert the laser beam into the waveguide on the return from the disk as in conventional pickups. In the SAW pickup, the laser beam on the detector moves due to deflection of the laser beam. We must take this problem into consideration when designing an optical pickup.

Figure 7 shows a five-segmented detector used in the experiment. A and C detectors are used for TR, and B_1–B_3 detectors are used for RF and AF signals. The laser beam moves transversely, but in this configuration the AF, TR, and RF signals are not interfered with by the laser spot motion on the detectors.

VI. Tracking Experiment Using a SAW Deflector

To perform the tracking, the three-spot method which makes use of a diffraction grating is employed. When the tracking servo is switched off [Fig. 8(a)] and on [Fig. 8(b)], the tracking error and RF signals are shown. When the servomechanism is off, the laser spot is fixed and the tracking error signal and the RF signal vary, because the laser spot crosses the data tracks due to decentering. When the tracking servomechanism is switched on, the continuous RF and suppressed tracking error signals are observed. The TR error signals are reduced to one-seventh of the peak value, ensuring that tracking using the SAW device is performed successfully. The fluctuation of the RF signal amplitude caused by variations in the diffraction efficiency that are due to the frequency change in the SAW deflector can be easily corrected with an automatic power control.

VII. Track Jump Experiment

Figure 9 shows the waveforms of the track jump experiment. Before jumping, stable reduced and con-

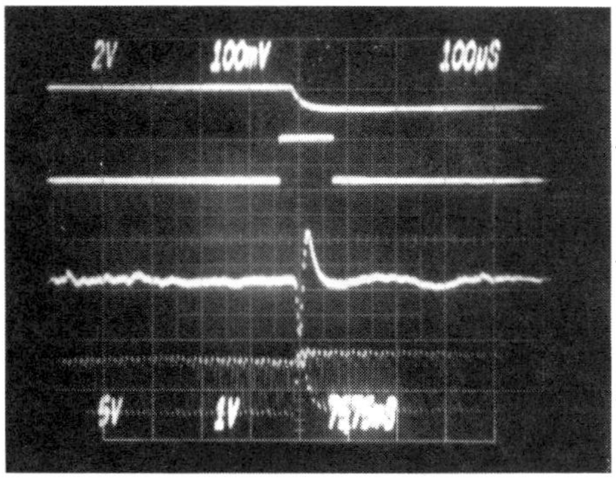

Fig. 9. Tracking jump experiment.

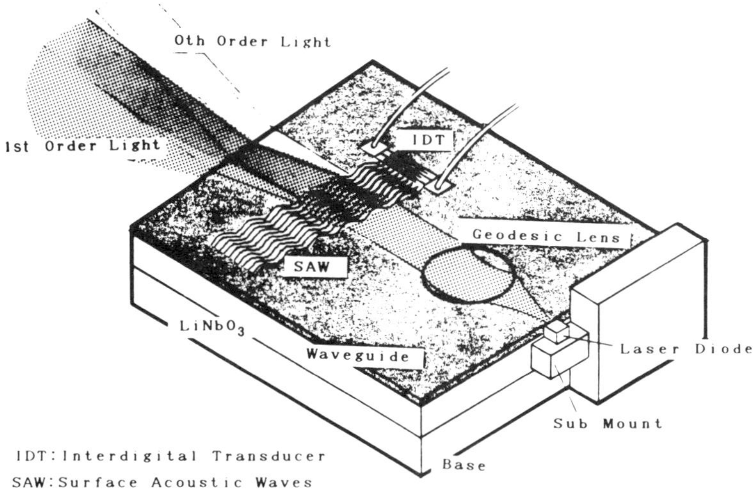

Fig. 10. Compact SAW deflector.

tinuous RF signals are observed. When the TR jump signal is applied and the SAW frequency is forced to change from one frequency to another, the TR error and RF signals fluctuate for a short time but soon stabilize. The horizontal axis unit is 100 µs, so a track jump can be done within 100 µs. This means that the fine access time can also be <100 µs.

VIII. Compact SAW Deflector

In the TR experiment, the optical system is complicated and impractical because optical transmittance is <10%. Figure 10 shows the next generation SAW deflector. A laser diode is attached and a geodesic lens is fabricated for beam collimation. This geodesic lens is chromatic aberration free, and it is quite important because the diode laser oscillates in multiple modes due to the mode-hopping and the feedback of the laser beam. More than 30% of the laser beam is inserted into the waveguide. The deflection was confirmed but tracking was not completed, mainly due to a fabrication error in the geodesic lens.

IX. Conclusion

To reduce access times and replace magnetic disk systems, elimination of mechanical actuators from an optical pickup is necessary. An electronically controlled SAW deflector with a fine access time of <100 µs was introduced. Further integration of the optical pickup, such as fully integrated pickup proposed by the Osaka University group,[5] will reduce the coarse access time.

References

1. M. Ishikawa, Y. Ohba, Y. Watanabe, H. Sugawara, and M. Yamamoto, "Room-Temperature cw Operation of Transverse-Mode-Stabilized 680-nm InGaAlP Visible Light Laser Diodes Grown by MOCVD," in *Technical Digest, Conference on Lasers and Electro-Optics* (Optical Society of America, Washington, DC, 1986), paper TUQ1.
2. M. Ojima, T. Nakao, H. Sukeda, N. Ohta, H. Yasuoka, T. Nishida, and M. Terao, "High Speed Overwritable Optical Disk," Proc. Soc. Photo-Opt. Instrum. Eng. **899,** 154–159 (1988).
3. A. Arimoto, K. Muraoka, T. Shimano, K. Senda, and S. Ishikawa, "Wave Guide Optical Deflector for Optical Disk Actuator Using Surface Acoustic Wave Device," in *Technical Digest, Conference on Integrated and Guided-Wave Optics* (Optical Society of America, Washington, DC, 1989), paper MBB5.
4. G. Bouwhuis, "Apparatus for Reading a Flat Record Carrier," U.S. Patent 3,876,842 (1975).
5. S. Ura, T. Suhara, H. Nishihara, and J. Koyama, "An Integrated-Optic Disk Pickup Device," IEEE/OSA J. Lightwave Technol. **LT-4,** 913–918 (1986).

Reprinted from *Optical Data Storage*, Proc. SPIE Vol. 1316, pp. 2-10 (1990).

An 86 mm magneto-optical disk drive with a compact and fast-seek-time optical head

Naoya Eguchi, Minoru Tobita, and Masataka Ogawa

Corporate Research Laboratories, Sony Corporation
1-7-4 Konan, Minato-ku, Tokyo 108, Japan

ABSTRACT

An average seek time less than 20 msec was achieved using a head with a 6 gr moving part with an 86 mm diameter magneto-optical disk drive.

1. INTRODUCTION

The magneto-optical (MO) disk is now being used as a data storage medium and its market is rapidly expanding. To encourage wider application of MO disk systems, these systems should be smaller, have a higher data transfer rate, a shorter seek time, and the ability to read read-only-memory (ROM) disks easily. To satisfy the above requirements, we have developed an 86 mm MO disk drive based on the sampled servo system which employs a compact optical head. In this paper we describe the structure of the disk, the optical head and the tracking system used.

2. STRUCTURE OF SAMPLED SERVO DISK

We modified a discrete-block-format[1] (DBF) to expand the disk capacity to more than 200 Mbytes using a variable data clock with high linear density and a narrow track pitch. Figure 1 shows the structure of a sampled servo disk. The total number of tracks is 12480 with a track pitch of 1.35 μm. Each track is divided into 1612 segments, and each segment subdivided into two areas, namely, servo area and data area. The servo area contains a grey code for detecting the relative track number, wobbled pits for tracking and a clock pit for clocking of time base. There are two clocks used in this system. One is a servo clock which has a fixed rate and the other is a data clock which has a rate proportional to linear velocity. The user data capacity is 13 kbytes at the most inner track and 20.5 kbytes at the most outer track. The disk rotates at 3600 rpm in order to achieve a high data transfer rate of 9.2 to 15.5 Mbps and a servo signal sampling rate of 96.72 kHz.

3. OPTICAL HEAD

Figure 2 shows the overall structure of the optical head. In order to achieve a faster seek time, the head is divided into three blocks: a slide motor with focus actuator, a galvano-motor and a fixed optical block. If it is to be used as a computer peripheral, an 86 mm-MO-disk drive must have a form factor of 41(H)×102(W)×147(D) mm^3. Since the height of the optical head is the major factor in the drive thickness, we developed a 15 mm-thick optical head employing optical components with a 3 mm-diameter pupil aperture, positioning all three blocks in one plane.

3.1. Slide motor

As is shown in Fig. 3, the slide motor consists of a focus actuator and a slide actuator for radial tracking. The focus actuator, an aspheric objective lens suspended by two leaf springs, is mounted on the slide actuator. The slide actuator is mounted on guides and moves with ball bearings. The guides act as yokes of the magnetic circuits which drive the actuators. By applying the magnetic flux commonly to the coils of the two actuators and by locating all the moving components except for the objective lens inside the coils, the moving part of the slide motor can weigh just 6 gr, which allows the parasitic resonances of the slide motor to be higher than 3 kHz. As the electromagnetic coil is made of rolled laminated film without a bobbin, more than 1.8 times ampere-turns is obtained as compared to the conventional wire coil. The small, light and bobbinless moving part of the slide motor allows the acceleration sensitivity to the drive current to be 12 G/A.

3.2. Galvano-motor

The galvano-motor is moving-coil (MC) type. A metal spring suspends the mirror to reduce the hysteresis of the galvano-mirror motion. The inertia of the galvano-mirror is 9 gr·mm^2 and the acceleration of the laser spot on the disk by the driving current is 100 G/A.

3.3. Optical block

Figure 4 shows the structure of the fixed optical block. Employing single-beam servo detection and a micro-prism detector[2] (MPD), a very small and simply structured optical block is built using only nine optical parts to provide a 3 mm-diameter pupil aperture.

4. TRACKING SYSTEM

Figure 5 is a block diagram of the tracking system. The system is composed of two feedback loops, namely, the position feedback loop and the velocity feedback loop. The former is a tracking servo loop using a tracking error signal and the latter is a seeking loop with the servo bandwidth of about 700 Hz using a velocity signal detected from the disk. A two-stage servo system[3,4] is employed for the tracking servo system of this optical head. The tracking error signal is divided into two frequency bandwidths, the higher frequency range for the galvano-motor and the lower frequency range for the slide motor. Figure 6 shows the open loop transfer characteristics of the tracking servo. The solid curve indicates the total gain of the two broken curves which is the gain of the two actuators. The entire tracking servo bandwidth is 4 to 5 kHz. The total loop gain of a two-stage servo system tends to decrease significantly near the crossing point of the gain curves of the two actuators. This tendency is minimized by matching the phase of the higher frequency servo signal and that of the lower frequency servo signal by adjusting the gain near the crossing point.

4.1. Detection of tracking error

In order to optimize the linearity of the tracking servo, a tracking error signal applying a new concept of an extended tracking error[5] is introduced to the present system. Figure 7 shows the relation between the conventional tracking error and the extended tracking error, and Fig. 8 shows the block diagram of the extended tracking error detection system.

The sampled signal difference of a pair of wobbled pits is employed as the conventional tracking error, TE (Fig. 8). Each one of a pair of wobbled pits is located a quarter track pitch from the track center, one inside the track and the other outside. A clock pit is located on the track between the wobbled pits (Fig. 1). The pair of wobbled pits are sampled and the signals added and the total halved. The difference between this value and signal from the clock pit is detected as a traverse signal, TS. The signals TE and TS are given by

$$TE = \sin(2\pi v/q),$$
$$TS = \cos(2\pi v/q),$$

where q denotes the track pitch and v the light spot position in the radial direction. The extended tracking error which is used here is given by the following formula.

$$\text{Extended Tracking Error} = \arctan(TE/TS),$$
$$= 2\pi v/q, \qquad (-q/2 \leq v < q/2),$$

This signal is the linear tracking error signal periodical for track pitch q (Fig. 7). Improvement of the ramp response of this extended tracking servo enables the light spot to locate a particular track at more than double the velocity of a conventional tracking servo. Figure 9 shows the overshoot of the extended and the conventional tracking servo ramp responses with the servo bandwidth of 4 kHz. The horizontal axis indicates the track crossing velocity and the vertical axis indicates the overshoot. In the conventional tracking servo, the light spot can follow a particular track at a velocity of less than 9 mm/sec[6]. However, in the extended tracking servo, the light spot can follow the track at as fast as 18 mm/sec, which enables the light spot to land on a particular track with much less deceleration. When deceleration is less then, of course, seek time is shortened. Figure 10 shows the ramp response of the tracking servo at a track crossing velocity of 14 mm/sec. It will be seen that the extended tracking servo system eliminates the two-track overrun which occurs in the conventional tracking servo.

4.2. Detection of track crossing velocity

The relative velocity between the disk and laser spot can be detected from a grey code or a tracking error. The grey code is a periodic code having 8 levels for every 2 tracks and is repeated every 16 tracks (Fig. 1). The velocity is calculated by difference of successively detected grey codes. The number of the crossed tracks can be calculated by accumulating the differences. On the other hand, the velocity from the tracking error is obtained from differentiation of the extended tracking error.

As to the measurable velocity range, the velocity detected by a grey code is 0 to 1.95 m/sec for a disk rotation of 3600 rpm and a track pitch of 1.35 μm. The velocity detected by the tracking error is 0 to 0.065 m/sec. As the phase delay of the velocity signal detected from the grey code increases as the velocity decreases, the velocity detected from the tracking error is preferred when the head slows down to reach the destination track. These two velocity signals for high speed and low speed are switched from one to the other according to the relative velocity between the disk and the laser spot.

5. ACCESS

Figure 11 shows the velocity signal and the tracking error in a 10000-track seek. In the present system, the acceleration, the deceleration and the maximum velocity of the moving head are 13 G, 7 G and 0.64 m/sec, respectively. Figure 12 indicates that the average seek time is less than 20 msec for 1000 random-address seeks. (This average seek time is equal to the seek time over 4000 tracks which are approximately one-third of full stroke.) The average access time, adding average latency (8.3 msec at 3600 rpm), therefore, is less than 30 msec. This short access time is achieved by a wide velocity servo bandwidth of about 700 Hz, which is made possible by making parasitic resonances of the slide motor higher than 3 kHz. It should be noted that in the sampled servo system, tracks are counted using a grey code and the accumulated track-counting-error is less than one track.

6.CONCLUSION

A compact, 86 mm-diameter magneto-optical disk drive in combination with thin lightweight optical head and sampled servo system was developed. The moving part of the head was designed to be 6 gr, which enabled wide servo bandwidth and high acceleration resulting with high track-seek speed. A sampled servo with a grey code was employed for precise track counting during track-seek, and a new extended tracking error detection system is used to locate the destination track at high speed. With the head and the servo system, an average seek time of less than 20 msec was achieved. A high density optical disk rotating at 3600 rpm achieved a transfer rate of 9.2 to 15.5 Mbps and a disk capacity of more than 200 Mbytes.

7. ACKNOWLEDGMENTS

The authors would like to thank S. Miyaoka and H. Ooki for their encouragement , and S. Kubota for his helpful advice and discussions.

8. REFERENCES

1. ISO JTC1/SC23/WG2 N10, Heidelberg, FGR, Jun. 20, 1989.

2. M. Oka, A. Fukumoto, K. Osato, and S. Kubota, "A New Focus Servo Method for Magneto-Optical Disk System." Jpn. J. Appl. Phys., Proc. Int. Symp. on Optical Memory, vol. 26 Supplement 26-4,, pp. 187-190, 1987.

3. H. Ishibashi, S. Tanaka, R. Shimizu, M. Kuwamoto, M. Imura, H. Yamamoto, and T. Shimamoto, "High speed accessing magneto-optical disk drive," SPIE, Optical Storage Technology and Application, vol. 899, pp. 8-5, 1988.

4 K. Koumura, F.Takizawa, T. Iwanaga, H. Inada, Y. Yamanaka, and K.Yoshida, "High Speed Accessing using Split Optical Head," SPIE, Optical Data Storage Topical Meeting, vol. 1078, pp. 239-243, 1989.

5. S. Kobayashi and N. Nishida, "Tracking Servo Uses Tangent Function to Determine Tracking Error Signal," United States Patent, US4853918, 1988.

6. N. Eguchi, "Tracking Technology of Optical Head," Japan Society of Precision Engineering, Seminar text-book of "New Positioning Technology" (in Japanese), pp. 81-91, 1987.

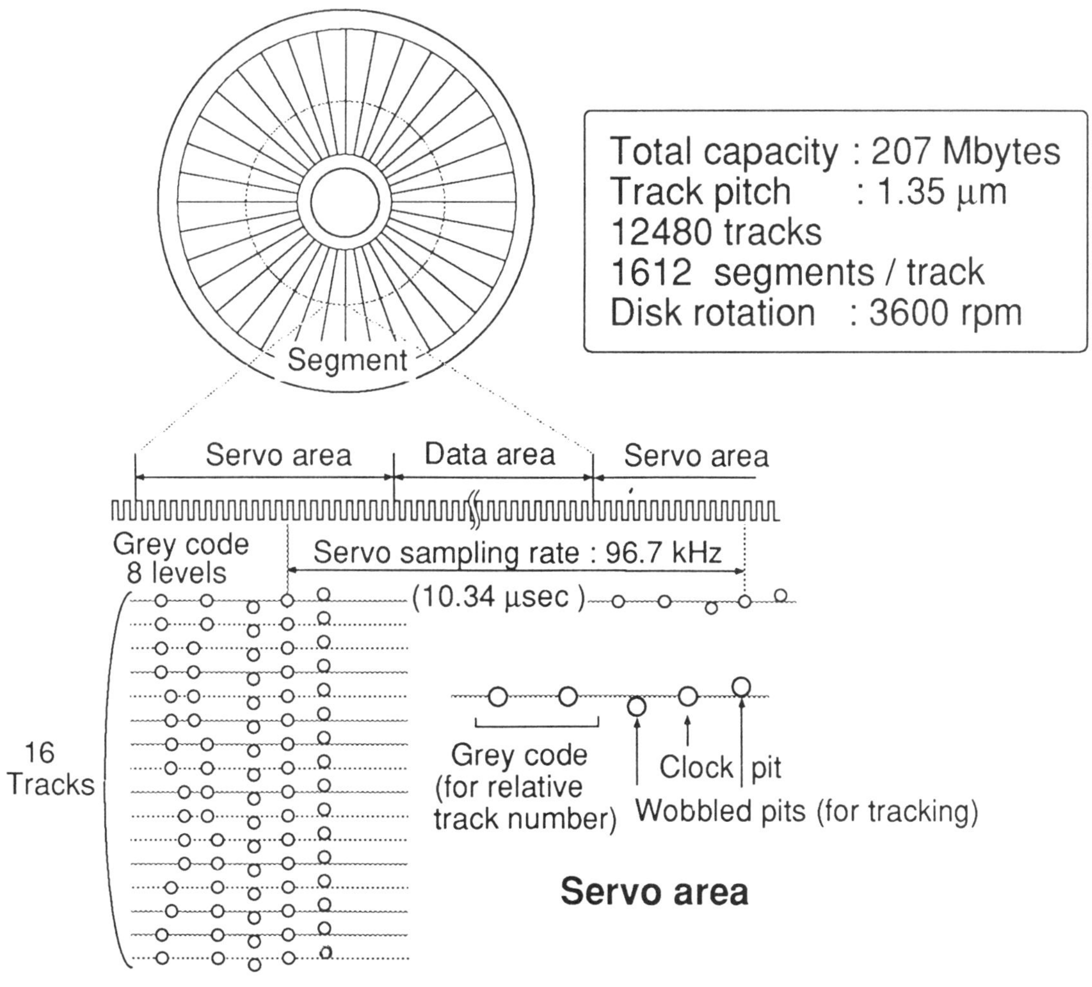

Fig. 1. Sampled servo disk

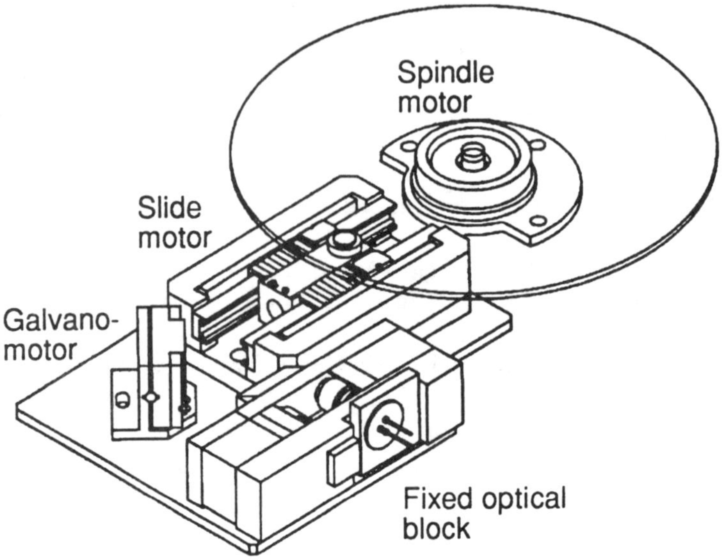

Fig. 2. Optical head

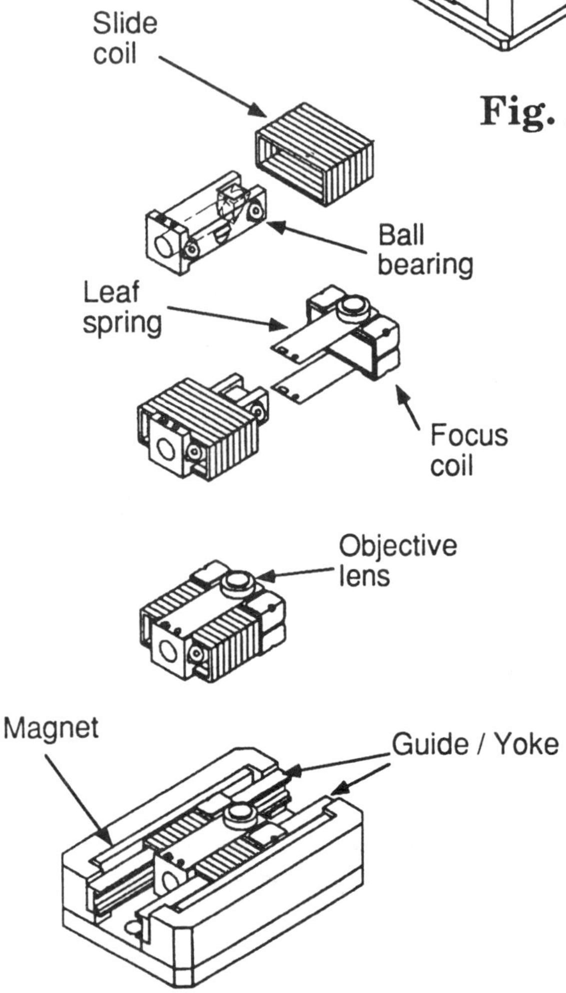

Fig. 3. Slide motor

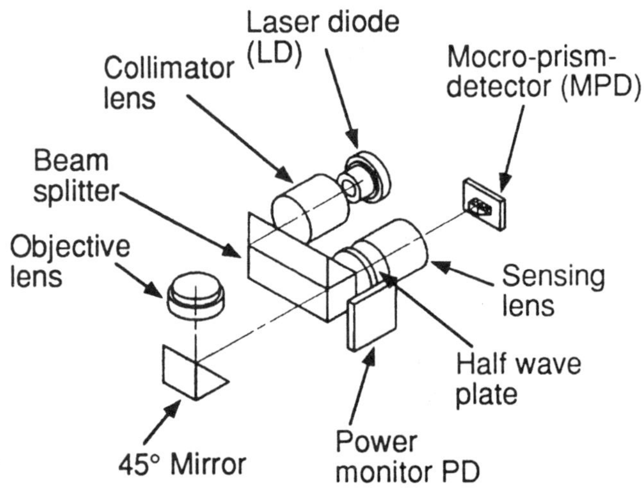

Fig. 4. Fixed optical block

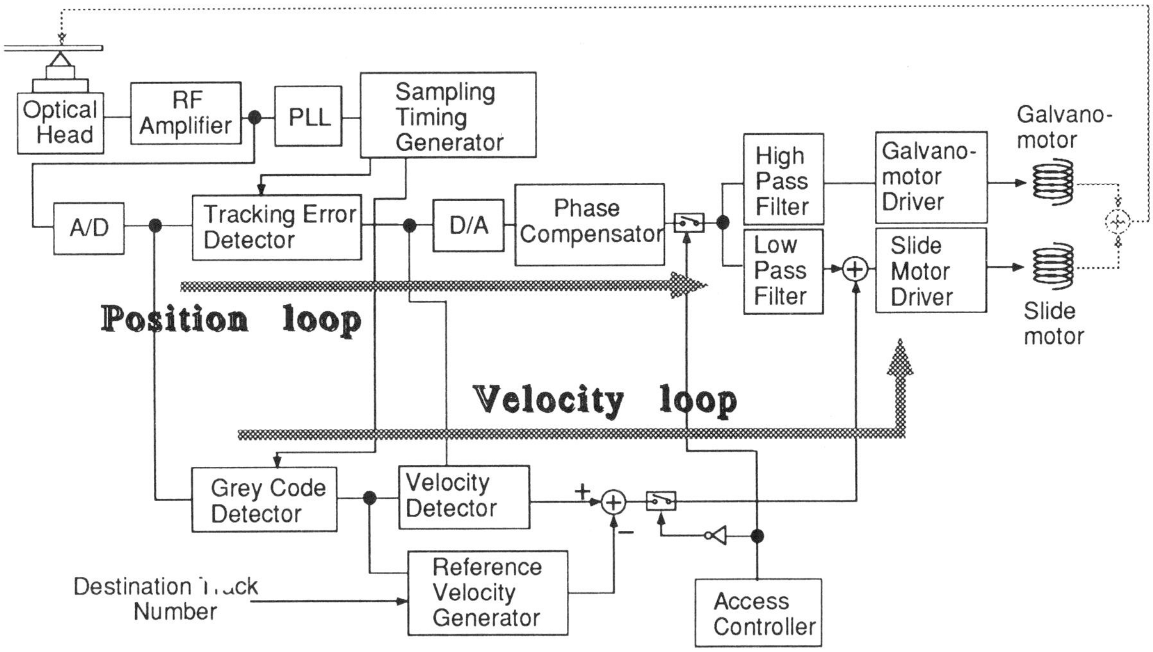

Fig. 5. Tracking system

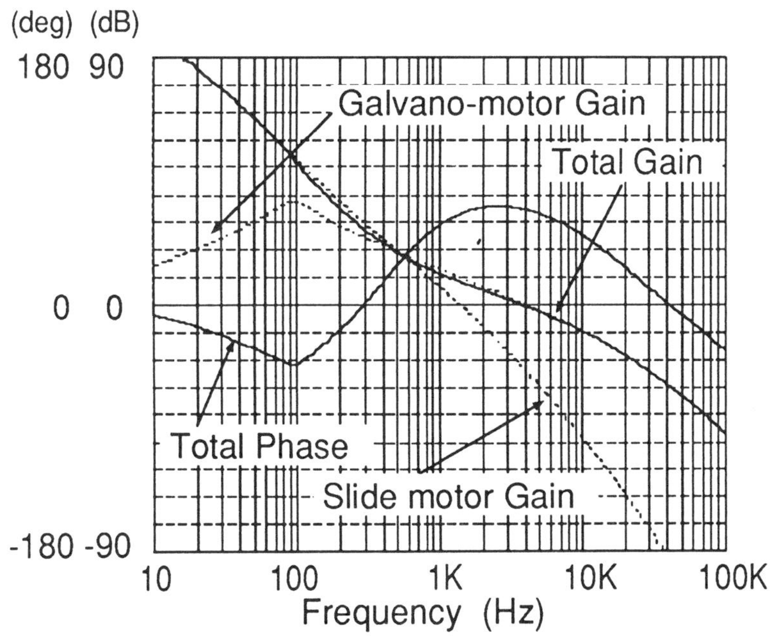

Fig. 6. Open loop transfer characteristics of two-stage servo

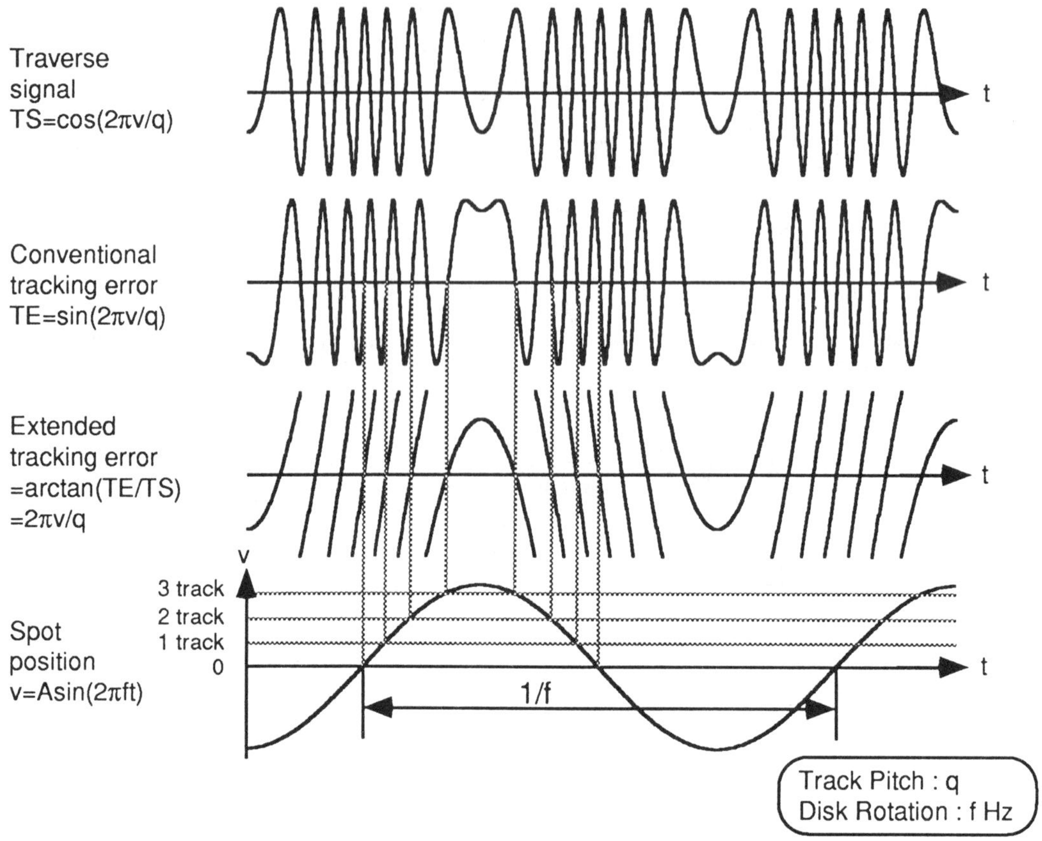

Fig. 7. Relation between conventional and extended tracking error

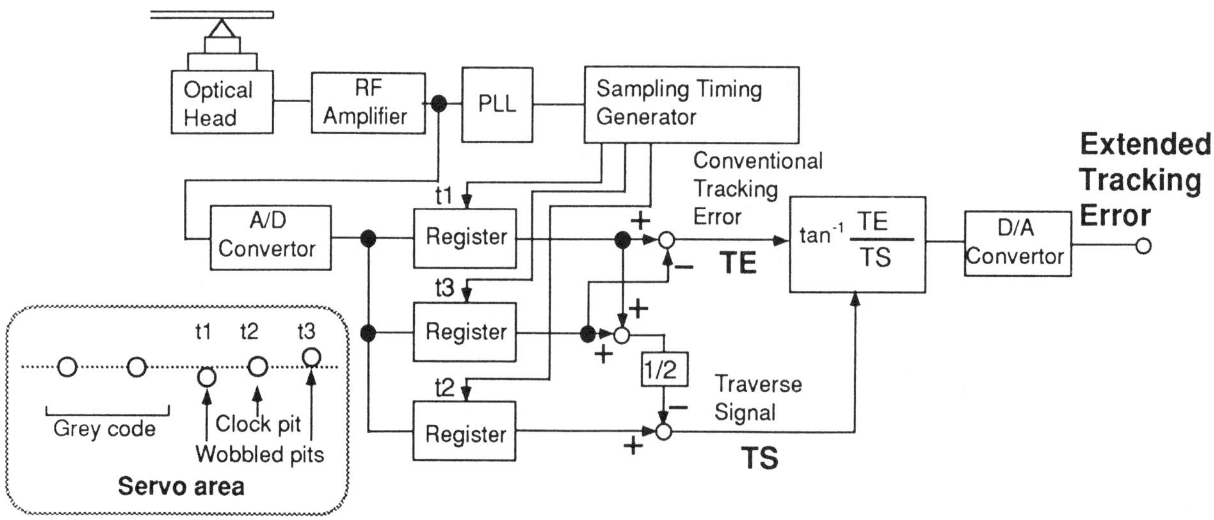

Fig. 8. Extended tracking error detection system

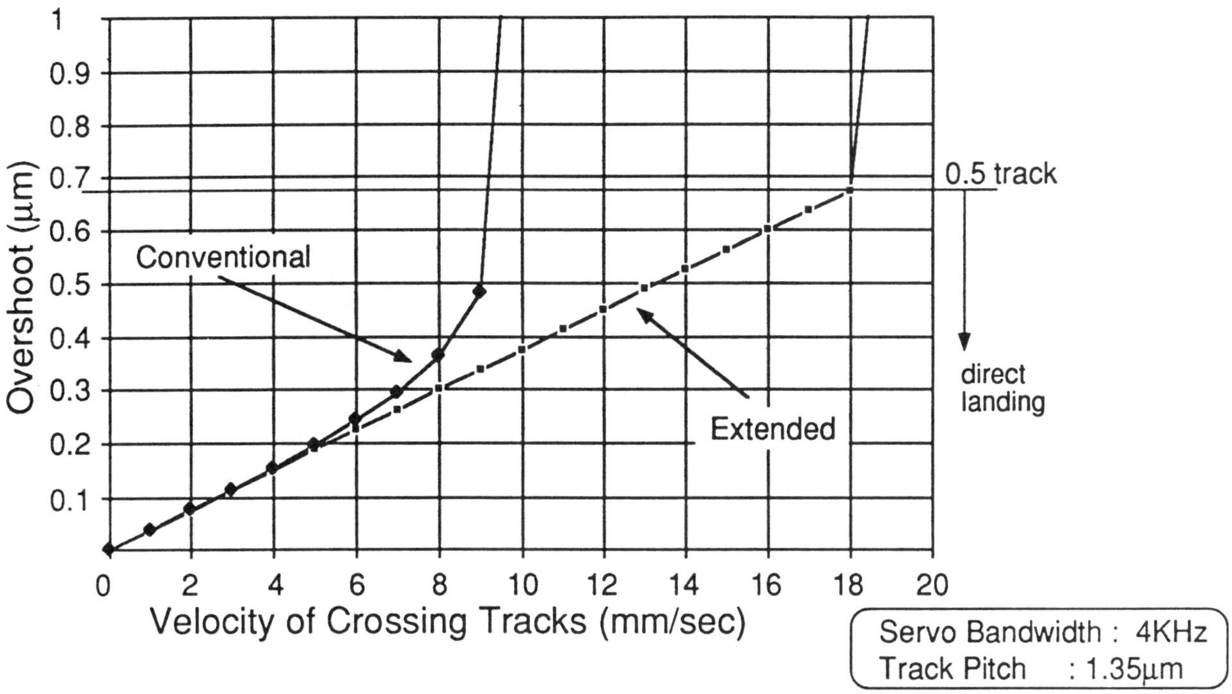

Fig. 9. Relation between overshoot and velocity of crossing tracks

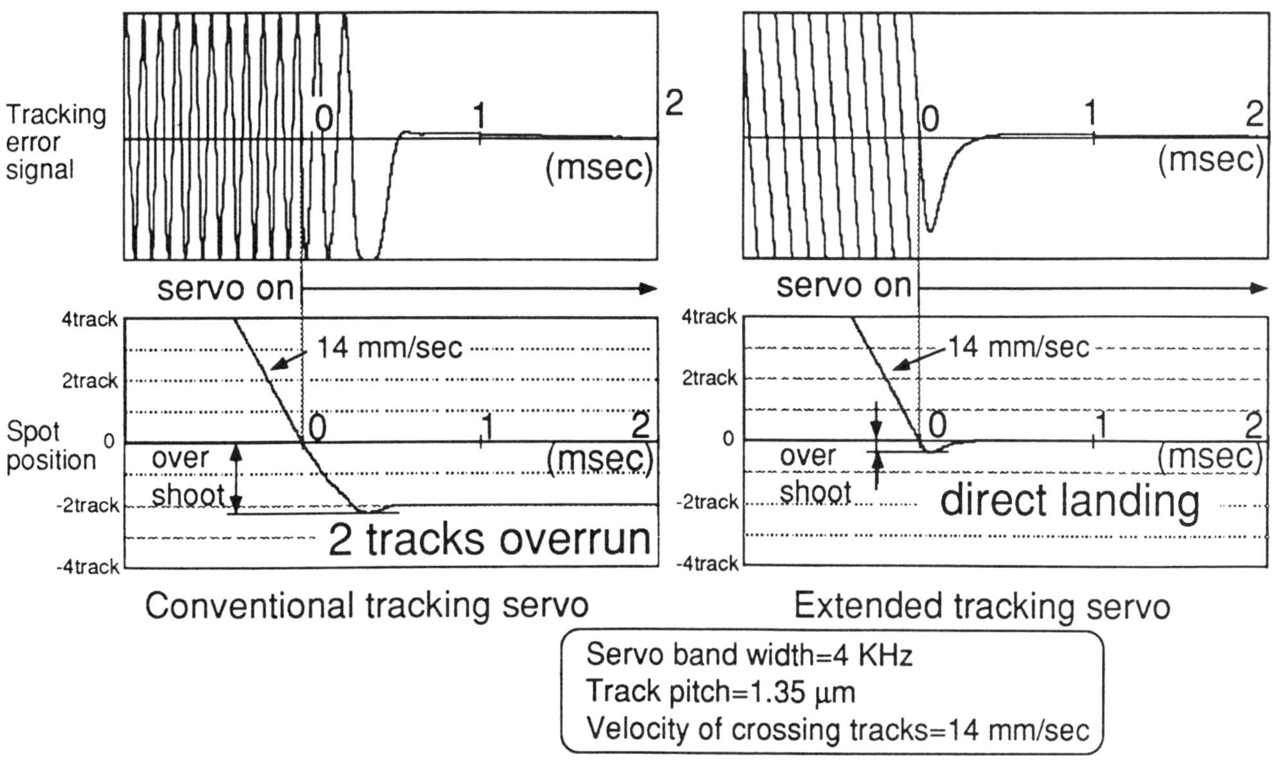

Fig. 10. Ramp responses of tracking servo

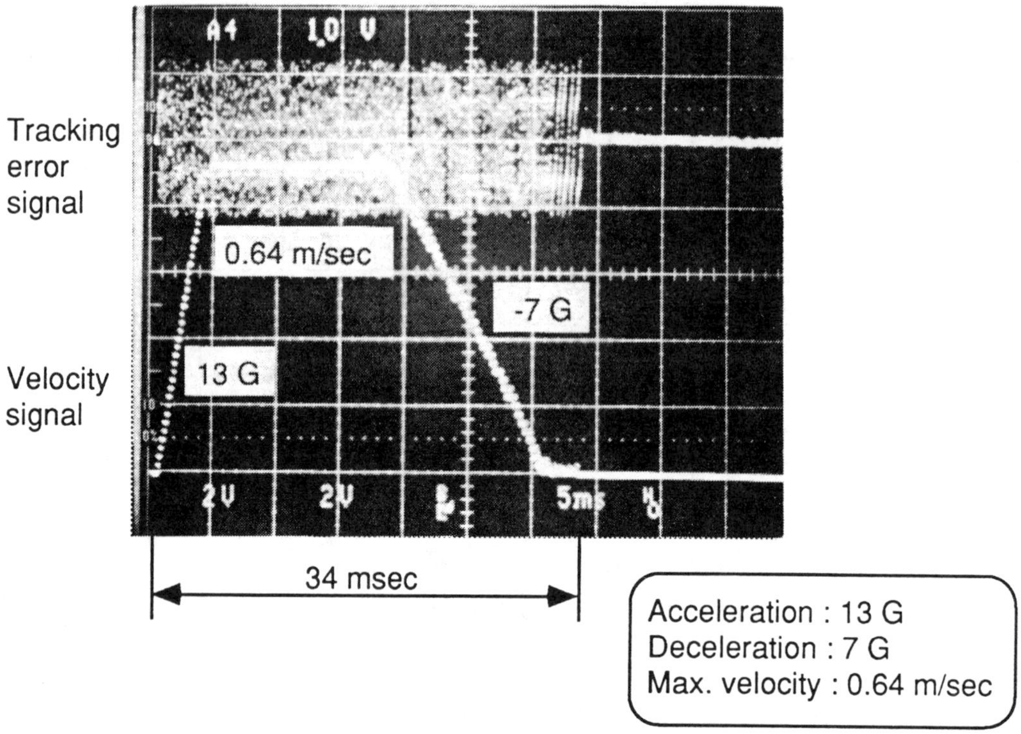

Fig. 11. 10000-track seek

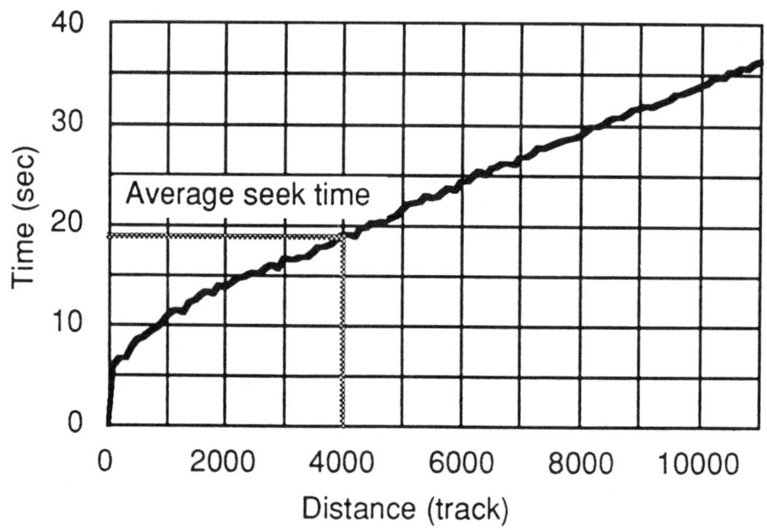

Fig. 12. Seek time

Section Eight
DRAW/DOW

Reprinted from *Optical Mass Data Storage II*, Proc. SPIE Vol. 695, pp. 182-186 (1986).

Two-beam Optical Head

K. Hamada, S. Arai, T. Iwamoto, K. Ogawa

FUJITSU LABORATORIES LTD., KAWASAKI, JAPAN

Abstract

Real time error checking is required in optical disk systems to improve data reliability and to increase data throughput to/from the computer. This paper describes a two-beam optical head which meets the requirement. The head is composed of two laser diodes (830 nm for write and servo operations, 780 nm for read operation), dichroic mirrors, and an achromatic objective lens.

New optical servo technologies have been developed and incorporated in this optical head. One is a new focusing technology (called the Wax-wane method) which needs only two optical components. Another is a new tracking technology (called the ILT method) which enables a tracking range of ±400 μm.

Introduction

A number of optical disk drives have been introduced in recent years. These were not always suitable for computer systems because of poor data throughput. In order to enable immediate verification of the data to guarantee the data reliability, the data was written at one revolution of the disk and verified at the next revolution. This recording process needed two or three times as much recording time as magnetic disk drives.

A two-beam optical head is developed. The preceding beam is used for write and servo operation and the following beam is used for read operation. This construction with two beams can realize direct-read-during-write and the same recording performance as magnetic disks.

The Two-beam Optical Head

Figure 1 is a photograph of the two-beam optical head. In spite of having many functions and advantages which are described below, smaller size of 60 x 50 x 18 mm and lighter weight of 90 grams are obtained.

Figure 2 is a photograph of the two-beam profile. The head contains two laser diodes (830 nm and 780 nm) and each beam is focused within 0.85 μm diameter FWHM (full width, half maximum) on the disk. The distance between the two beam spots is 10 ±1 μm in the track following direction and within ±0.05 μm in the radial direction.

The write beam is also used for detection of tracking/focusing error signals, because it is most important to prevent mis-writing or loss of recorded data (caused by off-tracking write), and to prevent decrease of write power margin (caused by defocused write beam).

Optical Systems

The schematic diagram of the optical head is shown in Figure 3.

Read System

Laser light at 780 nm from laser diode (LD1) is collimated using a lens (CL1) of comparatively small N.A. (0.15). Therefore no elements such as prisms or cylindrical lenses are needed to shape a collimated beam. The parallel light beam with circular cross-section passes through a polarizing beam splitter (PBS) and is reflected by a dichroic mirror (DM1). The beam reflected by DM1 goes back along the same path, and is reflected again by PBS because the beam passed through a quarter-wave plate (QWP1) twice. The circular-polarized light from the second quarter-wave plate (QWP2) comes to an objective lens (OL) via a galvano mirror (GM) and a mirror (M), and is focused to a micro-sized spot about 0.75 μm in diameter FWHM by the objective lens having an N.A. of 0.55.

The diffracted light including the information signal recorded on the disk returns along the same optical path in the opposite direction, and is polarized to P-component by the quarter-wave plate (QWP2). The light transitted from the PBS, a dichroic mirror (DM2) and a shortwave pass filter (SWF) is focused to a sutable spot size by a lens (L1), and falls on a PIN photodiode (PD1).

Write and Servo System

A high-power laser diode (LD2, 830 nm, 25 mW, CW) is used for write and servo operation. Laser light is collimated using a lens (CL2) having an N.A. of 0.30 and a collimation prism (PR1) to obtain highly efficient laser power. The collimated light beam is reflected by the dichroic mirror (DM2) and is combined with the read beam at the polarizing beam splitter (PBS). Passing along the same optical path as the read beam, the write beam is focused to a spot about 0.85 µm in diameter FWHM by OL. Light transmission efficiency of more than 45% is obtained.

The light reflected from the disk comes back to the polarizing beam splitter (PBS) after being polarized to S-component by the quarter-wave plate (QWP2). The light turns to the left in Figure 3, and reaches a quadrant PIN photodiode (PD2) via a detector lens (L2) and a longwave pass filter (LWF).

Objective Lens

Because the two laser lights are of different wavelengths, there could be some beam-displacement between the two spots, causing mis-reading and mis-writing. To prevent this, an achromatic objective lens is used in the head. A conventional non-achromatic objective lens causes beam-displacement of more than 1.5 µm on the disk in the radial direction, if the beams strike the objective lens 0.5 mm from its center. Our achromatic objective lens does not cause beam-displacement in the wide tracking range of ±0.5 mm.

Focusing

A new focusing technology (called the Wax-wane method *1)) is developed and incorporated in this head. The Wax-wane method requires only two components -- a very common lens (L2) and a detector (PD2) -- to detect the focusing signal. Because the optical detector is positioned not at the exact focal point of lens but at a far-field position, the placement of these components is not as critical as in conventional focusing technologies. As shown in Figure 4, the Wax-wane method is also very effective to reduce crosstalk.

Tracking

A new tracking technology (called the ILT method *2)) is also introduced in this optical head. The ILT method requires two separate actuators (a two-dimensional actuator and a galvano actuator) for track following. The two actuators are driven together, enabling a tracking range of ±400 µm.

Figure 5. shows the actuators used in the ILT method. The reflected beam, driven by the galvano mirror in the radial direction, is always kept passing through the back focal point of the objective lens.

To stabilize the linked movement of the two actuators, both actuators have tracking position sensors.

These position sensors are useful not only for the ILT method but also for other purposes. The sensors guarantee the neutral position for tracking movement, without springs, rubber, or other suspension for moving parts in the optical head. A second purpose of these sensors is electrical damping or electrical lock during long range access for quickly stopping mechanical vibration.

This ILT method greatly enlarges the track following range, which enables the application of a stepping motor as a coarse head carriage.

Performance of Two-beam Optical Head

The head had some worries about crosstalk between the write beam and the RF signal, and about stability against severe ambient temperature and mechanical impact. The spectral and phase characteristics of optical components are strictly controlled, and extremely low crosstalk of less than -50 dB is realized. The separate optical parts (shown in Figure 3) are built into one body, and the head is found to be very stable against severe conditions. Figures 5 and 6 show the accuracy of the relationship between the two beam spots in the radial direction under high temperature and mechanical impact.

Summary

A two-beam optical head for write-once-read-mostly (WORM) optical disk drives has been described. This head enables real time error checking because the preceding beam writes the data and provides signals for the track following servo, and the following beam reads (verifies) the written data. The ILT tracking method enables wider tracking range, and makes it possible for a rough pitch stepping motor such as in an FDD (floppy disk drive) to drive the optical head. The Wax-wane focusing method reduces the crosstalk between focusing and tracking error signals. The head is very stable against thermal and mechanical shock. The ILT construction using two actuators, costs only a little more. The ILT connected with a stepping motor provides a rather low cost optical disk drive.

As explained above, this two-beam optical head has many advantages and is suitable for optical disk drives of computer systems.

Moreover our two-beam optical head is also appropriate for a phase change optical disk, if a cylindrical lens is inserted into the optical path between the collimation lens (CL1, shown in Figure 3) and the polarizing beam splitter (PBS).

References

1) S. Arai, K. Hamada and K. Ogawa, "New Focusing Method for Draw-Type Optical Head", TECHNICAL DIGEST of TOPICAL MEETING ON OPTICAL DATA STORAGE, ThBB1-1, 1985.

2) S. Arai, T. Iwamoto, K. Hamada and K. Ogawa, "Interlinked Tracking Servo Technology", Proc. SPIE conference. 1986.

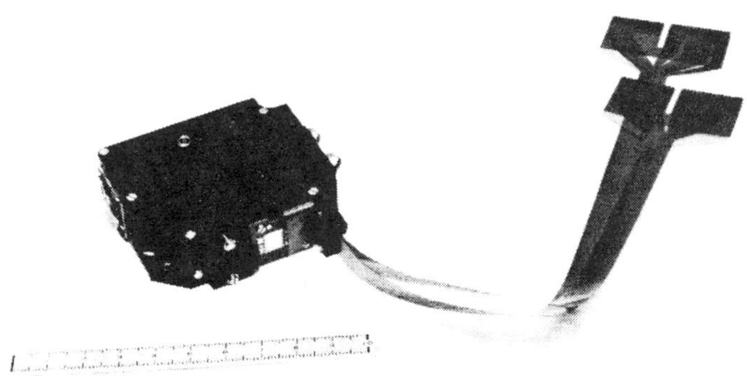

Fig. 1 Two-beam optical head

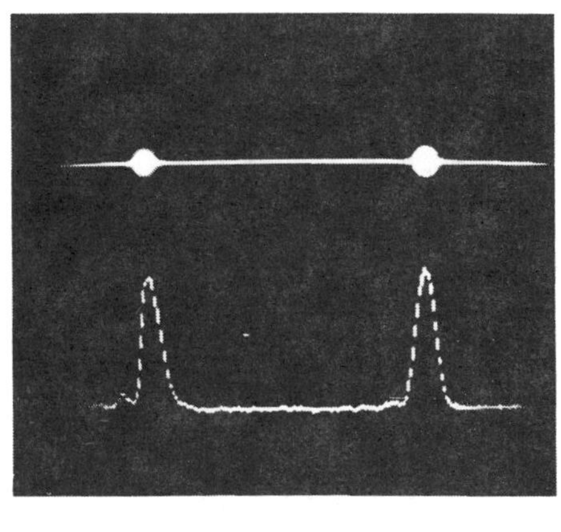

Fig. 2 Two-beam profile

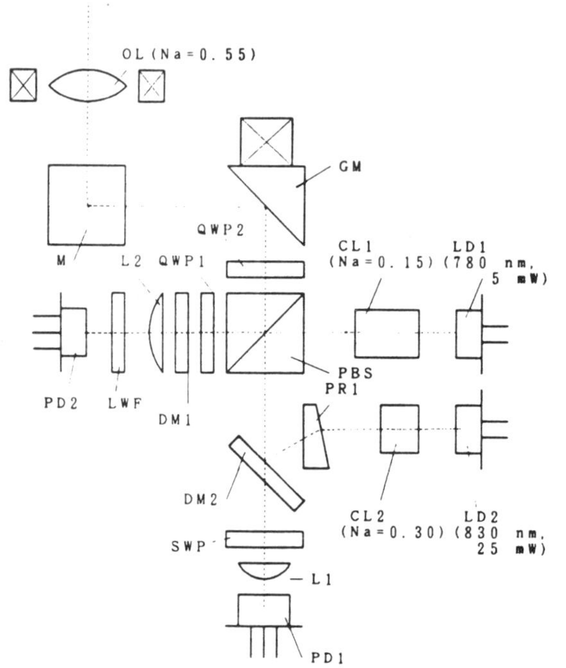

Fig. 3 Schematic diagram of two-beam optical head

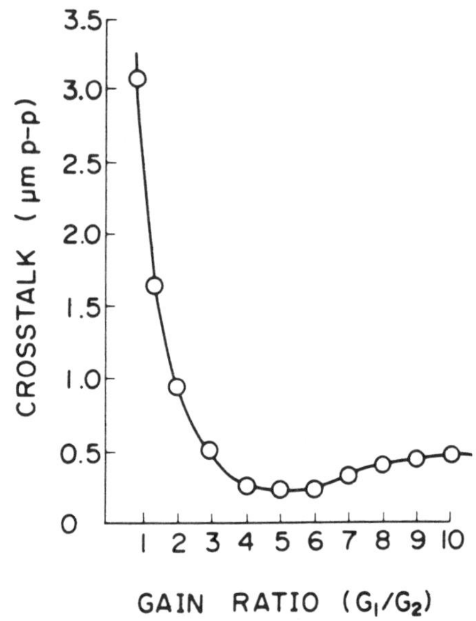

Fig. 4 Gain ratio vs crosstalk in the Wax-wane method *1)

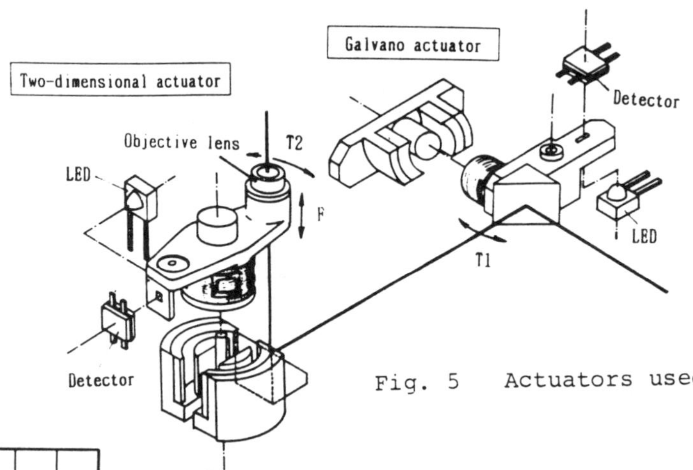

Fig. 5 Actuators used in the ILT method

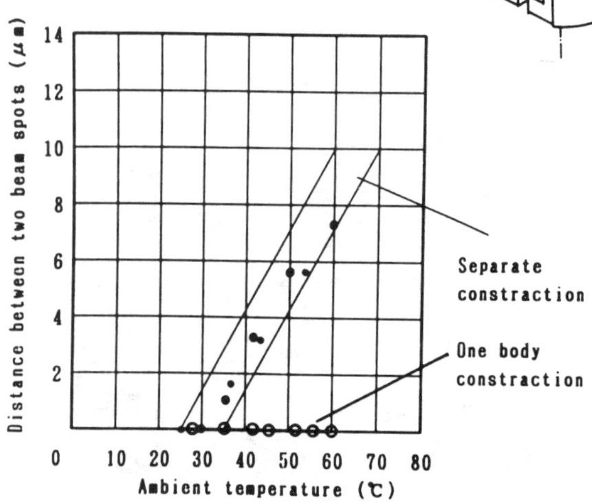

Fig. 6 Distance between two beam spots in the radial direction at high temperature

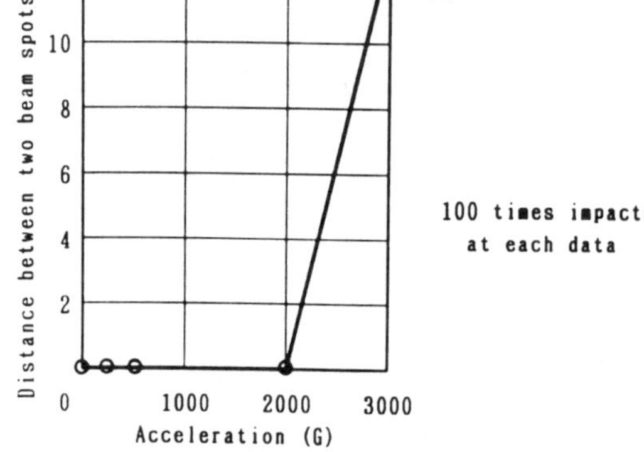

Fig. 7 Distance between two beam spots in the radial direction under mechanical impact

Reprinted with permission from *IEEE Transactions on Magnetics,* Vol. 24(1), pp. 666-669 (January 1988).

Magnetooptic Direct Overwrite Using a Resonant Bias Coil

DANIEL RUGAR

***Abstract*—A new method is described for achieving direct (single pass) overwrite in magnetooptical disk systems. In this method, a sinusoidal magnetic bias field is efficiently generated by incorporating the bias coil into a parallel resonant *LC* circuit. Magnetooptic domains of either sign can be written by timing laser pulses to coincide with either positive or negative peaks of the field. Continuous domains of variable length are written by overlapping the domains generated by successive laser pulses. Using a bias field with a fixed frequency of 4.5 MHz, carriers at various lower frequencies were written on a TbFe disk with 47-dB carrier-to-noise ratio. Direct overwrite resulted in erasure of old carriers by at least 34 dB. Suitable run length limited coding schemes are discussed and a new code type, called (*d*, *k*)-even 0, is introduced.**

I. Introduction

Magnetooptical disk recording is an emerging technology which offers the potential of high-density erasable data storage (see, e.g., [1]-[4]). One disadvantage of this recording method, however, is that most current schemes require two passes to overwrite old data with new data. Some preliminary work on single-pass (direct) overwrite of magnetooptic data can be found in [5]-[8]. In this paper we describe a novel technique which has been used to overwrite data at rates equivalent to 3 Mbit/s.

In principle, magnetooptic direct overwrite can be achieved if the magnetic bias field can be switched rapidly enough to follow the incoming data stream. Achieving megahertz field switching with a coil located millimeters away from the disk active layer is a difficult prospect, however. To understand this, consider the energy stored in a characteristic volume V due to a magnetic field of strength H. In Gaussian units, the stored energy in air is $E_s = (1/8\pi) H^2 V$. If the field is reversing at frequency f, then this energy must be supplied to the coil twice per cycle. Therefore, the powr required (ignoring resistive loss) is $P = 2E_s f = (1/4\pi) H^2 V f$. For a field strength of 200 Oe in a 0.1-cm^3 volume switching at 5 MHz, the required power is $P = 160$ W. This power is dissipated in the power supply, not in the coil itself (which was assumed lossless).

One way to avoid this large power requirement is to reduce the field volume as much as possible by using a small coil and submillimeter coil-disk spacing [5], [6]. An alternative approach is somehow to recycle the stored magnetic energy so that the power supply does not need to supply it anew for each reversal. If a sinusoidal magnetic field can be used as the bias, then energy recycling can be easily achieved using an *LC* resonant circuit. For a resonant circuit with quality factor Q, the energy lost per cycle is $E_L = 2\pi E_s/Q$, where E_s is the stored energy. This follows from the definition of Q. The power required to maintain E_s is $P = E_L f = 2\pi E_s f/Q$, which is smaller than the nonresonant case by the factor π/Q. Since Q values in the range of 10-100 are practical, the power savings can be significant. Alternatively, one can view the resonant approach as a method to increase coil-disk spacing while maintaining the same power requirement.

II. Overwrite Demonstration

To demonstrate direct overwrite using a sinusoidal magnetic field, we used the tapped inductor tank circuit shown in Fig. 1. The coil was wound on a 3.2-mm diameter ferrite rod and had an inductance of approximately 1.5 μH. The circuit had a center frequency of approximately 4.5 MHz, a Q of 50, and a real input impedance at resonance of 38 Ω. The coil was located approximately 2 mm from the disk active layer and was driven sinusoidally by a power amplifier with a 50-Ω output impedance. Typical peak voltage across the coil input terminals was 40 V (21 W average power), though satisfactory results could be obtained with voltages as low as 20 V (5 W average power). Peak field strength at the active layer was believed to be in the 100-200-Oe range, though this was not accurately measured.

To write on a disk with the sinusoidal applied field, the laser was pulsed with 70 ns pulses. By pulsing the laser in synchronism with the positive or negative peaks of the sinusoidal field, circular domains of either orientation can be written. If the disk velocity is sufficiently low (as was the case in this study), then successively written domains will overlap. By using the overlap of adjacent laser written marks, continuous variable-length domains can be written, as shown in Fig. 2. Domain transitions, which signify binary ones, are formed by shifting the timing of the laser pulse train 180° with respect to the phase of the sinusoidal field.

To demonstrate the overwrite technique, a series of carrier frequencies were written on a TbFe disk. The magnetic field frequency f_B was 4.5 MHz, the media velocity was 4.5 m/s, and the intrinsic written mark diameter was

Manuscript received July 1, 1987; revised August 31, 1987.
The author is with the IBM Almaden Research Center K67/802, 650 Harry Road, San Jose, CA 75120.
IEEE Log Number 8717711.

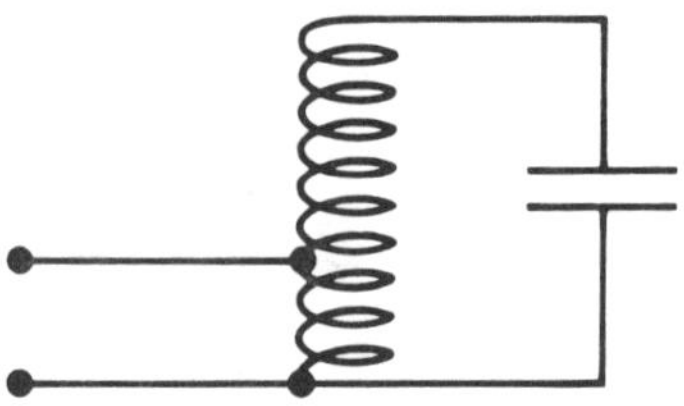

Fig. 1. Tapped inductor resonant circuit used to generate sinusoidal bias field.

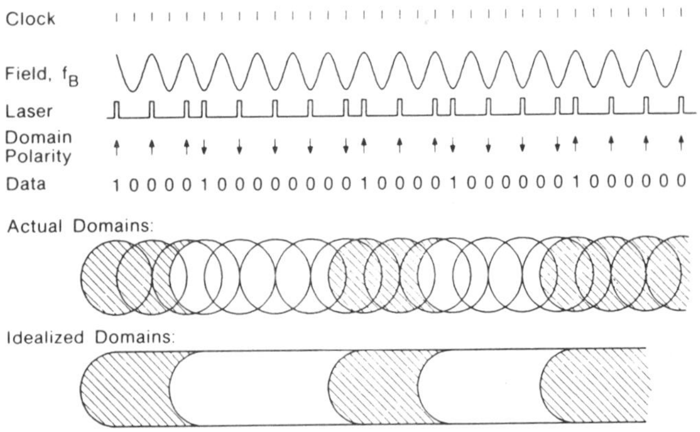

Fig. 2. Timing diagram for writing variable length domains. Laser is pulsed at either positive or negative field peaks. Circles show marks (domains) written by individual laser shots. Domains from individual laser shots overlap to form continuous domains. Binary ones are coded as domain transitions. Hatched and clear regions denote two domain polarities.

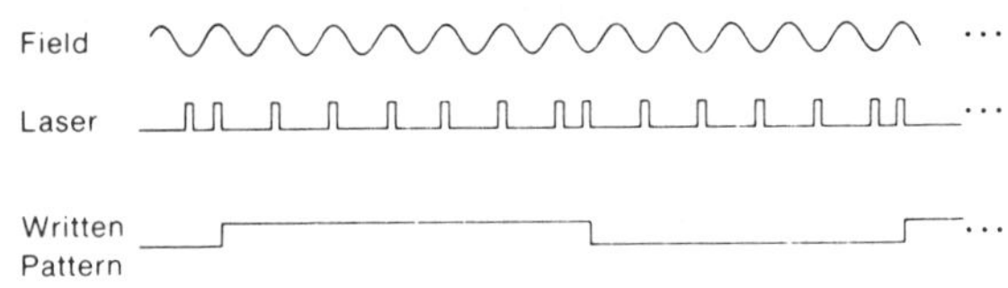

Fig. 3. Laser pulse sequence used to write 375 kHz carrier using 4.5 MHz bias field.

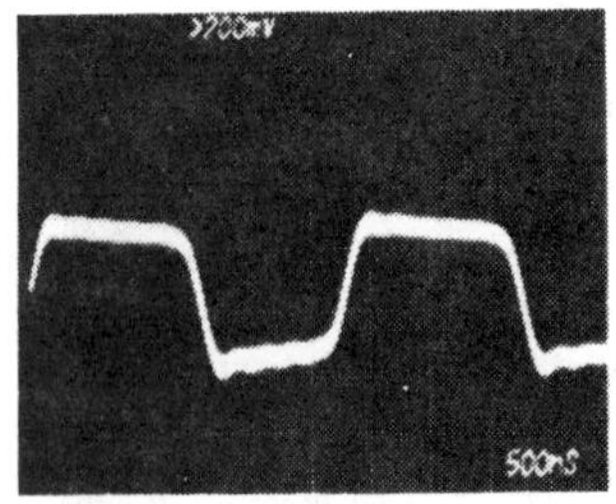

(a)

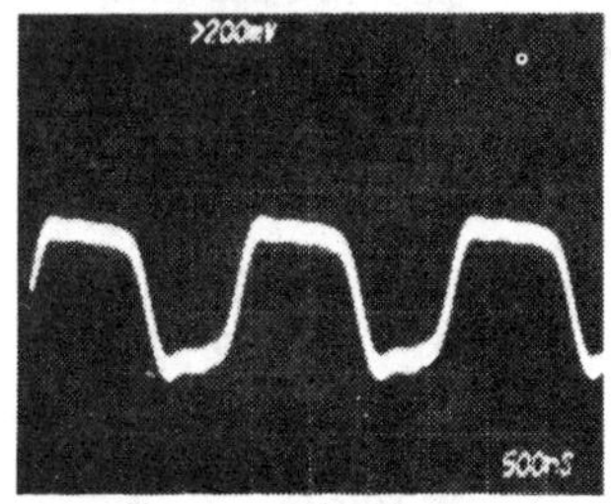

(b)

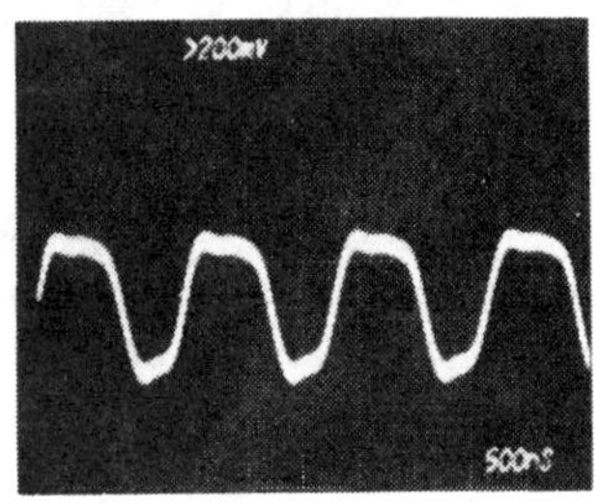

(c)

Fig. 4. Readback waveforms. (a) For carrier 1 at 375 kHz. (b) For carrier 2 at 563 kHz written over carrier 1. (c) For carrier 3 at 750 kHz written over carriers 1 and 2.

in the 1.5–2.0-μm range. Since the maximum distance between laser shots is 1 μm, good domain overlap is ensured. Carrier 1 at 375 kHz was recorded by firing the laser seven times in succession when the field was at its positive peak and six times when the field was negative. This sequence is shown in Fig. 3 and was repeated until an entire track was written. The resulting readback waveform is shown in Fig. 4(a) and the corresponding spectrum analyzer plot is shown in Fig. 5(a). Some ripple is evident on the waveform and is probably due to imperfect overlap of the successively written circular domains. The carrier-to-noise ratio (CNR) is 47 dB in a 30-kHz bandwidth. This is within 3 dB of the value obtained when the disk was written using the conventional static field method.

To demonstrate the overwrite capability, carrier 2 at 563 kHz was written over carrier 1 using the pattern of firing the laser five times when the field was at its positive peak and four times when it was negative. The resulting waveform and spectrum are shown in Figs. 4(b) and 5(b), respectively. The new carrier has a CNR of 47 dB and the level of carrier 1 has been reduced 34.8 dB, which is sufficient erasure for most digital applications.

Finally, a third carrier at 750 kHz was written, with results shown in Figs. 4(c) and 5(c). From the plot in Fig. 5(c), one sees that the signal from carrier 2 has been erased by 37 dB. Furthermore, residual signal from carrier 1 dropped an additional 11 dB (virtually into the noise) as the result of the second overwrite. This indicates that residual signals are noncumulative.

III. Immunity to Laser Power/Media Sensitivity Variation

A side benefit of this direct overwrite technique is that the timing between mark transitions is immune to slow variation of laser power or media sensitivity. Fig. 6 compares marks written conventionally with those written using the direct overwrite technique. Conventionally written marks are convex at both ends due to the circular shape of the writing beam. If, for example, the laser power increases or the media becomes more sensitive, then the radius of the convex ends will increase, resulting in increased mark length. This effect is shown in Fig. 6(a) and (b).

In the direct overwrite case, it is expected that the leading edge of the mark will be convex while the trailing end will be concave. This mark shape was verified by optical microscopy and was also observed in [5] and [6]. If the

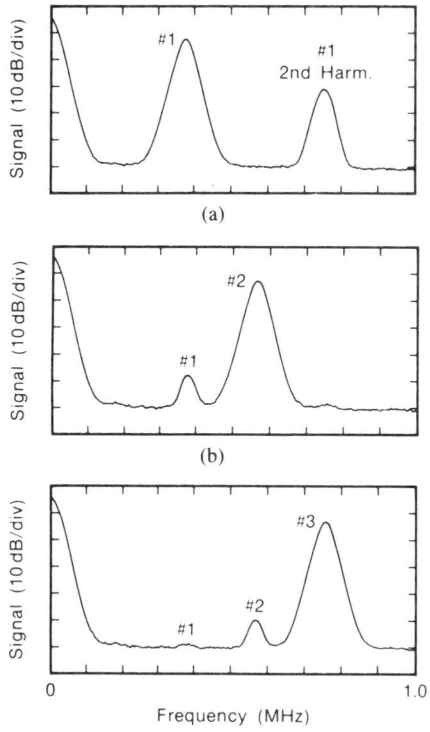

Fig. 5. Readback spectrum analyzer traces corresponding to three carriers shown in Fig. 4.

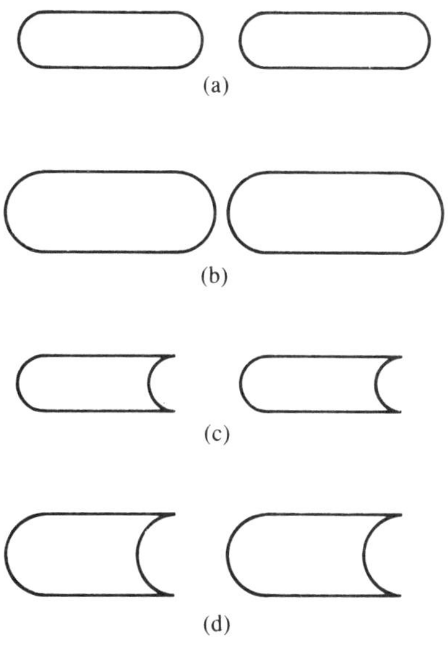

Fig. 6. (a), (b) How length and spacing of conventionally written marks are affected by change in laser power or media sensitivity. (c), (d) Immunity achieved using direct overwrite method.

laser power increases, then the mark position shifts slightly while the mark length remains unchanged (see Fig. 6(c) and (d)). This predicted immunity of mark length to laser power variation has been verified experimentally. Carriers were written using both the conventional and the direct overwrite method. When the laser power used for writing was 9.5 mW at the disk, the average readback pulsewidth as measured by a time interval analyzer was approximately 900 ns. When the laser power was increased ten percent, the average readback pulsewidth increased 22 ns for the conventional case and only 2 ns for the direct overwrite case.

IV. Run Length Limited Codes

Data for the sinusoidal field direct overwrite technique can be coded using a special run length limited pulse width modulation code. We start with the concept of (d, k) codes [9], [10], where d and k are the minimum and maximum number of successive zeros between binary ones, respectively. Binary ones are defined as the occurrence of a mark transition during a clock period, and binary zeros are the absence of a transition. To apply these concepts to the sinusoidal field direct overwrite scheme, we must first define the clock frequency. Since the laser can be pulsed during either the positive or negative halves of the sine wave field, a convenient definition for the clock frequency is $f_c = 2f_B$, where f_B is the frequency of the magnetic field.

With this definition of the clock, the concepts of (d, k) codes can be applied in a straight forward manner with the following exception: the number of successive zeros must be even. This constraint is a result of the fact that transitions can be written only when the field has reversed its sign. An opportunity for writing a transition occurs only for every other clock cycle as can be seen in Fig. 2. We designate this new class of codes as (d, k)–even 0.

The Shannon capacity of (d, k)–even 0 codes has been investigated. For example, it was found [11] that a (2, 6)–even 0 code has a Shannon capacity of 0.337. It thus seems likely that a practical code with rate 1/3 can be achieved. (A rate 1/3 code means that one data bit is translated into three channel bits.) Assuming that a rate 1/3 code can be constructed, then the decoded data rate would be $1/3 \times f_c$ or $2/3 \times f_B$. Thus our demonstration with a 4.5-MHz sinusoidal field would support a data rate of 3 Mbit/s.

V. Conclusion

We have demonstrated a magnetooptical direct overwrite method using a sinusoidal magnetic field and a pulsed laser beam. This technique is compatible with resonant coil configurations which can reduce power supply requirements at high data rates. Direct overwrite capability has been demonstrated with good erasability and immunity to laser power/media sensitivity variation. The (d, k)–even 0 code was introduced as a new type of code which is compatible with this overwrite technique. In the future, higher frequency operation and lower power consumption is expected to be possible if the field volume is reduced. This can be accomplished by using a smaller coil geometry and reducing the coil-media spacing below the current 2-mm value.

Acknowledgment

The author thanks V. Jipson, R. Lynch, and P. Siegel for encouragement and helpful discussion. A. Dewey, V.

Marrello, and U. Gerber designed and constructed the disk tester. A. Bell and L. Bynum fabricated the TbFe disk.

References

[1] N. Imamura and C. Ohta, "Experimental study on magneto-optical disk exerciser with laser diode and amorphous magnetic thin films," *Jpn. J. Appl. Phys.*, vol. 19, pp. L731-L734, Dec. 1980.

[2] N. Imamura, S. Tanaka, F. Tanaka, and Y. Nagao, "Magneto-optical recording on amorphous films," *IEEE Trans. Magn.*, vol. MAG-21, pp. 1607-1612, Sept. 1985.

[3] T. Deguchi *et al.*, "Digital magneto-optic disk drive," *Appl. Optics*, vol. 23, pp. 3972-3978, Nov. 1985.

[4] Y. Aoki, T. Ihashi, N. Sato, and S. Miyaoka, "A magneto-optic recording system using TbFeCo," *IEEE Trans Magn.*, vol. MAG-21, pp. 1624-1627, Sept. 1985.

[5] F. Tanaka, S. Tanaka, and N. Imamura, "Magneto-optic recording characteristics of TbFeCo media by magnetic field modulation method," *Jpn. J. Appl. Phys.*, vol. 26, pp. 231-235, Feb. 1987.

[6] M. Takahashi, T. Niihara, H. Sukeda, T. Maeda, and N. Ohta, "Study on magnetic properties and recorded domains for a very high density magneto-optical disc," presented at International Magnetics Conference, Tokyo, Japan, Apr. 1987.

[7] M. Hartmann, J. Braat, and B. Jacobs, "Erasable magneto-optical recording media," *IEEE Trans. Magn.*, vol. MAG-20, pp. 1013-1018, Sept. 1984.

[8] H. D. Shieh and M. H. Kryder, "Magneto-optic recording materials with direct overwrite capability," *Appl. Phys. Lett.*, vol. 49, pp. 473-4, 1986.

[9] J. Isailovic, "Codes for optical recording," in *Proc. Third Int. Conf. Optical Mass Data Storage*, R. A. Sprague, Ed., *Proc. SPIE*, vol. 529, pp. 161-168, 1985.

[10] P. Siegel, "Recording codes for digital magnetic storage," *IEEE Trans. Magn.*, vol. MAG-21, pp. 1344-1349, Sept. 1985.

[11] ——, personal communication.

Daniel Rugar received the B.A. degree in physics from Pomona College in 1975, and the Ph.D. degree in applied physics from Stanford University, Stanford, CA, in 1982.

From 1982 to 1984 he was a Research Associate at Stanford, where he worked on acoustic microscopy and phonon propagation in liquid helium. In 1984 he joined the IBM Almaden Research Center and currently works on magnetooptical recording.

Reprinted with permission from *IEEE Translation Journal on Magnetics in Japan*, Vol. 3(2), pp. 132-138 (February 1988). ©1988 IEEE.

Real-Time Recording and Erasure of Information by Optical Head with Laser Diode Array in Magneto-Optical Disk

K. Torazawa, S. Sumi, S. Murata, S. Minechika and Y. Ishii

Abstract--*The realization of an optical data system using magneto- optical disks remains an important subject for research. The functions of such a system would consist of real-time recording and erasure of information. We developed an optical head using a laser diode array with three elements, which successfully performs these functions. We also measured the real-time recording/erasure characteristics as a function of the composition of TbFeCo disks, using the optical head and six TbFeCo disks with different compositions. The characteristics were evaluated on the basis of the carrier-to-noise ratio and block error rate in Compact Disc formats. Real-time recording/erasure could be performed by applying a magnetic field of more than 400 Oe. But TbFeCo disks with compositions in the neighborhood of the magnetic compensation point delivered more stable performance with respect to deviations in the applied magnetic field and the recording power than disks with other compositions. We concluded that TbFeCo disks with 23 to 27 at% Tb are optimal for real-time recording/erasure by an optical head using a laser diode array.*

K. Torazawa, S. Sumi, S. Murata, S. Minechika, and Y. Ishii are with Sanyo Electric Co., Ltd.

From Journal of the Magnetics Society of Japan, Vol. 11, No. 2, 1987, Page 217-220.

I. Introduction

Magneto-optical disks have evolved to the point where they now have sufficiently good initial characteristics, with respect to S/N ratio, bit error rate and other performance parameters, to enable use as rewritable recording media, and recently, diverse applications have been reported. In future, to enable practical application of systems employing magneto-optical disk media, the addition of an overwrite function (a real-time recording/erasure function) will be essential. Methods which may be used to achieve overwriting include:

1) Direct magnetic field modulation
2) Head methods (a recording head plus an erase head)
3) Multi-beam methods

In the direct magnetic field modulation method, the inductance of the electromagnet presents an obstacle to high-speed recording. Two-head methods, on the other hand,

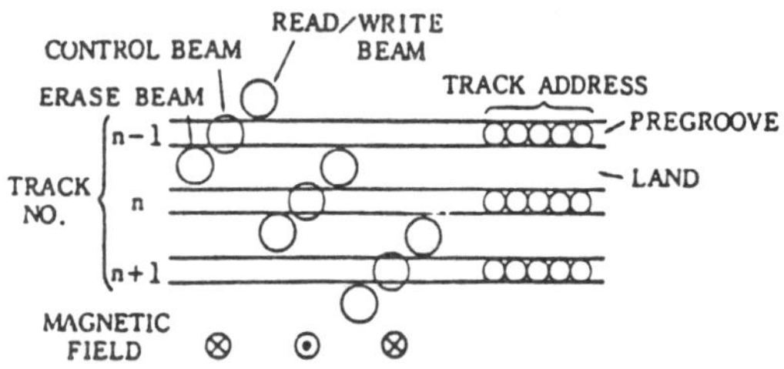

Fig. 1. Format for real-time recording/erasure.

Table 1. Specifications of the optical head.

OBJECTIVE LENS	f=3.9 NA=0.55
COLLIMATOR LENS	f=16 NA=0.17
LASER DIODE ARRAY	WAVELENGTH 8100Å POWER 25mW ELEMENTS 3 PITCH 150μm
FOCUS ERROR SIGNAL	ASTIGMATISM METHOD
TRACKING ERROR SIGNAL	PUSH-PULL METHOD
DIMENTION	H36×D55×W110

inevitably lead to greater system complexity and increased costs. However, M. Hartmann and coworkers have described the principle of an overwrite technique employing two beams of light [1]. The authors have themselves developed an optical head capable of overwriting, using a laser diode array [2]. Using this optical head, we investigated the overwrite characteristics, using magneto-optical disks of various compositions.

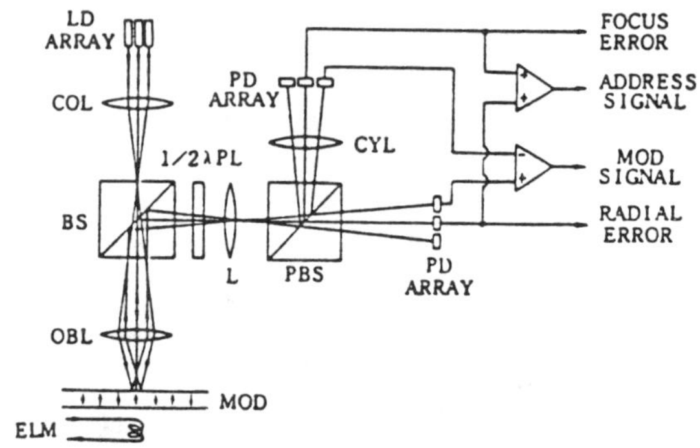

Fig. 2. Schematic diagram of optical head.

II. Experimental Procedure

Fig. 1 illustrates the overwrite format of the newly developed optical head. A laser diode array is used to generate three beams, for control, for reading/writing, and for erasing. The control beam is run along the pregrooves, while the read/write beam and the erase beam are swept along the lands, on which data is recorded. The erase beam runs one track ahead of the read/write beam. The polarities of the applied magnetic fields when the control beam runs along track (n-1) and when it runs along track (n) are opposite; hence the read/write beam can record on a track magnetized in one direction by the erase beam after one disk revolution. In this way, what is effectively direct overwriting of continuously transferred data is possible. The control beam is used for detecting focusing error signals, tracking error signals and track address signals.

Fig. 2 is a schematic diagram of the optical head construction. By positioning each element of a PIN photodiode array to correspond to the individual elements of the laser diode array, beams are spatially separated. Signals written on the disk are detected by differential detection of the PIN photodiode outputs. An electromagnet placed opposite the optical head is used to switch the polarity of the magnetic field applied to the

Table 2. Specifications of magneto-optical disks.

MATERIAL	TbFeCo
Hc	5KOe
θk	0.3
Tc	200 °C
R	30%
C/N f:1MHz Bw:30KHz	50 dB
BLER	0.2%

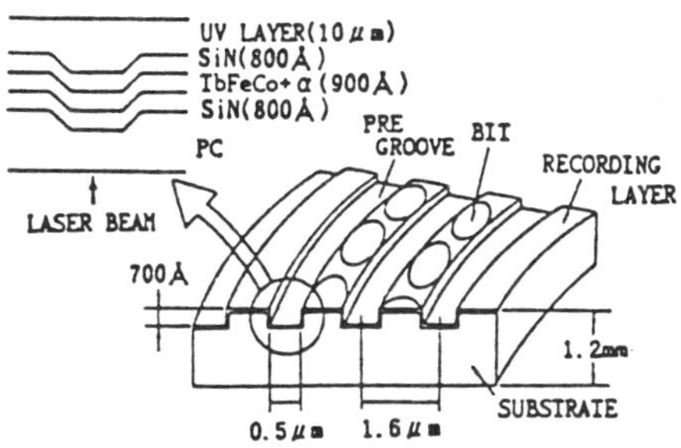

Fig. 3. Schematic diagram of magneto-optical disk construction.

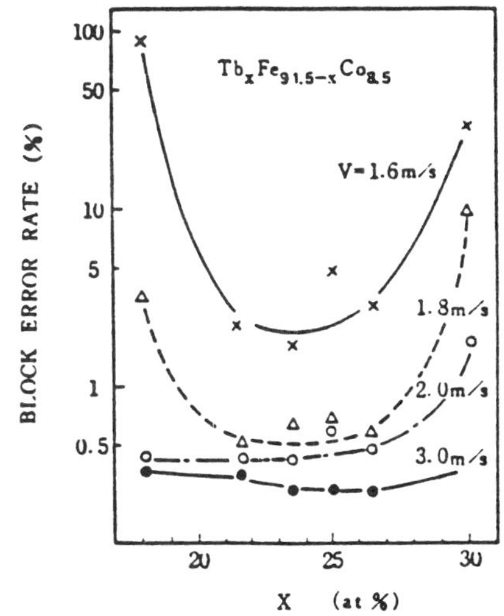

Fig. 4. Block error rate characteristics of magneto-optical disks in real-time recording-erasure, as function of medium composition.

disk. The average distance of the electromagnet from the disk is 0.5 mm, and the time for polarity switching is less than 50 μsec. The specifications of the optical head are given in Table 1. Because the head employs a laser diode array, it has such features as (1) a simple construction compared with optical heads adopting several separate laser diode elements, and (2) easy adjustment of the optical system.

The construction and basic specifications of the disks used in experiments are given in Fig. 3 and Table 2. Polycarbonate (PC) disks 120 mm in diameter and with pregrooved surfaces, produced by injection molding, were used as disk substrates. Film layers were fabricated by rf magnetron sputtering in the order SiN/TbFeCo/SiN on the PC substrate without breaking the vacuum, using a TbFeCo hybrid target for deposition of the magnetic film and a Si_3N_4 target to sputter-form the dielectric SiN layers.

Using six types of disk with different composition ratios, overwrit experiments were performed, and the dependences of the write and erase levels of carrier signals and of the block error rate (BLER) in Compact Disk format on the applied magnetic field, write power and erase power, were measured.

III. Experimental Results and Discussion

We began by measuring the read/write characteristics of the trial-fabricated disks. The C/N ratio was a constant 50 dB, regardless of the medium composition ratio. In Fig. 4 are shown the results for the BLER

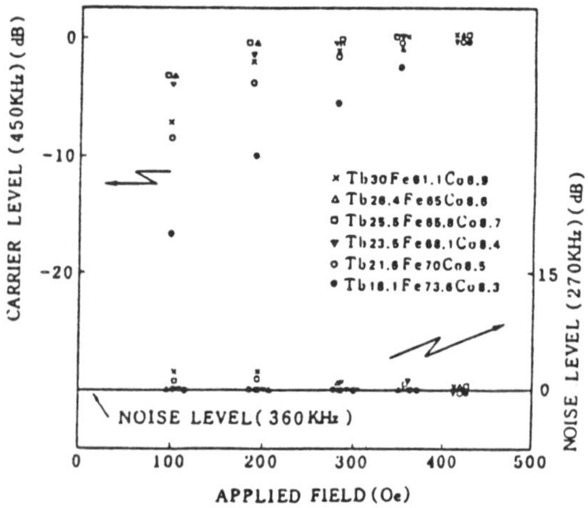

Fig. 5. Real-time recording/erasure characteristics (carrier signal), as function of applied field strength.

characteristic as a function of the Tb content, measured for various linear recording speeds. A single block consists of 588 bits, including data related to sync signals and to data error correction. The compensation composition of the medium is at 25 at% Tb. The figure indicates that high-density recording is difficult for disks with film compositions deviating considerably from the compensation composition, probably because, as the coercivity falls, the shapes of recorded bits tend to vary, and the jitter in the reproduced signals increases.

Experiments on overwrite characteristics were conducted, at a linear recording speed of 2 m/sec. Fig. 5 indicates the measured results for the levels of newly written carrier signals and erased carrier signals, as functions of the applied magnetic field. The applied magnetic field was measured at a point 0.5 mm from the tip of the electromagnet, using a Gaussmeter. In experiments, disks on which 270 kHz carrier signals were already recorded at a write power of 3 mW and an applied magnetic field of 500 Oe were used, and 450 kHz carrier signals were overwritten under a recording and erasing power of 3 mW. The figure indicates that, even for diverse medium composition ratios, signals could be completely rewritten at applied fields of 400 Oe and above. Under weak applied fields, disks with compositions near the compensation composition exhibited exceptionally good read/write characteristics; and the farther the medium composition deviated from the compensation composition, the more the written carrier level dropped, and the more difficult recording became. In general, when recording on a disk uniformly magnetized to saturation and with a composition deviating from the compensation composition, the demagnetizing field from material surrounding recorded bits makes it possible to write bits even without an applied magnetic field [3]. For instance, in our experiments using uniformly saturation-magnetized $Tb_{18.1}Fe_{73.6}Co_{8.3}$ disks, the C/N ratio when recording under a 100 Oe field is only 3 dB lower than when recording with a 500 Oe field applied. But in Fig. 5, a dramatic drop in C/N of about 18 dB results. This is attributed to the following two reasons:
(1) After erasure of carrier signals by the erase beam, the recording track is not uniformly magnetized to saturation, but is

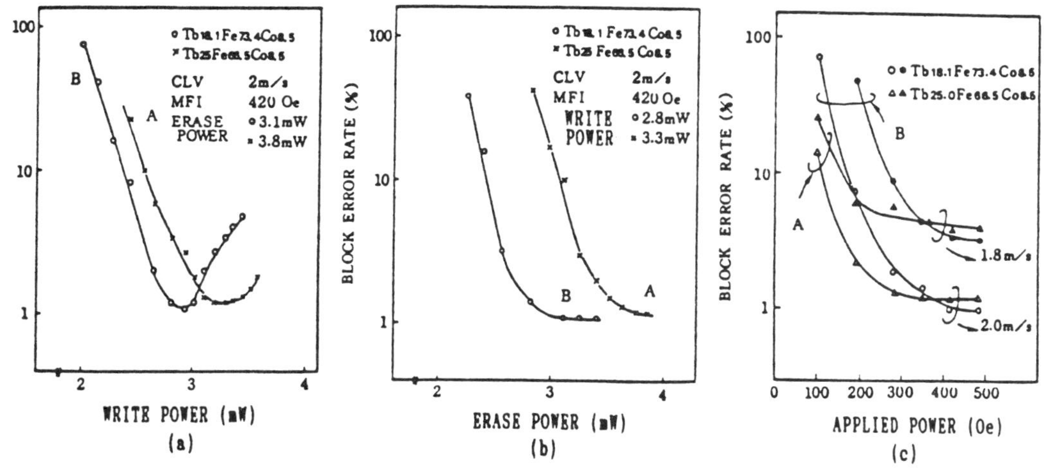

Fig. 6. Block error rate characteristics in real-time recording/erasure.

rather made up of maze-like magnetic domains. The demagnetizing field from areas surrounding recorded bits is small, and is not effective in shaping recorded bits. (2) Because of the small applied field and large magnetization in the recording film, the demagnetizin field within the written area itself, which appears during cooling after the write beam irradiation, tends to impede formation of the bit. Fig. 6 presents the results of measurements of BLER characteristics for the data overwrite process. Two disks were used--one near the compensation composition (with medium composition $Tb_{25}Fe_{66.5}Co_{8.5}$), hereafter called disk A, and one far from the compensation composition ($Tb_{18.1}Fe_{73.4}Co_{8.5}$), hereafter disk B.

Fig. 6(a) shows the BLER as a function of the write power. The BLER is smallest when the write power is 3.3 mW for disk A, and 2.8 mW for disk B. Although the magnetic layers of these disks have a Curie temperature near 200°C, the write power dependences of the BLER differ greatly for the two. This is thought to be because, for disk B, (1) the coercivity characteristic is more gentle near the Curie point, and (2) since the magnetization is large and the demagnetizing field from the material surrounding written areas is also large, larger recorded bits are formed even at the same write power [4]. Also, and probably for similar reasons, disk A, with composition near the compensation composition, has the better BLER characteristic with respect to the write power.

Fig. 6(b) shows the measured results for the BLER versus the erase power. Erase powers for which performance is satisfactory are 3.8 mW for disk A and 3.2 mW for disk B. Thus it is thought that the erase power should be set 10 to 20% above the write power.

In Fig. 6(c) are shown the BLER characteristics with

respect to the applied magnetic field. As with the data of Fig. 5, for media which deviates considerable from the compensation composition, decreasing the field strength leads to rapid deterioration of the BLER characteristic, whereas media close to the compensation composition show a more gradual decline in the BLER characteristic. Hence disks with a medium composition close to the compensating composition will suffer from less degradation in BLER due to changes in the applied magnetic field, which might be caused by fluctuations in the distance between the electromagnet and the disk due to disk deflection. Also, while the demagnetizing field from material surrounding the written area is in the same direction as the applied field during writing, during erase operations the demagnetizing field from the regions around the erased area is directed so as to reduce the applied field strength. Consequently, when two heads--an ordinary recording head and an erase head--are used, or when a single head is used to first erase data and then to write new data, the applied field is set higher during erasure than during writing. But in the present scheme, the strength of the applied field is the same during recording and during erasure; hence disks with media near the compensation composition, which have small magnetizations, are thought to be suited for use in data overwriting. A similar statement can be made about overwriting in direct magnetic field modulation methods [5]. The 0.5% increase in the error rates in Fig. 6 compared with the values in Fig. 4, is thought to arise primarily because of incomplete signal recording when the polarity of the electromagnet is switched.

IV. Conclusion

An optical head was developed for use in overwriting data, and the overwrite characteristics were measured for magneto-optical disks with different composition ratios. The overwrite performance of the optical head was verified, and the following results obtained.

(1) It is possible to rewrite carrier signals under applied magnetic fields of 400 Oe or more for disks with 18 at% $<$ Tb $<$ 30 at%; but as the medium composition deviates from the compensation composition, it becomes more difficult to record under weak applied fields.

(2) Setting the erase power 10 to 20% above the write power leads to a reduction in the block error rate.

(3) Magneto-optical disks with a magnetic film composition close to the compensation composition had comparatively stable block error rate characteristics with respect to fluctuations in the applied magnetic field and the write power. Hence it is concluded that disks with magnetic film close to the compensation composition, which suffer from less deterioration in performance due to changes in the applied field strength caused by, for instance, fluctuations in the disk-electromagnet distance with disk deflection, are suitable for use in data overwriting.

In closing, the authors wish

to express their heartfelt gratitude to Center Director Ohmori for continuous guidance and advice in the course of this work, and to those persons in the Optical Disk Division of Sanyo Electric who cooperated in developing the disk substrate.

References

[1] M. Hartmann, J. Braat and B. Jacobs, IEEE Trans. Magn., MAG-20-5, 1013, 1984.
[2] K. Torazawa, S. Sumi, S. Minechika, H. Hamada, K. Minaguchi and Y. Ishii, 47th Autumn 1986 Conf. Jpn. Soc. Appl. Phys., 29p-ZE-14, 1986.
[3] S. Tanaka and N. Imamura, Papers Tech. Meet. Magn., IEEJ, MAG-81-16, 1981.
[4] M. Takahashi, N. Ohta, S. Takayama and Y. Sugita, Papers Tech. Meet. Magn., IEEJ, MAG-86-84, 1986.
[5] F. Tanaka, S. Tanaka and N. Imamura, Spring 1986 Meeting Jpn. Soc. Appl. Phys. Rel. Socs., 2p-ZE-1, 1986.

Reprinted with permission from *Japanese Journal of Applied Physics,* Vol. 28, Suppl. 28-3, pp. 3-9 (1989).

Direct Overwrite in Magneto-Optical Recording Technology

M. H. KRYDER and M. D. SCHULTZ

Magnetics Technology Center Carnegie Mellon University
Pittsburgh, PA 15213-3890, USA

Progress on implementing direct overwrite in magneto-optical recording is reviewed.

Observations made with a 10 nsec exposure time magneto-optic camera system show that in single-layer media in which a constant bias field may be used for writing and erasing, circular domains erase by a collapse process. Observations made on stripe erasure show that erasure proceeds most easily from the end of the stripe. Experiments with high-speed scanning indicate that both circular domains and stripes may be overwritten at scanning speeds at least as high as 15 m/sec.

Three techniques for achieving direct overwrite with these single-layer media are described. The single-beam technique utilizes only a single laser beam with a modulated power level to achieve direct overwrite. Because stripes are easier to erase from the ends, this technique may require restricted modulation codes in which two ones are separated by no more than one zero such as the (1, 7) code. The erase-before-write technique uses two beams through a single lens. The first beam erases any previously written data on the track and the second writes the new data. This technique requires no restriction on modulation codes and also prevents interactions between the previously existing data and the new data. The read-before-overwrite technique requires that the previously written data be read and selectively altered to produce the new data pattern. Because of the added complexity of the electronics, this is the least favored technique.

The single-beam and erase-before-write techniques compare favorably with respect to approaches using field modulation or exchange-coupled multilayers, because both the head and media are relatively simple. On the other hand, data reported thus far in the literature indicate the best carrier-to-noise ratio is obtained with field modulation. Since research is still underway in a number of laboratories, the reported CNR, for all techniques are likely to improve, however.

§1. Introduction

The magneto-optical (MO) disk drives which have thus far appeared on the market use two passes of the laser beam over the media to overwrite data—blocks of data are erased on the first pass, then the new data are written on the second pass. This scheme more than doubles the latency time required to overwrite as compared to the direct overwrite technique used in conventional magnetic disk drives where no erasure before overwriting is necessary.

In order to improve the performance of future generation MO disk drives, techniques for achieving direct overwrite in MO recording are drawing an increasing amount of attention. Shieh and Kryder[1] first described a direct overwrite technique in 1986. Their approach utilized single-layer MO media and either zero or constant magnetic fields for both writing and erasure. Since this publication, a large number of theoretical[2,3,4] and experimental[5-15] papers on related research have been presented. In addition, several alternative approaches have been proposed.[16-22]

In the field modulation technique,[16,17] a continuous laser beam is swept over the media while a magnetic head, flying close to the media, is used to apply magnetic fields of alternating polarity corresponding to the data to be recorded. To achieve direct overwrite using the field modulation approach and large head-to-media spacings, a resonant bias coil technique has been demonstrated.[18]

Various techniques using multilayer exchange-coupled media have also been proposed for achieving direct overwrite in MO recording.[19-21] These techniques do not require modulation of the magnetic field, although some of them do utilize permanent magnets to initialize the recording media.

Recently, another multilayer media approach, which utilizes a high permeability layer to shield the recording layer from a dc magnetic field in the absence of a laser beam, but allows the magnetic field to penetrate when the high permeability layer is heated by a laser beam has been described.[22]

In this paper, primary attention will be given to the single-layer direct overwrite approach[1] and variations on it.[11] However, in the Discussion section, the advantages and disadvantages as well as the work yet to be done on a large variety of techniques will be discussed.

§2. Experimental Techniques

Single-layer direct overwrite materials have been made by both rf diode and dc magnetron sputtering. The systems used at Carnegie Mellon University include Leybold Heraeus Z400 and Z650 systems and a Perkin Elmer 2400J. All of these systems can be pumped down to the low 10^{-7} Torr range. Sputtering conditions were selected to give films with a high density featureless microstructure as determined by electron microscopy.[23]

The anisotropy of the deposited films was determined with a torque magnetometer; the magnetization, by a vibrating sample magnetometer; and the coercivity, compensation point, Curie temperature and saturation magneto-optic Kerr rotation by a Kerr hysteresis loop tracer with a temperature controlled stage.

Studies of the high-speed nucleation and erasure processes in the films were carried out with a high-speed camera system diagrammed in Fig. 1. The sample to be observed is placed in a Leitz Orthoplan polarized light microscope suitable for observing the domains by the polar Kerr magneto-optic effect. A diode laser is coupled

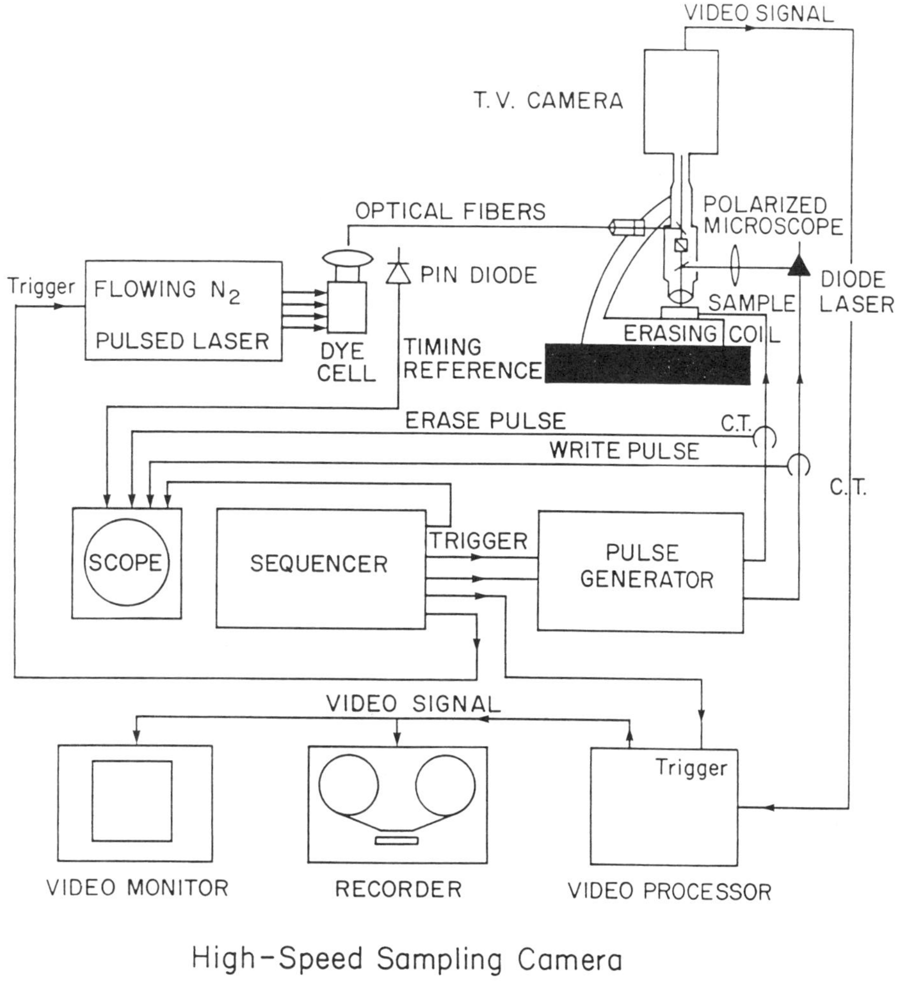

Fig. 1. A schematic diagram of the high-speed magneto-optic camera system.

into the objective lens by a dielectric mirror which is nearly totally reflecting at the 800 nm wavelength of the diode laser, but transparent in the visible spectrum. The diode laser is used to write and erase domains in the sample. A nitrogen-pulsed dye laser is used as the light source for high-speed (< 10 nsec) observations of the domain writing and erasure process. Both single shot and stroboscopic observations may be made with the system. The < 10 nsec nitrogen-pulsed dye laser pulse used for observation is synchronized to the order of ±1 nsec with the diode laser pulse used for writing and erasing the domains and may be positioned at any selected time before, during or after the diode laser pulse. A Perceptics model 9200 image processor allows background subtraction and/or averaging of several frames from repeated events to improve contrast and decrease noise.

In some experiments a scanning galvanometer system was used to scan the diode laser beam at velocities up to 15 m/sec across a sample in the microscope used for domain observation. A simplified schematic diagram of the system is shown in Fig. 2. The laser beam used to write and erase domains emanates from a 20 mw laser with corrective optics necessary to form a circular beam, passes through two lenses, is reflected off a galvanometer mirror and travels through an aperture in the microscope nose

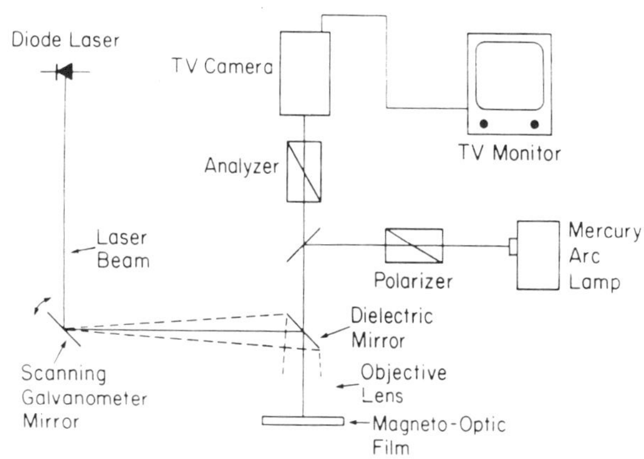

Fig. 2. A schematic diagram of the high-speed scanning galvanometer system.

piece. The galvanometer mirror is driven by a CX-660 control amplifier from General Scanning Inc. The galvanometer mirror can be oscillated at frequencies up to 220 cycles per second or can be triggered for only one oscillation. The amplitude of oscillation can be varied from zero up to a total angle of 40°.

The velocity of the laser beam scan on the sample is

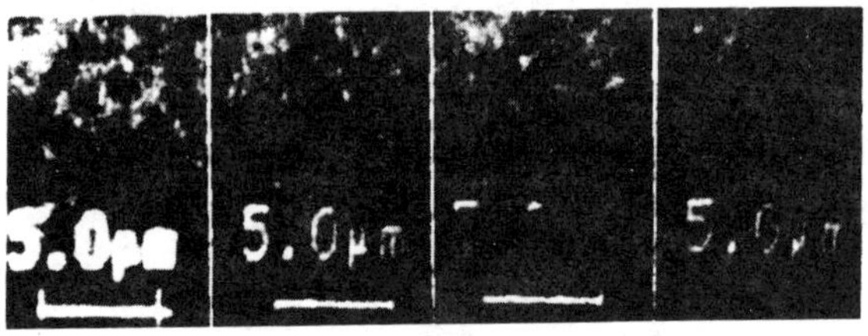

Fig. 3. A sequence of photographs illustrating writing and erasure of 1 mm domains with constant bias field applied.

limited by the laser beam velocity achieved on the back of the objective and the demagnification produced by the objective. Scanning velocities up to 15 m/sec have been achieved with this system.[11]

§3. Physics of the Single-Layer Direct Overwrite Phenomena

Single-layer direct overwrite phenomena were discovered by accident when a written domain was pulsed by a laser beam without a magnetic field applied. A sequence of photographs illustrating a direct overwrite phenomenon is shown in Fig. 3. The five domains on the left were written with an 8 mw laser beam 250 nsec in duration with no field applied. The center domain was then illuminated with a second laser pulse of 8 mw and 80 nsec duration, again with zero magnetic field, causing it to vanish. Another write pulse of 8 mw and 250 nsec duration caused the domain to be written again while a subsequent erase pulse of 8 mw and 80 nsec duration caused it to vanish again. Note that the writing and erasing of the micrometer sized domain spaced a micrometer from its neighboring domains has no influence on the neighboring domains. Note, also, that the erasure appears to be complete; there is no visible evidence of a remnant domain.

In the initial descriptions of this phenomenon,[1,5] it was hypothesized that the erasure occurred because a second domain was nucleated inside the first, forming a toroidally shaped domain which then collapsed. Hansen modeled such a process and predicted compositions under which it could occur.[2,3] This hypothesis was later proven to be incorrect when we used the high-speed camera system to photograph the domain during and after the erase pulse. High-speed photography showed that the domain shrunk and collapsed and that there was no evidence of nucleation of a second domain inside the first during the erase process.[8,10]

It was pointed out by Ruger *et al.*[7] that the gradient in domain wall surface tension produced by a Gaussian thermal profile in a film would provide a pressure which acted to collapse the domain. Savage *et al.*[4] subsequently developed a numerical model for this collapse process. They pointed out that forces from both domain wall surface tension and the gradient in the domain wall surface tension act in a direction to collapse the domain. Qualitatively their model seems to describe the observed erasure process.

Margins for the writing and erasure processes have been previously published.[5] They are reproduced in Fig. 4.

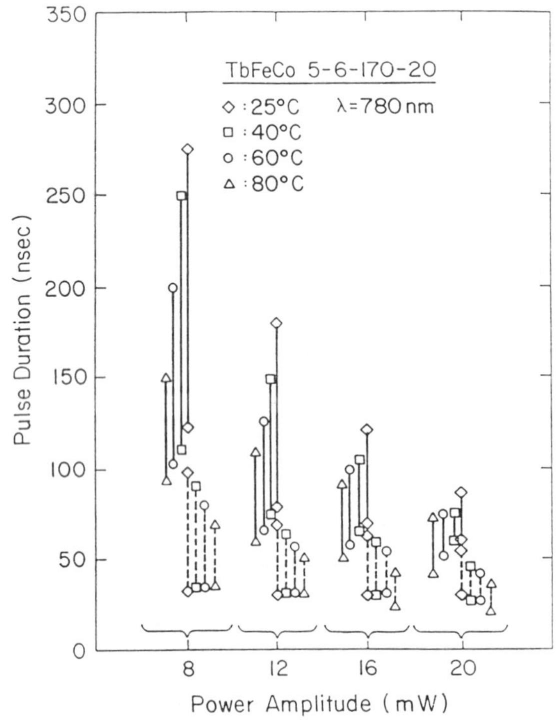

Fig. 4. Margins for writing domains which can be erased in zero magnetic field (solid lines), and for erasing those same domains in zero magnetic field (dashed lines).[5]

Note that it is possible to switch between writing and erasing either by varying the laser pulse width or the laser pulse amplitude. For example, to write a domain in the film used to gather the data in Fig. 4, a 70 nsec pulse of 20 mw amplitude could be used; whereas, to erase the domain the pulse width could be reduced to 35 nsec. Alternatively, the same 70 nsec, 20 mw pulse could be used for writing, and, instead of reducing the pulse width, the pulse amplitude could be reduced to 8 mw to erase the domain.

Although the erasure process with static media appeared to be quite reliable, there was a serious concern about the tolerance for misregistration of the erase pulse beam with respect to the center of a written domain during a dynamic erase. Experiments carried out by Schultz *et al.*[6] and Suzuki *et al.*[9] indicate that the erase beam can be quite far from the written domain center. Data is shown in Fig. 5 illustrating that a 0.8 mm diameter domain is reliably erased by a laser beam which is as much as 0.3 mm off-center.

Suzuki *et al.*[9] hypothesized that the erase process involved an initial expansion of the domain, which was followed by a collapse. They hypothesized this expansion to explain how the beam could be so far off-center of the domain and still cause erasure. However, in none of the high-speed photographs of the erasure process has any domain been observed to expand prior to erasure when a pulse with the energy required to erase the domain was applied.

Although data was initially presented only on the erasure of circular domains, it was later shown that elongated stripes could be written and erased with a

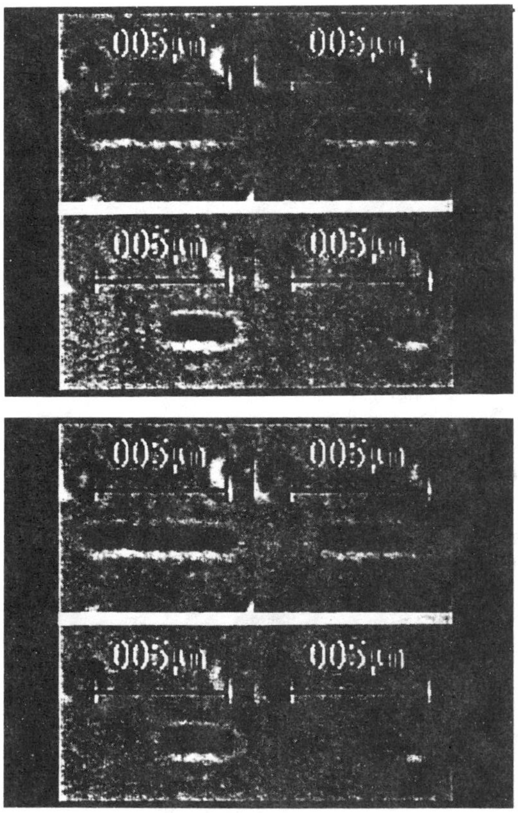

Fig. 5. A stripe domain written with a scanned laser beam is erased with a scanned and rapidly pulsed laser beam. Both the writing and erasure were done in the same constant magnetic field. The bar marker is 5 micro-meters long.

scanned laser beam which was either pulsed at a high rate with respect to the scanning speed or left on continuously.[6,9,11,12] A stripe domain written with a rapidly pulsed laser beam is shown in Fig. 5 at the top left. In the upper right and lower left and right photos of Fig. 5, are shown pictures of this same stripe as it is being erased with a scanned and rapidly pulsed beam. High-speed scanning experiments with the galvanometer mirror system shown in Fig. 2 have confirmed that this stripe erasure process works at least up to velocities of 15 m/sec.[11]

The mechanisms which cause the erasure of the stripe from one end, when it is illuminated with a laser beam are believed to be essentially the same as for a circular domain. Pressures, acting on the stripe domain wall, cause it to move inward as illustrated in Fig. 6.

Another experiment which was performed with stripe domains was to explore the possibility of erasing the center portion of a stripe. It was found that in some cases when a relatively high-power erase pulse was used, the center of a stripe could be erased. More typically, however, the region of the stripe which was pulsed with the laser beam was only slightly narrowed with respect to its original width.

The reason the center of a stripe is more difficult to erase than the ends is very likely due to the fact that, in the center of a stripe, there is no wall curvature. The domain wall surface tension, instead of providing an inward directed pressure on the wall, actually acts against

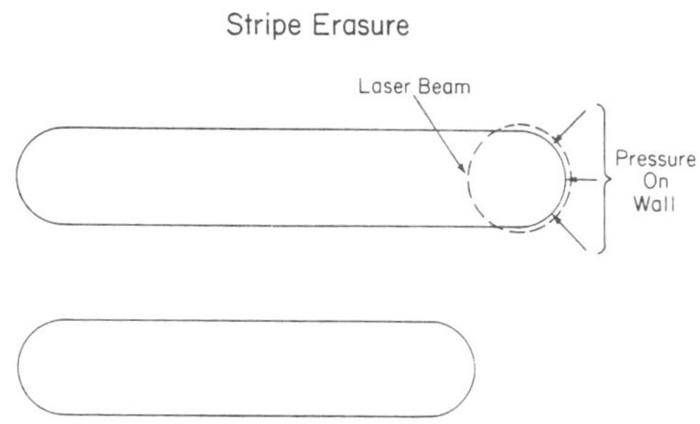

Fig. 6. Pressures from wall surface tension and a gradient in domain wall energy cause the stripe head to shrink when it is heated with a laser beam.

the collapse of the stripe center and tends to keep the walls straight.

§4. Direct Overwrite Techniques Utilizing Single-Layer Media

The above described domain writing and erasure phenomena can be utilized to achieve direct overwrite in MO recording. Three alternative methods of performing direct overwrite with these phenomena have been reported.[1,11] They are illustrated in Fig. 7.

The process which was first described1 is what is referred to as the read-before-overwrite process and is illustrated in Fig. 7(a). In this technique, two laser beams are focussed through a single lens onto different positions on the MO medium. The first beam reads any previously recorded data on the MO medium. The second beam is used to write data into the MO medium. Data read by the first beam, which is positioned a few bits ahead of the second beam on the MO material, are compared bit-by-bit to the data to be overwritten. If the bits match, the second beam is not pulsed. On the other hand, if the bits do not match, the second beam is pulsed with the appropriate energy to either write a reverse domain where there is none or erase one where it is not desired.

A second, more desirable direct overwrite technique is illustrated in Fig. 7(b) and is referred to as erase-before-write. In this technique, also, two laser beams are focussed through a single lens. The first is operated with its power adjusted to such a level that, at the rate it is scanned, it erases everything in a sector of the track. The second beam is then pulsed at the power level required for writing new domains in a saturated medium.

A third direct overwrite technique which utilizes only a single laser beam is illustrated in Fig. 7(c) and referred to as single-beam direct overwrite. In this technique, during overwrite the laser output is simply modulated between that required for writing and that required for erasing. Either a continuous beam or a beam pulsed at high frequency can be used.

Domains overwritten by this single-beam direct overwrite technique are shown in Fig. 8. In the upper image a stripe followed by a circular domain was written by scanning an amplitude modulated laser beam of the form shown in Fig. 9(b). In the lower image, the same region of the film is shown after it was subsequently scanned by an amplitude modulated laser beam of the form shown in

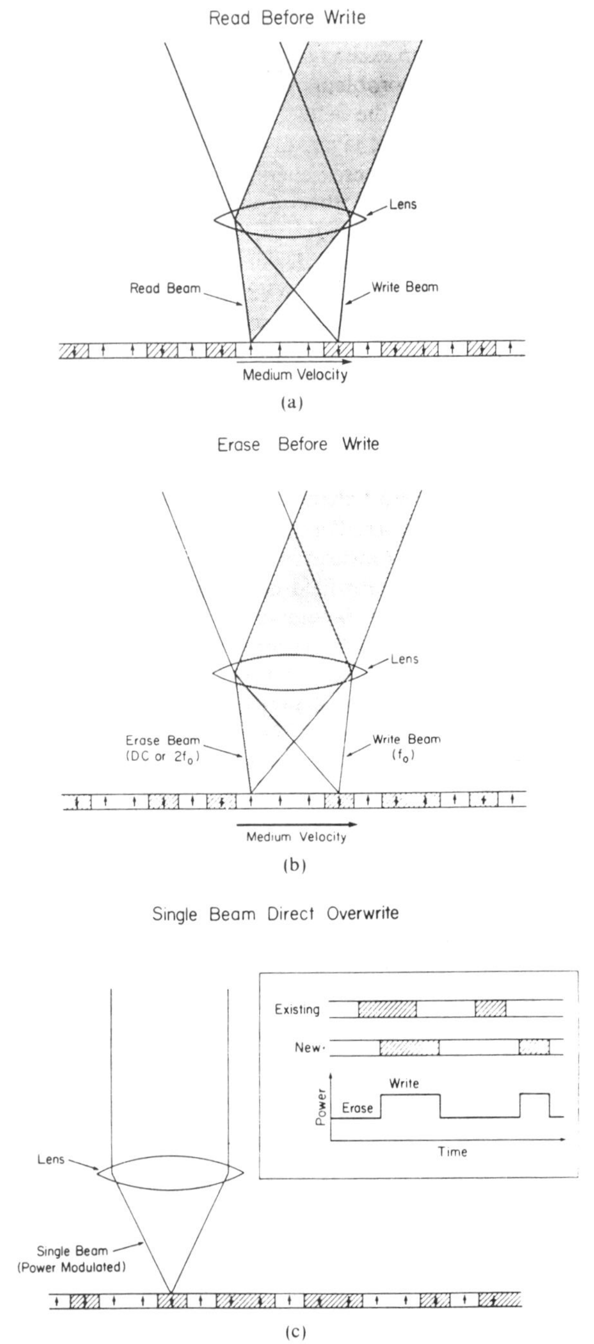

Fig. 7. Alternative single-layer direct overwrite techniques which have been demonstrated to work with constant bias field: (a) the read-before-overwrite technique, (b) the erase-before-write technique and (c) the single-beam direct overwrite technique.

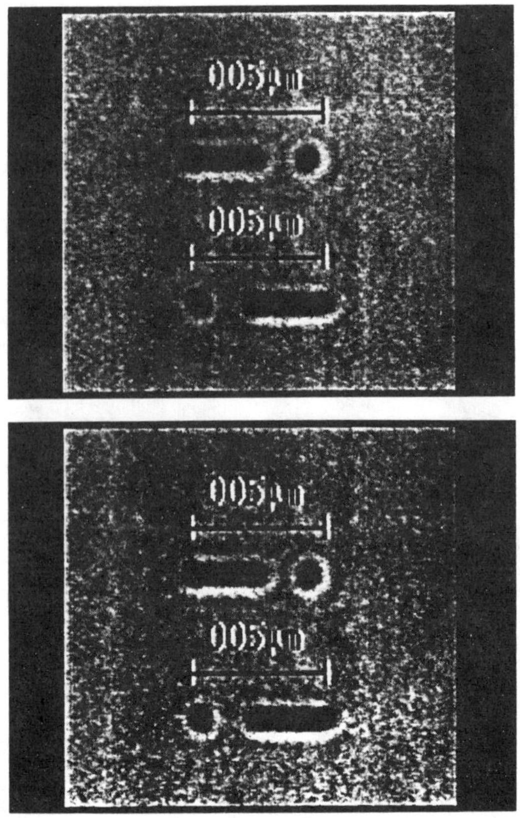

Fig. 8. The stripe-dot domain pattern in the upper photograph was directly overwritten by the dot-stripe pattern in the lower photograph using the single-beam overwrite technique with constant magnetic field and no intervening erasure of the original domains. The bar marker is 5 μm long.

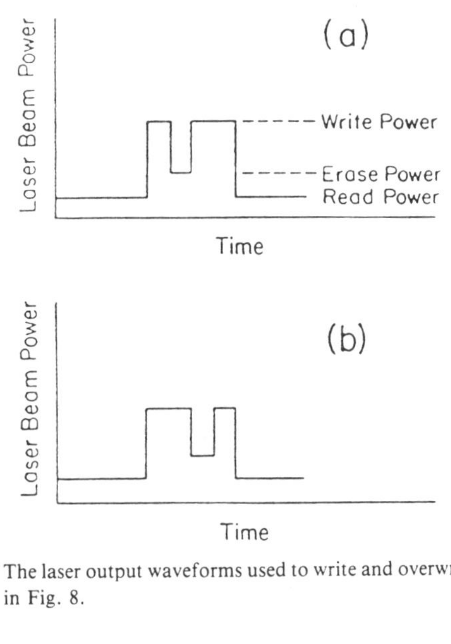

Fig. 9. The laser output waveforms used to write and overwrite the domains in Fig. 8.

Fig. 9(a), with no intervening erasure of the original domains.

§5. Discussion: A Comparison of Direct Overwrite Techniques

Of the single-layer direct overwrite techniques described above, the single-beam technique illustrated in Fig. 7(c) requires the least complex optical head and control electronics. A single-laser is adequate for both reading and overwriting. It is fully analogous to the direct overwrite process utilized in conventional magnetic recording.

The drawback of the technique is that, as mentioned above, the margins for erasing the center of a stripe have been found to be considerably narrower than for erasing a segment of a stripe from the end. This problem may be overcome as the erasure process becomes better understood and media are improved. Alternatively, a restricted set of modulation codes, in which stripe domains longer than some maximum length are not allowed, could be used. It is noted, for example, that the very popular (2, 7) block code used in magnetic recording does not allow more than two zeros between ones and therefore would allow at most 3:1 aspect ratio stripes. The also-popular (1, 7) code would allow at most 2:1 aspect ratio stripes. The high-speed overwriting of dot-dash patterns by dash-dot patterns shown in Fig. 8 indicates such narrow aspect ratio stripes may be directly overwritten. Hence, this single-layer, single-beam direct overwrite technique appears highly desirable.

The erase-before-write technique illustrated in Fig. 7(b) is also quite attractive, but requires a dual-laser optical head. The dual lasers could be achieved with either entirely separate lasers or with a laser diode array. Maintaining proper alignment in a head utilizing separate lasers could be a problem. Hence the laser diode array is preferred.

To compensate for the added complexity of the optical head, the erase-before-write technique allows fully unrestricted modulation codes to be utilized. Also, because the old data is fully erased before the new data is written, there is no concern about possible bit-shift caused by the interaction of the previously written data with the new data.

Of the three single-layer direct overwrite techniques, the read-before-write technique appears to be the least desirable. In this technique, not only is a dual-laser optical head required, but also there are added complexities in the electronics because of the necessity to compare the new data with the old. Furthermore, timing margins may be a problem because any overwrite must be properly timed with respect to the previously read data.

In comparison to other proposed direct overwrite techniques such as the field modulation technique and the various multilayer techniques, the single-layer techniques offer several advantages.

In comparison to field modulation, single-layer direct overwrite, even utilizing a dual-laser head to achieve erase-before-write, has a simpler head structure. The head is entirely on one side of the recording medium and there is no need to fly the head a few micrometers above the surface. Furthermore, the single-layer techniques can be performed on double-sided media, while, because of the required small magneticheadto-media spacing, when using field modulation, only single-sided media are possible. Additionally, when laser diode arrays become practical for writing and reading on parallel tracks, field modulation could not be used. The magnetic field from a field modulation optical head covers a large area and the fields on close-by tracks could not be individually modulated unless flying heights and heads similar to those in magnetic recording were utilized.

All the multilayer media techniques have the obvious disadvantage compared to the single-layer techniques that additional layers imply increased media cost and a larger number of defects. Since most of the multilayer media techniques rely not only upon the properties of the multiple layers themselves, but also on their interfacial exchange coupling, the problem is even more severe. Maintaining a uniform exchange coupling over large areas can be a significant problem. Furthermore, reports in the literature indicate the fields from the magnet used to initialize the reference layer, disturb the written bits. This forced one group of researchers to utilize three layers[20] instead of two and another group to use four layers.[21] With so many layers required, it is difficult to imagine that media cost or defect density can be low.

The approach proposed by Van den Berg[22] does not require exchange-coupling between the layers and does not appear to place unusual requirements on the drive. Little data has been reported for this technique; however, and it is therefore too early to assess its viability.

The single-layer direct overwrite techniques therefore appear to require the least complex drive and media. This makes them the most attractive approach, assuming they can be made to work. On the other hand, the carrier-to-noise ratio (CNR) measurements reported thus far are the most favorable for field modulation and multi-layer media and the least favorable for the single-layer approaches. Nakao *et al.*[17] reported a CNR of 50 dB using field modulation; Aratani *et al.*[20] reported a CNR of 50 dB using three layer exchange-coupled media; and the only reported measurement of CNR using single-layer approach is 31 dB.[14] It is our experience that the regularity of domains and the ease with which they may be overwritten in single-layer direct overwrite media is strongly dependent upon the media utilized. We therefore remain optimistic that the low CNR reported thus far will be significantly improved.

Acknowledgements

The authors wish to express their thanks to H-P. D. Shieh and W. H. Meiklejohn who contributed substantially to much of the work in the papers referenced and reviewed here.

References

1) H-P. D. Shieh and M. H. Kryder: Appl. Phys. Lett. **49** (1986) 473.
2) P. Hansen: Appl. Phys. Lett. **50** (1987) 356.
3) P. Hansen: J. Appl. Phys. **62** (1987) 216.
4) C. M. Savage et al.: Appl. Phys. Lett. **52** (1988) 1277.
5) H-P. D. Shieh and M. H. Kryder: IEEE Trans. Magnet. **MAG-23** (1987) 171.
6) M. D. Schultz, H-P. D. Shieh and M. H. Kryder: J. Appl. Phys. **63** (1988) 3844.
7) D. Rugar, J. C. Suits and C-J. Lin: Appl. Phys. Lett. **52** (1988) 1537.
8) M. Schultz and M. Kryder: Appl. Phys. Lett. **54** (1984) 1371.
9) Y. Suzuki, N. Ohta, M. Takahashi and S. Yonezawa: Appl. Phys. Lett. **55** (1989) 315.
10) M. D. Schultz and M. H. Kryder: "The Erase Process in Direct Overwrite Magneto-Optic Recording", Intermag Conference, Washington, DC, March 1989, to be published in IEEE Trans. Magnet. (1989).
11) M. H. Kryder, H-P. D. Shieh, W. H. Meiklejohn and R. E. Skoda: "Improved Methods of Direct Overwrite in Magneto-Optical Recording Media", Paper EA-10, Intermag Conference, Washington, DC, March 1989. To be published in IEEE Trans. Magnet., **MAG-25** (1989).
12) Y. Suzuki, M. Takahashi, N. Ohta and S. Yonezawa: Paper EA-6, Intermag Conference, Washington, DC, March 1989.
13) S. Takayama, T. Shimizu and N. Tanaka: "Direct Overwrite

Capability in TM-Rich Side Rare-Earth Transition Amorphous Films'', submitted to Magnetism and Magnetic Materials Conference, Boston, Nov. 28-Dec. 1, 1989.
14) Y. Hashimoto, N. Nakashima, M. Maeda, H. Inoue and S. Ogawa: ''Direct Overwriting Capability of Magneto-Optical Disk'', submitted to Magnetism and Magnetic Materials Conference, Boston, Nov. 28-Dec. 1, 1989.
15) Y. Suzuki, T. Niihara and N. Ohta: ''A New Mode of Thermomagnetic Direct Overwrite with a Fixed Bias Field'', submitted to Magnetism and Magnetic Materials Conf., Boston, Nov. 28-Dec. 1, 1989.
16) F. Tanaka, S. Tanaka and N. Imamura, Jap. J. Appl. Phys. **26** (1987) 231.
17) T. Nakao, H. Sukeda, M. Ojima, S. Taka and T. Nishiyama: ''High Speed Over-writing Characteristics of Magneto-Optic Disk by Magnetic Field Modulation Recording'', Paper EA-2, Intermag Conf., Washington, DC, March 1989.
18) D. Rugar: IEEE Trans. Magnet. **MAG-24** (1988) 666.
19) J. Saito, M. Sato, H. Matsumoto and H. Akasaka: Proc. Int. Symp. on Optical Memory, Tokyo, Japan 1987, Jpn. J. Appl. Phys. **26** (1987) Suppl. 26-4, p. 149.
20) K. Aratani, Y. Mutoh, M. Kaneko, K. Watanabe and H. Makino: Technical Digests of ''Topical Meeting on Optical Data Storage'', Los Angeles, CA, (1989) p. 154.
21) T. Fukami, Y. Nakaki, T. Tokunaga, M. Taguchi and K. Tsutsumi: ''Novel Direct Overwriting Technology for Magneto-Optical Discs by Exchange-Coupled RE-TM Quadri-Layer Films'', submitted to Magnetism and Magnetic Materials Conference, Boston, Nov. 28-Dec. 1, 1989.
22) H. A. M. van den Berg and R. Rockelein: ''A Direct Overwrite System for Magneto-Optical Storage Systems with a Soft-Magnetic Field Shield'', Paper HQ-09 Intermag Conference, Washington, DC, March 1989.
23) M. H. Kryder: ''The Design and Fabrication of Magneto-Optic Recording Materials'', presented at the Materials Research Society Meeting, San Diego, April 1989; to be published.

Section Nine
General Media

Reprinted with permission from *Science*, Vol. 154, pp. 1550-1551 (December 23, 1966). ©1966 American Association for the Advancement of Science.

Helium-Neon Laser: Thermal High-Resolution Recording

Abstract. *Scan-line recording by means of a moving laser spot has been achieved on metallic and organic thin films. Recording rates of the order of one million spots per second were obtained with a laser beam power of 38 milliwatts at the recording surface. Typical recorded line widths were of the order of 2 microns.*

Basic considerations indicated that laser recording of very high resolution (> 400 lines per millimeter) should be possible with what we now term heat-mode recording. We define laser heat-mode recording as any detectable physical or chemical change in the recording medium caused by a temperature rise due to absorption of energy from the laser beam. In general, such a recording is "real time" since the recorded data is immediately available for sensing and does not make use of a latent image effect requiring subsequent processing or treatment.

Our preliminary theoretical studies indicated that high-resolution, heat-mode recording should be possible on thin films with optics of reasonable numerical aperture (> 0.2) with a low divergence, uniphase laser beam with an output greater than 20 mw. Practical considerations dictated that the laser should be continuous wave. The particular wavelength of the laser was of minor concern in that we were free to select recording media that showed adequate absorption at the laser wavelength. We preferred a laser wavelength in the visible range because of the availability of suitable optics, the convenience of working in the visible portion of the spectrum, and the high resolution possible with visible radiation.

A Spectra-Physics model 125 helium-neon laser was used for our initial experiments. Its uniphase power output at 6328 Å exceeded 50 mw with a beam divergence that was within a factor of 2 of being diffraction-limited. Apparatus was constructed to move a focused spot at the recording medium through a series of parallel lines. The optical system was carefully designed to conserve power and to limit losses to unavoidable reflections. Although desirable, optical surfaces were not coated with antireflection films for the 6328-Å wavelength. The system could also operate without the scanning motion to record single spots.

The cross-sectional distribution of energy in the laser beam has been assumed to be gaussian. The radius of the beam is defined as the distance from the center of the beam to the point where the flux density has fallen to 0.135 of its peak value. When the laser beam is focused to a small spot, its distribution remains gaussian. The final recording lens had a 0.4 numerical aperture and produced a focused spot, as defined, which was 1.8 μ in diameter. The maximum power of the beam transmitted to the recording medium was 38 mw. This corresponds to an average power density of 1.5×10^6 watts/cm^2.

The recording media have been in the form of thin films on glass sub-

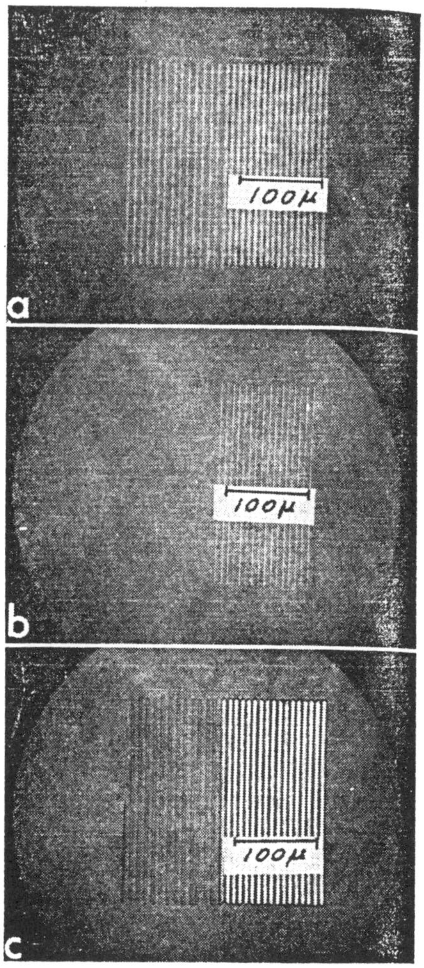

Fig. 1 High-resolution recording by laser thermal effects on thin films, illustrating (a) lead, (b) tantalum, and (c) a triphenylmethane dye in a plastic binder. Left half of a, b, and c at 0.1 msec per scan line, and right half at 1.0 msec per scan line; beam power 38 mw.

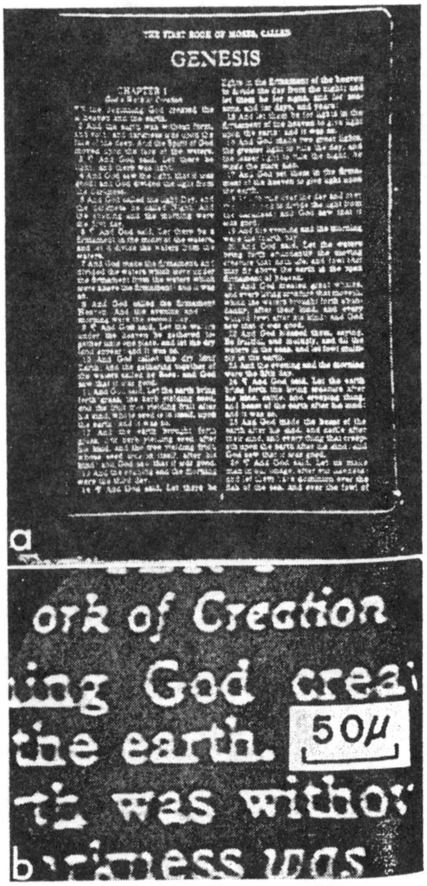

Fig. 2 A facsimile-type laser thermal recording of the first page of the Bible, showing (a) the complete page, and (b) a portion of the page to illustrate detail.

Table 1. Line widths and spot recording rates.

Fig.	0.1-Msec scan		1.0-Msec scan	
	Line width (μ)	Spots/sec	Line width (μ)	Spots/sec
1a	1.8	1.3×10^{6}	2.0	1.2×10^{5}
1b			2.0	1.2×10^{5}
1c	2.5	9.6×10^{5}	4.5	5.3×10^{4}

strates. Evaporated films were generally in the 500- to 1000-Å range, and the organic coatings 4 μ or less. Of the large number of media investigated, the discussion here will be limited to the following: a 500-Å evaporated coating of lead; a 500-Å evaporated coating of tantalum; a 1.0-μ coating of a triphenylmethane dye in a plastic binder.

In-focus line patterns were recorded on the three media with beam power, horizontal velocity, and vertical velocity being the independent variables. The recorded patterns are 240 by 240 μ. A beam power of 38 mw at the recording medium was used in each case. The left half of each pattern was recorded at 0.1 msec per vertical scan line, and the right half at 1.0 msec per vertical scan line. In the case of tantalum (Fig. 1b), beam power was insufficient to record the left half of the pattern. Measurements of the width of the clear portion of the scan lines for Fig. 1, a, b, and c, were made (Table 1). Knowledge of this line scan time permits calculation of the effective spot recording rate. The minimum line scan time was limited by the apparatus to 0.1 msec.

While the measurements on the dye-plastic sample (Fig. 1c) were made with respect to clear lines, this coating also exhibits a darkening effect on the edges of the cleared tracks. At higher scan rates, or lower beam powers, or both, this can lead to an overall image which is darker than the background, and need not be accompanied by a clear track at the line center. A similar effect, but resulting from different mechanisms, occurs with a number of evaporated films.

We recorded a page of the Bible by direct document input (by means of a facsimile-type optical scanner), and an ultrasonic diffraction-grating light shutter for beam modulation. The recording objective in this case had a numerical aperture of 0.2 with a useful flat-field diameter of 3 mm. The 3-mm field contained 1600 scan lines of which 1200 were actually used in recording the document. It should be noted that the resolution of the recording was limited by the optical resolution of the document scanner rather than by the laser recorder.

C. O. CARLSON
E. STONE
H. L. BERNSTEIN
W. K. TOMITA
W. C. MYERS

National Cash Register Company, Hawthorne, California

6 September 1966

Reprinted with permission from *Bell System Technical Journal* Vol. 50(6), pp. 1761-1789 (July-August 1971). ©1971 AT&T.

Micromachining and Image Recording on Thin Films by Laser Beams

By D. MAYDAN

(Manuscript received March 22, 1971)

A theoretical and experimental description of laser machining of thin metallic films is given. Calculations are carried out for the temperature rise of a thin film in response to a pulse of incident light energy and for the dependence of the temperature rise on the different thermal constants of the substrate and film and on the illumination conditions.

Experiments have been conducted on the pulsed machining of thin bismuth films with glass and mylar as substrates using a lowest-order transverse mode argon laser with an average power output capability of 20 milliwatts. A fast intracavity acoustooptical modulation system produced optical pulses with durations controllable from 25 ns to several ms. The pulse repetition rate could be varied from a single pulse up to several MHz. The peak power depended on the duty cycle but was limited to 2 watts for low duty cycles. Experiments were done with both front and back illumination. (In back illumination the laser beam is incident on the film through the substrate.) For optimal machining conditions in which the optical beam diameter is adjusted to produce the maximum diameter of transparent area for a given pulse energy, less than 50 percent of the removed material left the surface of the substrate. The remainder was displaced so as to leave some areas free of bismuth while the thickness of bismuth in other areas was increased. With a pulse duration of 25 ns and a 600-Å-thick bismuth film on mylar, the peak power required to machine a 6-μm-diameter spot was about 0.7 watt.

Variable amplitude light pulses produced by the intracavity modulation system with a 1-MHz repetition rate and duration of 25 ns were used to write images on a 600-Å-thick bismuth film, deposited on a mylar substrate, by deflecting the laser beam in raster fashion over the surface of the bismuth film. The average laser power output was 20 mW. The film area was 8 × 10 mm². Both positive and negative continuous-tone images were recorded. The images consisted of an array of 1200 by 2000 completely independent spots of varying diameter and a spot density of 4 × 10⁶/cm². Images of documents which were originally of size 8½ × 11 inches showed a limiting resolution of over 175 lines per inch when magnified to the original size.

I. INTRODUCTION

Images can be recorded on an opaque metallic film on a transparent substrate by using an amplitude-modulated laser beam to machine the metallic film. The modulated laser beam is scanned in raster fashion over the surface of the film, and the thermal energy dissipated in the film causes a local displacement or removal of the metallic material. In this way a permanent image suitable for immediate display or storage is created.

Experiments along this line have been reported[1,2] but reveal a number of shortcomings from the standpoint of practical application. Generally, there is no gray scale, only negative line images are reproduced, resolution is limited, and the required laser power is large.

A new laser machining technique is described in this paper which overcomes these shortcomings and permits the generation of high-quality, continuous-tone, permanent images using low-power gas lasers. The significant departure from previous work rests in the use of very short laser pulses to record the image in the form of an array of discrete holes in the metal film and in the modulation of the intensity of the laser pulses so as to vary the size of the holes. Each laser pulse serves to machine a single, nearly circular hole in the metal film by displacing and removing metal from the transparent substrate. In this way a transparency is created, and incandescent light can be directed through the film to display the image on a screen. The delivery of the laser energy to the metal film in short (20–30 nanosecond) pulses reduces the energy density required to raise the metal film to a given temperature by at least an order of magnitude over that which would be required for the same writing speed if a CW laser were used. Consequently, the average laser power required to displace or remove a given amount of metal per unit time from the transparent substrate is correspondingly reduced. In addition, the area of the transparent spot machined in the film varies in a nearly linear fashion with the pulse height, so that a good gray scale results. Equally good positive or negative images are produced, and successive, isolated machined spots have a transmission factor independent of adjacent spots or position along the scan line. The spots were machined at the rate of 10^6 spots/second, and the frames consisted of about 2000 lines of spots with about 1200 spots in each line.

The short laser pulses were generated with an argon ion laser in conjunction with an intracavity acoustooptic modulator[3,4]. The laser had an average power capability of 20 mW. The intracavity modulator served to deflect very short pulses from the laser cavity with much higher peak power but with an average power of approximately 20 mW.

In what follows, a model of the machining process will be developed based on electron micrographs of the machined spot. A calculation of the temperature rise of the film as a function of time is described and used to predict the laser power required to produce threshold machining for pulses of a given length.

II. DISCUSSION

2.1 *Temperature Rise of Thin Films Due to Laser Radiation*

A laser beam incident on a thin, optically absorbing film will cause the film temperature to rise.[5–7] Assuming a laser power density P_o transmitted through the film surface, the light intensity absorbed per unit thickness of the film at depth x may be described by

$$P_{abs}(x) = -\frac{d}{dx}(P_0 e^{-x/\delta}) = \frac{P_o}{\delta} e^{-x/\delta} \tag{1}$$

where δ is the skin depth of the film, that is, the depth at which the light intensity drops by a factor e from its initial value. If the laser beam diameter is very large compared with the film thickness, the heat flow may be treated as a one-dimensional problem. It will be assumed that the thermal and optical* properties of the absorbing material are independent of its temperature and phase.† Under these conditions and assuming no heat loss by radiation, the temperature rise of the thin film and the substrate is described by the differential equations of heat conduction:[8]

$$\frac{\partial^2 T_1}{\partial x^2} - \frac{1}{k_1}\frac{\partial T_1}{\partial t} = -\frac{1}{K_1}\frac{P_o}{\delta} e^{-x/\delta} \tag{2}$$

$$\frac{\partial^2 T_2}{\partial x^2} - \frac{1}{k_2}\frac{\partial T_2}{\partial t} = 0 \tag{3}$$

where T_1 is the film temperature, T_2 is the substrate temperature measured relative to the ambient temperature, k_1 and k_2 are the thermal diffusivity of the film and substrate respectively, $(k = K/\rho c)$, ρ is the density, c is the specific heat, K_1 and K_2 are the thermal conductivity of the film and substrate respectively, and t is time. The initial conditions, corresponding to the onset of illumination, are:

$$\text{at} \quad t = 0, \qquad T_1 = T_2 = 0.$$

The boundary conditions are:

$$\text{at} \quad t > 0 \quad \text{and} \quad x = 0,$$

$$\frac{\partial T_1}{\partial x} = 0;$$

$$\text{at} \quad t > 0 \quad \text{and} \quad x = d$$

(where d is the thickness of the metal film),

$$T_1 = T_2$$

and

$$K_1 \frac{\partial T_1}{dx} = K_2 \frac{\partial T_2}{\partial x}$$

(equal temperature and heat fluxes at film-substrate boundary). The substrate is assumed to be infinitely thick, so that $T_2 = 0$ at $x = \infty$. The temperature at depth x within the thin film can then be shown to be approximately equal to

$$\begin{aligned} T_1(x, t) \cong \frac{P_o}{K_1} \sum_{n=0}^{\infty} \Bigg\{ &+ 2\sqrt{k_1 t}\,\alpha^{n+1}\,\text{ierfc}\left(\frac{2(n+1)d \pm x}{2\sqrt{k_1 t}}\right) \\ &- 2\sqrt{k_1 t}\,e^{-d/\delta}\frac{\Lambda\alpha^n}{1+\Lambda}\,\text{ierfc}\left(\frac{(2n+1)d \pm x}{2\sqrt{k_1 t}}\right) \\ &+ 2\sqrt{k_1 t}\,e^{-d/\delta}\frac{\alpha^n}{1+\Lambda}\,\text{erfc}\left(\frac{(2n+1)d \pm x}{2\sqrt{k_1 t}}\right) \\ &- \delta e^{-x/\delta} + 2\sqrt{k_1 t}\,\text{ierfc}\left(\frac{x}{2\sqrt{k_1 t}}\right)\Bigg\} \end{aligned} \tag{4}$$

* The optical properties of thin bismuth films were measured and no significant change was detected below the vaporization temperature. The measurements were carried out by monitoring the amount of transmitted and reflected light from the bismuth film during a laser machining process.

† While this assumption clearly limits the accuracy of the calculations, the experimental results presented later show reasonable agreement with the calculations (See Fig. 14, for instance) and hence the error involved may not be very great.

in which erfc is the complementary error function and ierfc is its integral over the argument between limits x and ∞.

$$\alpha \equiv \frac{\Lambda - 1}{\Lambda + 1}$$
$$\Lambda \equiv \left(\frac{K_1}{K_2}\right)\sqrt{\frac{k_2}{k_1}}. \tag{5}$$

All the terms in equation (4) which have $\pm x$ should be repeated once with $+x$ and a second time with $-x$. The first term for example is equal to:

$$\sum_{n=0}^{\infty} 2\sqrt{k_1 t}\,\alpha^{n+1}\left\{\text{ierfc}\,\frac{2(n+1)d + x}{2\sqrt{k_1 t}} + \text{ierfc}\,\frac{2(n+1)d - x}{2\sqrt{k_1 t}}\right\}.$$

For the case of back illumination, where the laser beam is incident on a transparent substrate, the differential equations are the same as equations (2) and (3) with x replaced by $d - x$ in the exponential term. The boundary and initial conditions are the same as in the previous case. The temperature of the thin film in this case is approximately equal to

$$T_1(x, t) \cong \frac{P_o}{K_1}\sum_{n=0}^{\infty}\left\{-2\sqrt{k_1 t}\,e^{-d/\delta}\,\alpha^{n+1}\,\text{ierfc}\left(\frac{2(n+1)d \pm x}{2\sqrt{k_1 t}}\right) + 2\sqrt{k_1 t}\,\frac{\Lambda\alpha^n}{\Lambda + 1}\,\text{ierfc}\left(\frac{(2n+1)d \pm x}{2\sqrt{k_1 t}}\right) - 2\sqrt{k_1 t}\,e^{-d/\delta}\,\text{ierfc}\left(\frac{x}{2\sqrt{k_1 t}}\right) + \delta\,\frac{\alpha^n}{1+\Lambda}\,\text{erfc}\left(\frac{(2n+1)d \pm x}{2\sqrt{k_1 t}}\right) - \delta e^{-(d-x/\delta)}\right\}. \tag{6}$$

The model used above refers only to illumination by a step function of optical intensity. For comparison with experimental pulses having finite rise and fall times, a correction factor depending on the time dependence of the incident light pulse should be included in equations (4) and (6). For most practical cases in which $d > \delta$ all the terms in equations (4) and (6) which contain the expression $e^{-d/\delta}$ are smaller than the other terms and can be ignored. In addition, for the cases in which the illumination duration t is 10 ns or longer, $\sqrt{kt} \gg \delta$ so that all other terms with δ can be ignored. Numerical results based on equations (4) and (6) were obtained for the temperature distribution across the film and for the temperature dependence on the time from the onset of illumination. Of a number of metallic films examined, bismuth films required the least energy to machine a given size of spot at a given optical density. For this reason, the results presented herein all relate to bismuth films on glass and mylar substrates.

The thermal constants of bismuth, glass, and mylar are given in Table I. The light absorption in bismuth was measured, and δ was found to be ~140 Å for the 4880-Å wavelength of the argon ion laser. The calculated temperature rise of the bismuth film as a function of the light pulse duration, with the film thickness as a parameter, is shown in Figs. 1, 2, and 3. Figures 1 and 2 illustrate the case of front illumination with mylar and glass substrates, respectively. The results in Figs. 1 and 3 are given for bismuth film thicknesses of 330, 670, and 1000 Å and in Fig. 2 for film thicknesses of 330 and 1000 Å. The temperature rise is shown both at the bismuth-substrate boundary ($x = d$) and the bismuth-air ($x = 0$) surface. Figure 3, which refers to a bismuth film on mylar with the laser beam incident onto the mylar substrate, is seen to be essentially identical to Fig. 1. For long times, the temperature is proportional to $\sqrt{t}$ (slope equal to that of the broken line on the figures). This same proportionality is obtained for the surface temperature of a semi-infinite solid with constant flux of heat at the surface.[9]

As seen in Figs. 1, 2, and 3, for short times (<50 ns), the temperature rise approaches a linear dependence on time such as one might expect if the thermal loss to the substrate were small compared with the absorption of thermal energy by the thin metallic film. This is especially so with the mylar substrate due to the lower thermal conductivity of the mylar as compared to glass. For the same reason, the temperature rise of the bismuth film on glass is smaller than that of the same film on mylar.

The temperature distribution through the thickness of the film in the case of a 1000-Å-thick bismuth film on mylar is shown in detail in

TABLE I—THERMAL CONSTANTS

Substrate	Conductivity [cal/(s)(cm²)(°C/cm)]	Diffusivity [cm²/s]	α	Λ
Bismuth	2×10^{-2}	7×10^{-2}		
Glass	28×10^{-4}	58×10^{-4}		
Mylar	4×10^{-4}	1.1×10^{-3}		
Bismuth on glass			0.35	2.06
Bismuth on mylar			0.72	6.30

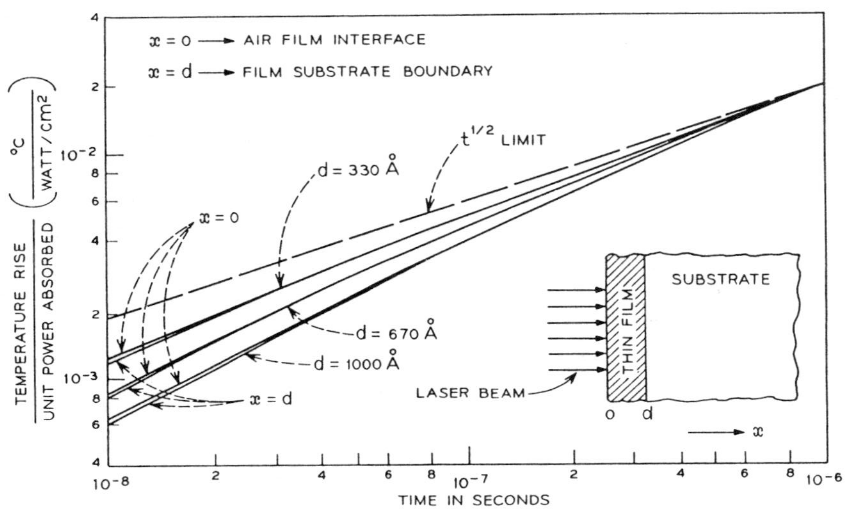

Fig. 1—Temperature rise of a bismuth film on a mylar substrate caused by front illumination.

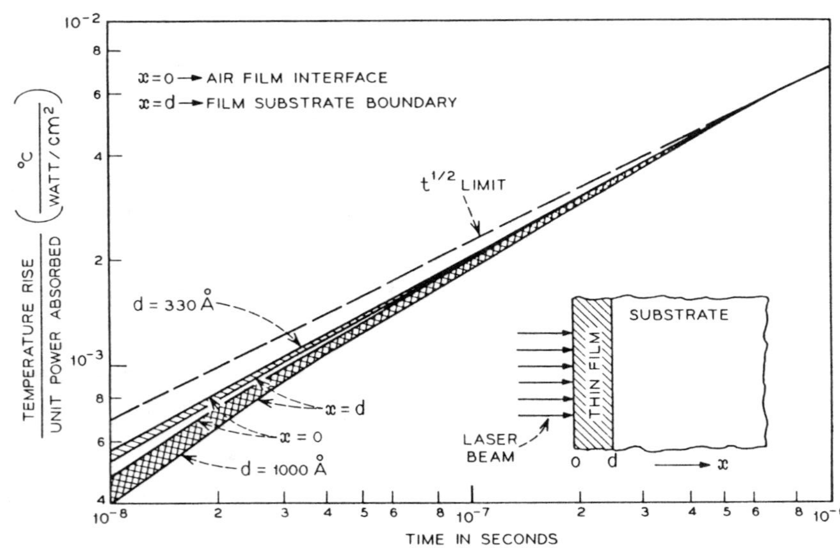

Fig. 2—Temperature rise of a bismuth film on a glass substrate caused by front illumination.

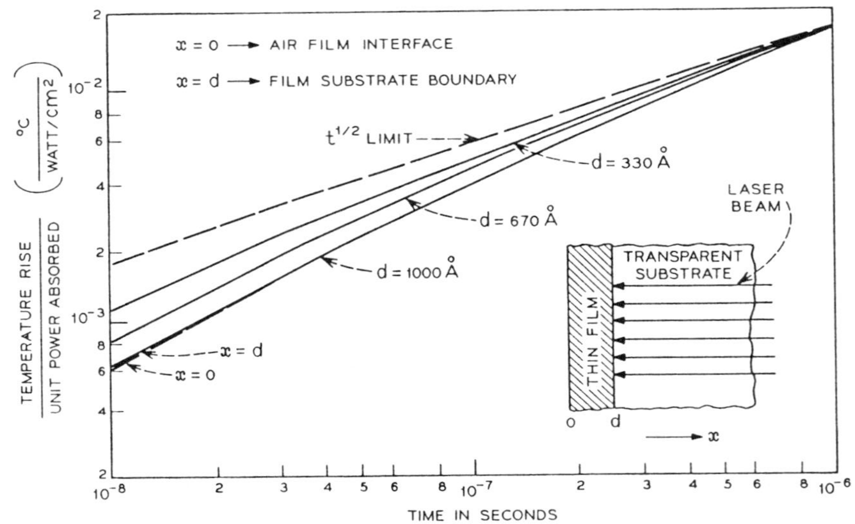

Fig. 3—Temperature rise of a bismuth film on a mylar substrate caused by back illumination.

Fig. 4a and b. The temperature distribution is given for the case of front illumination at the end of 10 ns (Fig. 4a) and at the end of 100 ns (Fig. 4b). The temperature gradient decreases with longer illumination, which can also be seen from the previous figures.

The temperature gradient across a 670-Å bismuth film on mylar, illuminated from the back, is shown in Fig. 5. Results are shown after 10 ns (Fig. 5a), after 100 ns (Fig. 5b), and after 1 μs (Fig. 5c). As expected, for short illumination duration, the temperature is higher near the substrate interface, where most of the laser intensity is absorbed. Because of loss of heat to the substrate, the location of maximum temperature shifts in time towards $x = 0$, and the temperature gradient decreases (Fig. 5c).

The temperature gradient across the film thickness is always very small (about 2 percent of surface temperature for the case described in Fig. 4b). For this reason, the model used for the foregoing calculations, which assumes that a single phase of the film is in existence at any instant, is adequate up to the vaporization temperature.

2.2 *Optimal Machining Considerations*

It is assumed that the laser beam power density absorbed by the thin film has an azimuthally symmetric Gaussian shape of the form

$$I(z, r) = I_o(z)e^{-(r/w(z))^2} \quad (7)$$

in which r is the radial distance from the axis, $I(z, r)$ is the laser intensity at a distance z from the focal plane along the optical axis, $I_o(z)$ is the laser intensity at $r = 0$, and $w(z)$ is the radius of the waist at the $1/e$ points of the intensity. For a beam propagating in the z direction, H. Kogelnik and T. Li[10] have shown that

$$\left(\frac{w(z)}{w_o}\right)^2 - 1 = \left(\frac{\lambda z}{2\pi w_o^2}\right)^2 \quad (8)$$

where λ is the light wavelength and w is the waist radius at the focal plane, $w_o \equiv w(o)$. Assuming a laser beam with a specified total power P, one can write

$$P = \int_0^\infty I_0 e^{-(r/w)^2} \cdot 2\pi r \, dr = \pi I_o w^2 \quad (9)$$

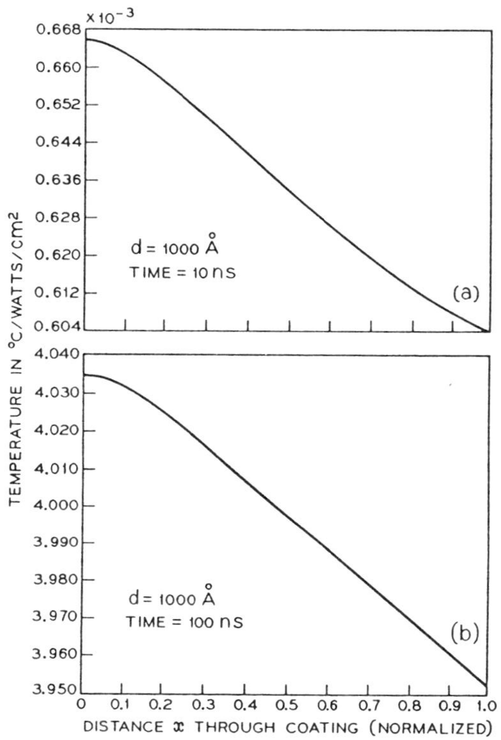

Fig. 4—Temperature distribution across **1000** Å of a bismuth film on mylar: (a) after **10** ns of front illumination; (b) after **100** ns of front illumination.

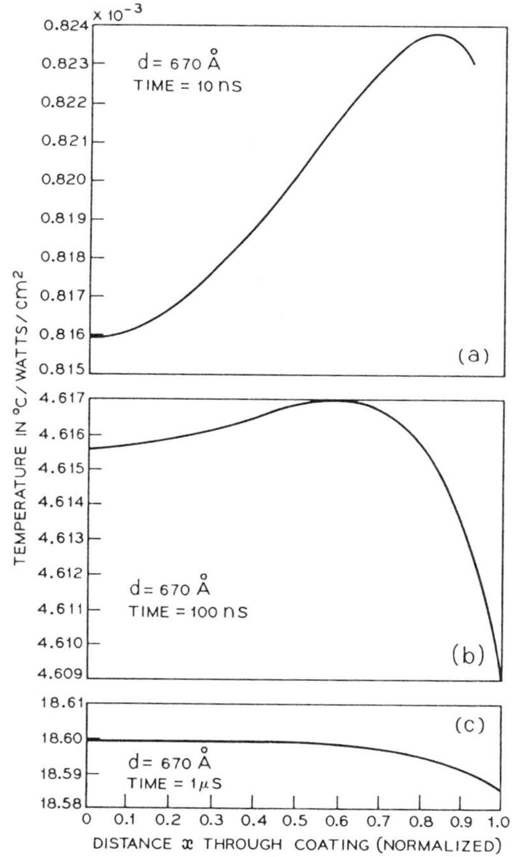

Fig. 5—Temperature distribution across 670 Å of a bismuth film on mylar after 10 ns (a), 100 ns (b), and 1 μs (c) of back illumination.

which defines

$$I_o(z) = \frac{1}{\pi}\frac{P}{w(z)^2}. \quad (10)$$

Machining or evaporation of the thin films with a spot diameter D and a given pulse duration requires a certain threshold intensity $I_T = I(z, D/2)$. At the focal plane ($z = 0$), it follows from equation (7) that

$$I_T = \frac{1}{\pi} P \frac{1}{w_o^2} e^{-(D/2w_o)^2} \quad (11)$$

or

$$P = \pi I_T w_o^2 e^{(D/2w_o)^2}$$

where D is the machined spot diameter. By taking the derivative of P with respect to w_o and setting $dP/dw_o = 0$, it can be shown that the laser power required to machine a spot of given diameter is minimum when

$$\frac{I_o}{I_T} = e. \tag{12}$$

It follows that the optimal beam waist diameter machining a spot of diameter D is

$$2w_{opt} = D. \tag{13}$$

As an example, consider the machining of a 1000-Å bismuth film with a mylar substrate. The laser power density absorbed in the film is assumed to have a Gaussian shape with a peak intensity I_o equal to 3×10^5 watts/cm². The melting temperature of bismuth is about 271°C. From Fig. 3 it is found that starting from room temperature the thin bismuth film will reach the melting temperature after 16 ns of illumination time. The latent heat of fusion of the bismuth is $L = 10.2$ cal/g, and the specific heat is $C = 0.03$ cal/g°C. The additional time Δt used in supplying the latent heat of fusion is estimated from the time required to raise the temperature by the amount $\Delta T = L/C =$ 300°C. From Fig. 3 this time is found to be $\Delta t \cong 20$ ns. Following that, the temperature will continue to rise toward the evaporation temperature. The time required to reach the evaporation temperature after supplying the latent heat of fusion is found from Fig. 3 to be equal to 124 ns. Assuming that machining is done under optimal conditions, as previously described, the temperature rise was calculated at different points across the diameter of the Gaussian beam. Results are shown in Fig. 6 for the temperature of the bismuth film after 160 ns illumination duration with a laser peak intensity of 3×10^5 watts/cm². The four lines shown correspond to the temperature at $a = 0$ (the center of the machined spot, at $a = 0.25D$, at $a = 0.50D$ (the edge of the machined spot), and at $a = 0.67D$, which is the radius where the film just reached the melting point.

When the laser beam is focused to a radius different from w_{opt}, the machined spot diameter is reduced. The ratio between the diameter of the machined spot and the optical beam diameter is found from equations (7), (10), and (12) to be:

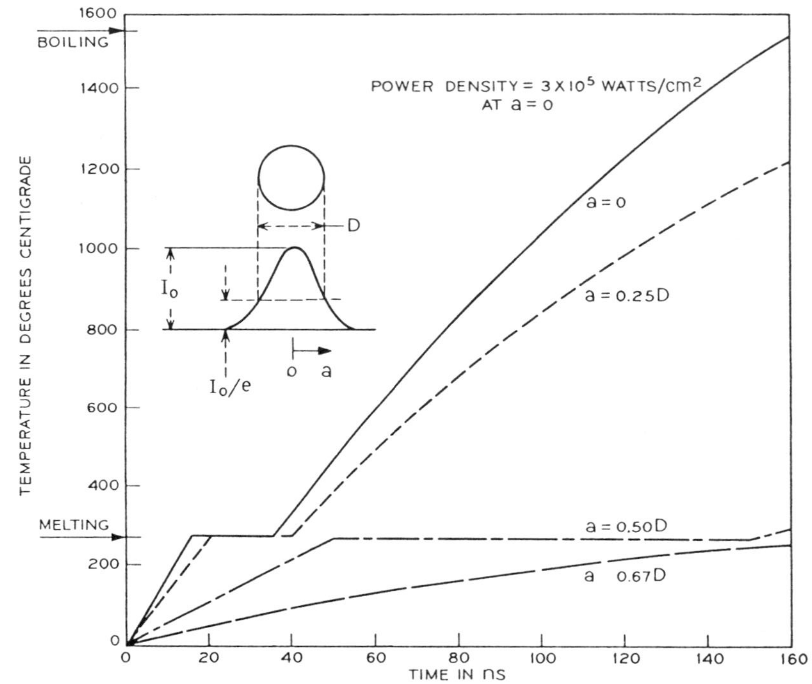

Fig. 6—Temperature distribution across the machine spot diameter.

$$\frac{D(z)}{2w(z)} = \sqrt{1 - 2 \ln (w(z)/w_{opt})}. \tag{14}$$

Suppose a beam is focused to a waist w_{opt} which is optimal for machining a spot of diameter $D_{opt} = 2w_{opt}$. Let the depth of field be the distance l from the focal plane for which the machining spot diameter is reduced by a factor of $\sqrt{2}$. It is found from equations (8) and (14) that

$$l = \pm\frac{1.7D_{opt}^2}{\lambda}. \tag{15}$$

As an example, to machine a spot diameter of 10μm, with an argon ion laser which has a wavelength of 0.488 Å, the field depth according to (15) is

$$l = \pm\frac{1.7D_{opt}^2}{\lambda} = \pm 348\mu\text{m}.$$

III. EXPERIMENTAL RESULTS

Machining experiments have been conducted on thin bismuth films using glass and mylar as substrate materials. These films were prepared by vacuum evaporation techniques in thickness of 200, 400, 500, 600, 800, and 1000 Å respectively. A folded-cavity, CW argon laser was used in conjunction with a fast acoustooptic modulator[3] to provide light pulses whose length could be varied from 25 ns to several ms. The repetition rate of the light pulses could be controlled and varied from a single pulse up to several million pulses/second. The focused spot could be swept in one dimension using deflectors, such as galvanometers,[11] rotating mirrors, etc. However, for most of the data reported in this section, the focused spot was kept fixed, and the film was moved in a direction perpendicular to the optical axis of the light pulses, as shown in Fig. 7. Tilting the film at a small angle α with respect to its direction of travel (shown in Fig. 7) causes the bismuth film to pass from inside the plane of focus to outside in its travel past the focused spot. With sufficient power in the light pulse a line of circular spots was machined.

Figure 8 shows four pairs of lines machined with four different light intensities. The spot diameters along these lines exhibit two maxima corresponding to the optimum case described in Section II [equations (12) and (13)]. The two maxima are at equal distances from the spot machined when the film is at the beam waist. The pulse repetition rate and film speed were adjusted to produce nonoverlapping but nearly contiguous spots along each line; i.e., each spot was formed by a single optical pulse. The film in this case was a bismuth film of 1000 Å thickness on a glass substrate. The optical beam was focused to a diameter $2w_o$ of about 4μm. The light pulse duration was 25 ns with peak intensities of 1.6, 0.8, 0.5, and 0.25 watts for the first, second, third, and fourth pairs of lines, respectively. The maximum spot diameter (which occurs at the brightest part of each line in Fig. 8) and the separation between the two maxima in each individual line increase with the intensity of the incident light. The machining obtained when the beam was in focus on the bismuth film is marked by the vertical line in the center. The disappearance of machining at the two ends of each line occurs where the peak power at the center of the beam falls below the machining threshold power I_T.

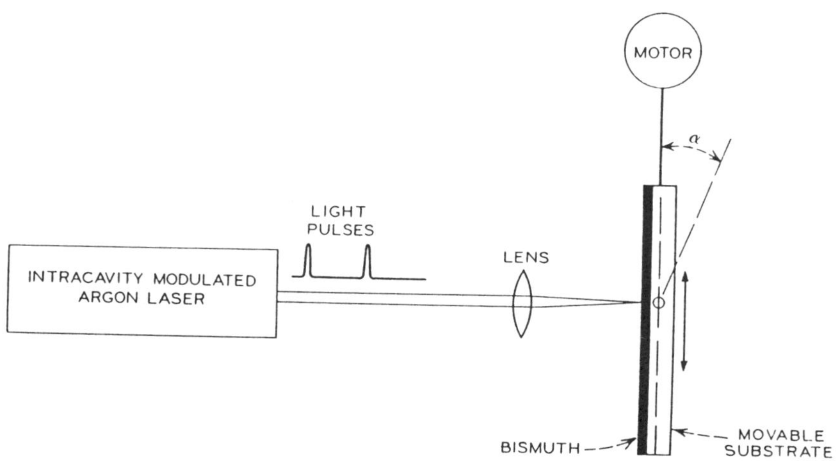

Fig. 7—Experimental setup.

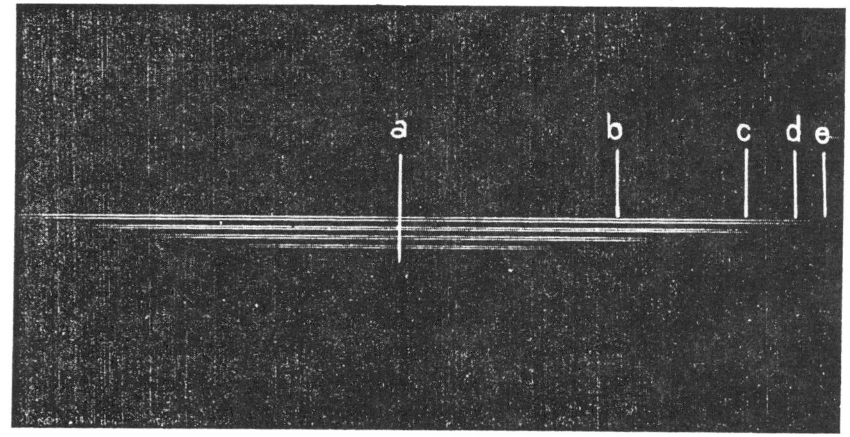

Fig. 8—Four sets of lines machined with different intensities.

A more detailed picture of some individual spots machined with the same laser intensity is given in Fig. 9. These photographs were obtained from a scanning electron microscope at 10,000 and 2000 times magnification. The approximate locations of these spots are indicated by the letters a, b, c, d, and e in Fig. 8. All spots shown in Fig. 9 are located on the top line of machined spots in Fig. 8. The individual machined spot magnified by 10,000 times in Fig. 9 is also marked by a small arrow on each 2000 magnification photograph. The machined spot obtained when the laser beam was focused in the plane of the bismuth film is shown in (a) of Fig. 9. Machining obtained with a power

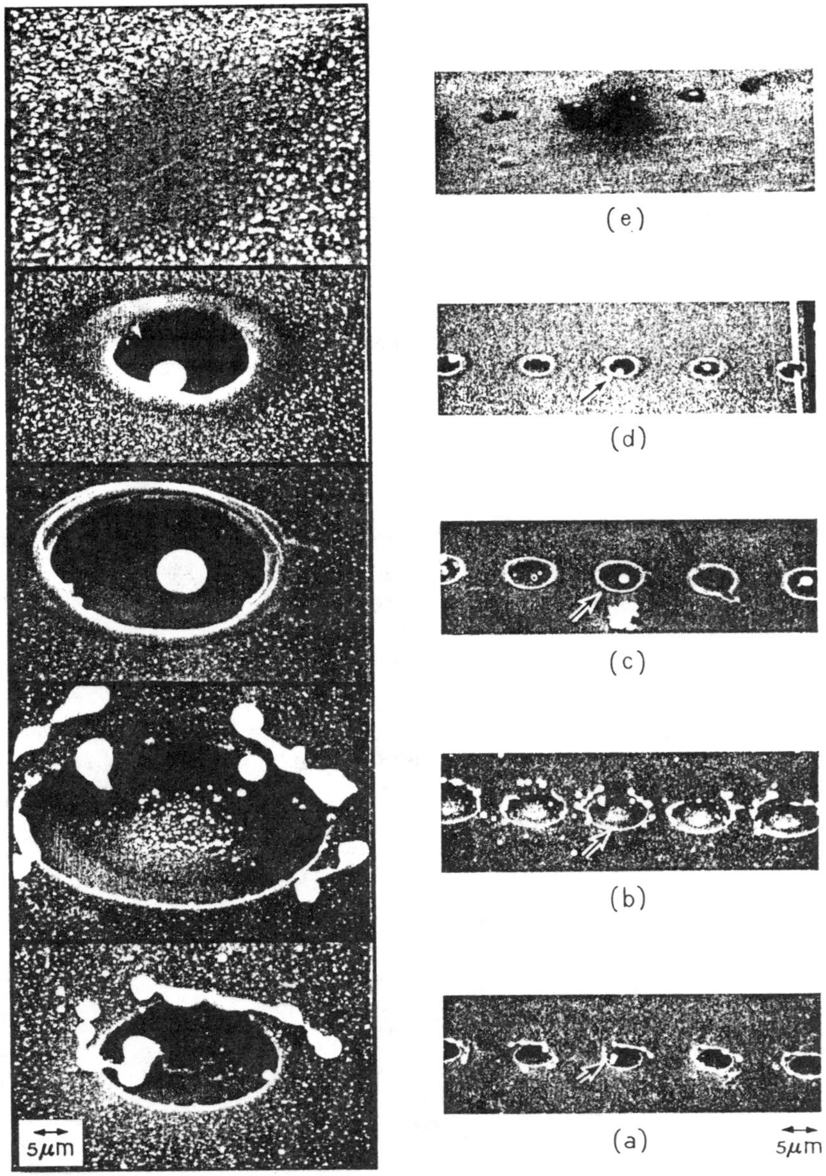

Fig. 9—Photographs of individual machining spots obtained from a scanning electron microscope.

just above threshold is shown in (d). The largest machined spot obtained is shown in (b). As clearly seen from these photographs, very little, if any, of the bismuth film was evaporated at the lower power densities, as shown in (c) and (d). Near the threshold power, the metal was only melted and the surface tension pulled the metal back from the center forming a crater, as shown in (c). In the machined spot at the center of (c), surface tension caused a large percentage of the liquid metal to resolidify into a sphere within the crater. In (e) the laser power was just below the threshold level, and while the metal appears to have been melted at each spot, only in a few cases were craters formed.

In (a) and (b) of Fig. 9 the optical beam intensity at the center of the spot is greater, and it is likely that some evaporation of material occurred in this region. The metal vapor pressure has forced some of the liquid metal away from the substrate, and surface tension has caused the remainder of the liquid bismuth to resolidify in the irregular form evident in (a) and (b). By estimating the volume of material accumulated around the crater in (b), the amount of material evaporated for the optimum machining is found to be less than 50 percent of the total metal displaced from the crater.

In order to measure the machining field depth, a similar experiment was carried out. Argon laser pulses with 25 ns duration and peak power of 0.75 watt were focused to a minimum diameter of 2μm. The bismuth sample was again moved perpendicular to the optical axis and was tilted at a small angle α with respect to its direction of travel. Assuming that the beam has a Gaussian shape, the optical beam diameter could be calculated for each machined spot. The relation between the machined spot diameter and the light beam diameter is given in Fig. 10. The results are given for an 800-Å-thick bismuth film on a glass substrate. The solid line is the theoretical line given by equation (14). As seen in this figure, the observed machining field depth is extended beyond that of the theoretical curve, especially at the lower power densities. This may result from the surface tension effect near the threshold power.

Figure 11 shows the effect of varying the laser intensity. A 400-Å-thick bismuth film on a mylar substrate was used in this case. The optical beam was in focus all along the machined line, and the machining was obtained by back illumination. As indicated on this figure, the beam intensity was varied from 100 percent to 24 percent, the 100-percent intensity corresponding to a light pulse of 1 watt peak intensity with 25 ns duration. The machining spots varied in diameter according

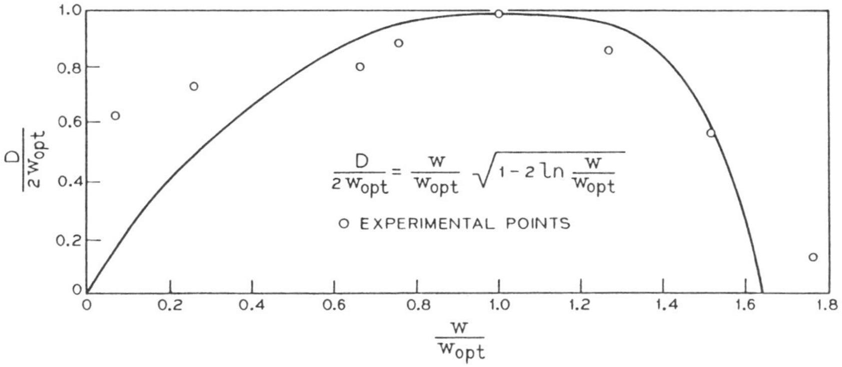

Fig. 10—Relation between machining spot and light beam diameters. The notation is that used in equations (13) and (14).

to the incident light intensity, from about 8 to 1μm. The optical beam diameter was 2 $w_o \approx 6\mu$m. As shown in Fig. 12, above threshold the measured machining area varied linearly with the light beam intensity for these particular machining conditions.

The variation of threshold machining power density with pulse duration was measured for pulse durations in the range from 25 to 4000 ns. The light pulse shapes for several durations are shown in Fig. 13. The results obtained with a 600-Å-thick bismuth film on mylar (illuminated from the substrate side) are given in Fig. 14. The solid line is the theoretical curve based on the results given in Section II. The small circles are the experimental results. Both the theoretical and the experimental results in Fig. 14 have been normalized so that they coincide at the 1-μs illumination period. The small deviations from the theoretical results for the short durations are not yet explained.

The peak pulse power required to machine 6-μm-diameter spots on 500-Å bismuth films on mylar with 25 ns pulse duration was found to be of the order of 0.7 watt. This corresponds to an average energy density of 0.06 joule/cm^2 over the area of the spot. As theory predicted (Figs. 1 and 3), the same power requirements were found for both back and front illumination.

Figure 15 shows the dependence of the energy density, required to machine 6-μm-diameter spots with laser pulses of 25 ns duration, on the film thickness. The energy density plotted in Fig. 15 is the peak pulse power times the pulse duration divided by the spot area. For optimal machining, this is also equal to the peak power density I_o at the center of the beam times the pulse duration [from equations (10)

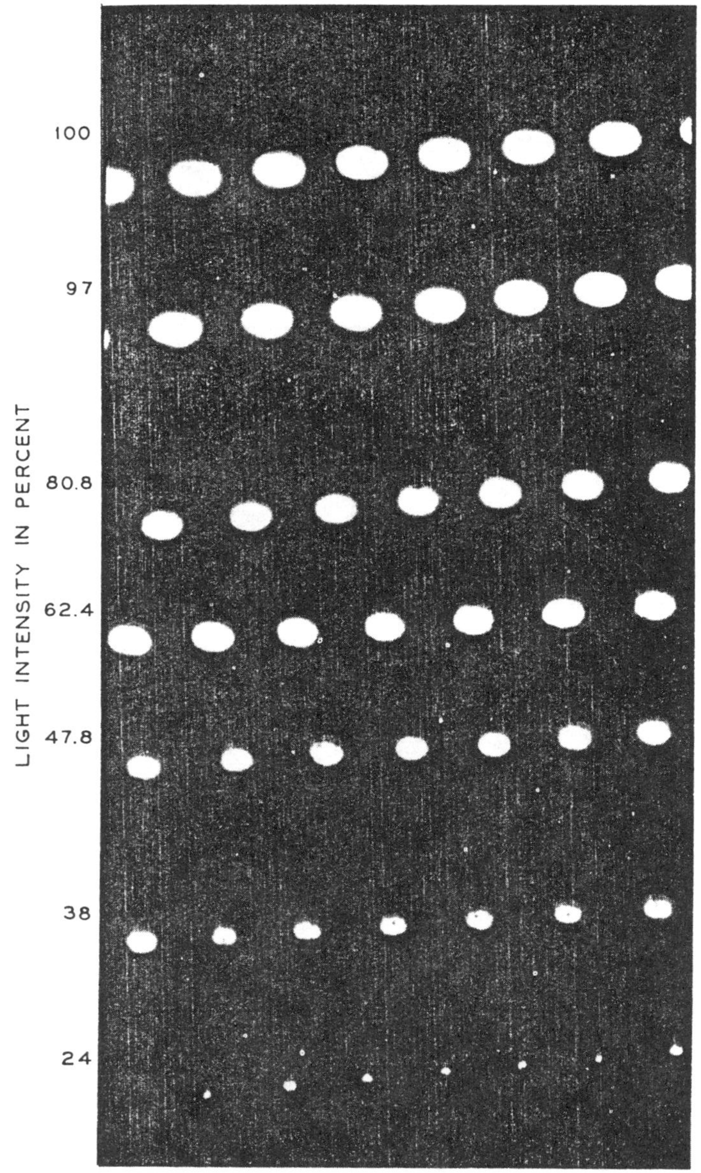

Fig. 11—Machining spots obtained with different light intensities. (The 100-percent value corresponds to a peak intensity of 1 watt with 25 ns pulse duration.)

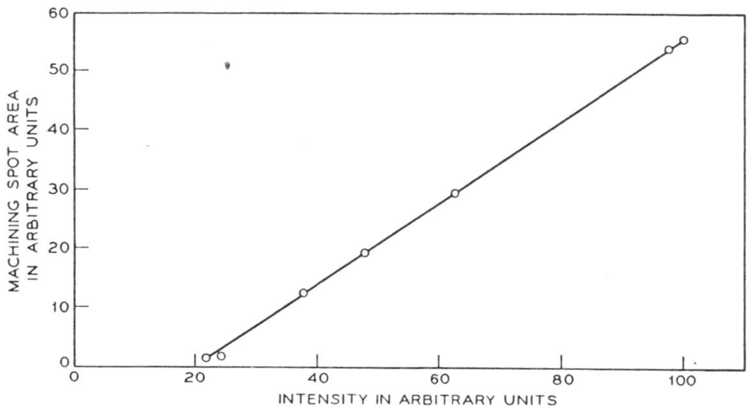

Fig. 12—Relation between machining spot area and illumination intensity.

and (13), $I_o = P/(\pi D^2/4)$]. The films were on mylar substrates and the machining was from the substrate side. As expected from the theoretical results presented in Figs. 1, 2, and 3, due to some loss of energy to the substrate, the required machining power increases relatively slowly with film thickness in the range from 200 to 1000 Å.

The power densities required to machine different thicknesses of bismuth films with different illumination duration can be estimated from Figs. 14 and 15. For the example given at the end of Section II, the peak power density required to machine a 1000-Å film with a 160-ns illumination duration, according to Fig. 14, is 4.5 times smaller than the peak power density with a 25-ns illumination duration. From Fig. 15 it is found that the energy density required to machine a 1000-Å film with a 25-ns pulse duration is 0.085 joule/cm^2. Assuming that due to reflectivity only 50 percent of the laser light is absorbed in the film, the peak power density I_o is then of the order of $0.5 \times 0.085/(4.5 \times 25 \times 10^{-9}) = 3.8 \times 10^5$ watts/cm^2. This is in agreement with the theoretical prediction presented in Fig. 6 and with the experimental results of Fig. 9 which indicate that only a small portion of the film is evaporated.

Back illumination machining experiments were carried out and compared for Mylar S* and Mylar D* substrates. Both of these materials have similar thermal constants. The Mylar S however has a better optical quality. The same power densities were required for machining in the two cases. As experimentally observed, due to the poorer optical

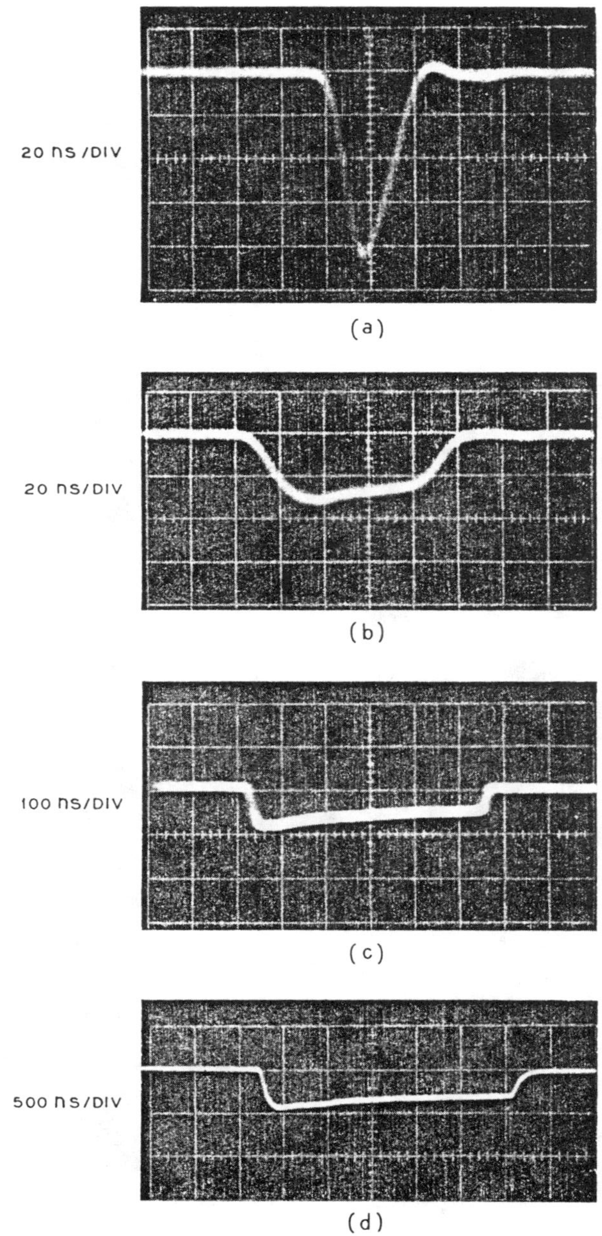

Fig. 13—Different light pulse durations obtained by the modulation system.

* Trademarks of DuPont.

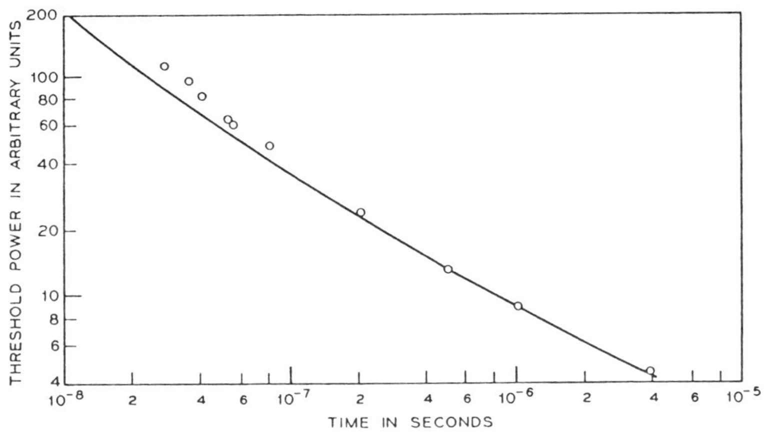

Fig. 14—Relation between threshold machining powers and illumination duration.

quality of the Mylar D, the optical beam shape was distorted causing a distortion in the shape of the machining. In Fig. 16a, b, and c, machining spot shapes are shown and compared for front and back illumination on Mylar D substrates and back illumination on Mylar S. As seen in Fig. 16a, the spot shape is distorted for the back illumination on Mylar D.

Many bismuth particles in the vicinity of the machining area can be observed in Fig. 16. These apparently were displaced as molten drops by the vapor pressure during machining.

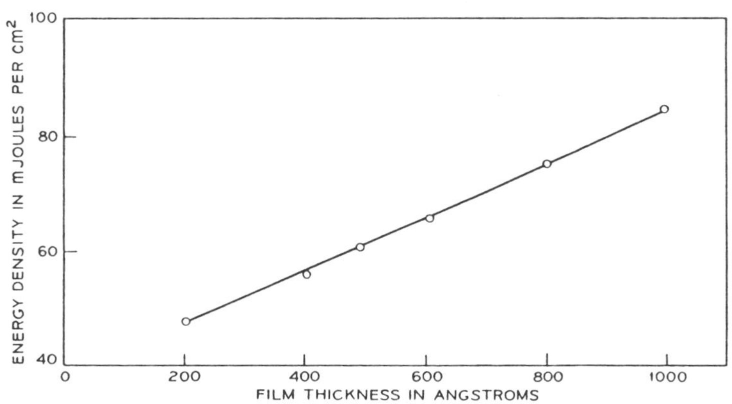

Fig. 15—Dependence of the energy densities, required to machine 6-μm-diameter spots, on the bismuth film thickness.

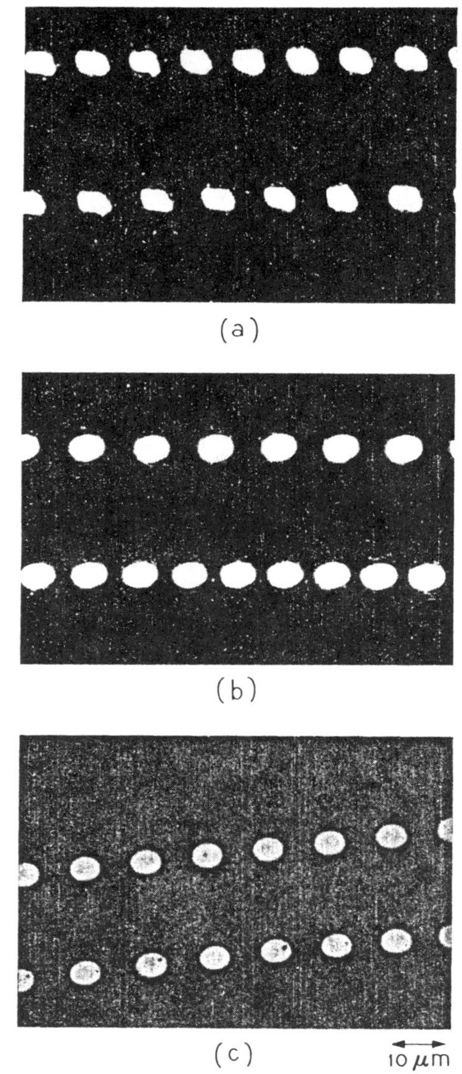

Fig. 16—Machining spot shapes with different substrates for front and back illumination: (a) Bismuth on Mylar D, back illumination. (b) Bismuth on Mylar D, front illumination. (c) Bismuth on Mylar S, back illumination.

IV. IMAGE RECORDING ON THIN BISMUTH FILM

The use of laser beams to record information on thin metallic film has been demonstrated by several authors.[1,2,12,13] C. O. Carlson, et al.[1] used a 38-mW CW He-Ne laser to write video information on various kinds of thin films on a glass substrate. The thin films used were 500 Å of lead and tantalum and 1μm of a triphenylmethane. A negative image of a page was formed by micromachining the thin film with the laser beam which was modulated at a video frequency. Machining lines had a width of 2μm with a total image size of 0.24 × 0.24 mm. The writing speed was equivalent to about 10^6 spots per second. The energy density required was 1.2 joules/cm^2.

A 100-Å-thick bismuth film on a glass substrate has been utilized to record holograms of various patterns with a pulsed ruby laser.[2]

In the present work, an intracavity modulated argon laser of average power capability, 20 mW, has been utilized to record images on a 600-Å-thick bismuth film on a mylar substrate. The recorded image size was 10 by 12 mm and the writing speed was 10^6 spots per second. Images of documents which were originally 8.5 × 11 inches have a resolution of over 175 lines/inch when magnified to the original size. The picture quality obtained is comparable with a good black and white photograph with 8 to 10 shades of gray. The energy density (given by the pulse peak energy divided by the spot area) had a maximum value of about 0.060 joule/cm^2.

The system is illustrated in Fig. 17. An acousto-optical intracavity modulation system was used to obtain 25-ns-duration light pulses from a CW argon laser producing 20 mW average power in the Gaussian mode. The modulation technique preserved the average output power of the laser. The video information was acquired by a He-Ne laser flying-spot scanner. A balanced mixer amplitude modulated a 5-mW 450-MHz RF signal with the video signal. By supplying a dc bias in addition to the video signal, the modulation depth could be adjusted to obtain the proper gray scale. A second balanced mixer, triggered with 30-ns-duration base band pulses at a 1-MHz repetition rate, was used to obtain the RF pulses required to dump the cavity. The video-modulated 30-ns-duration RF pulses, after being amplified by a 20-watt linear amplifier, were fed directly into the ZnO transducer sputtered on the fused silica modulator. As a result of a Bragg interaction between the acoustic pulses and the light beam, light pulses with the duration of 25–30 ns were dumped out of the cavity at a 1-MHz repetition rate. The pulses were intensity modulated at the video frequency. The light

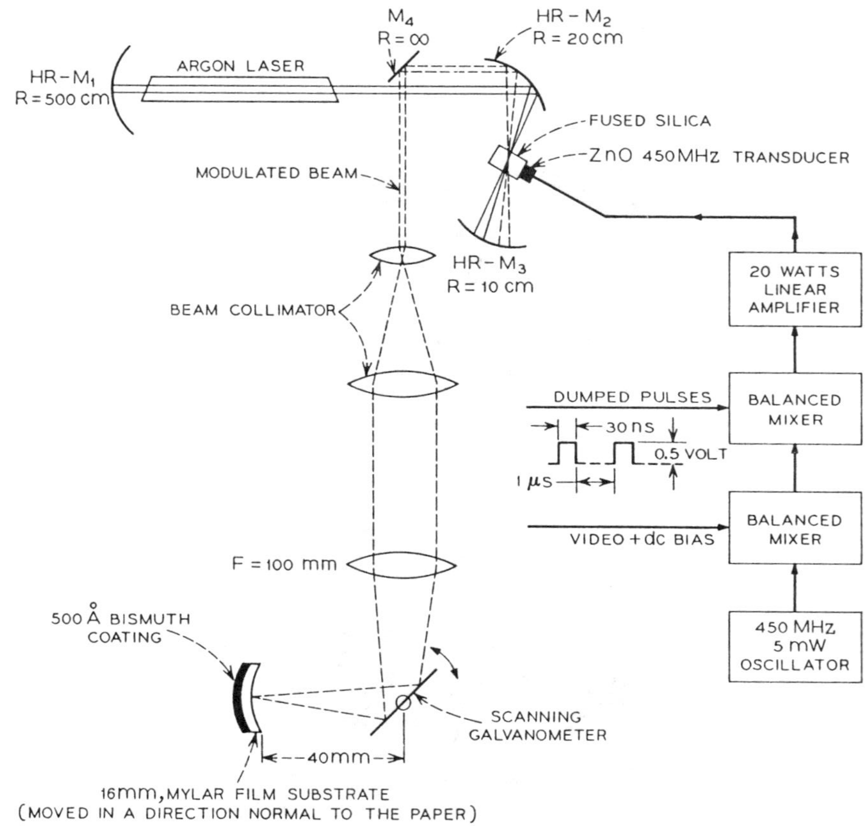

Fig. 17—The system setup for recording video information by laser machining.

pulses after being collimated were focused with an f/5.6 lens onto the bismuth film so as to machine 6-μm-diameter spots. The light pulses were deflected in the horizontal direction by a galvanometer mirror. The 16-mm mylar substrate with the bismuth coating moved at a constant speed in the vertical direction. To prevent the bismuth vapor from coating the optical components, the laser beam was incident on the thin film through the mylar substrate. The writing speed was 2 ms per line including 40-percent retrace time. The number of lines printed was about 2000 lines per frame (4-second frame with 2.4 × 10^6 total number of resolvable spots). Figure 18 shows a photograph of a micromachined copy of an IEEE facsimile test chart. As seen in this figure, about 8 to 10 shades of gray were obtained. A magnified

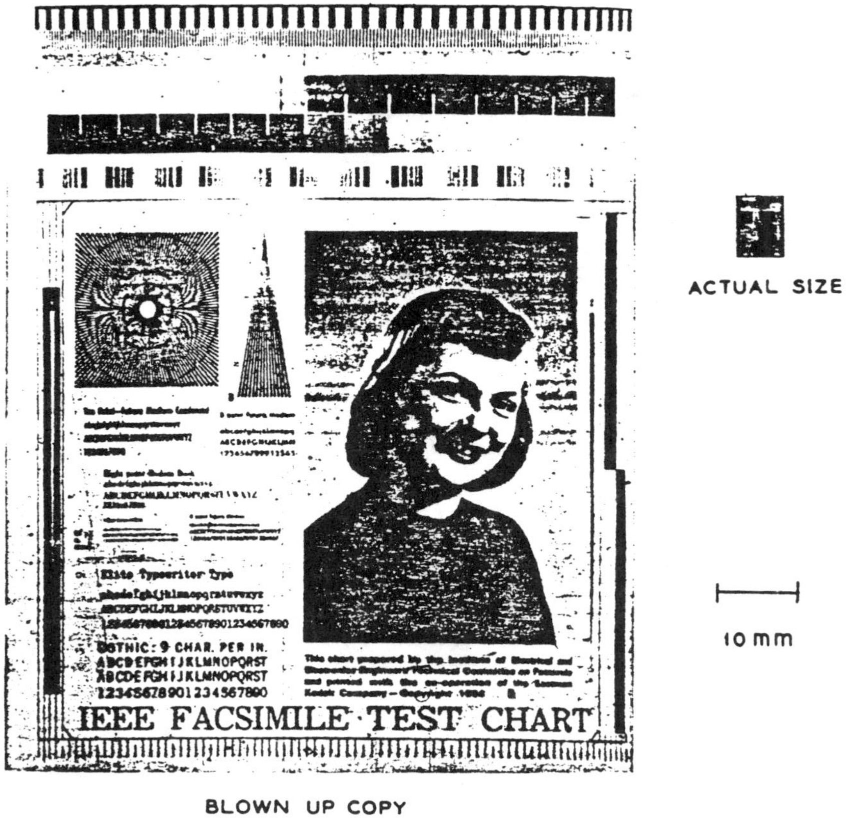

Fig. 18—Microfilm recording of a facsimile test chart.

region of an all-white area is shown in Fig. 19. A region of several gray scales (a detail of the eye in Fig. 18) is shown in Fig. 20.

By changing the polarity of the video signal (Fig. 17) either positive or negative images could be recorded. In Fig. 21, magnification of both positive and negative images of newspaper clippings are shown.

V. CONCLUSION

Calculations and experiments have been carried out for the machining of thin metallic films by a laser beam and for the associated temperature rise of the film. For a long illumination duration t, depending on the thermal constants of the film and substrate and on the film thickness, the temperature to which the film has risen by the end of the light pulse is proportional to $\sqrt{t}$. For a very short illumination duration,

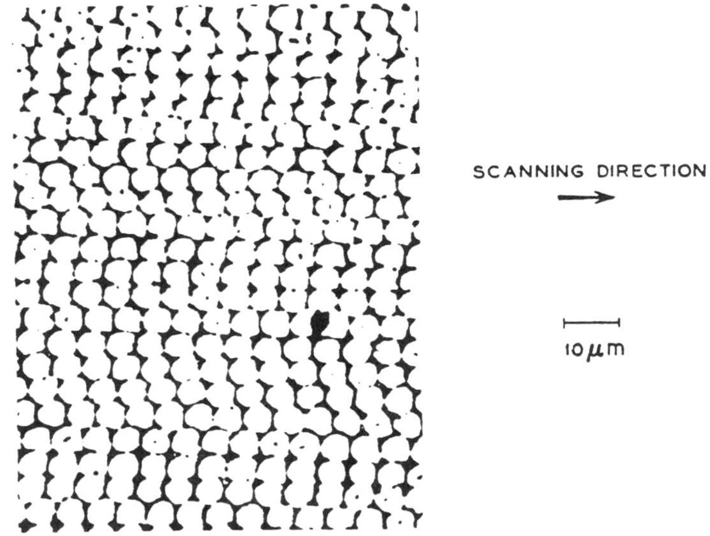

Fig. 19—1000 times magnification of a white area.

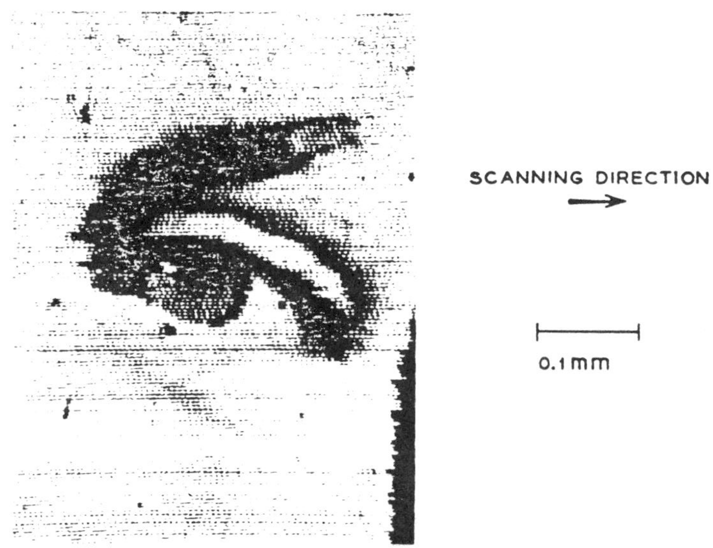

Fig. 20—Magnification of a gray scale area.

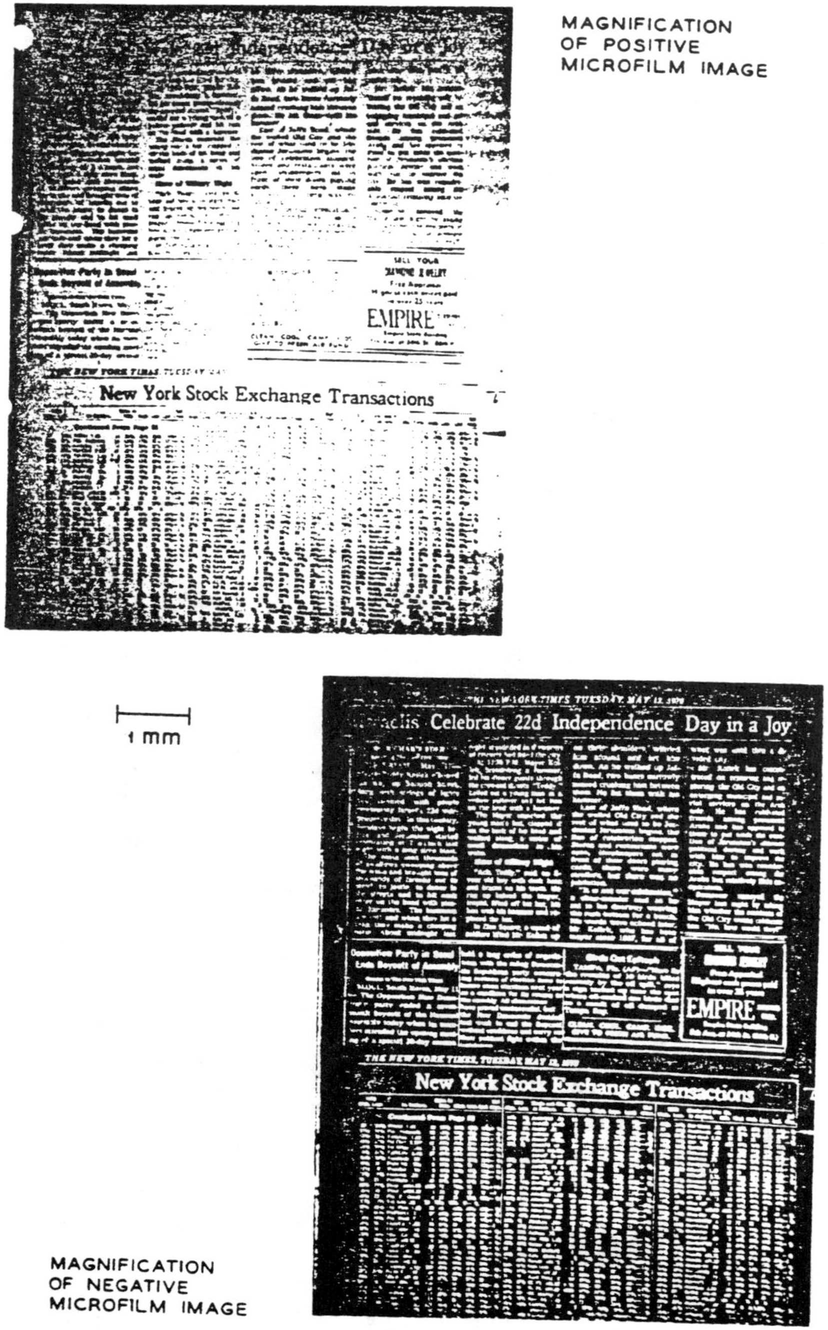

Fig. 21—Negative and positive microfilm recording of a newspaper.

however, the temperature rise approaches a linear dependence on time. For longer light pulses, a larger portion of the laser energy is lost to the substrate by heat conduction during the pulse. Machining of thin films on substrates of very low thermal conductivity is most efficiently obtained when the pulse duration is short enough that the temperature rise is linearly proportional to the pulse duration.

With optically transmitting substrates, machining could be obtained either by front illumination or by back illumination. The machining efficiency is approximately equal for the two methods. In many cases, however, back illumination is advantageous since the vapor from the surface is directed away from the focusing and deflecting optical components. By varying the laser pulse intensity, the machining spot area varies nearly linearly with the pulse intensity over a certain range above some threshold value. A better understanding of the dynamics of the machining process, however, is still required and further work is planned.

The use of very short laser pulses permits the machining of images of a given size and resolution in a given length of time with much lower average laser power than would be required with a CW laser. By varying the intensity of the laser pulses and hence the area of the machined spots, images with a continuous range of shades of gray were obtained. Images of printed material having a resolution of about 175 lines per inch of an original $8\frac{1}{2} \times 11$ inch document were machined.

VI. ACKNOWLEDGMENTS

The author wishes to acknowledge the expert collaboration of R. Sard in obtaining the electron microscope pictures of machined spots. He wishes also to acknowledge the invaluable help of R. G. Chemelli in the laboratory investigation of laser machining and of C. W. Van Hise in preparing the thin films and the acoustooptic transducers, and to express appreciation to W. J. Bertram and R. L. Field for the use of the laser scanner. Acknowledgments go to M. Feldman for many useful discussions, and to M. I. Cohen and R. E. Kerwin, who contributed some of the initial concepts which led to this work. The author also wishes to thank A. J. Chick for the numerical calculations, and R. C. Miller, H. A. Watson, and E. I. Gordon for encouragement and valuable discussions.

REFERENCES

1. Carlson, C. O., et al., "He-Ne Laser: Thermal High-Resolution Recording," Science, *154* (1966), pp. 1550–1551.
2. Harris, A. L., et al., "CW Laser Recording on Metallic Thin Film," Image Technology, *12*, No. 3 (April–May 1970), pp. 31–35.
3. Maydan, D., "Fast Modulator for Extraction of Internal Laser Power," J. Appl. Phys., *41*, No. 4 (March 1970), pp. 1552–1559.
4. Maydan, D., and Chesler, R. B., "Q-Switching and Cavity Dumping of Nd:YA1G Lasers," J. Appl. Phys., *42*, No. 3 (March 1971), pp. 1031–1034.
5. Ready, J. F., "Effects Due to Absorption of Laser Radiation," J. Appl. Phys., *36*, No. 2 (February 1965), pp. 462–468.
6. Cohen, M. I., Unger, B. A., and Milkosky, J. F., "Laser Machining of Thin Films and Integrated Circuits," B.S.T.J., *47*, No. 3 (March 1968), pp. 385–405.
7. Cohen, M. I., and Epperson, J. P., *Application of Lasers to Microelectronic Fabrication, Electron Beam and Laser Beam Technology*, New York: Academic Press, 1968, pp. 139–185.
8. Carslaw, H. S., and Jaeger, J. C., *Conduction of Heat in Solids*, 2nd Edition, New York: Oxford University Press, 1959, p. 10.
9. Ibid., p. 75.
10. Kogelnik, H., and Li, T., "Laser Beams and Resonators," Proc. IEEE, *54* (1966), pp. 1312–1329.
11. Feldman, M., private communication.
12. Amodei, J. J., and Mezrich, R. S., "Holograms in Thin Bismuth Films," Appl. Phys. Letters, *15*, No. 2 (15 July 1969), pp. 45–46.
13. Hamisch, H., "Aufzeichnung und Speicherung von Informationen mit Laserstrahlen," Clearinghouse, U.S. Department of Commerce, N 69-34646 (May 1969).

Reprinted with permission from *Optical Engineering*, Vol. 15(2), pp. 99-108 (March/April 1976). ©1976 Society of Photo-Optical Instrumentation Engineers.

Review and Analysis of Optical Recording Media

R. A. Bartolini, H. A. Weakliem, and B. F. Williams

RCA Laboratories
Princeton, New Jersey 08540

Abstract

Optical data storage systems employing large varieties of different recording materials have been investigated in great detail over the last few years. In this paper, we present a review of most of the known classes of optical recording media and describe a procedure for identifying materials with potential interest for specific applications. To provide a context for demonstrating this point, we pick as an example an optical disc recording and playback system. We feel the most useful application of this type of analysis is to help eliminate those materials from consideration which are unlikely to meet system requirements.

Introduction

Optical data storage systems employing large varieties of different recording materials have been investigated in great detail over the last few years. In this paper, we present a review of most of the known classes of optical recording media and describe a procedure for identifying materials with potential interest for specific applications. We will show that a detailed evaluation of the material properties which control sensitivity, resolution, and signal-to-noise ratio can provide sufficient information to reduce significantly the material options for a specific systems concept. To provide a context for demonstrating this point, we pick as an example an optical disc recording and playback system. By using the specific requirements of this system, we will identify materials and material classes which will satisfy these requirements. More importantly, we can eliminate a large number of seemingly interesting materials by paying strict attention to the system requirements.

In the following we will review our material classification system, describe in detail the requirements of an example optical disc recording playback system, and finally, present an analysis which identifies materials of interest for that system.

Material Classification

For convenience, we have organized optical recording materials into eleven different classes. These are: photographic films, photoresists, photopolymers, thermoplastics, photochromics, chalcogenide films, ablative thin films, magneto-optic, photoferroelectric, photoconductive/electro-optic, and electro-optic.

These eleven different classes of materials are listed in Table 1 along with a variety of descriptive properties. In the remaining tables, the most representative materials in

Table 1. Optical Recording Media

Medium	Processing	Recording Process	Recording Wavelength Range	Recording Sensitivity	Nominal Resolution	Storage Time	Readout Process	Recyclable	Replication
Photographic	Wet chemical or heat	Photochemical change	Panchromatic or orthochromatic	10^{-8}–10^{-1} J/cm^2	100-3000 c/mm	Permanent	Density or index change	No	Yes
Photoresists	Wet chemical or heat	Absorption & crosslinking	UV to 5500 Å	10^{-2} J/cm^2	1000-1500 c/mm	Permanent	Surface relief	No	Yes
Photopolymer	None or post exposure and post heating	Absorption & crosslinking	UV to visible	10^{-3}–100 J/cm^2	1500-5000 c/mm	Permanent	Index change	No	?
Ablative Thin Films	None	Ablation	Visible	~10^{-2} J/cm^2	1000–1500 c/mm	Permanent	Light amplitude	No	Yes
Thermoplastic	Pre-exposure corona charge and heat	Electric field produed deformation of heated material	Up to 6500 Å	10^{-4} J/cm^2	2000 c/mm	Permanent	Surface relief	Yes	Yes
Photochromic	None	Light induces new absorption bands	Entire visible range–most types most sensitive in UV or blue	3×10^{-3}–1 J/cm^2	300–>2000 c/mm	Minutes to months	Optical density change	Yes	Yes
Chalcogenide Films	None	Heat causes switching between amorphous and crystalline states	Visible to IR	10^{-2}–10^{-3} J/cm^2	1000 c/mm	Permanent	Different phases have different absorption & reflectivity	Yes	?
Magneto-optic	None	Thermomagnetic	Visible to IR	10^{-2} J/cm^2	100–2000 c/mm	Permanent	Faraday or Kerr effect	Yes	?
Photoferroelectric	None	Electric field & light induced birefringence change	Visible	10^{-2} J/cm^2	125-800 c/mm	Indefinite	Change in state of polarization of light in transmission or reflection	Yes	No
Photoconductive/Electro-optic	None	Photo-induced internal electric field redistribution	3500-6000 Å	5×10^{-6}–10^{-4} J/cm^2	1000 c/mm	Hours	Change in retardation induces change in polarization of light transmitted	Yes	No
Electro-optic	None (possibly heat to fix)	Light induced spatial modulation of refractive index	3500-5000 Å	10^{-1}–1 J/cm^2	1500 c/mm	Hours to indefinite	Diffraction of light	Yes	No

each class are listed with the most important properties of those materials. The experimental data given in these tables deal with the recording, readout, and erase-recycling processes of these materials.

Photographic Films – Table 2

Table 2 lists a variety of typical high resolution photographic materials along with their most important properties. Kodak Tri-X is also listed in this table for comparative purposes.

As can be seen from the table, these materials are prepared by emulsion coating techniques on a variety of rigid or flexible substrates in a usable thickness range of a few microns. A processing step is necessary for all these materials after its optical exposure. The recording process is photochemical in nature which results in an optical density change that modulates the intensity of a readout beam. In the case of dichromated gelatin, this density change can be converted into an index of refractive change by chemical bleaching. The recording sensitivity which extends over the entire visible range can be as high as 10^{-6} J/cm^2 for a resolution of >2500 cycles/mm (c/mm). There are no erase-recycling processes available for these materials. Replication can be accomplished by contact printing techniques. Most of these materials have been used as high density recording media for applications such as holography.

Photoresists – Table 3

Photoresists are light-sensitive organic materials that form imaged relief patterns upon exposure and development.

Table 2. Photographic Materials

Material	Usable Thickness	Preparation	Processing	Recording Processes	Recording Wavelength Range	Recording Sensitivity	Limiting Resolution	Readout Process	Erase Process	Replication	References
Kodak 649F	μm's	Coating technique	Wet chemical	Chemical change	Panchromatic	7×10^{-5} J/cm^2 for O.D. of 0.6	>3000 c/mm	Density change	None	Contact printing	[1,2]
Kodak 649GF	μm's	Coating technique	Wet chemical	Chemical change	Orthochromatic	5×10^{-5} J/cm^2 for O.D. of 0.6	>3000 c/mm	Density change	None	Contact printing	[1,2]
Agfa 8E75	μm's	Coating technique	Wet chemical	Chemical change	Panchromatic	2×10^{-5} J/cm^2 for O.D. of 0.6	>3000 c/mm	Density change	None	Contact printing	[1,3]
Agfa 8E56	μm's	Coating technique	Wet chemical	Chemical change	Orthochromatic	2×10^{-5} J/cm^2 for O.D. of 0.6	>3000 c/mm	Density change	None	Contact printing	[1,3]
Agfa 10E75	μm's	Coating technique	Wet chemical	Chemical change	Panchromatic	10^{-6} J/cm^2 for O.D. of 0.6	~2500 c/mm	Density change	None	Contact printing	[1,3]
Agfa 10E56	μm's	Coating technique	Wet chemical	Chemical change	Orthochromatic	10^{-6} J/cm^2 for O.D. of 0.6	~2500 c/mm	Density change	None	Contact printing	[1,3]
3M Dry Silver Film	μm's	Coating technique	Heat 115°C for ~20 sec	Chemical change	Panchromatic or orthochromatic	10^{-4}–10^{-5} J/cm for O.D. of 0.6	350 c/mm	Density change	None	Contact printing	[4]
Kodak Dry Silver Film	μm's	Coating technique	Heat 230°C for 4 sec	Chemical change	Panchromatic or orthochromatic	2×10^{-7} J/cm^2 for O.D. of 0.1	?	Density change	None	Contact printing	[2]
Dichromated Gelatin	μm's	Coating technique	Wet chemical	Chemical change	3500-5200 Å	5-10×10^{-2} J/cm^2 for $\Delta n \sim 10^{-2}$	200 c/mm	Index change	None	Contact printing	[5]
Technifax (Scott Graphics) D8-402 Diazo	μm's	Coating technique	Ammonia vapor	Chemical change	Up to 5000 Å	5×10^{-1} J/cm^2 for O.D. of 1.0	>2000 c/mm	Density change	None	Contact printing	[6]
Kalvar	μm's	Coating technique	Heat	Chemical change	3800 Å	3×10^{-2} J/cm^2 for O.D. of 1.0	1800 c/mm	Density change	None	Contact printing	[7]
Kodak Tri-X	μm's	Coating technique	Wet chemical	Chemical change	Panchromatic	10^{-8} J/cm^2 for O.D. of 0.6	100 c/mm	Density change	None	Contact printing	[1,2]

Table 3. Photoresists

Material	Usable Thickness	Preparation	Processing	Recording Processes	Recording Wavelength Range	Recording Sensitivity	Resolution	Readout Process	Erase Process	Replication	References
Shipley AZ-1350 (Pos)	μm's	Spin or spray coating	Wet chemical	Absorption & crosslinking	UV to 5000 Å	10^{-2} J/cm^2 at 4416 Å	>1500 c/mm	Surface relief	None	Embossing	[1,8,9]
Shipley AZ-111 (Pos)	μm's	Spin or spray coating	Wet chemical	Absorption & crosslinking	UV to 5000 Å	10^{-2} J/cm^2 at 4416 Å	<1000 c/mm	Surface relief	None	Embossing	[1,8]
Kodak KPR (Neg)	μm's	Spin or spray coating	Wet chemical	Absorption & crosslinking	UV to 4500 Å	10^{-2} J/cm^2 at 4000 Å	<1000 c/mm	Surface relief	None	Embossing	[1,2]
Kodak KTFR (Neg)	μm's	Spin or spray coating	Wet chemical	Absorption & crosslinking	UV to 4500 Å	10^{-2} J/cm^2 at 4000 Å	<1000 c/mm	Surface relief	None	Embossing	[1,2]
Kodak KMER (Neg)	μm's	Spin or spray coating	Wet chemical	Absorption & crosslinking	UV to 4500 Å	10^{-2} J/cm^2 at 4000 Å	<1000 c/mm	Surface relief	None	Embossing	[1,2]
Kodak KAR 3 (Pos)	μm's	Spin or spray coating	Wet chemical	Absorption & crosslinking	UV to 4500 Å	10^{-2} J/cm^2 at 4000 Å	>1500 c/mm	Surface relief	None	Embossing	[2]
GAF PR-115 (Pos)	μm's	Spin or spray coating	Wet chemical	Absorption & crosslinking	UV to 5000 Å	10^{-2} J/cm^2 at 4000 Å	~1000 c/mm	Surface relief	None	Embossing	[10]
Kodak KOR (Neg)	μm's	Spin or spray coating	Wet chemical	Absorption & crosslinking	UV to 5500 Å	10^{-2} J/cm^2 at 4880 Å	~1000 c/mm	Surface relief	None	Embossing	[1,2]
Horizons LHS7 (Neg)	μm's	Spin or spray coating	Heated air	Absorption & crosslinking	UV to 5500 Å	5×10^{-3} J/cm^2 at 4880 Å	>500 c/mm	Surface relief	None	Embossing	[11]

There are two types of photoresists – negative and positive. In negative-type photoresist, the exposed areas become insoluble as they absorb light so that upon development only the unexposed (soluble areas) are dissolved away. In positive-type photoresist the opposite is true; the exposed areas become soluble and dissolve away during the development process. The readout process for both types of photoresist is by phase detection due to the resulting surface relief after exposure and development. Spectral sensitivity is from the UV to the green and exposures of 10^{-2} J/cm^2 are necessary with a resolution capability of 1000-1500 c/mm. There is no erase-recycling process possible with these materials and replication is by embossing techniques. These materials have been used in optical data storage techniques such as holography and optical discs.

Photopolymers – Table 4

Many organic compounds are known to undergo photochemical reactions, some of which produce products having refractive indices substantially different from those of the starting material. A variety of media has been used to record optical data utilizing these refractive index changes.

While photographic and photoresist materials are commercially available the photopolymers are mostly in the experimental stage. Their spectral sensitivities can be varied by the addition of different sensitizers. The recording mechanism is generally permanent; there is no erase-recycling process and replication has been demonstrated only with the duPont material.

Ablative Thin Films – Table 5

Table 5 lists some metal and organic dye ablative films. The recording mechanism is essentially thermal in nature using the energy of the absorbed laser beam as an extermely localized and intense source of heat to either melt or actually ablate the material. Readout can be either reflection or transmission of light with the ablated holes acting as the modulated information. Metals are less sensitive than organic dyes but are more resistant to damage while handling. These are permanent recording materials and replication can be accomplished by embossing techniques.

Thermoplastic – Table 6

Table 6 describes the most widely used thermoplastic material which is a multilayer structure consisting of a substrate (glass or mylar), a thin conductive layer (Au, Ag, etc.), a photoconductor (polyvinyl carbazole sensitized with trinitro-9-fluorenone), and a thermoplastic (Staybelite Ester 10). The recording technique consists of first establishing a uniform charge on the surface of the thermoplastic usually with a corona-charging device. The voltage is capacitively divided between the photoconductor and thermoplastic layers. The next step is the optical exposure of the material which causes the photoconductor to conduct in the illuminated regions and thus discharges the voltage across it. However, the surface charge density on the thermoplastic surface is unchanged. To establish a force variation on the thermoplastic, the surface is recharged uniformly, thus adding charge to the illuminated regions. Finally, the thermoplastic is heated to its softening temperature and the electrostatic forces deform the surface into a relief pattern which corresponds to the optical recording information. When the thermoplastic is cooled, the relief pattern is permanently stored in the surface of the thermoplastic. In order to erase the information, the thermoplastic must be heated so that surface tension smooths out the relief variation. This material is one of the most promising for erasable optical memory applications. Spectral sensitivity can be varied depending on the choice of photoconductor. Resolution of 2000 c/mm has been achieved at exposures of 10^{-4} J/cm^2. Replicas can be made by embossing techniques.

Table 4. Photopolymers

Material	Usable Thickness	Preparation	Processing	Recording Processes	Recording Wavelength Range	Recording Sensitivity	Resolution	Readout Process	Erase Process	Replication	References
Dupont	mils	Coating Technique	Post Exposure	Absorption & crosslinking	UV to 5000 Å	2×10^{-2} J/cm^2	~1200 c/mm	Index change	None	Contact printing	[12]
PMMA	mm's	Cast	None	Absorption & crosslinking	UV	100 J/cm^2	5000 c/mm	Index change	None	?	[13,14]
PMMA (Q doped)	mils	Cast	None	Absorption & crosslinking	4880 Å	3 J/cm^2	5000 c/mm	Index change	None	?	[15]
Hughes	μm's	Coating Technique	UV Exposure	Absorption & crosslinking	6328 Å	10^{-3} J/cm^2	3000 c/mm	Index change	None	?	[16]
RCA	mm's	Cast	None or post heating	Absorption & crosslinking	5000 Å	1 J/cm^2	3000 c/mm	Index change	None	?	[17,18]

Table 5. Ablative Thin Films

Material	Usable Thickness	Preparation	Processing	Recording Processes	Recording Wavelength Range	Recording Sensitivity	Resolution	Readout Process	Erase Process	Replication	References
Metals											
Bismuth	100-1000 Å	Evaporation	None	Ablation	Visible	3×10^{-2} J/cm^2	~1000 c/mm	Light Amplitude	None	Embossing	[19,20]
Rhodium	100-1000 Å	Evaporation	None	Ablation	Visible	~10^{-1} J/cm^2	~1000 c/mm	Light Amplitude	None	Embossing	[21]
Dyes											
Fluorescein	~1000 Å	Evaporation or spinning	None	Ablation	4500 Å	10^{-2} J/cm^2	>1500 c/mm	Light Amplitude	None	Embossing	[22]
1,5 diamino-anthraquinone	~1000 Å	Evaporation or spinning	None	Ablation	5000 Å	10^{-2} J/cm^2	>1500 c/mm	Light Amplitude	None	Embossing	[22]
Rhodamine-B	~1000 Å	Evaporation or spinning	None	Ablation	5000-5600 Å	10^{-2} J/cm^2	>1500 c/mm	Light Amplitude	None	Embossing	[22]

Table 6. Thermoplastics

Material	Usable Thickness	Preparation	Processing	Recording Processes	Recording Wavelength Range	Recording Sensitivity	Resolution	Readout Process	Erase Process	Erase Time	Cycle Lifetime	Repli-cation	Refer-ences
A layered structure	μm's	Evaporation & spinning (multi-layer device)	Pre-exposure corona charge & heat	Electric field produced deformation of heated material	Up to 6500 Å	10^{-4} J/cm^2	2000 c/mm	Surface Relief	Heat	~1 sec	100	Embossing	[23,24]

Photochromic – Table 7

A photochromic material is one which may exist in two or more relatively stable states having different optical constants and which may be switched from one state to the other by photon irradiation. If the change of state is associated with different absorption spectra, the readout process is directly related to the change of amplitude of transmitted (or reflected) intensity either at the recording wavelength or at a different wavelength. If the photon irradiation causes a change in the refractive index, the information is recorded and read holographically. Recording in photochromics is accomplished entirely by the energy of the optical beam and so is less sensitive than processes which employ an additional source of energy for developing or fixing.

The examples given in Table 7 are classified as inorganic, silver halide doped glass, and organic. The inorganics are usually crystals doped with impurities which have electronic configurations having different absorption spectra. The electronic configurations which characterize the stable states correspond either to different oxidation states of the impurity or to localized states associated with impurity-defect combinations. The impurity concentrations are on the order of a few tenths of a percent, which requires relatively thick samples in order to obtain practically useful transmitted intensity differences between the different states.

The silver halide doped glasses have a thermally stable bleached state and a metastable, colored state having an absorption band centered in the visible. Absorption of UV causes the bleached state to transform to the colored state, which may, in turn, be switched back to the bleached state either by absorption of light in its visible band or by heat. The contrast and resolving power are also limited by thickness considerations in these cases.

There is a wide variety of photochromic effects in organic substances and they offer the advantage that they may be formed in a wide variety of physical configurations, such as polymeric films and deposits on discs. Most organic photochromics must be used diluted in an otherwise inactive host, have a stable colorless state, and are prone to fatigue, which is the result of competing chemical reactions, often with oxygen, resulting in a nonphotochromic product.

Two cases are particularly interesting for high contrast, high resolution recording media, photodimers, and spiropyranes with $RCBr_3$ added. The quantum efficiency for making and breaking the dimers is high and a strong absorption band in the blue is associated with the change. The absorption band of one of the states is in the UV, however, which is a distinct disadvantage. The spiropyranes have stable states with absorption bands in the blue to UV and have a strong visible band in the cleaved state. Permanent storage can be achieved by making use of a chemical reaction with a bromide salt and this has been done in a polymer film.

Most of the results on the organic photochromics are given in terms of the quantum efficiencies of conversion and the molar extinction (the absorption spectrum) of the two states. The results have most often been obtained in solution and often generalizing the results to a given solid configuration may only be done with much uncertainty. Concentration, solvent, and temperature effects are strong and not always known in advance. Some of the data in the tables (enclosed in brackets) are calculated and represent a best estimate based on available and photochemical data.

Table 7. Photochromatic

Material	Usable Thickness	Preparation	Processing	Recording Processes	Recording Wavelength Range	Recording Sensitivity	Resolution	Storage Time	Readout Process	Readout Wavelength Range	Erase Process	Erase Sensitivity	Erase Time	Cycle Lifetime	Repli-cation	Refer-ences
Inorganic																
CaF_2:Ce,Na $SrTiO_3$:Ni,Mo,Al $CaTiO_3$:Ni,Mo	mm to cm	Crystal	None	Color Center Trapping	UV to blue	10^{-1} J/cm^2 to give O.D. = 0.1	>2000 c/mm	Minutes to days	Amplitude	New absorption band created in visible	Visible light or heat	2-10 times less sensitivity than record to erase with light	[nsec]	Indef.	Contact printing	[25,26]
Glass Doped with Silver Halides																
Borosilicates	mm to cm		None	Reduction to Ag grains	UV to visible	3×10^{-2}-3×10^{-3} J/cm^2 to give O.D. = 0.2	>2000 c/mm	Days to months	Amplitude	Panchromatic	Heat or visible light	green-red 3×10^{-2}-6×10^{-2} J/cm^2	?	Indef.	Contact printing	[27]
Organic																
thio-indigo in epoxy	mm	Polymer films for matrix	None	Cis-tran isomerization	4900-5500 Å	0.2 J/cm^2 to give O.D. ~0.5	?	Minutes to months	Amplitude	New band 4200-4800 Å	Heat or absorb in new bands	0.1 J/cm^2 to give 70% of orig. state	?	Subj.to fatigue	Contact printing	[28,29]
salicylidene-anilene	2-10 μm	Polycrystalline films	Exposure at 3000-4000 Å first	Tautomerism	4500-5500 Å	0.2 J/cm^2 to give O.D. ~0.5	650 c/mm	Hours	Amplitude	4500-5500 Å	Visible light	2×10^{-2} J/cm^2 at 4880 Å	?	Can be repeatedly recycled	Contact printing	[30]
stilbene	?	Polymer films for matrix	None	Cis-tran isomerization	UV to blue	0.1 J/cm^2 for Δn = 0.3	2000 c/mm	?	Holographic	?	?	Low	?	?	?	[31]
Cyanamid dye #63-071	?	Film	None	Abs.spectrum change	4200-5000 Å	$3\cdot10^{-1}$ J/cm^2 for O.D. 0.2	?	Hours at room temp.	Amplitude	New band 5200-6500 Å	Heat or IR	?	?	Indef.	Contact printing	[32]
Anthracene	2-10 μm	Crystals or polymer films for matrix	None	Photodimerization	UV(2500-3000 Å)	0.1 J/cm^2(Δn ~10^{-3})	?	Days	Holographic	Visible	Heat 3200-3800 Å	0.2 J/cm^2 at 3200-3800 Å	[nsec]	Subj.to fatigue	No	[33]
Benzacridazinium	2-10 μm	Crystals or polymer films for matrix	None	Photodimerization	UV(2500-3000 Å)	0.1 J/cm^2(Δn 10^{-3})	?	Days	Holographic	Visible	4000-5000 Å	?	[nsec]	Subj.to fatigue	Contact printing	[33]
Spiropyrane	[1-10 μm]	Dilute solution in polymer films	None	Heterolytic cleavage	2200-3750 Å	[~0.5 J/cm^2] for O.D. 0.3	?	Min.-days strong temp. depend.	Amplitude	New band in visible	4400-6300 Å or heat	[~1 J/cm^2] in new band	?	Subj.to fatigue	Contact printing	[34]
Spiropyrane + $RCBr_3$	[1-10 μm]	Polymer films for matrix	None	Chemical reaction	2750 Å	~10^{-2} J/cm^2 for O.D. = 1	300 c/mm	Perm.in absence of UV	Amplitude	New band 5100-6200 Å	None	---	---	---	Contact printing	[35]

Chalcogenide – Table 8

Table 8 shows one of the most widely used chalcogenide materials. The recording mechanism is based on the reversible switching between the amorphous and crystalline states of these materials by heating with a laser. The readout in reflection or transmission is then provided by the difference in the optical properties between the crystalline and amorphous states. The chalcogenides are of particular interest because of the possibility for erasure and recycling and also because their high absorption coefficient at long wavelenghts allows for the use of GaAs injection lasers to record and readout.

Magneto-Optic – Table 9

While all magnetic-optic materials use the same mechanism for readout, i.e., Faraday effect for transmission readout and Kerr effect for reflective readout, the recording process can be by one of four different thermomagnetic effects. The most widely investigated recording effect is Curie Point writing. This is a method in which the temperature rise in the heated area caused by the optical information exceeds the Curie temperature of the material. Allowing the material to cool locally reverses the direction of magnetization, thereby storing the information. In ferrimagnetic materials, the two sublattice magnetizations are in opposite directions. At the compensation temperature, these sublattice magnetizations cancel out each other but a few degrees away from this temperature magnetization switching is possible. Thus, information from a laser heating beam can be stored in such materials by applying a switching field by operating the material at the compensation temperature. There are also other materials that have a temperature dependence of the coercivity that can be used to store information by applying a suitable field coincident with the laser heating. Thermally induced magnetization changes can also occur at a temperature below the Curie temperatures. The remanent magnetization M_R after such a thermal treatment is reduced from the saturation magnetization value M_S and under the influence of an external field, M_R can take on any value between $+M_S$ and $-M_S$. This effect is also sufficient to store optical information. The erase process for any of these recording techniques is the same as the recording process.

Table 8. Chalcogenide Films

Material	Usable Thickness	Preparation	Processing	Recording Processes	Recording Wavelength Range	Recording Sensitivity	Resolution	Readout Process	Erase Process	Erase Time	Cycle Lifetime	Replication	References
TeAsGe	500-1500 Å	Evaporation or sputtering	None	Heat causes switching between amorphous & crystalline states (absorption)	crystallizing: 6471 Å	3×10^{-2}–10^{-1} J/cm^2	1000 c/mm	Different phases have different absorption & reflectivity	Same as recording	μsec	Limited ~10	?	[36]
					crystallizing: 8500–9000 Å	3×10^{-2}–10^{-1} J/cm^2							
					amorphizing: 6471 Å	10^{-2} J/cm^2							
					amorphizing: 8500–9000 Å	2×10^{-2} J/cm^2							

Table 9. Magneto-optic Materials

Material	Usable Thickness	Preparation	Processing	Recording Processes	Recording Wavelength Range	Recording Sensitivity	Resolution	Readout Process	Erase Process	Erase Time	Cycle Lifetime	Replication	References
MnBi	300-700 Å	Evaporation	None	Curie point writing	6328 Å	10^{-2} J/cm^2	2000 c/mm	Faraday or Kerr effect (change in polarization)	Same as recording	~μsec	Indefinite	?	[37]
MnGaGe	300-700 Å	Evaporation	None	Curie point writing	6328 Å	?	?	Faraday or Kerr effect (change in polarization)	Same as recording	?	?	?	[38]
PtCo	μm's	Evaporation	None	Curie point writing	?-8500 Å	10^{-2} J/cm^2	200 c/mm	Faraday or Kerr effect (change in polarization)	Same as recording	?	?	?	[39]
GdlG	μm's	Evaporation	None	Compensation temperature writing	6328 Å	10^{-2} J/cm^2	300 c/mm	Faraday or Kerr effect (change in polarization)	Same as recording	?	?	?	[40, 41]
GdCo	μm's	Evaporation	None	Compensation temperature writing	4500-8200 Å	10^{-2} J/cm^2	?	Faraday or Kerr effect (change in polarization)	Same as recording	?	?	?	[42]
CrO_2	?	Evaporation	None	Thermo-remanent writing	5000 Å	10^{-2} J/cm^2	100 c/mm	Faraday or Kerr effect (change in polarization)	Same as recording	?	?	?	[43]
Co-P	200 Å	Evaporation	None	Coercivity temperature dependence	5000 Å	10^{-2} J/cm^2	200 c/mm	Faraday or Kerr effect (change in polarization)	Same as recording	?	?	?	[44]

Table 10. Crystalline Effects

Material	Usable Thickness	Preparation	Processing	Recording Processes	Recording Wavelength Range	Recording Sensitivity	Resolution	Storage Time	Readout Process	Erase Process	Erase Time	Cycle Lifetime	Replication	References
Photoferroelectric														
$Bi_4Ti_3O_{12}$	10-100 μm	Crystal sandwiched between PC & conductor	None	Electric field induced birefringence change	Depends on PC used	10^{-3} J/cm^2	800 c/mm	Indef.	Intensity changed, crossed polarizers	Same as record	msec	Indef.	No	[45, 46]
PLZT lead zirconate lead titanium lanthanum doped	mm	Ceramic between PC layer & conductor	Strain bias needed	Electric field induced birefringence change	Depends on PC used	10^{-2} J/cm^2	125 c/mm	Indef. with strain field on	Intensity change, crossed polarizers	Light & reversed polarity of applied field	msec	Fatigue a problem	No	[47]
Photoconductive-Electro-optic (Pockels Effect Device)														
$Bi_{12}SiO_{20}$	0.1-1 mm	Crystal sandwiched between conductor & insulating film	1-2 kV applied to record	Photo-induced internal electric field redistribution	3500-6000 Å	5×10^{-6}-10^{-4} J/cm^2 to produce 1/e intensity change	1000 c/mm	Hours	Polarized intensity change Readout using retardation induced by linear electro-optic effect, detected with polarizers	Similar to record	μsec	Indef.	No	[48]
Electro-Optic														
$LiNbO_3$	mm	Single crystal	160°C, 5 min	Index change	3500-5000 Å	1 J/cm^2	1500 c/mm	Indef.	Index change	Heat & same λ as record	Same as processing	Indef.	No	[49, 50, 51]
SBN strontium barium niobate	mm	Single crystal	None	Index change	3500-5000 Å	10^{-1} J/cm^2	1500 c/mm	Hours in the dark	Index change	Heat or same λ as record	Same as record	Indef.	No	[52]

Crystalline Effects – Table 10

Table 10 lists some of the most extensively investigated materials in the last three categories, i.e., photoferroelectric, photoconductive electro-optic, and electro-optic.

1. *Photoferroelectric.* A photoconductor-ferroelectric sandwich may be used to record information. The photoconductor absorbs the light and the resulting charge pattern induces switching between the stable remanent states of the ferroelectric. Particular orientations of the optical indicatrix and external polarizers must be used to obtain readout. Large area arrays, which are difficult to obtain with single crystals like bismuth titanate, may be obtained with the ceramic PLZT, which employs strain-biasing for orienting the optical indicatrix.

2. *Photoconductive Electro-optic.* A material that is both photoconductive and exhibits a linear electro-optic effect sandwiched between insulating layers having a voltage difference has been used as an information storing medium. Single crystals of $Bi_{12}SiO_{20}$ have been used in such a device. Photo-generated carriers in the crystal drift to the insulating layers and the field of this charge distribution produces a spatially varying retardation in the crystal *via* the linear electro-optic effect. The retardation is detected between crossed polarizers at a wavelength chosen to minimize photoconductivity. The image storage is several hours and the material can be reused indefinitely.

3. *Electro-optic.* Electro-optic materials such as $LiNbO_3$ rely on the generation of fields through photoexcitation and spatial rearrangement of electrons to cause changes in the refractive index of the material. The modulation of the refractive index stores the information to be retrieved at a later time. A practical limitation on optical storage in electro-optic crystals is the optical erasure of the information during readout. A thermal fixing process has been developed that converts the electronic charge information into optically stable ionic charge information which is permanently fixed. The ionic charge information can be erased by a later thermal treatment of the crystal.

System Description

As an example of how significant the details of the system requirements are for the choice of materials or material classes, we will consider materials for use in an optical disc record and playback system.

The recording process consists of focusing (to $< 1\mu$m) a modulated laser beam onto the recording material resulting in the storage of the information as a series of "spots" on a spiral track. For this system, we desire immediate playback so that no processing may be used to develop the exposed disc. The signal is readout from the disc by refocusing a less intense, unmodulated beam from the same laser on to the track. The track shape modulates the reflected light which is converted to the original signal by a photodetector. We will assume that we need a signal-to-noise ratio of 40 dB (peak-to-peak signal to RMS noise) in a bandwidth of 10 MHz readout from the disc to provide the recording we desire. The boundary conditions imposed by such a system are applied to the different possible optical recording materials below.

Materials Analysis

While the listing of all these materials in Tables 1 through 10 provides useful information, they are very difficult to compare for any specific application in this form. Thus, for this particular application of laser recorded optical disc, as a first step we compare these materials by plotting them on a graph of material resolution vs material sensitivity as shown in Figure 1.[21] By applying appropriate boundary conditions to this graph one can determine which materials are most likely to satisfy system requirements.

Since the optical disc recording system requires 1 μm (or smaller) spots, a boundary condition on material resolution of 1000 c/mm may be imposed on Figure 1. Materials with less resolution capability should not be considered for this application. We next consider exposure sensitivity. For a 1 μm spot recorded at 10 MHz and assuming that 1 mW of power is incident on the recording material, the exposure is 10^{-2} J/cm^2. A 1 mW power level is a practical value based

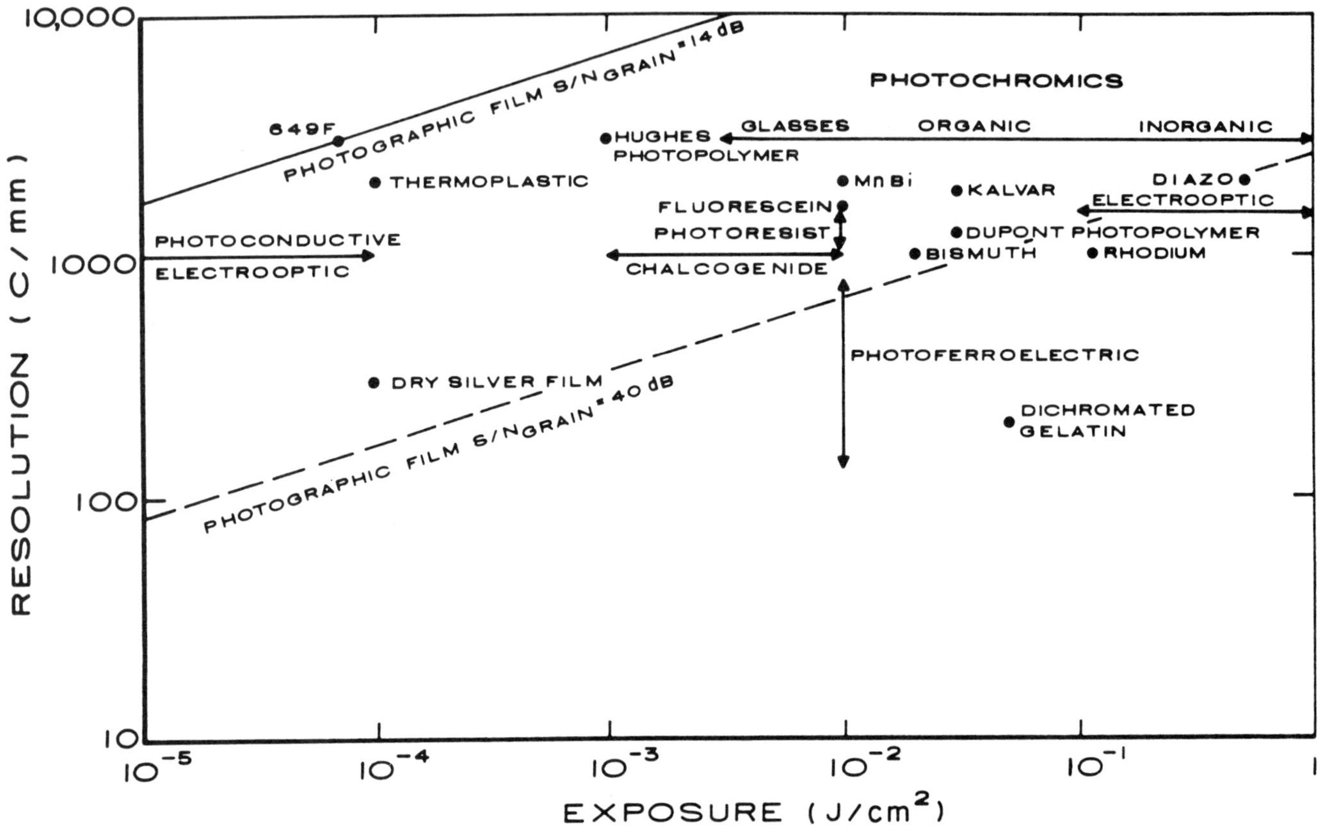

Fig. 1. Resolution (c/mm) vs exposure (J/cm²) for a variety of optical recording materials.

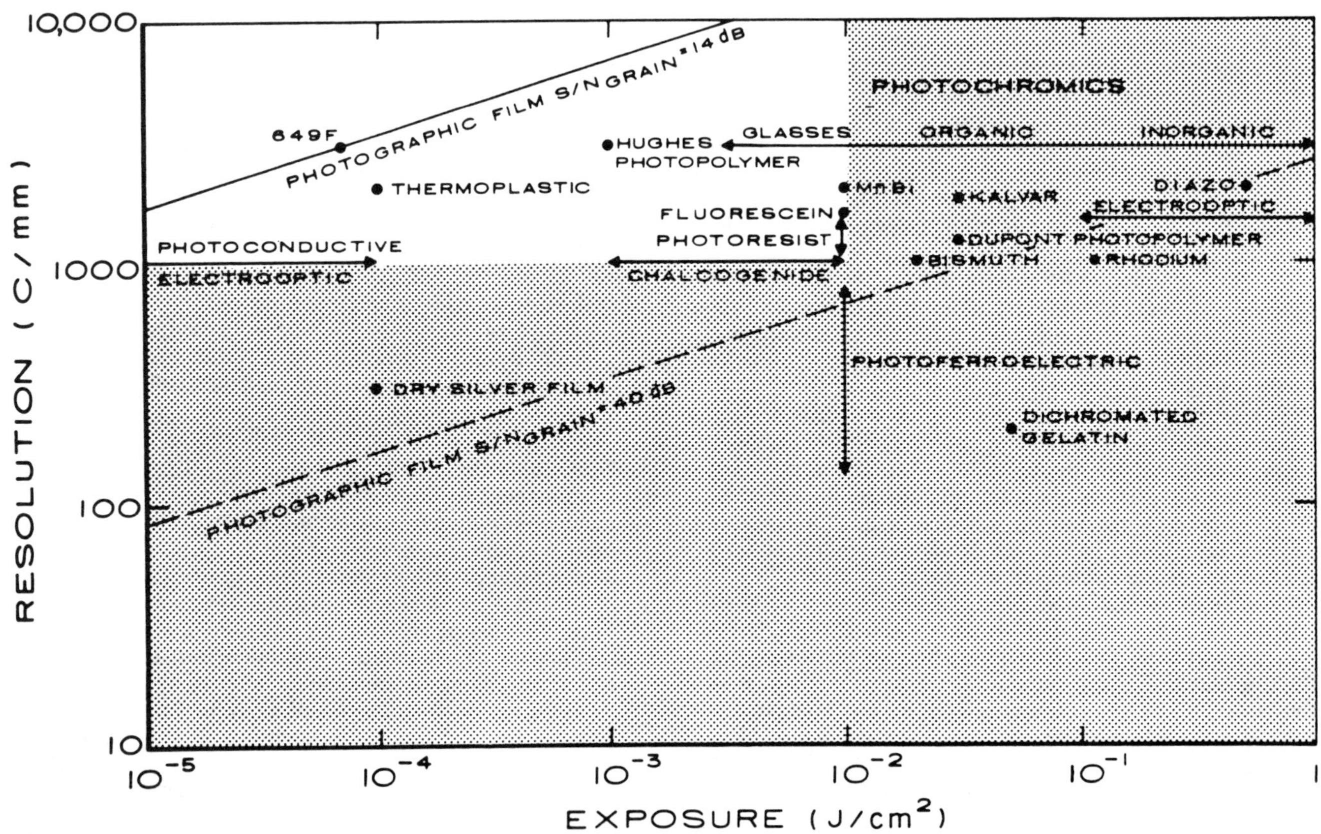

Fig. 2. Resolution (c/mm) vs exposure (J/cm²) for a variety of optical recording materials with restriction on recording resolution and recording exposure.

on existing lasers if one considers optical efficiencies, reflection, and other losses of the system. It should be clear that high power lasers are available and this will change the analysis, but our considerations will be restricted to lasers of moderate (∼10-100 mW) size.

The recording material sensitivity should correspond to 10^{-2} J/cm^2 to produce the desired optical effect. With these resolution and sensitivity restrictions, the number of useful materials is reduced as shown in Figure 2. Materials that are more sensitive than this 10^{-2} J/cm^2 would appear to be even more suitable. That this is not always the case can be seen by considering the energy required to read out the information from a signal-to-noise ratio (S/N) point of view.

During readout, the signal can be detected by either a photodiode or a photomultiplier tube. Neglecting system and material noises, the photodiode is limited by amplifier noise and it can be shown that for this case:[53]

$$(S/N)^2 \Delta f = \frac{[\eta_o \eta_q R\, R_1 Ce]^2 R_L}{(h\nu)^2 4kT} P_{in}^2 \tag{1}$$

where:

Δf = system bandwidth (Hz)
η_o = optical efficiency
η_q = quantum efficiency
R = material reflectivity
R_1 = optical efficiency loss due to tracking error
C = signal contrast
e = charge on electron (coulomb)
R_L = resistance of coupling resistor (ohms)
$h\nu$ = photon energy (J)
kT = thermal energy (J)
P_{in} = readout power onto disc (W).

Figure 3 shows a plot of $(S/N)^2 \Delta f$ vs P_{in} for the amplifier noise case for the parameters shown in the figure (all practical values). Since 40 dB S/N at 10 MHz is a system requiement, it requires 10^{-3} W of photons incident on the disc, corresponding to 10^{-2} J/cm^2 of exposures, for readout using a photodiode. Since this is also the exposure necessary for recording, there is no material which will be stable on readout and also sensitive enough to record on.

On the other hand, a shot noise limited detector, such as a photomultiplier, has better S/N performance in this regime. It can be shown that:[53]

$$(S/N)^2 \Delta f = \frac{\eta_o \eta_q R R_1 C^2}{2h\nu} P_{in}. \tag{2}$$

Figure 3 also shows a plot of $(S/N)^2 \Delta f$ vs P_{in} for this case for the parameters shown in the figure (all practical values). Using 40 dB and 10 MHz it requires 10^{-4} W of photons inci-

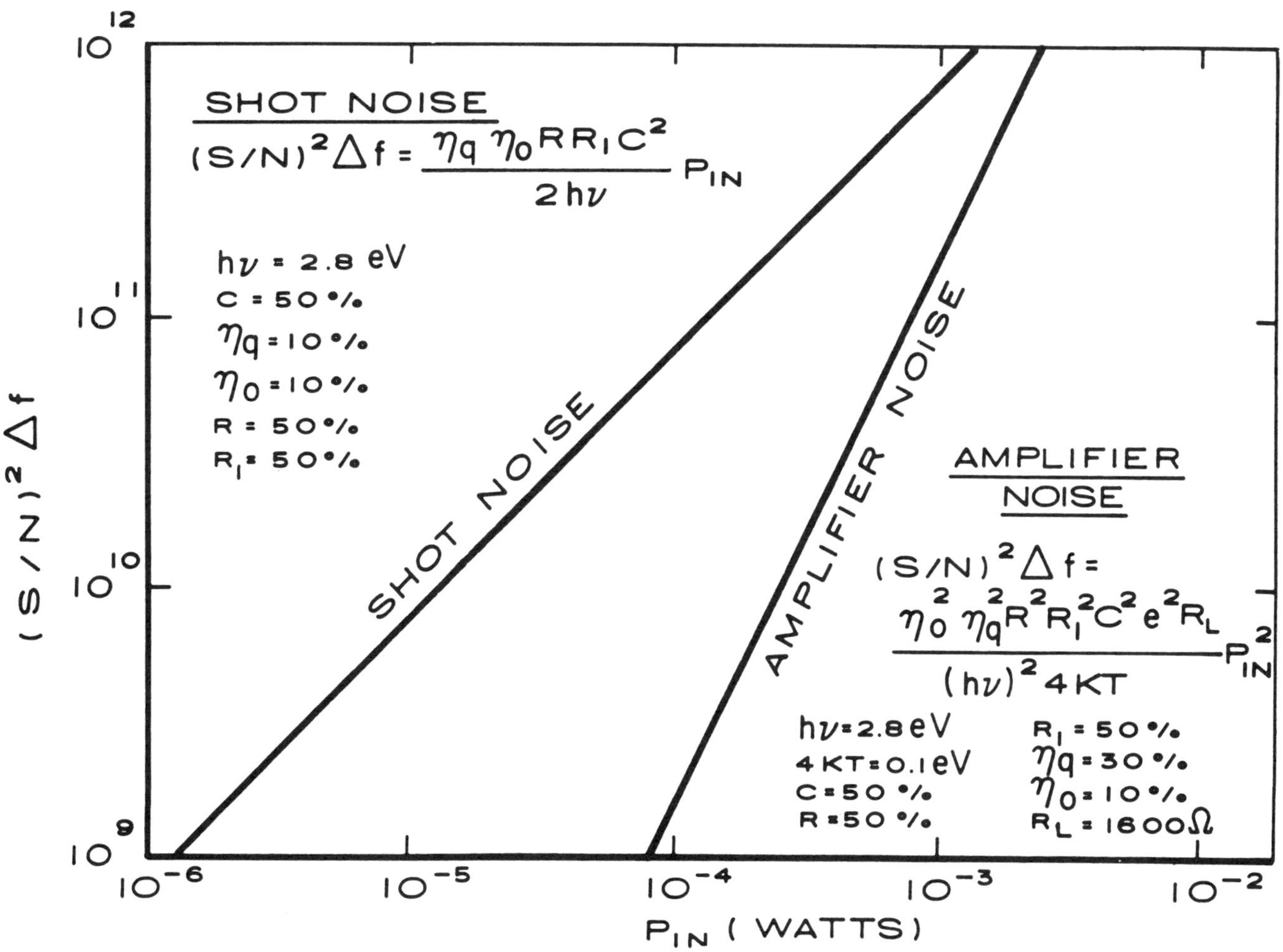

Fig. 3. (Signal-to-noise)^2 bandwidth [$(S/N)^2 \Delta f$] vs power onto recording material [P_{in}(W)] during readout of information for amplifier (photodiode) and shot noise (photomultiplier) limited cases.

dent on the disc, corresponding to 10^{-3} J/cm^2 exposure, for readout. Thus, the stable and unstable state exposure requirements are separated by an order of magnitude and non-destructive readout is possible using a photomultiplier tube. Destructive readout at higher powers may possibly be avoided for some classes of materials by using a different wavelength for readout but this requires two lasers, which in turn may also require different optics. This increase in system complexity will not be considered further.

With the above determined boundary condition (1000 c/mm, 10^{-2} J/cm^2 recording exposure and 10^{-3} J/cm^2 readout exposure) we can now determine which materials satisfy these conditions for an optical disc application.

The region of interest in Figure 4 includes materials from almost every class discussed above, however, many of the choices are not appropriate. Photographic film and photoresist require processing steps which are undesirable. What is surprising, however, is that photographic film does not even have sufficient resolution for this application. Notice that the region of film sensitivities depicted at the top of Figure 4 is for limiting resolution. If we assume that this corresponds to about 25 photographic gains per picture elements ($S/N_g = 5$ corresponds to 14 dB),[54] then we must consider a minimum picture element whose area is 400 times larger to get a signal-to-noise of 100 (40 dB). This reduction in resolution capability is shown by the dashed line in Figure 4. Thus, the grain nature alone of photographic film precludes its use for this application. The Hughes photopolymer appears to have limited shelf life. Of the reversible materials, the thermoplastics, the chalcogenides, and magneto-optic MnBi, only the latter can be ruled out, as is discussed below.

While the recording sensitivity of MnBi is just about ideal, the contrast on readout is quite low. In fact, for 400 Å thick film, which maximizes (contrast^2 x intensity) on readout, the intensity of the readout light must be increased by a factor of 10 over that shown in Figure 4. We have already seen that the readout power requirements for the parameters of Figure 3 require a material stable at 10^{-3} J/cm^2. MnBi must be stable at 10 times this on readout. This is nominally the power needed to write, so the material will be erased on reading. None of the other magneto-optic materials offers an improvement over MnBi on the basis of these considerations.

The thermoplastic materials and the chalcogenide films offer interesting opportunities for erasable recording in this system. It is worth noting that the reason thermoplastic materials can be considered here even though the sensitivity is in the restricted region is that they may be rendered insensitive for readout. This applies to all of the materials which are used in a photoconductive mode or with a photoconductor. However, the photoconductive electro-optic materials must be used in their single crystal forms and are not so useful in a disc format.

While the photochromic glasses exhibit sensitivities in the region of interest, the optical absorption is low so that the films must be very thick to achieve usable optical densities (mm's to give O.D.$\sim$0.1). They are located in the high resolution region of the chart because there is no grain and in a holographic mode, the high resolution may be realized. For the spot recording we are considering, these materials are not useful because of their thicknesses.

We now come to the final and, at the moment, most

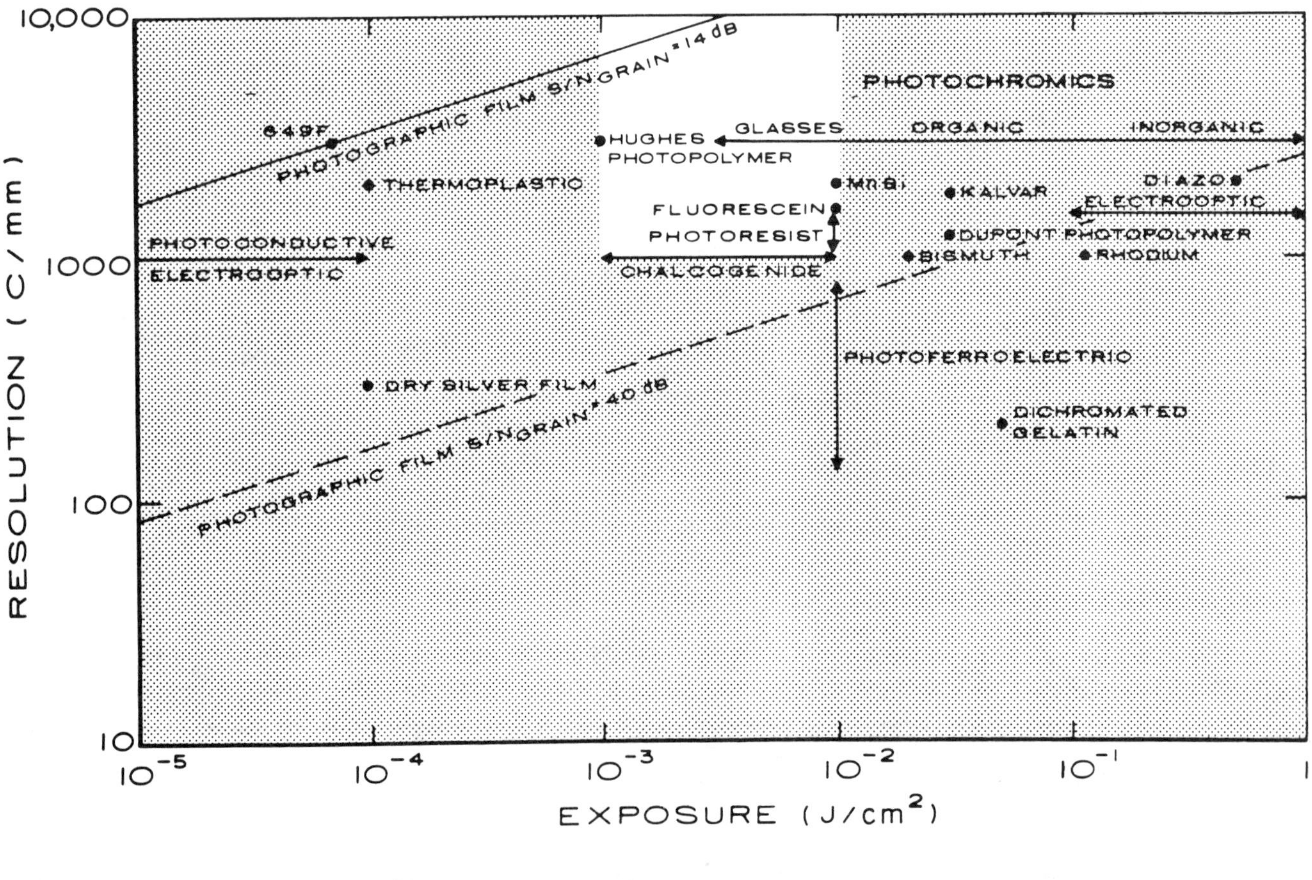

Fig. 4. Resolution (c/mm) vs exposure (J/cm^2) same as Fig. 2 with the addition of a restriction on readout exposure.

interesting class, the "ablative" thin film materials. This class, and the metal films in particular, has been the most extensively studied. For write once and read-only applications, these materials are practically ideal. The recording sensitivity is low enough so that the materials are stable on readout and high enough so that small lasers ($\sim$10 mW) may be used. For applications requiring erasure, the thermoplastics and chalcogenide films need to be investigated.

Conclusions

After presenting the characteristics of most of the known classes of optical recording materials, we have given a specific example of how the overall system requirements may be used to define areas of material research in some detail. We feel the most useful application of this type of analysis is to help eliminate those materials from consideration which are unlikely to meet system requirements.

References

1. K. S. Pennington, *Handbook of Lasers,* Chemical Rubber Co. Press, 1971.
2. Eastman Kodak Co., Rochester, New York.
3. Scientific Emulsions, Agfa Gevaert, Teterboro, New Jersey.
4. 3-M Corporation, St. Paul, Minnesota.
5. D. Meyerhofer, RCA Review *33*, 110 (1972).
6. K. O. Hill and G. W. Jull, Fall Meeting of Optical Society of America (October 1971).
7. Kalvar Corporation, New Orleans, Louisiana.
8. Shipley Co., Newton, Massachusetts.
9. R. A. Bartolini, Appl. Opt. *13*, 129 (1974).
10. GAF Co., New York, New York.
11. R. G. Zech, J. C. Dwyer, F. Fichter, and M. Lewis, Appl. Opt. *12*, 2822 (1973).
12. B. L. Booth, Appl. Opt. *11*, 2994 (1972).
13. W. J. Tomlinson, I. P. Kaminow, E. A. Chandross, R. L. Fork, an W. T. Silfvast, Appl. Phys. Lett. *16*, 486 (1970).
14. J. M. Moran and I. P. Kaminow, Appl. Opt. *12*, 1964 (1973).
15. F. P. Laming, Polym. Eng. Sci. *11*, 421 (1971).
16. J. A. Jenny, J. Opt. Soc. Am. *60*, 1155 (1970).
17. A. Bloom, R. A. Bartolini, and D. L. Ross, Appl. Phys. *Lett. 24,* 612 (1974).
18. R. A. Bartolini, A. Bloom, and H. A. Weakliem, Appl. Opt., submitted for publication.
19. J. J. Amodei and R. S. Mezrich, Appl. Phys. Lett. *15*, 45 (1969).
20. D. Mayden, Bell Syst. Tech. J. *50*, 1761 (1971).
21. D. Chen and J. D. Zook, Proc. of IEEE *63*, 1207 (1975).
22. R. S. Braudy, U. S. Patent 3,745,586.
23. T. L. Credelle and F. W. Spong, RCA Review *33*, 206 (1972).
24. W. S. Colburn and E. N. Tompkins, Appl. Opt. *13*, 2934 (1974).
25. R.C. Duncan, RCA Review *33*, 248 (1972).
26. J. J. Amodei, *Handbook of Lasers,* Chemical Rubber Co. Press, 1971.
27. G. K. Megla, Optics and Laser Technology, 61 (April 1974); and G. K. Megla, Appl. Opt. *5*, 945 (1966).
28. D. L. Ross, Appl. Opt. *10*, 571 (1971).
29. D. L. Ross and J. Blanc, *Photochromism,* Wiley and Sons, 1971.
30. D. S. Lo, Appl. Opt. *13*, 861 (1974).
31. R. P. Guzik, Electro-opt. Sys. Design, p. 22, June 1974.
32. A. Reich and G. M. Dorian, Opt. and Electro-Opt. Inf. Proc., MIT Press (1965).
33. W. J. Tomlinson, E. A. Chandross, R. L. Fork, C. A. Pryde, and A. A. Lamola, Appl. Opt. *11*, 533 (1972).
34. R. C. Bertelson, *Photochromism,* Wiley and Sons, 1971.
35. S. Maslowski, Appl. Opt. *13*, 857 (1974).
36. A. W. Smith, Appl. Opt. *13*, 795 (1974).
37. R. S. Mezrich, Appl. Opt. *9*, 2275 (1970); and R. W. Cohen and R. S. Mezrich, RCA Review *33*, 54 (1970).
38. D. Chen, Appl. Opt. *13*, 795 (1974).
39. D. Treves, J. T. Jacobs, and E. Sawatsky, J. Appl. Phys. *46*, 2760 (1975).
40. J. T. Chang, J. F. Dillon, Jr., and U. F. Gianola, J. Appl. Phys. *36*, 1110 (1965).
41. R. E. MacDonald and J. W. Beck, J. Appl. Phys. *40, 1429* (1969).
42. P. Chandhari, J. J. Cuombo, and R. J. Gambino, Appl. Phys. Lett. *22*, 337 (1973).
43. R. K. Waring, Jr., J. Appl. Phys. *42*, 1763 (1971).
44. D. Treves, R. P. Hunt, and B. Dickey, J. Appl. Phys. *40*, 972 (1969).
45. S. E. Cummins and L. E. Cross, Appl. Phys. Lett. *10*, 14 (1967).
46. S. E. Cummins Proc. IEEE *55*, 1536 (1967).
47. J. R. Maldonado and L. K. Anderson, IEE Trans. Elec. Dev. *ED-18*, 774 (1971).
48. S. G. Lipson and P. Nielson, Appl. Opt. *13*, 2052 (1974).
49. J. J. Amodei and D. L. Staebler, RCA Review *33*, 71 (1972).
50. W. Phillips, J. J. Amodei, and D. L. Staebler, RCA Review *33*, 94 (1972).
51. D. L. Staebler and W. Phillips, Appl. Opt. *13*, 788 (1974).
52. J. B. Thaxter and M. Kestigian, Appl. Opt. *13*, 913 (1974).
53. RCA Electro-Optic Handbook, 1974.
54. R. Rose, *Vision–Human and Electronic,* Plenum Press, p. 87, 1974.

Reprinted with permission from *IEEE Journal of Quantum Electronics*, Vol. QE-14(7), pp. 487-495 (July 1978).

Antireflection Structures for Optical Recording

ALAN E. BELL AND FRED W. SPONG

Abstract—Thin-film antireflection structures find a natural application in high density optical video recording systems. We describe the design and optimization of two new antireflection structures which are suitable for use in conjunction with a wide variety of recording media, including both organic dyes and metals. Our experimental results show that by using these new designs it is possible to achieve improvements in both the recording sensitivity and the playback signal-to-noise ratio of the optical video disk.

I. Introduction

OPTICAL data storage systems which utilize a highly focused laser beam to record and instantaneously play back information [1]-[5] offer extemely high density storage with rapid random access and potential archival properties. In Fig. 1 we show a schematic of an optical recording system which is suitable for recording video information onto disk substrates. An Ar^+ laser is used as the recording source (λ = 488 nm) and an electrooptic modulator is used to intensity modulate the light with the input FM video signal. The recording optics produce a diffraction limited focused spot on the surface of the disk which rotates at 1800 r/min. Where the incident intensity of the beam exceeds the threshold for recording, the surface of the recording medium is disturbed and a series of pits is formed along a track. For playback of the video information, the modulator is given a dc bias and the laser power is reduced so that the CW intensity incident on the recorded track in the focused playback beam does not exceed the threshold for recording. Due to the presence of the pits, the light reflected from the track is modulated in intensity with the original FM video signal. This reflected beam is directed via the polarizing beam splitter and the playback optics to the photodetector, where the electrical playback signal is generated.

While there exists a wide variety of possible mechanisms by which optical recording of information may be effected [6]-[7], the most intensively investigated materials have been those in which the recording proceeds via the purely thermal process [8]-[15]. In the simplest thermal recording process, a thin coating of the recording medium is deposited directly onto the substrate and the focused recording beam is used solely as an intense and highly localized source of heat

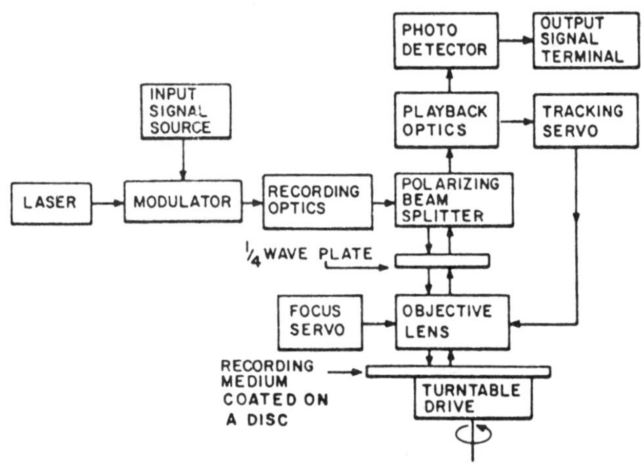

Fig. 1. Optical video disk recording system.

which elevates the temperature of the recording medium above some threshold value, wherepon a pit is formed. The pit formation process may be melting, vaporization, or some combination of these two effects. The presence of the pits causes the local reflectivity and transmissivity of the recording medium to change from its original value. In our system we have chosen the reflective mode of information retrieval since this approach readily permits the design of optical video disks on which information may be recorded independently on both surfaces.

The two most important parameters which characterize a material from the standpoint of high density optical recording are the sensitivity to recording laser beam and the signal-to-noise ratio (SNR) which may be achieved on playback of the information. For purely thermal recording processes, the threshold laser recording power is determined by the combined optical and thermal efficiencies of the recording structure. The optical efficiency is equal to that fraction of the incident light beam which is absorbed into the recording medium. The optical efficiency is therefore directly affected by the optical constants, particularly the extinction coefficient, of the recording medium as well as the design of the interference structure within which it is incorporated. The thermal efficiency is measured by the degree to which heat generated in the recording layer remains localized in the region of absorption during the pit ablation process. The thermal efficiency depends strongly on the thermal diffusivities of the recording medium and the substrate as well as

Manuscript received January 11, 1978.
The authors are with RCA Laboratories, David Sarnoff Research Center, Princeton, NJ 08540.

the duration of the ablation process. The latter is limited by the transit time for a point on the recording medium to pass through the focused laser beam.

Since the original video information is FM encoded on a carrier signal, any process which perturbs the instantaneously detected frequency (or timing) of the recorded pits along the track will represent a potential noise source in the demodulated playback video signal. There are two such noise components which are related directly to the choice of the recording medium and to the design of the recording structure. The first is the detailed nature of the recording process; i.e., melting, vaporizing, with or without an encapsulating layer [15]. The pit formation process will determine the precision with which the individual pits are machined into the recording medium [16], and any statistical variation in the physical location of the pits as they are recorded along the signal track will represent a "frozen in" or stationary type of materials-associated error. Such a noise source results in a synchronous and therefore stationary noise component in the playback signal of a single video frame which is generated by repeatedly playing out a single circular recorded track. Since the recorded information is detected optically, there must also be a photon shot noise component associated with the playback signal [7]. In order to minimize photon shot noise, the contrast between the disturbed (pit) and undisturbed portions of the track should be as high as possible, and this contrast will clearly be influenced by the interference properties of the recording structure.

Antireflection types of multilayer interference structures offer significant advantages over the conventional single layer approach [monolayer design in Fig. 2(a)] [17]. The optical efficiency of such structures can be very high since they eliminate the wasteful reflection of light at the top surface of the recording medium. In addition, if the structure is designed so that the effect of pit formation is to disrupt the antireflection condition, then the potential exists for very high signal contrast on playback.

In this paper we describe the design, optimization, and recording characteristics of two antireflection configurations which we have developed for optical recording. The first, Fig. 2(b), is a bilayer antireflection structure where a thin layer of material, which absorbs at the recording wavelength, is coated to an optimum thickness on a reflector layer. This structure is most suited to moderately light absorbing recording media such as organic dyes. The second design, Fig. 2(c), is a trilayer antireflection structure which is ideally suited to strongly absorbing recording materials such as metals. In the trilayer structure, the relatively thin absorber layer is separated from the reflector layer by a transparent dielectric layer.

II. Theory of Multilayer Structures

In order to calculate the reflection and transmission coefficients at normal incidence for a multilayer thin film stack, it is very convenient to use the matrix technique described by Heavens [18] and others [19]. At normal incidence, for a stack of $(m - 2)$ layers labeled with the convention shown in Fig. 3, the incident, reflected, and transmitted electric field amplitudes are related by the expression

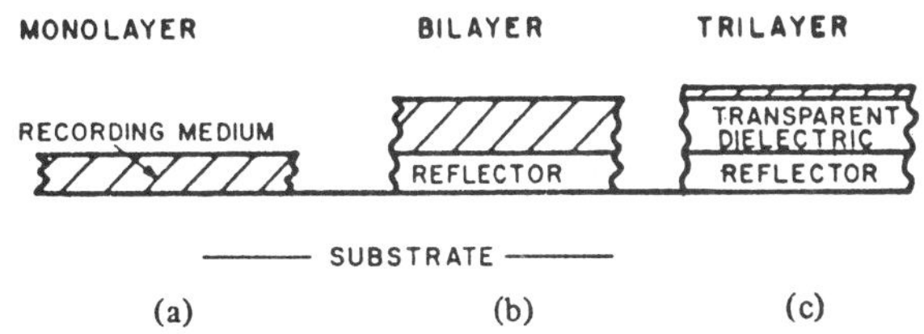

Fig. 2. Structures for optical recording.

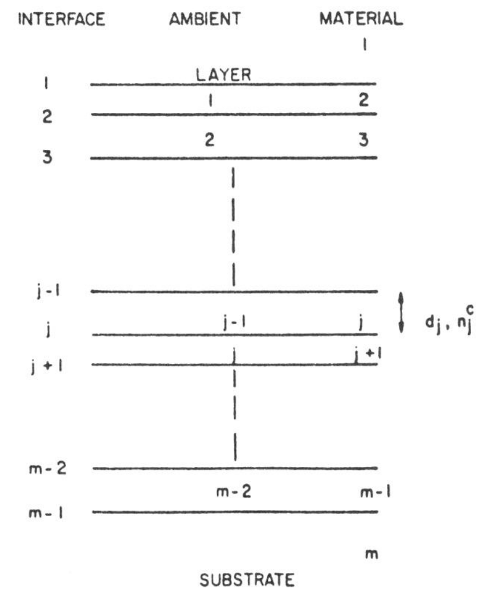

Fig. 3. Labeling convention for multilayer interference structures.

$$\begin{pmatrix} E_1^+ \\ E_1^- \end{pmatrix} = \frac{1}{t_1 t_2 \cdots t_j \cdots t_{m-1}} \begin{pmatrix} 1 & r_1 \\ r_1 & 1 \end{pmatrix} \cdot \prod_{j=2}^{m-1} \begin{pmatrix} e^{i\delta_j} & r_j e^{i\delta_j} \\ r_j e^{-i\delta_j} & e^{-i\delta_j} \end{pmatrix} \begin{pmatrix} E_m^+ \\ 0 \end{pmatrix} \tag{1}$$

where

$$\delta_j = \frac{2\pi}{\lambda} n_j^c d_j, \tag{2}$$

d_j = thickness of $(j - 1)$th layer and n_j^c is the complex refraction index of the jth material

$$n_j^c = n_j - ik_j. \tag{3}$$

r_j and t_j are, respectively, the Fresnel reflection and transmission coefficients at the boundary between the jth and the $(j + 1)$th material, i.e.,

$$r_j = \frac{n_j^c - n_{j+1}^c}{n_j^c + n_{j+1}^c} \tag{4}$$

and

$$t_j = \frac{2 n_j^c}{n_j^c + n_{j+1}^c}. \tag{5}$$

Equation (1) reduces to the form

$$\begin{pmatrix} E_1^+ \\ E_1^- \end{pmatrix} = \begin{pmatrix} a_{11} & a_{12} \\ a_{21} & a_{22} \end{pmatrix} \begin{pmatrix} E_m^+ \\ 0 \end{pmatrix} \tag{6}$$

so that the energy reflectance and transmission coefficients R and T are given by

$$R = \left|\frac{E_1^-}{E_1^+}\right|^2 = \left|\frac{a_{21}}{a_{11}}\right|^2 \tag{7}$$

$$T = \frac{n_m}{n_1}\left|\frac{E_m^+}{E_1^+}\right|^2 = \frac{n_m}{n_1}\left|\frac{1}{a_{11}}\right|^2. \tag{8}$$

By conservation of energy, the total absorption (A) in the multilayer stack is given by

$$A = 1 - R - T. \tag{9}$$

The matrix method of computation is particularly convenient since it allows the electric field to be calculated as a function of position through all the constituent layers of the stack. This allows one to calculate the Poynting vector as a function of position within the stack since [20]

$$P(z) = \text{Re}\,[E(z)\,H^*(z)]$$

and

$$H(z) = n_j^c E(z), \tag{10}$$

where $E(z)$ and $H(z)$ are the total electric and magnetic field amplitudes at point z in the jth material of the stack [$(j-1)$th layer]. With this knowledge of the spatial dependence of the energy flux, the individual absorption in each of the layers of the stack A_j can be computed, even when more than one absorbing material is present, and clearly

$$A = \sum_{j=2}^{m-1} A_j. \tag{11}$$

III. Design and Performance of Multilayer Structures

In this section, we discuss the suitability of the three structures shown in Fig. 2 for incorporating both strongly (e.g., metals) and weakly (e.g., organic dyes) absorbing recording materials with high optical efficiency and high contrast on readout.

A. Monolayer Structure

The optical characteristics of a thin film of material coated directly on a substrate depend on the optical constants of the material and the thickness of the coating. In Fig. 4, we show the calculated reflectance, transmittance, and absorbance at $\lambda = 488$ nm for a bismuth ($n^c = 1.6\text{-}i2.7$) [21] film coated onto a glass substrate ($n^c = 1.50$). We see that above 20 nm thickness there is very little further increase in the absorption, i.e., the optical efficiency, and only the reflectivity continues to increase as the transmission is reduced. At a film thickness of 25 nm, the absorption is 0.4 and the reflectivity is sufficiently high to provide good contrast on playback. Different metals will clearly optimize at different thicknesses, and our calculations show that for titanium ($n^c = 2.4\text{-}i3.3$) [22], the absorption reaches a value of 0.4 at a film thickness of 15 nm whereas in the case of rhodium ($n^c = 1.6\text{-}i4.6$) [23], the optical efficiency never exceeds 0.3, although a 10 nm thick film is sufficient to achieve this value.

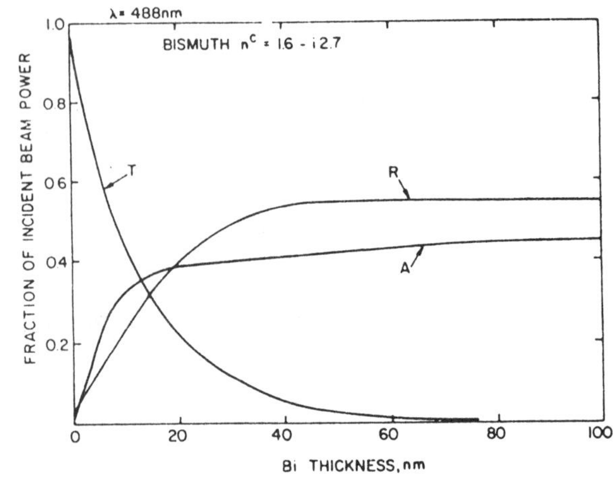

Fig. 4. Computed values of reflection, transmission, and absorption for a bismuth layer ($n^c = 1.6\text{-}i2.7$) layer on a glass substrate.

The duration of the pit formation process τ is limited by the transit time of the recording medium through the focused recording beam. For our system, having a diffraction limited spot of $1/e^2$ diameter $\sim$0.5 μm and a rotation rate of 1800 r/min, $\tau \approx 30$ ns when $r \approx 10$ cm. The thermal diffusion length $l = \sqrt{\kappa\tau}$ (where κ is the thermal diffusivity) for heat absorbed in a metallic film far exceeds the optimum film thicknesses discussed above, so that the thermal efficiency will be significantly limited by the conductive heat leak to the substrate. If $\kappa \gtrsim 0.1$ cm^{2}s^{-1} (which is the case for most metals), the thermal diffusion length becomes comparable to the beam diameter so that the radial spreading in the metal film will further reduce thermal efficiency.

In Fig. 5 we show the recording characteristics of a 20 nm layer of bismuth and a similar thickness of titanium, each coated onto glass substrates using standard vacuum techniques. In each case, the playback signal contrast $(S_1 - S_2)/(S_1 + S_2)$, where S_1 and S_2 are the light intensities reflected from the track and the pit areas, respectively, is plotted as a function of the incident laser beam power used to record the signal. The reduced melting point of bismuth results in a reduced threshold for recording compared to titanium; however, we note that in both cases, the playback video SNR is limited by the materials noise component to a value less than 40 dB (peak-to-peak signal to integrated rms noise in the 4.2 MHz video bandwidth). Microscopic examination of the recorded tracks in these metallic coatings [16] reveals recorded pits having irregular geometry resulting from the imperfect redistribution of molten material which takes place during the pit formation process.

Organic dyes represent an alternative class of light absorbing materials for use in optical recording. From the thermal point of view they appear attractive since, compared to metals, their low melting or vaporization temperature and low thermal diffusivities will work in favor of producing a low threshold for recording. In general, the absorption coefficients for organic dyes lie in the range $\alpha = 10^4\text{-}10^5$ cm^{-1} compared to $\alpha = 10^5\text{-}10^6$ cm^{-1} for most metals. The effect of the reduced absorption coefficient is shown in Fig. 6 where the computed reflectivity, absorption, and transmission are plotted versus

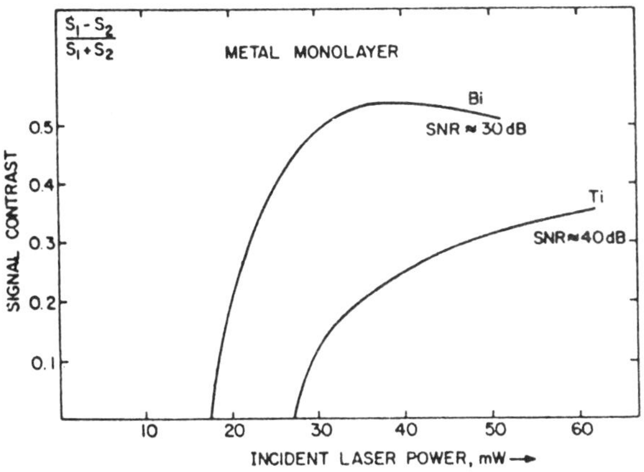

Fig. 5. Recording characteristics of 20 nm thick layers of bismuth and titanium coated onto glass substrates.

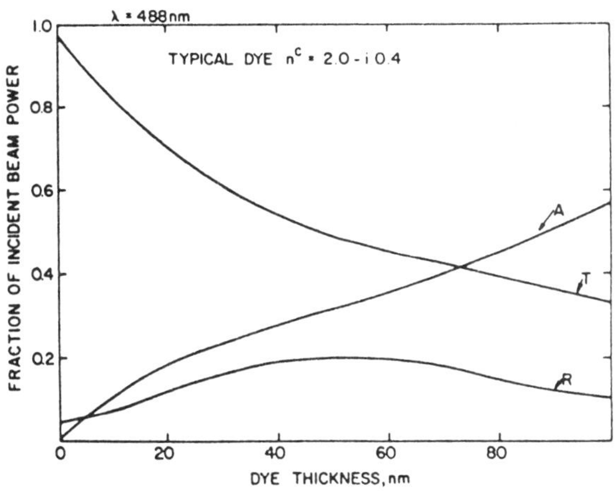

Fig. 6. Computed values of reflection, transmission, and absorption for a moderately absorbing ($n^c = 2.0\text{–}i0.4$) layer on a glass substrate.

the thickness of a fictitious material having optical constant $n^c = 2.0\text{–}i0.4$, typical of a relatively highly absorbing dye ($\alpha \approx 10^5$ cm^{-1}). We see that in the monolayer configuration, the effect of the reduced absorption coefficient is to limit both the absorption and also the reflectivity of the dye layer, thus lowering both the optical efficiency and also the contrast in playback in the reflective mode. A higher contrast can be obtained if playback is made in the transmissive mode and a relatively thick (>1000 Å) layer of dye is used, but then only one surface of the disk can be recorded upon. We conclude that the simplest monolayer approach, when used in the reflective playback mode, is really most suited to highly absorbing, generally metallic recording materials.

B. Bilayer Antireflection Structure

In the bilayer antireflection structure Fig. 2(b), we first coat the substrate with a layer which is reflective at the recording wavelength. This could be a layer of aluminum ($R \sim 0.9$ for a layer 30 nm thick) or it could equally well be a dielectric stack such as $TiO_2|MgF_2|TiO_2$ in quarter-wave thickness ($R \sim 0.8$). The absorptive recording material is then coated onto the reflective surface to approximately quarter-wave thickness, and the operation of this structure is shown schematically in Fig. 7. When the incident ray strikes the

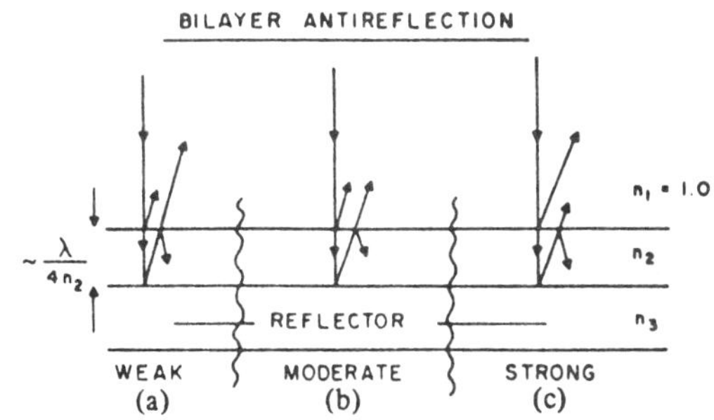

Fig. 7. Interference properties of the bilayer antireflection structure.

air/dye boundary, some fraction will be reflected, but most will be transmitted into the dye and continue to the dye/reflector interface where it is almost entirely reflected. When this reflected beam returns to the dye/air interface, it has traveled an extra $\lambda/2n_2$ path length and is out of phase with the initial reflected beam so that phase cancellation occurs, producing a minimum in the reflectivity. If the recording material is relatively weakly absorbing at the recording wavelength, Fig. 7(a), then although the phase condition is satisfied, the amplitudes differ in magnitude and the antireflection condition is only a weak minimum. For moderately absorptive material, Fig. 7(b), the phase condition is satisfied and the attenuation of the reflected wave is such that a very deep antireflection condition may be achieved. In Fig. 7(c) we see that, in the case of a strongly absorbing material, the reflectivity minimum will also be weak as a joint result of the high reflectivity at the now metal-like incident boundary coupled with the high attenuation of the beam which travels through the absorbing layer to the underlying reflective boundary. In the ideal case, where the boundary between the recording medium and the reflector is totally reflecting, all of the incident light which is not reflected from the bilayer structure must be coupled into the absorptive recording material, so that a low reflectivity implies a high optical efficiency.

The optimum value for the extinction coefficient is shown more quantitatively in Fig. 8, where we have computed the reflectivity of the bilayer structure as a function of the thickness of the absorptive layer for different values of the extinction coefficient of this layer. The real part of the refractive index of the absorptive layers is held constant so that the reflectivity minimum occurs at about the same thickness; however, the depth of the minimum, and therefore the optical efficiency, is greatest for $k = 0.5$. The computed reflectivity at the minimum together with the computed value for the absorption in the absorbing layer, is shown in Table I.

The advantage of the bilayer structure is now evident since with the moderately absorbing material, $n^c = 2.0\text{–}i0.5$, fully 0.84 of the incident beam is coupled into the recording layer at the antireflection thickness of 65 nm. One may regard the quarter-wave thick absorbing layer as a resonant cavity for light at the recording wavelength. The light entering the coating is reflected back and forth between the air/absorber interface and the absorber/reflector interface so that the reduced absorption coefficient of the material is offset by the extended path length. In the case of a dye layer, the

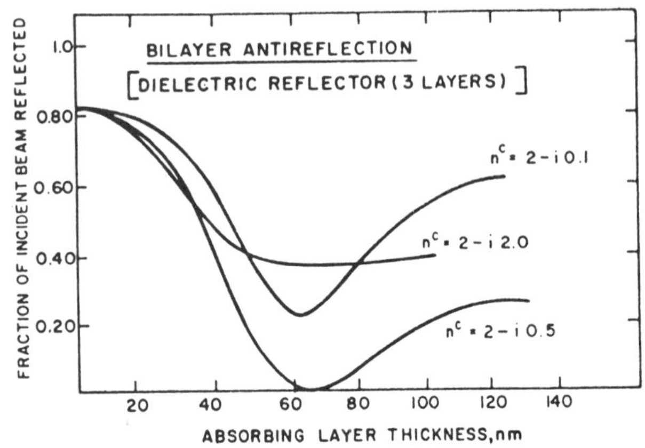

Fig. 8. Dependence of the reflectivity of the bilayer structure on the thickness of the absorber layer for layers having low, moderate, and high extinction coefficients.

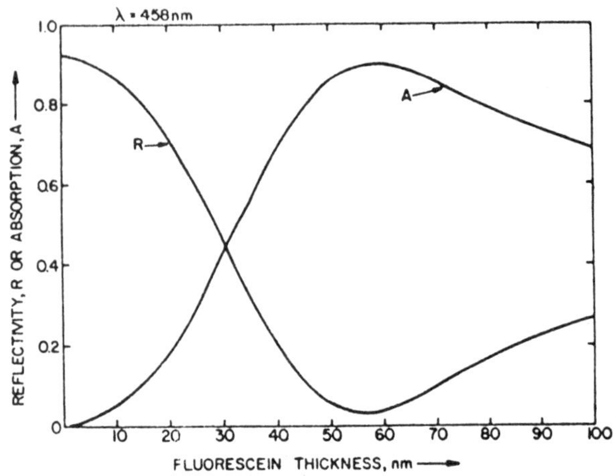

Fig. 9. Bilayer optical characteristics for a fluorescein layer deposited on an aluminum reflector layer as a function of the thickness of fluorescein.

TABLE I

BILAYER OPTIMIZATION AS A FUNCTION OF THE EXTINCTION COEFFICIENT OF THE RECORDING MATERIAL (λ = 488 nm)

n^c	R_{min}	A
2 - i0.1	0.22	0.38
2 - i0.5	0.00	0.84
2 - i2.0	0.36	0.62
2 - i3.0	0.55	0.45

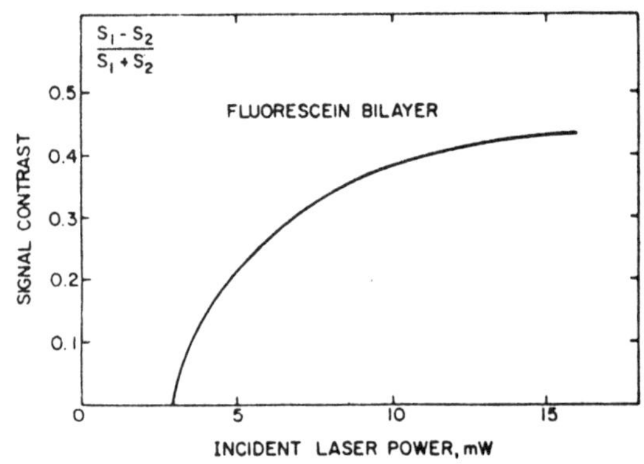

Fig. 10. Recording characteristics of a fluorescein bilayer disk.

absorbed light generates heat which melts or vaporizes the dye layer but leaves the reflector layer intact. The reflectivity of the pit is thus essentially that of the reflector layer which is exposed, and a strong positive going signal with high contrast will be obtained on playback.

The organic dye fluorescein is an example of a relatively strongly absorbing dye material which may be vacuum evaporated to produce thin amorphous and specular films. The spectral absorption is maximum near the λ = 458 nm emission line of the Ar^+ laser and we have determined the optical constants $n^c = 1.84\text{-}i0.36$ for fluorescein at this wavelength. In Fig. 9 we show the computed reflectivity and absorption at λ = 458 nm for a layer of fluorescein coated onto an aluminum reflector layer. We see that the antireflection condition is reached for a fluorescein thickness of 58 nm, where the absorption in the dye layer reaches 90 percent of the incident beam. In Fig. 10 we show the recording characteristics of this disk and note the high playback contrast as well as the considerably reduced recording threshold compared to the bismuth monolayer. The playback video SNR achieved a maximum value in excess of 45 dB, a significant improvement over the 30 dB observed in the case of the bismuth monolayer. The result is due mainly to the differing nature of the recording process, since in the case of fluorescein, the pit is formed cleanly by vaporization of the dye material whereas for bismuth a significant residue of molten material is left distributed nonuniformly and irregularly around the rim of the pit.

At this point, we have two distinct approaches available to us: 1) The monolayer, which is the simplest design but better suited to the most strongly absorbing (metallic) recording materials; and 2) The antireflection bilayer, which our experiments show is capable of producing both improved sensitivity and playback SNR when used in conjunction with the moderately absorptive organic dye materials.

Despite their inferior performance in the monolayer structures, the metals remain attractive because of their relative mechanical durability and adhesion to the substrate. Compared to most organic materials, the metals are also available in purer form and may be evaporated more rapidly to form the necessary specular layer. The low sensitivity of metallic layers is due in part to the low optical efficiency of the monolayer structure. The low SNR may be improved if the thickness of the metal layer could be reduced so that the volume of material, which is redistributed around the pit rim, is less. In the next section we describe a novel trilayer antireflection structure which allows relatively thin, highly absorbing materials to be incorporated with extremely high optical efficiency.

C. Trilayer Antireflection Structure

As we discussed in Section III-B, the failure of the bilayer configuration to achieve a strong antireflection minimum for highly absorbing materials is due to the imbalance in the amplitudes [Fig. 7(c)] of the ray reflected from the reflector layer boundary, despite the fact that the phase condition for de-

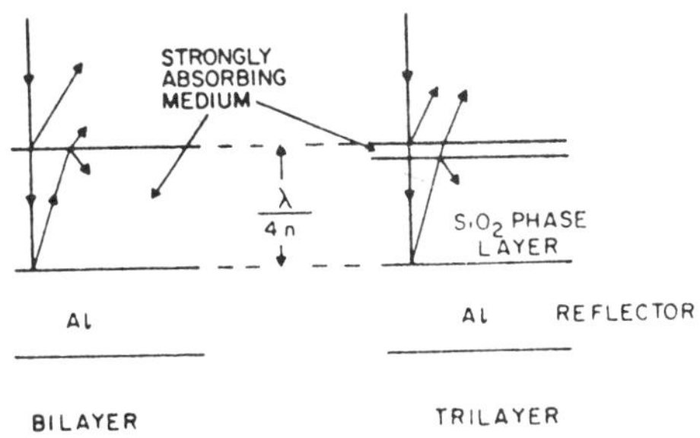

Fig. 11. Interference properties of the bilayer and the trilayer antireflection structures for strongly absorbing recording materials.

structive interference is correct. In Fig. 11 we show schematically how, by replacing some of the strongly absorbing material with a transparent dielectric layer, it is possible to reduce the attenuation of the ray which travels to the reflector boundary. The phase cancellation condition at the top surface of the trilayer is retained by appropriate choice of the thickness of the dielectric layer and a very effective antireflection condition can result. In Fig. 12 we show the variation of the reflectivity of a trilayer structure as a function of the thickness of a dielectric phase layer which has $n^c = 1.5 - i0.0$. The reflector layer is aluminum and the absorber layer is 9 nm of a material with $n^c = 2.0 - i2.0$. We observe that a very complete antireflection condition is achieved when the thickness of the dielectric layer is 70 nm. We also show the fraction of the incident beam which is absorbed into the 9 nm absorber layer. When the thickness of the dielectric layer produces the reflection minimum, the absorption is maximized and reaches 95.5 percent so that the optical efficiency of this configuration is extremely high. The contrast achieved on recording, i.e., ablation of a pit in the absorber layer, is shown in Fig. 13 where the reflectivity is plotted as a function of the absorber layer thickness. When the absorber layer is absent, the reflectivity approaches that of the reflector layer, since the presence of the transparent dielectric causes only a small reduction in reflectivity. This trilayer configuration is therefore capable of giving high signal contrast on playback, as well as extremely high optical efficiencies in conjunction with relatively thin absorbing layers.

The effect on the optical properties of the trilayer structure when the extinction coefficient of the absorber layer is varied from a high value to a moderate value is shown in Table II. In each case, the properties have been computed at $\lambda = 488$ nm for a dielectric with $n^c = 1.5 - i0.0$ and with aluminum as the reflector layer.

We see that in every instance it is possible to design a structure with both high optical efficiency and extremely low reflection; however, as the extinction coefficient is decreased, so a thicker and thicker absorber layer is required to achieve the optimum. As the thickness of the absorber layer increases, that of the dielectric layer is progressively reduced in order to maintain the phase cancellation condition at the top surface of the structure. The optical properties of the trilayer are periodic with the value of the dielectric thickness so that the values indicated in Table II represent those required for the first occurrence of the antireflection condition. Alternative dielectric thicknesses giving identical values of $R_{\min}$ and A are given by

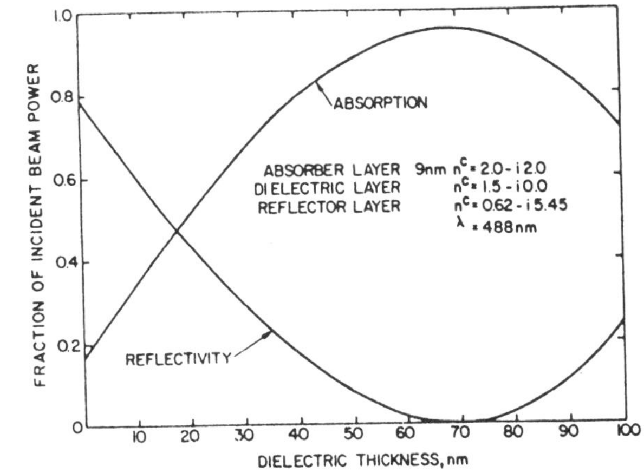

Fig. 12. Computed reflectivity and optical efficiency of a trilayer structure as a function of the thickness of the dielectric layer.

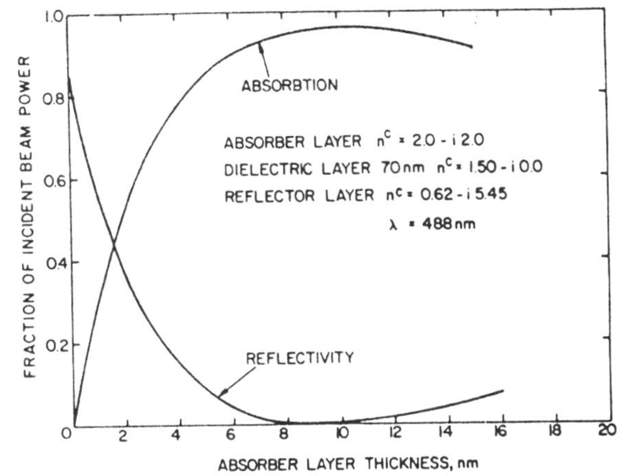

Fig. 13. Computed reflectivity and optical efficiency of a trilayer structure as a function of the thickness of the absorber layer.

TABLE II

TRILAYER OPTIMIZATION AS A FUNCTION OF THE EXTINCTION COEFFICIENT OF THE ABSORBER LAYER ($\lambda = 488$ nm)

n^c Absorber	Dielectric Thickness (nm)	Absorber Thickness (nm)	A_{max}	R_{min}
2 - i3.0	85.0	6.0	0.95	<0.01
2 - i2.0	70.0	9.0	0.96	<0.01
2 - i1.0	50.0	18.0	0.95	<0.01
2 - i0.5	0.0	54.0	0.93	<0.01

$$t_m = t_1 + (m-1)\frac{\lambda}{2n_D}, \quad m = 1, 2, \cdots \tag{12}$$

where n_D is the refractive index of the dielectric. In some instances it may be preferable to use the second- or third-order minimum since the dielectric layer thickness is then sufficient for this layer to act as a thermal barrier, reducing the heat flow between the absorber layer and the aluminum reflector, which acts as a heat sink to reduce thermal efficiency.

In Table III we compare the optical efficiency of the bilayer and the trilayer structure as a function of the extinction coefficient of the absorber. We see that for extinction coefficients in excess of $k = 0.5$ (the optimum value for the bilayer),

TABLE III
COMPARISON OF OPTIMIZED BILAYER AND TRILAYER STRUCTURES
(λ = 488 nm)

	Bilayer		Trilayer	
n^c	Absorber Thickness (nm)	A	Absorber Thickness (nm)	A
2 - i3.0	47.5	0.45	6.0	0.95
2 - i2.0	50.0	0.64	9.0	0.96
2 - i1.0	52.5	0.88	18.0	0.95
2 - i0.5	54.0	0.93	54.0	0.93

TABLE IV
TRILAYER OPTIMIZATION MAINTAINING 5 nm ABSORBER THICKNESS
(λ = 488 nm)

n^c	Dielectric Thickness (nm)	A	R_{min}
2 - i3.0	80.0	0.93	0.001
2 - i2.0	70.0	0.85	0.08
2 - i1.0	60.0	0.61	0.29
2 - i0.5	60.0	0.37	0.50

the optical efficiency of the trilayer exceeds that of the bilayer and, in addition, couples the light into a thinner absorbing layer.

In the case of recording materials (e.g., metals) where the mechanism of pit formation is essentially a melting process, the contribution to the playback noise due to fluctuations in the rim formation can be considerable. These fluctuations are reduced as the film thickness is reduced since the volume of molten material involved in the pit formation is decreased. For this reason it is desirable to use the thinnest absorber layer which is consistent with obtaining a homogeneous and continuous film. In Table IV we show the optical efficiency and minimum reflectivity of a trilayer structure when the absorbing layer is limited in thickness to 5 nm and the absorption coefficient is varied as before. For strongly absorbing materials, the 5 nm layer is sufficient to ensure high optical coupling and low reflectivity for the structure. For the two least absorbing materials, the reduction of the absorber layer thickness from the optimum value shown in Table IV has resulted in significant reduction of the optical coupling and potential playback contrast.

We have fabricated several trilayer structures where the reflector, dielectric, and absorber layer materials were chosen to be aluminum, silicon dioxide, and titanium, respectively. Titanium was chosen as the first candidate for incorporation into the trilayer structure since it is generally found to deposit as a continuous specular and amorphous layer at relatively low thickness (<10 nm). The high extinction coefficient at $\lambda = 488$ nm ($n^c = 2.3\text{-}i3.3$) makes it possible to obtain high optical efficiency in an absorber layer only 5 nm thick. Fig. 14 shows the variation of the reflectivity of the trilayer structure as a function of the silicon dioxide layer thickness for two fixed thicknesses, 2.5 and 5 nm, respectively, of the titanium layer. We note the particularly strong antireflection condition which exists in the case of the 5 nm titanium layer when the silicon dioxide thickness is set equal to 80 nm. When the structure is optimized at the antireflection condition fully 95 percent of the incident beam is absorbed by the thin titanium layer, a substantial improvement over that computed for titanium in the monolayer configuration. In Fig. 15 we show the effect of varying the titanium layer thickness on the optical properties of the trilayer structure when the silicon dioxide layer has the optimum value of 80 nm. The titanium thickness required to achieve a low reflectivity is not very critical since a range of thickness from ~3.5 to 7 nm still produces a substantially antireflective condition.

A titanium trilayer was fabricated on a glass disk substrate.

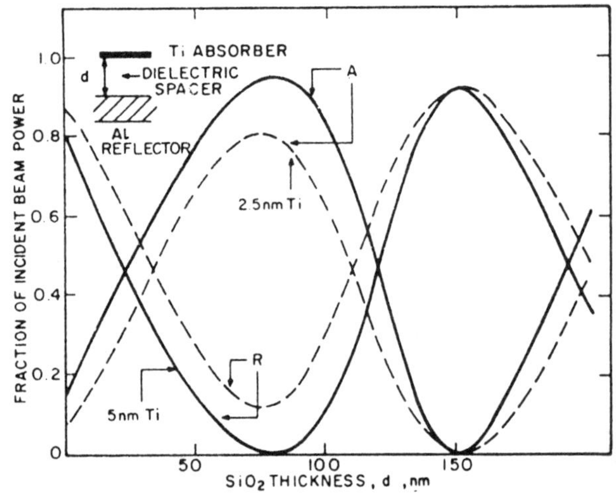

Fig. 14. Determination of the optimum thickness of the silicon dioxide phase layer for a titanium trilayer having either a 5 or a 2.5 nm titanium layer thickness. The optical efficiency in each case is also plotted.

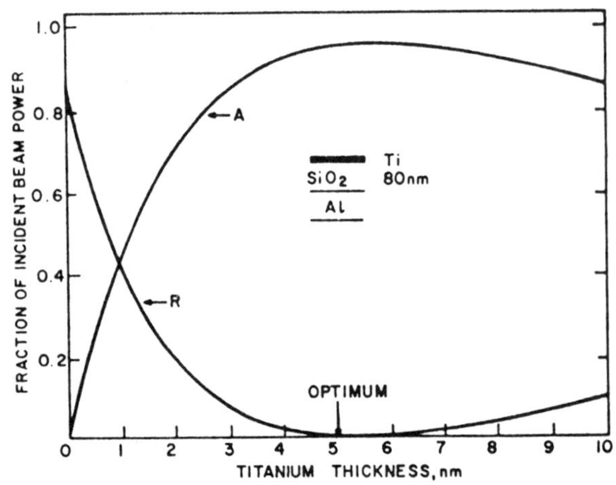

Fig. 15. Dependence of the optical properties of an optimized 5 nm titanium trilayer on variations in the thickness of the titanium layer.

The aluminum layer (~60 nm thick) was vacuum deposited using a resistance heated boat, while the silicon dioxide and titanium layers were deposited using an electron beam vacuum coating system. In order to allow for the formation of an oxide surface film on the titanium layer [22], a coating ~7.5 nm was applied and the final reflectivity of the structure at 488 nm was measured to be 3 percent. The recording characteristics of this disk are shown in Fig. 16 as a function of the incident laser power used to record the signal. The threshold for recording for titanium in the trilayer configuration is

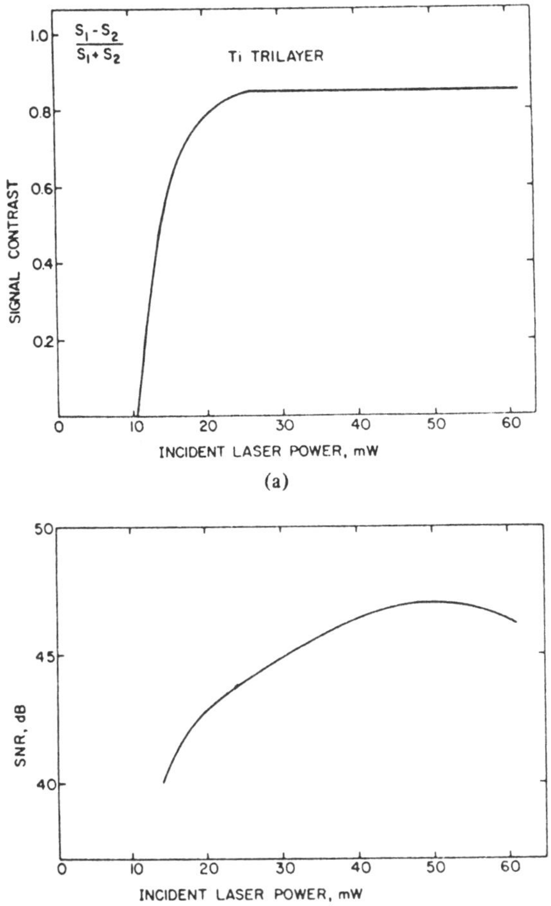

Fig. 16. Recording characteristics of a titanium trilayer. (a) Playback signal contrast versus recording laser beam power incident on the disk; and (b) playback SNR versus recording laser beam power incident on the disk.

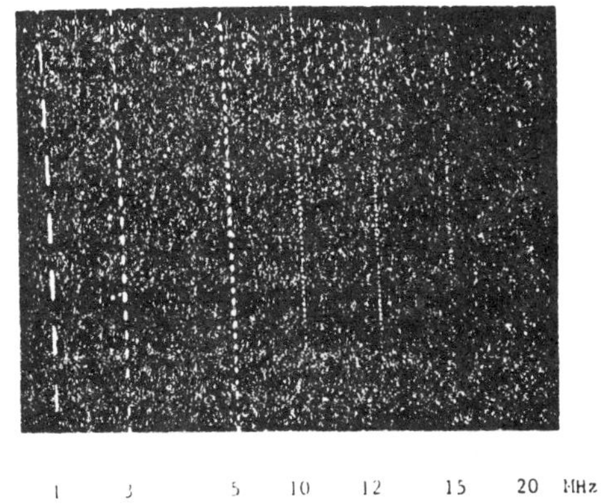

Fig. 17. Recorded pits in a titanium trilayer structure, magnification 500X. The signal frequencies range from 1 MHz to 20 MHz.

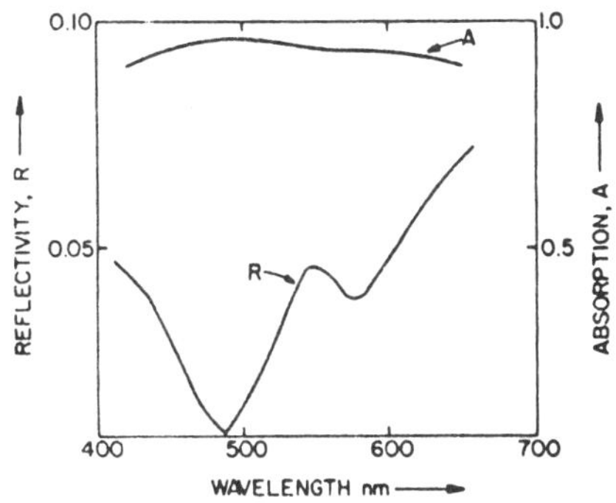

Fig. 18. Reflectivity of a titanium trilayer optimized for $\lambda = 488$ nm as a function of wavelength.

reduced considerably from that observed for the simple titanium layer shown in Fig. 5, a direct result of the improved optical efficiency of the antireflection design. Despite the high melting point of titanium (1668°C) the recording threshold in the trilayer configuration is only 10 mW of incident laser power. In Fig. 17 we show a photomicrograph (500X magnification) of tracks recorded on the trilayer disk over a range of signal frequencies from 1 to 20 MHz and we note the high signal contrast characteristic of this structure. The SNR is still limited by the materials noise associated with the fluctuations in the position of the rim of the melted pit, but the peak value of 47 dB is about 7 dB in excess of the best value achieved with a simple layer of titanium deposited directly onto a glass substrate. This improvement implies that the fluctuations in the position of the pit edges have been reduced for the titanium trilayer. This is consistent with the reduced thickness of the titanium layer, in turn reducing the volume of material melted and redistributed during the process of pit formation.

The reflectivity of a titanium trilayer optimized at $\lambda = 488$ nm is shown as a function of the wavelength in Fig. 18. We observe that although the interference properties produced the best antireflection condition at the chosen wavelength, the reflectivity remains below 10 percent across the entire visible spectrum. In fact, provided the absorber layer retains its high extinction coefficient, all that is necessary to reoptimize the structure to a new recording wavelength is to adjust the thickness of the dielectric spacer layer. The sensitivity of the recording process is then retained, except for the obvious adjustment due to the wavelength dependence of the diffraction limit on the diameter of the focused spot. A distinct advantage in this respect occurs for metallic recording media since in general their high absorption spans all of the useful laser recording wavelengths. Thus, a single material can be used in the trilayer configuration at a variety of recording wavelengths. On the other hand, the high absorption coefficient of most organic dyes occurs only over a limited spectral region so that a change in the recording laser wavelength will often require the development of a new recording medium.

IV. Conclusion

Antireflection structures find a natural application in the design of coatings for optical data storage. We have introduced and discussed two antireflection designs which can result in significant improvements in the performance of the broad classes of materials which are available for optical recording via the thermal processes of pit formation.

The bilayer antireflection structure allows effective use to be made of the low melting point and low thermal conductivity organic dyes which show only moderate absorption coefficients compared to metals. The result is an extremely low threshold configuration with high SNR on playback.

The trilayer antireflection structure allows high optical

efficiency to be achieved with extremely thin layers of strongly absorbing materials such as metals. This new design leads to a significant reduction in the laser threshold recording power as well as an increase in the playback SNR compared to the same recording medium used in the simple monolayer design.

ACKNOWLEDGMENT

The experiments were performed on an optical recording system developed by W. Barnette, R. E. Flory, R. W. Jebens, and M. Lurie. We wish to thank L. J. Levin and F. J. Tams for technical assistance in disk fabrication and R. A. Bartolini, A. Bloom, C. B. Oakley, and B. F. Williams for technical discussions.

REFERENCES

[1] D. Lou, F. Zernike, G. Kenney, A. Chan, P. Janssen, R. McFarlane, and J. Wagner, "A prototype optical disc recorder," in *Conf. Rec., 1977 IEEE/OSA Conf. Laser Engineering Applications*, Washington, DC.

[2] R. Pepperl and I. Sander, "High density direct read after write (DRAW) recording," *Opt. Acta*, vol. 24, pp. 427-431, 1977.

[3] E. E. Gray, "Laser mass memory system," *IEEE Trans. Magn.*, vol. MAG-8, pp. 416-420, 1972.

[4] H. R. Dell, "Design of a high density optical mass memory system," *Comput. Design*, pp. 49-53, Aug. 1971.

[5] K. McFarland and H. Hashiguchi, "Laser recording unit for high density permanent digital data storage," in *Conf. Rec., 1968 Fall Joint Computer Conference*, vol. 33, pp. 1369-1380.

[6] D. Chen and J. D. Zook, "An overview of optical data storage technology," *Proc. IEEE*, vol. 63, pp. 1207-1230, 1975.

[7] R. A. Bartolini, H. A. Weakliem, and B. F. Williams, "Review and analysis of optical recording media," *Opt. Eng.*, vol. 15, pp. 99-108, 1976.

[8] D. Maydan, "Micromachining and image recording on thin films by laser beams," *Bell Syst. Tech. J.*, vol. 50, pp. 1761-1789, 1971.

[9] R. J. Von Gutfeld and P. Chaudhari, "Laser writing and erasing on chalcogenide films," *J. Appl. Phys.*, vol. 43, pp. 4688-4691, 1972.

[10] A. W. Smith, "Injection laser writing on chalcogenide films," *Appl. Opt.*, vol. 13, pp. 795-798, 1974.

[11] L. N. Deryugin, V. A. Komotskii, and G. Kh. Fridman, "Characteristics of thin film metal arrays for laser beam information storage," *Sov. J. Quantum Electron.*, vol. 2, pp. 555-558, 1973.

[12] G. R. Levison and V. I. Smilga, "Mechanism of the damage of thin metal films by focused laser radiation," *Sov. J. Quantum Electron.*, vol. 3, pp. 220-223, 1973.

[13] D. B. Congleton, M. R. Smith, and A. S. Diamond, "A comparison of plastic vs metal coatings for real time laser recording," *J. Appl. Photographic Eng.*, vol. 3, pp. 97-105, 1977.

[14] J. Corcoran and H. Ferrier, "Melting holes in metal films for real time high density permanent digital data storage," in *Conf. Rec., SPIE 21st Int. Tech. Symp. Instrument Display*, San Diego, CA, Aug. 1977.

[15] H. J. Tolle and R. Memming, "Fast imaging process in PbI_2 films," *Appl. Phys. Lett.*, vol. 26, pp. 349-351, 1975.

[16] A. E. Bell and W. J. Burke, to be published.

[17] A. E. Bell, R. A. Bartolini, A. Bloom, and F. Spong, "Optical and thermal aspects of the design of laser recording structures," Invited Paper, presented at the Optical Soc. America 1977 Annu. Meeting, Toronto, Ont., Canada.

[18] O. S. Heavens, *Optical Properties of Thin Solid Films*. New York: Dover, 1955.

[19] Z. Knittl, *Optics of Thin Films*. New York: Wiley, 1976.

[20] M. Born and E. Wolf, *Principles of Optics*. New York: Pergamon, 1956.

[21] R. L. Aagard, "Optical constants of the constituents of manganese-bismuth films," *J. Opt. Soc. Amer.*, vol. 64, pp. 1456-1458, 1974.

[22] G. Haas and A. P. Bradford, "Optical properties and oxidation of evaporated titanium films," *J. Opt. Soc. Amer.*, vol. 47, pp. 125-129, 1957.

[23] G. A. Bolotin and T. P. Chukina, *Opt. Spectra*, vol. 23, pp. 333-335, 1967.

Reprinted with permission from *IEEE Spectrum*, pp. 20-28 (August 1978).

Optical disk systems emerge

Laser-powered recording systems with 'instant' playback could upstage magnetic tape as a medium for high-speed mass data storage

The needs of television broadcasters and other communications professionals to store and then rapidly retrieve extremely large quantities of video information (e.g., text or moving pictures) are growing rapidly. However, it now appears that further large improvements in magnetic recording will be increasingly difficult to obtain. An alternative high-speed, mass-storage medium is required that is economical, compact, and archival. Random-access capability will also be necessary.

A variety of different systems have been proposed to meet this challenge. Optical disk recording is attractive because it allows instantaneous playback, very fast random access, high recording density (better than magnetic recording), and archival storage. And a particularly promising technique is the optical disk system that uses laser recording to store encoded information at very high data rates with extremely high density. "Playback only" optical disk storage systems are available now and have been reviewed in detail elsewhere. To be described here are those optical disk systems under ongoing development for professional use that permit both recording and immediate playback of the stored information.

When presented as a functional block diagram (Fig. 1), the essential components and basic operation of an optical disk system can be quickly identified and understood. The actual hardware implementations, however, involve many intricate techniques to overcome focus errors, tracking errors, and signal-to-noise problems. Meanwhile, despite the fact that recording materials (disks) have been developed that offer some of the desirable features, ongoing laboratory experiments are bringing further improvements in the tradeoffs between features. Recording materials represent the key component of any data storage/retrieval system, and particular emphasis will be given here to the various types that are available.

Independent developments show similarities

The design of an acceptable video disk record/playback system is (mostly) well-defined by either the desired performance or fundamental physical limitations. It is not surprising then that systems developed independently at various locations, in several countries, show many similarities. The concept of a video disk system (Fig. 1) is relatively simple. A bright beam of laser light is focused onto a spinning disk to make a recording. The recorded information is retrieved by illuminating the disk with a somewhat less intense beam and then detecting the light reflected back from the disk. Information is recorded by turning the recording beam on and off at different rates, producing pit marks of varying lengths along a spiral or circular track on the disk. During playback, these marks will alter the amplitude of the reflected beam for different periods of time, producing a frequency-modulated (FM) signal. Other forms of encoding will produce different characteristic tracks.

Again referring to Fig. 1, note that the upper channel, labeled "record/play," contains all the essential optical components. In some systems, it is necessary to read two tracks simultaneously, thus requiring an extra "play only" channel that is identical to the "record/play" channel, except that it lacks a modulator (not needed for playback). With two play channels available, one can be playing while the other acquires a new track; this permits rapid electronic switching between signals after the electromechanical acquisition process is complete.

Making and playing an optical recording

Optical recording and playback begin with the laser output. The light is first divided, with most of it going to the channel that records. It then goes to a modulator (e.g., an electrooptic Pockels cell), which modulates the beam by rotating its polarization according to an applied voltage, and then passing it through a polarization analyzer. The attenuator following the modulator is used to reduce the light intensity during playback. (The modulator could also perform this function, but because of the potential for accidental erasure a passive, stable attenuator seems the prudent choice.)

Next, the beam is expanded to fill a large-aperture focusing lens, and then passes through a polarizing beam splitter and a quarter-wave plate. These two components transmit nearly all the incident light toward the disk, while directing nearly all the reflected light toward the detector. A "tracking" mirror is usually mounted on a galvanometer so that it can be deflected to follow an out-of-round track, or skip to a nearby track quickly.

Finally, the laser beam is concentrated onto the disk surface by the focusing lens (typically a 0.40 to 0.85 numerical aperture microscope objective). An actuator moves the lens perpendicular to the disk, maintaining the proper focal distance at all times. The laser light hitting the disk either records a track or is modulated, on playback, by the existing track and is reflected back along an optical path to the detector.

As an example, consider the method for optically recording a television picture. Details of the disk format are largely determined by the particular television system involved (NTSC, PAL, etc.). Recording an integral number of fields in each revolution of the disk simplifies implementation of such playback features as freeze-frame and fast or slow motion. The recording of one NTSC field (1/60 second) per revolution results in a disk speed of 3600 r/min. At that speed, it is very easy to achieve a wide

R. A. Bartolini, A. E. Bell, R. E. Flory, M. Lurie, F. W. Spong RCA Laboratories bandwidth and low video noise level. But a better compromise between bandwidth and recording time is obtained at half that speed (half the bandwidth and double the playing time). Most optical video disks described in the literature operate at two fields per revolution: 1800 r/min for NTSC, 1500 r/min for PAL. Disk speed can be further reduced to 900 r/min (or with special techniques, even to 450 r/min), and it will still reproduce a television image.

At 1800 r/min, the velocity along a track at an average radius of 10 cm is about 20 m/s. Using only intuitive calculations, we would expect a 0.5-μm-diameter spot of light just to produce (or resolve) a 0.5-μm-long mark on the disk, resulting in a cutoff frequency of 40 MHz with a useful upper frequency limit of about 20 MHz. The half-power diameter of the focused spot is more important in determining the maximum resolution than the null, or Airy pattern, diameter. It is not difficult to focus a coherent beam to a half-power diameter of about 0.5 μm. For example, consider a standard 40X microscope objective with numerical aperture of 0.65 using a wavelength of 488 nm. Diffraction theory shows that the half-power diameter ($d^{1/2}$) of the focused spot will be 0.4 μm, and the diameter of the first null (the Airy pattern dA) will be 0.9 μm.

The resulting track will be no more than 0.9 μm wide. Separation of the adjacent tracks by approximately the same distance results in a pitch of about 2 μm per track. With a maximum radius of 15 cm and a minimum radius of 5 cm, there is a useful recording band of 10 cm or 50 000 tracks. At 1800 tracks per minute, the disk will record or play continuously for 28 minutes. Thus we see that an 1800-r/min turntable, a standard microscope objective, a laser light source with modulator, and a 30-cm-diameter disk should be able to record a signal with 20-MHz bandwidth long enough to contain any typical "half-hour" TV program (sans commercials).

Overcoming mechanical errors

Several difficulties arise when implementing this type of video disk system. They are largely mechanical. As the disk spins, its surface will move axially in and out of focus, and recorded tracks will move back and forth radially across the lens axis, unless both the disk and its turntable are extraordinarily precise. Rather than maintain such precision, it is preferable to follow the disk's movements with servos (one for focusing and one for tracking). Mentioned here are just a few of the many focus and tracking servo systems that have been tried.

The depth of focus (range of focus distances that maintain a spot of acceptable size) for the resolution required is about $\pm$ 0.5 μm. The focus servo must respond to errors of that size, and move the lens to correct them. Suitably accurate mechanical spacing can be maintained by supporting the lens on an air film, or air bearing, over the disk, or conversely, by supporting a flexible disk on air over the lens. Both processes are commonly used in magnetic disk recording, but greater accuracy is required for the optical disk system. The air-support method seems limited to disks that are relatively flat or that are very flexible and can be flattened by the air bearings. Larger disk motions can be followed by moving the lens with a motor capable of producing the required high acceleration.

With an optical video disk turning at 1800 r/min, the surface undulations generate error frequencies that are multiples of 30 Hz. Since these frequencies are in the audio range, it is very effective to move the lens with a loudspeaker-type voice coil. A coil is wound around the lens and the entire unit is placed inside a permanent magnet. Current applied to the coil moves the lens just as it moves the cone in a loudspeaker. This is the most common type of focus servo. One laboratory has explored the use of a linear motor that can follow larger low-frequency excursions.

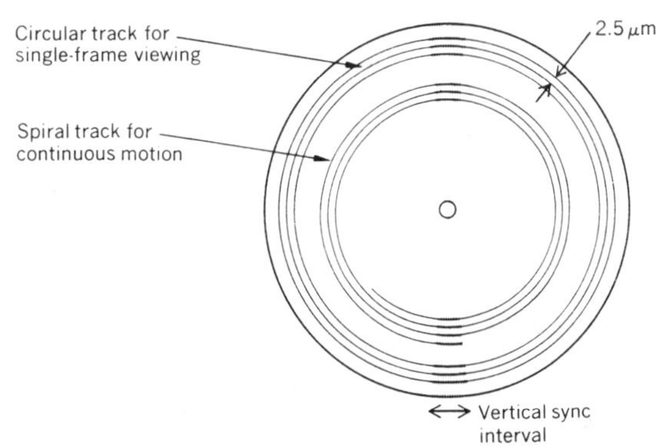

[1] In an optical video disk system, laser light both records and plays back from a single disk—without interim processing steps. If two "play" channels are available, rapid electronic switching between recovered signals is possible after the electromechanical acquisition process is complete.

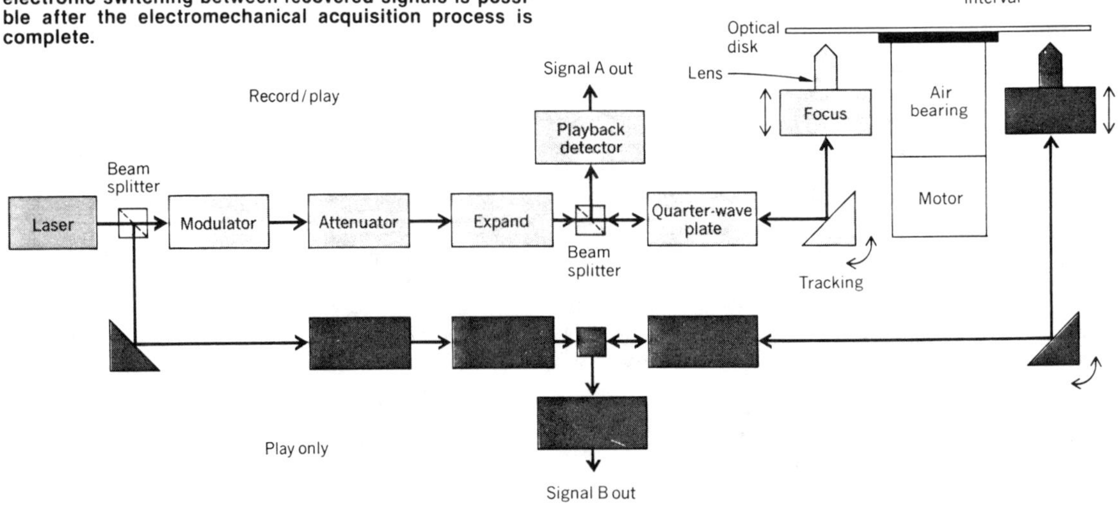

Focus errors can be sensed by measuring the distance etween the lens and the disk, for example, with a apacitance probe. But any method that measures iechanical separation to correct focus can cause prob-ms if a protective layer with nonuniform properties lies etween the disk surface and the lens, and this layer may ot provide a constant optical path length. When a pro-ctive layer is used, it is better to sense focus optically to erive correction signals for the focus adjustment. Several ptical methods have been described for this purpose. hey detect any change in size, shape, or position as the oot moves in and out of proper focus.

During playback, miscentering, bearing looseness, and sk distortion all cause the track to move across the field f the lens. This motion can be followed by deflecting the ot with a rotating mirror. Such a mirror, several illimeters in diameter and mounted on a galvanometer, in move a few degrees at 30 Hz, and through smaller igles at much higher frequencies. This is enough to llow miscentering errors spanning many track widths, id also permits rapid jumping across several tracks for calized high-speed random access.

Various approaches to sensing tracking errors have en tried. One uses three spots, two of which straddle the formation track so an increase in signal returned from ther spot shows that the track has moved. The third ot, centered between the other two, stays "on track," id reads the recorded information.

Another method uses a single spot that is "dithered" ick and forth across the information track at a speed uch higher than that of any expected tracking errors. ie phase of the resulting amplitude modulation (AM) of e detected signal indicates the tracking error. The AM is sily removed during limiting and FM decoding. The ther can be accomplished by superimposing a low-nplitude vibration of a few kilohertz on the tracking lvanometer, or with a second galvanometer or ezoelectric deflector designed specifically for high-equency use.

Though not usually considered a "component," dirt is navoidable in every optical video disk system. For exam-le, particles just a few micrometers in diameter can pre-nt the recording of many signal elements, or obscure em after recording. In either case, the signal is lost. epending on one's point of view, the problem may be onsidered mechanical or optical. However, the solution volves both factors. It is not reasonable to keep a disk rface perfectly free of all contamination—or to clean it peatedly. The most plausible solution is to manufacture e active disk surface under the cleanest possible condi-ons, and then immediately cover this sensitive surface ith a transparent layer. Recording and playing take ace through the protective layer, so dirt clinging to the sk exterior is always out of focus. Two approaches to oviding this protective layer have been described. In the hillips DRAW system, two recording surfaces are sembled into a sandwich, with their substrates facing it and an airspace in between (Fig. 2A). Thus, the bstrate itself becomes the protective layer. RCA coats e recording surface with a transparent material (Fig.

I. Comparison of candidate optical recording materials/media

Medium	Processing	Recording Process	Recording Wavelength Range, A	Recording Sensitivity, J/cm^2	Nominal Resolution, c/mm	Archival Storage Time	Readout Process	Recyclable?	Replication?
Photographic	Wet chemical or heat	Photochemical change	Panchromatic or orthochromatic	$10^{-8}-10^{-1}$	100–3000	Permanent	Density or index change	No	Yes
Photoresists	Wet chemical or heat	Absorption and cross-linking	Ultraviolet to 5500	10^{-2}	1000–1500	Permanent	Surface relief	No	Yes
Photopolymer	None or post-exposure and postheating	Absorption and cross-linking	Ultraviolet to visible	$10^{-3}-100$	1500–5000	Permanent	Index change	No	?
Ablative thin films	None	Ablation	Visible	$\sim 10^{-2}$	1000–1500	Permanent	Light amplitude	No	Yes
Thermoplastic	Preexposure corona charge and heat	Electric field produced deformation of heated material	Up to 6500	10^{-4}	2000	Permanent	Surface relief	Yes	Yes
Photochromic	None	Light induces new absorption bands	Entire visible range—most types most sensitive in ultraviolet or blue	$3\times10^{-3}-1$	300–>2000	Minutes to months	Optical density change	Yes	Yes
Chalcogenide films	None	Heat causes switching between amorphous and crystalline states	Visible to infrared	$10^{-2}-10^{-3}$	1000	Permanent	Different phases have different absorption and reflectivity	Yes	?
Magnetooptic	None	Thermomagnetic	Visible to infrared	10^{-2}	100–2000	Permanent	Faraday or Kerr effect	Yes	?
Photoferroelectric	None	Electric field- and light-induced birefrigence change	Visible	10^{-2}	125–800	Indefinite	Change in state of polarization of light in transmission or reflection	Yes	No
Photoconductive/Electrooptic	None	Photo-induced internal electric field redistribution	3500–6000	$5\times10^{-6}-10^{-4}$	1000	Hours	Change in retardation induces change in polarization of light transmitted	Yes	No
Electrooptic	None (possibly heat to fix)	Light-induced spatial modulation of refractive index	3500–5000	$10^{-1}-1$	1500	Hours to indefinite	Diffraction of light	Yes	No

2B). In both cases it is necessary to use optics designed for the appropriate transparent layer.

Accounting for noise

Even with all components idealized, including a defect-free recording surface, there will still be noise resulting from dark current and thermal effects in the light detector/amplifier, as well as shot noise due to photon "randomness" in the light beam. Thus, before going further, the concept of noise should be carefully defined. For instance, "carrier" refers to what is read from the disk, to distinguish it from "signal," which refers to the decoded information. The noise is best described as carrier-to-noise ratio (CNR) over the applicable bandwidth, and is defined as the root mean square (rms) carrier amplitude with no signal or noise divided by the rms noise amplitude with no signal. A carrier bandwidth of 20 MHz will be required here. Twenty times the log of this expression yields CNR in decibels.

To achieve a CNR greater than 40 dB, the light level must be so high that dark current can be neglected. At the lowest useful light levels, thermal noise dominates. With more light, thermal noise becomes less than shot noise. Of course, this is all based on maintaining a detector bandwidth of at least 20 MHz—basically a matter of placing a low load resistance across the detector so that the *RC* time constant formed with the detector capacitance is short compared with the frequencies being detected. Unfortunately, a low load resistance increases the effective thermal noise. Avalanche diode detectors, with a large internal gain, greatly reduce the effect of thermal noise.

Figure 3 shows the CNR obtainable from several detectors, each using a bandwidth of 20 MHz and typical values for the other parameters. At a detector light level of 250 μW or more, the CNR due to shot and thermal noise will exceed 59 dB. Notice that at such a high light level PIN diode detectors are superior to the avalanche

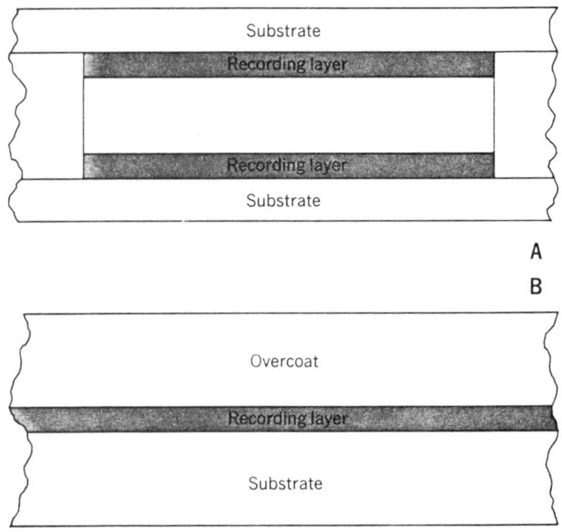

[2] Recording through a protective layer means dirt is out of the focal plane. Two approaches to providing this feature have been described. The Philips DRAW system (A) has two recording surfaces sandwiched together with substrates facing out. RCA simply coats the back recording surface with transparent material (B).

Experimental optical video disk apparatus can record studio-quality television in real time and play it back instantly. RCA demonstrated this equipment in 1976 at the National Association of Broadcasters Convention.

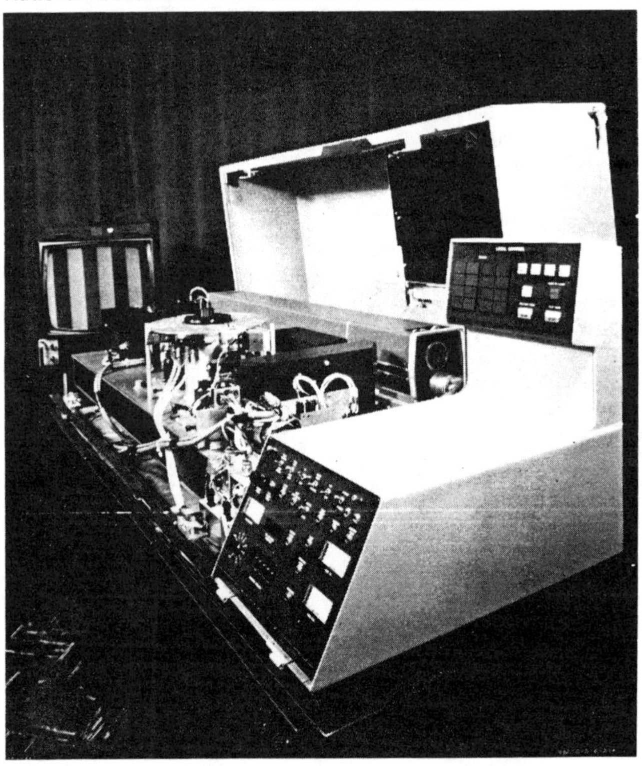

diode, because the gain mechanism in the avalanche diode is itself noisier than a good amplifier. At an input of approximately 25 μW the avalanche detector is superior, giving a CNR of 49 dB. As the noise is nearly all shot noise under these conditions, the value of the load resistor does not affect the CNR.

Optical path efficiency is typically 15 percent in either direction, and maximum record reflectivity is about 80 percent, so delivering 25 μW to the detector requires 208 μW at the record—which, in turn, requires 1.4 mW from the laser (a reasonable value). With a larger laser, and a recording medium that will not be damaged by reading with more than 208 μW, the CNR can be increased to more than 49 dB.

A typical FM format with preemphasis provides a video signal with a signal-to-noise ratio (SNR) about 10 dB greater than the CNR. For standard NTSC video signals, SNR is usually the peak-to-peak signal divided by the rms noise over a 4.2-MHz video bandwidth—which is the definition used here

Video disk apparatus has been constructed using many of these principles. Such systems can record studio-quality television in real time, and play it back instantly. A succession of stills or movies can be automatically accessed, then brought on the air without interruption. The equipment illustrated on page 23 was demonstrated by RCA at the 1976 National Association of Broadcasters Convention.

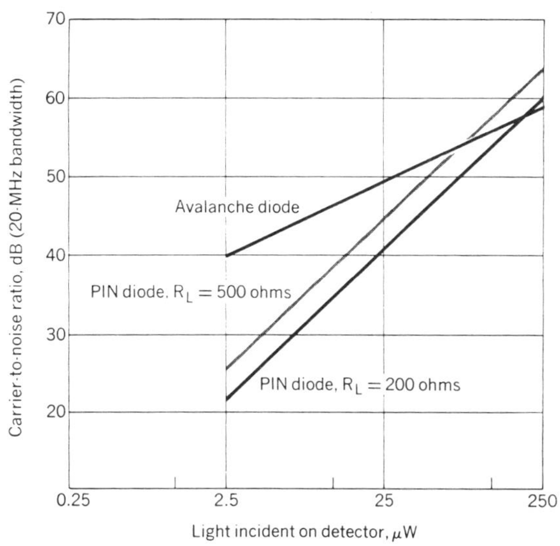

[3] PIN or avalanche diodes may be used as optical disk readout detectors; however, avalanche diode detectors, with a large internal gain, greatly reduce the effect of thermal noise. (Here bandwidth = 20 MHz; avalanche gain = 100; avalanche noise figure = 10; amplifier noise figure = 2. Over the range shown, load resistance does not affect signal-to-noise ratio for the avalanche diode.)

Recording material: the key component

Many classes of materials may undergo a change in their optical properties when absorbing light from a recording laser beam. Table I lists 11 such material categories, together with those properties that make direct record and playback possible. In some cases (photographic, photoresist, or photochromic), recording is actually a quantum process. In other instances (ablative thin films, magnetooptic materials, or chalcogenides), the recording mechanism is essentially thermal, using the absorbed energy from the recording laser beam simply as a highly localized, intense heat source that precipitates melting or vaporization, a change in magnetization, or a change in the degree of crystallinity respectively.

Since each category of Table I contains many candidates, the task of testing and evaluation in order to identify the optimum material may appear to be endless. But, by considering the overall system requirements, it is possible to eliminate *a priori* many obviously unsuitable materials. As a first step in this process, the various categories of materials are compared by plotting material resolution versus material sensitivity (Fig. 4). Acceptable material must allow the recording of 1-μm (or smaller) areas to satisfy the system requirement for high packing density. Materials with lesser resolution should not be considered further.

The exposure sensitivity of the recording medium is another very important parameter and should permit the use of moderate- to low-power recording lasers. On the other hand, the material must be sufficiently insensitive to allow repeated playback of the recorded information without causing deterioration. For a 1-μm spot recorded

[4] Candidate materials for optical disk recording are located within the unshaded "window." The open area summarizes requirements for high resolution coupled with a lower bound on exposure sensitivity due to shot-noise considerations.

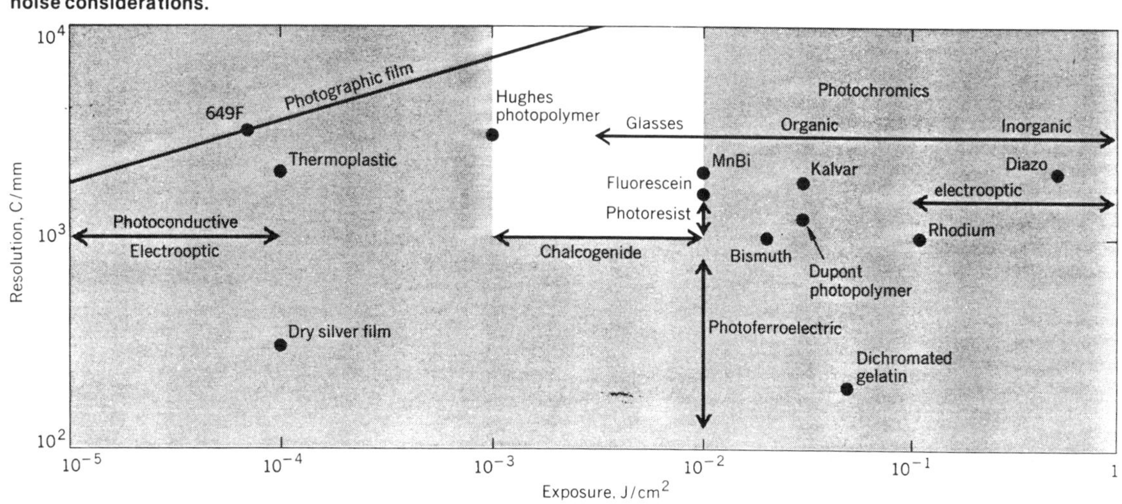

For further reading

- Bartolini, R.A., *et al.*, "In-contact encapsulation of a dye-based optical video disc," Conference on Laser and Electrooptical Systems (CLEOS), San Diego, Calif., Feb. 1978.
- Bell, A. E., *et al.*, "Fully encapsulated trilayer optical video disc," CLEOS, San Diego, Calif., Feb. 1978.
- Bell, A. E., *et al.*, "Optical and thermal aspects of the design of laser recording structures," Optical Society of America Annual Meeting, Toronto, Ont., Canada, Oct. 1977.
- Bell, A. E., and Spong, F. W., "Antireflection structures for optical recording," *J. Quantum Electronics*, vol. 14, p. 487, July 1978.
- Berger, K., and Hamisch, H. J., "Laser picture storage system for professional applications," *Femseh Aktuell*, p. 10, Feb. 1977.
- Bouwhuis, G., and Bergstede, P., "The optical scanning system of the Phillips VLP record player," *Phillips Tech. Rev.*, vol. 33, p. 186, 1973.
- Broadbent, K., "A review of the MCA Disco Vision system," J. SMPTE., vol. 83, p. 554, July 1974.
- Congleton, D. R., *et al.*, "A comparison of plastic vs. metal coatings for real time laser recording," *J. Appl. Photographic Eng.*, vol. 3, pp. 97–105, 1977.
- Corcoran, J., and Ferrier, H., "Melting holes in metal films for real time high density permanent digital data storage," SPIE 21st International Tech. Symposium and Instrument Display, San Diego, Calif., 1977.
- Firester, A. H., *et al.*, "Optical recording of the RCA VideoDisc," *RCA Rev.*, vol. 39, Sept. 1978. (This article refers to optical mastering of the RCA consumer video disk.)
- Gray, E. E., "Laser mass memory system," *IEEE Trans. Magnetics*, vol. MAG-8, pp. 416–420, 1971.
- Harris, A. L., "Continuous wave laser recording on metallic thin film," *Image Technology*, pp. 31–35, Apr./May 1970.
- Janssen, P. J. M., and Hay, P. E., "Control mechanisms in the Phillips VLP record player," *Phillips Tech. Rev.*, vol. 33, p. 190, 1973.
- Kenney, G., *et al.*, "Experimental optical disc data recorder," CLEOS, San Diego, Calif., Feb. 1978.
- Le Carvennec, F., "Professional applications of optical video disc technology," CLEOS, San Diego, Calif., Feb. 1978.
- Lurie, M., and Flory, R. E., "Developing a broadcast quality optical video disc system," CLEOS, San Diego, Calif., Feb. 1978.
- Maydan, D., "Micromatching and image recording on thin film by laser beams," *Bell System Tech. J.*, vol. 50, pp. 1761–1789, 1971.
- Naruse, Y., and Ohkoski, A., "A new optical video disc system with one hour playing time," IEEE Chicago Spring Conference on Consumer Electronics, June 1978.
- Palermo, P., *et al.*, "Video disc mastering and replication," *Optics and Laser Technology*, p. 169, Aug. 1977.
- Pepper, R., and Sander, I, "High density direct read and write (DRAW) recording," *Optica Acta*, vol. 24, p. 427, 1977.
- Spitz, E., "Optical video discs," Conference on Laser Engineering and Applications (CLEA), Washington, D.C., May 1975.

at a 10-MHz rate an exposure of about 10^{-2} J/cm^2 will be achieved if 1 mW of power is absorbed by the recording material. Allowing for reflection losses at the surface of the recording medium and the relatively low optical path efficiencies throughout the system, the 1-mW absorbed power will require 10–100 mW of laser output power. Lasers that operate in this range are commonly available and include HeNe, HeCd, and Ar+ ion types.

Once the information track has been recorded, it is played back by monitoring the modulation of either a reflected or a transmitted laser beam. In either case, since quanta are being detected, the fundamental statistics of counting must be considered as a source of noise. In fact, since the statistics of detection improve directly as the detected quantum flux increases, a certain minimum detected playback light level must be used to achieve a particular CNR in the playback signal. If this minimum playback beam intensity corresponds to an exposure greater than that required to effect a recording, the material is not useful. The interesting conclusion is that some materials can actually be too sensitive for a particular optical recording system.

As mentioned earlier, a playback beam power of about 200 μW incident on the disk limits the shot noise to an acceptable value. This corresponds to an exposure level of 10^{-3} J/cm^2 during playback, and places a lower bound on the exposure sensitivity of the recording material. Taken together with the requirements for high resolution and practical laser output power for recording, only the materials located within the unshaded "window" of Fig. 4 remain as candidates. However, this window still includes many different material categories. Several of the choices are inappropriate, and finer selection is called for.

Verification and rapid correction of recorded information require that the recording material need no processing before playback—which eliminates photographic and

[5] Antireflection structures in bilayer (B) and trilayer (C) configurations have been developed to increase the fraction of the recording beam that is absorbed by the recording medium. The multilayer designs also yield better contrast than is possible with a simple single layer (A).

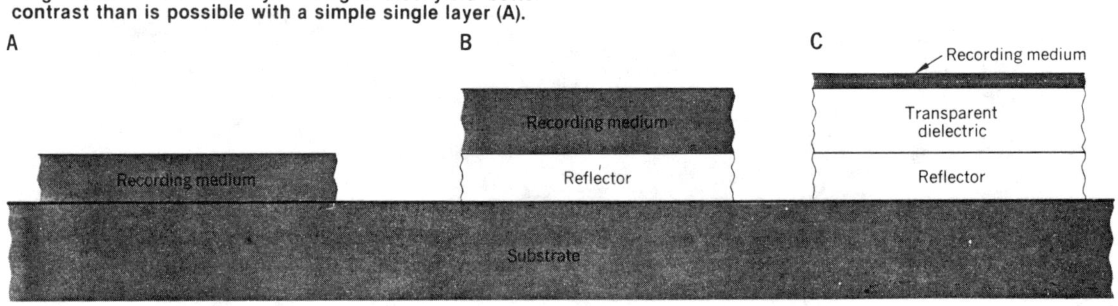

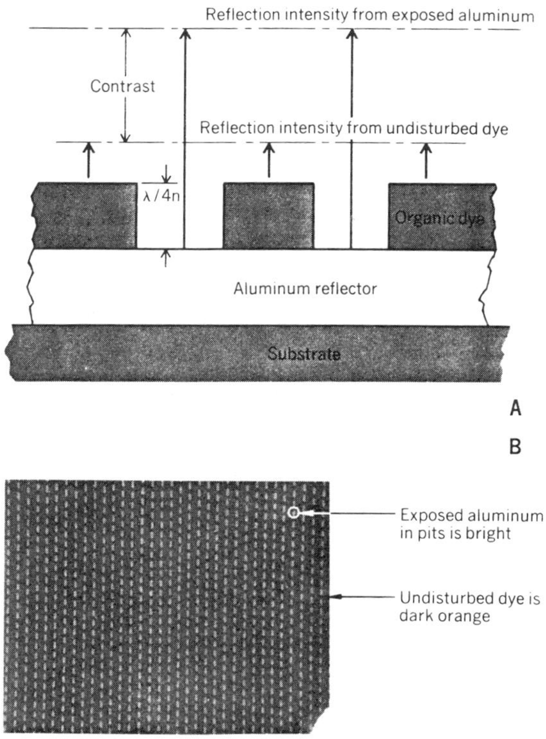

[6] Optimized for laser light at λ = 488 nm (A), characteristically orange organic dye in a bilayer structure (B) has a recording threshold of about 4-nW incident power.

[7] Almost 95 percent of the incident beam is coupled into a 5-nm titanium film in this trilayer structure (A). Though the structure is optimized for recording at λ = 488 nm, titanium is strongly light absorbing in all the visible spectrum (B).

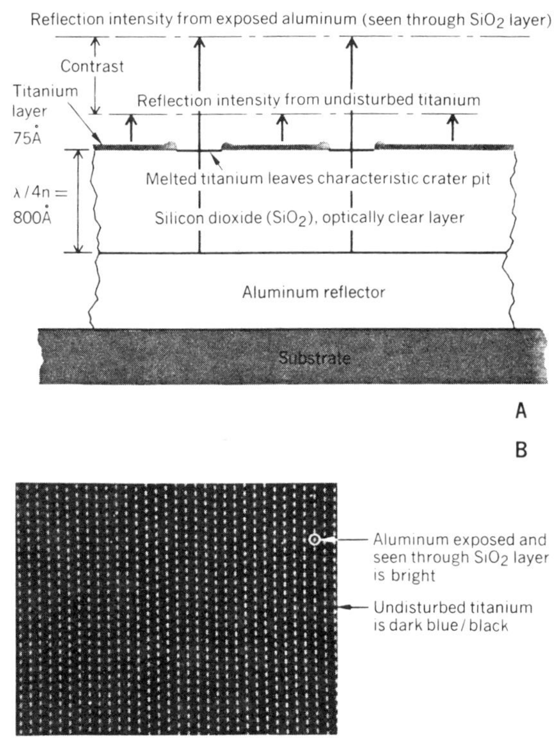

photoresist materials. And if a prolonged, repetitive readout is required, successive scans must not cause cumulative degradation of the recorded information. This implies that the recording process should possess a well-defined exposure threshold, below which no change occurs in the recording medium. For quantum types of recording processes such a threshold does not exist. From this perspective, many of the material categories in Table I are unattractive.

A careful look at thermal recording

For thermal recording, such as the ablation of pits in a thin layer of metal, the exposure threshold is well-defined; it is either the recording material's melting temperature or its vaporization temperature. If the heat from the playback beam is below the recording temperature, there is no damage, regardless of how often playback occurs. Of all the thermal recording processes, the ablation of thin film is the simplest and has been the most extensively investigated for optical recording. Thin coatings of many materials (inorganic and organic) can be easily vacuum-deposited onto the substrate disk. Relatively high contrast between the recorded pit and the adjacent undisturbed track is obtained in either the reflective or transmissive playback mode, and no supplementary magnetic fields or polarized detection schemes are needed.

The melting or vaporization temperature and the thermal diffusivity will largely determine the minimum laser power required to record in ablative thin-film materials. However, the threshold for recording is also significantly affected by the combined optical and thermal efficiencies of the surrounding structure that incorporates the recording material. Consider the simplest configuration, a thin metallic layer deposited directly onto a substrate. Only a portion of the incident beam is absorbed by the layer, with the remainder being reflected or transmitted into the substrate. That fraction of the incident beam absorbed into the metallic layer represents the optical efficiency of the layer.

The thermal efficiency is a measure of how much of the heat generated in the recording layer remains localized during the pit ablation process. Even for metals that conduct heat poorly the thermal diffusion length during pit exposure will considerably exceed the metallic recording layer thickness. Here, the heat loss from the absorption region is dominated by the diffusion of heat into the substrate disk. For this reason, it is important that the substrate material immediately adjacent to the recording material layer have low thermal diffusivity.

Radial heat loss into the surrounding metallic layer can also be significant if the thermal diffusivity of the metal is moderate to high. Low-melting-point metals that have relatively low thermal diffusivity include indium, lead, tellurium, and bismuth. Optical disks using tellurium as a recording medium have achieved a 40-dB CNR using 10-mW incident power and recording at 2 Mbit/s. Organic dyes also represent a broad class of materials with low melting points and low thermal diffusivities. Although less absorptive than metals, they are more difficult to work with since they tend to decompose rather than evaporate during vacuum deposition.

Advantages of multilayer materials

Antireflection multilayer coatings offer significant advantages over the single-layer approach just discussed. Optical efficiency can be very high since wasteful reflections from the top surface of the recording medium are eliminated. And if the pit-formation process in the recording medium is also designed to disrupt the antireflection condition, then high contrast can enhance playback quality (high SNR).

Antireflection structures have been developed in both bilayer and trilayer configurations (Fig. 5). The bilayer approach is most suited for use with moderately light-absorbing organic dye materials. The trilayer structure is most effectively used with strongly absorbing materials such as metals.

In the bilayer structure, a layer of dye is evaporated onto a thin reflective coating of aluminum and this, in turn, is deposited directly onto the disk substrate. If the thickness of the dye is maintained at a quarter of the recording-beam wavelength in the dye material, an antireflection condition is created at the upper dye surface. This happens because the ray reflected at the dye/aluminum interface has traveled an extra half-wavelength as compared with the ray reflected at the air/dye interface. Destructive interference will occur, producing low reflectivity. Actual dye thickness required to achieve antireflection is about 50 nm for a $\lambda = 488$-nm recording wavelength. For the optimized bilayer structure (dye layer consists of 4-phenylazo-1-napthylamine), reflectivities as low as 3 percent can be achieved with absorption in the dye layer, allowing optical efficiencies of 80–90 percent.

The light absorbed in the dye layer causes its temperature to rise until vaporization occurs and a pit is formed in the dye layer while the aluminum layer remains intact (Fig. 6A). Since the aluminum is revealed, the pit areas have a high reflectivity compared with the undisturbed bilayer coating (Fig. 6B). The organic dye (4-phenylazo-1-napthylamine) absorbs light best in the blue and green portions of the visible spectrum, and is relatively transparent to yellow and red. For this reason, the bilayer disk—optimized for recording at $\lambda = 488$ nm—displays a characteristically orange appearance.

The recording threshold for this disk is about 4-mW incident power. And recordings that give an SNR in excess of 50 dB on playback may be made with 15-mW incident power. As this dye material has minimal red absorption, reoptimizing the bilayer design for use with a HeNe laser or GaAs laser is not possible, even though the instrinsic sensitivity of the dye theoretically is high enough to allow recording with these sources. A limited spectral-absorption range precludes the possibility of using a single organic dye for recording with a wide range of laser sources.

On the other hand, most metallic recording materials absorb light throughout the visible (and infrared) spectrum. They are also more durable and easier to evaporate than organic dyes—but due to a relatively high absorption coefficient, metals are not suitable for achieveing a low-reflectivity bilayer disk structure. Instead, the trilayer antireflection structure has been developed (Fig. 5C), which is well suited for use with strongly absorbing recording materials. In the trilayer structure, the aluminum layer is first coated with a layer of transparent dielectric (the "spacer" layer) about a quarter-wavelength thick at the recording wavelength. This spacer can be a coat of silicon dioxide or any other transparent dielectric material that deposits well. A thin film of the metallic recording material is deposited on the dielectric spacer and, provided the thickness is chosen correctly, the interference conditions will produce a low-reflectivity trilayer structure. As an example, consider a trilayer structure with only 5 nm of titanium as the recording layer. When the silicon dioxide spacer is about 80 nm thick, the reflectivity will be less than 3 percent.

The trilayer principle may be better understood by analogy to the Fabry-Perot interferometer. Once the incident beam passes the uppermost low-reflectivity boundary of the structure, it is reflected back and forth between the air/absorber and the spacer/reflector interfaces. On each passage, some fraction of the beam is absorbed, so eventually almost 95 percent of the incident beam is coupled into a 5-nm titanium layer. The absorbed light melts a pit in the titanium, revealing the underlying aluminum (Fig. 7A). Thus, a series of highly reflective pits are recorded through the low-reflectivity trilayer coating (Fig. 7B). In contrast to the organic dye, titanium is strongly light absorbing in all of the visible spectrum. Even though the structure may be optimized for recording at $\lambda = 488$ nm, the reflectivity remains quite low across the whole visible spectrum so that the disk displays a characteristic deep blue-black appearance.

Although titanium has a reasonably low thermal diffusivity, its melting point is relatively high. Nevertheless, it was selected because, even in very thin layers (< 10 nm), the evaporated deposits are of excellent quality with

[8] Practical recording structures will use inexpensive plastic substrates (A). Playback signal-to-noise ratio (SNR) is ultimately limited by the quality of pit formation (B), regardless of increases in contrast (note 54 dB).

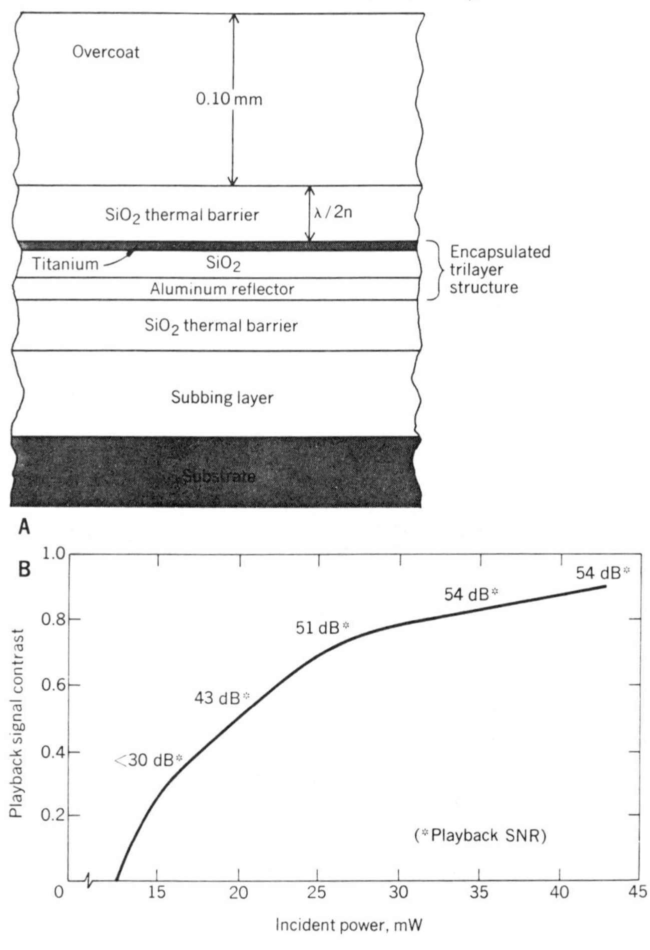

essentially no pinholes, texture, or graininess. Many low-temperature metals tend to agglomerate on deposition and the resulting texture can mar the playback signal quality. Despite the high melting point, it has been found that the recording threshold for the titanium trilayer configuration is only 10 mW, and that playback SNR of about 50 dB may be obtained with 30-mW incident beam power. Also, the structure can be easily reoptimized to a new recording wavelength simply by adjusting the spacer layer thickness (provided, of course, that titanium remains absorbing at the new wavelength).

Accepting and minimizing defects

No recording medium is perfect, and any discussion of the design of recording structures should consider the effect of defects on the playback signal. Since the density of information is so high, even 1-μm imperfections will result in a significant loss of information. Basically, there are three sources of defects that must be taken into account.

The first and foremost defect source is precipitated dust particles that accumulate on the disk surface. These particles may obscure recorded pits, or even prevent their formation. However, as already discussed, a transparent overcoat will prevent signal degradation due to dust by placing the particles out of the focal plane.

Defects can also lie in the evaporated coatings that form the recording structures. It is imperative that these coatings be very nearly pinhole-free, thus necessitating rigorous standards for cleanliness during the evaporation steps.

A third source of problems involves surface defects present on the glass substrate disk. Even the most carefully polished glass will have a finite density of cavities and microscratches on its surface. For any practical application of optical recording, both for reasons of economy and of bulk, it is evident that a plastic substrate will be required and the defect density is likely to be considerably higher than for glass. Plastic substrate surface defects can be hidden by covering the substrate with a thin layer of plastic solution. Applied by spin coating, the plastic solution dries and surface tension causes the uppermost surface to harden and form an almost defect-free platform for depositing the recording structure.

As one example, consider a fully encapsulated titanium trilayer structure based on an inexpensive plastic substrate disk (Fig. 8A). The titanium trilayer itself is sandwiched between two layers of silicon dioxide. These layers protect the overcoat and the subbing layers (both are low-temperature polymers) from excessive heat when the pits are melted into the titanium layer. Any thermal damage in the overcoat or subbing layer could reduce the quality of the playback signal. The overcoat layer is a spin-coat of silicone rubber that subsequently is heat cured.

Playback signal contrast plotted as a function of the incident recording-beam power illustrates the disk's recording characteristics (Fig. 8B). The threshold for recording is about 13 mW incident. Playback SNR>50 dB is achieved for recording powers of 25 mW or more. If reasonable care is taken to avoid contamination during fabrication the initial defect level in the playback signal can be less than 1 in 10^5 (about one defect per video frame). Note that encapsulation of the titanium layer has *not* led to a reduction in the playback SNR. This is due to the extreme thinness of the titanium layer and the fact that the pit-formation process is one of melting as opposed to vaporization.

What remains to be done

The advantages of an optical disk recorder product over other systems could be greatly increased if one (or two) significant events occur. These are

- Development of reusable and erasable recording materials
- Development of GaAs lasers for record and playback

The best performing recording materials currently used by all experimental optical systems are not erasable. But a number of possible recording materials are erasable (magnetooptic, thermoplastic, and chalcogenides). Most of these materials have not been developed to the point where they are usable in optical recording systems.

The small size and low power requirement of solid-state GaAs lasers make them very attractive as replacements for the relatively large and inefficient Ar^+ ion laser. And since GaAs lasers can be modulated directly, there is no requirement for an external modulator. However, a number of serious problems remain to be solved before these solid-state lasers become practical for optical record/playback applications. Such problems include insufficient output power, insufficient operating lifetime at high powers, and lack of modal stability and modal purity. ◆

Robert A. Bartolini (SM) joined the technical staff of RCA Laboratories in 1966 and has since been involved in the development of optical recording materials for relief-phase and volume-phase holography, integrated optic devices, and optical data storage. Dr. Bartolini is a member of the Optical Society of America, Eta Kappa Nu, Tau Beta Pi, and Sigma Xi.

Alan E. Bell became a member of the technical staff at RCA Laboratories in 1974, and has been mainly involved in the optical video recording research program, with particular interest in the design of recording structures. Dr. Bell is a member of the Optical Society of America, an Associate of the Royal College of Science, and a Diplomate of Imperial College.

Robert E. Flory (SM) joined RCA Laboratories in 1954, and has since been associated with video recording systems, TV standards conversion, ultra-high-speed computers, and various TV display projects. He participated in the earliest work on the RCA video disk, and he was team leader for video and sound on the HoloTape holographic video recording project. Mr. Flory is a member of the Society of Motion Picture and Television Engineers and a past editor of the IEEE Princeton Section Newsletter.

Michael Lurie (M) joined the RCA David Sarnoff Research Center in 1969 and has been involved in research on noise in holographic systems and on applying holography to video recording. Dr. Lurie was team leader for optics development on the HoloTape project. He is a member of the Optical Society of America, Eta Kappa Nu, and Tau Beta Pi.

Fred W. Spong joined the technical staff of the RCA Laboratories in 1964. His activities have included work on gas lasers, holography, and television recording by various means, including photography, holography, electron beam, and laser beam. Dr. Spong is a member of Phi Beta Kappa, Sigma Xi, and the American Physical Society.

Reprinted with permission from *Proceedings of the IEEE,* Vol. 70(6), pp. 589-597 (June 1982). ©1982 IEEE.

Optical Recording: High-Density Information Storage and Retrieval

ROBERT A. BARTOLINI, SENIOR MEMBER, IEEE

Abstract--Optical recording systems capable of information storage and retrieval, using a highly focused laser beam as the recording and playback source, have been developed to a point where product introduction in the 1983-1985 time frame is highly probable. The product potentials range from low-data-rate (10-20 Mbits/s), single disk (10^{10} to 10^{11} bits), random-access electronic office filing cabinets (augmenting existing word processing units) to high-data-rate (hundreds of megabits per second), multidisk (10^{13} bits on line "jukeboxes"), random-access archival mass memories. This article presents the basic features of these optical disk data storage and retrieval systems with particular emphasis on the recording source and recording media.

I. Introduction

IN TODAY'S SOCIETY there is an ever-increasing need to store and rapidly retrieve large amounts of information. This information may be in a variety of forms, from computer digital data to analog FM video signals. Storing this information requires an economical, compact, and high-speed mass storage medium. Retrieval of this information requires high-speed random access to selected segments of the data. Until recently, conventional storage techniques such as magnetic recording, through evolutionary developments, have been able to keep pace with the growth in information storage and retrieval applications. However, it now appears that a point has been reached where further improvement in magnetic recording is becoming increasingly difficult. Fortunately, an attractive new technique, the use of a highly focused laser beam for optical recording of data on a disk format is emerging as a solution to this problem. Capacities as high as 10^{11} bits per 30-cm-diameter optical disk, and data rates up to 50 Mbits/s for a single-channel system and 320 Mbits/s for a multichannel system have been reported [1], [2]. Playback-only optical-disk systems are commercially available now and have been described elsewhere [3], [4]. To be described in this article are optical disk data storage systems that permit both recording and playback of the stored information [5], [6].

In the following sections will be described the significant features of the optical disk recording and playback apparatus with particular emphasis on the two key elements of these types of systems; the recording laser source and the recording media.

II. Recording and Playback System

The advantages that optical disk recording systems provide are:

1) Very high information density (10^{10} to 10^{11} bits per disk have been demonstrated).

2) Very high data rates (up to 50 Mbits/s in a single-channel recorder have been demonstrated).

3) Rapid random access to the data ($\ll$1 s) has also been demonstrated.

4) The cost of the media has been projected to be 5×10^{-8} cent/bit.

5) The medium has the potential for archival storage (10 years or more) of the data.

For these reasons, many different organizations have been actively developing optical disk recording systems for a number of years [7]-[15]. Although there is a great difference between the various recording media that have been developed, the various systems, developed independently at different locations in several countries, show many similarities.

Shown in Fig. 1 is a block diagram of a typical optical disk recording and playback system. For recording, the output of the laser is directed through a modulator (both electrooptic and acoustooptic have been used) which intensity-modulates the light in response to an input electrical signal. The modulated laser beam is enlarged by the recording optics so that it will fill the aperture of the focusing lens. This lens focuses the laser beam into a small (<1-μm diameter) area on the recording medium so that the intensity modulation of the beam results in the information being stored as a series of "spots" along a track in the surface of the recording medium. For readout, an unmodulated, less intense (so that further recording will not take place) laser beam from the same laser follows the same path as the recording beam to the recording medium. The light is reflected from the recorded "spots" and passes back, through the focusing lens and a quarter-wave plate (not shown). The light, now rotated by 90° in polarization by the two passes through the quarter-wave plate, passes through a polarizing beam splitter (not shown) and is directed by the optics to a photodetector. The photodetector converts the reflected light beam (modulated by the recorded "spots") to an electrical output signal representative of the original input recording signal to the modulator. The radial position of the laser's spot on the disk is determined by the position of the motor-driven translation stage and the track mirror. The control system can move the translation stage so that either spiral or concentric circular tracks can be recorded. During readout, the translation stage is used to determine the approximate track location while the track mirror provides fine tracking control to lock on and read the desired information.

Several mechanical difficulties arise when implementing a system of this kind. As the disk spins, its surface will move in and out of focus, and a previously recorded track will move back and forth radially across the lens axis, unless both the disk and its turntable are extremely precise.

Manuscript received January 18, 1982.
The author is with RCA Laboratories, Princeton, NJ 08540.

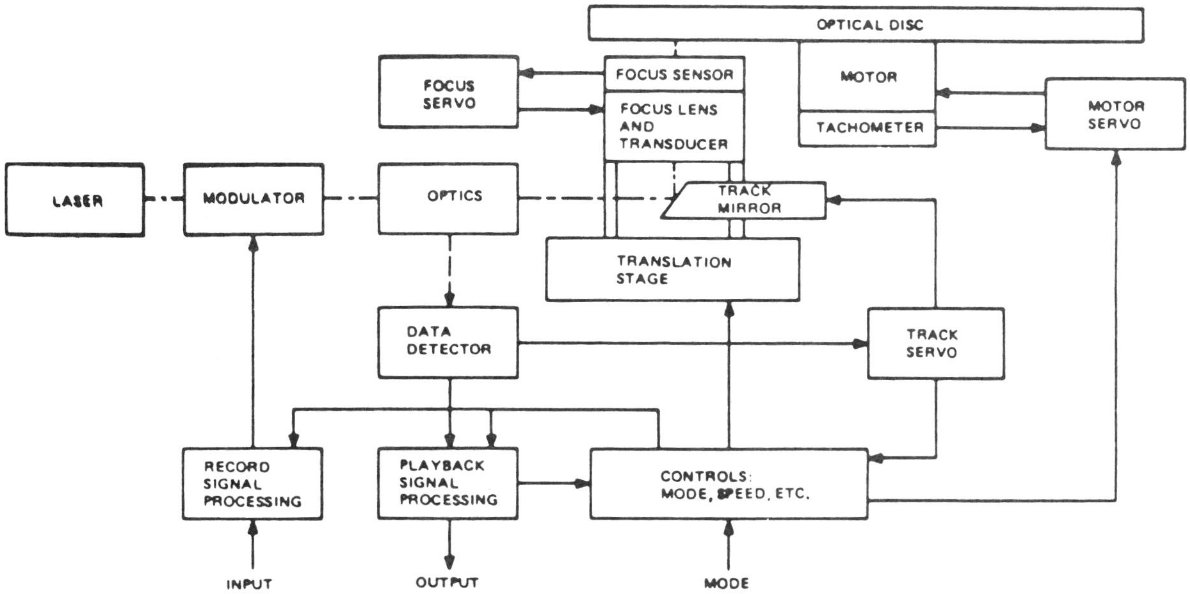

Fig. 1. Schematic of optical disk recording and playback apparatus.

Rather than attempt to maintain such precision in the mechanics, it is preferable to follow the disk movements in both dimensions with servos, one for focusing and one for tracking. In addition, a motor servo is required to maintain the disk rotational speed (S) constant which keeps the data transfer rate (R) fixed since

$$R = \frac{\#\,\text{bits}}{\text{revolution}}\,S. \tag{1}$$

The focus servo must detect focus errors comparable to the depth of focus of the recording lens (typically $\pm 0.5\ \mu$m), and move the lens accurately enough to correct for these errors. Suitably accurate mechanical spacing can be maintained by supporting the lens on an air film, or air bearing, over the disk, or conversely, by supporting a flexible disk on air over the lens [13]. Both processes are used commonly in magnetic disk recording. Greater accuracy is required in the optical system, so the air-support method is limited to disks that are relatively flat, or disks that are very flexible and can be flattened by the air bearings. Larger disk motions can be followed by moving the lens with some type of motor capable of producing the required high acceleration. With a disk turning at about 30 Hz, the surface undulations to be followed will lie in the range of 30 Hz to several times that frequency. Since those frequencies are in the audio range, it is very effective to move the lens with the mechanism used in loudspeakers. A coil (voice coil) is wound around the lens and the whole unit is placed inside a permanent magnet. Current applied to the coil moves the lens just as it moves the cone in a loudspeaker. This is the most common focus adjuster used today [11], [16].

Focus errors can be sensed by measuring the distance between the lens and the disk, for example, with a capacitance probe. That method, as well as the air bearing, has the weakness that optical path length is not measured. In the presence of a protective layer with nonuniform properties between the disk surface and the lens, mechanical separation or capacitance may not correspond to optical path length. It is better to optically sense focus to provide the correction signals to the focus actuator. Several optical methods have been described [16]. Generally, they detect a change in focused spot size or shape as the spot moves through focus.

During playback, disk miscentering, bearing looseness, and distortion of the disk all cause the track to move across the field of the lens. This motion can be followed by deflecting the laser spot with a rotating mirror. A mirror several millimeters in diameter mounted on a galvanometer can move through an angle of a few degrees at 30 Hz, and through smaller angles at much higher frequencies. That is enough to follow errors of many track widths, and also permits rapid jumping across several tracks for extremely high-speed local track access. Several approaches to sensing tracking errors have been used. One uses three laser spots [17]. Two of the spots straddle the track (one on each side) so an increase in signal returned from either spot shows that the track has moved to one side or the other. The third spot is centered on the track, and is used to read the recorded information. Another method uses a single laser spot that is "dithered" back and forth across the track at a speed much higher than any expected tracking errors [7]. The phase of the resulting amplitude modulation of the detected signal indicates the tracking error. These servos are shown schematically in Fig. 1.

When recording data, two important performance parameters are total data capacity of the storage media and the rate at which the data can be stored and retrieved. It is desirable to have both of these parameters as large as possible, e.g., 10^{10} to 10^{11} bits total capacity and 20- to 50-Mbit/s data recording rates.

The total data capacity of an optical disk depends on the number of data bits that can be stored on a single revolution (track) of the disk and the total number of revolutions (tracks). The primary factors that determine the bit density along the track are minimum recorded area size and modulation encoding scheme (i.e., number of bits per recorded area). The major factors that determine total number of tracks are size of the disk and track-to-track spacing which must be sufficient to reduce crosstalk below a maximum acceptable level.

Assuming a modulation-encoding scheme that requires 1 bit per recorded area and a track-to-track crosstalk limitation that requires the track density to be 1/3 the bit density along the track, then the total number of bits (C_0) that can be stored on an optical disk rotating at constant angular velocity is

$$C_0 = \frac{\pi}{6d_0^2}[D_oD_i - D_i^2] \tag{2}$$

where

d_0 recorded area diameter,
D_o disk outer recording diameter,
D_i disk inner recording diameter.

For C_0 to be a maximum, $D_i = D_o/2$ and assuming $D_o = 30.5$ cm, (2) reduces to

$$C_0 = \frac{1.2 \times 10^{10}}{d_0^2} \tag{3}$$

where the units of d_0 are in micrometers.

The data recording rate (R) [see (1)] is given by

$$R = \frac{\pi D_i}{d_o} S \tag{4}$$

where S is the disk rotation rate in revolutions per second. The disk rotational velocity is limited by mechanical properties of the disk substrate and focus and tracking servo performance. For many different substrates and the present servo systems, a disk rotational velocity of 30 r/s is practical, and with $D_i = D_o/2$, (4) reduces to

$$R = \frac{14.4 \times 10^6}{d_0} \text{ bits/s} \tag{5}$$

where the units of d_0 are in micrometers.

Equations (3) and (5) show the dependence of the disk data capacity and data rate on the recorded area size which, in turn, depends on the laser spot size. The laser spot size is a function of the laser wavelength (λ) and the numerical aperture (NA) of the recording focusing lens; i.e.,

$$d_{1/2p} = 0.56\left(\frac{\lambda}{\text{NA}}\right) \tag{6}$$

where $d_{1/2p}$ is the spot diameter at the 1/2 power points for a Gaussian beam truncated by the lens at the $1/e^2$ points [18]. The recorded area is typically of the order of this 1/2 power-spot diameter and thus (3) and (5) become

$$C_0 = 3.8\left(\frac{\text{NA}}{\lambda}\right)^2 \times 10^{10} \text{ bits} \tag{7}$$

and

$$R = 25.7\left(\frac{\text{NA}}{\lambda}\right) \times 10^6 \text{ bits/s.} \tag{8}$$

Focusing lens with NA's from 0.6 to 0.85 and lasers with wavelengths of 0.442 to 0.82 μm have been used in optical disk recording systems. Table I compares the disk capacity and data rate for these ranges.

As can be seen from Table I, the use of a longer wavelength laser requires a sacrifice in both disk capacity and data rate even when compared to a shorter wavelength system using an NA of 0.6. More will be said about this tradeoff in the next section.

Data access time is another parameter in which tradeoffs can be made [1]. Three separate operations must take place before data can be read from an optical disk. The first step involes slewing the translation stage to the appropriate track location. Two basic mechanisms can be used to perform this task: an air sled with a linear motor, or a ball screw with a rotary motor. The air sled is a translation table supported on a film of air to provide a smooth, almost frictionless motion. Typical air sleds move at rates of about 25 cm/s. The ball-screw design uses a precision threaded rotating screw to move the translation stage supported by a set of ball bearings. This system also moves at about 25 cm/s.

TABLE I
DISK CAPACITIES AND DATA RATES FOR SOME TYPICAL OPTICAL DISK RECORDING SYSTEMS

λ(μm)	NA	C_o (bits)	R (Mbits/sec)
0.442	0.6	7×10^{10}	35
0.442	0.85	1.4×10^{11}	50
0.82	0.6	2×10^{10}	19
0.82	0.85	4×10^{10}	27

TABLE II
OPTICAL RECORDING LASERS (CW, SINGLE MODE)

Type	λ(nm)	($d_{1/2p}$ (μm) NA=0.85)	P_{output} (max)
HeCd	442.	0.29	40mW
Ar^+	488.	0.32	1 - 4W
HeNe	633.	0.42	50mW
AlGaAs	820.	0.54	40mW

The second step in accessing data is to lock onto and read out the address of the track accessed mechanically to see if it is the desired track. If this is not the desired track, the tracking mirror moves the read beam to the estimated new location and repeats the address verification step and continues until the correct track is located. The limiting factor in this step is the time required to read the track address, and this depends on the data format. Multiple recording of the track address around a track reduces this time.

The third and final step in accessing data involves finding the desired data on a track once the proper track has been located. The time required to perform this step is dependent on the disk rotational speed. A worst case time is the once-around rate of the disk. Typical data access times are on the order of hundreds of milliseconds [1]. One technique which reduces access time is the use of multiple heads. With a two-head system, for example, one head can be accessing data while the other is reading out data, or one head could be assigned to half of the disk area and the other head to the other half.

III. RECORDING SOURCES

The recording source for all optical disk recording systems is a laser and the related optics. The optics must be capable of forming the necessary recording beams at the disk surface as well as bringing the beams to and from the disk surface and to the detector for readout.

A. Lasers

Table II lists some of the commercially available lasers (with relevant properties) that have been extensively used as recording sources in optical disk systems. Choosing the proper laser depends on the particular optical recording application. For example, the HeCd laser produces the shortest wavelength beam and, thus produces the smallest recording element which provides for the highest disk data capacity and highest single-channel recording rate. The Ar^+ laser, on the other hand, due to its high output power can have its output beam split into a number of sub-beams (up to 9 have been reported [2]), which can be independently modulated and used to simultaneously record data at *n* times the single-channel data rate, *n* being the number of sub-beams. Commercial availability and proven lifetime characteristics have made the HeNe laser an attractive optical disk recording source.

The HeCd, Ar^+, and HeNe lasers are all gas lasers and, until recently, were the only lasers used for optical recording applications. But the development of optical recording media with increased sensitivity, and the continued improvement of the modal characteristics and output power capabilities of the AlGaAs semiconductor diode lasers [19] has allowed recording of video [20], [21] and digital [20], [22] information with these semiconductor sources. This represents a major development in the field of optical recording since the diode laser permits the design of extremely compact systems with significant reductions in size and cost of the record and playback system. Because the output power of the diode laser can be modulated directly by the input signal, the external light modulator common to all gas-laser optical recording systems can be eliminated and diode-laser arrays will ultimately permit extremely high data rate (hundreds of megabits per second) systems. The high efficiency of the diode laser also permits the electrical input power to the system to be reduced drastically compared to gas-laser systems. Finally, the diode laser is potentially more reliable than its tube counterpart.

Diode-laser requirements for optical recording applications are operation at peak powers of 40 mW or more, at 50-percent duty cycle, and at rates up to 30 MHz. In addition, the laser should exhibit stable fundamental spatial mode operation in both transverse and lateral directions. "Lateral" and "transverse" designate planes respectively parallel and perpendicular to the junction plane, both being perpendicular to the direction of beam propagation, i.e., the longitudinal direction.

Recently, a new diode-laser structure (constricted double-heterojunction large optical cavity (CDH-LOC)) has been developed [23] which provides for high output power (up to 100 mW) in a single stable mode and the promise of good lifetime (more than 10 000 h to date at 40 mW, 50-percent duty cycle, and 10 MHz).

Lasers of this type (CDH-LOC) have typical characteristics illustrated by the lateral-mode far-field patterns shown in Fig. 2. This laser has very reproducible spatial mode characteristics over a wide range of drive conditions ranging from 8- to 100-ns pulsewidth, 10-percent duty cycle to CW, and power outputs ranging from 5 to 50 mW. Any changes in far-field pattern occur at the mode wings which are excluded from the optics when filling the recording lens aperture. These lasers produce nonastigmatic beams and provide diffraction-limited spot sizes at the disk surface.

B. Optics

The optical arrangement used for optical recording systems depends on a number of factors, which include the type of laser, type of applications (digital or analog), and number of recording beams. For illustration purposes we will describe an optical disk recording system which employs a diode laser as the recording source.

Fig. 3 shows the diode-laser optical recording system which begins with a microscope objective of sufficient NA to collect

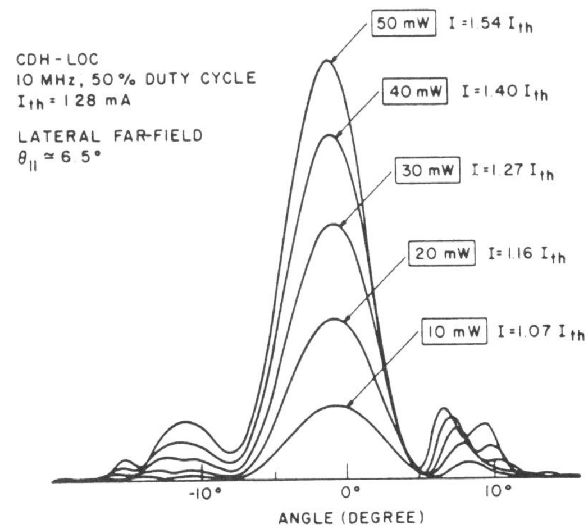

Fig. 2. Lateral mode far-field pattern for CDH-LOC laser.

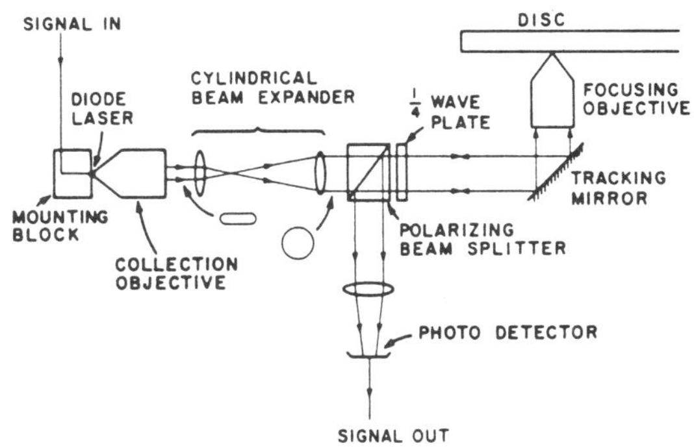

Fig. 3. Diode-laser optical-recording system.

most of the laser light (NA = 0.35). This is positioned to collimate the light, but further beam conditioning is required because of the asymmetric beam spreading in the lateral and transverse planes. This conditioning of the beam is accomplished by a combination of two simple plano-convex cylindrical lenses which function as a beam expander in one dimension. The positions of the cylindrical lenses are adjusted for common focus in both lateral and transverse directions. When the optical system is aligned, an approximately symmetrical beam just fills the focusing objective lens (NA = 0.85) producing a spot size of 0.6 μm (half-power diameter) in both lateral and transverse directions.

Also shown in Fig. 3 is the readout path and readout elements of the optical-disk system (i.e., tracking mirror, quarter-wave plate, polarizing beam splitter, and photodetector).

Since the data are written in real time with an instantaneous playback capability, a read-after-write (RAW) scheme can be used to verify that the data have been recorded without errors. A separate laser (lower power) can be used for this purpose as shown in Fig. 4. In addition to the diode laser and optics of Fig. 3, the HeNe in Fig. 4 is used to perform the function of focusing, tracking, read-after-write, and final data detection. Recording results using this system and gas-laser systems will be presented later in this article.

IV. Recording Media

The key component of the optical-disk data storage and retrieval system is the recording medium. Ideally, it should provide all the characteristics listed below:

1) High sensitivity—A few milliwatt incident onto the recording surface is a practical value based on existing lasers if

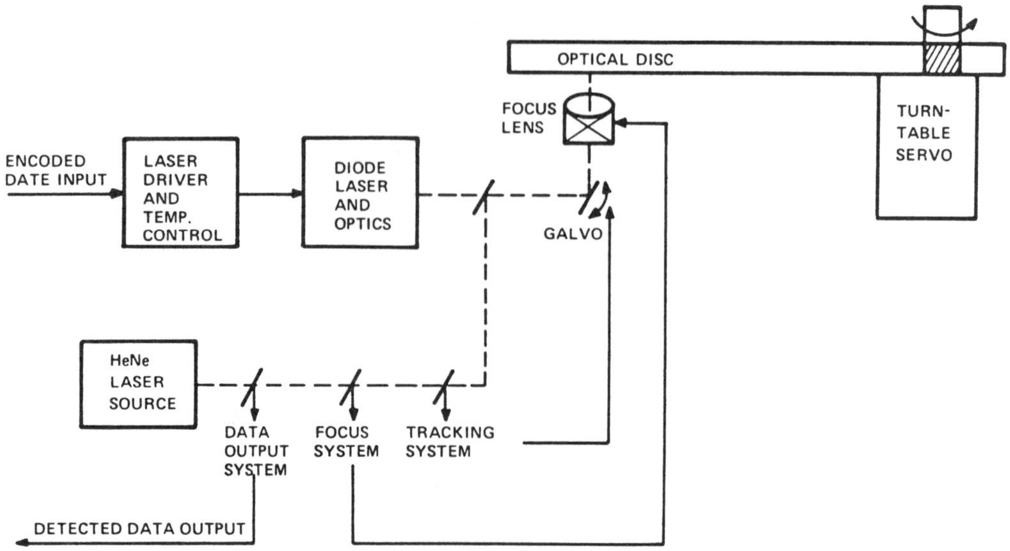

Fig. 4. Diode-laser recording and playback system.

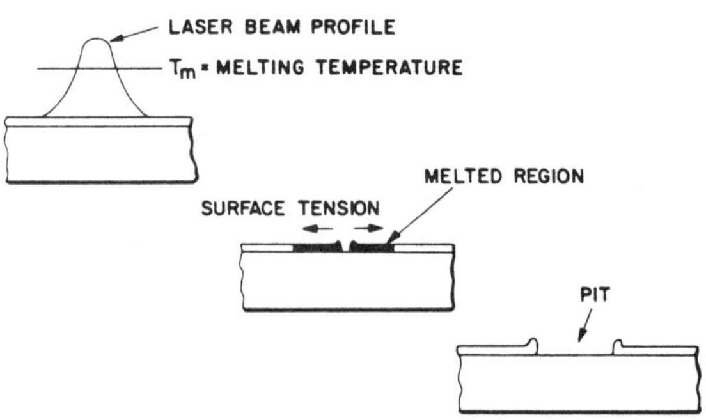

Fig. 5. Media thermal recording process.

one considers optical efficiencies, reflections, and other system losses.

2) High resolution—Since high-density optical recording systems require 1-μm (or smaller) spots, the material must have resolution capability >1000 lines pairs/mm.

3) High signal-to-noise ratio (SNR)—The signal to be useful must be recorded with sufficient SNR (the limit on SNR is dictated by system considerations; i.e., digital versus analog, data recording rate, disk capacity, etc.).

4) Real-time recording and instant playback—The material must have real-time recording characteristics and allow immediate retrieval of the stored information, i.e., there should be no processing steps required.

5) High immunity to defects—Material pinholes, dust, and other surface contaminants can prevent the recording of many signal elements, or obscure them after recording, thus the material must allow for protection from this source of "noise."

6) Archival storage—The material should allow permanent recording of the data and should not degrade under ambient conditions or prolonged readout.

There are many classes of materials in which changes in the optical properties may be induced by the absorption of light from the recording laser beam [24].

The types of materials that have been found to satisfy all of these conditions are ablative (melted or vaporized) thin films [25], [29]. The highly focused laser beam serves as an intense and extremely localized source of heat. This thermal recording process is depicted in Fig. 5, where a thin film is locally heated by the focused laser beam and surface tension is shown to open a hole (pit) in the melted area. To read the data, an unmodulated laser beam at an intensity below recording level is scanned across the pits by the rotating disk, and the change in reflectivity from pit to land is detected as the data output signal.

In general, for thermal types of recording media where the pit-formation process proceeds via melting of the layer of recording medium, the energy (E) which is required to form the pit and, therefore, the power required in the incident recording beam depend on three major parameters as shown in (9)

$$E \propto \frac{Q_m}{\eta_{\mathrm{OPTICAL}}\eta_{\mathrm{THERMAL}}}. \tag{9}$$

Q_m is the amount of energy which is required to melt the pit in the layer of recording medium, neglecting any heat losses to the substrate or overcoat layers. η_{OPTICAL} is the optical efficiency of the structure in which the recording medium is placed, i.e., the fraction of the incident recording beam which is coupled into and absorbed by the layer of the recording medium. η_{THERMAL} is the thermal efficiency of the recording process, being that fraction of the heat generated within the layer of recording medium that does not diffuse into the substrate or overcoat materials which are in contact with the layer of the recording medium.

The simplest ablative recording medium that has been used for optical recording is a single 30–50-nm vacuum-deposited layer of metal [25]–[30]. This type of optical recording structure reflects about 40 percent of the incident light and also absorbs about 40 percent. The remaining 20 percent is transmitted. The optical efficiency (η_{OPTICAL}) of this structure is, thus, 40 percent. A metallic optical recording structure that has been found to improve significantly on this factor is a thin-film multilayer antireflection (AR) structure called the trilayer optical recording structure [31].

The trilayer structure is shown schematically in Fig. 6. The substrate is first coated with a reflective layer of aluminum, followed by a layer of transparent dielectric material, and finally a very thin layer of the metallic recording medium. By choosing the thickness of the dielectric and the thin metal layer appropriately, it is possible to create an AR condition for light of the recording wavelength; this results in a very high fraction (up to 90 percent) of the incident beam being absorbed by the thin metallic layer. The optical coupling of the trilayer structure can, thus, be up to twice that which can be

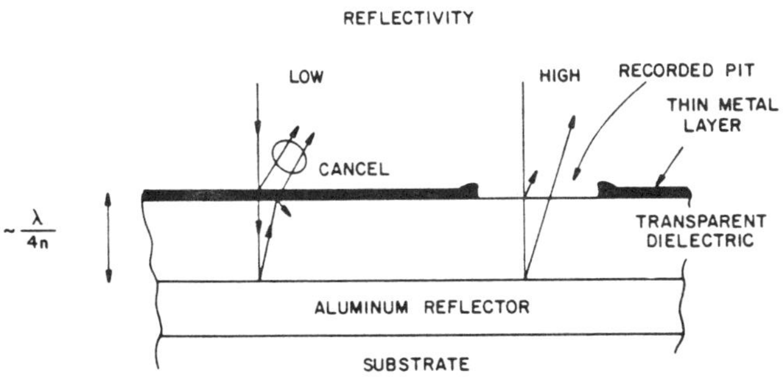

Fig. 6. Optical characteristics of the trilayer structure.

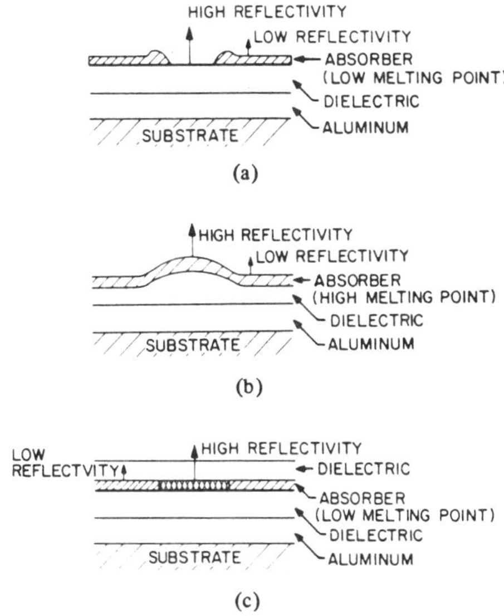

Fig. 7. Trilayer recording modes. (a) Pit-forming mode. (b) Bubble-forming mode. (c) Reversible mode.

TABLE III
RECORDING CHARACTERISTICS OF METAL TRILAYERS
(T_m: melting-point temperature; P_T: threshold incident recording power; SNR: playback signal-to-noise ratio)

Metal	T_m (°C)	P_T (mW)	SNR (dB)
Te	450	[illegible]	54
Pb	327	3	36
Bi	271	3	41
In	156	5	40

achieved using a single layer of the recording material due to the losses by reflection and partial transmission through the single layer. The trilayer design thus provides for the highest value for η_{OPTICAL} in (9).

The trilayer structure also provides an advantage over the single-layer approach during readout where the signal contrast is enhanced, because the pits formed during recording exhibit a high reflectivity due to the presence of the exposed aluminum reflection layer, while the AR condition is maintained in the unrecorded areas.

The amount of energy Q_m required to melt the pit in the metallic recording layer of Fig. 6, is heavily dependent on the melting point temperature of the metal itself. Of a large variety of low-melting-point metals tested, Te has been found to provide the highest playback SNR[1] (see Table III) [32].

The choice of Te as the absorbing layer in the trilayer structure, thus, provides for an optimum value for Q_m in (9).

The thermal efficiency, η_{THERMAL}, depends not only on the thermal properties of the recording medium and material in contact with this layer, but also on the duration of the pit-formation process, which is determined by system-related factors such as recorded signal bandwidth, laser-output duty cycle, turntable speed, and laser spot size.

The presence of the aluminum reflector layer in the trilayer structure represents a potential heat sink which may reduce the thermal efficiency of the recording process. The heat generated in the tellurium layer by the absorption of the incident recording beam will diffuse through the dielectric layer and, after some characteristic time dictated by the thickness and the thermal properties of the dielectirc layer, it will reach the aluminum reflector layer. Since aluminum is a good thermal conductor, once this layer is thermally coupled into the process of pit formation, the thermal efficiency of the recording can be substantially reduced, leading to a corresponding reduction in the recording sensitivity.

In general, in order to eliminate the deleterious effects of the heat losses to the aluminum layer, the parameters of the dielectric layer must be specified such that the thermal time constant for heat to diffuse from the layer of recording medium to the aluminum layer exceeds the duration of the pit-formation process. A polymer dielectric material, for example, was found [20] to provide a factor-of-two improvement in recording sensitivity over an SiO_2 dielectric layer. Thus the choice of materials for the dielectric layer can optimize the value of η_{THERMAL} in (9).

Besides the ablative pit-forming mode of optical recording (see Fig. 7(a)) in a trilayer structure, two new interesting modes of recording have been recently reported. The first is a bubble-forming mode [33], [34] (see Fig. 7(b)). In this mode, the absorbing layer is a high-melting-point metal (such as titanium) instead of the low-melting-point metal (such as tellurium), while the dielectric layer material is chosen to have a vaporization temperature well below the melting temperature of the absorber layer. Upon exposure to the highly focused laser beam, bubbles are formed as the top absorbed layer bulges due to the gas pressure from the vaporized dielectric layer beneath (see Fig. 7(b)). The bubble disrupts the AR condition and, thus, the readout is similar to the pit-forming mode (i.e., high-reflectivity recording regions and low-reflectivity unrecorded region). The other new mode is a reversible recording mode [35] (see Fig. 7(c)). In this mode, no topographical (pit or bubble) disruption of the trilayer structure occurs during recording. Instead, the optical properties of the recorded region are changed when the material switches between the crystalline and amorphous phases.

The pit-forming mode in this case is inhibited by the use of a dielectric capping layer (see Fig. 7(c)). The recorded region can be erased (restored to the original phase) by continuing high-power readout (but less power than the recording beam) of the recorded information. The area can then be rerecorded and erased many times (up to 50-record/erase cycles have been reported) [35].

Thus far in the discussion of the materials aspects of the design of the recording structure, the effects of defects on the quality of the playback signal have been neglected. Since the density of information is so high, even defects which have sizes in the range of 1 μm will result in a significant loss of

[1] SNR is the peak-to-peak signal to rms noise in a 4.2-MHz bandwidth.

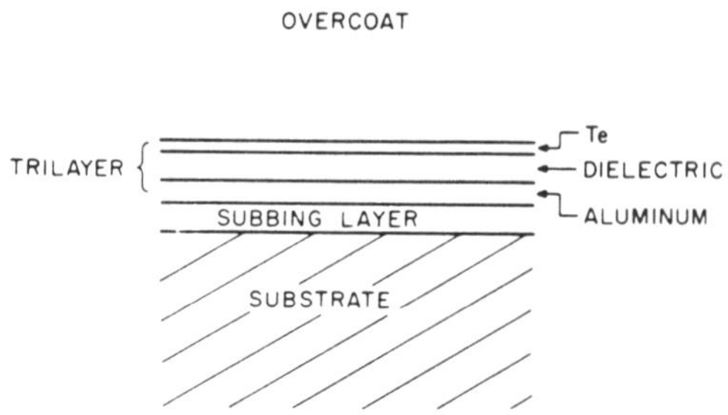

Fig. 8. Overcoated Te-trilayer structure with subbed substrate.

information. There are basically three sources of such defects and the overall design of the optical recording disk must take all of them into account.

The first and probably most important source of defects is that of precipitated dust particles which accumulate over a period of time on the surface of the recording medium. These particles may obscure a number of recorded pits, or even prevent their formation if the particle is present during the recording step. A relatively thick (180-μm) transparent overcoat or encapsulation layer will protect against signal degradation due to this source of defect by placing them out of focus [36]. Fig. 8 shows such an overcoat in contact with the Te-trilayer structure.

An alternate approach to providing protection from dust is to assemble two recording disks into a sandwich, with their transparent substrate facing out and recording surfaces facing in and an airspace in between. In this way, the substrate itself becomes the protective layer. The details of this Air Sandwich[2] system can be found in [6] and [9].

The second source of defects lies in the intrinsic quality of the evaporated coatings which form the recording structures. It is imperative that these coatings be pinhole free to a very high degree and this necessitates the most rigorous standards for cleanliness during the evaporation steps.

The third source of defects is due to intrinsic surface defects present in the substrate disk. Even the most carefully polished glass substrate will have a finite density of cavities and microscratches on its surface, and in the case of an inexpensive plastic substrate the defect density is likely to be considerably higher. The surface defects intrinisic to the substrate can be hidden relatively easily by precoating the substrate with a thin layer of plastic material applied by coating from a solution. Such a layer is applied by spin coating, and as it dries, surface tension causes the uppermost surface to harden and form an almost defect-free surface for coating with the recording structures. This surfacing or subbing layer is shown in Fig. 8.

Recording results using some of these structures are discussed in the following section.

V. Recording Results

Recording results using the diode laser in the recording arrangement of Fig. 3 and the optical disk structure of Fig. 6 are shown in Figs. 9 and 10.

Fig. 9 is a photograph of a TV monitor displaying the playback video signal which was recorded and read out using the diode-laser recording system [20]. The SNR is approximately 45 dB.

Fig. 10 shows digital recording results using the diode-laser recording system [20]. Fig. 10(a) shows the input bit pattern

[2]Registered Trademark of Philips Corporation.

Fig. 9. Photograph of TV monitor playback of video signal recorded and readout with a diode laser.

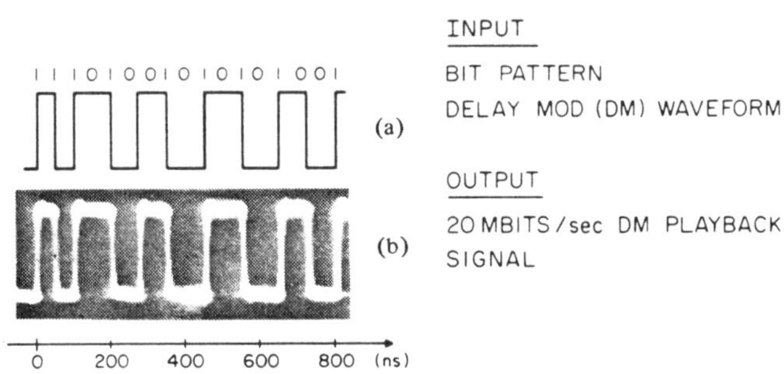

Fig. 10. Diode laser recording and playback of a 20-Mbit/s DM digital signal. (a) Input bit pattern and DM waveform. (b) Output DM signal.

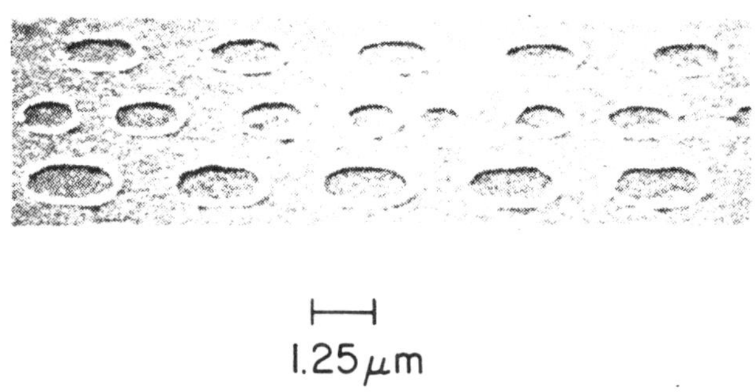

Fig. 11. SEM playback of three-channel (125-Mbit/s) optical disk recordings.

and corresponding delay modulation (DM) 20-Mbit/s digital encoding waveform which was used to directly modulate the diode laser. The output signal during playback is shown in Fig. 10(b).

A high-power (~1-W) Ar^+ laser whose output beam was divided into multiple beams which had been independently modulated has been used to simultaneously record digital data at rates of 125 to 320 Mbits/s. Fig. 11 shows an SEM photograph of three-channel 125-Mbit/s digital data recordings (41.67 Mbits/s per channel) made on the optical disk structure of Fig. 6. The track-to-track spacing is 1.25 μm with a track width of about 0.4 μm [2]. Fig. 12 shows a photomicrograph of nine-channel 320-Mbit/s digital data recordings (35.56 Mbits/s per channel) also made on the optical disk structure of Fig. 6. The data produce a pseudorandom delay modulation digital pattern [2].

Using the optical disk structure of Fig. 7(c), reversible video-

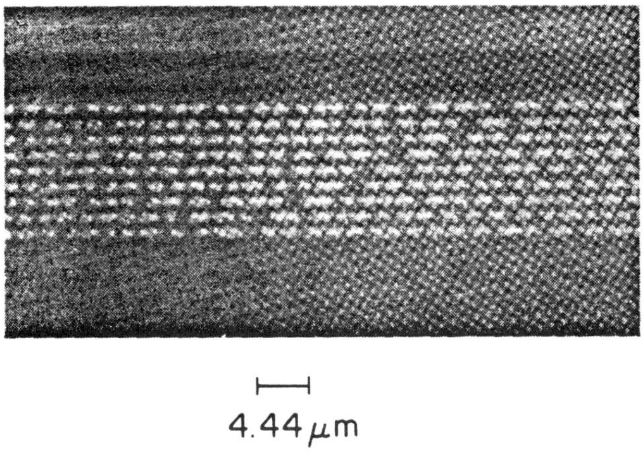

Fig. 12. Microscope photograph of nine-channel (320-Mbit/s) optical disk recordings.

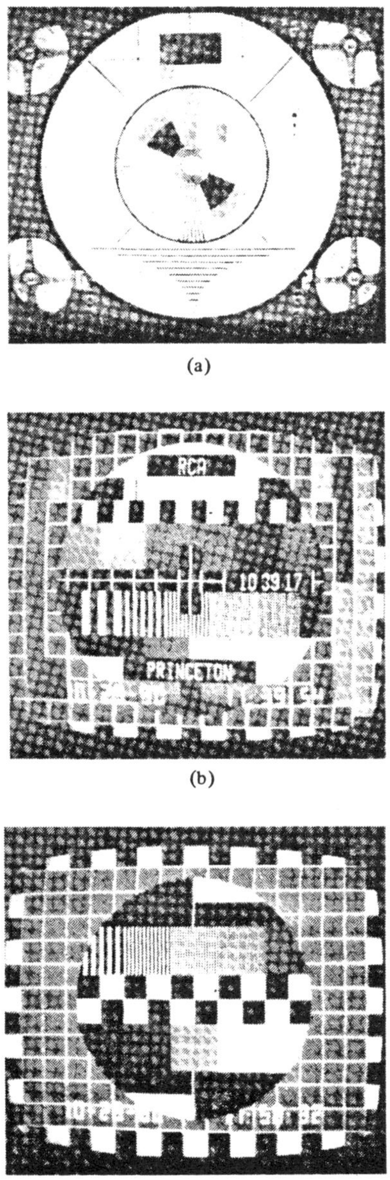

(a)

(b)

(c)

Fig. 13. Photograph of TV monitor playback of reversible recordings. (a) Recording #1. (b) Recording #30. (c) Recording #50.

optical recording has been demonstrated [35]. Fig. 13 shows photographs of the playback of video information displaying the 1st, 30th, and 50th successive recordings made on one track of a disk made during the record/erase/rerecord sequence.

VI. Conclusions

Optical data recording and playbacks systems have been talked about for well over ten years now. Recent developments, including high-power-single-mode diode lasers, multi-channel-high-data-rate systems, single-channel low-data-rate prototypes, as well as user needs, have increased the probability of product introduction in the 1983–1985 time frame. The product potentials range from low-data-rate (10–20-Mbit/s), single disk (10^{10} to 10^{11} bits), random-access electronic office filing cabinets (augmenting existing-word processing units) to high-data-rate (hundreds of megabits persecond), multidisk (10^{13} bits on line "jukeboxes"), random-access archival mass memories. The key to success in any of these application areas is the emergence of a useful recording media product.

This article has presented the basic features of optical disk data storage and retrieval systems with particular emphasis on the recording source and recording media. For more details on the technology and applications, the reader is directed to the references.

References

[1] G. J. Ammon, C. W. Reno, and R. J. Tarzaiski, "Digital optical video disc," Final Rep. Gov. Contract F30602-78-C-0050, Rome Air Development Center, Apr. 1979.

[2] G. J. Ammon and C. W. Reno, "Optics for multi-beam optical disc systems," *Proc. SPIE*, vol. 299, 1981.

[3] D. Mennie, "Television on a silver platter," *IEEE Spectrum*, vol. 12, 34–40, Aug. 1975.

[4] The July 1, 1978 issue of *Applied Optics* features seven articles on "Video Long Play Systems."

[5] R. A. Bartolini, A. E. Bell, R. E. Flory, M. Lurie, and F. W. Spong, "Optical disk systems emerge," *IEEE Spectrum*, vol. 15, pp. 20–28, Aug. 1978.

[6] G. C. Kenney, D.Y.K. Lou, R. McFarlane, A. Y. Chan, J. S. Nadan, T. R. Kohler, J. G. Wagner, and F. Zernike, "An optical disk replaces 25 mag tapes," *IEEE Spectrum*, vol. 16, pp. 33–38, Feb. 1979.

[7] K. Berger and H. J. Hamisch, "Laser picture storage system for professional applications," *Fernseh Aktuell*, pp. 10–16, Feb. 1977.

[8] K. Broadbent, "A review of the MCA disco vision system," *J. Soc. Motion Pict. Telev. Eng.*, vol. 83, pp. 554–559, July 1974.

[9] R. Pepperl and I. Sander, "High density direct read and write (DRAW) recording," *Opt. Acta*, vol. 24, pp. 427–431, 1977.

[10] G. Kenny, D. Lou, J. Wagner, F. Zernike, R. McFarlane, and A. Chan, "Experimental optical disc data recorder," presented at the Conference on Lasers and Electro-Optical Systems (CLEOS), San Diego, CA, Feb. 1978.

[11] M. Lurie and R. E. Flory, "Developing a broadcast quality optical video disc system," presented at the Conf. on Lasers and Electro-Optical Systems (CLEOS), San Diego, CA, Feb. 1978.

[12] Y. Naruse and A. Ohkoski, "A new optical video disc system with one hour playing time," presented at the IEEE Chicago Spring Conf. on Consumer Electronics, Chicago, IL, June 1978.

[13] E. Spitz, "Optical video discs," presented at the Conf. on Laser Engineering and Applications (CLEA), Washington, DC, May 1975.

[14] F. Le Carvennec, "Professional applications of optical video disc technology," presented at the Conf. on Lasers and Electro-Optical Systems (CLEOS), San Diego, CA, Feb. 1978.

[15] P. Palermo, A. Korpel, G. Dickinson, and W. Watson, "Video disc mastering and replication," *Opt. Laser Technol.*, pp. 169–174, Aug. 1977.

[16] P.J.M. Janssen and P. E. Hay, "Control mechanisms in the Philips VLP record player," *Philips Tech. Rev.*, vol. 33, pp. 190–193, 1973.

[17] G. Bouwhuis and P. Burgstede, "The optical scanning system of the Philips VLP record player," *Philips Tech. Rev.*, vol. 33, pp. 186–189, 1973.

[18] G. O. Olaofe, "Diffraction by Gaussian apertures," *J. Opt. Soc. Amer.*, vol. 60, pp. 1654–1657, 1970.

[19] D. Botez, F. W. Spong, and M. Ettenberg, "High-power constricted double-heterojunction AlGaAs diode lasers for optical recoding," *Appl. Phys. Lett.*, vol. 36, no. 1, pp. 4-6, 1980.

[20] R. A. Bartolini, A. E. Bell, and F. W. Spong, "Diode laser optical recording using trilayer structures," *J. Quantum Electron.*, vol. 17, no. 1, pp. 69-77, 1981.

[21] P. M. Asbeck, D. A. Cammack, J. J. Daniele, D. Lou, J.P.J. Heemskerk, W. J. Klouters, and W. H. Ophey, "High-density optical recording with (Ga, Al)As DH lasers," *Appl. Phys. Lett.*, vol. 34, no. 12, pp. 835-837, 1979.

[22] K. Bulthuis, M. G. Carasso, J.P.J. Heemskerk, P. J. Kivits, W. J. Kleuters, and P. Zalm, "Ten billion bits on a disk," *IEEE Spectrum*, vol. 16, pp. 26-33, Aug. 1979.

[23] D. Botez, "CW high power single mode operation of constricted double-heterojunction AlGaAs lasers with a large optical cavity," *Appl. Phys. Lett.*, vol. 36, no. 3, pp. 190-192, 1980.

[24] R. A. Bartolini, H. A. Weakliem, and B. F. Williams, "Review and and analysis of optical recording media," *Opt. Eng.*, vol. 15, pp. 99-108, 1976.

[25] A. L. Harris, "Continuous wave laser recording on metallic thin film," *Image Technol.*, pp. 31-35, Apr./May 1970.

[26] D. Maydan, "Micromatching and image recording on thin film by laser beams," *Bell Syst. Tech. J.*, vol. 50, pp. 1761-1789, 1971.

[27] E. E. Gray, "Laser mass memory system," *IEEE Trans. Magn.*, vol. MAG-8, pp. 416-420, 1971.

[28] D. R. Congleton, M. R. Smith, and A. S. Diamond, "A comparison of plastic vs metal coatings for real time laser recording," *J. Appl. Photographic Eng.*, vol. 3, pp. 97-105, 1977.

[29] J. Corcoran and H. Ferrier, "Melting holes in metal films for real time high density permanent digital data storage," presented at the SPIE 21st Int. Tech. Symp. add Instrument Display, San Diego, CA, 1977.

[30] M. Terau, K. Shigematsu, M. Ojima, Y. Taniguchi, S. Horigome, and S. Yonczawa, "Chalcogenide thin films for laser-beam recordings by thermal creation of holes," *J. Appl. Phys.*, vol. 50, no. 11, pp. 6881-6886, 1979.

[31] A. E. Bell and F. W. Spong, "Antireflection structures for optical recording," *J. Quantum Electron.*, vol. 14, pp. 487-495, 1978.

[32] A. E. Bell and R. A. Bartolini, "High performance te-trilayer for optical recording," *Appl. Phys. Lett.*, vol. 34, no. 4, pp. 275-276, 1979.

[33] J. Cornet, J. C. Lehureau, and F. Le Carvennec, "Characteristics and performance of a novel recording material," presented at the Conf. on Lasers and Electro-Optics (CLEO), Washington, DC, June 1981.

[34] W. Robbins, R. Freese, T. Smith, and R. Willson, "Bubble forming media for optical recording: A new approach," presented at the Conf. on Lasers and Electro-Optics (CLEO), Washington, DC, June 1981.

[35] A. E. Bell and F. W. Spong, "Reversible optical recording in trilayer structure," *Appl. Phys. Lett.*, vol. 38, pp. 920-921, 1981.

[36] A. E. Bell, R. A. Bartolini, and F. W. Spong, "Optical recording with the encapsulated titanium trilayer," *RCA Rev.*, vol. 40, pp. 345-362, 1979.

Section Ten
Magneto-Optic Media

Curie-Point Writing on Magnetic Films*

LUDWIG MAYER
Mechanical Division, General Mills, Inc., Minneapolis, Minnesota
(Received February 26, 1958)

WILLIAMS, *et al.*,[1] have recently shown that MnBi films having a uniaxial direction of easy magnetization normal to their surfaces are well suited for magnetic writing. Writing was accomplished by drawing the end of a very thin permanent-magnet wire across the MnBi film. For high contrast, the film must first be magnetized to saturation normal to the surface. The pattern drawn by the permanent-magnet pen becomes magnetized in the opposite direction and can be read-out optically by utilizing the Faraday effect.

It is the purpose of this brief communication to report first results of experimental work intended to determine the feasibility of a novel method of writing on magnetic films for which the over-all term *Curie-point writing* is proposed.

Phenomenologically, Curie-point writing means to record magnetic patterns permanently (but erasable at will) onto a magnetic film by elevating the film temperature locally above the Curie temperature or any suitable transformation temperature. Such a concept appears physically possible. A magnetic film magnetized in a direction normal to its surface and then heated above the Curie temperature at a given spot will first become nonmagnetic at this spot. Then, after the spot has cooled below the Curie point, the area of the spot will become magnetic again, and—this is important—will in general become magnetized in the opposite direction. This effect is simply caused by the fact that minimum magnetic energy requires magnetic flux closure through the temporarily nonmagnetic area, which in turn should result in a reversal of the direction of magnetization after the area of the spot has cooled enough to become magnetic again.

To check the feasibility of this concept for magnetic writing, MnBi films have been prepared after Williams *et al.*[1] For the purpose of Curie-point writing, these films were preheated to a temperature of about 260°C, and then a heated pen, mounted on a micromanipulator, was moved across the film without touching it. The heated pen was a platinum wire formed like a tiny hairpin and flattened at the tip to a width of about 0.06 mm. This Pt-hairpin pen was heated to a temperature of about 780°C; the distance between the tip of the pen and the film was kept at about 0.1 mm while the pen moved. Observation, either optically by utilizing the Faraday effect or electronically with an electron mirror microscope, showed that reversal of the direction of magnetization actually occurred beneath the path of the heated pen. As an example, Fig. 1 shows a micrograph of a cut-out area of such an MnBi film onto which, after magnetization to saturation in the direction normal to the film, two traces were written. The narrow dark trace stems from a permanent Cunife-magnet wire† of 0.12 mm diameter, and the wider dark trace crossing the narrow trace was written onto the film by Curie-point writing as described above. Both traces could be erased completely by subjecting the film to saturation magnetization. Curie-point writing and erasing could be repeated without leaving any permanent, i.e., not erasable, trace on the film. Too frequent and too prolonged preheating of the MnBi film in air resulted sometimes in discoloration of the film, which apparently was caused by surface oxidation. For this reason it proved advantageous to evaporate a thin protective coating of silica on the film.

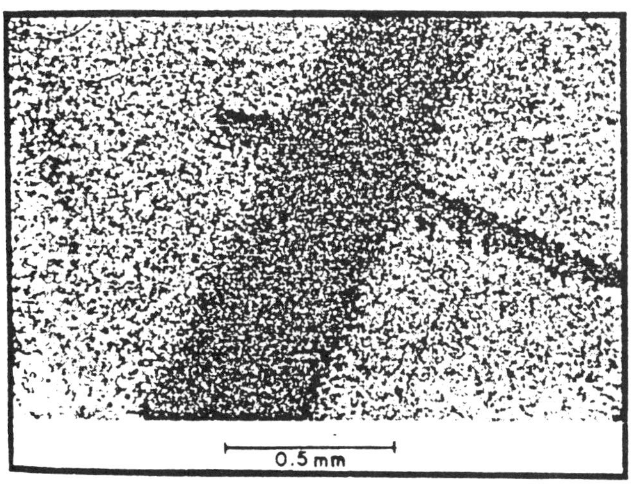

FIG. 1. Curie-point writing on MnBi film. The wide dark trace was written with a heated Pt-hairpin pen 0.1 mm above the preheated MnBi film. The narrow dark line was written with a Cunife permanent-magnet wire of 0.12 mm diameter.

Writing with a heated pen is only one way to accomplish the Curie-point writing. Of several other possible methods, Curie-point writing utilizing the dissipation energy of a focused electron beam appears to be one of the most interesting because it might permit an all-electronic magnetic momory system using a device based on the same basic principle as an electron mirror microscope[2] for electronic read-out. In this case electronic Curie-point read-in could, of course, be accomplished by the same device merely by switching the bias potentials on the different electrodes to values best suited for Curie-point read-in with the electron beam.

* This research was supported by the U. S. Air Force under Contract AF 33(616)-3852, monitored by the Aeronautical Research Laboratory, WCLJL, WADC.

[1] Williams, Sherwood, Foster, and Kelley, J. Appl. Phys. 28, 1181 (1957).

† Kindly supplied by Irvin L. Cooter, Magnetic Measurements Section, National Bureau of Standards.

[2] See, for example, Ludwig Mayer, J. Appl. Phys. 26, 1228 (1955); 28, 975 (1957); and 29, 658 (1958).

Reprinted with permission from *IEEE Transactions on Magnetics,* Vol. MAG-3(3), pp. 433-452 (September 1967).

Magnetic Films and Optics in Computer Memories

DONALD O. SMITH, MEMBER, IEEE

Abstract—The thermal cycle-time and read bandwidth are calculated for a magnetic film memory which is written by heating from focussed light and read by the same beam at reduced power. The heat flow model approximates the memory bit as a planar disc source and the cycle time T is determined by the condition that, for repetitive writing, the ratio of transient to dc temperature must be >1. Then it is found that the bit radius a must be $< \sqrt{\kappa T}$ = thermal diffusion length in time T (κ = thermal diffussivity of substrate); for $T < 1\mu s$ and an Ag substrate this gives $a \sim 5\mu$. For a magneto-optical reflection coefficient $k \sim 10^{-3}$ the shot-noise limited bandwidth is then found to be ~100 MHz for a temperature rise of 10 degrees in the memory bit. Surface noise due to imperfections will limit the bandwidth to less than the shot-noise limit unless some form of signal processing can be used to extract the magnetic information. For incident light polarized at, say, 45° to the plane of incidence, phase modulation of one of the components will generate a magneto-optical signal with a phase of 0 or π relative to the modulation, depending on the magnetic state of the memory bit. Additional topics which are considered include a review of magneto-optical enhancement using multilayer dielectric films, a discussion of possible memory materials, and a proposal for a content-addressed memory which is interrogated magneto-optically.

Manuscript received April 25, 1967.
The author is with the M.I.T. Lincoln Laboratory, Lexington, Mass., operated with support from the U. S. Air Force.

I. Introduction

Magnetic-film memory systems as currently conceived are sensed by the voltage induced during flux reversal. For cycle times ~1 μs the maximum bit density is then ~10^3 bits/cm^2, determined by a combination of considerations which include demagnetizing fields, drive currents, and thermal noise in the sensing channel. Hence a speed-density product $SD \sim 10^3$ bits/cm$^2\cdot\mu$s can be considered as the approximate upper limit of present magnetic film memory technology. In order to increase SD, a method of sensing must be used which can make more total signal energy available. Conceptually this can be done by magneto-optical sensing. Since now the sense signal is proportional to the intensity of the sensing light, which does not necessarily depend on bit density, in principle it should be possible to increase bit density without decreasing the cycle time. Furthermore, if writing can also be accomplished optically, the limiting bit diameter will be of the order of the diffraction limit for focussed light (~1 μ), corresponding to a bit density some 10^5 times greater than presently possible. Since there is no known direct interaction between light and electron spins which is strong enough to reverse the spin state, it is proposed to convert optical energy to heat, which in turn will interact with the electron spins to effect reversal in the presence of an applied magnetic field.

The calculations presented in this paper support the conception of a thermomagneto-optical (TMO) memory with an $SD \gtrsim 10^8$ bits/cm$^2 \cdot \mu$s.[1] The problems considered include: 1) Writing by thermo-optical effects which have a limiting spatial resolution $\sim 1\ \mu$ determined by optical diffraction; speed will be determined by heating and cooling time constants, which for available CW lasers and properly designed thermal structures should be $<1\ \mu$s. Memory cycle time is calculated from a simple model which considers the memory bit as a planar disc source of heat. 2) Reading by means of either the longitudinal or polar magneto-optical effect. As a specific example, reading at reduced power by the same beam used for writing is considered. Then the read beam is highly convergent since small spot size requires large numerical aperature. Such a beam cannot be extinguished by means of a crossed analyzer due to inhomogeneous ellipticity after reflection or transmission. Analysis of the optical noise then shows that some kind of modulation is required to extract the magnetic information from the beam. A method of phase modulating the incident light is described which converts the *one* and *zero* magnetic states into 0 or π phase information even in a convergent beam.

Additional topics which are considered include a review of magneto-optical enhancement using multilayer dielectric films, a discussion of possible memory materials, and a proposal for a content-addressed memory which is interrogated magneto-optically.

II. Thermo-Optical Writing

Thermal writing of information in a magnetic film appears to have been first suggested by Mayer,[1] who described an experiment in which a MnBi film initially magnetized normal to the film surface was locally heated by a Pt hair-pin pen to a temperature above the Curie point. Upon cooling this region was magnetized in the opposite direction due to the demagnetization fields from adjacent regions of the film. Mayer referred to this experiment as Curie point writing, although it was clear to him that temperatures exceeding the Curie point are sufficient, although perhaps not necessary, to achieve thermal writing. Furthermore, the use of a heated pen is clearly only of passing interest and any exploitation of thermal writing will surely use either electrons or photons to provide the energy for heating. Although, in principle, sufficient energy is available for heating by electrons, as was also demonstrated by Mayer,[2] no satisfactory read-out scheme using electrons was then or is yet available.[2] Since, in 1958, the laser did not exist, insufficient collimated optical energy was available for high-density Curie point writing. In principle, however, if enough optical energy were available for writing, magneto-optical effects in almost any material are large enough to provide a means for read-out. Consequently with the advent and rapid development of the laser, the whole subject of thermal writing in magnetic films has taken on new significance. It is probably not overoptimistic to say that the goal of a memory having the bit density of photographic film and the speed and flexibility of magnetic cores or films is now a realistic, even if somewhat distant, goal.

A. *Mathematical Model for Heat Flow*

It is desired to know the temperature of a memory bit as a function of power input to the bit. The most stringent conditions on cycle time arise in connection with random access memory, for which the most severe operating situation arises when one particular bit is written repetitively on each memory cycle. Then two questions must be answered, namely: 1) what is the incremental temperature rise at the end of a heating period; 2) what is the steady-state dc temperature level at the end of a cooling period. It is clear that for successful memory operation the incremental rise must exceed the dc level, a requirement which will be independent of the actual temperatures involved. Of course the actual temperatures are important when considering some particular memory material. Since considerable power is now available from CW lasers, it is likely that the cycle time of a thermally written memory will be determined by the previous condition on the incremental and dc temperatures, rather than the actual value of the temperature rise.

Consider the physical configuration shown in Fig. 1(a) in which a circular thin-film memory bit is heated by an optical beam and cooled by heat conduction to a boundary at which the temperature v is held at zero. The $v = 0$ boundary could be realized in different ways, e.g., as the edge of the thermal boundary layer of a flowing liquid, or at ∞ for a very thick substrate. A mathematical model for this physical configuration is shown in Fig. 1(b), in which the film is considered as a planar disc source of heat at $x = x'$ with the temperature $v = 0$ at the plane $x = 0$. For initial simplicity it is assumed that all materials in the structure have the same thermal properties.[3] Then Green's function for the problem can be obtained by introducing a sink in the image plane $x = -x'$. The energy input to the magnetic film is assumed to be pulses of width T_1 occurring with period T starting at $t = 0$. The transient and steady-state solution of this problem is given in Appendixes I, II, and III. Analytic expressions

[1] The problem of light deflection is not considered. It is assumed that the required technology will be developed.

[2] Use of Lorentz electron microscopy for read-out does not appear to be practical for a variety of reasons. A fundamental difficulty is due to the very small value of the Lorentz force.

[3] For the case in which the heat diffusivity κ_1 of the material in the region $x > x'$ is very much less than the diffusivity κ_2 in the region $x < x'$, a better model is obtained by taking $\kappa_1 = 0$, $\kappa_2 \neq 0$. Then the temperature distribution for this problem is twice that for the case for $\kappa_1 = \kappa_2 = \kappa$.

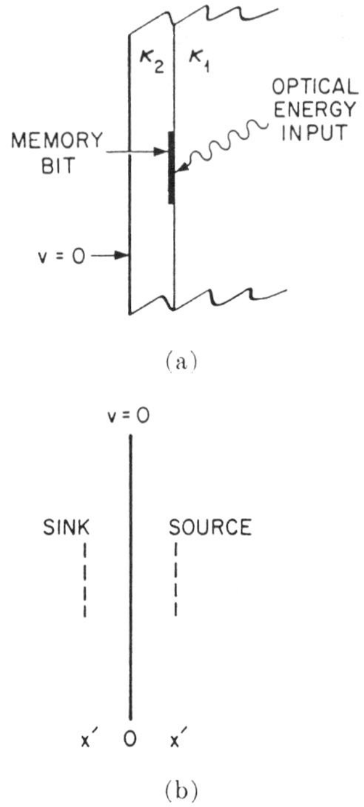

Fig. 1. Thermo-optically written memory bit. (a) Physical configuration. (b) Mathematical model: the film is considered as a planar disc source of heat at $x = x'$ and at the plane $x = 0$ the temperature $v = 0$; if the heat diffusivities κ_1 and κ_2 are assumed equal, a sink in the image plane $x = x'$ can be introduced to obtain Green's function for this problem.[3]

can only be obtained on the axis of the memory bit, which fortunately corresponds to the worst case.

The steady-state on-axis temperature v is given by

$$\begin{aligned} v &= v_\infty - v_a \\ &= v_0(I_\infty - I_a) \end{aligned} \tag{1}$$

where

v_∞	temperature due to infinite plane source
v_a	correction due to spreading from a disc source of radius a
v_0	normalizing temperature $= P_aD/KA$
P_a	power absorbed during the heating period
D	thermal diffusion distance during a period T ($D = \sqrt{\kappa T}$)
κ	thermal diffusivity
K	thermal conductivity
A	area of the source
I_∞, I_a	dimensionless integrals. (2)

In order to obtain the temperature at any time t numerical integration of I_∞ and I_a is required. However, at the times of most interest, i.e., at the end of a heating or cooling period, analytic expressions can be obtained.

Before writing the expressions for I_∞ and I_a, introduce the dimensionless times

$$\begin{aligned} \tau &= t/T \\ \tau_1 &= T_1/T \end{aligned} \tag{3}$$

in terms of which it is found that

$$I(\tau_1) = I(1) + I(\Delta). \tag{4}$$

In (4) the integral $I(1)$ represents the dc level present at the end of a cooling period $\tau = 1$, and $I(\Delta)$ is the incremental rise at the end of a heating period $\tau = \tau_1$. Also introduce the dimensionless distances

$$\begin{aligned} \xi_\pm &= (x \pm x')/D \\ \chi_\pm &= [(x \pm x')^2 + a^2]^{1/2}/D \end{aligned} \tag{5}$$

where

$$a = \text{radius of the memory bit.} \tag{6}$$

Then $I_\infty(1)$ and $I_a(1)$ are found to be

$$I_\infty(1) = \frac{\tau_1}{2\sqrt{\pi}} \sum_{n=1}^{\infty} \frac{1}{\sqrt{n}} (e^{-\xi_-^2/4n} - e^{-\xi_+^2/4n}) \tag{7a}$$

$$I_a(1) = \frac{\tau_1}{2\sqrt{\pi}} \sum_{n=1}^{\infty} \frac{1}{\sqrt{n}} (e^{-\chi_-^2/4n} - e^{-\chi_+^2/4n}). \tag{7b}$$

Note that the functional forms for $I_\infty(1)$ and $I_a(1)$ are identical, the expressions differing only in the arguments $\xi_\pm$ and $\chi_\pm$; a similar result is found for $I(\Delta)$. The expression for $I_\infty(\Delta)$ is

$$\begin{aligned} I_\infty(\Delta) = &-(1/2)\xi_-\left[1 - \phi\left(\frac{\xi_-}{2\sqrt{\tau_1}}\right)\right] + \\ &(1/2)\xi_+\left[1 - \phi\left(\frac{\xi_+}{2\sqrt{\tau_1}}\right)\right] + \sqrt{\frac{\tau_1}{\pi}}\,(e^{-\xi_-^2/4\tau_1} - e^{-\xi_+^2/4\tau_1}) \end{aligned} \tag{8a}$$

where

$$\begin{aligned} \phi(\eta) &= \text{probability integral} \\ &= \frac{2}{\sqrt{\pi}} \int_0^\eta e^{-u^2}\, du. \end{aligned} \tag{9}$$

$I_a(\Delta)$ is given by

$$I_a(\Delta) = \text{(8a) with } \xi_\pm \to \chi_\pm. \tag{8b}$$

B. *Memory Cycle Time*

Memory cycle time is determined by the ratio of the incremental to dc temperature at the memory bit, i.e., by

$$\begin{aligned} \nu &= \frac{v'(\Delta)}{v'(1)} \\ &= \frac{I'(\Delta)}{I'(1)}. \end{aligned} \tag{10}$$

The absolute minimum value of ν for successful memory operation is 1, while in practice a somewhat larger value will be required. To investigate the behavior of ν it is

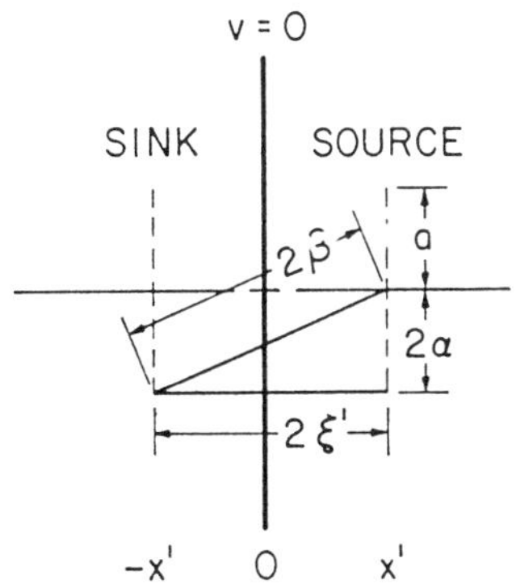

Fig. 2. Relationships between $\xi_\pm$ and $\chi_\pm$ at $x = x'$ ($\xi_\pm' = 2\xi'$, $\chi_+' = 2\beta$, $\chi_-' = 2\alpha$).

necessary to consider the values of $\xi_\pm$ and $\chi_\pm$ at $x = x'$ (Fig. 2),

$$\xi_+' = 2\xi'$$
$$\xi_-' = 0 \tag{11a}$$

$$\chi_+' = \frac{1}{D}\sqrt{(2x')^2 + a^2} = 2\beta$$
$$\chi_-' = \frac{a}{D} = 2\alpha. \tag{11b}$$

Note that (see Fig. 2)

$$\beta^2 = \alpha^2 + \xi'^2. \tag{12}$$

In addition to the diffusion length $D = \sqrt{\kappa T}$, it is useful to introduce a second diffusion length determined by the heating time according to

$$D_1 = \sqrt{\kappa T_1}. \tag{13}$$

From (7) and (8) it is seen that $I'(1)$ and $I'(\Delta)$ will be determined by x', a, and $\sqrt{(2x')^2 + a^2}$ normalized to D or D_1, respectively. Hence it is useful to define

$$\xi_1' = \frac{x'}{D_1}$$
$$2\beta_1 = \frac{1}{D_1}\sqrt{(2x)^2 + a^2}$$
$$2\alpha_1 = \frac{a}{D_1}. \tag{14}$$

For convenient reference $d = \sqrt{\kappa t}$ for representative materials and times is given in Table I, which shows a range of d from 0.003 to 10μm.

Next consider the values of a and x' which will be encountered in practice. The minimum value of a is determined by optical diffraction and for visible light will be $\sim$1 μm; in principle there is no upper limit to a. If x' is taken as the thickness of the boundary layer of a flowing liquid it is not a simple matter to determine practical values of x' although an order of magnitude estimate will be attempted as follows: For laminar flow over a flat plate, x' depends on the distance along the plate y and the Reynolds number based on y, R_y, according to[3]

$$x' = \frac{5y}{\sqrt{R_y}} \tag{15}$$

where

$R_y = \rho u y/\mu$
u — velocity of the main stream
ρ — density
μ — viscosity. (16)

Now x' cannot be made indefinitely small since turbulence[4] occurs for some critical value of the Reynolds number. Experimentally $R_{y\,\mathrm{crit}}$ depends on the extent to which the input flow is free of turbulence. The largest values of $R_{y\,\mathrm{crit}}$ which have been observed are in the range

$$R_{y\,\mathrm{crit}} \sim 10^6 \tag{17}$$

hence for a 1-cm memory plane

$$x' \sim 50\ \mu. \tag{18}$$

The preliminary nature of this estimate of x' should be emphasized.

It is clear that the longer the diffusion length the better the cooling. This suggests the use of a liquid metal as a cooling fluid. For a 1-μs cycle time, i.e., $T = 10^{-6}$ seconds and Hg as a coolant, Table I gives

$$D = 2\mu\mathrm{m}. \tag{19}$$

Hence from (18) and (19)

$$\xi' = 25. \tag{20}$$

Since the incident medium must be transparent, e.g., glass, the diffusion length in this medium will be much less than in the liquid metal coolant. Hence the model for heat flow should be enlarged to account for a layered structure. Heat flow in layered structures can be treated by using characteristic matrices[4] analogous to the use of such matrices in the theory of the optical behavior of

TABLE I

Thermal Diffusion Length $d = \sqrt{\kappa t}$ for Representative Materials and Times

	κ(cm²/s)	$d = \sqrt{\kappa t}$ (μm) $t = 10^{-6}$ seconds	10^{-8}	10^{-10}
Ag	1.7	13	1.3	0.13
Glass	0.006	0.8	0.08	0.008
Hg	0.04	2	0.2	0.02
Water	0.001	0 3	0.03	0.003

[4] It is assumed that turbulent flow would result in erratic cooling on the desired μsecond or less time scale of memory operation.

multilayer films, see Section III C. Although no conceptual difficulties are expected, the labor of such calculations will be considerable. For the present discussion the assumption $\kappa_2 = 0$ is certainly adequate, which merely introduces a factor of 2 in v_0.

The value of ν will depend on the relative magnitudes of x', a, D, and D_1. First note that $D > D_1$; then from Table I and the previous estimates of a and x' it is seen that it is always true that $D < x'$, while on the other hand $D < a$, $D_1 < a < D$, or $a < D_1$. These relationships are depicted graphically in Table II and designated as cases 1, 2, 3, and 4, respectively. It can be anticipated that case 3 will yield the most favorable value of ν, since the conditions $D_1 < a$ and $a < D$ provide for efficient incremental heating and dc cooling, respectively.

From Appendix II [see (129) and (128)]

$$I_\infty'(1) = \tau_1\left(\xi' - \frac{3}{2\sqrt{\pi}}\right), \ (\xi' > 1) \tag{21a}$$

$$I_a'(1) = I_a'(1, \beta) - I_a'(1, \alpha) \tag{21b}$$

where for α or $\beta > 1$ [(129)]

$$I_a'(1, \alpha) = \tau_1\left(\alpha - \frac{3}{2\sqrt{\pi}}\right)$$

$$I_a'(1, \beta) = \tau_1\left(\beta - \frac{3}{2\sqrt{\pi}}\right) \tag{22}$$

and for α or $\beta < 1$ [(130)]

$$I_a'(1, \alpha) = 1.49\ \tau_1\alpha^2$$

$$I_a'(1, \beta) = 1.49\ \tau_1\beta^2. \tag{23}$$

TABLE II

DEPENDENCE OF INCREMENTAL AND DC TEMPERATURES AND THE RATIO OF THESE TEMPERATURES ON THE RELATIVE MAGNITUDES OF THE PARAMETERS D_1, D, x', AND a. THE MOST FAVORABLE PARAMETER CONFIGURATION IS CASE 3. IN THIS CASE THE CONDITIONS $D_1 < a$ AND $D > a$ PROVIDE FOR EFFICIENT INCREMENTAL HEATING AND DC COOLING, RESPECTIVELY.*

Case	Distance Relationships	$I'(\Delta)$	$I'(1)$	ν
1	0 D_1 D x' a	$\sqrt{\frac{\tau_1}{\pi}}$	$\tau_1\xi'$	$\frac{1}{\sqrt{\pi}\xi'\sqrt{\tau_1}}$
2	0 D_1 D a x'	$\sqrt{\frac{\tau_1}{\pi}}$	$\tau_1\xi'$	$\frac{1}{\sqrt{\pi}\xi'\sqrt{\tau_1}}$
3†	0 D_1 a D x'	$\sqrt{\frac{\tau_1}{\pi}}$	$1.49\tau_1\alpha^2$	$\frac{1}{1.49\sqrt{\pi}\alpha^2\sqrt{\tau_1}}$
4†	0 a D_1 D x'	α	$1.49\tau_1\alpha^2$	$\frac{1}{1.49\alpha\tau_1}$

* The table entries are good approximations when $\xi' \gtrsim 1.5$, $\alpha \lesssim 0.5$.
† Multiply $I'(\Delta)$ and $I'(1)$ by 2 to obtain the temperature for the case $\kappa_2 = 0$.

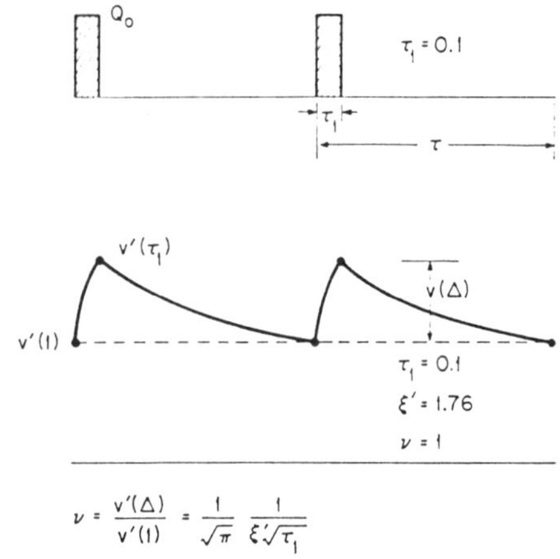

Fig. 3. Steady-state temperature at $x = x'$ vs. normalized time for $\nu = 1$.

Also from Appendix II [(123) and (122)]

$$I_\infty'(\Delta) = \xi'[1 - \phi(\xi')] + \sqrt{\frac{\tau_1}{\pi}}\,(1 - e^{-\xi_1'^2}) \tag{24a}$$

$$I_a'(\Delta) = -\alpha[1 - \phi(\alpha_1)] + \beta[1 - \phi(\beta_1) + \sqrt{\frac{\tau_1}{\pi}}\,(e^{-\alpha_1^2} - e^{-\beta_1^2}). \tag{24b}$$

Using these equations $I'(\Delta)$, $I'(1)$, and ν are readily evaluated and are given in Table II for the distance relationships discussed previously. The table entries are good approximations when $\xi' \gtrsim 1.5$, $\alpha \lesssim 0.5$. As an example of the use of the previous results consider case 1 or 2 of Table I. The steady-state temperature $v'(\tau)$ for $\nu = 1$ and $\tau_1 = 0.1$ is shown in Fig. 3; the resulting value of $\xi' = 1.76$ is large enough for the entries in Table II to be valid.

Cases 1 and 2 differ in a qualitative way from cases 3 and 4. Thus for cases 1 and 2 letting $\nu = 1$ gives a least upper bound on τ_1, which with $\xi' = 25$ [see (20)] is found to be

$$\tau_1 \lesssim 4 \times 10^{-4}\tau \tag{25}$$

which is a very short time for the heating pulse. It should be noted, however, that mode-locked lasers are available which can deliver pulses $\sim 10^{-10}$ seconds wide at a 10 MHz rate.[5]

Cases 3 and 4 both give values of $\nu \gg 1$ for conservative values of τ_1. Thus for $\alpha = 0.25$, $\tau_1 = 0.2$

$$\nu \sim 13 \tag{26}$$

for either case 3 or 4. As expected on physical grounds, case 3 for $\alpha \ll 1$ gives the greatest value of ν. It is important to note that for these cases ν does not depend on the distance to the boundary $v = 0$. This occurs because the radius of the memory bit is less than the diffusion length for cooling ($a < D$) so that the dc temperature does not have a chance to build up. The condition $a < D$ can always be satisfied if the period T is long enough;

however for $T \sim 1\ \mu s$ the largest $D(= 13\mu m)$ in Table I is for Ag so that $a \lesssim 6\mu$ is required.

This treatment of heat flow and cycle time has approximated the memory bit as a disc of zero thickness. As long as the thermal diffusion lengths D_1 and D and the radius a are all large compared to the thickness of the memory bit, this approximation should be a good one. Should this not be the case, it is straightforward to extend the calculation to take account of finite source thickness. Another improvement in the model would consider the dependence of absorption of the input optical energy on penetration depth into the magnetic film; for the case of very thin magnetic films on Ag energy will also be absorbed in the Ag. These and other possible refinements of the mathematical model are certainly important and interesting; however, it is to be expected that the principal qualitative features regarding memory cycle time have been elucidated by the present model.

III. Magneto-Optical Reading

A. *Signal-to-Noise Ratio*

1) Shot Noise

The SNR arising from magneto-optical signals and shot noise in a photo-detector has been considered by several authors.[6]–[10] It is most usual to consider an optical configuration in which the incident light is polarized nearly $\|$ or $\perp$ to the plane of incidence and an analyzer is used to transform the magneto-optical rotations of the plane of incidence into an amplitude modulation of the reflected light. However, an analyzer is not necessary if the incident light is polarized at a large angle to the plane of incidence, as was first shown in Lee et al.[6] and is considered in some detail in Section III B of this paper. In this case the conversion from magneto-optical rotation to amplitude modulation occurs at the reflecting surface. Whatever method is used to obtain the final magneto-optical amplitude modulation, the resulting signal power ΔP resulting from a reversal of the magnetization is always of the same order of magnitude and for the present purposes can be taken as[5]

$$\Delta P = P_i r k \tag{27}$$

where

P_i power in the incident light
k magneto-optical coefficient (Section III C)
r reflection coefficient (Section III C) $\simeq 1/\sqrt{2}$. (28)

[5] For example, when an analyzer is used a typical expression or the signal power is $\Delta P = P_r \psi \sin 2\theta$, where P_r = reflected power, ψ = magneto-optical angle $\simeq 10^{-3}$ radians, θ = analyzer setting from minimum transmittance [(69)]. Now $\psi \simeq k/r$ and $P_r = r^2 P_i$, so that $\Delta P = P_i rk \sin 2\theta$, which is (27) except for the $\sin 2\theta$-factor. This factor is not important for the present discussion, however, since the fact that often, in practice, the analyzer is set for $\theta \simeq 0$ is not due to shot-noise considerations. When an analyzer is not used the signal is given, for example, by (44) or (53) of Section III B. In this case the difference $r_\perp - r_\|$ can typically be taken as $\sim 1/\sqrt{2}$.

The average reflected power $\bar{P}$ is of the order of

$$\bar{P} \simeq P_i r^2. \tag{29}$$

The shot-noise current I_n in photo detectors is given by the well known formula[11]

$$I_n = (2\Delta f)^{1/2}(eq\bar{P})^{1/2} \tag{30}$$

where

Δf bandwidth
e electronic charge
q conversion coefficient of the photo cathode. (31)

The signal current is

$$I_s = q\Delta P \tag{32}$$

so that the resulting SNR is

$$\text{SNR} = I_s/I_n = \left(\frac{qP_i}{2e\Delta f}\right)^{1/2} k. \tag{33}$$

From Table II for case 3 the temperature at the end of the reading (heating) time Δv, multiplied by two to include the assumption $\kappa_2 = 0$, is

$$\Delta v = 2v_0 \sqrt{\frac{\tau_1}{\pi}},\ (\kappa_2 = 0). \tag{34}$$

Taking v_0 from (2) and using $P_a = P_i(1 - r^2)$ gives

$$P_i = \frac{\pi^{3/2}K}{2(1 - r^2)D_1}\, a^2 \Delta v \tag{35}$$

from which

$$\text{SNR} = 1/2\left[\frac{\pi\sqrt{\pi}qK}{e}\,\frac{1}{1 - r^2}\right]^{1/2} \times \left[\frac{T\Delta v}{D_1}\right]^{1/2} ak,\ (r \sim 1/\sqrt{2})$$
$$\simeq 5 \times 10^8 \left[\frac{T\Delta v}{\sqrt{T_1}}\right]^{1/2} ak \tag{36}$$

where $q = 3 \times 10^{-2}$ A/W, $e = 1.6 \times 10^{-19}$ C, $r = 1/\sqrt{2}$, and for an Ag heat sink $K = 4$ J/s·cm·degree, $\kappa = 1.7$ cm²/s. This expression for SNR is the same as that obtained by Treves[10] with $T^{1/2}/T_1^{1/4}$ replaced by $T^{1/4}$.

In the discussion of cycle time it was shown that for $\alpha = 0.25$, $T = 1\ \mu s$, $\tau_1 = 0.2$ the ratio $\nu = 13$ [see (26)] and $a \sim 6\ \mu m$. For these same parameter values

$$\text{SNR} \simeq 10^4\sqrt{\Delta v}k \tag{37}$$

so that for $\Delta v = 10$ degrees and $k = 10^{-3}$, the SNR $\simeq 50$. The power required to obtain the temperature Δv is

$$P_i \simeq 5 \times 10^{-3}\Delta v \text{ watts}. \tag{38}$$

Note that in principle for given allowed Δv the SNR can be made as large as desired by taking T_1 small enough; the effect is quite weak, however, depending only on $T_1^{-1/4}$. Physically this occurs because as T_1 decreases P_i

must increase in order to keep $v(\tau_1)$ constant. Then by (33), the SNR increases, an effect which has essentially been noted previously by Treves.[10] Perhaps advantage could be taken of the dependence of SNR on $T_1^{-1/4}$ by using a mode-locked laser which could deliver very short high-power pulses. However, it should be pointed out that the modulation scheme described in Section III B would no longer be practical.

It is interesting to compare the SNR for case 3 [(36)] with case 4. Calculation gives

$$\frac{(\text{SNR})_3}{(\text{SNR})_4} = \sqrt{\pi}\,\frac{a}{(D_1)_3} \tag{39}$$

and since $(a/D_1)_3 > 1$, the most favorable conditions for reading and writing are the same.

2) Other Noise

Treves[10] has discussed two additional sources of noise, namely: 1) surface noise due to imperfections (e.g., pin holes) in the magnetic film, and 2) laser noise. If an analyzer is used it is found that the SNR from surface noise $(\text{SNR})_S$ increases as the angle θ measured from extinction increases; conversely, the SNR from laser noise $(\text{SNR})_L$ decreases with θ. Furthermore, both of these noise sources are large compared to shot noise. Hence the optimum analyzer setting occurs when the $(\text{SNR})_S$ and $(\text{SNR})_L$ are approximately equal.

Methods of reducing surface noise have been described by Treves,[10] namely: 1) use of aperture stops to eliminate the diffracted depolarized light from small pin holes, and 2) split-beam dc signal processing in which the signal is additive in the two beams while the noise cancels. Although these methods are very powerful some difficulties remain, namely: 1) the use of an analyzer limits the extent to which a focussed beam can be used,[12] 2) perfect balance cannot be achieved. A signal processing technique is described in the next section which can, in principle, overcome these problems.

B. Incident-Light Modulation

To extract the magneto-optical signal from surface noise some modulation of the incident light is desired which will produce a signal which depends on the magnetization of the memory bit but which is independent of surface perfection. Furthermore, since laser noise is characterized by a high-frequency cut-off, such noise can be eliminated if the magneto-optical signal appears as modulation on a carrier of sufficiently high frequency. A final requirement is that the modulation scheme should work even for fully focussed light. The theory for a type of modulation satisfying these requirements is given next, followed by a short discussion of how such a scheme can be realized physically as well as some preliminary experimental results.

1) Theory

For simplicity, the following modulation theory will be given in terms of reflection effects; only slight changes in notation are required for the corresponding transmission effects. Magneto-optical signals without the use of an analyzer were first observed by Lee et al.[6]

a) Magneto-Optical Intensity Modulation Without an Analyzer: Reflection from a magnetic film can be described by means of a reflection matrix defined by [see (64)]

$$\begin{bmatrix} e_\perp \\ e_\parallel \end{bmatrix}^{(r)} = \begin{bmatrix} r_\perp & k_\perp \\ k_\parallel & r_\parallel \end{bmatrix} \begin{bmatrix} e_\perp \\ e_\parallel \end{bmatrix}^{(i)} \tag{40}$$

where the subscripts $\perp$ and $\parallel$ refer to components of field perpendicular and parallel to the plane of incidence, respectively.

Equation (40) is general and holds for any orientation of the magnetization $\vec{M}$, although when $\vec{M}$ is perpendicular to the propagation vector of the incident light (transverse orientation) the ks are equal to zero. For nonzero ks it is true in general that

$$k(\vec{M}) = -k(-\vec{M}). \tag{41a}$$

Also it is usually true that to a good approximation

$$k_\perp \simeq -k_\parallel. \tag{41b}$$

Finally, the crucial observation is that

$$r_\perp \neq r_\parallel \tag{41c}$$

either in phase or amplitude.

Now for simplicity, assume

$$r \text{ and } k = \text{real}$$

$$r_\parallel > 0$$

$$r_\perp = -r_\parallel. \tag{42}$$

Then for the incident $\vec{e}$ polarized at $+45°$ to the plane of incidence the vector diagram of Fig. 4(a) can be constructed; the subscripts 0 and 1 are used to denote oppositely directed magnetization states and the simplified notation

$$k \doteq k_\perp = -k_\parallel \tag{43}$$

is used. From Fig. 4(a) it is seen that the reflected optical electric fields, e_0^r and e_1^r, which are associated with the *zero* and *one* magnetic states, differ in amplitude. Hence a photodetector which responds to energy, i.e., $|e|^2$, can distinguish between these states without the use of an analyzer.

It follows from the vector diagram that for real r and k and to first-order in k

$$r^2 = r_\parallel^2 + r_\perp^2 + 2k(r_\perp - r_\parallel). \tag{44}$$

Also for real r and k it is known that k depends on the rs according to [see (77)]

$$k = (\text{const})[(1 - r_\perp^2)(1 - r_\parallel^2)]^{1/2}. \tag{45}$$

Then maximizing r^2 with respect to $r_\perp$ and $r_\parallel$ gives the conditions

$$r_\parallel = -r_\perp = \pm\sqrt{3} \tag{46a}$$

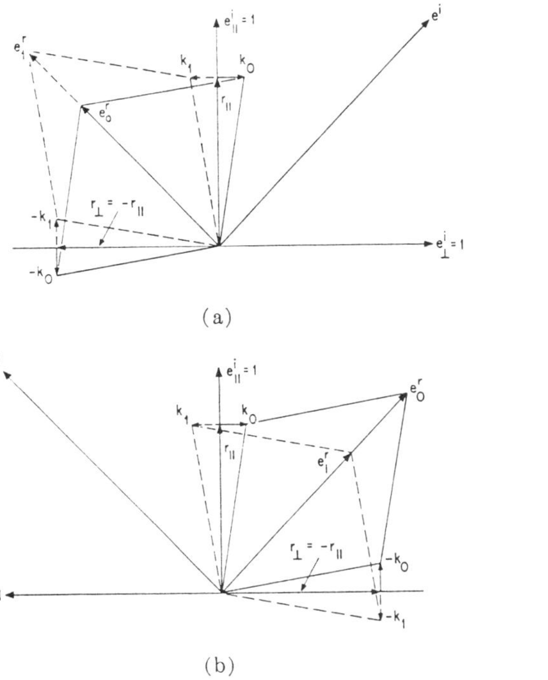

Fig. 4. Magneto-optical intensity modulation without the use of an analyzer. (a) One and zero output for incident $\vec{e}$ polarized at $+45°$ to the plane of incidence. (b) Incident $\vec{e}$ at $-45°$.

so that with k_0 = const in (45)

$$r_{\max}^2 = 2/3(1 \pm 2\sqrt{3}k_0). \tag{46b}$$

For the ferromagnetic metals $|k| \sim 10^{-2}$ to 10^{-3} so that the magnetic term in (44) can easily be obscured by fluctuations in the nonmagnetic term. For example, scanning the surface of a magnetic material with a light beam to determine the local magnetic state will usually result in fluctuations $r_{\|}$ and $r_\perp$ which are greater than k. Hence modulation schemes are necessary which will separate magnetic from surface effects. The simplest such modulation is to induce motion in $\vec{M}$ by means of external magnetic fields. In many situations this is not desirable or possible so that other methods would be of interest.

b) Phase Modulation of the Incident Light: Signals characterized only by the magnetic state can be obtained by phase modulation of the incident light. For example, assume that the phase of $e_\perp{}^i$ is shifted by 90° relative to $e_\|{}^i$ so that in the example of Fig. 4 the incident $\vec{e}$ is now at $-45°$ to the plane of incidence. Then by comparing the vector diagrams of Fig. 4(a) and 4(b) it is seen that the difference in reflected amplitudes for the *zero* and *one* magnetic states, i.e., the quantity $(r_0 - r_1)$, has changed sign. Hence, if the phase of $e_\perp{}^i$ is modulated sinusoidally at frequency ω_m it is anticipated that sinusoidal signals will be generated which have opposite phase for the *zero* and *one* magnetic states.

Consider the arrangement of Fig. 5 in which light is incident at an angle θ to a reflecting magnetic surface. After passing through a polarizer the light is linearly polarized with electric vector $\vec{e}$ at an angle ϕ measured from an axis $\perp$ to the plane of incidence. After passing

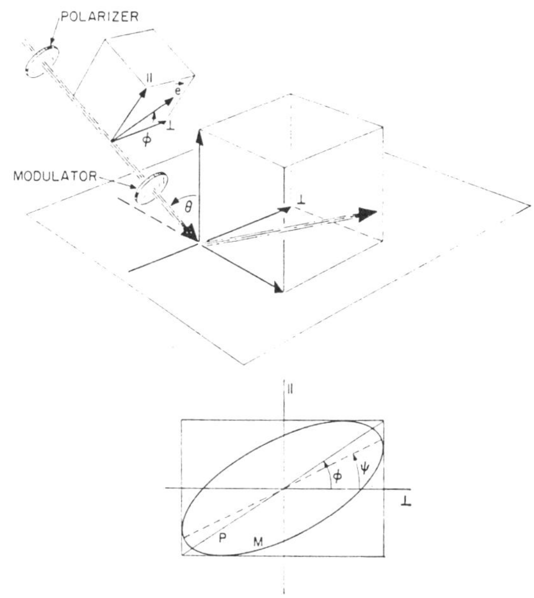

Fig. 5. Experimental arrangement for bit detection using phase modulation of one of the components of the incident light.

through the modulator a phase shift τ is introduced into the component $e_\perp{}^i$, resulting in elliptical light with major axis at an angle ψ. Hence the light incident at the reflecting surface has components

$$e_\|{}^i = \sin\phi \cos\omega t$$
$$e_\perp{}^i = \cos\phi \cos(\omega t - \tau) \tag{47}$$

where

ϕ — orientation of polarizer (Fig. 5)
ω — optical frequency
$\tau = \tau_0 \cos\omega_m t$
ω_m — modulation frequency. (48)

The reflected light then has components ($e_\|{}^i = e_\perp{}^i = 1$)

$$e_\|{}^r = r_\perp \cos\phi \cos(\omega t - \tau) + k_\perp \sin\phi \cos\omega t$$
$$e_\|{}^r = r_\| \sin\phi \cos\omega t + k_\| \cos\phi \cos(\omega t - \tau). \tag{49}$$

To first-order in the ks

$$|e_\perp{}^r|^2 = |r_\perp|^2 \cos^2\phi \cos^2(\omega t - \tau) + (k_\perp{}' r_\perp{}' + k_\perp{}'' r_\perp{}'') \sin 2\phi \cos\omega t \cos(\omega t - \tau) \tag{50a}$$

$$|e_\|{}^r|^2 = |r_\||^2 \sin^2\phi \cos^2\omega t + (k_\|{}' r_\|{}' + k_\|{}'' r_\|{}'') \times \sin 2\phi \cos\omega t \cos(\omega t - \tau) \tag{50b}$$

where

$$r = r' + ir''$$
$$k = k' + ik''. \tag{51}$$

Now since in practice $\omega \gg \omega_m$ the modulation τ may be considered constant over an optical period so that the average of $\cos^2(\omega t - \tau)$ over an optical period is unity, independent of τ, i.e.,

$$\langle \cos^2(\omega t - \tau) \rangle = 1. \tag{52a}$$

The average over an optical period of $\cos \omega t \cos (\omega t - \tau)$, however, does depend on τ and is given by

$$\langle \cos \omega t \cos (\omega t - \tau) \rangle = \cos \tau. \quad (52b)$$

Hence the phase modulation ac signal S_T is given by

$$S_\tau = \langle |e_{||}|^2 + |e_\perp| \rangle^2_{ac} = K \sin 2\phi \cos \tau \quad (53)$$

where

$$K = (k_\perp' r_\perp' + k_\perp'' r_\perp'') + (k_{||}' r_{||}' + k_{||}'' r_{||}'') \quad (54a)$$

or when $k = k_\perp = -k_{||}$

$$K = k'(r_\perp' - r_{||}') + k''(r_\perp'' - r_{||}''). \quad (54b)$$

The harmonic content of $\cos \tau$ is given by the Fourier expansion

$$\cos \tau = J_0(\tau_0) + 2\sum_1^\infty (-1)^n J_{2n}(\tau_0) \cos 2n\omega_m t \quad (55a)$$

where the Js are Bessel functions. For future reference also

$$\sin \tau = 2\sum_1^\infty (-1)^{n+1} J_{2n-1}(\tau)_0 \cos (2n - 1)\omega_m t. \quad (55b)$$

Note that $\cos \tau$ and $\sin \tau$ contain only even and odd harmonics, respectively.

It is important to note that for this type of modulation $\langle |e|^2 \rangle$ does not contain any terms independent of k, a fact which is due to (52a). However, if the modulation is not perfect, i.e., introduces modulation other than specified by (47), nonmagnetic signals will be generated. The most serious undesirable modulation is considered in the next section.

c) Orientation Modulation of the Incident Light: It is clear that magnetic signals can be generated by modulating the orientation of plane-polarized incident light. This method is not practical, however, due to the simultaneous generation of nonmagnetic signals. Since orientation modulation is the principal source of nonmagnetic signals in the experimental realization of phase modulation, the theory is briefly considered in the following.

Equations (50a) and (50b) for reflected energy contain the polarization orientation angle ϕ, which is now taken to depend on time according to

$$\phi = \phi_0 + \delta \quad (56)$$

where

$$\delta = \delta_0 \cos \omega_m t. \quad (57)$$

The nonmagnetic signal is contained in the factors $\sin^2\phi$ and $\cos^2\phi$, while a magnetic signal is generated by $\sin 2\phi$. Calculation then gives for the orientation-modulation ac signals S_ϕ,

$$S_\phi = (R \cos 2\phi_0 + K \sin 2\phi_0) \cos 2\delta - (R \sin 2\phi_0 - K \cos 2\phi_0) \sin 2\delta \quad (58)$$

where

$$R = (1/2)(|r_\perp|^2 - |r_{||}|^2). \quad (59)$$

The harmonic content of $\cos 2\delta$ and $\sin 2\delta$ is obtained from (55a) and (55b).

In the case of orientation modulation all harmonics of both the magnetic and nonmagnetic signals are present. Harmonic separation can be achieved by proper choice of ϕ_0, e.g., when $\phi_0 = \pi/4$

$$S_\phi = K \cos 2\delta - R \sin 2\delta \quad (60a)$$

or when $\phi = -\pi/2$

$$S_\phi = K \sin 2\delta + R \cos 2\delta. \quad (60b)$$

However in practice $R \gg K$ so that small experimental deviations from the simple theory will limit the usefulness of orientation modulation.

The nonmagnetic signals arising from orientation modulation present as a perturbation on phase modulation can be reduced by proper choice of ϕ_0. From (53) note that S_τ is maximum for $\phi_0 = \pi/4$ and furthermore contains only even harmonics; however, from (60a) the even harmonics of the nonmagnetic terms in S_ϕ are zero for this choice of ϕ_0.

2) Physical Realization

Longitudinal magneto-optical experiments with the incident light polarized at 45° to the plane of incidence have been performed using no analyzer and switching the magnetic film to generate an ac signal. The most interesting qualitative observation was the fact that signals are readily observable even in strongly focussed light such that the illuminated area was of the order of 2μ diameter. Of course, more definitive experiments require phase modulation of the incident light. A brief discussion of this problem is given, followed by a description of the use of a split beam in conjunction with phase modulation.

a) Crystal Modulators: Phase modulation of polarized light is commonly accomplished by means of a Kerr or Pockels electro-optical cell. In fact the simple theory of these cells predicts exactly the behavior required for the modulation scheme described before. However, preliminary experiments with a KDP crystal show that additional effects are present which result in large nonmagnetic signals. At present it is assumed that these effects are some form of orientation modulation. Nevertheless, by careful adjustment of the cell orientation and voltages applied to the cell, magneto-optical read-out of large area films has been accomplished. At this point there is no reason to believe that electro-optical phase modulation cannot be made to work satisfactorily, but the problems have not been carefully studied.

b) Moving Mirror: Another approach to optical phase modulation is to use a moving mirror. Incorporation of a moving mirror into an interferometer has been used[13] to accomplish amplitude modulation of light. Applied to the present problem the arrangement shown in Fig. 6 is proposed and for the most part is self-explanatory. If the moving mirror is a Z-cut quartz crystal electrically driven at some high harmonic of the fundamental longitudinal mode, frequencies of 10 to 100 MHz should be easily attainable.

c) Split Beam: The use of a split beam in conjunction with phase modulation is shown in Fig. 7. The planes of incidence for memories 1 and 2 are at right angles so that signals generated by spurious orientation modulation are out-of-phase and cancel in the difference amplifier. However, by appropriate choice of the directions of magnetization in the two memories, the magneto-optical signals can be made to add.

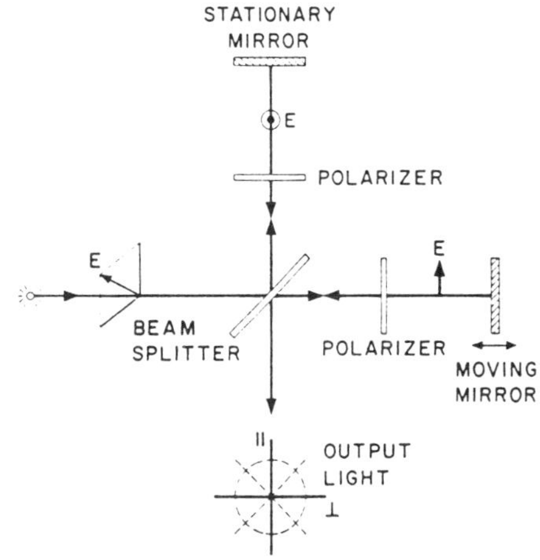

Fig. 6. Interferometer arrangement to obtain phase modulation of one of the components of polarized light by means of a moving mirror.

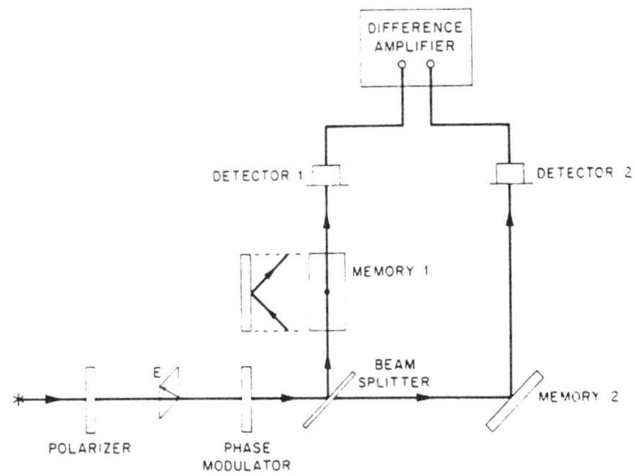

Fig. 7. Split beam used in conjunction with an imperfect phase modulator to cancel unwanted orientation modulation signals.

C. *Enhancement Theory*

A very considerable body of work has been done on the question of whether magneto-optical effects can be *enhanced* by use of auxiliary dielectric layers.[8], [9] [14]–[24] The theory of this problem is somewhat involved and the physical conclusions tend to be obscured by algebra. Hence a qualitative review will be given here which emphasizes physical results.

1) *Statement of the Enhancement Problem*

To formulate the enhancement problem consider the reflection or transmission of polarized light from a magnetic film [Fig. 8(a)]. The magnetization is oriented either in the plane of the film and simultaneously in the plane of incidence (longitudinal orientation), or normal to the film (polar orientation). Then with the incident light polarized in the plane of incidence magneto-optical interaction results in the appearance of a component of the reflected or transmitted light which is polarized perpendicularly to the plane of incidence, i.e., there has been magneto-optical mode conversion from the $\|$ to the $\perp$ mode of polarization; the converse effect occurs for the $\perp$ mode incident. Due to the presence of the converted mode in the output light, the plane of polarization of the output light is rotated by some angle ψ with respect to the incident light. This angle will be referred to as the magneto-optical angle ψ.[6] Experimentally ψ is easy to measure and hence it is natural to state the enhancement problem as follows: *To what extent can the magneto-optical angle ψ be increased by the use of general multilayer dielectric film structures above and/or below the magnetic film* [Fig. 8(b)].

However, consideration of the theory of these magneto-optical angles shows that this statement of the enhancement problem is not satisfactory. A better formulation will be given after a brief discussion of this theory.

[6] Historically a confusing terminology has been employed in referring to this angle. The difficulty stems from inconsistent labeling of special cases with special names. For example, transmission and reflection magneto-optical effects are often denoted as the Faraday and Kerr effect, respectively. However, some authors adopt the view that the Faraday effect is restricted to the polar orientation and the Kerr effect to the longitudinal orientation. Since there is no qualitative and only small quantitative differences in any of these special cases, the unspecific designation of *magneto-optical angle* will be used in this paper.

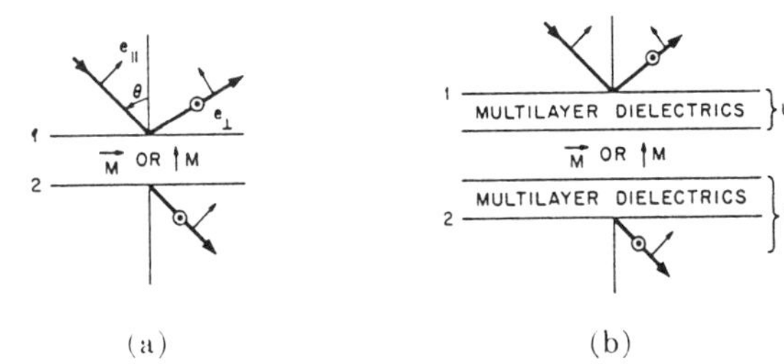

Fig. 8. General magneto-optical problem. (a) Longitudinal or polar magneto-optical mode conversion. (b) General *enhancement* problem.

A description of the magneto-optical effects under discussion can be obtained at the level of Maxwell's equations by introducing a skew-symmetric dielectric tensor

$$\epsilon = \epsilon \begin{bmatrix} 1 & -iq & 0 \\ iq & 1 & 0 \\ 0 & 0 & 1 \end{bmatrix} \tag{61}$$

where the magnetization $\vec{M}$ is taken along the z direction and in general ϵ and q are complex. The theoretical enhancement problem then is to express the output waves in Fig. 8(b) in terms of the input wave, the tensor ϵ, and the parameters describing the multilayer structures. As a first step in this problem it is usual to define the *scattering* coefficients which relate the amplitudes of the output waves to the input wave. An example of such a relation might be

$$e_{\perp}{}^{r} = r_{\perp}e_{\perp}{}^{i} + k_{\perp}e_{\|}{}^{i} \tag{62}$$

where

r	(ordinary) reflection coefficient
k	(reflection) magneto-optical coefficient
$\perp, \|$	refer to the electric field perpendicular and parallel to the plane of incidence, respectively
r, i	refer to the reflected and incident waves, respectively. (63)

The entire set of scattering coefficients required to describe the problem depicted in Fig. 8, generalized to include input light at the lower boundary and $\|$ polarization in the inputs, is conveniently defined by the matrix equation[7]

$$\begin{bmatrix} e_{\perp 1} \\ e_{\| 1} \\ e_{\perp 2} \\ e_{\| 2} \end{bmatrix}^{(s)} = \begin{bmatrix} r_{\perp} & k_{\perp} & t_{\perp}' & \kappa_{\perp}' \\ k_{\|} & r_{\|} & \kappa_{\|}' & t_{\|}' \\ t_{\perp} & \kappa_{\perp} & r_{\perp}' & k_{\perp}' \\ \kappa_{\|} & t_{\|} & k_{\|}' & r_{\|}' \end{bmatrix} \begin{bmatrix} e_{\perp 1} \\ e_{\| 1} \\ e_{\perp 2} \\ e_{\| 2} \end{bmatrix}^{(i)} \tag{64}$$

where

t	(ordinary) transmission coefficient
κ	(transmission) magneto-optical coefficient
s, i	refer to scattered (output) and incident (input) light, respectively
1, 2	refer to the top and bottom boundaries, respectively (see Fig. 8)
$'$	refers to the output due to input light impinging from below (boundary 2). (65)

The matrix of coefficients in (64) is commonly called the scattering matrix.

The principal labor in the theory is now to calculate the elements of the scattering matrix in terms of the dielectric tensor $\bar{\epsilon}$. A description of the mathematical procedure is given in Smith.[19] For the present purpose, however, it is required only to express the magneto-optical angle ψ in terms of k or κ and r. For simplicity, consider a reflection experiment. Then ψ, k, and the incident mode r are connected by the equations relating the parameters of a polarization ellipse to the orthogonal fields k and r, i.e., for an ellipse with ellipticity χ and major axis at an angle ψ to the plane of incidence[25]

$$\begin{aligned} \tan 2\psi &= \tan 2\rho \cos \tau \\ \sin 2\chi &= \sin 2\rho \sin \tau \end{aligned} \tag{66}$$

where

$$\begin{aligned} \tan \rho &= \left|\frac{k}{r}\right| \\ r &= r_0 e^{i\tau_r} \\ k &= k_0 e^{i\tau_k} \\ \tau &= \tau_k - \tau_r. \end{aligned} \tag{67}$$

For small ψ it follows that

$$\psi \simeq \left|\frac{k}{r}\right| \cos \tau \tag{68a}$$

$$\chi \simeq \left|\frac{k}{r}\right| \sin \tau. \tag{68b}$$

From (66), (67), and (68) it is clear that the magneto-optical angle ψ can be made as large as desired by using well-known antireflection dielectric coating techniques to make $r \to 0$. Therefore it is clear that a measurement of the resultant enhanced ψ does not elucidate the extent to which the basic magneto-optical interaction in the magnetic material has been enhanced, if at all, by the use of antireflection layers. The situation can be made more precise by considering the magneto-optical contrast and magneto-optical signal. Thus, for an analyzer set at an angle θ from the angle for minimum transmission, the difference in light intensity upon reversing the magnetization is

$$\begin{aligned} \Delta P_r &= P_r[\sin^2\theta - \sin^2(\theta - \psi)] \\ &\simeq P_r \psi \sin 2\theta \text{ (small } \psi) \\ &\simeq P_i|rk| \sin 2\theta \; (\cos \tau \simeq 1). \end{aligned} \tag{69}$$

Then the signal $= \Delta P$ and the contrast $= \Delta P/\bar{P}$ are given by

$$\text{signal} \simeq P_i|rk| \sin 2\theta \tag{70a}$$

$$\text{contrast} = 2\psi \operatorname{cotan} \theta. \tag{70b}$$

Hence if $r \to 0$ in order to make $\psi \to \pi/2$, the signal goes to zero,[8] while the contrast becomes maximum.[9]

A satisfactory statement of the enhancement problem can now be given in terms of the magneto-optical coefficients (k or κ), which provide a direct measure of the energy which is magneto-optically converted from the incident $\perp$ or $\|$ polarization to the opposite output polarization: *To what extent can the magneto-optical coefficients (k or κ) be increased by the use of general multilayer dielectric film structure above and/or below the magnetic film* [Fig. 8(b)].

2) Present Status of Enhancement

In discussing the answer to the k or κ enhancement problem it is first necessary to make a qualitative dis-

[7] The labeling of the ks and κs is arbitrary; the present choice of subscripts is opposite from that used elsewhere.[19]–[23] Considerable notational simplification occurs later in the theory with this new choice.

[8] The signal does not actually go to zero due to higher order terms in ψ.

[9] From (70b) for any $\psi \neq 0$ the contrast can be made as large as desired by taking θ small enough. In practice this does not occur due to imperfections in the optical system.

tinction between magnetic materials based on the relative magnitude of the magneto-optical constant q and the optical loss. A convenient and easily visualized comparison of this kind is obtained by defining the normalized Faraday rotation f as the ratio of the rotation of the plane of polarization of linearly polarized light propagating along M in an infinite medium, i.e., the Faraday rotation F,[6] to the absorption coefficient α. In terms of the elements of the dielectric tensor

$$\text{Faraday rotation} = \frac{\pi|nq|}{\lambda_0} \text{ rad/cm}$$

$$\text{Absorption coefficient} = \frac{2\pi n''}{\lambda_0} \text{ cm}^{-1} \quad (71)$$

where

$$\lambda_0 = \text{free space wavelength}$$

$$n = c\sqrt{\mu\epsilon} = n' + in'' \quad (72)$$

so that

$$f = \frac{|nq|}{2n''} \text{ radians.} \quad (73)$$

Values of f for various materials are given in Fig. 9 (in units of degrees).

a) Low f Materials: Most ferromagnetic materials have a value of f which is in the range of 10^{-2} to 10^{-3} radians, and for such materials the theory of enhancement yields the following remarkable relations between the magneto-optical and ordinary reflection coefficients:

$$|k|^2 = |k_0|^2(1 - |r_\perp|^2)(1 - |r_{||}|^2)$$
$$|\kappa|^2 = |\kappa_0|^2(1 - |r_\perp|^2)(1 - |r_{||}|^2). \quad (74)$$

Equations (74) are valid for any angle of incidence, any dielectric structures, and for either the longitudinal or polar orientation. A complete physical interpretation of these equations is still lacking, although the factor $(1 - |r|^2)$ involving r for the incident mode is clear enough. This factor accounts for the fact that any output light which is still in the original state of polarization can be thought of as not having interacted with the material and could in principle be recirculated and used as input again, resulting in a net increase in magneto-optical mode conversion. However, a satisfactory physical reason why a similar factor containing the converted mode r appears in k and κ has not yet been given.

Equations (74) contain considerably more information than is indicated by the abbreviated notation which has been used in this initial discussion. The derivation of (74) is discussed in Smith.[19] However, all that is needed for the present discussion is an understanding of some additional notation. Referring again to Fig. 8, the dielectric structure above the film is designated as the U structure, while everything below the film is labeled as the Y structure. Furthermore, the mathematical be-

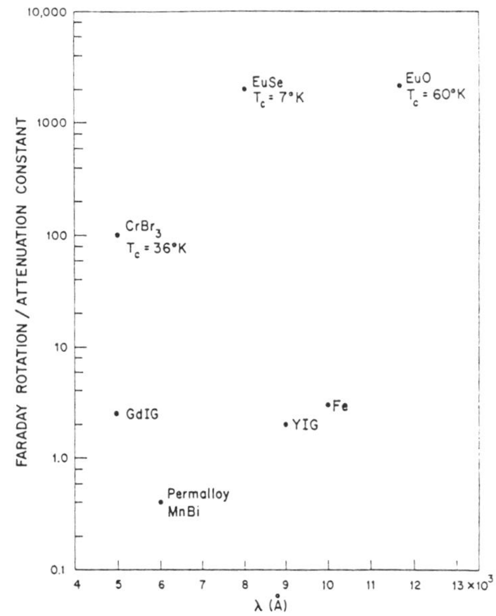

Fig. 9. Faraday rotation normalized to the absorption coefficient, $f = F/\alpha$ degree, for various materials. The data was obtained from the following references: EuO and EuSe;[27] $CrBr_3$;[28] Fe, YIG, and Permalloy;[22] GdIG;[29] MnBi.[30]

havior of these structures can be described by introducing 4×4 U and Y *characteristic* or *transfer* matrices. Now the U and Y structures are assumed to be isotropic, from which it follows that the U and Y matrices consist of two separate 2×2 matrices associated separately with the $\perp$ and $||$ polarization modes. For example, U may be written as

$$U = \begin{bmatrix} U_\perp & 0 \\ 0 & U_{||} \end{bmatrix} \quad (75)$$

where $U_\perp$ and $U_{||}$ are 2×2 matrices associated with $\perp$ and $||$ polarization, respectively. It is convenient to note at this time that a magnetic film (or semi-infinite substrate) is represented in this scheme by a 4×4 matrix of the form

$$\text{Magnetic transfer matrix} = \begin{bmatrix} \perp\text{-mode} & q\text{-coupling} \\ q\text{-coupling} & ||\text{-mode} \end{bmatrix} \quad (76)$$

where the off-diagonal 2×2 matrices are due to the off-diagonal elements iq and $-iq$ in the dielectric tensor $\bar{\epsilon}$ [(61)], and provide for the magneto-optical coupling between the otherwise uncoupled $\perp$ and $||$ modes.

Returning now to (74), which will be rewritten using the previous matrix notation to indicate how the various factors depend on the physical structure, we find

$$|k|^2 = |k_0(Y_I, Y_C)|^2[1 - |r_I(Y_I,U_I)|^2] \times [1 - |r_C(Y_C,U_C)|^2]$$

$$|\kappa|^2 = |\kappa_0(Y_I, U_C)|^2[1 - |r_I(Y_I,U_I)|^2] \times [1 - |r_C(Y_C, U_C)|^2] \quad (77)$$

where

I refers to incident mode
C refers to converted mode. (78)

Of course it is known beforehand that the ordinary reflection coefficients depend only on parameters associated with one or the other polarization modes so that inclusion of this information in (77) is merely for the sake of completeness. The new information lies in the dependence of k_0 and κ_0 on the Y and U structures. The two cases will be discussed separately in the following.

1) *Reflection:* For the reflection case note that k_0 does not depend on the upper dielectric structure at all. This fact yields an immediate and complete formal solution to the enhancement problem for reflection from lossy magnetic materials, namely: *For a Y structure chosen to maximize k_0, choose the U structure to make $r_I = r_C = 0$.*

The use of dielectric films as antireflection[10] coatings is well known in optical technology so that this part of the enhancement problem takes on a familiar aspect. Note that for complete enhancement r_I and r_C should simultaneously be zero, an unusual condition in the established optical art.

The maximum value that k_0 can attain has been calculated for the two limiting cases of a magnetic substrate (infinitely thick film) and a very thin magnetic film.[20] The results are given in Table III. Note that for the longitudinal case k_0 (thin film)/k_0 (thick film) $\simeq 2|n|^2/n''$, which in general is greater than one. Hence a modest improvement is predicted in the maximally enhanced magneto-optical reflection coefficient of a very thin film compared to a thick film of the same material. The reason for this effect is discussed in physical terms in Smith.[20]

Table III also shows that the maximum polar k_0 for a

TABLE III
MAXIMUM ENHANCED VALUE OF THE REFLECTION MAGNETO-OPTICAL COEFFICIENTS*

	$\lvert k_0 \rvert$ Longitudinal	$\lvert k_0 \rvert$ Polar
Magnetic substrate†	$\dfrac{\sin\theta\,\lvert q\rvert}{4\lvert\beta\rvert^2 n'}$	$\dfrac{\lvert nq\rvert}{4n'}$
Very thin film (on $EH_{\parallel}$ mirror)	$\dfrac{\lvert n\rvert^2\lvert\beta q\rvert}{4n'n''}$	—

* *Note:* β measures the direction of light in the magnetic film and usually $\beta \simeq 1$; θ is the angle of incidence of the input light (see Fig. 8).

† It may seem surprising that the limiting values of k_0 for thick films do not contain the optical loss and hence do not depend on the normalized Faraday coefficient [(73)]. This is an accident of the geometry in the sense that most of the converted wave is propagating in the incident direction to infinity (or extinction) and does not appear in the reflected output.

[10] In the language of electrical engineering the antireflection condition corresponds to impedance matching.

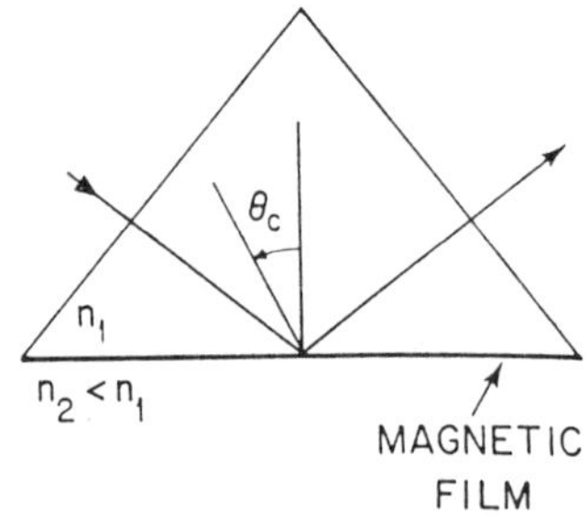

Fig. 10. Magnetic film on a prism.

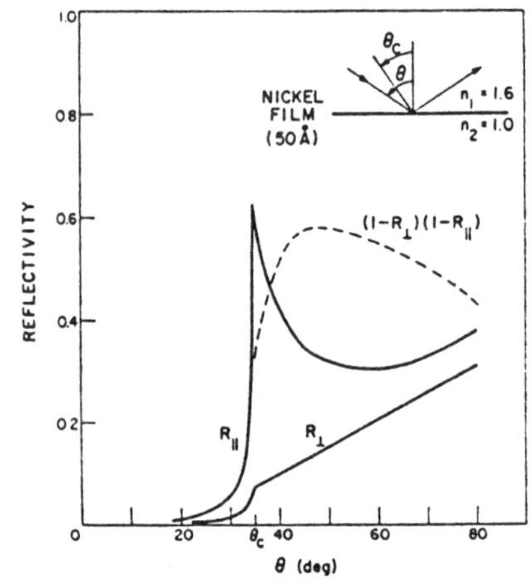

Fig. 11. Impedance matching to a film on a prism.

thick film is greater than the longitudinal k_0 by a factor ~ $|n|$. Physically this is due to the fact that in the magnetic material the light is refracted toward the normal, which in the polar case corresponds to the direction of $\vec{M}$, which in turn corresponds to the direction of maximum magneto-optical interaction according to the form of the dielectric tensor $\bar{\epsilon}$ [(61)]. The maximum polar k_0 has not yet been calculated and since there is no reason to expect any unusual behavior, the result should again yield a small improvement over the thick film case. The fundamental similarity, both qualitatively and quantitatively, between the polar and longitudinal reflection effects should be emphasized, since there is a widespread opinion that the polar effect is considerably superior to the longitudinal effect. These remarks apply equally well to both the reflection and transmission cases.

Most of the experimental work on enhancement has been concerned with antireflection coatings on thick metallic ferromagnets. A detailed review of the results will not be attempted here. In general, reasonable agreement with theory has been found. A somewhat unusual example in which the conditions $r \simeq 0$ can be realized without the use of dielectric coatings is encountered with very thin films when the incident light cannot propagate in the medium beyond the film due to total internal reflection (Fig. 10).[20],[26] Experimental values for $|r_\perp|^2$, $|r_\parallel|^2$, and the resulting calculated value of $(1 - |r_\perp|^2)(1 - |r_\parallel|^2)$ are shown in Fig. 11 for a 50 Å thick N film as a function of the angle of incidence. It is seen that for $\theta > \theta_c$ where θ_c is the critical angle for total internal reflection, very favorable values of $|r_\perp|^2$ and $|r_\parallel|^2$ are observed. Longitudinal magneto-optical reflection angles $\sim$ 1 degree have been observed in similar Permalloy films. A future publication on this work is planned.

2) *Transmisson:* We return now to the questions of transmission enhancement. For this case the problem is more difficult since from (77) it is seen that κ_0 depends on the upper dielectric structure through elements of the U_C matrix. Now $r_C = 0$ does not necessarily occur for the same U structure for which κ_0 is maximum. Note, however, that as far as the incident mode is concerned, enhancement is accomplished as before by merely impedance matching to make $r_I = 0$. The transmission enhancement problem with respect to the converted mode has not yet been rigorously solved. However, theoretical studies to date [21], [23] do not indicate that any direct and significant advantage results from the dependence of κ_0 on U_C.

However, the behavior of κ_0 is notable in a rather unexpected way, namely, it is found that κ_0 depends on $1/\cos\theta_t$ where θ_t is the angle of transmission of the output light (Fig. 8). Hence for small enough θ_t, κ_0 can be made to approach unity.[11] Numerically it is found that for materials such as Fe (see Fig. 9), θ_t must be within a few tenths of a degree of 90°, so that no practical use can be made of this effect.

Although it does not seem possible to make direct use of the effect $\kappa_0 \rightarrow 1$ as $\theta_t \rightarrow \pi/2$, the following indirect approach deserves investigation. Suppose that $\theta \rightarrow \pi/2$, even allowing the converted mode to be beyond the angle for total internal reflection so that propagation of the converted mode is not possible in the output medium. Then, since the transmitted mode is no longer accessible to the outside world, dissipate the converted mode energy within the magnetic film, with a consequent decrease in the reflected incident mode; then rotation of the magnetization to the transverse orientation will turn off the conversion, resulting in an increase in the reflected mode. Calculation of this effect is in progress and appears promising.

b) High f Materials: It remains to discuss enhancement in the case of magnetic materials with a high quality factor, since if $f \sim 1$, the approximation which leads to (74) and (77) is no longer valid. It can be shown by means of a general argument that the enhancement problem is now qualitatively different from an impedance matching problem and the term *conversion matching* has been coined to describe this case. The mathematical argument leading to this conclusion has been given in Smith[19] and will not be repeated here. The procedure is to show that the loss of energy from the incident mode by conversion to the output mode cannot be represented in an equivalent circuit as the real part of a magneto-optical impedance uniquely determined by the $\bar{\epsilon}$ tensor.

[11] Higher order terms in the theory keep $\kappa_0 \leq 1$ as $\theta_t \rightarrow 0$.

No detailed investigation of enhancement in high f material has yet been carried out. This is no doubt due to the fact that until very recently ferromagnetic materials with $f \gtrsim 1$ did not exist. The appearance of such materials makes it likely that the corresponding problem of enhancement will be investigated. It is a reasonable conjecture to suppose that in this case highly efficient magneto-optical mode conversion can be obtained using thin magnetic films and auxilliary dielectrics for any of the configurations of Fig. 8.

3) Relation of Enhancement to the Memory Problem

It seems clear that enhancement will always be very sensitive to the incidence angle of the input light. Hence, in a memory which is read by means of focussed light it is unlikely that enhancement could be used, even if it were fully successful in making k or $\kappa \rightarrow 1$. However, in any application in which it is necessary or desirable to read out with parallel light, enhancement becomes a possibility. One such application arises in connection with the content addressed memory described in Section V. Another application can be conceived if it is desired to reduce the demands on a light deflector by simultaneously illuminating some part of a random-access memory and using a conductor matrix for bit-selection and writing.

IV. Memory Material

The choice of magnetic material to be used in TMO memories will undoubtedly be a subject of considerable research in the coming years. In evaluating potential materials the Faraday constant normalized to the absorption, $f = F/\alpha$, discussed in Section III C, is an important guide. Accurate data for this parameter are not always readily available. Experimental results on new materials are often reported in terms of some observed angle of magneto-optical rotation, a parameter which by itself cannot be interpreted in terms of either f or the elements of the dielectric tensor $\bar{\epsilon}$. The present state of experimental knowledge is briefly summarized in Fig. 9. A complicating factor arises due to the strong dispersion of both F and α. For example, the real and imaginary parts of q for Fe and Ni vs. λ are shown in Fig. 12.[31] The sharp peaks observed in these materials are characteristic of the results for other materials. The entries in Fig. 9 are for the maximum values of f in the visible and near infrared, except for Permalloy and MnBi for which this information is not available. It is likely that f_{max} for Permalloy occurs near that for Ni, i.e., $\sim 5\ \mu$m from Fig. 12.

Of course f is not the only measure of material performance for a TMO memory. Other important factors include, e.g., coercive force, mechanism of thermal writing, thermal excursion for reading, and efficiency of heating. A comprehensive discussion of these factors will not be attempted here. However, a few brief observations will be made.

Coercive force: For bits magnetized in the plane the demagnetizing field $= H_D = 4\pi M$ (diameter/thickness).

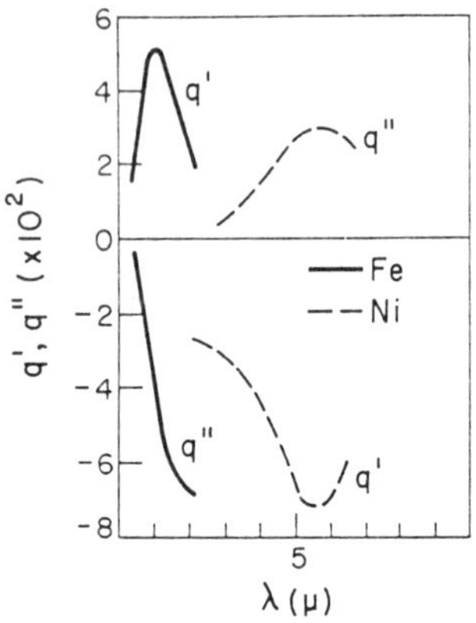

Fig. 12. q vs. λ for Fe and Ni (after Krinchik[31]).

For diameter $= 2\ \mu$m, thickness $0.2\ \mu$m, $H_D = 4\pi M \times 10^{-1}$ ~ 100 to 200 Oe for the ferromagnetic metals. Coercive forces of this order of magnitude can be obtained; bits of $\sim 4\ \mu$m diameter have been reported by Treves[10] in Ni-Co film a few hundred angstroms thick.

Recently an ingenious suggestion was made which eliminates demagnetizing fields in the storage state. It is proposed to use gadolinium iron garnet (GdIG) operating at the compensation temperature T_{comp} as the storage medium.[32] In this material T_{comp} is approximately at room temperature and, furthermore, since a compensation point is characterized by zero net magnetic moment, there are no demagnetizing fields. Heating above T_{comp} produces a net moment so that switching in an external field can be effected. At first sight the principal advantage appears to be that thick films can be used without loss of storage due to demagnetization. However, then only small temperature excursions are permitted in order not to generate too large an external moment, which will limit the SNR calculated in Section III A. Note that according to the data available in Fig. 9, the value of f for GdIG is not exceptional. In a high-density random access application the thickness $\ll$ radius $\sim 6\ \mu$m (see Section II B) in order to prevent thermal spreading from bit to bit.

Mechanism of thermal writing: The most obvious mechanism of thermal writing is to exceed the Curie (or compensation) point. For this mode of writing clearly a material like Fe ($T_c = 770°$C) is not feasible. However, it would seem likely that some suitable material can be found with $T_c < 100°$C. Of course it may not be necessary to exceed T_c to achieve writing. In this connection it should be noted that recently thermal writing on Permalloy was achieved by means of magnetostrictive effects.[33] It is not unreasonable to expect that other unusual mechanisms of writing will be discovered.

One curious feature of the writing problem should be pointed out when considering the random access problem. For a cycle time $\lesssim 1\ \mu$s the memory bit must be placed on a good heat conductor, say Ag (Section II B). Then if a low loss memory material is being considered a problem may arise in heating the bit for writing, since very little optical energy will be dissipated in the bit itself. An auxiliary metallic film on top of the bit might be used to achieve the best heating compromise between writing and reading.

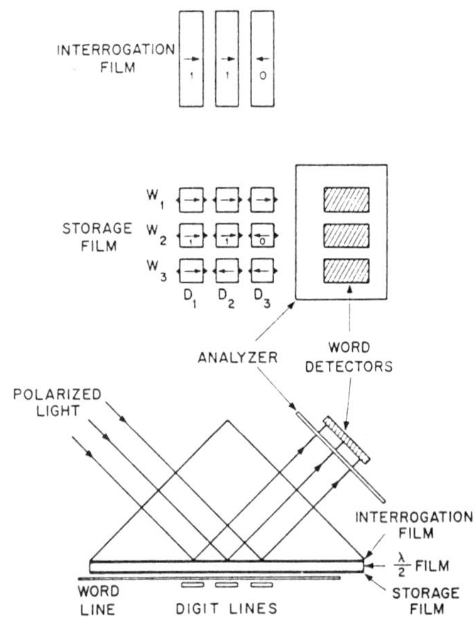

Fig. 13. Magneto-optically interrogated content-addressed memory.

V. Content-Addressed Memory

In 1956 Slade and McMahon[34] described a cryotron memory which was addressed by coincidence between an external key and the contents of a word in memory. Memories of this type are commonly called content-addressed memories (CAM). In order to be useful all words in a CAM memory must be interrogated at the same time. Hence, the existence of high-power cm lasers which could be used to illuminate many bits simultaneously suggests consideration of magneto-optical interrogation for CAM memories.[35],[36] A specific proposal for such a memory is shown in Fig. 13. This system makes use of two magnetic films separated by a $\lambda/2$ dielectric film, where λ is the wavelength of the interrogating beam. The $\lambda/2$ spacing makes the magnetic films optically anticongruent so that the total magneto-optical effect of the two films is the difference of the magneto-optical effects for the separate films. Now designate the bottom film as the storage or memory film, and the top film as the interrogation film. Information is written into the memory film in the conventional way,[37] while the interrogation key is written onto the interrogation film as bands parallel to the digits of the memory film. Then it follows that the detector corresponding to the word which contains the same word as the interrogation will have a null output. The length of a word which can be interrogated is determined by the contrast between matched and unmatched bits. The phase modulation scheme described in Section III B should be particularly useful in achieving high contrast.

To be able to write into the interrogation film without destroying the information in the memory film, it is required that the threshold for interrogation writing must be less than for memory writing. This can be achieved in a variety of ways, perhaps the easiest being to vary the composition of the magnetic films. For example, a high coercive force Ni-Co alloy could be used for the storage film and low coercive force Permalloy for the interrogation film; at the same time, the magneto-optical effects of the two films must be identical. The simultaneous achievement of these conditions should be possible by careful but straightforward experimentation.

Appendix I

Heat Flow From an Infinite Plane Source Periodic in Time

A mathematical model for the thermal response of a magnetic film which is written by optical heating is shown in Fig. 1. As a first step in solving the problem of a finite circular bit, consider the bit to be an infinite heat source at x', bounded on the right by a semi-infinite medium; on the left the temperature $v = 0$ at $x = 0$.

The temperature $v(x, t)$ in this structure is governed by the differential equation[38]

$$\frac{\partial^2 v}{\partial x^2} - \frac{1}{\kappa}\frac{\partial v}{\partial t} = -\frac{w(x,t)}{K} \tag{79}$$

where

K—thermal conductivity
$\kappa = K/\rho c$
ρ—density
c—specific heat
w—rate of heat supplied per unit area at point x and time t. (80)

By definition Green's function $u(x,t;\ x',t')$ is the temperature at x due to an instantaneous unit source[12] at x' at time t', with initial temperature $v(x, o) = 0$ and the boundary kept at $v(o, t) = 0$. The temperature $v(x, t)$ in an infinite solid due to a unit instantaneous plane source at x' and t' is known to be[39]

$$v = \frac{1}{2(\pi\kappa)^{1/2}(t-t')^{1/2}}\, e^{-(x-x')^2/4\kappa(t-t')}. \tag{81}$$

Then Green's function for the geometry of Fig. 1 can be obtained by introducing an instantaneous sink at the image point $-x'$, since by symmetry this insures the boundary condition $v(o, t) = 0$. The desired Green's function is therefore

$$u(x,t;\ x't') = \frac{1}{2(\pi\kappa)^{1/2}(t-t')^{1/2}}\left[e^{-(x-x')^2/4\kappa(t-t')} - e^{-(x+x')^2/4\kappa(t-t')}\right]. \tag{82}$$

[12] The heat liberated by a unit source is ρc.

Now assume that the heat source depends on time according to the general form

$$Q(t) \equiv 0,\ t < 0$$
$$Q(t) \not\equiv 0,\ t > 0 \tag{83}$$

then the temperature at time $t > 0$ is given by

$$v = \int_0^t Q(t')u(t-t')dt'. \tag{84}$$

The Laplace transform of v, $L\{v(t)\} = \bar{v} = \int_0^\infty e^{-pt}v(t)dt$, is

$$\bar{v} = L\left\{\int_0^t Q(t')u(t-t')dt'\right\} = \bar{Q}\bar{u} \tag{85}$$

using the superposition theorem for the Laplace transform.

Next assume the function $Q(t)$ to be pulses of width T_1 and magnitude Q_0 with period T starting at $t = 0$ (Fig. 3). Then $\bar{Q}$ and $\bar{u}$ are given by[40]

$$\bar{Q} = \frac{Q_0(1 - e^{-pT_1})}{p(1 - e^{-pT})} \tag{86}$$

and

$$\bar{u} = \frac{1}{2\kappa q}(e^{-qx_-} - e^{-qx_+}) \tag{87}$$

where

$$q = \sqrt{p/\kappa} \tag{88a}$$
$$x_- = x - x'$$
$$x_+ = x + x'. \tag{88b}$$

The temperature $v(t)$ is found from the inversion formula[41]

$$v(t) = \frac{1}{2\pi i}\int_{\gamma-i\infty}^{\gamma+i\infty} e^{pt}\bar{v}dp \tag{89}$$

where p is now taken as complex and γ is taken large enough that all the poles of the integrand lie to the left of the line Re $p = \gamma$. The singularities of $\bar{v}$ are: 1) branch point at $p = 0$, and 2) poles at $p = i\,(2\pi n/T)$, $n = \pm 1, \pm 2, \ldots$

Hence the integral of (89) can be evaluated by means of the contour shown in Fig. 14 according to

$$v = v_s + v_{tr} \tag{90}$$

where

$$v_s = \text{steady periodic function}$$
$$= \Sigma \text{ residues} \tag{91a}$$
$$v_{tr} = \text{transient function}$$
$$= -(I_{bc} + I_0)$$

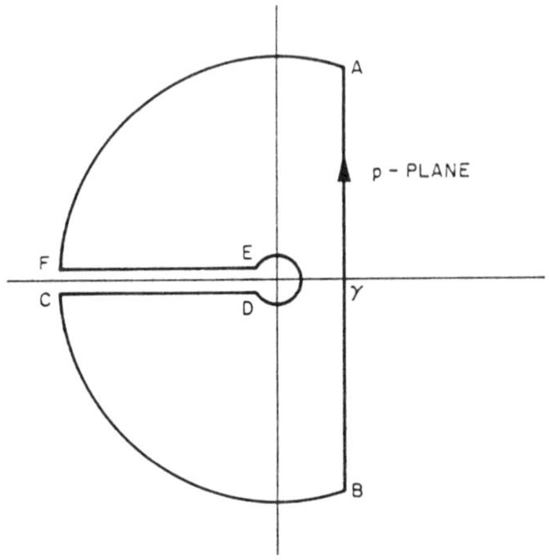

Fig. 14. Contour for evaluating $\int_{\gamma-i\infty}^{\gamma+i\infty} e^{pt}\,\bar{v}dp$.

$$I_{bc} = \text{integral along the branch cut (FE + DC)}$$

$$I_0 = \text{integral around the origin.} \tag{91b}$$

Along the branch cut take $|p| = \rho$ so that with

$$r = \sqrt{\rho/\kappa} \tag{92a}$$

the quantities p and q become

$$p = \rho e^{i\pi} = -\kappa r^2$$

$$q = i\sqrt{\rho/\kappa} = ir \text{ (along F} \rightarrow \text{E)} \tag{92b}$$

$$p = \rho e^{-i\pi} = -\kappa r^2$$

$$q = -i\sqrt{\rho/\kappa} = -ir \text{ (along D} \rightarrow \text{C)} \tag{92c}$$

$$dp = -2\kappa\, r\, dr \text{ (along both sides).} \tag{92d}$$

Then along the branch cut

$$I_{bc} = \frac{\kappa}{\pi i}\int_0^\infty e^{-\kappa r^2 t}\,\bar{Q}(-\kappa r^2)\,[\bar{u}(ir) - \bar{u}(-ir)]r\,dr \tag{93a}$$

$$= -\frac{1}{\pi}\int_0^\infty e^{-\kappa r^2 t}\,\bar{Q}(-\kappa r^2)C(rx)dr \tag{93b}$$

where

$$C(rx) = \cos rx_- - \cos rx_+. \tag{94}$$

Around the origin take $p = \rho e^{i\theta}$, so that

$$I_0 = \frac{1}{2\pi}\lim_{\rho\rightarrow 0}\int_\pi^{-\pi} e^{pt}\,\bar{Q}\bar{u}pd\theta \tag{95a}$$

$$= \frac{Q_0}{\kappa}\frac{T_1}{T}\left(\frac{x_- - x_+}{2}\right). \tag{95b}$$

Hence the transient is given by

$$v_{tr} = -\frac{Q_0}{\kappa}\left[\frac{T_1}{T}\left(\frac{x_- - x_+}{2}\right) + \frac{2}{\pi}\int_0^\infty \frac{e^{-\kappa r^2 t} - e^{-\kappa r^2(t-T_1)}}{1 - e^{\kappa r^2 T}}\,C(xr)\,\frac{dr}{r^2}\right]. \tag{96}$$

It is convenient at this point to introduce the dimensionless variables

$$\tau = t/T$$
$$\tau_1 = T_1/T$$
$$\rho = Dr$$
$$\xi = x/D \tag{97a}$$

where

$$D = \sqrt{\kappa T}. \tag{97b}$$

Also define

$$v_0 = \frac{Q_0 D}{\kappa} = \frac{P_a D}{KA} \tag{98}$$

where

$$P = \text{power absorbed during } T_1$$
$$A = \text{area.} \tag{99}$$

Using the previous notation v_{tr} becomes

$$v_{tr} = -v_0\left[\tau_1\left(\frac{\xi_- - \xi_+}{2}\right) + \frac{i}{\pi}\int_0^\infty \frac{e^{-\rho^2\tau} - e^{-\rho^2(\tau-\tau_1)}}{1 - e^{\rho^2}}\,C(\rho\xi)\,\frac{d\rho}{\rho^2}\right]. \tag{100}$$

The steady periodic part of v could be evaluated as a Fourier series from Σ residues. A more convenient form can be obtained as follows:[42] first expand $\bar{v}$ as

$$\bar{v} = \frac{Q_0}{p}\left(1 - e^{-pT_1} + e^{-pT} - e^{-p(T+T_1)} + \ldots\right)\bar{u} \tag{101}$$

then v is given in the range $0 < t < T_1$ by

$$v(0 < t < T_1) = \frac{Q_0}{2\pi i}\int_{\gamma-i\infty}^{\gamma+i\infty} e^{pt}\,\bar{u}\,\frac{dp}{p} \tag{102a}$$

while in the range $T_1 < t < T$[13]

$$v(T_1 < t < T) = v(0 < t < T_1) - \frac{Q_0}{2\pi i}\int_{\gamma-i\infty}^{\gamma+i\infty} e^{p(t-T_1)}\,\bar{u}\,\frac{dp}{p}. \tag{102b}$$

By the same methods used to evaluate $v(t)$, one finds, in terms of the dimensionless variables [(97a)],

$$v(0 < \tau < \tau_1) = -v_0\left[\left(\frac{\xi_- - \xi_+}{2}\right) + \frac{1}{\pi}\int_0^\infty e^{-\rho^2\tau}\,C(\rho\xi)\,\frac{d\rho}{\rho^2}\right]. \tag{103a}$$

[13] Recall that if $g(t) = H(t - T_1)f(t - T_1)$, where H is the unit step function applied at T_1, then $\bar{g} = e^{-pT_1}\bar{f}$.

One also finds

$$v(\tau_1 < \tau < 1) = -v_0 \frac{1}{\pi} \int_0^\infty (e^{-\rho^2\tau} - e^{-\rho^2(\tau-\tau_1)}) C(\rho\xi) \frac{d\rho}{\rho^2}. \tag{103b}$$

Now (100) for v_{tr} is good for all τ and hence when subtracted from (103a) must yield $v_s(0 < \tau < \tau_1)$; further, since v_s is periodic with period 1, the same result must hold for times given by $n < \tau < n + t$. Thus

$$v_s(n < \tau < n + \tau_1) = v_0 \left[(\tau_1 - 1) \left(\frac{\xi_- - \xi_+}{2} \right) + \frac{1}{\pi} \int_0^\infty F(\tau) C(\rho\xi) \frac{d\rho}{\rho^2} \right] \tag{104}$$

$$(0 < \tau < \tau_1)$$

where

$$F(\tau) = \frac{e^{-\rho^2(\tau-1)} - e^{-\rho^2(\tau-\tau_1)}}{1 - e^{\rho^2}}. \tag{105}$$

Similarly for the time intervals between $(n + \tau_1)$ and $(n + 1)$

$$v_s(n + \tau_1 < \tau < n + 1) = v_0 \left[\tau_1 \left(\frac{\xi_- - \xi_+}{2} \right) + \frac{1}{\pi} \int_0^\infty G(\tau)\, C(\rho\xi) \frac{d\rho}{\rho^2} \right] \tag{106}$$

$$(\tau_1 < \tau < 1)$$

where

$$G(\tau) = \frac{e^{-\rho^2(\tau-1)} - e^{-\rho^2(\tau-1-\tau_1)}}{1 - e^{\rho^2}}. \tag{107}$$

Using (100), (104), and (106), numerical integration could now be used to yield $v(\tau)$ for various values of the parameters τ and ξ'. However, for $\tau_1 \ll 1$ approximate analytic results at the beginning and the end of the heating period can be obtained, i.e., when $\tau = 1$ (or 0) and τ_1 with

$$\tau_1 \ll 1. \tag{108}$$

Since τ_1 and 1 (or 0) are the limits of the ranges of τ in (104) and (106), the question of possible discontinuities in v_s at these point arises.[14] However, it is found that $v(\tau)$ is in fact continuous at τ_1 and 1. Hence arbitrarily using (100) and (106)

$$v(\tau_1) = v_0 I(\tau_1)$$

$$v(1) = v_0 I(1) \tag{109}$$

[14] The differential equation (79) is the limiting form of the wave equation including dissipation when the velocity of propagation approaches ∞. Hence, since in the present problem the source is turned on and off instantaneously, it is perhaps not obvious that $v(\tau)$ will be continuous at τ_1 and 1.

where

$$I(\tau_1) = \frac{1}{\pi} \int_0^\infty G(\tau_1) C(\rho\xi) \frac{d\rho}{\rho^2} \tag{110a}$$

$$I(1) = \frac{1}{\pi} \int_0^\infty G(1) C(\rho\xi) \frac{d\rho}{\rho^2} \tag{110b}$$

$$G(\tau_1) = \frac{1 - e^{-\rho^2\tau_1}}{1 - e^{-\rho^2}}$$

$$= (1 - e^{-\rho^2\tau_1}) + (1 - e^{-\rho^2\tau_1}) \sum_{n=1}^\infty e^{-n\rho^2} \tag{111a}$$

$$G(1) = -\frac{e^{-\rho^2}(1 - e^{\rho^2\tau_1})}{1 - e^{-\rho^2}}$$

$$= -(1 - e^{\rho^2\tau_1}) \sum_{n=1}^\infty e^{-n\rho^2}. \tag{111b}$$

Now the factors $(1 - e^{\mp\rho^2\tau_1})$ in (111) can be expanded around $\rho = 0$ keeping only the first term, since the factor $e^{-\rho^2}$ in $\sum_{n=1}^\infty e^{-n\rho^2}$ will insure rapid convergence. Proceeding in this way leads to

$$I(\tau_1) = I(1) + I(\Delta) \tag{112}$$

where

$$I(\Delta) = \frac{1}{2} \int_0^\infty (1 - e^{-\rho^2\tau_1}) C(\rho\xi) \frac{d\rho}{\rho^2}. \tag{113}$$

Note that $I(\Delta)$ is the periodic temperature rise (normalized to v_0) above the steady-state dc level $I(1)$.

The integral $I(1)$ is evaluated in Appendix II and for $x = x'$ is found to be

$$I'(1) = \frac{\tau_1}{2\sqrt{\pi}} \sum_{n=1}^\infty \frac{1}{\sqrt{n}} (1 - e^{-\xi'^2 n}). \tag{114}$$

$I(\Delta)$ is also evaluated in Appendix II and for $x = x'$ is given by

$$I'(\Delta) = \xi' \left[1 - \phi\left(\frac{\xi'}{\sqrt{\tau_1}} \right) \right] + \sqrt{\frac{1}{\pi}} (1 - e^{-\xi'^2 \tau_1}) \tag{115}$$

where

$$\phi(\eta) = \frac{2}{\sqrt{\pi}} \int_0^\eta e^{-u^2 du}. \tag{116}$$

Appendix II further discusses these integrals.

Appendix II

Evaluation of Integrals

A. *Evaluation of* $I(\Delta)$

From (113)

$$I(\Delta) = \frac{1}{\pi} \times \int_0^\infty (1 - e^{-\rho^2\tau_1}) (\cos \rho\xi_- - \cos \rho\xi_+) \frac{d\rho}{\rho^2}. \tag{117}$$

Now the separate terms in the integral in (117) diverge at $\rho = 0$ but the difference remains finite. In order to evaluate $I(\Delta)$ introduce a parameter ϵ^2 as a term in the denominator and take the limit as $\epsilon \to 0$, i.e.,

$$I(\Delta) = \frac{1}{\pi} \times$$

$$\lim_{\epsilon \to 0} \int_0^\infty (1 - e^{-\rho^2 \tau_1})(\cos \rho\xi_- - \cos \rho\xi_+) \frac{d\rho}{\rho^2 + \epsilon^2}. \quad (118)$$

Using a table of integrals

$$\int_0^\infty e^{-\tau_1 \rho^2} \cos \rho\xi \frac{d\rho}{\rho^2 + \epsilon^2} = \frac{\pi e^{\tau_1 \epsilon^2}}{4\epsilon} [2 \cosh \epsilon\xi - e^{-\epsilon\xi}\phi(\eta_-) - e^{\epsilon\xi}\phi(\eta_+)] \quad (119)$$

where

$$\phi(\eta) = \text{erf}(\eta) = \frac{2}{\sqrt{\pi}} \int_0^\eta e^{-u^2} du$$

$$\eta_\pm = \epsilon\sqrt{\tau_1} \pm \frac{\xi}{2\sqrt{\tau_1}}; \quad (120)$$

also

$$\int_0^\infty \frac{\cos \rho\xi}{\rho^2 + \epsilon^2} d\rho = \frac{\pi}{2\epsilon} e^{-\epsilon\xi}. \quad (121)$$

Carrying out the indicated limiting process leads to

$$I(\Delta) = -(1/2)\,\xi_- \left[1 - \phi\left(\frac{\xi_-}{2\sqrt{\tau_1}}\right)\right] + (1/2)\,\xi_+ \left[1 - \phi\left(\frac{\xi_+}{2\sqrt{\tau_1}}\right)\right] + \sqrt{\frac{\tau_1}{\pi}}\,(e^{-\xi_-^2/4\tau_1} - e^{-\xi_+^2/4\tau_1}). \quad (122)$$

A case of particular interest occurs for $x = x'$; then $\xi_- = 0$, $\xi_+ = 2x'$, and

$$I(\Delta)_{x=x'} = I'(\Delta) = \xi'\left[1 - \phi\left(\frac{\xi'}{\sqrt{\tau_1}}\right)\right] + \sqrt{\frac{\tau_1}{\pi}}\,(1 - e^{-\xi'^2/\tau_1}). \quad (123)$$

B. Evaluation of I(1)

From (110b) and (111b) the integral $I(1)$ to first-order in τ_1 is

$$I(1) = \frac{\tau_1}{2\sqrt{\pi}} \sum_{n=1}^\infty \int_0^\infty e^{-n\rho^2} (\cos \rho\xi_- - \cos \rho\xi_+)\, d\rho \quad (124a)$$

$$= \frac{\tau_1}{2\sqrt{\pi}} \sum_{n=1}^\infty \frac{1}{\sqrt{n}} f(n) \quad (124b)$$

where

$$f(n) = (e^{-\xi_-^2/4n} - e^{-\xi_+^2/4n}). \quad (125)$$

When $\xi_\pm > 1$, the sum over n can be approximated as an integral for $n > N$, where N is some as yet unspecified integer, so that

$$I(1) \simeq \frac{\tau_1}{2\sqrt{\pi}} \left[\int_{N+1}^\infty f(n) \frac{dn}{\sqrt{n}} + \sum_{n=1}^N \frac{1}{\sqrt{n}} f(n)\right], (\xi_\pm > 1). \quad (126)$$

Then with $n = 1/u^2$ the integral in (126) can be written as

$$\int_{N+1}^\infty f(n) \frac{dn}{\sqrt{n}} = 2\left[\lim_{\epsilon \to 0} \int_0^\infty f\left(\frac{1}{u^2}\right) \frac{du}{u^2 + \epsilon^2} - \int_{1/\sqrt{N+1}}^\infty f\left(\frac{1}{u^2}\right) \frac{du}{u^2}\right]. \quad (127)$$

The final result for $I(1)$ is then found to be

$$I(1) = \tau_1 \left\{\left(\frac{\xi_+ - \xi_-}{2}\right) - \sqrt{\frac{N+1}{\pi}}\,(e^{-\xi_+^2/16(N+1)} - e^{-\xi_-^2/16(N+1)}) + \frac{\xi_-^2}{4}\left[1 - \phi\left(\frac{\xi_-^2}{4\sqrt{N+1}}\right)\right] - \frac{\xi_+^2}{4}\left[1 - \phi\left(\frac{\xi_+^2}{4\sqrt{N+1}}\right)\right] + \frac{1}{2\sqrt{\pi}} \sum_{n=1}^N \frac{1}{\sqrt{n}} f(n)\right\}. \quad (\xi_\pm > 1). \quad (128)$$

Taking $N = 1$ and $x = x'$ then gives for $\xi' > 1$

$$I'(1) \simeq \tau_1 \left[\left(\xi' - \frac{3}{2\sqrt{\pi}}\right) - \frac{1}{2\sqrt{\pi}} e^{-\xi'^2}\right], (\xi' > 1). \quad (129)$$

For small ξ', power series expansion around $\xi' = 0$ gives

$$I(1) = \frac{\zeta(3/2)}{\pi} \tau_1 \xi'^2 = 1.49 \tau_1 \xi'^2 \quad (130)$$

where

$$\zeta = \text{Riemann zeta function}. \quad (131)$$

Appendix III
Heat Flow From a Plane Disc Source Periodic in Time

A better model for thermo-optical writing in a magnetic film can be obtained by considering the film as a plane disc source of radius a (Figs. 1 and 3). Then for a unit such source located at x' at time t' the temperature $v(x, t)$ along the axis of the disc is[43]

$$v = \frac{1}{2(\pi\kappa)^{1/2}(t - t')^{1/2}} (1 - e^{-a^2/4\kappa(t-t')})\, e^{-(x-x')^2/4\kappa(t-t')}. \quad (132)$$

It is seen that the temperature generated by the disc source is given by the infinite plane source temperature minus a correction due to *spreading*, i.e.,

$$v = v_\infty - v_a \quad (133)$$

where

v_∞ temperature due to infinite plane source
v_a correction due to spreading from a disc source of radius a. (134)

Now v_∞ has already been calculated in Appendix I and, analogous to (82), Green's function for the spreading effect is

$$u_a(x,t;\,x',t') = \frac{1}{2(\pi\kappa)^{1/2}(t-t')^{1/2}}\left\{e^{-[(x-x')^2+a^2]/4\kappa(t-t')} - e^{-[(x+x')^2+a^2]/4\kappa(t-t')}\right\}. \quad (135)$$

Proceeding as in Appendix I it follows that all of the formulas derived there can be used to evaluate the spreading effect by merely replacing $\xi_\pm$ by $\chi_\pm$, where

$$\chi_\pm = [(x \pm x')^2 + a^2]^{1/2}/D. \quad (136)$$

As an example of the use of (133) and (136), the integral $I(\Delta)$ giving the incremental temperature at the end of a heating period is written as

$$I(\Delta) = I_\infty(\Delta) - I_a(\Delta) \quad (137)$$

where from (137)

$$I_\infty(\Delta) = -\frac{1}{2}\xi_-\left[1 - \phi\left(\frac{\xi_-}{2\sqrt{\tau_1}}\right)\right] + \frac{1}{2}\xi_+\left[1 - \phi\left(\frac{\xi_+}{2\sqrt{\tau_1}}\right)\right] + \sqrt{\frac{\tau_1}{\pi}}\left(e^{-\xi_-^2/4\tau_1} - e^{-\xi_+^2/4\tau_1}\right) \quad (138)$$

so that

$$I_a(\Delta) = -\frac{1}{2}\chi_-\left[1 - \phi\left(\frac{\chi_-}{2\sqrt{\tau_1}}\right)\right] + \frac{1}{2}\chi_+\left[1 - \phi\left(\frac{\chi_+}{2\sqrt{\tau_1}}\right)\right] + \sqrt{\frac{\tau_1}{\pi}}\left(e^{-\chi_-^2/4\tau_1} - e^{-\chi_+^2/4\tau_1}\right). \quad (139)$$

Acknowledgment

As often in the past, it is a pleasure to acknowledge many discussions with K. J. Harte.

References

[1] L. Mayer, "Curie-point writing on magnetic films," *J. Appl. Phys.*, vol. 29, p. 1003, June 1958.
[2] ——, "Magnetic writing with an electron beam," *J. Appl. Phys.*, vol. 29, pp. 1454–1456, October 1958.
[3] H. Schlichting, *Boundary Layer Theory*. New York: McGraw Hill, 1960, ch. I.
[4] H. S. Carslaw and J. C. Jaeger, *Conduction of Heat in Solids*. London: Oxford U. P., 1959, p. 326.
[5] D. A. Stetsen and A. J. DeMaria, "Optical spectra of ultra-short optical pulses generated by mode-locked glass: Nd lasers," *Appl. Phys. Letters*, vol. 9, pp. 118–120, August 1966.
[6] E. W. Lee, D. R. Callaby, and A. C. Lynch, "The use of the Kerr effect for studying the magnetization of a reflecting surface," *Proc. Phys. Soc.*, vol. LXXII, pp. 233–238, 1958.
[7] G. Fan, E. Donath, E. S. Barrekette, and A. Wirgin, "Analysis of a magneto-optic readout system," *IEEE Trans. Electronic Computers*, vol. EC-12, pp. 3–9, February 1963.
[8] P. H. Lissberger, "Kerr magneto-optic effect in nickel-iron films. I. Experimental," *J. Opt. Soc. Am.*, vol. 51, pp. 948–957, September 1961.
[9] P. H. Lissberger, "Kerr magneto-optic effect in nickel-iron films. II. Theoretical," *ibid.*, pp. 957–966.
[10] D. Treves, "Magneto-optic detection of high density recording," *J. Appl. Phys.*, 1967.
[11] V. K. Zworykin and G. A. Morton, *Television*. New York: Wiley, 1940, p.37.
[12] D. Treves, "Limitations of the magneto-optic Kerr technique in the study of microscopic magnetic domain structures," *J. Appl. Phys.*, vol. 32, pp. 358–364, March 1961.
[13] D. L. Fried, W. S. Read, and D. B. Pollock, "An interferometric optical modulator," *Appl. Opt.*, vol. 6, pp. 697–701, June 1964.
[14] J. Kranz and A. Hubert, "Application of the Kerr effect for observing magnetic domains," *Z. angew. Phys.*, vol. 15, pp. 220–232, 1963.
[15] J. Kranz and A. Kritikos, "Magnetooptische Bereichsbeobachtungen an dünnen Eisenschichten," *Naturwissenschaften*, vol. 51, pp. 378–379, 1964.
[16] P. H. Lissberger, "Modification of the longitudinal Kerr magneto-optic effect by dielectric layers," *J. Opt. Soc. Am.*, vol. 54, pp. 804–812, June 1964.
[17] ——, "Design of multilayer structure of magnetic and dielectric films for optimum longitudinal Kerr magneto-optic effect," *J. Opt. Soc. Am.*, vol. 56, pp. 192–197, February 1966.
[18] A. J. Kolk and M. Orlovic, "Increasing the Kerr magneto-optic effect in thin films," *J. Appl. Phys.*, vol. 34, pp. 1060–1061, April 1963.
[19] D. O. Smith, "Magneto-optical scattering from multi-layer magnetic and dielectric films, I. General theory," *Optica Acta*, vol. 12, pp. 13–45, January 1965.
[20] ——, "Magneto-optical scattering from multi-layer magnetic and dielectric films. II. Applications of the general theory," *Optica Acta*, vol. 12, pp. 193–204, April 1965.
[21] ——, "Maximizing the longitudinal magneto-optical transmission-scattering from multilayer magnetic and dielectric films," *Optica Acta*, vol. 13, pp. 121–138, 1966.
[22] ——, "Longitudinal Kerr effect using a very thin Fe film," *J. Appl. Phys.*, vol. 36, pp. 1120–1121, March 1965.
[23] D. O. Smith and K. J. Harte, "Improved method of optimizing longitudinal magneto-optical transmission-scattering in thin magnetic films," *Optica Acta* (in press).
[24] R. D. Hunt, *J. Appl. Phys.*, spring 1967.
[25] M. Born and E. Wolf, *Principles of Optics*. New York: Pergamon, 1959, p. 24.
[26] T. Turbadar, "Complete absorption of light by thin metal films," *Proc. Phys. Soc.*, vol. 73, pt. I, pp. 40–44, January 1959.
[27] C. D. Mee and G. J. Fan, "A proposed beam-addressable memory," *IEEE Trans. Magnetics*, vol. MAG-3, pp. 72–76, March 1967.
[28] J. F. Dillon, Jr., H. Kamimura, and J. P. Remeika, "Magneto-optical studies of chromium tribromide," *J. Appl. Phys.*, vol. 34, pp. 1240–1247, April 1963.
[29] J. F. Dillon, Jr., "Optical absorptions and rotations in the ferrimagnetic garnets," *J. Phys. Radium*, vol. 20, pp. 374–377, February-March 1959.
[30] D. Chen, Y. Gondo, and M. D. Blue, "Hall and Faraday effects in oriented MnBi films," *J. Appl. Phys.*, vol. 36, pp. 1261–1263, March 1965.
[31] G. S. Krinchik and R. D. Nuraliyeva, "Magneto-optic resonance in ferro-magnetics II. Near infra-red region," *Fiz. Metal. i Metalloved.*, vol. 5, pp. 694–698, 1959.
[32] J. T. Chang, J. F. Dillon, Jr., and U. F. Gianola, "Magneto-optical variable memory based upon the properties of a transparent ferrimagnetic garnet at its compensation temperature," *J. Appl. Phys.*, vol. 36, pp. 1110–1111, March 1965.
[33] H. J. Kump and P. T. Chang, "Thermostrictive recording on Permalloy films," *IBM J.*, pp. 255–260, May 1966.
[34] A. E. Slade and H. O. McMahon, "A cryotron catalogue memory system," *Proc. Eastern Joint Computer Conf.*, New York, N. Y., pp. 115–120, 1956.
[35] D. O. Smith and K. J. Harte, "Content-addressed memory using magneto- or electro-optical interrogation," *IEEE Trans. Electronic Computers (Correspondence)*, vol. EC-15, p. 123, February 1966.
[36] D. O. Smith and K. J. Harte, "Comment on 'content-addressed memory using magneto-optical interrogation'," *IEEE Trans. Electronic Computers (Correspondence)*, vol. EC-16, p. 372, June 1967.
[37] J. I. Raffel, T. S. Crowther, A. H. Anderson, and T. O. Anderson, "Magnetic film memory design," *Proc. IRE*, vol. 49, pp. 155–164, January 1961.
[38] H. S. Carslaw and J. C. Jaeger, *op. cit.* [4], p. 10.
[39] *Ibid.*, p. 259.
[40] *Ibid.*, p. 400.
[41] *Ibid.*, p. 302.
[42] *Ibid.*, p. 401.
[43] *Ibid.*, p. 260.

Reprinted with permission from *Journal of Applied Physics*, Vol. 39(8), pp. 3916-3927 (July 1968). ©1968 American Institute of Physics.

MnBi Thin Films: Physical Properties and Memory Applications

D. Chen, J. F. Ready, and E. Bernal G.
Honeywell Corporate Research Center, Hopkins, Minnesota
(Received 15 December 1967; in final form 15 February 1968)

Thin films o' MnBi possess many unusual physical properties and are particularly suitable for memory applications using laser Curie-point writing and magneto-optical readout. Films prepared on mica substrates have the easy direction of magnetization perpendicular to the film plane. They typically have a specific Faraday rotation of 5×10^5 deg/cm, and an absorption coefficient of 3.8×10^5 cm^{-1} for 6328 Å wavelength, permitting a readout rate of 10^9 bits/sec at 1 mW laser power level using available detectors. Curie-point writing and erasing of a spot a few microns in diameter in a 1000 Å thick film with a 4 μsec pulse from a 13 mW Gaussian laser beam have been achieved. For a 10^6 bits/cm^2 packing density, heating at adjacent spots is negligible. Calculations show that the writing magnetic field can be reduced to less than 10 Oe by using a memory plane consisting of discrete squares of MnBi. Analytical and experimental results as well as memory design considerations are presented.

I. INTRODUCTION

Thin-film MnBi possesses many interesting physical properties. It is a hard magnetic material with the easy direction of magnetization usually perpendicular to the film plane. The magnetic properties of these films can be readily studied by optical means because of their low optical density and large magneto-optic effect. MnBi has a Curie temperature of 360°C at which a first-order transition takes place. The properties of this material make it particularly well suited as a two-dimensional high-density magnetic memory medium. Writing on a MnBi film can be achieved by momentarily raising the temperature of a spot on the film to above the Curie temperature. When the material cools and regains its ferromagnetism, the direction of magnetization is determined by the direction of the magnetic field present at the spot. This so-called Curie-point writing technique has been demonstrated using a soldering iron,[1] and also using an electron beam.[2] In this paper, we propose the use of a laser beam for Curie-point writing (or more generally, thermomagnetic writing) on MnBi films. Laser heating offers extremely high packing density since the beam can be focused to a spot of dimensions comparable to the wavelength of the light. Furthermore, the same beam at reduced intensity can be used to nondestructively read the written bits of information by the magneto-optic technique.[3] The optical and magneto-optic properties of the MnBi film are well suited for this high-density optical-memory application. The optical-memory approach offers orders of magnitude increase in packing density over conventional mass memory techniques; together with

[1] L. Mayer, J. Appl. Phys. **29**, 1003 (1958).
[2] L. Mayer, J. Appl. Phys. **29**, 1454 (1958).
[3] Discussions on this subject have been given by a number of authors; see, for example, D. O. Smith, IEEE Trans. Mag. **MAG-3**, 433 (1967).

suitable high-speed beam deflectors and modulators, this type of memory system may be randomly addressed, erased, and read nondestructively.

Recently a number of transparent magnetic materials have been proposed for thermomagnetic writing, erasing, and magneto-optic readout using a laser beam.[4–6] In comparing with these potential materials for optical-memory applications, MnBi films show some unique advantages. In the following, we first describe the technique of sample preparation and the values of various relevant physical properties of MnBi films. We then analyze the processes involved in thermomagnetic writing, erasure, and magneto-optic readout techniques. Experimental results of high-speed writing and erasure experiments which support our analysis of the memory application are discussed.

II. MATERIAL PROPERTIES

A. Material Preparation

The technique for preparing thin films of MnBi was first described by Williams *et al.*[7] In their method a layer of Mn and a layer of Bi were vacuum deposited separately onto a glass substrate. These MnBi–glass films were then annealed in vacuum for 72 h at temperatures ranging from 225°–350°C. The resulting films were polycrystalline but oriented with the crystalline *c* axis perpendicular to the substrate surface. Subsequently, Mayer[8] found that the nucleation processes in films grown by this method were not controllable. As a result, films were not reproducible and uniform films of large area were difficult, if not impossible, to prepare.

The major improvement in our sample-preparation technique lies in the use of freshly cleaved mica as the film substrate for epitaxial growth.[9] The vacuum deposition and annealing procedure are essentially similar to Williams *et al.*[7] The choice of mica substrate is based on the fact that mica cleaves easily and has the same basal plane symmetry as that of MnBi. Films prepared in this manner, designated as MnBi–mica, were found to be single crystal in nature and extremely uniform, as confirmed by x-ray back-reflection measurements. The magnetic, optical, and magneto-optical properties of this type of film differ from those of MnBi–glass films. These properties will be discussed in a later section. We have also prepared films consisting of discrete 4 $\mu \times 4$ μ spots of MnBi separated by 12 μ. More than 90% of the spots on a particular film are ferromagnetic, indicating that the use of mica for the film substrate greatly reduces the problem of nucleation[8] because of the epitaxial nature of the crystal growth.

[4] J. F. Dillon, Jr., H. Kamimura, and J. P. Remeika, J. Appl. Phys. **34,** 1240 (1963).
[5] J. F. Dillon, Jr., J. Phys. Radium **20,** 374 (1959).
[6] C. D. Mee and G. J. Fan, IEEE Trans, Mag. **MAG-3,** 72 (1967).
[7] H. J. Willaims, R. C. Sherwood, F. G. Foster, and E. M. Kelley, J. Appl. Phys. **28,** 1181 (1957).
[8] L. Mayer, J. Appl. Phys. **31,** 384 *S* (1960).
[9] D. Chen, J. Appl. Phys. **37,** 1486 (1966).

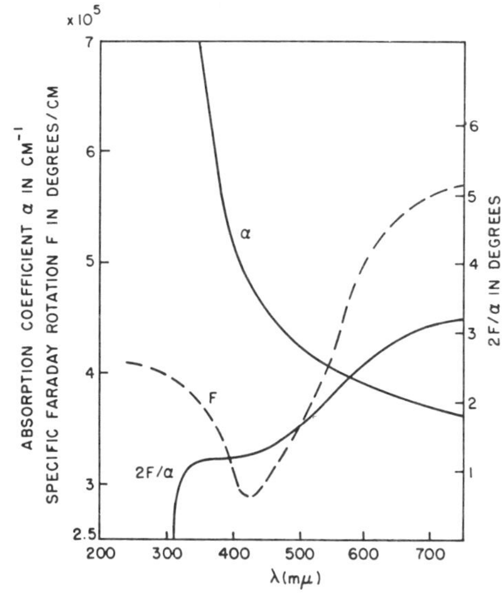

FIG. 1. Specific Faraday rotation F at magnetic saturation, absorption coefficient α, and performance factor $2F/\alpha$ for MnBi–mica as functions of wavelength λ. Rotation measured as magnetization goes from zero to saturation magnetization.

B. Magneto-optic Effect and Optical Absorption

For optical-memory applications using a magnetic film medium, beam addressable readout of stored information is accomplished by the magneto-optic effect.[3] The material figure of merit for readout purposes is proportional to the specific magneto-optic Faraday rotation F, and inversely proportional to the optical absorption. The specific Faraday rotation F is related to the saturation magnetization M_s by $\mathbf{F} = V\mathbf{M}_s$, where V is the Verdet constant. We define a performance factor as twice the Faraday rotation per neper optical intensity loss. (The utility of this definition will be apparent later.) Therefore, large specific Faraday rotation and low optical absorption are desirable for reading. On the other hand, efficient use of the laser beam for thermomagnetic writing on the magnetic film medium requires that the material have a large absorption coefficient at the laser wavelength. When one laser is used for both writing and reading, the wavelength of the beam must be properly selected to reach a compromise for the magnitudes of the material properties. In order to select the operating wavelength, we have measured the Faraday effect and the absorption coefficient of MnBi–mica films from the ultraviolet to the near infrared.

To measure the Faraday effect we used a photoelectric polarimeter in conjunction with a Faraday modulation cell. The measurement accuracy is better than 30 sec of rotation. A set of 50 Å bandwidth monopass filters was used with a tungsten lamp to obtain the required spectrally narrow lines for our Faraday effect measurements. The apparent absorption coefficient was measured on a Cary 14 spectrophotometer with a clear mica substrate in the reference beam. No corrections were made for the difference of the surface reflectivities

TABLE I. Values of $2F/\alpha$ in degrees for some transparent magnetic materials.

Material	MnBi–mica	MnBi–glass	GdIG	$CrBr_3$ (T_c=36°K)	EuO (T_c=60°K)	EuSe (T_c=7°K)
Wavelength λ (μ)	0.63	0.63	0.53	0.5	1.2	0.66
$2F/\alpha$ (deg)	2.85[a]	0.80	2.6	200	400[b]	300[b]
Reference	see text	11, 14	5	4	6	6

[a] On some recent samples $2F/\alpha$ as high as 3.6° has been obtained.
[b] These values reflect a reduction by a factor of 10 to compensate for the fact that the easy direction of magnetization for these films is in the plane of the film, whereas the incident beam is near normal to the film (see Ref. 6).

experienced by the main and reference beam. The apparent absorption data are useful for a practical device study. The determination of the true absorption coefficient requires reflectivity measurements, as well as transmission measurements, which are beyond the scope of the present work. The measured specific Faraday rotation F in deg/cm (at magnetic saturation), the apparent absorption coefficient α in cm^{-1}, and the performance factor $2F/\alpha$ in degrees are plotted in Fig. 1. The performance factor of MnBi–mica compares favorably with that of other known transparent magnetic materials except for materials requiring a cryogenic environment. This is shown in Table I.

C. Magnetic Properties

MnBi is an intermetallic compound of NiAs structure. The bonding in this compound is not entirely ionic, but is believed to be intermediate between ionic and metallic,[10] as indicated by the temperature dependence of the electrical conductivity.[11] The effective Bohr magnetons μ_B per Mn^{+3} cation is 3.95.[12] Present understanding of the ferromagnetism in this compound is that there exists a competitive $3d^4$–anion–$3d^4$ superexchange with quasistatic correlations.[13] The metallic conductivity in this case is apparently due to cation sublattice d-band formation made possible by a large number of interstitial cations. Hall effect measurement results support this picture.[14]

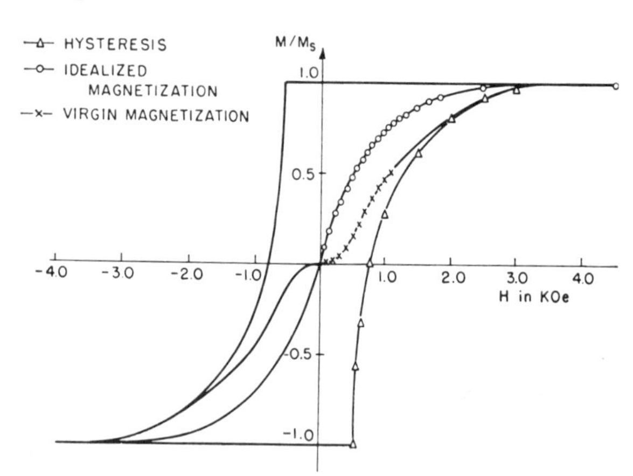

FIG. 2. Typical room-temperature magnetization curves for a 1500 Å thick MnBi–mica film. The idealized magnetization curve was measured with an ac magnetic field of magnitude exceeding H_c superimposed on the measuring field. The virgin magnetization curve was obtained from a freshly prepared film.

The MnBi–mica films possess a number of interesting magnetic properties. Because of the high specific Faraday rotation and relatively low optical density, these properties can be conveniently observed by magneto-optic techniques visually as well as photographically. In the following, we will summarize the static and dynamic magnetic properties of these films.

The easy direction of magnetization in MnBi films is almost entirely determined by the orientation of the c axis of the crystal; that is because of its extremely large crystalline anisotropy. If glass is used as a substrate, the orientation of the c axis of the crystallites cannot be readily controlled. It is possible to obtain films with the easy axis oriented either perpendicular or parallel to the film plane. For films grown epitaxially on mica substrates, the orientation of the growth assures that the easy axis is always perpendicular to the film plane. The easy axis magnetic hysteresis for a typical MnBi–mica film can be measured by either a torque magnetometer or by magneto-optic techniques. A typical room-temperature hysteresis curve is shown in Fig. 2. The coercive force of the films depends on their thickness. As film thickness increases, the coercive force decreases. For film thicknesses ranging from 200

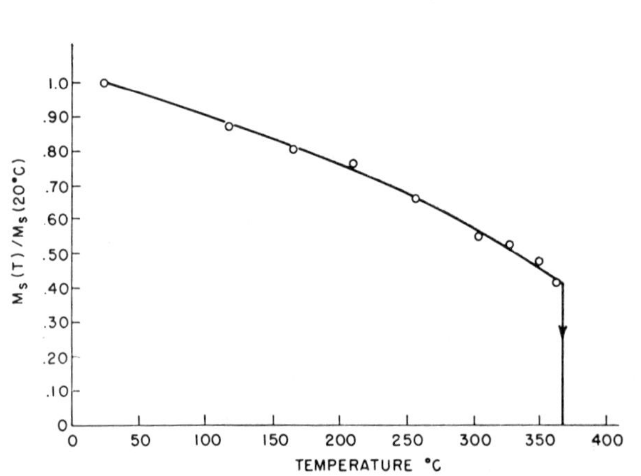

FIG. 3. Measured temperature dependence of magnetization of MnBi–mica films.

[10] K. Adachi, J. Phys. Soc. Japan **6**, 2187 (1961).
[11] D. Chen and Y. Gondo, J. Appl. Phys. **35**, 1024 (1964).
[12] R. R. Heikes, Phys. Rev. **99**, 446 (1955).
[13] J. B. Goodenough, A. Wold, R. J. Arnott, and N. Menyuk, Phys. Rev. **124**, 373 (1961).
[14] D. Chen, Y. Gondo, and M. D. Blue, J. Appl. Phys. **36**, 1261 (1965).

TABLE II. Physical properties of MnBi at 300°K.

Crystal structure	Uniaxial anisotropy constant $K=K_1+K_2$	Curie temp. T_c	Saturation magnetization $4\pi M_s$	Specific Faraday rotation[a] F	Absorption coefficient[a] α	Performance factor[a] $2F/\alpha$
NiAs type[b]	11.6×10^6 erg/cm^3	360°C	7700 G or 3.95 μ_B/Mn ion	5.4×10^5 deg/cm	3.8×10^5/cm	2.85°

[a] $\lambda=0.6328\ \mu$.
[b] $a_0=4.287$ Å, $c_0=6.126$ Å.

to 2900 Å, the coercive force varies from 2 kOe to 300 Oe. The thickness dependence of the coercive force in these films can be understood from simple domain theory calculations[15]; this dependence is of interest for applications of this material where adjustment of coercive force is required. On the other hand, the saturation field for manganese bismuth films is invariably near 4 kOe and insensitive to the thickness of the film (over the range studied here). The flux reversal mechanism in this type of film is believed to be mainly due to domain wall motion at fields higher than the coercive force.[15] However, in the region between the critical field value and the coercive force, the formation of doughnut-shaped domains takes place, partly because of domain reversal. The switching time[9] for this type of film was measured by the magneto-optical Faraday effect; a switching time of 0.067 Oe·sec was obtained,[9] indicating the switching mechanism is mainly due to domain wall motion.

The temperature dependence of the magnetization of these films is shown in Fig. 3; it differs only slightly from that measured on the bulk material.[12] The loss of magnetization at the magnetic transition occurs in our samples at 360°C; the magnetization recovers at 345°C during cooling back to room temperature. It should also be pointed out that at a temperature of 446°C, manganese bismuth decomposes to α-manganese in molten bismuth. Films heated above this temperature in air normally will not recover their magnetization.

The crystalline anisotropy of manganese bismuth is uniaxial in nature because of its nickel arsenide crystalline structure. The anisotropy constant has been measured by Guillaud[16] in the temperature range from liquid-nitrogen temperature up to about room temperature. The anisotropy constant is negative below 84°K and it increases monotonically to a positive value of 1.16×10^7 erg/cc at room temperature. This, together with the saturation magnetization value of 610 G, gives rise to an anisotropy field of 38 kOe. Measurements of the anisotropy constant between room temperature and the transition temperature of 300°C were attempted. It was found that over the entire temperature range of our measurements the anisotropy field never dropped below the 38 kOe level; this insures that the easy direction of magnetization for the films remains perpendicular to the film plane. Consequently, in the entire process of thermomagnetic writing with an applied magnetic field, we anticipate no complications due to the switching of the easy axis direction. The material properties of manganese bismuth films pertinent to optical mass memory applications are summarized in Table II.

III. WRITING TECHNIQUE

The Curie-point writing technique consists of raising the temperature of the bit to be written above the Curie temperature by laser heating and then allowing that bit to cool in the presence of a net magnetic field which determines the direction of magnetization. Therefore, in considering the feasibility of thermomagnetic writing in MnBi, one must answer two questions: The first question concerns the external field required to effect the desired changes in the direction of the magnetization; the second concerns the laser power and interval of time required to raise the temperature of the bit above the Curie temperature, as well as whether or not adjacent bits in the proposed array will be affected by this heating.

A. Writing Field

In the absence of an external magnetic field, a bit that has been heated with the laser beam and is cooling through the Curie temperature "sees" the magnetic field produced by all the magnetic material surrounding it; this is the so-called demagnetizing field. The direction of the demagnetizing field is fixed by the orientation of the magnetization of the magnetic material. To write the bit in a direction other than that determined by the demagnetizing field, one must apply an external field opposite to, and slightly larger than, the demagnetizing field when the bit cools through the Curie temperature. The worst possible case, i.e., the case requiring the largest external field, is that for which the entire memory plane is uniformly magnetized in one direction. We have calculated the demagnetizing field both for the case where the memory plane consists of a continuous film of MnBi and the case where the plane consists of discrete squares of MnBi deposited on a nonferromagnetic substrate. In the discrete spot case the results can be considered quantitative since the spots adjacent to the one being written on are essentially unaffected by the heating process and their magnetization can be considered to be the room-temperature magnetization. In the continuous film case both the boundary of the heated spot and the magnitude of the magnetization

[15] D. Chen, J. Appl. Phys. 38, 1309 (1967).
[16] C. Guillaud, thesis, Strasbourg (1943).

in that neighborhood are ill-defined; therefore, the demagnetizing field calculations can only be considered to be a qualitative guide to the fields to be expected.

The reason for considering both continuous and discrete film memory planes is that, while it is obvious that by increasing the separation between bits one loses packing density, it is also true that the demagnetizing field to be overcome by an external field will decrease as the separation between bits is increased in a discrete film plane. Since the demagnetizing field for a continuous film plane can be very large, one may be willing to sacrifice a certain amount of packing density in order to reduce the writing-field requirements. Hence, by calculating the demagnetizing field as a function of the separation between adjacent bits, we obtain its dependence on packing density and can decide whether or not the trade-off would be desirable.

For the purposes of this calculation we will assume that the bits consist of squares of side $2a$ with a center to center distance d. Thus, $d=2a$ corresponds to the continuous film case and $d>2a$ corresponds to discrete spots.

We first calculate the field produced at (x, y) by a single square of MnBi whose center is located at (x', y'), with sides parallel to the coordinate axes. As mentioned before, the uniaxial anisotropy in MnBi is such that it maintains the magnetization of the film perpendicular to the film plane at all temperatures between room temperature and the Curie temperature. Hence, we represent the film mathematically as a surface density of dipoles $\sigma=M_sh$, where h is the thickness of the film. We are interested only in the component of the demagnetizing field perpendicular to the film plane. This is simply given by

$$H_\perp(x, y)=\int (\sigma da'/|\mathbf{r}-\mathbf{r}'|^3). \quad (1)$$

Integrating over the square, and letting the unprimed coordinates be measured relative to the source point, one finds

$$H_\perp(x, y)=[\sigma/(x-a)]\{[(x-a)^2+(y+a)^2]^{1/2}/(y+a)-[(x-a)^2+(y-a)^2]^{1/2}/(y-a)\}$$
$$-[\sigma/(x+a)]\{[(x+a)^2+(y+a)^2]^{1/2}/(y+a)-[(x+a)^2+(y-a)^2]^{1/2}/(y-a)\}. \quad (2)$$

The field produced by the entire plane of discrete squares is then calculated as follows: Consider a two dimensional square lattice with spacing d. To each lattice point we assign one discrete square of MnBi and we sum the contribution of each spot to the field at the position of the nonmagnetic (heated) spot. We let the coordinates of an arbitrary point within the nonmagnetic square at the origin be defined by $x_k=ka$ and $y_l=la$, with $0\leq(k, l)\leq 1$; we also let the coordinates of the center of an arbitrary magnetic square with respect to the origin be given by $x_m'=md$, $y_n'=nd$, with $-\infty\leq(m, n)\leq\infty$. Summing contributions of the form (2) above for all the magnetic squares, one obtains the following expression for the magnetic field at the position of the nonmagnetic square

$$H_\perp(k, l)=\sum_{m,n=-\infty}^{+\infty}{}' H_{mn}(k, l), \quad (3)$$

where

$$H_{mn}(k, l)=[\sigma/(x_{mk}-a)]\{[(x_{mk}-a)^2+(y_{nl}+a)^2]^{1/2}/(y_{nl}+a)-[(x_{mk}-a)^2+(y_{nl}-a)^2]^{1/2}/(y_{nl}-a)\}$$
$$-[\sigma/(x_{mk}+a)]\{[(x_{mk}+a)^2+(y_{nl}+a)^2]^{1/2}/(y_{nl}+a)-[(x_{mk}+a)^2+(y_{nl}-a)^2]^{1/2}/(y_{nl}-a)\}$$

and

$$x_{mk}=x_m'-x_k$$
$$y_{nl}=y_n'-y_l.$$

The prime in the summation means that $m=n=0$ is not included in the sum.

The above expression can, in principle, be used to calculate the demagnetizing field at any point within the nonmagnetic spot. In practice it is found that the sum converges very slowly. Thus, for the actual numerical calculations, we first expanded the above expression in a Taylor series and then used a Fourier transform technique[17] that gives excellent convergence by summing over the nearest and next-nearest neighbors only. (The truncation error is less than one part in 10^6.)

The magnetic field produced at any point (k, l) within the nonmagnetic spot by an infinite continuous film of MnBi can be simply calculated by direct integration of Eq. (1). The result is

$$H_\perp(k, l)=(\sigma/a)[g(k+1, l+1)-g(k-1, l+1)$$
$$-g(k+1, l-1)+g(k-1, l-1)], \quad (4)$$

where

$$g(u, v)=(u^2+v^2)^{1/2}/uv.$$

At the center of the square, $k=l=0$, and

$$H_\perp(0, 0)=4\sqrt{2}\sigma/a. \quad (5)$$

Figure 4 shows the demagnetizing field at the center of the nonmagnetic square as a function of the fraction of the total area of the film plane that is filled with ferromagnetic material. The results are given for an infinite plane. The right side of the figure, where $d/2a=1$, corresponds to the continuous film case. The logarith-

[17] B. R. A. Nijboer and F. W. deWette, Physica **23**, 309 (1957).

mic slope of the curve is approximately 1.5. Thus, it can be seen that by reducing the packing density by one order of magnitude, the demagnetizing field can be reduced by approximately a factor of 30. Since the amount of power required to produce the writing field is proportional to the square of the required field, the power requirement for the writing field can be reduced by approximately three orders of magnitude. In fact, the reduction in the demagnetizing field in other parts of the heated square will be greater than we have estimated above. This can be clearly seen by examination of Figs. 5 and 6, which show the demagnetizing field for different positions on the nonmagnetic spot. It is

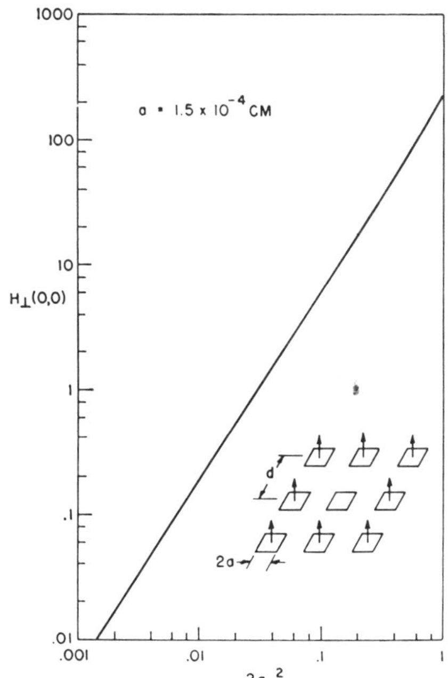

FIG. 4. Demagnetizing field at center of nonmagnetic spot as a function of the fraction of the memory-plane area filled with magnetic material.

evident from the comparison of those figures that the demagnetizing field increases much more rapidly toward the edges of the spot for the continuous-film case than for the discrete-film case. In fact, the rate of change of demagnetizing field across the dimensions of the spot depends strongly on the lattice spacing d. In the figures, the field is shown along a closed path in the square as identified in the inset. The upper scale corresponds to the upper branch of the closed curve, the lower scale of the lower branch which represents the field along the diagonal ($k=l$). The reason for going only half-way to the edge of the spot in the continuous film case (k, $l \leq 0.5$) is that the dipole approximation begins to break down if one gets very close to the source of the magnetic field.

The conclusion to be drawn from the above considerations is that there is a considerable gain to be obtained

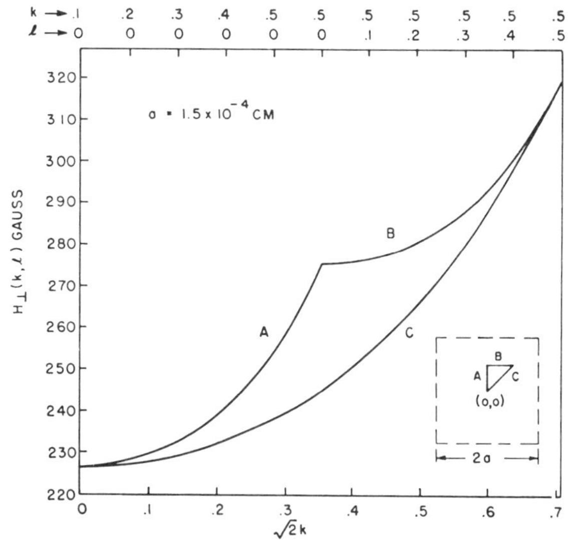

FIG. 5. Field produced by infinite plane of MnBi within a square hole of dimension $2a$ located at origin. $x=ka$ and $y=la$ are coordinates of points with hole $[0 \leq (k, l) \leq 1]$. See text.

by using a memory plane consisting of discrete spots separated by nonmagnetic material. This configuration can considerably decrease the amount of power required for the writing field with little loss in packing density. The loss of packing density is in fact not as great as it appears, since even if one were to use a continuous-film plane, the actual magnetic bits would have to be spaced by more than their dimension in order to avoid accidental reversal caused by thermal diffusion when a spot is written.

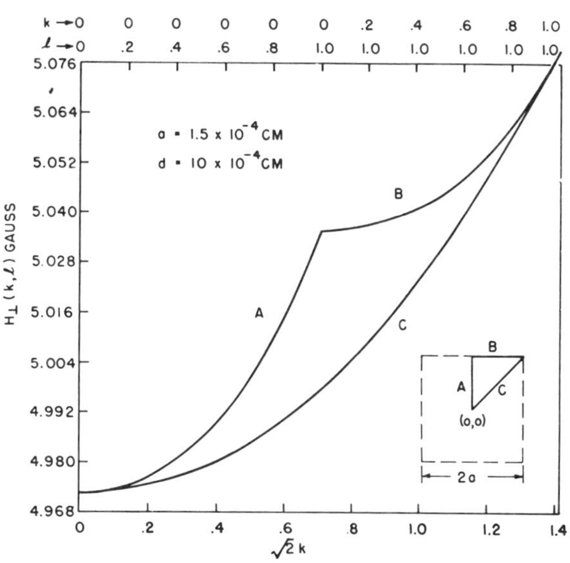

FIG. 6. Field produced by infinite array of discrete squares of MnBi within a nonmagnetic square located at origin. Squares of dimension $2a$ are separated by distance d between centers. $x=ka$ and $y=la$ are coordinates of points within square at origin $[0 \leq (k, l) \leq 1]$. See text.

B. Temperature-Rise Profile

In this section we consider temperature profiles in laser-irradiated MnBi films. The results show that for reasonable values of laser power, high writing rates with small spot areas can be attained. These results are based on calculations obtained by solving the heat-flow equation for various boundary conditions of interest, and indicate large temperature rises confined within small distances from where the laser energy is deposited.

Laser Curie-point writing has been experimentally demonstrated in micron-sized areas, as we shall describe below. It is still desirable to review the calculations, however, in order to consider the writing rates that should be available with present lasers to understand how the diffusion of heat affects the potential packing density and to investigate the relative merits of various geometries. For efficient Curie-point writing it is important that the heat not diffuse over too large an area. Generally the criterion is that the temperature of the spot which is being heated must rise above the Curie point while the temperature at a spot a short distance away (the distance corresponding to the position of the adjacent bit) must remain well below the Curie point. In order to achieve a high packing density, the temperature must fall rapidly in a radial direction from the center of the spot which is being heated.

First we will calculate the temperature profile in a thin film of MnBi with no substrate. This is not a realistic case physically, but the results are instructive as they relate to writing rate and packing density. We assume that the film is thermally thin, so thin that the temperatures of the front and back surfaces rapidly become equal. This is a reasonable assumption for the film thicknesses considered here ($\sim$1000 Å). The ratio of the temperature of the back surface to the temperature of the front surface when heat is applied in a rapid pulse at the front surface is approximately $\exp(-h^2/4\kappa t)$, where h is the thickness of the film, κ is the thermal diffusivity, and t is the time. For the films with which we will be concerned, the thickness is of the order of 10^{-5} cm and the diffusivity is of the order of 0.1 cm^2/sec. Thus, for times greater than about 10^{-8} sec, there is relatively little difference in temperature between the front and the back surfaces of this film. For times of the order of 1 μsec we can assume that the temperature is fairly uniform throughout the thickness of the film.

Under these conditions there is no heat flow in the direction perpendicular to the plane of the film. The situation is approximated by the heat flow near a continuous-line source. Such a source gives the temperature profile due to heat spreading out in the two dimensions perpendicular to the line. If the heat is supplied over a circular area, the incremental temperature at the point (x, y) due to the heat liberated by a line source through the area element centered at (x', y') is given by[18]

$$(1/4\pi\kappa)\int_0^t \phi(t')\exp[-r_1^2/4\kappa(t-t')](t-t')^{-1}dt', \qquad (6)$$

where $\phi(t)$ is equal to κ/Kh times the rate at which heat is liberated per unit time in the area element, κ and K are respectively the thermal diffusivity and thermal conductivity of the material, and

$$r_1^2=(x-x')^2+(y-y')^2.$$

Then the temperature at (x, y) may be calculated by integrating over the area of the circle in which the heat is being supplied with the proper weighting factor to take account of the distribution of energy within the circle. If one assumes that the function $\phi(t)$ is constant from time zero to the end of the pulse, then one obtains the following formula for the temperature T at a distance r from the center of the illuminated spot and at time t less than or equal to the pulse length t_p.

$$T(r,t)=\frac{P_0}{2\pi Kh}\int_0^\pi d\theta'\int_0^\infty r'dr'f(r') \times E_1\left(\frac{r^2-2rr'\cos\theta'+r'^2}{4\kappa t}\right);\ t\le t_p \qquad (7)$$

For $t>t_p$, the temperature is given by $T(r, t)-T(r, t-t_p)$. In this formula P_0 is the laser flux density at the center of the laser spot, and $f(r')$ is the radial distribution of energy within the spot normalized to unity. E_1 is the exponential integral function. This expression has been evaluated by numerical integration on a computer.

A typical result is shown in Fig. 7 for the case where the total incident power is 2.2 mW, the pulse length is

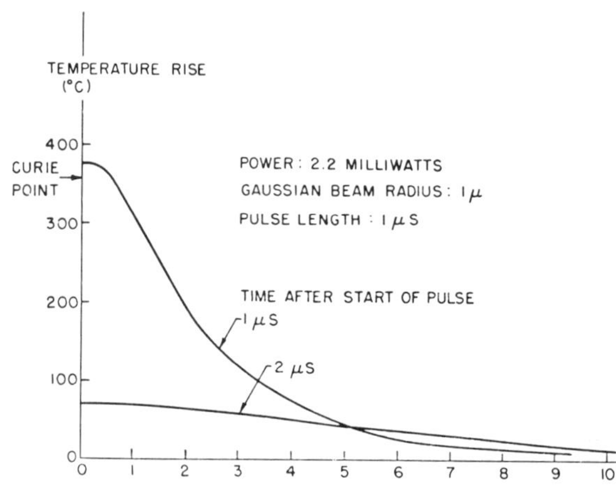

FIG. 7. Calculated temperature rise produced in an unbacked MnBi film by a pulsed laser beam with a Gaussian spatial profile as a function of radial distance from the beam center.

[18] H. S. Carslaw and J. C. Jaeger, *Conduction of Heat in Solids* (Oxford University Press, Oxford, 1959), Chap. X.

1 μsec, and the heat is deposited with a Gaussian spatial profile with a radius (e^{-2}) of 1 μ. The 2.2 mW is heat actually absorbed in the MnBi film. This does not take into account the fact that there may be losses because of reflection or transmission of the incident light through the film. Transmission is relatively small since even a 1000 Å film has an optical thickness greater than unity. The results show that near the center of the spot we can indeed produce relatively high temperatures even in a very short pulse with low-power beams provided the focusing is good. The radius of the area raised above the Curie point is less than 1 μ. We also see that at a distance a few microns from the center of the spot, the temperature never becomes high. In this particular figure the temperature profile is shown immediately after the end of the pulse as well as 1 μsec later, i.e., at 2 μsec after the beginning of the pulse. The temperature profile rapidly flattens after the end of the laser pulse and the temperature at the center of the irradiated spot drops quickly. This means that the heat remains localized and a spot a few microns away which would be meant to store another bit of information would be unaffected by the heating of the first spot. Therefore, the writing rate and packing density can be high.

However, an isolated film with no substrate has no mechanical stability. In practice the MnBi film will be deposited on insulating substrates. This will affect the temperature profiles in the material considerably. For our particular case we have deposited MnBi film on mica substrates, the thermal conductivity of which is considerably lower than that of the MnBi; if this were not so, the temperature at the surface would drop rapidly and it would be very difficult to heat above the Curie point with low laser powers.

We now consider the effect of the backing on the temperature profile. The treatment of the surface distribution in a two-layer system produced by a heat source at the surface involves solution of two heat-flow equations in the two media by means of a Laplace transform with respect to the time variable. One can obtain the following infinite series[19] for the temperature at the surface of the MnBi film for times t less than the pulse length at a distance r from the center of the illuminated area:

$$T(r,t)=\frac{P_t}{\pi r_0^2}\sum_{k=1}^{\infty}\frac{\exp(-\nu_k^2 r_s^2/4r_0^2)\,J_0(\nu_k r/r_0)}{J_1^2(\nu_k)} \times\left(\left\{\frac{\exp(h'\nu_k/r_0)-\exp(-h'\nu_k/r_0)}{\exp(h'\nu_k/r_0)+\exp(-h'\nu_k/r_0)}\frac{r_0}{\nu_k K_2}-\sum_{n=1}^{\infty}\frac{2\exp(-\mu_{nk}^2 t)\cos\{[(\mu_{nk}^2/\kappa_1)-(\nu_k^2/r_0^2)]^{1/2}h\}}{\mu^2_{nk}\cos\{[(\mu_{nk}^2/\kappa_2)-(\nu_k^2/r_0^2)]^{1/2}h\}F_{nk}}\right\}\right) \tag{8}$$

with

$$F_{nk}=((K_1h/\kappa_1)+(K_2h'/\kappa_2)\cos^2\{[(\mu_{nk}^2/\kappa_1)-(\nu_k^2/r_0^2)]^{1/2}h\}\sin^{-2}\{[(\mu_{nk}^2/\kappa_2)-(\nu_k^2/r_0^2)]^{1/2}h'\}).$$

In this doubly infinite series the ν_k are the roots of the zeroth-order Bessel function. r_s is the radius of the Gaussian profile in which the laser energy is deposited, and P_t is the total power in the beam. K_1 and κ_1 are the thermal conductivity and thermal diffusivity of the MnBi; K_2 and κ_2 are the thermal conductivity and diffusivity of the mica. The thicknesses of the MnBi and mica layers are h and h', respectively. In the boundary conditions the surface temperature is matched to be zero at $r=r_0$. Thus r_0 must be chosen to be much larger than the dimensions of interest in the problem. However, we find that as r_0 becomes large, the series converges very slowly. Therefore, one must make a careful choice of r_0 so that it is small enough to give reasonable convergence of the series and at the same time large enough so that the temperature profile for small values of r is not affected. Of course, for $r \ll r_0$, the series converges to a value that is independent of the choice of r_0. For the problems with which we are concerned, the choice of r_0 in the region of 100 to 200 μ is satisfactory. The values μ_{nk}^2 are the roots of the equation

$$K_1[(\mu_{nk}^2/\kappa_1)-(\nu_k^2/r_0^2)]^{1/2}\tan\{[(\mu_{nk}^2/\kappa_1)-(\nu_k^2/r_0^2)]^{1/2}h\}+K_2[(\mu_{nk}^2/\kappa_2)-(\nu_k^2/r_0^2)]^{1/2} \times\cot\{[(\mu_{nk}^2/\kappa_2)-(\nu_k^2/r_0^2)]^{1/2}(h-h')\}=0. \tag{9}$$

Equation (8) for the case of a 0.1 μ thick MnBi film deposited on a 25 μ thick mica substrate has been evaluated by summing numerically using a computer routine. Typical results are shown in Fig. 8, where the result is compared with that for the case of a MnBi film with no substrate as evaluated for similar laser-pulse conditions using the method described earlier. The conditions here are that the total laser power was 5 mW in a Gaussian beam of radius 10 μ and a pulse length of 10 μsec. The temperature profiles shown here are for the time at the end of the laser pulse. The radius of 10 μ is somewhat larger than what would be desirable in practice, but convergence of the series is so slow for $r\approx 1$ μ that an unreasonably large amount of computer time would be required. The results for $r=10$ μ show roughly the influence of the mica backing. We see that the presence of the mica backing does indeed cut down the temperature at the center of the irradiated spot as compared to the case of an isolated film with no backing by approximately a factor of two. It will still be possible

[19] N. R. Rykalin *et al.*, Dokl. Akad. Nauk SSSR **169**, 565 (1966) [Sov. Phys.—Dokl. **11**, 632 (1967)].

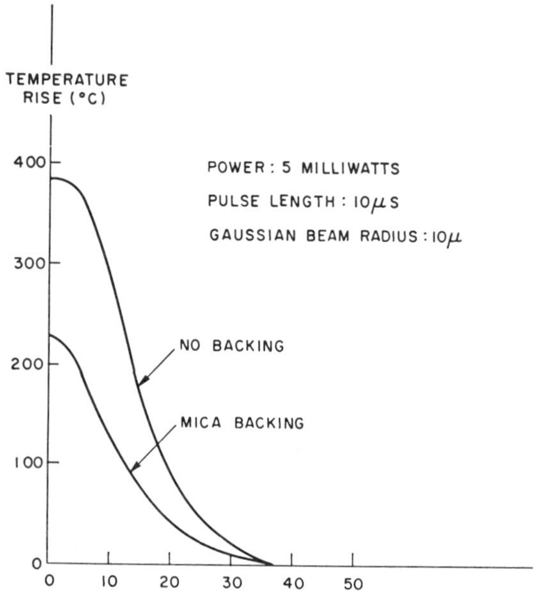

FIG. 8. Comparison of temperature profiles produced in mica-backed and unbacked MnBi films at the end of the laser pulse for a beam with a Gaussian spatial profile as a function of radial distance from the beam center.

to obtain a temperature rise suitable for Curie-point writing with modest laser powers even with the presence of the backing which conducts heat away from the surface. This indicates that the Curie-point writing will be feasible although more power will be required than is indicated by calculations done without including the presence of the backing. We see also that the radial distance at which the temperature has fallen to one-half the value at the center is not appreciably changed by the presence of the substrate. A similar problem has been treated by an alternative method by Smith.[20]

Now let us consider the case in which isolated spots of MnBi are deposited on a mica substrate. Each spot will represent one bit of information. The breaking up of the MnBi layer into these discrete spots will allow for more efficient heating of a given spot without conduction of heat to a neighboring spot which is meant to store a separate bit of information. This case is one of considerable interest as we described in Sec. III.A, where we calculated the switching field required for isolated MnBi spots. As the results of Fig. 8 indicate, a large fraction of the heat flow at a radial distance of several microns from the center of the illuminated spot occurs by conduction through the higher conductivity but thinner layer of MnBi. If this layer is broken up, the heat conducted to the surface of the mica some distance away from the spot in which the heat is being deposited will be greatly reduced. To carry out this calculation, we assume a uniform disk source of radius a. If the actual profile is Gaussian, this will lead to results which are inexact for $r<a$, but for $r>a$, the high ratio of the thermal conductivity of MnBi to that of mica should make the results almost independent of the exact profile of the laser beam. At some distance from the MnBi disk in which the energy is being deposited, the relatively high thermal conductivity of the MnBi makes it appear as if the source were uniform. We also assume that the thermal capacity of the MnBi is relatively small compared to that of the more massive mica substrate beneath it so that the adjacent MnBi spots deposited on the surface will assume approximately the temperature of the mica substrate which is heated by conduction through the mica. Under these conditions the temperature at the surface at a radial distance r from the center of the spot at a time t is given by the following expression[18]:

$$T(r,t)=(P_0a/2K)\int_0^\infty J_0(\lambda r)\,J_1(\lambda a)\times\operatorname{erf}[\lambda(\kappa t)^{1/2}]\lambda^{-1}d\lambda. \quad (10)$$

In this equation, a is the radius of the uniform disk source, i.e., the radius of one of the isolated MnBi layers, and P_0 is the power per unit area in the laser beam. Once again the above expression has been evaluated by numerical integration on a computer. A typical result is shown in Fig. 9 for a total laser power of 4 mW deposited in a 1.5 μ radius spot of MnBi. The pulse length is 1 μsec. For this case we see that there is indeed very little radial flow of heat. The temperature at a distance of 4 μ from the center of the spot is drastically reduced since the only way in which heat can reach that position is by the relatively inefficient mechanism of conduction through the mica substrate. At later times, after the laser radiation ceases, the profile flattens rapidly, and there is little temperature rise at a neighboring spot. The temperature at the center of the illuminated spot is indeed somewhat lower than would be the case of an unbacked continuous film of MnBi. However, with a reasonable laser power it should be

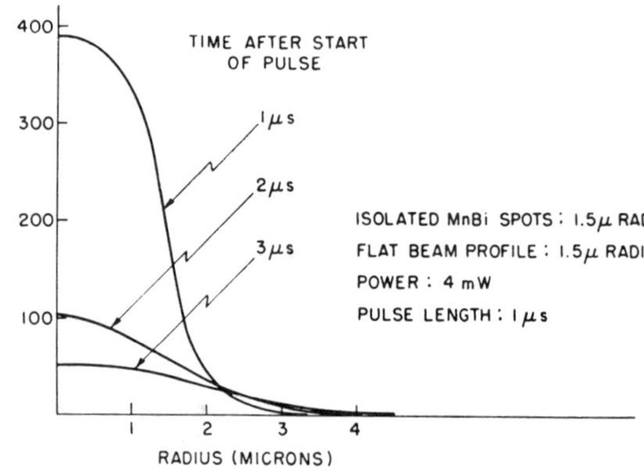

FIG. 9. Calculated temperature profiles for isolated MnBi spots on a mica substrate for a pulsed laser beam with a flat profile as a function of radial distance from the beam center.

[20] D. O. Smith, IEEE Trans. Mag. **MAG-3,** 433 (1967).

easily possible to achieve sufficient heating for Curie-point writing. Thus, this method of depositing isolated spots of MnBi on the surface of the substrate gives us the additional advantage of very little interaction between adjacent spots.

In a later section we will consider experimental results for the case of microsecond duration pulses of laser radiation. It is of interest to see how the temperature rise develops as a function of time for a pulsed source. Because of spreading of heat by conduction and losses by convection, the temperature at the center of the spot will approach a steady-state value. For microsecond duration pulses, convection losses are negligible, but they should be included in the millisecond region. The fraction $F(t)$ of the steady-state temperature that is reached at the end of a pulse of length t for a stationary source in a thin film when convection is included is given by[18]

$$F(t)=\left\{\frac{P_0}{4\pi Kh}\int_0^t \frac{dt'}{(t-t')}\exp[-\kappa k^2(t-t')-r^2/4\kappa(t-t')]\right\}\Big/\left\{\frac{P_0}{4\pi Kh}\lim_{t\to\infty}\int_0^t \frac{dt'}{(t-t')}\exp[-\kappa k^2(t-t')-r^2/4\kappa(t-t')]\right\}$$

$$=\left\{\int_0^t \frac{dt'}{(t-t')}\exp[-\kappa k^2(t-t')-r^2/4\kappa(t-t')]\Big/ 2K_0(rk)\right\}, \tag{11}$$

where $k^2=2H/Kh$, H is the convective heat-transfer coefficient, h is the film thickness, r is the radial distance from the point source, K_0 is the modified Bessel function of order zero, and the other notation is as before.

The integration was carried out numerically on a computer and results are shown in Fig. 10. These results show how the temperature rise would develop as a function of time for a pulse with a fixed power level as pulse length is varied. The results indicate that an appreciable fraction of the steady-state temperature is reached in a pulse in the microsecond regime. In the millisecond regime, the temperature approaches steady state and is insensitive to pulse length. The trade-off between power and pulse length is such that one gains by going to short pulse lengths because of smaller convection losses and spreading by conduction in the microsecond regime. The results above are not exact because of the assumptions of point source and no film backing made in the interest of computational simplicity, but they do serve to show the relation between results expected for millisecond pulses and microsecond pulses.

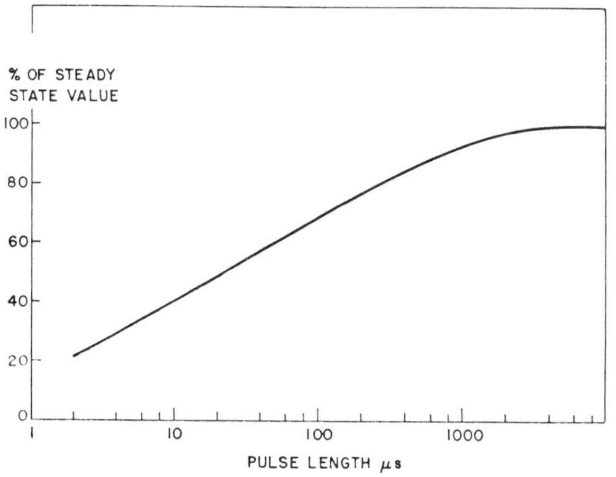

FIG. 10. Fraction of steady-state temperature rise 3 μ from a point source in an unbacked MnBi film as a function of pulse length.

The value of the thermal conductivity used in the calculations described above is of some interest. Because of the difficult geometry involving a very thin film in contact with a thicker substrate, we have not yet measured the thermal conductivity of MnBi films accurately. From the Wiedemann–Franz law and the electrical conductivity[11] we would obtain $K=0.028$ W/cm °C. This is a lower limit since it falls in the range where phonon conductivity could be expected to make a significant contribution in addition to electronic conductivity. The value actually used in the calculations was 0.17 W/cm °C. This value is consistent with the analysis of the experimental data on spreading of the heated spots in the millisecond regime. It is also a conservative value, since if it is too large, the calculated temperatures are underestimated and diffusion of the heat is overestimated. Thus, the main conclusions about the availability of large packing densities and high writing rates would be unaffected. The conclusion of this analysis is that Curie-point writing with microsecond duration pulses of modest laser output with high bit-packing densities and good isolation of adjacent bits is feasible.

IV. READING TECHNIQUE

A. Beam Intensity

To read a written bit of information on a MnBi film, the magneto-optical Faraday effect in the film is used. The laser-beam intensity is first adjusted to a suitable level so that no appreciable temperature rise will occur on the spot being read. The incident beam is linearly polarized, and after passing through the film it is analyzed by a polarizer and falls on a detector. Assuming the input optical field is

$$\begin{bmatrix} E_x \\ 0 \end{bmatrix},$$

the output field in such a system will be

$$\begin{bmatrix} E_x' \\ E_y' \end{bmatrix} = \begin{bmatrix} \cos^2\theta & \sin 2\theta/2 \\ \sin 2\theta/2 & \sin^2\theta \end{bmatrix} \begin{bmatrix} \cos\phi_f & -\sin\phi_f \\ \sin\phi_f & \cos\phi_f \end{bmatrix} \begin{bmatrix} \exp(-\alpha h/2) & 0 \\ 0 & \exp(-\alpha h/2) \end{bmatrix} \begin{bmatrix} E_x \\ 0 \end{bmatrix}. \quad (12)$$

The resulting intensity I is given by

$$I = (c/8\pi)[|E_x'|^2 + |E_y'|^2] = I_0 e^{-\alpha h} \cos^2(\theta - \phi_f), \quad (13)$$

where $I_0 = cE_x^2/8\pi$, $\phi_f = Fh$ = Faraday rotation, h = optical path length in the sample, and α = absorption coefficient. θ is the angle of the analyzer with respect to the polarization vector of the incoming radiation I_0. The signal intensity is directly proportional to the difference in the intensity of the beam passing through regions of opposite magnetization, or

$$\Delta I = I_0 e^{-\alpha h}[\cos^2(\theta - \phi_f) - \cos^2(\theta + \phi_f)]. \quad (14)$$

In the case where the analyzer is set to cross out the unreversed magnetization direction $(\theta = \phi_f + \pi/2)$, ΔI is maximized when $h = 2/\alpha$; then $\Delta I_{max} = I_0 e^{-2} \sin^2(2F/\alpha)$. For MnBi, at the helium–neon laser wavelength of 6328 Å, $\Delta I/I_0$ goes through a broad maximum exceeding 10^{-3} in the range of thickness from 300 Å to 1200 Å. Light leakage results from the imperfect analyzer, scattering in the film medium, and the fact that mica is biaxial; it has in practice been reduced to 1/50 of ΔI.

B. Reading Speed

To calculate the number of photons available at the detector for reading-rate considerations, we start with the intensity of the reading laser beam. From previous calculations in Sec. III we know that, when a 1 mW laser is used, heating of the film is negligible for rates much greater than 1 MHz. We assume that reading beam of 1 mW is attenuated by a factor of 10 to take into account losses in the optical components. It will be attenuated further, by a factor of 1000, by the MnBi film before it reaches the photodetector. From this we arrive at a total number of 10^{11} photons/sec available at the detector. Using a silicon detector peaked at 0.8 μ with a D_λ^* value of 10^{12} cm(Hz)$^{1/2}$/W and a detector area of 10^{-2} cm^2, the noise level will be about 3×10^9 photons/sec at 10^9 Hz bandwidth. Therefore, a signal to noise ratio of 30 appears feasible at a readout rate of 10^9 bits/sec. The limiting noise arises from the detector rather than from the imperfections in the system which were described in the preceding paragraph. In addition, methods for reducing the noise arising from the system imperfections have been described elsewhere.[20]

V. EXPERIMENTAL RESULTS

Thermomagnetic writing on continuous thin films of MnBi has been achieved using a laser as the heating source. Spots of dimensions approaching the diffraction limit of the beam have been written with a helium–neon laser operating at 6328 Å wavelength. The beam was focused by an f/0.65 microscope objective to a focal spot size less than 1.5 μ in diameter. A signal level difference of 50 between the written and unwritten spot was observed using a photomultiplier. This indicates that even when the spot size approaches the diffraction limit of the optics, detection by electronic means is still feasible.

We have demonstrated Curie-point writing with microsecond duration pulses from a He–Ne laser on an 800 Å thick film initially at room temperature. The beam switching was achieved by applying a voltage pulse to a Kerr cell between crossed polarizers interposed between the laser and the focusing lens. Some results are shown in Fig. 11(a) for spots written with laser pulse lengths in the microsecond regime. Eleven spots are visible: The three larger ones at the left end of the line are burned spots which serve as a reference to locate the area of interest in the MnBi film; the other eight are written spots. The pulse lengths were 3.8 μsec at half-power. For the four written spots at the left, the peak laser power was approximately 31 mW incident on the film surface; the diameters of the written spots are about 1.75 μ. For the four written spots on the right, the peak laser power incident on the film was about 13 mW; the spot diameters are about 1.25 μ. The distance between centers is about 12 μ. The film was moved between spots using a micrometer. Figure 11(b) shows the same area after erasure by the application of a 6 kOe magnetic field at room temperature.

In these experiments erasure was also accomplished by heating the written spot to above the Curie temperature and applying an external field slightly larger than the demagnetizing field during cooling. In other experiments we have achieved writing in the microsecond regime with a laser power as low as 3.5 mW. The writing

TABLE III. Erasure magnetic field (Oe).

Spot diameter (μ)	Erasure field (Oe) Calc.[a]	Expt.
~1.25	>543	950
~1.75	>388	700
~3	>226	400

[a] Values given correspond to calculated demagnetizing field at the center of heated spot.

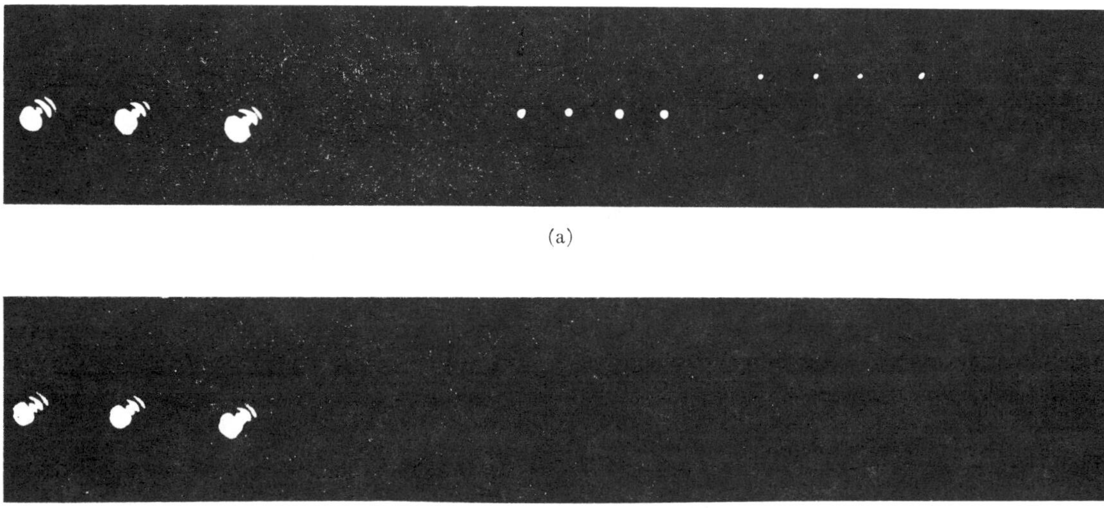

(a)

(b)

FIG. 11. Photograph of Curie-point writing on MnBi–mica using a focused He–Ne laser beam. (a) Observation of the written spots by the Faraday effect. The three large spots at the left are burned spots which identify the area of interest on the film. The eight written spots have diameters ranging from 1.25 to 1.75 μ. The spacings between centers are approximately 12 μ. See text for more details. (b) Same area after erasure by application of a magnetic field.

and erasure can be repeated on the same spot without deterioration.

The results described above clearly establish the feasibility of laser Curie-point writing in the microsecond regime with modest laser powers. At the present time it is not possible to make a detailed comparison with the results of Sec. III.B because of uncertainty about the exact diameter of the focused laser beam at the film plane and consequent lack of knowledge about the flux density. In addition, feedback by reflection from the MnBi film into the laser momentarily changes the laser output power when the shutter is opened. As we go from the millisecond pulse length regime to the microsecond regime, the increase in laser power required for writing is relatively small, in qualitative agreement with the results shown in Fig. 10.

Table III shows the experimental Curie-point erasure field requirements and the demagnetizing field calculated for the center of the heated spot. It is clear that the erasure field is inversely proportional to the diameter of the spots as predicted by Eq. (5) of Sec. III.A. The experimental value of the erasure field is, of course, expected to be larger than the demagnetizing field at the center of the spot. The experimental measurements show that spot dimensions of 3 μ with a spacing of 10 μ from center to center are certainly realizable, and that the switching field requirement will be approximately that of the calculated demagnetizing field.

VI. CONCLUSION

Thin films of MnBi have been prepared by vacuum-deposition technique on mica substrates. These films are extremely uniform and easily reproducible. Their optical, magneto-optical, and magnetic properties have been measured. The results of these measurements indicate that this material is particularly suitable for optical mass memory applications.[4,6,20] Information can be written by the Curie-point writing technique and read by the magneto-optic method, using a single laser beam. An analysis of the requirements for the writing laser beam and magnetic field, as well as for the readout signal-to-noise ratio, has been performed. The results indicate that on a continuous thin MnBi–mica film, writing of 3 μ spots can be achieved at a rate of 10^6 bits/sec, using a 10 mW He–Ne laser of 6328 Å wavelength. The required magnetic field for writing on a continuous film at 10^6 bits/cm^2 packing density is 265 Oe. This can be considerably reduced if films of discrete spots are used. The readout rate can be as high as 10^9 bits/sec, with a signal-to-noise ratio of 30, using a conventional silicon detector. Experimental results on the Curie-point writing technique are in general agreement with the analysis.

ACKNOWLEDGMENTS

The authors gratefully wish to thank R. Honebrink and G. N. Otto for their diligent and valuable contribution on the crystal growth and various experimental measurements of this work. They are indebted to F. Schmit for the results of the microsecond writing experiment, and to L. D. McGlauchlin for his initial contribution on the concept of the laser Curie-point writing on MnBi films.

Reprinted with permission from *Applied Physics Letters*, Vol. 22(7), pp. 337-339 (April 1, 1973).

Amorphous metallic films for magneto-optic applications

P. Chaudhari, J.J. Cuomo, and R.J. Gambino
IBM Thomas J. Watson Research Center, Yorktown Heights, New York 10598
(Received 28 December 1972)

It has been demonstrated that perpendicular uniaxial anisotropy and compensation points in the vicinity of room temperature can be obtained in amorphous thin films of rare-earth—transition-metal alloys. The magneto-optic properties of these films are reported, and it is shown that a remarkably good signal-to-noise ratio can be obtained in a thermomagnetically written film.

We have shown that amorphous Gd-Co and Gd-Fe films can be prepared with uniaxial magnetic anisotropy.[1] These materials were found to be suitable for bubble domain devices. It was also shown that the saturation magnetization at room temperature varied over large values by changing the composition of the sputtered films. This change in magnetization with composition was interpreted in terms of antiferromagnetic coupling between the rare-earth and transition-metal spin systems. In this note we show that this antiferromagnetic coupling leads to compensation points which vary with composition from approximately 40 to 500 °K. Compositions which lead to compensation points in the vicinity of room temperature suggest that thermomagnetic writing and erasing can be carried out in these films. The advantage of using amorphous films over oriented polycrystalline films lies in the large signal-to-noise ratio achievable in the absence of grain noise.[2] Grain size, if any, in amorphous films is less than 25 Å, which is orders of magnitude smaller than the wavelength of optical lasers.

The films were prepared by rf and dc sputtering from arc-melted targets onto water-cooled substrates. The details are described in Ref. 1. The magnetization was measured (as a function of temperature) in an inert atmosphere by means of a force balance. The data for a film containing 78.5 at.% Co are shown in Fig. 1.

The dependence of the compensation point on cobalt concentration in the Gd-Co system is shown in Fig. 2. Note that the compensation temperature is a simple function of the cobalt concentration. One of the major advantages of amorphous magnetic materials for this application is the ability to vary composition without the constraints of equilibrium phases. Within broad limits it is thus possible to select a composition with any compensation point desired.

The fact that these magnetization minima are indeed compensation points was confirmed by an examination of a film by electron microscopy. The Lorentz contrast indicated a reversal in the sign of magnetization on crossing a compensation point. This is shown in Fig. 3. By using Kerr contrast the change in stripe domain pattern was followed as a function of temperature (Fig. 4). With increasing temperature the width of stripe domains increased as the compensation point was approached. At still higher temperatures the strip width decreased again. This is in conformity with a decrease in saturation magnetization. The variation of strip width with magnetization can be explained on the basis of change in saturation magnetization.[3] However, the more interesting observation here is not the strip width but the shape of the strip. At room temperature the stripes are relatively smooth, but as the compensation point is approached the stripes are found to be nonuniform, indicative of increasing influence of coercive force. Faraday loop measurements of coercivity as a function of temperature show a sharp maximum at the compensation point. The temperature dependence of the coerciv-

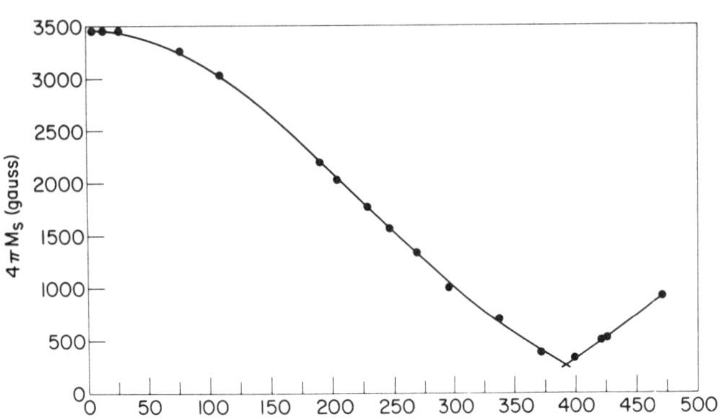

FIG. 1. Magnetization vs temperature of sputtered amorphous Gd-Co alloy containing 78.5 at.% cobalt.

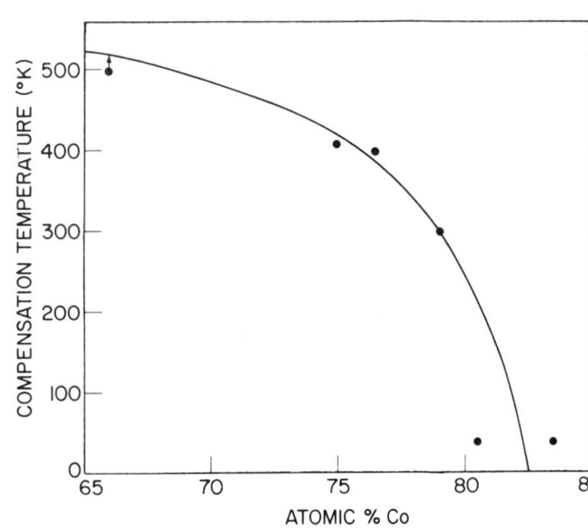

FIG. 2. Compensation point vs cobalt concentration in amorphous Gd-Co alloys. Compositions were determined by x-ray fluorescence analysis.

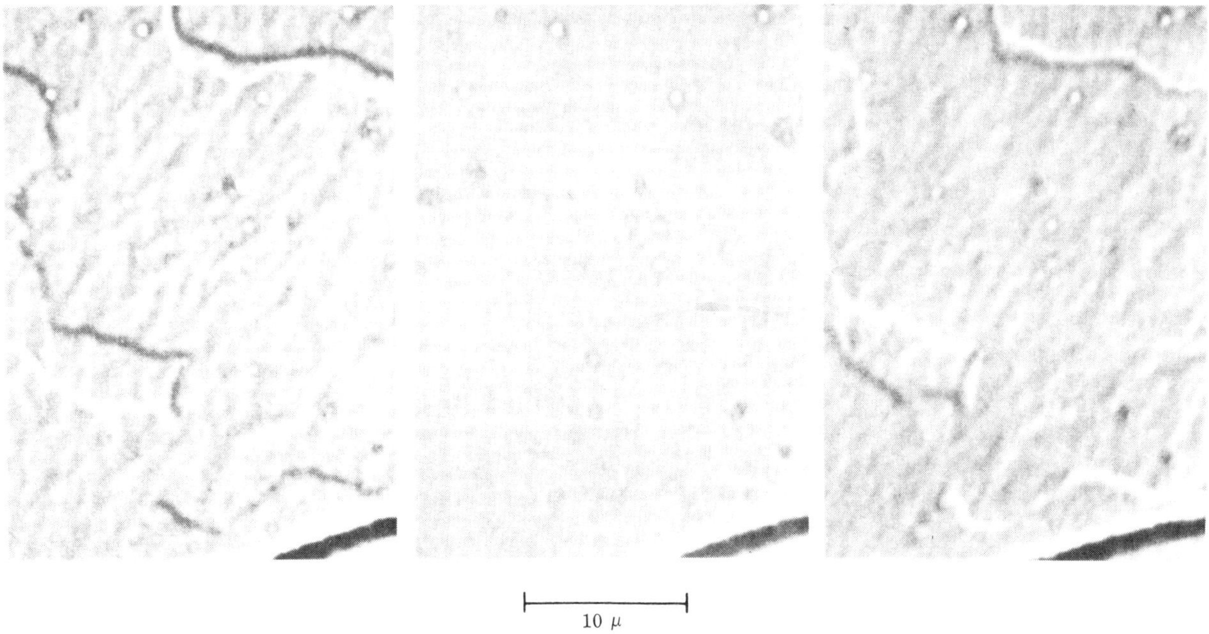

FIG. 3. Reversal of the sign of the magnetization through the compensation point as shown by Lorentz contrast. (Photographs courtesy of S. Herd, IBM Research.)

ity is characteristic of a ferrimagnet at its compensation point.[4] The value of coercivity was found to be influenced by the method of fabrication and by the impurity level in the sputtering working gas.

The Faraday rotation was determined to be 1.8×10^5 deg/cm, and the Kundt constant ~ 5000 deg/cm. Over the wavelength examined (4500–8200 Å) the linear absorption coefficient was essentially independent of wave-

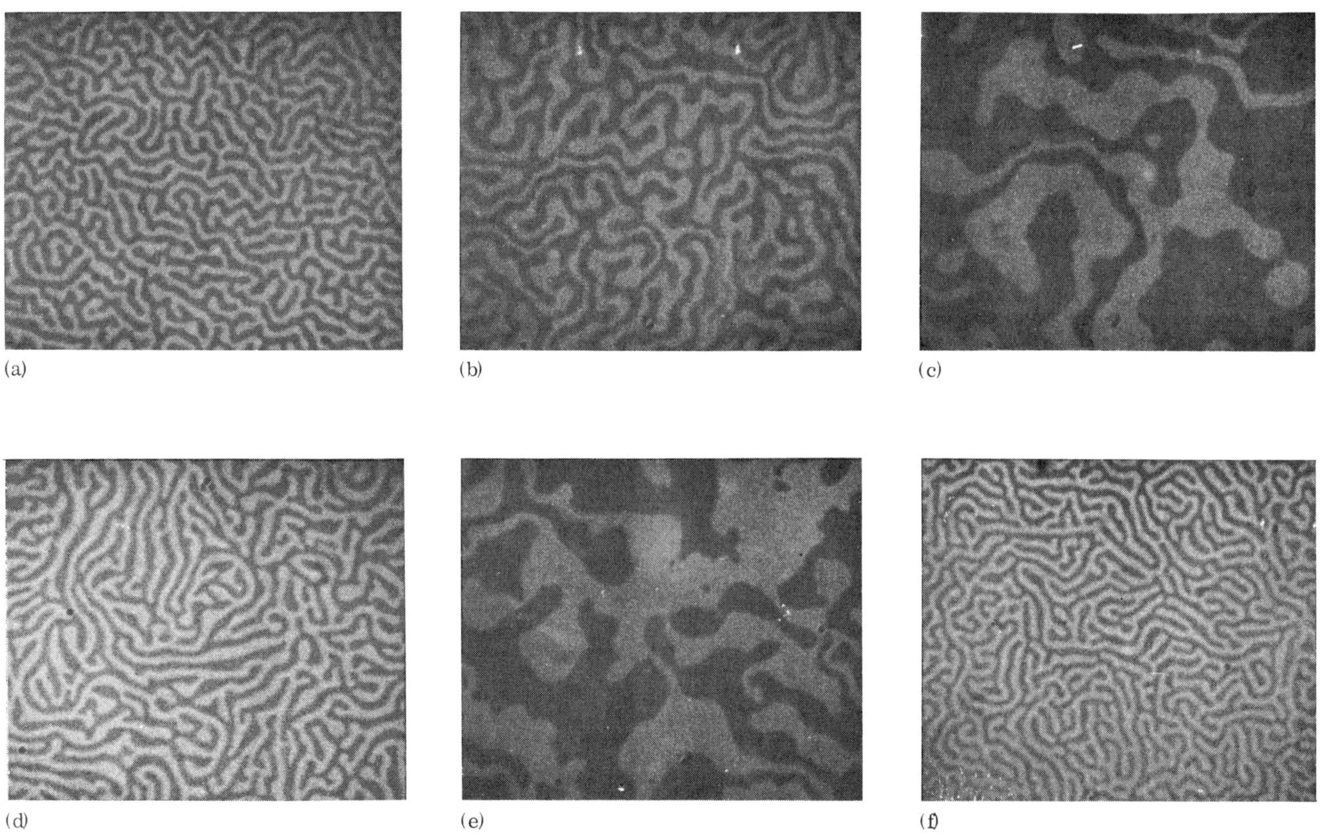

FIG. 4. Zero-field stripe domain pattern as a function of temperature. The conditions shown are (a) room temperature before heating, (b) 70 °C, (c) 100 °C, very near the compensation temperature, (d) 225 °C, (e) 90 °C during cooling, and (f) room temperature after heating. Note the irregular domain pattern at the compensation point. Kerr contrast. Magnification, 200×.

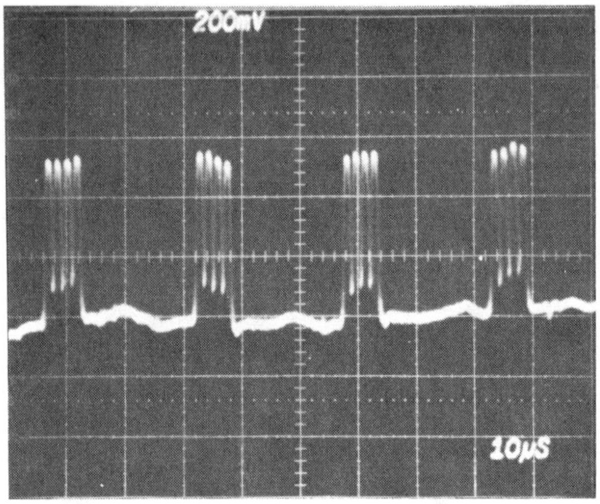

FIG. 5. Signal from laser-written bits on an 800-Å film of amorphous Gd-Co alloy. Note the absence of grain noise. Each division on the x axis corresponds to 100 μ, and each division on the y axis corresponds to a rotation of 0.2 deg. (Photograph courtesy of B. Brown, IBM Research.)

length and of the order of 8×10^5 cm^{-1}. The details of these measurements will be described elsewhere.

Films were examined for magneto-optic writing and erasing. Compensation point writing was achieved by using a dye laser or a low-temperature GaAs laser in conjunction with a bias field of up to several hundred G. Typically a laser power density of less than 0.5 nJ/μ^2 was used in conjunction with a 1-nsec pulse width to write over a few-micron spot. Erasure of the written spot was achieved both by applying a large bias field at room temperature and by a smaller bias field in combination with a laser pulse with similar requirements as the write cycle. Writing and erasing were carried out on the same spot many times in air without degradation of the film. A typical signal-to-noise ratio on a film moved at a relative speed of 1000 cm/sec with respect to the read beam is shown in Fig. 5. We note that the signal-to-noise ratio is remarkably good.

The authors are grateful to many colleagues for helpful discussions. In particular they should like to thank B. E. Argyle, F. Cardone, D. Cronemeyer, E. Da Silva, J. Dempsey, S. Herd, H. Lilienthal, W. Molzen, L-J. Tao, R. J von Gutfeld, and B. Welber of the IBM Yorktown Research Laboratory, and B. Brown, D. Franich, B. G. Huth, E. Sawatzky, and C. Stratton of the IBM San Jose Research Laboratory for data prior to publication.

---

[1]P. Chaudhari, J.J. Cuomo, and R.J. Gambino, IBM J. Res. Dev. 17, 66 (1973).
[2]R.P. Hunt, IEEE Trans. Magn. 5, 700 (1969).
[3]J.A. Cape and G.W. Lehman, J. Appl. Phys. 42, 5732 (1971).
[4]C.D. Mee, IBM J. Res. Dev. 11, 468 (1967).

Reprinted with permission from *IEEE Transactions on Magnetics*, Vol. MAG-12(6), pp. 779-781 (November 1976). ©1976 IEEE.

MAGNETIC PROPERTIES AND CURIE POINT WRITING IN AMORPHOUS METALLIC FILMS

Y. Mimura, N. Imamura, and T. Kobayashi
KDD Research and Development Laboratory,
1-23, Nakameguro, Tokyo, Japan

ABSTRACT

The Curie point writing in the amorphous ferrimagnetic films are described. The bits written on the films with relatively large values of $4\pi M_s$ are unsaturated. We could write and erase the saturated bits on the film with low values of $4\pi M_s$ with a low external field. The magnetic reversal profile inside the bits can be explained by the "self" demagnetizing field originating within the reversed region itself.

INTRODUCTION

Thermomagnetic writing of information on magnetic films is becoming an important technique for high density storage. Amorphous magnetic films are advantageous for this technique, since the large signal-to-noise ratio can be achieved in the absence of grain noise. A Gd-Co[1] sputtered film is the first amorphous magnetic films proposed for this technique.

We have shown that amorphous Tb-Fe[2] and Dy-Fe[3] films are suitable for applications to thermomagnetic writing because of the high coercivity and the low Curie temperature. We have also shown that the bits written thermomagnetically on these films without external field is generally nonsaturated.[4] The magnetic reversal process within the heated region is influenced by the demagnetizing field and therefore thermal behavior of the magnetization. Amorphous Tb-Fe and Dy-Fe films are ferrimagnetics in the sense that the Tb or Dy moments couple antiferromagnetically with the Fe moments. The demagnetizing field and the thermal behavior of the magnetization vary with composition in these films. Such variation of the magnetic behavior with composition will lead to the various magnetization profiles within the thermomagnetically written region. The main purpose of this paper is to describe the magnetic reversal process within the thermomagnetically written region in the amorphous ferrimagnetić films.

FILM PREPARATION

The films used in this study were prepared by rf cosputtering technique from the iron target on which small Tb or Dy plates were distributed onto water-cooled substrate. The sputtering pressure was 4.0 x 10^{-2} torr Ar. The distance between a target and a substrate was 50 mm. The sputtering rate was about 300 Å/min.

MAGNETIC PROPERTIES

Some magnetic properties of the amorphous Tb-Fe and Dy-Fe films such as the Curie temperature, compensation temperature, the room temperature magnetization, and B-H loops are shown in Fig. 1. For films with large values of Ms (I, V), the preferred direction of magnetization is in-plane, but it changes from in-plane to normal to the film plane for the films with low values of Ms (II III IV). However, the films in the composition region (II IV) are not suitable for thermomagnetic writing, because the shape of B-H loops are not rectangular under the influence of the large demagnetizing field, and hence a single domain can not be held without applied field. Ultimately, the films in the region III can be used for thermomagnetic writing. Within the region III, the Curie temperature in these films is about 130 and 70 °K for Tb-Fe and Dy-Fe films, respectively, and its compositional dependence is very small. The small compositional dependence of Tc is favorable to produce the film with a uniform Curie temperature.

DEMAGNETIZING FIELD AND THERMAL BEHAVIOR OF MAGNETIZATION

There appears to be general concensus that the demagnetizing field plays an important role in determizing the direction of the magnetization within the region heated above the Curie temperature upon cooling. Two types of the demagnetizing field have been reported. One is the usual demagnetizing field which originates from regions outside the Curie heated region, and it is denoted by H_{od}(OD). The other is the "self" demagnetizing field[5] which originates within the reversed region itself, and it is denoted by H_{sd}(SD). Since the source of the SD field is physically so close to the unmagnetized region, it can exert a disproportionally strong influence on the magnetization of this region. In particular, the $4\pi M_s$ of the present ferrimagnetic films is about one order of magnitude smaller than that of ferromagnetic films such as MnBi[6], accordingly it is considered that OD effect is comparatively small in the present films. As a result, we can expect that the SD effect is more important in these ferrimagnetic films than in ferromagnetic films.

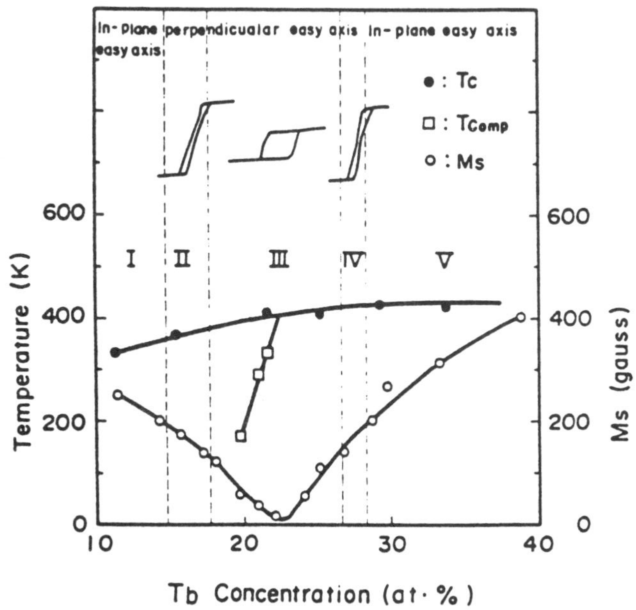

Fig.1(a) Curie temperature, compensation temperature, saturation magnetization and B-H loops vs. composition for amorphous Tb-Fe films.

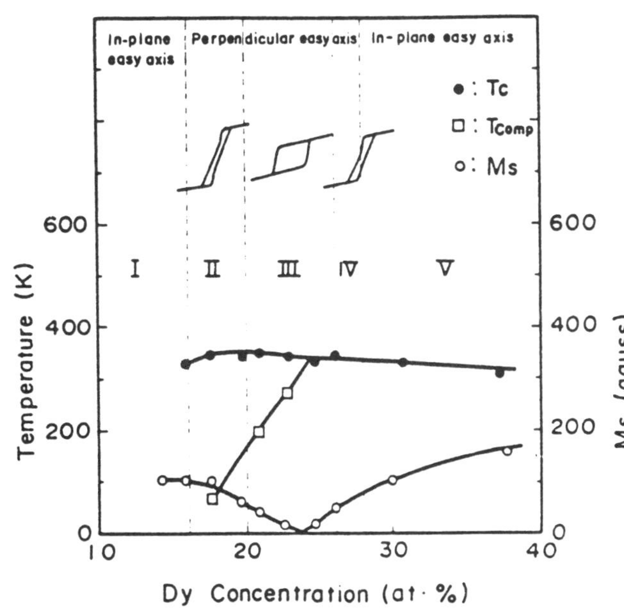

Fig.1(b) Curie temperature, compensation temperature, saturation magnetization and B-H loops vs. composition for amorphous Dy-Fe films.

In order to investigate the influence of these demagnetizing field effects on the writing process, we require a knowledge of the thermal behavior of the magnetization in the films. Fig.2 shows three typical magnetization vs. temperature curves observed in the present ferrimagnetic films. The A curve is observed typically for the Fe moment dominant films, and the C curve for the Tb moment dominant films. The B curve is observed for the films with the compensation point near room temperature. From these M-T curves, we can deduce the instantaneous distribution of the magnetization in the films, when the Gaussian distributed heat pulse is turned off, and it is shown schematically in Fig.3. The region (a) in Fig.3 is room temperature and (b) is heated above the Curie point. The magnitude of Hod in the A and B films should be same, if these films have almost the same values of $4\pi M_s$. However, Hsd in the A film should be stronger than that in the B film, because the magnetization in the periphery of the unmagnetized region is larger in the A film than in the B film. In the C film, $4\pi M_s$ in (a) is approximately 0 gauss and the magnetization in (b) is also small, therefore it is considered that the both of H_{od} and H_{sd} are small in this film.

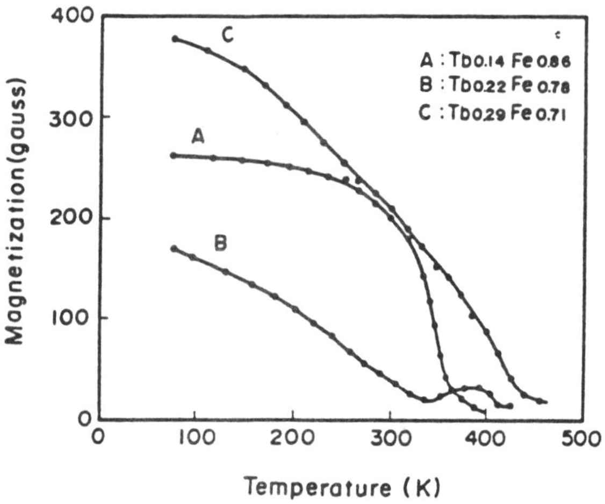

Fig.2 Magnetization vs. temperature for amorphous Tb-Fe films.

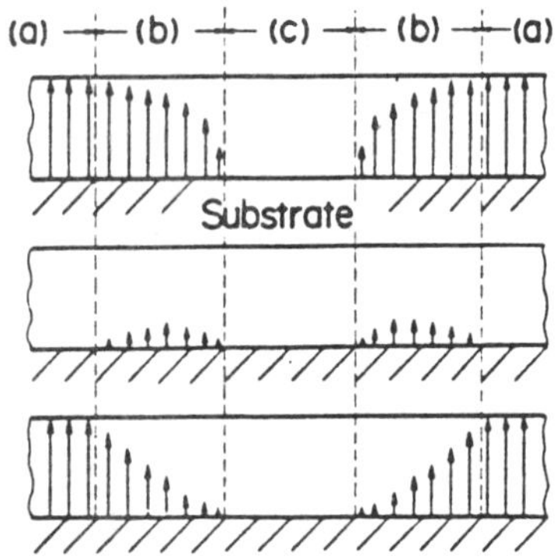

Fig.3 Magnetization distribution in the film at the instance the heat pulse is turned off. A: Fe moment dominant film. B: Film with the compensation temperature near room temperature. C: Tb moment dominant film.

THERMOMAGNETIC WRITING

The Curie point writing and erasing were achieved by using a He-Ne laser beam which was focused by an object lens onto the film mounted on an X-Y translation stage. According to the demand, the laser beam was pulsed by an electro-optic modulator, and the external magnetic field was applied to the film. The laser power was measured on this side of the object lens. The written domains were observed under a polarizing microscope using the polar Kerr magneto-optic effect. Six Tb-Fe and five Dy-Fe films were used for this study. The magnetic properties of these films are listed in Table 1. Fig.4 shows the microphotographs of the domain configurations written thermomagnetically on the Tb-Fe films without bias field. Firstly, the film was magnetically saturated in one direction, which we will denote by ⊖, and then a straight line was written magnetically using the magnetized Ni needle whose edge was sharpened to about 5 μm, which we will denote by ⊕. The laser beam was scanned across the ⊕ line written magnetically. As seen in Fig.4 (a) (b) (g) the stripes written thermomagnetically on the films with relatively large values of $4\pi M_s$ show a neutral tone between ⊖ and ⊕ region, indicating that the magnetization of this region is not saturated, but is demagnetized. It is also observed that the tone of the stripes is uniform over both region of ⊖ and ⊕, that is, the domain configurations within the heated area is not influenced by the direction of the magnetization of the region out-

TABLE 1

Magnetic properties of amorphous Tb-Fe and Dy-Fe films used for thermomagnetic writing.

composition (Tb at%)	thickness (Å)	H_c (KOe)	4πMs (gauss)	T_c (K)	T_{comp} (K)	dominant moment at room temperature
19.5	1460	1.6	860	400	100	Fe
21.7	1400	3.8	410	410	310	Fe
22.0	1390	7.0	~0	410	320	Tb
25.0	1460	2.1	790	415	~415	Tb
26.2	1300	1.1	1570	420	~420	Tb
27.1	1420	0.6	1900	425	~425	Tb
(Dy at%)	(Å)	(Oe)	(gauss)	(K)	(K)	
21.0	1110	500	550	335	200	Fe
24.5	1330	3000	~0	335	300	Dy
24.9	1170	760	210	330	~330	Dy
25.8	1450	410	480	340	~340	Dy
26.6	1800	210	700	340	~340	Dy

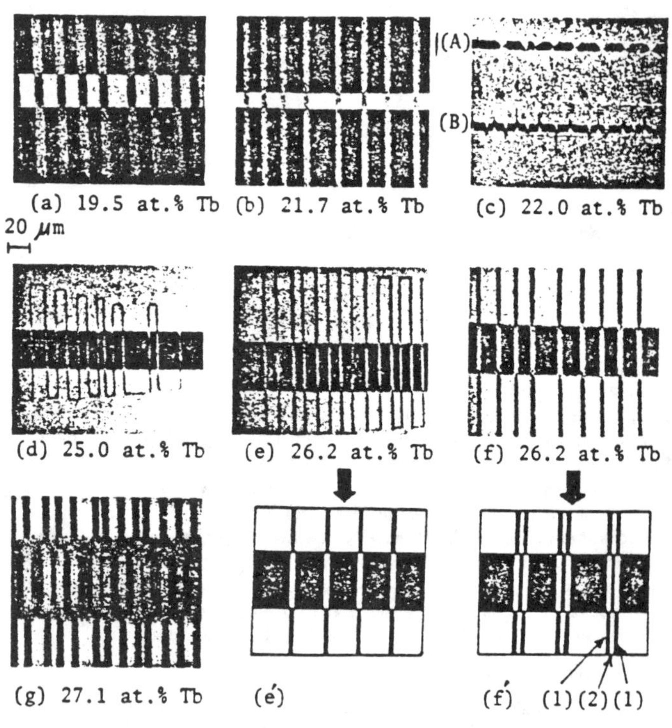

Fig.4 Microphotographs and schematics of the domain configurations written on amorphous Tb-Fe films without the external field. ⊖ region is dark and ⊕ region is light for (a)(b), and the reverse for (c)(g). Laser power is 7 mW for (a)(b)(d)(f)(g), 15 mW for (c)(A) and 5 mW for (c)(B) and (e).

side the Curie heated region. These observations show that the SD effect is more important in the magnetic reversal process within the heated region than the OD effect in these films.

The total energy in the films magnetized normal to the film plane is written as[8)]

$$E_t(\text{total}) = E_w(\text{wall}) - E_d(\text{magnetostatic}) + E_H(\text{applied}) \qquad (1)$$

The width of the reversed domain is determined so as to minimize E_t. Since $E_w \propto \sigma_w S$ and $E_d \propto M_s^2$, where σ_w is wall energy density and S is an area of domain wall, the domain width tends to spread with decreasing $4\pi M_s$ to reduce the wall energy, and a single domain become stable for the film with $4\pi M_s \approx 0$, under zero applied field. Such behaviors of the magnetic domain were observed in the present study. As shown in Fig.4 (a)→(b) and (g)→(f), a size of divided domain inside the stripes written thermomagnetically tends to spread with decreasing $4\pi M_s$. We could not write any reversed domain thermomagnetically on the films with $4\pi M_s \approx 0$ without the external field, and we could only erase the domain which had been written magnetically by the magnetized needle. When the laser beam of large power of 15 mW was applied to the reversed region, the erasing was possible without the external field, as shown in Fig.4 (c) (A), and when the laser beam of low power 5 mW was scanned across the reversed region, it was observed that the domain wall moved along with the temperature gradient as shown in Fig.4 (c) (B).

For the films with relatively small $4\pi M_s$ not 0, it was observed that both the SD and OD affected the magnetic reversal process. As shown in Fig.4 (f) (f'), when a comparatively wide area is heated above the Curie temperature, firstly the periphery of the unmagnetized region (A) is reversed to ⊕ by the demagnetizing field originated from the region outside the unmagnetized region (OD effect), and then the center of the heated area (B) is reversed to ⊖ by the demagnetizing field originated from (A) (SD effect). Fig.4 (e) (e') shows the reversed domain profile, when a narrow region is heated in the same film. The magnetization of the entire area of the heated region turns the opposite direction from that in the surrounding region, because the Hod affects not only the periphery of the heated region, but also the center of the heated region.

A similar behavior of the magnetic reversal profile was also observed for the Dy-Fe films, and it is shown in Fig.5.

Fig.6 shows the reversed bits written on Tb-Fe and Dy-Fe films by moving the film on an X-Y stage using the pulsed laser beam of various frequency under the external field of 50 Oe oriented parallel to the demagnetizing field. The bits written on 26.2 at.%Tb and 25.8 at.%Dy are saturated, and we could erase these bits under the external field less than 50 Oe oriented antiparallel to the demagnetizing field. On the other hand, the bits written on 19.5 at.%Tb and 21.0 at.%Dy films are nonsaturated, and an external field of more than 100 Oe was necessary to erase these bits. These observations coincide with the consideration that the SD effect is stronger in the Fe moment dominant films than in the Tb or Dy moment dominant films.

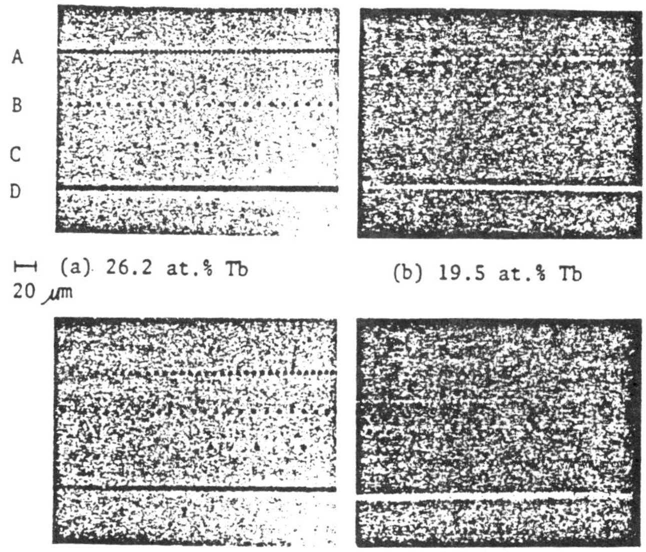

(a) 26.2 at.% Tb (b) 19.5 at.% Tb

(c) 25.8 at.% Dy (d) 21.0 at.% Dy

Fig.6 Bits written on amorphous Tb-Fe and Dy-Fe films using the pulsed laser beam of 10 µsec under the external field of 50 Oe. Laser power is 5 mW for Tb-Fe films and 2 mW for Dy-Fe films. A: 1 KHz B: 500 Hz C: 100 Hz D: continuous beam.

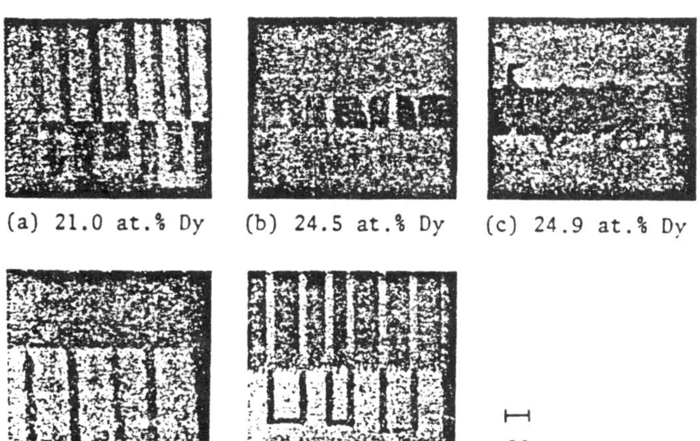

(a) 21.0 at.% Dy (b) 24.5 at.% Dy (c) 24.9 at.% Dy

(d) 25.8 at.% Dy (e) 26.6 at.% Dy

Fig.5 Microphotographs of the domain configurations written on amorphous Dy-Fe films without the external field. ⊖ region is dark and ⊕ region is light for (a) and the reverse for (b)(e). Laser power is 3 mW for (a)(c)(d)(e) and 5 mW for (b).

CONCLUSION

The Curie point writing in amorphous ferrimagnetic films was studied. The bits written on the films with relatively large values of $4\pi M_s$ are unsaturated because of the strong SD effect. However, the SD effect reduces with decreasing $4\pi M_s$, and therefore we can write and erase the saturated bits on the films with low values of $4\pi M_s$ under the low external field.

ACKNOWLEDGEMENT

The authors wish to thank H. Ogasawara andY. Hayashi for their technical assistance.

The authors also wish to thank Drs. S. Oshima, N. Sasamoto, Y. Nakagome and K. Morooka for their encouragement.

REFERENCES

(1) P. Chaudhari, J. J. Cuomo, and R. J. Gambino, Appl. Phys. Lett. 22, 337 (1973).

(2) Y. Mimura and N. Imamura, Appl. Phys. Lett. 28, 746 (1976).

(3) Y. Mimura and N. Imamura, Japan J. appl. Phys. 15, 937 (1976).

(4) Y. Mimura, N. Imamura, and T. Kobayashi, Japan J. appl. Phys. 15, 933 (1976).

(5) H. Wieder and R. A. Burn, Appl. Phys. Lett. 22, 188 (1973)

(6) D. Chen and R. L. Augard, J. Appl. Phys. 41, 2530 (1970).

(7) N. Imamura, Y. Mimura, and T. Kobayashi, Japan J. appl. Phys. 15, 183 (1976).

(8) C. Kooy and U. Enz, Phylips Res. Repts. 15, 7 (1960).

Reprinted with permission from *Journal of Applied Physics*, Vol. 54(9), pp. 4794-4798 (September 1983).

Thermal aspects of magneto-optical recording

M. Mansuripur and G. A. N. Connell
Xerox Palo Alto Research Center, Palo Alto, California 94304

(Received 3 December 1982; accepted for publication 11 May 1983)

We discuss the influence of thermal response on the magneto-optic readout performance of optically similar but thermally different quadrilayer media, using newly developed computational techniques for examining laser-induced heat flow in moving multilayer media. We first emphasize the importance of time scale for differentiating the write and read processes. We then show that the optimum design is a function of the medium velocity and available laser power for writing and is not normally one that provides maximum signal-to-noise ratio in the limit that laser heating during readout can be ignored. Finally, we describe a systematic approach to the selection of the best quadrilayer design for a given set of system conditions.

I. INTRODUCTION

In earlier work, we considered how a differential detection system and magneto-optical medium could best be configured for readout by the polar Kerr effect.[1,2] It was found that the generic optimum structure of the medium itself was closely related to the trilayer metal-dielectric structures that have been long used in optics when absorption in a thin metal layer is required.[3,4] In particular, Bell and Spong[5] have pointed out that this structure is most useful for ablative optical recording because the maximum light absorption occurring in the thin metal film results in a maximum writing sensitivity. In our case, we made use of the structure somewhat differently. Its characteristics were no longer designed for maximum light absorption, but rather were optimized along with the detection system parameters to give a maximum signal-to-noise ratio for magneto-optical readout by the polar Kerr effect. Typically the structure of the medium is then quite different from that for maximum light absorption. Indeed the signal-to-noise ratio for the latter is usually very small with a differential detection system.

While heating of the medium during readout was recognized in our earlier work[2] as an important parameter in the selection of an optimized system, it was included only inasmuch as media requiring different write powers can obviously withstand different read powers. This feature of the problem is closely related to earlier work done on the thermal aspects of ablative recording,[6,7] but now it is no longer important to maximize the writing sensitivity by preventing heat flow from the light absorbing layers, but rather to encourage it, subject to thermomagnetic writing remaining possible. This then allows a maximum of power to be used in readout. The approach we took earlier therefore represents a reasonable beginning to the solution of the complete readout problem, but it omits an important quantitative difference between the write and read processes, namely time scale. The consequences of this are demonstrated in Fig. 1 by calculations that simulate the thermal write process for a stationary quadrilayer medium. (The details of the layer structure are given in the figure caption.) In this example, the medium is irradiated for 50 ns by a 5-mW laser pulse focussed in a Gaussian beam to an e^{-1} radius of 500 nm. The point to note is that, independent of the temperature rise needed for writing the magnetic film, the writing process itself is completed prior to the establishment of thermal equilibrium. In contrast, readout is done in steady state, and our example therefore implies that readout must be carried out with even less power than might straightfowardly be deduced from the write power of the medium alone. The remainder of this paper will be concerned therefore with the problem of the selection of magneto-optical interference structures that provide best overall system operation when these differences between reading and writing are properly included. Moreover, the effects of relative motion of the laser beam and magneto-optic medium on the selection process are fully explored.

The paper is organized as follows. The computational procedures are outlined in Sec. II and then used in Sec. III to discuss the influence of thermal response on the read-write performance of optically similar quadrilayer media. Finally, the conclusions of the work are presented in Sec. IV.

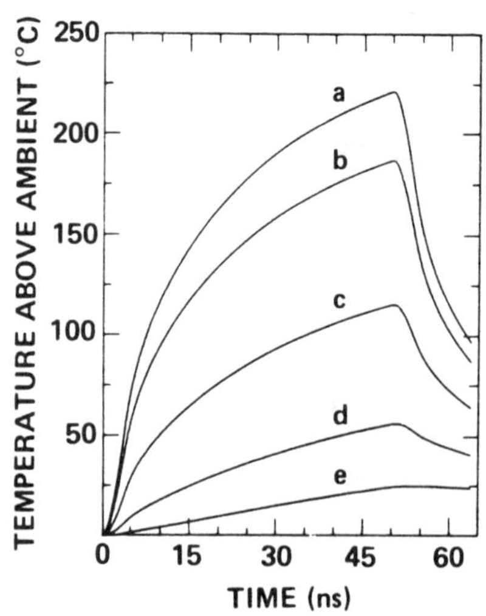

FIG. 1. Increase of temperature vs time at radii of (a) 0, (b) 250, (c) 500, (d) 750, and (e) 1000 nm that occurs in the magnetic film of a quadrilayer medium. The quadrilayer consists of a 50-nm-thick aluminum layer, and 80-nm-thick quartz intermediate layer, a 15-nm-thick magnetic metal, and a 120-nm-thick quartz overlayer. The thermal and optical parameters are given in Table II.

II. BRIEF OUTLINE OF THE COMPUTATIONAL PROCEDURES

We describe the technique fully elsewhere,[8] but the basis of the method is readily understood. We assume that the medium is moving with constant linear speed V and then approximate this motion by a step-like progression in which the medium jumps a distance $V\Delta t$ at the end of each time interval Δt. During each Δt, however, the medium is stationary and the static technique, used earlier[9] for the calculation of temperature distributions generated in multilayer media by the incidence of a Gaussian laser beam, is applicable to the problem. Moreover, for the specific case of temperature independent optical and thermal parameters, the absorption and diffusion equations become linear, and the moving medium problem may be solved by the superposition of these static solutions at progressive time intervals.

In the present case, we are mostly interested in the laser power required to heat the magnetic medium to a certain temperature for different values of V in two situations: when the laser pulse length is r_0/V, with r_0 being the e^{-1} radius of the Gaussian laser spot, and when the pulse is long enough for thermal equilibrium to be established. The two cases, therefore, relate to that of writing at fixed resolution (for a given magnetic field) and reading under fixed illumination, respectively. The results obtained will provide the required information for the complete optimization of the medium in a magneto-optical recording system. We also show the isotherms obtained during the read process for various media at the typical medium velocity of $V = 20$ m/S. These results then provide the basis for an intuitive understanding of the different responses of the various media structures considered, and indicate why the achievable bit density for a given magnetic material may be greater in one design than another.

III. RESULTS AND DISCUSSION

When the read power available is sufficiently low for thermal effects to be unimportant, we showed in earlier work[1] that an optimum quadrilayer structure provided maximum signal-to-noise ratio at a fixed read-beam power during readout in a magneto-optical system. The design of the device is shown in Fig. 2. The magnetic thin film lies between two transparent dielectric films that together rest on an opaque reflector, and the whole is supported by a glass substrate. In later work,[2] we took the first steps toward determining how the structure could be adjusted to make best use of the laser power available for the read and write processes and showed that considerable gains in signal-to-noise could be expected by detuning the optimum device in such a way that all the laser power available would be used for recording. The effects of detuning on the signal-to-noise were then more than made up for by the extra power that could be used during readout. In the following we will put this result on a quantitative basis.

The details of the media designs considered and the optical and thermal parameters used in the calculations for a wavelength of 840 nm are shown in Tables I and II, respectively. The quadrilayer designs themselves fall into two categories: the first representing optimum designs, in the sense of providing maximum signal-to-noise ratios at fixed laser power when heating during readout can be ignored, and the second representing sub-optimum situations but providing what we shall see are attractive thermal characteristics. The signal-to-noise ratio at the same fixed, small laser power, however, is only slightly different in the two cases (Δ SNR < 0.5 dB) and therefore any major differences in performance in a magneto-optical recording system will derive directly from different thermal characteristics. It therefore follows that the power required for a fixed temperature rise under continuous illumination will be a direct relative measure of the readout signal-to-noise for a given quadrilayer structure. Similarly, the power required for a fixed temperature rise under pulsed illumination will be a direct relative measure of the writing sensitivity at fixed applied magnetic field.

Figures 3 and 4 then show respectively the power required to raise the temperature of the magnetic medium by 100 °C during readout and writing versus the velocity of the medium for each of the structures. At the simplest level, the behavior for each structure is as expected—more power is needed for readout at fixed maximum temperature and for writing at fixed resolution as the medium velocity is increased. However, at the next level, an interesting effect can be seen, namely the rate of power increase is larger for writing than for reading, a result of considerable value when selecting a medium for a particular system. Moreover, different structures are seen to require very different power levels. In the remainder of this section, we will trace the origin of these effects and establish the conditions under which a particular structure would provide the best performance in a magneto-optical system.

The structure with greatest writing sensitivity (structure a) is an optimum quadrilayer with dielectric reflector. However, it also tolerates least power in readout, and therefore its selection as the medium of choice in a magneto-optical recording system would be appropriate in the case of an insensitive magnetic material or of severely limited laser power. For magnetic materials of current interest[10] and in view of the present availability of power from solid-state lasers, it is likely that neither of these conditions will pertain. It is important to note, however, that the thermal characteristics of the equivalent bilayer structure, i.e., a 20-nm-thick magnetic film on a glass substrate overcoated with a 100-nm-thick dielectric film, are obtained directly from these particular quadrilayer results by correcting for the smaller absorptance at the bilayer. In this case, there is an increased in the writing power by a factor of about 2. While there is a corresponding renormalization of the readout curve, the signal-to-noise actually achieved at the increased read beam power allowed remains unchanged because of the lower intrinsic signal-to-noise ratio of the bilayer devices. Nevertheless, these results do demonstrate that this simplest of realizable magneto-optic structures might be a sufficient magneto-optical medium, particularly in lower performance memories.

When the dielectric reflector is replaced by an aluminum one, the power levels for the readout and writing curves

TABLE I. Design of quadrilayers.[a]

Quadrilayer label	Over	Layer thickness (nm) Magnetic	Intermediate	Reflecting
Optimum:				
a	100	20	60	Thick (Dielectric)
b	100	20	60	50 (Aluminum)
c	310	20	60	50 (Aluminum)
d	100	20	60	Thick (Aluminum)
Detuned:				
e	100	20	30	50 (Aluminum)
f	100	20	30	Thick (Aluminum)

[a] All the optimum quadrilayers have the same signal-to-noise ratio for fixed system parameters. Under equivalent circumstances the detuned structures have a signal-to-noise that is at most 0.5 dB lower.

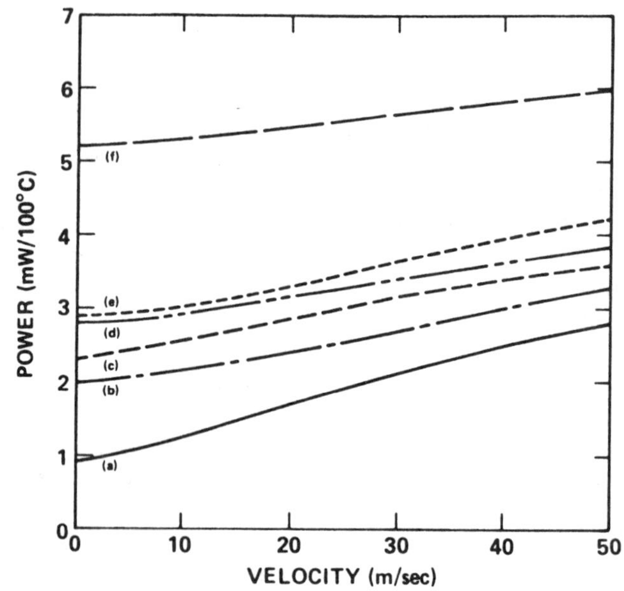

FIG. 3. Power (cw) required during readout to heat the magnetic film by 100 °C (peak of temperature distribution) vs medium velocity for the quadrilayer structures of Table I.

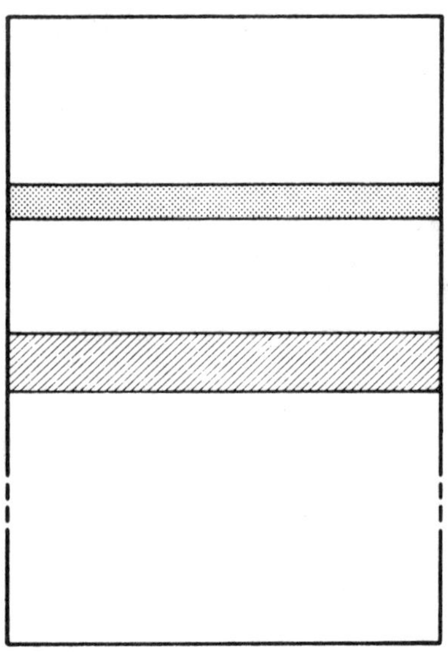

FIG. 2. A schematic representation of the quadrilayer configuration. The thicknesses of the reflector (hatched), intermediate layer, magnetic film (shaded), and overlayer may be adjusted to match the write sensitivity to a given magneto-optical system and thus obtain best readout signal-to-noise.

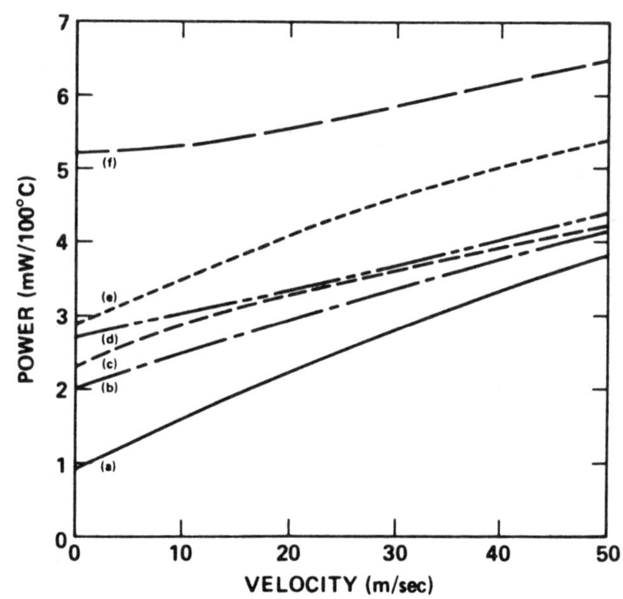

FIG. 4. Power (pulsed with duration r_0/V) required during writing to heat the magnetic film by 100 °C (peak of temperature distribution) vs medium velocity for the quadrilayer structures of Table I.

TABLE II. Optical and thermal parameters of materials.

Material		Refractive index 840 nm	Specific heat ($J/cm^3/deg$)	Heat conductivity (J/cm/deg/sec)
Glass substrate		1.5	2.0	0.015
Magnetic layers		$3.67 + 3.85i$	3.2	0.4
Over/intermediate layers		2.0	2.0	0.015
Reflector:	Aluminum	$2 + 7.1i$	2.7	2.4
:	Dielectric	$2 + 7.1i$[a]	2.0	0.015

[a] It is sufficient to use an effective refractive index for the dielectric reflector in this work since we are mainly concerned with the effects of heat flow toward the substrate.

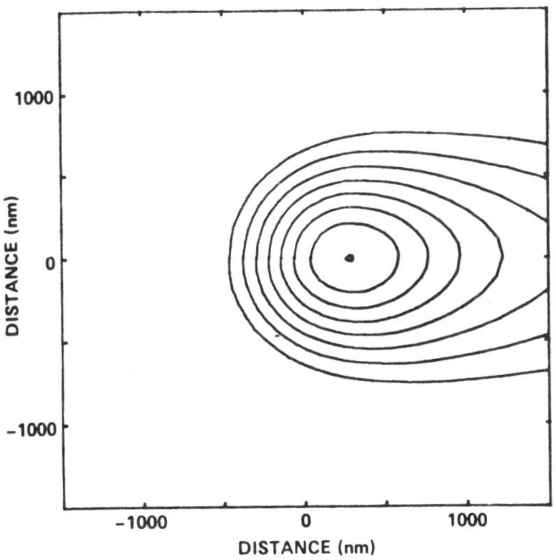

FIG. 5. Steady-state isotherms in the magnetic film of quadrilayer (a) of Table I when the medium moves at 20 m/S from the left to right relative to the beam. The central maximum is at 57.6 °C/mW, and each successive contour represents a temperature that is lowered by 10% of the maximum.

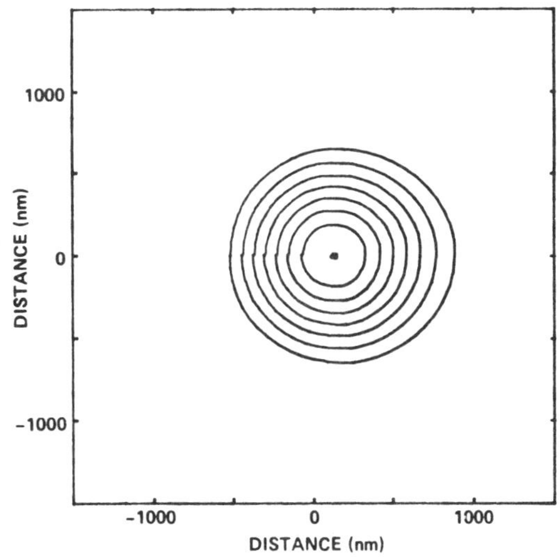

FIG. 7. Steady-state isotherms in the magnetic film of quadrilayer (d) of Table I when the medium moves at 20 m/S from left to right relative to the beam. The central maximum is at 31.4 °C/mW, and each successive contour represents a temperature that is lowered by 10% of the maximum.

increase progressively with aluminum thickness to provide one satisfactory way of adjusting the quadrilayer structure to the pertaining system writing constraints. For example, at a medium speed of 20 m/S, a decrease of writing sensitivity by a factor of about 1.5 leads to an increase in readout signal-to-noise at the read beam power now allowed of about 3 dB when the dielectric mirror is replaced by thick aluminum. These effects of course are caused by increased heat flow to the aluminum sink, and the isotherms for the magnetic layer shown in Figs 5, 6, and 7 make this very apparent. Namely, when heat flow to the substrate is minimized by using the dielectric reflector, a significant part of the heat generated in the magnetic film must be lost by lateral flow. Thus is Fig. 5, the maximum temperature in the magnetic film is displaced downstream from the center of the laser spot, and a long tail is apparent in the temperature distribution. However, as heat flow toward the substrate is increased in Figs. 6 and 7 by replacing the dielectric mirror with an aluminum layer and by increasing the aluminum layer thickness, the temperature distributions tend to move toward the original Gaussian intensity distribution of the laser spot. Note in particular the disappearance of the downstream tail and the movement of the temperature maximum toward the center of the laser beam. This is an important result because it demonstrates that heat coupling toward the substrate can be made large enough to overcome the enlarging effects of lateral heat flow on bit size that may occur during writing. For high performance memories, this factor may be of considerable value.

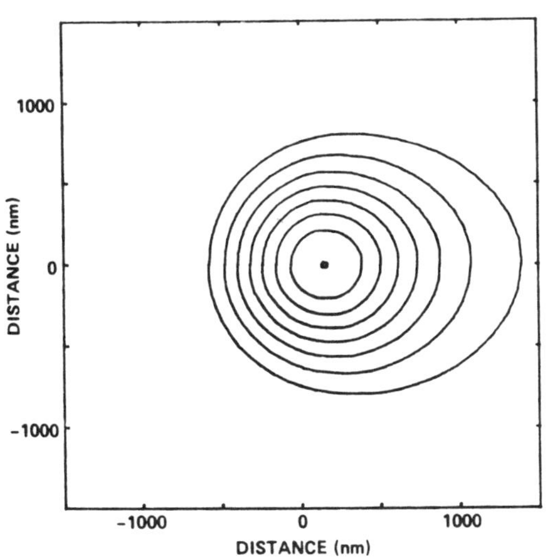

FIG. 6. Steady-state isotherms in the magnetic film of quadrilayer (b) of Table I when the medium moves a 20 m/S from left to right relative to the beam. The central maximum is at 41.5 °C/mW, and each successive contour represents a temperature that is lowered by 10% of the maximum.

Further control of the thermal characteristics of the quadrilayer structure may be obtained by adjusting the overlayer thickness. If this is increased discretely in half-wave steps, the quadrilayer maintains its optimum response but now requires more power for writing. Figures 3 and 4 show the effects of a half-wave increase in the overlayer thickness when a 50-nm-thick aluminum layer is used as reflector (structure c). Fortuitously, the writing and reading results are now very similar to those for the structure with a 100-nm-thick overlayer and relatively thick aluminum reflector. A combination of overcoat and reflector thickness adjustments can therefore yield optimum results under certain circumstances.

Finally, structures requiring the highest write powers and therefore providing the largest signal-to-noise in readout are obtained by decreasing the intermediate layer thickness to create a detuned design. Two examples using an intermediate layer thicknesses of 30 nm and different aluminum reflecting layer thickness are shown in Figs. 3 and 4 as curves e and f. As before, increases in write power and readout signal are evident when the aluminum layer thickness is increased, but now the media and system can be matched over a much larger range. For example, if sufficient laser power is available for writing when the medium travels at 20 m/S, the signal-to-noise attainable with the increased read-beam power now allowed for the detuned quadrilayer structure that employs a thick aluminum reflector, is about 2 dB greater than that possible for the equivalent optimum structure. In this case, the write power required at fixed applied magnetic field is about 1.7 times greater.

There is a common link in all of these trends. As the thermal coupling to the aluminum heat sink is improved, the time constant for attaining the pseudosteady state decreases. The temporal difference between the writing and reading processes therefore gets smaller and indeed has almost disappeared in the most detuned structure (cf. curves f in Figs. 3 and 4). If thermomagnetic writing can still be achieved satisfactorily, in this condition then both the signal-to-noise in readout and the domain resolution are simultaneously maximized.

IV. CONCLUSIONS

We have described how the basic optimum quadrilayer structure can be adjusted to match its recording needs to those of a particular system in such a way as to maximize the readout signal-to-noise and obtain maximum recording density. In all of this, we assumed that recording was carried out in a fixed applied magnetic field. While this is reasonable for the discussion presented, it is an unlikely constraint in practice. That is, the writing characteristics of a given material depend both on laser power and applied magnetic field such that two media structures with different thermal characteristics could, for example, be written as well with the same laser power and different applied magnetic fields, as with different laser powers and the same applied magnetic field. A system designer, therefore, should first consider the least thermally sensitive quadrilayer design prior to any other because of its highest signal-to-noise in readout and then compensate for lack of writing sensitivity by using a higher applied magnetic field. If this approach proved technologically unfeasible or economically unsound with the available magnetic material, then he can follow the systematic approach outlined here to obtain the media structure that provides the best compromise between the writing and reading constraints of the magneto-optical system. While this is a long procedure which must be done at each spin speed, for instance, it nevertheless represents a completely tractable problem.

ACKNOWLEDGMENT

Thanks are due to David Treves for many helpful discussions on the subject.

[1]M. Mansuripur, G. A. N. Connell, and J. W. Goodman, J. Appl. Phys. **53**, 4485 (1982).

[2]M. Mansuripur, G. A. N. Connell, and J. W. Goodman, SPIE **329**, 215 (1982).

[3]L. N. Hadley and D. M. Dennison, J. Opt. Soc. Am. **37**, 451 (1947).

[4]B. H. Billings, J. Phys. (Paris) **11**, 407 (1950).

[5]A. E. Bell and F. W. Spong, IEEE J. Quantum Electron. **QE-14**, 487(1978).

[6]P. Kivits, R. de Bont, and P. Zalm, Appl. Phys. **24**, 273 (1981).

[7]A. E. Bell, R. A. Bartolini, and F. W. Spong, RCA Rev. **40**, 345 (1979).

[8]M. Mansuripur and G. A. N. Connell, Appl. Opt. **22**, 666 (1983).

[9]M. Mansuripur, G. A. Connell, and J. W. Goodman, Appl. Opt. **21**, 1106 (1982).

[10]See, for example, the Proceedings of the 10th International Colloquium on Magnetic Films and Surfaces (Yokohama 1982), J. Magn. Magn. Mater **35** (1983).

Reprinted with permission from *Applied Optics*, Vol. 23(22), pp. 3983-3989 (November 15, 1984). ©1984 Optical Society of America.

Reflectivity, rotation, and ellipticity of magnetooptic film structures

Gerard J. Sprokel

A general relation is derived between the incident, reflected, and transmitted waves for a multilayer structure containing one or more layers characterized by a magnetooptic dielectric tensor. The method consists of formulating the characteristic matrix for each layer and evaluating the matrix products numerically. From the resulting matrix, desired quantities such as reflectivity, transmission, rotation, and ellipticity are obtained.

I. Introduction

Analysis of stratified media is discussed in textbooks such as Born and Wolf[1] and Heavens[2] and more exhaustively in review papers in Refs. 3 and 4. Take a coordinate system so that the boundary planes are normal to the Z axis and the plane of incidence is the Y-Z plane. At each interface, the tangential components of $\mathbf{E}$ and H and the normal components of D must be continuous. Thus at each interface there is a set of linear equations in $\mathbf{E}_x$, H_y, $\mathbf{E}_y$, H_x, and D_z. From the Maxwell relation between H_x and D_z, it follows that the tangential component of the wave vector is constant throughout the stack regardless of the form of the dielectric tensor. However, the relations between H_x and $\mathbf{E}_y$ and H_y and $\mathbf{E}_x$ involve the dielectric tensor. One can eliminate H and write the result as a matrix relation between the $\mathbf{E}$ vectors on both sides of the interface. On combining the equations for the two boundaries of one medium, one arrives at the relation between the incident and reflected waves on both sides of the layer. This matrix is commonly referred to as the characteristic matrix for that medium. Repeating the procedure for each layer one arrives at the optical response of the stack.

The form of the characteristic matrix is largely determined by the form of the dielectric tensor. If the tensor is symmetrical or if it is reduced to a scalar (e.g., isotropic material), the dispersion equation is of the second order, and the matrix separates into two 2×2 matrices describing, respectively, p- or s-polarized light. For a magnetooptic medium, the dielectric tensor has skew–symmetry; the dispersion relation is of the fourth order, and a 4×4 matrix results. This complicates the linear algebra part of the calculation, but it is unnecessary to resort to extraneous assumptions.

The author is with IBM Research Laboratory, San Jose, California 95193.

Received 20 April 1984.
0003-6935/84/223983-07$02.00/0.

Dielectric tensor relations may be found in Ref. 5. Landau and Lifshitz[6] derive the form of the tensor for magnetooptic materials from first principles for the polar case. Judy[7] discussed the form of the tensor for the longitudinal, the transverse, and the polar cases. However, Walter's review paper[3] does not include the derivation of the characteristic matrix for a skew–symmetrical dielectric tensor.

The procedure outlined above greatly simplifies numerical computation. Other schemes are possible, e.g., Mansuripur *et al.*[8] briefly indicate a method which leads to a recursive relation between the partial solutions for each layer. However, the matrix method has found numerous applications for isotropic and birefringent multilayer stacks.[9,10] Its use in magnetooptics probably was pioneered by Smith[11] and Judy.[7]

This paper is organized as follows: First, the dispersion and impedance relations are derived. Next, the characteristic matrices are formulated, and the final matrix relation is written. Calculation of reflectivity, transmission, rotation, and ellipticity is then straightforward. Computer-generated response curves exemplify the optical response of a stack.

II. Dispersion and Impedance Relations

The dispersion relation is the relation between wave vector $\mathbf{k}$ and permittivity ε of the medium. It can always be derived from the Maxwell equations and a material equation for the permittivity appropriate to the medium. The Maxwell equations can be written as

$$\left(\frac{\omega}{c}\right)^2 D = (\mathbf{k} \cdot \mathbf{k})\mathbf{E} - \mathbf{k}(\mathbf{k} \cdot \mathbf{E}); \tag{1}$$

the material equation for a magnetooptic medium is a tensor equation

$$D = [\varepsilon]\mathbf{E}. \tag{2}$$

If the medium is not birefringent and the magnetization is in the Z direction (polar), the appropriate form of the ε matrix is[12]

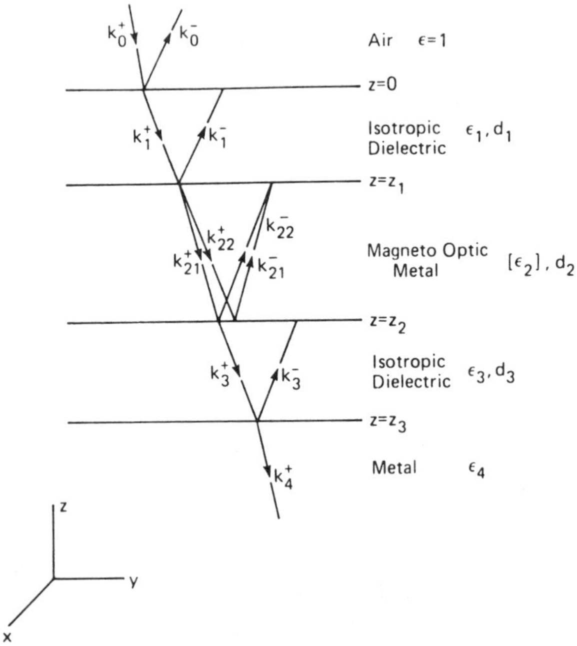

Fig. 1. Four-layer film structure containing one magnetooptic layer. Subscript index denotes the layer, superscript + is the incident direction, and superscript − is the direction of the reflected beam.

$$\varepsilon \cdot \begin{bmatrix} 1 & -ig & 0 \\ ig & 1 & 0 \\ 0 & 0 & 1 \end{bmatrix}. \tag{3}$$

This same form was used by Smith[11] and by Judy.[7] ε and g are, in general, complex quantities. Their effects on reflectivity and rotation will be discussed later. Elimination of D between Eqs. (1) and (2) leads to

$$\begin{bmatrix} \mathbf{k}^2 - \mathbf{k}_x^2 - \left(\frac{\omega}{c}\right)^2 \varepsilon & -\mathbf{k}_x\mathbf{k}_y + ig\left(\frac{\omega}{c}\right)^2 \varepsilon & -\mathbf{k}_x\mathbf{k}_z \\ -\mathbf{k}_x\mathbf{k}_y - ig\left(\frac{\omega}{c}\right)^2 \varepsilon & \mathbf{k}^2 - \mathbf{k}_y^2 - \left(\frac{\omega}{c}\right)^2 \varepsilon & -\mathbf{k}_y\mathbf{k}_z \\ -\mathbf{k}_x\mathbf{k}_z & -\mathbf{k}_y\mathbf{k}_z & \mathbf{k}^2 - \mathbf{k}_z^2 - \left(\frac{\omega}{c}\right)^2 \varepsilon \end{bmatrix} \times \begin{pmatrix} \mathbf{E}_x \\ \mathbf{E}_y \\ \mathbf{E}_z \end{pmatrix} = 0, \tag{4}$$

which has a nontrivial solution only if the determinant is zero. This leads to a fourth-order equation in k_z,

$$\left[\mathbf{k}^2 - \left(\frac{\omega}{c}\right)^2 \varepsilon\right]^2 + \left[\mathbf{k}^2 - \mathbf{k}_z^2 - \left(\frac{\omega}{c}\right)^2 \varepsilon\right] g^2 \left(\frac{\omega}{c}\right)^2 \varepsilon = 0 \tag{5}$$

written out explicitly, the four roots are

$$\mathbf{k}_{z1}^{+} = \kappa_0 \sqrt{1 + g\,\frac{\frac{\omega}{c}\sqrt{\varepsilon}}{\kappa_0}} \quad k_{z2}^{+} = \kappa_0 \sqrt{1 - g\,\frac{\frac{\omega}{c}\sqrt{\varepsilon}}{\kappa_0}}$$

$$k_{z1}^{-} = -\kappa_0 \sqrt{1 - g\,\frac{\frac{\omega}{c}\sqrt{\varepsilon}}{\kappa_0}} \quad k_{z2}^{-} = -\kappa_0 \sqrt{1 + g\,\frac{\frac{\omega}{c}\sqrt{\varepsilon}}{\kappa_0}}, \tag{6}$$

where

$$\kappa_0^2 = \left(\frac{\omega}{c}\right)^2 \varepsilon - (\mathbf{k}_x^2 + \mathbf{k}_y^2). \tag{7}$$

Note that

$$\mathbf{k}_{z1}^{-} = -\mathbf{k}_{z2}^{+} \quad \mathbf{k}_{z2}^{-} = -\mathbf{k}_{z1}^{+}. \tag{8}$$

The impedance relations are derived from the Maxwell relations and the dielectric tensor. The results are listed here for reference:

$$H_{x1} = \mathbf{E}_{y1} \cdot \mathbf{k}_{z1} \cdot \frac{-\frac{\omega}{c}\varepsilon}{\kappa_0^2},$$

$$H_{x2} = \mathbf{E}_{y2} \cdot \mathbf{k}_{z2} \cdot \frac{-\frac{\omega}{c}\varepsilon}{\kappa_0^2},$$

$$H_{y1} = \mathbf{E}_{x1} \cdot \frac{c}{\omega}\,\mathbf{k}_{z1},$$

$$H_{y2} = \mathbf{E}_{x2} \cdot \frac{c}{\omega}\,k_{z2}. \tag{9}$$

The relations between the tangential components of E for the two waves are:

$$\mathbf{E}_{x1} = \mathbf{E}_{y1} \cdot \frac{-i\frac{\omega}{c}\sqrt{\varepsilon}}{\kappa_0},$$

$$\mathbf{E}_{x2} = \mathbf{E}_{y2} \cdot \frac{i\frac{\omega}{c}\sqrt{\varepsilon}}{\kappa_0}. \tag{10}$$

III. Characteristic Matrix

The characteristic matrix is derived for a four-layer structure as illustrated in Fig. 1. The figure could represent a reflecting trilayer structure with an antireflecting coating. However, by appropriate choice of the various ε and d, the structure can also represent a single magnetooptic layer on a substrate with or without an antireflection layer or even a hypothetical structure such as a thin film in air. The X-Y planes form the boundary planes at $z = 0, z_1, z_2$, etc. The boundary conditions at each plane are a set of four linear equations in $\mathbf{E}_y$, H_x, $\mathbf{E}_x$, and H_y. The impedance relations are then used to transform the boundary conditions to a Jones matrix. The choice of the coordinate system simplifies the relations but they remain general.

At $z = 0$, the boundary equations are

$$\mathbf{E}_{0y}^{+} + \mathbf{E}_{0y}^{-} = \mathbf{E}_{1y}^{+} + \mathbf{E}_{1y}^{-},$$

$$H_{0x}^{+} + H_{0x}^{-} = H_{1x}^{+} + H_{1x}^{-},$$

$$\mathbf{E}_{0x}^{+} + \mathbf{E}_{0x}^{-} = \mathbf{E}_{1x}^{+} + \mathbf{E}_{1x}^{-},$$

$$H_{0y}^{+} + H_{0y}^{-} = H_{1y}^{+} + H_{1y}^{-}. \tag{11}$$

Eliminating the magnetic vectors and solving for $\mathbf{E}_0$ one obtains the matrix relation

$$\begin{pmatrix} \mathbf{E}_{0y}^{+} \\ \mathbf{E}_{0y}^{-} \\ \mathbf{E}_{0x}^{+} \\ \mathbf{E}_{0x}^{-} \end{pmatrix} = [\mathrm{MA1}] \times \begin{pmatrix} \mathbf{E}_{1y}^{+} \\ \mathbf{E}_{1y}^{-} \\ \mathbf{E}_{1x}^{+} \\ \mathbf{E}_{1x}^{-} \end{pmatrix}. \tag{12}$$

The matrix MA1 is

$$\begin{bmatrix} \dfrac{u_0+u_1}{2u_0} & \dfrac{u_0-u_1}{2u_0} & 0 & 0 \\ \dfrac{u_0-u_1}{2u_0} & \dfrac{u_0+u_1}{2u_0} & 0 & 0 \\ 0 & 0 & \dfrac{\mathbf{k}_{0z}+\mathbf{k}_{1z}}{2\mathbf{k}_{0z}} & \dfrac{\mathbf{k}_{0z}-\mathbf{k}_{1z}}{2\mathbf{k}_{0z}} \\ 0 & 0 & \dfrac{\mathbf{k}_{0z}-\mathbf{k}_{1z}}{2\mathbf{k}_{0z}} & \dfrac{\mathbf{k}_{0z}+\mathbf{k}_{1z}}{2\mathbf{k}_{0z}} \end{bmatrix} \cdot \quad (13)$$

As was mentioned earlier, for an isotropic medium the matrix separates into two 2×2 matrices, and it is often written as such. The 4×4 format is retained here to retain generality. The ratio H/E is denoted as u, the index indicates the medium.

At $z = z_1$, the boundary equations have the form

$$\mathbf{E}_{1y}^{+} \exp - i\mathbf{k}_{1z}^{+}z_1 + \mathbf{E}_{1y}^{-} \exp i\mathbf{k}_{1z}^{+}z_1 = \mathbf{E}_{2y1}^{+} \exp - i\mathbf{k}_{2z1}^{+}z_1 + \mathbf{E}_{2y1}^{-} \exp i\mathbf{k}_{2z2}^{+}z_1 + \mathbf{E}_{2y2}^{+} \exp - i\mathbf{k}_{2z2}^{+}z_1 + \mathbf{E}_{2y2}^{-} \exp i\mathbf{k}_{2z1}^{+}z_1 \quad (14)$$

and similar expressions for H_x, $\mathbf{E}_x$, and H_y. Equations (8) were used to express the various k_{2z}. Using the relations between H_x and $\mathbf{E}_y$, H_y and $\mathbf{E}_x$, and $\mathbf{E}_x$ and $\mathbf{E}_y$ [Eqs. (9) and (10)], the set of boundary equations can be combined into one matrix relation between the $\mathbf{E}$ vectors in mediums 1 and 2 at boundary z_1. One obtains

$$\begin{pmatrix} \mathbf{E}_{1y}^{+} \\ \mathbf{E}_{1y}^{-} \\ \mathbf{E}_{1x}^{+} \\ \mathbf{E}_{1x}^{-} \end{pmatrix} = [\mathrm{MA21}] \times \begin{pmatrix} \mathbf{E}_{2y1}^{+} \exp - i\mathbf{k}_{2z1}^{+}z_1 \\ \mathbf{E}_{2y1}^{-} \exp + i\mathbf{k}_{2z2}^{+}z_1 \\ \mathbf{E}_{2y2}^{+} \exp - i\mathbf{k}_{2z2}^{+}z_1 \\ \mathbf{E}_{2y2}^{-} \exp + i\mathbf{k}_{2z1}^{+}z_1 \end{pmatrix}, \quad (15)$$

[MA21]:

$$\begin{bmatrix} \dfrac{\exp i\mathbf{k}_{1z}z_1}{2u_1} & (u_1+u_{21} & u_1-u_{22} & u_1+u_{22} & u_1-u_{21}) \\ \dfrac{\exp - i\mathbf{k}_{1z}z_1}{2u_1} & (u_1-u_{21} & u_1+u_{22} & u_1-u_{22} & u_1+u_{21}) \\ \dfrac{i\frac{\omega}{c}\sqrt{\varepsilon}\exp i\mathbf{k}_{1z}z_1}{2k_{1z}\kappa_0} & (-\mathbf{k}_{1z}-\mathbf{k}_{2z1}^{+} & -\mathbf{k}_{1z}+\mathbf{k}_{2z2}^{+} & \mathbf{k}_{1z}+\mathbf{k}_{2z2}^{+} & \mathbf{k}_{1z}-\mathbf{k}_{2z2}^{+}) \\ \dfrac{i\frac{\omega}{c}\sqrt{\varepsilon}\exp - i\mathbf{k}_{1z}z_1}{2k_{1z}\kappa_0} & (-\mathbf{k}_{1z}+\mathbf{k}_{2z1}^{+} & -\mathbf{k}_{1z}-\mathbf{k}_{2z2}^{+} & \mathbf{k}_{1z}-\mathbf{k}_{2z2}^{+} & \mathbf{k}_{1z}+\mathbf{k}_{2z2}^{+}) \end{bmatrix} \quad (16)$$

$$u_{21} = -\frac{\omega}{c}\varepsilon\frac{\mathbf{k}_{2z1}}{\kappa_0^2},$$

$$u_{22} = -\frac{\omega}{c}\varepsilon\frac{\mathbf{k}_{2z2}}{\kappa_0^2}.$$

Combining Eqs. (12) and (15) results in

$$\begin{pmatrix} \mathbf{E}_{0y}^{+} \\ \mathbf{E}_{0y}^{-} \\ \mathbf{E}_{0x}^{+} \\ \mathbf{E}_{0x}^{-} \end{pmatrix} = [\mathrm{MA1}] \times [\mathrm{MA21}] \times \begin{pmatrix} \mathbf{E}_{2y1}^{+} \exp - i\mathbf{k}_{2z1}^{+}z_1 \\ \mathbf{E}_{2y1}^{-} \exp + i\mathbf{k}_{2z2}^{+}z_1 \\ \mathbf{E}_{2y2}^{+} \exp - i\mathbf{k}_{2z2}^{+}z_1 \\ \mathbf{E}_{2y2}^{-} \exp + i\mathbf{k}_{2z1}^{+}z_1 \end{pmatrix}. \quad (17)$$

At $z = z_2$, the boundary conditions have the form

$$\mathbf{E}_{2y1}^{+} \exp - i\mathbf{k}_{2z1}^{+}z_2 + \mathbf{E}_{2y1}^{-} \exp - i\mathbf{k}_{2z2}^{+}z_2 + \mathbf{E}_{2y2}^{+} \exp - i\mathbf{k}_{2z2}^{+}z_2 + \mathbf{E}_{2y2}^{-} \exp + i\mathbf{k}_{2z1}^{+}z_2 = \mathbf{E}_{3y}^{+} \exp - i\mathbf{k}_{3z}z_2 + \mathbf{E}_{3y}^{-} \exp i\mathbf{k}_{3z}z_2$$

and similar equations for H_x, E_x, and H_y. Note that the two beams in medium 2 combine to a single beam in medium 3 since there is only one value for

$$\mathbf{k}_{3z} = \sqrt{\left(\frac{\omega}{c}\right)^2 \varepsilon_3 - \mathbf{k}_{0y}^2}\,.$$

Proceeding as before, the matrix equation at z_2 is

$$[\mathrm{MA3}] \times \begin{pmatrix} \mathbf{E}_{2y1}^{+} \exp - i\mathbf{k}_{2z1}^{+}z_2 \\ \mathbf{E}_{2y1}^{-} \exp + i\mathbf{k}_{2z2}^{+}z_2 \\ \mathbf{E}_{2y2}^{+} \exp - i\mathbf{k}_{2z2}^{+}z_2 \\ \mathbf{E}_{2y2}^{-} \exp + i\mathbf{k}_{2z1}^{+}z_2 \end{pmatrix} = \begin{pmatrix} \mathbf{E}_{3y}^{+} \exp - i\mathbf{k}_{3z}z_2 \\ \mathbf{E}_{3y}^{-} \exp + i\mathbf{k}_{3z}z_2 \\ \mathbf{E}_{3x}^{+} \exp - i\mathbf{k}_{3z}z_2 \\ \mathbf{E}_{3x}^{-} \exp + i\mathbf{k}_{3z}z_2 \end{pmatrix}. \quad (18)$$

MA3 has the same form as MA2, but u_1 is replaced by u_3 and $\mathbf{k}_{1z}$ by $\mathbf{k}_{3z}$. Inverting Eq. (18) and including the phase vector in the matrix, one arrives at

$$\begin{pmatrix} \mathbf{E}_{0y}^{+} \\ \mathbf{E}_{0y}^{-} \\ \mathbf{E}_{0x}^{+} \\ \mathbf{E}_{0x}^{-} \end{pmatrix} = [\mathrm{MA1}] \times [\mathrm{MA21}] \times [\mathrm{MA31}] \times \begin{pmatrix} \mathbf{E}_{3y}^{+} \exp - i\mathbf{k}_{3z}z_2 \\ \mathbf{E}_{3y}^{-} \exp + i\mathbf{k}_{3z}z_2 \\ \mathbf{E}_{3x}^{+} \exp - i\mathbf{k}_{3z}z_2 \\ \mathbf{E}_{3x}^{-} \exp + i\mathbf{k}_{3z}z_2 \end{pmatrix}. \quad (19)$$

[MA31] is the inverse of [MA3] multiplied by the phase vector.

At $z = z_3$, the boundary equations take the form

$$\mathbf{E}_{3y}^{+} \exp - i\mathbf{k}_{3z}z_3 + \mathbf{E}_{3y}^{-} \exp + i\mathbf{k}_{3z}z_3 = \mathbf{E}_{4y}^{+} \exp - i\mathbf{k}_{4y}z_3 + \mathbf{E}_{4y}^{-} \exp + i\mathbf{k}_{4z}z_3 \quad (20)$$

and equivalent expressions for H_x, E_x, and H_y. The matrix equation takes the form

$$\begin{pmatrix} \mathbf{E}_{3y}^{+} \exp - i\mathbf{k}_{3z}z_3 \\ \mathbf{E}_{3y}^{-} \exp + i\mathbf{k}_{3z}z_3 \\ \mathbf{E}_{3x}^{+} \exp - i\mathbf{k}_{3z}z_3 \\ \mathbf{E}_{3x}^{-} \exp + i\mathbf{k}_{3z}z_3 \end{pmatrix} = [\mathrm{MA4}] \times \begin{pmatrix} \mathbf{E}_{4y}^{+} \exp - i\mathbf{k}_{4z}z_3 \\ \mathbf{E}_{4y}^{-} \exp + i\mathbf{k}_{4z}z_3 \\ \mathbf{E}_{4x}^{+} \exp - i\mathbf{k}_{4z}z_3 \\ \mathbf{E}_{4z}^{-} \exp + i\mathbf{k}_{4z}z_3 \end{pmatrix}. \quad (21)$$

Matrix MA4 is identical with MA1 if u_0 is replaced by u_3, u_1 by u_4, $\mathbf{k}_{0z}$ by $\mathbf{k}_{3z}$, and $\mathbf{k}_{1z}$ by $\mathbf{k}_{4z}$. Again, the phase part is incorporated in MA4, and the result is called [MA41]. Upon combining with Eq. (19), the result is

$$\begin{pmatrix} \mathbf{E}_{0y}^{+} \\ \mathbf{E}_{0y}^{-} \\ \mathbf{E}_{0x}^{+} \\ \mathbf{E}_{0x}^{-} \end{pmatrix} = [\mathrm{MA1}] \times [\mathrm{MA21}] \times [\mathrm{MA31}] \times [\mathrm{MA41}] \times \begin{pmatrix} \mathbf{E}_{4y}^{+} \exp - i\mathbf{k}_{4z}z_3 \\ \mathbf{E}_{4y}^{-} \exp + i\mathbf{k}_{4z}z_3 \\ \mathbf{E}_{4x}^{+} \exp - i\mathbf{k}_{4z}z_3 \\ \mathbf{E}_{4x}^{-} \exp + i\mathbf{k}_{4z}z_3 \end{pmatrix}. \quad (22)$$

z_3 is the last interface, but it is clear that the method can be expanded if there are more interfaces.

IV. Evaluation of the Matrix Product

The product of the characteristic matrices is itself a 4×4 complex matrix

$$M = [a_{ij}] \quad i,j = 1,2,3,4.$$

If there is no wave incident on the last surface, it is seen that

$$\begin{pmatrix} \mathbf{E}_{0y}^{+} \\ \mathbf{E}_{0x}^{+} \end{pmatrix} = \begin{bmatrix} a_{11} & a_{13} \\ a_{31} & a_{33} \end{bmatrix} \times \begin{pmatrix} \mathbf{E}_{4y}^{+} \\ \mathbf{E}_{4x}^{+} \end{pmatrix} \tag{23}$$

for $\mathbf{E}_{4y}^{-} = \mathbf{E}_{4x}^{-} = 0$.

Taking the inverse

$$\begin{pmatrix} \mathbf{E}_{4y}^{+} \\ \mathbf{E}_{4x}^{+} \end{pmatrix} = \frac{1}{D} \begin{bmatrix} a_{33} & -a_{13} \\ -a_{31} & a_{11} \end{bmatrix} \times \begin{pmatrix} \mathbf{E}_{0y}^{+} \\ \mathbf{E}_{0x}^{+} \end{pmatrix}, \tag{24}$$

D is the determinant of the first matrix. Equation (24) describes the beam transmitted through the last surface in terms of the incident beam at the first surface. Similarly,

$$\begin{pmatrix} \mathbf{E}_{0y}^{-} \\ \mathbf{E}_{0x}^{-} \end{pmatrix} = \begin{bmatrix} a_{21} & a_{23} \\ a_{41} & a_{43} \end{bmatrix} \times \begin{pmatrix} \mathbf{E}_{4y}^{+} \\ \mathbf{E}_{4x}^{+} \end{pmatrix},$$

$$\begin{pmatrix} \mathbf{E}_{0y}^{-} \\ \mathbf{E}_{0x}^{-} \end{pmatrix} = \frac{1}{D} \cdot \begin{bmatrix} a_{21} & a_{23} \\ a_{41} & a_{43} \end{bmatrix} \cdot \begin{bmatrix} a_{33} & -a_{13} \\ -a_{31} & a_{11} \end{bmatrix} \times \begin{pmatrix} \mathbf{E}_{0y}^{+} \\ \mathbf{E}_{0x}^{+} \end{pmatrix}. \tag{25}$$

Equation (25) describes the reflected beam at the front surface.

To continue, one must specify the incident beam. Assume that the incident beam is polarized in the incident plane (p polarized), then $\mathbf{E}_{0x}^{+} = 0$.

Abbreviating the matrix product in Eq. (25) as $[b_{ij}]$, there results

$$\begin{pmatrix} \mathbf{E}_{0y}^{-} \\ \mathbf{E}_{0x}^{-} \end{pmatrix} = \begin{bmatrix} b_{11} & b_{12} \\ b_{21} & b_{22} \end{bmatrix} \times \begin{pmatrix} \mathbf{E}_{0y}^{+} \\ 0 \end{pmatrix}$$

or

$$\begin{aligned} \mathbf{E}_{0y}^{-} &= b_{11}\mathbf{E}_{0y}^{+} \\ \mathbf{E}_{0x}^{-} &= b_{21}\mathbf{E}_{0y}^{+} \end{aligned} \qquad p \text{ polarized.} \tag{26}$$

Similarly, if the incident beam is s polarized, $\mathbf{E}_{0y}^{+} = 0$, and

$$\begin{aligned} \mathbf{E}_{0y}^{-} &= b_{12}\mathbf{E}_{0x}^{+} \\ \mathbf{E}_{0x}^{-} &= b_{22}\mathbf{E}_{0x}^{+} \end{aligned} \qquad s \text{ polarized} \tag{27}$$

the coefficients b_{ij} are in general complex; thus Eqs. (26) and (27) describe the two plane-polarized components of the reflected wave. The resultant will be an elliptically polarized wave. To determine the rotation and ellipticity for p-polarized light, write Eqs. (26) in the form

$$\mathbf{E}_{0y}^{-} = |\mathbf{E}_y| \exp i(\omega t + \theta),$$

$$\mathbf{E}_{0x}^{-} = |\mathbf{E}_x| \exp i(\omega t + \phi),$$

$$|\mathbf{E}_y| = \frac{b_{11}'}{\cos\theta} \quad \theta = \arctan\frac{b_{11}''}{b_{11}'},$$

$$|\mathbf{E}_x| = \frac{b_{21}'}{\cos\phi} \quad \phi = \arctan\frac{b_{21}''}{b_{21}'} \tag{28}$$

and define

$$\tan\alpha = \frac{|\mathbf{E}_x|}{|\mathbf{E}_y|}, \tag{29}$$

the angle of rotation ψ is calculated from

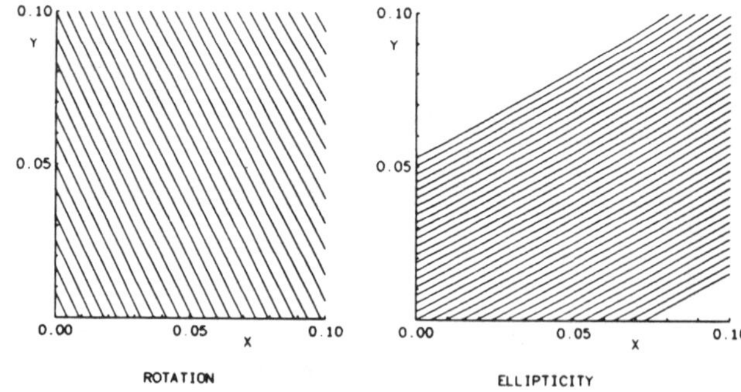

Fig. 2. Rotation and ellipticity as a function of g ($g = x + iy$). Step size for x and y is 0.01. The contour levels for rotation are 0.2–7° in 0.2° steps; for the ellipticity, −0.03 to +0.04 in 0.002 steps. Magnetooptic film thickness is 200 Å.

$$\tan 2\psi = \tan 2\alpha \cos(\phi - \theta), \tag{30}$$

and the ellipticity is calculated from

$$\sin 2\chi = -\sin 2\alpha \sin(\phi - \theta), \tag{31}$$

where the axis ratio is $\tan\chi$.

For s-polarized light, replace b_{11} with b_{12} and b_{21} with b_{22}. For details of the transformation refer to the Appendix. The reflectivity is then given by

$$R = |\mathbf{E}_y|^2 + |\mathbf{E}_x|^2. \tag{32}$$

V. Numerical Results and Discussion

In this section numerical results are presented to exemplify the properties of the matrix product. In principle, the observable quantities, reflectivity, rotation, and ellipticity calculated from the matrix product should depend on both ε_2 and g. However, for known materials, g is of the order of 10^{-2} or less. Numerical results show that, for small g, the reflectivity is independent of g (to the fourth decimal place), while rotation and ellipticity, although small, are proportional to g. This may be seen by evaluating matrix MA2, which is tantamount to reducing the stack to just one interface: air/magnetooptic metal film. For a p-polarized beam at normal incidence, the result is, to the first power in g,

$$\frac{\mathbf{E}_{0x}^{-}}{\mathbf{E}_{0y}^{-}} = \frac{ig\sqrt{\varepsilon_2}}{1 - \varepsilon_2} + \Theta(g^2), \tag{33}$$

which is of the order of g, and

$$\frac{\mathbf{E}_{0y}^{-}}{\mathbf{E}_{0y}^{+}} = \frac{1 - \sqrt{\varepsilon_2}}{1 + \sqrt{\varepsilon_2}} + \Theta(g^2), \tag{34}$$

which is of the order of 1.

Equation (34) is, of course, the Fresnel equation for this interface. On expanding Eqs. (30) and (31), it is seen that the rotation and ellipticity are proportional to $\mathbf{E}_{0x}^{-}/\mathbf{E}_{0y}^{-}$ and thus to g.

On the other hand, the reflectivity is given by

$$R = \frac{|\mathbf{E}_{0x}^{-2}| + |\mathbf{E}_{0y}^{-2}|}{|\mathbf{E}_{0y}^{+2}|} = \frac{|\mathbf{E}_{0y}^{-2}|}{|\mathbf{E}_{0y}^{+2}|}[1 + \Theta(g^2)], \tag{35}$$

which is nearly independent of g.

Turning now to numerical results for the four-layer stack in Fig. 1, in Fig. 2, rotation and ellipticity are plotted vs g. In general, g is complex, $g = x + iy$. Reflectivity, rotation, and ellipticity are calculated while varying x and y in steps of 0.01. Rotation and ellipticity are plotted as contour levels from 0.2 to 7° for the rotation and −0.03 to +0.04 for the ellipticity. From experimental values for rotation and ellipticity, g is determined by the intercept of the two contour levels. The calculation shows that rotation and ellipticity are proportional to g over a range far larger than encountered experimentally. For these parameters the reflectivity is ~0.05 and varies not more than 0.0007 with g. Thus the reflectivity is independent of g to better than 1%.

In Figs. 3, 4, and 5 the reflectivity, rotation, and ellipticity of the structure in Fig. 1 is plotted vs the thickness of the magnetooptic film. The numerical values used in the calculation are: For the magnetooptic film $n - i\mathbf{k} = 2.96 - 3.4i$ and $g = 0.001 - 0.025i$. For the reflector $1.3 - 7.1i$. The antireflection film thickness and the dielectric film thickness are set to $n \cdot \lambda/2$ or $(n + 1/2) \cdot \lambda/2$. The last is the antireflection condition, which will minimize the reflectivity of the stack. The first forms a convenient check on the numerical program: If the various ε are assumed real, the reflectivity for $n \cdot \lambda/2$ must be identical with that for the stack with the layer absent. It is seen that, for thick metal films, the dielectric film thickness has no effect, curves 1 and 3 and curves 2 and 4 converge at ~1000 Å, then the antireflection layer determines the reflectivity. Both curves 1 and 2 show steep minimum reflectivities: 100 Å for curve 1 and 50 Å for curve 2. For dielectric thickness of 1/4λ, the reflectivity is much lower than for 1/2λ. Figure 4 shows that the rotation increases sharply near a reflectivity minimum. $|\mathbf{E}_x|$ in Eq. (28) is of the order of 10^{-3}. $|\mathbf{E}_y|$ is much larger than $|\mathbf{E}_x|$, except near the minimum. At the reflectivity minimum, the sign of $|\mathbf{E}_y|$ reverses. Thus $\tan\alpha$ in Eq. (29) is small everywhere, except in the neighborhood of the reflectivity minimum, where $|\mathbf{E}_y|$ and $|\mathbf{E}_x|$ are of the same magnitude.

Figure 5 is a plot of $\tan\chi$ in Eq. (31). The term $\sin(\phi - \theta)$ changes sign slightly beyond the reflectivity minimum, and thus the ellipticity changes sign.

The angle of incidence enters the calculation through k_y [Eq. (5)]. For normal incidence $\mathbf{k}_y = 0$, and the equations for s- and p-polarized light are identical. In Fig. 6 reflectivity and rotation are plotted vs the angle of incidence for s and p polarization. The data were calculated for a 1000-Å magnetooptic film. The effect of the thickness of the dielectric film has largely disappeared (curves 1 and 3 and 2 and 4 practically coincide), and thus the structure resembles a thick magnetooptic film with an antireflection coating. For a 1/4λ AR coat (curves 1 and 3) and p-polarized light the reflectivity is nearly constant to ~45° incidence and increases thereafter as the structure becomes detuned. But the rotation increases steadily in this region. The data show that $|\mathbf{E}_x|$ increases much faster than $|\mathbf{E}_y|$, while $\cos(\phi - \theta)$ remains about constant. Hence ψ increases [see Eq. (30)]. For 1/2λ AR coating $|\mathbf{E}_y|$ decreases, $|\mathbf{E}_x|$ increases, and $\cos(\phi - \theta)$ is again nearly constant. Thus ψ increases faster, but it remains below

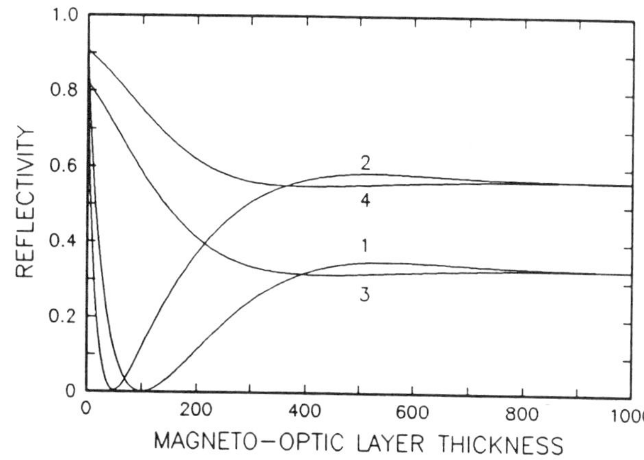

Fig. 3. Reflectivity vs magnetooptic layer thickness for the structure in Fig. 1 calculated in 5-Å steps. Normal incidence. Parameters, thickness of antireflection and dielectric layers (Dar,Ddi); curve 1, Dar 1/4λ Ddi:1/4λ curve 2, Dar 1/2λ Ddi 1/4λ; curve 3, Dar 1/4λ Ddi:1/2λ; curve 4, Dar 1/2λ Ddi 1/2λ.

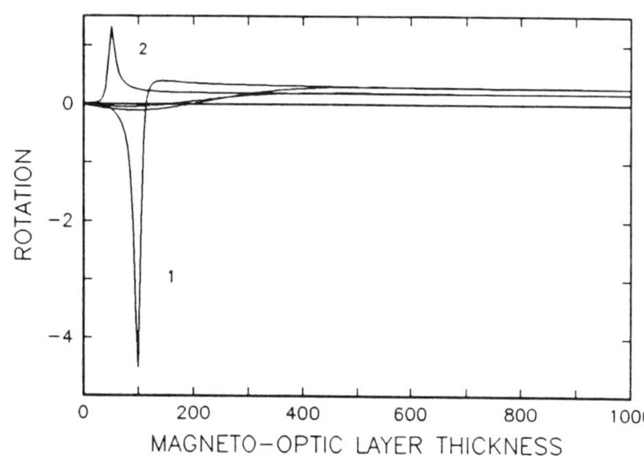

Fig. 4. Rotation vs magnetooptic layer thickness for the structure in Fig. 1 calculated in 5-Å steps as Fig. 3. Parameters, thickness of antireflection and dielectric layers (Dar,Ddi); curve 1, Dar 1/4λ Ddi:1/4λ curve 2, Dar 1/2λ Ddi 1/4λ; curve 3, Dar 1/4λ Ddi:1/2λ; curve 4, Dar 1/2λ Ddi 1/2λ. Curves 1 and 3 and 2 and 4 coincide for thick m.o. films.

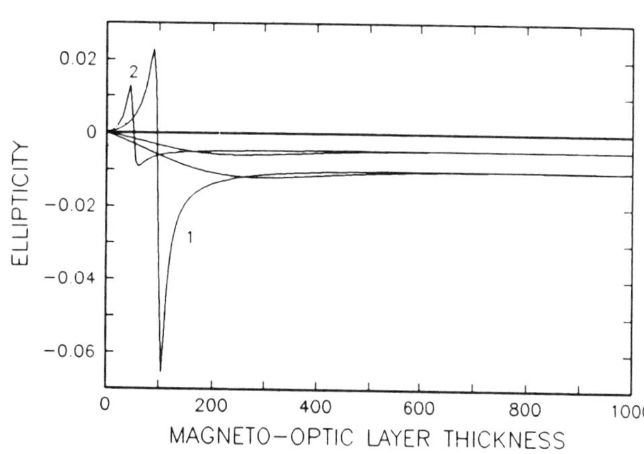

Fig. 5. Ellipticity vs magnetooptic layer thickness for the structure in Fig. 1 calculated in 5-Å steps as Fig. 3. Parameters, thickness of antireflection and dielectric layers (Dar,Ddi); curve 1, Dar 1/4λ Ddi:1/4λ curve 2, Dar 1/2λ Ddi 1/4λ; curve 3, Dar 1/4λ Ddi:1/2λ; curve 4, Dar 1/2λ Ddi 1/2λ. Curves 1 and 3 and 2 and 4 coincide for thick m.p. films.

the value for the antireflection case.

For thin magnetooptic films (200 Å), the rotation cannot be increased substantially above the values for curve 1 in Fig. 3. The reflectivity is determined mainly by the metal film and does not change much up to ~45°.

VI. Summary

A matrix relation is derived between the **E** vectors of the reflected and incident light for a multilayer structure containing a magnetooptic layer. The properties of the matrix are exemplified for a four-layer reflective structure. The dielectric function ε of a magnetooptic film can be determined from the reflectivity as for any other metal film. The off-diagonal term in the permittivity tensor g follows from the rotation and ellipticity. Rotation and ellipticity are proportional to g, the reflectivity is essentially independent of g. If the stack is designed to minimize the reflectivity, rotation and ellipticity will be maximum near the reflectivity minimum.

I thank M. Chen for critically reviewing the text and Dick Dunbar of the Research Computing facility for his operation.

Appendix: Rotation and Ellipticity

Equations (28) define an ellipse on the Y,X coordinates. They can be transformed to a new set of axes Y',X', rotated over an angle ψ with respect to Y,X so that Y' and X' are the major and minor axes of the ellipse. Thus

$$\mathbf{E}_y = |\mathbf{E}_y| \exp i(\omega t + \theta), \quad \mathbf{E}_x = |\mathbf{E}_x| \exp i(\omega t + \phi) \tag{A1}$$

must transform to

$$\mathbf{E}'_y = |\mathbf{E}'_y| \exp i\omega t, \quad \mathbf{E}'_x = |\mathbf{E}'_x| \exp i\left(\omega t + \frac{\pi}{2}\right) = i|\mathbf{E}'_x| \exp i\omega t \tag{A2}$$

according to

$$\begin{pmatrix}\mathbf{E}'_y\\ \mathbf{E}'_x\end{pmatrix} = \begin{bmatrix}\cos\psi & \sin\psi\\ -\sin\psi & \cos\psi\end{bmatrix} \times \begin{pmatrix}\mathbf{E}_y\\ \mathbf{E}_x\end{pmatrix}. \tag{A3}$$

Substitute Eqs. (A1) and (A2) in (A3) and separate the real and imaginary parts:

$$|\mathbf{E}'_y| = |\mathbf{E}_y| \cos\psi \cos\theta + |\mathbf{E}_x| \sin\psi \cos\phi \text{ (a)}, \quad 0 = |\mathbf{E}_y| \cos\psi \sin\theta + |\mathbf{E}_x| \sin\psi \sin\phi \text{ (b)}, \tag{A4}$$

$$|\mathbf{E}'_x| = |\mathbf{E}_x| \cos\psi \sin\phi - |\mathbf{E}_y| \sin\psi \sin\theta \text{ (c)}, \quad 0 = |\mathbf{E}_x| \cos\psi \cos\phi - |\mathbf{E}_y| \sin\psi \cos\theta \text{ (d)}, \tag{A5}$$

define $\tan\alpha = |\mathbf{E}_x|/|\mathbf{E}_y|$, then from (b) and (d)

$$\tan\psi = \frac{\tan\alpha \cos\phi}{\cos\theta} = -\frac{\sin\theta}{\tan\alpha \sin\phi}, \tag{A6}$$

$$\tan^2\alpha \sin\phi \cos\phi = -\sin\theta \cos\theta. \tag{A7}$$

Equation (A6) is a solution adequate for numerical calculations, but θ and ϕ are not easily measured. To obtain a more useful expression use

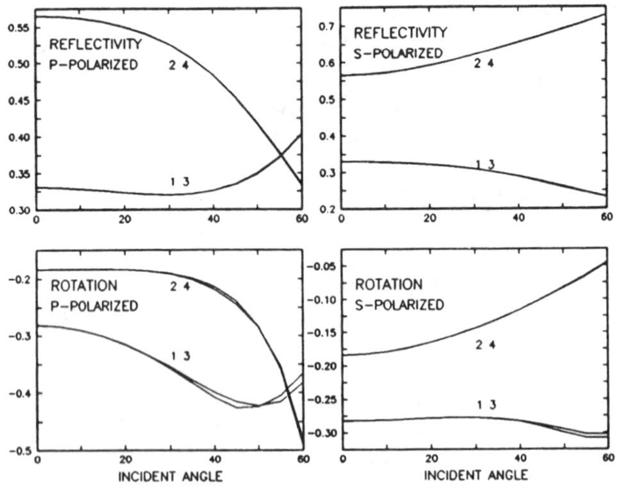

Fig. 6. Reflectivity and rotation vs angle of incidence for p and s polarization. The magnetooptic layer is 1000 Å, essentially a thick FeTb film with a $1/4\lambda$ or a $1/2\lambda$ antireflection layer.

$$\tan 2\psi = \frac{2\tan\psi}{1 - \tan^2\psi}.$$

Substitute Eq. (A6) and use Eq. (A7):

$$\tan 2\psi = \frac{2\tan\alpha \sin\phi \cos\phi}{\sin(\theta + \phi)}.$$

Substitute

$$2\tan\alpha = \tan 2\alpha(1 - \tan^2\alpha)$$

and use Eqs. (A6) and (A7) again. The expression becomes

$$\tan 2\psi = \tan 2\alpha \cdot \cos(\phi - \theta). \tag{A8}$$

The ellipticity follows from (a) and (c):

$$\frac{|\mathbf{E}'_x|}{|\mathbf{E}'_y|} = \frac{\tan\alpha \sin\phi - \tan\psi \sin\theta}{\cos\theta + \tan\alpha \tan\psi \cos\phi}. \tag{A9}$$

Substitute Eq. (A6) and use Eq. (A7):

$$\frac{|\mathbf{E}'_x|}{|\mathbf{E}'_y|} = \tan\alpha \frac{\sin\phi}{\cos\theta}. \tag{A10}$$

Equation (A10) has the same drawback as had Eq. (A6). Define

$$\frac{|\mathbf{E}'_x|}{|\mathbf{E}'_y|} = \tan\chi,$$

then

$$\sin 2\chi = \frac{2\tan\chi}{1 + \tan^2\chi} = \frac{2\tan\alpha \sin\phi \cos\phi}{\cos(\theta + \phi)}.$$

Substitute $2\tan\alpha = \sin 2\alpha\,(1 + \tan^2\alpha)$ and use Eq. (A7) again. The result is

$$\sin 2\chi = -\sin 2\alpha \sin(\phi - \theta).$$

Synopsis:

$$\tan\alpha = \frac{|\mathbf{E}_x|}{|\mathbf{E}_y|},$$

$$\tan\chi = \frac{|\mathbf{E}'_x|}{|\mathbf{E}'_y|},$$

$$\tan 2\psi = \tan 2\alpha \cos(\phi - \theta),$$

$$\sin 2\chi = -\sin 2\alpha \sin(\phi - \theta).$$

References

1. M. Born and E. Wolf, *Principles of Optics* (Pergamon, New York, 1975).
2. O. S. Heavens, *Optical Properties of Thin Solid Films* (Dover, New York, 1965).
3. H. Walter, "Optik Dunner Schichten," in *Handbuch der Physik, Vol. 25/1* (Springer, Berlin, 1961).
4. R. Jacobsson, "Light Reflection from Films of Continuously Varying Refractive Index," Prog. Opt. **5,** 249 (1966).
5. G. N. Ramachandran and S. Ramangeshan, "Crystal Optics," in Handbuch der physik, *Vol. 25/1* (Springer, Berlin, 1966).
6. L. D. Landau and E. M. Lifshitz, *Electrodynamics of Continuous Media* (Pergamon, New York, 1960).
7. J. H. Judy, "Magnetic Optics Theory," in *Proceedings, Conference on Advances in Magnetic Recording*, N.Y. Acad. of Science **189,** 239 (1972).
8. M. Mansuripur and G. A. N. Connell, "Signal and Noise in Magneto-Optical Readout," J. Appl. Phys. **53,** 4485 (1982).
9. D. W. Berreman, "4 × 4-Matrix Formulation for Anisotropic Media," J. Opt. Soc. Am. **62,** 502 (1972).
10. G. J. Sprokel, "The Reflectivity of a Liquid Crystal Cell in a Surface Plasma on Experiment," Mol. Cryst. Liq. Cryst. **68,** 39 (1981).
11. D. O. Smith, "Magneto-Optical Scattering from Multilayer Magnetic and Dielectric Films," Opt. Acta **12,** 18 (1965).
12. M. Freiser, "A Survey of Magneto-Optic Effects," IEEE Trans. Magn. **MAG-4,** 152 (1968).
13. M. Mansuripur, G. A. N. Connell, and J. W. Goodman, "Signal to Noise in Magneto-Optic Storage," Proc. Soc. Photo-Opt. Instrum. Eng. **329,** 215 (1982).
14. D. Cheng, D. Treves, and T. Chen, "Static Tests of TbFe Films for Magneto-Optical Recordings," Proc. Soc. Photo-Opt. Instrum. Eng. **329,** 223 (1982).

Reprinted with permission from *Applied Physics Letters*, Vol. 49(7), pp. 419-421 (August 18, 1986).

Lorentz microscopy of micron-sized laser-written magnetic domains in TbFe

J. C. Suits, R. H. Geiss, C. J. Lin, D. Rugar, and A. E. Bell
IBM Almaden Research Center, San Jose, California 95120-6099

High resolution images of laser-written magnetic domains in thin sputtered films of terbium iron have been obtained using Lorentz microscopy. The domains, written with a GaAs diode laser, are generally circular in shape with diameters in the range 0.9–2.2 μm. Conditions necessary for the observation of highly detailed images are described. It was found that with certain values of laser power and applied magnetic field, smaller irregular domains appear within the principal domains.

Amorphous magnetic films of the rare-earth/transition metal type (RE/TM) have been found to hold promise for use in reversible, high density optical memory devices.[1–3] Magnetic domains are written in these films by locally heating the film with a focused laser beam in the presence of a magnetic field. Readout of these domains of reversed magnetization is accomplished using the same laser beam with reduced intensity and is based on the Kerr effect.

In order to understand the domain formation process, and to optimize the magneto-optical material composition, it is extremely helpful to obtain sharp images of the detailed magnetic structure of the thermomagnetically written domains. Previously, observations of written domains in the 1–10 μm size range have been accomplished using optical microscopy.[4–6] However, in practical devices, where the domain diameter of interest is about 1 μm, the limitations of optical resolution preclude the clear observation of structure either at the domain boundary or within the domain. In this letter we describe a technique based on Lorentz microscopy which we have used to observe laser-written magneto-optical domain structure as small as 0.2 μm.

Films of TbFe or TbFeCo were prepared by dual source magnetron sputtering. The system was pumped out to 8×10^{-8} Torr before introduction of the argon gas to a pressure of typically 5×10^{-3} Torr. Sputtering rates were about 1 Å/s. The films are 50 nm thick. Film compositions were measured by x-ray fluorescence which was calibrated against pure elements. To facilitate preparation of samples for the transmission electron microscope (TEM) the films were deposited on mica substrates which had been precoated with 25 nm of carbon.

Magnetization measurements by vibrating sample magnetometer (VSM), as well as Kerr rotation measurements, both showed that the films predominantly exhibited an easy axis of anisotropy perpendicular to the plane. As discussed later, there may also be a small in-plane component of magnetization. For TbFe films the composition exhibiting a room-temperature compensation temperature was 21 at. % Tb.

The Lorentz microscopy,[7] in which magnetic domains or walls are revealed directly in thin films as a result of the deflection of the incident electron beam by internal magnetic fields, was done in a Phillips 400T transmission electron microscope using an accelerating voltage of 120 keV. The objective lens was turned off in order to provide a field-free environment for the sample. Images were taken using the Fresnel (phase contrast) mode of Lorentz microscopy.

After deposition, films which were to be examined in the TEM were subjected to a saturating field of 15 kOe applied perpendicular to the film surface. Domains of reverse magnetization were then written by focusing a GaAs laser with a FWHM beam width of 0.85 μm on the film (film side incident) and simultaneously applying a magnetic field of up to 500 Oe perpendicular to the film. Laser power varied up to 7 mW as measured at the sample. The writing pulse width was 500 ns. The films were then floated off the mica substrates in water and picked up using 200 mesh copper grids and examined in the TEM by Lorentz microscopy.

Figure 1 shows an example of domain observation on a film of TbFeCo which was prepared by the above method. Although one can see that written domains are present, the overall magnetic structure is complex. (The written domains are the large scale regular pattern superimposed on a fine scale ripple background.) One of the properties of the RE/TM systems is that they are susceptible to surface oxidation and the Tb-Fe system in particular corrodes in room air by developing a thin oxidized surface layer which is magnetic and exhibits in-plane anisotropy. This corrosion product has been shown recently to contain mostly metallic Fe in addition to oxides of Tb.[8] Since Lorentz microscopy depends for the most part on deflection of the incident electrons by in-plane components of magnetization in the film, it seemed possible that the major features in Fig. 1 were due to this surface layer with in-plane magnetization.

A series of films was fabricated with composition Tb_xFe_{100-x} where x varied from 14 to 32 at. %. These films were deposited with 10 nm of SiO_2 on both the top and bottom surfaces for protection against atmospheric oxidation. VSM measurements made on these films showed that no in-plane component of magnetization was present. VSM measurements were also done after immersion of a film in water for several minutes to ensure that no in-plane magnetization

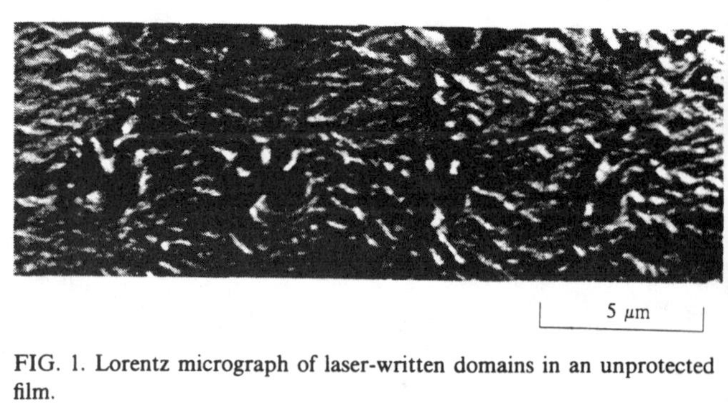

FIG. 1. Lorentz micrograph of laser-written domains in an unprotected film.

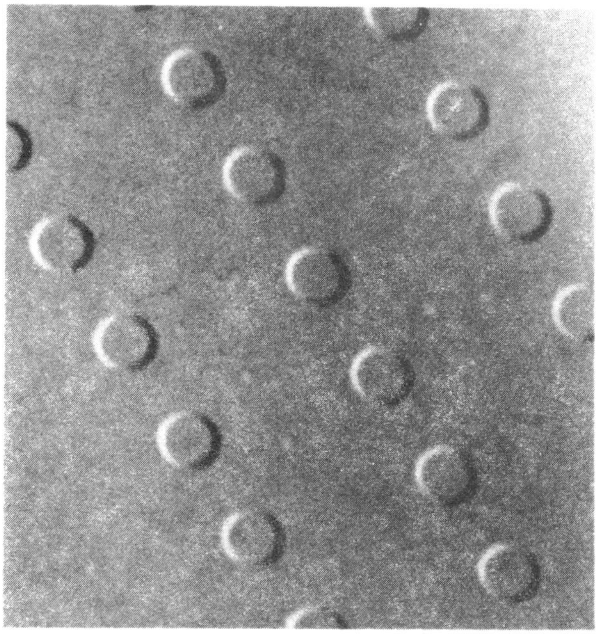

FIG. 2. Lorentz micrograph of laser-written domains in a protected film. The center-to-center spacing of the nearest domains is 5 μm. The laser power is 5.7 mW at the film surface and the applied magnetic field is 400 Oe.

was introduced during floatoff. Typical written domain images for a $Tb_{19}Fe_{81}$ film are shown in Fig. 2. No trace of the complex structure seen in Fig. 1 is evident. One requirement, therefore, for obtaining detailed TEM images of written domains is the prevention of the occurrence of a surface layer with in-plane magnetization. All films to be discussed below had 10 nm of SiO_2 on both surfaces.

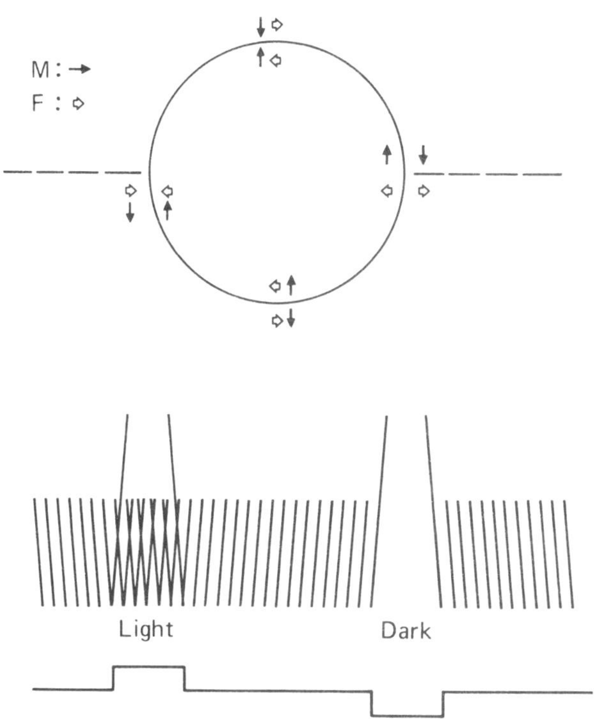

FIG. 3. Image formation in tilted film. Upper: looking down on a circular domain rotated about horizontal dashed line. M is the direction of in-plane magnetization resulting from this rotation and F is the direction of Lorentz force on an electron beam incident from above. Lower: side view of electrons incident on the fluorescent screen being deflected into either overlap or nonoverlap regions.

Figure 2 also demonstrates that magnetic domains may be laser written in RE/TM type materials with extremely regular and reproducible magnetic structure. This is important in order to obtain high signal/noise in practical devices.

In one method of doing Lorentz microscopy on magnetic domains in media with perpendicular magnetization, the incident electron beam is aligned normal to the film and a deflection of the electrons is caused by the in-plane component of the magnetic spins in the domain walls. This deflection allows the walls to become visible. In the case of TbFe, however, it was found that when the incident electron beam was perpendicular to the film surface, the domain contrast was extremely weak. Setting the angle between the incident beam and the film normal to about 20°–30° resulted in the best contrast. By tilting the sample in this way, one creates a component of magnetization which is perpendicular to the electron beam and which deflects the electrons to form an image. Figure 3 shows a schematic of the magnetization components and the contrast to be expected (light crescents on one side of the domain and dark crescents on the other). Rotating the angle of incidence through the normal direction causes the black crescents to go to white, and the white crescents to go to black.

The exchange stiffness A was calculated for $Tb_{19}Fe_{81}$ using the expression derived by Gangulee and Kobliska[9] and found to give $A = 2\times 10^{-7}$ erg/cm. Combining this with recent measurements of the anisotropy constant,[10] gives a domain wall width at 20 °C of approximately $\pi\sqrt{A/K} = 8$ nm. Due to the the very large K and the small A in these materials, the domain wall width is very small. This may be one reason that contrast is very weak unless the film is tilted with respect to the electron beam.

To illustrate the effect of laser power and magnetic field, H, applied during writing, an array of domains was written with power varying in one direction and H varying in the other direction. The result is shown in Fig. 4 for a film of composition $Tb_{32}Fe_{68}$. In one direction this pattern was written with fields ranging from $+200$ to -100 Oe. The orthogonal direction was written with constant magnetic field but with laser power varying in the range 1–7 mW as measured incident upon the film.

The largest, most regular, and saturated domains are written with the highest laser power and with the magnetic field in the direction which will aid in domain formation, i.e., in the direction of the reversed magnetization within the interior of the domain (see upper left-hand corner region of Fig. 4). As one reduces the laser power at constant bias field (top to bottom along a vertical column), the domains become smaller, as expected, and eventually disappear as the power falls below the threshold for recording. As one reduces the bias field at constant laser power (left to right along a horizontal row) subdomains appear within the principal domain, which itself remains approximately the same diameter. From observation of the black and white domain crescents one can determine that these subdomains have a direction of magnetization which is the same as that of the

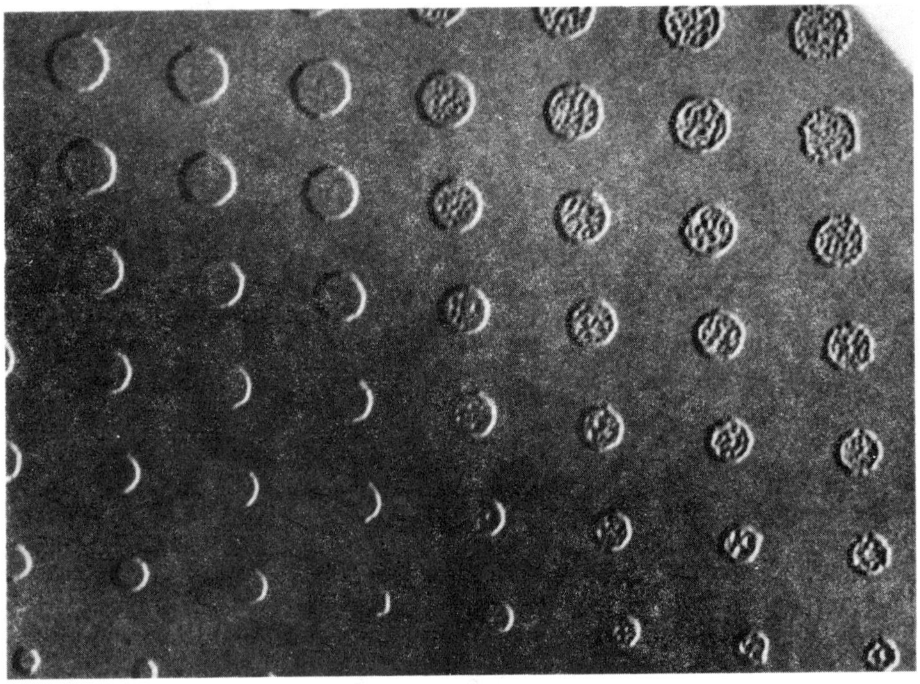

FIG. 4. Lorentz micrograph of film of composition $Tb_{32}Fe_{68}$. The minimum center-to-center domain spacing is 5 μm. Laser power decreases from top to bottom, and applied field in the writing direction decreases from left to right.

original film before writing. This unsaturated domain structure is expected when the net magnetic field inside the domain during its formation is dominated by the magnetostatic demagnetization field. It is believed that this is the first time that such interior structure has been observed with clear detail within a micron-sized written domain.

In summary, we have described a method for observation of highly detailed images of laser-written micron and submicron sized domains in the RE/TM system of magneto-optical materials. One requirement for observation of true domain images is the avoidance of a surface oxidation layer with in-plane magnetization. A second requirement for high contrast domain observation is tilting of the sample in the TEM. An example is shown where the observed domains are highly regular and saturated under some conditions of laser power and magnetic field, but are highly irregular and unsaturated under other conditions.

The authors wish to thank Lois Bynum for her technical assistance in film fabrication.

[1] P. Chaudhari, J. J. Cuomo, and R. J. Gambino, Appl. Phys. Lett. **22**, 337 (1973).

[2] Y. Mimura, N. Imamura, and T. Kobayashi, Jpn. J. Appl. Phys. **15**, 933 (1976).

[3] Y. Mimura, N. Imamura, T. Kobayashi, A. Okada, and Y. Kushiro, J. Appl. Phys. **49**, 1208 (1978).

[4] K. Sunago, S. Matsushita, and Y. Sakurai, IEEE Trans. Magn. **MAG-12**, 776 (1976).

[5] S. Honda, J. Hirookane, M. Ohkoshi, and T. Kusuda, J. Mag. Magn. Mater. **35**, 208 (1983).

[6] H. D. Shieh and M. H. Kryder, 1985 Topical Meeting on Optical Data Storage, IEEE Catalog No.85CH2234-3.

[7] P. S. Alexopoulos and R. H. Geiss, IEEE Trans. Magn. (to be published).

[8] R. B. van Dover, E. M. Gyorgy, R. P. Frankenthal, M. Hong, and J. Siconolfi, J. Appl. Phys. **59**, 1291 (1986).

[9] A. Gangulee and R. J. Kobliska, J. Appl. Phys. **49**, 4896 (1978).

[10] R. B. Van Dover, M. Hong, E. M. Gyorgy, J. F. Dillon, Jr., and S. D. Albiston, J. Appl. Phys. **57**, 3897 (1985).

Reprinted with permission from *Proceedings of the IEEE*, Vol. 74(11), pp. 1570-1581 (November 1986). ©1986 IEEE.

Magnetooptics: A Thermomagnetic Recording Technology

W. H. MEIKLEJOHN

Invited Paper

Erasable magnetooptic (MO) recording is viewed as a possible successor to electromagnetic recording that is used today in high-performance disk drives for mainframe computers, hard disks, and floppy disks for personal computers and magnetic tape drives. A large number of rare earth–transition metal alloys have been investigated for use as the magnetic recording material. The properties of the films are very sensitive to the film deposition processing parameters. Accelerated life tests show that there are magnetooptic materials that do have all of the necessary properties and stability for use in computer digital storage.

Introduction

Erasable magnetooptic (MO) recording is viewed as a possible successor to electromagnetic recording that is used today in high-performance disk drives for mainframe computers, hard disks, and floppy disks for personal computers and magnetic tape drives. MO offers bit densities ($\sim 10^8$ bits/cm^2) ten times that of high-performance disk drives and 50 to 100 times the density of the low-end disk drives.

The optical recording head is at least 1 mm above the disk surface. This arrangement eliminates the possibility of head crashes and also allows the disk to be removed. Rigid plastic disks with grooved recording tracks on 1.6-μm centers can be rotated at speeds as high as conventional Winchester drives. Since bit density along the track is comparable to today's Winchester technology, data rates can be the same.

There have been several reviews of MO recording [1]–[3] that extol the promise and describe the critical areas that require improvement. A 12-inch diameter MO prototype drive has been announced by Sony [4] that has a user capacity of 2.1 Gbytes. The data transfer rate is 360 kbytes/s and the seek time for $\frac{1}{3}$ full stroke is 80 ms. A 3.5-inch diameter prototype disk drive has been announced by Verbatim that is intended to serve the personal computer market. The MO drive contains 40 Mbytes of formatted data with overwrite capability of 1 Mbit/s. The $\frac{1}{3}$ stroke seek time is 20 ms and the track-to-track access time is 1 ms.

Manuscript received December 1, 1985; revised May 1, 1986.

The author is with the Department of Electrical and Computer Engineering, Carnegie-Mellon University, Pittsburgh, PA 15213, USA.

Several of the features of the highly developed "write once" optical disk storage technology, such as tracking and focussing, are directly applicable to the MO storage technology.

In this paper we describe the construction and operation of a typical MO drive, the status of MO recording materials, the expected bit density and signal-to-noise ratio (S/N), as well as the material stability and future developments.

Magnetooptic Recording Technology

Thermomagnetic recording [5] is used in the MO recording technology. The material is heated above a critical temperature by a diode laser beam and the heated spot is allowed to cool in a magnetic field. The direction of the magnetic field, which is normal to the disk surface, determines the direction of the magnetization in the recording. The critical temperature may be the Curie temperature (T_c) or a temperature above the compensation temperature (T_{comp}) where H_c is less than the bias field.

The rare earth–transition metal (RE-TM) magnetic materials are amorphous ferrimagnetic materials. There is usually a compensation temperature (T_{comp}) at which the magnetizations of the two subnetworks are equal. T_{comp} is about 60°C for the material shown in Fig. 1.

The insert is a schematic representation of the magnitude and direction of magnetization in the transition metal (TM) subnetwork, in the rare earth metal (RE) subnetwork and the net magnetization. AT T_{comp} the magnetizations are equal and opposite and M is zero. Below T_{comp}, the RE magnetization is greater while above T_{comp} the TM magnetization is greater. Note that the magnetic field lines up the greater subnetwork magnetization, and because of the negative exchange interaction between the two subnetworks, the other subnetwork has the opposite direction of magnetization.

The coercive force (H_c) tends toward infinity at T_{comp}. This behavior is due to H_c being inversely proportional to M which goes to zero at T_{comp}. This coercive force is not due to the magnetization having to mount a potential barrier, but rather that the torque (T) on the net magnetization,

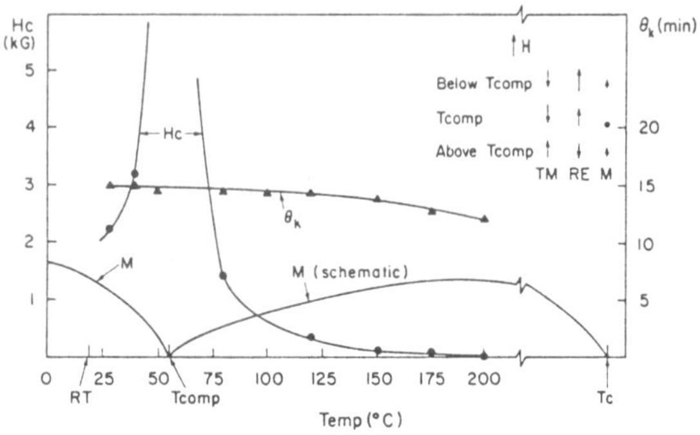

Fig. 1. Coercive force (H_c), and Kerr rotation angle (θ_k) as a function of temperature for a GdTbCo magnetic recording film, and a schematic representation of the temperature dependence of the magnetization (M). The insert shows a schematic representation of the subnetwork magnetization in the film at the compensation temperature (T_{comp}) and for temperatures below and above T_{comp}.

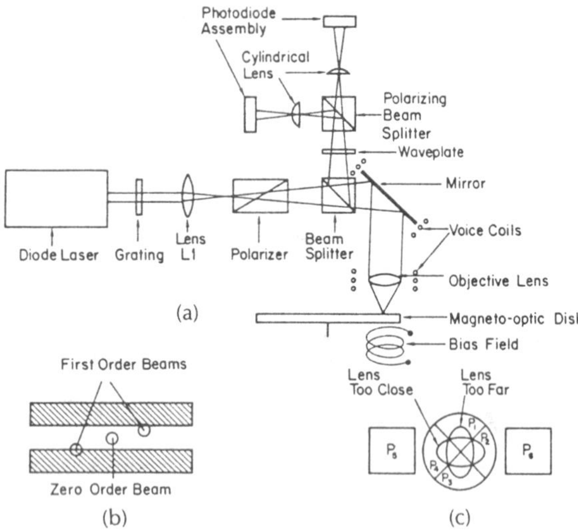

Fig. 2. (a) A schematic drawing of the optical system for a magnetooptic drive. (b) The tracking laser beam spots (first order) and the read–write laser beam spot (zero order) are shown on the disk groove. (c) Shown are the photodiodes (P_5) and (P_6) for tracking and the photodiode (P_1, P_2, P_3, and P_4) for focussing and for the signal.

given by

$$T = \overline{M}_{net} \times \overline{H} \tag{1}$$

goes to zero at T_{comp}. In the determination of H_c, the H in (1) is applied perpendicular to the plane of the film, but it is not precisely collinear with the net magnetization.

A compensation temperature is not necessary in magnetooptic materials used in Curie point recording. When the material cools from the paramagnetic state above T_c to the ferromagnetic state below T_c, the magnetization direction is the same as that of the magnetic field present during cooling.

Writing

The magnetic properties of a material that may be used for compensation point recording are shown in Fig. 1. The coercive force of the film at room temperature (RT) is greater than 5 kOe. To write thermomagnetically, a spot on the magnetic film is heated by the laser beam. The coercive force at the heated spot first increases as T_{comp} is approached and then decreases along the H_c curve until it gets below an applied bias magnetic field at some temperature above 100°C. The direction of the bias field determines the direction of the magnetization after the laser pulse is turned off. The bias field does not need to be present until the spot reaches room temperature, but only until a blocking temperature is reached. The blocking temperature is several degrees below the temperature at which H_c becomes greater than the bias field.

A material similar to that shown in Fig. 1, i.e., a material with a compensation temperature, can be used for Curie point recording. The composition of the material would be adjusted so that T_c is in the neighborhood of 150°C. A terbium iron (TbFe) alloy is frequently used for Curie point recording.

A schematic diagram of an optical assembly to record information and read the recorded information is shown in Fig. 2. There are a number of optical system designs [4], [6]–[9]. It is not possible to discuss all of these designs in this paper, but instead a specific example has been chosen to provide the reader with enough information to understand the issues.

The diode laser provides the heat source for writing. The beam passes through a polarizer and a beam splitter to a reflecting movable mirror and is focused on the disk by an objective lens as shown in Fig. 2(a).

The bias magnetic field mentioned earlier is provided by the coil shown in Fig. 2(a). The magnetic field generated in the coil by a current pulse is in one direction for writing and in the opposite direction for erasing the domain.

The Reading Process

There are regions of up and down magnetization in the recorded track due to the recording process. Assume that down magnetization corresponds to a "1" and up magnetization corresponds to a "0." We can now detect these directions of magnetization by means of the polar Kerr magnetooptic effect. When a linearly polarized light beam is reflected from a vertically magnetized magnetic material there is a rotation of the plane of polarization. The sense of the rotation depends upon the direction of magnetization in the magnetic film.

The light beam incident upon the magnetic film during the reading of the information is of a much reduced intensity compared to that during the write process. Thus the film's temperature increase and corresponding coercivity decrease are small enough that the direction of magnetization does not change. The laser beam is first linearly polarized by the polarizer. Then it passes through a beam splitter and is reflected by the mirror through a lens to the disk surface. The beam reflected from the disk has its plane of polarization rotated about 0.25° to 0.5° due to the polar Kerr effect. The rotation angle is designated θ_k. We will assume, for discussion purposes, that the rotation is in the positive direction for a written "1," and in the negative direction for a "0." The reflected beam passes through the lens and is reflected by the mirror to the beam splitter where some of the beam is again reflected through the waveplate to the polarizing beam splitter. If the polarizer between lens $L1$ and the beam splitter is set at a 45° angle with respect to the P axis of the polarizing beam splitter then signals will occur in the photodiode assembly that can be subtracted to give the desired signal without the common mode noise. Hence the pulse pattern of the recording laser is reproduced in the output of the photodetectors.

Consideration of the S/N ratio of presently available recording materials [2] shows that MO systems could have a bandwidth during reading of 100 MHz. MO systems with even greater bandwidth may be possible as discussed in the section entitled "Possible Future Developments."

Tracking and Focusing

The laser beam also passes through a grating and a lens before reaching the linear polarizer as shown in Fig. 2(a). The zero-order beam is used for writing and for reading when reduced in intensity.

The two first-order beams from the optical grating are used for tracking, as shown schematically in Fig. 2(b). The cross hatched region represents the grooves in the disk. The area between the grooves, called the land, is where the recordings are made. The grooves are about 0.07 μm deep and usually have a pitch of 1.6 μm. The three beams are focused on the disk so that one first-order beam leads and the other first-order beam lags the zero-order beam. The two first-order tracking beams are slightly displaced from the track in opposite directions. Only the portion of each first-order beam that falls on the land is reflected to a photodiode due to the sloping sides of the grooves. The reflected first-order beams fall on separate photodiodes P_5 and P_6 shown in Fig. 2(c). The difference signal from these photodiodes is maintained at zero by the movement of a mirror shown on the right side of Fig. 2(a), keeping the zero-order write-read beam on the track. The mirror transducer causes all three beams to move radially. Radial movement of the objective lens can be used in place of the mirror.

There is also an objective lens, shown in Fig. 2(a), which focusses the beam on the magnetooptic recording material. This lens is mounted on a voice coil–permanent magnet arrangement to allow movement of the lens normal to the surface of the disk. The photodetector consists of four pie shaped quadrants as shown in Fig. 2(c). The sum of the output of these four quadrants is the signal. The cylindrical lens shown between the polarizing beam splitter and the photodiode assembly in Fig. 2(a) causes an elongated beam to appear on the photodiode when the recording material is out of focus, the direction of elongation depending on whether the lens is too far or too close to the recording material. The beam is circular when the recording material is in focus. The difference signal of the sums of the opposite sectors is fed to the voice coil to maintain the focus.

The Disk Assembly

One method of making a magnetooptic disk substrate is to inject a liquid plastic material into a die that contains a grooved mold. The spiral grooves are about 0.07 μm deep, 0.7 μm wide, and are spaced about 1.6 μm apart. This plastic material must have a very low birefringence since the laser beam passes through this material. Birefringence causes a depolarization of the light beam and causes a decrease in the S/N.

The next step in fabricating the simplest type of MO disk involves the deposition of the magnetooptic magnetic film. The film is deposited on the groove side of the substrate by sputtering or evaporation of the metal alloy to a thickness of about 0.1 μm. A protective coating is then deposited over the metal storage magnetic film.

A cross-sectional view of a portion of this type of disk is shown in Fig. 3. This type of disk could be used in the magnetooptic recording system shown schematically in Fig. 2.

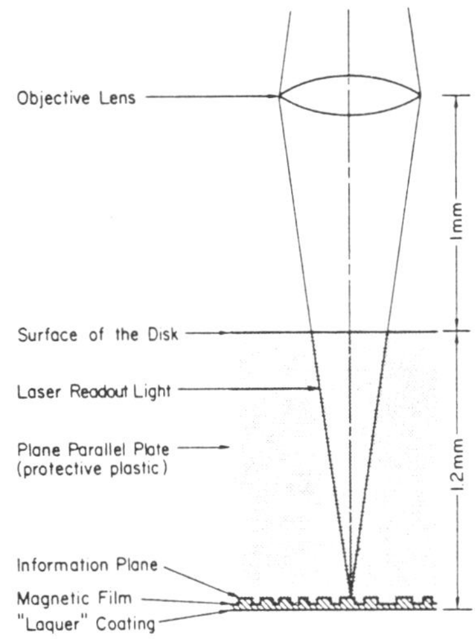

Fig. 3. A schematic representation of a portion of the disk substrate showing the grooves, magnetic film, and the aluminum reflection coating.

If the Faraday magnetooptic effect is used to read the disk, then the deposited magnetic recording film is only about 0.02 μm thick so that the laser beam can pass through the film to the detecting system. This system can use only one side of the disk for recording of information. Verbatim used this type of system in their previously mentioned drive demonstration.

To make a disk that can be used to record information on both sides, two of the disks having 0.1 μm thick films are placed back-to-back so that the grooved sides are adjacent to each other. The disks are then bonded together.

More complicated disk structures are made in order to enhance the signal-to-noise ratio of the system. These more complicated structures use magnetic film thicknesses similar to the films discussed above for Faraday detection of the recorded information. These structures are discussed under the heading "Signal-to-Noise Ratio."

Magnetooptic Recording Materials

An MO material should have a large value of intrinsic uniaxial perpendicular anisotropy energy density (K_u) whose origin lies in pair ordering or a columnar structure, and not in film strain due to the substrate. Films that derive K_u from strain may lose data due to loss of perpendicular magnetization in the film which can result from a change of stress in the film due to a temperature change. A change of stress may also arise from a change in the adherence of the film to the substrate which in the worst case would be the lifting of the film from the substrate.

K_u is usually derived from torque measurements made on the film in a saturating magnetic field, coupled with a measurement of the saturation magnetization M_s. K_u can also be obtained from the measurement of the anisotropy magnetic field (H_k), again coupled with a measurement of M_s.

Ferromagnetic resonance measurements (FMR) also yield the value of K_u.

In the case of both compensation point recording and Curie point recording, the MO material should probably not be heated above about 150°C during the recording process. The MO materials are in the amorphous state and repeated heating above this temperature may cause crystallization and the loss of magnetooptic properties. Repeated (10^7 times) heating at this temperature does not cause crystallization as discussed under "Material Stability." Heating the film at higher temperatures is also undesirable since more powerful lasers are required.

During the reading process, the laser power should be as great as possible without causing a loss of the recorded information, since the shot limited signal-to-noise ratio (S/N) is proportional to the read power. However, as shown in the section "Signal-to-Noise Ratio," the S/N is also proportional to the square of the Kerr angle (θ_k^2) for small values of θ_k. Since the Kerr angle is nearly proportional to the transition metal magnetization [10], θ_k tends to fall off sharply in the neighborhood of T_c. In the case of Curie point recording there may be a net loss of S/N at high laser read power, even though the stored information may not be destroyed. On the other hand, a compensation point recording material may have a very high T_c resulting in a constant θ_k up to the writing temperature of the film. For this reason, compensation point recording is preferred over Curie point recording. The compensation point recording material (GdTbCo) shown in Fig. 1 has a relatively substantially constant θ_k up to 200°C. However, from a manufacturing point of view, the composition of the compensation point recording material must be more rigidly controlled than for the Curie point recording material. The Curie temperature is not a strong function of the material composition while a 1 at % change in composition can shift T_{comp} by a large amount. As shown by Fig. 1, a small shift in T_{comp} will drastically change H_c at room temperature.

High coercive force (1-2 kOe) at RT is desirable so that stray magnetic fields do not disturb the recording of the data. The product $M \times H_c$ should be large [11], [12] so that stable submicrometer domains can be recorded as is discussed in greater detail under the heading "Recording Density." The Kerr angle should be about $\frac{1}{3}$ of a degree or larger in order to have a satisfactory S/N.

In some cases, all of the features may not be obtained in a single film. Double-layer films that are exchange coupled have been investigated to improve the properties [13]-[15].

Alloys of rare earth metals and transition metals are used as magnetooptic recording materials. Gadolinium terbium cobalt (GdTbCo) and gadolinium terbium iron (GdTbFe) alloys are typical MO recording materials.

These RE-TM alloys have two desirable features. Thin metal films of RE-TM alloys can be produced in the desired amorphous state by several deposition techniques, such as sputtering and evaporation. Unlike semiconductors which are usually in the form of single-crystal wafers of limited diameter, the MO film material can be deposited on a 14-inch diameter plastic substrate by vapor deposition techniques.

A second important property of these amorphous alloys is that they are ferrimagnetic. Although the films have only very short range atomic order, the transition metal magnetization is directed opposite to that of the rare earth metal magnetization. We do not speak of sublattices as in long range ordered ferrimagnetic materials, but rather of subnetworks. These two subnetworks have exchange coupled, oppositely directed magnetizations and have different temperature dependence of magnetization such that the magnetization curves cross each other. The temperatures at which the curves cross is called the compensation temperature (T_{comp}) as discussed earlier in this paper.

The easy direction of magnetization of these amorphous films is purposely made normal to the film surface by creating a large K_u. The exact origin of this anisotropy energy is not clear, although it is believed to be due to pair ordering, a columnar structure, or anisotropically shaped microstructural regions in the materials that are useful for MO recording. The "Origin of K_u" is discussed later in this paper. In some of the RE-TM films this anisotropy energy is the result of strain between the substrate and the magnetic film as has been shown for TbFe [16]-[18]. Films with large strain anisotropy due to stress induced by the substrate should not be used for MO recording because the magnetic properties will be sensitive to the temperature of the film and there is a possibility that the film may lose its adherence to the substrate in certain areas with the resultant loss of data.

The demagnetizing field (H_d) of recorded bits is about $4\pi M_s$ since the bit diameter is about 1 μm while the film thickness is about 0.1 μm. The demagnetizing energy is given by

$$E_d = \int_0^{M_s} 4\pi M dM = 2\pi M_s^2.$$

In order to have the magntization of the recording normal to the surface, the anisotropy energy K_u must be greater than $2\pi M_s^2$. This relationship can be expressed in terms of a quality factor (Q) by the equation

$$Q = K_u/2\pi M_s^2 > 1. \tag{2}$$

The origin of K_u in the MO material is very important. The most desirable sources of K_u are pair ordering and columnar growth. The least desirable source of K_u is due to strain in the magnetic film [16]-[19] since K_u will be temperature sensitive and the film may lift from the substrate. Either case may cause the condition specified by (2) to be violated with the resultant loss of data.

Pair ordering [20], [21] is believed to result from resputtering of atoms from the film during the deposition process which causes some alignment of atom pairs perpendicular to the plane of the film, i.e., in the direction of the incident atom flux. This resputtering is caused by argon ions hitting the film due to a negative bias on the substrate.

Columnar structure [22], [23] in sputtered amorphous films is mostly found in alloys containing cobalt. It has been proposed that this amorphous columnar structure is due to argon (~10 percent) incorporation in the film [24]-[26]. More recent studies [27] show that films of GdCo with only 1-percent argon in the film can have a large $K_u = 2.3 \times 10^5$ ergs/cm^3, suggesting that argon may not be necessary in the film. The columnar axis is believed to be the easy direction of magnetization due to compressive strain caused by the preferential oxidation of the rare earth atoms along the column boundaries through the film thickness [22], [28]. In the case of GdCo films the columns become cobalt rich.

The amount of stress in a film depends upon the film deposition conditions in some alloys. For GdFe alloys the stress in the film is highly dependent upon the argon pressure

Table 1 Properties of Magnetooptic Recording Materials

Material	Compensation Composition (at %)	Type of Recording	T_c (°C)	Kerr Angle (degrees)	H_c (kOe)	K_u 10 (ergs/cm)	Q	R (%)
	Ref. [ ]	Ref. [ ]	Ref. [ ]	Ref. [ ]	Ref. [ ]	Ref. [ ]	Ref. [ ]	Ref. [ ]
TbFe	Tb-22 [32]	Curie [31]-[33]	80-150 [34]	0.25	2-10 [32]			0.4
GdCo	Gd-21	Comp. [35]	~400	0.33 [38]	1-2 [37]	6.8 [39]	3.5	
GdFe	Gd-26	Curie [16], [18], [19], [40]	180 [41], [42]	0.5 [42], [43]	0.25 [13], [41]	2.5 [42]		
GdTbFe	(GdTb)	Curie [42], [50]	165 [7]	0.35 [7]	1-3 [43]	3.5 [42]		
GdTbCo		Comp.	~380	0.33 [55]	2 [53]	10		
GdFeCo		Curie [10]	220-270 [10]	0.42 [10]				
TbFeCo		Curie [43], [57], [59], [61]-[63]	180 [59]	0.3-0.4 [43], [62], [63], [65]		10-40 [43], [61]		

during deposition. The stress in sputtered GdFe films changes from compressive stress for an argon pressure of 6×10^{-2} torr to tensile stress for an argon pressure of 10×10^{-2} torr [18].

In the following discussion of alloys I have emphasized those experiments that show undesirable characteristics obtained on some films, as well as those experiments that show good characteristics of films. The intent is to give a new investigator a basis to choose the better alloy system to begin his (or her) investigation.

The magnetic and optical properties of some binary and ternary alloys are listed in Table 1. Specific compositions are not listed but rather the range of magnetic and optical properties obtained from a range of compositions are listed. Properties of specific compositions can be found in several publications [29], [30].

Binary Alloys

Terbium iron (TbFe) alloys have been prepared by evaporation and sputtering and have been extensively investigated [31]-[34] during the last decade. The origin of K_u is thought to be due to the Tb ion anisotropy. Some films have shown a large contribution to K_u due to stress between the film and the substrate [18], depending on the composition. The main drawback to using this material is that θ_k falls rapidly with increased reading power since the Fe magnetization is dropping rapidly near the Curie temperature. Hence, the *S/N* cannot be enhanced by higher reading power.

Gadolinium cobalt (GdCo) was one of the first materials suggested [35] for a thermomagnetic recording. GdCo films have shown magnetic domain instability [36] unless there is a perpendicular composition gradient introduced into the film during the deposition process [37].

The origin of K_u in GdCo is not clear, since it has been reported as due to a columnar structure [22], [23] and as due to pair ordering. However, there is also evidence [17] that as much as 30 percent of K_u comes from strain in the film, particularly for Co rich films.

Gadolinium iron (GdFe) alloys have been extensively investigated. A large contribution to K_u has been shown to be the result of stress in the film [16], [18], [19]. H_c values measured for GdFe films are lower than desired for a good MO material.

Double layer exchange coupled films of GdFe/TbFe have been investigated to get the higher θ_k of GdFe for reading and the higher H_c of TbFe for recording [13], [14].

Other Binary Alloys

Many other binary alloys such as DyFe, [14], [44], [45]; HoCo, [46]-[48]; HoFe, [46]; and GdY [49] have been investigated. None of these binary alloys appears to be useful for MO recording.

Ternary Alloys

Gadolinium terbium iron (GdTbFe) films show more promise [6], [7], [42] as a viable MO material than the binary alloy TbFe. The origin of K_u has not been reported for this alloy. Vacuum evaporated films do show some changes in K_u with aging tests performed at 160°C [42].

Gadolinium terbium cobalt (GdTbCo) appears to be one of the better MO materials. The origin of K_u has not been reported but it is probably due to a columnar structure or pair ordering [23], [51] because of the cobalt in the film. The stability of domains written in this material is excellent [52]-[54]. The material has a large value of $M \times H_c$ (8×10^4 Oe^2) [53] which is necessary to have small stable domains [11], [12].

Gadolinium iron cobalt (GdFeCo) alloys are similar to GdFe alloys when they contain only a small amount of cobalt. Since Curie point recording is used in this material the cobalt addition must be kept small since it increases T_c [10].

GdFeCo alloys that contain only a small amount of iron are similar to GdCo alloys and can be used for compensation point recording [55]. The addition of small amounts of bismuth (Bi) increases θ_k to 0.48° [56].

Terbium iron cobalt (TbFeCo) alloys have received a great deal of attention in recent years. The origin of K_u has not been reported for this alloy. Experiments show that TbFeCo is better than TbFe or GdTbFe for video recording [57].

Films have been prepared by evaporation [58] and sputtering [57], [59]-[61]. Stability tests on TbFeCo films indicate a life greater than 5 years [47], [58] when overcoated for protection from oxidation. Even unprotected films that are prepared by sputtering in a low argon pressure are quite stable [61], [62]. C/N values of 50 and 55 dB have been obtained with a bandwidth of 30 kHz [4], [57], [58], [63], [64].

Other Ternary Alloys

Several other ternary alloys have been investigated such as GdFeBi [66], GdFeSn [66], [67], GdFeY [41], GdFeB and GdCoB [68], and DyFeCo [63]. Only GdFeSn [67] and DyFeCo [63] have been evaluated in a drive.

Quaternary Alloys

Quaternary alloys GdTbFeGe [69], GdTbFeCo [42], [55], and GdFeCoBi [56] have been investigated. There is not sufficient published data on these materials to indicate their viability as an MO material.

Origin of K_u

The origin of the anisotropy energy density (K_u) in magnetooptic films is clearly not the result of one mechanism, nor is there one dominant mechanism. A discussion and summary of the present situation is contained in some recent papers [30], [70]. There is irrevocable experimental evidence [17], [18] that magnetostriction plays a major role in many MO films deposited on substrates such that there is a strain applied to the MO film by the substrate. It is clear that this source of K_u must be eliminated or evaluated in any experiment designed to determine other sources of K_u. One way to eliminate magnetostrictive effects is to remove the film from the substrate [18]. The magnitude of the magnetoelastic coupling and its effect on K_u can be determined by applying additional strain to the film in a torque magnetometer [18].

Many papers [24], [25], [71] have been written about the experimental correlation of K_u with the argon content in sputtered films. It is now clear that the effect of the argon is indirect, possibly increasing the strain in the film and thus influencing K_u through its magnetostrictive properties. MO films are being made with a very low argon content by dc magnetron sputtering [72] and by RF sputtering in a low argon pressure [30]. The films have $Q > 1$ and a large value of K_u perpendicular to the film.

The magnetic properties, particularly K_u, are very dependent upon the method of film deposition, such as evaporation, RF sputtering, or dc magnetron sputtering, and on the deposition parameters, such as argon pressure and bias voltage. As a consequence, the values of K_u measured by different investigators sometimes yield diametrically opposite results.

Pair ordering of the transition metal (TM) atoms has been proposed [20], [73] as the source of K_u in some MO alloys. Calculations [74] of the magnitude and the temperature dependence of K_u due to pair ordering show good agreement with the measured values of a GdCoMoAr alloy. Several attempts [75], [76], using large-angle X-ray scattering, have been made to observe this atomic scale ordering without success.

There is experimental evidence [70] of microstructural anisotropic magnetic regions in MO materials. These elongated regions are usually oriented with the axis perpendicular to the film. The regions are considered to arise due to shadowing during deposition. Calculations [74] of K_u due to a dipolar interaction of such regions having a slight compositional fluctuation did not yield an acceptable result to account for the magnetic anisotropy.

Another possible mechanism for the origin of K_u is exchange anisotropy [77]. Fig. 4 shows a schematic representation of microstructural anisotropic ferromagnetic regions surrounded by an antiferromagnetic structure in a material of the compensation composition. We consider a binary

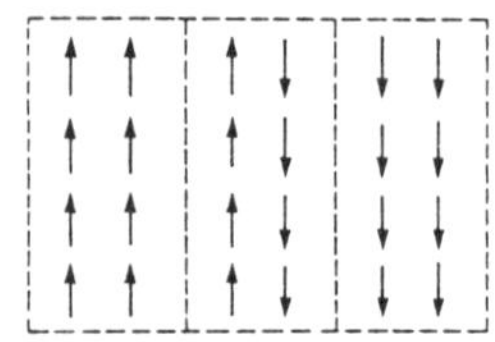

Fig. 4. A schematic representation of microstructural regions in a magnetooptic recording film which have positive exchange interaction surrounded by an antiferromagnetic region at the compensation composition.

alloy of A and B atoms with a positive exchange interaction between identical atoms and a negative exchange interaction between unlike atoms. The ferromagnetic regions are exchange coupled at the surface to the antiferromagnetic region which we assume to have a large anisotropy. K_u of such a system should have a maximum value at the compensation composition.

The anisotropy energy density for the structure shown in Fig. 4 is given by

$$K_u = V^{-1}\Big\{\sum_i (\tfrac{1}{2}N_{AA}M_{AA}^2V_{AA} + K_{EAA}V_{AA})_i + \sum_j (\tfrac{1}{2}N_{BB}M_{BB}^2V_{BB} + K_{EBB}V_{BB})_j + \sum_k (\tfrac{1}{2}N_{AB}M_{AB}^2V_{AB})_k\Big\} \quad (3)$$

where the sum over i, j, and k yields the total contribution due to the separate AA, BB, and AB coupled regions, respectively. Assuming that the microstructural regions are ellipsoidal in shape, then $N_{AA} = N_b - N_a$, where N_b is the demagnetizing factor of the minor axis and N_a is the demagnetizing factor of the major axis. M_{AA} is the magnetization of the AA coupled regions of volume V_{AA}. K_{EAA} is the exchange anisotropy energy density contribution to the anisotropy energy density of the ferromagnetic regions due to the surrounding ferrimagnetic regions. The other notations have similar meanings for the BB and the AB coupled regions.

In order to have $Q > 1$, the terms K_{EAA} and K_{EBB} must make an appreciable contribution to K_u since a strictly columnar structure has a $Q < 1$. The maximum value of K_u and Q should occur at the compensation composition where the AB coupled regions are antiferromagnetic and $M_{AB} = 0$.

We examine the exchange anisotropy energy density terms as a function of composition through the anisotropy field equation, $K_u = H_kM_{AB}/2$. We assume that H_k has the same dependence as a function of composition as H_c since they both depend upon a torque as a result of the magnetic field and the net magnetization as given by (1). In the determination of H_k the magnetic field H is applied in the plane of the film. We note in Fig. 1 that H_c drops very rapidly on each side of T_{comp}. M is shown schematically to increase almost linearly on each side of T_{comp}, as is also true in reality. Since $H_u = H_k M/2$, then K_u must decrease on each side of T_{comp} and have its maximum value at T_{comp}. Although Fig. 1 gives H_c and M as a function of temperature, a similar behavior is obtained for H_c and M as a function of composition, where T_{comp} represents the compensation composition.

If we consider K_u measurements on films that have been removed from the substrate we find that GdCo and TbFe exhibit a peak in K_u versus composition near the compensation composition, as shown in Fig. 5(a) and (b). These curves are from the work of Tsunashima *et al.* [17] and Takagi *et al.* [18], respectively. However, GdFe films [18] show a minimum near the compensation composition as shown in Fig. 5(c).

TbFe films of very low argon content prepared by dc magnetron sputtering show a peak in K_u near the compensation composition as shown in Fig. 5(d). This curve is from the work of van Dover *et al.* [72]. However, a TbFe film of unknown argon content prepared by RF sputtering shows a minimum in K_u [32] at the compensation composition.

Evidence of exchange coupling of TbFe to a ferromagnet on a macroscopic scale has been presented [78]. A shifted hysteresis loop was obtained on a double-layer film consisting of TbFe deposited on a permalloy film in the presence of an in-plane magnetic field.

We have found a shifted hysteresis loop in a double-layer film of TbCo on permalloy as shown in Fig. 6. The TbCo was sputtered onto a permalloy film (19 at % Ni) in the presence of an in-plane magnetic field. The hysteresis loop is shifted in the opposite direction to that of the magnetic field direction during the sputtering of the materials. Both films are about 400 Å thick.

The small shift in the hysteresis loop is believed to be due to creating a domain wall near the TbCo-permalloy interface. The microstructural regions that exist in the MO films are undoubtedly single-domain and hence the ferromagnetic regions will be more strongly coupled to the surrounding *AB* structure since a domain wall cannot exist in a single-domain region.

Film Deposition

The film morphology and the magnetic properties [61], [62], [79], [80] are sensitive to the processing parameters. The selection of the proper type of film deposition equipment and the proper film deposition parameters is just as important as selecting the proper MO material.

Vacuum evaporation can be used to prepare many MO films. The films, although columnar, have the easy direction of magnetization in the plane of the film ($Q < 1$) [40], [74], [81]. It is believed necessary to have some oxygen in the system during evaporation to have $Q > 1$. The oxygen is believed to oxidize the rare earth metal ions along the columns and produce a compressive stress on the columns which makes $Q > 1$, particularly for evaporated GdCo films. Evaporated alloys of TbFe and TbFeCo do not develop $Q > 1$ by this treatment.

It is necessary to use oblique incidence of the metal vapors in order to obtain a columnar structure [82]. The angle

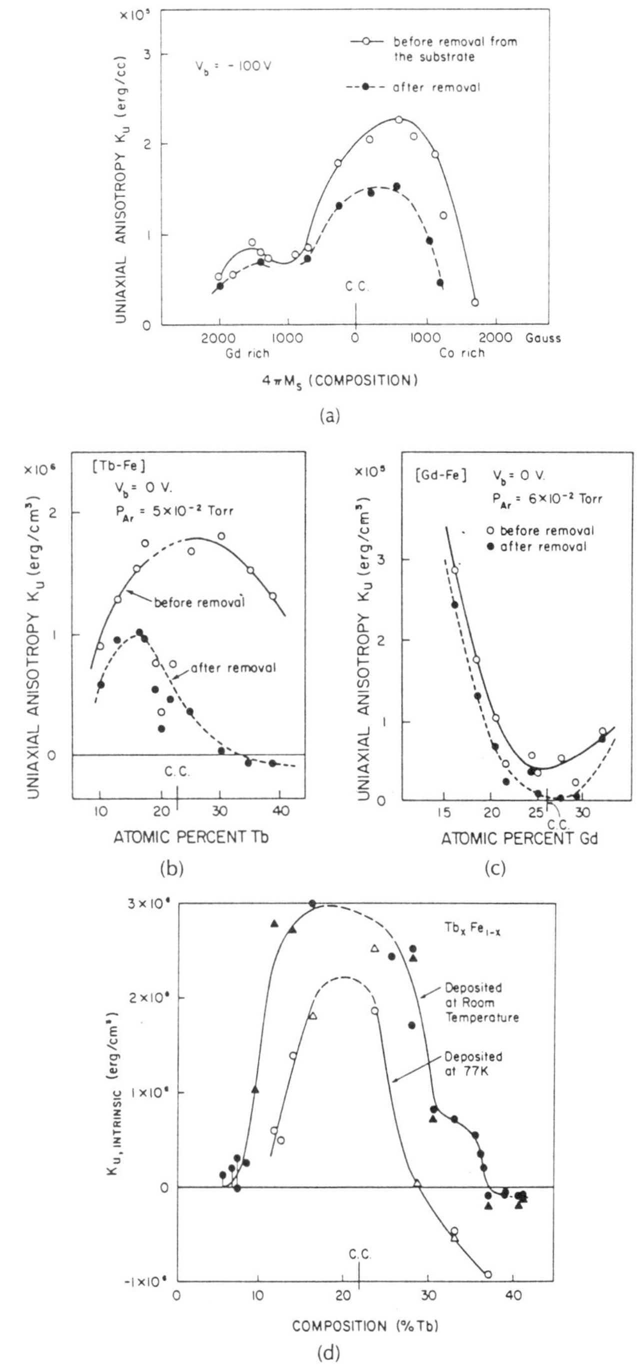

Fig. 5. Perpendicular anisotropy of magnetic films before and after removal from the substrate as a function of composition in (a) GdCo, (b) TbFe, and (c) GdFe. (d) The intrinsic anisotropy of TbFe thin films as a function of composition. Triangles represent measurements by torque magnetometry and circles by using the VSM. Solid symbols represent films deposited at room temperature, while open symbols denote films deposited on a liquid-nitrogen-cooled substrate.

of the columns (β) with respect to the substrate normal is related to the angle of incidence (α) of the metal vapor by the relation [40] $\tan \beta = (\frac{1}{2}) \tan \alpha$. The columnar structure is believed to result from self-shadowing under conditions of limited atom mobility. The resultant metal film has a very large void structure [74], [81]. Rotation of the substrate during oblique incident evaporation produces a perpendicular

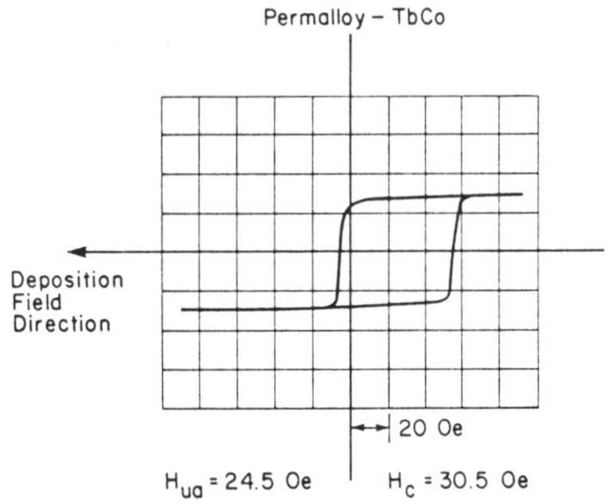

Fig. 6. A shifted hysteresis loop measurement of a double-layer film showing exchange anisotropy. A permalloy film was deposited on a glass substrate followed by the deposition of a TbCo film. Both films were deposited by RF sputtering and each film is about 400 Å thick.

columnar structure. Vacuum evaporation has been used to prepare MO films that have good stability and good performance [42], [63].

Radio frequency (RF) sputtering is a widely used method for depositing MO films. The RF electric field ionizes the argon gas that is usually used in the system at a pressure of 1 to 10 × 10^{-2} torr. The target may be an alloy of the metals desired in the film or it may be a composite target made by fixing the rare earth metals of the proper area percentages to the transition metal target plate. The target is biased using about −1 kV with respect to ground. The ionized argon ions hit the target and transfer their energy to atoms or clusters of atoms, some of which gain sufficient energy to exceed the work function of the metal, and are deposited in the system, which includes the substrate.

The substrate is placed in this flux of metal atoms and may also be biased by 0 to −200 V with respect to ground. The composition of the film and K_u of the film are dependent on the substrate bias due to the resputtering action.

Magnetron sputtering is also extensively used to deposit MO films. A magnetic field is used to increase the path length of the electrons and hence increase the ionization of the argon gas. Sufficient ionized argon gas can be obtained at much lower argon pressures than in the case of RF sputtering. There is evidence [61], [80] that MO films prepared by sputtering in low argon pressures have better film morphology and are more resistant to oxidation. Magnetron sputtering equipment is a good type of equipment for the preparation of MO films since the substrate is not part of the electrical circuit. As a result, the substrate is not bombarded by electrons and remains at a lower temperature than in the case of RF sputtering where the substrate is heated due to the electron bombardment.

Recording Density

In order to assess the maximum recording density we first examine the conditions that determine the minimum stable magnetic domain size [12]. We make use of Thiele's treatment [83] of the stability of bubble domains. We add a coercivity term (H_c) to his expression for the normalized forces per unit area (ϕ) of the domain wall of a cylindrical domain and get

$$\phi = -\sigma/4\pi M^2 h + (H_c - H)/4\pi M \times (d/h) + F(d/h) \quad (4)$$

where σ is the energy per unit area of the domain wall around the domain, M is the magnetization, h is the magnetic film thickness, and d is the diameter of the recorded bit. H is the externally applied magnetic field experienced by the magnetic domain and is directed opposite to the magnetization.

The first term on the right in (4) describes the normalized force per unit area on the domain wall which arises from the surface tension provided by the domain wall energy. The second term describes the normalized force per unit area which arises from the applied field and is corrected for the coercive force of the material. The third term describes the normalized force per unit area arising from the demagnetizing fields.

The minimum stable domain diameter (d_0) is determined by setting $\phi = 0$. After applying the conditions of a small value of M, $d/h \ll 1$, and $H \ll H_c$, which are applicable to MO recorded domains, we obtain

$$d_0 \sim \sigma/MH_c. \quad (5)$$

To obtain the minimum recorded bit size, the domain wall energy, $\sigma = 4\sqrt{AK_u}$, should be made as small as possible while still maintaining the condition $K_u > 2\pi M_s^2$. A is the exchange constant involved in the exchange interaction energy between the magnetic spins. The product MH_c should be made as large as possible. The validity of (5) was determined by data taken on the collapse field for recorded magnetic domains [12].

We now examine the consequence of packing the recorded domains close together to obtain the highest domain density. The placement or separation of the domain influences the domain stability through the net magnetic field (H) experienced by the recorded domain. This magnetic field does not appear in (5), because we have assumed that $H \ll H_c$. The magnitude and variation of the magnetic field at a given domain site due to other recorded domains can be determined [84].

Experimentally, it has been shown that 0.5-μm-diameter domains on 1.0-μm centers are stable. This arrangement yields a density of 10^8 domains/cm^2. The bit density is a function of the particular encoding scheme and may be two times the domain density.

There are many other factors that determine the minimum domain diameter. One of these factors is the diffraction-limited diameter of the laser beam used to heat the film. Assuming that the laser beam has a Gaussian intensity distribution, the first region to have the magnetization changed or reversed will be the center region of the laser beam due to the greater beam intensity in this region.

The nucleation and growth of thermomagnetically recorded domains has been examined [54]. Domains were formed in GdTbCo films using both 7.5- and 15-mW diode laser pulses of 200- to 1000-ns duration [54]. The nucleation and growth of these domains was observed with stroboscopic light pulses of 10-ns duration from a nitrogen/dye laser. Domain nucleation occurs within the first 10 to 20 ns of the beginning of the write pulse. Using the lower power the domain initially grows at a rate of about 4 μm/μs, with increased power producing only a slightly higher growth rate. The experiments suggest that the rate of domain growth is limited not by thermal considerations, but rather by the domain wall motion process. By proper adjustment of the laser beam pulse length and its intensity, along with the conductivity of the magnetic film and substrate, domains smaller than the diffraction-limited diameter of the beam can be recorded. The reading of the recorded domains may require a shorter wavelength laser.

Signal-to-Noise Ratio

A signal-to-noise ratio (*S/N*) of about 25 dB at 20-MHz bandwidth is considered more than adequate for an MO system [85], [86]. We first consider an ideal MO system where the limiting source of noise is the shot noise in the differential photodiode detectors. The shot-noise-limited signal-to-noise ratio is then

$$(S/N)_{shot} = 10 \log [2\eta P_0 R \sin^2 \theta_k / eB] \qquad (6)$$

where η is the sensitivity of the photodiodes in amperes per watt of illumination, P_0 is the average light intensity (watts) incident on the magnetic material during the read process, R is the reflectivity of the magnetic material, θ_k is the Kerr rotation angle, e is the electronic charge in coulombs, and B is the bandwidth of the system in hertz. If typical values [2] are inserted in (6), we get $(S/N)_{shot} = 38$ dB.

In order to obtain the maximum *S/N* we need to make P_0 as large as possible without disturbing the written domain. Compensation-point magnetooptic recording materials with a high Curie temperature can accommodate a large value for P_0.

The reflectivity is a function of the electronic properties of the magnetic film and its morphology and should be as large as possible.

The Kerr angle θ_k, which is the angular rotation of the plane of polarization of the incident laser light should also be large. This angle in most amorphous MO materials is about $\frac{1}{3}$ of a degree.

The origin of the Kerr effect is also found in the electronic structure of the MO film. Usually the Kerr effect is caused by incomplete cancellations of left and right polarized electronic excitations. In the ferromagnetic material PtMnSb we have the situation that one type of excitation is absent and a Kerr angle of 2.5° is obtained at room temperature with a 720-nm light source [87], [88]. PtMnSb is reported to be a poor MO recording material [89] because the magnetization is in the plane of the film instead of normal to the film.

Equation (6) has been derived for the case of light intensity P_0 falling on the MO film with P_0R being reflected to the photodetector. Placing a nonreflecting coating on the opaque MO film to increase the absorption, and hence the number of electronic transitions, can increase *S/N* by 4 dB [33]. The *S/N* can be increased even more by capturing more of the radiation from the electronic excitations. The distribution of radiation from the electronic transition is spherical. Some of this radiation can be redirected back to the photodetector by using a quadrilayer structure [33] shown in Fig. 7. The MO film is not opaque as in the case of the nonreflecting coating. Proper adjustment of thicknesses of the dielectric overlayer, the magnetic film, and the intermediate dielectric layer can result in an additional 4.5-dB gain in *S/N*. The total enhancement of the *S/N* of the system by use of the quadralayer structure is about 8.5 dB. The more complicated quadrilayer structure may require better manufacturing control of the material processes in order to achieve the same yield of usable disks as the simpler nonreflecting coating structure.

Signal-to-noise ratio (*S/N*) at operating bandwidths is not usually measured directly, but rather calculated from the measured carrier-to-noise ratio (*C/N*). The usual bandwidth used in the *C/N* measurements is 30 kHz. High-end storage systems usually operate in the tens of megahertz range. In order to convert the *C/N* measurement into a value for *S/N* at a higher frequency, a bandwidth correction must be made. The signal is well within the bandwidth of the *C/N* measurement and will be the same. The noise will increase with bandwidth. Using (6) we get

$$(S/N)_{f_s} = (C/N)_{f_c} - 10 \log f_s/f_c \qquad (7)$$

where f_s is the operating frequency desired for *S/N* and f_c is the measurement frequency of *C/N*. If $f_c = 30$ kHz and $f_s = 20$ MHz, the value of $10 \log (f_s/f_c) = 27$ dB. *C/N* values of 55 dB have been reported by several investigators [85], [86] for MO disks measured at 30 kHz. The calculated *S/N* at 20 MHz is $55 - 27 = 28$ dB. This value of *S/N* is quite adequate for an MO storage system.

There are other sources of noise besides the shot noise that bring the calculated value of $S/N = 38$ dB down to the experimental value of $S/N = 28$ dB. Some of these are laser noise, electronic noise, and media noise. One possible media noise source is the irregularity of the written domain. If the periphery of the domain is ragged instead of circular there will be some noise introduced into the system as well as bit shift. An examination of this potential problem [54] shows that some precaution must be taken. In order to obtain circular domains the MO material must have a large domain wall stiffness. The desirable material parameters are large domain wall energy density (σ), a small magnetization (M_s), and a small ratio of domain diameter to film thickness (d/h).

Material Stability

The magnetic films used in the MO recording systems are less than 0.1 μm in thickness and must be protected from oxygen and other destabilizing materials. The magnetic films are amorphous and must remain in that state for suc-

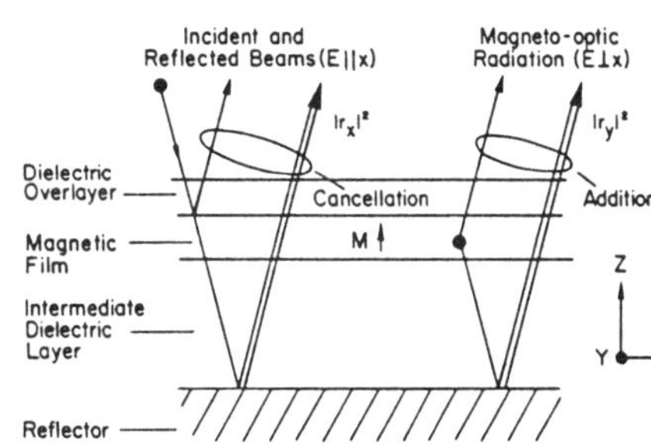

Fig. 7. A schematic representation of the quadrilayer configuration is shown. The incident beam is plane polarized with E parallel to x. The beam reflected with E parallel to x can then be considered to consist of two beams—one from the magnetic film, the other from a superposition of all beams having at least one bounce from the reflector. Similarly, the polar Kerr radiation, which has E parallel to x, can be considered to consist of the back-emitted radiation plus radiation having at least one bounce from the reflector. The trilayer parameters are adjusted for appropriate interference of these beams. (The difference between the behavior of r_x and r_y stems from the phase jump of r_x at the top surface of the sample layer which, of course, is absent for r_y.) (After M. Mansuripur and G. A. N. Connell.)

cessful operation of the MO system. Studies [71] of the crystallization of MO recording materials and similar composition materials in ribbon form show that it will take 100 years for the films to begin crystallization when exposed to a temperature of 100°C. The activation energy and the life time of the amorphous films were determined from time–temperature studies of the crystallization process. Since the crystallization process is well understood, these results are considered to be reliable.

The stability of MO recording materials against oxidation has been shown to be intimately related to the morphology of the films [62]. Smooth dense films prepared in low Ar pressure have shown much greater corrosion resistance than the films produced in high Ar pressure. The decrease in the magnitude of perpendicular anisotropy due to aging is much slower in denser films.

The determination of the chemical stability of the amorphous magnetic films over a long period of time is difficult. Accelerated life tests are usually performed at high temperatures in various gas atmospheres. The stability of the magnetic film is mainly determined by the packaging. It has been reported [6] that aluminum nitride films protect the magnetic films from changes in magnetic properties (H_c) much better than SiO_2 films. H_c of SiO_2 coated magnetic films of GdTbFe alloy exposed to 80°C in air change by more than 50 percent in 15 days. Aluminum-nitride-coated films show essentially no change in H_c under the same test conditions.

It is fortuitous that overcoating and undercoating of these magnetic films can enhance their performance (*S/N*) and at the same time provide protection against deterioration. Although these stability test results are not sufficient to predict that aging in MO films is not a problem, neither are they discouraging. Life tests [85], [86] on an undisclosed MO material were conducted at 50°C and 85-percent relative humidity on MO disks to determine the changes in *C/N*, bit error rate (BER), sensitivity, reflectivity, coercivity, Kerr angle, read signal amplitude, background noise level, chemical composition, and the mechanical parameters such as vertical and radial runout. In all cases, an increase in the BER can be detected before any other change in the media becomes apparent. Data obtained from these and other tests were used to predict a lifetime greater than 8 years.

The increase of "hard" and "soft" bit errors due to aging must also be determined. The use of the direct-read-after-write (DRAW) capability of MO systems could be used to determine the location of the "hard" bit errors due to aging and adjust for them. When the "hard" bit errors increase beyond an acceptable limit in a particular sector of the disk, the information can be written in a good sector.

Possible Future Developments

One of the most promising prospects for the improvement of the MO system is an increased data rate. Integrated arrays of diode laser have been fabricated [90], [91] that can be independently modulated. These arrays are so close together that they can be focussed within the field of view of a single lens. Each laser in the array writes and reads information in different tracks on the MO disk. The use of such arrays can increase the data rate to greater than 20 Mbytes/s.

A concept under development, for a high data rate, high-capacity optical disk buffer has been presented [92], for which the projected performance is a 24 disk capacity of 1.25×10^{11} bytes and a data rate of 200 Mbytes/s obtained by reading 12 disk surfaces in parallel. Each disk has an array of 9 laser diodes which are focussed through the same optical system. The center diode is used for tracking and the detection of track identification data that are contained in a permanent pilot track. The remaining eight diodes record four tracks of data on the "land" on each side of the pilot track. The data are recorded on tracks that are 1.4 μm apart. One module contains twelve 14-inch diameter double-sided disks.

An increase of the bit density in MO recording is possible even without any improvement (decrease) in the wavelength of diode lasers. The track spacing can probably be decreased to 1.1 μm [1] which will provide a 20-percent increase in bit density.

Conclusions

What are the prospects for MO recording technology? That question is best answered by comparing it to current Winchester technology and magnetic tape technology and by examining its prospects for manufacture and unique market niches. The salient issues are the following:

- MO technology offers a factor-of-ten increase in data storage density over the highest density Winchester drives, and a factor of 50 to 100 increase over the drives used in personal computers. MO drives should be smaller and less expensive storage systems per megabyte of stored information.
- Since the optical head is a few millimeters away from the disk surface, head crashes are eliminated and the disks are removable. Backup data can be stored on the removable disks by direct disk-to-disk copy.
- MO technology offers a greatly increased data rate over the Winchester drives, since several tracks can be written and read in parallel. This can be accomplished by the use of arrays of laser diodes that are focussed onto the disks through one optical system.
- MO technology is just emerging from the laboratory where it was evaluated for data storage, digital video recording, and digital audio recording. It is very encouraging that the first announced product is for data storage, where a very low bit error rate is required. Since this is a much more demanding application than digital audio or video, announcement of these products should be forthcoming. Since the MO drives can be made compatible with pressed digital audio disks (DAD), there should not be any impediment to their introduction into the DAD market.
- The greater access speed of the MO drive is certain to result in the replacement of tape systems with MO systems.
- The large storage density and short access time of the MO technology makes it the best candidate for a multipurpose recording system in the home, used to access and record library material, local store sales information, the morning newspaper, television programs, or digital audio from a satellite.

Acknowledgement

I am indebted to M. H. Kryder and H. D. Shieh for helpful discussions and a critical review of this paper. I am also indebted to other Professors and graduate students at Carnegie Mellon University as well as many other people for helpful suggestions. I wish to thank D. Shieh for Fig. 1 and B. Cain for Fig. 6.

References

[1] A. E. Bell, "Critical issues in high density magnetic and optical data storage," *Proc. SPIE—Opt. Data Storage*, vol. 382, pp. 2-15, Jan. 1983.
[2] M. H. Kryder, "Magneto-optic recording technology," *J. Appl. Phys.*, vol. 57, no. 1, pp. 3913-3918, Apr. 15, 1985.
[3] G. A. N. Connell, "Magneto-optics and amorphous metals: An optical storage revolution, *J. Magn. and Magn. Mat.*, vol. 54-57, pp. 1561-1566, 1986.
[4] Y. Aoki, T. Ihashi, N. Sato, and S. Miyaoka, "A magneto-optic recording system using TbFeCo," *IEEE Trans. Magn.*, vol. MAG-21, no. 5, pp. 1624-1628, Sept. 1985.
[5] A. E. Berkowitz and W. H. Meiklejohn, "Thermomagnetic recording: Physics and materials," *IEEE Trans. Magn.*, vol. MAG-11, no. 4, pp. 996-1017, July 1975.
[6] T. Deguchi *et al.*, "Digital magnetooptic disk drive," *Appl. Opt.*, vol. 23, no. 22, pp. 3972-3978, Nov. 1984.
[7] N. Imamura and C. Ota, "Experimental study on magneto-optical disk exerciser with the laser diode and amorphous magnetic thin films," *Jap. J. Appl. Phys.*, vol. 19, no. 12, pp. L731-L734, Dec. 1980.
[8] A. M. Earman, "Optical focus servo for optical disk mass data storage system application," *SPIE—Opt. Disk Technol.*, vol. 329, pp. 89-93, 1982.
[9] K. Yoshizumi, S. Goto, H. Inoue, and R. Imanaka, "Fast access actuator for optical disk memory," presented at the OSA Topical Meeting on Optical Data Storage, Washington, DC, paper no. ThAA5, Oct. 1985.
[10] S. Tsunashima, S. Masui, T. Kobayashi, and S. Uchiyama, "Magneto-optic Kerr effect on amorphous Gd-Fe-Co films," *J. Appl. Phys.*, vol. 53, no. 11, pp. 8175-8177, Nov. 1982.
[11] B. G. Huth, "Calculations of stable domain radii produced by thermomagnetic writing," *IBM J. Res. Devel.*, vol. 18, pp. 100-109, Mar. 1974.
[12] M. H. Kryder, W. H. Meiklejohn, and R. E. Skoda, "Stability of perpendicular domain in thermomagnetic recording materials," *Proc. SPIE—Opt. Storage Media*, vol. 420, pp. 236-241, June 1983.
[13] A. Takenouchi, E. Shichi, T. Kato, S. Tsunashima, S. Uchiyama, and N. Imamura, "Magneto-optical characteristics of double-layered amorphous RE-Fe films by RF-sputtering," *J. Appl. Phys.*, vol. 55, no. 6, pp. 2164-2166, Mar. 1984.
[14] S. Tsunashima, H. Fsuji, T. Kobayashi, and S. Uchiyama, "Thermomagnetic writing on charge-coupled amorphous rare-earth iron double-layer films," *IEEE Trans. Magn.*, vol. MAG-17, no. 6, pp. 2840-2842, Nov. 1981.
[15] Y. Yokoyama, S. Tsukahara, and T. Tanaka, "Thermomagnetic recording characteristics of amorphous TbFe/$(BiY)_3(FeGa)_5(O)_{12}$," *IEEE Trans. Magn.*, vol. MAG-20, no. 5, pp. 1039-1041, Nov. 1981.
[16] T. Katayama, M. Hirano, Y. Koizumi, and K. Kawanishi, "Different origin of the perpendicular anisotropy in amorphous Gd-Fe from Gd-Co films," *IEEE Trans. Magn.*, vol. MAG-13, no. 5, pp. 1603-1605, Sept. 1977.
[17] S. Tsunashima, H. Takagi, K. Kamegaki, T. Fuji, and S. Uchiyama, "Magnoelestic contribution to perpendicular anisotropy in amorphous GdCo and GdFe films," *IEEE Trans. Magn.*, vol. MAG-14, no. 5, pp. 844-846, Sept. 1978.
[18] H. Takagi, S. Tsunashima, and S. Uchiyama, "Stress induced anisotropy in amorphous GdFe sputtered films," *J. Appl. Phys.*, vol. 50, no. 3, pp. 1642-1644, Mar. 1979.
[19] S. H. Gill, M. Chen, and J. H. Judy, "Electron microscopy of in-situ ion beam sputtered amorphous GdFe alloy thin films," *J. Appl. Phys.*, vol. 49, no. 3, pp. 1741-1743, Mar. 1978.
[20] R. J. Gambino, J. Chaudhari, and J. J. Cuomo, "Amorphous magnetic materials," *AIP Conf. Proc.*, vol. 18, no. 18, pp. 578-592, 1973.
[21] Y. Nishihara, T. Katayama, Y. Yamaguchi, S. Ogawa, and T. Tsushima, "Anisotropic distribution of atomic pairs induced by the preferential resputtering effect in amorphous Gd-Fe and Gd-Co films," *Jap. J. Appl. Phys.*, vol. 17, no. 6, pp. 1083-1088, June 1978.
[22] S. Yasugi, S. Honda, M. Ohkoshi, and T. Kusuda, "Cross-sectional structures and depth profiles in bias sputtered GdCo," *J. Appl. Phys.*, vol. 52, no. 3, pp. 2298-2300, Mar. 1981.
[23] T. Kusuda, S. Honda, and M. Ohkoshi, "Perpendicular anisotropy of bias-sputtered GdCo film," *J. Appl. Phys.*, vol. 53, no. 3, pp. 2338-2340, Mar. 1982.
[24] C. T. Burilla, W. R. Bekebrede, M. Kestigian, and A. B. Smith, "Influence of inert gas incorporation on uniaxial anisotropy of sputtered GdCo films," *J. Appl. Phys.*, vol. 49, no. 3, pp. 1750-1752, Mar. 1978.
[25] S. Esho and S. Fujiwara, "Growth induced anisotropy in sputtered GdCo films," *AIP Conf. Proc.*, vol. 34, no. 34, pp. 331-333, 1976.
[26] F. E. Luborsky, "Kinetics for changes in anisotropy, coercivity, and argon content of transition metal-rare earth films," *J. Appl. Phys.*, vol. 57, no. 1, pp. 3592-3594, Apr. 1985.
[27] D. D. Bacon, M. Hong, E. M. Gyorgy, P. K. Gallagher, S. Nakahara, and L. C. Feldman, "Magnetic anisotropy in dc-diode getter sputtered GdCo films—How important is the argon content in the films?" *Appl. Phys. Lett.*, vol. 48, no. 11, pp. 730-732, Mar. 1986.
[28] H. Hoffman, A. J. Owen, and F. Schrope, "Electron microscopy of evaporated and sputtered Gd/Co and Ho/Co films," *Phys. Status Solidi (A)*, vol. 52, pp. 161-174, 1979.
[29] T. Numata and Y. Sakurai, "Amorphous magnetic films for magneto-optical memory," *JARECT, Recent Magnetics for Electronics*, vol. 15, pp. 91-108, 1984.
[30] H. Heitmann, M. Hartmann, M. Rosenkrantz, and H. J. Tolle, "Amorphous rare earth-transition metal films for magneto-optical storage," *Jour. Phys.*, vol. 46, no. C6, pp. 9-18, Sept. 9, 1985.
[31] Y. Mimura and N. Imamura, "Magnetic properties of amorphous TbFe thin films prepared by RF sputtering," *Appl. Phys. Lett.*, vol. 28, no. 12, pp. 746-748, June 1976.
[32] T. Chen, D. Cheng, and G. B Charlan, "An investigation of amorphous TbFe thin films for magneto-optic memory application," *IEEE Trans. Magn.*, vol. MAG-16, no. 5, pp. 1194-1196, Sept. 1980.
[33] M. Mansuripur and G. A. N. Connell, "Signal and noise in magneto-optical readout," *J. Appl. Phys.*, vol. 53, no. 6, pp. 4485-4494, June 1982.
[34] S. Honda, J. Hirokane, M. Ohkoshi, and T. Kusada, "Static and dynamic properties of thermomagnetic writing in amorphous TbFe films," presented at the ICMF-10, Yokohama, Japan, paper no. 13a5, Sept. 1982.
[35] P. Chaudhari, J. J. Cuomo, and R. J. Gambino, "Amorphous metallic films for magneto-optic applications," *Appl. Phys. Lett.*, vol. 22, no. 7, pp. 337-339, Apr. 1973.
[36] B. R. Brown, "Optical data storage potential of six materials," *Appl. Opt.*, vol. 13, no. 4, pp. 761-766, Apr. 1974.
[37] Y. Togami, "Stability of small bits written in amorphous GdCo thin films," *Appl. Phys. Lett.*, vol. 32, no. 10, pp. 673-675, May 1978.
[38] Y. Togami, "Magneto-optic storage," *IEEE Trans. Magn.*, vol. MAG-18, no. 6, pp. 1233-1237, Dec. 1982.
[39] R. Hasegawa, "Static bubble domain properties of amorphous GdCo films," *J. Appl. Phys.*, vol. 45, no. 7, pp. 3109-3112, July 1974.
[40] A. G. Dirks and H. J. Leamy, "Microstructure and magnetism in amorphous rare earth-transition metal thin films," *J. Appl. Phys.*, vol. 49, no. 3, pp. 1735-1737, Mar. 1978.
[41] Y. Mimura, N. Imamura, and T. Kobayashi, "Thermomagnetic writing on GdFe and GdFeY amorphous film," *Jap. J. Appl. Phys.*, vol. 17, no. 8, pp. 1365-1369, Aug. 1978.
[42] M. Hartmann, J. Bratt, and B. Jacobs, "Erasable magneto-optic recording media," *IEEE Trans. Magn.*, vol. MAG-20, no. 5, pp. 1013-1018, Sept. 1984.
[43] N. Sato, Y. Aoki, and S. Miyaoka, "Magneto-optical recording on TbFe based thin films," *IEEE Trans. Magn.*, vol. MAG-20, no. 5, p. 1022, Sept. 1984.
[44] Y. Mimura, N. Imamura, and T. Kobayashi, "Curie point writing in amorphous magnetic films," *Jap. J. Appl. Phys.*, vol. 15, no. 5, pp. 933-944, 1976.
[45] Y. Mimura and N. Imamura, "Magnetic properties of amorphous DyFe alloy films," *Jap. J. Appl. Phys.*, vol. 15, no. 5, pp. 937-938, 1976.
[46] A. Sarkis, "Magnetic anisotropy of Ho_2Co_{17} and Ho_2Fe_{17}," *J. Appl. Phys.*, vol. 53, no. 3, pp. 2365-2366, Mar. 1982.
[47] T. Suzuki, H. Ichinose, and E. Aoyagi, "Microstructure and magnetic properties of sputter-deposited HoCo alloy thin

films with perpendicular magnetic anisotropy," *Jap. J. Appl. Phys.*, vol. 23, no. 5, pp. 585-592, May 1984.

[48] Y. Sakurai, "Preparation and properties of RE-TM amorphous films," presented at the ACMFS-10, Yokohama, Japan, paper 13al-1, Sept. 1982.

[49] J. Kwo, E. M. Gyorgy, D. B. McWhan, M. Hong, F. J. DiSalvo, C. Vettier, and J. E. Bower, "Magnetic and structural properties of single crystal rare earth Gd/Y superlattices," *Phys. Rev. Lett.*, vol. 55, p. 1402, 1985.

[50] P. Dekker, "Manganese bismuth and other magnetic materials for beam addressable memories," *IEEE Trans. Mag.*, vol. MAG-12, no. 4, pp. 311-327, July 1976.

[51] S. Honda, M. Ohkoshi, and T. Kusuda, "Dependence of compositional variation on sputtering condition in GdCo and GdFe amorphous films," in *Ferrites: Proc. Int. Conf.*, Japan, Sept. 1980.

[52] H. Shieh and M. H. Kryder, "Direct observation of thermomagnetic writing," presented at the OSA Topical Meeting on Optical Data Storage, Monterey, CA, paper no. Thc-c3, Apr. 1984.

[53] Y. Togami, A. Yoshihara, T. Tamaki, and K. Sato, "Magnetic properties of amorphous GdTbCo film," *J. Mag. and Mag. Mat.*, vol. 31-34, pp. 1497-1498, 1983.

[54] H. P. Shieh and M. H. Kryder, "Reproducibility of thermomagnetically written domains," presented at the Topical Meeting on Optical Data storage, Washington, DC, Optical Society of America, paper no. WAA4, Oct. 1985.

[55] Y. Togami and R. Sato, "Magneto-optical Kerr effect in amorphous RE-CoFe films," *Jap. J. Appl. Phys.*, vol. 24, no. 1, pp. 106-107, 1985.

[56] S. Masui, T. Kobayashi, S. Tsunashima, and S. Uchiyama, "Magnetic and magneto-optic properties of amorphous GdFeCo films," *IEEE Trans. Magn.*, vol. MAG-20, no. 5, pp. 1036-1038, Sept. 1984.

[57] N. Sato, Y. Aoki, and S. Miyaoka, "Magneto-optical recording on TbFe based thin films," *IEEE Trans. Magn.*, vol. MAG-20, no. 5, pp. 1022-1024, Sept. 1984.

[58] N. Imamura, S. Tanaka, F. Tanaka, and Y. Nagao, "Magneto-optical recording on amorphous films," *IEEE Trans. Magn.*, vol. MAG-21, no. 5, pp. 1607-1612, Sept. 1985.

[59] H. Tsujimoto, K. Saiki, S. Inokuchi, and Y. Sakurai, "Effect of bias voltage on Kerr rotation angle in amorphous TbFeCo films prepared by RF-sputtering," *IEEE Trans. Magn.*, vol. MAG-20, no. 5, pp. 1027-1029, Sept. 1984.

[60] S. Takayama, T. Niihara, T. Kaneko, and Y. Sugita, "Magnetic and magneto-optical properties of TbFeCo amorphous film," presented at ICMFS-11, Asilomar, CA, paper no. AB-17, Sept. 1985.

[61] M. Hong, E. M. Gyorgy, R. B. van Dover, S. Nakahara, D. D. Bacon, and P. K. Gallagher, "DC magnetron and diode sputtered polycrystalline Fe and amorphous Tc(FeCo) films morphology and magnetic properties," *J. Appl. Phys.*, vol. 59, no. 2, pp. 551-556, Jan. 2, 1986.

[62] M. Hong, D. D. Bacon, R. B. van Dover, E. M. Gyorgy, J. F. Dillon, and S. D. Albiston, "Aging effects on amorphous Tb-transition-metal films prepared by diode and magnetron sputtering," *J. Appl. Phys.*, vol. 57, no. 15, pp. 3900-3902, Apr. 1985.

[63] F. Tanaka, Y. Nagao, and N. Imamura, "Dynamic read/write characteristics of magneto-optical TbFeCo and DyFeCo disk," *IEEE Trans. Magn.*, vol. MAG-20, no. 5, pp. 1033-1035, Sept. 1984.

[64] T. C. Anthony, J. Brug, S. Naberhuis, and H. Birecki, "Thermal stability of magneto-optic quadrilayers," *J. Appl. Phys.*, vol. 59, no. 1, pp. 213-217, Jan. 1986.

[65] H. Tsujimoto, Y. Sakurai, A. Saito, and S. Matsushita, "Magnetic properties and magnetic Kerr rotation of amorphous TbFeCo and TbFeCr films," *IEEE Trans. Magn.*, vol. MAG-19, no. 5, pp. 1757-1759, Sept. 1983.

[66] M. Hartman and T. R. McGuire, "Relationship between Faraday rotation and Hall effect in amorphous rare-earth-transition-metal alloys," *Phys. Rev. Lett.*, vol. 51, no. 13, pp. 1194-1197, Sept. 1983.

[67] M. Urner-Wille, P. Hansen, and K. Witter, "Magnetic, magneto-optic and switching properties of amorphous GdFeSn alloys," *IEEE Trans. Magn.*, vol. MAG-16, no. 5, pp. 1188-1193, Sept. 1980.

[68] R. C. Taylor and A. Gangulee, "Magnetic properties of amorphous GdFeB and GdCoB alloys," *J. Appl. Phys.*, vol. 53, no. 3, pp. 2341-2343, Mar. 1982.

[69] Y. Kudo, A. Muto, N. Hara, M. Birukawa, and H. Kawabata, "Preparation and properties of MnCuBi, GdTbFe, and GdTbFeGe films for magneto-optical memories," *Nat. Tech. Rep. (Jap.)*, vol. 28, no. 6, pp. 33-42, Dec. 1982.

[70] H. J. Leamy and A. G. Dirks, "Microstructure and magnetism in amorphous rare-earth-transition-metal thin films. II Magnetic anisotropy,"' *J. Appl. Phys.*, vol. 50, no. 4, pp. 2871-2882, Apr. 1979.

[71] F. E. Luborsky, "Crystallization kinetics of amorphous Co-Gd ribbons and films," *J. Non-Cryst. Solids*, vol. 62-62, p. 829, 1984.

[72] R. B. van Dover, M. Hong, E. M. Gyorgy, J. F. Dillon, and S. D. Albiston, "Intrinsic anisotropy of TbFe films prepared by magnetron co-sputtering," *J. Appl. Phys.*, vol. 57, no. 1, pp. 3897-3899, Apr. 1985.

[73] D. C. Cronemeyer, "Perpendicular anisotropy in GdCo amorphous films prepared by RF sputtering," *AIP Conf. Proc.*, vol. 18, no. 18, pp. 85-89, 1973.

[74] C. N. J. Cargill and T. Mizoguchi, "Dipolar mechanisms for magnetic anisotropy in amorphous ferrimagnetic alloys," *J. Appl. Phys.*, vol. 49, no. 3, pp. 1753-1755, Mar. 1978.

[75] C. N. J. Cargill, "Ferromagnetism in amorphous solids," *AIP Conf. Proc.*, vol. 24, no. 24, pp. 138-144, 1975.

[76] C. N. J. Wagner, in *Rapidly Quenched Metals*, N. J. Grant and B. C. Giessen, Eds., 1966.

[77] W. H. Meiklejohn, "Exchange anisotropy—A review," *J. Appl. Phys. Suppl.*, vol. 33, no. 3, pp. 1328-1335, Mar. 1962.

[78] F. Hellman, R. B. von Dover, and E. M. Gyorgy, "Anomalous unidirectional anisotropy in Tb-Fe/permalloy bilayer films," *Bull. Amer. Phys. Soc.*, no. 31, 1986.

[79] F. E. Luborsky, J. T. Furey, R. E. Skoda, and B. C. Wagner, "Relation between some process parameters and properties of transition metal-rare earth films for magneto-optic recording," *Proc. SPIE—Opt. Mass Data Storage*, vol. 529, pp. 55-59, Jan. 1985.

[80] S. Nakahara, M. Hong, R. B. van Dover, E. M. Gyorgy, and D. D. Bacon, "Microstructures of thin sputtered amorphous TbFe and polycrystalline Fe films," *J. Vac. Soc. A Tech 1986*, to be published.

[81] A. Brunsch and J. Schneiderer, "Perpendicular magnetic anisotropy in evaporated amorphous GdCo films," *J. Appl. Phys.*, vol. 48, no. 6, pp. 2641-2643, June 1977.

[82] J. M. Nieuwenhuizen and H. B. Haanstra, "Microfractography of thin films," *Philips Tech. Rev.*, vol. 27, no. 3/4, pp. 87-91, 1966.

[83] A. A. Thiele, "The theory of cylindrical magnetic domains," *Bell Syst. Tech. J.*, vol. 48, no. 10, pp. 3287-3335, Dec. 1969.

[84] W. F. Druyvesteyn, D. L. A. Tjaden, and J. Dorleijn, "Calculation of the stray field of a magnetic bubble, with application to some bubble problems," *Philips Res. Rep.*, vol. 27, pp. 7-27, 1972.

[85] S. C. Webster, R. N. Gardner, R. P. Freese, T. A. Rinehart, and J. C. Ellis, "High performance magneto-optic recording materials," presented at INTERMAG 1985, session DC.

[86] R. P. Freese and T. Takahashi, "High performance magneto-optic recording media," *Jap. J. Mag.*, vol. MAG-85-88, pp. 81-88, 1985.

[87] P. G. van Engen and K. H. J. Buschow, "Magnetic properties and magneto-optical spectroscopy of Heusler alloys based on transition of metals and Sn," *J. Mag. and Mag. Mat.*, vol. 30, no. 30, pp. 374-382, 1983.

[88] R. A. de Groot, F. M. Mueller, P. G. van Engen, and K. H. J. Buschow, "Half-metallic ferromagnets and their magneto-optical properties," *J. Appl. Phys.*, vol. 55, no. 6, pp. 2151-2154, Mar. 1984.

[89] R. Oyama, J. Abe, and K. Matsubara, "Preparation of PtMnSb thin films and magneto-optical properties," *IEEE Trans. J. Magn. In Japan*, vol. TJMJ-1, no. 1, p. 122, Apr. 1985.

[90] J. P. Van der Ziel, R. A. Logan, and R. M. Mikulyak, "A closely spaced (50 micrometer) array of 16 individually addressable buried heterostructure GaAs lasers," *Appl. Phys. Lett.*, vol. 41, no. 1, pp. 9-11, July 1982.

[91] D. Botez, J. C. Connolly, D. B. Gilbert, M. G. Harvey, and M. Ettenberg, "High-power individually addressable buried heterostructure GaAs lasers," *Appl. Phys. Lett.*, vol. 41, no. 11, pp. 1040-1042, Dec. 1982.

[92] M. L. Levene, "A high data rate, high capacity optical disk buffer," in *Proc. 7th Symp. on Mass Storage Systems* (Nov. 4-7, 1985), pp. 17-21.

Reprinted with permission from *IEEE Transactions on Magnetics,* Vol. MAG-23(5), pp. 2263-2265 (September 1987).

SUBMICRON DOMAINS FOR HIGH DENSITY MAGNETO-OPTIC DATA STORAGE

D. Rugar, C.-J. Lin, and R. Geiss
IBM Almaden Research Center
650 Harry Rd.
San Jose, CA 95120 USA

ABSTRACT - This paper discusses the potential of using submicron magneto-optic domains for high density data storage applications. In TbFe, domains as small as 200 nm were controllably written and domains down to 70 nm diameter were observed to be magnetically stable. In order to demonstrate the use of submicron domains for data storage, a disk tester operating at 488 nm wavelength with 0.5 μm spot size was constructed. Using a TbFeCo disk, an areal density of 1.8×10^8 transitions/cm^2 was demonstrated with 44 dB carrier-to-noise ratio (CNR). Reasons for decreased CNR at short wavelengths are discussed.

I. INTRODUCTION

Magneto-optic recording has received considerable attention lately as a means for high density data storage. The ultimate areal bit density that can be achieved by this technique depends, in part, on the ability of the media to support small, well defined domain structures. Most magneto-optic data storage experiments reported to date have concentrated on the recording performance of approximately 1 μm size domains thermomagnetically written by diode lasers with focused spot sizes in the 0.8 - 1 μm range[1]. The purpose of this paper is to investigate the potential of MO recording at significantly higher areal bit densities. First, in section II of this paper, we demonstrate the existence and magnetic stability of domains with diameters in the 70-300 nm range. These domains are thermomagnetically written on coupons of TbFe. In section III, we present high areal density recording results from a magneto-optic disk tester operating at 488 nm wavelength with an optical spot size of 0.5 μm. Finally, in section IV, we discuss how signal-to-noise ratio scales with wavelength for the shot noise limited case.

II. SUBMICRON DOMAINS IN TbFe

TbFe and related alloys, such as TbFeCo, are good materials for high density data storage because of their large perpendicular anisotropy and adequate Kerr rotation. With proper choice of composition the coercivity can rise steeply below the Curie temperature, which allows the stable formation of small domains. In this section we present small domain writing results for two alloys: $Tb_{19}Fe_{81}$ and $Tb_{32}Fe_{68}$. Preparation methods for these films have been described previously[2]. Briefly, the films were sputter deposited on carbon coated mica substrates and then laser written using a static coupon tester. The films were floated off the substrates and subsequently imaged by Lorentz-mode transmission electron microscopy[2]. Some magnetic properties of the two films are given in Table I.

TABLE I

Properties of Tb_xFe_{1-x}

x	T_{curie} (K)	M (300K) (emu/cm^3)	H_c (300K) (kOe)
.19	397	109	5.1
.32	405	250	1.4

A. Domain Stability

Stability of small cylindrical domains in a perpendicular magnetic medium has been considered previously by several authors[3,4,5]. Briefly, the domain wall feels a collapsing force arising from the domain wall energy, an expanding force due to the magnetostatic energy and a "frictional" force due to the coercivity.

Consider a thin magnetic film of thickness h, coercivity H_c, magnetization M and wall energy density σ. If there is no applied field and if magnetostatic energy can be ignored, then the minimum stable domain diameter is given by[4,5]

$$d_o = \frac{\sigma}{MH_c}. \tag{1}$$

This formula is valid when $d_o >> h$, which is often the case for low anisotropy films such as GdCo[6] or GdFe[7].

If d_o is comparable to the film thickness, then the magnetostatic energy should be taken into account since there can be significant demagnetizing field acting on the wall. The demagnetizing field helps to stabilize the domain and allows domains to be smaller than predicted by equation (1). To analyze this case, one can use the graphical method of Bobeck[3]. Alternatively, Callen and Josephs[8] found that the effective driving field H_{eff} acting on the domain wall could be accurately approximated by[8,9]

$$H_{eff} = \frac{\sigma}{Md} - \frac{4\pi Mh}{h+3d/4}. \tag{2}$$

Since domains will be stable if $|H_{eff}| < H_c$, then the minimum stable domain diameter, d_o, is found by solving the equation $H_{eff} = H_c$. The solution is

$$d_o = A + \left(A^2 + \frac{4\sigma h}{3MH_c}\right)^{1/2}, \tag{3}$$

where

$$A = \frac{\sigma}{2MH_c} - \frac{2h}{3} - \frac{8\pi Mh}{3H_c}. \tag{4}$$

In the limit $h \rightarrow 0$, the formula in (3) reduces to equation (1), as it should.

Using the room temperature magnetic properties of our films from Table I and an estimated value of 3 erg/cm^2 for σ, we find from (3) that d_o = 47 and 35 nm for $Tb_{19}Fe_{81}$ and $Tb_{32}Fe_{68}$, respectively. In the case of $Tb_{32}Fe_{68}$, the effect of magnetostatic energy is significant. If magnetostatic energy is ignored and equation (1) is used instead of (3), then d_o = 82 nm is obtained.

B. Generating Small Domains

The above calculation shows that, at least theoretically, domains smaller than 100 nm should be stable if a means can be found for writing them. We have used three techniques to thermally write small domains, even though the laser diameter (FWHM) was around 0.9 μm.

The first method is to write with an applied bias field which is insufficient to form a saturated mark. For certain compositions, the resulting written mark will consist of a collection of microdomains. Fig. 1 shows the result of writing $Tb_{32}Fe_{68}$ with a bias field of -100 Oe, where the minus sign indicates that the field was in the direction which opposes writing. Domains as small as 70 nm are visible in the figure. The domains are apparently quite robust in their magnetic stability since the domains were subjected to a -400 Oe field subsequent to writing. This field would act to collapse marginally stable domains.

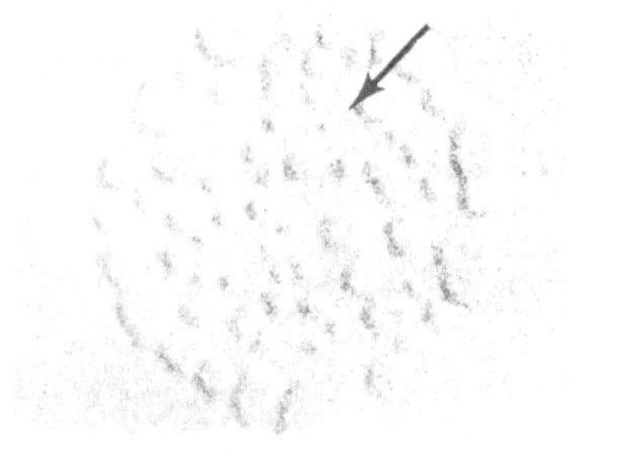

Fig. 1. A laser written mark on $Tb_{32}Fe_{68}$ consisting of microdomains. Mark diameter is approximately 2.2 μm. The arrow points to a domain approximately 70 nm in diameter.

Although the above method was successful in generating domains smaller than 100 nm, it fails to make single domain marks. In lieu of using a smaller focused laser beam, small isolated marks can be written by reducing the laser power close to the writing threshold. By doing this, only a small area in the center of the focused beam reaches a temperature high enough to write. Figure 2 shows the results of a near-threshold writing experiment performed on $Tb_{19}Fe_{81}$. The writing power at the left of the figure was approximately 6 mW and was successively reduced by 0.1 mW for each mark to the right. The write pulse length was 20 nsec and the bias field was 500 Oe, which guaranteed that the written marks were saturated (single domain). The smallest domain is at the right of the figure and is approximately 200 nm in diameter.

The final method used to generate small domain structures requires two laser pulses. The first laser pulse is fired while the bias field is in the writing direction and, as a result, a domain approximately 1.5 μm in diameter is formed. The sample is then translated slightly and the laser is pulsed a second time, but with the bias field switched to the erasing direction[10]. The result is that all but a thin sliver of the original domain is erased. Figure 3 shows an example of the crescent shaped domains which can be written using this method. A domain narrower than 300 nm can be observed in the figure.

Fig. 2. Laser written domains in $Tb_{19}Fe_{81}$. The smallest domain is approximately 200 nm in diameter. Spacing is 2 μm.

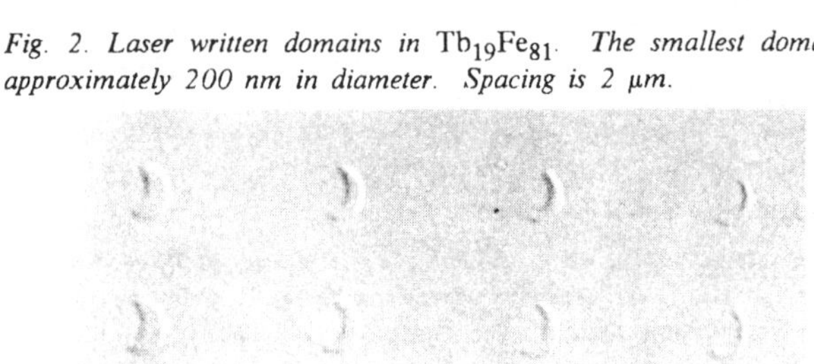

Fig. 3. Crescent shaped domains obtained by partial erasure of circular domains. Width of narrowest crescent is less than 300 nm. Domain spacing is 6 μm.

III. DISK RECORDING WITH SUBMICRON DOMAINS

The examples of small domains presented in section II show that magneto-optic media are capable of supporting bit densities significantly greater than currently demonstrated in disk systems. To fully exploit the storage potential of submicron domains, the focused laser spot diameter should be as small as possible. This is essential for good readback performance and helps the writing performance as well. To achieve small focused spot size, some combination of shorter wavelengths and higher numerical aperture can be used. We have constructed a dynamic disk tester which uses an argon laser operating at 488 nm wavelength and a 0.6 numerical aperture (NA) objective lens. The resulting focused spot diameter is approximately 0.5 μm FWHM.

The system is shown schematically in Fig. 4. A fast acousto-optic modulator controls the intensity of the incident beam. A spatial filter ensures good wavefront quality. The focusing lens is a Leitz 32x, 0.6 NA microscope objective and is attached to a piezoelectric positioner for focus control. Knife edge detection is used to derive a focus error signal which is used to control the lens position in a servo loop. Since no track servoing was done, an air bearing spindle was used to minimize random radial runout. The disk used in these experiments was unpatterned and consisted of a TbFeCo active layer with a single layer anti-reflection coating. The disk was recorded with laser light incident on the film side of the disk.

Figure 5 shows a spectrum analyzer trace from a 5 MHz carrier written at a linear velocity of 10 m/s. The written marks, approximately 1.0 μm in diameter with 2.0 μm period, were intentionally written "large" so that the maximum carrier-to-noise (CNR) was obtained. As can be seen in the figure, the CNR was approximately 50 dB in a 30 kHz bandwidth.

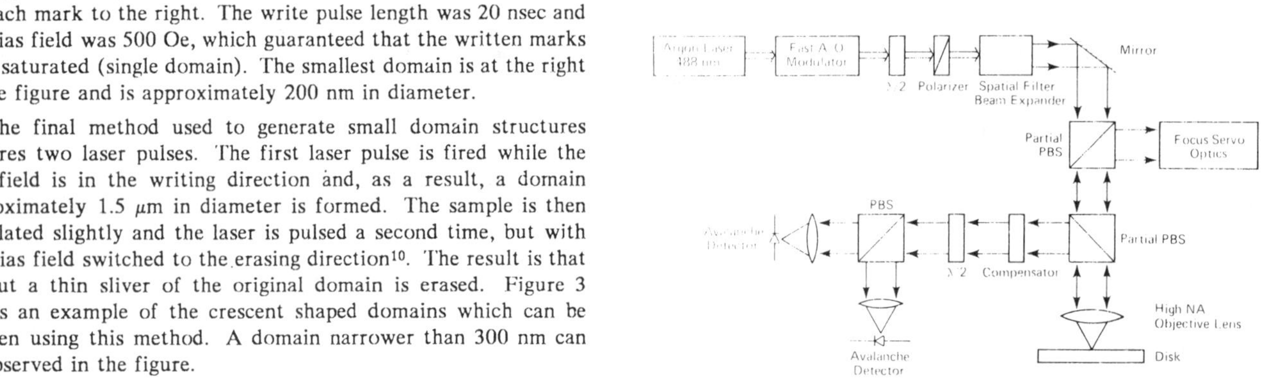

Fig. 4. Block diagram of disk tester operating at 488 nm. The readback signal is taken from the difference of the two avalanche detectors.

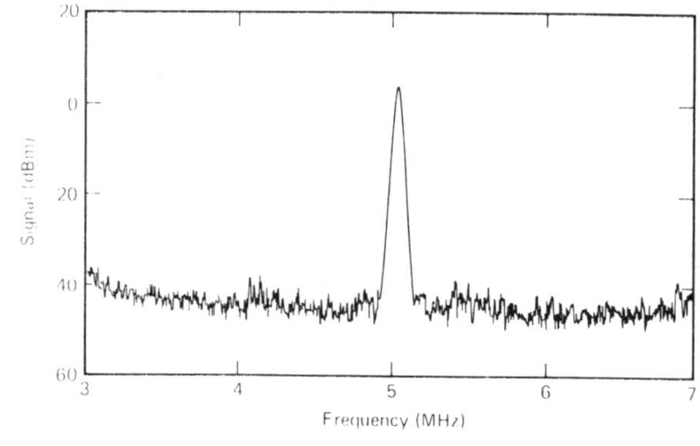

Fig. 5. Spectrum analyzer trace of carrier written with 1.0 μm marks and 2 μm period.

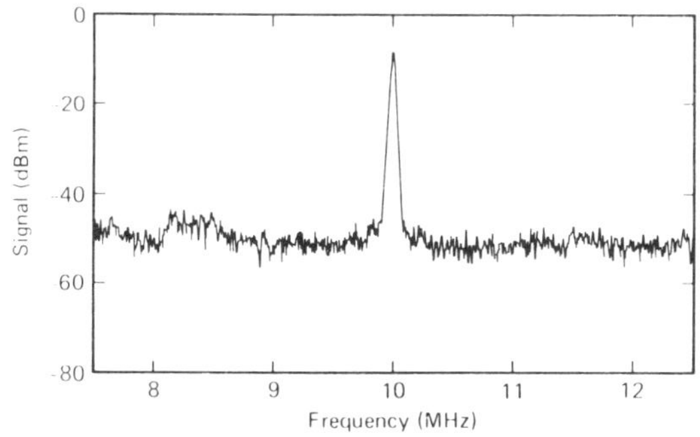

Fig. 6. Spectrum analyzer trace of carrier written with 0.5 μm marks and 1 μm period.

Higher density recording performance is shown in Fig. 6. This spectrum is from a 10 MHz carrier written with 0.5 μm marks and 1.0 μm period. A CNR of approximately 44 dB is observed. The writing pulse width was short, approximately 15 nsec, in order to prevent elongation of the marks due to disk motion and to minimize widening of the thermal spot profile due to thermal diffusion. Tracks could be written as close together as 1.1 μm with crosstalk of approximately -35 dB.

Using the parameters associated with Fig. 6 (1.1 μm track pitch and 1.0 μm mark period) and noting that each written mark results in two transitions, the resulting areal density is 1.8×10^8 mark transitions per cm^2 or 1.2×10^9 mark transitions per square inch. We use "mark transitions" as the unit of measure rather than "bits" in order to be independent of the encoding scheme. (Encoded bit density in most pulse width modulation coding schemes would be somewhat higher than the mark transition density.) Although results at comparable densities have been reported for write-once optical recording[11], most reported magneto-optic recording results have had areal densities less than half that reported here.

IV. SCALING OF CNR WITH WAVELENGTH

As was shown in Fig. 5, the CNR obtained from a carrier with large marks was approximately 50 dB. This is about 10 dB lower than would be expected from an analogous recording made on an analogous disk at 830 nm wavelength. To understand the reason for the poorer performance at 488 nm, consider how the shot noise limited CNR should scale with wavelength.

Following the analysis of Dewey,[12] the shot noise limited voltage CNR for big marks (i.e., negligible optical transfer function roll-off) is given by

$$\mathrm{CNR} = \left(\frac{S\,P\,R\,L}{e\,\Delta f} \right)^{1/2} \Theta_k, \tag{5}$$

where S is detector responsivity (A/W), P is the optical readback power at the disk, R is disk reflectivity, Θ_k is Kerr rotation angle, e is electronic charge and Δf is the detection bandwidth. L is a factor accounting for optical losses. Detector responsivity is related to wavelength λ according to $S = (e/hc)\lambda\eta$, where η is the detector quantum efficiency, h is Planck's constant, and c is speed of light. Thus (5) becomes

$$\mathrm{CNR} = \left(\frac{\lambda\,\eta\,P\,R\,L}{h\,c\,\Delta f} \right)^{1/2} \Theta_k. \tag{6}$$

The effects of various terms in (6) are summarized in Table II for 830 and 488 nm. The values used for R and Θ_k are for TbFe.[13,14] TbFeCo data is expected to be similar.[14] The values used for η are typical for the avalanche detectors used in the present experiment.[15] Table II shows that even if the readback power is the same at the two wavelengths, CNR loss of 9.2 dB is

TABLE II

CNR loss at 488nm compared to 830nm due to various parameters in equation (6).

Parameter	Value @ 830nm	Value @ 488nm	CNR Loss
λ	830nm	488nm	2.3 dB
η	0.77	0.38	3.1 dB
R	0.61	0.53	0.6 dB
Θ_k	0.45°	0.31°	3.2 dB
Total			9.2 dB

expected. This is consistent with the reduced CNR observed experimentally.

V. CONCLUSION

We have presented experimental evidence for the stability of submicron domains in TbFe. Domains as small as 200 nm have been controllably written and domains down to 70 nm have been observed to be magnetically stable. If such small domains could be used in a recording system, more than an order of magnitude increase in areal density would result.

As a first step toward harnessing submicron domains for magneto-optic recording, a disk tester operating at 488 nm was constructed. An areal density of 1.8×10^8 transitions/cm^2 was demonstrated on a TbFeCo disk with a CNR of 44 dB. This density is a factor of 2-3 higher than is usually associated with magneto-optic recording.

ACKNOWLEDGEMENT

The authors are grateful to A. Bell, L. Bynum, A. Dewey, V. Jipson, C. Stratton and J. Suits for their contributions to this work.

REFERENCES

1. Many possible references exist. See for example: H. D. Shieh and M. H. Kryder, J. Appl. Phys. (1986); M. Takahashi, Y. Sugita, M. Yoshihiro and K. Shigematsu, IEEE Trans. Magn. MAG-22, 931 (1986); F. Tanake, Y. Nagao and N. Imamura, IEEE Trans. Magn. MAG-20, 1033 (1984).
2. J. C. Suits, R. H. Geiss, C. J. Lin, D. Rugar and A. E. Bell, Appl. Phys. Lett. 49, 419 (1986).
3. A. H. Bobeck, Bell System Tech. J. 46, 1901 (1967).
4. A. A. Thiele, J. Appl. Phys. 41, 1139 (1970).
5. W. D. Doyle, G. K. Goldberg and W. E. Flannery, IEEE Trans. Magn. MAG-6, 548 (1970); M. H. Kryder, W. H. Meiklejohn and R. E. Skoda, Proc. SPIE 420, 236 (1983).
6. Y. Togami, Appl. Phys. Lett. 32, 673 (1978).
7. Y. Mimura, N. Imamura and T. Kobayashi, Jap. J. Appl. Phys. 17, 1365 (1978).
8. H. Callen and R. M. Josephs, J. Appl. Phys. 42, 1977 (1971).
9. S. Honda, K. Ueda and T. Kusuda, J. Appl. Phys. 52, 2295 (1981).
10. S. Araki, A. Asayama and H. Kobayashi, IEEE Transl. J. Magn. Jpn. TJMT-1, 691 (1985).
11. G. J. Ammon, R. F. Kenville, A. A. Litwak, M. S. Nigro and C. W. Reno, Proc. SPIE 392, 33 (1982).
12. A. Dewey, Proc. SPIE 695, 72 (1986).
13. G. A. N. Connell, Appl. Optics 22, 3155 (1983).
14. G. Sprokel, personal communication.
15. RCA Model C30950 Avalanche detector module

Reprinted with permission from *Journal of Applied Physics*, Vol. 64(1), pp. 252-261 (July 1, 1988).

Thermomagnetic writing in Tb-Fe: Modeling and comparison with experiment

J. C. Suits, D. Rugar, and C. J. Lin
IBM Research, Almaden Research Center, 650 Harry Road, San Jose, California 95120-6099

Computer model calculations and experimental comparison are presented for thermomagnetic writing in TbFe media. Two models are discussed which are appropriate for two observed types of writing behavior. A "bubble" type writing model is found to give good experimental agreement when wall motion processes dominate and the laser-written marks exhibit single-domain behavior. This model accurately accounts for the writing behavior of $Tb_{0.23}Fe_{0.77}$. The gradient of the domain-wall energy is found to be an important factor in the writing process and can cause nucleation to be delayed until some cooling has occurred. This results in written domain sizes which are a strong function of the applied bias field. For compositions which exhibit multiple domain writing behavior, a "nucleation field" model is found to be appropriate. This model predicts the observed weak dependence of mark size on applied bias field. In addition, the nucleation field model is used to calculate the bias fields which correspond to the onset of writing (onset field) and single-domain mark formation (saturation field).

I. INTRODUCTION

Optical storage of information by thermomagnetic recording is of increasing interest today because of its potential for higher bit density and increased head-disk spacing compared to conventional magnetic recording.[1] The thermomagnetic writing process is critical to this new recording technology since it is this process that determines written domain size and therefore ultimate density. The magneto-optical storage media of greatest current interest today is the family of Tb-Fe type alloys such as Tb-Fe-Co.

Computer modeling of the thémomagnetic writing process has been considered by a number of authors.[2–8] The first modeling of a rare-earth–transition–metal alloy, GdCo, was done by Huth[3] who used an approach similar to that used for magnetic bubbles modified by the inclusion of both a laser heating source and temperature-dependent parameters. Although Huth showed two examples of domain size as calculated by this model, he did not make a comparison with experimentally measured domain sizes.

Modeling by Mansuripur and Connell[4] considered the energetics of circular domains during writing and took wall width into account. They showed a sample calculation for a typical TbFe film, but no comparisons with experiment were given. In their modeling calculations, Takahashi *et al.*[5] do show a comparison between experimental and calculated domain sizes for two compositions of TbFeCo. However, we were not able to extend this method of calculation to compute domain sizes over a wide range of experimental conditions. Another model, based on an array of small elements, has also been applied to thermomagnetic writing in TbFe, but no detailed comparison of domain diameter with experimental results has been shown.[6,7]

In this paper, we present computer model calculations with experimental comparison for thermomagnetic writing in TbFe media. The experiments explore writing behavior for a variety of compositions, laser pulse lengths, and applied bias fields. Two theoretical models are used to explain the results. The input parameters to the calculations include laser parameters, as well as thermal and magnetic properties of the materials to the extent that they are known. Once these inputs are determined, no parameter adjustment (in most cases) is used to obtain reasonable agreement between experiment and the model calculations. We present comparison with experiment for mark size, onset field (the bias field at which writing first occurs), and saturation field (the field required to ensure single-domain marks).

II. EXPERIMENT

Thin films of 500-Å-thick Tb_xFe_{1-x} were deposited on glass or carbon-coated mica coupons by dual-source dc magnetron sputtering. Samples were made with Tb fractions of $x = 0.14, 0.19, 0.23, 0.28$, and 0.32. Protective layers of 100-Å SiO_2 were deposited above and below the TbFe film to protect against oxidation and corrosion. Films deposited on the mica coupons could be floated off for transmission electron microscopy (TEM) Lorentz-mode imaging.[9,10] The samples were characterized by x-ray florescence for composition, VSM for room-temperature magnetization, torque magnetometry for room-temperature anisotropy, and Kerr loops for Kerr rotation and coercivity as a function of temperature. Thermomagnetic writing of domain patterns was accomplished using an automated static tester.[10] The full width half-maximum (FWHM) diameter of the focused laser beam was approximately 0.85 μm.

Figure 1 shows photomicrographs of laser-written magnetic domains for three different compositions of Tb_xFe_{1-x}. These images were taken with a polarizing optical microscope using a high-resolution oil-immersion objective lens (1.25 NA). The domains in Fig. 1 were all formed under identical laser conditions (power and pulse width). Results for three different bias fields applied during writing are shown. Domains written with bias fields close to the onset

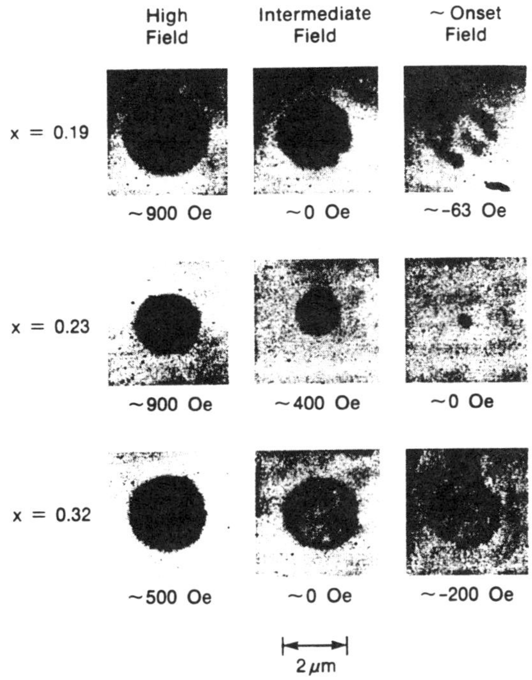

FIG. 1. Optical images of laser-written domains in Tb_xFe_{1-x} for $x = 0.19$, 0.23, and 0.32. Domains written for three different values of applied field are shown.

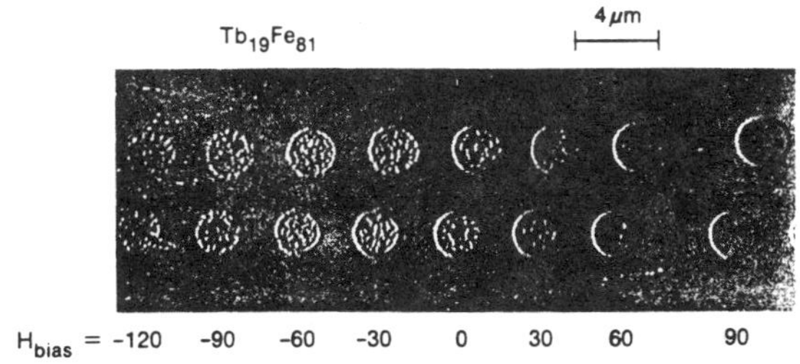

FIG. 2. Lorentz micrograph of domains written in $Tb_{0.19}Fe_{0.81}$ using 500-ns laser pulses. Bias field values are in Oe.

field are shown at right. Intermediate field results are shown at center, and fields large enough to ensure magnetic saturation are shown at the left.

Important differences in writing behavior are evident in Fig. 1. First, one may note that the dependence of mark size upon bias field is quite weak for the $x = 0.19$ and 0.32 compositions, whereas a strong field dependence is observed in the $x = 0.23$ sample. Another characteristic of the $x = 0.23$ sample is that the laser-written mark remains saturated (single domain) even down to fields close to the onset field. In contrast, the other two compositions show unsaturated (multiple domain) behavior at lower fields. These differences in writing behavior are believed to be due to differences in the relative importance of magnetostatic energy and domain-wall energy. In the $x = 0.23$ sample, the compensation temperature is above room temperature and, consequently, the magnetization in the temperature range of interest is quite small. Thus, the importance of magnetostatic energy is diminished relative to that of domain-wall energy. The greater relative importance of domain-wall energy promotes single-domain behavior.

Domain structure can be seen with high resolution in the Lorentz micrograph shown in Fig. 2. The sample is $Tb_{0.19}Fe_{0.81}$ and the field was varied during writing from -120 to $+90$ Oe in steps of 30 Oe. (Positive field is defined to be in the direction to promote writing.) No writing was observed to occur for fields less than -120 Oe. At -120 Oe, which is approximately equal to the onset field, the written pattern is observed to consist of a small number of microdomains which appear to have nucleated independently. As the field is increased, the nucleation density increases until, at about 0 Oe, the rim of the mark becomes essentially continuous. However, the mark interior retains substantial microdomain structure. At $+30$ Oe a well-formed circular mark is seen, but the mark interior is still stippled with a small number of microdomains which did not reverse during writing. For field values of $+90$ Oe and above, the marks in Fig. 2 are saturated (single domain).

III. THERMOMAGNETIC WRITING MODELS

In this section we present two models for the thermomagnetic writing process. The first model, which we call the "bubble" model, is essentially the same as the model developed by Huth.[3] This model is valid when the writing process can be described by the expansion or contraction of a single circular domain wall. This model is obviously not applicable, however, when the writing process gives rise to multidomain behavior, as observed in our 19% Tb and 32% Tb samples. To account for the cases where many independent nucleation sites appear to be active, we introduce a different model which we call the "nucleation field" model.

A. Bubble model

Assume one has a circular domain of radius r written in a thin film of thickness h. Following the bubble model, a wall energy σ may be assigned to the domain wall where σ is the energy per unit area of the domain wall. This energy is given by $4\sqrt{AK}$, where A is the exchange stiffness and K is the uniaxial anisotropy constant. In addition to wall energy, the domain will have two other energy terms—one due to demagnetizing field, and one due to the applied field. One may write the total energy of the domain (relative to no domain at all) as

$$E = 2\pi rh\sigma - E_H - 4\pi hH_a \int_0^r M(r')r'\,dr' , \tag{1}$$

where E_H is the demagnetizing energy, H_a is the applied field, and $M(r)$ is the magnetization of the film as a function of radial distance r from the center of the domain. Huth[3] has shown that this equation may be differentiated with respect to r to obtain the total radial force per unit area acting on the wall

$$F_t = \left(-\frac{1}{2\pi rh}\right)\left(\frac{\partial E}{\partial r}\right) = -\frac{\sigma}{r} - \frac{\partial \sigma}{\partial r} + 2MH_d + 2MH_a \, , \quad (2)$$

where the first two terms come from the derivative of the wall energy, the third from the derivative of the E_H term, and the last term from the derivative of the last term in Eq. (1). The sign of the force is such that a positive force will expand the domain.

The demagnetizing field H_d may be calculated from[11]

$$H_d(r,\theta,z) = 4\int_0^\infty \frac{M(r')(z-h)E(\xi)r'\,dr'}{[(r+r')^2+(z-h)^2]^{1/2}[(r-r')^2+(z-h)^2]} - 4\int_0^\infty \frac{M(r')zE(\xi')r'\,dr'}{[(r+r')^2+z^2]^{1/2}[(r-r')^2+z^2]} \, , \quad (3)$$

where

$$\xi = 4rr'/[(r+r')^2+(z-h)^2] \, , \quad \xi' = 4rr'/[(r+r')^2+z^2] \, ,$$

and $E(x)$ is the complete elliptic integral of the second kind.

In order to investigate the stability of a circular domain using this model one compares the total force F_t acting on the wall with the force F_c exerted on the wall by the coercive force, i.e., for wall motion one requires

$$|F_t| \geqslant F_c = 2MH_c \, , \quad (4)$$

where H_c is a coercive force which resists wall motion. In our calculations we have used experimental values of coercive force as a function of temperature as measured by Kerr loops. If Eq. (4) is satisfied, the wall will move in the direction of the total force F_t, i.e., a positive force will move the domain in the positive r direction; otherwise the frictional force, F_c, acting on the wall will keep the wall stationary.

B. Nucleation field model

In the bubble model just discussed, the writing process for the bubble occurs simply by the expansion or contraction of a circular domain wall. Another possible mode for writing is by nucleation of microdomains at many sites within the laser-heated region. As indicated previously, evidence for the existence of multiple nucleation sites within a single mark can be seen in Fig. 2. For negative applied field, the nucleation density is low and domains as small as 70 nm have been observed.[12] As the field is increased in the positive (writing) direction, the nucleation density increases until the entire mark becomes a single micron-size domain. This phenomenon was observed in all of our samples except for the 23% Tb sample, which exhibited single-domain behavior even at low bias fields.

Additional evidence for magnetization reversal via nucleation has been observed by polarized light microscopy at room temperature. We have observed that for some compositions of Tb-Fe when an external field is applied at room temperature to a previously written domain, the domain does not reverse by wall motion of that domain. Instead, one sees a gradual grey scale change inside the domain. It appears that the domain is reversing in very tiny regions which are below the resolution of the microscope, but which give the appearance of a gradual uniform change of magnetization.[13]

To account for domain writing when nucleation behavior is more important than wall motion, we have developed the following "nucleation field" model. The criterion we have chosen here for the initiation of magnetization reversal is a comparison of total internal field, H_t, which consists of demagnetizing field plus applied field, with a nucleation field H_n:

$$H_t = H_d + H_a > H_n \, . \quad (5)$$

For reversal of a small region to occur, the total field in that region must be at least as large as H_n, which is the field required to nucleate reversal. It is not necessary to explicitly account for domain-wall energy in this model since this is implicit in the definition of the nucleation field. H_n is a strong function of temperature and is determined experimentally using the relation

$$H_n = H'_n + 4\pi M \, . \quad (6)$$

H'_n is the applied field where magnetization reversal first starts on a B-H or Kerr loop. The extra term, $4\pi M$, must be added because this is the value of the demagnetizing field which is acting to help nucleation when the loop measurement is made. Inclusion of this term is essential for accurate modeling results. Because the experimental loops are quite square, we approximate H'_n in our calculations by H_c, where H_c is the experimental coercive force.

It is important to note that even though the resulting equation, Eq. (5), is similar to the bubble model [Eqs. (2) and (4)] with the wall-energy terms dropped, the reversal mechanism in the two cases is entirely different. As mentioned above, the nucleation field model does not ignore domain-wall energy, but, instead, contains it implicitly in H_n.

IV. TEMPERATURE PROFILE

The first step in a domain-writing calculation is to determine the temperature profile generated by the laser pulse. For this purpose a computer program written by Wieder *et al.*[14] was used which solves the heat diffusion equation in

TABLE I. Thermal parameters.

	Thermal conductivity (cal/s cm deg)	Specific heat (cal/deg cm^3)	Ref.
SiO_2	0.0037	0.40	15
Tb-Fe	0.095	0.76	16
Glass	0.0025	0.55	17

three dimensions by an explicit finite difference method. The geometrical configuration used in the experiments and in the thermal modeling was a three-layer film structure on top of a glass substrate (Corning 7059 glass). The top layer was 100 Å of SiO_2, the second layer was 500 Å of Tb-Fe, and the third layer was another layer of 100 Å of SiO_2, followed by a thick glass substrate. The thermal parameters used in the model are shown in Table I (Refs. 15–17).

Power incident on the sample was measured to be 7.9 mW ($\pm 10\%$). The measured reflectivity and film transmission were measured to be 0.63 and 0.015 ($\pm 5\%$). To obtain the best fit to domain-diameter measurements it was necessary to assume the maximum laser power and minimum reflectivity within these uncertainty ranges. Examples of calculated thermal profiles are shown in Fig. 3 illustrating cooling after a 50-ns duration laser pulse.

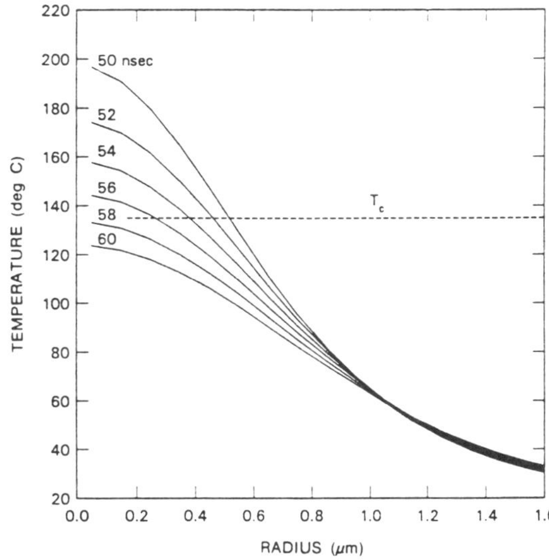

FIG. 3. Temperature profiles of the Tb-Fe layer at the end of a 50-ns laser pulse and during cooling. Approximate value of the Curie temperature is shown for reference.

V. MAGNETIC PROPERTIES

After modeling the temperature profile, the next step in the calculation is to model the magnetic properties of the medium in which the domains are to be written. For the calculation of magnetization we use a mean field model described by Mansuripur.[18] In this model TbFe is represented by three subnetworks. One subnetwork consists of terbium atoms and the other two consist of iron atoms. All the iron moments are coupled antiferromagnetically to the Tb moments. A certain fraction, α, of the iron is assigned to a second network in which the Fe-Fe coupling is antiferromagnetic. The third network contains iron which has ferromagnetic Fe-Fe coupling. The parameter α, the spin on the iron atom S_{Fe}, and the composition, at. % Tb, are all adjustable parameters.[18] The at. % Tb is adjustable within the experimental uncertainty of about ± 1 at. %.

Films were made with the following measured percentages of terbium: 14, 19, 23, 28, and 32 at. %. The foregoing adjustable parameters were then varied for each composition to obtain a good fit to experimentally measured values of Curie temperature T_C, compensation temperature T_{co} if available, room-temperature magnetization $M(T = 293$ K), and at. % Tb.

Table II shows a comparison of experimental and calculated values of these parameters. The values of the fitting parameters α and S_{Fe} are shown at the bottom of the table. It was found possible to use the same exchange integrals for Fe-Fe, Fe-Tb, and Tb-Tb exchange as used in Ref. 18 (with the one exception that the Fe-Tb exchange was reduced by 20%). Figure 4 shows the temperature dependence of the magnetization of our samples calculated in this way.

TABLE II. Experimental and calculated material parameters. The values of M, K, H_c, and A are given for 20 °C.

	14% Tb	19% Tb	23% Tb	28% Tb	32% Tb
%Tb(exp)	13.8	19.4	23.1	27.7	32.4
%Tb(calc)	13.9	20.8	23.2	27.7	31.0
M(exp) (emu/cm^3)	217	109	− 40	− 162	− 250
M(calc) (emu/cm^3)	217	27	− 37	− 162	− 255
T_{co}(exp) (K)	···	258	346	···	···
T_{co}(calc) (K)	···	258	345	···	···
T_C(exp) (K)	378	397	406	413	405
T_C(calc) (K)	381	398	406	415	407
K(exp) (10^6 erg/cm^3)	3.5	1.7	···	3–3.5	3.7
H_c(exp) (kOe)	0.4	5.1	···	3.4	1.4
A(calc) (10^{-7} erg/cm)	1.42	1.83	2.02	2.10	2.00
α(%)	47	36	30	35	33
S_{Fe}	0.97	0.97	0.97	1.05	1.05

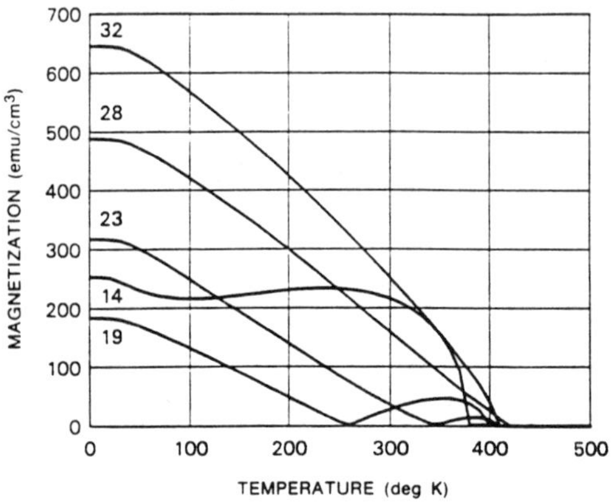

FIG. 4. Calculated magnetization vs temperature for $x = 0.14$, 0.19, 0.23, 0.28, and 0.32 in Tb_xFe_{1-x}.

Uniaxial anisotropy K at 20 °C for each composition was measured by torque balance and the results are also shown in Table II. The room-temperature value for 23 at. % Tb is uncertain due to the inability to saturate that sample in the available field. We were not able to measure the temperature dependence of K for our samples. For fitting purposes we have assumed the temperature dependence of the anisotropy to obey

$$K \propto [m_{Tb}(T)]^n , \qquad (7)$$

where $m_{Tb}(T)$ is the relative magnetization of the terbium subnetwork. The exponent n is expected to be 2 (Ref. 19) or to vary in the range 2–3 for simple molecular-field theory.[20]

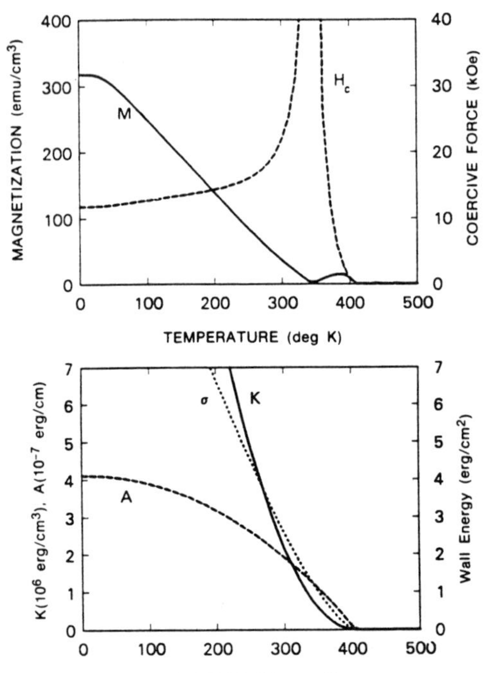

FIG. 5. Calculated media properties vs temperature for the 23 at. % Tb composition. The properties are magnetization, coercive force, anisotropy K, exchange stiffness A, and wall energy σ.

The exchange stiffness A was calculated from an equation by Gangulee and Kobliska.[21] The A parameter is not an adjustable parameter. The wall energy as a function of temperature may then be calculated from

$$\sigma(T) = 4\sqrt{A(T)K(T)} . \qquad (8)$$

The coercive force for irreversible wall displacement may be calculated from $H_c = (1/2M) \times (\partial\sigma/\partial x)$, where $\partial\sigma/\partial x$ is the maximum gradient in wall energy as a function of wall displacement.[22] Assuming this gradient is proportional to σ, then

$$H_c \propto \sigma/M . \qquad (9)$$

It was found that this expression gives the correct general behavior of $H_c(T)$ but fails in detail; for the calculations in this paper, Eq. (9) was fitted to experimental coercive force values, or the experimental values of coercive force were used directly. An example of the temperature dependence of these material parameters is shown in Fig. 5 which gives the results of the above calculations for the temperature dependence of $M(T)$, $H_c(T)$, $K(T)$, $A(T)$, and $\sigma(T)$ for the $x = 0.23$ TbFe composition. The $K(T)$ curve was calculated using $n = 4$ which will be discussed in more detail later.

VI. MARK SIZE

The previous two sections have described how the temperature profiles and temperature-dependent magnetic

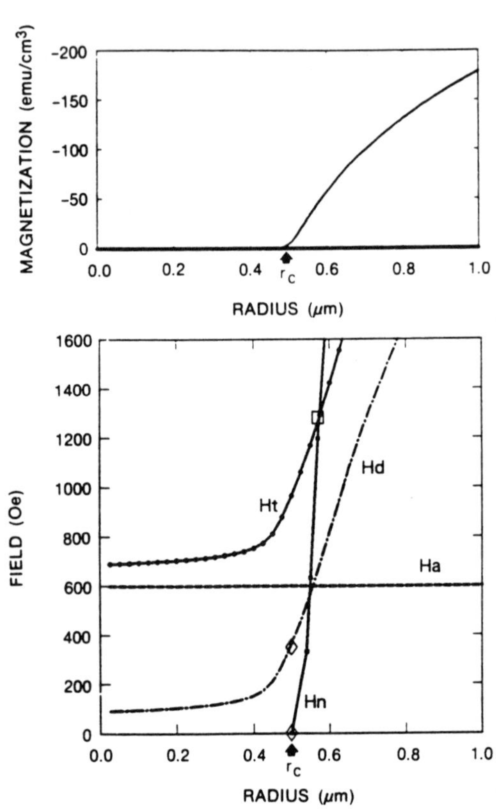

FIG. 6. Upper: radial dependence of magnetization in $Tb_{0.32}Fe_{0.68}$ at the end of a 50-ns, 8.7-mW laser pulse. Lower: radial dependence of demagnetizing field H_d, applied field H_a (600 Oe), total field H_t, and nucleation field H_n.

properties were determined. In this section and the two that follow, we use the thermal and magnetic media modeling results as input for the thermomagnetic writing models. We start with a calculation of written mark size.

Figure 6 illustrates the result of a calculation of mark size using the nucleation field model [Eq. (5)] for the 32% Tb composition. As mentioned previously, this is the appropriate model to use for this composition (at least at low fields) because of the multidomain behavior observed in Fig. 1. The laser pulse is assumed to be 50 ns. Calculation of the temperature profile $T(r)$ at the end of the pulse gives the profile shown as the upper curve in Fig. 3. Combining $T(r)$ with the $M(T)$ curve for the 32% Tb composition shown in Fig. 4 gives the magnetization profile $M(r)$ shown in the upper part of Fig. 6. The Curie temperature (where $M \to 0$) is located at a "Curie radius" of about $r_C = 0.5\,\mu$m. Figure 6 (lower) shows the radial dependencies of the various terms in Eq. (5). The demagnetizing field H_d is calculated from the magnetization profile shown in Fig. 6 (upper). It is assumed that no domain walls are present. As expected, H_d is small at the interior of the heated region ($r \sim 0$) and rises rapidly where the magnetization increases at the cooler temperatures near the edge of the heated region. For this calculation the applied field H_a is assumed to be 600 Oe (applied in the writing direction). The total field H_t acting on a small region is the sum of H_d and H_a, as given in Eq. (5).

The region over which nucleation can occur is found by comparing the total field with the field H_n required to nucleate magnetization reversal. The two fields H_t and H_n intersect at $r = 0.58\,\mu$m (open square in Fig. 6). For radii smaller than this and larger than r_C, the total field is greater than the nucleation field and magnetization reversal takes place. For radii larger than 0.58 μm the total field is not strong enough for nucleation to occur. Therefore, the written region will extend out to a diameter of $2 \times 0.58 = 1.16\,\mu$m. The measured diameter for these experimental conditions is 1.32 μm. Note that implicit in the nucleation field model is the assumption that once nucleation occurs, the resulting magnetization reversal will remain stable during cooldown.

Mark diameters calculated in the same way as above are shown in Fig. 7 for four different TbFe compositions, two laser pulse lengths differing by an order of magnitude, and five different applied field values. Comparisons with the experimental values are also shown. With the exception of the 23% Tb sample, the agreement is excellent for the 50-ns cases and reasonably good for the 500-ns cases. Note that the nucleation field model successfully predicts the observed weak dependence of mark size on applied field which is most evident in the 19% and 32% samples.

For the 23% Tb composition, the agreement with experiment is quite poor, especially at low bias fields. The experimental domain sizes are seen to decrease with decreasing applied field much more rapidly than do the calculated values (see also the experimental results in Fig. 1). As noted earlier, the writing behavior for this composition, the one with a compensation temperature above room temperature, is markedly different from the other compositions. (The 28% Tb composition seems to be an intermediate case, showing good agreement at high fields and poor agreement at low fields.)

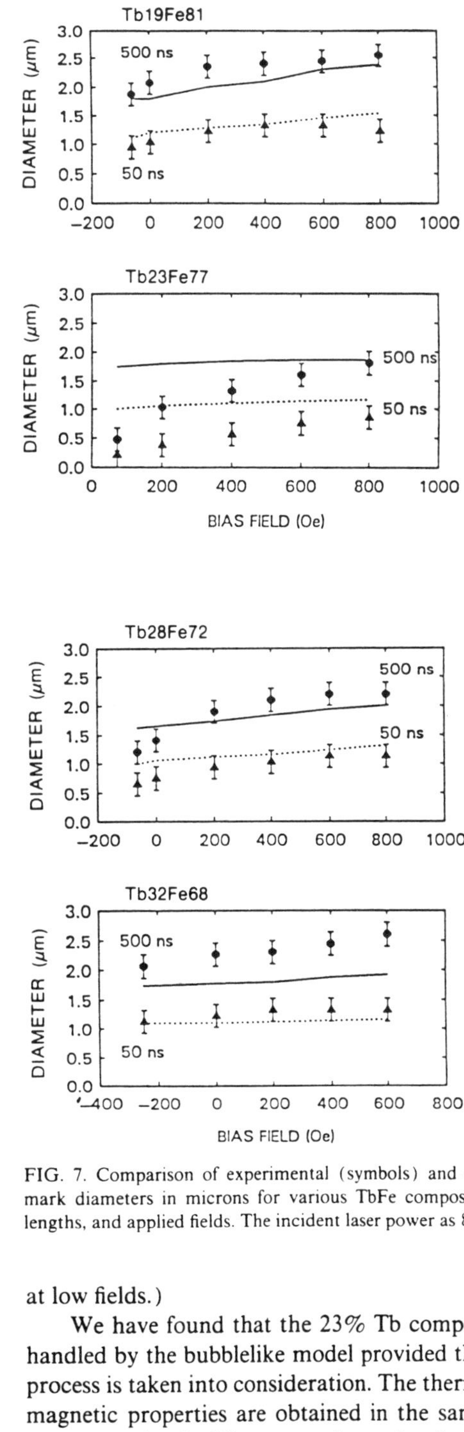

FIG. 7. Comparison of experimental (symbols) and calculated (lines) mark diameters in microns for various TbFe compositions, laser pulse lengths, and applied fields. The incident laser power as 8.7 mW.

We have found that the 23% Tb composition may be handled by the bubblelike model provided that the cooling process is taken into consideration. The thermal profile and magnetic properties are obtained in the same way as discussed previously. We assume that a circular domain wall is present and use Eqs. (2) and (4) to calculate the forces acting on the wall. Figure 8 (upper) shows $M(r)$ and wall energy $\sigma(r)$ calculated just at the end of a 50-ns laser pulse. Figure 8 (lower) shows the various forces as a function of the wall position. As might be expected from the close proximity of the compensation temperature, the force on the wall due to the demagnetizing field, $F_d = 2MH_d$ is essentially zero. The force due to the applied field, $F_a = 2MH_a$, is calculated in this case for a bias field of 400 Oe. The force due to the first wall energy term $F_{w1} = -\sigma/r$ is not especially

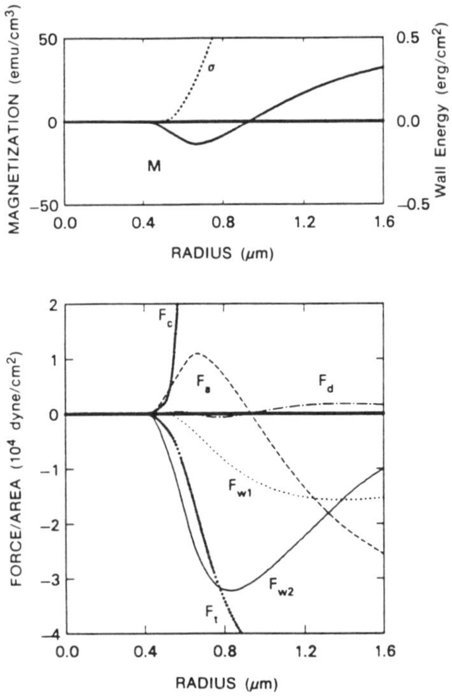

FIG. 8. Upper: radial dependence of magnetization (solid line) and wall energy (dotted line) at the end of 50-ns laser pulse for $Tb_{0.23}Fe_{0.77}$. Lower: radial dependence of force/area acting on a circular domain wall: F_a is the force due to the applied field (400 Oe); F_d is due to the demagnetizing field; F_{w1} is the first wall force $-\sigma/r, F_{w2}$ is the second wall force $-\partial\sigma/\partial r$, F_t is the total field; and F_c is the force due to the coercive force.

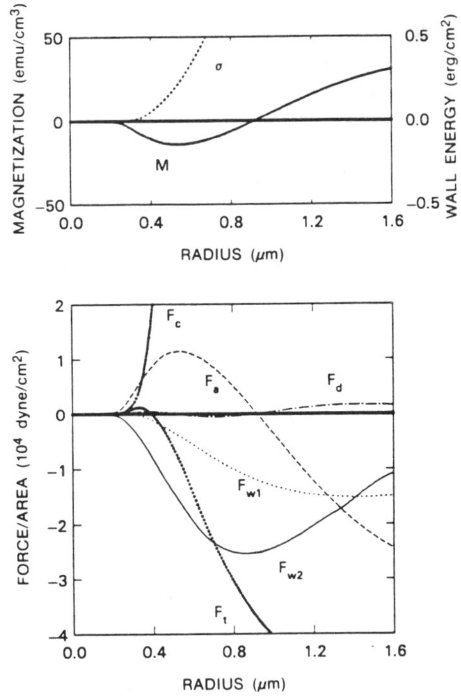

FIG. 9. Same as Fig. 8 except at 6 ns after the end of a 50-ns laser pulse.

large, but the force due to the second wall energy term, $F_{w2} = -\partial\sigma/\partial r$, is large and is significant. The net result is that the total force on the wall, F_t, is negative or zero in Fig. 8 for all values of radii. This suggests that even if a wall could nucleate, it would shrink and disappear.

We now extend this calculation to times after the laser has been turned off. The results at a time 6 ns after the end of the 50-ns laser pulse are shown in Fig. 9. In contrast to Fig. 8, after some cooling has occurred, and for some values of r very close to r_C where F_c is small, the total force F_t on the wall is both positive and greater than F_c.

Thus, Eq. (4) is satisfied and a domain wall can expand out to where $F_t = F_c$, yielding a domain radius of 0.3 μm. The results of similar calculations for a variety of fields and pulse lengths for this 23% Tb sample are shown with measured values in Fig. 10. Agreement is excellent. In contrast with the nucleation field model, this bubblelike model, when calculated during cooling, predicts the observed strong dependence of diameter with applied field.

The reason for the delayed domain nucleation is as follows: Right at the end of the laser pulse, the thermal profile is quite steep and, therefore, the wall energy term involving the gradient of wall energy, $F_{w2} = -\partial\sigma/\partial r$, is large near T_C. This wall energy gradient is illustrated by the slope of the dotted curves in the upper parts of Figs. 8 and 9. Since F_{w2} tends to inhibit domain formation, no domain is formed at this time. As the heated region cools down, the thermal gradient in the vicinity of T_C becomes much smaller due to thermal diffusion (see Fig. 3). As a result, F_{w2} is no longer large enough to inhibit domain formation and a domain is nucleated. The amount of cooling that needs to take place before domain formation depends on the strength of F_a. The smaller the bias field, the smaller the wall energy gradient that can be tolerated, and therefore, more cooling must take place before domain formation can occur. This is why the mark size depends so strongly on bias field.

Choosing the proper value for anisotropy is a matter of concern. As noted earlier, we could not measure K at room temperature for this composition by the torque method. However, it seems likely that K is similar in value to that of its neighboring compositions,[19] so we have chosen the average of its two neighbors, 2.5×10^6 ergs/cc, as K(20 °C) for this composition. However, it should be noted that at least two examples appear in the literature where K shows a sharp

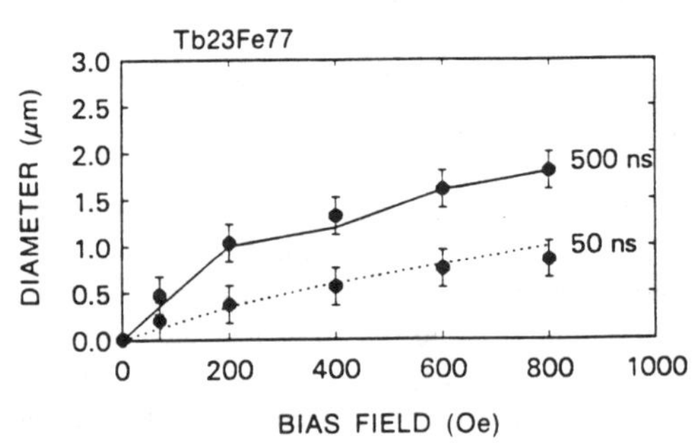

FIG. 10. Experimental (symbols) and calculated (lines) domain diameters for two different laser pulse lengths as a function of applied field. The TbFe composition is 23 at. % Tb, and the laser power is 8.7 mW. The calculation uses the bubble model with cooling.

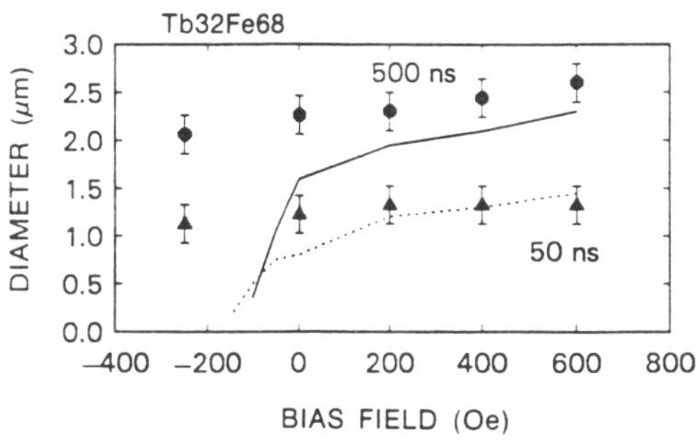

FIG. 11. Same as Fig. 10 except the composition is 32% Tb.

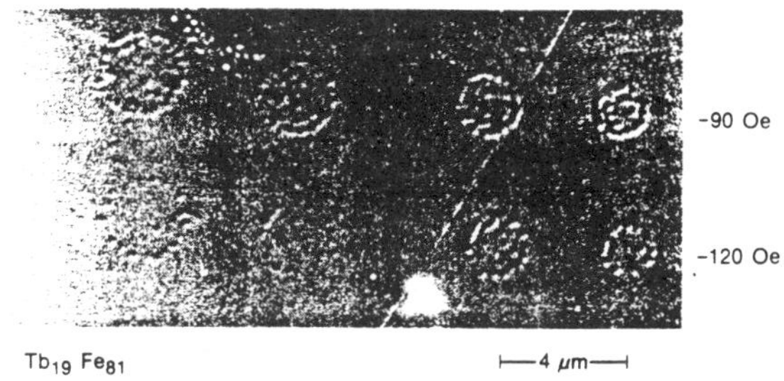

FIG. 12. Lorentz micrograph of marks written with bias fields close to the onset field. Note the preferential nucleation near the edges of the heated regions.

dip at the composition for which the compensation temperature is near room temperature.[23,24]

We are not aware of any measurements of the temperature dependence of K for TbFe. However, $K(T)$ for GdCoMo (Ref. 20) has been measured and the data indicate that the best fit to Eq. (7) would require $n = 4$, which is the value we have chosen here in this study At this time there does not appear to be any theoretical justification for choosing $n = 4$; we are using n here simply as a fitting parameter. We have found that a "poor to fair" fit to the data of Fig. 10 could be obtained by assuming $n = 2$, but in that case it was necessary to reduce the value of room-temperature K by about a factor of 5 below what we have assumed above.

The good agreement between calculation and experiment in Fig. 10 shows that the bubble model accurately accounts for the writing behavior of the 23% Tb composition, where the written marks are single domain for all values of bias field. One might argue that the bubble model should be equally applicable to the other compositions at high fields, where the written marks are also observed to be single domain. To check this, we have made a calculation of mark diameter using the bubble model for the 32% Tb composition. The results of this calculation are shown in Fig. 11. It may be seen that at high fields the bubble model does give mark diameters which are in reasonable agreement with experiment and which are similar to the predictions of the nucleation field model (see Fig. 7). For fields less than about 100 Oe, however, the bubble model fails badly. This is in accord with the observation that, at low bias fields, this composition produces fragmented marks, in violation of the basic assumptions of the bubble model.

TABLE III. Onset field.

Pulse width (ns)	Tb (%)	Exp (Oe)	Calc (Oe)
500	19	125	100
	28	118	100
	32	262	205
50	19	172	107
	28	180	140
	32	394	350

VII. ONSET FIELD

As the applied field during writing is changed in a direction so as to promote domain writing, the field at which magnetization reversal first starts is called the onset field. In this section we use the nucleation field model to calculate the onset field for a variety of writing conditions.

Figure 6 (lower) shows the fields acting at the end of a 50-ns laser pulse for the 32% Tb composition. For $H_a = 0$ the total field is just the demagnetizing field, labeled H_d in the figure. Since H_d is larger than H_n in the region $0.50 < r < 0.56$, magnetization reversal can occur in this region even when the applied field is zero. To suppress this reversal, an external field H_a must be applied such that the total field $H_a + H_d < H_n$ for all r [see Eq. (5)]. Inspection of Fig. 6 (lower) will show that the maximum difference between H_d and H_n occurs at the Curie radius, r_C. Since $H_n(r_C) = 0$, the field required to just suppress nucleation (the onset field) is simply

$$H_a = -H_d(r_C)\,. \tag{10}$$

This is shown as the vertical distance between the two open diamonds in Fig. 6 (lower) and corresponds to a value of -350 Oe.

Onset field calculations based on Eq. (10) were made for the other samples and for both 50- and 500-ns laser pulse durations. Comparisons with experiment is shown in Table III. Agreement is, in most cases, about 20%.

The calculation for onset field described above was for the time just at the end of the laser pulse. This will give the true value of the onset field only if that field is also sufficient to overcome the demagnetizing field present during cooldown. Calculations of $H_d(r_C)$ were made for times during the cooling period and it was verified that the largest value occurs just at the end of the pulse. Physically, this result is due to the fact that the temperature gradient in the vicinity of r_C is greatest just at the end of the pulse. Consequently, this is also the time when the gradient in magnetization will be maximum. This is the situation which will generate the largest demagnetizing field in the vicinity of the Curie radius.

Further evidence that the greatest driving force for nucleation occurs just at the end of the pulse can be seen in Fig. 12. This figure shows a Lorentz micrograph of domains written in $Tb_{0.19}Fe_{0.81}$ with a bias field just slightly less negative than the onset field. As a result, nucleation driven by the demagnetizing field occurs. However, notice that the nucleation showed a strong preference for the outer circumference of the marks. Few nucleation sites are seen in the mark interiors. This supports the idea that the driving force for nucleation peaks in the vicinity of r_C just at the end of the laser pulse. Note that this result is in sharp contrast with the delayed nucleation observed for the case of $Tb_{0.23}Fe_{0.77}$.

VIII. SATURATION FIELD

As the writing field is increased in the positive (writing) direction, the field at which the written mark is fully saturated is called the saturation field. This field may be calculated using the nucleation field model by assuming a domain wall has been written and is fixed in location. The demagnetizing field within this wall is then calculated as the domain cools. As cooling occurs the region of reversed magnetization inside this wall increases and so the negative contribution to the demagnetizing field by this region also increases. If this demagnetizing field achieves a large enough negative value, it will cause demagnetization inside the initial domain wall. This demagnetization results in the stippled appearance of the laser-written domains in Figs. 1 and 2.

Figure 13 shows an example of a calculation of the growth of this negative demagnetizing field H_d as cooling occurs after the end of a 500-ns pulse. A wall is assumed to be located at $r = 0.96\,\mu$m. It may be seen that during the cooldown period, the Curie radius shrinks and the region of written (positive) magnetization grows. The growing positive magnetization region leads to a growing negative H_d. If H_d becomes sufficiently negative such that the absolute value of $H_d + H_a$ becomes larger than H_n, then the nucleation of

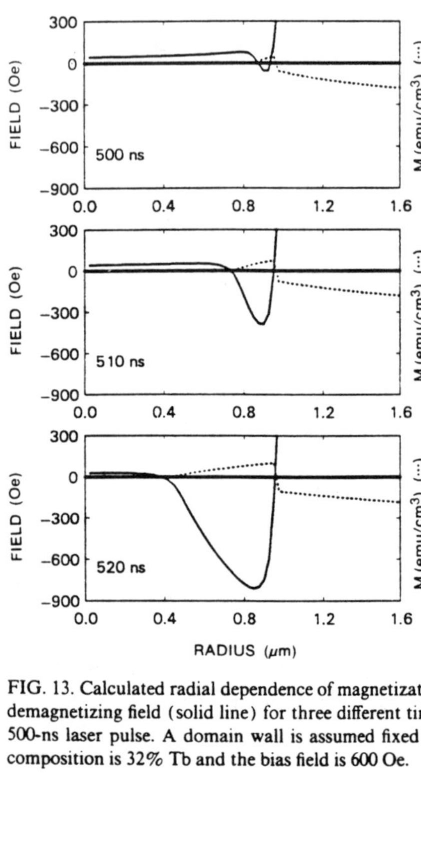

FIG. 13. Calculated radial dependence of magnetization (dotted line) and demagnetizing field (solid line) for three different times after the end of a 500-ns laser pulse. A domain wall is assumed fixed at $r = 0.96\,\mu$m. The composition is 32% Tb and the bias field is 600 Oe.

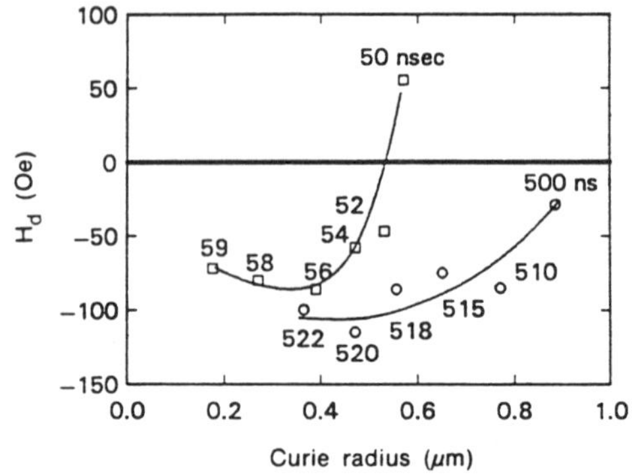

FIG. 14. Calculation of demagnetizing field at the Curie radius during cooldown. The upper curve is for $Tb_{0.19}Fe_{0.81}$ using a 50-ns laser pulse and with a wall assumed fixed at $r = 0.61\,\mu$m radius. The lower curve is for $Tb_{0.32}Fe_{0.68}$ using a 500-ns laser pulse and with a wall assumed fixed at $r = 0.96\,\mu$m. Lines are drawn through the calculated points for ease of viewing.

negative magnetization within the positive domain can occur. Since $H_n \to 0$ at $r = r_C$, the necessary condition to prevent this negative nucleation is to require $H_t(r_C) > 0$ or $H_a > -H_d(r_C)$ during the cooldown period. Although this condition appears to be the same as Eq. (10), in this case H_d is calculated with an already nucleated domain wall fixed in place. Figure 14 shows the calculated values of $H_d(r_C)$ for the 32% Tb sample wth a 500-ns pulse, and also for the 19% Tb sample with a 50-ns pulse. For the first case, the calculation shows that the largest negative value of $H_d(r_C)$ is approximately -110 Oe and it occurs about 20 ns after the end of the laser pulse. The saturation field is thus predicted to be $+110$ Oe, which agrees well with the 120-Oe experimental value. The second case in Fig. 14 is for 19% Tb and a 50-ns laser pulse. The saturation field is calculated to be 80–90 Oe and the experimental value obtained by Lorentz microcopy is 60–90 Oe.

IX. SUMMARY

Measurements of laser-written mark diameter have been made on TbFe films as a function of film composition, laser pulse length, and applied magnetic field during writing. It was found that the dependence of diameter on field is quite strong in the case of the low magnetization composition (23% Tb), and weak for the other compositions (19%, 28%, and 32% Tb). In addition, the 23% Tb sample remained single domain for all values of bias field. In contrast, the other compositions showed multidomain writing behavior at low bias fields.

Calculations of domain diameter for the 23% Tb composition gave good agreement with experiment provided a bubblelike model is used. The thermally induced domain-wall energy gradient was found to play an important role in the writing process and to act to delay domain formation until cooling has occurred. For the other compositions, a nucleation field model gave reasonable agreement with experiments. It was found that the field for the initiation of magnetization reversal (onset field) and the field required to write single-domain marks (saturation field) could be calculated in reasonable agreement with experiment.

We have pointed out a distinction between two types of writing behavior. In the 23% Tb sample, the compensation temperature is within 60 °C of the Curie temperature, which leads to relatively low magnetization values in the vicinity of the Curie temperature. Consequently, the magnetostatic energy is low and the wall energy is relatively more important. This dominance of wall energy is believed to be responsible for the observed single-domain behavior and good agreement with the bubble writing model. In the other compositions studied, the magnetostatic energy is larger, a situation which favors multidomain behavior and the activity of multiple nucleation sites. This is the situation which is suitable for the nucleation field model.

ACKNOWLEDGMENTS

The authors wish to thank A. E. Bell and R. Kant for helpful discussions, and L. Bynum for assistance with film preparation.

[1]A. E. Bell, Proc. SPIE **382**, 2 (1983).
[2]W. D. Doyle, G. K. Goldberg, and W. E. Flannery, IEEE Trans. Magn. **MAG-6**, 548 (1970).
[3]B. G. Huth, IBM J. Res. Develop. **18**, 100 (1974).
[4]M. Mansuripur and G. A. N. Connell, J. Appl. Phys. **55**, 3049 (1984).
[5]M. Takahashi, N. Ohta, S. Takayama, Y. Sugita, M. Yoshiro, and K. Shigematsu, IEEE Trans. Magn. **MAG-22**, 931 (1986).
[6]M. Mansuripur, J. Appl. Phys. **61**, 3334 (1987).
[7]T. W. McDaniel and M. Mansuripur, IEEE Trans. Magn. **MAG-23**, 2943 (1987).
[8]P. Hansen, J. Appl. Phys. **62**, 216 (1987).
[9]J. C. Suits, R. H. Geiss, C. J. Lin, D. Rugar, and A. E. Bell, Appl. Phys. Lett. **49**, 419 (1986).
[10]J. C. Suits, R. H. Geiss, C. J. Lin, D. Rugar, and A. E. Bell, J. Appl. Phys. **61**, 3509 (1987).
[11]See Eq. (A6) of Ref. 3. An error has been corrected in that equation.
[12]D. Rugar, C. J. Lin, and R. Geiss, IEEE Trans. Magn. **MAG-23**, 2263 (1987).
[13]C. J. Lin, J. C. Suits, and R. H. Geiss, J. Appl. Phys. **63**, 3835 (1988).
[14]H. Wieder, S. S. Lavenberg, G. J. Fan, and R. A. Burn, J. Appl. Phys. **42**, 3458 (1971).
[15]*Handbook of Thermophysical Properties of Solid Materials*, edited by A. Goldsmith, T. E. Waterman, and H. J. Hirschborn (Macmillan, New York, 1961).
[16]*Thermophysical Properties of Matter*, edited by Y. S. Touloukian (IFI/Plenum, New York, 1970). The values shown are estimated from the pure elements.
[17]Courtesy of Corning Glass Works, Corning, New York.
[18]M. Mansuripur and M. F. Ruane, IEEE Trans. Magn. **MAG-22**, 33 (1986).
[19]R. B. van Dover, M. Hong, E. M. Gyorgy, J. F. Dillon, Jr., and S. D. Albiston, J. Appl. Phys. **57**, 3897 (1985).
[20]P. Chaudhari and D. C. Cronemeyer, AIP Conf. Proc. **29**, 113 (1976).
[21]A. Gangulee and R. J. Kobliska, J. Appl. Phys. **49**, 4896 (1978). See their Eq. (27).
[22]S. Chikazumi and S. H. Charap, in *Physics of Magnetism* (Wiley, New York, 1964).
[23]Y. Mimura, N. Imamura, T. Kobayashi, A. Okada, and Y. Kushiro, J. Appl. Phys. **49**, 1208 (1978).
[24]R. Hasegawa, J. Appl. Phys. **45**, 3111 (1974).

Section Eleven
Phase Change Media

Reprinted with permission from *Journal of Non-Crystalline Solids*, Vol. 2, pp. 99-106 (1970).

AN INTRODUCTION TO OVONIC RESEARCH

S. R. OVSHINSKY

Energy Conversion Devices, Inc., 1675 W. Maple, Troy, Michigan 48084, U.S.A.

Reversible switching in amorphous materials depends upon localized structural considerations. In the case of the threshold switch, elements are chosen to provide several functions, among which are the inhibition of crystallization and the introduction, by their chemical bonds, of large amounts of localized states bridging the valence and conduction bands. Memory materials are chosen which allow for reversible structural changes. They are therefore balanced between ordered and disordered states. A brief history of previous switching work is given as well as a description of the direction of some of ECD's current research.

This paper will briefly review our past work in amorphous switching and discuss some of the operational concepts underlying our continued effort. In the past, periodicity of atomic structure has been the basis of the understanding of solid state physics. Part of the problem in understanding the amorphous state is the terms themselves, for the words "disordered" and "amorphous" for non-crystalline materials have misleading connotations as they project an image of a material with lack of specific structure. In this paper we will describe some of our working hypotheses which pay particular attention to spatial relationships of the localized structural units as a means of developing unusual electronic and electrical changes in amorphous materials.

In the late 1950's, the well-known Soviet physicist, A. F. Ioffe, considered the possibility that short-range order[1]) could have a direct bearing on semiconducting properties and he, in conjunction with a notable group of scientists led by B. T. Kolomiets, started investigations into the characteristics of amorphous materials[2,3]). This Soviet group studied amorphous materials, measured their properties at low electric fields, and found them, in fact, to be rather uninteresting in the sense that they behaved like intrinsic crystalline semiconductors. Since intrinsic semiconductivity is common, one would not expect exciting new phenomena from such materials. Other conventional measurements important in crystalline materials were also discouraging when made on amorphous materials. For example, the Hall mobility is very low, almost non-existent, in most amorphous semiconductors. There appeared to be very few available carriers with little ability to move freely in an electric field[4]).

At approximately the same time, I independently started an investigation of the electrical properties of disordered materials with the avowed purpose of finding switching effects. My interest in disordered materials arose from an interest in neurophysiology. I had posed to myself the question of what is the physical basis of intelligence, that is, how do nerve cells store, switch, alter, and transfer information in an organism.

We divided the problem into two parts, one having to do with the traffic problems of communication, that is, threshold switching events, and the other, the storage and alteration of information, which is memory. By building models which simulated biological action we could develop insights into possible biological mechanisms and at the same time develop a superior switch for technological use. This bionic approach has been a particular interest of both my wife's, who is a biochemist, and mine. We, therefore, worked on changing a film reversibly between a high and a low impedance (the threshold) and on storing and interrogating an energy event in a film by altering the atomic configuration of the material (the memory). Regarding the first, we originally expected a strongly disordered material to have many traps and recombination centers utilized for conduction processes by a sufficiently high electric field. As to the second type of switching, we viewed the expected change of atomic configuration as a change in the degree of local order in the disordered polymeric material. Many disordered systems were explored. Both threshold and memory switching were found and we proceeded with our original purpose of making nerve models and developing commercially usable switches.

In June of 1958, I succeeded in making a switch based on an amorphous material. I used a valve metal, the transition metal tantalum, which had an amorphous layer of tantalum oxide about 900 Å thick, and an electrolyte solution formed a portion of the control electrode. Switching occurred when the film was sufficiently polarized. This work was widely reported in technical journals such as Electronics[5]) and Control Engineering[6]) in the summer of 1959, and discussed in the book *Anodic Oxide Films* by Young[7]). A suitable metallized electrode led to a memory version of this switch which was demonstrated at a lecture given at the Detroit Physiological Society Meeting in 1959[8]). A pulse of one polarity set the memory, and a pulse of the opposite polarity shut it off.

We have been demonstrating switching in amorphous materials since 1958 and our group at Energy Conversion Laboratories (now Energy Conversion Devices, Inc.) has been working continuously since 1960 with great commitment to glassy semiconducting switching. Our work, originally kept secret for commercial reasons, has led to many patents, granted and in process. The switches[9]) that were developed can be divided into three types, all of them based on disordered materials[10]) and fabricated by a common technology. The devices commercially utilized were made of chalcogenide materials. The electrical characteristics of three types of devices (threshold, memory and adaptive memory) will be described in detail in subsequent papers.

In the early 1960's, Dewald, Northover, and Pearson[11]) at Bell Labs, were working with glasses. They also found amorphous materials which had memory action, and could be kept in either a conducting or a non-conducting condition by pulses of opposite polarity. They encountered problems of material stability and reproducibility, and their original project was eventually dropped. Movies of the memory device of the Bell Telephone Labs group recently shown[12]) and a description of materials and electrodes used by them clearly show chemically unstable thermodynamic processes taking place.

It is important to caution that there are artifact switching and memory actions which are neither reproducible nor stable, nor long-lived, and can cause confusion among the uninitiated. These artifact switching effects can be summarized briefly. Placing two narrowly separated electrodes in a high electric field can cause the growth or throwing of a metallic bridge from the electrodes that can be destroyed by passing a high enough electrical current through it. This is an artifact because it uses up the electrodes and it is certainly not a reproducible or long-lived element. Even if the metal throwing takes place penetrating a solid film, it will still fall into the same category. Another common artifact is the use of chemically unstable material between two electrodes. Here chemical decomposition takes place, resulting in irreproducible and irreversible phenomena. This type can be characterized as using up the material. A third type draws upon the environment for its mechanism and concerns itself with forms of oxidation and reduction which are thermally induced. A form of this occurs with some oxides wherein a destructive dielectric breakdown can sometimes be healed by utilizing the oxygen either inside or surrounding the material to repair the oxide. Another type of artifact has to do with contacts that can react with the switching materials used. Non-stable switching can be seen as alloying takes place.

In contrast to this, ovonic threshold and memory switches are completely reversible. Their mechanisms do not involve chemical reactions with electrodes, materials, or environment.

We will now briefly describe ovonic switching. The Ovonic Threshold Switch is a device which requires a sustaining current to keep it in the conducting condition. When the sustaining current is not present, the device reverts to its high resistance state. These devices have been cycled 8×10^{12} times without significant change of parameters, at which point the test was discontinued. The mechanism is considered to be primarily an electronic one and does not involve structural change. Thermal effects can be cooperative with the electronic mechanism[13–15].

The Ovonic Memory Switches can remain indefinitely either in the ON or OFF state without any sustaining energy input, that is the information is stored in an available stable structure of the material and one must alter the structure of the material in order to change its electrical resistance. Electronic threshold switching initiates the change in conductance from high to low. The change of conductivity, however, is stored in the structure. In essence, this switch is based on reversible phase changes. A recent test run was stopped after 3×10^{8} switching operations were observed without failure.

Let us now differentiate between the Ovonic Threshold Switch and the Ovonic Memory Switch. The threshold switch is made up of materials that are specifically chosen not to have structural changes, either at elevated temperatures or when slowly cooled. However, the memory material is chosen so that the cross-link elements, which shorten the polymeric chains and tie the bonding sites, allow for some polymeric movement that can include bond rotation and switching, chain shortening and lengthening and very limited spatial diffusion of atoms. Such atomic movement need not extend further than a few angstroms. The new structural configuration develops towards a different stable state which involves more order, therefore it tends toward crystallinity. This more conducting configuration is connected with lowering the local satisfaction of valence requirements of the amorphous state.

There is the possibility that as soon as positional and compositional disorder starts to change towards more order, the Fermi level moves away from the center of the gap and the material becomes a low resistivity semiconductor with perhaps a smaller band gap. Since this involves the changing of atomic overlap, it may be related to a Mott transition. It was Mott who also conceived of the ability of amorphous materials to accommodate changes in local structure which result in satisfying bonding requirements[16]).

Memory action involves the transformation of a material from one stable state to another. This transformation is affected by either the taking in or giving up of energy in the transitions. Fritzsche, who has been a most valued and early collaborator, has been utilizing the DTA to elucidate the properties of memory materials[17,18]) and will report some of his findings at this meeting[19]). Electronic processes initiated by the electric field create the energetic conditions which provide for the limited atomic motion that allows the structural reorganization to take place. The changes of electrical resistance are, therefore, associated with ordering and disordering processes. Electric field, current density and thermal gradients may affect the filamentary ordering processes since they are anisotropic in nature[9, 13–15]. Ridley has shown that in S-shape breakdown, filamentary conduction should be expected[20].

A. Bienenstock, at this meeting, will show that, by using X-ray diffraction and spectrographic techniques, the transitions identified with DTA are associated with ordering through the formation of small crystallites[21].

A third device which has interesting possibilities is a variation of the ovonic memory device called the Ovonic Adaptive Memory. Some aspects of this action will be discussed by Evans and Helbers[22]. The adaptive memory is deliberately designed so as to have a very large number or continuous range of resistance values between the OFF and ON conditions. It is energy-controlled in that the amount of energy put into the materials determines the resistive value which then retains the information state without the need for a holding energy and yet it can be altered. We consider this effect to be associated with a varying degree of structuring related to energy deposited in the material.

Amorphous materials are based upon positional and, in the case of alloys, also compositional disorder. However, paradoxically, it is the understanding of the role of short-range order in the form of local bonds and the steric architecture involved which is essential for both the understanding and the design of amorphous switching devices.

Various single elements can be condensed on cold substrates as amorphous films and yet they are usually not preferable for device usage. The spatial relationship of one atom to another can be altered by the increase in mobility of the atoms with temperature, for example, and therefore new local structures can be formed with different relative positions and changes of overlapping orbitals[23]. This is most pronounced in chain forming materials, but holds true for most polymer-forming single elements. Obviously this is a form of memory action, for change of structure can be reflected in change of electrical resistance, but unfortunately single element memory is not very stable. For most memory work, one finds that once a memory made only of a single element crystallizes, the energy necessary to disrupt the crystal and revert it back to its amorphous state becomes high and difficult to control. This leads to destructive breakdown. It is useful to add atoms anchoring single element structures to prevent creep and aging, which are the reflections of configurational changes.

For improved memory action, material compositions are chosen which are allotropic in that they can have an ordered or disordered phase, that is they can be either crystalline or amorphous, for example, germanium telluride.

The introduction of suitable crosslinking materials such as arsenic, phosphorus, etc. provides an additional chemical bonding distortion which favors the amorphous state. Under suitable energy conditions, at least one of the species in the mixture can become ordered and can become a nucleation point for other such arrangements. Disordering is accomplished by passing sufficient energies through the structure, causing randomization once more. Amorphousness can be assured by chemical bonding of cross-linking atoms.

The threshold materials are chosen with different structural principles in mind. The preferred materials are stabilized by alloying, so as to prevent atoms from moving and changing their spatial relationships. This is done by fixing the local order desired through the addition of compatible elements that can bind all available sites. These elements are chosen for two reasons:

one, to provide interfering additives to inhibit crystallization, and two, to provide, through the processes of chemical bonds of different energies, a large density and a wide spectrum of localized states which bridge the valence and conduction bands. The band gap model of Cohen, Fritzsche and Ovshinsky[24–26]) takes into account the importance of the localized states for the establishment of a high density of charged and neutral trapping centers. Fagen has reported[27, 28]), and will report further at this meeting[29, 30]), the large density of recombination centers that do exist at the Fermi energy. In our model, the almost complete local satisfaction of valence requirements yields the observed intrinsic behavior of the conductivity. Another possible explanation of intrinsic behavior will be described by our colleague, Böer[31]). In a material which has a high density of localized states, injecting processes can be considered to be a dominating factor in the transformation of the material into the conducting state, for if traps can be made to be inoperative, then clearly the material can become highly conducting until the current is decreased beneath a minimum holding current which is necessary to sustain the injecting process. Henisch is considering space charge effects in this kind of model[32]).

The field of disordered materials is an exciting one since it is a relatively new and expanding area in physics. Conventional concepts are going to be challenged and new techniques will be developed that will allow us to understand electronic processes better. As with the early transistor, one wipes out the distinction between pure and applied physics, for in order to see high field phenomena, one must really make a device. The device and the phenomena interesting to physicists become indistinguishable. Neale, an early collaborator, will describe the development of the manufacturing techniques that mate science and technology[33]).

What are some of the problems that we face as we develop amorphous switching? There is continuous material development, a portion of which is devoted to raising the glass transition temperature. This will minimize the problem of straight dc operation associated with some threshold switching. Another important part of our effort is to make sure that catalytic agents which enhance structural change are not introduced either during manufacturing or operation. We are also concerned with possible reactions between contact materials and semiconducting switching materials. Fleming has contributed greatly to our development with his materials work as well as directing the electroluminescent activities reported at this Conference[34]).

There is much work being done on circuit applications by the group at ECD under the direction of Nelson[35]). Various aspects of this work have been reported[18, 36, 37]). Programs in display and memory circuits are being actively pursued with some aspects of memory utilization being investigated by Sie. Electrical characterization of devices continues as a major project as exemplified by Shanks' paper at this meeting[38]).

The size of this meeting illustrates the great interest in amorphous switching and gives us great satisfaction. It has been a long journey from 1958 to the present. We have been fortunate to witness the increase of understanding in this area of physics and look forward to a rapid growth of this field. The work that will be reported here by our group reflects the progress being made.

References

1) This idea guided Mooser and Pearson to predict semiconducting behavior in a large variety of crystalline materials. See E. Mooser and W. B. Pearson, Can. J. Phys. **34** (1956) 1369.

2) For references to their work see A. F. Ioffe and A. R. Regel, Non-Crystalline, Amorphous, and Liquid Electronic Semiconductors, in: *Progress in Semiconductors*, Vol. 4, Ed. A. F. Gibson (Wiley, New York, 1960) pp. 237–291.
3) For references to their work see B. T. Kolomiets, Phys. Status Solidi **7** (1964) 359, 713.
4) For a discussion of electronic properties of amorphous materials see J. Tauc, Science **158** (1967) 1543.
5) S. R. Ovshinsky, Electronics **32** (1959) 76.
6) S. R. Ovshinsky, Control Engineering **6** (1959) 121.
7) L. Young, *Anodic Oxide Films* (Academic Press, New York, 1961) pp. 147–149.
8) S. R. Ovshinsky, The Physical Base of Intelligence – Model Studies, presented at Detroit Physiological Society, Dec. 17, 1959.
9) S. R. Ovshinsky, Symmetrical Current Controlling Device, U.S. Patent No. 3, 271, 591.
10) S. R. Ovshinsky, Phys. Rev. Letters **21** (1968) 1450.
11) A. D. Pearson, W. R. Northover, J. F. Dewald and W. F. Peck, Jr., Chemical, Physical, and Electrical Properties of Some Unusual Inorganic Glasses, in: *Advances in Glass Technology* (Plenum Press, New York, 1962) pp. 357–365.
12) A. D. Pearson, in: *Proc. Symp. on Instabilities in Semiconductors*, IBM Watson Research Center, Yorktown Heights, New York, March 1969.
13) S. R. Ovshinsky, The Ovonic Switch as an Amorphous Switching Device, presented at IV Symposium on Vitreous Chalcogenide Semiconductors, Acad. Sci., USSR, Leningrad, 1967.
14) S. R. Ovshinsky, Ovonic Switching Devices, presented at Intern. Colloq. on Amorphous and Liquid Semiconductors, Acad. Soc. Repub. Rumania, Bucharest, 1967.
15) S. R. Ovshinsky, Ovonic Switching Devices, Proc. of 1968 Electronic Components Conf., Washington, D.C., 1968, p. 313.
16) N. F. Mott, Advan. Phys. **16** (1967) 49.
17) H. Fritzsche, Physics of Instabilities in Amorphous Semiconductors, Symp. on Instabilities in Semiconductors, IBM Watson Research Laboratory, Yorktown Heights, New York, March 1969.
18) H. Fritzsche, Bull. Am. Phys. Soc. [II] **14** (1969) 342.
19) H. Fritzsche and S. R. Ovshinsky, J. Non-Crystalline Solids **2** (1970) 148.
20) B. K. Ridley, Proc. Phys. Soc. (London) **82** (1963) 996.
21) A. Bienenstock, F. Betts and S. R. Ovshinsky, J. Non-Crystalline Solids **2** (1970) 347.
22) E. J. Evans, J. H. Helbers and S. R. Ovshinsky, J. Non-Crystalline Solids **2** (1970) 334.
23) Effects of this nature were observed in the form of an increase in the resistivity of evaporated films of Ge and Si as a result of heat treatment. For a review see P. A. Walley and A. K. Jonscher, Thin Solid Films **1** (1968) 367.
24) M. H. Cohen, H. Fritzsche and S. R. Ovshinsky, Phys. Rev. Letters **22** (1969) 1065.
25) H. Fritzsche and S. R. Ovshinsky, J. Non-Crystalline Solids **2** (1970) 393.
26) M. H. Cohen, J. Non-Crystalline Solids **2** (1970) 432.
27) E. A. Fagen, S. R. Ovshinsky and H. Fritzsche, Bull. Am. Phys. Soc. [II] **14** (1969) 311.
28) H. Fritzsche, E. A. Fagen and S. R. Ovshinsky, Bull. Am. Phys. Soc. [II] **14** (1969) 311.
29) E. A. Fagen and H. Fritzsche, J. Non-Crystalline Solids **2** (1970) 170.
30) E. A. Fagen and H. Fritzsche, J. Non-Crystalline Solids **2** (1970) 180.
31) K. W. Böer, J. Non-Crystalline Solids **2** (1970) 444.
32) H. K. Henisch, personal communication.
33) R. G. Neale, J. Non-Crystalline Solids **2** (1970) 558.
34) G. R. Fleming, J. Non-Crystalline Solids **2** (1970) 540.
35) D. L. Nelson, J. Non-Crystalline Solids **2** (1970) 528.
36) S. R. Ovshinsky and D. L. Nelson, Ovonic Switches and Their Application, Proc. IEEE Intern. Convention, New York, March 1969.
37) S. R. Ovshinsky, E. J. Evans, D. L. Nelson and H. Fritzsche, Radiation Hardness of Ovonic Devices, IEEE Trans. Nuclear Science (Dec. 1968), p. 304.
38) R. R. Shanks, J. Non-Crystalline Solids **2** (1970) 504.

Reprinted with permission from *Applied Physics Letters,* Vol. 46(8), pp. 734-736 (April 15, 1985).

Reversibility and stability of tellurium alloys for optical data storage applications

M. Chen, K. A. Rubin, V. Marrello, U. G. Gerber, and V. B. Jipson
IBM Research Laboratory (K47/282), 5600 Cottle Road, San Jose, California 95193

We demonstrate for the first time that $Te_{87}Ge_8Sn_5$ films, which are amorphous as deposited, can be optically switched between the crystalline and amorphous states more than 10^6 times. The measured reflectivity changed from 40% to 60% and the transmission changed from 3% to 1.5%, respectively, between the amorphous and crystalline states. The crystallization temperature of the cycled spots is ≈ 75 °C and these spots are observable after > 20 weeks. It is found that the crystallization temperature of cycled spots is typically about 20 °C lower than that of the unwritten film. Increasing the Ge concentration leads to increased crystallization temperature and increased minimum crystallization time without affecting reversibility.

Optical data storage based on marking of micron-sized spots on a disk using a laser is an area with on-going research activity.[1] Optical disk data storage has the combined advantages of high areal storage density, disk removability, and large head-disk working distance. In recent years, write-once optical storage products have been commercially introduced and more research effort is being devoted to reversible optical storage materials.[2]

Early work on Te-based alloys suggested several potential problems which need to be addressed. These problems include limited reversibility,[3] due potentially to phase segregation,[4,5] and varying recording conditions with cycling.[6] More recently, films of TeGeAs have been successfully cycled between the amorphous and the crystalline states for more than 10^4 cycles.[7] However, these films contain very little Ge and As so that the data retention time of the amorphous phase is limited to about three weeks. In fact, the crystallization temperatures of these films are sufficiently low so that they crystallize either during deposition or rapidly (hours?) thereafter at room temperature. Te alloy films have also been reported to be capable of cycling more than 500 times, to have fast erasure times, to have good signal-to-noise ratios, and to have > 100 °C transition temperatures.[8] It is not clear in Ref. 8 whether all these properties were obtained for the same alloy.

In this letter we report static testing results for two TeGeSn alloy films. We examined their limits of cycling, their optical properties as a function of the number of cycles, and the crystallization temperatures of the cycled amorphous spots as well as the unwritten film.

We designed our experimental setup to rapidly and flexibly study reversible phase change optical recording materials. Two lasers are used in the experiment. These two laser beams are combined using a dichroic mirror and directed onto the sample through a commercial Olympus lens with a numerical aperture of 0.5. One of the lasers is a GaAs laser which is used to allow continuous automatic focusing of the lens on the sample under test. Focus error is derived using a knife edge and a split detector.[9] The focus error is fed to servoelectronics which drive the actuator in the lens assembly to maintain focus. To ensure that the focus laser beam does not affect the spot under test, the focus laser spot is offset slightly from a second Kr-ion laser spot. The Kr-ion laser is used to write and erase spots on the sample; it is also used to measure changes in the optical properties of the tested spots. Chromatic aberration of the lens is compensated for by incorporating suitable additional lenses in the GaAs laser beam path. The size of the focused Kr-ion laser on the sample is about 0.8 μm full width at half-maximum. An acousto-optic modulator is used to produce Kr-ion laser pulses of various intensities and durations. The desired sequence of write, erase, and read pulses from the Kr-ion laser is obtained by using three pulse generators, a summing amplifier, a 200-MHz radio frequency (rf) oscillator, a rf amplifier, and a rf switch. The reading is done both in reflection and in transmission using two Si photodiode detectors with associated preamplifiers; simultaneously a third detector and preamplifier combination measures the instantaneous laser power to allow correction for laser power fluctuations. Each of the preamplifiers is connected to a sample-and-hold and an analog-to-digital converter. The reading pulses are 10 μs long. The samples are mounted on stepper-motor-driven stages and the entire experiment is controlled through an IBM S/1 computer. On this experimental setup, only relative changes in reflectivity (R) and transmission (T) are measured. A calibration of the relative R and T was done by separately measuring the absolute R and T for an amorphous and a crystalline sample.

The Te alloy films were prepared by thermal evaporation from multiple sources onto 1 in. $\times$ 1 in. glass substrates. The film thickness is 750 Å and the deposition rate used was 300 Å/min. The samples were then placed in another vacuum system where SiO_x ($x \approx 2$) overcoats were deposited by thermally evaporating SiO in the presence of 2×10^{-4} Torr of O_2 at a rate of 2 Å/min. The overcoat thickness is about 3000 Å. Overcoats are necessary to suppress hole formation which reduces the laser power margin for the amorphization process.[10]

Typical reversible cycling data for a $Te_{87}Ge_8Sn_5$ film are shown in Fig. 1(a). We have plotted R of the spot under test against the cycle number. Odd-numbered pulses of 1.7-mW power and 30-μs duration were used to erase the spot. Even-numbered pulses of 24-mW power and 50-ns duration were used to write the spot. The erase pulses increased R and decreased T while the write pulses decreased R and increased T. The effect of the erase pulse on R and T was

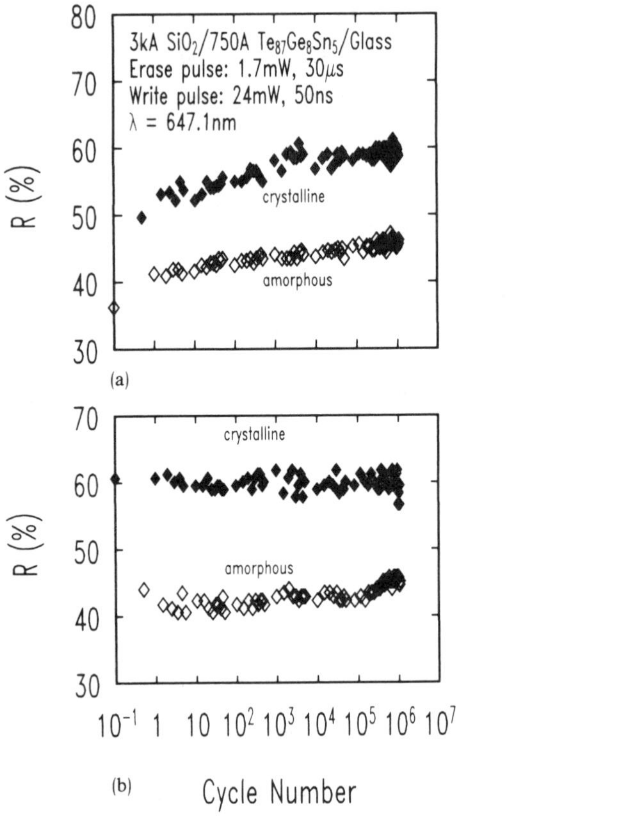

FIG. 1. Reflectivity vs cycle number for a $Te_{87}Ge_8Sn_5$ film: (a) as deposited and (b) after thermal annealing at 150 °C for about 10 min. For illustration purpose, the initial reflectivity of the sample is shown on the *y* axes (at cycle number = 0.1).

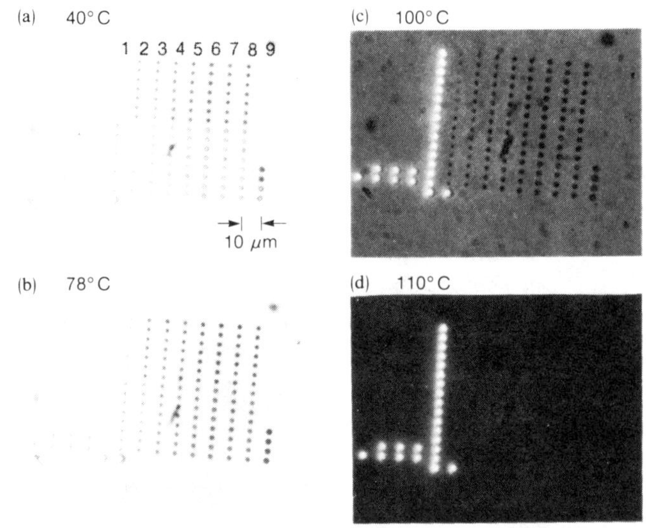

FIG. 2. Transmissive optical micrographs of an array of cycled spots for as-deposited $Te_{87}Ge_8Sn_5$ film at (a) 40 °C, (b) 78 °C, (c) 100 °C, and (d) 110 °C. All four photographs were taken with the same exposure. The large bright spots to the left of each photo are ablated holes used for identification. The nine nonablated rows are identified by row numbers. From bottom to top, row 1 consists of one ablated spot, eight spots which were written with one erase pulse only, and seven spots which were written with one erase and one write pulse (one cycle). Rows 2 through 8 each consists of eight spots which have been cycled an integral number of times and seven spots which have been cycled the same number of times plus one additional erase pulse. The numbers of cycles for rows 2 to 8 were 5, 50, 500, 5 K, 50 K, 50 K, and 50 K, respectively. The time delay between the erase and the write pulses was varied in rows 6 to 8 and no difference is observed in the spots, indicating that the spots cooled sufficiently between pulses. Row 9 consists of two spots which have been put through $> 1.1 \times 10^6$ cycles and two spots which have been put through $> 1.1 \times 10^6$ cycles plus an additional erase pulse. In every case when a spot has been cycled an integral number of times, the written region is surrounded by a crystalline ring. The width of this ring increases with the number of cycles.

similar to what we obtained by heating the sample on a heating stage to 150 °C (2 °C/min heating rate). We conclude, therefore, that the as-deposited film is amorphous, the erase pulses crystallized the spot under test, and the write pulses amorphized the spot under test. A clear change in R, between $\approx 40\%$ and $\approx 60\%$, and in T, between $\approx 1.5\%$ and $\approx 3\%$, is observed following each erase and write pulse for more than one million cycles. Laser amorphized spots are surrounded by crystalline rings whose size grows with cycling (see Fig. 2). These rings are detected by the skirt of the read beam and contribute towards R and T measured using the focused beam. Thus, the gradual change of R and T with cycling is explained by the presence of these rings. Preannealing the sample so that these rings cannot be differentiated from the background (crystalline) film would yield better reversible writing characteristics. Shown in Fig. 1(b) is the reversible cycling data for the sample after annealing to 150 °C for about 10 min. In this case, as expected, R was found to be independent of the number of cycles for both the written and the erased states. The number of cycles achieved at this spot was limited to about 1 million due to eventual hole formation, as indicated by the down turn in R after $\gtrsim 1$ million cycles.

For erase pulses, power can be traded off against duration. However, a minimum erase pulse width of 30 μs was found to be necessary to allow reversible cycling of the film. Shorter erase pulses were found to be unreliable in erasing, independent of power. Using this minimum erase time of 30 μs, we found acceptable erase powers to range from 1.5 to 2.1 mW.

The crystallization temperatures of the cycled spots as well as of the unwritten film were measured. Arrays of spots which were cycled various numbers of times, using minimum erase pulse width, 50-ns write pulse width, and optimum write and erase powers, were made on the films. The film was then placed in a heating stage and attached to an optical microscope. Data taken using transmitted light are shown in Fig. 2. A constant exposure was used for all photographs. A heating rate of 2 °C/min was used. Figure 2(a) is taken at 40 °C and shows the array of spots which have not changed from their room-temperature state. A spot which is left in the erased state (i.e., after an integral number of cycles was performed, an additional erase pulse was applied) is simply a crystalline dot and appears as a dark spot. A spot which is left in the written state (i.e., cycled through an integral number of times) consists of a light, amorphous central spot surrounded by a dark, crystalline ring. The size of the crystalline area increased with cycling. At 78 °C, as shown in Fig. 2(b), all cycled spots have crystallized; however, the background film has not. Further, the crystallization temperature of an amorphous spot which had undergone previous cycling did not depend on the number of cycles. In Fig. 2(c), which was taken at a sample temperature of 100 °C, the background film had begun to crystallize; cycled spots are still observable as darker spots. The entire film crystallized

TABLE I. Summary of cycling and crystallization temperature data.

Material	Phase (as deposited)	Cycles	Erase time (μs)	T_x data[a] (°C)	T_x film[a] (°C)
Te	crystalline	$>10^6$	<1	$\lesssim 20$	$\lesssim 20$
$Te_{87}Ge_8Sn_5$	amorphous	$>10^6$	$\gtrsim 30$	75	100
$Te_{79}Ge_{16}Sn_5$	amorphous	$>5\times10^5$	$\gtrsim 200$	95	120

[a] Based on visual observation under an optical microscope.

at 110 °C, as shown in Fig. 2(d). Due to the resolution limit of the microscope, it was not possible to determine if the crystallization rate was significantly higher next to the crystalline ring. Therefore, it is possible that the apparently lower crystallization temperature of the cycled spots is simply due to growth of the crystalline ring and not a result of a real change in the crystallization temperature. Spots on the annealed film which was kept under normal ambient conditions were monitored periodically using a transmissive optical microscope. Although a dark region developed in the center of each spot after about two weeks, no further change was observed and the spots remained clearly observable after more than 20 weeks.

To study the effect of varying the Ge concentration, we also carried out measurements for a $Te_{79}Ge_{16}Sn_5$ film as well as a pure Te film. The results are summarized in Table I. Based on the cycling data, we conclude that the addition of up to 16 at. % of Ge to a TeSn alloy film does not significantly affect the ability of the film to be reversibly cycled (up to half a million times). Phase segregation of Te and TeGe, which likely took place when the sample was laser crystallized[4] and almost certainly when the sample was annealed on a heating stage,[5,11] does not appear to be a limiting factor for reversibility. We believe that when a spot is melted during the amorphization pulse, a mixing of the segregated phases takes place. The spot is then quenched in a mixed, amorphous state; resulting in a crystallization temperature which is substantially higher than that of pure Te. Higher Ge concentration does lead to an alloy with a substantially longer required erase time as well as a more stable amorphous phase, as evidenced by a higher crystallization temperature. A long erase time is undesirable for optical recording applications because it would require either a long transit time under the erase laser beam (elongated erase spot) or multiple passes under a micron-sized laser spot.

In summary, we draw four major conclusions from the data presented. (1) Reversibility is not limited by phase segregation but by ablation, and up to 16 at. % substitution of Ge for Te in a TeSn film does not change the reversibility (0.5–1 million cycles) of the film, (2) The required minimum crystallization time increases rapidly with Ge concentration, (3) The apparent crystallization temperature of a cycled spot is approximately 20 °C lower than that of the as-deposited material, (4) Crystallization temperature is increased by increasing the Ge concentration in the film. We have demonstrated, for the first time, that Te alloy films which are amorphous as deposited, suggesting high crystallization temperatures and hence potential long data retention time, can be reversibly cycled for more than half a million times. For these films to be considered for optical disk applications, it would be desirable to reduce the minimum erasure time and to increase the crystallization temperature.

It is a pleasure to acknowledge Dr. A. E. Bell, Dr. W. Y. Lee, and Dr. R. Barton for many stimulating discussions. We would like to thank R. N. Payne and W. C. McChesney for the preparation of the samples. We are grateful to E. Alger and R. W. Martin for providing S/1 programming support, and to Dr. J. R. Salem and R. D. Tom for assistance in the SiO_x deposition.

[1]See, for example, numerous articles and references in Proc. Soc. Photo-Opt. Instrum. Eng. **420** (1983).
[2]See, for example, Technical Digest of Topical Meeting on Optical Data Storage, Monterey, CA, April 1984.
[3]B. R. Brown, Appl. Opt. **13**, 761 (1974).
[4]P. Chaudhari and S. R. Herd, J. Non-Cryst. Solids **8-10**, 56 (1972).
[5]J. Cornet, Ann. Chim. **10**, 239 (1975).
[6]A. W. Smith, Appl. Opt. **13**, 795 (1974).
[7]P. C. Clemens, Appl. Opt. **22**, 3165 (1983).
[8]D. Strand and D. Adler, Proc. Soc. Photo-Opt. Instrum. Eng. **420**, 200 (1983).
[9]C. Bricot, J. C. Lehureau, and C. Puech, IEEE Trans. Consumer Electron. Nov. 304 (1976).
[10]M. Chen and V. Marrello, J. Vac. Sci. Technol. **18**, 75 (1981).
[11]See, for example, the phase diagram for Te-Ge in *Constitution of Binary Alloys*, 2nd ed., edited by M. Hansen (McGraw–Hill, NY, 1958), p. 776.

Reprinted with permission from *Applied Physics Letters*, Vol. 49(9), pp. 502-504 (September 1, 1986). ©1986 American Institute of Physics.

Compound materials for reversible, phase-change optical data storage

M. Chen, K. A. Rubin, and R. W. Barton
IBM Almaden Research Center (K67/802), 650 Harry Road, San Jose, California 95120-6099

Results of rapid (laser induced) and slow (heating stage induced) crystallization studies on $Te_{1-x}Ge_x$, $0 \leqslant x \leqslant 0.6$, are reported. The time it takes to laser crystallize varies with x by more than four orders of magnitude. Films with stoichiometric compositions, Te and GeTe, can be crystallized using laser pulses of less than 100 ns duration. Unlike Te, which spontaneously crystallizes at room temperature, GeTe has a crystallization temperature of > 150 °C. From these results we argue that, in general, compound materials allow realization of fast-switching, reversible, phase-change optical recording media. Furthermore, this fast-switching capability, from the amorphous to the crystalline state and back, can be attained simultaneously with long term data (amorphous phase) stability, provided the melting temperature of the compound is sufficiently high.

Reversible optical information storage based on the amorphous to crystalline phase transformation was first studied in the early 1970s.[1] A particular "phase-change" technique, termed reverse-mode recording, records data by quenching amorphous spots in the material with short laser pulses and erases data by recrystallizing with long laser pulses.[2] A number of initial studies were performed which concentrated mainly on Te alloys.[3,4] Significant problems cited in these studies were limited reversibility[5] and aging effects which required adjustment of recording and erasing powers with cycling.[4] Progress has been reported recently in solving these problems for Te alloys as well as alloys based on other elements.[6] However, to date, no alloys have been shown to possess both a short crystallization time and a long data retention time—requirements that appear to be mutually exclusive. The need for a fast crystallization process is a result of the short time ($\lesssim 300$ ns) which a given spot on an optical disk spends under the laser beam during each revolution. Takenaga *et al.*[7] have reported a medium with $\approx 1\ \mu s$ erase speed and > 1 yr data stability. This medium is more complex than Te alloys and consists of small (≈ 20 Å) TeGeSn alloy particles embedded in a TeO_2 matrix. Very recently, Sb_2Se has been found to possess both fast erase speed and data stability.[8]

In this letter we present data on the speed with which laser-induced crystallization of $Te_{1-x}Ge_x$ films can be accomplished. The crystallization temperatures of the these films, measured on a heating stage, are also reported. The results can be interpreted and generalized based on familiar concepts in material science. As a consequence, we can develop guidelines for the searching of media with both fast crystallization speed and long data retention capability. We will also briefly address the reversibility issue.

The search for potential materials for phase-change optical recording has mostly been based on one approach. Starting with a group V A or VI A element in the periodic table, attempts were made to improve or to achieve desired properties through alloying with other elements. A familiar example of this approach is the alloying of Ge with Te to circumvent a major limitation of Te, i.e., instability of the amorphous state at room temperature. The addition of elements such as Ge and As to Se is known to crosslink the Se chains and to cause an increase in the stability of the amorphous phase.[9] This crosslinking model is expected to apply to Te-Ge alloys as well. The improvement in stability, however, is at the expense of greatly increased erase time.[10] The reason for the slowing down of the crystallization speed upon alloying is apparent upon examination of published phase diagrams. For example, the phase diagram of the Te-Ge binary system, shown in Fig. 1(a), indicates that the stable crystalline phases are Te, Ge, and GeTe.[11] Amorphous alloy films can be prepared with a wide range of com-

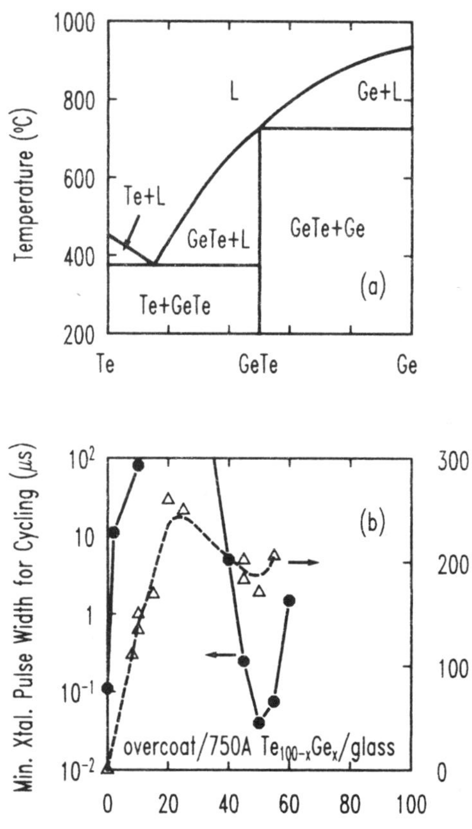

FIG. 1. (a) Phase diagram for the binary Te-Ge system, obtained from Ref. 11. (b) Minimum crystallization pulse width and crystallization temperature vs composition. The minimum crystallization pulse width is defined as the shortest laser pulse width required for crystallization in order to obtain phase-change cycling with saturated contrast. The crystallization temperature is determined by measuring the optical transmission of the sample as a function of the sample temperature. The heating rate used was 23 °C/min.

positions, but upon crystallization these films are subject to phase separation. Phase separation in laser-crystallized alloy $Te_{0.88}Ge_{0.07}As_{0.05}$ has been reported.[3] We have also confirmed phase separation in laser-crystallized films of $Te_{0.95}Ge_{0.05}$ and $Te_{0.85}Ge_{0.15}$ using transmission electron microscopy (TEM). Phase partitioning during crystallization must be accompanied by diffusion of atomic species, which is slow and severely limits the crystallization speed. In contrast, when films are prepared with compositions which allow congruent melting, crystallization should proceed much faster. Stoichiometric compounds (or pure elements) are sufficient, though not necessary, to guarantee this. We present in Fig. 1(b) experimental data for the Te-Ge binary system which demonstrate the large sensitivity of erase speed to alloy composition. The minimum erase pulse width required to allow cycling of an alloy on our static tester[10] is plotted versus alloy composition in Fig. 1(b). The data show clearly that partitionless crystallization can be very fast, i.e., erase pulses of 100 and < 30 ns can crystallize pure Te and GeTe, respectively. The erase speed decreases progressively as the composition deviates from the single phase compositions. A 5 at. % departure from the stoichiometric compositions increases the required erase pulse duration by a factor of 2–10. The minimum required erase pulse width for films with composition in the middle of the two phase region is more than four orders of magnitude longer than that for films with stoichiometric compositions.

TEM observations of laser-crystallized GeTe films confirm the existence of the face-centered-cubic GeTe structure with no evidence of either second phase Te crystals or remnant amorphous material. In general, laser crystallization of both Te and GeTe films produces very large grains, often on the order of 0.5 μm in diameter. In most instances, these grain sizes are limited by the size of the spot heated by the laser beam. Large grains are indicative of a high ratio of crystal growth rate, unencumbered by phase separation, to nucleation rate. Similar large grain sizes were reported for laser-crystallized metallic films.[12]

Several other stoichiometric chalcogen-based compounds that can be laser crystallized with pulse widths less than 1 μs were found to be Sb_2Se_3, Sb_2Se, and Sb_2Te_3. However, GeSe and $GeSe_2$ required much longer time to crystallize. Of these latter two compounds, at least $GeSe_2$ is known to possess an unusually high viscosity near its melting point (≈ 6 P).[13] In contrast, the fast crystallizing compounds have viscosities in the range of 0.01 P upon melting.[14] Materials which form very viscous liquids (such as SiO_2 or organic polymers) are known to be slow crystallizers. We conclude, in general, that fast crystallizing media should be based on materials which stay single phase through the amorphous to crystalline phase transition and which have low melting point viscosities.

In our usual sample structure consisting of a polymethyl methacrylate substrate on top of which is deposited a 750-Å active layer, sandwiched between two 1-kÅ ZnS layers, we were unable to amorphize GeTe with laser pulses as short as 30 ns. Laser amorphization was possible only after the sample structure was modified by incorporating an additional layer of Al next to the top dielectric layer. This Al layer increased the cooling rate of the sample sufficiently to allow GeTe to be quenched into the amorphous state. Optimization of optical disks which incorporate a fast-crystallizing phase-change active layer, therefore, requires careful balancing of the thermal design of the disk structure with the cooling requirements of the active material.

The stability of recorded data in various media was investigated. The crystallization temperatures (T_x) of various Te-Ge alloys, determined by measuring optical transmission changes at a heating rate of 23 °C/min, are shown in Fig. 1(b). In contrast to the minimum erase speed, the dependence of T_x on alloy composition is not very dramatic. The change of T_x as Ge is added to pure Te is attributable to crosslinking of Te chains by Ge atoms. For films near the GeTe composition, only minor changes of T_x versus composition were observed. Our measured T_x for GeTe of 170 °C is in reasonable agreement with that reported previously.[15] Isothermal annealing of materials similar to GeTe, with $T_x \approx 200$ °C, was also carried out. Written data (amorphous spots) were found to be stable for between 7 and 8 months at an elevated temperature of 110 °C. If one assumes an Arrhenius behavior with an activation energy of 0.5 eV, the extrapolated data retention time at 50 °C would be > 5 yr. Thus, using a compound material as a reversible medium does not require compromising either data stability or erase speed.

There are several factors which lead to the stability of the amorphous phase of GeTe near room temperature. First, since the melting temperature of GeTe is 725 °C, the glass transition temperature (T_g) of GeTe is expected to be considerably higher than room temperature.[16] Thus, self-diffusion rates of Te or Ge in amorphous GeTe should be typical of those for solid state diffusion, $\approx 10^{-8}$–10^{-14} cm^2/s.[17] This slow diffusion limits the crystallization rate and allows recorded amorphous spots to be stable at room temperature. In contrast, during erasing with a laser beam, crystallization occurs at temperatures well above T_g where the diffusion rates are more typical of those in liquids and are expected to be 10^{-4}–10^{-5} cm^2/s.[17] A second factor contributing to the stability of amorphous GeTe may be the density difference between the amorphous and crystalline states, which is about 10%.[14] Strain energy of crystallites nucleated in an amorphous matrix can affect the growth rate of the crystalline phase.

Correlation between T_g and the melting temperature (T_m) has been noted.[16] It is often found that T_g is approximately 2/3 of T_m, when both temperatures are absolute temperatures. However, materials which are difficult to amorphize may have T_g as low as 0.5 T_m. If a glass transition temperature of 200 °C is desired, and assuming T_g equal to 0.5 T_m, the required T_m is about 700 °C. This temperature is higher than typical melting temperatures of Te alloys (≈ 450 °C) which implies that the more stable amorphous compound materials will be less sensitive in the writing process, which requires melting, than Te alloys.

We found that reversibility of our GeTe-based phase-change media is limited not by the phase transformation process but by the tendency for pinhole formation in the active material after repetitive cycling. A probability of 10^{-4}–10^{-5}

of pinhole formation translates into more errors on a disk than can be handled by reasonable error correction capabilities of an optical disk drive and limits the reversibility of the material in a practical application. Currently, weak spots on disks which are more prone to pinhole formation limit reversibility to about 10^3 cycles.

In summary, we have shown that the compound GeTe has an erase speed of <30 ns along with a T_x of 170 °C. Furthermore, these results may be generalized, based on familiar material science arguments, into principles which are critical to the selection of materials for phase-change optical recording. Choice of compounds, which do not require phase separation upon crystallization, can result in sub-100 ns erase speeds. In conjunction with this fast crystallization, the amorphous phase is stabilized by using compounds with high glass transition temperatures.

We thank A. G. Dewey, C. R. Davis, G. Lim, R. E. Geiss, and E. E. Marinero for performing various experiments which aided our understanding of the materials studied. Helpful discussions with V. B. Jipson, C. J. Lin, and D. Rugar are gratefully acknowledged. We thank W. G. McChesney for the preparation of samples used in this study.

[1]J. Feinleib, J. deNeufville, S. C. Moss, and S. R. Ovshinsky, Appl. Phys. Lett. **18**, 254 (1971).
[2]R. J. von Gutfeld, Appl. Phys. Lett. **22**, 257 (1973).
[3]R. J. von Gutfeld and P. Chaudhari, J. Appl. Phys. **43**, 4688 (1972); K. Weiser, R. J. Gambino, and J. A. Reinhold, Appl. Phys. Lett. **22**, 48 (1973); J. Cornet, Ann. Chim. **10**, 239 (1975).
[4]A. W. Smith, Appl. Opt. **13**, 795 (1974).
[5]B. R. Brown, Appl. Opt. **13**, 761 (1974).
[6]P. C. Clemens, Appl. Opt. **22**, 3165 (1983); M. Terao, T. Nishida, Y. Miyauchi, T. Nakao, T. Kaku, S. Horigome, M. Ojima, Y. Tsunoda, Y. Sugita, and Y. Ohta, Proc. Soc. Photo-Opt. Inst. Eng. **529**, 46 (1985).
[7]M. Takenaga, N. Yamada, S. Ohara, K. Nishiuchi, M. Nagashima, T. Kashihara, S. Nakamura, and T. Yamashita, Proc. Soc. Photo-Opt. Inst. Eng. **420**, 173 (1983).
[8]R. W. Barton, C. R. Davis, K. A. Rubin, and G. Lim, Appl. Phys. Lett. **48**, 1255 (1986).
[9]P. H. Gaskell, J. Phys. C **12**, 4337 (1979).
[10]M. Chen, K. A. Rubin, V. Marrello, U. G. Gerber, and V. B. Jipson, Appl. Phys. Lett. **46**, 734 (1985).
[11]M. Hansen, ed., *Constitution of Binary Alloys* (McGraw–Hill, New York, 1958).
[12]C. J. Lin and F. Spaepen, in *Chemistry and Physics of Rapidly Solidified Materials*, edited by B. J. Berkowitz and R. Scattergood (TMS–AIME, New York, 1983), p. 273.
[13]A. Laugier, G. Chaussemy, and J. Fornazero, J. Non-Cryst. Solids **23**, 419 (1977).
[14]V. M. Glasov, S. N. Chizhevskaya, and N. N. Glagoleva, *Liquid Semiconductors* (Plenum, New York, 1969).
[15]D. J. Sarrach, J. P. deNeufville, and W. L. Haworth, J. Non-Cryst. Solids **22**, 245 (1976).
[16]D. Turnbull, Contemp. Phys. **10**, 473 (1969).
[17]G. H. Geiger and D. R. Poiriev, *Transport Phenomena in Metallurgy* (Addison–Wesley, Menlo Park, CA, 1973).

Reprinted with permission from *Thin Solid Films,* Vol. 181, pp. 129-139 (1989).

PROGRESS AND ISSUES OF PHASE-CHANGE ERASABLE OPTICAL RECORDING MEDIA

KURT A. RUBIN AND MARTIN CHEN

IBM Research Division, Almaden Research Center, 650 Harry Road, San Jose, CA 95120-6099 (U.S.A.)

Recent progress in realizing practical media for phase-change reversible optical recording is described. Key attributes required of the media and the material science which underlie them are discussed. These include: (1) the kinetics of crystallization and amorphization, (2) chemical and data stability, (3) reversibility, and (4) readout signal-to-noise ratio. The crystallization kinetics of recording media is discussed using two examples: $(GeTe)_{0.85}Sn_{0.15}$ and the pseudo-binary InSb–GaSb stoichiometric compound system. A discussion of the recorded signal strength, sources of media noise and resulting signal-to-noise ratio is presented. Non-uniform optical index of refraction was shown to increase media noise. Methods to reduce the effect of media noise are presented.

1. INTRODUCTION

Erasable optical recording is an emerging technology for achieving high-density data storage. The two leading contenders are magneto-optic recording and phase-change recording. Magneto-optic recording detects small polarization rotations of light reflected from magnetic domains. Phase-change recording uses differences of reflected light intensity to distinguish recorded data bits. While magneto-optic recording is presently the most mature technology, phase-change optical storage has several attributes which are attractive. These include an optical head with fewer components, which simplifies alignment. The lack of polarization-preserving components makes future development of a compact integrated optical head much more feasible. Finally, the magnitude of the phase-change signal, which is several orders of magnitude greater than that from magneto-optic media, provides the potential for much higher signal-to-noise ratio (SNR).

The principle underlying optical recording using phase-change media is the controlled, reversible switching of a spot on the disk between two states, usually the amorphous and crystalline states. The written amorphous state is created by heating the material above the melting temperature with a laser beam and then letting it cool rapidly. Erasure is accomplished by heating the spot above the crystallization temperature for sufficient time to cause complete recrystallization of the amorphous spot. Changes in electronic structure accompany these phase transformations and result in significant changes in optical properties such as reflectivity. The reflectivity difference is detected using a low-power, continuous laser beam to read the data. Presently, many materials are being investigated. Tellurium-based materials have been studied most extensively because glass formation is easy for a wide range of compositions.

This paper describes several of the key attributes required of phase-change reversible recording media. These include: (1) fast writing and erasing, (2) adequate number of reversible cycles, (3) stable written and erased states, (4) large signal together with low noise. Attribute (1) is necessary for one very important feature of phase-change media, which is single-beam direct overwrite, *i.e.* the ability to use a circular beam to write new data without having to erase the old data[1].

2. HISTORY

Research of phase-change recording materials is now more than two decades old. It began with the work of Ovshinsky *et al.*[2]. Early works dealt mostly with tellurium-based alloys[3–9]. As the glassy state was most easily generated in materials with compositions near eutectic points, these exploratory studies invariably used alloys; and coupon samples rather than disks were typically used.

Progress in this technology accelerated beginning in 1983 with the report by Clemens[10] that 40 000 cycles between the amorphous and crystalline states were achieved. In the same year, Takenaga *et al.* claimed 10^6 cycles on a disk with 55 dB carrier-to-noise ratio (CNR) using a tellurium-oxide-based active layer[11]. The crystallization process of tellurium-oxide based write-once materials was shown to take up to several minutes to complete[12]. This is also a potential problem for reversible media. Also, the minimum erase time was 1 μs which necessitated a complicated two-laser-beam head, with an elliptically shaped erase beam and a circular write beam. The alignment of the optical head was undoubtedly very challenging.

In 1985, Chen *et al.* reported that erase times of less than 30 ns could be achieved without compromising data stability by using media based on stoichiometric compounds[13]. This allowed a much simpler, single-laser-beam optical head to be used for both writing and erasing, and the practicality of a phase-change recording system was greatly enhanced.

Other significant developments, reported in 1989, include Sato *et al.*[14], who reported 10^5 reversible cycles using In–Sb–Te; Terao *et al.*[15], who obtained more than 10^5 cycles using Ge–Sb–Te–Co; and Ohta *et al.*[16], who reported 2×10^6 reversible cycles in GeTeSb-based films, using the most stringent criterion, namely bit error rate, to monitor failure. Furthermore, all of these studies used single-beam direct overwrite of old data, with more than 28 dB of overwrite ratio.

3. KINETICS OF PHASE TRANSFORMATIONS

There are several factors which control the transformation kinetics of the media between the amorphous and the crystalline phases. The active layer has an intrinsic, composition-dependent crystallization time. Also, the temperature–time history which the layer experiences is important. The latter is affected by the laser power and pulse width and by the thermal properties of the disk structure. The laser pulse conditions dictate whether melting occurs or if adequate time for crystallization is provided. The disk structure controls the cooling rate of the media which is important for amorphization.

We have consistently observed that the minimum time to crystallize tellurium-based compounds and alloys depends on the history of the film. For example, the minimum time to crystallize 1 μm diameter as-deposited, amorphous (vapor-quenched) spots and laser-amorphized (liquid-quenched) spots of $(GeTe)_{85}Sn_{15}$ are approximately 700 ns and 200 ns, respectively. The reason for this difference is suggested by Fig. 1, which contains TEM micrographs of laser-crystallized spots in as-deposited films of $(GeTe)_{0.85}Sn_{0.15}$[1]. Spots irradiated with 7.4 mW of laser power for 300, 500 and 700 ns durations are shown in Figs. 1(a), 1(b) and 1(c), respectively. The onset of crystallization occurred with pulse lengths slightly less than 300 ns. Using a 500 ns pulse, crystallization of the irradiated spot was about half complete. Finally, using a 700 ns pulse, crystallization was almost complete. The 300 ns minimum time to initiate crystallization was roughly independent of the applied laser power. The microstructure resulting from higher or lower power 500 ns pulses all showed a similar ring-like structure while the outer diameter of the transformed regions increased monotonically with laser power. Laser irradiation with pulses shorter than 300 ns and powers up to 9.0 mW (causing melting at the center of the spot) did not cause crystallization to occur. This suggests that a finite incubation time, about 300 ns, was needed for nucleation. Following nucleation, crystal growth then proceeded. Since laser-quenching (quench-rate about 10^9 to $10^{10}\,K\,s^{-1}$) is not as rapid as vapor-quenching (quench rate about 10^{12} to $10^{14}\,K\,s^{-1}$), the laser-amorphized spots might have contained some residual crystal nuclei. This would have allowed the incubation period for nucleation of crystallites in laser-amorphized spots to be by-passed, resulting in significantly

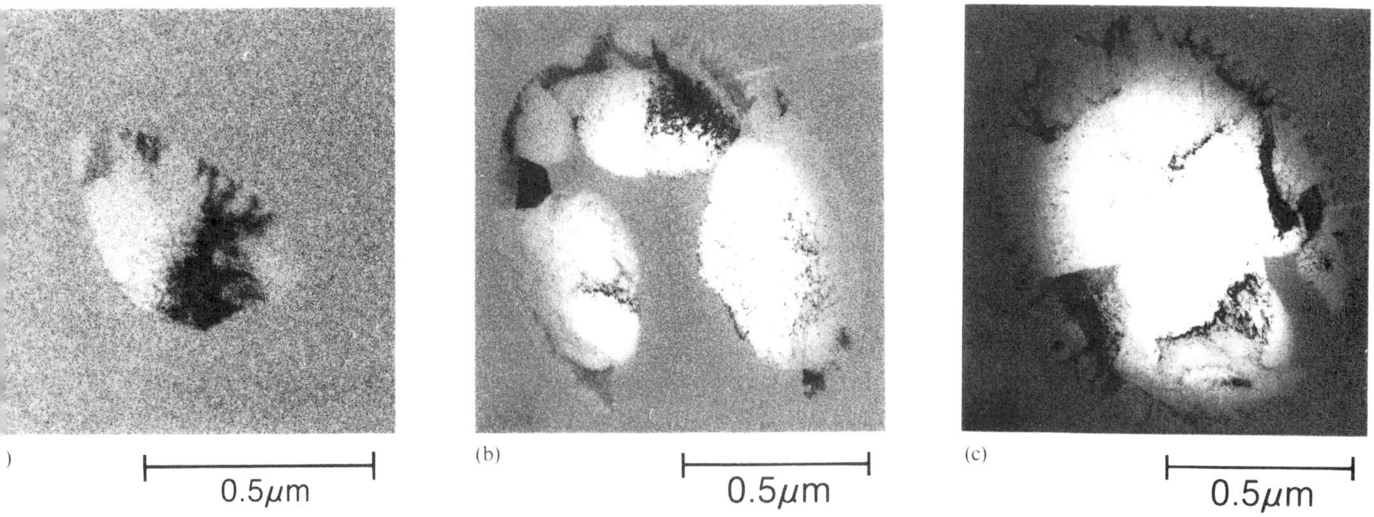

Fig. 1. TEM micrographs showing progression of crystallization *vs.* laser irradiation time. The 647 nm laser light was focused to an approximately 1 μm diameter spot on three separate regions of the sample. The sample was a 750 Å $(GeTe)_{0.85}Sn_{0.15}$ amorphous as-deposited film which was floated off carbon-coated mica after the irradiation. (a) Irradiation by a 300 ns laser pulse caused a single crystalline grain to form. Comparison of grain position with adjacent fully crystallized spots showed that the grain growth initiated near the edge of the laser spot. (b) A 500 ns laser pulse caused more grains to nucleate at the spot edge. Further growth of the crystallites toward the center of the spot can be seen. (c) Crystallization was almost complete after a 700 ns laser pulse. The presence of only a few grains within the spot was typical and showed that crystallization was limited by nucleation and not growth.

faster crystallization. In disk applications, as-deposited amorphous media would be initialized, *i.e.* appropriately crystallized first. Thereafter, only crystallization of laser-amorphized spots would be required.

A complete characterization of the crystallization of a sample can be represented by a phase transformation kinetics diagram which displays the effect of laser power and pulse width on the transformations of a 1 μm spot[17]. The results for films of $(In_{0.9}Ga_{0.1})_{0.5}Sb_{0.5}$ and GaSb are shown in Figs. 2 and 3, respectively. The most important parameter obtainable from each figure is the minimum time, irrespective of laser power, needed to crystallize a spot. This minimum crystallization time is seen to be approximately 170 ns and 30 ns for $(In_{0.9}Ga_{0.1})_{0.5}Sb_{0.5}$ and GaSb, respectively. In the latter case, the measured crystallization time was actually limited by the rise time of the acousto-optic modulator in our experimental setup and not by the material itself.

Whereas compounds can crystallize rapidly, alloys generally crystallize too slowly to be used for recording media. This has been clearly illustrated in the binary Ge–Te system[13]. The solubility of excess germanium or tellurium in GeTe is quite small, so alloys not close to the stoichiometric composition GeTe crystallize slowly. Another example of a compound system which exhibits fast erase is the pseudo-binary InSb–GaSb system. This $(InSb)_{1-x}(GaSb)_x$ system is continuously miscible for the entire range $0 \leqslant x \leqslant 1$, as shown by the phase diagram in Fig. 4(a)[18]. This indicates that no phase separation will occur during crystallization. The minimum crystallization time *vs.* the fraction of GaSb is shown in Fig. 4(b). There was a monotonic decrease in crystallization time from 170 to 30 ns as the amount of GaSb increased. The power which gave the minimum crystallization time is also shown in Fig. 4(a). The required power increased as the GaSb content of the thin film increased. There are two reasons for this: the first is that shorter crystallization times required more laser power to reach the same temperature; secondly, the crystallization temperature of the film with greater GaSb content was higher. Experiments with alloy InSb films showed that, as with the GeTe system, the minimum crystallization time was significantly longer[19].

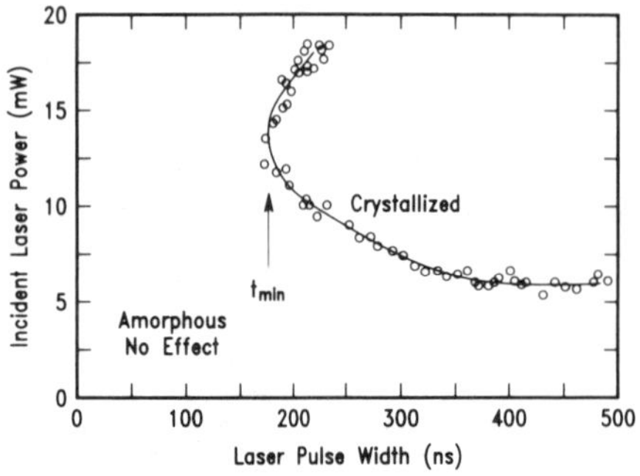

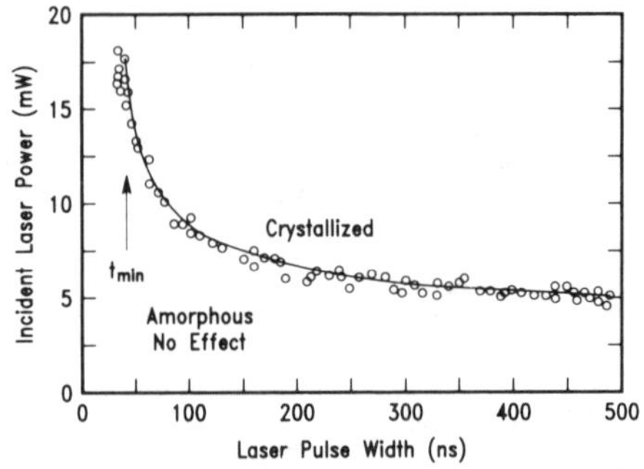

Fig. 2. Phase transformation kinetics plot for crystallization of an approximately 1.0 μm diameter region of as-deposited amorphous 750 Å $(In_{0.9}Ga_{0.1})_{0.5}Sb_{0.5}$/glass sample using substrate-incident 647 nm krypton ion radiation. The reflectivity difference between the amorphous and crystalline state was used to measure the degree of crystallization. The plotted line corresponds to 75% of the maximum measured reflectivity decrease. The minimum time to crystallize this pseudo-binary composition irregardless of laser power is indicated by the arrow.

Fig. 3. Same experiment as in Fig. 2 for amorphous as-deposited 750 Å GaSb/glass sample.

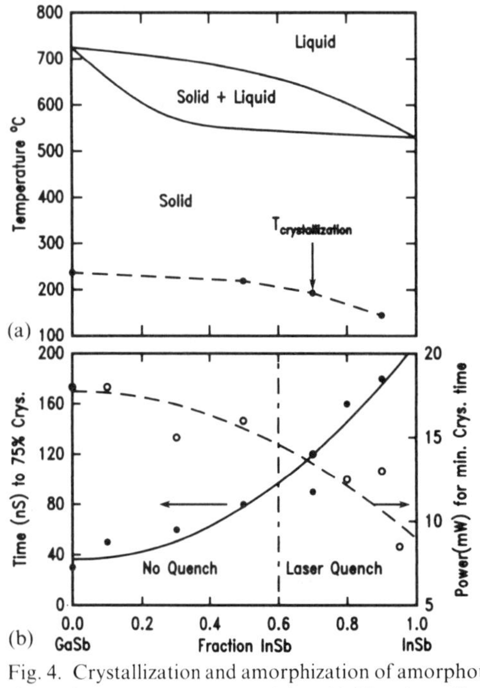

Fig. 4. Crystallization and amorphization of amorphous as-deposited 750 Å $(InSb)_x(GaSb)_{1-x}$ films on glass substrates versus composition. (a) The phase diagram of InSb–GaSb pseudo-binary showing full solid- and liquid-phase miscibility for all x.[18] The crystallization temperatures for amorphous as-deposited films heated at 23 °C min^{-1} are indicated by a dashed line. (b) Minimum pulse width for crystallization (solid line) of an approximately 1.0 μm diameter spot versus composition as determined by phase transformation kinetic measurements. These pulse widths correspond to 75% of the maximum reflectivity decrease associated with crystallization. The short crystallization times are consistent with the single-phase nature of this pseudo-binary compound system. The power required to achieve minimum crystallization time (dashed line) is higher for GaSb-rich compositions. Only films with 60% or more InSb (vertical dot-dash line) could be laser quenched in this structure.

Re-amorphization of laser-crystallized spots was examined for $(InSb)_x(GaSb)_{1-x}$. There was a crossover point, at $x = 0.6$ mole fraction InSb, which separated compositions that could and could not be reamorphized. This crossover composition is indicated by the vertical dashed line in Fig. 4(b). The reason for this is that GaSb-rich compositions can recrystallize during cool-down from the molten state, *i.e.* the critical quench rate required for amorphization, which we estimate to be of the order of 10^{10} K s^{-1}, could not be achieved. To obtain such cooling rates requires a very good heat sink layer such as aluminum to be placed near the active layer. However, such a heat sinking layer would also be operative during writing, decreasing the writing sensitivity of the media.

The search for compounds which are fast erasing is not restricted to the already known stable compounds. Laser irradiation of phase-change media has produced some new metastable compounds. For example, Sb_2Se was found to be a metastable compound with an erase time less than 200 ns[20]. Another example of a metastable system is $(GeTe)_x(Sb_2Te_3)_{1-x}$, for which Akahira *et al.* have reported fast erase times over the entire range of x[21]. X-ray diffraction data from each crystallized film revealed only a face-centered cubic phase, implying that a metastable crystal was formed for films with compositions intermediate between the five reported stable compounds[22]. $(GeTe)_{0.85}Sn_{0.15}$ is also believed to crystallize into a metastable state[1].

4. AMORPHOUS PHASE AND CHEMICAL STABILITY

Two aspects of stability are important for phase-change media. First, the stability of the amorphous state against crystallization is important in determining the data retention capability of the media. Second, chemical stability in operating environments containing moisture and pollutants, always an issue for thin films, is crucial for archival stability of the disk.

The stability of the vapor-quenched amorphous films has been examined by Yagi *et al.*[23] for the Sb–Te binary system, by Matsushita *et al.*[24] for the Ge–Te binary system, by Yamada *et al.*[22] for the GeTe–Sb_2Te_3 pseudobinary system, and by Sarrach and deNaufville[25] for the Ge–Te–Se and the Ge–Te–Sb ternary systems. While there are compositions, especially tellurium-rich films, that have low crystallization temperatures, it is not difficult to find compositions with crystallization temperatures exceeding 150 °C and with activation energies exceeding 1.5 eV. For compounds, which have well defined melting temperatures, there is a correlation between the melting temperature and the crystallization temperature[1]. The crystallization temperature is about half the melting temperature of the active layer. Thus, there is a trade-off between amorphous phase stability and writing sensitivity. In assessing data stability, it is important to determine the resistance against crystallization of laser-amorphized spots. This has been done in some instances and the results indicate no significant problems[1,22].

Extensive studies have been done to examine the environmental stability of write-once disks on phase-change materials. With suitably designed packaging, these disks have been judged to be sufficiently stable to be introduced into commercial products, with greater than five year guarantees. Reversible disks, using elements similar to those in write-once disks, are expected to have similarly dependable reliability. In the only published study on this subject, Ohta *et al.* have shown that GeTeSb-based disks meet this expectation[16].

5. REVERSIBLE AMORPHOUS-TO-CRYSTALLINE CYCLING

Limited reversibility has been and continues to be a problem for phase-change media. High temperature encountered during amorphization is one of the major causes of cycling failure. For example, it was found that more than 1 million static cycles could be achieved with low melting point tellurium alloys[26]. Since then, more than 10^7 static cycles have been demonstrated without any sign of failure with pure tellurium which melts at 452 °C[27]. However, materials with higher melting temperatures, such as GeTe, are required to guarantee long-term structural stability of the amorphous state[1]. These high melting temperature materials generally do not cycle as many times. The melting temperature of GeTe is 725 °C. To amorphize a spot with a diameter equal to the FWHM of the beam, the temperature at the FWHM diameter of the Gaussian laser beam must be raised to 725 °C, and the center of the spot will reach more than 1450 °C. This high temperature may cause vaporization of GeTe or delamination of the thin film layers from each other or from the substrate through stress build-up, followed by flow of the molten GeTe. Experimentally, pinholes in the active layer are observed after a limited number of reversals.

Evaluation at a disk level, typically involving recording and erasing of a track containing about 10^5 bits, yielded even more limited reversibility. Because of the high reliability demands of digital data recording, bit error rates must be kept below about 10^{-5}. Thus, the weakest spots on the disk, which can fail up to ten times earlier than average, has limited cyclability to 500–1000 cycles[1]. One clear source of weak spots are pre-existing pinholes in the active layer which have been observed to promote pinhole formation. The defective spots of the disks were found to grow in length along the groove direction as the number of cycles was increased. Despite these problems, Sato *et al.*[14], Terao *et al.*[15] and Ohta *et al.*[16] have recently reported achieving more than 10^5, 10^5 and 2×10^6 cycles, respectively. Achieving this high number of cycles required accurate control of the laser power. It is believed that the addition of a metallic reflection layer, introduced for optical purposes, also plays a role in extending cyclability by controlling the thermal response of the disk structure during and following laser irradiation. Beyond an existence proof of more than 10^6 cycles, a better understanding of how improved cyclability is achieved with a "cooling" layer may eventually eliminate cyclability as a major concern for phase-change media.

6. READBACK SIGNAL-TO-NOISE RATIO

One of the most important features of properly selected phase-change recording media is the very large achievable signal level. For example the reflectivities of the amorphous and crystalline states of $(GeTe)_{0.85}Sn_{0.15}$ at 633 nm are 37% and 62%, respectively. Combined with a suitably designed optical enhancement layer, it is possible to obtain reflectivities of 10–20% and 70–80% in the amorphous and crystalline states, respectively, for $(GeTe)_{0.85}Sn_{0.15}$. Very high reflectivity of the crystalline state, with an associated low absorption, has a negative impact on recording sensitivity. There are also undesirable consequences if the reflectivity in the amorphous state is made very large and the crystalline state is made very small. Thus, it is unlikely that the signal level for phase-change media can be made significantly higher than what has already been achieved.

The SNR which is currently achieved is typically more than 25 dB in a 10 MHz bandwidth with grooved phase change media. This is quite adequate for readback of recorded data. Further increases of the SNR would allow higher recorded data densities to be realized. A SNR of more than 60 dB in a 10 MHz bandwidth could be achieved with a 1 mW readback intensity if the only contribution to noise was fundamental shot noise. Thus, further gains in SNR may be achievable with understanding and reduction of media noise.

There are several contributions to the media noise of a phase-change disk. These include the noise from (1) tracking grooves, (2) as-deposited amorphous films, (3) crystal grains, and (4) writing non-uniform spots. Groove noise currently limits attainable SNR of phase-change disks. Using grooveless substrates for laboratory studies, we have found that the media noise, while present in the as-deposited amorphous thin films, can be made lower than groove noise. Groove noise is certainly not a fundamental limitation. Better fabrication process can lead to smoother grooves and reduced noise. It is also possible to avoid grooves altogether using sampled tracking servo schemes. Because of the statistical nature of the thin film growth process, the film does not grow at exactly the same rate uniformly[28]. Rather, there are small peaks and valleys which cause surface roughness. Also, there often are columnar type microstructures which accompany most physical vapor deposition processes. This leads to the presence of voids which will contribute to variations of the optical properties of the thin films. Noise due to crystal grains can be significantly reduced by using materials which crystallize into a cubic structure[1]. This criterion applies equally to any dielectric layer which is crystalline. Further reduction of both amorphous and crystalline thin film media noise requires research on the growth of the films. It is possible that techniques such as ion-beam-assisted deposition, which have reduced light scatter of laser mirrors,

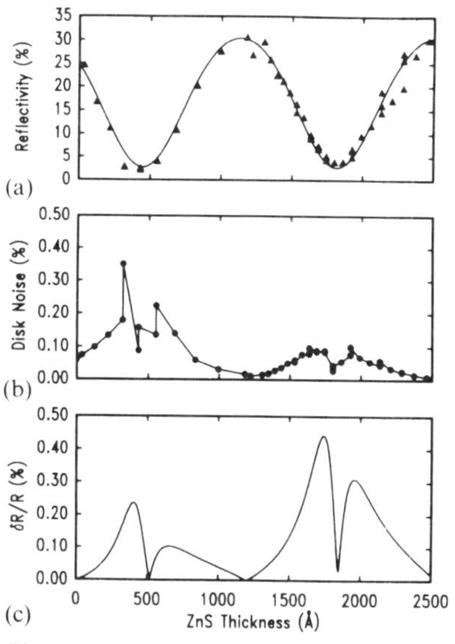

Fig. 5. Measurement and calculation of reflectivity and noise of a disk with the structure: 2500 Å ZnS/750 Å $(GeTe)_{0.85}Sn_{0.15}$/X-ZnS/PMMA. Light of wavelength 647 nm was incident through the disk substrate. The thickness, $0 < x < 2500$ Å, of the ZnS film was calculated from the geometry of the deposition mask and by fitting measurements of the relative reflectivity to calculations based on optical interference. The media speed was $4\,m\,s^{-1}$. (a) Reflectivity *vs.* ZnS thickness. Triangles are relative reflectivity, scaled to calculated reflectivity (solid line). (b) Measured noise at 2.5 MHz, in a 30 kHz bandwidth, normalized to reflectivity. (c) Calculated noise, $\Delta R/R$ *vs.* ZnS thickness for index of refraction (n) and $\delta n = 0.0005$ of ZnS of 2.36 and 0.0005, respectively.

may also help to reduce optical recording media noise. Currently, writing noise from a disk is well below the other sources of noise discussed above.

Dielectric layers are often placed between the plastic disk substrate and the active layer. They are designed to accomplish two purposes. One is to increase the sensitivity of the media by performing an anti-reflection function which optically couples more laser light into the active layer. Second, they thermally insulate the plastic substrate (with typical glass-transition temperature of about 150 °C) from the high temperatures reached in the active layer during writing. These dielectric layers can significantly increase media noise. This is shown by an experiment in which media noise was measured *vs.* the thickness of dielectric ZnS. The measured and calculated reflectivity *vs.* ZnS thickness are shown in Fig. 5(a). The variation in reflectivity was caused by optical interference of light reflected from the substrate/ZnS and from the ZnS/active-layer interfaces. The amount of noise depended on the ZnS thickness, as shown in Fig. 5(b). The disk noise was inversely proportional to the disk reflectivity except that sharp decreases occurred at approximately 500 Å and 1700 Å ZnS thicknesses, where the reflectivity was minimum. Optical modelling gave a strong indication of the origin of the noise. The sensitivity of normalized reflectivity variation, $\Delta R/R$, to refraction index and thickness variations of the active layer, the ZnS layer and the substrate was calculated. Qualitative explanation of the data could only be obtained if a variation of the index of refraction (n) of ZnS existed. Using a variation δn of 0.0005 for the modelling, we obtained the results shown in Fig. 5(c). This variation is consistent with published results[29] which showed that the mean index of refraction of ZnS increased with film thickness. The simultaneous requirement of high writing sensitivity and low media noise can be achieved by designing the dielectric layer thickness to fall at the center of the sharp dips. This will require precise process control.

There are ways to reduce the effect of media noise without modifying the media. One can use the fact that media noise is fixed on the disk. If the local media reflectivity is measured, it becomes deterministic. This is illustrated by the data in Fig. 6, which is a plot of the recorded peak height *vs.* the reflectivity at the same

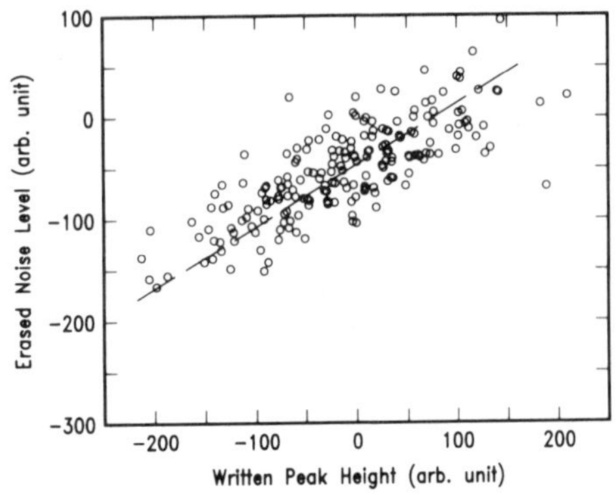

Fig. 6. Correlation (dashed line) of peak height of recorded signal *vs.* noise level of same region of the track before recording. Data were obtained by digitizing the reflectivity signal of a rotating grooved disk. The d.c. component was subtracted so only the reflectivity variations, which constitute media noise, are plotted.

location of the rotating disk before recording. Variations of the reflectivity from one location to another constitute media noise. A clear correlation was seen between the reflectivity before and after recording. The height of the recorded peak could in principle be made more constant by using a two-beam optical head[30]. A read beam would precede the write beam. The reflectivity measured by the read beam would provide a control signal which may be used to modify the power of the writing beam. Similarly, variations in recorded peak position could be minimized using the read beam signal to delay or advance timing of the write pulse. This write compensation scheme would reduce the effect of the media noise.

7. CONCLUSION

Recent progress in realizing operational phase change erasable media was reviewed. Improved understanding of the kinetics of crystallization has led to the development of many materials which can be switched between the amorphous and crystalline states in less than 100 ns. This fast-switching capability has been exploited to attain single-beam direct overwrite of data. Long-term amorphous state stability has been demonstrated and, while more detailed studies still need to be done, chemical stability is not expected to be a major issue. Rapid progress is being made in increasing the number of reversible cycles of phase-change media. Over one million cycles has been demonstrated recently. Finally, good readout SNR has been achieved. With increased understanding of the noise sources, even higher SNR may be possible in the near future.

ACKNOWLEDGMENTS

We thank W. McChesney for sample preparation. The assistance of R. Geiss with TEM and R. Lynch with media noise analysis is appreciated. Critical reading of the manuscript by E. Engler, J. Hurst, M. Libera, C. Ortiz, D. Rugar, J. Suits, and T. Suzuki and useful discussions with C. J. Lin are acknowledged.

REFERENCES

1 M. Chen and K. A. Rubin, *Proc. Soc. Photo-Opt. Inst. Eng.*, *1078* (1989) 150.
2 S. R. Ovshinsky, *IEEE Trans. Electron Devices*, *ED-20* (1973) 91.
3 D. Strand and D. Adler, *Proc. Soc. Photo-Opt. Inst. Eng.*, *420* (1983) 200.
4 J. Feinleib, J. deNeufville, S. C. Moss and S. R. Ovshinsky, *Appl. Phys. Lett.*, *18* (1971) 254.
5 R. J. von Gutfeld, *Appl. Phys. Lett.*, *22* (1973) 257.
6 R. J. von Gutfeld and P. Chaudhari, *J. Appl. Phys.*, *43* (1972) 4688.
7 K. Weiser; R. J. Gambino and J. A. Reinhold, *Appl. Phys. Lett.*, *22* (1973) 48.
8 J. Cornet, *Ann. Chim.*, *10* (1975) 239.
9 A. W. Smith, *Appl. Opt.*, *13* (1974) 795.

10 P. C. Clemens, *Appl. Opt. Lett.*, *22* (1983) 3165.
11 M. Takenaga, N. Yamada, O. Phara, K. Nisiuchi, M. Nagashima, T. Kashihara, S. Nakamura and T. Yamashita, in A. Bell and A. Jamberdino (eds.), *Proc. Soc. Photo-Opt. Inst. Eng.*, *420* (1983) 173.
12 W. Y. Lee, H. Coufal, C. R. Davis, V. Jipson, G. Lim, W. Parrish and F. Sequeda, *J. Vac. Sci. Technol. A*, *4* (1986) 2988.
13 M. Chen, K. A. Rubin and R. W. Barton, *Appl. Phys. Lett.*, *49* (1986) 502.
14 Y. Sato, Y. Maeda, H. Andoh, I. Ikuta, M. Nagai, N. Tsuboi, H. Minemura and M. Ishigaki, *Proc. Soc. Photo-Opt. Inst. Eng.*, *1078* (1989) 11.
15 M. Terao, Y. Miyauchi, K. Andao, H. Yasuoka and R. Tamura, *Proc. Soc. Photo-Opt. Inst. Eng.*, *1078* (1989) 2.
16 T. Ohta, M. Uchida, K. Yoshioka, S. Furukawa and K. Kotera, *Proc. Soc. Photo-Opt. Inst. Eng.*, *Techn. Digest Ser.*, *1* (1989) 14.
17 K. A. Rubin, R. W. Barton, M. Chen, V. B. Jipson and D. Rugar, *Appl. Phys. Lett.*, *50* (1987) 1488.
18 I. E. Gorshkov and N. A. Goryunova, *Zh. Neorgan. Khim.*, *3* (1958) 650.
19 J. Solis, K. A. Rubin and C. Oritiz, to appear in *J. Mater. Res.*
20 R. W. Barton, C. R. Davis, K. A. Rubin and G. Lim, *Appl. Phys. Lett.*, *48* (1986) 1255.
21 N. Akahira, N. Yamada, K. Kimura and M. Takao, in D. Carlin, Y. Tsunoda and A. Jamberdino (eds.), *Proc. Soc. Photo-Opt. Inst. Eng.*, *899* (1988) 188.
22 N. Yamada, M. Takao and M. Takenaga, in R. P. Freese, M. DeHaan and A. Jamberdino (eds.), *Proc. Soc. Photo-Opt. Inst. Eng.*, *695* (1986) 79.
23 S. Yagi, S. Fujimori and H. Yamazaki, *Techn. Digest of Int. Symp. on Optical Memory*, Tokyo, Japan, Sept. 1987.
24 T. Matsushita, A. Suzuki, T. Aoki, T. Nakau and M. Okuda, *Optical Memory Symp. (Domestic)*, Japan, 1988, Tokyo.
25 D. J. Sarrach and J. P. deNeufville, *J. Non-Cryst. Solids B*, 1 (1983) 229.
26 M. Chen, K. A. Rubin, V. Marrello, U. G. Gerber and V. B. Jipson, *Appl. Phys. Lett.*, *46* (1985) 734.
27 K. A. Rubin, unpublished.
28 See articles in *Modelling of Optical Thin Films*, in M. R. Jacobson (ed.), *Proc. Soc. Photo-Opt. Inst. Eng.*, *821* (1987).
29 *Proc. 6th Int. Vacuum Congr. 1974*, *Jpn. J. Appl. Phys.*, *Suppl. 2*, *Pt. 1* (1974) 769.
30 M. Chen and R. T. Lynch, Jr., *IBM Techn. Disclosure Bull.*, *29* (11) (1987) 4987.

Section Twelve
Substrates

Reprinted from *Optical Disk Technology*, Proc. SPIE Vol. 329, pp. 236-241 (1982).

Developments in fabrication of optical disks

Yoshihiro Okino, Kazuhiko Sano, Toshiaki Kashihara
First Development Division, Matsushita Electric Industrial Co., Ltd.
22-6, Motomachi, Kadoma, Osaka 571, Japan

Abstract

To establish the high density disc system, the most important point is how discs are produced stably on high rate of yield. From the point of this view we developed a new laser cutting machine of master disc and a novel duplicator of replica disc. This cutting machine which is compactly designed has a improved optical system which are stabel against temperature fluctuation and other environmental disturbance. And we can produce high quality discs at high rate of index time by our disc duplicator using liquid photo-polymer resin which is cured by ultra-violet ray.

Introduction

In several high density recorded disc system, the biggest technical problem which will determine whether the system is good or not is to fabricate discs in stable and high yield rate and at low cost, that is how to produce soft-hard.

Disc fablication has two major processes. One is mastering process which consist of recording of the glass master and production of the metal mold from the glass master. The another is replication process which produce discs from the mold which should be handed to final users.

The main problem of the former process is how to record the recording medium. There are some methods for that, for example mechnical recording method in which recording medium (for example copper plate) is recorded by use of the stylus which is vibrated by the piezo-electric device, and electron beam method in which electron beam is exposed to the recording medium. And each has its own characteristic. For optical disc, optical cutting system is generally used in which modulated laser beam is exposed to the recording medium.

As for the replication process, there have been three methods. The first one is compression method which follow the conventional audio record manufucturing process. The second one is injection method which was introduced as something like a new technique. The third one is injection-compression method which took the merits of the both methods aforementioned. Presently the injection method using acrylic resin is generally used. But recently, 2P method using the resin cured by ultra-violet light is being developed. And this methos is thought to be promising in future because of its simple equipment and extremely high fidelity replication.

In this paper, technical problem on optical disc fablication is reviwed, and the cutting machine in mastering process and the 2P duplicator in replication process which were developed by us are introduced.

Mastering process

Manufacturing process from glass master to metal mold (nickel stamper) is called mastering process. Mastering process has several steps as shown in Fig. 1. First, glassmaster which is made of green glass is prepared and then polished to highly precise surface. Then, photo resist is coated on its surface 1000 Å - 2000 Å thickness. In most cases, to improve the adhesiveness between glass master and photo resist, silane coupling is coated as primer in advance. This resist coated glass master is exposed by the laser beam modulated by information signal in the cutting machine. After the exposure, the glass master is developed and loses the exposed portions and makes the relief pattern. This is resist master. Then, nickel is deposited by sputtering 500 - 1000 Å thickness on the surface of this resist master. In this step the surface of the resist master become conductive and by making it cathode electroforming is done. By following steps as shown in Fig. 1 similar to the conventional audio record manufacturing processes, master, mother and stamper can be made.

By replicating master to make mother andstamper, a pretty amount of decrease in S/N ratio is unavoidable. That means discs in best quality can be made from master. Mastering in optical system is a struggle against dust, and especially in the steps from resist coating to recording extremely high cleaness of atomosphere is required.

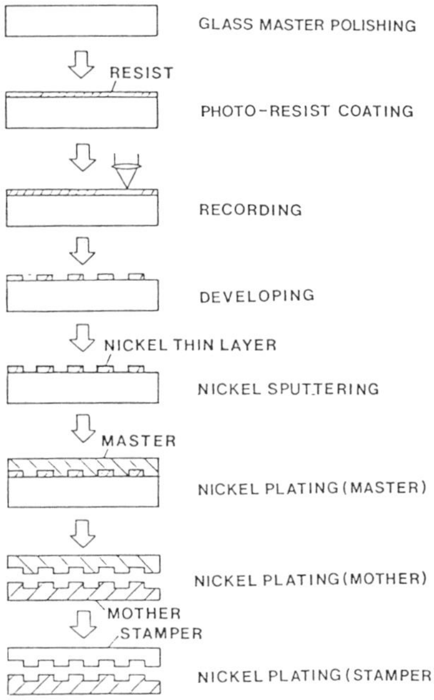

Fig. 1 Mastering process

Laser cutting Machine

In mastering processes, recording process is very important part and recording equipment used in this process -laser cutting machine- is one of the most important equipments. The following is the cutting machine which we developed for optical PCM digital audio disc system.

Constitution

This cutting machine is a system of which turn table moving with a glass master on it, and Optics system does not move. Each unit, laser for cutting, Optics bench on which optical system are arranged, and turn table feed mechanism, are independently fixed on a steel base which has a large mass and mutual influences of vibration can be restrained. For the spindle of the turn table, Air spindle is used. The steel base is supported by air servo mount (self leveling system) and is prevented from vibration from the floor. Optics bench is coverd by transparent yellow acryl plate which absorbs only blue light of laser beam and protects optical system from air disturbance.

This optics cover is made to be opened in every faces and makes the optics alignment easy. The external appearance is shown in Fig. 2.

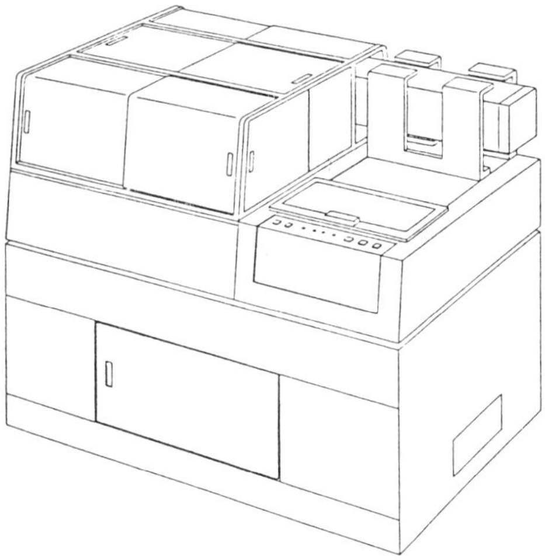

Fig. 2 Cutting machine

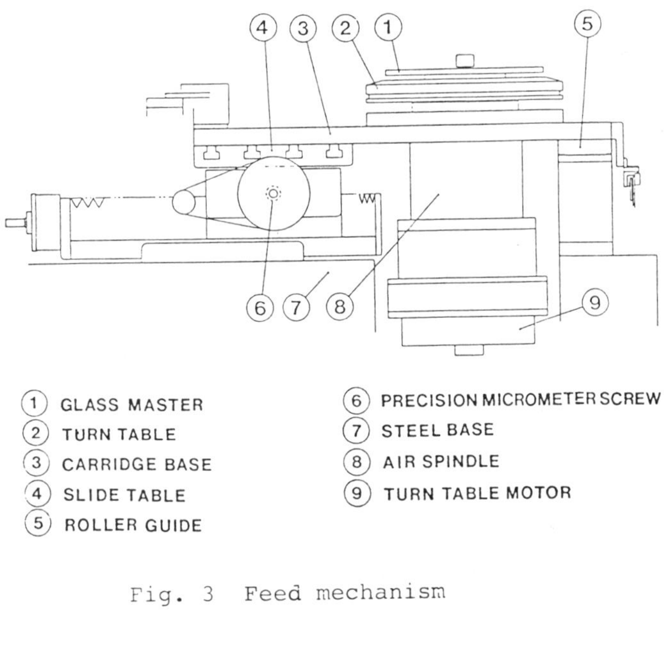

Fig. 3 Feed mechanism

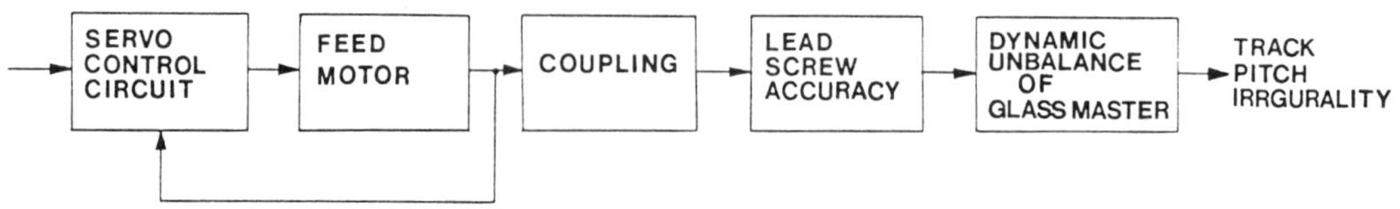

Fig. 4 Factors of track pitch irregularity

Feed mechanism

The main part of the feed mechanism is shown in Fig. 3. The carridge base is supported on one side by the slide table which has a precision micrometer screw and precision bearing face guide and by frictionless cross-roller guide on the other side. For feeding the carridge base the precision micrometer screw is driven by motor. The guide of the slide table is a face to face contact type with thin lubricant layer in between. So, the guide has high dynamic rigidity and very stable for vibration. In such a system, the vibration caused by the dynamic unbalance of the glass master become one of the causes of track pitch irregularity. But this machine is also very stable for such a vibration.

The causes of track pitch irregularity are shown in Fig. 4, most of them are mechanical factors. The data of track pitch irregularity mainly caused by mechanical factors is shown in Fig. 5 which was measured when the feed mortor was driven with the constant revolution.

In Fig. 5, one wavelength means the feed of $\lambda/2$ = 3164 Å and 5 wavelengths is nearly equivalent to 1.6 µm which is track pitch in this machine. Accordingly, the deviation of the point of the fifth wave, is considered to be the track pitch irregularity. In this measurement track pitch irregularity is about ± 3 %, this is ± 0.05 µm for 1.6 µm.

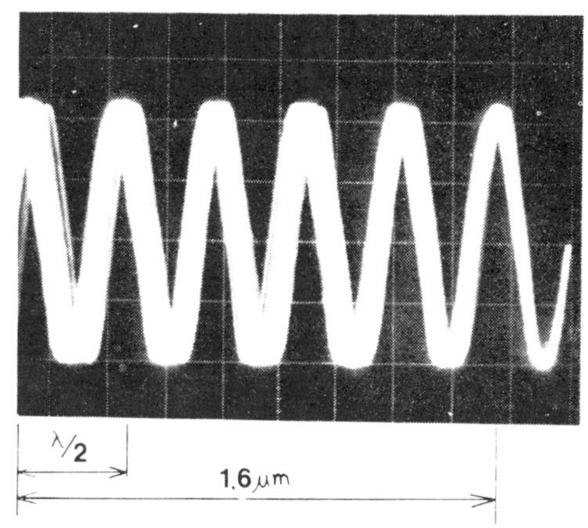

Fig. 5 Track pitch irregularity

Optical system

The optical system of this cutting machine is shown schematically in Fig. 6. The beam from the He - Cd laser, 4416 Å wave length, is set to pass through the two Acoust Optical Modulator. Modulator A01 is used for laser beam power control and reduces the noise which will occur in the laser cavity. A02 is for PCM signal modulation and has 13 MHz (-3 dB down) bandwidth characteristic and 1-order beam is available. BS4 is a optical parallel. The beam which impinges on BS4 is reflected both on the front and the back surface and makes interference image on the monitor plate. To make the interval of these fringes maximum the beam can be made to be parallel. And by the same way using

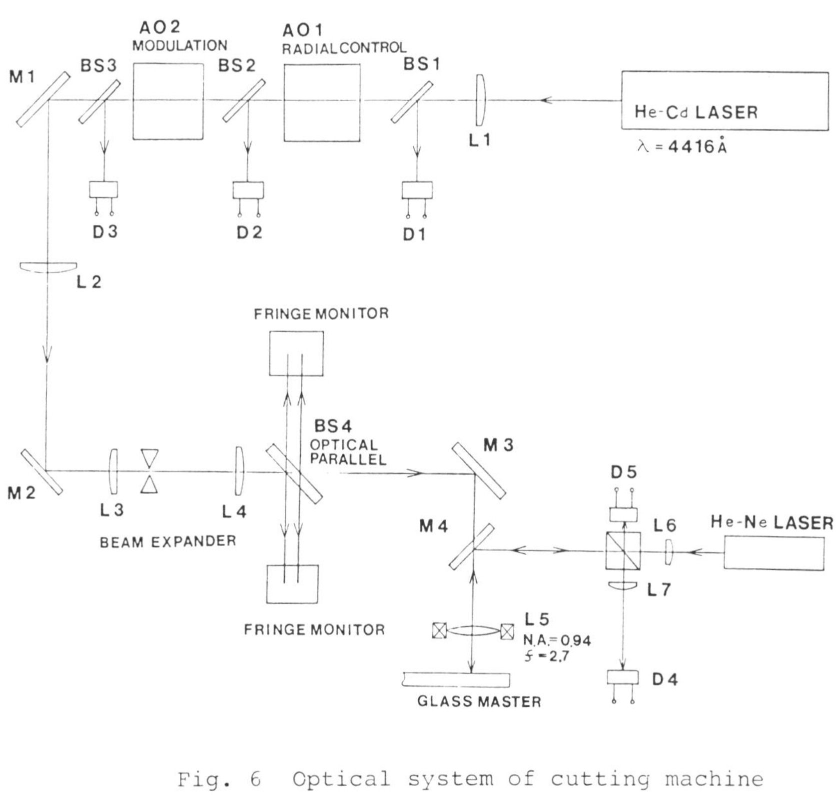

Fig. 6 Optical system of cutting machine

the reflected beam from the glass master, the objective lense can be brought into focus. In the focus servo system, the change of the distance between the glass master and the objective lense is detected by the detector D-4 using the He-Ne laser beam, and the objective lense is controled to follow the glass master by means of the voice coil type actuator. The focus servo gain is about 55 dB.

Replication process

In the process to make replica discs which should be handed to the users from the nickel mold which is replicated from the original master by electroforming, 2P process which has high efficiency and high yield rate is mentioned below.

In this process photo polymer resin is used which is cured by the irradiation of ultraviolet light, and the steps to make disc is shown in Fig. 7. The first step is the preparation of the mold to replicate and the base plate which become the disc substrate.

For this base plate acrylic resin or polycarbonate is used. In the second step, liquid of photo polymer resin is filled up or coated between the nickel mold and the base plate. In the third step ultraviolet light is irradiated to the 2P resin through the base plate. The 2P resin which is irradiated ultraviolet light is cured in several ten minutes by photopolymerization. In the fourth step 2P resin with base plate is peeled off the mold and the replication is finished.

These four steps are the fundamental process. Coating of the reflection layer onto the relief pattern of the replica surface by sputtering, coating of the resin which will protect the aluminum reflection layer, pasting two replicas together with the protection layers facing each other, and trimming of it are done after that as the following process.

The characteristics of the 2P replica disc are as follows.

1) High fidelity replication.
   Because of liquid it permeates into details and has smooth surface.

2) Simplicity of menufacturing equipment.
   Clean plant layout is possible and no environmental pollution such as noise, heat and smog.

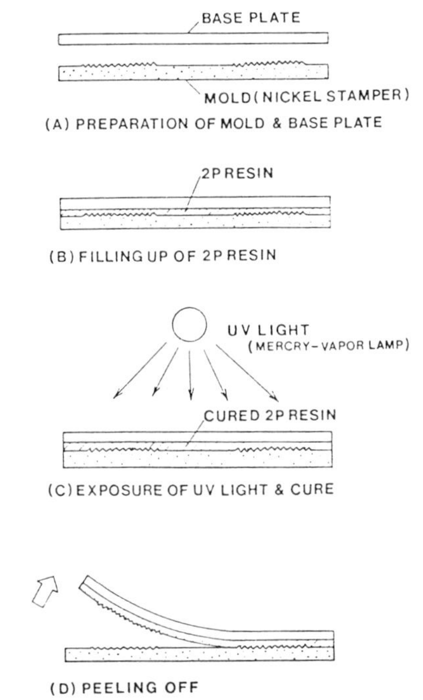

Fig. 7 Disc replication by 2P process

3) Almost no deterioration to the mold, it is possible to make a lot of discs from the nickel master.

4) As plastic resin is liquid, it is possible to remove extraneous substances by filtration, and quality control is simple.

2P disc duplicator

The disc duplicator which contains the fundamental process shown in Fig. 7 in it was developed. As mentioned befor the duplication of replica disc has four steps fundamentally.

(1) preparation of base plate.
(2) filling up 2P resin.
(3) exposure by ultraviolet light.
(4) peeling off from the stamper.

So as shown in Fig. 8, four stampers are set on the turn table and four processes above mentioned are set to each position. When the turn table rotates in the direction of center arrows, base plate is supplied and after each process replica disc is made.

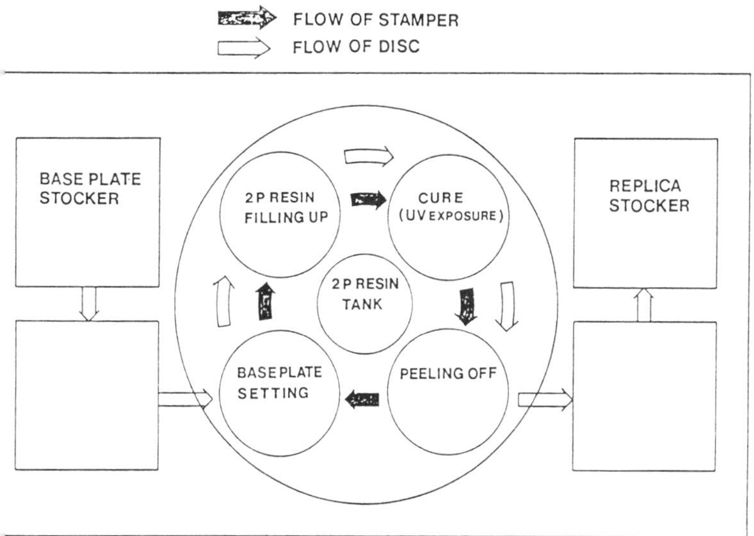

Fig. 8 Schematic diagram of 2P duplicator

Fig. 9 Replica disc made by 2P duplicator

In our duplicator, one mercury-vapor lamp of which required electric power is 500 W is enough to cure the resin completely, and index time is under 25 seconds. The replica made by this machine is shown in Fig. 9. This become the complete disc after the coating of the aluminum layer and so forth.

Method for filling up 2P resin

The most significant problem in filling up the photopolymer resin between the nickel stamper (matrix) and base plate is invasion of air bubble. To solve this problem center inject nozzle shown in Fig. 10 was developed. As shown in Fig. 10, the center pin is set pushed up from the stamper and then base plate is placed on it. While applying the prepressure P_0 to the center axis portion, a fixed quantity of photo polymer resin is injected from the center nozzle. In this manner the invasion of air bubble can be restrained. After the injection, keeping the pre-pressure P_0, pressure P_1 is applied to the whole base plate surface and make the distance between the stamper and the base plate smaller to the 10 - 15 μm thickness layer and complete this pressure coating process. By controling the spreading speed of the resin, it is possible to restrain the invation of air bubble in this pressure coating process.

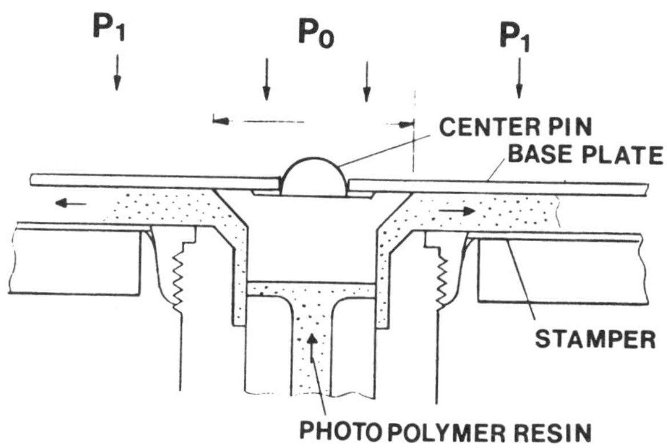

Fig. 10 Center inject nozzle

Summary

We have described the novel processes of optical disc fabrication, which include the mastering process to make matrix (nickel stamper) and duplication process to produce disc. Recording and replication are most significant problem in the former process and in the latter respectively. From these points of view we developed the laser cutting machine and the 2P duplicator as facility for stable and high-yield production. These are for PCM audio disc of 12 cm diameter, but it's easy to apply same technique to optical disc of all other system.

Reprinted from *Third International Conference on Optical Mass Data Storage*, Proc. SPIE Vol. 529, pp. 62-68 (1985).

Developments in optical disk mastering

Jan H.T. Pasman, Hans F. Olijhoek and Bill Verkaik

Philips,
Optical Disk Mastering Department,
Eindhoven, The Netherlands

Abstract

This paper describes the principles of mastering as a means of preformatting recordable optical data storage disks with sector heading marks and tracking grooves. It is shown from calculations based on vectorial diffraction theory that optimum read-out signals are obtained from grooves wider than pits on read-only optical disks. Two methods to realize such wide grooves are indicated.

Introduction

Over the last years optical disk systems have left the research stage behind and entered the market. In two of such systems, i.e. the LaserVision (LV) video disk system and the Compact Disc (CD) digital audio system and its offspring CD ROM, the information is fully prerecorded in the mastering stage on a master disk in the form of a sequence of pits of well-defined dimensions. It is this coded height profile which makes mass replication relatively simple and cheap, imperative for consumer products. The mastering of such prerecorded disks now is a standard and well-controlled process [1,2,3].

In a third application of optical disks, the Digital Optical Recording (DOR) data storage system[5], it is the user who writes the information by locally changing the reflectivity of the disk, and therefore most of the disk should be blank when sold. Yet such disks also require some sort of prerecording, or rather preformatting, for addressing, tracking and synchronization purposes. At present two different methods are used, the choice being determined mainly by whether the system designer's background is in magnetic or optical recording. In a first method, which we will not further discuss, each individual disk is formatted by servo writing. In the method presented in this paper a master disk is produced, the surface profile of which, modulated by pregrooves, heading pits and clock signals, can easily be replicated by standard optical disk replication methods.

Although the physical and chemical principles of DOR mastering do not differ from those now standard for fully prerecorded disks there are some requirements, in particular concerning the geometry of the height profile, which are specific to DOR mastering. In this paper we briefly describe the mastering process and next consider some aspects of DOR mastering, in particular the relation between pregroove dimensions and read-out requirements.

Mastering process

In the mastering process [1,2,3] a laser beam recorder (LBR) is used to record the pit and groove structure onto the surface of the master disk, a polished glass substrate coated with a photo sensitive resist layer of thickness of 0.1 to 0.2 micron, the exact value depending on the read-out wavelength. Figure 1 shows a schematic diagram of the LBR. The beam of an Ar-Ion laser operating at a wavelength of 457.9 nm is modulated in intensity by a fast optical modulator. Unlike LV and CD mastering where only on/off modulation is required DOR mastering also uses intermediate intensity levels for the recording of pregroove and clock. After expansion by a telescope the beam passes through a high quality objective lens of high numerical aperture (NA), to produce a focussed spot which exposes the resist layer on the rotating master. When the objective lens moves in the outward direction with constant speed or step by step, a spiral track or a series of concentric tracks is exposed.

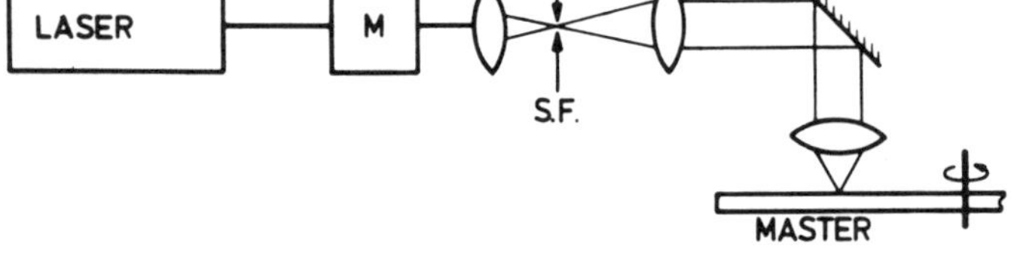

Figure 1. Schematic diagram of LBR.

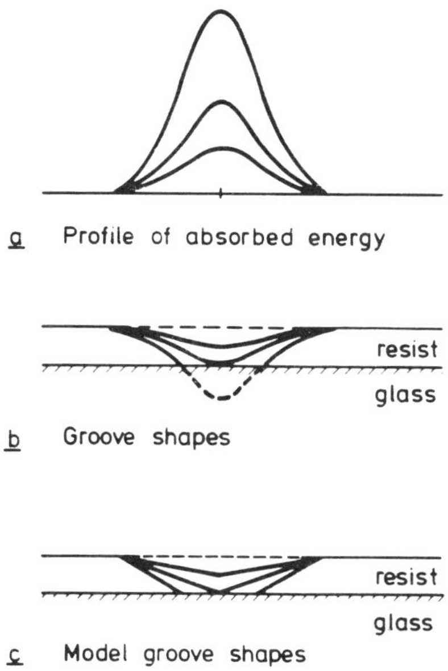

Figure 2. Groove formation.

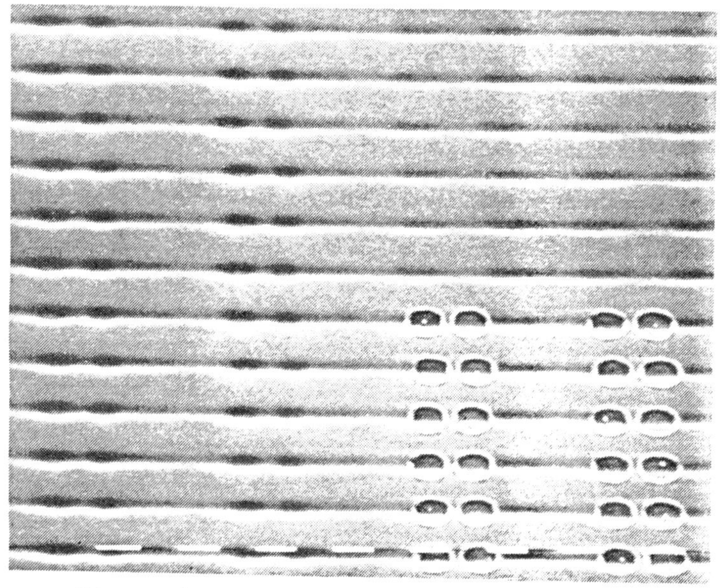

Figure 3. SEM photograph of a DOR replica.

The absorption of radiant energy by the photo resist causes a local increase in the solubility. The desired relief pattern is obtained by removing the exposed resist in a development process (Figure 2). The groove dimensions can be modified by changing the intensity (groove depth) and NA (groove width) of the recording spot while keeping the development parameters constant. After evaporation of a silver layer the master is inspected. Then in an electroplating process stampers are made for further replication.

Geometry of DOR format

On DOR disks the preformatted tracks are divided in the tangential direction into sectors, each consisting of a relatively short sequence of pits, the header, containing address and synchronization patterns followed by an empty user area. Both in the header and user area a shallow pregroove is recorded for tracking purposes. The groove is modulated in depth to provide a clock signal for synchronization of disk and drive. Figure 3 shows a SEM photograph of a DOR replica with heading pits (left) and pregrooves (right). On the bottom right hand side user data are written in the form of holes in a reflective, Te based, layer.

The opto-electronic read-out signal from the prerecorded heading pits is obtained in essentially the same way as from pits on LaserVision or Compact Disc disks by scanning with a focussed spot of full width at half maximum (FWHM) of 0.8 micron. This implies that DOR pits also should be about 0.6 micron wide. Because of the longer wavelength (820 nm) of the DOR semiconductor laser as compared to the LV HeNe laser (633 nm) DOR pits should be deeper than LV pits, resulting in optimum detection of pits about 0.15 micron deep.

For the choice of the pregroove dimensions two read-out signals are of major importance, the push-pull (PP) radial tracking servo signal and the central aperture (CA) reflection signal. The third signal, the clock, will not further be considered here.

Read-out

At read-out of a disk the beam of an AlGaAs laser of wavelength λ = 820 nm is focussed by the read-out lens of numerical aperture NA = 0.52 onto the groove structure on the disk. This structure acts like a reflection grating and produces diffraction orders [6], the direction of which is fully determined by the geometrical parameters wavelength, grating period (track pitch) and direction of incidence, in accordance with the grating law (Figure 4). The relative amplitudes and phases of the diffracted orders will be seen to depend on the dimensions of the grooves. Only part of the diffracted light passes through the entrance pupil of the read-out lens, which is imaged onto a detector.

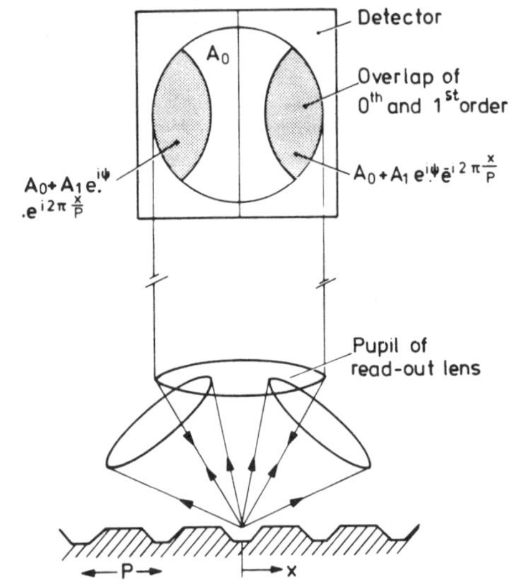

Figure 4. Read-out arrangement.

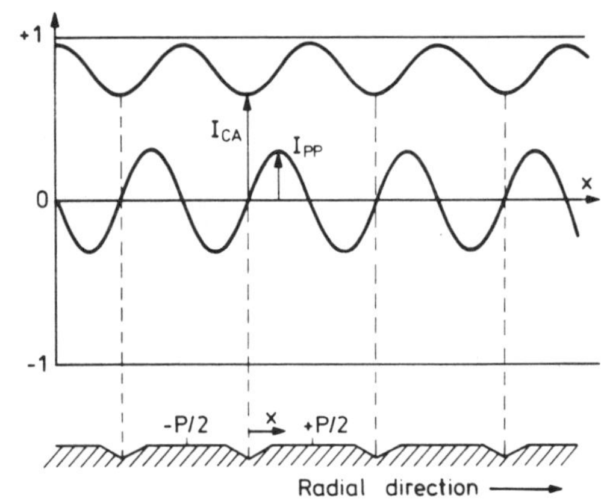

Figure 5. CA and PP signals.

At the detector the complex amplitudes of the various orders interfere in regions of overlap and the resulting intensity distribution is integrated by the detector to give the read-out signal. When a dual element photo diode is used as a detector we may add (CA detection) or subtract (PP detection) the signals from both radially directed diode halves. When only first and zeroth orders are considered both signals are given in terms of the absolute values of the amplitudes A_0, A_1 and the difference in phase Ψ_{01} of zeroth and first order, the relative area of overlap S on one detector half and the radial position x of the spot relative to the tracks of period p:

$$i_{CA}(x) = A_0^2 + 2 \cdot S \cdot A_1^2 + 4 \cdot S \cdot A_0 \cdot A_1 \cdot \cos \Psi_{01} \cdot \cos(2\pi x/p) \quad (1)$$

$$i_{PP}(x) = 4 \cdot S \cdot A_0 \cdot A_1 \cdot \sin \Psi_{01} \cdot \sin(2\pi x/p) \quad (2)$$

In general the CA signal is minimum when the spot is centred on the groove and maximum between grooves (Figure 5). Since the user data are to be written as local changes in reflectivity in the groove, the available signal space may be defined as the CA signal amplitude I_{CA} at the groove centre. The PP signal, being a differential signal, equals zero when the spot position is at points of mirror symmetry in the radial direction, i.e. at the groove centre and midway between grooves. The PP signal reaches a maximum I_{PP} when the spot is displaced over 1/4 of a track distance. Clearly optimum read-out is achieved when grooves produce maximum signal amplitudes of both the PP signal I_{PP} and the groove reflection signal I_{CA}, a situation which cannot occur as will be seen later. Yet, as is the essence of this paper, this situation can be approached by an appropriate choice of the pregroove dimensions.

If header pits are simply considered to be only parts of a continuous groove Eq. 1 indicates that they are best detected in CA when they produce Ψ_{01} values of 180°, as is the case on LV disks.

Now we have come to the stage where, if we want to calculate read-out signals in order to design an optimum groove shape, we need to know how the amplitudes and phases of the reflected diffraction orders depend on the dimensions of the grooves. For this purpose diffraction theory is used.

Vectorial diffraction calculations

One has to apply vectorial diffraction theory, based on Maxwell's Equations, rather than scalar theory[6] because the groove dimensions are of the same order as the wavelength of the read-out light giving rise to different results for different states of polarization of the incident light[4,7,8]. For a description of the vectorial theory we refer to the literature[4,7,8], here only results are presented.

First we have to define the electromagnetic diffraction problem. We consider the read-out parameters of the Philips - OSI DOR system[5]: light of wavelength λ = 820 nm is focussed by a lens of numerical aperture NA = 0.52 through a glass substrate of refractive index n_0 = 1.5 onto a Te based layer of complex refractive index $\tilde{n}_1$ = 5.3 + 1.8i. The effect of read-out through the substrate is to scale down the effective wavelength by a factor 1.5 to

547 nm and to reduce the optical contrast of the glass-Te interface. The reflection of a plane interface then becomes 36 %.

We approximate the cross-section of the groove by straight line segments resulting in a triangular or trapezoidal (truncated triangular) shape, depending on whether after development the tip of the groove has reached the resist-glass interface (Figure 2). In all calculations we assume the maximum groove depth, which in practice is determined by the thickness of the photo resist layer on the master, to be 145 nm (Figure 6). A constant track pitch p = 1600 nm is taken, resulting in a normalized overlap area S of 0.397. The width W at the top of the groove is taken as a parameter.

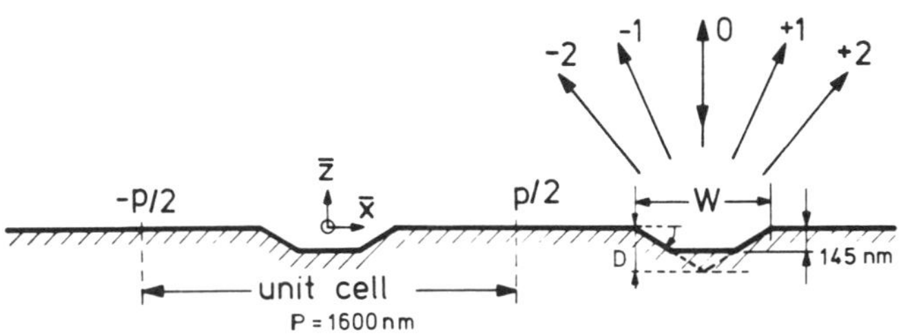

Figure 6. Diffraction geometry.

We have calculated the amplitudes and phases of zeroth, first and second reflected orders produced when a plane wave of unit amplitude is normally incident. The direction of polarization of the electric vector is either parallel (E-polarization) or perpendicular (H-polarization) to the grooves.

The absolute values of the amplitudes A_1 and A_2 of first and second orders, together with the phase difference Ψ_{01} between first and zeroth order have been plotted in Figure 7 as a function of the depth D of the deepest point in the groove for two values of W. Values of D larger than 145 nm represent the virtual depth of the truncated tip of the triangle.

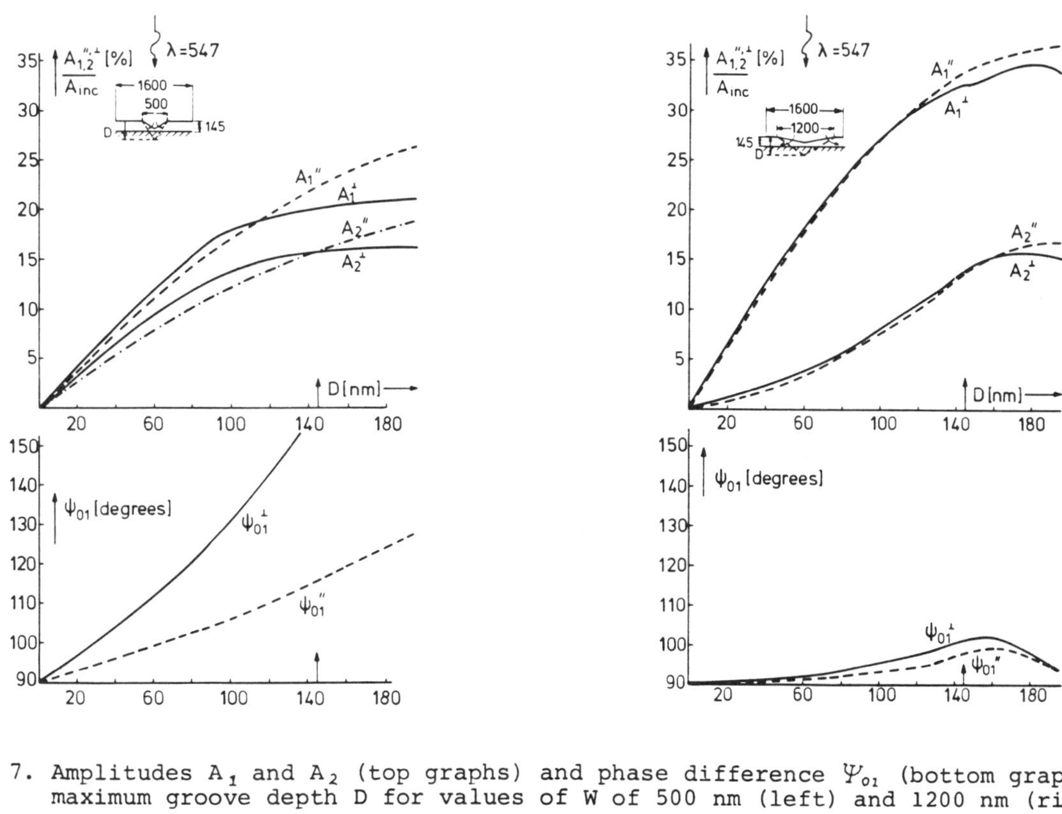

Figure 7. Amplitudes A_1 and A_2 (top graphs) and phase difference Ψ_{01} (bottom graphs) versus maximum groove depth D for values of W of 500 nm (left) and 1200 nm (right).

The left-hand curves, where W = 500 nm, more or less represent the case when a standard sized LaserVision mastering spot is used. Because of the nonlinear development rate of the photo resist the lower recording intensity which will appear to be necessary for DOR grooves results in grooves narrower than 0.6 micron. Such grooves of width smaller than the wavelength generally exhibit polarization effects in such a way that the groove appears to be narrower and shallower in parallel polarization, as explained in References 4 and 7. We note that the polarization effect in particular is present in the phase difference Ψ_{01}. Very shallow grooves produce a phase difference Ψ_{01} equal to or somewhat larger than 90° resulting in a large sin Ψ_{01} term in the PP amplitude (cf. Eq. 2). The amplitude A_1 then however is very small, so the groove must be deeper, which again results in lower values of sin Ψ_{01}. As a result we expect a maximum in the PP amplitude at a certain optimum depth. The CA amplitude then will be reduced due to the negative cos Ψ_{01} term.

When we compare the case W = 1200 nm with the previous W = 500 nm case, several observations can be made:

- The differences in amplitudes and phases for the two states of polarization are much smaller since now W is much larger than the wavelength.
- The second order, compared to the first order is much weaker, an effect quite analogous to the well-known fact that in periodic signals second harmonic components disappear when the duty cycle of the signal approaches ½.
- For equal depths the first order amplitude is larger for wider grooves, as can easily be understood to be caused by the larger volume of the groove.
- Maybe the most important observation is that Ψ_{01} remains very close to 90°, even for larger depths, once the width W approaches the track pitch p. This is an example of the more general statement that grooves of radial duty cycle of ½ of which the average depth is not larger than $\lambda/4$ produce Ψ_{01} values equal to 90°. As a result we expect larger PP and CA signal amplitudes from wider grooves.

To verify this qualitative idea both CA and PP amplitudes have been calculated from equations of the form of Eqs. 1 and 2, but now including the, though small, effect of second diffracted orders. The signal amplitudes are normalized with respect to the CA reflection signal from a plane disk with no grooves present. The values of A_0, A_1, A_2 and Ψ_{01} to be substituted have been calculated for normal incidence only, but we expect this to be a good approximation of the real situation where the maximum angle of incidence amounts to only 20°. Also the intensity distribution of the incident light over the pupil is taken to be uniform. It is interesting to note that in this case the maximum possible PP amplitude amounts to 56 %, a value which is reached when $\Psi_{01} = 90°$ and the reflected light is equally divided over the zeroth order and the two first orders.

Figure 8 shows the calculated CA and PP amplitudes as a function of groove depth for different values of W. In all cases the PP amplitude indeed shows a maximum, reached when the maximum groove depth D is about 100 to 120 nm, implying that the groove should not reach the resist-glass interface. The maximum value is largest in parallel polarization, owing to the smaller Ψ_{01} value, although for larger width this polarization effect decreases. Wide grooves produce the largest PP signal amplitudes which are seen to approach the largest possible value. The CA amplitude decreases when, with increasing depth, the groove becomes more pronounced until a minimum is reached. In practice a reduction in groove reflection of 20 % is just allowed, which implies that the corresponding PP signal amplitudes will be off-optimum. However, for wider grooves I_{CA} falls off less rapidly, which again pleads for wider grooves. In Figure 9 these results are summarized. The maximum PP signal amplitude (a), averaged over the two states of polarization, together with the average PP amplitudes corresponding to average CA amplitudes of 80 % (b) and 90 % (c) are plotted versus the width W. Clearly grooves of width W larger than about 1200 nm give best results. In that case the average groove width approaches half the track pitch.

Realization

In mastering such wide and shallow pregrooves have been realized by broadening the recording spot through a reduction of the recording NA to a value of about 0.3 and by reducing the exposure level. However one should take into account that the heading pits are to be recorded by the same spot and, as stated before, these pits are best detected when they are about 0.6 micron wide. This can be seen from Figure 8 where wider pits, approximated by grooves of depth larger than 145 nm, do not modulate the CA reflection as deep as narrower pits do. As a consequence a compromise must be made between narrow heading pits and wide pregrooves. At present standard DOR masters for OSI are recorded with a NA of about 0.4 to 0.5 which give satisfactory pit and groove signals : a groove reflection I_{CA} of 90 % which is reduced by heading pits to about 25 % and PP amplitudes of 20 to 25 %. This PP amplitude is on the low side as compared to our calculated values because at read-out the pupil of the read-out lens is not uniformly filled, which reduces the normalized PP signal.

A more fundamental solution to the problem of conflicting requirements of wide grooves and narrow pits is to record pits and grooves simultaneously but independently by two recording spots of different width and both independently modulated in intensity. The use of such two recording spots yields the extra advantage that by a radial displacement of one spot the header pits, and consequently the user information bits when the disk is written, can in a simple way be located midway between the pregrooves. Here, as shown in Figure 5, the CA reflection signal is maximum rather than minimum and the groove can so be designed as to give maximum PP amplitude. Figure 10 shows a SEM photograph of such a structure recorded by a dual spot LBR.

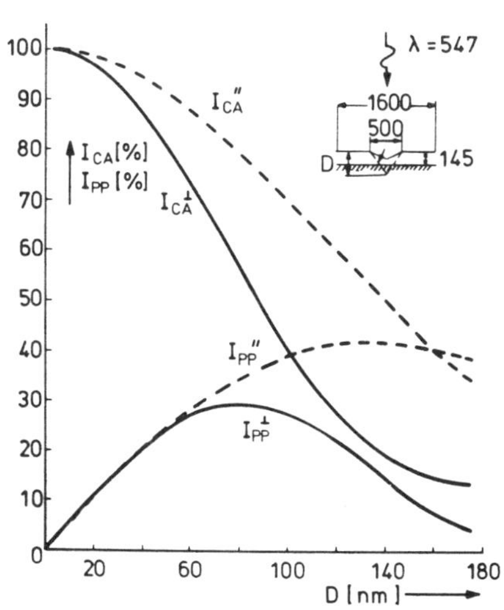

Figure 8 a. W = 500 nm

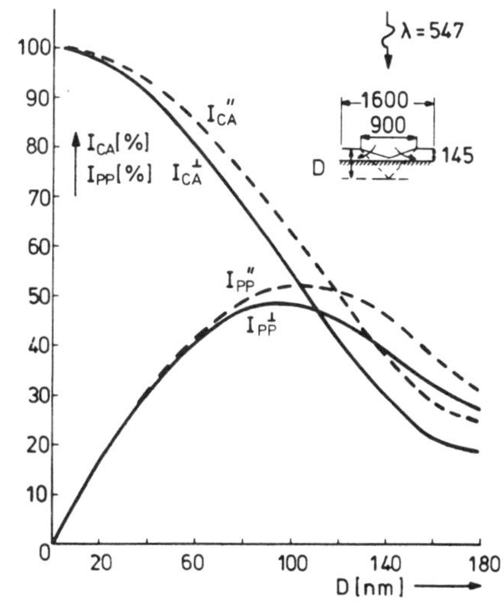

Figure 8 b. W = 900 nm

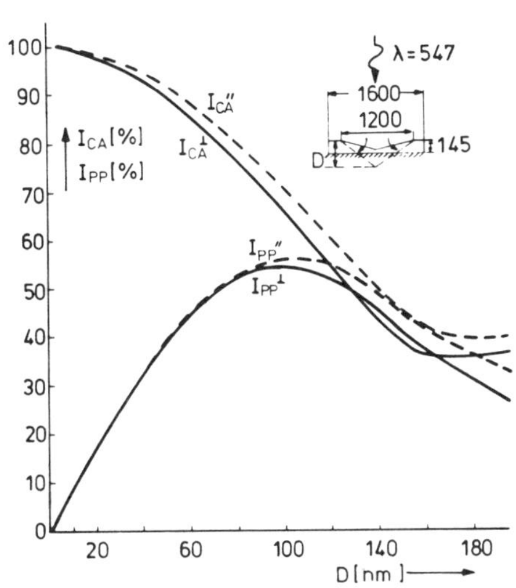

c. W = 1200 nm.

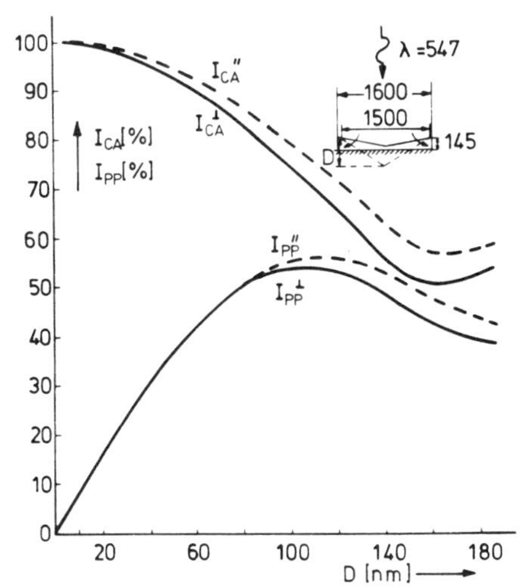

d. W = 1500 nm.

Figure 8. Calculated CA and PP signal amplitudes versus groove depth for grooves of width W = 500 nm, 900 nm, 1200 nm and 1500 nm.

Conclusion

In this paper we have described how the mastering required for the preformatting of DOR data disks deviates in several aspects from the now standard video and audio disk mastering. An important difference is the presence of pregrooves which are recorded by a spot of intermediate intensity and of width larger than usual on prerecorded disks. Such pregrooves produce optimum read-out signals when the average groove width approaches half the track pitch.

Within Philips' Optical Disk Mastering Department we have started development of new laser beam recorders, both single and dual spot machines, especially dedicated to the preformatting of recordable optical disks. These master recorders not only perform the above-mentioned functions, but other non-conventional demands can also be satisfied, such as radial wobble of the track, recording of concentric grooves rather than a spiral track and combinations of preformat and prerecorded information. In this way we aim to maintain our capability to supply the masters and mastering systems to the rapidly evolving market for optical disk systems.

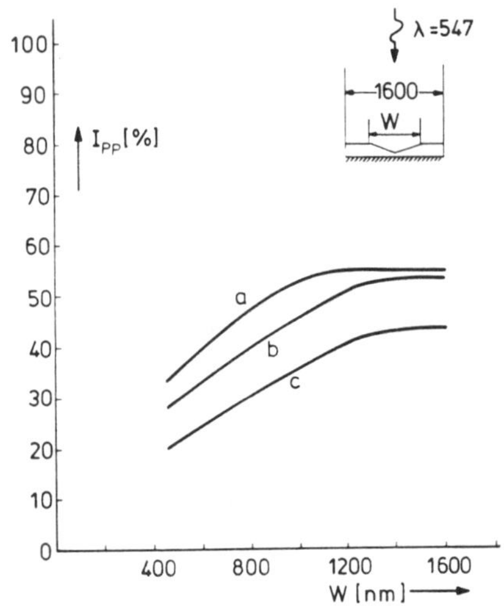

Figure 9. Average PP signal amplitudes versus groove width W. Maximum values (a) and amplitudes when I_{CA} = 80 % (b) and I_{CA} = 90 % (c).

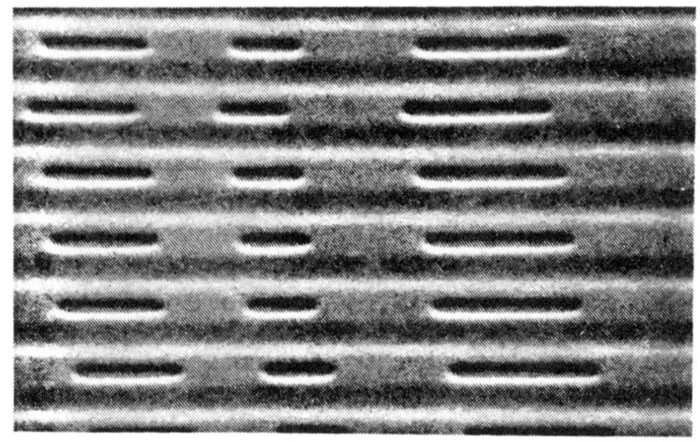

Figure 10. SEM photograph of pregrooves pregrooves recorded between heading pits.

Acknowledgements

The authors wish to express their thanks to E. Penning de Vries and D. Verhaart for their cooperation and fruitful discussions.

References

1. Olijhoek, J.F, Peek, Th.H. and Wesdorp, C.A., Mastering Technology for the Philips Optical Disk systems, Video Disk Technology Overview, Electro '81, New York.
2. Verkaik, W., Compact Disc Mastering - An Industrial Process, AES Premiere Conference, Rye, New York, 1982.
3. Pasman, J.H.T., Chapt. 5 in "Optical Disc Systems", ed. G. Bouwhuis, to be published in 1985 by Adam Hilger Ltd.
4. Pasman, J.H.T., ibid., Chapt. 3.
5. Rooijmans, C.J.M. and Verhoeven, J.A.Th., Characteristics of the Optical Media Laboratory disk, SPIE Proc., Vol. 420, Optical Storage Media Conference, Arlington, 1983.
6. Hopkins, H.H., Diffraction theory of laser read-out systems for optical video disks, J. Opt. Soc. Am., Vol. 69, No. 1, 1979.
7. Dil, J.G. and Jacobs, B.A.J., Apparent size of reflecting polygonal obstacles of the order of one wavelength, J. Opt. Soc. Am., Vol. 69, No. 7, 1979.
8. Pasman, J.H.T., Rigorous diffraction theory applied to video disk geometries, SPIE Proc., Vol. 369, 1983, The Max Born Centenary Conference, Edinburg, 1982.

Reprinted with permission from *Applied Optics*, Vol. 25(22), pp. 4031-4034 (November 15, 1986). ©1986 Optical Society of America.

Optical pregroove dimensions: design considerations

Manabu Yamamoto, Akinori Watabe, and Hiroo Ukita

This paper describes the designing pregroove dimensions to achieve stable data read-out. The spot size detection method and the push-pull method are used for focusing and tracking error signal detection, respectively. Experimental and theoretical results show that pregroove crosstalk and decrease in light intensity due to groove diffraction become smaller as the edge gradient of the pregroove decreases. The permissible range of the depth for a triangular pregroove is 40% larger than that of a rectangular pregroove. The triangular pregroove depth to attain the maximum tracking signal efficiency is about 100 nm for a refractive index of 1.5.

I. Introduction

Pregrooved optical disks have the advantage of making the tracking servo easy when writing data and enabling the disk to be exchanged. In pregrooved disks, push–pull tracking servo signals are detected by a change in the intensity distribution of the diffracted beam. Both tracking and focusing servo signals can be obtained by the same photodiode placed in a far field. However, interference between tracking error signals and focusing error signals makes the focusing servo unstable,[1,2] and the decrease in light intensity due to groove diffraction causes a reduction in signal amplitude. These problems depend not only on the width and depth but also on the gradient of the edge of a pregroove.

In this paper the relationship between these pregroove dimensions and the readout requirements needed to achieve stable data readout is considered.

II. Experiment

Figure 1 shows a schematic diagram of the experimental optical head[3] and the six-element photodiode used for tracking and focusing error signal detection. The spot size detection method is used for focusing error signal detection because of the small number of optical components and easy adjustment compared with other methods, such as the astigmatic or knife-edge. The tracking error signal is derived via the well-known push–pull method. The laser beam passes through the objective lens $L3$ having a 0.5 N.A. and is focused on the pregrooved disk. The pregroove acts like a reflection grating and produces beam diffraction. Only part of the reflected beam passes through the entrance pupil of the objective lens $L3$. This part is projected onto the photodiode $PD2$ as shown by the dotted arrows. The intensity distribution of the reflected beam on the photodiode is shown by the solid and dotted lines in the figure. We can obtain tracking and focusing error signals by subtracting the output from the diode halves which are composed of the diode elements shown in Fig. 1. The light intensity P_0 is obtained by adding the output of the six photodiode elements. The beam of an AlGaAs laser of wavelength 830 nm is focused by a lens with 0.5 N.A. through a glass substrate of refractive index $n = 1.5$. The intensity profile of the laser beam is Gaussian, and the spot diameters (full width at half-maximum) in both radial and tangential directions are 0.85 μm.

We used the pregrooves whose width W and gradient of the edge θ are changed, but depth D is constant (70 nm). These parameters were measured by an interference microscope and a contact profilometer.

Figure 2 shows the waveform of the tracking and focusing error signals as well as the definition of width, depth, and gradient. ΔF is the level of the pregroove crosstalk on the focus signal. The pregroove crosstalk is produced by the disturbance of the light intensity distribution due to groove diffraction. It makes the servo gain fluctuate and the focusing offset vary. When the edge gradient θ is ~10°, the pregroove crosstalk on the focus signal has little effect on the tracking signal, because the focusing offset error produced by the pregroove crosstalk corresponds to a defocusing of less than ±1 μm. On the other hand, when θ is ~40°, a large focusing offset error is caused by the pregroove crosstalk, and the tracking signal is disturbed. Figure 3 shows the pregroove crosstalk on focus signal ΔF as well as the tracking signal efficiency ΔT as a function of the gradient θ. In Fig. 3 the dotted lines are the calculated results discussed later, and the solid line is a curve through the experimental points. The triangle is used to separate tracking error data from focusing error data. As the gradient is increased, the ΔF amplitude increases correspondingly. In our experiment using the spot size detection method the

---

The authors are with NTT Electrical Communication Laboratories, 3-9-11, Midori-cho, Musashino-shi, Tokyo 180, Japan.

Received 4 February 1986.
0003-6935/86/224031-04$02.00/0.

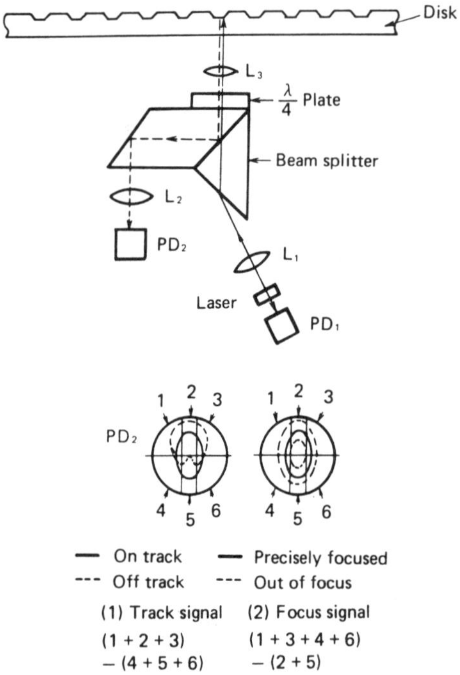

Fig. 1. Schematic diagram of optical head.

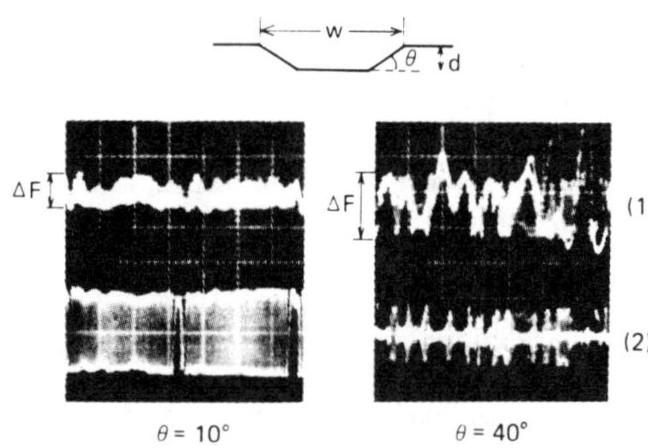

Fig. 2. Waveform: (1) focusing error signal, (2) tracking error signal.

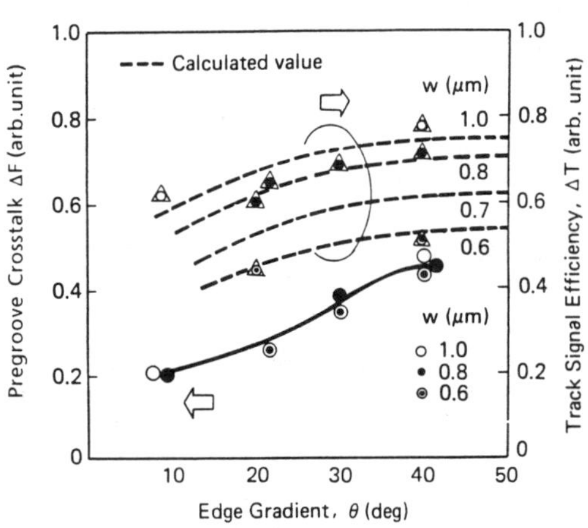

Fig. 3. Focus servo error and track signal efficiency vs edge gradient.

focusing offset error produced by the pregroove crosstalk is lower than $\pm(1\text{–}2)$ μm for a gradient of less than ~25°, and the tracking signal is not disturbed in this range.

The tracking signal efficiency ΔT is defined by

$$T = (P_a - P_b)/P_0. \tag{1}$$

Here P_a and P_b are the output of the diode halves. P_0 is the sum of P_a and P_b. Figure 3 shows that as the gradient θ increases from 10 to 40°, ΔT amplitude increases gradually but does not increase as much as the focusing servo error. It can also be seen that the ΔT amplitude depends mostly on width W.

Figure 4 shows the reduction of P_0 in a pregroove as a function of the gradient θ for different values of the width W. The light intensity P_0 is normalized by the light intensity reflected from the flat surface disk. The reduction of P_0 is caused by an eclipse of the diffracted beam out of the objective entrance pupil. The experimental result indicates that (1) as the gradient θ decreases from 40 to ~10°, the P_0 amplitude increases in each groove width, and (2) the reduction in light intensity depends more on the gradient of the edge than on the groove width.

III. Theoretical Consideration

We calculated the intensity distribution of the reflected light by the scalar diffraction theory.[4] This theory provides reasonable agreement with the vector theory[5] when the N.A. is smaller than 0.5.[4,6] The light amplitude in the far field can be expressed by

$$\begin{aligned} U(u) &\propto \frac{1}{\lambda d} \int f(x) \exp[j\phi(x)] \exp\left(-j\frac{2\pi}{\lambda} r\right) dx, \\ r &= \sqrt{(x-u)^2 + d^2}, \\ f(x) &= \exp[-x^2(A + Bi)], \\ 1/A &= w_0^2\left[1 + \left(\frac{\lambda d}{\pi w_0^2}\right)^2\right], \\ 1/B &= \frac{\lambda d}{\pi}\left[1 + \left(\frac{\pi w_0^2}{\lambda d}\right)^2\right]. \end{aligned} \tag{2}$$

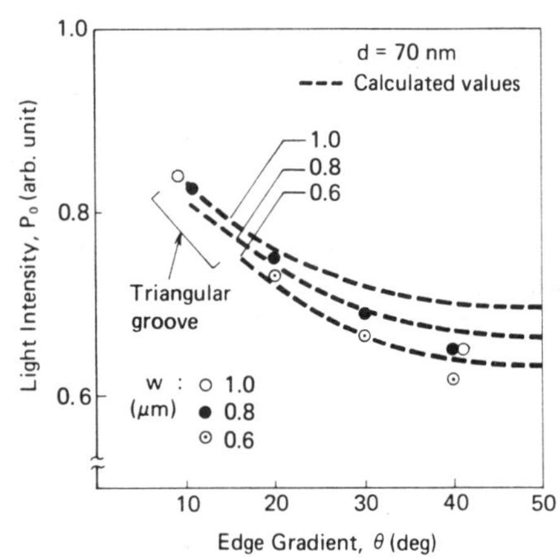

Fig. 4. Light intensity vs edge gradient.

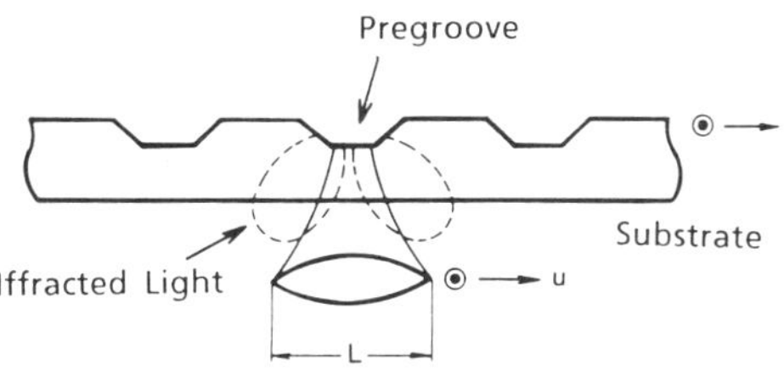

Fig. 5. Definition of the coordinate.

Here $U(u)$ is the light amplitude reflected back from a pregroove; $\phi(x)$ is the cross section of a pregroove; $f(x)$ is the incident beam distribution; w_0 is the spot radius of the beam waist ($1/e^2$); λ is the wavelength of the laser beam; and d is the focal length of the objective lens. The coordinate x lies on the medium surface, and the coordinate u lies in the far field as shown in Fig. 5. We approximated the cross section of the groove by straight line segments resulting in a triangular or trapezoidal shape, assuming that the pregroove imposes only a phase change on the light.

The tracking signal efficiency ΔT and light intensity P_0 are described as follows:

$$\Delta T = \left\{\int_0^{L/2} |U(u)|^2 du - \int_{-L/2}^{0} |U(u)|^2 du\right\} \Big/ \int_{-\infty}^{\infty} |U(u)|^2 du; \qquad (3)$$

$$P_0 = \int_{-L/2}^{L/2} |U(u)|^2 du. \qquad (4)$$

Here L is the lens diameter.

The dotted lines in Figs. 3 and 4 show the calculated ΔT and P_0 amplitude as a function of the edge gradient for different values of width W. The result indicates that (1) as the gradient θ increases, ΔT amplitude increases gradually and saturates when the gradient θ is wider than ~40°, and (2) the light intensity P_0 becomes maximum at a triangular shape for each width. Figure 6 shows the light intensity P_0 and tracking signal efficiency as a function of width W and depth D for a triangular shape. As the depth is increased, light intensity P_0 decreases for each width. The groove depth of the maximum tracking signal effiency is ~100 nm.

IV. Discussion and Conclusion

From the experimental and theoretical results, we evaluated the permissible pregroove dimensions for the stable readout requirement. The factors which determine pregroove dimensions are (1) tracking signal efficiency, (2) light intensity reflected from a pregroove, and (3) pregroove crosstalk on the focus signal.

Figure 7 shows a permissible range of θ for each W. The depth is 70 nm. As the groove width increases, the permissible range of the gradient also increases. Lines a and b are obtained from the conditions of the pregroove crosstalk on the focus signal and the light intensity P_0, respectively. The condition of line a is that the range of the gradient for a stable focusing servo is lower than ~25°. In our experiment using the spot size

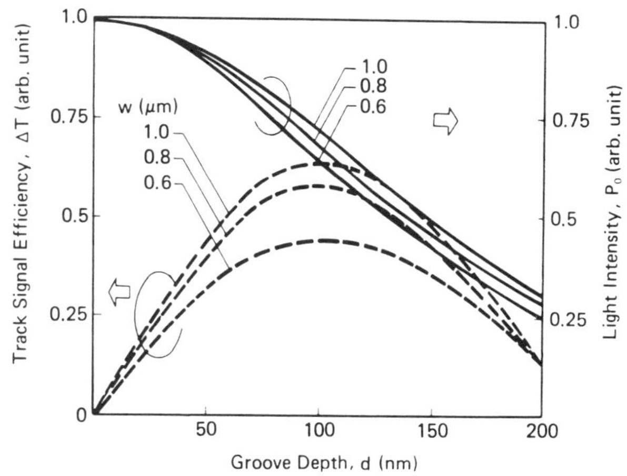

Fig. 6. Calculated results for a triangular shape.

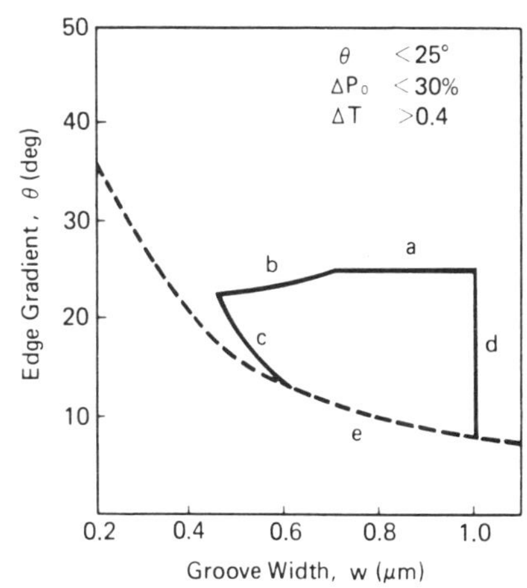

Fig. 7. Permissible range of edge gradient for each width.

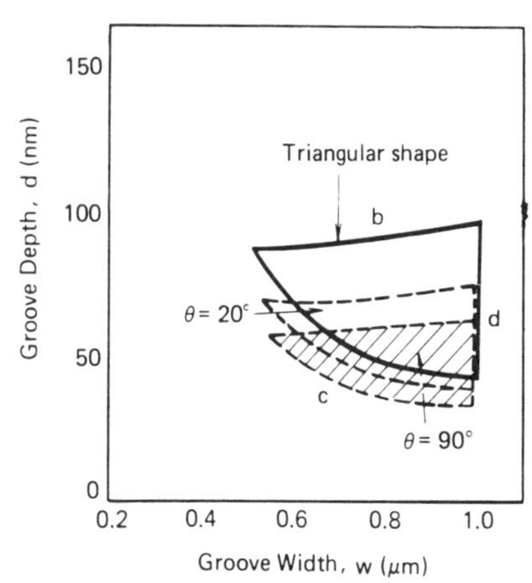

Fig. 8. Permissible range of groove depth for each width.

detection method the pregroove crosstalk is comparatively small within the range in which the tracking signal is not disturbed. The condition of line b is that the intensity reduction of the light in a pregroove is lower than 30% compared with the flat surface. Line c is obtained when the tracking signal efficiency must be higher than 0.4 according to compact disk (CD) standards. Line d will result when the condition of the crosstalk from an adjacent track is satisfied. However, in our experiment there is no noticeable difference between the different edge gradients when the pit diameter is 0.8 μm and the tracking offset error is less than ± 0.1 μm. According to the experimental result, line d, which does not depend on the gradient, has been obtained. Dashed line e will result when the condition for the triangular shape is satisfied. Figure 8 shows the permissible range of depth D and width W in various pregroove shapes. Lines b, c, and d are obtained with the same conditions as in Fig. 7. The shaded area and dotted line show the permissible range of a rectangular and trapezoidal shape, respectively. The solid line shows that of a triangular shape. The permissible range of the depth is large for this shape.

Factors (1) and (2) mentioned above do not depend on the focusing signal detection method, but (3) depends on the focusing signal detection method.

For example, the knife-edge method, which places a detector in the focal plane, does not have the focusing offset error produced by the pregroove crosstalk. In principle, this is because the crosstalk is generated when the detector is placed in a far field. The astigmatic method using a cylindrical lens has a small offset error produced by the pregroove crosstalk. This is because the pregroove crosstalk can be canceled by the differential detection in this method. In practice, there is pregroove crosstalk in these methods because of the adjustment error or deviation in the detector position. However, if these methods are used, the permissible range of the pregroove dimensions can be large compared with the spot size detection method described here.

The authors thank K. Itao and S. Hara for encouragement. They also thank S. Fukunishi and K. Katoh for experimental help.

References

1. Y. Tsunoda, S. Horigome, and Z. Tsutsumi, "Optical Digital Data Storage Technologies with Semiconductor Laser Head," Proc. Soc. Photo-Opt. Instrum. Eng. **382,** 24 (1983).
2. O. Ito, K. Yoshimoto, I. Watanabe, and T. Kubo, "Improvement of Focusing and Tracking Servos in Optical Disk Memory," Soc. Photo-Opt. Instrum. Eng. **529,** 145 (1985).
3. H. Ukita, A. Watabe, and K. Katoh, "MALS Head for Optical Recording," in *Technical Digest, Conference on Lasers and Electro-Optics* (Optical Society of America, Washington, DC, 1983), paper WR2.
4. H. H. Hopkins, "Diffraction Theory of Laser Read-out Systems for Optical Video Discs," J. Opt. Soc. Am. **69,** 4 (1979).
5. J. H. T. Pasman, H. F. Olijhok, and B. Verkaik, "Developments in Optical Disk Mastering," Proc. Soc. Photo-Opt. Instrum. Eng. **529,** 61 (1985).
6. T. Murakami, "Design of the Pregroove-Shape on an Optical Disk Based on the Light Diffraction Analysis," Trans. Jpn. IECE **J67-C,** 2 (1984) (in Japanese).

Reprinted from *Optical Storage Technology and Applications*, Proc. SPIE Vol. 899, pp. 123-128 (1988). ©1988 Society of Photo-Optical Instrumentation Engineers.

Novel stamper process for optical disc

Shinkichi Horigome, Yoshinori Miyamura, Keizo Katou, Masaru Itou, and Norio Ohta

Central Research Laboratory, Hitachi Ltd
Kokubunji, Tokyo 185

ABSTRACT

Two new types of stampers have been developed in order to improve the quality of optical discs and to simplify the process for manufacturing stampers. One is a plastic stamper which consists of a PMMA substrate and a photo-polymer resin layer having fine patterns such as pits and grooves on its surface. The other is a quartz stamper which has the pits and grooves with different depth formed on the surface of the quartz substrate by reactive ion etching process. Both these stampers can be made in a short processing time, and their quality is found to be higher than that of the conventional one.

These stampers are transparent for ultraviolet light, so that they may also be used for making a replica disc with non-transparent substrate such as an Al substrate.

1. INTRODUCTION

In an optical disc, grooves and pits are needed for tracking and for adressing, respectively. There are two methods widely used for forming these fine patterns on an optical disk, namely, the 2p(photo-polymerization) method and the injection method.[1] For both of the methods, stampers are necessary. The conventional process for manufacturing stampers is as follows. Grooves and pits are formed by laser irradiation onto photoresist coated on a glass substrate followed by developing process. Then, a layer of nickel is deposited on the surface of the patterned photoresist by electroplating, and is separated from the glass substrate. It is used as a metal stamper. An electroplating process requires a number of steps and takes a long time. Therefore, it is desirable to introduce a new method for manufacturing a stamper in a more simplified process without electroplating. In addition, improving the quality of the stamper is also required. An Al substrate is supposed to be more promising than a glass or plastic substrate for a higher data rate optical disc. In this case, the stamper transparent for ultraviolet light is necessary to form grooves and pits on an Al substrate by a 2P method. Furthermore, the transparent stamper is expected to play an important role in the photochemically casting process newly developed by us for molding replica discs.[2] This paper describes new methods for fabricating a stamper to meet such requirements.

2. PLASTIC STAMPER

2.1. Manufacturig process

The process for a plastic stamper is shown in Fig. 1. A photoresist layer was coated on a glass substrate with a thickness of 10mm, and then laser beam exposure and development were carried out to form the groove and pit pattern by a conventional mastering method. A glass substrate having a patterned photoresist is called a glass master. Then, a fluorine containing polymer film or metal film was deposited on the surface of the patterned photoresist by an RF sputtering process. This polymer film having low surface energy serves as a layer to improve releasability of a plastic stamper from the glass master. The liquid of a photo-polymer resin was dropped onto the patterned surface. Then, a plate of PMMA having a thickness of 2mm was placed on this liquid. The surface of the PMMA plate was given a pretreatment to enhance the adhesive strengh to the photo-polymer resin layer. The photo-polymer resin was cured by irradiation of ultraviolet light through the PMMA plate. After this, the plastic stamper was separated from the glass master as shown in

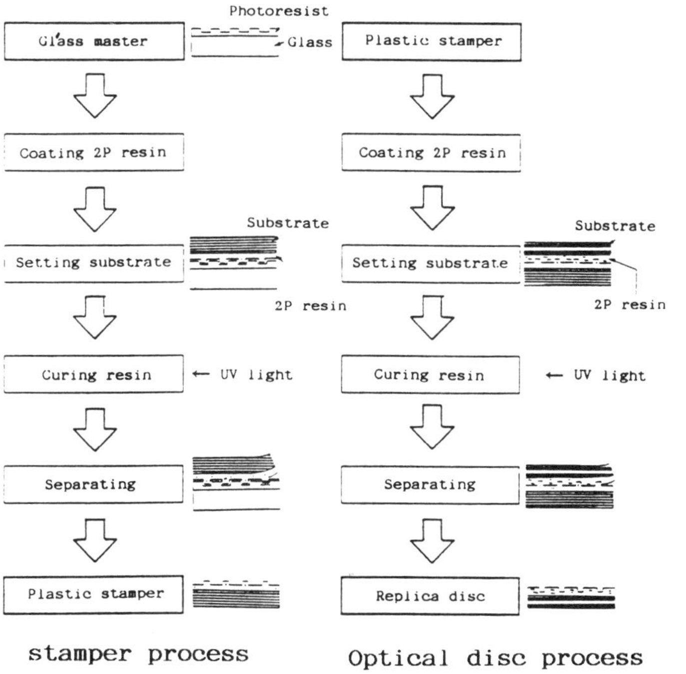

Fig. 1 Optical disc process using plastic stamper

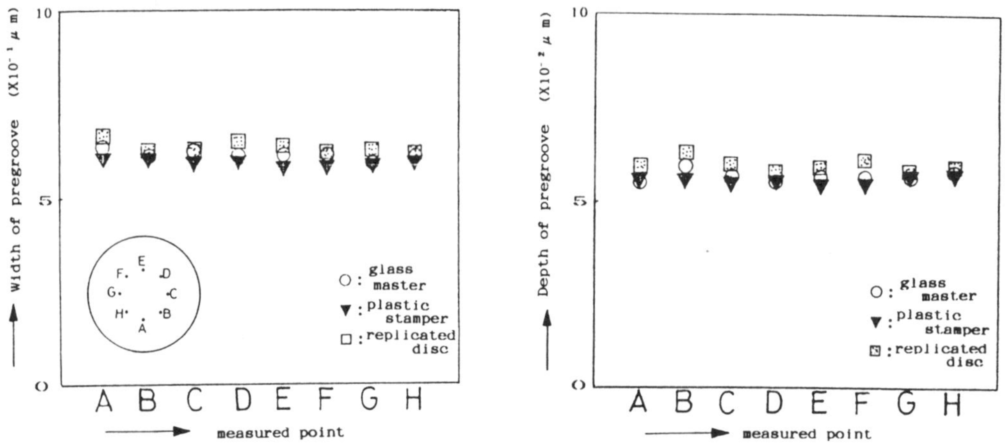

(1) Width of pregroove (2) Depth of pregroove

Fig. 2 Depth and width of pregroove measured by diffraction method

Fig. 1. It should be noted that, by this method, over 20 pieces of plastic stampers could be obtained from a single piece of glass master without deterioration in the quality of the fine patterns. A plastic stamper can be manufactured in a much shorter time, than a conventional electroplated stamper, because it is directly reproduced from the master in fewer steps. The plastic stamper, then, was baked for being cured completely. A separation layer was deposited on the surface of the photo-polymer resin. Its thickness is about 50nm. By use of this stamper, a 5.25 inch replica disc of glass substrate was made by a conventional 2P method as shown in Fig. 1. The glass substrate was chemically reinforced and 1.15mm in thickness. The photo-polymer resin used here was selected for its hardness and heat resistance which are necessary to avoid deformation of its patterned surface due to the stress in the recording layer and the heat during the following processes.

2.2. Evaluation

The quality of the replica discs obtained from the plastic stamper was evaluated .

The depth and width of the grooves were derived from the measurement of the intensities of diffracted beams when the light was incident on the disc. These results are shown in Fig 2. It is clear that there is little difference between the patterns on the plastic stamper and on the glass master. Therefore, the groove shape of the glass master is accurately transfered to the replica disc by this new process. Moreover, the circumferential uniformity of the groove shape is found to be good as shown in Fig. 2. Fig. 3 is SEM photographs showing the groove shape on a plastic stamper and on the replica disc. For this

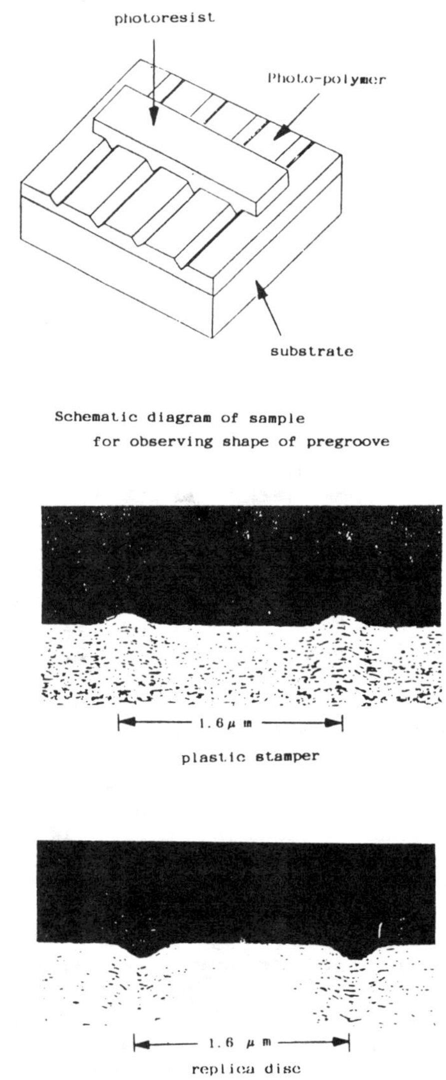

Fig. 3 Shape of groove of plastic stamper and replica disc

observation, a rectanglar photoresist layer was formed on the grooves by photolithographic technique as shown in Fig. 3. These photographs also indicate that the grooves on the stamper were transfered to the replica disc with high fidelity. The photographs in Fig. 4 show the pits and grooves of the 1st disc and the 100th disc replicated from the same plastic stamper. From these photographs, it can be concluded that it is possible to make several hundred pieces of replica discs from a single plastic stamper.

Next, the mechanical properties of these replica discs were measured and the results are shown in Fig. 5. The skew angles of the discs were less than 0.3°, and the vertical acceleration of the rotating disc at 1800rpm was less than 0.8G. These results mean that the degree of the flatness of the replica disc is sufficient for practical use. The surface noise of the replica disc on which a magneto-optical recording layer sandwiched between protective layers was deposited was measured by using an optical disc drive and a spectrum analyser. The results are shown in Fig. 6. The noise level in this new process is slightly lower on groove and on land than that in the conventional process. Therefore, the roughness of the pattern surface does not increase throughout the processing.

The plastic stamper process is particularly suited for the case in which many kinds of optical discs having different information have to be produced at a low cost.

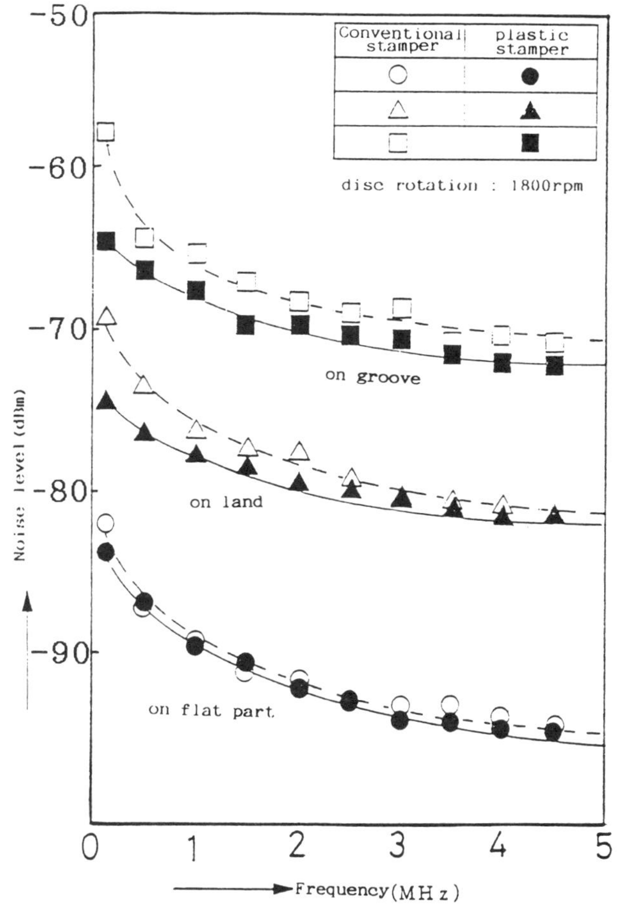

Fig. 6 Noise level comparison of optical disc replicated from plastic stamper and the conventional disc

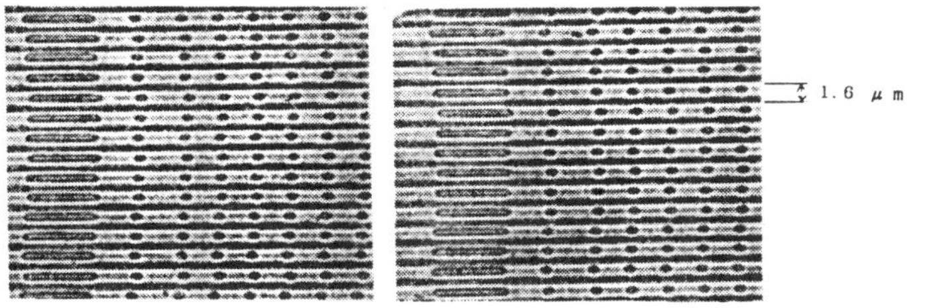

Fig. 4 Pits and grooves pattern of 1st disc and one hundredth disc replicated same plastic stamper

Fig. 5 Mechanical properties of 5.25inch optical glass disc replicated from plastic stamper

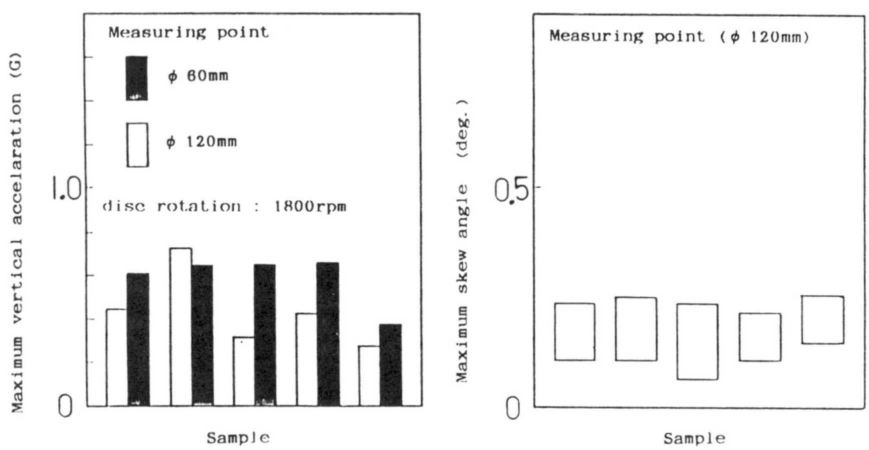

(1) Maximum vertical accelaration (2) Maximum skew angle

3. A QUARTZ STAMPER

3.1. Manufacturing process

It has already been reported[3,4] that the groove alone can be formed in a glass surface or a SiO_2 layer by reactive ion etching(RIE) process. In this study, however, a new approach for fabricating the groove and pit with different depth has been tried. The fabrication process is shown in Fig. 7. Initially, photoresist was coated on a quartz substrate which was 200mm in diameter and 5mm in thickness. The thickness of the photoresist was 140nm. The surface of the quartz substrate was fine polished beforehand. Then, grooves and pits were formed on the photoresist on the quartz substrate by a mastering process. The depth of the pit was equal to the thickness of the photoresist, and the depth of the groove was about half that of the pit. Then, the quartz substrate was etched through the photoresist pattern used as a mask by reactive ion etching in pure CF_4 gas. Finally, the residual photoresist was removed by plasma ashing in oxygen. For forming the grooves in the surface of the quartz substrate, it is necessary to etch the residual photoresist initially and then the quartz surface in order to maintain the groove shape. On the other hand, no photoresist was left at the pits, so that these parts of the quartz surface were directly etched from beginning. Fig. 8 is the schematic model showing the relation between the pit depth and the groove depth during the etching process. The thickness of the photoresist is d_1, and thus, the pit depth in the photoresist is also d_1. The depth of the groove as defined in Fig. 8 is d_2. The depth of the pits and grooves etched in the surface of the quartz is d_1' and d_2', respectively. When t is an etching time, the following relation is obtained:

$$t=\frac{d_1'}{Eq} \quad \ldots(1) \qquad Es=\frac{d_1'-d_2'}{d_1-d_2} \quad \ldots(2)$$

where Eq is the etching rate of the quartz and Es is the etching selectivity which is the etching rate ratio of the quartz to the photoresist. Eq and Es are depend on the etching conditions such as RF power, CF_4 gas flow rate, and so on. Generally, an RIE process has a tendency to damage surface smoothness of a substrate to be etched by the plasma particles. Therefore , the suitable etching conditions were chosen in order to obtain the lowest noise level of the etched surface. The noise level of the quartz surface etched under these conditions is shown in Table 1. The noise level was measured after the quartz surface had been deposited with an Al layer. The value for the unetched part of the quartz surface is also given in Table 1. The noise level of the etched surface is 2dBm higher than that of the unetched surface at 1MHz, but there is no difference between them at 5MHz.

Under the etching condition as described above, Es was reproducible and its value was 1.4. When d_1 and d_1' are 140nm, the depth of the groove d_2 in the photoresist should be 97nm from the equation(2) in order to obtain the groove depth etched d_2' of 80nm. The pattern in the photoresist which had the same d_1, d_2 as obtained above was formed on a quartz surface.
Then, this was etched through the patterned photoresist. The depth of the grooves and the pits formed on the quartz stamper was 80nm and 140nm, respectively.

3.2. Evaluation

Fig. 9 is a SEM image of the pits and grooves on the quartz surface. Fig. 10 is a TEM image showing the groove shape. The SEM image shows that the surface roughness is larger in the groove than in the pits and in the land which has the original quartz surface. Then, the noise level was measured and the results are shown in Table 2 with that of a replica disc from a conventional nickel stamper. As shown in Table 2, the noise level on land of the

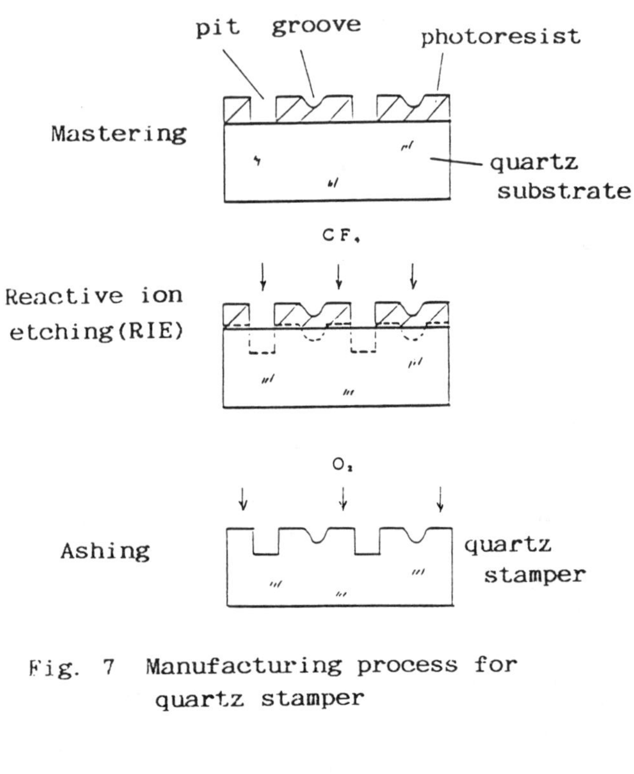

Fig. 7 Manufacturing process for quartz stamper

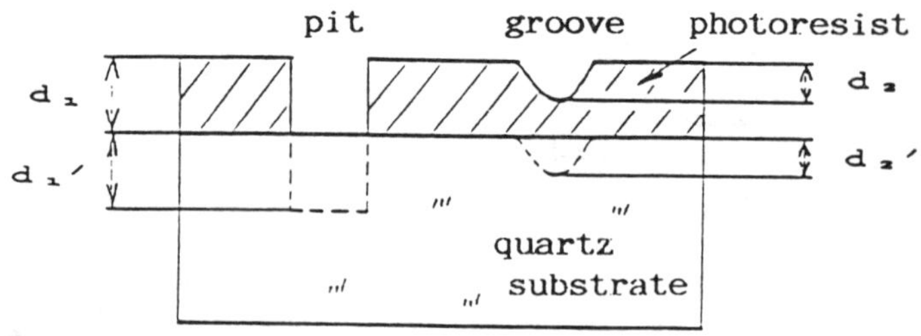

Fig. 8 Relation between pit depth and groove depth during etching

quartz stamper is about 4dBm lower and that on groove is 2~3dBm higher than these of a conventional disc. This high noise level on groove is consistent with the surface roughness shown in Fig. 9. This roughness of the groove surface could be attributed to the development of the photoresist on the grooves to nearly half the original thickness. Furthermore, this roughness increased in proportion to the etching selectivity in the RIE process.

The noise level on land must be essentially the same as that on the unetched surface shown in Table 1. However, the noise level on land is 3~4dBm higher than that of the unetched part. This can be explained as follows. Since the diameter of the laser beam for reading is slightly larger than the width of the land, a small part of the laser beam is reflected from the rough surface of the grooves. The noise level on land of the quartz stamper is -78dBm, which is sufficiently low for plactical use. The processing conditions, however, must be investigated to decrease the roughness of the groove surface. Fig. 11 shows the readout signal of the pits and grooves of the quartz stamper.

This quartz stamper is expected to be applicable not only to duplication of discs by the 2P method, but also to fabrication of a metal stamper by the electroplating process.

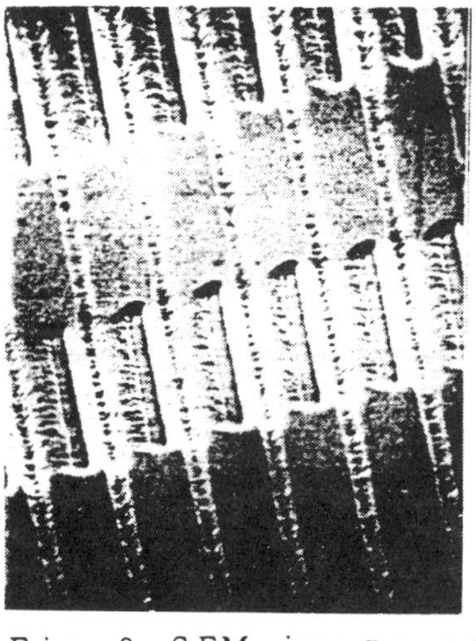

Fig. 9 SEM image of pits and grooves on quartz stamper

Fig. 10 TEM image of cross section of grooves on quartz stamper

Table 1. Noise Level of Etched and Unetched Quartz Surface

process	noise level (dBm)	
	1MHz	5MHz
unetched	-83.0	-86.7
etched	-81.0	-86.7

rotation 1800rpm

Table 2. Noise Level Comparison of Quartz Stamper and Conventional Stamper

stamper	noise level (dBm)			
	on groove		on land	
	1MHz	5MHz	1MHz	5MHz
conventional Ni stamper	-70.6	-77.5	-73.3	-79.4
quartz stamper	-68.9	-74.5	-77.8	-83.6

rotation 1800rpm

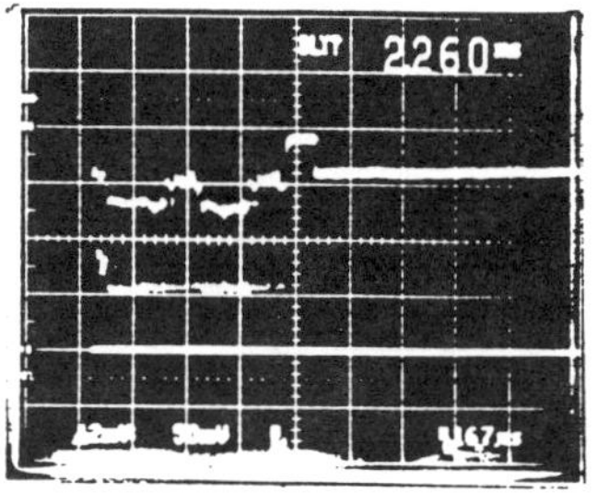

Fig. 11 Readout signal of pits and grooves

4. CONCLUSION

New methods for manufacturing stampers have been studied with the aim of improving the quality of optical discs and simplifying the process and the following results are obtained.

(1) A plastic stamper made of a PMMA substrate and photo-polymer resin layer with pits and grooves can be manufactured in a greatly reduced time. Therefore, this process is very convenient for producing many kinds of optical discs having different information at a low cost.

(2) A quartz stamper having pits and grooves with different depth can be made by reactive ion etching through patterned photoresist. The noise level on land is sufficiently low, but that on groove must be improved.

5. ACKNOWLEDGMENT

The authors wish to express their thanks to Dr. Yoshito Tsunoda for his encouragement and helpful suggestions and Miss Mari Ichikawa for her invaluable assistance.

6. REFERENCES

1. H. C. Haverekorn van Risewijk, P. E. J. Legierse and G. E. Thomas, Philips Tech. Rev. 40, No. 10, 289 (1982)
2. R. Sudou, H. Miwa, T. Tajima, S. Horigome, E. Koyama and N. Ohta, This proceedings, 899-20 (1988)
3. K. Tanii, T. Kashihara, N. Tsuboi and Y. Okino, Proceedings of Symposium on Dry Process, II-5, Tokyo, 53 (1986)
4. M. Miyagi, A. Iwasawa and H. Yamazaki, International Symposium on Optical Memory, Tokyo, Japan, 105 (1987)

Section Thirteen
Other Storage Media

Reprinted with permission from *Japanese Journal of Applied Physics,* Vol. 28, Suppl. 28-3, pp. 221-227 (1989).

Photon-Gated Persistent Spectral Hole-Burning

W. E. MOERNER

IBM Research Division Almaden Research Center,
650 Harry Road San Jose, California 95120, U.S.A.

(Received September 6, 1989)

This article reviews recent progress in the area of photon-gated persistent spectral hole-burning, in which one photon selects absorbers in an inhomogeneously broadened line and a second ''gating'. photon of a different wavelength completes the excitation necessary to produce a spectral hole. This phenomenon provides a crucial threshold in the hole formation process, allowing reading with the first wavelength alone to be nondestructive. Examples of photon-gating in both inorganic and organic materials are summarized, with emphasis on the organic materials.

§1. Introduction

Frequency domain optical storage, in which digital information in encoded as the frequency location of persistent spectral holes,[1] shows promise of high areal densities in the 10^9–10^{11} bits/cm^2 range while affording fast random access by beam deflection as well as high data rates by laser frequency tuning and optical parallelism. Persistent spectral hole-burning (PSHB) occurs in inhomogeneously broadened optical transitions of impurities in transparent solids at low temperatures when optical excitation of resonant impurity centers causes a long- lived transformation to a new nonabsorbing ground state. In addition to the interest in PSHB for possible storage applications,[2] this phenomenon has proven to be extremely useful in the study of the statics and dynamics of absorbing centers in a wide variety of amorphous and crystalline solids.[3]

A material in order to be useful in a practical frequency-domain optical storage system must simultaneously show the ability to form deep holes in short (nanosecond) burning times and yet allow nanosecond reading at high signal-to-noise ratios with tightly focused beams. These requirements place several well-defined constraints on the dynamical properties of the hole-burning mechanism. For the case of single-photon materials in which the photoinduced change proceeds with a certain fixed probability after the absorption of one photon by each absorbing center, a thorough analysis of the coupled reading-writing problem[4] shows that these materials can provide sufficient SNR only in a limited range of absorption cross sections and quantum efficiencies. The essential problem with single-photon processes is that there is no threshold or hysteresis in the hole formation mechanism, as opposed to the situation with all other successful schemes for long-term information storage such as magnetic, magneto-optic, ferroelectric, and phase change recording.

One way around the destructive reading problem with single-photon materials is to consider two-step spectral hole formation mechanisms, called gated mechanisms (see Fig. 1). A gated mechanism has the property that irradiation with the wavelength λ_1 alone causes essentially no photoinduced changes (writing), i.e., after the absorption event takes the center from the ground state 1 to the excited state 2, the center simply returns to the original ground state 1 after the excited state lifetime. However, when the center has been placed in the excited state 2 by absorption of a photon at λ_1 and an external ''gating field'' is also present, a photoinduced change occurs and the center enters a new ground state or permanent reservoir that no longer absorbs at λ_1. The ''dip'' or region of reduced absorption that is left behind in the inhomogeneous line is the resulting spectral hole that can be detected nondestructively with λ_1 alone. In this manner, gated mechanisms add a crucial threshold to the hole formation process, which allows the writing and reading processes to be uncoupled. The external field may be a second photon of a different wavelength or the gating could perhaps be achieved by any other external field, such as electric field, magnetic field, stress field, and the like. When the gating field is a second light field,[5] the mechanism is said to be *photon-gated.*

The importance of photon-gated processes for scientific studies of PSHB and for frequency domain optical

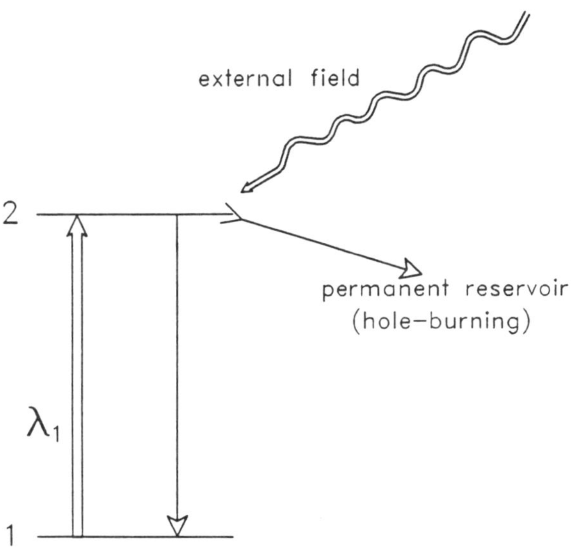

Fig. 1. Illustration of basic scheme for gated, two-step PSHB. The line pointing toward ''permanent reservoir'' depicts the transformation of the center to a new ground state that produces hole-burning.

storage has stimulated much recent research to discover new photon-gated materials. This paper will first present schematic energy level schemes for photon-gating appropriate for 3-level and 4-level systems (§2). In most cases, the mechanism is two-step photoionization, although other biphotonic photochemical reactions have also produced photon-gating (see §3.2). In §3, recent progress in the search for photon-gated systems in inorganic materials as well as in organic materials will be briefly reviewed. Some emphasis will be placed on the organic materials since a review of the inorganic materials has appeared recently.[6] This work concludes (§4) with a description of the optimal cross section, quantum efficiency, and number density for an optimal photon-gated material for the case of frequency domain readout in focused spots.

§2. Generalized Photon-Gated Mechanisms

Figure 2 illustrates some general requirements on three-level and four-level photon-gated PSHB mechanisms in order to achieve high overall efficiency and high gating ratio, where the gating ratio is a measure of the efficiency of two-color hole production divided by the efficiency of one-color hole production with λ_1 alone. While the exact materials requirements depend strongly upon the actual system configuration and signal-to-noise requirements, some general comments about the required level structure can be made.

The left side of Fig. 2 illustrates the three-level system which is the common configuration for many inorganic materials. A first requirement is that λ_1 should be larger than λ_2, so that the site-selecting beam cannot also easily act as the gating beam. The absorption in the circular regions should be small to reduce one-color burning (by three λ_1 photons, case a) or bleaching by the gating light (by two λ_2 photons, case b). The lifetime τ of the intermediate state 2 should be long enough to allow large populations to build up in this state for further excitation to the reactive levels 3. In cases where τ is much larger than the data access time the storage material does not have to be exposed to both photon energies simultaneously. Further, there is no fundamental need for a frequency-selective narrow-band transition from level 2 to level 3; however, in certain instances, narrow-band levels 3 involving transitions with high peak cross sections may be preferable to a continuum absorption such as a conduction band. The microscopic yield η from the reactive

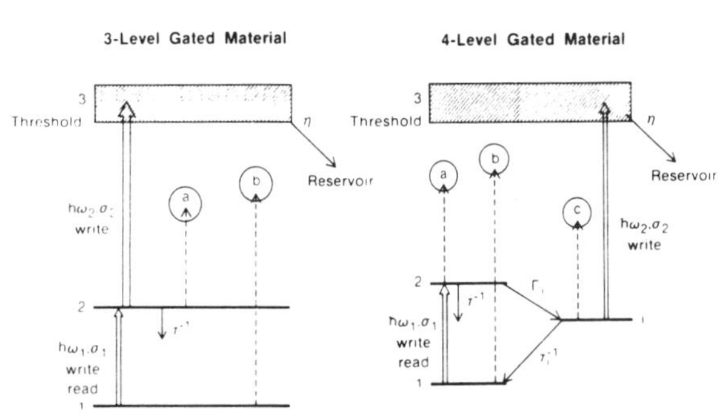

Fig. 2. Two schemes for photon-gated PSHB. Left side: Three-level system. Right side: Four-level system. The optical absorption should be small in the circled regions.

levels 3 should be as large as possible for high overall gating efficiency.

In principle, systems with $\lambda_1 = \lambda_2$ can exhibit gated PSHB in the sense that the hole-burning yield is nonlinear with laser intensity.[7] For example, at low powers the hole formation rate may scale quadratically with laser power, whereas at high powers level 2 saturates and the hole formation rate scales linearly with laser power. However, in actual situations, signal-to-noise requirements often require that the reading laser power be so large that the quadratic regime cannot be utilized. In addition, the requirement of non-destructive reading makes it desirable to use systems with $\lambda_1 \neq \lambda_2$ permitting complete decoupling of reading and writing processes.

The right half of Fig. 2 illustrates optimized photon-gated PSHB with a four-level system. Here the intermediate state i is distinct from state 2, which allows independent optimization of the level lifetimes. This level structure occurs quite commonly for organic materials in which states 1 and 2 are part of the singlet manifold and level i is the lowest triplet state. The required lifetime τ of level 2 will often be determined by a combination of the required hole width, data rate, and absorption strength. For example, if τ is too short, the holes will be necessarily broad. For efficient gated PSHB the $(2 \rightarrow i)$ rate Γ_i should be as large as possible consistent with the requirement that the lifetime of level 2 not be too short. Further, a long intermediate state lifetime τ_i would be advantageous in achieving a large population in level i. It is evident that absorptions $2 \rightarrow$(a), $1 \rightarrow$(b), and $i \rightarrow$(c) involving photons of frequency ω_1, ω_2, and ω_1, respectively, should not be large in order to prevent undesired bleaching and inefficient excitation of interfering levels. Of course, the microscopic yield η from the photoreactive levels 3 should be as large as possible for efficient photon-gating.

§3. Examples of Photon-Gating

3.1 *Inorganic materials*

The first observation of photon-gating resulted from experiments on Sm^{2+} ions in BaClF crystals.[8] This materials is in the class of 3-level materials (Fig. 2, left side), and the specific energy levels of interest are shown in Fig. 3. Despite the discovery of several other inorganic systems showing photon-gating, this material continues to be of interest due to the large gating ratio and other properties to be described below. The first photon near 690 nm excites the system from the $^7F_0$ ground state (level 1) to the $^5D_0$ level (level 2). Extended irradiation at 690 nm produces essentially no hole production, but brief periods of simultaneous irradiation with $\lambda_2 = 514$ nm or shorter produces spectral holes at λ_1. The second photon excites the ion from $^5D_0$ to the conduction band or to an autoionizing level above the conduction band edge and the liberated electron is subsequently trapped in the host matrix leaving behind a Sm^{3+} ion.

Figure 4 shows examples of the spectral holes for the Sm^{2+}:BaClF system. The gating ratio, or the ratio of gated hole depth to single-color hole depth for equal λ_1 burning conditions, has been observed to be quite large, on the order of 10^4. Another novel property of this system is room temperature cyclability of the written in-

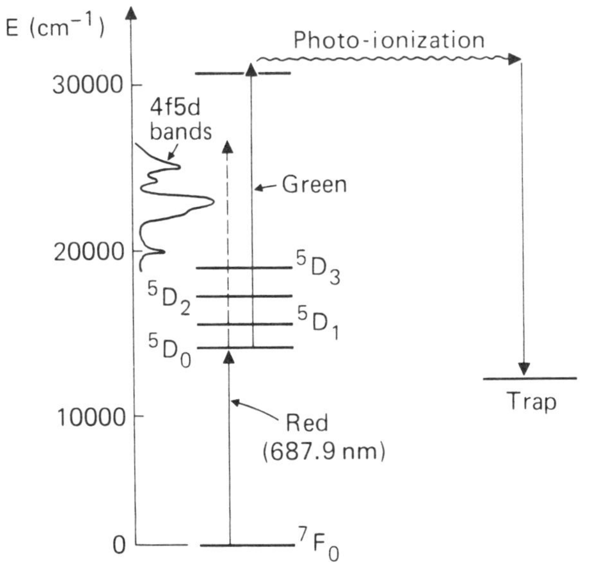

Fig. 3. Energy level diagram of BaClF:Sm^{2+} showing the two-step photoionization process responsible for photon gating.

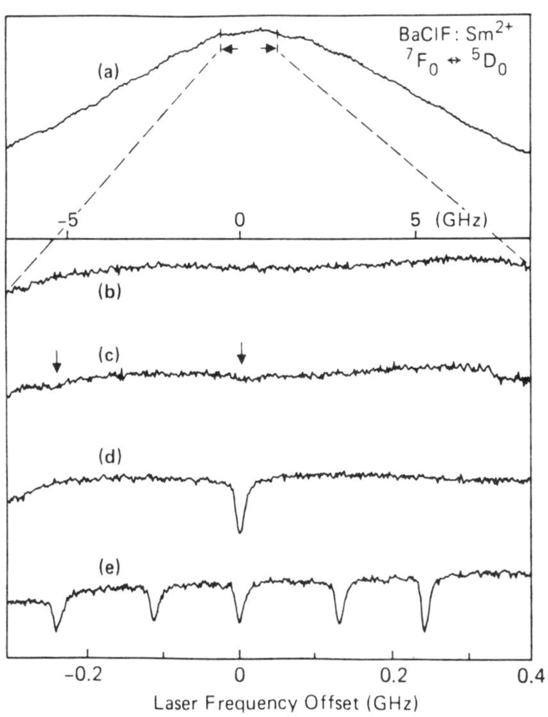

Fig. 4. Photon gated hole-burning in the $^7F_0 \rightarrow {}^5D_0$ transition of BaClF:Sm^{2+}. (a) the 13 GHz wide inhomogeneous line profile at 6879 Å, (b) a 700 MHz section of the line before hole-burning. (c) after burning at 0 MHz and −220 MHz with 2 W/cm^2 for 2000 s, (d) a single hole burned at 0 MHz in 3 s by adding 20 W/cm^2 of 5145 Å gating light, (e) multiple holes burned 110 MHz apart.

formation.[9] For example, after burning holes at liquid helium temperatures, the sample can be warmed up to room temperature for 24 hours, and then upon recooling to helium temperature, the previously written holes can still be easily detected.

Since the hole-burning process appears to involve conversion of Sm^{2+} to Sm^{3+} with trapping of the ejected electron at Sm^{3+} ions, the act of hole-burning simply redistributes absorption strength within the inhomogeneous absorption of the Sm^{2+} ions. The holes can be erased by irradiation in the homogeneously broadened transitions in the blue, which redistributes the ions between the divalent and trivalent states. One drawback of this material, however, is the extremely small oscillator strength of the λ_1 transition, which makes high SNR experiments in thin films difficult.

Table I. Inorganic materials showing photon-gated persistent spectral hole-burning.

Material System	λ_1, nm	λ_2, nm	Mechanism[a]	Reference
Sm^{2+} in BaClF	630,690	514–454	two-step PI	8–10
Sm^{2+} in SrClF	690	350	two-step PI	6
Sm^{2+} in CaF_2	690	514	two-step PI	6
Sm^{2+} in $CaSO_4$	690	351–514	two-step PI	11
Sm^{2+} in BaF(Cl, Br)	562	562	two-step PI	12
Co^{2+} in $LiGa_5O_8$	660	673	two-step PI	13
Cr^{3+} in $SrTiO_3$	790	790–1060	two-step PI	14

[a]PI-photoionization.

Table I lists the inorganic photon-gated materials studied to date. For full detail, the reader is encouraged to consult the references. As can be seen, the divalent Sm ion shows gating in a number of host crystals. In addition, the transition from the ground state to the $^5D_1$ state shows photon-gating.[10] Recently, using a mixed crystal of BaClF and BaBrF to increase the inhomogeneous broadening, photon gating was observed at 77 K by Wei *et al.*[12]

One possible approach to higher absorption cross sections in the inorganics lies in the study of multivalent transition metal ions in crystals (see the last two entries in Table I). For example, Co^{2+} in an inverse spinel crystal, $LiGa_5O_8$ has shown photon-gating.[13] This inorganic material utilizes 660 nm for λ_1 and longer wavelengths for λ_2 leading to a gating ratio of $\simeq 20$. The mechanism is similar to that for the Sm^{2+}:BaClF material: two-step photoionization and trapping of the ejected electron in the host crystal. Cr^{3+} in $SrTiO_3$ has also shown photon-gating,[6] but with a reduced gating ratio.

3.2 *Organic materials*

Spectral hole-burning nonlinear in the burning intensity was observed quite early in the history of hole-burning (see ref. 7 for details), but the importance of photon-gating with two-color excitation was not realized until much later, after the SNR analysis of single-photon materials.[4] Table II lists organic materials show in two-color photon-gated PSHB to date. In these materials, the $1 \rightarrow 2$ transition is usually within the singlet-single manifold, thus the absorption cross section is often quite large. Optical densities of 0.3 or more are relatively easy to achieve in thin films. The level scheme most often applicable to these systems is shown in the right half of Fig. 2.

The first example of two-color photon-gated PSHB in organic materials was provided by carbazole molecules in boric acid glass.[15] Upon excitation in the single-singlet origin with $\lambda_1 = 335$ nm, the molecule undergoes intersystem crossing with a high yield to build up a large

Table II. Organic materials showing photon-gated persistent spectral hole-burning.

Material System[a]	λ_1, nm	λ_2, nm	Mechanism[b]	Reference
Carbozole in boric acid glass	335	360–514	two-step PI	15
Zn-TBP′ in PMMA	628	337	?	16
Anthracene-tetracene photoadducts in PMMA	326	442	two-step PD	17, 18
TZT with halomethanes in PMMA	630	350–800	two-step DA ET.	19–21
TMT with halomethanes in PMMA	630	350–800	two-step DA ET	19, 20
TZT in PVC	630	514	two-step DA ET	24
Carbazole in PMMA	337	441–514	two-step NHS	25
Perylene in boric acid glass	441	441	two-step PI	26

[a]PMMA-poly(methylmethacrylate), PVC-poly(vinyl chloride), TBP′-tetrabenzoporphyrin derivative, TZT-meso-tetra-(p-tolyl)-Zn-tetrabenzoporphyrin, TMT-meso-tetra-(p-tolyl)-Mg-tetrabenzoporphyrin.
[b]PI-photoionization, PD-photodissociation, NHS-H-H bond scission, DA ET-donor-acceptor electron transfer.

metastable population in the lowest triplet state, T_1. In the presence of $\lambda_2=360$–514 nm, holes are formed at the singlet excitation wavelength, λ_1, due to photoionization of the molecule in the triplet manifold and trapping of the ejected electron in the boric acid glass matrix. The observed gating ratio for this material is near 400. The boric acid glass matrix plays an important role: in comparison to less polar hosts like PMMA, boric acid glass lowers the ionization potential of the carbazole guest from the gas phase value by 2.3 eV. The difficulty with this material is the short wavelength of the site-selecting transition: λ_1 alone is sufficient to cause photoionization via a biphotonic process.

The group of Korotaev *et al.*, reported two-color hole-burning for a derivative of Zn-tetrabenzoporphyrin in uncharacterized PMMA[16] with excitation at 628 and 337 nm. The mechanism was not determined, and the authors left that a photoionization process was unlikely in this system.

The organic material composed of photoadducts of anthracene and tetracene (A-T) in PMMA[17] provides a quite different type of photoreaction leading to photon-gating: two-step photodissociation of the adduct. The gating ratio for this material is small, however, due to one-color photoreactions caused by the absorption of two photons of $\lambda_1=326$ nm. The one-color and two-color hole-burning kinetics for this system show both transient and photochemical saturation effects.[18]

Studies of the molecule TZT (meso-tetra-(p-tolyl)-Zn-tetrabenzoporphyrin) in ultrapure primary standard PMMA using toluene as a solvent yielded only single-photon photophysical hole-burning.[19] However, by utilizing halomethane solvents, photon-gating was observed by a particularly interesting mechanism: donor-acceptor electron transfer from the excited porphyin donor to a nearby halomethane acceptor.[19] The TZT donor (or its Mg analog) was found to exhibit photon gating in

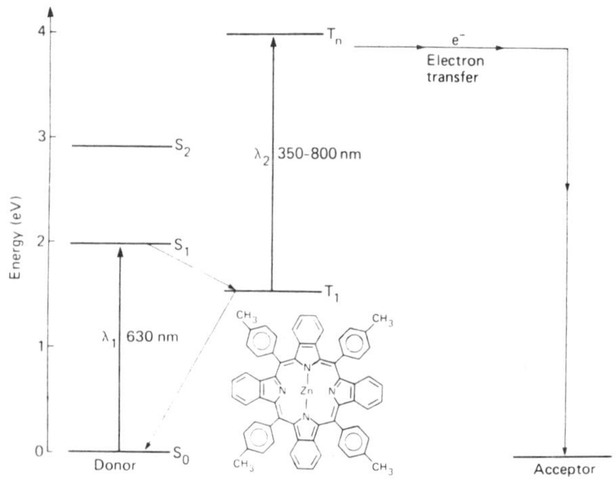

Fig. 5. Level diagram for photon-gated PSHB via donor-acceptor electron transfer. The structure of the donor chromophore TZT is shown in the inset.

well-characterized PMMA thin films in the presence of a variety of halomethane acceptors, such as chloroform ($CHCl_3$), methylene chloride, or methylene bromide.[20] The schematic mechanism for hole production is shown in Fig. 5. The frequency-selecting photon $\lambda_1 \simeq 630$ nm excites the lowest singlet-singlet transition, and tripletes are produced efficiently by virtue of the high triplet yield (0.8). The triplet lifetime of 40 ms (for the Zn compound) facilitates the buildup of a large population in T_1 When the gating light at $\lambda_2=350$–800 nm is present, the electron is excited to an upper triplet T_n from which it tunnels to a nearby halomethane acceptor. This mechanism has been confirmed by time-delayed two-color hole burning and measurements of the spectrum of the porphyrin cation photoproduct.[20]

This first example of photon-gating via a clearly defined donor-acceptor electron transfer mechanism has the useful property of allowing the gating effect to be controlled by varying the acceptor concentration and/or acceptor electron affinity (by choosing a different acceptor). However, due to subsequent reaction of the reduced acceptor with the matrix, reversal of the electron transfer is somewhat difficult. This lack of reversibility is not a fundamental problem because one can envision directly coupled donor-acceptor molecules in which the electron can be reversibly shuffled back and forth between the two halves of the molecule in a fashion analogous to the proton tautomerization reaction for the free-base porphyrins and phthalocyanines.

Aside from the specific disadvantage of irreversibility, the TZT/$CHCl_3$/PMMA material and its analogs with Mg and with other halomethanes have a further useful property: the overall efficiency of gated hole production is high: i.e., fairly deep ($\simeq 1\%$ in transmission) holes can be burned with a *single* 8 ns pulse at λ_1 followed immediately by a 200 ms gating pulse at λ_2.[21] This is a result of several factors: the large absorption cross section at λ_1, the large triplet yield, the reasonably long triplet lifetime, and the large intrinsic probability for electron transfer from the high triplet levels to the nearby acceptor.

Figure 6 illustrates a further advantage of this system;

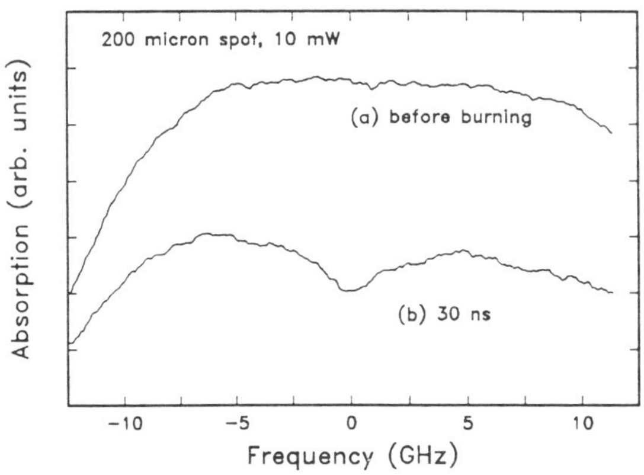

Fig. 6. Fast burning in small spots for the TZT/$CHCl_3$/PMMA material. (a) baseline before burning. The curvature in this trace is due to weak Fabry-Perot resonances in the optical system. The zero of absorption is off-scale at the bottom. (b) After burning with a 10 mW beam at λ_1 for 30 ns, followed immediately with a 200 ms gating beam at 488 nm of 17 mW.

with the TZT/$CHCl_3$/PMMA material, fast burning in small laser spots can be achieved. This property has not been observed in any single-photon material to date mostly due to the monophotonic nature of the hole-burning process. As was demonstrated earlier for a reasonably efficient single-photon material,[22] in order to detect the shallow hole burned on a ns time scale, very large reading laser spots are required to limit the reading intensity. By contrast, Fig. 6 shows that for the donor-acceptor photon-gated system, a clearly detectable spectral hole can be produced with a 30 ns pulse at λ_1 followed immediately by a 200 ms gating pulse at λ_2. The λ_1 spot diameter was 200 μm (limited by the mechanical stability of the cryostat), and the gating beam was unfocused. The 200 ms gating light pulse acts as a "developer" for centers that were placed in the lowest triplet state by the site-selecting λ_1 pulse. If many bits were to be written in the frequency domain in this material, one would simply inject short pulses at all the λ_1 wavelengths desired and then irradiate with the long λ_2 pulse to make the spectral holes permanent. The observation directly illustrates the superiority of photon-gated over single-photon mechanisms for frequency-domain optical storage applications.

One measure of the gating ratio is the inverse ratio of burning times for shallow holes of constant depth, assuming constant burning intensity at λ_1. This measure, called G_i, is approximately equivalent to the ratio of the slopes of hole growth curves for two-color versus one-color irradiation. The value of G_i for the TZT/$CHCl_3$/PMMA system is $\simeq 100$ at 1.5 K.[20] In recent measurements, G_i was observed to *increase* to a value near 10^4 when the temperature was raised to 20 K.[23] This surprising result can be understood if the one-color hole-burning at low temperatures is due to a photophysical rearrangement of two-level systems of the nearby host polymer over shallow barriers. At higher temperatures, any one-color holes due to transitions among the same set of shallow barriers thermally erase immediately after burning. To form a detectable one-color hole at higher temperatures, much longer burning times are required because barriers must be surmounted that are large compared to kT.

Returning to Table II, another variation of gated hole-burning material formed from the TZT molecule may be constructed by providing the halomethane acceptor as an intrinsic part of the host. Toward these ends, photon-gating has been observed for TZT in PVC (poly(vinyl chloride)) thin films.[24] However, since the effective concentration of chlorine atoms close to the TZT dopant molecules is reduced, this material shows relatively small gating ratios.

In recent studies, the group of Alshits *et al.*, have observed two-color photon-gating for carbazole in PMMA[25] and nonlinear one-color hole production for perylene in boric acid glass.[26] For the former system, the mechanism was determined to be N-H bond scission as was observed in previous two-photon holography experiments.[27] In the latter system, photoionization was observed upon the absorption of two quanta at 440 nm without participation of the triplet states, thus, this material may be regarded as a limiting example of 3-level photon gating shown in the left half of Fig. 2.

§4. Optimal Photon-Gating with Frequency-Domain Readout

We now turn to a discussion of the *optimal* materials parameters for practical gated recording media. In order to identify the required characteristics of a practical photon-gated storage medium, some assumptions must be made about the configuration of the frequency domain optical storage system, specifically, the detection technique used. Possible detection methods include transmission, fluorescence detection, holographic methods, electric-field readout, and time-domain readout.[1] In addition, the level of parallelism during writing and reading affects the power and time available for hole formation and detection. The configuration that has been treated in some detail that is assumed here is the case of quantum-limited transmission detection with frequency-domain readout with a single beam.[28] Some aspects of two-step spectral hole-burning with short pulses have been analyzed separately.[29]

For the case of frequency domain readout, a plausible set of constraints are: 30 ns reading and writing per bit, shot-noise-limited transmission detection with a least 26 dB wideband signal-to-noise ratio, and 10 μm diameter laser spots. A useful quantity for modeling purposes is the effective hole-burning yield for photon-gated hole-burning, η_e, which is defined as the relative absorption change $\eta_e = (\Delta\alpha/\alpha)$, that is produced by the best-case two-color hole-burning during the writing time of 30 ns. Due to the complex nature of gated PSHB processes, the value of η_e depends critically on the specific microscopic properties of the gated PSHB mechanism as well as on the writing conditions.

A thorough SNR analysis for this situation[28] seeks appropriate values for the absorption cross section σ_1 at λ_1, the density of centers within a homogeneous width of the laser frequency, N_ω, sample length L, and yield η_e which result in SNR ≥ 26 dB in a 16 MHz bandwidth. Since the values of L and N_ω are not intrinsic material properties and can be controlled when fabricating the storage

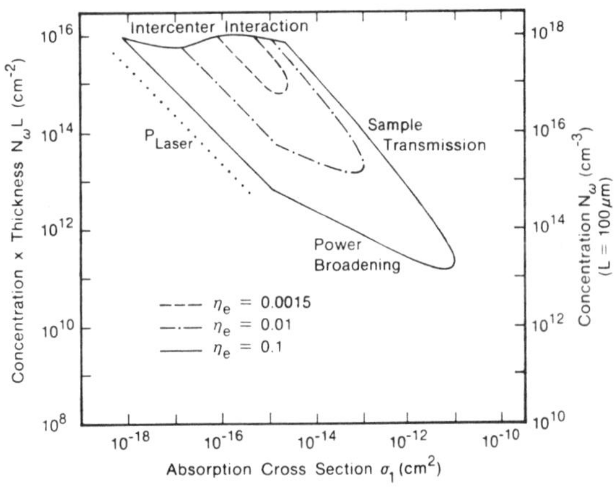

Fig. 7. Materials constraints for gated PSHB materials in order to achieve practical SNR. The various regions and symbols are defined in the text. To illustrate the effect of increased laser power, the dotted line shows the $\eta_e=0.1$ boundary for 100 mW reading power.

medium, it is helpful to classify materials by a concentration-thickness product $N_\omega L$, as long as $L \leq 100\ \mu m$.

Figure 7 shows the results of the analysis. Materials within the shown boundaries for particular values of η_e permit data readout with SNR equalt o or greater than the required value with a maximum of 10 mW reading power. For high cross sections σ_1 and large values $N_\omega L$ the optical absorption becomes so strong as to prevent sufficient light from reaching the detector. For high σ_1 and low values $N_\omega L$ the achievable signal-to-noise is limited by power broadening of the detected holes. For $\sigma_1 \leq 10^{-15}$ cm^2 the available laser power of 10 mW is lower than the saturation power; therefore the available reading power defines the S/N limit. At the top of figure, undesirable interactions between the optically active centers, *e.g.* spectral diffusion and cooperative excited-state quenching, can limit the usable concentrations of centers.

As always, spectral broadening of the produced hole, either by saturation of the transition or by excessive hole-burning, imposes an upper limit on yield η_e. It is reasonable to restrict the hole-burning yield to $\eta_e \leq 0.1$. The $\sigma_1 - N_\omega L$ parameter space shrinks rapidly as η_e decreases. In order to successfully implement a frequency-domain optical storage system based on gated mechanisms, it is critical to find a material that permits gated PSHB with very high efficiency in the short writing times required for fast data transfer rates.

For example, with the inorganic material composed of Sm^{2+} in BaClF, $N_\omega L \simeq 10^{14}$ cm^{-2} and $\sigma_1 \simeq 10^{-18}$ cm^2, and the value of η_e has not been measured. The achievable SNR in this material is insufficient without more than 100 mW of reading laser power. For the case of the organic TZT/$CHCl_3$/PMMA material, $N_\omega L \simeq 5 \times 10^{12}$ cm^{-2}, $\sigma_1 \simeq 5 \times 10^{-12}$ cm^2, and $\eta_e \simeq 0.01$.[20] This material would be barely acceptable if the overall efficiency were higher or if the absorption cross section were smaller (and if the mechanism were reversible).

The use of a data access method other than frequency scanning in a focused spot will clearly alter the exact materials requirements for practical information storage using photon gating. Other access methods that have been considered are time-domain storage,[30] holographic storage,[31] and electric-field access,[32-34] and a similar SNR analysis should be performed for each of these access methods. In each case, the presence of a threshold in the writing process improves the performance, so photon-gated materials are expected to continue to be important for information storage schemes using persistent spectral hole-burning, no matter what the access mechanism.

§5. Conclusions

The number of examples of gated spectral hole-burning continues to grow, further widening this novel class of mechanisms for PSHB. Considering the limitations on single-photon materials for frequency domain optical storage applications, the search for gated mechanisms should be an important area for future study. This is especially true because photon-gating directly resolves a paradox that limits single-photon materials: how to achieve high efficiency photochemistry for writing and yet have low efficiency during reading. The practical realization of technological applications of PSHB presents a stimulating challenge for interdisciplinary research: identification and characterization of gated hole-burning mechanisms in materials that will satisfy all the requirements for frequency domain optical storage as well as other useful applications of persistent spectral hole-burning.

Acknowledgements

The author acknowledges stimulating prior collaborations on photon-gated materials with H. W. H. Lee, M. Gehrtz, T. P. Carter, Chr. Bräuchle and W. P. Ambrose and the author thanks R. M. Macfarlane for permission to use Figs. 3 and 4. This research has been supported in part by the U.S. Office of Naval Research.

References

1) See W. E. Moerner, W. Lenth and G. C. Bjorklund: "Frequency Domain Optical Storage and Other Applications of Persistent Spectral Hole-Burning," Chapter 7 of *Persistent Spectral Hole-Burning: Science and Applications*, W. E. Moerner, editor, Topics in Current Physics Vol. 44 (Springer, Berlin, Heidelberg, 1988), and references therein.
2) W. E. Moerner: J. Molec. Elec. **1** (1985) 55.
3) See Chapters 1 through 6 of ref. 1 for a review of many photophysical and photochemical mechanisms for persistent spectral hole-burning; also see J. Friedrich and D. Haarer: Angew. Chem. Int. Ed. Engl. **23** (1984) 113; R. Jankowiak and G. J. Small: Science **237** (1987) 618.
4) W. E. Moerner, M. D. Levenson: J. Opt. Soc. Amer. B: Optical Physics **2** (1985) 915.
5) D. M. Burland, F. Carmona, G. Castro, D. Haarer and R. M. Macfarlane: *IBM Tech. Discl. Bull.* **21** (1979) 3770.
6) R. M. Macfarlane: J. Lumin. **38** (1987) 20.
7) As for example with dimethyl-s-tetrazine. See D. M. Burland and D. Haarer: IBM J. Res. Devel. **23** (1979) 534, and references therein.
8) A. Winnacker, R. M. Shelby, R. M. Macfarlane: Opt. Lett. **10** (1985) 350.
9) A Winnacker, R. M. Shelby, R. M. Macfarlane: J. de Phys. Colloq. C7, Suppl. 10, **46** (1985) C7-543.
10) R. M. Macfarlane, R. M. Shelby and A. Winnacker: Phys. Rev. B **33** (1986) 4207.
11) R. J. Danby, K. Holliday and N. Manson: J. Lumin. **42** (1988) 83.
12) C. Wei, S. Huang and J. Yu: J. Lumin. **43** (1989) 161.
13) R. M. Macfarlane and J. C. Vial: Phys. Rev. B **34** (1986) 1.

14) A. J. Silversmith, W. Lenth and R. M. Macfarlane: IQEC-87 Technical Digest (1987) 190.
15) H. W. H. Lee, M. Gehrtz, E. Marinero and W. E. Moerner: Chem. Phys. Lett. **118** (1985) 611.
16) O. N. Korotaev, E. I. Donskoi and V. I. Glyadkovskii: Opt. Spektrosk. **59** (1985) 492.
17) M. A. Iannone, G. W. Scott, D. Brinza and D. R. Coulter: J. Chem. Phys. **85** (1986) 4863.
18) M. A. Iannone and G. W. Scott: J. Chem. Phys. **89** (1988) 2640.
19) T. P. Carter, C. Bräuchle, V. Y. Lee, M. Manavi and W. E. Moerner: Opt. Lett. **12** (1987) 370.
20) T. P. Carter, C. Bräuchle, V. Y. Lee and W. E. Moerner: J. Phys. Chem. **91** (1987) 3998.
21) W. E. Moerner, T. P. Carter and C. Bräuchle: Appl. Phys. Lett. **49** (1987) 430.
22) M. Romagnoli, W. E. Moerner, F. M. Schellenberg, M. D. Levenson and G. C. Bjorklund: J. Opt. Soc. Am. B: Optical Physics **1** (1984) 341.
23) W. P. Ambrose and W. E. Moerner: in preparation for submission to Chem. Phys. Lett.
24) T. P. Carter, W. P. Ambrose and W. E. Moerner: unpublished results.
25) E. I. Alshits, B. M. Kharlamov and R. I. Personov: Opt. Spectrosc. **65** (1988) 173.
26) E. I. Alshits, B. M. Kharlamov and R. I. Personov: Opt. Spectrosc. **65** (1988) 326.
27) G. C. Bjorklund, C. Bräuchle, D. M. Burland and D. C. Alvarez: Opt. Lett. **6** (1981) 159.
28) W. Lenth and W. E. Moerner: Opt. Commun. **58** (1986) 249.
29) I. K. Rebane: Phys. Stat. Sol. (b) **145** (1988) 749.
30) T. W. Mossberg: Opt. Lett. **7** (1982) 77.
31) A. Renn, A. J. Meixner, U. P. Wild and F. A. Burkhalter: Chem. Phys. **93** (1985) 157.
32) G. Castro, R. H. Dicke and D. Haarer: IBM Tech. Discl. Bull. **21** (1979) 3333.
33) U. Bogner, P. Schätz, R. Seel and M. Maier: Chem. Phys. Lett. **102** (1983) 267.
34) U. P. Wild, S. E. Bucher and F. A. Burkhalter: Appl. Opt. **24** (1985) 1526.

Reprinted from *Optical Data Storage,* Proc. SPIE Vol. 1316, pp. 65-68 (1990).

An Optical Tape Recorder Using Linear Scanning

Dan Gelbart, CREO Products, Inc.
110 - 3700 Gilmore Way
Burnaby, B.C. Canada V5G 4M1

1. Specifications

The CREO Optical Tape Recorder (OTR) has the following key specifications:

1. Capacity: 1 TByte on a 30cm (12") reel of 75 micron (3 mil) optical tape. Capacity can be increased to 4 TByte on a 35cm (14") reel of 25 micron (1 mil) tape.
2. Data Rate: 3 MByte/sec read and write. Any data rate from 0 to 3 MB/sec can be used. Data rate can be different for read and write.
3. Access time: 30 sec average, 60 sec worst case in 1 TByte.
4. Size: 60cm W x 190cm H x 75cm D (24" x 75" x 30").
5. Power: Powered by standard 120V 15A 60Hz outlet (or 220 - 240V 50Hz).
6. Media: Write once, non formatted (ICI 1012 optical tape or eq.).
7. Interfacing: SCSI, serial and others.
8. Directory & Indexing: An auxiliary longitudinal address track, readable during high speed search, is provided.

2. Scanning Method

The OTR uses linear scanning across the tape. The tape is stationary during the scan and stepped rapidly at the end of each scan. The scanning is bi-directional and the position of the scanning mirror is measured and controlled by an interferometer. The position is interpolated down to 1 nm using a digital frequency synthesizer. Other details of the system are shown in Fig 3. The advantages of this scanning method over rotary scanning or AO scanning are:

A. Allows the use of large N.A. lenses.
B. Requires focus only along a straight line, not an area.
C. Does not involve rotation of the optical image (important in multi-spot recording).
D. Has a high duty cycle without using multiple facets or lenses.
E. Unlike AO scanning, number of spots per line is only limited by scan length.

3. Writing Method and Format

The OTR writes 33 tracks in parallel: 32 data tracks and a servo track. Since the tape used is unformatted the servo track is written at the same time as the data. The writing is done by an array of 33 discrete laser diodes each one delivering 30mW at 830nm. The arrangement of this array is shown in Fig 4. Since the lasers are staggered the data has to be delayed electronically to align the 33 spots on the recorded tape. The written spots are about 1 micron in diameter and the spacing in both directions is 1.5 microns.

The advantage of this writing methods is that large arrays of laser diodes can be manufactured from discrete lasers without having to wait for the development of monolithic arrays. Also, each laser can be replaced individually, thus the array can be repaired if needed. Fig 5 shows the written data magnified about 600X.

4. Reading Method

All data tracks are tracked and read simultaneously using a CREO custom I.C. This I.C. contains all photodetectors, amplifiers, comparators, data recovery and tracking logic.

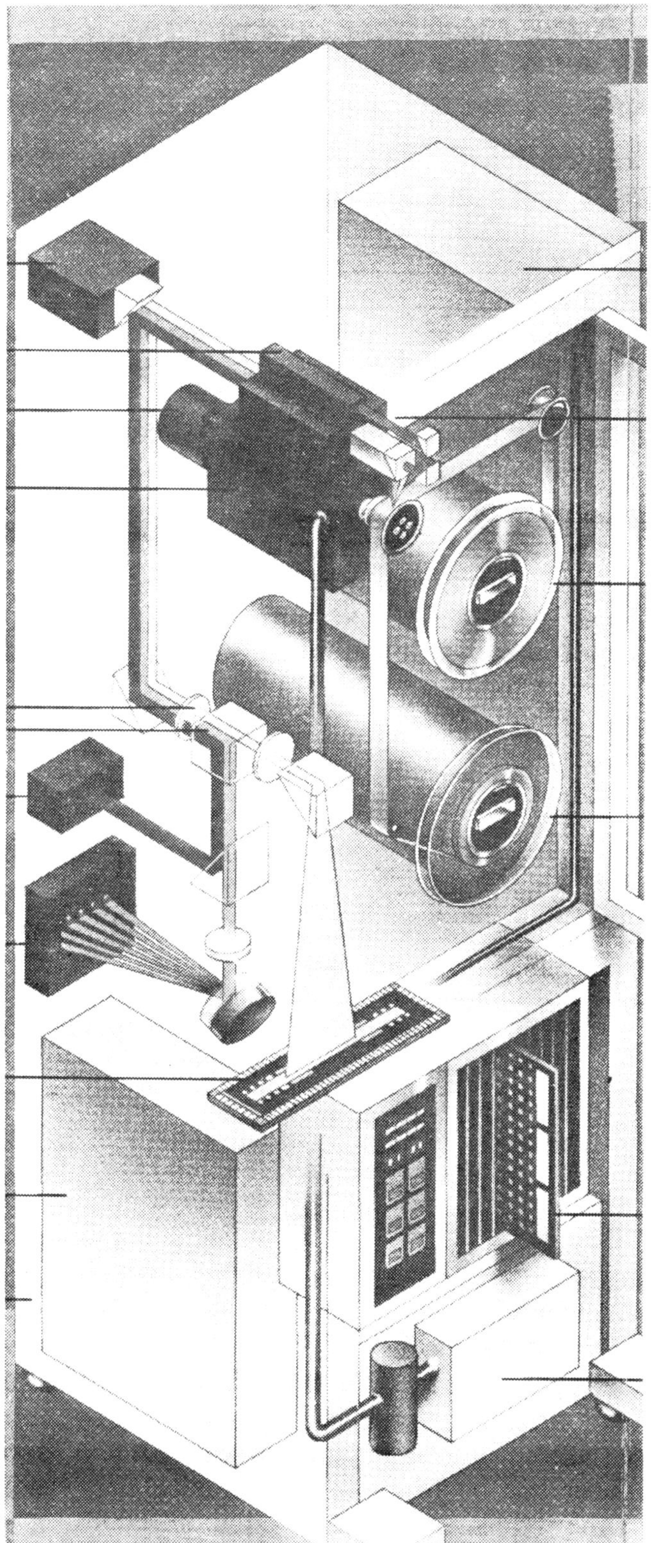

FIG.1 The inside

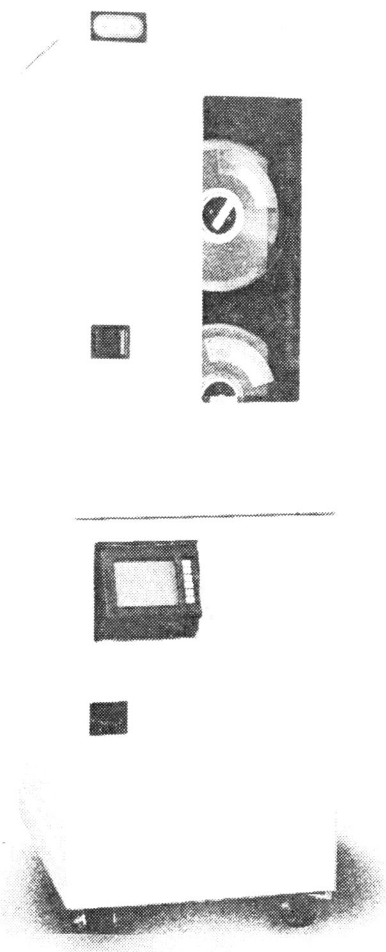

FIG.2 The outside

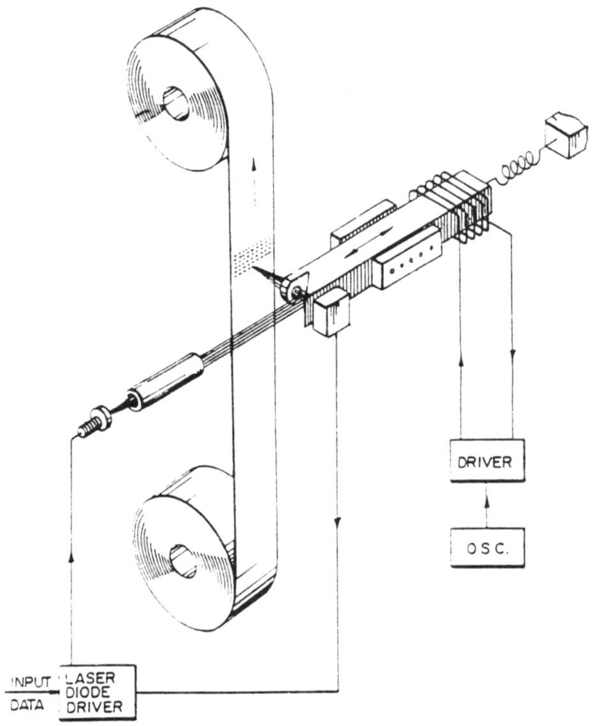

FIG.3 Linear scanning

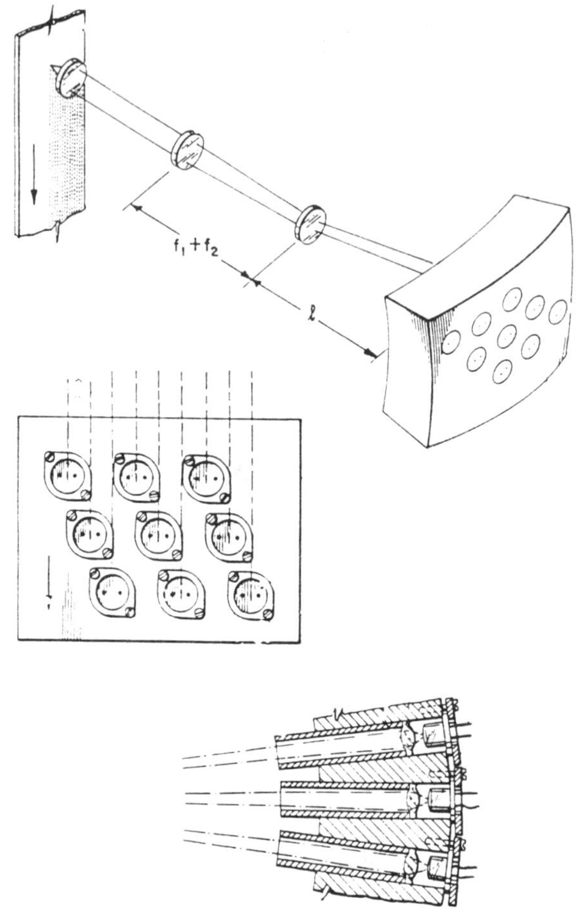

FIG.4 Laser diode array

FIG.5 Written data

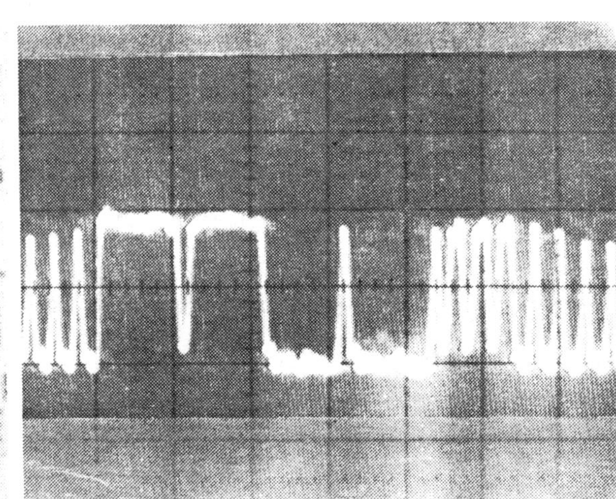

FIG.6 Read waveform

In order to avoid a mechanical tracking servo the data is oversampled by a ratio of 4:1 and the correct sample is selected electronically. A focus servo is holding the scanning lens within one micron during read and write. The analog read waveform from one of the data channels is shown in Fig 6. This waveform is only present inside the read I.C. - all outputs of the I.C. are digital. The OTR is a true read-after-write machine, namely it writes, waits a few microseconds for all thermal effects to die out and then reads and checks the data.

5. Modulation and Error Correction Coding

No modulation codes are used - the data is recorded in NRZ format. An interleaved Reed-Solomon code is used for EC. The overhead is 20%. Each record across the tape is 80KB, out of which 64KB are user data and 16KB used for EC. The EC will correct a burst error equivalent to a 3mm wide scratch and will handle random error to 10-3 while correcting to 10-12.

6. Transport

The OTR uses an innovative transport capable of search speeds in excess of 15 m/s (600 ips). The same transport can step the tape to a 2 micron accuracy within 5 mS. The transport uses 3 motors controlled by a digital servo using a combination of predictive (feed-forward) and closed loop systems. The reel motors are 2 HP brushless and slotless DC motors controlled by a sinewave output three phase amplifier. The tape compartment is kept as a class 100 clean room using recirculating air. The transport is designed to prevent tape damage not only during operation, but during extreme situations, such as power loss during high speed search.

7. Diagnostics

The OTR contains an unusual amount of built-in diagnostics which is helpful in new and complex equipment. Some of the unusual diagnostic features provided in the OTR:

A. BER measurements and error distribution maps (before EC). This mode can also be used for tape testing and certification.

B. A built-in scanning laser microscope for a close-up look at data bits and defects at sub-micron resolution.

C. A built-in test pattern generator and various levels of loop-back testing. The OTR can be fully tested in stand alone mode.

8. Project Status

The first batch of these units is being built at CREO. A working prototype exists since 1989. Deliveries are scheduled this year.

9. Media

The only qualified media at this moment is ICI type 1012 optical tape. Other tapes are being sampled; the most noteworthy is Dow Chemical's hardcoated metal alloy tape.

Reprinted from *Optical Computing,* Proc. SPIE Vol. 963, pp. 35-45 (1989).

Data Storage in Photorefractives Revisited

Steve Redfield and Lambertus Hesselink

Microelectronics and Computer Technology Corporation (MCC)
P.O. Box 200195, Austin, TX 78720, USA
and
Stanford University
Dept. of Aeronautics and Astronautics, Durand Bldg, Stanford, CA 94034 USA

ABSTRACT

The results of a fresh look at holographic data storage in photorefractive materials is reported. This work concludes that advances in photorefractive material, particularly SBN, AO deflector technology, CCD technology, and laser technology could make a very interesting low latency mass storage device possible. Two advances in this technology, a technique for a highly asymmetric write/read cycle and a method for building artificial crystal shapes, are described.

1.INTRODUCTION

The slow access time of conventional mass storage devices has been found to be a bottleneck for data intensive architectures currently under development at MCC. This has lead us to take a fresh look at volume holographic storage in photorefractive materials. This paper describes the results obtained from a working device that is called Bobcat. We present several innovations conceived of during this experimentation to overcome problems encountered, and the architecture of a very promising system, that utilizes these improvements.

Photorefractive volume holographic storage (PVHS) has been investigated in the past, but has been unsuccessful. We believe that this in part has been caused by an emphasis on capacity, while we think that the real advantage of this technology is speed. In the mean time electrooptic technology has also advanced considerably and some of the past problems have been overcome.

2.CONSTRUCTION

The optical configuration built is fairly standard where data is in the form of a 2-d array of spots, called a page [1],[2],[3],[4]. A hologram of the Fourier transform of this array is stored. The Fourier transform is chosen because it is a convenient way to get a small recording size and it provides some immunity to material defects. Data is placed in the photorefractive crystal as a 2-d array of stacks of the pages.

The source of coherent light is a 35 mw argon ion laser tuned to the green 514 nm wavelength The output beam of this laser is reflected through a shutter which provides control over exposure times. The shutter was initially a mechanical shutter capable of times as short as 2 ms. It has been replaced with an A-O modulator which can produce exposure times as short as 300 ns. After passing thru the shutter the beam is reflected off the x-y deflector. This is simply a mirror with 2 digital micrometers to adjust its tilt. The mirror moves the beam before it is split so both the reference and object beam track with respect the deflection produced. Two folding mirrors are placed next in the path. They allow the x-y deflector to work with the small angles that an A-O deflector would produce. This would ultimately be the deflector in a final system. After the folding mirrors the beam passes thru a collimator. Then it is split by a HOE into the reference beam and the object beam.

Tracing out first the reference beam from the beam splitter, a flat rotatable plate is encountered. This plate shifts the beam laterally. It is controlled by a stepper motor which in turn is controlled by a microprocessor. The microprocessor can remember page positions. After the plate another HOE is encountered. This HOE is equivalent to a lens and a prism. It deflects the beam onto the region in the crystal where a page is to be read or written. It always directs the beam to the same spot (assuming all beams parallel in all positions) so it translates the lateral displacement produced by the plate into an angular change. Different positions of the plate then give rise to different reference angles.

Now tracing out the object beam, the first item encountered is a HOE called the "fly's eye" which corresponds to an array of small lenses, one for each stack. Next is the transform lens and SLM. The transform lens is placed immediately in front of the SLM. It focuses through the SLM onto the region in the crystal where a page is to be written. It then forms the Fourier transform of the SLM pattern at that place in the crystal. The SLM is the data input device and several different types. One is a 35 mm camera body which is used as a film holder. Another is just a holder for single frames. Another is a cannibalized Radio Shack liquid crystal TV set.

The crystal is mounted in a holder which has a thermopile at its base. This permits heating and cooling of the crystal. To apply various electric fields across the crystal, high voltage leads from a 6 Kv power supply are attached to it.

On the output side of the crystal is an imaging lens which images the output pattern onto the detector array. Several different types of detectors are used including a CCD TV camera, a 35 mm camera body and a Polaroid film holder. The Polaroid is used for taking pictures of test patterns that are recorded. Another is a pin photodiode. It is used for precise monitoring of the magnitude or strength of reads. A photomultiplier tube is added to perform high speed measurements.

The entire setup is interfaced to an IBM PC through a measurement interface card permitting computer control of exposures, of the sequencing of reads and writes, of exposures by the Polaroid plates, and of sampling by the photodiode or photomultiplier tube.

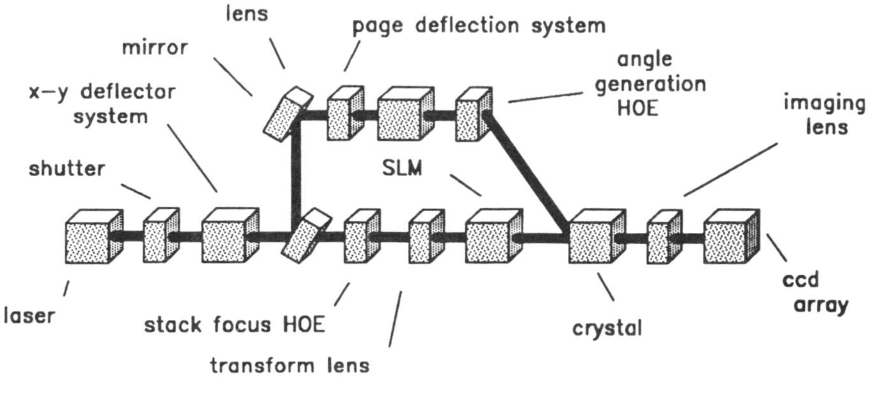

figure 1 optical configuration

3.CAPABILITIES MEASURED

Measurement results are shown in Table 1. Shown are the targets which need to be met for a competitive mass storage device and what is felt achievable in light of the results. It should be noted the results in table 1 represent independent measurements. In an actual system a compromise is needed between these characteristics in order to optimize performance, for example read speed verses quality (SNR).

The capacity is going to be a close one to call, even though Bobcat did very well in its demonstration of capacity, because the signal-to-noise-ratio, SNR, will be small at high storage densities. This increase in error rate may be compensated for by using some of the additional density for more redundancy. Theory says that the number of pages will be strongly limited by crosstalk. If one assumes that the composite error rate (all the errors from various sources) is approximately Gaussian in distribution, theory says that a SNR of about 18 in needed to get an error rate less than 10^{-13}.

	BOBCAT TARGET	BSO	SBN
SPEED	1 ms page wr	1.5 ms page wr	2 ms page wr
	200 ms stack wr		300 ms stack wr
	100 us page rd	10 us page red	
DENSITY	50K bits/page	79K bits/page	99K bits/page
	10 pages/stack	4 pages/stack	10 pages/stack
	200 stacks	400 stacks	1000 stacks
	----------	----------	
	100 M bits	100 M bits	1 G bits
STABILITY	20 rds of stack		20 rds of stack
	100 flash rds	5K flash reads	
	24 hr retention		1/64 at 24 hr
QUALITY	some loss @ edge	satisfact photo	satisfact photo
	satisfact images	of 140x140 dot	of 280x280 dot

table 1 bobcat goals/results

Speed is the advantage of this technology. The speed targets look easily obtainable, in fact should be exceeded for readout. No significant read speed limitations are observed or are expected for the basic holographic reconstruction of a recorded image. Important read speed limitations do occur in the optical detection system which converts the optical signal to electronic form. These limitations are, however, compatible with targets. Write speed for the material depends on power and can meet the SLM speed.

Where the volatility hurts the most is in recording multiple pages. The write of a new page into a stack area is far more damaging to existing pages in that area, than reads are. Stability is a problem.

Permanent recording can be achieved however through the utilization of electrical fixing and the non destructive readout technique.

The page experiments made it very clear that some sort of fixing, restoring, or desensitizing mechanism has to be part of any pvhs device. Write damage severely limited the number of pages that can be overlaid in both BSO and SBN. For BSO the number achieved using a ratioing strategy is just 4 while for SBN it is 10. The equal write strategy yields 6 for SBN. In all cases the first pages written are severely damaged by the writing of the later pages and only 2 to 3 reads can be made of them before their recording is in the noise. It should be noted that the reads are for 10 to 100 ms with power densities between .2 to .5 mw/mm^2 which is 4 to 5 orders of magnitude more energy than what it is envisioned a final pvhs device system would use for a read.

The quality measurements are the most difficult to make and the most subjective in interpretation. Within the limited capabilities of Bobcat there appear to be no fundamental difficulties which would lead one to worry about a quality problem. While no quality problems are encountered, the first round of experiments did not quantitatively address error rates. New experiments are now being set up to do this. Speckle is a problem. It did not degrade the photographs because they are taken over a relatively long time. Speckle is visible in the real time CCD camera readout. It is found that the use of diffusers and cross gratings helped somewhat. Some cleverness with these will probably be needed in a final system.

4.NON DESTRUCTIVE READOUT

A major difficulty with data storage using holograms recorded in photorefractive media has been the destructive readout which limits the number of reads that can be made before the signal-to-noise ratio becomes too low.

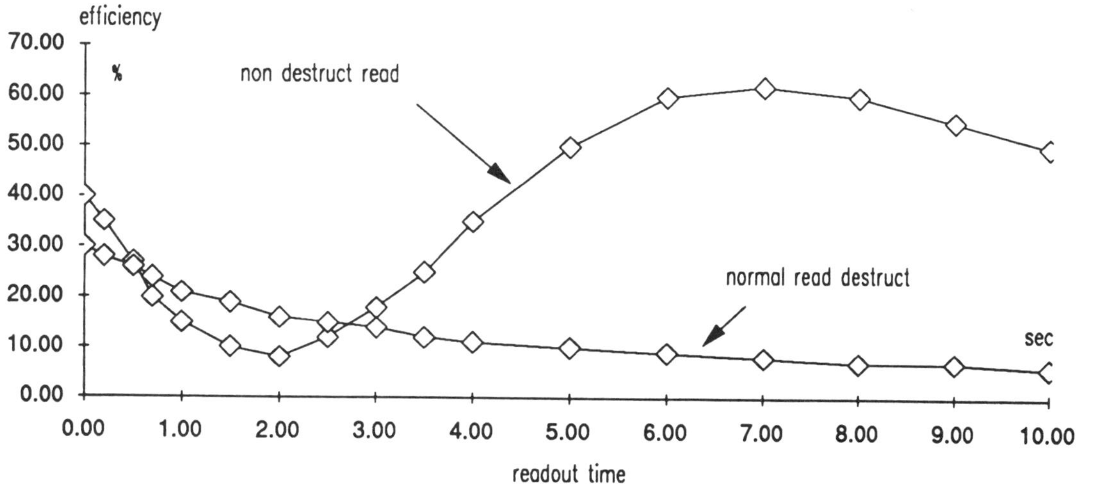

figure 2 read destruct curves

A non destructive readout technique was discovered which provides a new method to promote construction/reconstruction asymmetry [5]. The technique can, under certain conditions, produce an enhanced reconstruction that is substantially nondestructive over a very large number of reconstruction cycles. The key new ingredient is utilization of degrees of freedom in polarization of the reconstruction beam. The procedure, in its optimum or preferred form, involves first recording at a spatial frequency of around 200 lines/mm for a particular length of time with a high applied electric field, around 6 Kv/cm, and ordinary polarized beams. The reconstruction is then carried out with the applied electric field reduced to around 1 Kv/cm and the polarization of the reconstruction beam rotated 90 degrees. The reconstructed beam first drops in intensity, but subsequently grows in strength above the starting value, approaching 100% efficiency in some cases. The reconstruction is almost nondestructive with erasure times exceeding 3 hours of continuous readout. This equates to over 1 billion 10 us readouts with signal-to-noise ratios exceeding 20 db due to the high efficiency. Recording exposures aree approximately 0.3 mJ/cm^2.

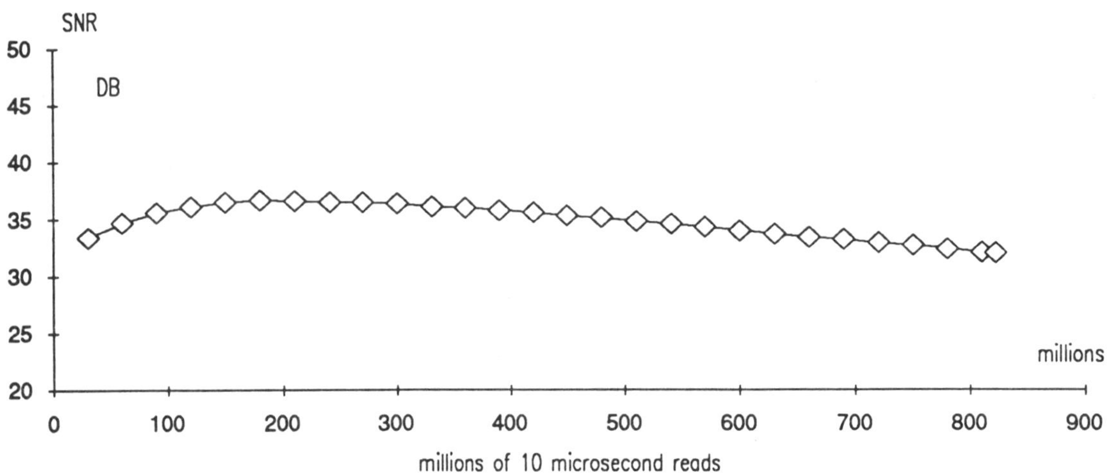

figure 3 nondestructive reads

The improved readout process may be understood, at least qualitatively, by considering first the recording process. In the presence of an applied field and for ordinary beams, the space charge field builds up in an oscillatory manner. Under these conditions the light intensity pattern and the space charge field are in phase and little energy coupling between the writing beams occurs. During readout with extraordinary light and no applied field, the readout beam gives rise to a strong diffracted beam producing an interference pattern which is out off phase with the space charge field by $\pi/2$. As a result strong energy coupling between the readout beam and the diffracted beam occurs. In this case the fringes are no longer straight and the new space charge field reinforces the already present one at certain locations in the crystal. The diffracted beam grows in amplitude until the beam ratio between the readout and the diffracted beam becomes so unbalanced, that the space charge fields are erased. As a result the diffracted beam decays exponentially. The magnitude of the decay time constant is substantially larger than when reading the hologram with ordinary light. In that case no energy coupling between the readout and diffracted beam occurs in this crystal, and the space charge field decays rapidly. The improved readout characteristics are not fully understood, but we believe that ionic charge patterns may be responsible for the prolonged readout times. An ionic pattern is erased with the dielectric relaxation time for ions, which can be orders of magnitude larger than for electrons. The exact nature cf these ionic charge patterns are not identified at present. It should be noted here, that the mechanism described appears to be similar to the fixing process described by Amodei [6], but in fact this is not so. In the earlier experiments, either temperature or electric based fixing procedures have been described, whereas here we are exploiting polarization properties of ferroelectrics. Since the energy transfer between the readout and diffracted waves is dependent on the magnitude of the space charge field, the diffracted beam decays exponentially when the writing process is stopped early on, before the peak in the space charge field is reached.

5.CRYSTALLYTES

Previous work with photorefractive materials has been carried out with bulk crystals on the order of about one centimeter by one centimeter on the face and half a centimeter in depth. It is difficult to grow larger crystals of good optical quality and, in fact, it is difficult to grow electro-optically superior crystals such as $BaTiO_3$ and SBN to even half this size. As a consequence of this difficulty in making large bulk crystal, wide spread application of the bulk photorefractive technology has not occurred, despite an initial surge of development in the decade of the 1970's.

Other limitations with a bulk crystal are the inability to selectively control regions of bulk material, and the restriction of bulk photorefractive materials to linear optical effects (because of the inability to sustain high enough energy densities)

An alternative to bulk crystals which we propose is the fabrication of artificial crystals out of bundles of fibers or matrixes of small crystals, which we refer to as crystallyte configurations [7]. They potentially have several significant advantages over monolithic crystal approaches. The promise of using these materials in such new architectures under improved conditions appears to warrant further research in this area.

The crystallyte idea is to use a composite array of small, isolated photorefractive recording volumes which may be either cubits (small crystals) or fibers which are assembled in a matrix to synthesize a larger volume. They are used in place of a bulk crystal. Typically, the cross-section of each crystallyte is less than one millimeter and in the case of fibers can be very long, presently of the order of ten's of centimeters in length. The array geometry can be tailored to specific applications and may take on different configurations. Isolation may be achieved by the refractive index differences, coatings on the sides, or the intervention of a substrate material.

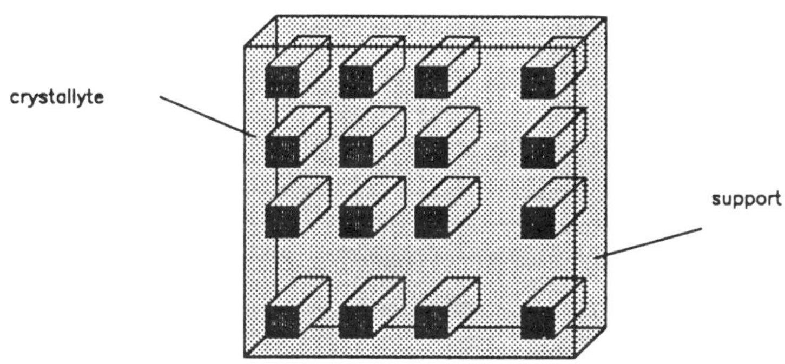

figure 4 generic crystallytes

One of the advantages of this approach is the prospect of using new materials. Some of the most promising materials such as SBN, KTN and $KNbO_3$ may be grown in fiber form, but not in large bulk sizes. To obtain the equivalent of a large sized crystal, these fiber may be composed into a matrix of a suitable host material. It is important to note here, that design of such a fiber array needs to be carefully considered. Application of an electric field for efficiency enhancement and fixing often needs to be applied along the c-axis, which is also the axis along which the fibers are most commonly grown. Thus, in order to achieve large angular sensitivity, fiber length may be several mm, which still requires applied voltages on the order of kV/cm. Perhaps, by sandwiching a few layers of this matrix, applied voltages of a few hundred volts may be sufficient to perform enhancement and fixing procedures. The total information content stored in such structures needs to be evaluated, and is different from the capacity obtained in conventional bulk holograms using the same recording configuration [8].

Another advantage is the increased region of interference between the reference and object beams. This has the potential of much stronger recordings. Also from the standpoint of geometry, the confinement of the beams provided by a crystallyte allows much larger angles between the beams to be used in recording. This allows higher spatial frequencies and a wider range of angles for pages.

In using a crystallyte approach, it is important to carefully study the required orientation of the material in relation to the recording beams and the direction of the applied field. For SBN, for example, the grating vector is usually aligned with the c-axis and in the direction of the applied field. This requires that the recording beams are incident more or less transverse with respect to the c-axis, or counter propagating to it. In either case, the applied voltage may be large when working with cm long crystallytes. The magnitude of the required applied voltage is directly proportional to the thickness of the crystallytes. Thick crystals are usually needed to obtain a large angular sensitivity in bulk materials, but for data storage in fibers multimode recording may alter the length requirement as compared with results obtained from Kogelnik theory for bulk materials. This subject needs to be investigated before a full assessment of the architecture can be made [8]

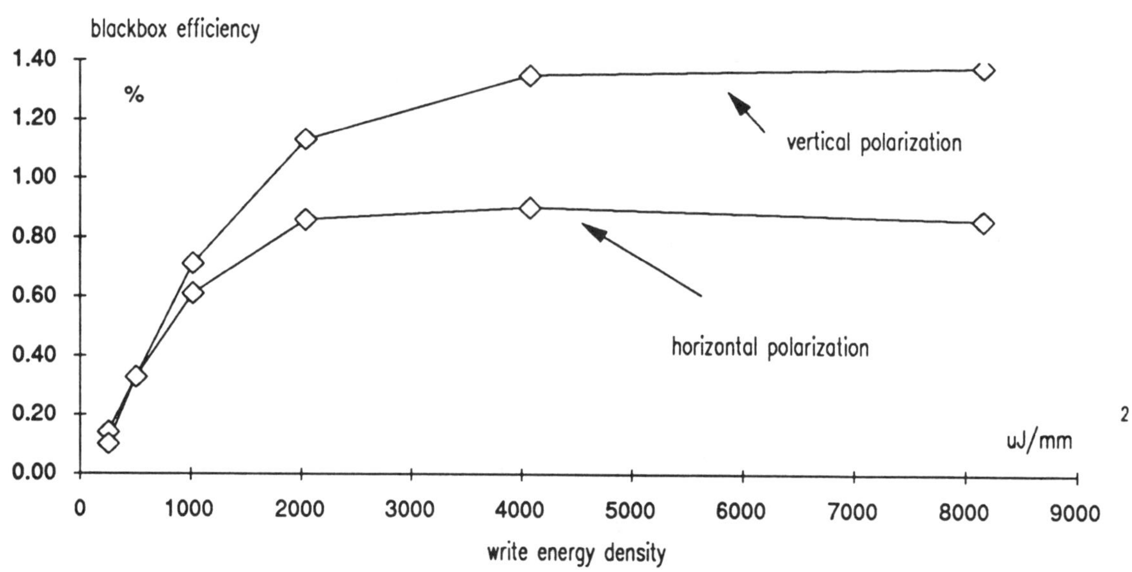

figure 5 fiber saturation curves

Recently, we have achieved, to our knowledge, the first photorefractive holographic recordings in a fiber. A plane wave reference and an image carrying object beam are used; the beams are incident on the fiber in a reflection geometry with the object beam propagating along the bore. The image recorded is an AirForce RES 1 test pattern. To date best resolution has been group 3 set 2 which gives 9 l/mm. The limitation seems to be the roughness on the fiber surfaces and improvement is expected, in particular for smaller fibers. In these experiments the angular sensitivity ranged from +/-.3 degrees off Bragg for on axis recordings to +/-.15 degrees for 20 degree off axis recordings.

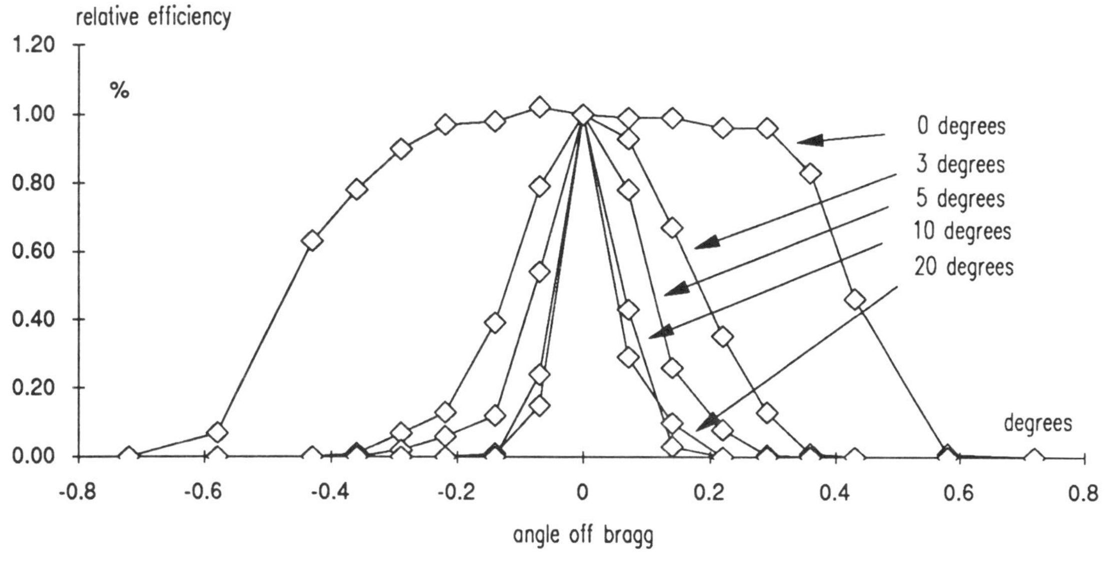

figure 6 fiber bragg angle sensitivity

These experiments are carried out for two relatively short 4 mm and 10 mm multimode 1 mm diameter fibers. The results suggest that an array of fibers might favorably replace bulk materials for certain computer and signal processing applications.

6.ARCHITECTURE

Results from the Bobcat experiments coupled with the non destructive readout technique and the crystallyte idea gives confidence that a reasonable volume holographic storage device can be built. An optical architectural approach similar to Bobcat appears most suitable. Some refinements will need to be made to this system which we will call Bobcat II. It incorporates acoustic optic deflectors for no moving part addressing, has a to-be-identified transmissive SLM with 10,000 or more pixels, use an array of photorefractive fibers or cubits for the storage media, and utilize a CCD array for output. It has electrical fixing on a stack by stack basis exploiting the fiber array isolation properties. The non destructive technique is used for readout. Bobcat II is be designed to fill a systems architecture role of a mass storage "gap filler".

Bobcat II is organized as a page oriented device with 10,000 bit pages (the current SLM technology forcing this), 10 pages per stack and 1000 stacks. The laser used is a laser diode pumped YAG with sufficient power for 1 ms writes and 10 us reads of the pages.

SBN is chosen as the photorefractive material. Cerium doped SBN:60 which is now becoming available has good sensitivity and also reasonably slow dark decay. In place of a bulk photorefractive crystal an array of cubits is used, each holding a single stack. This permits a large storage volume and allows selective control over applied electric field on a stack by stack basis.

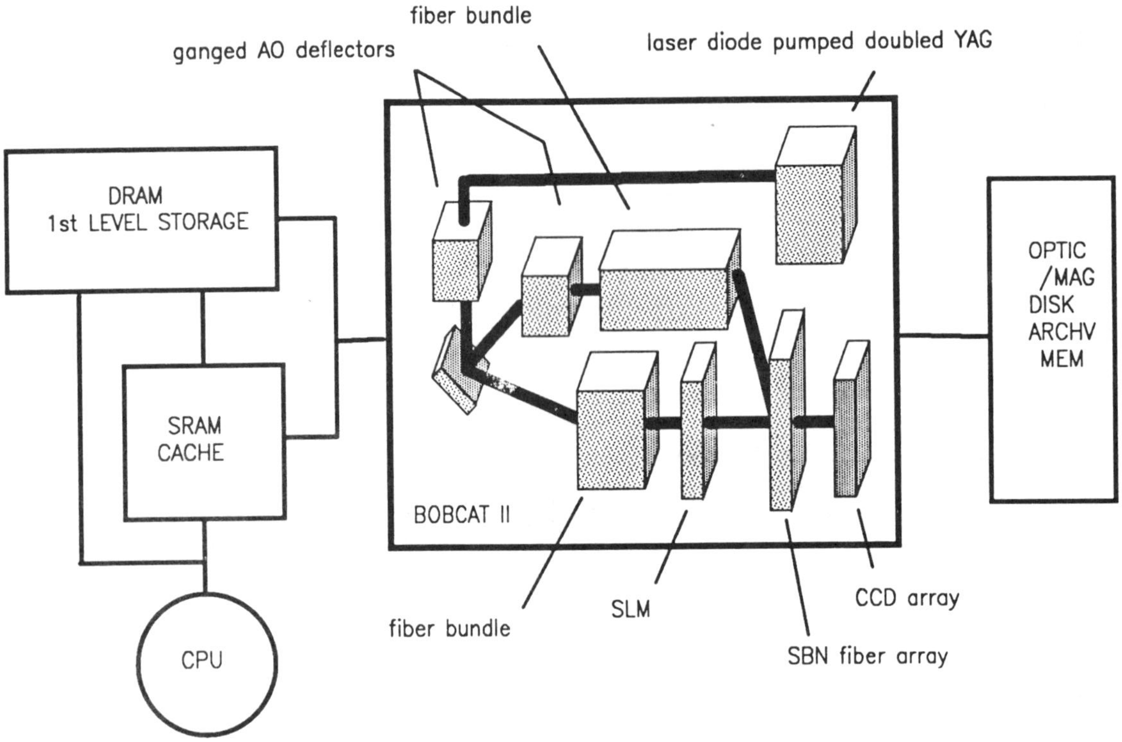

figure 7 Bobcat II configuration

SBN has the additional advantage in that it can be electrically fixed [9]. Electrical fixing requires a procedure by which the hologram is first poled to align all polarization domains. A hologram is recorded and then fixed by applying an electric field of 1.2 Kv/cm for .5 seconds antiparallel to the original poling field. Polarization reversal occurs at those locations where the sum of the space charge field plus applied field is above the coercive field. The result is a replication of the original trapped charge pattern. This pattern masks the holographic space charge field and the efficiency of the reconstruction is initially quite small. Upon illumination with the reference beam the charges redistribute themselves revealing the domain pattern. The ultimate efficiency is quite high (often higher than the original efficiency before the switching field is applied), which is attributed to over cancellation of the space charge field by local polarization switching. Stacks would be electrically fixed as soon as all their pages are written. A update strategy is chosen, which is common for database systems, which does not write in place but creates a new version. Old versions are eventually garbage collected.

To route the object and reference beams precisely and in a compact fashion, a fiber bundle approach is taken. There is a fiber bundle for the reference beam with a separate fiber for each stack position and page Bragg angle; that is a fiber for each page or 10,000 fibers. It is hoped that this fiber bundle approach for beam directing makes it possible to fit Bobcat II into a 5-1/4 inch form factor.

Ganged acoustic optic deflectors with a lens between them will be the x-y deflection system which achieves stack addressing by moving both reference and object beam in tandem. The lateral deflector needed for page addressing is accomplished by another pair of AO deflectors. Each of these pairs of deflectors would launch their beam into a particular fiber of a fiber bundle. The pairing of the deflectors is arranged to compensate for any optical frequency shift.

7.OPPORTUNITY

The 5 orders of magnitude difference that exists between the access time of DRAM and magnetic/optical disks creates a distortion in storage hierarchies. The DRAM and Optical Disk columns in table 2 illustrate this gap. Bobcat II is designed to provide a storage level which fits into this gap. It supplies large blocks of data to the DRAM first level storage in a system. Because of its high bandwidth it can also be used to supply megablocks to a CPU cache. This might be an attractive way to set up a working set in a cache for a new job, or to save the results of one job before another is started. Bobcat II has an optical disk for archival purposes. The optical disk can now be designed for its optimum characteristic, namely high capacity, and Bobcat II masks its slow latency.

A final Bobcat II system where the SLM is not a limiting factor can be projected to have the following capabilities as a "gap filler".

	DRAM	BOBCAT II	OPTICAL DISK
cost per Mbyte	$15.00 to $25.00	$2.00 to ...$6.00	$0.50 to ...$2.00
latency	50 nanosec to 100 nanosec	1 microsec to 10 microsec	5 millisec to 20 millisec
transfer bandwidth	40 Mbyte/s to 160 Mbyte/s	5 Gbyte/s to 50 Gbyte/s	6 Mbyte/s to 15 Mbyte/s
transfer unit	32 bits to 128 bits	10,000 bits to 100,000 bits	1 bit to 8 bits
capacity	10 Mbytes to 100 Mbytes	200 Mbytes to 1 Gbyte	500 Mbytes to 3 Gbyte

table 2 system performance

8.CONCLUSIONS

Some of the experimental work described here verifies what has been reported in the literature [10],[11],[12],. The number of pages, the read and write dynamics, and the volatility agree with earlier work. The recording speed of the Ce SBN crystal is surprisingly high being faster than predicted. This is important because SBN has attractive fixing characteristics. One exception to past results is the dark decay time of a recording. For some unknown reason, this is much shorter than reported for both crystals.

We believe this experimental work has overcome a major roadblock with our new prolonged readout procedure. The crystallyte technique should also improve storage capacity and open up new materials.

At this juncture PVHS devices still appear to have the potential for orders of magnitude better latencies than magnetic or optical disk. With respect to density, however, because of difficulties in getting true volumetric or 3-d storage, it looks like these devices may only have the potential of matching the densities of magnetic or optical disk. To achieve these potentials, the biggest technology problem is the SLM and the biggest material problem is stability.

Richard Rallison is acknowledged for his expert help in setting up Bobcat and Keith Daniels for his help with the experiments. We also wish to thank Ratnakar Neurgaonkar at Rockwell Science Center for supplying SBN material used for some of the experiments.

REFERENCES

[1] Alphonse, G. A. and W. Phillips, "Iron-Doped Lithium Niobate as a Read-Write Holographic Storage Medium," *RCA Review*, vol. 37, 1976.

[2] d'Auria, L., J.P. Huignard, C. Sleazak, and E. Spitz, "Experimental Holographic Read-Write Memory Using 3-d Storage," *Appl Opt.*, vol. 13, 1974.

[3] Reddersen, B. R. and L. M. Ralston, "Digital Optical Data Storage and Retrieval," *Opt Eng*, vol. 19, no. 2, March/April 1980.

[4] Weaver, J. E. and T. K. Gaylord, "Evaluation Experiments on Holographic Storage of Binary Data in Electro-optic Crystals," *Opt. Eng.*, vol. 20, 1981.

[5] Redfield, Steve, and L. Hesselink, "Enhanced Nondestructive Holographic Readout in SBN", to be published, *Optics Letters*, Oct 1988

[6] Amodei, J. J., "Analysis of Transport Processes During Holographic Recording in Insulators," *RCA Review*, vol. 32, 1971, p. 185.

[7] Hesselink, Lambertus. and S. Redfield, "Photorefractive Holographic Recording in SBN Fibers", to be published, Optics Letters, Oct 1988

[8] Jannson, T., "Information Capacity of Bragg Holograms in Planar Optics," *J Opt. Soc. Am.*, vol. 71, no. 3, March 1981, p.342.

[9] Micheron, F. and G. Bismuth, "Electrical Control of Fixation and Erasure of Holographic Patterns in Ferroelectric Materials," *App. Phys. Lett.*, vol. 20, 1972.

[10] Gaylord, Thomas K., Digital Data Storage," *Handbook of Optical Holography*, Academic Press, 1979.

[11] Glass, A. M., "The Photorefractive Effect," *Opt Eng*, vol. 17, 1978.

[12] Huignard, J. P. and F. Micheron, "High Sensitivity Read-Write Volume Holographic Storage in BSO and BGO Crystals.," *Appl. Phys. Lett.*, vol. 29, no. 9, November 1976.

Author Index

Subject Index